FINITE MATHEMATICS WITH CALCULUS:
A MODELING APPROACH

FINITE MATHEMATICS WITH CALCULUS: A MODELING APPROACH

Richard Bronson
Fairleigh Dickinson University

Gary Bronson
Fairleigh Dickinson University

Brooks/Cole
Thomson Learning™

Australia • Canada • Denmark • Japan • Mexico • New Zealand • Philippines • Puerto Rico • Singapore
Spain • United Kingdom • United States

Sponsoring Editor: *Margot Hanis*
Marketing Team: *Karin Sandberg, Laura Hubrich*
Production Editor: *Keith Faivre*
Manuscript Editor: *Susan Gerstein*
Permissions Editor: *Mary Kay Hancharick*
Interior Design: *Lois Stanfield, Lightsource Images*
Cover Design: *Roy R. Neuhaus*

Cover Image: *Aurio Sectio*
Art coordinator: *Lisa Torri*
Interior Illustration: *Tech-Graphics*
Photo Researcher: *Terry Powell*
Print Buyer: *Vena Dyer*
Typesetting *UG/GGS Information Services, Inc.*
Printing and Binding: *West Publishing Company*

For more information, contact:
BROOKS/COLE
511 Forest Lodge Road
Pacific Grove, CA 93950 USA
www.brookscole.com

For permission to use material from this work, contact us by
Web: www.thomsonrights.com
fax: 1-800-730-2215
phone: 1-800-730-2214

Printed in United States of America

10 9 8 7 6 5 4 3 2 1

Library of Congress Cataloging-in-Publication Data

Bronson, Richard
    Finite mathematics with calculus: a modeling approach / Richard Bronson, Gary Bronson.
      p.   cm.
    Includes index.
    ISBN 0-534-35652-4
    1. Mathematical models. I. Bronson, Gary J. II. Title.
QA401.B688 2000
511'.8—dc21   99-38894

To the many wonderful teachers
who showed us why and how,
especially
Mary M. Brown
Mary Mannion
Nicholas J. Rose

> *Philosophy is written in this grand book—I mean the universe—which stands continually open to our gaze, but it cannot be understood unless one first learns to comprehend the language and interpret the characters in which it is written. It is written in the language of mathematics.*
>
> GALILEO

> *Mathematics, rightly viewed, possesses not only truth, but supreme beauty.*
>
> BERTRAND RUSSELL

CONTENTS

PREFACE

The world changed dramatically in the twentieth century, and as the world became more complex, so too did its problems. Issues dealing with solid waste disposal, global warming, international finance, pollution, and nuclear proliferation are relatively new, and their solutions challenge even our best technologies and most advanced mathematics.

To better understand the dynamics of complex systems, analysts often simplify them into more manageable parts through a process called *modeling*. They then study the model to gain a better understanding of the real system they need to manage. Basic arithmetic skills, by themselves, are inadequate for much of this work. New problems require new methodologies for analyzing them.

In this book, we introduce readers to many of the newest ideas in mathematical thought. We describe in great detail the process of mathematical modeling, and we develop techniques for analyzing models. The theme of this book is modeling, and each chapter focuses on a class of models in common use today. The intent is to develop an understanding of modern mathematics that will assist managers of systems—be they social, political, commercial, or ecological—with their decision-making processes. Through such understanding, both the beauty and relevance of mathematics becomes apparent.

In this book, new concepts are carefully motivated from the experiences of most readers. Academic transcripts, road maps, and organizational charts are used to illustrate models in Chapter 1. Matrices are introduced through inventory levels in a department store. Financial decisions begin with a bank savings account and move later to installment loans. Functions are introduced as input–output relationships and then applied to design and production problems. Rates of change are introduced through automobile odometer readings, and definite integration follows naturally from area calculations. The definite integral evolves naturally from calculating areas, beginning with two-dimensional grids for irregularly shaped regions—a unique feature of this book—and moving through Riemann sums as the regions themselves become more structured.

Chapter 1 introduces the concept of a model and develops the rudimentary skills of modeling, from identifying important factors to establishing relationships among those factors. Great care is taken to explain the process of modeling because that skill is central to all that follows. Chapters 2 and 3 develop a distinct class of models that are grouped into four broad categories. Deterministic models include linear models, linear programming models, financial models, and presentation models. Probabilistic models include basic probability models, Markov chain models, and probability distributions. Calculus models focus on functions, derivatives, and integrals.

The progression of chapters is logically consistent, although there is great flexibility in the order that topics may be studied. Of course, some topics by necessity must follow others, as shown in the diagram on the following page.

The order of topics may be changed to suit individual goals. Financial models can be addressed at any point after Chapter 1. Calculus models can be introduced at any point after Chapter 2. Although the material on financial models from Chapter 4 is not required for any of the material in calculus, some of the models developed in Chapter 4 are yield rich examples for the exponential functions introduced in Section 10.3. While Section 5.1 on sets and Section 5.4 on tree diagrams are essential to Chapter 6 on probability, the other sections in Chapter 5 may be skipped if one wants to get to probability topics quickly. One may also move directly from Chapter 6 to Chapter 8.

Combinatorics and Bernoulli trials appear at the end of Chapter 6 because we prefer to introduce all the concepts of probability first, and not to complicate those concepts with the strictly mechanical procedures for counting. Others may prefer to move these two sections to the end of Chapter 5, following tree diagrams, or even to the end of Chapter 8.

There is some material in this book that is unique to our presentation, including the focus on modeling, the inclusion of simulation, and the sources for many of our examples

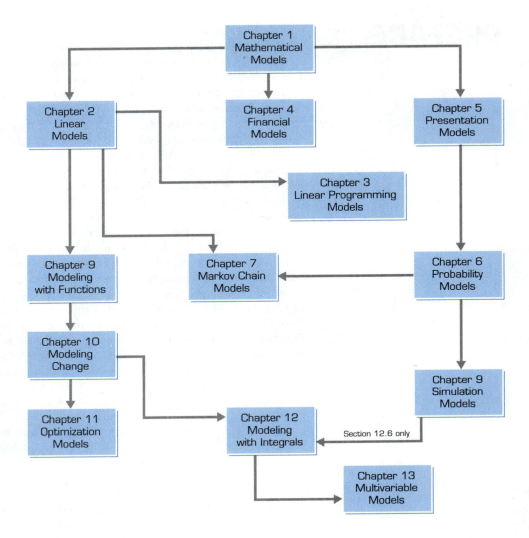

and problems. To emphasize the relevance of mathematics to everyday concerns, we have taken the information for many of our problems from newspapers, magazines, professional journals, government agencies, the General Social Survey, and books such as the *Economist's Book of Vital World Statistics*.

LEARNING AIDS

To make this book accessible to the largest possible audience, we included a variety of learning aids.

Highlight boxes

Highlight boxes are interspersed throughout the text to showcase significant material. These boxes emphasize new ideas, useful procedures, and notable models.

Illustrative examples

New methods are coupled with a large number of completely worked-out problems. We are convinced that such examples, when presented in detail, are essential to promoting real understanding, and we use such examples liberally.

Visual aids

Over 800 figures, tables, and photographs are included to help readers visualize the material featured in this book. Reprints from newspapers and magazines are included to show the relevance of certain models to everyday life.

Chapter keys

At the end of each chapter, a chapter summary in the form of chapter keys provides a concise review. These summaries list new terminology, important concepts, and key formulas and procedures.

Chapter tests

Each chapter ends with a set of problems that readers can use to test themselves on the material in that chapter. The even-numbered problems in each set may be used as one test, the odd-numbered problems as a second test.

EXERCISES

Because mathematics is not a spectator sport, one must experience mathematics to understand it. Consequently, this book is rich in problems. The exercises are separated into six distinct categories, each designed to support a different type of skill mastery or cognitive learning process.

Improving Skills consists of problems on manipulation. Readers are asked to apply mechanical operations on very basic information. The purpose of these problems is to develop competency in the basic steps of various methods.

Creating Models asks readers to construct mathematics models from descriptive problems. Its function is to develop the modeling capabilities of the reader.

Exploring in Teams encourages readers to expand their horizons and develop material beyond what is explicitly covered in the book. The problems are exploratory, and as such are ideal for group learning environments where readers can use the dynamics of a team to forge new understandings among its members.

Exploring with Technology directs readers to use existing technologies, most notably graphing calculators and electronic spreadsheets, to explore more deeply the relationships in this book. Technology can enhance understanding, and the problems in this section are designed to do just that. Technology, when used indiscriminately, can also obscure underlying concepts. Therefore, we do not ask readers to use calculators or computers in those sections where the use of technology is inappropriate.

Reviewing Material is designed for review and reinforcement of material from previous chapters. These problems first appear in Chapter 2 where they are limited to a single problem per section from Chapter 1. There are two problems in each section of Chapter 3, one on material from each of the previous two chapters; three problems in each section of Chapter 4; and four problems in each section of Chapter 5. From Chapters 6 through 13, the number of review problems in each section is capped at four. Review problems are never the focus of a section, so the number of such problems in each section is strictly limited.

Recommending Action asks readers to recommend actions, in writing, for addressing particular problems that are presented in writing to the problem solver. While problem-solvers generally use mathematical techniques to analyze systems, ultimately any recommendation must be communicated in an acceptable form (usually in writing).

PEDAGOGICAL INITIATIVES

Group learning, writing-across-curriculum, and enabling technologies are now common features of many mathematics courses. This book supports all of three innovations.

Group learning

Any problem or set of problems in the exercise sections can be assigned to students for solution by a team. However, the problems in *Exploring in Teams* are particularly good for this purpose. These problems take students into unchartered waters where group dynamics provide effective mediums for exploration and discovery.

Writing across curriculum

Any problem or set of problems in the exercises can be assigned as part of a report to be typed and submitted. However, the problems in *Recommending Action* are designed expressly for this purpose. They require the reader to synthesize the material in each section into a memo to assist someone else in addressing a problem.

Enabling technologies

It is assumed that all readers have access to calculators. A good scientific calculator, one that performs the basic arithmetic operations, raises numbers to a power, and creates random numbers, is sufficient for the material in the first eight chapters. Raising numbers to nonintegral powers is essential in Chapter 4 dealing with financial models; a simple random number generator is needed for the probability distribution models in Chapter 8. A good graphing calculator is essential for the material in the last five chapters. No other technology is needed within the main body of the text or for most of the problems. The only exceptions are the problems under the heading *Exploring with Technology*, which exploit the power of a high-powered graphing calculator or computer software to enhance basic understanding and, on occasion, to further develop some of the concepts introduced in the book.

SUPPLEMENTS

Solutions manuals

The *Instructor's Solutions Manual* (ISBN 0-534-36839-5) provides complete solution to all *Exploring in Teams* and *Recommending Action* problems and to the even-numbered *Improving Skills, Exploring with Technology,* and *Creating Models* problems. The *Student's Solutions Manual* (ISBN 0-534-36840-9) contains complete solutions to all *Reviewing Material* and *Testing Yourself* problems and to odd-numbered *Improving Skills, Exploring with Technology,* and *Creating Models* problems. Both manuals were prepared by Alison Paradise and Rosemary Hirschfelder of the University of Puget Sound.

Printed test items

Written by Leslie Cobar, the *Printed Test Items* (ISBN 0-534-36895-6) contain approximately 100 test questions per chapter, of which 70% are multiple choice.

Thomson Learning *Testing Tools*

This fully integrated suite of test creation, delivery, and classroom management tools includes *Testing Tools Test, Testing Tools Online,* and *Testing Tools Manager. Testing Tools* allows professors to deliver tests via print, floppy, hard drive, LAN, or Internet. With these tools, professors can create cross-platform exam files from published files, create and edit questions, and provide their own feedback to objective test questions, enabling the system to work as a tutorial or an examination. In addition, professors can generate questions algorithmically, creating tests that include multiple-choice, true/false, and matching questions. Professors can also track the progress of an entire class or an individual student. Testing and tutorial results can be integrated into the class management tool, which offers scoring, grade book, and reporting capabilities.

ACKNOWLEDGMENTS

We are particularly indebted to Deans Dario Cortes and Paul Lerman for their substantial support and encouragement. We could not have moved as expeditiously as we did without their backing.

Our editors at Brooks/Cole were a joy to work with and made significant contributions to the scope and format of this book. Special thanks go to our sponsoring editor, Margot Hanis, and our production editor, Keith Faivre, for their expertise and constant support; we are most grateful to both of them. We are also indebted to our manuscript editor, Susan Gerstein, for the magnificent job she did in shapinng many of the chapters into their current form.

We also gratefully acknowledge the helpful suggestions made by our many colleagues who reviewed and critiqued various portions of our manuscript. They include:

Jeff Allbutten, Middle Tennessee State University

Judy Barclay, Cuesta College

James Balch, Middle Tennessee State University

Steve Blasberg, Western Valley College

Robert Bradshaw, Ohlone College

James Carpenter, Iona College

Robert Chaney, Sinclair Community College

Darrah Chavey, Beloit College

Charles Cleaver, The Citadel

Ted Clinkenbeard, Des Moines Area Community College

Leslie Cobar, University of New Orleans

Jerry Davis, Lyndon State College

Maureen Dion, San Joaquin Delta College

Richard Easton, Indiana St. University

Michael Ecker, Penn State University–Wilkes-Barre

Matthew Gould, Vanderbilt University

Frances Gulick, University of Maryland

Joe Guthrie, University of Texas–El Paso

Jim Hatton, College of the Siskiyous

Sue Henderson, DeKalb College

Yvette Hester, Texas A&M University

John Hill, Lincoln College

Myron Hood, California State Polytechnical University, San Luis Obispo

Robert Horvath, El Camino College

Claire Krukenberg, Eastern Illinois University

Joyce Lindstrom, St. Charles Community College

James Loats, Metropolitan State College of Denver

Giles Wilson Maloof, Boise State University

Charles Miller, Foothill College

Cynthia Miller, Christian Brothers University

Michelle Mosman, Des Moines Area Community College

John Muzzey, Lyndon State College

Donald Myers, University of Arizona

Richard Nadel, Florida International University

Karla Neal, Louisiana State University

Alison Paradise, University of Puget Sound

Patrick Perkins, University of Washington

Matt Pickard, University of Puget Sound

Dave Ponick, University of Wisconsin

Elaine Russell, Angelina College

Sandra Savage, Orange Coast College

Harry Smith, University of New Orleans/Loyala University

Louis Talman, Metropolitan State College of Denver

Anthony Vance, Austin Community College

Gary Van Velsir, Anne Arundel Community College

Giovanni Viglino, Ramapo College of New Jersey

Richard Werner, Santa Rosa Junior College

Charles Zimmerman, Robert Morris College

1

MATHEMATICAL
MODELS

THE MODELING PROCESS

There are two reasons for studying mathematics: First, mathematics provides a framework for solving real-world problems, and this, in turn, helps people prosper. Second, mathematics itself is an art and science anchored on reason and logic. In a world often filled with irrational behavior and haphazard events, the lure of an ordered world governed solely by rational thought has an irresistible appeal to some—the mathematicians.

For most people, the appeal of mathematics is directly proportional to its usefulness in solving problems. Many individuals are employed to solve problems—problems in government, economics, ecology, and commerce, to mention just a few areas. As the world becomes more complex, so do its problems, and any advantage gained in the ability to solve some of those problems is an advantage to be cherished.

Years ago, the world was simpler and so was the mathematics needed to be effective. Proficiency in arithmetic was often sufficient. Such skills are still essential, but they are not adequate for solving many of today's problems. One approach to this situation is to develop more sophisticated mathematical techniques; a second approach is to simplify problems so that they can be solved with existing methodologies. We shall develop both strategies in this book.

A model is a representation of a particular situation.

The technique of replacing a complex problem with a simpler one is called *modeling*. A *model* is a representation of a particular situation. The college transcript in Figure 1.1 is a model of a person's academic achievement, the map in Figure 1.2 is a model of a road network, and the organizational chart in Figure 1.3 is a model of a company's management structure. Many situations, especially dynamic ones that change with time, can be modeled by mathematical equations.

Very few models are ever complete; that is, models do not reveal every aspect of the situation they represent. A college transcript does not indicate the ease with which the grades were achieved, a road map does not include weather conditions, and an organizational chart does not describe the personalities of the people filling the positions.

FIGURE 1.1

A college transcript as a model of academic achievement.

PERMANENT RECORD

Sem. End Mo. Yr.	Cat. No.	Course Title	Grade
–12/94	82EL102*	ENGL SEC LANG II	A
–12/94	88MG505	MGMT THRY & PRAC	B+
–12/94	88CS525	INTR COMP SCIENCE	A
–12/94	88CS550	BUSI PROG TECH	A
–		4PT SEM GPR 3.82	12.0
–		4PT CUM GPR 3.77	9.0
–05/95	87AC610	MGRL ACCT APPL	A
–05/95	87FN506	FIN ANALYSIS	A
–05/95	88CS603	COMPUTER ARCHITECTURE	A
–05/95	88CS720	MGMT INFO SYS	A
		4PT SEM GPR 4.00	12.0
		4PT CUM GPR 3.90	21.0

EXAMPLE 1

A patient's medical record in a doctor's office is a model of that patient's health. List some of the factors that this model would include and some that it may neglect.

Solution

This model would include the patient's vital medical statistics, such as age, weight, and blood pressure, from the patient's last physical examination. It should contain the patient's

FIGURE 1.2
A map as a model of a road network.

SOURCE: ©AAA reproduced by permission.

medical history, including major illnesses and operations, as well as relevant family history on genetically linked diseases. It would also list the patient's allergies and reactions to medicines and contain copies of all reports from laboratory tests and recommended consultations with other doctors.

The model may not include the patient's financial condition, which could be a cause of stress or the current state of the patient's relationships (antagonistic, distant, harmonious) with friends and family members. It might also neglect behavior patterns, such as a passion for tanning, that could affect health, and it would not include information the patient omitted either purposefully or through forgetfulness. In addition, the model will not include information on emerging conditions that are not yet noticeable, such as microscopic tumors.

Models are built for a particular purpose that often requires representing only part of a situation. A flowchart for a computer program (see Figure 1.4) may be an adequate model

FIGURE 1.3
An organizational chart as a model of corporate structure.

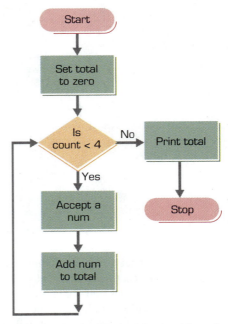

SOURCE: Bronson, Gary and Stephen Menconi. *A First Book of C: Fundamentals of C Programming,* West Publishing Company, 1988, p. 141.

A model is adequate if it meets the needs of the people using it.

for analyzing the logic of that program. An academic transcript may be a suitable model for an employer making a hiring decision. If models fit the needs of the people using them, they are good models. The criterion on which models are judged is, Does the model adequately meet the needs of the person using the model?

The word *adequate* is important. What is considered adequate by one person may not be adequate to someone else. What is considered adequate in one situation may not be adequate in another. There are no absolutes in modeling. The usefulness of a model is relative to its purpose.

Why build models? Why not analyze the real situation itself? The answer often involves a savings of either time, money, lives, or a combination of all three.

Models can save time, money, or lives.

The Sahel is a region of Africa with limited water, previously populated by nomadic tribes with their small herds in a state of equilibrium. As technology improved, deep wells were dug for additional water and humans and animals were inoculated against diseases. As a result, both human and animals populations grew. Unfortunately, larger herds soon depleted the available grasslands, and many animals starved. As the herds diminished, the larger number of nomads also starved. A good model could have predicted the effects of larger animal populations on the land and avoided the human tragedy that followed.

EXAMPLE 2

A power company has three competing designs for a new nuclear reactor it wants to build. What are the advantages of modeling in this situation?

Solution

The company could build all three, run all three, and then have hard data to determine the safest design, but this approach is too expensive. Furthermore, if one design is not safe and if it is actually built and operated, then the consequences could be disastrous to people in the surrounding communities. Mathematical models based on the design specifications of each plant are cheaper and safer. If the model's behavior implies that a power plant will explode, a disaster can be avoided with little damage.

EXAMPLE 3

A land developer wants to convert 30,000 acres of wetlands into a planned community. The developer has convincing arguments for why such a move would improve the economic base of the surrounding region, but public officials worry about the effect of the development on the ecology of the region. What are the advantages of modeling this situation?

Solution

Once the developer converts the wetlands into a planned community, any effects on the ecology and the economy would be apparent. It would, however, be too late to avoid any adverse effects. An ecological model could forecast changes in the ecology without actually disturbing it. If the changes are manageable, then the development plan might be good; if the changes are disastrous either to wildlife or, in terms of flooding, to humans, then changes the proposed development are appropriate. One also could construct an economic model, which would be quite different from an ecological model, to test the developer's hypothesis that a planned community would improve the region economically.

We can experiment on models; often, we cannot experiment on real processes.

A model allows us to analyze a real situation through the model. If the model reacts badly to a decision and if the model is an adequate representation of the real situation, then we expect the real situation also to react badly to the same decision. If the model reacts well to a proposed action, then we expect the real situation to react similarly. In this way, we can test various actions and decisions and identify the best ones. We can experiment on the model! Often, we cannot experiment on the real situation.

The various stages in using models to help make real-world decisions are shown schematically in Figure 1.5. When an action or decision is required on real-world behavior, a

FIGURE 1.5

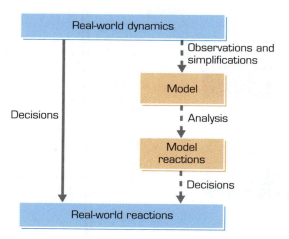

decision can be made directly on the actual system, as represented by the solid line in Figure 1.5. The underlying system will then improve or deteriorate. Alternatively, decision makers can build a model of the real-world behavior, test various decisions on the model, observe the reaction of the model to each decision, and then apply the most promising solution to the real-world situation with the expectation that real-world behavior will mimic the model's behavior. This approach is represented by the three dashed arrows in Figure 1.5.

IMPROVING SKILLS

In Problems 1 through 10, describe advantages of modeling the given situations.

1. A proposed space mission to place astronauts on Mars.
2. The number of toll booths to place at one end of a suspension bridge currently under construction.
3. A mass transit plan to build a parking lot and a commuter bus station in a suburb of a large metropolitan city.
4. The placement of roads in an undeveloped park to make the park accessible to people without harming the animal and plant life.
5. A national policy to end a long-standing recession by choosing among a tax cut, deficit spending, and lower interest rates.
6. Training airline pilots to fly safely through changing weather conditions and aircraft malfunctions.
7. The amount of inventory a department store needs for each item it carries.
8. New fishing regulations to protect both the interest of the fishing industry and the stability of the fish population.
9. The economic consequences of a supermarket chain building a sixth store in a city.
10. The number of tellers and teller positions needed at a new branch of a bank.

In Problems 11 through 17, describe some of the factors that each of the given models neglect.

11. A baseball player's statistics on the back of a baseball card as a model of that player's abilities.
12. A monthly statement from a bank as a model of an individual's financial position.
13. A schematic diagram of the electronic circuit boards in a television set as a model of that set.
14. A diorama of a city as a model of that city.
15. A fiberglass airplane model that replicates perfectly the exterior and shape of the aircraft.
16. A flight plan, listing the route an aircraft will follow, as a model of that flight.
17. A model railroad replica of a rail yard as a model of that yard.

EXPLORING IN TEAMS

18. A couple planning to buy a house is considering two choices. Table 1.1, taken from their realtor's multiple-listing guide, summarizes the most important features of each house and serves as a model of both. The realtor suggests that for each feature the couple assign a point to the house holding the advantage and then bid on the house with the most points. Under this system, which house is the most attractive? Are there any disadvantages to this ranking system?

TABLE 1.1

FEATURES	HOUSE #1	HOUSE #2
Rooms	5	6
Bathrooms	2	2.5
Fireplace	1	none
Garage	2-car	1-car
Lot size	0.5 acre	0.3 acre
Air-conditioning	central	none
Kitchen	old	modern
Basement	finished	semifinished
Price	$270,000	$250,000

19. A friend of the couple suggests another ranking scheme. Each of the nine features is ranked from one (least important) to nine (most important), and the house having the advantage in each category is assigned the same number of points as the ranking. To use this system, the couple created the ranking shown in Table 1.2. Which house is the most attractive under this system? Are there are disadvantages to this ranking system?

TABLE 1.2

FEATURES	RANK
Rooms	4
Bathrooms	7
Fireplace	5
Garage	1
Lot size	6
Air-conditioning	3
Kitchen	8
Basement	2
Price	9

20. Construct a ranking system of your own that differs from the ones given above.

21. What are some of the factors that have been neglected in using just the data available from the realtor's guide?

RECOMMENDING ACTION

22. Respond by memo to the following request:

MEMORANDUM

Subject: J. Doe Reader

From: Superintendent's Office

Date: Today

Subject: **Academic Transcripts**

Parents are pressuring us to include more information about each high school student on academic transcripts. These parents contend that colleges and prospective employers need a student profile rather than just an academic transcript, because the transcript alone does not present an adequate model of the student.

I must report on this matter later this month at the next Board of Education meeting, and I would like your opinions. Do you feel that our academic transcripts are adequate? If not, what changes do you suggest?

1.2 MODELING WITH EQUALITIES

A model is a representation of a particular situation. A model can be a chart or a map, as we saw in Section 1.1, or it can be a schematic or a graph. Figure 1.6 is a line-drawing model of an inspection station for new television sets. Televisions arrive at the inspection station and wait in line until one of the two inspectors is free. If a set passes inspection, it goes to the packing department for shipment; if a set fails inspection, it is sent to an adjuster for repairs, then it is inspected again. Figure 1.7 is a bar-graph model showing the voter turnout in U.S. presidential elections. Because there are different ways to represent various situations, there are different types of models. Many situations, especially dynamic ones that change with time, are modeled by mathematical equations.

Constraints are limitations on what can be accomplished. For example, there is a limit to the amount of money a company can commit to advertising; a constraint is the money the company controls plus the money it can borrow. There is a limit to the number of redwood trees that can be harvested; a constraint is the number of such trees alive in the world. There is a limit to the time any individual lives. Every process is constrained in one fashion or another, hence the equations and relationships that model those processes must account for the constraints.

Most problems are a collection of statements that describe a process and a problem in detail. To model the process, we model each of the statements describing the process. We begin the modeling process, therefore, by modeling individual statements.

Statements that relate the factors in a process to each other are constraints, because each relationship limits the flexibility of those factors. The most useful statements compare factors numerically. To say that Candidate A has more support than Candidate B is imprecise. There is, after all, a significant difference between leading by 2 percentage points

> *Constraints are limitations on what can be accomplished.*

FIGURE 1.6

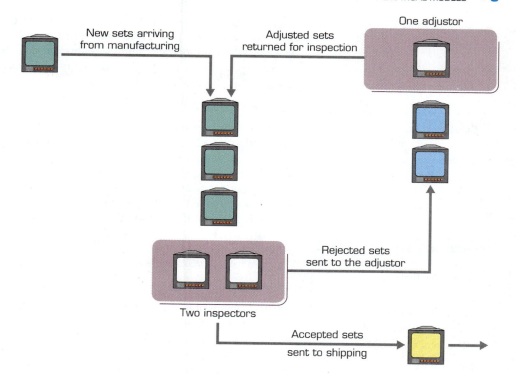

and leading by 20 percentage points. To say that industrial output exploded during the industrial revolution is imprecise. Did it double, increase 10-fold, or increase by 100-fold? Numbers add relevance.

A powerful modeling procedure involves identifying the important factors in a statement and then writing an equation or a set of equations that relate the factors to each other.

FIGURE 1.7
U.S. presidential elections

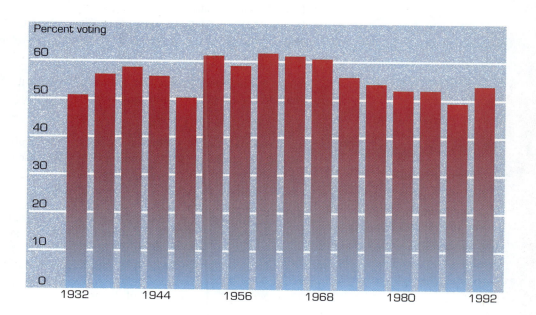

For convenience, we often use descriptive letters to label the factors. Common designations are C for cost, P for profit, n for number of items, s for sales, R for revenue, and d for demand. When appropriate, multiple letters can be used, such as TC for total cost and NI for new inventory. It takes less time to write a letter than a word, and in a complicated equation this shorthand notation increases clarity.

EXAMPLE 1

Write equations that model the following statement: The daily income of a salesperson is a base salary of $20 plus commissions of 5% on all sales.

Solution

Three important factors present themselves immediately: daily income, base salary, and commissions, which we designate by the letters DI, BS, and C, respectively. The daily income is the sum of the base salary plus commissions, so

$$DI = BS + C \tag{1}$$

The base salary is $20, hence

$$BS = 20 \tag{2}$$

Finally, commissions are 5% of sales, but we are not given total sales. No matter. Total sales is an important factor of the process, even if its exact value is unknown at present; we designate total sales by TS. If total sales are $100, then the commission is 5% of $100 or $5; if total sales are $450, then the commission is 5% of $450 or $22.50. The commission C based on total sales of TS is modeled by the equation

$$C = 0.05 \cdot TS \tag{3}$$

Equations (1) through (3) constitute a model of the original statement and are depicted in Model 1.1.

MODEL 1.1

Important factors

DI = daily income ($)
BS = base salary ($)
C = commission ($)
TS = total sales ($)

Constraints

$DI = BS + C$
$BS = 20$
$C = 0.05 \cdot TS$

Model 1.1 is *one* model for the given statement. Other models for the same statement can be created by algebraically manipulating Equations (1) through (3) or by using different labels for the important factors. A good model is one that adequately represents a system for our needs, and any model that meets this criterion is an acceptable model. In particular, Model 1.1 is one of many acceptable models for the statement in Example 1.

EXAMPLE 2

Write equations that model the following statement: A manufacturer's revenues come solely from the sale of lamps that are currently priced at $8 each.

Solution

Two important factors present themselves immediately: total revenue (i.e., income to the manufacturer) and price of a lamp; we designate these factors by the letters R (for revenue) and p (for price). The price (in dollars) is

$$p = 8$$

The total revenue is the price of each lamp, p, times the number of lamps sold. We are not given the number of lamps sold, but it is clearly an important factor, so we designate it by the letter n. It follows that

$$R = p \cdot n$$

Thus, Model 1.2 is one model of the original statement.

MODEL 1.2

Important factors

R = total revenue ($)
p = price per lamp ($)
n = number of lamps produced

Constraints

$p = 8$
$R = p \cdot n$

EXAMPLE 3

Write equations that model the following statement: A public relations firm has $350,000 to spend on radio messages, which cost $2,000 per minute, and newspaper advertisements, which cost $1,000 per quarter page.

Solution

Three important factors are total advertising expenditures, radio advertising costs, and newspaper advertising costs, which we designate by the letters T (for total), R (for radio) and N (for newspaper), respectively. We are given that

$$T = 350,000 \text{ (dollars)}$$

and that this amount will be distributed between radio and newspaper advertising. Therefore,

$$T = R + N$$

The initial statement includes two other pieces of numerical information on the costs of advertising: a minute of radio time costs $2,000, and each quarter page of newspaper space costs $1,000. This data suggest two other important factors, radio time purchased and newspaper space purchased, which we designate by the letters t (for time) and s (for space), respectively. Using units suggested by the original statement, we measure t in terms of minutes and s in terms of quarter-page units. Thus, $s = 4$ refers to four quarter-page ads while $s = 10$ refers to ten quarter-page ads.

The values of t and s are unknown. Still, we know that each minute of radio time costs $2000, hence t minutes of purchased radio time will cost (in dollars) $2,000 \cdot t$. Therefore,

$$R = 2,000 \cdot t$$

We also know that each quarter page of newspaper advertising costs $1,000, so the purchase of s quarter pages will cost (in dollars) $1,000 \cdot s$ and

$$N = 1,000 \cdot s$$

Model 1.3 is one model of the original statement.

MODEL 1.3

Important factors

T = total advertising cost ($)

R = radio advertising costs ($)

N = newspaper advertising costs ($)

t = radio time purchased (minutes)

s = newspaper space purchased (quarter pages)

Constraints

$T = 350,000$

$T = R + N$

$R = 2,000t$

$N = 1,000s$

A set of statements is modeled by modeling each statement individually and then collecting together the separate models, one for each statement, to form a larger model.

EXAMPLE 4

Write equations that model the following statements: A mining company operates two mines in West Virginia. The first mine produces 5,000 tons of high-grade ore and 1,000 tons of low-grade ore each day it is in operation. The second mine produces 2,000 tons of high-grade ore and 9,000 tons of low-grade ore for each day of operation.

Solution

We model the collection of statements by modeling each statement individually.

Statement: A mining company operates two mines in West Virginia.

This statement establishes the type of process under investigation. There is nothing much to model here, but it is clear that the process involves two mines, which we identify as mines A and B, respectively.

A set of statements is modeled by modeling each statement individually.

Statement: The first mine produces 5,000 tons of high-grade ore and 1,000 tons of low-grade ore each day it is in operation.

Two important factors are the amount of high-grade ore and the amount of low-grade ore produced from mine A; we designate these amounts by the letters HA and LA, respectively. The first letter denotes the type of ore (H for high, L for low), while the second letter identifies the mine. Each amount is in units of tons and depends on the number of days mine A is in operation. Although we are not given this number, it is an important component; we designate the number of days mine A is in operation by d_A, using the letter d for days and subscript A (written slightly below d) to designate mine A. It follows that

$$HA = 5{,}000 \cdot d_A$$

and

$$LA = 1{,}000 \cdot d_A$$

Statement: The second mine produces 2,000 tons of high-grade ore and 9,000 tons of low-grade ore for each day of operation.

Two important factors are the amount of high-grade ore and the amount of low-grade ore produced from mine B; we designate these amounts by the letters HB and LB, respectively. The first letter again denotes the type of ore, and the second letter denotes the mine. We denote the number of days mine B is in operation by d_B, using the letter d for days and the subscript B to designate mine B. It follows that

$$HB = 2{,}000 \cdot d_B$$

and

$$LB = 9{,}000 \cdot d_B$$

Collecting together the individual parts, we generate Model 1.4 as a model of the given statements.

MODEL 1.4

Important factors

HA = amount of high-grade ore taken from mine A (tons)
LA = amount of low-grade ore taken from mine A (tons)
HB = amount of high-grade ore taken from mine B (tons)
LB = amount of low-grade ore taken from mine B (tons)
d_A = number of days mine A is in operation
d_B = number of days mine B is in operation

Constraints

$HA = 5{,}000d_A$
$LA = 1{,}000d_A$
$HB = 2{,}000d_B$
$LB = 9{,}000d_B$

> *The answers to a question about the underlying real system become the objective of a model.*

The purpose of a model is to enhance our ability to answer a question about the underlying real system. Therefore, the question is of prime importance, and an answer to the question is the *objective* of any model. The objective is the first item that needs to be identified, and it immediately defines some of the important factors in the model. A complete mathematical model consists of three blocks: the objective, a list of other important components, and constraints.

Many questions are of the form, How many . . . such that . . . ? or Determine the number of . . . if. . . . The quantity we seek the number of or how many of is clearly an important factor in the model. We label this factor with an appropriate letter or letters and include it in the model as the objective. The statement following *such that, in order to,* or *if* is simply another constraint and should be modeled as such.

EXAMPLE 5

Identify the objectives in the following problem: How many hours must Mary keep her store open next week in order to make a profit of $300?

Solution

The question is the number of hours that Mary must keep her store open. If we label the answer *n,* then *n* becomes an important component and finding *n* is the objective of the problem. That is, under the heading of important factors, we write

$$n = \text{number of hours Mary must keep her store open}$$

and under the heading objective, we write

$$\text{Find } n.$$

The fact that Mary must make a profit of $300 is a constraint. Thus, we add *P* for profit to our list of important components and the equation

$$P = 300$$

to our list of constraints. A model for this problem is shown in Model 1.5.

MODEL 1.5

Important factors

n = number of hours Mary must keep her store open

P = profit ($)

Objective

Find n.

Constraints

$P = 300$

EXAMPLE 6

Identify the objective in the following problem: How many weeks will it take for Rachel's savings to equal Eric's savings?

Solution

The question is how many weeks it will take *for* something to happen. That something, the phrase following the word *for,* is a constraint. The answer to the question is the number of weeks. If we label the answer *w,* then *w* becomes an important component and finding *w* is the objective of the problem. That is, under the heading for important factors, we write

$$w = \text{number of weeks}$$

and under the heading for the objective, we write

$$\text{Find } w.$$

Rachel's savings *RS* is also an important component, as is *ES,* Eric's savings, so we add them both to our list of important factors. The fact that Rachel's savings must equal Eric's savings leads to the equation

$$RS = ES$$

as a constraint. A model for this problem is shown in Model 1.6.

MODEL 1.6

Important factors

w = weeks

RS = Rachel's savings ($)

ES = Eric's savings ($)

Objective

Find w.

Constraints

$RS = ES$

Some questions take the form, How many of each type...? The solution to such a problem involves multiple answers, one for each type, and each answer is part of the objective. To formulate an objective, one must be clear on the number of answers needed to solve a particular problem and include each answer in that objective.

EXAMPLE 7

MODEL 1.7

Important factors

nr = number of regular adult tickets sold

ns = number of senior citizens' tickets sold

nc = number of children's tickets sold

R = receipts ($)

Objective

Find nr, ns, and nc.

Constraints

$R = 1,200$

A model is incomplete if the constraints do not include all the factors in the objective.

Identify the objective in the following problem: A movie house sells regular adult tickets, senior citizens' tickets, and children's tickets at different prices. How many tickets of each type did the movie house sell yesterday if their box office receipts totaled $1,200?

Solution

The question asks for the number of tickets of each type sold, and there are three different types of tickets involved: regular, senior citizens', and children's. We require the number of each type of ticket, so each is an important factor and must become part of the objective. Another important factor is box office receipts, but information on it becomes part of the constraints. Model 1.7 is a model for the problem as stated.

A model is *incomplete* if the constraints do not include all of the variables in the objective. In this case, additional constraints must be identified! Perhaps the modeler has overlooked constraints that were stated but not modeled. Perhaps additional information is needed. Either way, the original problem cannot be solved using the model until every factor in the objective also appears in at least one constraint. Model 1.7 is incomplete because the constraints do not inclue nr, ns, and nc, which are all part of the objective. More information is required, such as the price of each ticket, before we can solve this problem. Model 1.6 is also incomplete because the constraints do not include the factor w from the objective.

With a well-defined objective, we can judge the simplicity of the model. The simplest and most elegant models are those having the fewest number of constraints involving the factor or factors in the objective. Any model that adequately represents all the constraints is a valid model. However, sometimes constraint equations can be combined to form fewer equations. If this is done without eliminating the factors of interest in the objective, then the resulting model is simpler and is preferred.

EXAMPLE 8

Model the following: A mining company operates two mines in West Virginia. The first mine produces 5,000 tons of high-grade ore and 1,000 tons of low-grade ore each day it is in operation. The second mine produces 2,000 tons of high-grade ore and 9,000 tons of low-grade ore for each day of operation. Determine the number of days that each mine should be in operation if the company needs to produce exactly 50,000 tons of high-grade ore and 80,000 tons of low-grade ore.

Solution

The first three statements are identical to Example 4 and are modeled in Model 1.4. In terms of the notation used there, the objective is to find d_A and d_B. We are asked to find the number of days that each mine should operate. There are two mines, so we require two answers, the number of days mine A should operate and the number of days mine B should operate.

We are also presented with two additional constraints. First, the company must produce

50,000 tons of high-grade ore, so in terms of the notation defined in Model 1.4, we add the constraint

$$50,000 = HA + HB$$

In addition, the company must produce 80,000 tons of low-grade ore, so we also have the constraint

$$80,000 = LA + LB$$

With these additions to Model 1.4, we generate Model 1.8 as a model for the given set of statements. Note that both objectives, d_A and d_B also appear in constraints.

MODEL 1.8

Important factors

HA = amount of high-grade ore taken from mine A (tons)

LA = amount of low-grade ore taken from mine A (tons)

HB = amount of high-grade ore taken from mine B (tons)

LB = amount of low-grade ore taken from mine B (tons)

d_A = number of days mine A is in operation

d_B = number of days mine B is in operation

Objective

Find d_A and d_B.

Constraints

$$HA = 5,000d_A$$
$$LA = 1,000d_A$$
$$HB = 2,000d_B$$
$$LB = 9,000d_B$$
$$50,000 = HA + HB$$
$$80,000 = LA + LB$$

The model in Model 1.8 can be simplified. If we substitute into the last two constraint equations the values for HA, HB, LA, and LB, we reduce the constraint equations to two, both involving the factors specified in the objective. The equation

$$50,000 = HA + HB$$

becomes

$$50,000 = 5,000\ d_A + 2,000\ d_B$$

and the equation

$$80,000 = LA + LB$$

becomes

$$80,000 = 1,000\ d_A + 9,000\ d_B.$$

The result is Model 1.9 which is the simplest model for the given statements. The constraints are represented as two equations in two unknowns, which can be solved algebraically to yield $d_A \approx 6.74$ days and $d_B \approx 8.14$ days. To meet production goals, the company should operate mine A for 6.74 days and mine B for 8.14 days.

MODEL 1.9

Important factors

d_A = number of days mine A is in operation

d_B = number of days mine B is in operation

Objective

Find d_A and d_B.

Constraints

$$5,000d_A + 2,000d_B = 50,000$$
$$1,000d_A + 9,000d_B = 80,000$$

EXAMPLE 9

Model the following: A consortium of automobile dealerships sells 20,000 vehicles each year with no advertising and has data to indicate that every million dollars in advertising expenditures would result in an additional 5,000 vehicles being sold. How much money should the consortium spend on advertising if it wants to sell 60,000 cars, the maximum number it can receive from the manufacturer in a year?

Solution

The question is what the consortium's advertising expenditures should be. We designate this factor as E, whereupon the objective becomes *Find E*. Another important factor is the number of vehicles sold, which we designate by the letter n. Using units suggested by the original statements, we measure advertising expenditures in terms of millions of dollars and the number of vehicles sold in actual sales. The consortium will sell 20,000 vehicles without any advertising and this number will increase with advertising. Therefore,

$$n = 20{,}000 + \text{additional sales from advertising}$$

Each million dollars in advertising results in 5,000 additional sales, hence E million dollars spent on advertising will increase sales by $5{,}000 \cdot E$. Thus,

$$n = 20{,}000 + 5{,}000 \cdot E \tag{4}$$

We are also told that the number of cars to be sold is 60,000. Therefore, we have the constraint

$$n = 60{,}000 \tag{5}$$

By substituting Equation (5) into Equation (4), we create the model shown in Model 1.10. The constraints are represented by a single equation in one unknown factor, the factor appearing in the objective. Solving this equation algebraically, we find that $E = 8$. Thus, the consortium should spend 8 million dollars on advertising.

MODEL 1.10

Important factors

E = advertising expenditures (in millions of dollars)

Objective

Find E.

Constraints

$20{,}000 + 5{,}000E = 60{,}000$

The process of modeling statements with mathematical equations is summarized in Figure 1.8.

CREATING MODELS

In each of the following problems, write equations that model the given statements. Do not solve the equations.

1. Last week, a waiter earned $722 from a base salary of $80 a week plus 15% of the cost of all the food and drinks he served to his customers.
2. Data from a telephone-marketing firm shows that it sells one magazine for every 20 calls it places to homeowners.
3. A manufacturing company discards as defective one of every ten units produced.
4. A conservation policy requires loggers to plant two saplings for every tree harvested.
5. An aspirin producer believes that every thousand dollars spent on advertising will increase sales by 5,000 bottles, although 200,000 bottles would be sold even with no advertising.
6. Candidate A leads Candidate B by ten percentage points.
7. Candidate A receives twice the votes of Candidate B.
8. A mail-order firm expects to sell 5,000 skirts for every 20,000 catalogs it distributes.
9. A rent-a-car company estimates that its automobiles depreciate (*i.e.*, lose value) at the rate of $10 a day.

FIGURE 1.8

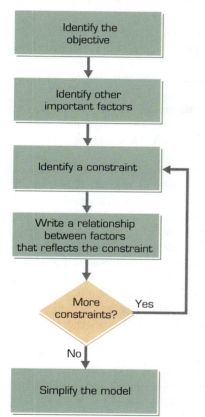

10. A machine was purchased for $100,000 and is expected to depreciate at the rate of $5,000 per year.

11. A movie theater has gross receipts of $2,856 from the sale of $8 adult tickets and $5 children's tickets.

12. A woodworking firm manufacturers tables and chairs and allots three hours for its carpenters to cut and assemble each table and two hours to cut and assemble each chair.

13. The woodworking firm in the previous problem also allots 1 hour for its decorators to paint each table and $1\frac{1}{2}$ hours to decorate each chair.

14. A mattress company produces two different styles of mattresses. The deluxe style uses 50 springs per mattress while the regular style uses 35 springs per mattress.

15. The mattress company in the previous problem also uses ten pounds of padding in each deluxe mattress but only eight pounds of padding in each regular mattress.

16. It takes a company twice as long to reupholster a love seat as it takes to reupholster a wing chair.

17. A calculator manufacturer determines that the cost of producing a calculator is $1.50 per unit; each calculator is then sold for $3.00.

18. A maker of tree swings and tree-house ladders has 14,000 feet of rope and wants to use all of this rope in making its two products. Each swing requires 10 feet of rope, and each ladder requires 25 feet of rope.

19. A recent concert brought in $37,200 from the sale of seats at $35 and seats at $20. The total number of seats sold was 1,250.

20. A customer is billed $64.20 for one day's rental from a rent-a-car company that charges $29 per day and 10¢ a mile for each mile driven.

21. A moving company charges $100 (to cover the cost of traveling between the client's locations and the home-base of the moving company) plus $59 per hour for the time actually needed to complete the move (*i.e.*, a client pays only for the time used).

22. A bookbinding company uses glue and cardboard and has two binding jobs to complete. Each book in the first job requires half a unit of cardboard, while books in the second job require one unit of cardboard. Each book in the first job also requires two ounces of glue, while those in the second job require three ounces.

23. A bookbinding company has 20,000 units of cardboard in inventory and wants to deplete it between two jobs. Each book in the first job requires two units of cardboard, while each book in the second job requires three units of cardboard.

24. An upholstery shop reconditions sofas and love seats. Each sofa requires eight yards of material and 12 pounds of padding. Each love seat requires six yards of material and 10 lbs of padding.

25. A pipe company manufactures two types of metal joints, each of which must be molded and threaded. Type I joints require 60 seconds on the molding machine and five seconds on the threading machine. Type II joints require 50 seconds on the molding machine and seven seconds of the threading machine.

26. A pipe company has available to it 500 hours of machine time on a molding machine, and this time is to be completely allocated between the production of two types of metal joints. Type I joints require 60 seconds on the molding machine, while Type II joints require 50 seconds of molding.

27. A mining company operates two mines. The first mine produces 4,000 tons of high-grade ore, 7,000 tons of medium-grade ore, and 2,000 tons of low-grade ore each day it is in operation. The second mine produces 3,000 tons of high-grade ore, 12,000 tons of medium-grade ore, and 6,000 tons of low-grade ore for each day of operation.

28. A mining company operates three mines. The first mine produces 2,000 tons of high-grade ore and 9,000 tons of low-grade ore each day it is in operation. The second mine produces 4,000 tons of high-grade ore and 4,000 tons of low-grade ore for each

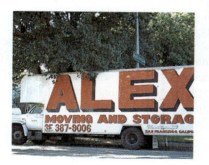

day of operation. The third mine produces only low-grade ore at the rate of 10,000 tons for each day of operation. How many days should the company operate each mine to produce 100,000 tons of high-grade ore and 90,000 tons of low-grade ore?

29. A rent-a-car company charges $32 per day and 9¢ a mile for each mile driven. A woman rents a car in the morning and returns it in the afternoon of the following day. She is charged $78.58. How many miles did she drive?

30. A telephone-answering service begins operations with an expectation of answering 500 calls a day. The sales force is expected to generate new clients who will increase this number of calls at the rate of 10 a day. How many days will it take for the company to reach its capacity of 2,000 calls per day?

31. A salesperson earns $200 per week plus 5% of all her sales. Determine her total sales for a week in which her salary was $630.

32. An automobile was purchased for $15,000 and depreciates at the rate of $200 per month. How many months will it take for this automobile to depreciate to half of its new-car value?

33. A woman lends her nephew $40,000 as part of a down payment on a house. The nephew agrees to repay his aunt $59,000 after five years to repay the initial loan with interest. Interest is computed by applying an annual interest rate to the full amount of the loan for each year the loan is outstanding. What interest rate did the aunt and nephew agree on?

34. A bank lends an entrepreneur $20,000 and charges interest at the rate of 12% a year on the initial amount for as long as the loan is outstanding. The entrepreneur signs a loan agreement that requires her to repay $31,000 in principal and interest. How long will she have the loan?

35. A mattress company produces two different styles of mattresses. The deluxe style uses 50 springs and 12 pounds of padding per mattress, while the regular styles uses 35 springs and 10 pounds of padding per mattress. The company has in inventory 16,900 springs and 4,400 pounds of padding. How many mattresses of each type should the company produce if it wants to use up its entire inventory of springs and padding?

36. A wood-cabinet manufacturer produces cabinets for television consoles and frames for grandfather clocks, both of which must be assembled and decorated. Each television cabinet requires three hours to assemble and five hours to decorate. In contrast, each grandfather clock frame requires ten hours to assemble and eight hours to decorate. The firm has 6,300 hours of assembly time and 6,600 hours of decorating time available. How many television cabinets and grandfather clock frames should it produce if it wants to use all the available time?

37. Assume that the wood-cabinet manufacturer described in the previous problem also produces lamp bases, each of which takes one hour to assemble and one hour to decorate. How many television cabinets, grandfather clock frames, and lamps should the manufacturer produce if it wants to use all the available time?

38. A pipe company has available 220 hours of machine time on a threading machine that will be completely allocated between the production of two types of metal joints. Type I joints require 10 seconds on the threading machine, while Type II joints require 15 seconds of threading. How many joints of each type should the company thread if it wants to utilize all available machine time?

39. Continental Motors has two plants that manufacture taxis and airport limousines for fleet operators. Plant A produces 75 taxis and 6 limousines daily, while plant B produces 50 taxis and 2 limousines each day. To fulfill contractual obligations to fleet owners in November, the company must manufacturer 3125 taxis and 206 limousines. How many days should the company operate each plant if it wants to meet its contracts?

EXPLORING IN TEAMS

40. One equation in the model developed in Example 3 relates radio advertising costs to minutes of radio time purchased by the formula $R = 2,000 \cdot t$. Here R is in units of dollars, and t is in units of minutes. What units are associated with the multiplier 2,000? What units are associated with the product $2,000 \cdot t$?

41. One equation in the model developed in Example 4 relates the amount of high-grade ore taken from mine A to the number of days that the mine is in operation by the formula $HA = 5,000 \cdot d_A$. Here HA is in units of tons, while d_A is in units of days. What units are associated with the multiplier 5,000? What units are associated with the product $5,000 \cdot d_A$?

EXPLORING WITH TECHNOLOGY

42. Graph the first constraint equation in Model 1.9 on a graphing calculator. For convenience, replace the label d_A with x and the label d_B with y. What type of curve do you get?

43. Now graph the second equation constraint equation in Model 1.9 on the same coordinate system. By adjusting the range on both the horizontal and the vertical axes, locate a window that shows the two curves intersecting Then use the zoom feature on your calculator along with the cursor to estimate this point of intersection. What is its significance?

RECOMMENDING ACTION

44. Respond by memo to the following request:

MEMORANDUM

To: J. Doe Reader

From: Mike

Date: Today

Subject: **Help!!!**

I'm in trouble, and I hope you can help. My job may by on the line!

My boss wanted to know how many centimeters are in a foot. I checked some books and found the following conversion factors:

2.54 centimeters/inch   and   12 inches/foot

I then divided 2.54 into 12 and gave my boss the answer. My boss responded with anger, claiming I should have know the answer was wrong just from the units. What did he mean by that?

1.3 MODELING WITH PROPORTIONALITY

Television screens come in many sizes, from two-inch hand-held models to screens that fill an entire wall. The widths of television screns vary by model, as do the heights, yet all standard screens share an interesting property: the ratio of height to width is 0.75. That is,

$$\frac{\text{height of a television screen}}{\text{width of a television screen}} = 0.75 \tag{6}$$

Two positive quantities are *directly proportional* if their quotient is a constant. The constant is called *a constant of proportionality*. If we denote two quantities by the letters A and B and their constant of proportionality by the letter k, we say that A is directly proportional to B if

$$\frac{A}{B} = k \tag{7}$$

> *Two positive quantities are directly proportional if their quotient is a constant.*

Multiplying both sides of Equation (7) by B, we have

$$A = kB, \tag{8}$$

so it is equally correct to say that two quantities are directly proportional if and only if one is a constant multiple of the other. Equation (6) has the form of Equation (7) with the height of a television screen replacing A, the width of a television screen replacing B, and

Standard television screens have a constant height-to-width ratio.

k equal to 0.75, so we say that the height of a television screen is directly proportional to its width.

$$\text{height of a television screen} = 0.75 \cdot (\text{width of a television screen})$$

EXAMPLE 1

The sales tax on restaurant meals in a popular convention city is 8%. Show that the amount of money collected as sales tax on meals in restaurants is directly proportional to the amount of money restaurants charge for meals consumed by their customers.

Solution

The formula for calculating sales tax revenues is

$$\text{sales tax revenues} = 0.08 \cdot (\text{restaurant charges})$$

This equation has the form of Equation (8) with sales tax revenues replacing A, restaurant charges replacing B, and 8% in decimal form as the constant of proportionality k. Dividing both sides of this formula by the restaurant charges, we have

$$\frac{\text{sales tax revenues}}{\text{restaurant charges}} = 0.08$$

which is in the form of Equation (7).

Many models are replicas of real systems drawn to *scale*, that is, measurements on the model are directly proportional to measurements on the real system. Road maps are models of highway systems drawn to a smaller scale, perhaps one inch of map for every 500 miles of roadway. The distance between any two points on the map is directly proportional to the actual distance between the same two points on the highway. The constant of proportionality is the ratio of the scales. In particular, a map scaled so that each inch represents 500 miles of roadway has a constant of proportionality of $1/500 = 0.002$ in units of inches per mile, or $500/1 = 500$ in units of miles per inch.

EXAMPLE 2

Model the following statements: The distance between the cities of Deerfield and Oyster Bay is 2.37 inches on a road map scaled so that one-fourth inch represents 50 miles. What is the actual distance between these two cities?

Solution

The objective is to find the actual distance between Deerfield and Oyster Bay. The proportionality relationship is

$$\frac{\text{distance between points on the map}}{\text{distance between points on the road}} = k$$

A model is drawn to scale when measurements on the model are directly proportional to measurements on the real system.

with the constant of proportionality k defined by the ratio of the scales. Each quarter inch on the map represents 50 miles. When the distance between two points on the map is one-fourth inch, the actual distance between the same two points on the road is 50 miles, hence

$$\frac{\frac{1}{4} \text{ inch}}{50 \text{ miles}} = k$$

or $k = 1/200 = 0.005$ inch per mile. The proportionality relationship becomes

MODEL 1.11

Important factors

d = mileage between Deerfield and Oyster Bay

Objective

Find d.

Constraints

$$\frac{2.37 \text{ inches}}{d} = 0.005 \, \frac{\text{inch}}{\text{mile}}$$

$$\frac{\text{distance between points on the map}}{\text{distance between points on the road}} = 0.005 \, \frac{\text{inch}}{\text{mile}} \qquad (9)$$

The two points of interest are Deerfield and Oyster Bay, and in terms of these cities, Equation (9) becomes

$$\frac{\text{distance between Deerfield and Oyster Bay on the map}}{\text{distance between Deerfield and Oyster Bay on the road}} = 0.005 \, \frac{\text{inch}}{\text{mile}}$$

We know that the distance between these cities on the map is 2.37 inches. If we let d denote the actual distance between these two cities, we have

$$\frac{2.37 \text{ inches}}{d} = 0.005 \, \frac{\text{inch}}{\text{mile}}$$

as a constraint. Solving this equation for d, we have

$$\frac{2.37 \text{ inches}}{0.005 \, \dfrac{\text{inch}}{\text{mile}}} = d$$

or $d = 474$ miles as the actual distance between Deerfield and Oyster Bay. A complete mathematical model for this problem is displayed in Model 1.11.

EXAMPLE 3

Model the following statements: The floor plan shown in Figure 1.9 is scaled so that three-quarters of an inch represents 10 feet. What are the actual dimensions of the garage?

FIGURE 1.9

Solution

The objective is to determine the actual dimensions of the garage. The garage has two dimensions, length and width, so the objective is to find both the length and the width of the garage. The proportionality relationship is

$$\frac{\text{measurements on the floor plan}}{\text{measurements on the real house}} = k$$

with the constant of proportionality k defined by the ratio of the scales. When a measurement on the floor plan is $\frac{3}{4}$ inch, the comparable measurement on the house is 10 feet, hence

$$\frac{(3/4) \text{ inch}}{10 \text{ feet}} = k$$

or $k = \frac{3}{40} = 0.075$ inches per foot. The proportionality relationship becomes

$$\frac{\text{measurements on the floor plan}}{\text{measurements on the real house}} = 0.075 \frac{\text{inch}}{\text{foot}} \qquad (10)$$

One measurement of interest is the length of the garage, and in terms of this measurement Equation (10) becomes

$$\frac{\text{length of garage on the floor plan}}{\text{length of real garage}} = 0.075 \frac{\text{inch}}{\text{foot}} \qquad (11)$$

Using a ruler, we find that the garage measures $1\frac{5}{8} = 1.625$ inches on the floor plan. If we denote the length of the real garage by l, then Equation (*11*) becomes

$$\frac{1.625 \text{ inches}}{l} = 0.075 \frac{\text{inch}}{\text{foot}}$$

as one constraint.

The second measurement of interest is the width of the garage, and in terms of this measurement Equation (*10*) becomes

$$\frac{\text{width of garage on the floor plan}}{\text{width of real garage}} = 0.075 \frac{\text{inch}}{\text{foot}} \qquad (12)$$

Using a ruler, we find that the garage measures $1\frac{3}{8} = 1.375$ inches wide on the floor plan. Designating the width of the real garage by w, we can rewrite Equation (*12*) as

$$\frac{1.375 \text{ inches}}{w} = 0.075 \frac{\text{inch}}{\text{foot}}$$

for another constraint.

Solving each constraint equation for the single variable that appears in it, we find that

$$\frac{1.625 \text{ inch}}{0.075 \frac{\text{inch}}{\text{foot}}} = l$$

$l \approx 21.667$ feet or 21 feet and 8 inches, while

$$\frac{1.375 \text{ inches}}{0.075 \frac{\text{inch}}{\text{foot}}} = w$$

MODEL 1.12

Important factors

l = length of garage (feet)
w = width of garage (feet)

Objective

Find l and w.

Constraints

$$\frac{1.625}{l} = 0.075$$

$$\frac{1.375}{w} = 0.075$$

Model trains in O scale (right), HO scale (center), and N scale (left).

$w \approx 18.333$ feet or 18 feet and 4 inches. The real garage measures 21′8″ long by 18′4″ wide. A complete mathematical model for this problem is displayed in Model 1.12.

EXAMPLE 4

Model the following statements: HO gauge model trains have a scale of 1 inch to 87 inches. How long is a real train car if its HO model measures 8 inches?

Solution

The objective is to find the length of a real train car. The proportionality relationship is

$$\frac{\text{length of model train car}}{\text{length of real train car}} = \frac{1 \text{ inch}}{87 \text{ inches}} = \frac{1}{87}$$

The length of the model train car is 8 inches. If we denote the length of the real train car by L, then the proportionality relationship becomes

$$\frac{8 \text{ inches}}{L} = \frac{1}{87}$$

Solving this equation for L, we find that the length of the real train car is $L = 8(87) = 696$ inches or 58 feet. A mathematical model for this problem is displayed in Model 1.13.

MODEL 1.13

Important factors

L = length of real passenger train car (inches)

Objective

Find L.

Constraints

$$\frac{8}{L} = \frac{1}{87}$$

EXAMPLE 5

The revenues of a firm that manufactures a single product are directly proportional to sales. Last month the company booked $2,460,000 on a total sales of 2 million items. Next month it hopes to increase sales by 100,000 items. How much revenue can the firm expect next month if its sales projections are met?

Solution

The objective is to determine the firm's revenue for the next month. The proportionality hypothesis is modeled by the equation

MODEL 1.14

Important factors

R = total revenue ($)

Objective

Find R.

Constraints

$$\frac{R}{2,100,000} = 1.23$$

$$\frac{\text{revenue}}{\text{sales}} = k \tag{13}$$

Last month, *sales* were 2,000,000 items and yielded a *revenue* of $2,460,000, so this proportionality relationship becomes

$$\frac{2,460,000 \text{ dollars}}{2,000,000 \text{ items sold}} = k$$

or $k = 1.23$ in units of dollars per item sold. Thus, Equation (*13*) can be rewritten as

$$\frac{\text{revenue}}{\text{sales}} = 1.23 \frac{\text{dollars}}{\text{items sold}} \tag{14}$$

Next month, *sales* will equal 2,100,000, and we want to find the corresponding revenue. Substituting this value into Equation (*14*), we have

$$\frac{\text{revenue}}{2,100,000 \text{ items sold}} = 1.23 \frac{\text{dollars}}{\text{item sold}}$$

Finally, if we label *revenue* as R, we have the model shown in Model 1.14. Solving the single constraint equation for R, we determine next month's revenue to be

$$\text{revenue} = R = \left[1.23 \frac{\text{dollars}}{\text{item sold}}\right] \cdot [2,100,000 \text{ items sold}] = \$2,583,000$$

IMPROVING SKILLS

FIGURE A

1. Using Equation (*6*), determine the height of a television screen that measures 24 inches wide.
2. Using Equation (*6*), determine the width of a television screen that measures 24 inches high.
3. A proposed standard for high-definition television sets the ratio of screen height to screen width at 9/16. Determine the height of such a television screen if its width is 24 inches.
4. A second proposed standard for high-definition television sets the ratio of screen height to screen width at 0.6. Determine the width of such a television screen if its height measures 12 inches.
5. Using the data provided in Example 1, find the total of all restaurant charges for food and drink in a quarter year in which a restaurant paid $13,400 in sales taxes.
6. Using the data provided in Example 2, find the distance between Deerfield and the state line, if that distance is 0.75 in. on the map.
7. Use the floor plan shown in Figure 1.9 and the data in Example 3 to estimate the dimensions of (*a*) the family room, (*b*) the foyer, and (*c*) the outside dimensions of the house *without* the garage.
8. An architect will add an additional room to the floor plan shown in Figure 21. If the actual length of the room is to be 15.25 feet, what would its length be on the floor plan?
9. Use the floor plan shown in Figure A to determine the apartment's overall dimensions if the plan is drawn to a scale of $\frac{3}{4}$ to 10 feet.
10. Solve Problem 9 if the scale is 1 inch to 15 feet.

11. *N*-gauge model trains have a scale of 1 inch to 160 inches. What is the length of a model passenger car in this scale if the real train car measures 58 feet?

12. *O*-gauge model trains have a scale of 1 inch to 48 inches. What is the length of a model passenger car in this scale if the real train car measures 58 feet?

13. A locomotive in an O-gauge train set measures 9 inches in length. Using the data in Problems 11 and 12, determine the length of the same type of locomotive in an *N*-gauge train set.

14. *A* is directly proportional to *B* with a constant of proportionality of 7. Determine whether (*a*) *A* is directly proportional to 2*B*, (*b*) 2*A* is directly proportional to 3*B*, (*c*) *A* is directly proportional to *AB*, (*d*) A^2 is directly proportional to B^2, (*e*) A^3 is directly proportional to B^3, and (*f*) *A* is directly proportional to A^2. Determine the constants of proportionality for those quantities that are directly proportional.

15. Show that if *A* is directly proportional to *B*, then *B* is directly proportional to *A*. What is the relationship between their constants of proportionality?

16. Show that if *A* is directly proportional to *B* with constant of proportionality k_1 and if *B* is directly proportional to *C* with constant of proportionality k_2, then *A* is directly proportional to *C*. What is the constant of proportionality between *A* and *C*?

17. The Greeks noted that the area of a circle was always directly proportional to the square of its radius. (*a*) Express this relationship as a formula. (*b*) Find the constant of proportionality if a circle of radius 5 in. has an area of 78.54 in^2.

18. Some people believe that the weight of a fish is directly proportional to the cube of its length, for fish of the same species. (*a*) Express this relationship as a formula. (*b*) Find the constant of proportionality if a fish of length 12.5 in. weighs 17 oz.

CREATING MODELS

Model and solve Problems 19 through 34.

19. Property tax revenues expected by a municipality are directly proportional to the assessed value of those properties. The property tax rate is 0.6%, and the municipality expects to receive 18 million dollars next year from this tax. What is the total assessed value of the properties in this municipality?

20. An architectural drawing is scaled so that 1 inch represents 10 feet. How long is the front of a building if it measures 12.5 inches on the drawing?

21. A model airplane is scaled so that 0.75 inch represents 5 feet. How long is the plane's wingspan if it measures 8 inches on the model?

22. A model airplane is scaled so that 1 inch represents 18 feet. Determine the length of the model if the length of the plane is 320 ft.

23. Interest paid on a savings account is directly proportional to the balance in the account. A person receives a quarterly interest payment of $805 at a bank that offers a quarterly interest rate of 2%. How much was in the account when interest was paid?

24. A person received $805 in interest on a savings account balance of $9,200. (*a*) How much interest is due on a balance of $10,500? (*b*) What is the significance of the constant of proportionality?

25. A person receives $1,332 in interest on a certificate of deposit with a balance of $14,400. How much interest would have been paid on an account balance of $12,000?

26. Interior designers traditionally use a scale of $\frac{1}{4}$ inch to a foot when they draw diagrams of a room. What is the actual length of a wall that measures 8.3 inches on such a diagram?

27. After a drug is injected into a patient, doctors estimate that the amount of the drug removed each day by the patient's body chemistry is directly proportional to the amount in the body at the beginning of the day. One cubic centimeter of the drug is injected into a patient, and a day later a blood test shows that 0.92 cc remains. (*a*) How much of the drug will be in the body after two days? (*b*) After three days?

28. The number of left-handed people in a population is hypothesized to be directly proportional to the total population with a constant of proportionality of 0.18. Using this proportionality hypothesis, estimate the number of left-handed people in a city of 41,000 residents.

29. Using the data provided in Problem 28, determine whether the number of right-handed people is directly proportional to the total population and, if so, find its constant of proportionality.

30. Using the data provided in Problem 28, determine whether the number of right-handed people is directly proportional to the number of left-handed people in a given population, and if so, find the constant of proportionality.

31. The number of births in a country in a year is proportional to the population of that country at the beginning of the year. A country with two million people at the beginning of a year records 460,000 births by the end of the year. How many births are expected the following year if the population at the beginning of that year is 2,200,000 (taking into account all births and deaths)?

32. If one end of a spring is attached to a ceiling, its elongation is directly proportional to the weight attached to the other end of the spring. How much will a 5-ounce weight stretch a spring if a 2-pound weight stretches it 4 inches?

33. The velocity of a free-falling object in a vacuum is directly proportional to its time in flight. How fast will such an object be traveling after 10 seconds if its velocity after one second is 32 ft/sec?

34. In a culture with ample nutrients and space, the number of new strands of bacteria created in an hour is directly proportional to the total number of strands at the beginning of the hour. A culture initially contains 1,000 strands, and after one hour it has grown to 1,050. How many strands would be expected after two hours?

EXPLORING IN TEAMS

35. We want to test the proportionality hypothesis of Problem 28 by randomly selecting 1,400 people and asking each whether he or she is left-handed. What conclusions can you draw if that sample contains 256 left-handed people? How many left-handed people would you expect if the hypothesis is true? Do you really expect to obtain that *exact* number?

36. A coin is tossed many times. A common supposition is that the number of times a head appears is directly proportional to the number of throws. Five people are asked to toss a coin 200 times each and count the number of heads they get. The results are tabulated in the following table. What do you conclude?

Number of heads	96	99	101	102	102
Number of throws	200	200	200	200	200

EXPLORING WITH TECHNOLOGY

37. The proportionality equation in Example 1 is

$$\text{sales tax revenues} = 0.08 \cdot (\text{restaurant charges})$$

Designate *sales tax revenues* by *y* and *restaurant charges* by *x*, and then graph the resulting equation on a graphing calculator. What type of curve do you get?

38. The proportionality equation in Example 2 is

$$\frac{\text{distance between points on the map}}{\text{distance between points on the road}} = 0.005 \; \frac{\text{inch}}{\text{mile}}$$

Let *distance between points on the map* be *y* and *distance between points on the road* be *x*, and then graph the resulting equation on a graphing calculator. What type of curve do you get?

39. The proportionality equation in Example 4 is

$$\frac{\text{length of an HO model train car}}{\text{length of a real train car}} = \frac{1}{87}$$

Designate *length of an HO model train car* by *y* and *length of a real train car* by *x*, and then graph the resulting equation on a graphing calculator. What type of curve do you get?

40. Use the results of Problems 37 through 39 to formulate a theorem about direct proportionality equations and their graphs. Test your theorem on other direct proportionality equations in this section.

RECOMMENDING ACTION

41. Respond by memo to the following request:

MEMORANDUM

To: J. Doe Reader

From: Regional Sales Office

Date: Today

Subject: **Sales**

At our last national sales meeting, Audrey Tsai made a persuasive argument for linking national sales to advertising. If I remember correctly, she showed that sales were directly proportional to the amount of money spent on television advertising. I wonder whether the same is true for regional sales.

The advertising budgets for our region for the last three years were $1.2 million, $1.5 million, and $1.4 million. Sales for the region were $4.8 million, $5.9 million, and $5.7 million, respectively. Can we transfer Audrey's conclusions about national sales to our region?

MODELING WITH GRAPHS

Graphs have visual impact and are a powerful medium for presenting data and relationships.

Graphs have visual impact and are a powerful medium for presenting data and relationships. They are used by the Census Bureau to display national statistics, by newspapers to show trends, and by social scientists to demonstrate correlations between social behaviors. Corporations use graphs to illustrate growth, and speakers routinely use graphs to support their conclusions. Graphs cut across all disciplines as the most widely used model for displaying real-world behavior.

Graphs are most often displayed on a *Cartesian coordinate system*, which is created by drawing two intersecting perpendicular lines called *axes*. The intersection of these two axes is the *origin*, which is the reference point of the system. The horizontal line is often called the *x-coordinate axis* (or just the *x-axis*); the vertical line is often called the *y-coordinate axis* (or *y-axis*). Although the letters *x* and *y* are the most widely used generic symbols for the axes, other letters such as *s* for supply and *d* for demand or *AE* for advertising expenditures and *TS* for total sales often replace *x* and *y* when they are more meaningful in a particular problem.

Tick marks divide each axis into fixed units of length. Units to the right of the origin on the *x*-axis and units above the origin on the *y*-axis are assigned positive values. Units to the left of the origin on the *x*-axis and units below the origin on the *y*-axis are assigned negative values. Arrowheads are affixed to the positive portions of the *x*- and *y*-axes to indicate the directions of increasing values of *x* and *y*. Successive tick marks must be equally spaced; the units between successive tick marks must be the same, although the units or *scale* on the horizontal axis can, and frequently does, differ from the scale used on the vertical axis.

Every point on the plane is uniquely determined by an ordered pair.

Every point on the plane is uniquely determined by an ordered pair of the form (x, y), called the *coordinates* of the point. The first number in an ordered pair is often called the *x-coordinate*, and the second number, the *y-coordinate*. The coordinates of point *P* in Figure 1.10 are $(6, 4)$; its *x*-coordinate is 6 and its *y*-coordinate is 4. The coordinates of point *A* in Figure 1.10 are $(5, -3)$; its *x*-coordinate is 5 and its *y*-coordinate is -3.

Geometrically, the set of *all* the solutions of one equation in two unknowns is a curve on the plane. The form of this curve is often determined by looking at a plot of some of the solution points and then making an educated guess about the pattern of all solutions.

FIGURE 1.10

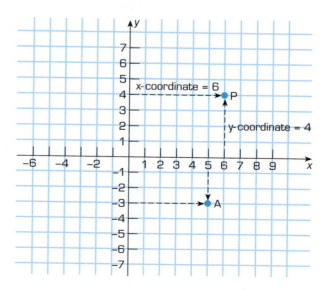

This procedure is called *graphing*, and the final curve is the *graph of the equation* under consideration. Graphing many points is always better than graphing too few.

EXAMPLE 1

Find the graph of the equation $P = 2Q^2 + 1$, putting Q on the horizontal axis and P on the vertical axis.

Solution

We identify some solutions to this equation by arbitrarily assigning values to Q and then finding the corresponding values for P. For example,

$$\text{when } Q = 0, \qquad P = 2(0)^2 + 1 = 1$$

$$\text{when } Q = -2, \qquad P = 2(-2)^2 + 1 = 9$$

Continuining in this manner, we generate Table 1.3.

TABLE 1.3

Q	−3	−2	−1	0	1	2	3
P	19	9	3	1	3	9	19

Plotting these points on a cartesian coordinate system and then connecting the points with a curve, we obtain Figure 1.11, which is the graph of the equation $P = 2Q^2 + 1$.

FIGURE 1.11

FIGURE 1.12

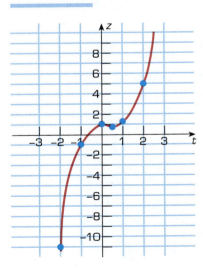

When solving an equation in two unknowns, the solver decides which variable to assign values to and which variable to calculate; the choice is dictated by personal preference and the ease with which the second unknown can be found. For the equation $P = 2Q^2 + 1$, it is a bit easier to select values for Q and then solve for P rather than the reverse. Given the equation $y^5 - 2y^2 = x + 1$, it is simple to pick values of y and solve for x but quite difficult to pick values of x and then solve for y.

EXAMPLE 2

Graph the equation $z = t^3 - t^2 + 1$ with t on the horizontal axis.

Solution

With this equation, it is easier to select values for t and then solve for z, rather than the reverse. In particular,

when $t = 1$, $\quad z = (1)^3 - (1)^2 + 1 = 1$

when $t = -2$, $\quad z = (-2)^3 - (-2)^2 + 1 = -8 - 4 + 1 = -11$

Continuining in this manner, we generate Table 1.4.

TABLE 1.4

t	-2	-1	0	0.5	1	2
z	-11	-1	1	0.875	1	5

We plot these points on a cartesian coordinate system, and then connect the points, as illustrated in Figure 1.12, to obtain the graph of the equation $z = t^3 - t^2 + 1$.

Graphing calculators and graphical software packages provide new tools for automating the evaluation of equations and the plotting of points. We can use these tools to produce the solutions to Examples 1 and 2 simply and quickly. When such technology is available, its use is highly recommended.

EXAMPLE 3

Graph the equation $x^2 + y^2 = 25$.

Solution

Either we select values for y and solve the given equation for x, or we select values for x and solve the given equation for y. Both paths are equal in difficulty, so we arbitrarily take the first, and rewrite the equation as

$$x = \pm\sqrt{25 - y^2}$$

When $y = -4$,

$$x = \pm\sqrt{25 - (-4)^2} = \pm\sqrt{25 - 16} = \pm 3$$

Continuing in this manner, we generate Table 1.5.

TABLE 1.5

y	−5	−4	−3	0	3	4	5
x	0	±3	±4	±5	±4	±3	0

Plotting these points, we obtain Figure 1.13 as the graph of the given equation.

FIGURE 1.13

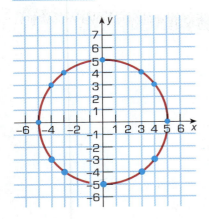

Exponential-type growth is characterized by a curve that begins as a nearly horizontal line and later changes into a nearly vertical line.

Graphs are to data what pictures are to words: a visual model. As another example, we will graph the data in Table 1.6, which records the effect of compound interest on a savings account over a two-hundred-year period.

The account was started with a deposit of $1,000 and then left to accumulate interest at an annual rate of 5%, without any additional deposits or withdrawals. We shall see how to generate such a table in Chapter 4, but that is not our concern here.

If we plot each point in Table 1.6 on a cartesian coordinate system and then connect the plotted points with a curve, as shown in Figure 1.14, it appears that the account balance grows very slowly at first, following a straight-line pattern with very little rise in the vertical direction until year 100, when the curve shows rapid growth in the positive vertical direction. Such behavior is commonly called *exponential-type growth*. A graphical model of exponential-type growth is characterized by a curve that begins as a nearly horizontal line and later (that is, further along the positive horizontal axis) changes into a nearly vertical line.

Figure 1.14 illustrates the effect that scaling can have on a graph. It appears from this graph that there is no growth over the first 50 years. This is not true, of course, as we can see from Table 1.6. At time zero, the balance is actually $1000, and after 50 years the balance has grown to over $11,000. The visual effect of this growth is lost on the graph because of the scale used for the vertical axis. Because the account balance grows from $1,000 to over 17 *million* dollars over the 200-year period, the vertical axis was scaled in two million dollar units to accommodate this growth. With such scaling, half the distance between two tick marks is a million dollars, a tenth of the distance between two tick marks

TABLE 1.6

YEAR	BALANCE	YEAR	BALANCE
0	$1,000.00	110	$214,201.69
10	$1,628.89	120	$348,911.99
20	$2,653.30	130	$568,340.86
30	$4,321.94	140	$925,767.37
40	$7,039.99	150	$1,507,977.50
50	$11,467.40	160	$2,456,336.44
60	$18,679.19	170	$4,001,113.23
70	$30,426.43	180	$6,517,391.84
80	$49,561.44	190	$10,616,144.55
90	$80,730.37	200	$17,292,580.82
100	$131,501.26		

FIGURE 1.14

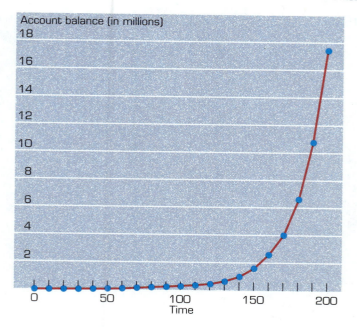

is $200,000 and a thousandth of the distance between two tick marks is $2,000. Thus, $1,000 is $\frac{1}{2000}$ the distance between the 0 tick mark and the 2 million dollar tick mark, so it's very close to zero. Eleven thousand dollars is approximately $\frac{11}{2000}$ the distance between the same two tick marks, again very close to zero.

A point that represents a few hundredths of the distance between tick marks on a graph is difficult to differentiate from the tick mark itself, and such differentiations are generally lost on a graph. Once a scale is chosen, points on a graph are essentially rounded, for graphing purposes, to one-tenth the distance between two successive tick marks. If the difference between two tick marks on the vertical axis is two million dollars, as in Figure 1.14, then one-tenth of that is $200,000, so all points are essentially rounded to the nearest $200,000. Thus, $1,000 is rounded to zero, for graphing purposes, as are $2,000, $4,000, and even $30,000. In contrast, $131,501 is rounded to $200,000 for graphing purposes and is graphed approximately $\frac{1}{10}$ the distance between the 0 tick mark and the 2 million dollar tick mark, which is still very near the horizontal axis.

To alter the effects of scaling, we change the scale. In Figure 1.14, the distance between successive tick marks on the vertical was two million dollars. We could have chosen a different scale. If we separate successive tick marks by a quarter of an inch and let the distance between two tick marks represent $1,000, we create a scale of a quarter inch for each thousand dollars. How long must the vertical axis be to display 18 million dollars? This is a proportionality problem, which can be solved by the methods developed in Section 1.3. The answer is 4,500 inches (see Problem 7) or 125 yards! Much too large for this book!

A better technique for altering the scale is to *zoom in* on a smaller portion of the graph and magnify it for analysis. If we graph just the first 100 years of Table 1.6, where the balance grows from $1,000 to just over $130,000, we can accommodate all the points in this portion of the graph by scaling in the thousands rather than in the millions. The result is shown in Figure 1.15. The same type of exponential-growth is observed but it is now clear that there is some growth in the first 50 years.

> *To alter the effects of scaling, change the scale.*

FIGURE 1.15

Zooming-in to feature a small part of a graph is an effective way to accent interesting portions of a graph. The result is a *window* containing a magnified portion of the graph. The window in Figure 1.15 includes all points on the curve with a horizontal coordinate between 0 and 100 years and a vertical coordinate between 0 and 140 thousand dollars. In contrast, the window in Figure 1.14 includes all points with a horizontal coordinate between 0 and 200 years and a vertical coordinate between 0 and 18 million dollars.

EXAMPLE 4

Table 1.7 contains data provided by the Census Bureau on the population of the United States. What type of growth does it exhibit?

TABLE 1.7

YEAR	POPULATION (MILLIONS)	YEAR	POPULATION (MILLIONS)
1790	3.9	1900	76.0
1800	5.3	1910	92.0
1810	7.2	1920	105.7
1820	9.6	1930	122.8
1830	12.9	1940	131.7
1840	17.1	1950	151.3
1850	23.2	1960	179.3
1860	31.4	1970	203.3
1870	39.8	1980	226.5
1880	50.2	1990	248.7
1890	62.9		

Solution

We take time to be the horizontal axis and population to be the vertical axis. We have no data for years before 1790, the first census year, so we need not show any part of the time axis prior to 1790. There is no data provided past 1990, so we stop our graph at that point. The population figures are between 3.9 and 248.7 million, so that is the interval of interest on the vertical axis. To allow for convenient labeling of the vertical tick marks, we enlarge this interval to include 0 and 250 million. We therefore zoom in on the window containing points with horizontal coordinates between 1790 and 1990 and vertical coordinates between 0 and 250 million. Plotting the points from Table 1.7 in this window, we generate Figure 1.16. The graph appears to exhibit exponential-type growth, although there is room for doubt. The lower-than-expected growth around 1940 can be explained by the effects of the Great Depression on family size.

FIGURE 1.16

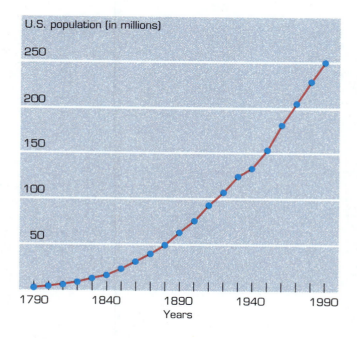

Exponential-type growth assumes the ability to grow without bound. This is a reasonable assumption for savings accounts, but not for populations. Space and food availability place constraints on the size of a population and limits its growth. When a population begins to approach its limiting size, growth moderates and the population begins to level. The result is a curve that resembles the letter *S*.

In the social sciences and in many business situations, curves that resemble the letter *S* are called *S-shaped curves*. The term *S*-shaped curve is not a mathematical name, but we shall use it in this book because it is so widely accepted in other fields to which mathematics is applied. A graphical model of *S*-shaped growth is characterized by a curve that begins as a nearly horizontal line, later changes into a nearly vertical line, and later still changes back to a nearly horizontal line. An *S*-shaped curve looks much like an exponential-type growth curve until it gets close to its limiting value. The population of the United States probably has a limiting value; the graph in Figure 1.16 suggests it has not yet come close to reaching that limit.

S-shaped growth is modeled by a curve that begins as a nearly horizontal line, later changes into a nearly vertical line, and later still changes back to a nearly horizontal line.

EXAMPLE 5

Table 1.8 contains data taken at a downtown parking garage with a maximum capacity of 400 cars. The number of cars in the garage was recorded every fifteen minutes from 5:00 a.m., when the garage opened until 10:00 a.m., when peak usage was expected. What type of growth does the data exhibit?

TABLE 1.8

TIME	CARS	TIME	CARS
5:00	0	7:45	247
5:15	2	8:00	275
5:30	5	8:15	312
5:45	10	8:30	342
6:00	21	8:45	364
6:15	36	9:00	385
6:30	54	9:15	391
6:45	88	9:30	394
7:00	116	9:45	396
7:15	154	10:00	397
7:30	205		

Solution

We take time to be the horizontal axis and the resident car population to be the vertical axis. Since no point can have a second coordinate greater than 400, the limit of the garage, we shall not show any part of the vertical axis beyond 400. Since all the data fall between 5:00 a.m. and 10:00 a.m., the horizontal axis will contain only that portion of time. Plotting the points in Table 1.8, we generate the window shown in Figure 1.17. The graph exhibits the classic form of *S*-shaped growth.

FIGURE 1.17

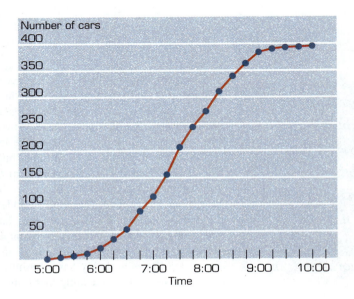

TABLE 1.9

YEAR	NEW HOME PRICE (×$1000)	YEAR	NEW HOME PRICE (×$1000)
1970	23.4	1981	68.9
1971	25.2	1982	69.3
1972	27.6	1983	75.3
1973	32.5	1984	79.9
1974	35.9	1985	84.3
1975	39.3	1986	92.0
1976	44.2	1987	104.5
1977	48.8	1988	112.5
1978	55.7	1989	120.0
1979	62.9	1990	123.0
1980	64.6		

EXAMPLE 6

Table 1.9 shows U.S. Department of Commerce data on the price of new, single-family homes in the United States. What type of growth does the data exhibit?

Solution

We take time to be the horizontal axis and the price of homes (in thousands of dollars) to be the vertical axis. Plotting the points from Table 1.9 and limiting our graph to a small region that contains all the data, we generate the window shown in Figure 1.18. The graph exhibits almost *straight-line growth*, that is, growth that closely follows a straight line.

FIGURE 1.18

FIGURE 1.19

U.S. population (in millions)

Years

FIGURE 1.20

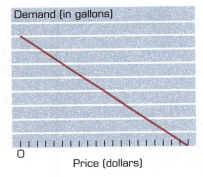

Demand (in gallons)

0   Price (dollars)

FIGURE 1.21

Demand (in gallons)

0   Price (dollars)

All of the preceding graphs were *quantitative graphical models*, that is, they associate with each point on a graph a set of numerical coordinates and they have axes with well-defined tick marks. We can locate one coordinate of a point on the graph once the other coordinate is specified. For example, to estimate the population of the United States in 1965, we locate 1965 on the horizontal time axis in Figure 1.19. We then follow along a straight line segment parallel to the vertical axis until that line segment intersects the curve. From the point of intersection on the growth curve, we follow another straight line segment parallel to the horizontal axis until we reach the vertical axis. The point of intersection on the vertical axis is approximately 190 million; this point is the vertical coordinate of the point on the growth curve associated with 1965, and it represents an estimate of the population in 1965. The dashed line segments in Figure 1.19 illustrate this process.

To estimate when the U.S. population reached 100 million, we follow the lower set of dotted lines in Figure 1.19 as indicated by the arrowheads, beginning with 100 million on the vertical axis and reaching the horizontal axis between 1910 and 1920, at approximately 1914. Thus, 1914 is the horizontal coordinate of the point on the growth curve associated with a vertical coordinate of 100 million in population, and 1914 becomes our estimate of the time when a population of 100 million was reached.

Sometimes we have a good idea about the shape of a curve but we don't know the appropriate scales for the coordinate axes. Consider, for example, the relationship between the demand for gasoline and its price. As price (in dollars) increases, demand (in gallons) decreases, so we might expect the straight-line relationship shown in Figure 1.20. The window in Figure 1.20 is limited to nonnegative values for demand and price. (A negative price would be equivalent to gasoline companies paying customers to take gasoline, a theoretical possibility that has no practical application, and negative demand would be equivalent to customers returning gasoline to the stations, another situation of no practical interest.)

According to Figure 1.20 there would be a finite demand for gasoline even when the price is zero. This is indeed reasonable, because there is a limit to the amount of gasoline any one driver can use and store, and when these individual limits are summed, we have a limit for the country. However, we do not know what this limit is, so we cannot label the tick marks on the vertical axis. Figure 1.20 also indicates that at some price the demand for gasoline will drop to zero. We do not know what this limiting price is, so we cannot label the tick marks on the horizontal axis.

With some additional reflection, we might modify our hypothesized relationship between the price of gasoline and demand and produce a curve like the one in Figure 1.21. Now we reason that there will be very little change in demand as the price increases but remains low. The demand for gasoline at 10¢/gallon will not differ much from the demand at 35¢/gallon. At some point, however, the price will begin to bite, and the demand will drop in proportion to price increases. Nevertheless, some people such as people running delivery services and traveling salespeople, need gasoline at almost any price, and they will purchase at high prices and then pass the cost along to their customers. Thus, we expect the demand never to fall to zero.

Figures 1.20 and 1.21 are examples of *qualitative graphical models*, that is, graphs that lack a scale for one or both of their coordinate axes. Quantitative models are always preferred to qualitative models, but in the absence of a quantitative model, a good qualitative model can provide a decision maker with significant information about the underlying process and related real-world behavior.

> *A qualitative graphical model is a graph that lacks a scale on one or both of its coordinate axes.*

EXAMPLE 7

Hypothesize a relationship between the life cycle of a successful product and its cumulative profits over its lifetime.

Solution

Once a product is conceived, it must be developed and brought to market. This process involves research and development and perhaps investment in new machinery to manufacture the product, all of which take time and cost money. These investments yield no immediate return and result in initial losses or negative profits. Once the product is brought to market and begins to sell, it makes money, but because it takes time for a company to recoup initial costs, there is an interval during which cumulative profits remain negative but move towards the break-even point. Past the break-even point, a successful product generates positive cumulative profits.

All products eventually lose their customer base as they are replaced by other products from competitors or from the same company. At some point, the original product becomes a drain on company profits and will lose money if kept on the market. Thus, we expect the relationship between time and cumulative profit to be given by Figure 1.22. The time

FIGURE 1.22

required to bring a product to market, the initial losses that are sustained, and the profits eventually realized are all product-dependent and will vary from company to company and product to product, so we cannot label the tick marks on the axes. Nevertheless, we might expect all successful products to follow the general trends exhibited by Figure 1.22.

Figure 1.22 is a combination of two interesting shapes. Between the time a product is conceived and the time it becomes profitable, the graph in Figure 1.22 has the general shape of the letter *U*. Curves that resemble the letter *U* are frequently called *U-shaped curves*. From the time the product becomes profitable through the remainder of its life cycle, the graph in Figure 1.22 has the general shape of an inverted *U*. Curves that resemble the letter *U* turned upside down are frequently called *inverted U-shaped curves*. Although the terms *U*-shaped and inverted *U*-shaped are not mathematical names, they are used frequently in business and the social sciences.

IMPROVING SKILLS

1. A graph is to contain points with vertical coordinates ranging from just over one million to just under five million. Determine a reasonable scale for the tick marks on the vertical axis of a window that will display all the points of this graph.
2. A graph is to contain points with vertical coordinates ranging from 8,250 to 99,170. Determine a reasonable scale for the tick marks on the vertical axis of a window that will display all the points of this graph.
3. A graph is to contain points with horizontal coordinates ranging between 5,900 and 6,400. Determine a reasonable scale for the tick marks on the horizontal axis of a window that will display all the points of this graph.
4. Create a window for graphing data with horizontal coordinates ranging between 0 and 1,000 and vertical coordinates ranging between 0 and 1 million. Plot in this window the points (*a*) (5, 1000), (*b*) (200, 200), (*c*) (500, 500), (*d*) (700, 200,000).
5. Create a window for graphing data with horizontal coordinates ranging between 130 and 150 and vertical coordinates ranging between 29 and 31. Plot in this window the points (*a*) (130, 30.5), (*b*) (135, 30.2), (*c*) (140, 30), (*d*) (148, 29).
6. Create a window for graphing data with horizontal coordinates ranging between 5,000 and 9,000 and vertical coordinates ranging between −3,000 and −2,000. Plot in this window the points (*a*) (5,000, −2,995), (*b*) (5,010, −3,000) (*c*) (5,500, −2,500), (*d*) (5,525, −2,490).
7. A window for a graph is to contain points with vertical coordinates ranging between zero and 18 million dollars. How high must the window be if tick marks are placed a quarter inch apart and if the distance between successive tick marks represents $1,000?
8. A window for a graph is to contain points with horizontal coordinates for population ranging between 0 and 400,000. How long must the window be if tick marks are placed an eighth of an inch apart and if the distance between successive tick marks represents 5,000 people?
9. A window for a graph is to contain points with vertical coordinates ranging between 10 million and 12 million dollars. How high must the window be if tick marks are placed a half inch apart and if the distance between successive tick marks represents $200,000?
10. A window for a graph is to contain points with horizontal coordinates ranging between 10,000 and 90,000 units. How wide must the window be if tick marks are placed a

quarter inch apart and if the distance between successive tick marks represents 500 units?

11. (a) What is the second coordinate of every point on the x-axis? (b) What is the first coordinate of every point on the y-axis?

12. Construct a cartesian coordinate system and draw a line parallel to the x-axis. What do all points on this line have in common?

13. Construct a cartesian coordinate system and draw a line parallel to the y-axis. What do all points on this line have in common?

In Problems 14 through 29, plot the graphs of the given equations.

14. $y = 2x$
15. $y = -2x$
16. $y = x^2/2$
17. $y = -x^2/2$
18. $y = 2x^2$
19. $y = x^2 + 1$
20. $y = x^2 - 2$
21. $2x - 3y = 5$
22. $N = t^2 - 3$, with t on the horizontal axis
23. $N = t^2 - 3$, with N on the horizontal axis
24. $p = d^3 - 2d^2 + d$, with d on the horizontal axis
25. $y = \sqrt{x}$
26. $y = 1/x$
27. $S = 1/E^2$, with E on the horizontal axis
28. $x^2 + y^2 = 4$
29. $x^2 - y^2 + 4 = 0$

30. Figure A is a graph of the life expectancy of U.S. citizens born between 1975 and 1989, based on data provided by the U.S. National Center for Health Statistics. Estimate (a) the life expectancy of someone born in 1977, (b) the life expectancy of someone born in 1985, (c) the date when life expectancy rose to 74 years, and (d) the date life expectancy rose to 75 years.

31. What is the general trend of the curve in Figure A?

FIGURE A

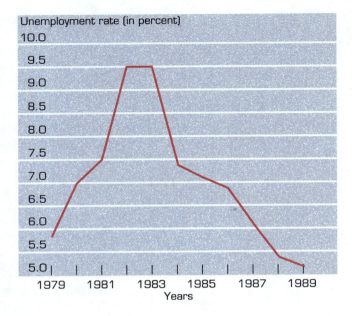

32. Figure B is a graph of the unemployment rate in the United States between 1979 and 1989 as reported by the U.S. Bureau of Labor Statistics. Estimate (*a*) the unemployment rate in 1983, (*b*) the unemployment rate in 1988, (*c*) the dates when the unemployment rate reached 8%, and (*d*) the dates of maximum unemployment.
33. What is the general trend of the curve in Figure B?
34. Figure C is a graph of mortage rates for new homes in the United States between 1980 and 1990, as reported by the Board of Governors of the Federal Reserve System. Estimate (*a*) the date when mortgage rates reached 12%, (*b*) the dates when mortgage rates reached 15%, (*c*) the mortgage rate on January 1, 1983, and (*d*) the mortgage rate on July 1, 1987.

FIGURE C

FIGURE D

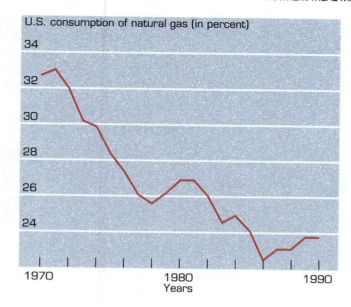

U.S. consumption of natural gas (in percent)

35. Figure D, which is based on data provided by the U.S. Energy Information Admin-istration shows U.S. consumption of natural gas between 1970 and 1990 as a per-centage of the total energy consumption. Estimate (*a*) the consumption in 1973, (*b*) the consumption in 1985, (*c*) when consumption reached its maximum percentage during this period, and (*d*) when consumption reached its minimum percentage.

36. Table 1.10 contains data provided by the U.S. Bureau of Economic Analysis on the Gross National Product (GNP) of the United States. The GNP is reported in billions of dollars, adjusted to constant 1982 dollars to account for the effects of inflation. (*a*) Plot this data on a graph and connect the points with an appropriate curve. (*b*) Determine the general trend of the curve.

TABLE 1.10

Year	1940	1950	1960	1970	1980
GNP	100.4	288.3	515.3	1,015.5	2,732.0

37. The data in Table 1.11 was compiled by the U.S. National Aeronautics and Space Administration and shows the percentage of the Federal Space Program budget allo-cated to NASA.

TABLE 1.11

Year	1960	1965	1970	1975	1980	1985
NASA's Percent	43.3	73.9	66.4	59.3	53.9	34.3

(*a*) Plot the data from 1965 through 1980 (omitting 1960 and 1985) on a graph and connect the points with an appropriate curve. (*b*) Determine the general trend of the curve. (*c*) Add to the graph the data from 1960 and 1985 and determine the general trend of the expanded curve.

38. Table 1.12 shows data reported by the U.S. Department of Commerce on the total horsepower capacity (in millions) of all prime movers in the United States and the percentage attributed to motor vehicles. (*a*) Plot the data on horsepower capacity against time (i.e., horsepower on the vertical axis and years on the horizontal axis) and connect the points with an appropriate curve. (*b*) Determine the general trend of the curve.

TABLE 1.12

YEAR	HORSEPOWER (MILLIONS)	PERCENT IN MOTOR VEHICLES	YEAR	HORSEPOWER (MILLIONS)	PERCENT IN MOTOR VEHICLES
1975	25,100	94.6	1983	31,337	94.7
1976	25,732	94.6	1984	31,819	94.7
1977	26,469	94.5	1985	32,529	94.7
1978	27,379	94.6	1986	32,660	94.6
1979	28,162	94.5	1987	33,266	94.7
1980	28,922	94.6	1988	34,200	94.8
1981	29,507	94.6	1989	36,570	95.0
1982	30,495	94.6			

39. Using the data in Table 1.12, plot the percentage in motor vehicles against time (i.e., percentage on the vertical axis and years on the horizontal axis) in a window whose vertical axis is scaled between 94.4% and 95.0% in units of one-tenth of a percent between successive tick marks. Determine the general trend of the curve.

40. Redo Problem 39, changing the scale on the vertical axis to range between 90% and 100% in units of a whole percent between successive tick marks.

41. Table 1.13 contains data compiled by the U.S. Weather Service on the annual precipitation in Baltimore from 1940 through 1955. (*a*) Plot this data on a graph and connect the points with an appropriate curve. (*b*) Determine the general trend of the curve.

TABLE 1.13

Year	1940	1941	1942	1943	1944	1945	1946	1947	1948	1949	1950	1951	1952	1953	1954	1955
Rainfall (inches)	44.3	34.7	46.0	36.8	45.5	46.6	37.6	46.2	54.7	37.7	44.0	46.9	55.9	49.3	30.5	47.9

CREATING MODELS

42. Construct a qualitative graphical model that relates the percentage of homes in the United States that have at least one computer (vertical axis) to time (horizontal axis) over the last 35 years.

43. Construct a qualitative graphical model that relates the profits of a large department store (vertical axis) to the months of the year (horizontal axis).

44. Construct a qualitative graphical model that relates the tolls collected (vertical axis) on a bridge into a major metropolitan city to the hours of the day (horizontal axis). Assume that the toll is the same in both directions.

45. Construct a qualitative graphical model that relates world population (vertical axis) to years (horizontal axis) from 4000 B.C. to the present.

46. Construct a qualitative graphical model that relates profits from a product (vertical axis) to the number of products produced each year (horizontal axis).

47. Construct a qualitative graphical model that relates the number of violators of a law (vertical axis) to the severity of punishment (horizontal axis).

48. Construct a qualitative graphical model that relates the *total* number of cases of chicken pox reported in a city (vertical axis) to time (horizontal axis) over the period from late winter to mid-spring, when the disease is most prevalent.

49. Redo Problem 48 making the vertical axis the number of *new* cases.

50. Every Monday, a gasoline station has its reserves of unleaded regular gasoline refilled to maximum capacity by trucks from the gasoline company. Construct a qualitative graphical model that relates the supply of unleaded regular gasoline at the station (vertical axis) to time (horizontal axis).

51. Construct a qualitative graphical model that relates temperature on a typical summer day (vertical axis) to time (horizontal axis) over a 24-hour period.

EXPLORING IN TEAMS

52. Match each of the following descriptions with *one* of the graphs displayed in Figures E through K. (*a*) An airplane flies from Phoenix to Houston. (*b*) An airplane flies from Houston to Phoenix. (*c*) An airplane flies from Houston to Phoenix, remains at Phoenix overnight, and then flies back to Houston. (*d*) An airplane circles the airfield at Houston. (*e*) An airplane circles the airfield at Houston and then lands.

FIGURE E

FIGURE F

FIGURE G

FIGURE H

FIGURE I

FIGURE J

FIGURE K

EXPLORING
WITH
TECHNOLOGY

53. Using a graphing calculator, graph the following two equations on the same coordinate system.

$$6x - y = -3$$
$$5x - y = 2$$

Using the cursor on the calculator, identify the point of intersection. Substitute the coordinates of this point into the two equations. What conclusions do you make?

54. Repeat the procedure described in Problem 53 for the following equations, but now use the zoom-in feature of your calculator to estimate your answer to two decimal places.

$$y = 5x^2 - 2$$
$$y = x + 3$$

55. Repeat the procedure described in Problem 53 for the following two equations:

$$y = x^2 + 1$$
$$y = -x^2$$

What can you say about solutions to this set of equations?

RECOMMENDING
ACTION

56. Respond by memo to the following request:

MEMORANDUM

To: J. Doe Reader

From: Head of Planning and Budget

Date: Today

Subject: **Figure 1.22**

I am confused about Figure 1.22, which relates cumulative profit to time. I do not understand how one can say the a product *loses* money just when the graph reaches its highest point. Isn't that when a product makes the most money? It seems to me the graph is wrong. Should I have the graphics department redraw the graph, or am I missing something?

1.5 LINEAR EQUATIONS AND STRAIGHT LINES

The graphing process described in Section 1.4 can be shortened considerably if we already know the shape of the curve. This is the case whenever we deal with a *linear equation*. An equation is *linear* in two variables x and y if it can be written in the form

$$Ax + By = C \qquad (15)$$

where A, B, and C are known numbers with A and B not both zero (to avoid equations of the form $0 = C$). The variables x and y can be replaced by any other convenient letters. In particular, if x and y are replace by m and n, respectively, then Equation (15) becomes the linear equation $Am + Bn = C$ in the variables m and n.

EXAMPLE 1

Show that each of the following equations is linear:

(a) $-x + 7y = 0$ (b) $y = 0$
(c) $2x + 5y + 4 = 8y$ (d) $\frac{1}{2}N - \frac{3}{4}P = 1.7$

Solution

(a) This equation has the form $Ax + By = C$ with $A = -1$, $B = 7$, and $C = 0$.
(b) This equation has the form $Ax + By = C$ with $A = C = 0$ and $B = 1$.
(c) This equation can be rewritten as $2x - 3y = -4$ which has the form $Ax + By = C$ with $A = 2$, $B = -3$, and $C = -4$.
(d) This is a linear equation in the variables N and P. It has the form of Equation (15) with N and P replacing x and y, respectively. Here, $A = \frac{1}{2}$, $B = -\frac{3}{4}$, and $C = 1.7$.

EXAMPLE 2

Determine whether the equation $x^2 + y^2 = 4$ is linear.

Solution

No, it is not linear. Here the variables x and y are squared, whereas Equation (15) requires x and y to appear only to the first power multiplied only by known numbers. No algebraic manipulations will transform the given equation into the form specified by Equation (*15*).

EXAMPLE 3

Determine whether the equation $1/x + 2y = 0$ is linear.

Solution

Here x appears in the denominator, which is not the form required by Equation (15). We can rewrite the equation as

$$x^{-1} + 2y = 0$$

but now the variable x is raised to the -1 power, not the first power, as required. Alternatively, we can multiply the original equation by x, obtaining

$$1 + 2xy = 0$$

but again this is not in the form of Equation (*15*). Now both variables are raised to the first power, as required, but they multiply each other, which is not allowed. Only a known number, not another variable, can multiply a variable in a linear equation. No algebraic manipulations will transform the given equation into the form of Equation (15), so the equation is not linear.

The graph of a linear equation in two variables is a straight line.

The graph of a linear equation in two variables is a straight line. A straight line is uniquely determined by two distinct points, so to graph a linear equation in two variables, plot *two* solutions to the equation and then draw the straight line through those points. Points with one coordinate set to zero, that is, points of the form $(0, b)$, or $(a, 0)$, are the points where the line crosses each of the coordinate axes, and these points are often the easiest to locate. The point where the line crosses the horizontal axis is the *horizontal intercept* or, in an x-y coordinate system, the x-intercept; the point where the line crosses the vertical axis is the *vertical intercept* or, in an x-y coordinate system, the y-intercept.

To check for errors, it is good practice to plot a third point when graphing a linear equation. If the third point does not fall on the straight line defined by the first two points, then at least one of the three points has been identified or plotted incorrectly.

EXAMPLE 4

Graph the linear equation $p + 100q = 1,500$ with q on the horizontal axis.

Solution

Setting $p = 0$ and solving for q, we obtain

$$0 + 100q = 1,500$$

or $q = 15$. The q-intercept is located at $(15, 0)$. Setting $q = 0$ and solving for p, we obtain

$$p + 100(0) = 1,500$$

or $p = 1{,}500$. The p-intercept is located at $(0, 1{,}500)$. To locate a third point, we set $q = 10$, because it is similar in magnitude to the q-values of the two intercepts. Solving for p we obtain

$$p + 100(10) = 1{,}500$$

or $p = 500$, thus a third point on the graph is $(10, 500)$. These three solutions are plotted in Figure 1.23 along with the straight line that contains them; this line is the graph of the given equation.

FIGURE 1.23

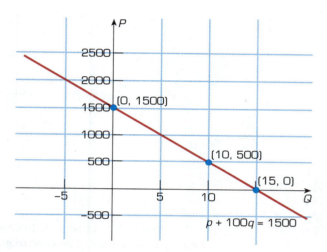

EXAMPLE 5

Graph the equation $\frac{1}{2}N - \frac{3}{4}P = 1.7$ with N on the horizontal axis.

Solution

Replacing each fraction by its decimal equivalent, we rewrite the linear equation (see part (d) of Example 1) as

$$0.5N - 0.75P = 1.7$$

Setting $N = 0$, we have

$$0.5(0) - 0.75P = 1.7$$

or $P = 1.7/(-0.75) \approx -2.27$, rounded to two decimals for graphing purposes. The point $(0, -2.27)$ is the vertical intercept or, in this coordinate system, the P-intercept. Setting $P = 0$, we obtain

$$0.5N - 0.75(0) = 1.7$$

or $N = 1.7/(0.5) = 3.4$. The point $(3.4, 0)$ is the horizontal intercept or, in this coordinate system, the N-intercept. To identify a third point on the graph, we arbitrarily set $N = 1$, because it is similar in magnitude to N-values of the two intercepts. Solving for P, we obtain

$$0.5(1) - 0.75P = 1.7$$

$$-0.75P = 1.2$$

or $P = 1.2/(-0.75) = -1.6$. Thus $(1, -1.6)$ is a third point on the graph. These three solutions are plotted in Figure 1.24 along with the straight line that contains them; this line is the graph of the given equation.

FIGURE 1.24

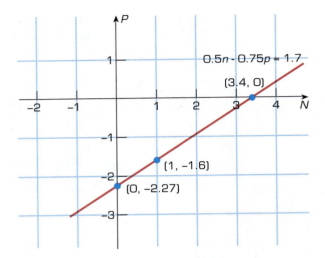

Sometimes it is not possible to construct the straight-line graph of a linear equation by locating both its vertical and horizontal intercepts. These situations occur when one of the coefficients A, B, or C in Equation (15) is zero. If $A = 0$, Equation (15) reduces to $0x + By = C$, $By = C$, or more simply

$$y = k \qquad (16)$$

where $k = C/B$. The graph of Equation (16) is a horizontal straight line parallel to the x-axis and consisting of all points having a y-coordinate equal to k. Graphs of Equation (16) for $k = 2$ and $k = -1.5$ are displayed in Figure 1.25. When k is positive, the corre-

FIGURE 1.25

FIGURE 1.26

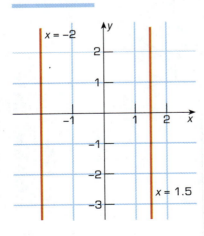

sponding horizontal line is above the x-axis; when k is negative, the corresponding horizontal line is below the x-axis. Such lines have no horizontal intercepts. If $k = 0$, the line is the x-axis.

If $B = 0$, Equation (15) reduces to $Ax + 0y = C$, $Ax = C$, or more simply

$$x = h \tag{17}$$

where $h = C/A$. The graph of Equation (17) is a vertical straight line parallel to the y-axis and consists of all points having an x-coordinate equal to h. Graphs of Equation (*17*) for $h = 1.5$ and $h = -2$ are displayed in Figure 1.26. When h is positive, the corresponding vertical line is to the right of the y-axis; when h is negative, the corresponding vertical line is to the left of the y-axis. Such lines have no vertical intercepts. When $h = 0$, the line is the y-axis.

If $C = 0$, Equation (15) reduces to $Ax + By = 0$, $By = -Ax$, or more simply

$$y = mx \tag{18}$$

where $m = -A/B$. Setting $y = 0$, we find $x = 0$, and vice-versa, so both intercepts are the origin. To locate a second and a third point, we must select *nonzero* values for either x or y and then solve the linear equation for the other variable. Graphs of Equation (18) for selected values of m are displayed in Figure 1.27.

FIGURE 1.27

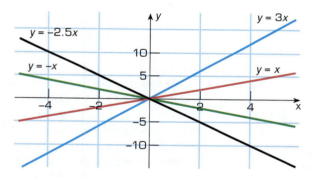

A defining characteristic of a line is its steepness, which is often measured by its slope. Let $P_1 = (x_1, y_1)$ and $P_2 = (x_2, y_2)$ denote two distinct points on the same straight line (or, alternatively, satisfying the same linear equation in two variables). The *run* from P_1 to P_2 is the difference $x_2 - x_1$, the *rise* from P_1 to P_2 is the difference $y_2 - y_1$, as illustrated

FIGURE 1.28

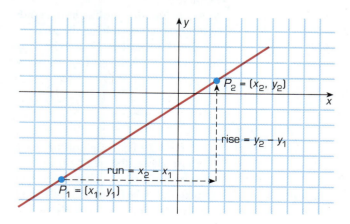

in Figure 1.28. If the run is *not* zero, then the *slope* of the line through these two points, denoted by the letter m, is the quotient of the rise over the run, that is,

$$m = \frac{\text{rise}}{\text{run}} = \frac{y_2 - y_1}{x_2 - x_1} \tag{19}$$

providing $x_1 \neq x_2$. If the run is zero, then the slope is undefined.

EXAMPLE 6

> *The slope of a line through two points is the quotient of the rise over the run between those points.*

Find the slope of a straight line that contains the two points (1, 2) and (3, 5).

Solution

If we take the first point to be (1, 2), so that $x_1 = 1$ and $y_1 = 2$, and the second point to be (3, 5), so that $x_2 = 3$ and $y_2 = 5$, then

$$m = \frac{y_2 - y_1}{x_2 - x_1} = \frac{5 - 2}{3 - 1} = \frac{3}{2} = 1.5$$

If, instead, we take the first point to be (3, 5), so that $x_1 = 3$ and $y_1 = 5$, and the second point as (1, 2), so that $x_2 = 1$ and $y_2 = 2$, then

$$m = \frac{y_2 - y_1}{x_2 - x_1} = \frac{2 - 5}{1 - 3} = \frac{-3}{-2} = 1.5$$

Either way, the slope is 1.5.

We can multiply both the numerator and the denominator of Equation (19) by -1 without changing the equality. Since this multiplication has the effect of reversing the order of the two points in the formula, we see that when calculating slope it does not matter which point on a straight line is designated as the first point and which as the second point. Furthermore, x and y in Equation (15) can be replaced by any two letters of convenience, as long as the variables in the numerator refer to the vertical coordinates and the variables in the denominator refer to the horizontal coordinates of their respective points.

EXAMPLE 7

Find the slope of the straight line defined by the linear equation $v + 200t = 1,500$ in a coordinate system with t on the horizontal axis.

Solution

Points on the straight line have coordinates (t, v). To calculate slope, we need two points. Setting $t = 0$ and solving the given equation for v, we obtain $v = 1,500$, so one point on the line is located at (0, 1,500). Setting $v = 0$ and solving the given equation for t, we find that $t = 7.5$, hence (7.5, 0) is a second point on the line. The slope of this line is

$$m = \frac{v_2 - v_1}{t_2 - t_1} = \frac{0 - 1,500}{7.5 - 0} = \frac{-1,500}{7.5} = -200$$

Figure 1.29 is a graph of the linear equation $v + 200t = 1,500$ discussed in Example 7; it also shows the two points $P_1 = (0, 1,500)$ and $P_2 = (7.5, 0)$ used to calculate the slope of the line. To move from (0, 1,500) to (7.5, 0), one must travel a distance of 7.5 units to the right from (0, 1500) and then down a distance of 1,500 units, as indicated by the arrows. The rise from P_1 to P_2 is $0 - 1,500 = -1,500$, where the negative sign

FIGURE 1.29

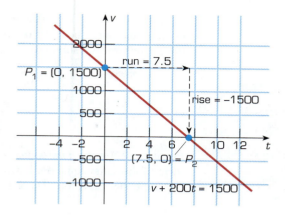

signifies movement in negative or downward direction on the graph. The run from P_1 to P_2 is $7.5 - 0 = 7.5$, and the ratio of rise to run is the slope.

A nonvertical straight line has a nonzero run between any two points on the line. Slopes are defined for such lines, and the slope of any line is independent of which two points on the line are used to calculate slope. That is, the slope of a line is a constant. We know from Equation (19) that

$$\frac{\text{rise}}{\text{run}} = m \qquad (20)$$

The rise in a nonvertical straight line is directly proportional to the run with a constant of proportionality given by the slope.

From this relationship we conclude that the rise is directly proportional to the run with a constant of proportionality given by the slope. Starting from any point on a nonvertical line, the slope is the change in the vertical direction needed to remain on the line for each unit change in the positive horizontal direction.

Figure 1.30 shows the line through the two points defined in Example 6. The slope of the line is 1.5, so if we start at any point on the line and move 1 unit in the positive horizontal direction, we must also move 1.5 units in the positive vertical direction if we wish to remain on the line. The slope of the line in Figure 1.29 is -200. If we start at any point on that line and move 1 unit in the positive horizontal direction, we must also move -200 units in the positive vertical direction (or 200 units in the negative vertical direction) if we wish to remain on the line.

FIGURE 1.30

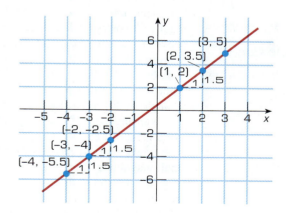

Let $P_1 = (x_1, y_1)$ be a fixed point on a straight line and let $P_2 = (x, y)$ denote any other point on the line. The rise from P_1 to P_2 is $y - y_1$, the run is $x - x_1$, and as long as the run is nonzero, Equation (20) becomes

$$\frac{y - y_1}{x - x_1} = m \qquad (21)$$

or

$$y - y_1 = m(x - x_1) \qquad (22)$$

Equation (22) is the *point-slope equation* for a straight line. Equation (21) is an alternative form of the same equation. If we know the slope of a line and one point on the line, we can use either Equation (21) or Equation (22) to obtain the equation of the line.

EXAMPLE 8

Find the equation of the line passing through the point $(-8, 70)$ with slope -5.

Solution
Using Equation (22) with $x_1 = -8$, $y_1 = 70$, and $m = -5$, we have

$$y - 70 = -5\,(x - (-8)) \qquad \text{or} \qquad y - 70 = -5(x + 8)$$

which can be simplified to $5x + y = 30$.

EXAMPLE 9

Find the equation of the line that passes through the two points $(1, 2)$ and $(3, 5)$.

Solution
To use Equation (22), we must have a point on the line and the slope of the line. We know from Example 6 that the slope is $m = 1.5$. With $(x_1, y_1) = (1, 2)$, the point-slope equation becomes

$$y - 2 = 1.5(x - 1) \qquad \text{or} \qquad 1.5x - y = -0.5$$

Had we taken instead $(x_1, y_1) = (3, 5)$, the point-slope equation would be

$$y - 5 = 1.5(x - 3) \qquad \text{or, again} \qquad 1.5x - y = -0.5$$

The slope is the quotient of the rise to the run. If a line has a *positive* slope, then a positive run requires a corresponding positive rise. Such a line slants upward to the right. Thus, a line has positive slope if the angle between the line and the positively directed horizontal axis is betwen 0° and 90°. Similarly, if a line has *negative* slope, a positive run requires a corresponding negative rise. Therefore, a line has negative slope if the angle between the line and positively directed horizontal axis is between 90° and 180° (see Figures 1.31 and 1.32).

A line with a large positive slope goes up more steeply, moving from left to right, than a line with a smaller positive slope. In particular, a line intersecting the horizontal axis at an angle of 30° has a slope of $m \approx 0.577$, while a steeper line intersecting the horizontal axis at an angle of 45° has a slope of $m = 1$. An even steeper line intersecting the horizontal axis an angle of 80° has a slope of $m \approx 5.67$. A line with negative slope of large magnitude goes down more steeply, moving from left to right, than a line with a negative slope of small magnitude.

A line parallel to the x-axis has zero slope, because any two points on such a line have the same y-coordinates and thus yield a zero numerator in Equation (19). A line parallel to the y-axis does not have a slope.

FIGURE 1.31

FIGURE 1.32

The general form of the equation of a straight line is

$$Ax + By = C \qquad \text{(15 repeated)}$$

If B is not zero, which means the graph of the equation is not a vertical line, we can rewrite the equation as

$$By = -Ax + C$$

or

$$y = (-A/B)x + (C/B)$$

Setting $m = -A/B$ and $b = C/B$, we have

$$y = mx + b \qquad (23)$$

which is the *slope-intercept equation* of a straight line.

The slope and vertical intercept of a line can be read directly from the slope-intercept equation of that line.

An advantage to the slope-intercept equation is that the slope and y-intercept can be read directly from the equation. The slope is m, the coefficient of the x-term and the y-intercept is at $(0, b)$. Both facts are easy to verify. If (x_1, y_1) and (x_2, y_2) are any two points on the straight line graph of Equation (23), then those points must satisfy the equation. Therefore, $y_1 = mx_1 + b$ and also $y_2 = mx_2 + b$. It now follows that

$$\text{slope} = \frac{y_2 - y_1}{x_2 - x_1} = \frac{(mx_2 + b) - (mx_1 + b)}{x_2 - x_1}$$

$$= \frac{mx_2 - mx_1}{x_2 - x_1} = \frac{m(x_2 - x_1)}{x_2 - x_1}$$

$$= m$$

The y-intercept is the point where the line crosses the y-axis. At such a point, $x = 0$. Setting $x = 0$ in (23), we have $y = b$; the y-intercept is located at $(0, b)$.

EXAMPLE 10

Find the slope and the y-intercept of the line $3x + 4y = 10$.

Solution

Solving the equation for y, we obtain the slope-intercept form of the equation as

$$y = -0.75x + 2.5$$

with $m = -0.75$ and $b = 2.5$. The slope is -0.75 and the y-intercept is located at $(0, 2.5)$.

EXAMPLE 11

Find the slope of the straight line defined by the linear equation $v + 200t = 1{,}500$ in a coordinate system with t on the horizontal axis.

Solution

Solving the equation for v, the variable on the vertical coordinate axis, we obtain the slope-intercept form of the equation as

$$v = -200t + 1{,}500$$

with $m = -200$ and $b = 1{,}500$. The slope is -200. Compare this solution procedure with one demonstrated in Example 7.

Any nonvertical line passing through the origin has its y-intercept at $(0, 0)$. Thus, $b = 0$ and Equation (23) reduces to

$$y = mx \qquad \text{(18 repeated)}$$

Two quantities are directly proportional if and only if their graph is a straight line through the origin.

Recall from Section 1.3 that two quantites (variables) A and B are directly proportional if there exists a constant k such that

$$A = kB \qquad \text{(8 repeated)}$$

The resemblance between Equation (8) and Equation (18) is striking! We conclude that two quantites are directly proportional if and only if their graph is a straight line through the origin. The slope of that line is the constant of proportionality.

EXAMPLE 12

Table 1.14 is the result of years of data collecting by the American Citrus Corporation. Find an equation relating the number of orange trees that bear fruit, F, to the number of orange trees planted, N.

Solution

These four data points are plotted in Figure 1.33, and they fall on a straight line. Therefore, the variables F and N are related by a linear equation. Two points on the line are given by the first two points in Table 1.14 as $N_1 = 120$, $F_1 = 114$ and $N_2 = 140$, $F_2 = 133$. The slope of the line is

$$m = \frac{F_2 - F_1}{N_2 - N_1} = \frac{133 - 114}{140 - 120} = \frac{19}{20} = 0.95$$

The units of m are the units of F divided by the units of N, or trees that bear fruit per tree planted. Using the point-slope equation for a line with the letters F and N replacing y and x, respectively, we obtain

$$F - 114 = 0.95(N - 120)$$

or more simply $F = 0.95N$. This is the equation of a straight line through the origin, even though this fact is not apparent from the window displayed in Figure 1.33. It follows that

TABLE 1.14

Number of trees planted	120	140	160	180
Number that bear fruit	114	133	152	171

FIGURE 1.33

Two lines are parallel if they have the same slope; two lines are perpendicular if their slopes are negative reciprocals of one another.

the number of orange trees that bear fruit is directly proportional to the number of orange trees planted with a constant of proportionality equal to

$$0.95 = \frac{\text{trees that bear fruit}}{\text{trees planted}}$$

Thus, 95% of all orange trees planted will bear fruit.

Two lines are parallel if they have the same slope. Two lines are perpendicular if their slopes are negative reciprocals of one another. That is, if one line has slope m, then all lines parallel to it will have slope m and all lines perpendicular to it will have slope equal to $-1/m$, providing $m \neq 0$.

EXAMPLE 13

Find the equation of the line that passes through the point $(2, -500)$ and is (a) parallel to the line defined by the linear equation $6x + 3y = 450$, and (b) perpendicular to the line defined by $6x + 3y = 450$.

Solution

Solving the equation for $6x + 3y = 450$ for y, we generate the slope-intercept equation

$$y = -2x + 150$$

The slope of the line defined by this equation is the coefficient of the x term, namely $m = -2$.

(a) A line parallel to the graph of $6x + 3y = 450$ has the same slope, $m = -2$. Using the point-slope equation, we determine the equation of the parallel line passing through $(2, -500)$ as

$$y - (-500) = -2(x - 2)$$

or

$$y = -2x - 496.$$

(b) A line perpendicular to the graph of $6x + 3y = 450$ has slope $m = 1/2$, which is the negative reciprocal of $m = -2$. Using the point-slope equation, we determine the equation of the perpendicular line passing through $(2, -500)$ as

$$y - (-500) = \tfrac{1}{2}(x - 2)$$

or, $y = \tfrac{1}{2}x - 501$.

IMPROVING SKILLS

1. Determine which of the following equations are linear:
   (a) $2x = y$ (b) $2x = 1/y$ (c) $xy = 4$ (d) $x = 4$
   (e) $2x - 3y = 0$ (f) $y = 4x$ (g) $y = 4x^2$ (h) $x - 2 = 3y$
   (i) $1/x + 1/y = 2$ (j) $x = y$

In Problems 2 through 13, graph the given equations placing the first variable that appears in each equation on the horizontal axis.

2. $2x + 3y = 6$ 8. $x = y$
3. $-2x + 3y = 6$ 9. $-x + 200y = 5,000$
4. $2x - 3y = 6$ 10. $-N + 2M = 10$
5. $2x + 3y = -6$ 11. $20N - 150M = -80,000$
6. $3x + 2y = 6$ 12. $P = 1 + 2Q$
7. $10x - 5y = 50$ 13. $5r - 2s = 0$

In Problems 14 through 23, graph the equations on traditional x-y coordinate systems.

14. $x = 7$ 19. $y = -4000$
15. $x = -7$ 20. $y = x/2$
16. $x = 1.5$ 21. $y = 2x$
17. $y = 4$ 22. $y = -2x$
18. $y = 400$ 23. $y = -4x$

In Problems 24 through 33, find the slopes of the straight lines passing through the given pairs of points:

24. $(1, 2)$ and $(2, 5)$ 29. $(5, 50)$ and $(-1, -130)$
25. $(7, -3)$ and $(-1, -8)$ 30. $(1, 1000)$ and $(2, -1000)$
26. $(-1, 2)$ and $(4, 2)$ 31. $(400, 400)$ and $(190, 800)$
27. $(1, 0)$ and $(0, 1)$ 32. $(100, 200)$ and $(190, 200)$
28. $(2, -1)$ and $(2, 4)$ 33. $(100, 200)$ and $(190, 400)$

In Problems 34 through 51, find the slopes and the vertical intercepts for the straight lines defined by each equation.

34. $2x + 3y = 6$ 40. $x = y$
35. $-2x + 3y = 6$ 41. $-x + 200y = 5000$
36. $2x - 3y = 6$ 42. $x = 7$
37. $2x + 3y = -6$ 43. $y = 7$
38. $3x + 2y = 6$ 44. $x = -7$
39. $10x - 5y = 50$ 45. $y = -7$
46. $-p + 2q = 10$, with p on the horizontal axis
47. $20y - 150t = -80,000$, with t on the horizontal axis

48. $x = 1 + 2t$, with t on the horizontal axis
49. $5y - 2x = 0$
50. $r = s/2$, with s on the horizontal axis
51. $y = 2x$
52. Which of the equations given in Problems 34 through 51 model proportionality relationships?

In Problems 53 through 62, find the equation of the straight line that passes through the given point and has the given slope.

53. $(1, 0)$, $m = 5$
54. $(1, 0)$, $m = 3$
55. $(0, 1)$, $m = 3$
56. $(1, 0)$, $m = -3$
57. $(0, 1)$, $m = -3$
58. $(0, 1)$, $m = 0$
59. $(-1, 2)$, $m = 2$
60. $(90, 40)$, $m = -1$
61. $(8, 8)$, $m = 40$
62. $(8, 6{,}000)$, $m = 15$

In Problems 62 through 72, find the equation of the straight line that passes through the given pairs of points.

63. $(1, 0)$ and $(4, 1)$
64. $(-1, -2)$ and $(1, 2)$
65. $(100, 0)$ and $(0, 200)$
66. $(2, 5)$ and $(8, 5)$
67. $(700, 1)$ and $(700, 10)$
68. $(700, 700)$ and $(500, 500)$
69. $(-1, 2)$ and $(3, 2)$
70. $(90, 40)$ and $(40, 90)$
71. $(8, 8)$ and $(8, 20)$
72. $(8, 8)$ and $(19, 8)$

73. Determine which of the following equations have graphs that are parallel to the graph of $y = 3x + 7$, which have graphs perpendicular to it, and which have graphs that are neither perpendicular nor parallel.
    (a) $y + 3x = 8$ (b) $3y + x = 8$ (c) $2y - 6x = 8$
    (d) $6x - 2y = 8$ (e) $2x = -6y + 1$ (f) $2x = 6y + 1$

74. Find the equation of a line that passes through $(1, 1)$ and is perpendicular to the line defined by the equation $3x + 2y = 0$.
75. Find the equation of a line that passes through $(20, 70)$ and is perpendicular to the line defined by the equation $3x + 2y = 0$.
76. Find the equation of the line that passes through $(10, 20)$ and is parallel to the line defined by the equation $2x + 4y = -900$.
77. Find the equation of the line that passes through $(-50, 200)$ and is parallel to the line defined by the equation $2x + 4y = -900$.
78. Find the equation of the line that passes through $(-1, 50)$ and is perpendicular to the line defined by the equation $y = 4$.
79. Find the equation of the line that passes through $(100, 200)$ and is parallel to the line defined by the equation $x = 40$.
80. Sales data from a large appliance center indicates that n, the number of clock radios of a particular model sold on any given day, is governed by the equation

$$n = 182 - 7p$$

where p represents the price of each clock radio (in dollars). Graph this equation. Use your graph to determine (a) the number of clock radios sold on a day when they are priced at \$20 each, (b) the price on a day when ten clock radios were sold, and (c) the physical significance of the p-intercept.

81. It has been determined by the Chubby Cat Food Corporation that A, the amount of cat food sold on any given day in Newark, New Jersey, is governed by the equation

$$A = 2{,}000 - 50p$$

where p represents the price of each can (in nickels). Graph this equation. Use your graph to determine (a) the number of cans of cat food the company can expect to sell if it prices each can at 50¢, (b) the number of cans the company can sell at $1.00 per can, (c) the physical significance of the p-intercept.

82. A company's annual sales S are related to its advertising expenditures E by the equation

$$S = 2E + 3.5$$

where both S and E are in millions of dollars. Graph this equation. Use your graph to determine (a) the amount of money that must be committed to advertising to realize annual sales of ten million dollars, (b) the annual sales expected with an advertising budget of two million dollars, and (c) the physical significance of the S-intercept.

83. Based on depreciation schedules, the current value V of a building is related to the building's age in years t by the equation

$$V = -21,000t + 630,000$$

Graph this equation. Use your graph to determine (a) the value of the building when it is four years old, (b) the age of the building when it becomes worth exactly $300,000 and (c) the physical significance of the V-intercept.

84. Temperature is traditionally measured in either Celsius (C) or Fahrenheit (F). The two are related by the formula

$$F = \frac{9}{5}C + 32$$

Graph this equation with Celsius on the horizontal axis and Fahrenheit on the vertical axis. Use your graph to determine (a) the temperature in Celsius when it is reported as 90° F, (b) the temperature in Fahrenheit when it is reported as 20° C, (c) the physical meanings of the vertical and horizontal intercepts.

85. A rent-a-car company uses the formula

$$C = 0.12m + 35$$

where m denotes the miles driven, to calculate the charges C (in dollars) to its daily customers. Graph this equation. What are the physical meanings of the horizontal and vertical intercepts?

86. Figure A illustrates the cumulative monthly attendance at a local amusement park for the past year. Find an equation relating attendance A to time t (in months).

FIGURE A

FIGURE B

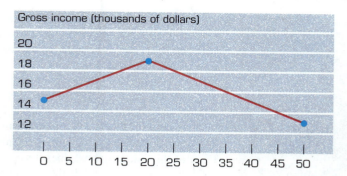

Gross income (thousands of dollars)

87. Figure B illustrates the gross income (*i.e.*, weekly sales receipts) of a supermarket over the past year. Find (*a*) the equation relating gross income *I* to time *t* for the first 20-week period and (*b*) the equation relating *I* to *t* for the last 32 weeks of the year.

CREATING MODELS

88. A rent-a-car company charges $28 per day plus 9¢ a mile for every mile driven. Show that an equation relating total daily charge *C* to *m*, the number of miles driven, is linear.

89. A waitress receives $2 per hour in salary plus 15% of the total charges billed to her customers. Show that her pay *P* is related to *b*, the total charges billed to her customers, by a linear equation.

90. A large television manufacturer has determined that the number of television sets sold annually, denoted by *N*, is related directly to the amount of money spent on advertising. Every million dollars in advertising expenditures results in an additional 50,000 television sets being sold, although 10,000 sets would be sold with no advertising. Let *E* denote the amount of money (in millions of dollars) committed to advertising. Show that an equation relating *N* to *E* is linear.

91. A car dealer determines that *C*, the daily cost of operating a showroom, can be separated into a fixed cost of $630, which includes insurance, rent, lighting, and so on, plus a variable salary cost of $40 per day for each salesperson. Let *n* denote the number of salespeople employed each day (*n* can change daily). Show that an equation relating *C* to *n* is linear.

92. After carefully studying used-car price lists, Mr. Rheza has determined that his particular model car, purchased yesterday for $20,000 will depreciate $1800 per year. Let *V* denote the value of Mr. Rheza's car at any given time and let *t* denote the age of that car. Show that an equation relating *V* to *t* is linear.

93. The White-All Bleach Company believes that the number of bottles of its bleach purchased by people with coupons is directly proportional to the number of coupons distributed through the mail. A recent mailing of 50,000 coupons resulted in 2,000 subsequent purchases with those coupons. Graph the relationship between purchases with coupons, *P*, and the number of coupons mailed *C*. What is the equation of the line?

94. Quality control tests on light bulbs have shown that the number of defective bulbs is directly proportional to the number produced. A sample of 25,000 bulbs was tested, and 75 were found to be defective. Graph the relationship between the number of defective bulbs, *d*, and the number of bulbs produced, *n*. What is the equation of the line?

95. The pressure on deep-sea divers increases linearly with the depth to which they submerge. The pressure at sea level is 15 lb/in^2, and it rises to 27 lb/in^2 at a depth of 20 ft. Graph the relationship between pressure (p) on the vertical axis and depth (d) on the horizontal axis. What is the equation of the line?

96. A salesman's weekly pay increases linearly with the amount of sales he records. One week he booked $62,000 in sales and received a salary of $1,240. Another week he booked $105,000 in sales and received a salary of $2,100. Graph the relationship between his salary (s) on the vertical axis and the amount of sales booked (S) on the horizontal axis. (*a*) What is the equation of the line? (*b*) Where is the vertical intercept of this line and what does it say about his salary?

97. Repeat Problem 96 if the salesman's salary is $970 on sales of $62,000 and $1400 on sales of $105,000.

EXPLORING IN TEAMS

98. The slope of a line is a measure of the line's steepness, yet steepness is a visual concept that can be altered by changes in scale. Consider the graph of the equation $y = 5x$ on a traditional x-y coordinate system in a window that is restricted to the first quadrant and has 11 equally spaced tick marks on each axis, with the first tick mark at the origin. (*a*) Graph this equation if the scale on both the x- and y-axes is one unit per tick mark. (*b*) Graph this equation if the scale on the x-axis is one unit per tick mark but the scale on the y-axis is two units per tick mark. (*c*) Graph this equation if the scale on the x-axis is one unit per tick mark but the scale on the y-axis is five units per tick mark. (*d*) Graph this equation if the scale on the x-axis is five units per tick mark but the scale on the y-axis is one unit per tick mark. Compare each of these graphs, paying particular attention to the "steepness" of each line and the slope of the line. What conclusions do you make?

EXPLORING WITH TECHNOLOGY

99. Graph each of the lines in Problems 34 through 51 on a graphing calculator. Place the cursor arbitrarily on each line and then move the cursor one unit to the right. If the line is not horizontal, the cursor will no longer be on the line. Now move the cursor vertically so that the cursor is once more on the line. How many units did you have to move the cursor in each case? How does this distance relate to slope?

100. (*a*) Set a graphing calculator's viewing window to $[-10, 10]$ on both the horizontal and vertical axes and then graph the equation $y = x$.
   (*b*) Graph $y = 3x$ in the same viewing window and note the larger angle that this line makes with the positive x-axis. Is this expected?
   (*c*) Experiment with different viewing windows, leaving the horizontal range at $[-10, 10]$ but changing the range on the vertical axis, and find a viewing window in which the graph of the equation $y = 3x$ *looks* identical to the graph you got in Part (*a*). Does this mean that the slope of a line changes with the viewing window?

RECOMMENDING ACTION

101. Respond by memo to the following request:

> MEMORANDUM
>
> To: J. Doe Reader
>
> From: President's Office
>
> Date: Today
>
> Subject: **Making Us Look Good**
>
> Sales have been relatively flat this year, $2.17 million compared with $2.14 million last year. Not something to brag about at our stockholders' meeting next week. I'd like to avoid the whole topic, but I am sure the stockholders will not allow me to do so. Is there some truthful way to *present* our profit picture so that it appears better than it is?

1.6 GRAPHING SYSTEMS OF EQUATIONS

A solution to a system of linear equations is a set of numbers, one for each variable, that satisfies all the equations in the system.

It is not likely that a mathematical model of a real-world system will contain only one constraint. The real world is rarely that simple. It is equally unlikely that such a mathematical model will have only one or two variables. Present-day problems routinely involve hundreds of constraints and thousands of variables. A mathematical model for the movement of goods through a Sears or K-Mart distribution warehouse, for example, could have separate variables for each type of radio, each brand of paint, each size of blanket and every other individual item number in the facility!

The next chapter focuses on systems with many constraints and many variables. For now we limit ourselves to systems of linear equations in two variables because we can analyze these systems with graphs. An example of a system with two linear equations is

$$2x + y = 1 \tag{24}$$
$$3x - y = 4$$

A *solution* to such a system of linear equations is a set of numbers, one number for each variable, that satisfy all the equations in the system.

EXAMPLE 1

Determine whether the pair of numbers $x = 1$ and $y = -1$ is a solution to the system of equations in Equation (*24*).

Solution

Substituting the proposed solution into the left side of the first equation, we obtain $2(1) + (-1) = 1$, which does equal the right side of that equation. Substituting these same values into the left side of the second equation, we obtain $3(1) - (-1) = 4$, which also equals the right side of that equation. Since the pair $x = 1$ and $y = -1$ satisfies all the equations of the system, it is a solution to the system.

EXAMPLE 2

Determine whether the pair of numbers $x = 2{,}000$ and $y = 4{,}000$ is a solution to the system

$$5x + 8y = 42{,}000$$
$$3x + 10y = 33{,}000$$

Solution

Substituting the proposed solution into the left side of the first equation, we obtain

$$5(2{,}000) + 8(4{,}000) = 10{,}000 + 32{,}000 = 42{,}000$$

which equals the right side of that equation, so the pair $x = 2{,}000$ and $y = 4{,}000$ is a solution of the first equation. Substituting these same values into the left side of the second equation, we obtain

$$3(2{,}000) + 10(4{,}000) = 6{,}000 + 40{,}000 = 46{,}000$$

which does *not* equal 33,000, the right side of the second equation. Consequently, the pair $x = 2{,}000$ and $y = 4{,}000$ is not a solution of the second equation and not a solution of the system. A solution of a system must satisfy *every* equation in the system, not just some.

A solution to a system of two linear equations in two variables is the pair of coordinates of the point of intersection of the two lines defined by those equations.

Examples 1 and 2 demonstrate how to check whether a proposed solution is indeed a solution. They do not show how to locate solutions, but the graphing techniques introduced earlier do. Recall that the graph of a linear equation in two variables is a straight line and that the pair of coordinates for each point on that line is a solution to the equation. If we have a system of two such equations, then both their graphs are straight lines. The coordinates of a point are a solution to the first equation if the point is on the line defined by the first equation. Similarly, these same coordinates are a solution to the second equation if the point is also on the line defined by the second equation. Being a solution to both equations is equivalent to being a point on both lines! That is, the coordinates of the point of intersection of the two lines defined by a system of two linear equations in two variables are a solution to that system of equations.

EXAMPLE 3

Find a solution to the system

$$x + y = 6$$
$$2x - y = 0$$

Solution

Each equation is graphed in Figure 1.34. There is one point of intersection, and its coordinates, read directly from the graph as $x = 2$ and $y = 4$, are a *proposed* solution. To conclude that these coordinates are the solution, we substitute $x = 2$ and $y = 4$ into each equation and verify that both equations are satisfied.

EXAMPLE 4

Find a solution to the system

$$5x + 8y = 42{,}000$$
$$3x + 10y = 33{,}000$$

FIGURE 1.34

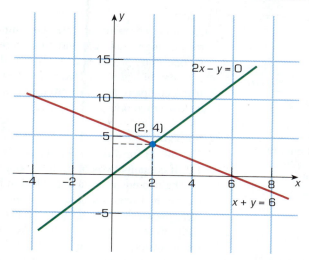

Solution

Each equation is graphed in Figure 1.35. There is one point of intersection and its coordinates, read directly from the graph as $x = 6{,}000$ and $y = 1{,}500$, are a *proposed* solution. Substituting these coordinates into the equations, we find that the pair $x = 6{,}000$ and $y = 1{,}500$ makes both equations true and is, therefore, the solution to the given system.

Any solution read from a graph is only an estimate.

As we noted in Section 1.4, coordinates on a graph are accurate only to approximately one-tenth of the distance between successive tick marks. Consequently, when we read points from a graph, we are only approximating their location. We read the point of intersection in Figure 1.35 as $x = 6{,}000$ and $y = 1{,}500$. It could just as well have been $x = 6{,}002$ and $y = 1{,}495$ or $x = 6{,}005$ and $y = 1{,}504$. It is impossible for us to distinguish between such points on Figure 1.35. Until we substitute the coordinates back into the equations of interest, we cannot be certain that we have a solution to a system.

FIGURE 1.35

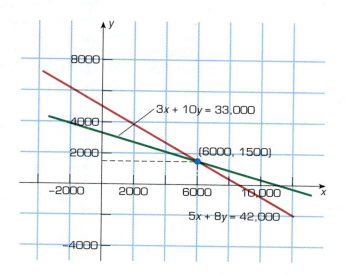

EXAMPLE 5

MODEL 1.9

Important factors

d_A = number of days mine A
      is in operation

d_B = number of days mine B
      is in operation

Objective

Find d_A and d_B.

Constraints

$5{,}000d_A + 2{,}000d_B = 50{,}000$

$1{,}000d_A + 9{,}000d_B = 80{,}000$

FIGURE 1.36

A mining company operates two mines in West Virginia. The first mine produces 5,000 tons of high-grade ore and 1,000 tons of low-grade ore each day it is in operation. The second mine produces 2,000 tons of high-grade ore and 9,000 tons of low-grade ore for each day of operation. Determine the number of days that each mine should be in operation if the company needs to produce exactly 50,000 tons of high-grade ore and 80,000 tons of low-grade ore.

Solution

The model for this problem is Model 1.9, which was developed in Example 8 of Section 1.2. It contains two constraints, both linear equations in the variables d_A and d_B.

The two equations are graphed in Figure 1.36 with d_A on the horizontal axis and d_B on the vertical axis; the horizontal and vertical intercepts were first plotted and then lines were drawn through those intercepts. There is one point of intersection, but to identify it with any precision, we need a different graph. Note that much of Figure 1.36 is in the fourth quadrant, where values for d_B are negative. This region is not relevant, since d_B denotes the number of days mine B is in operation. Realistically, both d_A and d_B must be nonnegative.

Days operating mine A

Let us redraw the graph, zooming in on a tighter window in the first quadrant. We rescale the tick marks and plot a smaller area of the graph to see the region of interest in more detail. Since the first constraint equation generates negative values for d_B whenever d_A is greater than ten, we shall limit our window to values of d_A between zero and ten. For the first constraint equation we plot the points (0, 25) and (8, 5) and for the second we plot the points (0, 8.9) and (8, 8). Drawing lines through these pairs of points, we generate Figure 1.37. The horizontal component of the point of intersection is approxi-

FIGURE 1.37

Days operating mine A

mately 6.8. The vertical component is just above the plotted point (8, 8), so we estimate it to be 8.1. Thus a proposed solution is the pair of values $d_A = 6.8$ and $d_B = 8.1$.

To determine whether the proposed solution is indeed a solution, we substitute it into each constraint equation. The left side of the first constraint equation becomes

$$5000(6.8) + 2000(8.1) = 50,200$$

The left side of the second constraint equation becomes

$$1000(6.8) + 9000(8.1) = 79,700$$

Neither constraint is satisfied, so the proposed solution is *not* a solution. It is close, however, and is probably as good an *estimate* to the solution as we can get from this graph.

To obtain greater precision, we must zoom in and magnify the region around the point of intersection. We restrict our viewing window (graphing region) to [6.5, 7.5] on the horizontal axis and [7.5, 8.5] on the vertical axis. The result is a window very much like Figure 1.38. With a graphing calculator, we would use the cursor to estimate the position of the point of intersection. In Figure 1.38, we use the dotted lines emanating from the point of intersection to *estimate* the solution as the pair of values $d_A \approx 6.74$ and $d_B \approx 8.14$. The actual solution is the pair $d_A = 290/43 \approx 6.744186$ and $d_B = 350/43 \approx 8.1395349$, but to find it we need the techniques of the next chapter.

FIGURE 1.38

Systems of two equations in two variables are useful in business for analyzing the relationships between costs and income from sales or *revenues*. Every manufacturing process involves costs, which often are separated into fixed costs and variable costs. *Fixed costs* include rent, insurance, property taxes, and other expenses that occur regardless of the number of items produced. *Variable costs* are those expenses directly attributable to the manufacture of the items themselves, such as labor, raw materials, and packaging. Variable costs depend directly on the number of items produced; the more items, the higher the variable costs.

Over any extended time period, inflation, revised labor contracts, and other macroeconomic trends change the cost and price of any product. Over the short run, generally a year or less, these economic conditions have little influence, and over short intervals both the price of a product and the costs associated with its production are constant.

If we designate the total cost by C, the direct cost associated with manufacturing a single item by a, the number of items being manufactured by n, and the fixed cost by FC, we see that the variable cost is $a \cdot n$ and the total cost is the sum of the variable cost and the fixed cost or

$$C = a \cdot n + FC \qquad (25)$$

Over the short run, the costs a and FC are fixed and known, so Equation (25) is a linear equation in the variables n and C. As n, the number of items produced, changes, so will total cost C.

EXAMPLE 6

A company manufacturing a single product has recently signed contracts with its suppliers and the labor unions representing its employees. For the duration of these contracts, the material and labor costs for manufacturing each item will be $300. The company has fixed monthly costs totalling $100,000. Determine the company's total cost per month.

Solution

Using Equation (25) with $a = 300$ and $FC = 100,000$, we have as the total monthly cost $C = 300n + 100,000$. In particular, if 500 items are produced, the total monthly cost is $C = 300(500) + 100,000 = \$250,000$. If no items are produced, the total monthly cost is $C = 300(0) + 100,000 = \$100,000$, which is the fixed cost.

If we continue to restrict ourselves to the short run, the price of each item produced and sold is also a constant. The sales revenue is simply the number of items sold times the price of each item. Designating the sales revenue by R, the price per item by p, and the number of items sold by N, we have

$$R = p \cdot N \qquad (26)$$

With p constant over the short run, Equation (26) is a linear equation in the two variables R and N.

EXAMPLE 7

The manufacturing company described in Example 6 has priced its product at $400 per item. Determine the sales revenue the company can expect.

Solution

Using Equation (26) with $p = 400$, we have as the estimated revenue $R = 400N$. In particular, if the company sells 500 items, it will generate $R = 400(500) = \$200,000$ in revenues. If the company sells no items, its revenue will be $R = 400(0) = 0$.

We see from Examples 6 and 7 that a production level of 500 items per month will cost the company $250,000 and generates sales revenue of $200,000, leaving the company with a net loss of $50,000 for the month. Such embarrassing situations can be predicted with *break-even analysis*. As the name suggests, break-even analysis determines the level of production at which a company's costs equal its income. This production level is called the *break-even point*. Below the break-even point, a company loses money; above the break-even point, a company makes a profit.

FIGURE 1.39

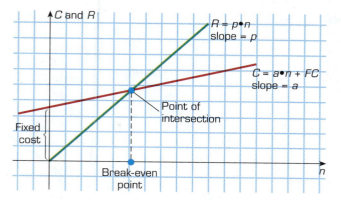

In determining a break-even point, it is assumed that all items produced are sold. With N, the number of items sold, equal to n, the number of items produced, Equation (26) becomes

$$R = p \cdot n \qquad (27)$$

The break-even point occurs when total cost equals sales revenue, that is, when C in Equation (25) equals R in Equation (27), or when

$$a \cdot n + FC = p \cdot n \qquad (28)$$

> *A break-even point occurs when total cost equals income.*

With a, FC, and p specified, Equation (28) is one equation in the single variable n, so it can be solved algebraically for n.

A graphical solution to the break-even problem can be obtained by graphing Equations (25) and (27) on the *same* coordinate system, as is done in Figure 1.39. The horizontal axis is n, the number of items produced and presumably sold. The *vertical axis, however, has two designations*: it is cost when we plot Equation (25) and revenue when we plot Equation (27). Since both equations are linear, their graphs are straight lines. *Their point of intersection defines the break-even point.* The production level required to reach the break-even point is the horizontal coordinate of the point of intersection. The vertical coordinate of the point of intersection is the total cost C at the break-even point, and it is also the total revenue R, because cost and revenue are equal at this point.

We can say even more about the geometry of the lines in Figure 1.39. The slope of the line defined by Equation (27) is the unit price, p, while the slope of the line defined by Equation (25) is the direct cost associated with manufacturing each item, a. The y-intercept of the graph of Equation (25) is the fixed cost, FC. The line defined by Equation (27) passes through the origin, indicating that revenue is directly proportional to the number of products sold.

EXAMPLE 8

Use graphical methods to determine the break-even point for the manufacturing process described in Examples 6 and 7, assuming that the company can sell all the items it produces.

Solution

We first plot the cost equation $C = 300n + 100{,}000$ from Example 6 on a cartesian coordinate system with n as the horizontal axis (see Figure 1.40). The price of each item is \$400, so it follows from Equation (27) that total revenue is $R = 400n$; this equation is

FIGURE 1.40

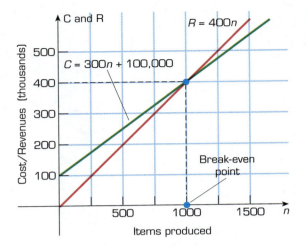

also drawn in Figure 1.40. The point of intersection is read directly from the graph as (1000, 400,000). We can verify by direct substitution that $C = R = 400,000$ when $n = 1,000$, thus the break-even point is 1,000 items. At this point the total cost and the revenue from sales are both $400,000.

The information that graphs provide about solutions of systems of equations is not readily obtained from the equations themselves. A solution of a set of two linear equations in two variables is the pair of coordinates of the points of intersection of the graphs of those equations. Since the equations are linear, their graphs are straight lines. If two distinct straight lines are parallel, they do not intersect, and their defining system of equations has no solution. An example of a system that has no solution is

$$x - y = 1$$
$$x - y = -1$$

which is graphed in Figure 1.41.

It is also possible for two linear equations to have the same straight line graph. An example is

$$2x + 3y = 6$$
$$4x + 6y = 12$$

FIGURE 1.41

FIGURE 1.42

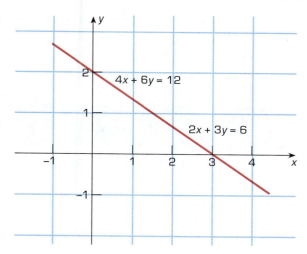

which is graphed in Figure 1.42. Every point on the graph of the first equation is also a point on the graph of the second equation, so the lines have infinitely many points of intersection, and consequently, the equations that define these lines have infinitely many solutions.

Two straight lines can intersect at a single point, as we saw in Figures 1.34, 1.35, 1.37, 1.40, at infinitely many points, or at no points. It then follows that a system of two linear equations in two variables has either one solution, infinitely many solutions, or no solution.

The same type of graphical analysis can be applied to the solution of a system of *three* linear equations in two unknowns. Each equation graphs as a straight line. If all three lines intersect at the same point, as shown in Figure 1.43, then the underlying system has a solution, which is the coordinates of the point of intersection.

> *A system of linear equations has either one solution, infinitely many solutions, or no solution.*

FIGURE 1.43

FIGURE 1.44

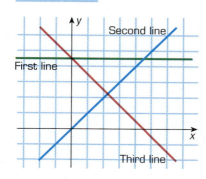

If there is no common point of intersection for all three lines, then the underlying system of equations has no solution. Consider the situation depicted in Figure 1.44. Each line intersects with the other two lines, but there is no point common to all three lines. Therefore, the system of equations that defined these lines does not have a solution.

Finally, if the graphs of all three equations appear as the same line, then there are infinitely many points of intersection and infinitely many solutions to the system of equations that defined these lines. We conclude, as in the case with two equations, that a system of three linear equations in two variables has either exactly one solution, infinitely many solutions, or no solution.

IMPROVING SKILLS

1. Determine whether any of the following proposed solutions solve the system

$$3x + 8y = 12,600$$

$$4x + 3y = 7,600$$

(**a**) $x = 600, y = 1,350$ (**b**) $x = 4,200, y = 0$
(**c**) $x = 1,000, y = 1,200$ (**d**) $x = 1,150, y = 1,000$

2. Determine whether any of the following proposed solutions solve the system

$$2x - y = 4$$

$$x + y = 14$$

(**a**) $x = 3, y = 2$ (**b**) $x = 10, y = 16$
(**c**) $x = 7, y = 7$ (**d**) $x = 6.1, y = 8.2$

In Problems 3 through 20, use graphical methods to estimate solutions to the given systems of linear equations, then determine whether the estimates are actually solutions.

3. $2x - y = 4$
    $x + y = 14$

4. $2x + y = 7$
    $x + y = 2$

5. $4x - 3y = 0$
    $2x - y = 6$

6. $x + 2y = 2$
    $2x + 3y = -1$

7. $2x - y = -5$
    $-4x + 2y = 10$

8. $2x - y = 40$
    $3x - 2y = 30$

9. $x + 2y = 4$
    $2x - y = 4$

10. $2x - y = 4$
    $6x - 3y = 4$

11. $2x + y = 100$
    $x + 2y = 50$

12. $3x - y = 15$
    $-3x + y = 9$

13. $3x - y = 0$
    $2x + 3y = 0$

14. $3x + 4y = 6$
    $6x + 8y = 12$

15. $3x - y = 6$
    $2x - 3y = 12$

16. $3x + 2y = 0$
    $4x + 3y = 10$

17. $2x + 4y = 2{,}000$

   $4x + 3y = 3{,}000$

   $5x + 5y = 4{,}000$

18. $2x + 4y = 2{,}000$

   $4x + 3y = 3{,}000$

   $5x + 5y = 8{,}000$

19. $3x - y = 4$

   $2x + 3y = 6$

   $x + y = -2$

20. $2x - 4y = 6$

   $x - 2y = 3$

   $-3x + 6y = -9$

CREATING MODELS

In Problems 21 through 29, model the given problems and then use graphs to solve them.

21. A manufacturer of electric can openers currently has a fixed cost of $21,000 per day and determines that the cost of producing each can opener is $5.50. Each can opener is sold for $9.00. Find the number of can openers that must be produced to have total cost equal total revenue.

22. A bakery currently has a fixed cost of $120,000 per month and estimates its variable cost to be 23¢ per loaf of bread. Loaves are sold at wholesale for 55¢ per loaf. Determine the number of loaves that need to be produced so that total cost equals total revenue.

23. A publisher of a current economics textbook determines that the manufacturing costs directly attributable to each book are $10 and that the fixed costs are $20,000. Determine the break-even point if the publisher sells each book for $22.

24. A manufacturer of light bulbs determines that each bulb costs 15¢ to make and that the business incurs weekly costs of $15,000 independent of production runs. Determine the break-even point if each bulb sells for 25¢.

25. After a new labor contract is signed, it costs the manufacturer in the previous problem 5¢ more in labor costs to make a bulb. The company therefore raises its selling price by the same amount. What is the new break-even point?

26. A manufacturer of Lucite pipe holders currently has fixed costs of $220,000 per month. The labor and material costs associated with the production of the pipe holders are $28 per item. Determine the break-even point if each pipe holder sells for $42.

27. A manufacturer of staplers determines that the manufacturing costs directly attributable to the production of each stapler are $2 and that fixed costs are $15,000 per month. Determine the break-even point if each stapler sells for $3.

28. Redo the previous problem if each stapler sells for $4.

29. A new factory lease increases the fixed monthly cost to the manufacturer in the previous problem by $30,000 per month. The company does not feel it can increase the price of its product and remain competitive. What is the new break-even point?

EXPLORING IN TEAMS

30. The graphical procedures described in this section can be extended to solving any two equations in two variables, not just linear equations. Of course, if the equations are

not linear, their graphs will not be straight lines. Graph the following two equations on the same coordinate system:

$$y = x^2$$

$$y = x + 2$$

Locate the points of intersection of the two graphed curves, and show that their coordinates are solutions to the given system of equations.

31. Use graphical methods to solve the system of equations

$$y = x^3$$

$$y = -x^2 + 2$$

32. Use graphical methods to show that the system of equations

$$y = x^4 + 2$$

$$y = x^2$$

has no solution.

33. A manufacturer of fire alarm systems determines that its total cost is given by

$$C = n^2 + 500n + 100,000$$

(Note that this equation is not linear.) Determine the break-even point if each system sells for $1,200.

EXPLORING WITH TECHNOLOGY

34. Use a graphing calculator to produce a more precise estimate of the solutions to Example 5 by setting your viewing window to [6.7, 6.8] on the horizontal axis and [8.1, 8.2] on the vertical axis. Continue to zoom in on the point of intersection. How close can you come to the true answer?

35. Change the right side of the equations in Example 4 to 42,010 and 32,950, respectively, and then use a graphing calculator to estimate to two decimal places the solution to the new system of equations:

$$5x + 8y = 42,010$$

$$3x + 10y = 32,950$$

36. Use a graphing calculator to estimate to two decimal places the solution to the system of equations given in Problem 9.

37. Use a graphing calculator to estimate to two decimal places the solution to the system of equations given in Problem 15.

38. Use a graphing calculator to estimate to two decimal places the solution to the system of equations

$$y = x^2 - 1$$

$$y = x + 2$$

39. Use a graphing calculator to find the solution to the system of equations

$$y = x^2 - 1$$
$$y = x^3 + 1$$

RECOMMENDING ACTION

40. Respond by memo to the following request:

> **MEMORANDUM**
>
> To: J. Doe Reader
>
> From: President's Office
>
> Date: Today
>
> Subject: **Hocus Pocus**
>
> I loved the graph you prepared for me last week on profit. By changing the scale, you made it *look* as though profits were increasing dramatically while, in fact, they were flat. This got me thinking about our break-even point: Is the break-even point on our graphs for cost and revenue also dependent on the scale? That is, if we change the scale on our graphs, do we change the break-even point?

CHAPTER 1   KEYS

KEY WORDS

adequate model (p. 4)
break-even analysis (p. 68)
break-even point (p. 68)
Cartesian coordinate system (p. 29)
constant of proportionality (p. 20)
constraints (p. 8)
coordinates (p. 29)
exponential-type growth (p. 32)
fixed costs (p. 67)
graph of an equation (p. 30)
horizontal intercept (p. 48)
incomplete model (p. 14)
linear equation (p. 47)
model (p. 2)
modeling process (p. 16)
objective (p. 13)
origin (p. 29)

parallel lines (p. 57)
perpendicular lines (p. 57)
proportional (p. 20)
qualitative graphical models (p. 39)
quantitative graphical models (p. 38)
revenue (p. 68)
rise (p. 51)
run (p. 51)
S-shaped curve (p. 35)
scale (p. 29)
slope (p. 52)
solution (p. 63)
straight lines (p. 48)
system of linear equations (p. 63)
variable costs (p. 67)
vertical intercept (p. 48)

KEY CONCEPTS

1.1 The Modeling Process

- A model is a representation of a real situation.

- If a model fits the needs of the people using it, the model is adequate.

- We can experiment on a model; often, we cannot experiment on the real situation.

1.2 Modeling with Equalities

- Constraints are limitations on what can be accomplished.

- Mathematical modeling involves identifying the important factors in a statement and then writing equations that relate those factors to each other.

- A set of statements is modeled by modeling each statement individually.

- The purpose of a model is to enhance our ability to answer questions about the underlying real system. Therefore, the questions define the objective of any model.

- A model is incomplete if the constraints do not include all the factors in the objective.

1.3 Modeling with Proportionality

- Two positive quantities are directly proportional if their quotient is a constant. The constant is called a constant of proportionality.

- Many models are replicas of real systems drawn to scale, that is, measurements on the model are directly proportional to measurements on the real system. The constant of proportionality is the ratio of the scales.

1.4 Modeling with Graphs

- Graphs have visual impact and are a powerful medium for presenting data and relationships.

- Every point on a cartesian coordinate system is uniquely determined by an ordered pair of numbers. By convention, the first number is a value on the horizontal axis while the second number is a value on the vertical axis.

- Successive tick marks must be equally spaced, and the units between equally spaced tick marks must be the same. However, the scale on the horizontal axis can differ from the scale on the vertical axis.

- Geometrically, the set of all solutions of one equation in two unknowns is a curve on the plane.

- Graphing too many points is always better than graphing too few.

- Scaling can affect the presentation of a curve. Zooming in to feature a small part of a graph is effective for minimizing the effects of scaling and accenting interesting portions of a graph.

- Once a scale is chosen, numbers are rounded to one-tenth the distance between successive tick marks for graphing purposes.

1.5 Linear Equations and Straight Lines

- The graph of a linear equation in two variables is a straight line.

- The slope of a straight line is the same regardless of which two points on the line are used to calculate the slope.

- Starting from any point on a line, the slope is the change in the vertical direction (the rise) needed to accompany a unit change in the positive horizontal direction (the run) if the resulting point is to be on the line.

- A line with a positive slope slants upward to the right; a line with a negative slope slants downward to the right.

- A line parallel to the x-axis has zero slope; a line parallel to the y-axis does not have a slope.

- Two quantities are directly proportional if and only if their graph is a straight line through the origin. The slope of that line is the constant of proportionality.

- Two lines are parallel if they have the same slope. Two lines are perpendicular if their slopes are negative reciprocals of one another.

1.6 Systems of Linear Equalities

- A solution to a system of linear equations is a set of numbers, one for each variable, that together satisfy all the equations in the system.

- The coordinates of the point of intersection of the two lines defined by a system of two linear equations in two variables are the solution of the system of equations (if such a point exists).

- Any solution read from a graph is only a proposed solution. If the proposed solution solves the underlying system of equations, the proposed solution is indeed a solution. If not, the proposed solution is only an approximate solution.

- A system of two linear equations in two variables has one solution, infinitely many solutions, or no solution.

- A break-even point is the level of production at which a company's costs equal its income.

KEY FORMULAS

- Quantities A and B are directly proportional if there exists a constant k such that

$$\frac{A}{B} = k$$

or

$$A = kB$$

- The slope of a straight line passing through the points (x_1, y_1) and (x_2, y_2) is

$$m = \frac{y_2 - y_1}{x_2 - x_1}$$

- The point-slope formula for a nonvertical straight line passing through the point (x_1, y_1) and having slope m is

$$y - y_1 = m(x - x_1)$$

- The slope-intercept formula for a nonvertical straight line with slope m and $(0, b)$ as its vertical intercept is

$$y = mx + b$$

- The equation of a nonvertical line through the origin with slope m is

$$y = mx$$

- The graph of a horizontal straight line parallel to the x-axis satisfies the equation

$$y = k$$

where k is a constant.

- The graph of a vertical straight line parallel to the y-axis satisfies the equation

$$x = h$$

where h is a constant.

- Let C denote total cost, a the cost associated with manufacturing a single item, n the number of items being manufactured, and FC the fixed cost. Then the variable cost is $a \cdot n$, and the total cost is

$$C = a \cdot n + FC$$

- Let R denote the sales revenue, p the price per item, and N the number of items sold. Then

$$R = p \cdot N$$

KEY PROCEDURES

- To model a problem:

  Step 1   Identify the objective.

  Step 2   Identify other important factors.

  Step 3   Identify the constraints.

  Step 4   For each constraint, construct a relationship between factors that reflects the constraint.

  Step 5   Combine constraint relationships if it can be done without deleting factors that are part of the objective.

- To solve a system of linear equations in two variables by graphing:

  Step 1   Graph both equations on the same coordinate system.

  Step 2   Locate the point of intersection common to all of the straight lines obtained in Step 1. If no such point exists, the system has no solution; otherwise the pair of coordinates of the point of intersection is the solution to the given system.

TESTING YOURSELF

Use the following problems to test yourself on the material in Chapter 1. The odd problems comprise one test, the even problems a second test.

1. Model the following problem but do not solve: A company announces that it will cut its work force by 5%.
2. Graph the equation $S = 3/(E + 1)$ with E on the horizontal axis.
3. New car dealers give their customers promotional brochures describing the features of their new models. Such a brochure is a model of the car. List some factors that this model includes and some factors that it neglects.
4. Model the following problem, but *do not solve*: A population is half male and half female, and each year 60% of the females give birth to a single offspring.
5. Model the following problem, but do not solve: The owner of an ice-cream cart sells only ice-cream cones. Single scoops cost $1.50, and double scoops cost $2.50. Yesterday, the cart owner took home $500.50. How many cones of each type were sold if 265 cones were used from inventory?
6. Graph the equation $x = -5.5$.
7. Graph the equation $y = 3x$.
8. A photograph of a baseball stadium is a model of that stadium. List some of the factors that this model includes and some factors that it neglects.
9. Model the following problem, but do not solve: A caterer charges clients $200 for the rental of equipment plus $65 for each meal served.
10. An architectural drawing is scaled so that $\frac{1}{4}$ inch represents 8 feet. How big is a structure that measures $2\frac{1}{2}$ inches by $3\frac{1}{8}$ inches on the drawing?
11. Use graphical methods to estimate the solution to the following system of equations:

$$3x + 7y = 210$$
$$5x + 2y = 180$$

12. Model the following problem, but do not solve: A manufacturer announces that it will increase the price of its computers 3%.
13. The number of reported injuries in a marathon is directly proportional to the number of participants. Last year, 11,600 people ran in a marathon, and 174 injuries were reported. How many injuries can be expected this year if 12,000 runners are entered in the marathon?
14. Model the following problem, but do not solve: A restaurant serves only complete dinners and offers two types: a tourist special for $12 and a gourmet delight for $23. Last night the restaurant served 130 customers. How many dinners of each type were served if total income from food was $2,077?
15. Graph the equation $y = x^2 - 2x + 2$.
16. Use graphical methods to estimate the solution to the following system of equations:

$$40x - 30y = 24,000$$
$$10x + 50y = 30,000$$

2

LINEAR

MODELS

2.1 ELEMENTARY ROW OPERATIONS

Equations are the bedrock of mathematical modeling because they represent limitations on what can be accomplished in a given situation. Most real-world systems involve multiple constraints and many unknowns, hence mathematical models of such systems contain multiple equations in many variables.

In Section 1.6, we used graphical methods to estimate solutions to systems of two linear equations in two variables. Graphical methods are not applicable when constraint equations involve more than three variables, and they are difficult to use even with three variables. Other methods are needed, and one of the very best is based on *elementary row operations*.

In Section 1.5 we defined a linear equation in the two variables x and y as an equation of the form

$$Ax + By = C \tag{1}$$

where A, B, and C are known numbers and A and B are not both zero. We now extend this definition to include equations in three or more variables. An equation in the three variables x, y, and z is linear if it has the form

$$Ax + By + Cz = D \tag{2}$$

where A, B, C, and D are known numbers and A, B, and C are not all zero (to avoid equations of the form $0 = D$).

EXAMPLE 1

Determine which of the following equations are linear equations in the three variables x, y, and z:
(a) $2x + 2y - 3z = 0$
(b) $y - z = -7$
(c) $x + 3y^2 + 4z = 2$
(d) $3x + 2y - xz = 1$

Solution

(a) This equation is linear because it is in the form of Equation (2) with $A = B = 2$, $C = -3$, and $D = 0$.
(b) This equation has form of Equation (2) with $A = 0$, $B = 1$, $C = -1$, and $D = -7$, so it is linear.
(c) This equation is *not* linear because the y term appears to the second power; the first power is required by Equation (2).
(d) This equation is *not* linear because one variable is multiplied by another in the xz term; in a linear equation, only known numbers can multiply variables.

A solution to a system of equations is a set of numbers, one for each variable, that satisfies all the equations in the system.

As always, the letters x, y, and z in Equation (2) can be replaced by any other letters that are more convenient or more descriptive for a particular problem. The form of an equation, not the letters, makes the equation linear. Equation (2) requires the variables to be raised only to the first power and to be multiplied only by known numbers, regardless of the letters used to denote those variables.

An equation in the four variables x, y, z, and w is linear if it has the form

$$Ax + By + Cz + Dw = E \tag{3}$$

where A, B, C, D, and E are known numbers and A, B, C, and D are not all zero. The pattern for linear equations in five or more variables is similar.

A system of linear equations is a set of two or more linear equations. As in Chapter 1, a solution to a system of equations is a set of numbers, one for each variable, that makes *all* the equations true.

EXAMPLE 2

Determine whether the set of values $R = 1$, $S = -1$, and $U = V = 2$ is a solution to the system

$$2R + S - U + V = 1$$
$$R + S + 2U - V = 2$$
$$R + 2S + U + V = 3$$
$$R + S + U + 2V = 4$$

Solution

Each of these four equations is linear because each has the form specified by Equation (*3*) with the variables R, S, U, and V replacing x, y, z, and w, respectively. Substituting the proposed solution into each equation sequentially, we find that the first three equations are satisfied, but the left side of the last equation is

$$(1) + (-1) + (2) + 2(2) = 6$$

which does *not* equal 4, the right side of the last equation. Since the proposed solution does not satisfy *all* the equations, it is *not* a solution.

Our goal is to find solutions to systems of linear equations. As we might expect, some systems of linear equations are easy to solve, while others are not. For example, it is not at all clear what the solutions are to the system

$$x + y - 2z = -3$$
$$-x - y + z = 0 \tag{4}$$
$$2x + 3y - 2z = 2$$

while we can quickly solve the system

$$x + y - 2z = -3$$
$$y + 2z = 8 \tag{5}$$
$$z = 3$$

The last equation in Equation (5) establishes that $z = 3$. Substituting this value of z into the second equation, we obtain

$$y + 2(3) = 8 \quad \text{or} \quad y = 2$$

Then substituting $y = 2$ and $z = 3$ into the first equation of System (5), we find

$$x + (2) - 2(3) = -3 \quad \text{or} \quad x = 1$$

Back-substitution begins with the last equation of a system and moves up through the system sequentially, solving each equation for one variable.

Together, $x = 1$, $y = 2$, and $z = 3$ are a solution to System (5). We confirm this by substituting these values into each equation and verifying that the equations are satisfied. The process of solving a system of equations by beginning with the last equation, sequentially moving up through all the equations, and using each equation to solve for one variable at a time is known as *back-substitution*.

EXAMPLE 3

Use back-substitution to solve the system

$$x + 2y + 6z = 24.4$$
$$y + 4z = 9.6 \tag{6}$$
$$z = 2.857$$

Solution

The last equation establishes that $z = 2.857$. Substituting this result into the second equation, we have

$$y + 4(2.857) = 9.6 \quad \text{or} \quad y = = -1.828$$

Finally, substituting $z = 2.857$ and $y = -1.828$ into the first equation of System (6), we obtain

$$x + 2(-1.828) + 6(2.857) = 24.4 \quad \text{or} \quad x = 10.914$$

The solution is $x = 10.914$, $y = -1.828$, $z = 2.857$.

FIGURE 2.1

Original system of equations

Transform without changing solutions

Simple system of equations

Back-substitutions

Identify solutions

Back-substitution is effective for solving only special types of systems. In particular, back-substitution is *not* well suited to solving System (4). Note, however, that Systems (4) and (5) have the *same* solution, $x = 1$, $y = 2$, and $z = 3$, which suggests an interesting strategy: Transform a system of equations that is difficult to solve, such as System (4), into another system of equations that is easy to solve, such as System (5). Do this *without* altering the solution of the original system (nothing is gained by transforming one system into another if the solution of the simpler system is not also the solution of original system). This strategy is depicted schematically in Figure 2.1.

There are three operations on equations that change the look of the equations (and therefore the look of the system) but *do not* change the solution. They are:

E_1 Interchange the positions of any two equations.

E_2 Multiply an equation by a nonzero number.

E_3 Add to one equation a nonzero constant times another equation.

To illustrate these operations, consider again the equations

$$x + y - 2z = -3$$
$$-x - y + z = 0 \tag{4 repeated}$$
$$2x + 3y - 2z = 2$$

which has as its solution $x = 1$, $y = 2$, and $z = 3$. If we apply operation E_2 to System (4) by multiplying the first equation by -2, the first equation becomes

$$-2(x + y - 2z) = -2(-3)$$

or

$$-2x - 2y + 4z = 6$$

The new system is

$$-2x - 2y + 4z = 6$$
$$-x - y + z = 0 \tag{7}$$
$$2x + 3y - 2z = 2$$

which *looks* different from System (4) but still has the same solution: $x = 1$, $y = 2$, and $z = 3$.

If we apply operation E_3 to System (4) by adding to the second equation 3 times the third equation (note that we are not changing the third equation, only the second equation), the second equation becomes

$$(-x - y + z) + 3(2x + 3y - 2z) = 0 + 3(2)$$

or

$$5x + 8y - 5z = 6$$

The new system is

$$x + y - 2z = -3$$
$$5x + 8y - 5z = 6 \tag{8}$$
$$2x + 3y - 2z = 2$$

which again *looks* different from System (4) but still has the same solution.

Operations E_1 through E_3 affect only the coefficients in the equations, not the variables. Each equation in Systems (4), (7), and (8) contains an x, a y, and a z, and as we moved from System (4) to System (7) or from System (4) to System (8), those variables remained intact. Therefore, we direct our attention to the parts of a system of equations that can change under operations E_1 through E_2: the coefficients.

For each system of linear equations, we create a table of coefficients, enclosed in brackets. The rows correspond to the individual equations in the system, row one for equation one, row two for equation two, and so on. The columns represent the variables in the equations, with the first column holding the coefficients of the first variable, the second column holding the coefficients of the second variable, and so on. The last column of the table is reserved for the numbers on the right side of the equality sign. To differentiate this column from the others, a line is drawn between the last two columns of the table. The table associated with

$$x + y - 2z = -3$$
$$-x - y + z = 0 \tag{4 repeated}$$
$$2x + 3y - 2z = 2$$

An augmented matrix for a system of equations is a table of numbers enclosed in brackets, where the rows represent the equations, all columns but the last column hold the coefficients of the variables in the equations, and the last column is the right side of the equations.

is

$$
\begin{array}{ccc}
x & y & z \\
\end{array}
$$
$$
\left[\begin{array}{ccc|c}
1 & 1 & -2 & -3 \\
-1 & -1 & 1 & 0 \\
2 & 3 & -2 & 2
\end{array}\right]
\begin{array}{l}
\textit{first equation} \\
\textit{second equation} \\
\textit{third equation}
\end{array}
\tag{9}
$$

Table (9) has many names, including schedule (such as a train schedule), spreadsheet, tableau, and augmented matrix. We shall adopt the term *augmented matrix* (plural, augmented matrices), because later we will manipulate these objects in ways that are not generally associated with either spreadsheets or tables.

EXAMPLE 4

Construct augmented matrices for the systems of equations defined by (7) and (8).

Solution

System (7) is

$$-2x - 2y + 4z = 6$$
$$-x - y + z = 0$$
$$2x + 3y - 2z = 2$$

The variables are x, y, and z, and we will leave them in this order. Column 1 will hold the coefficients of x, column 2 the coefficients of y, and column 3 the coefficients of z. The augmented matrix for System (7) is

$$
\begin{array}{ccc}
x & y & z
\end{array}
$$
$$
\left[\begin{array}{ccc|c}
-2 & -2 & 4 & 6 \\
-1 & -1 & 1 & 0 \\
2 & 3 & -2 & 2
\end{array}\right]
\begin{array}{l}
\textit{first equation} \\
\textit{second equation} \\
\textit{third equation}
\end{array}
$$

Similarly, the augmented matrix for the system of equations

$$x + y - 2z = -3$$
$$5x + 8y - 5z = 6 \qquad\qquad \text{(8 repeated)}$$
$$2x + 3y - 2z = 2$$

is

$$
\begin{array}{ccc}
x & y & z
\end{array}
$$
$$
\left[\begin{array}{ccc|c}
1 & 1 & -2 & -3 \\
5 & 8 & -5 & 6 \\
2 & 3 & -2 & 2
\end{array}\right]
\begin{array}{l}
\textit{first equation} \\
\textit{second equation} \\
\textit{third equation}
\end{array}
$$

EXAMPLE 5

Determine the system of equations in the variables a and b represented by the augmented matrix

$$
\begin{array}{cc}
a & b
\end{array}
$$
$$
\left[\begin{array}{cc|c}
1 & \frac{1}{2} & \frac{1}{2} \\
0 & 1 & -1 \\
0 & 0 & 0
\end{array}\right]
\begin{array}{l}
\textit{first equation} \\
\textit{second equation} \\
\textit{third equation}
\end{array}
$$

Solution

The system of equations is

$$1a + \tfrac{1}{2}b = \tfrac{1}{2}$$
$$0a + 1b = -1$$
$$0a + 0b = 0$$

which can be simplified into the system

$$a + \tfrac{1}{2}b = \tfrac{1}{2}$$
$$b = -1$$
$$0 = 0$$

EXAMPLE 6

Model the following problem, and then construct an augmented matrix for the constraint equations. The National Marrow Donor's Program lists the blood types of over one million volunteers. When a match exists between a volunteer and a person needing healthy marrow, a marrow transplant is performed. The characteristics of marrow are inherited, so matches generally involve people of similar racial origins. From 1987 through April 1994, there were 145 transplants to patients of African-American and Hispanic origin, and there were 41 more African-American recipients than Hispanic recipients. How many patients of each type received marrow transplants during this period?

Solution

The objective of the problem is to find the number of patients in each category who received transplants. There are two categories of interest here, African-American and Hispanic, so there are two factors to determine. We label the number of African-Americans who received transplants by A and the number of Hispanics who received transplants by H. The objective is to find both A and H.

There were a total of 145 transplants to African-Americans and Hispanics, hence one constraint is

$$A + H = 145$$

Forty-one more African-Americans than Hispanics received transplants, so the difference between the two groups is 41 and a second constraint is

$$A - H = 41$$

A complete model for this problem is shown in Model 2.1.

The constraints form the system of equations

$$A + H = 145$$
$$A - H = 41$$

in the variables A and H. Column 1 will hold the coefficients of A and column 2 the coefficients of H. The augmented matrix for this system is

$$\begin{array}{cc} A & H \end{array}$$
$$\left[\begin{array}{cc|c} 1 & 1 & 145 \\ 1 & -1 & 41 \end{array}\right] \begin{array}{l} \textit{first equation} \\ \textit{second equation} \end{array}$$

MODEL 2.1

Important factors

A = number of African-American transplant recipients

H = number of Hispanic transplant recipients

Objective

Find A and H.

Constraints

$A + H = 145$

$A - H = 41$

When we rephrase operations E_1 through E_3 into words appropriate to augmented matrices, we generate the three *elementary row operations*:

ELEMENTARY ROW OPERATIONS:

1. Interchange the positions of any two rows.

2. Multiply one row by a nonzero number.

3. Add to one row a nonzero constant times another row.

As an illustration of these elementary row operations, let us apply each one to the augmented matrix

$$\begin{bmatrix} 1 & 1 & -2 & | & -3 \\ -1 & -1 & 1 & | & 0 \\ 2 & 3 & -2 & | & 2 \end{bmatrix}$$

(9 repeated)

If we interchange the positions of the second and third rows (an example of the first elementary row operation), we obtain

$$\begin{bmatrix} 1 & 1 & -2 & | & -3 \\ 2 & 3 & -2 & | & 2 \\ -1 & -1 & 1 & | & 0 \end{bmatrix}$$

We symbolize this operation by $R_2 \leftrightarrow R_3$, where R denotes a row and the subscript the row number. The notation $R_2 \leftrightarrow R_3$ signifies that row 2 and row 3 are being interchanged and therefore replacing one another in the augmented matrix.

If we multiply the first row of Augmented Matrix (9) by -2 (an example of the second elementary row operation), we obtain

$$\begin{bmatrix} -2 & -2 & 4 & | & 6 \\ -1 & -1 & 1 & | & 0 \\ 2 & 3 & -2 & | & 2 \end{bmatrix}$$

We symbolize this operation by $-2R_1 \rightarrow R_1$, to signify that we are multiplying row 1 by -2 and then using the result as the new row 1. The arrow is read as "becomes the new." Here -2 times row 1 becomes the new row 1.

If we add to the second row of Augmented Matrix (9) three times the third row, as an example of the third elementary row operation, then 3 times row 3 is added to row 2 and the result becomes the new row 2. We symbolize this operation by $R_2 + 3R_3 \rightarrow R_2$. For Augmented Matrix (9)

$$\begin{aligned}
R_2 + 3R_3 &= [-1 \ -1 \ 1| \ 0] + 3[2 \ 3 \ -2| \ 2] \\
&= [-1 \ -1 \ 1| \ 0] + [6 \ 9 \ -6| \ 6] \\
&= [5 \ 8 \ -5| \ 6] \rightarrow R_2
\end{aligned}$$

and we obtain as the new augmented matrix

$$\begin{bmatrix} 1 & 1 & -2 & | & -3 \\ 5 & 8 & -5 & | & 6 \\ 2 & 3 & -2 & | & 2 \end{bmatrix}$$

Our strategy for solving a system of linear equations will be to transform that system into another system that has the same solution as the original system and is easy to solve by back-substitution. We must therefore know the types of systems that are amenable to solution by back-substitution. Consider again the system

$$\begin{aligned}
x + 2y + 6z &= 24.4 \\
y + 4z &= 9.6 \\
z &= 2.857
\end{aligned}$$

(6 repeated)

from Example 3, and its associated augmented matrix

$$\begin{bmatrix} 1 & 2 & 6 & | & 24.4 \\ 0 & 1 & 4 & | & 9.6 \\ 0 & 0 & 1 & | & 2.857 \end{bmatrix} \tag{10}$$

Note, in particular, the positions of the ones and zeros in Augmented Matrix (*10*). These values and the positions they hold are crucial.

We define the *main diagonal* of an augmented matrix as the set of all augmented matrix entries that lie to the left of the vertical line and have identical row and column positions. The main diagonal thus consists of the element in the first row and first column, the element in the second row and second column, the element in the third row and third column, and so on, for as many rows and columns in which matches exist. Note that in Augmented Matrix (*10*) all of the main diagonal elements are ones.

An augmented Matrix is *upper triangular* if all numbers below the main diagonal are zero. Augmented Matrix (*10*) is upper triangular. So too are

$$\begin{bmatrix} 1 & 2 & 3 & | & 11 \\ 0 & 4 & 5 & | & 12 \\ 0 & 0 & 6 & | & 13 \end{bmatrix}, \quad \begin{bmatrix} 1 & 2 & | & -3 \\ 0 & 3 & | & 6 \\ 0 & 0 & | & 0 \\ 0 & 0 & | & 0 \end{bmatrix}, \quad \text{and} \quad \begin{bmatrix} 1 & 0 & 2 & 0 & | & 8 \\ 0 & 0 & 1 & 3 & | & 0 \end{bmatrix}$$

(the main diagonal is shaded). Upper triangular form places no restrictions on the numbers on or above the main diagonal; they can be anything, even zero. The only restriction is that the elements below the main diagonal must be zero.

A system of equations is ready for solution by back-substitution when its augmented matrix is upper triangular and when all nonzero elements on the main diagonal are ones. Thus our strategy for solving systems of equations becomes the path of solid lines and arrows in Figure 2.2, which is just a fleshed-out version of Figure 2.1. The direct path from a given system to its solution, shown by the dotted line in Figure 2.2, is the preferred route, but we have no general methods for implementing it. Instead we execute Steps 1 through 4, because we know how to implement each step.

Step 1 is the construction of an augmented matrix from a system of linear equations; this is a very simple procedure, as illustrated in Example 4. Step 3 is the reverse process—going from an augmented matrix to the corresponding system of linear equations—and it is equally simple, as illustrated in Example 5. Step 4 is back-substitution, a bit more time

FIGURE 2.2

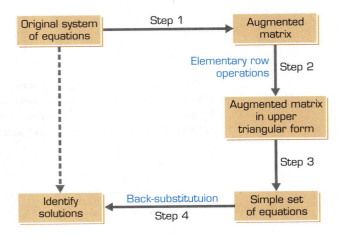

consuming than Steps 1 and 3 but almost as simple, as illustrated in Example 3. Most of the work in the *algorithm* (that is, the sequence of steps) shown in Figure 2.2 occurs in Step 2: using elementary row operations to transform an augmented matrix into upper triangular form. We will show how to do this step in the next section.

IMPROVING SKILLS

In Problems 1 through 8, determine whether the given systems are linear.

1. $x - 2y + 3z - w = 4$
$2x \quad\quad - z + 4w = 8$

2. $2a - 3b + 4c = 2$
$a + b - 2c = 0$
$-a + 2b + 5c = 1$

3. $2a + ab = 3$
$-a + 2b = 1$

4. $2^3a + b + c = 3$
$4a + 2b - c = 2$

5. $2a^3 + b + c = 3$
$4a + 2b - c = 2$

6. $2(3)^3 + b + c = 3$
$4a + 2b - c = 2$

7. $2x + \dfrac{1}{y} = 3$

$3x - \dfrac{2}{y} = 7$

8. $r - 2s = 1$
$r + 2s = -1$
$2r - s = 4$
$2r - 4s = 2$

9. Determine whether the given values of the variables x, y, and z are solutions to the system

$$x - 2y + 2z = 0$$
$$3y - 2z = 0$$
$$-8x + y + 2z = 0$$

(*a*) $x = 1, y = 2, z = 3$

(*b*) $x = 1, y = 0, z = -1$

(*c*) $x = 0, y = 0, z = 0$

10. Determine whether the given values for the variables x, y, z, and w are solutions to the system

$$2x + y - z + w = 0$$
$$3x - 2y + z + 2w = 2$$
$$x - y - z - w = 1$$
$$4x + 5y + z + 3w = 2$$

(*a*) $x = 0, y = -\frac{1}{2}, z = 0, w = \frac{1}{2}$
(*b*) $x = 0, y = -\frac{7}{8}, z = -\frac{1}{2}, w = -\frac{3}{8}$
(*c*) $x = 1, y = 0, z = 1, w = -1$

In Problems 11 through 17 write augmented matrices that correspond to the given systems of equations (this is Step 1 in Figure 2.2).

11. $2a + 3b = -4$
$5a - 6b = 7$

12 $-p + 11q = 100$
$2p + 25q = 500$

13. $23a - 37b = 0$
$66a - 72b = 0$

14. $-2N - 5C = 0.75$
$3N + 16C = 0.43$

15. Use the system in Problem 1.

16. Use the system in Problem 8.

17. Use the system in Problem 4.

In Problems 18 through 32, write out the system of equations that corresponds to the given augmented matrix (this is Step 3 in Figure 2.2).

18.
$$\begin{array}{cc} x & y \\ \left[\begin{array}{cc|c} 11 & 12 & 20 \\ 13 & 14 & 30 \end{array}\right] \end{array}$$

19.
$$\begin{array}{cc} y & x \\ \left[\begin{array}{cc|c} 11 & 12 & 20 \\ 13 & 14 & 30 \end{array}\right] \end{array}$$

20.
$$\begin{array}{cc} a & b \\ \left[\begin{array}{cc|c} 11 & 12 & 20 \\ 13 & 14 & 30 \end{array}\right] \end{array}$$

21.
$$\begin{array}{cc} x & y \\ \left[\begin{array}{cc|c} -8 & 2 & 0 \\ 0 & 3 & 1 \end{array}\right] \end{array}$$

22.
$$\begin{array}{cc} y & z \\ \left[\begin{array}{cc|c} -8 & 2 & 0 \\ 0 & 3 & 1 \end{array}\right] \end{array}$$

23.
$$\begin{array}{cc} x & y \\ \left[\begin{array}{cc|c} -1 & 0 & 3 \\ 0 & 2 & 4 \end{array}\right] \end{array}$$

24.
$$\begin{array}{cc} r & s \\ \left[\begin{array}{cc|c} 1 & 4 & 0 \\ 0 & 0 & 0 \end{array}\right] \end{array}$$

25.
$$\begin{array}{ccc} x & y & z \\ \left[\begin{array}{ccc|c} 1 & 2 & 3 & 10 \\ 2 & 3 & 4 & 20 \\ 3 & 4 & 5 & 30 \end{array}\right] \end{array}$$

26.
$$\begin{array}{ccc} x & y & z \\ \left[\begin{array}{ccc|c} 1 & 0 & 1 & 4 \\ 1 & 1 & 0 & 5 \\ 0 & 1 & 1 & 6 \end{array}\right] \end{array}$$

27.
$$\begin{array}{ccc} A & B & C \\ \left[\begin{array}{ccc|c} -1 & 2 & -3 & 5 \\ 2 & -6 & 6 & -7 \\ -3 & 0 & 4 & 0 \end{array}\right] \end{array}$$

28.
$$\begin{array}{ccc} r & s & t \\ \left[\begin{array}{ccc|c} 1 & 4 & 5 & 0.15 \\ 0 & 1 & 6 & 0.25 \\ 0 & 0 & 1 & 0.60 \end{array}\right] \end{array}$$

29.
$$\begin{array}{ccc} x & y & z \\ \left[\begin{array}{ccc|c} 1 & -4 & \frac{1}{2} & 2 \\ 0 & 1 & \frac{1}{2} & -3 \\ 0 & 0 & 0 & 0 \end{array}\right] \end{array}$$

30.
$$\begin{array}{ccc} N & E & S \\ \left[\begin{array}{ccc|c} \frac{1}{2} & \frac{1}{4} & 3 & -5 \\ 0 & \frac{1}{3} & \frac{3}{4} & -8 \\ 0 & 0 & 4 & \frac{3}{4} \end{array}\right] \end{array}$$

31.
$$\begin{array}{cccc} x & y & z & w \\ \left[\begin{array}{cccc|c} 1 & -2 & 0 & 5 & 100 \\ 0 & 1 & -1 & 4 & 150 \\ 0 & 0 & 1 & 3 & 200 \\ 0 & 0 & 0 & 1 & 250 \end{array}\right] \end{array}$$

32.
$$\begin{array}{cccc} w & x & y & z \\ \left[\begin{array}{cccc|c} 1 & 3 & 3 & 0 & 0 \\ 0 & 1 & 0 & -2 & 0 \\ 0 & 0 & 1 & -5 & 0 \\ 0 & 0 & 0 & 1 & 0 \end{array}\right] \end{array}$$

33. Use elementary row operations on the augmented matrix

$$\left[\begin{array}{cc|c} 1 & 2 & 5 \\ 3 & 4 & 6 \end{array}\right]$$

to perform each of the following operations and give the symbolic notation for each operation.
(a) Interchange the first and second rows.
(b) Multiply the second row by -2.
(c) Add to the second row 4 times the first row.
(d) Convert the number 3 to 1.

34. Use elementary row operations on the augmented matrix.

$$\begin{bmatrix} 0 & 1 & 2 & | & 0.2 \\ 3 & 4 & 5 & | & -0.6 \\ 6 & 7 & 8 & | & 0.3 \end{bmatrix}$$

to perform each of the following operations and give the symbolic notation for each operation.

(**a**) Interchange the first and third rows.
(**b**) Add to the second row 2 times the first row.
(**c**) Add to the third row -2 times the second row.
(**d**) Convert the number 3 to 1.
(**e**) Convert the number 6 to 1.

In Problems 35 through 38, perform the specified elementary row operations sequentially, beginning with the given matrix and applying each successive row operation to the matrix obtained from the previous operation. Give the symbolic notation for each operation.

35.
$$\begin{bmatrix} 1 & -5 & | & 8 \\ 3 & 3 & | & 6 \end{bmatrix}$$

(**a**) Add to the second row -3 times the first row.
(**b**) Multiply the second row by $\frac{1}{18}$.

36.
$$\begin{bmatrix} 2 & -4 & | & -2 \\ -5 & 9 & | & 3 \end{bmatrix}$$

(**a**) Multiply the first row by $\frac{1}{2}$.
(**b**) Add to the second row 5 times the first row.
(**c**) Multiply the second row by -1.

37.
$$\begin{bmatrix} 1 & 2 & 3 & | & 10 \\ 4 & 9 & 6 & | & 20 \\ 0 & 1 & 4 & | & 30 \end{bmatrix}$$

(**a**) Add to the second row -4 times the first row.
(**b**) Add to the third row -1 times the second row.
(**c**) Multiply the third row by 0.1.

38.
$$\begin{bmatrix} 2 & 4 & 6 & | & 0 \\ 8 & 10 & 12 & | & 0 \\ 14 & 16 & 18 & | & 0 \end{bmatrix}$$

(**a**) Multiply the first row by 0.5.
(**b**) Add to the second row -8 times the first row.
(**c**) Add to the third row -14 times the first row.
(**d**) Multiply the second row by $-\frac{1}{6}$.
(**e**) Add to the third row 12 times the second row.

In Problems 39 through 50, solve the given systems by back-substitution (this is Step 4 in Figure 3).

39. $x + 2y = 4$
$\qquad\quad y = 8$

41. $11x - 8y = 0$
$\qquad\quad -y = 0$

40. $2x - 3y = -1$
$\qquad\qquad 2y = 5$

42. $7x - 11y = 0$
$\qquad\qquad 3y = 1$

43.
$$x + 2y - z = 1$$
$$y + 3z = 2$$
$$z = 6$$

44.
$$x - y - 3z = 0$$
$$2y + 3z = 6$$
$$z = 4$$

45.
$$x - 8y + 7z = 0$$
$$y - 8z = 0$$
$$z = -2$$

46.
$$2x - z = 1$$
$$3y + z = 8$$
$$2z = 5$$

48.
$$3x + 4y - 5z = 2$$
$$6y + 7z = 1$$
$$8z = 9$$

47.
$$2x + 3y = -1$$
$$3y - 4z = 0$$
$$3z = -1$$

49.
$$x - y - 2z + w = 0$$
$$y + 5z - 2w = 1$$
$$z + 8w = 5$$
$$w = 2$$

50.
$$x + 4y - z + 5w = 10$$
$$y + 6z - 7w = 20$$
$$z + 8w = 30$$
$$w = 40$$

51. Show that each of the systems of equations in Problems 39 through 50 has an upper triangular augmented matrix.

CREATING MODELS

52. A Christmas tree vendor carries two types of trees, real evergreens priced at $40 each and artificial trees priced at $65 each. One day, 62 trees are sold and total sales are $2,930. The vendor wants to know how many real trees R and how many artificial trees A were sold. Show that the solution to this problem is a solution to two equations in R and A.

53. A movie theater sells 1,201 tickets for a show and has a total box-office take of $8,510. There are two types of tickets, adult tickets costing $8 and youth tickets costing $5. The theater's manager needs to determine how many adult tickets A and how many youth tickets Y were sold. Show that the solution to this problem is a solution to two equations in A and Y.

54. A movie theater sells 206 tickets for a matinee and has a total box-office take of $1,252. There are three types of tickets, adult tickets for $8, youth tickets for $5, and senior citizen's tickets for $4. Twice as many senior citizen's tickets as youth tickets were sold. The theater's manager wants to determine how many adult tickets a, how many youth tickets y, and how many senior citizen's tickets s were sold. Show that the solution to this problem is a solution to three equations in a, y, and s.

55. An airline sells 207 seats on a flight from Miami to Boston. Some of the seats are coach seats priced at $250 each, and the rest are first-class seats priced at $490 each. The flight generated a total income of $54,150 from the sale of seats. Now the airline wants to know how many coach seats c and how many first-class seats f were sold for the flight. Show that the solution to this problem is a solution to two equations in c and f.

56. An airline sells 270 seats on a flight from Boston to Miami. Coach seats are priced at $250 each, business-class seats are priced at $340 each, and first-class seats are priced at $490 each. The flight generated a total income of $76,680 from the sale of seats. The airline wants to know how many coach seats C, business-class seats B, and first-class seats F were sold for the flight. Show that a model of this problem based on the given information involves two equations in C, B, and F.

57. The Heartland Cereal Company produces JumpStart, a ready-to-eat breakfast food made from a blend of a fortified cereal and a dehydrated dairy product. Each serving of JumpStart is guaranteed to contain exactly 40 units of protein and 60 units of calcium. Heartland's supplier certifies that every ounce of fortified cereal contains 30 units of protein and 5 units of calcium, while every ounce of dehydrated dairy product contains 10 units of protein and 30 units of calcium. Heartland wants to know how much fortified cereal C and dehydrated dairy product D should be blended into each serving of JumpStart to meet the guarantee. Show that the solution to this problem is a solution to two equations in C and D.

58. A caterer must prepare 200 gallons of a fruit punch that is 5% cranberry juice, 20% grapefruit juice, and 75% pineapple juice. This punch will be made from a blend of three commercial drinks that the caterer keeps in stock in very large quantities. Commercial drink A contains 50% pineapple juice and 50% grapefruit juice. Commercial drink B is 15% pineapple juice, 45% grapefruit juice, and 40% cranberry juice, while commercial drink C is 100% pineapple juice. Show that the amount of each commercial drink to be blended into the fruit punch is a solution to three equations in three variables.

EXPLORING IN TEAMS

59. We discovered in Section 1.6 that systems of linear equations can have one solution, no solutions, or infinitely many solutions. Describe real-life situations in which one solution to a problem is more advantageous to people solving the problem than the other two alternatives, and then describe other situations where having many solutions is more advantageous. Based on your examples, which alternative do you prefer?

EXPLORING WITH TECHNOLOGY

60. Most graphing calculators include menus for automatically performing the three elementary row operations. Using this technology reproduce the illustrations on page 87.

61. Solve Problems 33 through 38 using a graphing calculator with menus for elementary row operations.

REVIEWING MATERIAL

62. (Section 1.4) Graph the equation $y = x^3$. Why is this curve not a straight line?

RECOMMENDING ACTION

63. Respond by memo to the following request:

MEMORANDUM

To: J. Doe Reader

From: Payroll

Date: Today

Subject: **Bonuses**

As you know, our company's end-of-the-year employee bonus pool is 10% of profits after city and state taxes have been paid. The city tax is 3% of taxable income, while the state tax is 8% of taxable income with credit allowed for the city tax as a pretax deduction. This year, our profit was $800,000, and taxable income is this profit less the end-of-the-year bonus.

Calculating the bonus is complicated, because taxable income depends on the bonus, the bonus depends on the taxes, and taxes depend on the taxable income. Everything depends on everything else! As a result, Payroll spends days determining these amounts, using a trial and error procedure. Is there a better way?

2.2 GAUSSIAN ELIMINATION

We can solve systems of simultaneous linear equations using the four-step procedure outlined in Figure 2.2 of Section 2.1: (1) construct an augmented matrix for the system, (2) transform this augmented matrix into upper triangular form, (3) write the simplified system of equations associated with the upper triangular augmented matrix, and (4) use back-substitution to solve the simplified system. The technique of using elementary row operations to transform an augmented matrix into upper triangular form is called *Gaussian elimination*.

In an upper triangular augmented matrix, all the entries below the main diagonal are zeros. Most augmented matrices do not have zeros in these positions, so the goal of Gaussian elimination is to place zeros below the main diagonal when they are not there initially. This is done using elementary row operations, beginning with column one and proceeding one entry at a time. After all the elements below the main diagonal in column one are transformed into zeros, we move to column two and work with the entries in this column, one at a time. After all the entries below the main diagonal in column two are transformed into zeros, we move to column three. We move sequentially through the columns until all the columns have zeros where we want them.

The only operations at our disposal are the elementary row operations, because those operations do not alter solutions. For reference, these operations are:

> *Gaussian elimination is the process of using elementary row operations to place zeros below the main diagonal of an augmented matrix.*

A pivot is a nonzero element in an augmented matrix that is used to transform elements directly below it to zero.

In general, *we use each entry on the main diagonal to transform all of the entries directly below it to zero.* A *pivot* is an element in an augmented matrix that is used to transform all elements directly below it to zero. Pivots must be nonzero. If an element on the main diagonal is zero when we need it as a pivot, we use the first elementary row operation to interchange the row containing the zero diagonal entry with any row below it that has a nonzero entry in the same column as the zero diagonal entry.

EXAMPLE 1

Use elementary row operations to transform the augmented matrix

$$\left[\begin{array}{cc|c} 0 & 2 & 4 \\ 1 & 3 & -1 \end{array}\right]$$

into upper triangular form.

Solution

The main diagonal in this augmented matrix consists of the elements connected by the shaded screen, and we want all entries below this line to be zero. We begin with the first column; its diagonal element is the entry in the first row and first column, and it is zero. We check to see if there are any nonzero elements below it in the first column. There is such an element, and it sits in the second row. Therefore, we use the first elementary row operation to interchange the first and second rows. Doing so, we obtain

$$\left[\begin{array}{cc|c} 0 & 2 & 4 \\ 1 & 3 & -1 \end{array}\right] \rightarrow \left[\begin{array}{cc|c} 1 & 3 & -1 \\ 0 & 2 & 4 \end{array}\right] \begin{cases} \text{by interchanging the} \\ \text{first and second rows} \\ R_1 \leftrightarrow R_2 \end{cases} \tag{11}$$

The last augmented matrix in (*11*) is in upper triangular form.

Computations are simplified if we require that pivots always equal one. Thus as a final step in Gaussian elimination, we also require that the first nonzero element in any nonzero row (a row that is not all zero) be one. If an element is not one when we want it to be one, we use the second elementary row operation to transform it to one by multiplying the entire row in which the element sits by the reciprocal of that element.

The final augmented matrix in (*11*) has a one as the first nonzero element in the first row but a two as the first nonzero element in the second row. We transform the two to a one by multiplying entire second row by $\frac{1}{2}$, the reciprocal of two. In particular,

Pivots are transformed into the number one.

Then each nonzero entry k directly below a pivot is transformed to zero by adding −k times the row containing the pivot to the row in which k sits.

$$\begin{bmatrix} 1 & 3 & | & -1 \\ 0 & 2 & | & 4 \end{bmatrix} \rightarrow \begin{bmatrix} 1 & 3 & | & -1 \\ 0 & 1 & | & 2 \end{bmatrix} \begin{cases} \text{by multiplying the} \\ \text{first row by } 1/2 \\ \frac{1}{2} R_2 \rightarrow R_2 \end{cases}$$

Now the resulting matrix is upper triangular, and the first nonzero element of each nonzero row is one.

Once a pivot is one, we use the third elementary row operation to transform to zero all nonzero entries below the pivot. To transform a nonzero entry k to zero, we add $-k$ times the row containing the pivot to the entire row containing k. In the examples that follow, we use a shaded screen to identify the main diagonal, a hat or caret (ˆ) to mark the *position* of the current pivot, a circle around an element to target it as next number we want transformed to zero, and arrows to denote the transformations.

EXAMPLE 2

Use elementary row operations to transform the augmented matrix

$$\begin{bmatrix} 1 & 1 & -2 & | & -3 \\ -1 & -1 & 1 & | & 0 \\ 2 & 3 & -2 & | & 2 \end{bmatrix}$$

into upper triangular form.

Solution

$$\begin{bmatrix} \hat{1} & 1 & -2 & | & -3 \\ \text{\textcircled{-1}} & -1 & 1 & | & 0 \\ 2 & 3 & -2 & | & 2 \end{bmatrix} \rightarrow \begin{bmatrix} \hat{1} & 1 & -2 & | & -3 \\ 0 & 0 & -1 & | & -3 \\ \text{\textcircled{2}} & 3 & -2 & | & 2 \end{bmatrix} \begin{cases} \text{by adding to the second} \\ \text{row 1 times the first row} \\ R_2 + R_1 \rightarrow R_2 \end{cases}$$

$$\rightarrow \begin{bmatrix} 1 & 1 & -2 & | & -3 \\ 0 & \hat{0} & -1 & | & -3 \\ 0 & 1 & 2 & | & 8 \end{bmatrix} \begin{cases} \text{by adding to the third row} \\ -2 \text{ times the first row} \\ R_3 + (-2)R_1 \rightarrow R_3 \end{cases}$$

$$\rightarrow \begin{bmatrix} 1 & 1 & -2 & | & -3 \\ 0 & 1 & 2 & | & 8 \\ 0 & 0 & -\hat{1} & | & -3 \end{bmatrix} \begin{cases} \text{by interchanging the} \\ \text{second and third rows} \\ R_2 \leftrightarrow R_3 \end{cases}$$

$$\rightarrow \begin{bmatrix} 1 & 1 & -2 & | & -3 \\ 0 & 1 & 2 & | & 8 \\ 0 & 0 & 1 & | & 3 \end{bmatrix} \begin{cases} \text{by multiplying the} \\ \text{third row by } -1 \\ (-1)R_3 \rightarrow R_3 \end{cases}$$

EXAMPLE 3

Use elementary row operations to transform the augmented matrix

$$\begin{bmatrix} 2 & 1 & | & 1 \\ 3 & -1 & | & 4 \\ 2 & 2 & | & 0 \end{bmatrix}$$

into upper triangular form.

Solution

$$
\begin{bmatrix} \hat{2} & 1 & | & 1 \\ 3 & -1 & | & 4 \\ 2 & 2 & | & 0 \end{bmatrix} \rightarrow \begin{bmatrix} \hat{1} & \frac{1}{2} & | & \frac{1}{2} \\ ③ & -1 & | & 4 \\ 2 & 2 & | & 0 \end{bmatrix} \left\{ \begin{array}{l} \text{by multiplying the first row} \\ \text{by } \frac{1}{2} \text{ to make the pivot 1} \\ \frac{1}{2}R_1 \rightarrow R_1 \end{array} \right.
$$

$$
\rightarrow \begin{bmatrix} \hat{1} & \frac{1}{2} & | & \frac{1}{2} \\ 0 & -\frac{5}{2} & | & \frac{5}{2} \\ ② & 2 & | & 0 \end{bmatrix} \left\{ \begin{array}{l} \text{by adding to the second} \\ \text{row } -3 \text{ times the first row} \\ R_2 + (-3)R_1 \rightarrow R_2 \end{array} \right.
$$

$$
\rightarrow \begin{bmatrix} 1 & \frac{1}{2} & | & \frac{1}{2} \\ 0 & -\frac{5}{2} & | & \frac{5}{2} \\ 0 & 1 & | & -1 \end{bmatrix} \left\{ \begin{array}{l} \text{by adding to the third row} \\ -2 \text{ times the first row} \\ R_3 + (-2)R_1 \rightarrow R_3 \end{array} \right.
$$

$$
\rightarrow \begin{bmatrix} 1 & \frac{1}{2} & | & \frac{1}{2} \\ 0 & \hat{1} & | & -1 \\ 0 & ① & | & -1 \end{bmatrix} \left\{ \begin{array}{l} \text{by multiplying the second row} \\ \text{by } -2/5 \text{ to make the pivot 1} \\ (-2/5)R_2 \rightarrow R_2 \end{array} \right.
$$

$$
\rightarrow \begin{bmatrix} 1 & \frac{1}{2} & | & \frac{1}{2} \\ 0 & 1 & | & -1 \\ 0 & 0 & | & 0 \end{bmatrix} \left\{ \begin{array}{l} \text{by adding to the third row} \\ -1 \text{ times the second row} \\ R_3 + (-1)R_2 \rightarrow R_3 \end{array} \right.
$$

EXAMPLE 4

Use elementary row operations to transform the augmented matrix

$$
\begin{bmatrix} 5 & 10 & 30 & | & 122 \\ 10 & 30 & 100 & | & 340 \\ 30 & 100 & 354 & | & 1156 \end{bmatrix} \tag{12}
$$

into upper triangular form.

Solution

$$
\begin{bmatrix} \hat{5} & 10 & 30 & | & 122 \\ 10 & 30 & 100 & | & 340 \\ 30 & 100 & 354 & | & 1156 \end{bmatrix} \rightarrow \begin{bmatrix} \hat{1} & 2 & 6 & | & 24.4 \\ ⑩ & 30 & 100 & | & 340 \\ 30 & 100 & 354 & | & 1156 \end{bmatrix} \left\{ \begin{array}{l} \text{by multiplying the first row} \\ \text{by } 1/5 \text{ to make the pivot 1} \\ (1/5)R_1 \rightarrow R_1 \end{array} \right.
$$

$$
\rightarrow \begin{bmatrix} \hat{1} & 2 & 6 & | & 24.4 \\ 0 & 10 & 40 & | & 96 \\ ㉚ & 100 & 354 & | & 1156 \end{bmatrix} \left\{ \begin{array}{l} \text{by adding to the second row} \\ -10 \text{ times the first row} \\ R_2 + (-10)R_1 \rightarrow R_2 \end{array} \right.
$$

$$
\rightarrow \begin{bmatrix} 1 & 2 & 6 & | & 24.4 \\ 0 & 1\hat{0} & 40 & | & 96 \\ 0 & 40 & 174 & | & 424 \end{bmatrix} \left\{ \begin{array}{l} \text{by adding to the third row} \\ -30 \text{ times the first row} \\ R_3 + (-30)R_1 \rightarrow R_3 \end{array} \right.
$$

$$
\rightarrow \begin{bmatrix} 1 & 2 & 6 & | & 24.4 \\ 0 & \hat{1} & 4 & | & 9.6 \\ 0 & ㊵ & 174 & | & 424 \end{bmatrix} \left\{ \begin{array}{l} \text{by multiplying the second row} \\ \text{by } 1/10 \text{ to make the pivot 1} \\ (1/10)R_2 \rightarrow R_2 \end{array} \right.
$$

$$\rightarrow \begin{bmatrix} 1 & 2 & 6 & | & 24.4 \\ 0 & 1 & 4 & | & 9.6 \\ 0 & 0 & 14 & | & 40 \end{bmatrix} \quad \begin{cases} \text{by adding to the third row} \\ -40 \text{ times the second row} \\ R_3 + (-40)R_2 \rightarrow R_3 \end{cases}$$

$$\rightarrow \begin{bmatrix} 1 & 2 & 6 & | & 24.4 \\ 0 & 1 & 4 & | & 9.6 \\ 0 & 0 & 1 & | & 2.857 \end{bmatrix} \quad \begin{cases} \text{by multiplying the third row by 1/14} \\ \text{and rounding the result to three} \\ \text{decimal places } (1/14)R_3 \rightarrow R_3 \end{cases} \qquad (13)$$

We now have developed all the steps we need to solve systems of linear equations using the algorithm in Figure 2.2.

EXAMPLE 5

Solve the system

$$5x + 10y + 30z = 122$$
$$10x + 30y + 100z = 340$$
$$30x + 100y + 354z = 1156$$

Solution

The augmented matrix for this system is Augmented Matrix (*12*), which can be transformed into the upper triangular matrix (*13*), as we did in Example 4. The system of equations corresponding to Augmented Matrix (*13*) is

$$x + 2y + 6z = 24.4$$
$$y + 4z = 9.6$$
$$z = 2.857$$

The solution to this last system was found in Example 3 of Section 2.1: $x = 10.914$, $y = -1.828$, $z = 2.857$. This is also the solution (rounded to three decimal places) of the original system of equations.

EXAMPLE 6

Solve the system

$$x + y - 2z = -3$$
$$-x - y + z = 0$$
$$2x + 3y - 2z = 2$$

Solution

The augmented matrix for this system is

$$\begin{array}{ccc} x & y & z \end{array}$$
$$\begin{bmatrix} 1 & 1 & -2 & | & -3 \\ -1 & -1 & 1 & | & 0 \\ 2 & 3 & -2 & | & 2 \end{bmatrix}$$

which, as we saw in Example 2, can be transformed into the upper triangular matrix

$$\begin{array}{ccc} x & y & z \end{array}$$
$$\left[\begin{array}{ccc|c} 1 & 1 & -2 & -3 \\ 0 & 1 & 2 & 8 \\ 0 & 0 & 1 & 3 \end{array}\right]$$

The system of equations corresponding to this augmented matrix is

$$x + y - 2z = -3$$
$$y + 2z = 8$$
$$z = 3$$

We found in Section 2.1 that the solution to this last system is $x = 1$, $y = 2$, $z = 3$, which is also the solution to the original system of equations.

EXAMPLE 7

Solve the system

$$2p + q = 1$$
$$3p - q = 4$$
$$2p + 2q = 0$$

Solution

The augmented matrix for this sytem is

$$\begin{array}{cc} p & q \end{array}$$
$$\left[\begin{array}{cc|c} 2 & 1 & 1 \\ 3 & -1 & 4 \\ 2 & 2 & 0 \end{array}\right]$$

which, as we saw in Example 3, can be transformed into the upper triangular matrix

$$\begin{array}{cc} p & q \end{array}$$
$$\left[\begin{array}{cc|c} 1 & \frac{1}{2} & \frac{1}{2} \\ 0 & 1 & -1 \\ 0 & 0 & 0 \end{array}\right]$$

The system of equations corresponding to this augmented matrix is

$$p + \tfrac{1}{2}q = \tfrac{1}{2}$$
$$q = -1$$
$$0 = 0$$

The solution to this last set of equations is easily determined by back-substitution. Clearly $q = -1$. Substituting this value into the first equation, we obtain $p + \frac{1}{2}(-1) = \frac{1}{2}$ or $p = 1$. The solution to both this last set of equations and the original system is $p = 1$, $q = -1$.

Most of the work in solving each of the last three examples involves using elementary row operations to transform an augmented matrix into upper triangular form. Fortunately, these operations are automated on most graphing calculators and in specialized software

> *Elementary row operations can replace the computations associated with back-substitution.*

packages designed for matrix manipulations. Keystrokes differ from one machine to the next, so a user's manual must be consulted for the correct sequence. Such technology greatly reduces the work required to solve systems of linear equations, and this technology should be used whenever it is available.

We can use elementary row operations to replace the computations associated with back-substitution. This is particularly relevant if we have access to technology that automates the elementary row operations, because we can use that technology to automate back-substitution. Looking again at Example 6, we derived the system

$$x + y - 2z = -3$$
$$y + 2z = 8$$
$$z = 3$$

Solving by back-substitution we immediately have $z = 3$. Substituting this value of z into the second equation, we *eliminate z from the second equation*, obtaining $y = 2$. Then, we substitute these values of y and z into the first equation, *eliminating y and z from the first equation*, obtaining $x = 1$. Notice that eliminating variables from equations is equivalent to placing zero coefficients into an augmented matrix!

A system of equations is ready for solution by back-substitution when its augmented matrix is upper triangular and when all nonzero elements on the main diagonal are ones. If we now use elementary row operations to place zeros *above* each one on the main diagonal, we effectively eliminate variables from such equations and eliminate the need for back-substitution. This procedure is called *Gauss-Jordan reduction*.

GAUSS-JORDAN REDUCTION:

STEP 1 Use Gaussian elimination to transform an augmented matrix [**A** | **b**] into the augmented matrix [**C** | **d**] where **C** is an upper triangular matrix with all nonzero elements on the main diagonal set to one.

STEP 2 Beginning with the last column of **C** having a one on its main diagonal and progressing backward sequentially to the second column, use only the third elementary row operation to transform all elements in **C** above the main diagonal to zero. Complete all work on one column before moving to another column, and apply all operations to the entire augmented matrix.

EXAMPLE 8

Use Gauss-Jordan reduction to solve the system

$$x + y - 2z = -3$$
$$-x - y + z = 0$$
$$2x + 3y - 2z = 2$$

Solution
The augmented matrix for this system is

$$
\begin{array}{ccc}
x & y & z
\end{array}
$$
$$
\begin{bmatrix}
1 & 1 & -2 & -3 \\
-1 & -1 & 1 & 0 \\
2 & 3 & -2 & 2
\end{bmatrix}
$$

which, as we saw in Example 2, can be transformed into the upper triangular matrix

$$
\begin{array}{ccc}
x & y & z
\end{array}
$$
$$
\begin{bmatrix}
1 & 1 & -2 & -3 \\
0 & 1 & 2 & 8 \\
0 & 0 & 1 & 3
\end{bmatrix}
$$

This completes Step 1 of Gauss-Jordan reduction. We now begin with column 3 and use elementary row operations to place zeros *above* the one on the main diagonal. Once this is done, we move to column 2 and place a zero *above* the one on the main diagonal in that column. Sequentially, we have

$$
\begin{bmatrix}
1 & 1 & -2 & -3 \\
0 & 1 & 2 & 8 \\
0 & 0 & 1 & 3
\end{bmatrix}
\rightarrow
\begin{bmatrix}
1 & 1 & -2 & -3 \\
0 & 1 & 0 & 2 \\
0 & 0 & 1 & 3
\end{bmatrix}
\begin{cases}
\text{by adding to the second} \\
\text{row } -2 \text{ times the third row} \\
R_2 + (-2)R_3 \rightarrow R_2
\end{cases}
$$

which eliminates z from the second equation, and

$$
\rightarrow
\begin{bmatrix}
1 & 1 & 0 & 3 \\
0 & 1 & 0 & 2 \\
0 & 0 & 1 & 3
\end{bmatrix}
\begin{cases}
\text{by adding to the first} \\
\text{row 2 times the third row} \\
R_1 + 2R_3 \rightarrow R_1
\end{cases}
$$

which eliminates z from the first equation.

$$
\rightarrow
\begin{bmatrix}
1 & 0 & 0 & 1 \\
0 & 1 & 0 & 2 \\
0 & 0 & 1 & 3
\end{bmatrix}
\begin{cases}
\text{by adding to the first row} \\
-1 \text{ times the second row} \\
R_1 + (-1)R_2 \rightarrow R_1
\end{cases}
$$

The last step eliminates y from the first equation. The system of equations corresponding to this last augmented matrix is $x = 1$, $y = 2$, and $z = 3$, which also gives the solution to the original system without requiring any back-substitution.

A system of linear equations has no solution if one of the equations associated with the augmented matrix in upper triangular form is false.

Each of the previous examples in this section dealt with systems having a unique (only one) solution. As we know, a system does not always have a unique solution. For example, we showed in Figure 1.41 of Section 1.6 that the graphs of the equations

$$
\begin{aligned}
x - y &= 1 \\
x - y &= -1
\end{aligned}
\tag{14}
$$

are parallel straight lines with no points of intersection and that this system of equations therefore has no solution. Let us see how our general method for solving systems of equations handles System (*14*). The augmented matrix for System (*14*) and its transformation to upper triangular form are as follows:

FIGURE 2.3

FIGURE 2.4

FIGURE 2.5

FIGURE 2.6

$$\begin{bmatrix} \hat{1} & -1 & | & 1 \\ ① & -1 & | & -1 \end{bmatrix} \rightarrow \begin{bmatrix} 1 & -1 & | & 1 \\ 0 & 0 & | & -2 \end{bmatrix} \quad R_2 + (-1)R_1 \rightarrow R_2$$

The equations corresponding to the last augmented matrix are

$$x - y = 1$$
$$0 = -2$$

The second equation is clearly false! We conclude that a system of linear equations has *no* solution if the equations associated with its augmented matrix in upper triangular form contain a false equation such as $0 = -2$.

Graphical methods are useful for representing equations when we can assign a different axis to each variable. If a problem involves two variables, we need only two axes and the equations can be graphed on a plane. If a problem involves three variables, we need three axes and the equations must be graphed in space. The graph of a linear equation in three variables is a plane in space. A system of three linear equations in three variables is represented graphically by a set of three planes, one plane for each equation. A solution to such a system is a point in space that is simultaneously on all three planes.

Analyzing graphs of planes should therefore reinforce our understanding of the solutions that can exist for different types of systems. Figure 2.3 shows an example in which three planes intersect at a single point; it represents a system of three linear equations in three variables that has a unique solution. The linear systems graphed in Figures 2.4 and 2.5 have no solutions. In Figure 2.4, the three planes are parallel, and there is no point that lies on any two planes, much less on all three. In Figure 2.5, there are infinitely many points of intersection between any two of the three planes, but there is no point that lies on all three planes. Figure 2.6 is a graph of a linear system with infinitely many solutions, all with coordinates on the line common to the three planes. Another example of a system with infinitely many solutions can be obtained by collapsing the three planes in Figure 2.4 on to the same plane, that is, by taking two of the planes to be copies of the third. If Gaussian elimination and Gauss-Jordan reduction are to be truly general, then they must identify multiple solutions in situations such as these.

Observe that using back-substitution for a system of equations associated with an augmented matrix in upper triangular form solves the equations sequentially, from the bottom up, by solving each equation for the *first* variable that has a coefficient of one. The same is true for Gauss-Jordan reduction, except there we no longer require back-substitution to solve the resulting system. In both methods, we solve each equation for the *first* variable that appears in that equation. These are the variables associated with the coefficients of one on the main diagonal. Variables that do not appear first in any equation can be assigned any numbers we choose. Other variables (variables that appear first in some other equation) are specified in terms of known numbers and these arbitrary variables.

EXAMPLE 9

Solve the system

$$x + 2y + z = 3{,}000$$
$$2x + 5y + 3z = 7{,}000$$
$$3x + 7y + 4z = 10{,}000$$

Solution

We use Gauss-Jordan reduction to eliminate the need for back-substitution. First, we must transform the augmented matrix for this system into upper triangular form. Doing so, we obtain

$$\left[\begin{array}{ccc|c} \hat{1} & 2 & 1 & 3{,}000 \\ ② & 5 & 3 & 7{,}000 \\ 3 & 7 & 4 & 10{,}000 \end{array}\right] \rightarrow \left[\begin{array}{ccc|c} \hat{1} & 2 & 1 & 3{,}000 \\ 0 & 1 & 1 & 1{,}000 \\ ③ & 7 & 4 & 10{,}000 \end{array}\right] R_2 + (-2)R_1 \rightarrow R_2$$

$$\rightarrow \left[\begin{array}{ccc|c} 1 & 2 & 1 & 3000 \\ 0 & \hat{1} & 1 & 1000 \\ 0 & ① & 1 & 1000 \end{array}\right] R_3 + (-3)R_1 \rightarrow R_3$$

$$\rightarrow \left[\begin{array}{ccc|c} 1 & 2 & 1 & 3000 \\ 0 & 1 & 1 & 1000 \\ 0 & 0 & 0 & 0 \end{array}\right] R_3 + (-1)R_2 \rightarrow R_3$$

Next, we use the third elementary row operation to place zeros above the ones on the main diagonal. The last column having a one on its main diagonal is the second column, so we begin with it, obtaining

$$\left[\begin{array}{ccc|c} 1 & ② & 1 & 3000 \\ 0 & \hat{1} & 1 & 1000 \\ 0 & 0 & 0 & 0 \end{array}\right] \rightarrow \left[\begin{array}{ccc|c} 1 & 0 & -1 & 1000 \\ 0 & 1 & 1 & 1000 \\ 0 & 0 & 0 & 0 \end{array}\right] R_1 + (-2)R_2 \rightarrow R_1$$

The equations corresponding to this last augmented matrix are

$$x \qquad - z = 1000$$
$$y + z = 1000$$
$$0 = 0$$

The last equation is true but uninteresting, so we ignore it. We use the first equation to solve for x and the second equation to solve for y, because in these two equations they are the first variables to appear with a coefficient of one. There is no equation in which z appears as the first variable, so we may assign an arbitrary value to z and then solve for y and x in terms of z. The first equation becomes $x = 1000 + z$, and the second equation becomes $y = 1000 - z$. The solution is $x = 1000 + z$, $y = 1000 - z$, and z is arbitrary.

Since z is arbitrary, we can assign it any value we please. If we set $z = 1{,}500$, then $x = 2{,}500$ and $y = -500$; this set of three values (for the three variables) comprises one solution. Setting $z = -400$, we obtain $x = 600$ and $y = 1{,}400$ as a second set of values that solves the original set of equations. In fact, every choice of z leads to a different solution to the original system. To emphasize this choice, we leave the solution in terms of arbitrary z.

EXAMPLE 10

Solve the system

$$a + 2b - c \qquad = 0$$
$$2a + b - c + 3d = 0$$

Solution

We use Gauss-Jordan reduction to eliminate the need for back-substitution. Transforming the augmented matrix for this system into upper triangular form, we obtain

$$\left[\begin{array}{ccc|c} \hat{1} & 2 & -1 & 0 \\ ② & 1 & -1 & 3 \end{array}\middle| \begin{array}{c} 0 \\ 0 \end{array}\right] \rightarrow \left[\begin{array}{ccc|c} 1 & 2 & -1 & 0 \\ 0 & -3 & 1 & 3 \end{array}\middle| \begin{array}{c} 0 \\ 0 \end{array}\right] \quad R_2 + (-2)R_1 \rightarrow R_2$$

$$\rightarrow \left[\begin{array}{ccc|c} 1 & 2 & -1 & 0 \\ 0 & 1 & -\frac{1}{3} & -1 \end{array}\middle| \begin{array}{c} 0 \\ 0 \end{array}\right] \quad (-\tfrac{1}{3})R_2 \rightarrow R_2$$

Next, we use the third elementary row operation to place zeros above the ones on the main diagonal. The last column having a one on its main diagonal is the second column, so we begin with it, obtaining

$$\left[\begin{array}{ccc|c} 1 & ② & -1 & 0 \\ 0 & \hat{1} & -\frac{1}{3} & -1 \end{array}\middle| \begin{array}{c} 0 \\ 0 \end{array}\right] \rightarrow \left[\begin{array}{ccc|c} 1 & 0 & -\frac{1}{3} & 2 \\ 0 & 1 & -\frac{1}{3} & -1 \end{array}\middle| \begin{array}{c} 0 \\ 0 \end{array}\right] \quad R_1 + (-2)R_2 \rightarrow R_1$$

The equations corresponding to the last augmented matrix are

$$a \qquad - \tfrac{1}{3}c + \quad 2d = 0$$

$$b - \tfrac{1}{3}c - \quad d = 0$$

We use the first equation to solve for a and the second equation to solve for b, because they are the variables that first appear with a coefficient of one in these equations. There are no equations in which c or d appears as the first variable with a coefficient of one, so c and d are arbitrary, and we solve for a and b in terms of c and d. The complete solution is $a = \tfrac{1}{3}c - 2d$, $b = \tfrac{1}{3}c + d$, with both c and d arbitrary.

IMPROVING SKILLS

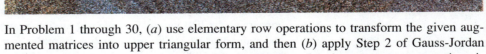

In Problem 1 through 30, (a) use elementary row operations to transform the given augmented matrices into upper triangular form, and then (b) apply Step 2 of Gauss-Jordan reduction to each result. If technology for automating the elementary row operations is available, use it.

1. $\left[\begin{array}{cc|c} 1 & 2 & 1 \\ 3 & 7 & -1 \end{array}\right]$

2. $\left[\begin{array}{cc|c} 1 & -1 & 2 \\ 4 & -3 & 3 \end{array}\right]$

3. $\left[\begin{array}{cc|c} 1 & 1 & 0 \\ -1 & 0 & 2 \end{array}\right]$

4. $\left[\begin{array}{cc|c} 1 & 2 & 7 \\ -5 & -9 & 0 \end{array}\right]$

5. $\left[\begin{array}{cc|c} 0 & 1 & 10 \\ 1 & 2 & 20 \end{array}\right]$

6. $\left[\begin{array}{cc|c} 1 & 2 & 7 \\ -5 & -8 & 0 \end{array}\right]$

7. $\left[\begin{array}{cc|c} 0 & 1 & 3 \\ 2 & 1 & 4 \end{array}\right]$

8. $\left[\begin{array}{cc|c} 0 & 3 & -9 \\ -2 & 5 & -9 \end{array}\right]$

9. $\begin{bmatrix} 1 & 1 & | & 1 \\ -2 & 3 & | & 1 \end{bmatrix}$

10. $\begin{bmatrix} 1 & -2 & | & 1 \\ -2 & 1 & | & 1 \end{bmatrix}$

11. $\begin{bmatrix} 2 & 4 & | & 8 \\ -3 & 1 & | & 3 \end{bmatrix}$

12. $\begin{bmatrix} 2 & 4 & | & 5 \\ 3 & -4 & | & 0 \end{bmatrix}$

13. $\begin{bmatrix} 1 & 2 & 3 & | & -2 \\ 4 & 5 & 6 & | & 0 \end{bmatrix}$

14. $\begin{bmatrix} 4 & 4 & 3 & | & -12 \\ -2 & 1 & -1 & | & 10 \end{bmatrix}$

15. $\begin{bmatrix} 1 & 3 & | & -1 \\ 2 & 5 & | & -1 \\ 3 & 5 & | & 0 \end{bmatrix}$

16. $\begin{bmatrix} 1 & 3 & | & 2 \\ 1 & -1 & | & 1 \\ 4 & 1 & | & 2 \end{bmatrix}$

17. $\begin{bmatrix} \frac{1}{2} & \frac{1}{4} & 3 & | & -5 \\ 0 & \frac{1}{3} & \frac{3}{4} & | & -8 \\ 0 & 0 & 4 & | & \frac{3}{4} \end{bmatrix}$

18. $\begin{bmatrix} 1 & 0 & 1 & | & 4 \\ 1 & 1 & 0 & | & 5 \\ 0 & 1 & 1 & | & 6 \end{bmatrix}$

19. $\begin{bmatrix} \frac{1}{2} & 1 & 3 & | & -1 \\ 1 & \frac{1}{2} & 2 & | & 0 \\ 0 & 0 & 1 & | & 1 \end{bmatrix}$

20. $\begin{bmatrix} 1 & 0 & 1 & | & 4 \\ 0 & 1 & 0 & | & 5 \\ 1 & 1 & 1 & | & 6 \end{bmatrix}$

21. $\begin{bmatrix} 1 & 0 & 1 & | & 4 \\ 0 & 2 & 0 & | & 5 \\ 1 & 1 & 1 & | & 6 \end{bmatrix}$

22. $\begin{bmatrix} 1 & 0 & 1 & | & 4 \\ 1 & 1 & 0 & | & 5 \\ 1 & 1 & 1 & | & 6 \end{bmatrix}$

23. $\begin{bmatrix} 1 & 1 & 1 & | & 1 \\ 1 & 1 & 0 & | & 0 \\ 2 & 1 & 0 & | & -1 \end{bmatrix}$

24. $\begin{bmatrix} 1 & 1 & 1 & | & -1 \\ 2 & 1 & 0 & | & 1 \\ 1 & 0 & 1 & | & -2 \end{bmatrix}$

25. $\begin{bmatrix} 2 & 1 & 1 & | & 3 \\ 3 & 0 & 4 & | & 2 \\ 4 & -1 & 3 & | & 0 \end{bmatrix}$

26. $\begin{bmatrix} 2 & 1 & 0 & | & 7 \\ 0 & 3 & 1 & | & 1 \\ 5 & 0 & 1 & | & 2 \end{bmatrix}$

27. $\begin{bmatrix} 1 & 2 & 3 & | & 0 \\ 4 & 5 & 6 & | & 0 \\ 7 & 8 & 9 & | & 0 \end{bmatrix}$

28. $\begin{bmatrix} 1 & 2 & 3 & | & 0 \\ 4 & 5 & 6 & | & 0 \\ 7 & 8 & 9 & | & 1 \end{bmatrix}$

29. $\begin{bmatrix} 5 & 1 & 3 & | & -200 \\ 1 & 2 & -1 & | & -100 \\ 1 & 1 & 1 & | & -500 \end{bmatrix}$

30. $\begin{bmatrix} -2 & 0 & 1 & | & 40{,}000 \\ 3 & 1 & 1 & | & 20{,}000 \\ 2 & 1 & 0 & | & 30{,}000 \end{bmatrix}$

In Problems 31 through 50, use either Gaussian elimination or Gauss-Jordan reduction to solve the given systems of linear equations. Use any available technology for automating the elementary row operations.

31. $\begin{aligned} x + y &= 3 \\ 2x + 5y &= 7 \end{aligned}$

32. $\begin{aligned} x + 2y &= 3 \\ 2x + 6y &= 7 \end{aligned}$

33. $\begin{aligned} 4a + 2b &= 1 \\ 7a - 3b &= 4 \end{aligned}$

34. $\begin{aligned} 5p - 10q &= 80 \\ -p + 2q &= -16 \end{aligned}$

35. $\begin{aligned} 3N + 4S &= 12 \\ 2N + S &= 6 \end{aligned}$

36. $\begin{aligned} 2x + 4y &= -8 \\ 2x + 6y &= 10 \end{aligned}$

37. $\begin{aligned} 2.1s - 15t &= 3.7 \\ 4s + 7.4t &= 4 \end{aligned}$

38. $\begin{aligned} \tfrac{1}{2}x + 2y &= 0 \\ 3x - \tfrac{1}{2}y &= 0 \end{aligned}$

39. $5x + 7y = 3$
$4x + 2y = 5$
$6x + 12y = 1$

40. $5x + 7y = 3$
$4x + 2y = 5$
$6x + 12y = 7$

41. $x + 2y - z = 1$
$2x + 3y - 2x = 4$
$3x - y + 5z = 0$

42. $3a + 4b - 4c = 0$
$a - 2b + 3c = 1$
$-2a + b - c = 2$

43. $x + y - 2z = 3$
$2x + 4y - 3z = 4$
$x + 3y - z = 1$

44. $u + 2v - w = 1$
$u - 3v + w = 2$
$u + v - 4w = -1$

45. $x + y - 2z = 3$
$2x + 4y - 3z = 4$
$x + 3y - z = 0$

46. $u - v + 2w = 0$
$2u + 3v - w = 0$
$3u + 2v + w = 0$

47. $2N + 3S - 2E = 300$
$2N + 4S - 3E = 300$
$4N - 3S + 2E = 300$

48. $u - v + 2w = 0$
$2u - 2v + 4w = 0$
$3u - 3v + 6w = 0$

49. $x + 2y - z + w = 2$
$-2x - 3y - z - w = 4$
$4x + 2y - 4z + 5w = 0$
$-3x - 3y + 2z - 7w = 14$

50. $2x + 3y - z + 2w = 1$
$4x + 6y - 2z + 4w = 2$
$3x + y - 3z + w = 2$
$5x + 4y - 3z + w = 3$

CREATING MODELS

Model and solve Problems 51 through 60.

51. A mattress company produces two different styles of mattresses. The deluxe style uses 50 springs and 12 lbs of padding per mattress, while the regular style uses 35 springs and 10 lbs of padding per mattress. The company has in inventory 16,900 springs and 4,440 lbs of padding. How many mattresses of each type should the company produce if it wants to deplete its inventory?

52. A Christmas tree vendor carries two types of trees, real evergreens priced at $40 per tree and artificial trees priced at $65 a tree. One day, the vendor sells 62 trees and takes in $2,930 in total sales. How many trees of each type were sold?

53. A movie theater sells 1,201 tickets for a show with a total box-office take of $8,510. There are two types of tickets, adult tickets that cost $8 and youth tickets that cost $5. How many tickets of each type were sold for that show?

54. A movie theater sells 206 tickets for a matinee and has a total box-office take of $1,252. There are three types of tickets: adult tickets cost $8, youth tickets cost $5, and senior citizen's tickets cost $4. Twice as many senior citizen's tickets as youth tickets were sold. How many tickets of each type were sold?

55. An airline sells 207 seats on a flight from Miami to Boston. Some of the seats are coach seats priced at $250 each, and the rest are first-class seats priced at $490 each. How many seats of each type were sold if the flight generated a total ticket revenue of $54,150?

56. Continental Motors manufactures taxis and airport limousines at its two plants. Plant *A* produces 75 taxis and 6 limousines daily, while plant *B* produces 50 taxis and 2 limousines each day. To fulfill its contractual obligations for November, the company must manufacture 3,125 taxis and 206 limousines. How many days should the company operate each plant if it wants to exactly meet its contracts?

57. Continental Motors retools its two plants to produce vans as well as taxis and airport limousines. Plant *A* now produces 75 taxis, 12 vans, and 6 limousines daily, while plant B produces 50 taxis, 5 vans, and 2 limousines each day. To fulfill its contractual obligations for December, the company must manufacture 3,125 taxis, 384 vans, and 206 limousines. How many days should the company operate each plant if it wants to exactly meet its contracts?

58. Continental Motors has purchased an additional plant that manufactures vans and airport limousines. Plant A produces 75 taxis, 12 vans, and 6 limousines daily, while plant B produces 50 taxis, 5 vans, and 2 limousines each day, and plant C produces 30 vans and 15 limousines each day but no taxis. To fulfill its contractual obligations for January, the company must manufacture 2,485 taxis, 934 vans, and 915 limousines. How many days should the company operate each plant if it wants to exactly meet its contracts?

59. A gasoline company has 180,000 gallons of low-octane gasoline and 120,000 gallons of high-octane gasoline in inventory, and it mixes them to produce regular, plus, and supreme blends for its gasoline stations. Each gallon of regular consists of 70% low-octane gasoline and 30% high-octane gasoline. Each gallon of plus consists of 60% low-octane gasoline and 40% high-octane gasoline, while each gallon of supreme consists of 55% low-octane gasoline and 45% high-octane gasoline. How many gallons of each blend should the company produce if it wants to deplete all its inventory?

60. A cabinet manufacturer produces cabinets for television consoles, lamp bases, and frames for grandfather clocks, all of which must be assembled and decorated. Each television cabinet requires 3 hours to assemble and 2 hours to decorate, each grandfather clock frame requires 10 hours to assemble and 7 hours to decorate, and each lamp base requires 1 hour to assemble and 1 hour to decorate. The firm has 5,900 hours of assembly time and 4,300 hours of decorating time available. How many television cabinets, lamp bases, and grandfather clock frames should it produce if it wants to use all the available time?

EXPLORING IN TEAMS

For a few systems of equations (which appear rarely in practice) Gaussian elimination and Gauss-Jordan reduction must be modified slightly before they can be used to generate solutions. Solve the systems given in Problems 61 through 63, modifying as needed.

61. $x + 2y + 3z = 4$
    $2x + 4y + 2z = 4$
    $3x + 6y + 8z = 8$

62. $x + 2y + 3z + w = 4$
    $2x + 4y + 2z - 2w = 4$
    $3x + 6y + 8z + 4w = 4$

63. $x + 2y + 3z = 4$
    $2x + 4y + 6z = 8$
    $2x + 4y + 5z = 4$

EXPLORING WITH TECHNOLOGY

64. The constraint equations in Example 6 of Section 2.1 were

$$A + H = 145$$
$$A - H = 41$$

(*a*) Using a graphing calculator, graph each of these equations in the same viewing window and then use the zoom feature on the calculator to find the point of intersection.

(*b*) Solve this system of equations by Gauss-Jordan reduction, using a graphing calculator to perform the required elementary row operations.

(*c*) Many graphing calculators have a menu for solving systems of equations directly. Use this feature, if it is available, to solve the given system of equations.

Which of these methods do you prefer?

65. Apply each of the three solution procedures described in Problem 64 to the system of equations

$$4x + 3y = 800$$
$$3x + 4y = 500$$

All solutions should be accurate to at least three decimal places. Which method do you prefer?

REVIEWING MATERIAL

66. (Section 1.5) Find the equation of the straight line that passes through the two points (100, 200) and (100, 500), and then find the equation of the straight line that passes through the point (50, 60) and has slope $-\frac{1}{2}$. Are these two lines perpendicular?

RECOMMENDING ACTION

67. Respond by memo to the following request:

MEMORANDUM

To: J. Doe Reader

From: Brett

Date: Today

Subject: **Solutions**

I was speaking with Harvey today, and he mentioned that any system of linear equations with more unknowns than equations must have infinitely many solutions. According to Harvey, a system of two equations in three variables has infinitely many solutions, as does a system of three equations in five variables. The only requirement is to have fewer equations than unknowns.

I don't see why. I asked Harvey, but he said it was obvious from Gaussian elimination. Well, you know Harvey. To him everything is obvious. Not to me, though. Is Harvey correct, and if so, why?

2.3 MODELING DATA WITH STRAIGHT LINES

As we noted in Section 1.4, data sometimes follows a pattern. When trends in data are apparent, they often reflect a significant relationship between important factors in a real-world environment, and these relationships can be modeled. Figure 2.7 graphs data gathered by the U.S. National Center for Health Statistics between 1975 and 1987 on the median age of women at the time of their first marriage. There is clearly a trend here, and it looks very much like a straight-line relationship!

> *Variations from expected patterns due to outside forces are called noise.*

Keep in mind when analyzing data that distortions are expected. For example, suppose you ask a friend to drive a car down the middle of a straight road. To assist your friend, you paint a line down the middle of the road. Now you attach to the back of the car a paint pot that intermittently releases drops of paint. If your friend drives flawlessly, all the paint spots left by the car will land on the line you painted—but that is not likely to happen! Gusts of wind, muscle twitches in the hands on the steering wheel, road imperfections, distractions such as a cat darting across the road or a bird flying towards the windshield, and a host of other random events will all conspire to move the car slightly from its intended path. In fact, the paint spots might look very much like the data points in Figure 2.7—close to a straight line but with variations. Variations from expected patterns due to external forces are called *noise*, and data affected by noise is called *noisy data*.

FIGURE 2.7

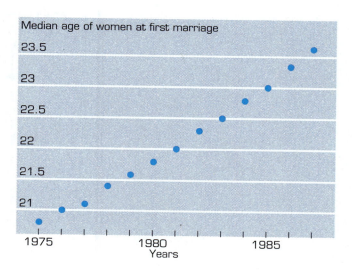

EXAMPLE 1

Discuss some influences that might contribute to noise in the data displayed in Figure 2.7.

Solution

Economic recessions, which have a tendency to delay marriages, high interest rates, which make borrowing more difficult and tend to delay marriages, the threat of war, which tends to accelerate marriage plans, and sudden changes in tax codes, which alter the financial benefits of marriage.

All data are noisy, some much more than others. When data seems to follow a pattern, we attempt to identify the pattern, most often by constructing the curve that "best fits" the data. We then take the position that the curve is the real pattern and that *the data reflect the system's attempt to follow the curve in the presence of noise.* Because the data in Figure 2.7 appear to follow a straight line, we will find the straight line that best fits the data, and the equation of that straight line will be our model for the behavior of the system. To create such a model, we must determine what is meant by "best fit" and then develop a method to produce curves that satisfy this criterion.

In this section, we will concentrate on data that follow straight-line patterns, which have equations of the form $y = mx + b$ (see Section 1.5). A straight line has one y value for each value of x, and each y value may or may not agree with a data point. Thus, for values of x for which data are available, we generally have two values of y: one value from the data and a second value from the straight-line approximation to the data (see Figure 2.8). The *residual* at each x, designated by $r(x)$, is the y value of the data point less the y value obtained from the straight-line approximation. Thus, $r(1)$ is the residual at $x = 1$, $r(2)$ is the residual at $x = 2$, and $r(1975)$ is the residual at $x = 1975$. If a data point falls above the line, its residual will be positive; data that fall below the line have negative residuals.

> *The residual at each x is the y value of the data point less the y value obtained from the straight-line approximation.*

FIGURE 2.8

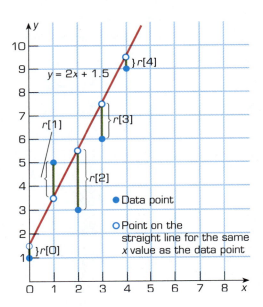

EXAMPLE 2

Calculate the residuals generated by approximating the data given in Figure 2.8 with the line $y = 2x + 1.5$.

Solution

The line and the given data points are plotted in Figure 2.8. There are residuals at $x = 0$, $x = 1$, $x = 2$, $x = 3$, and $x = 4$, which are the only x values for which data are supplied. We read the y-values for each data point directly from Figure 2.8, and the y values are shown with their corresponding x values in Table 2.1.

TABLE 2.1

x	0	1	2	3	4
y (from data)	1	5	3	6	9
y = 2x + 1.5	1.5	3.5	5.5	7.5	9.5

Also included in Table 2.1 are the y values obtained from evaluating the equation $y = 2x + 1.5$ at each x value where we have data. It now follows that

$$r(0) = 1 - 1.5 = -0.5$$
$$r(1) = 5 - 3.5 = 1.5$$
$$r(2) = 3 - 5.5 = -2.5$$
$$r(3) = 6 - 7.5 = -1.5$$
$$r(4) = 9 - 9.5 = -0.5$$

The sum of squares error is the sum of the squares of the residuals.

We seek a measure of the overall error in the fit of the straight-line approximation with the actual data points. We could define the overall error to be the sum of the residuals, but this would not be a good idea. Residuals can be both positive and negative (as we saw in Example 2), hence their sum could be zero even when the residuals are quite large. We want to reserve zero error for a straight line that fits the data perfectly, that is, a line that passes through every data point. There are a number of ways to define overall error; the most popular involves squaring the residuals. By squaring each residual, we force the results to be nonnegative, and the sum of nonnegative terms is zero only when each term is zero. Therefore, we define *the sum of squares error E* as the sum of the squares of the residuals.

EXAMPLE 3

Calculate the sum of squares error for the straight line approximation used in Example 2.

Solution

$$E = [r(0)]^2 + [r(1)]^2 + [r(2)]^2 + [r(3)]^2 + [r(4)]^2$$
$$= (-0.5)^2 + (1.5)^2 + (-2.5)^2 + (-1.5)^2 + (-0.5)^2$$
$$= 0.25 + 2.25 + 6.25 + 2.25 + 0.25$$
$$= 11.25$$

A *least-squares straight line* is the line having the smallest sum of squares error for a given set of data. Minimizing sum of squares error will be our criterion for "best fit." In the social sciences and in statistical analysis, fitting data with a least-squares straight line is often called *linear regression*. Least-squares straight lines are generated by solving a system of two linear equations involving summations. For notational convenience, we let

> *A least-squares straight line is the line having the smallest sum of squares error for a given set of data.*

N = the number of data points

S_x = the sum of all the x data values

S_{x^2} = the sum of the *squares* of all the x data values

S_y = the sum of all the y data values

S_{xy} = the sum of each x data value multiplied by its corresponding y data value.

(15)

It can be shown that the equation of the least-squares straight line $y = mx + b$ satisfies the following two equations for m and b:

$$(N)b + (S_x)m = S_y$$
$$(S_x)b + (S_{x^2})m = S_{xy}$$

(16)

The derivation of System (*16*) comes from the calculus, which is beyond the scope of this book. We can, however, use (*16*) with Gaussian elimination to find least-squares straight lines when we believe such lines are appropriate for modeling our data.

EXAMPLE 4

Find the least-squares straight line for the data in Figure 2.8.

Solution
The data is given in the first two lines of Table 2.1. There are five data points, hence $N = 5$. To evaluate the other summations in (*15*), we construct Table 2.2.

TABLE 2.2

x	y	x^2	xy
0	1	0	0
1	5	1	5
2	3	4	6
3	6	9	18
4	9	16	36
$S_x = 10$	$S_y = 24$	$S_{x^2} = 30$	$S_{xy} = 65$

With the column sums of Table 2.2, System (*16*) becomes

$$5b + 10m = 24$$
$$10b + 30m = 65$$

Solving this system by Gaussian elimination, we find $b = 1.4$ and $m = 1.7$, thus the equation of the least-squares straight line ($y = mx + b$) for this data is $y = 1.7x + 1.4$. Its graph and the original data points are shown in Figure 2.9.

FIGURE 2.9

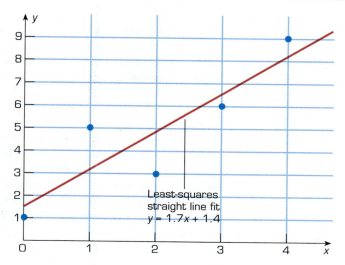

Least-squares
straight line fit
$y = 1.7x + 1.4$

EXAMPLE 5

Find the sum of squares error for the line found in Example 4.

Solution

Table 2.3 lists the data and the y values obtained from evaluating the equation $y = 1.7x + 1.4$ at each x-value where we have data.

TABLE 2.3

x	0	1	2	3	4
y (from data)	1	5	3	6	9
$y = 1.7x + 1.4$	1.4	3.1	4.8	6.5	8.2

It now follows that

$$r(0) = 1 - 1.4 = -0.4$$
$$r(1) = 5 - 3.1 = 1.9$$
$$r(2) = 3 - 4.8 = -1.8$$
$$r(3) = 6 - 6.5 = -0.5$$
$$r(4) = 9 - 8.2 = 0.8$$

and

$$E = [r(0)]^2 + [r(1)]^2 + [r(2)]^2 + [r(3)]^2 + [r(4)]^2$$
$$= (-0.4)^2 + (1.9)^2 + (-1.8)^2 + (-0.5)^2 + (0.8)^2$$
$$= 0.16 + 3.61 + 3.24 + 0.25 + 0.64$$
$$= 7.90$$

Observe that this error is *smaller* than the error for the line drawn in Examples 2 and 3, as it must be if $y = 1.7x + 1.4$ is the equation of the least-squares line, that is, the line with the smallest sum of squares error.

EXAMPLE 6

In a March, 1994, article in the *Journal of Interpersonal Violence*, Karen McCurdy and Deborah Daro relied on the data in Table 2.4 to analyze the number of maltreated children in the United States over an eight-year period. Construct a least-squares straight line for the number of such cases.

Solution

The data appear to follow a straight line trend, as Figure 2.10 shows, so a least-squares straight line fit is appropriate.

We are given 8 data points, hence $N = 8$. To evaluate the other summations in (*15*), we construct Table 2.5. Here x represents the years and y the estimated number of maltreated children, written in units of millions to keep the numbers in the table from becoming too large. With the column sums of Table 2.5, System (*16*) becomes

$$8b + 15{,}908m = 19.078$$

$$15{,}908b + 31{,}633{,}100m = 37{,}942.440$$

TABLE 2.4

YEAR	ESTIMATED NUMBER OF MALTREATED CHILDREN	NUMBER PER 1,000 U.S. CHILDREN
1985	1,919,000	30
1986	2,086,000	33
1987	2,157,000	34
1988	2,265,000	35
1989	2,435,000	38
1990	2,557,000	40
1991	2,723,000	42
1992	2,936,000	45

FIGURE 2.10

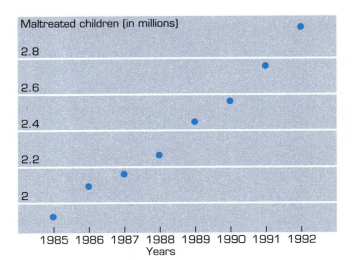

TABLE 2.5

x	y	x^2	xy
1985	1.919	3,940,225	3,809.215
1986	2.086	3,944,196	4,142.796
1987	2.157	3,948,169	4,285.959
1988	2.265	3,952,144	4,502.820
1989	2.435	3,956,121	4,843.215
1990	2.557	3,960,100	5,088.430
1991	2.723	3,964,081	5,421.493
1992	2.936	3,968,064	5,848.512
$S_x = 15,908$	$S_y = 19.078$	$S_{x^2} = 31,633,100$	$S_{xy} = 37,942.440$

Solving this system by Gaussian elimination, we find $b \approx -273.96940$ and $m \approx 0.13897619$. Thus the equation of the least-squares straight line, $y = mx + b$, for this data is $y = 0.13897619x - 273.96940$. Its graph and the original data points are displayed in Figure 2.11.

FIGURE 2.11

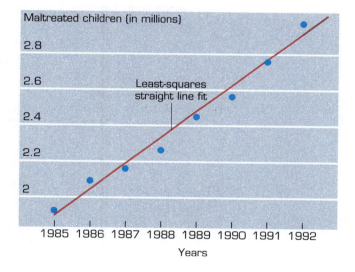

Data are frequently given for equally spaced time intervals, such as in Example 6 or Figure 2.7, or for equally spaced x-values, as in Example 2. The numbers associated with the equally spaced values can be large, particularly when they represent years, and hand calculations can become tedious, as we saw in Example 6. A useful trick when dealing with such data is to code the years as follows: If the total number of years in the data is odd, code the middle year as year 0; if the total number of years in the data is even, code the midpoint of the two years closest to the middle as year 0. Code all other years in the data according to the number of intervals, either positive or negative, by which those years *follow* year 0. This coding technique yields an augmented matrix for System (16) that is in upper triangular form and easy to solve. It also keeps the numbers in System (16) relatively small.

Coding simplifies hand calculations of least-squares straight lines for data recorded over equally spaced intervals.

EXAMPLE 7

Find the least-squares straight line for the data given in Table 2.6, which is the data displayed graphically in Figure 2.7.

TABLE 2-6

Year	1975	1976	1977	1978	1979	1980	1981	1982	1983	1984	1985	1986	1987
Median Age	20.8	21.0	21.1	21.4	21.6	21.8	22.0	22.3	22.5	22.8	23.0	23.3	23.6

Solution

There are 13 data points, hence $N = 13$. The middle year is 1981, so we code it as $x = 0$. All other years are coded relative to 1981 according to the number of years by which they follow 1981. For example, 1982 is one year after 1981 and is coded as $x = 1$, 1980 is one year before 1981 and is coded as $x = -1$, 1983 is two years after 1981 and is coded as $x = 2$, 1975 is six years before 1981 and is coded as $x = -6$. To evaluate the summations in (15) in this coding scheme, we construct Table 2.7. With the column sums of Table 2.7, System (16) becomes

$$13b + \quad 0m = 287.2$$

$$0b + 182m = \quad 42.4$$

This is a particularly easy system to solve, and it illustrates one of the primary advantages of coding. From the first equation, we find $b = 287.2/13 \approx 22.092$, and from the second equation $m = 42.4/182 \approx 0.23297$. The equation of the least-squares straight line

TABLE 2.7

x	y	x^2	xy
1975 → −6	20.8	36	−124.8
1976 → −5	21.0	25	−105.0
1977 → −4	21.1	16	−84.4
1978 → −3	21.4	9	−64.2
1979 → −2	21.6	4	−43.2
1980 → −1	21.8	1	−21.8
1981 → 0	22.0	0	0.0
1982 → 1	22.3	1	22.3
1983 → 2	22.5	4	45.0
1984 → 3	22.8	9	68.4
1985 → 4	23.0	16	92.0
1986 → 5	23.3	25	116.5
1987 → 6	23.6	36	141.6
$S_x = 0$	$S_y = 287.2$	$S_{x^2} = 182$	$S_{xy} = 42.4$

$y = mx + b$ for this data is $y = 0.23297x + 22.092$, where x denotes the year *relative to 1981* and both numbers are rounded to five significant digits. Its graph and the original data points are displayed in Figure 2.12.

FIGURE 2.12

Median age of women at first marrage

EXAMPLE 8

Find the least-squares straight line for the data in Example 6 after coding the years.

Solution

The data covers an eight-year period from 1985 through 1992, inclusive, so we have an even number of data years. The middle two years are 1988 and 1989, so we code the midpoint, 1988.5, as time 0. All years are then coded relative to 1988.5 according to the number of years by which they *follow* 1988.5. In particular, 1989 is one-half year after 1988.5 and is coded as $x = 0.5$. Similarly, 1988 is one-half year before 1988.5 and is coded as $x = -0.5$. With this coding scheme, we generate the entries in Table 2.8.

TABLE 2.8

x	y	x^2	xy
1985 → −3.5	1.919	12.25	−6.7165
1986 → −2.5	2.086	6.25	−5.2150
1987 → −1.5	2.157	2.25	−3.2355
1988 → −0.5	2.265	0.25	−1.1325
1989 → 0.5	2.435	0.25	1.2175
1990 → 1.5	2.557	2.25	3.8355
1991 → 2.5	2.723	6.25	6.8075
1992 → 3.5	2.936	12.25	10.2760
$S_x = 0$	$S_y = 19.078$	$S_{x^2} = 42.00$	$S_{xy} = 5.8370$

Least-squares models are used to estimate the values of a variable at points where data are not provided.

With the column sums of Table 2.8, System (*16*) becomes

$$8b + 0m = 19.078$$

$$0b + 42m = 5.837$$

This is a particularly easy system to solve, again illustrating one of the advantages of our coding scheme. From the first equation, we find $b = 19.078/8 = 2.38475$ and from the second equation, $m = 5.837/42 \approx 0.138976$. Now the equation of the least-squares straight line is $y = 0.138976x + 2.38475$, where x denotes the year *relative to 1988.5*.

Least-squares models can also be used to estimate values of a variable at points where data are not provided. Such points may be either within the range of the data (*interpolation*) or outside the given range (*extrapolation*). If data follow a straight-line trend, we find the least-squares straight line that best fits the data and then use the equation of that line to estimate values for the underlying system at points where data do not exist.

EXAMPLE 9

Use the results of Example 8 to estimate the number of maltreated children in the United States in 1994.

Solution

We require an estimate for a time that is outside of the range of data, so this is an example of extrapolation. The equation of the least-squares straight line is

$$y = 0.138976x + 2.38475$$

where x denotes the year relative to 1988.5. Since 1994 is 5.5 years after 1988.5, it corresponds to $x = 5.5$ in this coding scheme. Substituting this value of x into the least-squares equation, we obtain

$$y = 0.138976(5.5) + 2.38475 \approx 3.149118 \text{ million children}$$

If the trend from 1985 through 1992 continues, then we estimate the number of maltreated children in the United States in 1994 to be 3,149,000, rounded to the nearest thousand. We round to the nearest thousand because the original data in Table 2.4 were rounded to the nearest thousand.

A least-squares straight line has utility as a model only if the data follow a straight-line pattern.

Using a least-squares straight line to estimate values of a variable at points where data is not given assumes that the data follows a straight-line trend and that the trend will remain intact. Thus, one always has more confidence in interpolation, estimating between known data values, than in extrapolation, estimating outside the range of the existing data. Consider, in particular, the data on the age of women when they first marry (see Figure 2.7). It is clear, we cannot extend the least-squares line backwards too many years, because eventually it will go to zero, which is clearly an unreasonable result. Nor can we extend the least-squares line forward too many years, because eventually it will extend past 100. It is more likely that the median age of women at first marriage is represented by an *S*-shaped curve that was flat for many years, is now growing, and will flatten out again at some point. The data from 1975 through 1987 is probably part of the middle section of the *S*-shaped curve, where a straight-line approximation to the growth is reasonable. As long as we expect this straight-line growth to continue, we can use its equation to predict future values.

We close by noting that we can fit a least-squares straight line to *any* data, even when it is meaningless to do so. A least-squares straight line has utility as a model only if the data follow a straight-line pattern. If data follow some other pattern, or no pattern at all, then a least-squares straight line is worthless and should not be calculated. The first step in this modeling process is therefore to plot the data to determine whether they follow a straight-line pattern, as we did in Figure 2.10 for Example 6. If a straight-line pattern is apparent, then and *only* then is a least-squares straight line an appropriate model.

IMPROVING SKILLS

1. Calculate the sum of squares error that results from fitting the data in Table 2.5 of Example 6 with a line whose equation is $y = 0.14x - 276$.
2. Calculate the sum of squares error associated with fitting the data in Table 2.5 to the least-squares straight line determined in Example 6. Compare this answer with the one found in Problem 1.
3. Calculate the sum of squares error in fitting the data

x	−2	−1	0	1	2
y	10	13	11	15	14

with the line whose equation is $y = x + 12$.
4. Calculate the sum of squares error in fitting the data in the Problem 3 with a line whose equation is $y = 0.9x + 12.5$.
5. Show that the equation of the least-squares straight line for the data in Problem 3 is $y = x + 12.6$, and then calculate the sum of squares error for this line. Compare this answer to the answers for Problems 3 and 4.
6. Plot the data in the following table and determine whether a straight-line approximation is reasonable.

x	0	1	2	3	4	5	6	7
y	100	98	105	110	126	158	205	293

7. Plot the data in the following table and determine whether a straight-line approximation is reasonable.

x	10	20	30	40	50
y	64	43	30	48	60

8. Complete the following table and then use the results to write System (*16*) for the data. (Do not solve).

x	y	x^2	xy
0	−5		
1	−3		
2	0		
3	1		
4	4		
$S_x =$	$S_y =$	$S_{x^2} =$	$S_{xy} =$

9. Complete the following table and then use the results to write System (*16*) for the data. (Do not solve).

x	y	x²	xy
−2	60		
−1	70		
0	90		
1	110		
2	120		
$S_x =$	$S_y =$	$S_{x^2} =$	$S_{xy} =$

10. Code the years given in the following table and then complete all table entries using your coded values for x. Use your results to write System (*16*) for the data.

x	y	x²	xy
1985 →	−2		
1987 →	−1		
1989 →	1		
1991 →	2		
1993 →	4		
1995 →	5		
$S_x =$	$S_y =$	$S_{x^2} =$	$S_{xy} =$

11. Code the years given in the following table, and then complete all table entries using your coded values for x. Use your results to write System (*16*) for the data.

x	y	x²	xy
1950 →	1		
1960 →	3		
1970 →	6		
1980 →	8		
1990 →	10		
$S_x =$	$S_y =$	$S_{x^2} =$	$S_{xy} =$

CREATING MODELS

12. A large population of white mice was divided into groups that were fed different levels of the same poison. The data in the following table relate the percentage that died in each group to the level of poison they received.

x (dosage in mg)	4	6	8	10	12	14
y (% that died)	5	12	23	34	46	56

Find the equation of the least-squares straight line that fits this data, and then use the result to estimate the percentage of mice that will die from a dosage of 9 mg.

13. A well-known weight-loss center provides a new client with the following data on other clients. The data relate the number of weeks each client has been on the center's weight-loss program with the total weight in pounds that client lost to date.

x (weeks on diet)	1	2	3	4	7	9	10	12
y (weight loss)	2	6	10	8	20	30	38	34

Find the equation of the least-squares straight line that fits this data, and then use the result to estimate the number of weeks the new client should expect to stay on the weight-loss program to lose 25 pounds.

14. Solve Problem 13 if instead the data is given by the following table:

x (weeks on diet)	2	2	2	3	3	4	4	4
y (weight loss)	7	7	8	10	15	15	17	18

Observe that now there are multiple y values for each x value. Create the required summations for (15) as before by using each data point as a separate entry.

15. Using the data in Table 2.4 of Example 6, find the equation of the least-squares straight line for the number of maltreated children per 1,000 U.S. children (column 3 of the table) as it relates to time. Use the result to estimate the number in 1993.

16. In 1978, the bald eagle was listed as an endangered species in the 48 contiguous states. The following table presents information from the Department of the Interior on the number of adult bald eagle pairs occupying nesting areas in those states.

Year	1981	1984	1986	1988	1990	1993
Nesting Pairs	1,188	1,757	1,875	2,475	3,020	4,016

Find the equation of the least-squares straight line that fits this data and then use the result to estimate the number of nesting pairs in 1991.

17. The net sales reported by *Toys "Я" Us* were $4 billion in 1988, $4.8 billion in 1989, $5.5 billion in 1990, $6.1 billion in 1991, $7.2 billion in 1992, and $7.9 billion in 1993. Find the equation of the least-squares line that fits this data, and use the result to estimate total sales in 1994.

18. *Coca-Cola* reported profits of $311 million in 1986, $748 million in 1988, $878 million in 1990, and $1,369 million in 1992. Find the equation of the least-squares straight line that fits this data, and use the result to estimate Coca-Cola's 1989 profits.

19. The U.S. National Center for Health Statistics provides yearly data on the average life expectancy of U.S. residents.

Year	1980	1981	1982	1983	1984	1985	1986
Life expectancy (years)	73.7	74.2	74.5	74.6	74.7	74.7	74.8

Find the equation of the least-squares straight line that fits this data, and then use the result to estimate life expectancies in 1990 and 1970.

20. Solve Problem 19 again after adding one more data point: a life expectancy of 75.0 in 1987.

21. The following table, provided by the U.S. National Center for Health Statistics, lists the median age of men at the time of their first marriage.

Year	1980	1981	1982	1983	1984	1985	1986
Median Age	23.6	23.9	24.1	24.4	24.6	24.8	25.1

Find the equation of the least-squares straight line that fits this data, and then use the result to estimate the median age of men at the time of their first marriage for 1990 and 1975.

22. Solve Problem 21 again after adding one more data point: In 1987 the median age of men at the time of first marriage was 25.3 years.

23. The U.S. Bureau of Justice supplied the following data on the number of inmates in federal and state prisons per 100,000 people in the United States:

Year	1982	1983	1984	1985	1986
Prisoners	170.6	178.5	188.0	200.6	216.0

Find the equation of the least-squares straight line that fits this data, and then use the result to estimate the number of prisoners in 1992 and 1975.

24. Solve Problem 23 again after adding one more data point: 229.0 prisoners per 100,000 people in 1987.

25. The U.S. Bureau of the Census and the U.S. Department of Housing and Urban Development provided the following data on the median price of a new one-family house in the United States.

YEAR	PRICE (× $1000)	YEAR	PRICE (× $1000)	YEAR	PRICE (× $1000)
1970	23.4	1977	48.8	1984	79.9
1971	25.2	1978	55.7	1985	84.3
1972	27.6	1979	62.9	1986	92.0
1973	32.5	1980	64.6	1987	104.5
1974	35.9	1981	68.9	1988	112.5
1975	39.3	1982	69.3	1989	120.0
1976	44.2	1983	75.3	1990	123.0

Find the equation of the least-squares straight line that fits this data, and then use the result to estimate the price of new homes in 2000 and in 1965.

EXPLORING IN TEAMS

26. Least-squares analysis can be extended to finding the U-shaped curve that best fits data, that is, the U-shaped curve that has the smallest least squares error. The equation of such a curve has the form

$$y = ax^2 + bx + c$$

where a, b, and c satisfy the following set of equations:

$$(N)c + (S_x)b + (S_{x^2})a = S_y$$
$$(S_x)c + (S_{x^2})b + (S_{x^3})a = S_{xy}$$
$$(S_{x^2})c + (S_{x^3})b + (S_{x^4})a = S_{x^2y}$$

where N, S_x, S_{x^2}, and S_{xy} are as in (15) and

S_{x^3} = the sum of the cubes of all the x data values

S_{x^4} = the sum of the fourth powers of all the x data values

S_{x^2y} = the sum of each x data value squared multiplied by its corresponding y data value

Using these equations, find the least squares U-shaped curve that best fits the following data:

x	0	1	2	3	4
y	10	14	18	30	50

27. Use the equations given in Problem 26 to find the least-squares U-shaped curve that best fits the data in Problem 7.

EXPLORING WITH TECHNOLOGY

28. Many calculators and all the popular electronic spreadsheets include menus for calculating the equations of least-squares straight lines, most often under the heading of linear regression. Using this technology, reproduce the equation for the least-squares straight line found in Example 4.

29. Using a calculator or spreadsheet that supports linear regression, reproduce the equation for the least-squares straight line found in Example 6.

30. Using a calculator or spreadsheet that supports linear regression, reproduce the equation for the least-squares straight line found in Example 8. Compare your results to the solution of Problem 29, and then discuss the advantages of coding.

31. Solve Problems 12 through 25 using a calculator or spreadsheet that supports linear regression.

REVIEWING MATERIAL

32. (Section 1.4) A woman drives from her home to visit an uncle. She stays at her uncle's house for a few hours and then returns home. On the way home, she makes one stop at a local grocery store to buy a few items for dinner. Construct a qualitative graphical model that relates the woman's distance from her house (vertical axis) to the passage of time (horizontal axis), from the time she leaves home until she returns with her groceries.

RECOMMENDING ACTION

33. Respond by memo to the following request:

> MEMORANDUM
>
> To: J. Doe Reader
>
> From: Studio Head
>
> Date: Today
>
> Subject: **Marketing Movies**
>
> Did you see the numbers in *Variety* (March 14–20, 1994) for marketing a movie? In 1989, marketing costs per movie averaged $9.3 million, but they skyrocketed to $12.0 million in 1990, $12.1 million in 1991, $13.5 million in 1992, and $14.1 million in 1993. I am signing contracts now for movies that will not appear until 1996 or later. How much should I budget for marketing?

 MATRICES

In Sections 2.1 and 2.2, we wrote systems of linear equations as augmented matrices, because augmented matrices captured all the pertinent information clearly and concisely. Presenting information clearly and concisely is an important aspect of communications in all fields—this suggests that matrix representations have much wider applicability than just to systems of equations.

Consider a shoe store that carries one style of handmade men's sandals in whole sizes and in three different widths. Each evening, the store's manager must prepare a report on the current inventory. In one such report, the paragraph dealing with these sandals reads as follows:

> **Current stock of men's sandals**
> Two pairs of size eight narrow, one pair of size eight medium; five pairs of size nine narrow, two pairs of size nine medium; one pair of size ten narrow, three pairs of size ten medium, and one pair of size ten wide; four pairs of size eleven medium and six pairs of size eleven wide; three pairs of size twelve wide.

This report contains all the pertinent information, but it is not an easy format to analyze and it is not easy to identify individual pieces of information. To determine the number of pairs of size nine wide sandals in current stock, one has to read the entire report. In contrast, the following chart or matrix captures the same information more clearly:

$$
\begin{array}{ccccc}
 & & \text{sizes} & & \\
8 & 9 & 10 & 11 & 12
\end{array}
$$

$$
\begin{bmatrix}
2 & 5 & 1 & 0 & 0 \\
1 & 2 & 3 & 4 & 0 \\
0 & 0 & 1 & 6 & 3
\end{bmatrix}
\begin{array}{l}
narrow \\
medium \\
wide
\end{array}
$$

Now one can read at a glance the number of sandals of each type in stock. In particular, one can readily see that the store currently has no sandals in size nine wide.

A *matrix* is a rectangular array of elements enclosed in brackets and arranged in horizontal rows and vertical columns. One example is the stock matrix for sandals, where the rows pertain to sandal widths and the columns to sizes. Still other examples are:

$$A = \begin{bmatrix} 2 & 1 & -1 & 0 \\ -3 & 0 & -2 & 4 \end{bmatrix} \qquad B = \begin{bmatrix} 2 & 4 \\ 0 & \sqrt{5} \\ 1.7 & 1 \end{bmatrix}$$

$$C = \begin{bmatrix} 1 & 2 & 3 \\ 4 & 5 & 6 \\ 7 & 8 & 9 \end{bmatrix} \qquad D = \begin{bmatrix} 1 & 0 & 0 & 1 & 0 & 0 \\ 0 & 1 & 1 & 0 & 0 & 1 \\ 0 & 0 & 0 & 0 & 1 & 0 \end{bmatrix}$$

(17)

Column vectors are matrices having a single column, such as

$$a = \begin{bmatrix} 9 \\ 22 \end{bmatrix}, \qquad b = \begin{bmatrix} 100 \\ 50 \\ 200 \end{bmatrix}, \qquad \text{and} \qquad c = \begin{bmatrix} 21 \\ 0 \\ 0 \\ 5 \\ 0 \end{bmatrix}$$

(18)

Row vectors are matrices having a single row, such as

$$p = [21{,}000 \quad 14{,}000 \quad -7000 \quad 16000]$$

and

$$s = [100 \quad 50 \quad 200]$$

(19)

Augmented matrices, as defined in Section 2.1, are matrices in which a vertical line has been added between the last two columns.

We designate matrices by letters in boldface type. We will use uppercase boldface type for most matrices, as illustrated in (*17*), and reserve lowercase boldface type for row vectors and column vectors, as in (*18*) and (*19*), to emphasize their special characteristics. We can label the rows and columns of a matrix, as we did with the sandals, but when the meanings of the rows and columns are clear from context or are not needed, we will omit the labels.

By convention, the number of rows is given before the number of columns when discussing the size or *order* of a matrix. Matrix **A** in (*17*) has two rows and four columns, so we say it has order 2×4, read "two by four." Matrix **B** has three rows and two columns, so it has order 3×2, read "three by two." The orders of **C** and **D** are 3×3 and 3×6, respectively. The column vector **c** in (*18*) has order 5×1, while the row vector **p** in (*19*) has order 1×4. Because row vectors have only one row and column vectors have but a single column, it is common to specify their sizes just by listing the number of columns for a row vector or the number of rows for a column vector. This number is known as the *dimension of a vector*. Thus, **p** in (*19*) is a 4-dimensional row vector, while **c** in (*18*) is a 5-dimensional column vector. A matrix is *square* if it has the same number of rows and columns. Matrix **C** in (*17*) is square; **A**, **B**, and **D** are not.

The entries in a matrix are called *elements*. Because it is often confusing to identify elements by their numerical values—after all, which element do we mean if we refer to the zero element of

$$D = \begin{bmatrix} 1 & 0 & 0 & 1 & 0 & 0 \\ 0 & 1 & 1 & 0 & 0 & 1 \\ 0 & 0 & 0 & 0 & 1 & 0 \end{bmatrix}$$

we identify elements by their location in a matrix. We designate an element in a matrix by a lowercase letter (generally the same letter used for the matrix itself) with two subscripts that pinpoint its location. The first subscript specifies the row position and the second subscript the column position of the element. Accordingly, a_{24}, read "a sub two four," represents the element of **A** located in the second row and fourth column, whereas

d_{32}, read "d sub three two" designates the element in **D** located in the third row and second column. For a matrix **A** with p rows and n columns, we write

$$A = \begin{bmatrix} a_{11} & a_{12} & a_{13} & \cdots & a_{1n} \\ a_{21} & a_{22} & a_{23} & \cdots & a_{2n} \\ \vdots & \vdots & \vdots & \ddots & \vdots \\ a_{p1} & a_{p2} & a_{p3} & \cdots & a_{pn} \end{bmatrix}$$

In particular, if

$$A = \begin{bmatrix} 2 & 1 & -1 & 0 \\ -3 & 0 & -2 & 4 \end{bmatrix}$$

with two rows and four columns, then $a_{12} = 1$ and $a_{23} = -2$. If we have

$$B = \begin{bmatrix} 2 & 4 \\ 0 & \sqrt{5} \\ 1.7 & 1 \end{bmatrix}$$

with three rows and two columns, then $b_{22} = \sqrt{5}$ and $b_{31} = 1.7$. Here b_{13} does not exist because **B** does not have a third column. Double subscripting remains correct when identifying elements in row vectors and column vectors, but such subscripting is superfluous because these matrices have only a single row or a single column. Only a single subscript is needed to locate elements of a row or column vector precisely. Thus, p_1 denotes the first element in a vector **p** while c_4 denotes the fourth element in a vector **c**.

Two matrices are equal if they have the same order and if their corresponding elements are equal.

Two matrices are *equal* if they have the same order and if their corresponding elements are equal. In particular, the matrices

$$\begin{bmatrix} 1 & 2 \\ 3 & 4 \end{bmatrix} \quad \text{and} \quad \begin{bmatrix} 1 & 2 & 0 \\ 3 & 4 & 0 \end{bmatrix}$$

are not equal because they do not have the same order. The matrices

$$\begin{bmatrix} 1 & 2 \\ 3 & 4 \end{bmatrix} \quad \text{and} \quad \begin{bmatrix} 2 & 3 \\ 1 & 4 \end{bmatrix}$$

have the same order, both 2×2, but they are not equal because not all of their corresponding elements are equal. The 1-1 element (the element in row one and column one) of the first matrix is 1, while the corresponding 1-1 element of the second matrix is 2.

EXAMPLE 1

Find a, b, x and y if

$$\begin{bmatrix} 1 & 2y \\ x & 0 \end{bmatrix} = \begin{bmatrix} a & 6 \\ 2 & b \end{bmatrix}$$

Solution
Both matrices have the same order. Equating corresponding elements, we have $a = 1$, $x = 2$, $b = 0$, and $2y = 6$ or $y = 3$.

Earlier, we constructed a stock matrix for men's sandals:

$$S = \begin{bmatrix} 2 & 5 & 1 & 0 & 0 \\ 1 & 2 & 3 & 4 & 0 \\ 0 & 0 & 1 & 6 & 3 \end{bmatrix} \begin{matrix} narrow \\ medium \\ wide \end{matrix}$$

sizes: 8 9 10 11 12

Let

$$
D = \begin{array}{c} \\ \end{array}
\begin{array}{ccccc}
\text{sizes} \\
8 \quad 9 \quad 10 \quad 11 \quad 12 \\
\begin{bmatrix}
0 & 0 & 6 & 5 & 0 \\
4 & 1 & 3 & 1 & 3 \\
2 & 4 & 6 & 0 & 0
\end{bmatrix}
\begin{array}{l} narrow \\ medium \\ wide \end{array}
\end{array}
$$

be a delivery matrix that lists the number of pairs of new sandals shipped to the store overnight. By adding the new shipment of sandals to the existing stock, we obtain the new inventory. Thus, it makes sense to write

The sum of two matrices of the same order is the matrix obtained by adding together the corresponding elements of the original two matrices.

$$
S + D = \begin{array}{c} \end{array}
\begin{array}{ccccc}
\text{sizes} \\
8 \quad 9 \quad 10 \quad 11 \quad 12 \\
\begin{bmatrix}
2 & 5 & 7 & 5 & 0 \\
5 & 3 & 6 & 5 & 3 \\
2 & 4 & 7 & 6 & 3
\end{bmatrix}
\begin{array}{l} narrow \\ medium \\ wide \end{array}
\end{array}
$$

as the new stock matrix.

More generally, we define the *sum of two matrices of the same order* as a matrix obtained by adding together the corresponding elements of the original two matrices. Addition is not defined for matrices with different orders.

EXAMPLE 2

Find $A + B$, $C + D$, and $A + C$ for

$$
A = \begin{bmatrix} 2 & 1 & -1 \\ 0 & 1 & 2 \end{bmatrix} \qquad B = \begin{bmatrix} -1 & 0 & -1 \\ 0 & 2 & \frac{1}{2} \end{bmatrix},
$$

$$
C = \begin{bmatrix} 1 & 2 \\ 3 & 4 \end{bmatrix} \qquad D = \begin{bmatrix} -1 & 2 \\ 0.5 & -2.1 \end{bmatrix}
$$

Solution

$$
\begin{aligned}
A + B &= \begin{bmatrix} 2 & 1 & -1 \\ 0 & 1 & 2 \end{bmatrix} + \begin{bmatrix} -1 & 0 & -1 \\ 0 & 2 & \frac{1}{2} \end{bmatrix} \\
&= \begin{bmatrix} 2 + (-1) & 1 + 0 & -1 + (-1) \\ 0 + 0 & 1 + 2 & 2 + \frac{1}{2} \end{bmatrix} \\
&= \begin{bmatrix} 1 & 1 & -2 \\ 0 & 3 & 2.5 \end{bmatrix}
\end{aligned}
$$

$$
\begin{aligned}
C + D &= \begin{bmatrix} 1 & 2 \\ 3 & 4 \end{bmatrix} + \begin{bmatrix} -1 & 2 \\ 0.5 & -2.1 \end{bmatrix} \\
&= \begin{bmatrix} 1 + (-1) & 2 + 2 \\ 3 + 0.5 & 4 + (-2.1) \end{bmatrix} \\
&= \begin{bmatrix} 0 & 4 \\ 3.5 & 1.9 \end{bmatrix}
\end{aligned}
$$

The sum $A + C$ is not defined, because the matrices have different orders.

A zero matrix is a matrix whose elements are all equal to zero.

Because matrix addition is a collection of ordinary additions, matrix addition shares many of the properties of addition of real numbers. For matrices **A**, **B**, and **C** of identical order, the associative law of addition

$$A + (B + C) = (A + B) + C$$

is valid, as is the commutative law of addition

$$A + B = B + A$$

*The difference **A** − **B** of two matrices of the same order is the matrix obtained by subtracting from the elements of **A** the corresponding elements of **B**.*

A *zero matrix*, **0**, is a matrix whose elements are all equal to zero. In particular,

$$\begin{bmatrix} 0 & 0 \\ 0 & 0 \end{bmatrix} \quad \text{and} \quad [0 \ \ 0 \ \ 0 \ \ 0 \ \ 0]$$

are the 2×2 zero matrix and the 1×5 zero matrix, respectively. If **A** and **0** have the same order, then

$$A + 0 = A \tag{20}$$

Matrix subtraction is defined analogously to addition: the matrices must have the same order and the subtractions are performed on all the corresponding elements.

EXAMPLE 3

Find **A** − **B** and **A** − **C** for the matrices given in Example 2.

Solution

$$\begin{aligned} A - B &= \begin{bmatrix} 2 & 1 & -1 \\ 0 & 1 & 2 \end{bmatrix} - \begin{bmatrix} -1 & 0 & -1 \\ 0 & 2 & \frac{1}{2} \end{bmatrix} \\ &= \begin{bmatrix} 2 - (-1) & 1 - 0 & -1 - (-1) \\ 0 - 0 & 1 - 2 & 2 - \frac{1}{2} \end{bmatrix} \\ &= \begin{bmatrix} 3 & 1 & 0 \\ 0 & -1 & 1.5 \end{bmatrix} \end{aligned}$$

The difference **A** − **C** is not defined, because the matrices have different orders.

EXAMPLE 4

The stock of men's sandals at the beginning of a business day is given by the matrix

$$\begin{array}{c} \quad\quad\quad\quad\quad\quad \text{sizes} \\ \quad\quad 8 \quad 9 \quad 10 \quad 11 \quad 12 \\ S = \begin{bmatrix} 2 & 5 & 7 & 5 & 0 \\ 5 & 3 & 6 & 5 & 3 \\ 2 & 4 & 7 & 6 & 3 \end{bmatrix} \begin{array}{l} narrow \\ medium \\ wide \end{array} \end{array}$$

During the day, the store sells one pair of size 9 wide, three pairs of size 10 medium, two pairs of size 10 wide, one pair of size 12 medium and 2 pairs of size 12 wide. What will the inventory be at the end of the day?

Solution

Purchases for the day can be tabulated into the matrix

$$
\text{sizes}
$$

$$
\mathbf{P} = \begin{array}{cc} & \begin{array}{ccccc} 8 & 9 & 10 & 11 & 12 \end{array} \\ & \begin{bmatrix} 0 & 0 & 0 & 0 & 0 \\ 0 & 0 & 3 & 0 & 1 \\ 0 & 1 & 2 & 0 & 2 \end{bmatrix} \begin{array}{l} narrow \\ medium \\ wide \end{array} \end{array}
$$

Then

$$
\text{sizes}
$$

$$
\mathbf{S} - \mathbf{P} = \begin{array}{cc} & \begin{array}{ccccc} 8 & 9 & 10 & 11 & 12 \end{array} \\ & \begin{bmatrix} 2 & 5 & 7 & 5 & 0 \\ 5 & 3 & 3 & 5 & 2 \\ 2 & 3 & 5 & 6 & 1 \end{bmatrix} \begin{array}{l} narrow \\ medium \\ wide \end{array} \end{array}
$$

> *The product of a number c and a matrix A is the matrix obtained by multiplying every element of A by c.*

A matrix $\mathbf{A}$ can always be added to itself, generating the sum $\mathbf{A} + \mathbf{A}$. If $\mathbf{A}$ represents, for example, a stock matrix, then $\mathbf{A} + \mathbf{A}$ represents a doubling of the stock and we would like to write

$$
\mathbf{A} + \mathbf{A} = 2\mathbf{A} \tag{21}
$$

The right side of (*21*) is a number times a matrix, a product known as *scalar multiplication*, which is, as yet, undefined. Scalar multiplication is an important concept, but it must be defined in such a way that (*21*) remains true. Thus we define $c\mathbf{A}$, the product of a number c and a matrix $\mathbf{A}$, as the matrix obtained by multiplying every element of $\mathbf{A}$ by c.

EXAMPLE 5

Find $-3\mathbf{B}$ and $\frac{1}{2}\mathbf{C}$ for

$$
\mathbf{B} = \begin{bmatrix} -1 & 0 & -1 \\ 0 & 2 & \frac{1}{2} \end{bmatrix} \quad \text{and} \quad \mathbf{C} = \begin{bmatrix} 1 & 2 \\ 3 & 4 \end{bmatrix}
$$

Solution

$$
-3\mathbf{B} = -3 \begin{bmatrix} -1 & 0 & -1 \\ 0 & 2 & \frac{1}{2} \end{bmatrix} = \begin{bmatrix} 3 & 0 & 3 \\ 0 & -6 & -\frac{3}{2} \end{bmatrix}
$$

$$
\tfrac{1}{2}\mathbf{C} = \tfrac{1}{2} \begin{bmatrix} 1 & 2 \\ 3 & 4 \end{bmatrix} = \begin{bmatrix} \frac{1}{2} & 1 \\ \frac{3}{2} & 2 \end{bmatrix}
$$

EXAMPLE 6

Find $2\mathbf{C} - 3\mathbf{D}$ for the matrices given in Example 2.

Solution

$$
\begin{aligned}
2\mathbf{C} - 3\mathbf{D} &= 2 \begin{bmatrix} 1 & 2 \\ 3 & 4 \end{bmatrix} - 3 \begin{bmatrix} -1 & 1 \\ 0.5 & -2.1 \end{bmatrix} \\
&= \begin{bmatrix} 2 & 4 \\ 6 & 8 \end{bmatrix} - \begin{bmatrix} -3 & 6 \\ 1.5 & -6.3 \end{bmatrix} \\
&= \begin{bmatrix} 2 - (-3) & 4 - 6 \\ 6 - 1.5 & 8 - (-6.3) \end{bmatrix} \\
&= \begin{bmatrix} 5 & -2 \\ 4.5 & 14.3 \end{bmatrix}
\end{aligned}
$$

With our definition of scalar multiplication, we are assured that if c and d are any two numbers and $\mathbf{A}$ and $\mathbf{B}$ are any two matrices of the same order, then

$$c(\mathbf{A} + \mathbf{B}) = c\mathbf{A} + c\mathbf{B}$$

and

$$c(d\mathbf{A}) = (cd)\mathbf{A}$$

In other words, matrix addition, matrix subtraction, and scalar multiplication have the same arithmetic properties as real numbers.

The similarities between operations on real numbers and operations on matrices end when we move to matrix multiplication, because this operation is *not* defined as one might expect. The purpose of proposing a new construct, such as the matrix, is to use it to model real-world behavior and to solve problems that could not be modeled or solved easily without this new construct. If we expect to solve problems differently, then some of the characteristics of our new construct must be different, and one of the major differences between matrices and real numbers is in the way they multiply.

> *The product of two matrices* $\mathbf{AB}$ *is defined if and only if the number of columns in* $\mathbf{A}$ *equals the number of rows in* $\mathbf{B}$.

The *product* $\mathbf{AB}$ is defined if and only if the number of columns in $\mathbf{A}$ is equal to the number of rows in $\mathbf{B}$. Thus the product

$$\mathbf{AB} = \begin{bmatrix} 0 & 1 & 2 \\ 3 & 4 & 5 \end{bmatrix} \begin{bmatrix} 6 & 7 \\ 8 & 9 \\ -1 & -2 \end{bmatrix} \tag{22}$$

is defined, because $\mathbf{A}$ has three columns and $\mathbf{B}$ has three rows. In contrast, the product

$$\mathbf{AB} = \begin{bmatrix} 3 & 1 & 0 \\ -3 & 4 & 2 \end{bmatrix} \begin{bmatrix} 1 & 1 & 1 \\ -1 & 1 & 2 \end{bmatrix}$$

is *not* defined, even though both matrices have the same order, because the number of columns in $\mathbf{A}$ (three) is diffferent from the number of rows in $\mathbf{B}$ (two).

A simple schematic for determining whether two matrices can be multiplied is to write their respective orders next to each other and note whether the abutting numbers match. For the product in (22), we write

$$[2 \times \underline{3}][\underline{3} \times 2] \tag{23}$$

where the abutting numbers are underlined. If the abutting numbers match, as they do here, then the multiplication can be performed; if the abutting numbers do not match, then the multiplication is not defined. Moreover, when the abutting numbers match, we obtain the order of the product by deleting the matching numbers and collapsing the two $\times$ symbols into one. For example, if we delete the abutting threes in (23), we are left with 2×2, which is the order of the product $\mathbf{AB}$ in (22).

> *To calculate the i-j element of* $\mathbf{AB}$, *multiply the elements in the ith row of* $\mathbf{A}$ *by the corresponding elements in the jth column of* $\mathbf{B}$ *and sum the results.*

Knowledge of the order of a product is useful in computing the product. We know that

$$\mathbf{AB} = \begin{bmatrix} 0 & 1 & 2 \\ 3 & 4 & 5 \end{bmatrix} \begin{bmatrix} 6 & 7 \\ 8 & 9 \\ -1 & -2 \end{bmatrix} = \begin{bmatrix} -_{11} & -_{12} \\ -_{21} & -_{22} \end{bmatrix} \tag{24}$$

hence this product will have a 1-1 element, a 1-2 element, and 2-1 element, and a 2-2 element. To calculate the *i-j* element (where *i* is the row and *j* is the column) of $\mathbf{AB}$, multiply the elements in the *i*th row of $\mathbf{A}$ by the corresponding elements (the first with the first, the second with the second, and so on) in the *j*th column of $\mathbf{B}$ and sum the results.

Consider again the product **AB** in (*24*). The 1-1 element ($i = 1, j = 1$) is obtained by multiplying the elements in the first row of **A** by the corresponding elements in the first column of **B** and summing the results. Thus the 1-1 element is

$$[0 \quad 1 \quad 2 \ ] \begin{bmatrix} 6 \\ 8 \\ -1 \end{bmatrix} = 0(6) + 1(8) + 2(-1) = 6$$

The 1-2 element ($i = 1, j = 2$) is obtained by multiplying the elements in the first row of **A** by the corresponding elements in the second column of **B** and summing the results. Thus the 1-2 element is

$$[0 \quad 1 \quad 2] \begin{bmatrix} 7 \\ 9 \\ -2 \end{bmatrix} = 0(7) + 1(9) + 2(-2) = 5$$

The 2-1 element ($i = 2, j = 1$) is obtained by multiplying the elements in the second row of **A** by the corresponding elements in the first column of **B** and summing the results. Thus the 2-1 element is

$$[3 \quad 4 \quad 5] \begin{bmatrix} 6 \\ 8 \\ -1 \end{bmatrix} = 3(6) + 4(8) + 5(-1) = 45$$

Finally, the 2-2 element ($i = 2, j = 2$) is obtained by multiplying the elements in the second row of **A** by the corresponding elements in the second column of **B** and summing the results. Thus the 2-2 element is

$$[3 \quad 4 \quad 5] \begin{bmatrix} 7 \\ 9 \\ -2 \end{bmatrix} = 3(7) + 4(9) + 5(-2) = 47$$

Filling the blanks in (*24*), we have

$$\mathbf{AB} = \begin{bmatrix} 0 & 1 & 2 \\ 3 & 4 & 5 \end{bmatrix} \begin{bmatrix} 6 & 7 \\ 8 & 9 \\ -1 & -2 \end{bmatrix} = \begin{bmatrix} 6 & 5 \\ 45 & 47 \end{bmatrix}$$

EXAMPLE 7

Find **CD** and **DC** when $\mathbf{C} = \begin{bmatrix} 1 & 0 \\ -3 & 4 \end{bmatrix}$ and $\mathbf{D} = \begin{bmatrix} 2 & 6 \\ -1 & 5 \end{bmatrix}$

Solution

Both matrices have order 2×2, so in both cases the number of columns in the first matrix (two) equals the number of rows in the second matrix. For both products, our schematic is

$$[2 \times \underline{2}][\underline{2} \times 2]$$

The abutting numbers match, and by deleting them we see that both products are defined and are again 2×2 matrices.

$$\mathbf{CD} = \begin{bmatrix} 1 & 0 \\ -3 & 4 \end{bmatrix} \begin{bmatrix} 2 & 6 \\ -1 & 5 \end{bmatrix}$$

$$= \begin{bmatrix} 1(2) + 0(-1) & 1(6) + 0(5) \\ -3(2) + 4(-1) & -3(6) + 4(5) \end{bmatrix}$$

$$= \begin{bmatrix} 2 & 6 \\ -10 & 2 \end{bmatrix}$$

$$\mathbf{DC} = \begin{bmatrix} 2 & 6 \\ -1 & 5 \end{bmatrix} \begin{bmatrix} 1 & 0 \\ -3 & 4 \end{bmatrix}$$

$$= \begin{bmatrix} 2(1) + 6(-3) & 2(0) + 6(4) \\ -1(1) + 5(-3) & -1(0) + 5(4) \end{bmatrix}$$

$$= \begin{bmatrix} -16 & 24 \\ -16 & 20 \end{bmatrix}$$

Matrix multiplication is not, in general, commutative.

In Example 7, note that $\mathbf{CD} \neq \mathbf{DC}$! *Matrix multiplication is not, in general, commutative.* This is the first major difference between matrices and real numbers.

EXAMPLE 8

Find $\mathbf{Ab}$ and $\mathbf{bA}$ when

$$\mathbf{A} = \begin{bmatrix} 0 & 1 & 0 \\ -3 & 4 & 2 \end{bmatrix} \quad \text{and} \quad \mathbf{b} = \begin{bmatrix} 6 \\ 7 \\ 8 \end{bmatrix}$$

Solution

$\mathbf{A}$ has three columns and the column vector $\mathbf{b}$ has three rows, so the product $\mathbf{Ab}$ is defined.

$$\mathbf{Ab} = \begin{bmatrix} 0 & 1 & 0 \\ -3 & 4 & 2 \end{bmatrix} \begin{bmatrix} 6 \\ 7 \\ 8 \end{bmatrix}$$

$$= \begin{bmatrix} 0(6) + 1(7) + 0(8) \\ -3(6) + 4(7) + 2(8) \end{bmatrix}$$

$$= \begin{bmatrix} 7 \\ 26 \end{bmatrix}$$

Our schematic for the product $\mathbf{bA}$ is

$$[3 \times \underline{1}][\underline{2} \times 3]$$

The abutting numbers do not match, so the product is not defined. In this case $\mathbf{Ab}$ does not equal $\mathbf{bA}$ because the latter product does not exist!

EXAMPLE 9

Find $\mathbf{AB}$ when

$$\mathbf{A} = \begin{bmatrix} 2 & 4 & -1 \\ -4 & -8 & 2 \\ -2 & -4 & 1 \end{bmatrix} \text{ and } \mathbf{B} = \begin{bmatrix} 4 & 1 & -1 \\ 1 & 1 & 1 \\ 12 & 6 & 2 \end{bmatrix}$$

Solution

A has three columns and B has three rows, so the product **AB** is defined.

$$\mathbf{AB} = \begin{bmatrix} 2 & 4 & -1 \\ -4 & -8 & 2 \\ -2 & -4 & 1 \end{bmatrix} \begin{bmatrix} 4 & 1 & -1 \\ 1 & 1 & 1 \\ 12 & 6 & 2 \end{bmatrix}$$

$$= \begin{bmatrix} 2(4) + 4(1) + (-1)(12) & 2(1) + 4(1) + (-1)(6) & 2(-1) + 4(1) + (-1)(2) \\ -4(4) + (-8)(1) + 2(12) & -4(1) + (-8)(1) + 2(6) & -4(-1) + (-8)(1) + 2(2) \\ -2(4) + (-4)(1) + 1(12) & -2(1) + (-4)(1) + 1(6) & -2(-1) + (-4)(1) + 1(2) \end{bmatrix}$$

$$= \begin{bmatrix} 0 & 0 & 0 \\ 0 & 0 & 0 \\ 0 & 0 & 0 \end{bmatrix}$$

In Example 9, $\mathbf{AB} = \mathbf{0}$, yet neither **A** nor **B** is the zero matrix. This is another difference between matrices and real numbers. If the product of two real numbers is zero, then at least one of the numbers must be zero. The analogous property for matrices is not true!

Thankfully, some of the properties of multiplication with real numbers do carry over to matrices. If **A**, **B**, and **C** are matrices of compatible orders (so that the indicated operations are defined), then the associative law of multiplication

$$\mathbf{A(BC)} = \mathbf{(AB)C}$$

is valid, as are the left and right distributive laws

$$\mathbf{A(B + C)} = \mathbf{AB + AC}$$

and

$$\mathbf{(A + B)C} = \mathbf{AC + BC}$$

*An identity matrix **I** is a square matrix in which all main diagonal elements are equal to one and all other elements are equal to zero.*

Like the main diagonal of an augmented matrix, the *main diagonal* of a general matrix consists of every element whose row position equals its column position. One very important matrix is the *identity matrix*, designated by the letter **I**, defined as a square matrix in which all main diagonal elements are equal to one and all other elements are equal to zero. The 2×2, 3×3, and 4×4 identity matrices are, respectively.

$$\begin{bmatrix} 1 & 0 \\ 0 & 1 \end{bmatrix}, \quad \begin{bmatrix} 1 & 0 & 0 \\ 0 & 1 & 0 \\ 0 & 0 & 1 \end{bmatrix}, \quad \text{and} \quad \begin{bmatrix} 1 & 0 & 0 & 0 \\ 0 & 1 & 0 & 0 \\ 0 & 0 & 1 & 0 \\ 0 & 0 & 0 & 1 \end{bmatrix}$$

If **A** and **I** have compatible orders so that their product is defined, then

$$\mathbf{AI} = \mathbf{A}$$

and

$$\mathbf{IA} = \mathbf{A}$$

The same identity matrix will serve for both the above products if **A** is square. If **A** is not square, then a different identity matrix is needed for multiplication on either side of **A**. In particular, if **A** has order 2×3, then **I** must have order 3×3 for the product **AI** to be defined and order 2×2 for the product **IA** to be defined. In each case, however, the product is **A**. Thus the identity matrix is the matrix counterpart of the number 1 for real numbers.

Let's look at a commercial application of matrix multiplication. Consider the production requirements of a furniture manufacturer who produces only television cabinets and frames for grandfather clocks. Each item must be assembled by carpenters, decorated by artists, and then packed for shipping by the quality control inspector.

The times required to complete each operation on each item are constraints on the manufacturing process. A constraint matrix might be

$$
\mathbf{C} = \begin{bmatrix} \overset{\text{Assembly}}{\underset{\text{time}}{1}} & \overset{\text{Decorating}}{\underset{\text{time}}{2}} & \overset{\text{Packing}}{\underset{\text{time}}{\frac{1}{4}}} \\ 5 & 3 & \frac{1}{2} \end{bmatrix} \begin{matrix} \textit{Television cabinets} \\ \textit{Grandfather clock frames} \end{matrix} \tag{25}
$$

In particular, it takes five hours to assemble each grandfather clock frame and one-quarter hour to pack each television cabinet.

Assume that wages are \$20/hr for a carpenter, \$18/hr for a decorating artist, and \$15/hr for an inspector. We represent these wages by the wage vector

$$
\mathbf{w} = \begin{bmatrix} 20 \\ 18 \\ 15 \end{bmatrix} \tag{26}
$$

Then the product

$$
\begin{aligned}
\mathbf{Cw} &= \begin{bmatrix} 1 & 2 & \frac{1}{4} \\ 5 & 3 & \frac{1}{2} \end{bmatrix} \begin{bmatrix} 20 \\ 18 \\ 15 \end{bmatrix} \\
&= \begin{bmatrix} 1(20) + 2(18) + \frac{1}{4}(15) \\ 5(20) + 3(18) + \frac{1}{2}(15) \end{bmatrix} \\
&= \begin{bmatrix} 59.75 \\ 161.50 \end{bmatrix}
\end{aligned}
$$

represents the total labor costs for manufacturing each television cabinet and each grandfather clock frame, respectively.

IMPROVING SKILLS

1. Determine whether values of p and q exist so that
$$
\begin{bmatrix} p & 3 \\ 4 & q \end{bmatrix} = \begin{bmatrix} 4 & 3 \\ 5 & 1 \end{bmatrix}
$$

2. Determine values of a, b, c, and d so that
$$
\begin{bmatrix} 1 & a & 3 & b \\ 2 & 1 & 0 & -1 \\ c & 1 & 4 & d \end{bmatrix} = \begin{bmatrix} 1 & 2 & 3 & 4 \\ 2 & 1 & 0 & -1 \\ 4 & 1 & 4 & -5 \end{bmatrix}
$$

3. Determine values of x and y so that
$$
\begin{bmatrix} 2x + y \\ y - 7 \end{bmatrix} = \begin{bmatrix} x \\ 3 \end{bmatrix}
$$

4. Determine values of x, y, and z so that

$$2\begin{bmatrix} x \\ 3 \\ 4 \end{bmatrix} + 3\begin{bmatrix} -1 \\ y \\ 0 \end{bmatrix} = 4\begin{bmatrix} 0 \\ 4 \\ z \end{bmatrix}$$

5. Determine the orders of the following matrices:

$$\mathbf{A} = \begin{bmatrix} 1 & 2 \\ -3 & 1 \end{bmatrix} \qquad \mathbf{B} = \begin{bmatrix} 11 & -1 \\ 4 & 3 \end{bmatrix} \qquad \mathbf{u} = \begin{bmatrix} 1 \\ 7 \end{bmatrix}$$

$$\mathbf{C} = \begin{bmatrix} 2 & 1 & 3 \\ 4 & 1 & 7 \end{bmatrix} \qquad \mathbf{D} = \begin{bmatrix} -1 & 0 & 4 \\ 3 & 1 & 1 \end{bmatrix} \qquad \mathbf{v} = \begin{bmatrix} 3 \\ -2 \end{bmatrix}$$

$$\mathbf{E} = \begin{bmatrix} 1 & -1 \\ -2 & 0 \\ 1 & -2 \end{bmatrix} \qquad \mathbf{F} = \begin{bmatrix} -5 & 2 \\ -1 & 3 \\ 4 & -4 \end{bmatrix} \qquad \mathbf{w} = \begin{bmatrix} 2 \\ 1 \\ -2 \end{bmatrix}$$

$$\mathbf{G} = \begin{bmatrix} 1 & 0 & 1 \\ 3 & -1 & -2 \\ 0 & 2 & -1 \end{bmatrix} \qquad \mathbf{H} = \begin{bmatrix} 1 & 2 & 3 \\ 2 & 3 & 1 \\ 3 & 1 & 2 \end{bmatrix}$$

$$\mathbf{x} = \begin{bmatrix} 2 & -1 & 1 & -2 \end{bmatrix} \qquad \mathbf{y} = \begin{bmatrix} -3 & 2 & 1 & -6 \end{bmatrix}$$

6. Identify the 1-2 and 3-2 elements, if they exist, for matrices **A** through **H** in Problem 5.

In Problems 7 through 57, perform the indicated operations on the matrices defined in Problem 5.

7. $\mathbf{A} + \mathbf{B}$	**8.** $\mathbf{A} + \mathbf{C}$	**9.** $\mathbf{C} + \mathbf{D}$	**10.** $\mathbf{u} + \mathbf{v}$
11. $\mathbf{x} + \mathbf{y}$	**12.** $\mathbf{u} + \mathbf{y}$	**13.** $\mathbf{E} + \mathbf{F}$	**14.** $\mathbf{F} + \mathbf{E}$
15. $\mathbf{H} + \mathbf{G}$	**16.** $\mathbf{A} - \mathbf{B}$	**17.** $\mathbf{B} - \mathbf{A}$	**18.** $\mathbf{C} - \mathbf{D}$
19. $\mathbf{E} - \mathbf{F}$	**20.** $\mathbf{F} - \mathbf{E}$	**21.** $\mathbf{F} - \mathbf{G}$	**22.** $\mathbf{G} - \mathbf{H}$
23. $\mathbf{x} - \mathbf{y}$	**24.** $\mathbf{v} - \mathbf{u}$	**25.** $2\mathbf{A}$	**26.** $\frac{1}{2}\mathbf{A}$
27. $-3\mathbf{u}$	**28.** $100\mathbf{w}$	**29.** $-5\mathbf{C}$	**30.** $0.3\mathbf{E}$
31. $\mathbf{A} - 2\mathbf{B}$	**32.** $2\mathbf{u} + 3\mathbf{v}$	**33.** $2\mathbf{D} - \mathbf{F}$	**34.** $3\mathbf{E} + 5\mathbf{F}$
35. $2\mathbf{y} - \mathbf{x}$	**36.** $2\mathbf{G} - 3\mathbf{H}$	**37.** $\mathbf{AB}$	**38.** $\mathbf{BA}$
39. $\mathbf{uv}$	**40.** $\mathbf{CD}$	**41.** $\mathbf{DC}$	**42.** $\mathbf{ux}$
43. $\mathbf{EF}$	**44.** $\mathbf{FE}$	**45.** $\mathbf{Au}$	**46.** $\mathbf{Bu}$
47. $\mathbf{Cu}$	**48.** $\mathbf{Cw}$	**49.** $\mathbf{Gw}$	**50.** $\mathbf{uy}$
51. $\mathbf{vy}$	**52.** $\mathbf{yv}$	**53.** $\mathbf{Gy}$	**54.** $\mathbf{GE}$
55. $\mathbf{GC}$	**56.** $\mathbf{GH}$	**57.** $\mathbf{HG}$	

58. Using the matrices in Problem 5, find the orders of the following products, but do not find the products.

(*a*) $\mathbf{ACE}$ (*b*) $\mathbf{Auy}$ (*c*) $\mathbf{CFv}$ (*d*) $\mathbf{Bvx}$
(*e*) $\mathbf{ABCD}$ (*f*) $\mathbf{ABCE}$ (*g*) $\mathbf{GEAu}$ (*h*) $\mathbf{HECG}$
(*i*) $\mathbf{Gwx}$ (*j*) $\mathbf{CECE}$ (*k*) $\mathbf{BBBC}$ (*l*) $\mathbf{CCEC}$

59. Find **M** if $\mathbf{M} + 2\mathbf{A} = \mathbf{B}$, when **A** and **B** are defined as in Problem 5.

60. Find **N** if $2\mathbf{E} - 3\mathbf{N} = 4\mathbf{F}$, where **E** and **F** are defined as in Problem 5.

61. Verify that $\mathbf{AB} = \mathbf{AC}$, but that $\mathbf{B} \neq \mathbf{C}$, for

$$\mathbf{A} = \begin{bmatrix} 2 & 4 \\ 1 & 2 \end{bmatrix}, \qquad \mathbf{B} = \begin{bmatrix} 1 & 2 \\ 1 & -1 \end{bmatrix}, \qquad \text{and} \qquad \mathbf{C} = \begin{bmatrix} 5 & 6 \\ -1 & -3 \end{bmatrix}$$

Thus, the cancellation law is *not* valid in matrix multiplication.

62. Find two matrices **A** and **B** having the property that $\mathbf{AB} = \mathbf{BA}$.

CREATING MODELS

63. The inventory of the Village Appliance Store can be modeled by a 1 × 5 vector in which the first entry is the number of television sets, the second entry is the number of air conditioners, the third entry is the number of refrigerators, the fourth entry is the number of stoves, and the fifth entry is the number of dishwashers.

(**a**) Determine the January 1 inventory given by the vector [25 5 7 10 5].

(**b**) January sales are given by [5 0 1 2 1]. What is the inventory on February 1 if no new appliances are added to stock?

(**c**) February sales are given by [4 1 2 1 1]. New stock deliveries to the store in February are given by [6 8 0 0 3]. What is the inventory on March 1?

64. The daily gasoline supply of a local service station is given by a 1 × 3 vector in which the first entry is gallons of super, the second entry is gallons of plus, and the third entry is gallons of regular.

(**a**) Determine the supply given at the close of business on Monday by [6,000 10,000 2,000].

(**b**) Tuesday's sales are given by [2,000 2,500 1,000]. What is the supply when the station opens on Wednesday morning?

(**c**) On Wednesday, the station receives its weekly delivery of gas given by [15,000 20,000 6,000]. Determine the station's supply at the close of business on Wednesday if sales for that day were [1,500 1,700 2,500].

(**d**) Thursday's and Friday's sales are reported to the owner as [1,700 2,000 3,200] and [2,500 3,200 1,800], respectively. Why should the owner be upset?

65. The damage matrix below gives the number of damaged cases delivered by the Homestead Chocolate Company from its plants in New Jersey, Delaware, and Georgia during the past year.

	Chocolate Bars	One-pound Boxes	Novelty Items	
D =	50	30	40	NJ
	150	70	80	DE
	40	20	80	GA

(**a**) What will the damage matrix be next year if the company meets it goal of reducing damaged shipments by 10%?

(**b**) What will the damage matrix be next year if, instead, the company reduces the number of damaged cases of chocolate bars shipped from New Jersey by 50%, the number of damaged cases of one-pound boxes shipped from Delaware by 30%, and the number of damaged cases of all items from its Georgia plant by 5%?

66. Jane and Tom sell two types of computers for a manufacturer, and each month the manufacturer summarizes their total dollar sales for the month in a 2 × 2 matrix. January's sales are given by

	Low-end computers	High-end computers	
J =	30,100	62,500	Jane's sales
	45,600	50,800	Tom's sales

while February sales are given by

$$\mathbf{F} = \begin{bmatrix} 24{,}300 & 50{,}700 \\ 36{,}400 & 35{,}300 \end{bmatrix} \begin{matrix} \textit{Jane's sales} \\ \textit{Tom's sales} \end{matrix}$$

with column headers *Low-end computers* and *High-end computers*.

(a) What is the significance of $\mathbf{F} - \mathbf{J}$?

(b) Construct a commission matrix if Jane and Tom earn a 2% commission on all sales.

(c) The manufacturer wants the sale's matrix in March to be the average of those from the previous two months. What matrix models this goal?

67. The price schedule for a Chicago-to-Miami flight is given by

$$\mathbf{p} = [630 \quad 490 \quad 350]$$

where the vector elements pertain, respectively, to first-class tickets, business-class tickets, and coach tickets. The number of tickets purchased in each category for a recent flight is given by

$$\mathbf{n} = \begin{bmatrix} 8 \\ 11 \\ 103 \end{bmatrix}$$

Find the product $\mathbf{pn}$ and determine its significance.

68. The closing prices for the stocks in Ms. Dolin's stock portfolio during the past week are collected into the following portfolio matrix

$$\mathbf{P} = \begin{bmatrix} 20 & 52\frac{1}{2} & 7 \\ 20\frac{1}{2} & 52\frac{5}{8} & 7 \\ 21\frac{1}{4} & 52\frac{3}{4} & 6\frac{3}{4} \\ 21\frac{1}{4} & 52\frac{3}{8} & 6\frac{3}{4} \\ 21\frac{3}{4} & 52\frac{5}{8} & 6\frac{3}{8} \end{bmatrix} \begin{matrix} \textit{Monday} \\ \textit{Tuesday} \\ \textit{Wednesday} \\ \textit{Thursday} \\ \textit{Friday} \end{matrix}$$

with column headers *BTD*, *ACC*, *EME*.

where the columns contain the prices of stock in Breslin Tool and Die (*BTD*), American Citrus Corporation (*ACC*), and Eagle Mining Enterprises (*EME*), respectively. Ms. Dolin's holdings (that is, the number of shares she owns) in each of these companies are given by

$$\mathbf{h} = \begin{bmatrix} 100 \\ 200 \\ 550 \end{bmatrix}$$

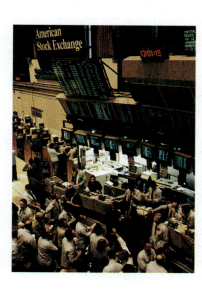

American Stock Exchange

Find $\mathbf{hP}$ and $\mathbf{Ph}$ and determine their meanings.

69. Refer to the data for the cabinet manufacturer given by matrices (25) and (26).

(a) What is the new wage vector $\mathbf{w}$ if each worker receives a $1/hr across the board salary increase?

(b) Determine the new product $\mathbf{Cw}$.

(c) Determine the percentage increase in the labor costs for manufacturing each item as a result of this new wage package.

70. (a) Construct a matrix of basic ingredients $\mathbf{B}$ for a specialty shop that bakes only sourdough bread and cherry pies. Each loaf of sourdough bread requires 3 units of flour, 1 unit of shortening, and 1 unit of yeast. Each cherry pie requires 1 unit of flour, 2 units of shortening, and 5 units of fruit.

(*b*) Each unit of flour costs 50¢, each unit of shortening 30¢, each unit of yeast 15¢, and each unit of fruit 60¢. Construct a materials cost vector **m** in such a way that the product of **B** with **m** is meaningful, and then calculate that product.

(*c*) Assume that the shop sells 1,000 loaves of bread and 400 pies daily. Construct a demand vector **d** so that the product of **d** and the product from part (*b*) is meaningful. Calculate this product and determine its significance.

71. (*a*) Construct a transportation cost matrix **C** for a manufacturer with two factories in the cities of Townsand and Rockland and two distribution warehouses in Evertsville and Foundling. It costs $5 to ship an item from Townsand to Evertsville and $8 to shop an item from Townsand to Foundling. Shipping costs from the Rockland factory are $4 an item to Evertsville and $10 an item to Foundling.

(*b*) The warehouse at Evertsville needs 800 items while Foundling needs 1000 items. Construct a demand vector **d** in such a way that the product of **C** with **d** is meaningful, and then calculate that product. What is its significance?

EXPLORING IN TEAMS

72. Rewrite the constraint matrix **C** given by (25) so that the rows hold the times required to assemble, decorate, and package each item, respectively, and the columns pertain to television sets and grandfather clock frames. What configuration must the wage vector **w** have if the product of **C** and **w** (in some sequence) is to again represent the labor costs of manufacturing various items?

73. Rewrite the portfolio matrix **P** given in Problem 68 so that the rows pertain to the companies in which Ms. Dolin holds stock and the columns pertain to the days of the week. How should the vector **h** of stock holdings be reconfigured if the product of **P** and **h** (in some sequence) is to be meaningful?

EXPLORING WITH TECHNOLOGY

74. Many graphing calculators have menus for performing matrix operations. Using such a calculator, reproduce the computations in Examples 2 through 9.

75. Solve Problems 7 through 57 using a calculator that supports matrix operations.

REVIEWING MATERIAL

76. (Section 1.6) Graph the following three equations on the same coordinate system:

$$3x + y = 0$$
$$x - y = 4$$
$$x + y = 1$$

From these graphs, what can you say about a solution that simultaneously satisfies all three equations?

RECOMMENDING ACTION

77. Respond by memo to the following request:

> MEMORANDUM
>
> To: J. Doe Reader
>
> From: Purchasing Department
>
> Date: Today
>
> Subject: **Inventory Reports**
>
> Our accountants recommend adopting a standard format for reporting daily inventory in each department in our store. Certainly standardization will enhance our ability to review purchasing policies throughout the store, but more importantly it will simplify accounting and lower those costs.
>
> Our shoe department favors the matrix approach they adopted last year, listing all inventory in a single stock matrix. I sent you a sample of such a matrix earlier for review. The dress department feels that a stock matrix approach is unwieldy and opposes it. They do not carry all items in every size and would be reporting a lot of zeros with a matrix approach. I would like your opinion on this matter before making a final decision.

 # 2.5 MATRIX INVERSION

MODEL 2.2

Important factors

s = number of
    Snack-packs to make

b = number of
    Big-packs to make

Objective

Find s and b.

Constraints

$3s + 5b = 265$

$s + 2b = 100$

The motivation for defining matrix multiplication as we did in Section 2.4 comes from our need to solve systems of linear equations. Equations represent constraints on real-world behavior, and if we want to analyze models of real-world behavior to solve real-world problems, then we must take into account all the constraints associated with that behavior.

As an example, consider the situation of a shopkeeper who sells specialty foods, including loose peanuts and cashews from bulk containers. At the end of each week, the shopkeeper takes any unsold nuts and combines them into small packages of mixed nuts, either a Snack-pack containing three ounces of peanuts and one ounce of cashews or a Big-pack containing five ounces of peanuts and two ounces of cashews. These packages are vacuum packed by the shopkeeper and sold during the following weeks. New shipments of loose nuts are delivered by a supplier at the beginning of each week, so that the nuts that the shopkeeper offers from open bulk containers are always fresh. One week the shopkeeper is left with 265 ounces of peanuts and 100 ounces of cashews. The problem is to determine how many packages of each type to make so that all the remaining peanuts and cashews are used. A model for the shopkeeper's problem is shown in Model 2.2. To solve the problem, the shopkeeper need only solve a system of two equations in two variables.

Recall how much easier it is to solve one linear equation in one variable than to solve many linear equations for many variables. A linear equation in one variable has the general form $Ax = B$, or

$$[\text{constant}] \cdot [\text{variable}] = \text{constant} \tag{27}$$

We solve for the variable in (27) by dividing the entire equation by the multiplicative constant on the left. The solution to $Ax = B$ with $A \neq 0$ is $x = B/A$.

We want to mimic this process for many equations in many variables! Ideally, we want a single master equation of the form

$$\begin{bmatrix} \text{package} \\ \text{of} \\ \text{constants} \end{bmatrix} \cdot \begin{bmatrix} \text{package} \\ \text{of} \\ \text{variables} \end{bmatrix} = \begin{bmatrix} \text{package} \\ \text{of} \\ \text{constants} \end{bmatrix} \tag{28}$$

that we can divide by the package of constants on the left to solve for all the variables at one time. To do this, we need an arithmetic of "packages," first to write systems of linear equations in the form of (28), which involves a multiplication of such "packages," and then to divide by "packages" to solve for the unknowns. Much of this arithmetic was developed many years ago! "Packages" were called matrices, and matrix multiplication, as we now know it, was just the operation needed to write systems of linear equations in the form of (28).

Let us see, as an example, how to write the constraint equations in Model 2.2

$$3s + 5b = 265$$
$$s + 2b = 100 \tag{29}$$

in form (28) using matrices. We package all the variables (s and b) into the column vector

$$\begin{bmatrix} s \\ b \end{bmatrix}$$

we package all the coefficients of these variables into the matrix

$$\begin{bmatrix} 3 & 5 \\ 1 & 2 \end{bmatrix}$$

and we package the constants on the right side of each equation into their own column vector

$$\begin{bmatrix} 265 \\ 100 \end{bmatrix}$$

We then form the matrix equation

$$\begin{bmatrix} 3 & 5 \\ 1 & 2 \end{bmatrix} \begin{bmatrix} s \\ b \end{bmatrix} = \begin{bmatrix} 265 \\ 100 \end{bmatrix} \tag{30}$$

Matrix equation (30) has the form of (28), and we can verify that Equation (30) is equivalent to System (29) by using our definition of matrix multiplication. Performing the multiplication on the left side of Equation (30), we obtain

$$\begin{bmatrix} 3 & 5 \\ 1 & 2 \end{bmatrix} \begin{bmatrix} s \\ b \end{bmatrix} = \begin{bmatrix} (3s + 5b) \\ (1s + 2b) \end{bmatrix}$$

whence (*30*) becomes

$$\begin{bmatrix} (3s + 5b) \\ (1s + 2b) \end{bmatrix} = \begin{bmatrix} 265 \\ 100 \end{bmatrix}$$

Every system of linear equations can be written in the form of the matrix equation **Ax** = **b**.

It follows from our definition of matrix equality that this last matrix (vector) equation is equivalent to System (*29*).

The same process can be used to convert any system of linear equations into one matrix equation. Collect all the variables into a column vector **x**, collect all the numerical coefficients of the variables into a matrix **A**, and collect the constants on the right side of each equation into a column vector **b**. Then the system of linear equations will be equivalent to the matrix equation

$$\mathbf{Ax} = \mathbf{b} \tag{31}$$

The matrix **A** is called a *coefficient matrix*, **x** is a *vector of variables*, and **b** is a *constant vector*. For System (*29*) (the constraint equations in Model 2.2),

$$\mathbf{A} = \begin{bmatrix} 3 & 5 \\ 1 & 2 \end{bmatrix}, \quad \mathbf{x} = \begin{bmatrix} s \\ b \end{bmatrix}, \quad \text{and} \quad \mathbf{b} = \begin{bmatrix} 265 \\ 100 \end{bmatrix}$$

EXAMPLE 1

Write the system of equations

$$x + 2y + 3z = 10$$
$$4x - 5y \quad\;\; = 20$$

in the matrix form **Ax** = **b**.

Solution

Set

$$\mathbf{A} = \begin{bmatrix} 1 & 2 & 3 \\ 4 & -5 & 0 \end{bmatrix}, \quad \mathbf{x} = \begin{bmatrix} x \\ y \\ z \end{bmatrix}, \quad \text{and} \quad \mathbf{b} = \begin{bmatrix} 10 \\ 20 \end{bmatrix}$$

The 2-3 element in **A** is zero because the coefficient of z in the second equation is zero, and the 1-1 element in **A** is one because the coefficient of x in the first equation is one. The negative sign in the second equation is incorporated into the coefficient of the y-term which is equivalent to rewriting the equation as

$$4x + (-5)y = 20.$$

The original system of equations can be written as the matrix equation

$$\begin{bmatrix} 1 & 2 & 3 \\ 4 & -5 & 0 \end{bmatrix} \begin{bmatrix} x \\ y \\ z \end{bmatrix} = \begin{bmatrix} 10 \\ 20 \end{bmatrix}$$

Whenever we convert a system of linear equations into matrix form, we must be scrupulous in placing only the coefficients of the first variable appearing in column vector **x** into the first column of **A**, only the coefficients of the second variable in **x** into the second column of **A**, and only the coefficients of the last variable in **x** into the last column of **A**. In Example 1, the elements in **x** were ordered x first, y second, and z third. Consequently, the first column of **A** contains the coefficients of x, the second column of **A** contains the coefficients of y, and the third column of **A** contains the coefficients of z.

EXAMPLE 2

Transform the system of equations

$$\begin{aligned} a + 3b + 2c &= 1 \\ b &= 4 \\ 2a - b - 3c &= 2 \end{aligned}$$

into the matrix equation $\mathbf{Ax} = \mathbf{b}$.

Solution
Take

$$\mathbf{A} = \begin{bmatrix} 1 & 3 & 2 \\ 0 & 1 & 0 \\ 2 & -1 & -3 \end{bmatrix}, \qquad \mathbf{x} = \begin{bmatrix} a \\ b \\ c \end{bmatrix}, \qquad \text{and} \qquad \mathbf{b} = \begin{bmatrix} 1 \\ 4 \\ 2 \end{bmatrix}$$

---

*A square matrix **B** is the inverse of a square matrix **A** if and only if their product is an identity matrix.*

The intent in writing systems of linear equations in the matrix form $\mathbf{Ax} = \mathbf{b}$ is to divide each side by **A**, leaving the variables isolated on one side of the equation. This sounds like a good idea, but we cannot implement it yet, because matrix division remains an undefined operation!

Fortunately, there is another operation that is just as good. To solve the linear equation $Ax = B$ for x we can either divide both sides of this equation by A or multiply both sides by the *reciprocal* of A. To solve the equation $2x = 5$, we can either divide both sides by 2 or multiply both sides by 0.5.

A number b is the reciprocal of a number a if their product is 1, that is, if $ab = 1$. The number 1 is significant because of its defining property that $1 \cdot a = a$ for every real number a. The matrix counterpart of the real number 1 is an identity matrix **I**, because $\mathbf{IA} = \mathbf{A}$ for an identity matrix of suitable order. In matrices, we use the term *inverse* instead of reciprocal, and we say that a matrix **B** is the *inverse* of a square matrix **A** if and only if

$$\mathbf{AB} = \mathbf{BA} = \mathbf{I} \tag{32}$$

Both products in (*32*) are defined and equal only if **A** and **B** are square matrices of the same order. It is shown in more advanced treatments of matrices that the equality $\mathbf{AB} = \mathbf{I}$ implies the equality $\mathbf{BA} = \mathbf{I}$, and vice-versa, when **A** and **B** are square matrices. Therefore, we need not verify all parts of (*32*), but only whether $\mathbf{AB} = \mathbf{I}$.

EXAMPLE 3

Determine whether **B** is an inverse of **A** when

$$\mathbf{A} = \begin{bmatrix} 3 & 5 \\ 1 & 2 \end{bmatrix} \qquad \text{and} \qquad \mathbf{B} = \begin{bmatrix} 2 & -5 \\ -1 & 3 \end{bmatrix}$$

Solution

We form the product **AB** and check whether it is the identity matrix. Here

$$\mathbf{AB} = \begin{bmatrix} 3 & 5 \\ 1 & 2 \end{bmatrix} \begin{bmatrix} 2 & -5 \\ -1 & 3 \end{bmatrix} = \begin{bmatrix} 1 & 0 \\ 0 & 1 \end{bmatrix} = \mathbf{I}$$

so **B** is an inverse of **A**.

EXAMPLE 4

Determine whether **B** is an inverse of **A** when

$$\mathbf{A} = \begin{bmatrix} 1 & 3 & 2 \\ 0 & 1 & 0 \\ 2 & -1 & 3 \end{bmatrix} \quad \text{and} \quad \mathbf{B} = \begin{bmatrix} 3 & -1 & 3 \\ 0 & 1 & 0 \\ 2 & -1 & 1 \end{bmatrix}$$

Solution

$$\mathbf{AB} = \begin{bmatrix} 1 & 3 & 2 \\ 0 & 1 & 0 \\ 2 & -1 & -3 \end{bmatrix} \begin{bmatrix} 3 & -1 & 3 \\ 0 & 1 & 0 \\ 2 & -1 & 1 \end{bmatrix} = \begin{bmatrix} 7 & 0 & 5 \\ 0 & 1 & 0 \\ 0 & 0 & 3 \end{bmatrix} \neq \mathbf{I},$$

so **B** is *not* an inverse of **A**.

The inverse of a square matrix **A** *is unique, when it exists, and is denoted by* $\mathbf{A}^{-1}$.

If a matrix **A** has an inverse, it has only one (see Problem 60); that is, an inverse, if it exists, is unique. Once we find an inverse, we have found *the* inverse. We denote the unique inverse of a matrix **A** as $\mathbf{A}^{-1}$. It follows from (*32*) that

$$\mathbf{AA}^{-1} = \mathbf{A}^{-1}\mathbf{A} = \mathbf{I} \tag{33}$$

Although (*32*) is a test for determining whether one matrix is the inverse of another matrix, it does not provide a method for finding inverses. Calculating inverses is called *inversion*, and an efficient procedure for this process is based on elementary row operations and the concept of a partitioned matrix. We introduced the concept of a partitioned matrix in Section 2.1, although we did not use that name at the time.

The augmented matrix for the system of equations $\mathbf{Ax} = \mathbf{b}$ can be written compactly as $[\mathbf{A}|\mathbf{b}]$ by combining **A** and **b** into a single matrix. The vertical line is only for convenience and serves to distinguish the elements of **A** from those of **b**. We can extend this notation to join any two matrices having the same number of rows. If

$$\mathbf{A} = \begin{bmatrix} 3 & 5 \\ 1 & 2 \end{bmatrix} \quad \text{and} \quad \mathbf{I} = \begin{bmatrix} 1 & 0 \\ 0 & 1 \end{bmatrix}$$

then

$$[\mathbf{A}|\mathbf{I}] = \left[\begin{array}{cc|cc} 3 & 5 & 1 & 0 \\ 1 & 2 & 0 & 1 \end{array} \right]$$

Such combinations, called *partitioned matrices*, are used in the following procedure for inverting matrices.

AN INVERSION ALGORITHM:

STEP 1 Form the partitioned matrix [**A** | **I**], where **I** is the identity matrix having the same order as **A**.

STEP 2 Use elementary row operations to transform **A** into upper triangular form, applying each operation to the entire partitioned matrix. Denote the result as [**C**|**D**], where **C** is in upper triangular form with one as the first nonzero element in each nonzero row.

STEP 3 Check whether **C** has any zeros on its main diagonal. If it does, stop; **A** does *not* have an inverse. Otherwise continue.

STEP 4 Beginning with the last column of **C** and progressing backward sequentially to the second column, use only the third elementary row operation to transform all elements above the main diagonal of **C** to zero. Complete all work on one column before moving to another column, and apply all operations to the entire partitioned matrix.

STEP 5 At the conclusion of Step 4, the partitioned matrix will have the form [**I** | **B**], with **B** = **A**$^{-1}$.

EXAMPLE 5

Find the inverse of

$$\mathbf{A} = \begin{bmatrix} 3 & 5 \\ 1 & 2 \end{bmatrix}$$

Solution

We construct the augmented matrix

$$[\mathbf{A} \mid \mathbf{I}] = \left[\begin{array}{cc|cc} 3 & 5 & 1 & 0 \\ 1 & 2 & 0 & 1 \end{array}\right]$$

Next, we use elementary row operations to transform **A** to upper triangular form, being mindful to apply each operation to the entire partitioned matrix. Accordingly,

$$\left[\begin{array}{cc|cc} 3 & 5 & 1 & 0 \\ 1 & 2 & 0 & 1 \end{array}\right] \rightarrow \left[\begin{array}{cc|cc} 1 & 2 & 0 & 1 \\ 3 & 5 & 1 & 0 \end{array}\right] \quad \begin{cases} \text{by interchanging the first and} \\ \text{second rows to make the pivot 1} \\ R_1 \leftrightarrow R_2 \end{cases}$$

$$\rightarrow \left[\begin{array}{cc|cc} 1 & 2 & 0 & 1 \\ 0 & -1 & 1 & -3 \end{array}\right] \quad \begin{cases} \text{by adding to the second row} \\ -3 \text{ times the first row} \\ R_2 + (-3)R_1 \rightarrow R_2 \end{cases}$$

$$\rightarrow \left[\begin{array}{cc|cc} 1 & 2 & 0 & 1 \\ 0 & 1 & -1 & 3 \end{array}\right] \quad \begin{cases} \text{by multiplying the} \\ \text{second row by } -1 \\ (-1)R_2 \rightarrow R_2 \end{cases}$$

The last transformation is required to convert the first nonzero element in the second row to one. If we denote this last partitioned matrix as [**C** | **D**], then **C** is in upper triangular form with one as the first nonzero element of each nonzero row. Both elements on the main diagonal of **C** are one and none are zero, so **A** has an inverse. Continuing with the reduction of the left side of our partitioned matrix by Step 4 of the inversion algorithm, we have

$$\rightarrow \begin{bmatrix} 1 & 0 & 2 & -5 \\ 0 & 1 & -1 & 3 \end{bmatrix} \begin{cases} \text{by adding to the first row} \\ -2 \text{ times the second row} \\ R_1 + (-2)R_2 \rightarrow R_1 \end{cases}$$

The matrix on the left side of this partition is **I**, so the matrix on the right side is the inverse of **A**, that is,

$$\mathbf{A}^{-1} = \begin{bmatrix} 2 & -5 \\ -1 & 3 \end{bmatrix}$$

To ensure against errors, we check whether $\mathbf{AA}^{-1} = \mathbf{I}$. Hence

$$\mathbf{AA}^{-1} = \begin{bmatrix} 3 & 5 \\ 1 & 2 \end{bmatrix} \begin{bmatrix} 2 & -5 \\ -1 & 3 \end{bmatrix} = \begin{bmatrix} 1 & 0 \\ 0 & 1 \end{bmatrix} = \mathbf{I}$$

so we have found the inverse of the given matrix.

EXAMPLE 6

Find the inverse of

$$\mathbf{A} = \begin{bmatrix} 1 & 3 & 2 \\ 0 & 1 & 0 \\ 2 & -1 & -3 \end{bmatrix}$$

Solution

We construct the augmented matrix

$$[\mathbf{A} \mid \mathbf{I}] = \begin{bmatrix} 1 & 3 & 2 & 1 & 0 & 0 \\ 0 & 1 & 0 & 0 & 1 & 0 \\ 2 & -1 & -3 & 0 & 0 & 1 \end{bmatrix}$$

Next, we use elementary row operations to transform **A** to upper triangular form, being mindful to apply each operation to the entire partitioned matrix. Accordingly,

$$\begin{bmatrix} 1 & 3 & 2 & 1 & 0 & 0 \\ 0 & 1 & 0 & 0 & 1 & 0 \\ 2 & -1 & -3 & 0 & 0 & 1 \end{bmatrix} \rightarrow \begin{bmatrix} 1 & 3 & 2 & 1 & 0 & 0 \\ 0 & 1 & 0 & 0 & 1 & 0 \\ 0 & -7 & 7 & -2 & 0 & 1 \end{bmatrix} R_3 + (-2)R_1 \rightarrow R_3$$

$$\rightarrow \begin{bmatrix} 1 & 3 & 2 & 1 & 0 & 0 \\ 0 & 1 & 0 & 0 & 1 & 0 \\ 0 & 0 & -7 & -2 & 7 & 1 \end{bmatrix} R_3 + (7)R_2 \rightarrow R_3$$

$$\rightarrow \begin{bmatrix} 1 & 3 & 2 & 1 & 0 & 0 \\ 0 & 1 & 0 & 0 & 1 & 0 \\ 0 & 0 & 1 & \frac{2}{7} & -1 & -\frac{1}{7} \end{bmatrix} (-\tfrac{1}{7})R_3 \rightarrow R_3$$

If we denote this last partitioned matrix as $[\mathbf{C} \mid \mathbf{D}]$, then **C** is in upper triangular form with one as the first nonzero element of each nonzero row. All elements on the main diagonal of **C** are one and none are zero, so **A** has an inverse, although we are still a few steps away from finding it. Continuing with the reduction of the left side of our partitioned matrix using Step 4 of the inversion algorithm, we have

$$\rightarrow \begin{bmatrix} 1 & 3 & 0 & \frac{3}{7} & 2 & \frac{2}{7} \\ 0 & 1 & 0 & 0 & 1 & 0 \\ 0 & 0 & 1 & \frac{2}{7} & -1 & -\frac{1}{7} \end{bmatrix} R_1 + (-2)R_3 \rightarrow R_1$$

$$\rightarrow \begin{bmatrix} 1 & 0 & 0 & \frac{3}{7} & -1 & \frac{2}{7} \\ 0 & 1 & 0 & 0 & 1 & 0 \\ 0 & 0 & 1 & \frac{2}{7} & -1 & -\frac{1}{7} \end{bmatrix} R_1 + (-3)R_2 \rightarrow R_1$$

The matrix on the left side of this partition is **I**, so the matrix on the right side is the inverse of **A**, that is,

$$\mathbf{A}^{-1} = \begin{bmatrix} \frac{3}{7} & -1 & \frac{2}{7} \\ 0 & 1 & 0 \\ \frac{2}{7} & -1 & -\frac{1}{7} \end{bmatrix}$$

To check against errors, we calculate the product $\mathbf{AA}^{-1}$:

$$\mathbf{AA}^{-1} = \begin{bmatrix} 1 & 3 & 2 \\ 0 & 1 & 0 \\ 2 & -1 & -3 \end{bmatrix} \begin{bmatrix} \frac{3}{7} & -1 & \frac{2}{7} \\ 0 & 1 & 0 \\ \frac{2}{7} & -1 & -\frac{1}{7} \end{bmatrix} = \begin{bmatrix} 1 & 0 & 0 \\ 0 & 1 & 0 \\ 0 & 0 & 1 \end{bmatrix} = \mathbf{I}$$

so we have the required inverse.

EXAMPLE 7

Find the inverse of

$$\mathbf{A} = \begin{bmatrix} 1 & 2 \\ 2 & 4 \end{bmatrix}$$

Solution

$$\begin{bmatrix} 1 & 2 & 1 & 0 \\ 2 & 4 & 0 & 1 \end{bmatrix} \rightarrow \begin{bmatrix} 1 & 2 & 0 & 1 \\ 0 & 0 & -2 & 1 \end{bmatrix} R_2 + (-2)R_1 \rightarrow R_2$$

If we denote this last partitioned matrix as [**C** | **D**], then **C** is in upper triangular form and the first nonzero element in each nonzero row of **C** is 1. Note that the second row in **C** is a zero row. Thus, one of the elements on the main diagonal of **C** (the 2-2 element) is zero, and **A** does *not* have an inverse.

If a coefficient matrix **A** has an inverse, we can use that inverse to solve systems written in the matrix form

$$\mathbf{Ax} = \mathbf{b} \tag{31 repeated}$$

Multiplying both sides of Equation (*31*) by $\mathbf{A}^{-1}$ on the left, we have

$$\mathbf{A}^{-1}(\mathbf{Ax}) = \mathbf{A}^{-1}\mathbf{b}$$

$$(\mathbf{A}^{-1}\mathbf{A})\mathbf{x} = \mathbf{A}^{-1}\mathbf{b} \quad \text{(associative law of multiplication)}$$

$$\mathbf{Ix} = \mathbf{A}^{-1}\mathbf{b}, \quad \text{(Eq. (33))}$$

and, because the identity matrix has the property that $\mathbf{Ix} = \mathbf{x}$,

$$\mathbf{x} = \mathbf{A}^{-1}\mathbf{b} \tag{34}$$

*If **A** has an inverse, then the unique solution to the matrix equation* **Ax** = **b** *is* **x** = **A**$^{-1}$**b**.

Equation (*34*) is a matrix representation of the solution of any system of linear equations, providing the inverse of the coefficient matrix **A** exists. The solution vector is the product of $\mathbf{A}^{-1}$ with the known vector **b**.

EXAMPLE 8

Solve the problem modeled in Model 2.2, that is, solve the system

$$3s + 5b = 265$$
$$s + 2b = 100$$

for s, the number of Snack-packs, and b, the number of Big-packs to be made from the existing stock of peanuts and cashews.

Solution

We showed earlier (see Equation (*30*)), that these constraint equations are equivalent to the matrix equation $\mathbf{Ax} = \mathbf{b}$ with

$$\mathbf{A} = \begin{bmatrix} 3 & 5 \\ 1 & 2 \end{bmatrix}, \qquad \mathbf{x} = \begin{bmatrix} s \\ b \end{bmatrix}, \qquad \text{and} \qquad \mathbf{b} = \begin{bmatrix} 265 \\ 100 \end{bmatrix}$$

Using the results of Example 5, we have

$$\begin{bmatrix} s \\ b \end{bmatrix} = \mathbf{x} = \mathbf{A}^{-1}\mathbf{b} = \begin{bmatrix} 2 & -5 \\ -1 & 3 \end{bmatrix}\begin{bmatrix} 265 \\ 100 \end{bmatrix} = \begin{bmatrix} 30 \\ 35 \end{bmatrix}$$

Two column vectors are equal if and only if their components are equal; thus, $s = 30$ and $b = 35$ is the solution to the original system of equations. To use existing inventory, the shopkeeper should make 30 Snack-packs and 35 Big-packs.

EXAMPLE 9

One week later, the shopkeeper in Example 8 has 256 ounces of peanuts and 90 ounces of cashews left over. How many Snack-packs and Big-packs should the shopkeeper make to again fully deplete all the existing inventory of loose peanuts and cashews?

Solution

A model for this problem is almost the same as Model 2.2. Here, however, the constraint equations are

$$3s + 5b = 256$$
$$s + 2b = 90$$

This system is equivalent to the matrix equation $\mathbf{Ax} = \mathbf{b}$ with

$$\mathbf{A} = \begin{bmatrix} 3 & 5 \\ 1 & 2 \end{bmatrix}, \qquad \mathbf{x} = \begin{bmatrix} s \\ b \end{bmatrix}, \qquad \text{and } \mathbf{b} = \begin{bmatrix} 256 \\ 90 \end{bmatrix}$$

The coefficient matrix remains unchanged from Example 8, as does its inverse. It follows that

$$\begin{bmatrix} s \\ b \end{bmatrix} = \mathbf{x} = \mathbf{A}^{-1}\mathbf{b} = \begin{bmatrix} 2 & -5 \\ -1 & 3 \end{bmatrix}\begin{bmatrix} 256 \\ 90 \end{bmatrix} = \begin{bmatrix} 62 \\ 14 \end{bmatrix}$$

Therefore, $s = 62$ and $b = 14$. To use all the remaining inventory of peanuts and cashews, the shopkeeper should now make 62 Snack-packs and 14 Big-packs.

EXAMPLE 10

Solve the system

$$a + 3b + 2c = 1$$
$$b \qquad = 4$$
$$2a - b - 3c = 2$$

Solution

This system is equivalent to the matrix equation $\mathbf{Ax} = \mathbf{b}$ with

$$\mathbf{A} = \begin{bmatrix} 1 & 3 & 2 \\ 0 & 1 & 0 \\ 2 & -1 & -3 \end{bmatrix}, \qquad \mathbf{x} = \begin{bmatrix} a \\ b \\ c \end{bmatrix}, \qquad \text{and} \qquad \mathbf{b} = \begin{bmatrix} 1 \\ 4 \\ 2 \end{bmatrix}$$

Using the results of Example 6, we have

$$\begin{bmatrix} a \\ b \\ c \end{bmatrix} = \mathbf{x} = \mathbf{A}^{-1}\mathbf{b} = \begin{bmatrix} \frac{3}{7} & -1 & \frac{2}{7} \\ 0 & 1 & 0 \\ \frac{2}{7} & -1 & -\frac{1}{7} \end{bmatrix}\begin{bmatrix} 1 \\ 4 \\ 2 \end{bmatrix} = \begin{bmatrix} -3 \\ 4 \\ -4 \end{bmatrix}$$

The solution is $a = -3$, $b = 4$, and $c = -4$.

We now have two methods for solving systems of linear equations: matrix inversion and Gaussian elimination. Both are powerful methods and each has its advantages.

One advantage of matrix inversion is that the inversion algorithm is available as a preprogrammed option on many calculators and electronic spreadsheets. Thus, calculating an inverse, when an inverse exists, is reduced to a few keystrokes. Since many coefficient matrices are square and most square matrices have inverses, using technology to calculate inverses is an efficient and simple way to solve those systems.

A second advantage of matrix inversion becomes apparent when we solve different sets of equations with the same coefficient matrix, as we did in Examples 8 and 9. The only difference between these two sets of equations is their right sides. The solution to both equations is $\mathbf{x} = \mathbf{A}^{-1}\mathbf{b}$ with the same inverse! Thus, we calculate the inverse only once, and we solve the two sets of equations by multiplying each new $\mathbf{b}$ by the same $\mathbf{A}^{-1}$. The shopkeeper in Examples 8 and 9 will probably have a different amount of peanuts and cashews left each week, but determining what to do with the leftover nuts is reduced to a simple matrix multiplication involving the same $\mathbf{A}^{-1}$ and a new inventory vector.

One advantage of Gaussian elimination is its applicability to all systems, while matrix inversion is limited to systems with coefficient matrices that have inverses. For example, the system

$$p + 2q = 4$$
$$2p + 4q = 8$$

is equivalent to the matrix equation $\mathbf{Ax} = \mathbf{b}$ with

$$\mathbf{A} = \begin{bmatrix} 1 & 2 \\ 2 & 4 \end{bmatrix}, \qquad \mathbf{x} = \begin{bmatrix} p \\ q \end{bmatrix}, \qquad \text{and} \qquad \mathbf{b} = \begin{bmatrix} 4 \\ 8 \end{bmatrix}$$

but we showed in Example 7 that $\mathbf{A}$ has no inverse, so matrix inversion cannot be used to solve this system. Using Gaussian elimination, however, we obtain the solution $p = 4 - 2q$ with q arbitrary. Similarly, we showed in Example 1 that the system

$$x + 2y + 3z = 10$$
$$4x - 5y \qquad = 20$$

is equivalent to the matrix equation $\mathbf{Ax} = \mathbf{b}$ with

$$\mathbf{A} = \begin{bmatrix} 1 & 2 & 3 \\ 4 & -5 & 0 \end{bmatrix}, \qquad \mathbf{x} = \begin{bmatrix} x \\ y \\ z \end{bmatrix}, \qquad \text{and} \qquad \mathbf{b} = \begin{bmatrix} 10 \\ 20 \end{bmatrix}$$

Here $\mathbf{A}$ is not square, so by definition it has no inverse. Again, matrix inversion is not applicable, while Gaussian elimination is.

Matrix inversion is also the basis of a simple scheme to encode and decode sensitive messages for transmission. Initially, each letter in the alphabet is assigned an integer between 1 and 26 as illustrated in Figure 2.13. Then each letter in a message is replaced by its integer counterpart with words separated by zeros.

FIGURE 2.13

A B C D E F G H I J K L M N O P Q R S T U V W X Y Z
1 2 3 4 5 6 7 8 9 10 11 12 13 14 15 16 17 18 19 20 21 22 23 24 25 26

The message

$$I \quad AM \quad AMY$$

is encoded

$$9 \quad 0 \quad 1 \quad 13 \quad 0 \quad 1 \quad 13 \quad 25 \qquad (35)$$

and placed into a matrix, one column at a time, two elements to a column. The unscrambled message matrix $\mathbf{M}$ for the above code is

$$\mathbf{M} = \begin{bmatrix} 9 & 1 & 0 & 13 \\ 0 & 13 & 1 & 25 \end{bmatrix}$$

$\mathbf{M}$ is scrambled by multiplying it on the left by a 2×2 matrix $\mathbf{A}$ that has an inverse. The matrix

$$\mathbf{A} = \begin{bmatrix} 3 & 5 \\ 1 & 2 \end{bmatrix}$$

from Example 5 will do nicely. We form the scrambled matrix $\mathbf{S}$

$$\mathbf{S} = \mathbf{AM} = \begin{bmatrix} 3 & 5 \\ 1 & 2 \end{bmatrix} \begin{bmatrix} 9 & 1 & 0 & 13 \\ 0 & 13 & 1 & 25 \end{bmatrix} = \begin{bmatrix} 27 & 68 & 5 & 164 \\ 9 & 27 & 2 & 63 \end{bmatrix}$$

and send the scrambled message

$$27 \quad 9 \quad 68 \quad 27 \quad 5 \quad 2 \quad 164 \quad 63$$

One effect of scrambling is to the send the letter A, which appears twice in the message I AM AMY, first as 68 and then as 2, and the letter M is sent first as 27 and then as 164. The scrambled message sends the integer 27 twice, but the first time it represents the letter I and the second time it represents the letter M. A well-scrambled message is more difficult for an unauthorized party to decipher than the unscrambled one, and therefore it is more secure. Anyone possessing the key to the message, the matrix $\mathbf{A}$, can unscramble the message by placing the transmitted code back into the matrix $\mathbf{S}$ and then multiplying $\mathbf{S}$ on the left by the inverse of $\mathbf{A}$. Using the result of Example 5, we have

$$\mathbf{M} = \mathbf{A}^{-1}\mathbf{S} = \begin{bmatrix} 2 & -5 \\ -1 & 3 \end{bmatrix} \begin{bmatrix} 27 & 68 & 5 & 164 \\ 9 & 27 & 2 & 63 \end{bmatrix} = \begin{bmatrix} 9 & 1 & 0 & 13 \\ 0 & 13 & 1 & 25 \end{bmatrix}$$

from which the original message is easily deciphered using Figure 2.13.

We can also scramble a message using a 3×3 matrix by placing the integer counterparts of the message into a matrix one column at a time with *three* elements to a column. The message matrix $\mathbf{M}$ for the code (*35*) is

$$\mathbf{M} = \begin{bmatrix} 9 & 13 & 13 \\ 0 & 0 & 25 \\ 1 & 1 & 0 \end{bmatrix}$$

where we used zeros to fill the last column when necessary, here just for the 3-3 element. **M** is scrambled by multiplying it on the left by a 3×3 matrix **A** that has an inverse. The matrix

$$A = \begin{bmatrix} 1 & 3 & 2 \\ 0 & 1 & 0 \\ 2 & -1 & -3 \end{bmatrix}$$

from Example 6 will do, although the scrambling is better if all the elements of **A** are nonzero. The scrambled matrix **S** becomes

$$S = AM = \begin{bmatrix} 1 & 3 & 2 \\ 0 & 1 & 0 \\ 2 & -1 & -3 \end{bmatrix} \begin{bmatrix} 9 & 13 & 13 \\ 0 & 0 & 25 \\ 1 & 1 & 0 \end{bmatrix} = \begin{bmatrix} 11 & 15 & 88 \\ 0 & 0 & 25 \\ 15 & 23 & 1 \end{bmatrix}$$

which is transmitted as

$$11 \quad 0 \quad 15 \quad 15 \quad 0 \quad 23 \quad 88 \quad 25 \quad 1$$

This message is unscrambled by placing the transmitted code back into the matrix **S** and then multiplying **S** on the left by the inverse of **A**. Using the results of Example 6, we have

$$M = A^{-1}S = \begin{bmatrix} \frac{3}{7} & -1 & \frac{2}{7} \\ 0 & 1 & 0 \\ \frac{2}{7} & -1 & -\frac{1}{7} \end{bmatrix} \begin{bmatrix} 11 & 15 & 88 \\ 0 & 0 & 25 \\ 15 & 23 & 1 \end{bmatrix} = \begin{bmatrix} 9 & 13 & 13 \\ 0 & 0 & 25 \\ 1 & 1 & 0 \end{bmatrix}$$

from which the original message is easily deciphered using Figure 2.13.

IMPROVING SKILLS

In Problems 1 through 20, find the inverses of the given matrices, if they exist.

1. $\begin{bmatrix} 1 & 3 \\ 3 & 8 \end{bmatrix}$ **2.** $\begin{bmatrix} 1 & 2 \\ 4 & 9 \end{bmatrix}$ **3.** $\begin{bmatrix} 2 & 3 \\ 3 & 4 \end{bmatrix}$ **4.** $\begin{bmatrix} 3 & -5 \\ 2 & -3 \end{bmatrix}$

5. $\begin{bmatrix} 3 & 2 \\ 2 & 2 \end{bmatrix}$ **6.** $\begin{bmatrix} 3 & 6 \\ 8 & 16 \end{bmatrix}$ **7.** $\begin{bmatrix} 0 & 1 \\ 1 & 0 \end{bmatrix}$ **8.** $\begin{bmatrix} 2 & 1 \\ 3 & 5 \end{bmatrix}$

9. $\begin{bmatrix} 3 & 1 \\ -4 & 1 \end{bmatrix}$ **10.** $\begin{bmatrix} 20 & 5 \\ 10 & 3 \end{bmatrix}$ **11.** $\begin{bmatrix} 0 & 0 & 1 \\ 0 & 1 & 0 \\ 1 & 0 & 0 \end{bmatrix}$ **12.** $\begin{bmatrix} 0 & 1 & 0 \\ 0 & 0 & 1 \\ 1 & 0 & 0 \end{bmatrix}$

13. $\begin{bmatrix} 1 & 0 & 0 \\ 2 & 1 & 0 \\ 0 & -3 & 1 \end{bmatrix}$ **14.** $\begin{bmatrix} 1 & 0 & 4 \\ 0 & 1 & 0 \\ 4 & 0 & 1 \end{bmatrix}$ **15.** $\begin{bmatrix} 1 & 0 & -2 \\ 2 & 1 & 0 \\ 0 & 0 & 1 \end{bmatrix}$

16. $\begin{bmatrix} 1 & 0 & 3 \\ -2 & 1 & 1 \\ 0 & 1 & 0 \end{bmatrix}$ **17.** $\begin{bmatrix} 1 & 1 & 3 \\ 1 & 2 & -1 \\ 1 & 1 & 1 \end{bmatrix}$ **18.** $\begin{bmatrix} 1 & 2 & 3 \\ 0 & 1 & 1 \\ 2 & 3 & 4 \end{bmatrix}$

19. $\begin{bmatrix} 1 & 2 & 3 \\ 4 & 5 & 6 \\ 7 & 8 & 9 \end{bmatrix}$ **20.** $\begin{bmatrix} 0 & 5 & 2 \\ 1 & 1 & -3 \\ 1 & -1 & -3 \end{bmatrix}$

21. Write each of the following systems in the matrix form $\mathbf{Ax} = \mathbf{b}$, and then solve using matrix inversion. Note that each system has the same coefficient matrix.

(**a**) $x + 3y = 8$
      $2x + 7y = 3$

(**b**) $x + 3y = -1$
      $2x + 7y = 4$

(**c**) $x + 3y = 100$
      $2x + 7y = 150$

(**d**) $p + 3q = -1$
      $2p + 7q = 4$

22. Write each of the following systems in the matrix form $\mathbf{Ax} = \mathbf{b}$, and then solve using matrix inversion. Note that each system has the same coefficient matrix.

(**a**) $2x - 4y = 2$
      $3x - 7y = -2$

(**b**) $2C - 4N = 3$
      $3C - 7N = 0$

(**c**) $2x - 4y = 10,000$
      $3x - 7y = 40,000$

(**d**) $2r - 4s = -1$
      $3r - 7s = -8$

In Problems **23** through **42**, write each system of equations in the matrix form $\mathbf{Ax} = \mathbf{b}$, and solve using matrix inversion when possible.

23. $x + 2y = 3$
      $2x + 5y = 7$

24. $x + 2y = 3$
      $2x + 6y = 7$

25. $4a + 2b = 1$
      $7a - 3b = 4$

26. $5p - 10q = 80$
      $-p + 2q = -16$

27. $3N + 4S = 12$
      $2N + S = 6$

28. $2x + 4y = -8$
      $3x + 6y = 10$

29. $2.1s - 15t = 3.7$
      $4s + 7.4t = 4$

30. $\frac{1}{2}x + 2y = 0$
      $3x - \frac{1}{2}y = 0$

31. $5x + 7y = 3$
      $4x + 2y = 5$
      $6x + 12y = 1$

32. $5x + 7y = 3$
      $4x + 2y = 5$
      $6x + 12y = 7$

33. $x + 2y - z = 1$
      $2x + 3y - 2z = 4$
      $3x - y + 5z = 0$

34. $3a + 4b - 4c = 0$
      $a - 2b + 3c = 1$
      $-2a + b - c = 2$

35. $x + y - 2z = 3$
      $2x + 4y - 3z = 4$
      $x + 3y - z = 1$

36. $u + 2v - w = 1$
      $u - 3v + w = 2$
      $u + v - 4w = -1$

37. $x + y - 2z = 3$
      $2x + 4y - 3z = 4$
      $x + 3y - z = 0$

38. $u - v + 2w = 0$
      $2u + 3v - w = 0$
      $3u + 2v + w = 0$

39. $2N + 3S - 2E = 300$
      $2N + 4S - 3E = 300$
      $4N - 3S + 2E = 300$

40. $u - v + 2w = 0$
      $2u - 2v + 4w = 0$
      $3u - 3v + 6w = 0$

41. $x + 2y - z + w = 2$
      $-2x - 3y - z - w = 4$
      $4x + 2y - 4z + 5w = 0$
      $-3x - 3y + 2z - 7w = 14$

42. $2x + 3y - z + 2w = 1$
      $4x + 6y - 2z + 4w = 2$
      $3x + y - 2z + w = 2$
      $5x + 4y - 3z + w = 3$

43. Scramble the message I AM ME using Figure 2.13 and the matrix

$$A = \begin{bmatrix} 1 & 2 \\ 3 & 4 \end{bmatrix}$$

44. Solve Problem 43 using the matrix

$$A = \begin{bmatrix} 2 & 3 \\ -4 & 5 \end{bmatrix}$$

45. Scramble the message LLOYD IS HERE using Figure 2.13 and the matrix given in Problem 44.

46. Solve Problem 45 with the matrix

$$A = \begin{bmatrix} 1 & 2 & 3 \\ 2 & 3 & 4 \\ 3 & 4 & 5 \end{bmatrix}$$

47. Decode the message

$$22 \quad 56 \quad 45 \quad 91 \quad 5 \quad 15 \quad 44 \quad 102 \quad 23 \quad 69$$

which was transmitted using Figure 15 and the coding matrix **A** given in Problem 43.

48. Decode the message

$$53 \quad -51 \quad 60 \quad 12 \quad 57 \quad 95 \quad 75 \quad 15 \quad 28 \quad -56$$

which was transmitted using Figure 2.13 and the coding matrix **A** given in Problem 44.

CREATING MODELS

49. A movie theater charges $8 per adult and $5 per child. (*a*) How many adults and children attended the first show if 240 tickets were sold and the box-office receipts were $1,356? (*b*) How many adults and children attended the second show if 188 tickets were sold and box office receipts were $1,477?

50. A political candidate contracts with a survey company to conduct a public opinion poll on issues for the next election. The company will survey 1,200 voters, some through in-person interviews and some through telephone interviews. It takes, on average, 10 minutes to conduct a telephone interview and 15 minutes to conduct an in-person interview, and the company will allocate 230 hours of interviewing time to this contract. How many interviews of each type should the company schedule to generate exactly 1200 interviews using all the time available to it?

51. (*a*) How would the model in the previous problem change if instead the company contracted to survey *at least* 1200 interviews *within* the time allocated to the project?
(*b*) Under what real physical conditions would an equality time constraint be more fitting that an inequality constraint?

52. A retired couple needs to earn $15,000 a year from their $200,000 life savings. They have decided on two investments, bank certificates of deposits paying 6% interest per year with no risk and corporate bonds paying 10% per year with moderate risk. How much money should they allocate to each investment to achieve their financial objective?

53. An investor has $50,000 to invest in two opportunities. One pays 14% per year and is highly risky, the second pays $8\frac{1}{4}$% per year and is moderately risky. How much should be invested in each opportunity if the total return on the two investments must be $10\frac{1}{2}$%?

54. A fertilizer company produces two lawn fertilizers, regular and deluxe. Each bag of deluxe fertilizer contains three pounds of active ingredients and seven pounds of inert substances, while each bag of regular fertilizer contains two pounds of active ingredients and eight pounds of inert substances. How many bags of each product should the company produce if it wants to deplete its entire inventory of 15,700 pounds of active ingredients and 52,300 pounds of inert ingredients?

55. The Air Ride Bus Company manufactures buses at two plants. Plant *A* has a fixed cost of $400,000 and a variables cost of $80,000 per bus. Plant *B* has a fixed cost of $600,000 and a variable cost of $75,000 per bus. The company has a contract for 142 buses. How should it divide the production of these buses between its plants if it wants to equalize the total cost at the two plants?

56. A company must pay both city and state taxes on all profits. The city tax is 2% of profits after state taxes are deducted as a pretax expense. The state tax is 8% of profits after city taxes are deducted as a pretax expense. How much tax must the company pay the city and the state on a pretax profit of $200,000?

57. Solve the previous problem if the company must also pay a federal tax of 20% on profits after city and state taxes are deducted as pretax expenses. Neither the city nor state, however, allows any deductions from the federal tax.

58. Yummy Nuts has an inventory of 65,000 ounces of peanuts, the same amount of almonds, and 46,000 ounces of cashews, from which it will manufacture three nut mixes. Its regular mix contains 10 oz of peanuts, and 3 oz each of almonds and cashews. Its special mix contains 6 oz of peanuts, 6 oz of almonds, and 4 oz of cashews, while its deluxe mix contains no peanuts, 10 oz of almonds, and 6 oz of cashews. How many items of each type nut mix should the company produce if it wants to deplete its entire inventory?

59. A mattress company produces three styles of mattresses. Each deluxe mattress contains 50 springs, 12 lbs of padding, and 10 yards of material. Each regular mattress contains 40 springs, 10 lbs of padding, and 10 yards of material, while each economy mattress contains 35 springs, 9 lbs of padding, and again 10 yards of material. The company has in inventory 105,000 springs, 26,100 lbs of padding, and 25,500 yards of material. How many mattresses of each type should the company produce if it wants to deplete its inventory?

EXPLORING IN TEAMS

60. Assume that **B** and **C** are *two* inverses of a matrix **A**. Explain why each of the following equalities is true.

$$\mathbf{B} = \mathbf{IB} = (\mathbf{CA})\mathbf{B} = \mathbf{C}(\mathbf{AB}) = \mathbf{CI} = \mathbf{C}$$

61. Assume that **A** and **B** are square matrices of the same order and that both have inverses. Explain why each of the following equalities is true.

$$(\mathbf{B}^{-1}\mathbf{A}^{-1})(\mathbf{AB}) = \mathbf{B}^{-1}[\mathbf{A}^{-1}(\mathbf{AB})] = \mathbf{B}^{-1}[(\mathbf{A}^{-1}\mathbf{A})\mathbf{B}] = \mathbf{B}^{-1}[\mathbf{IB}] = \mathbf{B}^{-1}\mathbf{B} = \mathbf{I}$$

What does this say about the inverse of the product **AB**?

62. Solve Example 9 if the shopkeeper has 280 ounces of peanuts and 75 ounces of cashews. What conclusions do you draw?

63. Under what conditions will the shopkeeper in Example 9 use all existing inventory? What can you say about the relationship between the amount of peanuts and the amount of cashews left in inventory at the end of the week.

EXPLORING WITH TECHNOLOGY

64. Many graphing calculators and all the popular electronic spreadsheets have menus for performing matrix inversion. Using such technology, reproduce the results of Examples 5, 6, and 7.
65. Solve Problems 1 through 20 using technology that supports matrix inversion.
66. Using technology that supports both matrix inversion and matrix multiplication, reproduce the results of Example 8, 9, and 10.
67. Solve Problems 23 through 42 using technology that supports matrix inversion and matrix multiplication.

REVIEWING MATERIAL

68. (Section 1.5) Figure A shows the relationship between corporate profits P and years t. Find an equation relating profit P to time t.

FIGURE A

RECOMMENDING ACTION

69. Respond by memo to the following request:

> ### MEMORANDUM
>
> To: J. Doe Reader
>
> From: Research and Development
>
> Date: Today
>
> Subject: **Our Next-Generation Calculator**
>
> Too many new features have been proposed for our next-generation calculator. We simply do not have the memory or available keys to support all of them. Either we eliminate some of these features or we replace some of the current features. One proposal is to remove matrix operations. I think we need to keep matrix inversion and matrix multiplication, but I also think we can eliminate matrix addition, subtraction, and scalar multiplication. What do you think?

CHAPTER 2   KEYS

KEY WORDS

augmented matrix (p. 84)
back-substitution (p. 82)
coefficient matrix (p. 141)
column vector (p. 125)
dimension of a vector (p. 125)
element (p. 125)
elementary row operations (p. 86)
equal matrices (p. 126)
extrapolation (p. 118)
Gauss-Jordan reduction (p. 100)
Gaussian elimination (p. 94)
identity matrix (p. 133)
interpolation (p. 118)
inverse (p. 142)
inversion (p. 143)
least-squares straight line (p. 111)

linear regression (p. 111)
main diagonal (p. 88)
matrix (p. 125)
noise (p. 109)
order (p. 125)
partitioned matrix (p. 143)
pivot (p. 95)
residual (p. 110)
row vector (p. 125)
scalar multiplication (p. 129)
solution (p. 82)
square matrix (p. 125)
sum of squares error (p. 111)
upper triangular form (p. 88)
vector of variables (p. 141)
zero matrix (p. 128)

KEY CONCEPTS

2.1 Elementary Row Operations

- A solution to a system of equations is a set of numbers, one for each variable, that makes all the equations true when substituted back into the equation of the system.

- Back-substitution is the process of solving a system by beginning with the last equation, sequentially moving up through all the equations, and using each equation to solve for one variable at a time.

- An augmented matrix for a system of linear equations is a matrix in which the rows represent the equations, all columns except the last column hold the coefficients of the variables in each equation, and the last column contains the constants on the right side of the equations. A vertical line between the last two columns is added to indicate the placement of the equality signs in the equations.

- The three elementary row operations are (1) interchange the positions of any two rows, (2) multiply one row by a nonzero number, and (3) add a nonzero constant times one row to another row.

- The main diagonal of an augmented matrix is the set of all matrix entries, that lie to the left of the vertical line, whose row position is the same as their column position. An augmented matrix is in upper triangular form if all numbers below its main diagonal are zero.

2.2 Gaussian Elimination

- In Gaussian elimination, an augmented matrix for a system of linear equations is transformed into upper triangular form using elementary row operations, then the corresponding set of equations is solved through back-substitution.

- Back-substitution is replaced by a sequence of elementary row operations in Gauss-Jordan reduction.

- A system of linear equations has no solution if the equations associated with its augmented matrix in upper triangular form include an absurd statement.

- A system of linear equations has infinitely many solutions if the equations associated with its augmented matrix in upper triangular form contain a variable that does not appear first in any of these equations.

2.3 Modeling Data with Straight Lines

- Data is expected to contain some distortions due to noise. If a trend is apparent, we assume that that trend is the underlying pattern and that the data reflect the system's attempt to follow the trend in the presence of noise.

- The residual at each (x, y) data point is the y value of the data point less the y value obtained from a straight line approximation to the graph of the data. The sum of squares error is the sum of the squares of the residuals at each data point.

- A least-squares straight line (or linear regression line) is the line having the smallest sum of squares error for a given set of data.

- A least-squares straight line has utility as a model only when the data follow a straight-line pattern.

2.4 Matrices

- A matrix is a rectangular array of elements arranged in horizontal rows and vertical columns. Column vectors are matrices having a single column; row vectors are matrices having a single row.

- By convention, the number of rows is given before the number of columns when discussing the order of a matrix; the row number is also listed first when giving the location of an element in a matrix.

- Two matrices are equal if they have the same order and if their corresponding elements are equal.

- The sum of two matrices of the same order is another matrix obtained by adding together the corresponding elements of the original two matrices.

- The difference of two matrices of the same order is another matrix obtained by subtracting the corresponding elements of the original two matrices.

- The product of a number c and a matrix A is another matrix obtained by multiplying each element of A by c.

- The product AB is defined if and only if the number of columns of A is equal to the number of rows of B. To calculate the i-j element of AB, multiply the elements in the ith row of A by the corresponding elements in the jth column of B and sum the results.

- Multiplication of two matrices is not, in general, commutative.

2.5 Matrix Inversion

- Matrix division is not a defined operation.

- A matrix B is the inverse of a square matrix A if and only if $AB = BA = I$, in which case we write $B = A^{-1}$.

KEY FORMULAS

- A linear equation in the three variables x, y, and z has the form

$$Ax + By + Cz = D$$

where A, B, C, and D are known numbers, with A, B, and C not all zero. An equation in the four variables w, x, y, and z is linear if it has the form

$$Aw + Bx + Cy + Dz = E$$

where A, B, C, D, and E are known numbers, with A, B, C, and D not all zero.

- The equation of the least-squares straight line $y = mx + b$ satisfies the following two equations for m and b:

$$(N)b + (S_x)m = S_y$$
$$(S_x)b + (S_{x^2})m = S_{xy}$$

where

    N = the number of data points

    S_x = the sum of all the x data values

    S_{x^2} = the sum of the *squares* of all the x data values

    S_y = the sum of all the y data values

    S_{xy} = the sum of each x data value multiplied by its corresponding y data value.

- The unique solution of the matrix equation $\mathbf{Ax} = \mathbf{b}$ is

$$\mathbf{x} = \mathbf{A}^{-1}\mathbf{b}$$

providing $\mathbf{A}$ has an inverse.

KEY PROCEDURES

- If the total number of years in a data set over equal intervals of time is odd, code the middle year as year 0; if the total number of years in the data is even, code the midpoint of the two years closest to the middle as year 0. Code all other years in the data by the number of intervals, either positive or negative, by which these years follow year 0.

- A system of linear equations can be transformed into the matrix equation $\mathbf{Ax} = \mathbf{b}$ by collecting all the variables into a column vector $\mathbf{x}$, collecting all the numerical coefficients of the variables into a matrix $\mathbf{A}$, and collecting the constants on the right side of each equation into a column vector $\mathbf{b}$. The coefficients of the first variable in $\mathbf{x}$ are placed into the first column of $\mathbf{A}$, the coefficients of the second variable in $\mathbf{x}$ are placed into the second column of $\mathbf{A}$, and the coefficients of the last variable in $\mathbf{x}$ are placed into the last column of $\mathbf{A}$.

- A matrix is transformed into upper triangular form by:

Step 1   Using elementary row operations to transform all elements below the main diagonal to zero, proceeding one element at a time and one column at a time beginning with column 1 and moving through the columns sequentially.

Step 2   Using the first elementary row operation to place a nonzero element (a pivot) on the main diagonal in the current column of interest whenever that element is zero and there is a nonzero element below it in the same column.

Step 3   Using the second elementary row operation to transform all pivots to unity.

Step 4   Using the third elementary row operation with a pivot to transform all nonzero elements below the pivot to zero.

- To effect a Gauss-Jordan reduction on an augmented matrix.

Step 1   Use elementary row operations to transform the augmented matrix into an augmented matrix $[\mathbf{C} \mid \mathbf{d}]$ in upper triangular form, with all nonzero elements on the main diagonal equal to one.

Step 2   Use only the third elementary row operation to transform all elements in $\mathbf{C}$ above the main diagonal to zero, proceeding one element at a time and one column at a time, beginning with the last column of $\mathbf{C}$ having a 1 on its main diagonal and then progressing backward sequentially to the second column, being sure to apply all operations to the entire partitioned matrix.

- To calculate the inverse of a square matrix **A**:

Step 1 Form the partitioned matrix [**A** | **I**], where **I** is the identity matrix having the same order as **A**.

Step 2 Use elementary row operations to transform **A** into upper triangular form, applying each operation to the entire partitioned matrix. Denote the result as [**C** | **D**], where **C** is in upper triangular form and has one as the first nonzero entry in every nonzero row.

Step 3 Check whether **C** has any zeros on its main diagonal. If it does, stop—**A** does *not* have an inverse. Otherwise continue.

Step 4 Beginning with the last column of **C** and progressing backward sequentially to the second column, use only the third elementary row operation to transform all elements in **C** above the main diagonal to zero. Complete all work on one column before moving to another column, and apply all operations to the entire partitioned matrix.

Step 5 At the conclusion of Step 4, the partitioned matrix will have the form [**I** | **B**], with $\mathbf{B} = \mathbf{A}^{-1}$.

TESTING YOURSELF

Use the following problems to test yourself on the material in Chapter 2. The odd problems comprise one test, the even problems a second test.

1. Use Gaussian elimination to solve the system

$$x + 2y - z = 5$$
$$3x + 4y + 2z = 3$$
$$3y + z = 1$$

2. Use *matrix inversion* to solve the following system:

$$x + 3z = 15$$
$$x + y + 4z = 20$$
$$2x + z = 20$$

3. Use elementary row operations to reduce the following augmented matrix to upper triangular form (do **not** solve).

$$\begin{bmatrix} 1 & 2 & -1 & 4 \\ 3 & 8 & 0 & 0 \\ 2 & 9 & 1 & 1 \end{bmatrix}$$

4. The Tourism Authority in Thailand reported the following revenue from tourism (Baht is the Thai currency):

Year	1976	1978	1980	1982	1984
Revenue (in millions of Baht)	3,990	8,894	17,765	23,879	27,317

Find the equation of the least-squares straight line for this data and then use the result to estimate tourism revenues in 1981.

5. Use Gaussian elimination to solve the systems of equations expressed as the following augmented matrices for x, y, and z.

(a) $\left[\begin{array}{ccc|c} 1 & 1 & -2 & 4 \\ 0 & 1 & 3 & 4 \\ 0 & 0 & 0 & 0 \end{array}\right]$

(b) $\left[\begin{array}{ccc|c} 1 & 0 & 4 & 0 \\ 0 & 1 & -1 & 4 \\ 0 & 0 & 0 & 4 \end{array}\right]$

6. Use elementary row operations to reduce the following augmented matrix to upper triangular form (do **not** solve).

$$\left[\begin{array}{ccc|c} 1 & 2 & -1 & 3 \\ 3 & 8 & 0 & 1 \\ 5 & 9 & 2 & 0 \end{array}\right]$$

7. Let

$$\mathbf{A} = \begin{bmatrix} 1 & 2 \\ 0 & 1 \\ 1 & 1 \end{bmatrix}, \quad \mathbf{B} = \begin{bmatrix} 2 & -2 \\ 5 & -3 \\ 1 & 8 \end{bmatrix}, \quad \text{and} \quad \mathbf{C} = \begin{bmatrix} 2 & 1 & 1 \\ 3 & -1 & 2 \end{bmatrix}$$

Find (a) $\mathbf{A} + \mathbf{B}$ (b) $\mathbf{B} - \mathbf{C}$ (c) $\mathbf{B} - 2\mathbf{A}$
    (d) $\mathbf{AB}$ (e) $\mathbf{BC}$

8. Let

$$\mathbf{A} = \begin{bmatrix} 2 & -1 & 0 \\ 5 & 1 & 1 \end{bmatrix}, \quad \mathbf{B} = \begin{bmatrix} 0 & -2 \\ 3 & -3 \end{bmatrix}, \quad \text{and} \quad \mathbf{C} = \begin{bmatrix} 1 & 0 & 4 \\ 1 & -1 & 2 \end{bmatrix}$$

Find (a) $\mathbf{A} - \mathbf{B}$ (b) $\mathbf{A} - \mathbf{C}$ (c) $\mathbf{A} + 3\mathbf{C}$
    (d) $\mathbf{AB}$ (e) $\mathbf{BA}$

9. Find the inverse of

$$\mathbf{A} = \begin{bmatrix} 2 & 2 \\ 5 & 10 \end{bmatrix}$$

10. Use Gaussian elimination to solve the systems of equations, expressed as the following augmented matrices for x, y, and z.

(a) $\left[\begin{array}{ccc|c} 1 & 1 & -1 & 4 \\ 0 & 1 & 3 & 4 \\ 0 & 0 & 0 & 1 \end{array}\right]$

(b) $\left[\begin{array}{ccc|c} 1 & 1 & -1 & 4 \\ 0 & 1 & 3 & 4 \\ 0 & 0 & 1 & 0 \end{array}\right]$

11. Use *matrix inversion* to solve the system:

$$x + 2y + 3z = 17$$
$$y - 2z = -22$$
$$2x + 4y + 8z = 50$$

12. Use Gaussian elimination to solve the system in Problem 11.

13. In response to a survey, six waiters at an expensive continental restaurant reported their salaries for the past week to a sociologist who matched those figures against the total bills written by each waiter for the week. The results were compiled into the following table.

x (total billings)	$5,250	$5,870	$4,590	$5,320	$5,520	$4,810
y (salary)	$1,025	$1,195	$980	$1,090	$1,070	$950

Find the equation of the least-squares straight line that fits this data. (*a*) What is the significance of *m* and *b* in this situation? (*b*) Is salary directly proportional to total billings?

14. Find the inverse of

$$A = \begin{bmatrix} 5 & 2 \\ 4 & 2 \end{bmatrix}$$

3

LINEAR PROGRAMMING
MODELS

3.1 MODELING WITH INEQUALITIES

Although some constraints can be appropriately modeled by equalities, many constraints are not that binding. For example, if a person saves $4,000 to buy a used car, the constraint is not that the car cost *exactly* $4,000 but rather that the car cost *no more than* $4,000. Purchasing an acceptable car for less money is all to the good. If a bank requires borrowers to make 20% down payments on new homes, the constraint is not that a down payment be *exactly* 20% but rather that a borrower put down *at least* 20%. A borrower may always opt for a higher down payment and a smaller loan. Thus equalities are not always the appropriate relationship for the constraints we will model. Many real-world problems are modeled more accurately by inequalities.

> *The statement "X is no more than Y" is modeled as $X \le Y$; the statement X is at least as great as Y is modeled as $X \ge Y$.*

Two real numbers a and b are related by one of *three* possibilities: a is less than b, written $a < b$; a equals b, written $a = b$; or a is greater than b, written $a > b$. For example, $5 < 7$, $-3 < -2$, $4 = 16/4$, $8 > 5$, and $0 > -5$.

The inequalities $a < b$ and $a > b$ are called *strict inequalities*. Equally useful are the inequalities $a \le b$ and $y \ge x$, read, respectively, "a is less than or equal to b" and "y is greater than or equal to x." The word *or* is important. The statement $a \le b$ is true if either a is less than b or a equals b. It is as correct to write $7 \le 8$ as it is to write $7 < 8$, although the latter statement is stronger. It is as correct to write $4 \ge 16/4$ as it is to write $4 = 16/4$, although again the latter statement is the stronger one. To say that "X cannot exceed Y" or that "X is no more than Y" is the same as saying $X \le Y$. To say that "X is at least as great as Y" is equivalent to saying $X \ge Y$.

EXAMPLE 1

Model the following statements: A small company manufactures professional scissors for commercial use and standard scissors for home use; both models are popular and sell as quickly as the company can produce them. Each professional scissors requires 5 units of steel to produce and sells for $12. Each standard scissors requires 2 units of steel to produce and sells for $4. The company has 10,000 units of steel in inventory, with another delivery due the following day. How many items of each type should the company schedule for production if it wants its sales to at least cover its daily payroll of $12,000?

Solution

The objective is to determine the number of professional scissors and standard scissors to schedule for production, so these are important components. Let P denote the number of professional scissors to produce and S the number of standard scissors. There are two limitations to any production run: First, the production process cannot use more than the 10,000 units of steel available to the company, and second, the items produced must generate at least $12,000 in income.

We handle the steel constraint first. Each professional scissors requires 5 units of steel, so P professional scissors use $5 \cdot P$ units of steel. Each standard scissors requires 2 units of steel, so S standard scissors use $2 \cdot S$ units of steel. The total amount of steel used in manufacturing both products is

$$5P + 2S \qquad \text{(steel used)}$$

The amount of steel available to the company is

$$10,000 \qquad \text{(steel available)}$$

The amount of steel used cannot exceed (i.e., must be less than or equal to) the amount of steel available, hence

$$5P + 2S \leq 10,000$$

Next, we tackle the revenue constraint. Each professional scissors generates $12 in sales, so P professional scissors generate revenues of $12 \cdot P$ dollars. Each standard scissors generates $4 in sales, so S standard scissors generate revenues of $4 \cdot S$ dollars. The total amount of revenue generated from sales of all products is

$$12P + 4S \qquad \text{(dollars of revenue)}$$

The amount of dollars needed to cover payroll is

$$12,000 \qquad \text{(dollars needed)}$$

Sales revenues must be at least as much as (i.e., greater than or equal to) the payroll, hence

$$12P + 4S \geq 12,000$$

This model is shown in Model 3.1.

MODEL 3.1

Important factors

P = number of professional scissors to produce

S = number of standard scissors to produce

Objective

Find P and S.

Constraints

$5P + 2S \leq 10,000$

$12P + 4S \geq 12,000$

For many processes of interest, the important components are the items to be produced, transported, or consumed. The production, shipment, or consumption of these items is often limited by the availability of resources, and these limitations are the constraints. Sometimes information about each item can be summarized in a table similar to Table 3.1, which we call a *constraint table*. In general, the data to the right of the first column in each row of the table yields an inequality constraint. The number of rows and columns is adjusted to meet the conditions of the process being modeled. Table 3.2 is a constraint table for the process described in Example 1.

Some constraints, called *hidden conditions*, are so obvious that they are not stated explicitly. Hidden conditions include nonnegativity conditions, when they exist, for important components. In many processes, negative amounts are meaningless. In Example 1, it is understood that the manufacturing company cannot produce negative amounts of scissors. The conditions $P \geq 0$ and $S \geq 0$ are not explicitly stated—they are just understood—but these conditions reflect reality and must be added to Model 3.1 if the model is to be complete.

Hidden conditions are constraints, such as nonnegativity conditions, that are so obvious they are not explicitly stated.

TABLE 3.1

	ITEMS IN THE OBJECTIVE FUNCTION				OVERALL LIMITATIONS
	First	**Second**	**Third**	**Fourth**	
Amount of each item					
First requirement per item					
Second requirement per item					
Third requirement per item					

TABLE 3.2

| | ITEMS IN THE OBJECTIVE FUNCTION | | |
	Professional Scissors	Standard Scissors	OVERALL LIMITATIONS
Amount of each item	p	s	
Steel requirement per scissors	5	2	10,000
Sales revenue per scissors	12	4	12,000

EXAMPLE 2

Model the following: A wood cabinet manufacturer produces cabinets for television consoles and frames for grandfather clocks, both of which must be assembled and decorated. Every television cabinet requires 3 hours to assemble and 5 hours to decorate, while each grandfather clock frame requires 10 hours to assemble and 8 hours to decorate. The manufacturer has 33,000 hours available each week for assembling these products (825 assemblers working 40 hours per week) and 42,000 hours available each week for decorating (1050 decorators working 40 hours per week). How many of each product can the manufacturer schedule for production each week?

Solution

The objective is to determine the number of television consoles and grandfather clock frames to produce, so these quantities are important components. We set

$$x = \text{the number of television cabinets to be produced}$$

$$y = \text{the number of grandfather clock frames to be produced}$$

whereupon the objective is to determine x and y. Table 3.3 is a constraint table for this process.

Using the row labeled *Assembly time*, we generate the inequality

$$3x + 10y \leq 33,000 \qquad \text{(assembly constraint)}$$

which mdoels the condition that the amount of assembly time used to produce both products, $3x + 10y$, cannot exceed the 33,000 hours of assembly time available to the company.

MODEL 3.2

Important factors

x = number of television cabinets to produce

y = number of grandfather clock frames to produce

Objective

Find x and y.

Constraints

$3x + 10y \leq 33,000$

$5x + 8y \leq 42,000$

$x \geq 0$

$y \geq 0$

TABLE 3.3

| | ITEMS IN THE OBJECTIVE FUNCTION | | |
	Television Cabinet	Grandfather Clock Frame	AVAILABLE TIME
Amount of each item	x	y	
Assembly time (hours) per item	3	10	33,000
Decorating time (hours) per item	5	8	42,000

The last row in Table 3.3 yields the inequality

$$5x + 8y \leq 42{,}000 \qquad \text{(decorating constraint)}$$

which models the condition that the amount of decorating time used to produce both products, $5x + 8y$, cannot exceed the 42,000 hours of decorating time available to the company. Clearly, the manufacturer will not produce negative amounts of cabinets or clock frames, so we also have the hidden conditions $x \geq 0$ and $y \geq 0$.

Collecting our results, we generate the model in Model 3.2.

It is important to realize that a constraint table may not accommodate all constraints. In particular, had the process in Example 2 involved a contract requiring the company to produce at least 2,000 grandfather clock frames, then this constraint would be modeled by the inequality $y \geq 2{,}000$. The conditions of this contract, like the hidden conditions, are not easily written in a constraint table.

EXAMPLE 3

Model the following: Continental Motors manufactures taxis and airport limousines at its two plants. Plant A produces 75 taxis and 6 limousines daily, while plant B produces 50 taxis and 2 limousines each day. To fulfill contractual obligations for March, the company must manufacture 3,000 taxis and 186 limousines. How many days should each plant operate in March to fulfill these contractual obligations?

Solution

The objective is to determine the number of days the company should operate Plant A and Plant B in March. We set

$$x = \text{the number of days Plant } A \text{ will operate in March}$$

$$y = \text{the number of days Plant } B \text{ will operate in March}$$

whereupon the objective is to determine x and y. Table 3.4 is a constraint table for this process.

Using the row dealing with taxi production, we generate the inequality

$$75x + 50y \geq 3{,}000 \qquad \text{(taxi constraint)}$$

which models the condition that the number of taxis produced at the two plants must be greater than or equal to the 3,000 taxis the company must deliver to meet contractual obligations. The last row of Table 3.4 yields the inequality

$$6x + 2y \geq 186 \qquad \text{(limousine constraint)}$$

which models the condition that the number of limousines produced at the two plants cannot be less than the 186 limousines needed to fulfill the contract.

MODEL 3.3

Important factors

x = number of days Plant A will operate in March

y = number of days Plant B will operate in March

Objective

Find x and y.

Constraints

$75x + 50y \geq 3000$

$6x + 2y \geq 186$

$x \geq 0, \quad y \geq 0$

$x \leq 31, \quad y \leq 31$

TABLE 3.4

	ITEMS IN THE OBJECTIVE FUNCTION		GUARANTEED AMOUNTS
	Plant **A**	Plant **B**	
Days in operations	x	y	
Production of taxis per day	75	50	3,000
Production of limousines per day	6	2	186

Neither plant can operate a negative number of days, so we also have the hidden conditions $x \geq 0$ and $y \geq 0$. Furthermore, March has only 31 days, which places an upper bound on operations. We model these hidden conditions as $x \leq 31$ and $y \leq 31$. Collecting our results, we have the model in Model 3.3.

EXAMPLE 4

Describe how the model for Example 3 changes if the company has a labor policy prohibiting either plant from operating more than five days longer than the other plant.

Solution

The labor policy requires that $x - y$, the number of days Plant A operates less the number of days that plant B operates, be no more than five days. Therefore,

$$x - y \leq 5$$

The policy also requires that the number of days Plant B operates less the number of days that plant A operates be no more than five days. Thus, we also have the constraint

$$y - x \leq 5$$

These two new constraints are not easily written into the constraint table developed in Example 3, yet they must be added to Model 3.3 to model the company's labor policy.

As we work with models to increase our understanding of real-world processes, we must observe the arithmetic properties of inequalities when they are present.

> *Multiplying or dividing both sides of an inequality by a negative number changes the sense of the inequality.*

Property 1 Multiplying both sides of an inequality by a *positive* number does not change the sense of the inequality; less than remains less than, and greater than remains greater than.

We have $5 < 7$, and $5(4) < 7(4)$. Similarly, $20 \geq 3$, and $20(10) \geq 3(10)$. Also $-7 \leq -4$, and $-7(600) \leq -4(600)$.

Property 2 Multiplying both sides of an inequality by a negative number changes the sense of the inequality; less than becomes greater than, and greater than becomes less than.

We have $5 < 7$, but $5(-4) > 7(-4)$. Similarly, $20 \geq 3$, but $20(-10) \leq 3(-10)$. Also $-7 \leq -4$, but $-7(-600) \geq -4(-600)$. Division follows the same pattern as multiplication.

Property 3 Dividing both sides of an inequality by a negative number changes the sense of the inequality; dividing by a positive number does not change the sense of the inequality.

In contrast to multiplication and division, the operations of addition and subtraction never change the sense of inequalities.

Property 4 Adding or subtracting a positive or negative number on both sides of an inequality does not change the sense of the inequality.

Since $5 < 7$, it follows that $5 + 4 < 7 + 4$ and $5 - 4 < 7 - 4$. Similarly, $20 \geq 3$, so $20 + 30 \geq 3 + 30$ and $20 - 30 \geq 3 - 30$.

We will use the properties of inequalities to solve a single inequality in one variable. The techniques we will use are identical to those used to solve equalities in one variable.

EXAMPLE 5

Find x if $15 - 2x \leq 31$.

Solution

We add -15 to both sides of the inequality, and noting from Property 4 that the sense of the inequality remains unchanged, we have

$$-2x \leq 16$$

If we divide both sides of this last inequality by -2, it follows from Property 3 that the sense of the inequality is *changed* and

$$x \geq -8$$

Any value of x greater than or equal to -8 satisfies the original inequality. One solution is $x = -8$, but other solutions are, for example, $x = -7$, $x = 0$, $x = 25$, and $x = 103.44$.

EXAMPLE 6

Find x if $-10 + 5x < 7 + 2x$.

Solution

$$10 + (-10) + 5x < 10 + 7 + 2x$$
$$5x < 17 + 2x$$
$$5x - 2x < 17 + 2x - 2x$$
$$3x < 17$$
$$\frac{3x}{3} < \frac{17}{3}$$
$$x < \frac{17}{3}$$

Any value of x less than $\frac{17}{3}$ satisfies the original inequality. Here, $x = \frac{17}{3}$ is *not* a solution, but $x = 5.66$ is a solution, as are $x = 5.65$, $x = 5.64$, $x = 3$, and $x = -22$.

EXAMPLE 7

Find x if $66 \geq 2(3 - 4x) > -14$.

Solution

We can separate the inequalities into the two inequalities

$$66 \geq 2(3 - 4x) \qquad \text{and} \qquad 2(3 - 4x) > -14$$

and then solve as we did in Examples 4 and 5, or, more simply, we can apply the properties of inequalities to the statement as given, in which case we have

$$66 \geq 2(3 - 4x) > -14$$
$$66 \geq 6 - 8x > -14$$
$$66 - 6 \geq 6 - 6 - 8x > -14 - 6$$
$$60 \geq -8x > -20$$
$$60/(-8) \leq (-8x)/(-8) < (-20)/(-8)$$
$$-7.5 \leq x < 2.5$$

Any value of x satisfying *both* conditions—x is greater than or equal to -7.5 *and also* less than 2.5—is a solution of the original inequality. One solution is $x = -5$, a second solution is $x = 0$, and a third solution is $x = 2.36$. The value $x = 4$ is *not* a solution, because it violates the condition $x < 2.5$; similarly, $x = -10$ is not a solution because it violates the condition $x \geq -7.5$.

An inequality in one variable usually has infinitely many solutions.

We conclude from the last three examples that an inequality in one variable usually has infinitely many solutions. The models developed in Examples 1 through 4 contain inequalities in two variables, and we will show how to solve them in the next section. We shall see, more often than not, that such models also admit infinitely many solutions.

IMPROVING SKILLS

In Problems 1 through 29, solve for the variable in each inequality.

1. $\frac{1}{2}x - 19 < 3$
2. $2y + 14 \geq 5$
3. $-2N + 14 \geq 5$
4. $3x > 4x + 5$
5. $x + 5 \leq 2x - 7$
6. $3z - 19 < -5z + 7$
7. $-\frac{1}{2}t + 2 \geq -\frac{1}{3}t + \frac{1}{2}$
8. $-3x \leq 0$
9. $-5x > 0$
10. $17 \geq 9 - 4N$
11. $250 \leq 500 + 20S$
12. $250 \leq 500 - 20S$
13. $4 \leq x + 3 \leq 9$
14. $4 \leq x - 3 \leq 9$
15. $8 < 2y < 12$
16. $-2 < 2y \leq 12$
17. $2 < -2y \leq 12$
18. $6 \geq -3y \geq -30$
19. $10 \leq 2z + 4 \leq 20$
20. $10 \leq 2z - 4 \leq 20$
21. $-3 \geq -2x - 1 > -7$
22. $z < 2z + 8 < z + 10$
23. $z < 3z - 8 < z + 10$
24. $z + 3 < 2z + 8 < z + 10$
25. $N > \frac{1}{3}N + 10$
26. $N + 3 \leq \frac{1}{3}N + 10$
27. $3p - 3 \leq \dfrac{p}{4} + 10$
28. $3p - 3 \leq \dfrac{p + 10}{4}$
29. $3p - 3 \geq \dfrac{10 - 40p}{4}$

CREATING MODELS

In Problems 30 through 49, model each problem but do not solve it.

30. A manufacturer of sofas and wing chairs has 2,000 pounds of stuffing in inventory with no access to more stuffing until the following week. Each wing chair requires 5 pounds of stuffing and each sofa requires 11 pounds of stuffing to produce. How many wing chairs and sofas can the company manufacture during the week?

31. A manufacturer of sofas and wing chairs sells each wing chair for $200 and sells each sofa for $500. How many wing chairs and sofas must the company manufacture if it must at least cover its weekly costs of $70,000?

32. An upholstery shop has 300 yards of material in inventory for reconditioning sofas and love seats. Each sofa requires 9 yards of material and each love seat requires 5 yards of material. How many items of each type can the company upholster?

33. Model Problem 32 if the company also has a contract with a department store to provide the store with a minimum of ten sofas.

34. A metal pipe company has 500 hours of machine time available on a molding machine that will be allocated between the production of two types of metal joints. Type I joints require 60 seconds on the molding machine, while Type II joints require 50 seconds of molding. How many joints of each type can the company mold?

35. Model Problem 34 if the company also has contracts to provide customers with 100 Type I joints and 200 Type II joints.

36. A mattress company produces two different styles of mattresses. The deluxe style requires 50 springs and 12 pounds of padding per mattress, while the regular style requires 35 springs and 10 pounds of padding per mattress. The company has in inventory 17,000 springs and 5,000 pounds of padding. How many mattresses of each type can the company produce?

37. Model Problem 36 if, in addition, company policy requires that it always produce at least twice as many regular mattresses as deluxe mattresses.

38. A wood cabinet manufacturer produces cabinets for television consoles and lamp bases, both of which must be assembled and decorated. Each television cabinet requires 2 hours to assemble and 3 hours to decorate. In contrast, each lamp base requires 30 minutes to assemble and 45 minutes to decorate. The firm has 6,000 hours of assembly time and 6,500 hours of decorating time available to it each week. How many television cabinets and lamp bases can it produce weekly?

39. Model Problem 38 if, in addition, the company has a contract to deliver 5,000 lamp bases each week to one of its customers.

40. Model Problem 38, if, in addition, the company knows it cannot sell more than 4,000 lamp bases.

41. A caterer must prepare 500 gallons of a fruit punch that contains at least 20% orange juice. The caterer has ample stocks of two types of drinks that will be blended together to prepare the punch. Drink A is 100% orange juice, while Drink B is 5% orange juice. How much of each type drink should the caterer use?

42. Model Problem 41 if the caterer has an ample supply of Drink B but only 70 gallons of Drink A in inventory.

43. A pet store prepares its house brand of rabbit food from a mixture of two commercial feeds. Feed I contains five units of fat per pound, while Feed II contains ten units of fat per pound. How much of each feed should the pet store blend into each pound of its house brand, which is guaranteed to contain no more than seven units of fat per pound?

44. Model Problem 40 if, in addition, Feed II is much tastier than Feed I, so that the pet store always uses at least twice as much Feed II as Feed I in its house brand.

45. A pet store prepares its house brand of rabbit food from a mixture of two commercial feeds. Feed I contains 20 units of protein per pound, while Feed II contains 5 units of protein per pound. How much of each feed should the pet store blend into each pound of its house brand, which is guaranteed to contain at least 10 units of protein per pound?

46. Model Problem 45 if, in addition, Feed I costs 80¢ per pound, Feed II costs 30¢ per pound, and the pet store wants to keep the cost of its house brand under 50¢ per pound.

47. A wood cabinet manufacturer produces cabinets for television consoles, frames for grandfather clocks, and lamp bases, all of which must be assembled and decorated. Each television cabinet requires 5 hours to assemble and 3 hours to decorate, each grandfather clock frame requires 10 hours to assemble and 8 hours to decorate, and each lamp base takes 1 hour to assemble and 2 hours to decorate. The firm has 10,000 hours of assembly time and 12,000 hours of decorating time available. How many items of each type can be produced?

48. Anew Motors customizes taxis and airport limousines at its three plants. Plant *A* can customize 75 taxis and 6 limousines daily, while plant *B* can customize 50 taxis and 2 limousines each day. Plant *C* has facilities only for limousines, and it can customize 20 a day. The company has contractual obligations in November for 3,000 taxis and 300 limousines. How many days should the company operate each plant?

49. Model Problem 45 if the pet store has four types of commercial feed available for its house mixture, the two feeds described in Problem 45 plus Feed III, which contains 15 units of protein per pound, and Feed IV, which contains 3 units of protein per pound.

EXPLORING IN TEAMS

Break-even analysis (Section 1.6) focuses on production levels at which total revenue *R* equals total cost *C*. If a company is to make a profit, however, total revenue must exceed total cost. The analysis is analogous to that described in Section 1.6, but instead of requiring $R = C$, the constraint becomes $R > C$. Use this last inequality to solve Problems 50 through 52.

50. A manufacturer of electric can openers currently has a fixed cost of $20,000 per day. The cost of producing each can opener is $5.38, and each can opener is sold for $9.00. How many can openers must the firm produce to make a profit?

51. A bakery currently has a fixed cost of $120,000 per month and estimates its variable cost to be 28¢ per loaf of bread. Loaves are sold wholesale for 56¢ per loaf. How many loaves must the bakery sell each month to show a profit?

52. A publisher of a current economics textbook determines that the manufacturing costs directly attributable to each book are $16.23 and that the fixed costs are $200,000. How many books must the publisher sell to show a profit if (*a*) each book sells for $32.50, and (*b*) each book sells for $39?

REVIEWING MATERIAL

53. (Section 1.4) On an island inhabited by only rabbits and foxes, the rabbits eat vegetation and the foxes eat the rabbits. Construct a qualitative graphical model that relates the rabbit population (vertical axis) to the fox population (horizontal axis).

54. (Section 2.2) Use Gaussian elimination to solve the following system for *s* and *p*:

$$2s + 3p = 3,000$$
$$8s + 7p = 2,000$$

RECOMMENDING ACTION

55. Respond by memo to the following request:

MEMORANDUM

To: J. Doe Reader

From: Manufacturing

Date: Today

Subject: **Scissors Production**

I have your model for the number of professional scissors P and standard scissors S that I should schedule for production, taking into account both the amount of steel we have in inventory and our budget requirements. Your constraints are

$$5P + 2S \leq 10{,}000$$

$$12P + 4S \geq 12{,}000$$

Can't I just replace the inequalities with equalities and then solve the equations? It is certainly easier, and what is lost?

3.2 GRAPHING LINEAR INEQUALITIES

The graphing techniques developed in Section 1.6 for solving linear equalities in two variables can be extended to solving linear inequalities in two variables, with some interesting distinctions. Recall that a system of linear equalities has infinitely many solutions only under highly unlikely circumstances: the graphs of every equation in the system must be the *same* straight line. In contrast, systems of linear inequalities generally have infinitely many solutions.

An inequality in one or more variables is *linear* if the equation obtained by replacing the inequality with an equality is linear (see Sections 1.5 and 2.1). For example, the inequality $2x - y > 0$ in the two variables x and y is linear because the equation $2x - y = 0$ is linear. In contrast, the inequality $x^2 + y^2 \leq 4$ is *not* linear because the equation $x^2 + y^2 = 4$ is not linear. The inequality $3r + 5s - 2t \geq 1{,}000$ in the three variables r, s, and t is linear because $3r + 5s - 2t = 1{,}000$ is a linear equation.

> *An inequality in one or more variables is linear if the equation obtained by replacing the inequality with an equality is linear.*

The graph of a linear equation in two variables is a straight line, and *every straight line in a plane divides the plane into two regions* (see Figures 3.1 and 3.2): the region above the line and the region below the line or, depending on the slope of the line, the region to left of the line and the region to the right of the line.

The general form of a linear equation in two variables is

$$Ax + By = C \tag{1}$$

where A, B, and C are known numbers with A and B not both zero. If the x and y coordinates of a point in the plane do *not* satisfy Equation (*1*), then that point is *not* on the straight

FIGURE 3.1

FIGURE 3.2

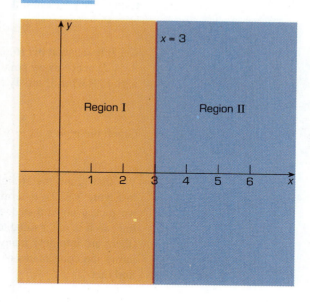

line graph of Equation (*1*) and it therefore must be in one of the regions on either side of the line. Furthermore, if $Ax + By \neq C$, then either

$$Ax + By < C \qquad (2)$$

or

$$Ax + By > C \qquad (3)$$

> *The graph of a linear inequality in two variables is the entire region to one side of a straight line.*

Points in the plane that satisfy (2) lie on one side of the line defined by Equation (*1*); points that satisfy (3) lie on the other side of the line. To determine which region satisfies which inequality, pick a test point in one of the regions and determine by direct substitution whether its x and y coordinates satisfy (2) or (3). The inequality satisfied by the test point is the inequality that is true for the coordinates of all points in the region containing the test point. Points on the other side of the straight line satisfy the reverse inequality. The simplest test point is the origin ($x = y = 0$) when the origin is in one of the two regions of interest.

EXAMPLE 1

Determine the inequality satisfied by all points in Region I of Figure 3.1, and then determine the inequality satisfied by all points in Region II.

Solution

The straight line that separates the two regions in Figure 3.1 is defined by the linear equation

$$2x + 3y = 6 \qquad (4)$$

Thus the coordinates of points in Regions I and II will satisfy either

$$2x + 3y < 6 \qquad (5)$$

or

$$2x + 3y > 6 \qquad (6)$$

The origin is *not* on the line defined by Equation (*4*), so it is a suitable test point. Substituting $x = y = 0$ into the left side of either (*5*) or (*6*), we have

$$2(0) + 3(0) = 0 < 6$$

so (*5*) is satisfied but (*6*) is not. The origin is in Region II and satisfies the inequality $2x + 3y < 6$, hence every point in Region II satisfies the same inequality. All points in the other region, Region I, satisfy the reverse inequality, $2x + 3y > 6$.

EXAMPLE 2

Graph the region $2x - y > 0$.

Solution

We first graph the straight line defined by the linear equation $2x - y = 0$, as is done in Figure 3.3. To determine which side of the line corresponds to the given inequality, we pick a point that does *not* lie on the line as a test point. Here the origin is on the line, so it is *not* a suitable test point. We see from Figure 3.3 that $(0, 2)$ is not on the line (there are infinitely many other points not on the line, and all are equally good). Substituting the coordinates of $x = 0$ and $y = 2$ into the left side of the given inequality, we find that

$$2(0) - 2 = 0 - 2 = -2$$

which is *less than* 0, the right side of our inequality. The point $(0, 2)$ satisfies the inequality $2x - y < 0$, as does every other point on the same side of the line as $(0, 2)$. The opposite region, which is shaded in Figure 3.3, satisfies the reverse inequality, $2x - y > 0$, and is the one of interest.

FIGURE 3.3

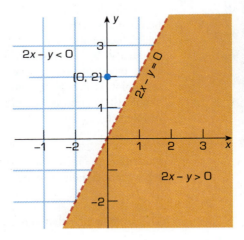

The inequality in Example 2 is a strict inequality that is *not* satisfied by the points on the line $2x - y = 0$. We emphasize that a line is *not* part of a region by graphing it as a dashed line, as is done in Figure 3.3. If a line is part of a region of interest, we show it as a solid line.

EXAMPLE 3

Graph the region $x + y \leq 6$.

Solution

The line defined by the equation $x + y = 6$ is graphed in Figure 3.4. Since any solution of the equality $x + y = 6$ also satisfies the inequality $x + y \leq 6$, the line is part of the region of interest and is drawn as a solid line. The origin is not on the line and is a suitable test point. Substituting its coordinates $x = y = 0$ into the left side of the given inequality, we find that $0 + 0 = 0$, which is less than 6 and *does* satisfy the given inequality. Thus, every point on the same side of the line as the origin (the shaded side in Figure 3.4) also satisfies the inequality $x + y \leq 6$.

FIGURE 3.4

EXAMPLE 4

Graph the region $y \geq -1$.

Solution

The line defined by the equation $y = -1$ is graphed in Figure 3.5. Since any solution of the equality $y = -1$ also satisfies the inequality $y \geq -1$, the line is part of the region of

FIGURE 3.5

interest and is drawn as a solid line. The origin is not on the line and is a suitable test point. Substituting its coordinates into the left side of the given inequality, we obtain 0 (because $y = 0$ at the origin), which is greater than -1, the right side of the given inequality. Thus, every point on the same side of the line as the origin (the shaded side in Figure 3.5) also satisfies the inequality $y \geq -1$.

> *To find the feasible region for a system of linear inequalities, first locate solutions to each inequality in the system and then identify the region common to all of them.*

A *solution* to a system of linear inequalities in two variables is a point whose coordinates satisfy *all* the inequalities in the system. To find solutions to a system of linear inequalities, first locate solutions to the individual inequalities in the system and then identify the region common to all of them. This region is called the *feasible region*, and every point in the feasible region is a solution to the given system. If there is no region common to all constraints, then the system has no solution and the feasible region is said to be empty. For clarity and for future reference, it is good practice when working with systems of linear inequalities to label each graphed line with its equation.

EXAMPLE 5

Find the feasible region defined by the system

$$2x - y > 0$$

$$x + y \leq 6$$

Solution

The regions satisfying these individual constraints were found in Examples 2 and 3. We place these regions on the same graph (see Figure 3.6), but since they are not the final answer, we use shading only to show which sides of the individual lines satisfy the individual inequalities.

For clarity, we also label each line with the equation that defines it. Points in the feasible

FIGURE 3.6

FIGURE 3.7

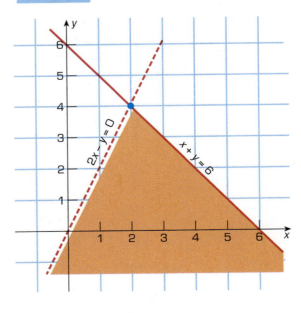

region must satisfy both inequalities; these points lie to the right of the line $2x - y = 0$ and also to the left of the line $x + y = 6$. This area is shaded in Figure 3.7; it is the feasible region for the given system of linear inequalities.

The feasible region in Example 5 is bounded by two lines with a single point of intersection, which is circled in Figure 3.7. This point lies on the straight lines defined by

A corner point is a point on the boundary of a feasible region where two lines bounding the feasible region intersect.

$$x + y = 6$$

and

$$2x - y = 0$$

The coordinates of the point of intersection is a solution to both equations; we can find by Gaussian elimination that it is $(2, 4)$. A *corner point* of a feasible region is a point on the boundary of the feasible region where two lines bounding the feasible region intersect. The feasible region in Example 5 has the single corner point $(2, 4)$.

EXAMPLE 6

Find the feasible region and all the corner points defined by the system

$$x + y \leq 2$$
$$2x + y \geq -1$$
$$y \geq -3$$

Solution

We first find the regions that satisfy the individual constraints and place those regions on the same Cartesian coordinate system, as we have done in Figure 3.8, shading only the sides of the lines that define each region. We also label each line with the equation that defines it.

FIGURE 3.8

FIGURE 3.9

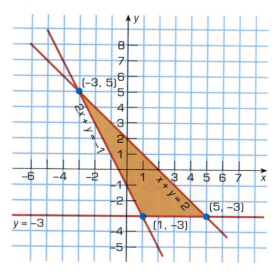

The origin $(0, 0)$ is a suitable test point for all three inequalities, and because its co-ordinates satisfy each inequality, the regions of interest include the origin. The feasible region of the system consists of points to the left of the line $x + y = 2$ *and* to the right of the line $2x + y = -1$ *and* above the line $y = -3$. The feasible region is shaded in Figure 3.9.

This feasible region has three corner points, which are circled in Figure 3.9. One corner point satisfies the system of equations

$$x + y = 2$$
$$y = -3$$

Solving this system by back-substitution, we find that the coordinates of this corner point are $x = 5$ and $y = -3$. A second corner point satisfies the system

$$2x + y = -1$$
$$y = -3$$

and has the coordinates $x = 1$ and $y = -3$, again found by back-substitution. The third corner point satisfies the system

$$x + y = 2$$
$$2x + y = -1$$

Using Gaussian elimination we find the coordinates of this point to be $x = -3$ and $y = 5$.

EXAMPLE 7

Find the feasible region and all corner points defined by the system

$$3x + 10y \le 33{,}000$$
$$5x + 8y \le 42{,}000$$
$$x \ge 0$$
$$y \ge 0$$

Solution

The inequalities $x \ge 0$ and $y \ge 0$ limit the feasible region to the first quadrant, so we shade region to the right of the y-axis and above the x-axis, as in Figure 3.10.

We see from Figure 3.10 that $(0, 0)$ is a suitable test point for the other two inequalities

FIGURE 3.10

FIGURE 3.11

because the origin is not on the lines defined by $5x + 8y = 42,000$ and $3x + 10y = 33,000$. Substituting $x = y = 0$ into the left sides of the first two inequalities of the given system, we find that the origin satisfies both, so the regions satisfying these inequalities lie on the sides of lines that contain the origin. The feasible region lies in the first quadrant below the line $5x + 8y = 42,000$ and below the line $3x + 10y = 33,000$. It is the shaded area in Figure 3.11.

This feasible region has four corner points, which are circled in Figure 3.11. Three of these points $(0, 0)$, $(0, 3,300)$ and $(8,400, 0)$ are on the coordinate axes. The other corner point satisfies the system of equations

$$3x + 10y = 33,000$$

$$5x + 8y = 42,000$$

Using Gaussian elimination we find the coordinates of this point to be $x = 6,000$ and $y = 1,500$.

FIGURE 3.12

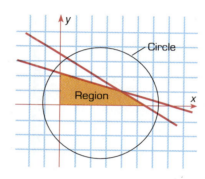

A feasible region is bounded if it can be enclosed in a circle.

The feasible region in Example 7 can be enclosed in a circle (see Figure 3.12) and is therefore conceptually different from the feasible region in Figure 3.7 of Example 5, which cannot be so contained.

We say that a feasible region is *bounded* if it can be enclosed within a (possibly very large) circle; regions that are not bounded are called *unbounded*. The feasible regions of Example 6 and 7 are bounded; the feasible regions of Examples 2 through 5 are unbounded.

IMPROVING SKILLS

In Problems 1 through 12, graphically find the feasible regions of the given linear inequalities.

1. $2x + 5y \leq 10$
2. $2x - 5y \leq 10$
3. $-2x + 5y \leq 10$
4. $-2x - 5y \leq 10$
5. $2x + 5y \leq -10$
6. $x - 3y \geq -6$
7. $y \geq 0$
8. $2x - 3y > 12$
9. $2x + 5y < 0$
10. $x + 12y \leq 0$
11. $x \geq 0$
12. $y < 3$

In Problems 13 through 38, find the feasible regions and corner points for the given systems of linear inequalities.

13. $x + 2y \geq 4$
    $2x - y \leq 3$

14. $x + 2y \geq 4$
    $2x - y \geq 3$

15. $3x - 4y < 12$
    $3x + 4y > 12$

16. $3x - 4y \geq 12$
    $3x + 4y \geq 12$

17. $2x + 5y \geq 1{,}000$
    $5x + 2y \geq 1{,}000$

18. $2x + 5y > 1{,}000$
    $5x + 2y < 1{,}000$

19. $2x - 3y < 12$
    $x \geq 5$

20. $2x + 7y \leq 14{,}000$
    $y < -1{,}000$

21. $x > 7$
    $y < 2$

22. $x \geq -2$
    $x \geq 3$

23. $x + 3y \leq 9$
    $x \geq 0$
    $y \geq 1$

24. $x - 3y \leq 9$
    $x \geq 0$
    $y \leq 2$

25. $5x + 4y \leq 20$
    $x \geq 0$
    $y \geq 0$

26. $5x - 4y \geq 20$
    $x \geq -1$
    $y \geq 3$

27. $5x - 4y \geq 20$
    $x \geq -1$
    $y \leq 3$

28. $5x + 3y \leq 15$
    $x \geq 5$
    $y \geq 0$

29. $3x + y \geq -3$
    $x - y \leq 4$
    $y \leq 2$

30. $10x + 8y \leq 80$
    $x - 4y \leq 12$
    $x \geq 0$

31. $10x + 8y \leq 80$
    $x - 4y \leq 12$
    $x \leq 0$

32. $10x + 8y \leq 80$
    $x - 4y \leq 12$
    $y \geq -3$

33. $6x + y \geq 12$
    $2x + 3y \geq 12$
    $x + y \leq 6$

34. $x + y \leq 10$
    $x - y \geq 0$
    $y \geq 1$

35. $5x + 7y \leq 35$
    $7x - 5y \leq 35$
    $-x - 5y \leq 10$
    $x - y \leq 5$

36. $5x + 7y \leq 35$
    $7x - 5y \geq 35$
    $-x - 5y \leq 10$
    $x - y \leq 5$

37. $5x + y \leq 10$
    $7x + 8y \leq 40$
    $x \geq 0$
    $y \geq 0$

38. $7x + 4y \leq 140$
    $x + 2y \leq 40$
    $x \geq 0$
    $y \geq 0$

CREATING MODELS

In Problems 39 through 47, identify the feasible regions for the given processes.

39. Problem 30 of Section 3.1.
40. Problem 31 of Section 3.1.
41. Problem 32 of Section 3.1.
42. Problem 33 of Section 3.1.
43. Problem 36 of Section 3.1.
44. Problem 37 of Section 3.1.
45. Problem 38 of Section 3.1.
46. Problem 39 of Section 3.1.
47. Problem 40 of Section 3.1.

EXPLORING IN TEAMS

48. Assume that x and y in Example 7 denote, respectively, the number of sofas and the number of wing chairs that an upholstery shop is capable of producing in a month, subject to the production constraints modeled by the inequalities in that example. Identify six different solutions in the feasible region. Determine the total sales revenues generated by each of these six solutions if a sofa sells for $400 and a wing chair for $200. Which solution generates the most revenue?

49. Assume that x and y in Problem 38 denote, respectively, the number of people from the east coast and the number of people from the west coast (both in thousands) who visit a state park in South Dakota and that the inequalities in that problem reflect constraints on advertising expenditure. Identify six different solutions in the feasible region and determine the total number of people who visit the park for each solution. Which solution generates the most visitors?

REVIEWING MATERIAL

50. (Section 1.5) Find the slope of a line parallel to the line defined by $2x + 8y = 9$.

51. (Section 2.4) Find x and y if

$$\begin{bmatrix} 1 & x \\ y & -2 \end{bmatrix} \begin{bmatrix} -1 \\ 5 \end{bmatrix} = \begin{bmatrix} 2 \\ -2 \end{bmatrix}$$

RECOMMENDING ACTION

52. Respond by memo to the following request:

> **MEMORANDUM**
>
> To: J. Doe Reader
>
> From: Dora
>
> Date: Today
>
> Subject: **A Little Help, Please**
>
> I have a model with just two inequalities. The first is $y < 5$ and the second is $y < x + 5$. Does this mean that $x = 0$? Clearly, this would be the case if I was dealing with equalities, but inequalities always make me a bit nervous. Can you help?

OPTIMIZING WITH GRAPHS

Having many solutions gives decision-makers choices, and choices provide opportunities.

Linear programming problems optimize a linear objective function subject to constraints modeled by linear equalities and/or linear inequalities.

Mathematical models consisting of systems of linear equations (Sections 1.6 and 2.2) or linear inequalities (Section 3.2) may have no solution, exactly one solution, or infinitely many solutions. "No solution" is definitive: the modeled constraints cannot be satisfied; for example, there is no solution if a person who wants to buy a new car, has twenty dollars in savings, and refuses to go into debt.

A single solution is equally definitive, because there is only one way to satisfy the modeled constraints. This situation may seem attractive, but a model with only one solution limits choice, and choices provide opportunities. If a menu of solutions is available, decision-makers can select the one most advantageous to themselves or to society according to some secondary criterion, such as maximizing efficiency or minimizing fuel consumption or waste.

Linear programming problems are problems that optimize a quantity—the *objective function*—defined by a linear equation, subject to constraints modeled by linear equalities and/or linear inequalities. The word *programming* in linear programming does *not* refer to computer programming but rather to a well-defined solution procedure that we shall describe shortly. Sometimes we shall want to maximize an objective function such as productivity; sometimes we shall want to minimize an objective function such as the time to complete a project. Rather than always saying "maximize or minimize," we will just say "optimize." We often let the letter z denote the quantity being optimized, but, as always, we are free to replace z with other letters, such as P for profit or C for cost, when those letters are more meaningful.

An example of a linear programming problem is

$$\text{Maximize:} \quad z = 2x + y \tag{7}$$
$$\text{subject to:} \quad 3x + 10y \leq 33{,}000$$
$$5x + 8y \leq 42{,}000$$
$$x \geq 0$$
$$y \geq 0$$

We seek values of x and y that satisfy all the constraints and also maximize the objective z, which is defined by the linear equation $z = 2x + y$. If the feasible region is empty, so that there is no solution that satisfies the constraints, then there is no solution to the linear programming problem. If the feasible region contains only one solution, then the coordinates of that point are the solution to the linear programming problem. If, however, the feasible region contains many points, then we want to find a point in the feasible region whose coordinates yield the best value for z.

In Example 7 of Section 3.2, we found that the feasible region for the constraints defined by System (7) was the shaded portion of Figure 3.13. This region contain infinitely many points, so we have many options! We can ask for more and we do! We want a point in the feasible region whose x and y coordinates yield the greatest value of the objective function $z = 2x + y$.

The point (2000, 1000) is in this feasible region, and it yields

$$z = 2(2{,}000) + 1{,}000 = 5{,}000$$

so we are assured of achieving a z of at least 5,000. The point (1000, 3000) is also in the feasible region, and it also yields

$$z = 2(1{,}000) + 3{,}000 = 5{,}000$$

FIGURE 3.13

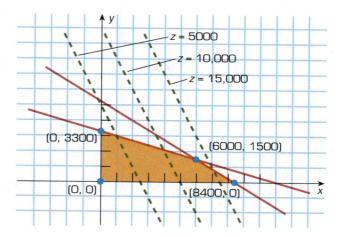

In fact, if we set $z = 5{,}000$, then the objective function in (7) becomes

$$5{,}000 = 2x + y$$

which is the equation of a straight line. This line is graphed in Figure 3.13, where it is labeled as $z = 5{,}000$, and it passes through the feasible region at many points; thus, there are many points in the feasible region that yield $z = 5{,}000$, including the two points (1000, 3000) and (2000, 1000).

Knowing we can attain $z = 5{,}000$, we ask, Can we do better? Can we achieve a value of the objective function equal to 10,000? We set $z = 10{,}000$, whereupon the objective function in (7) becomes

$$10{,}000 = 2x + y$$

This line ($z = 10{,}000$) is graphed in Figure 3.13, and because it passes through the feasible region, $z = 10{,}000$ is obtainable. Any point on the line $z = 10{,}000$ and in the feasible region satisfies all the constraints and yields the value of 10,000 for z. One such point is (5000, 0).

Can we do better still and achieve a value of the objective function equal to 15,000? With $z = 15{,}000$, the objective function in (7) becomes

$$15{,}000 = 2x + y$$

This line ($z = 15{,}000$) is also graphed in Figure 3.13, and because it passes through the feasible region, $z = 15{,}000$ is obtainable.

Note that the three z lines in Figure 3.13 are parallel, and as the lines move to the right in the increasing x direction, their corresponding z values increase. Thus, the largest attainable z will occur on a line that is parallel to the three z lines drawn in Figure 3.13 and is as far to the right as possible while still intersecting the feasible region. Such a line must pass through the corner point $x = 8{,}400$ and $y = 0$ and will have a corresponding maximum of

$$z = 2(8{,}400) + 0 = 16{,}800$$

The z line corresponding to any larger value of z will fall outside the feasible region and will not be attainable by points satisfying the constraints.

As the z lines in Figure 3.13 move to the left in the decreasing x direction, their corresponding z values decrease. Thus, the smallest attainable z will occur on a line that is parallel to the three z lines drawn in Figure 3.13 and is as far to the left as possible while

The maximum and minimum of a linear programming model occur at corner points of the feasible region when the feasible region is nonempty and bounded.

still intersecting the feasible region. The minimum value of the objective function in (7) therefore occurs at the corner point $x = y = 0$ where $z = 2(0) + 0 = 0$.

Applying the same graphical analysis to any linear programming problem in two variables with a bounded feasible region shows that the optimal value of the objective function (either the maximum or the minimum) occurs at a corner point. Consequently, there is no need to graph z lines; we only need to find the corner points of the feasible region and determine the one that yields the best value of the objective function. We can therefore use the following four-step procedure or algorithm for solving a linear programming problem in two variables:

GRAPHICAL SOLUTION OF LINEAR PROGRAMMING PROBLEMS:

STEP 1 Find the feasible region that satisfies the constraints. If there is no feasible region, stop: the problem has no solution; otherwise, continue.

STEP 2 Determine whether the feasible region is bounded. If not, stop: this method is not applicable; otherwise, continue.

STEP 3 Locate the coordinates of all corner points.

STEP 4 Substitute the coordinates of each corner point into the objective function, and determine the one that optimizes the objective function.

Although Step 2 opens a potential loophole, it is not one of practical importance, and we shall close it shortly.

EXAMPLE 1

$$\text{Maximize:} \quad z = 7x + 22y$$
$$\text{subject to:} \quad 3x + 10y \leq 33{,}000$$
$$5x + 8y \leq 42{,}000$$
$$x \geq 0$$
$$y \geq 0$$

Solution

This linear programming problem is different from the problem in (7) because it has a different objective function. The constraints are identical, however, so the feasible region

TABLE 3.5

CORNER POINTS		OBJECTIVE FUNCTION
x	y	$z = 7x + 22y$
0	0	$z = 7(0) + 22(0) \qquad = 0$
0	3300	$z = 7(0) + 22(3300) \qquad = 72{,}600$
6000	1500	$z = 7(6000) + 22(1500) = 75{,}000 \checkmark$
8400	0	$z = 7(8400) + 22(0) \qquad = 58{,}800$

is still the shaded portion of Figure 3.13, which is bounded. We could draw new z lines associated with the new objective function, but there is no need to. The maximum value of the objective function must occur at a corner point of the feasible region. Evaluating the objective function at each corner point, we complete Table 3.5 and see that z attains a maximum of 75,000 when $x = 6,000$ and $y = 1,500$.

EXAMPLE 2

$$\text{Minimize:} \quad z = 8x + 5y$$

$$\text{subject to:} \quad x + y \leq 2$$
$$2x + y \geq -1$$
$$y \geq -3$$

Solution

The feasible region for this set of linear constraints was found in Example 6 of Section 3.2 and is the shaded portion of Figure 3.14. This region contains many solutions and is bounded. The minimum value of the objective function must occur at a corner point of the feasible region. Evaluating the objective function at each corner point, we complete Table 3.6 and see that z attains a minimum of -7 when $x = 1$ and $y = -3$. We also note from Table 3.6 that the maximum of $z = 8x + 5y$, subject to the same set of constraints, is 25, and it occurs when $x = 5$ and $y = -3$.

FIGURE 3.14

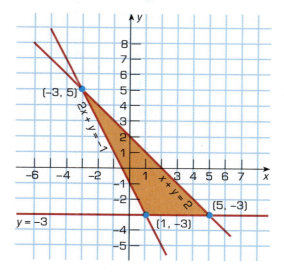

TABLE 3.6

CORNER POINTS		OBJECTIVE FUNCTION
x	y	$z = 8x + 5y$
-3	5	$z = 8(-3) + 5(5) = \quad 1$
5	-3	$z = 8(5) + 5(-3) = \quad 25$
1	-3	$z = 8(1) + 5(-3) = \quad -7 \checkmark$

EXAMPLE 3

The Lawn Care Company produces two different lawn fertilizers, Green Power Regular and Green Power Deluxe, both of which are highly popular and sell without difficulty. The profit on each bag of Regular is 75¢, while the profit on each bag of Deluxe is $1.10. Each bag of Regular contains 3 pounds of active ingredients (a combination of nitrogen, phosphoric acid, and potash) and 7 pounds of inert substances (essentially fillers). In contrast, each bag of Deluxe contains 4 pounds of active ingredients and 6 pounds of inert substances. Due to limited warehouse facilities, the company can stock only 7,400 pounds of active ingredients and 14,100 pounds of inert substances. How many bags of each product should the Lawn Care Company produce if its objective is to maximize profit?

Solution

We seek to maximize total profit P. The profit on each bag of Regular is 75¢, and the profit on each bag of Deluxe is $1.10, so if we set

$$x = \text{the number of bags of Regular to be produced}$$

$$y = \text{the number of bags of Deluxe to be produced}$$

we can express the objective function (in dollars) as

$$\text{Maximize } P = 0.75x + 1.10y$$

Table 3.7 is a constraint table for this process. The row dealing with active ingredients yields the inequality

$$3x + 4y \leq 7,400 \text{ (active ingredient constraint)}$$

The row dealing with inert ingredients yields the inequality

$$7x + 6y \leq 14,100 \text{ (inert ingredient constraint)}$$

The manufacturer will not produce negative bags of either fertilizer, so we add the hidden conditions $x \geq 0$ and $y \geq 0$. Combining these constraints, we have the linear programming model

$$\text{Maximize:} \quad P = 0.75x + 1.10y$$

$$\text{subject to:} \quad 3x + 4y \leq 7,400$$

$$7x + 6y \leq 14,100$$

$$x \geq 0$$

$$y \geq 0$$

TABLE 3.7

	ITEMS IN THE OBJECTIVE FUNCTION		
	Regular Fertilizer	Deluxe Fertilizer	AMOUNT IN STOCK
Amount of each item	x	y	
Active ingredients (pounds per bag)	3	4	7,400
Inert ingredients (pounds per bag)	7	6	14,100

Using the methods developed in Section 3.2, we identify the feasible region for this set of linear constraints, the shaded portion of Figure 3.15.

This feasible region contains many solutions and is bounded. The region has four corner points, which are circled in Figure 3.15. Evaluating the objective function at each corner point, we complete Table 3.8 and see that maximum profit is $2,035 when $x = 0$ and $y = 1850$.

FIGURE 3.15

TABLE 3.8

CORNER POINTS		OBJECTIVE FUNCTION
x	y	$P = 0.75x + 1.10y$
0	1850	$P = 0.75(0) + 1.10(1850)$ $= 2,035$ ✓
1200	950	$P = 0.75(1200) + 1.10(950) = 1,945$
2014.29	0	$P = 0.75(2014.29) + 1.10(0) = 1,510.72$
0	0	$P = 0.75(0) + 1.10(0)$ $= 0$

EXAMPLE 4

Solve Example 3 if the profit for Regular is 85¢ per bag and the profit for Deluxe is $1 per bag.

Solution

The only change from the linear programming model in Example 3 is in the objective function, which becomes

$$\text{Maximize } P = 0.85x + 1.00y$$

to reflect the new profit figures. The inequality constraints are identical, so the feasible region and corner points remain as shown in Figure 3.15. Evaluating this new objective function at all four corner points, we complete Table 3.9. The maximum value of P is now $1,970, and it occurs when $x = 1200$ and $y = 950$.

TABLE 3.9

CORNER POINTS		OBJECTIVE FUNCTION
		$P = 0.85x + 1.00y$
x	y	
0	1850	$P = 0.85(0) + 1.00(1850) = 1,850$
1200	950	$P = 0.85(1200) + 1.00(950) = 1,970$ ✓
2014.29	0	$P = 0.85(2014.29) + 1.00(0) = 1,712.15$
0	0	$P = 0.85(0) + 1.00(0) = 0$

> *An unbounded feasible region indicates the existence of additional hidden conditions.*

Step 2 of our graphical procedure for solving linear programming problems requires us to check whether the feasible region is bounded. All feasible regions associated with real-world problems are bounded, because all real-world processes are constrained by finite resources: inventories are finite, labor is finite, demands are finite. Even time is effectively finite, because we require solutions within specified time periods. An unbounded feasible region indicates the existence of additional hidden constraints that, when identified, bound the region.

EXAMPLE 5

The Heartland Cereal Company produces All-Pro, a ready-to-eat breakfast food made from a blend of fortified cereal and dehydrated dairy products. Each serving of All-Pro is guaranteed to contain at least 30 grams of protein, 50 units of calcium, and 20 grams of carbohydrates. Heartland's supplier certifies that every ounce of fortified cereal contains exactly 20 grams of protein, 10 units of calcium, and 30 grams of carbohydrates. Each ounce of the dehydrated dairy product contains exactly 10 grams of protein, 100 units of calcium, and 20 grams of carbohydrates. The fortified cereal and dehydrated milk product cost 3¢ and 2¢ per ounce, respectively. How much of each ingredient should be used in each serving of All-Pro to minimize supply costs while still maintaining the guaranteed nutritional composition?

Solution

We seek to minimize cost C, which is 3¢ times the number of ounces of fortified cereal plus 2¢ times the number of ounces of dehydrated milk product used in each serving of All-Pro. The actual amounts of each ingredient in All-Pro are unknown but clearly important components. We set

x = ounces of fortified cereal used in each serving of All-Pro

y = ounces of dehydrated milk product used in each serving of All-Pro

Then the cost (in dollars) is $C = 0.03x + 0.02y$, and the objective function is to

$$\text{Minimize } C = 0.03x + 0.02y$$

Table 3.10 is a constraint table for this process. The last three columns yield, respectively, the inequalities

$$20x + 10y \geq 30 \qquad \text{(protein constraint)}$$
$$10x + 100y \geq 50 \qquad \text{(calcium constraint)}$$
$$30x + 20y \geq 20 \qquad \text{(carbohydrate constraint)}$$

TABLE 3.10

	ITEMS IN THE OBJECTIVE FUNCTION		
	Fortified Cereal	**Dehydrated Milk Product**	GUARANTEED AMOUNTS
Amount of each item	x	y	
Protein content (grams per serving)	20	10	30
Calcium content (units per serving)	10	100	50
Carbohydrate content (grams per serving)	30	20	20

Each inequality is of the $\geq$ type because the amounts of the three nutrients in each serving of All-Pro cannot be less than the guaranteed minimums. Furthermore, Heartland cannot use negative amounts of fortified cereal or dehydrated milk products, so we add the hidden conditions $x \geq 0$ and $y \geq 0$. Combining these results, we generate the linear programming model

$$\text{Minimize:} \quad C = 0.03x + 0.02y$$

$$\begin{aligned} \text{subject to:} \quad 20x + 10y &\geq 30 \\ 10x + 100y &\geq 50 \\ 30x + 20y &\geq 20 \\ x &\geq 0 \\ y &\geq 0 \end{aligned}$$

The feasible region for the constraint inequalities is the shaded portion of Figure 3.16. Note that this region is not bounded, suggesting the existence of additional hidden conditions.

The Heartland Cereal Company will not use more ingredients than are absolutely nec-

FIGURE 3.16

essary in each serving of All-Pro, because that would just add expense. If All-Pro were made from just fortified cereal, the company would need $1\frac{1}{2}$ ounces to meet the protein requirement, 5 ounces to meet the calcium requirement, and $\frac{2}{3}$ ounce to meet the carbohydrate requirement. Therefore, 5 ounces of fortified cereal would meet all the requirements. Five ounces of cereal supplies $5(20) = 100$ grams of protein, $5(10) = 50$ units of calcium, and $5(30) = 150$ grams of carbohydrates—this is too much protein and carbohydrates but exactly the required amount of calcium. Thus, Heartland would never use more than 5 ounces of fortified cereal in any serving of All-Pro, so $x \leq 5$ is a hidden condition. If All-Pro were made from just the dehydrated milk product, the company would need 3 ounces to meet the protein requirement, $\frac{1}{2}$ ounce to meet the calcium requirement, and 1 ounce to meet the carbohydrate requirement. Therefore, 3 ounces of the dehydrated milk product would meet all the requirements and is thus an upper bound on the amount of dehydrated milk product in a serving of All-Pro, so another hidden condition is $y \leq 3$.

With these two additional inequalities, the feasible region in Figure 3.16 contracts to the shaded area in Figure 3.17. This region is bounded, and it has four corner points, which are circled in Figure 3.17. Evaluating the objective function at each corner point, we complete Table 3.11 and see that the minimum cost C is 4.7¢ per serving when $x = 1.316$ and $y = 0.368$. The Heartland Cereal Company should therefore mix 1.316 ounces of fortified cereal with 0.368 ounces of dehydrated milk product for each serving of *All-Pro*; this combination of ingredients will meet the guaranteed nutritional composition of the breakfast food at a minimum cost of 4.7¢ per serving.

FIGURE 3.17

TABLE 3.11

CORNER POINTS		OBJECTIVE FUNCTION	
x	y	$C = 0.03x + 0.02y$	
0	3	$C = 0.03(0) + 0.02(3)$	$= 0.06$
5	3	$C = 0.03(5) + 0.02(3)$	$= 0.21$
5	0	$C = 0.03(5) + 0.02(0)$	$= 0.15$
1.316	0.368	$C = 0.03(1.316) + 0.02(0.368)$	$= 0.047$ ✓

Observe that three of the inequality constraints in Example 5 ($x \geq 0$, $y \geq 0$, and $30x + 20y \geq 20$) are redundant; the feasible region does not change if they are deleted. Such inequalities are called *superfluous*. They are satisfied automatically by satisfying the other constraints in the system.

IMPROVING SKILLS

1. Consider the feasible region defined by the shaded portion of Figure 3.14 with corner points at $(-3, 5)$, $(5, -3)$ and $(1, -3)$. Determine a point in this feasible region that will
   (a) maximize $z = y - x$
   (b) maximize $z = x - y$
   (c) minimize $z = y - x$
   (d) minimize $z = x - y$

2. Consider the feasible region defined by the shaded portion of Figure 3.15 with corner points at $(0, 1850)$, $(1200, 950)$, $(2014.29, 0)$, and $(0, 0)$. Determine a point in this feasible region that will
   (a) maximize $z = x + y$
   (b) maximize $z = 2x$
   (c) maximize $z = 3y$
   (d) maximize $z = 3x + y$

3. Consider the feasible region defined by the shaded portion of Figure 3.17 with corner points at $(0, 3)$, $(5, 3)$, $(5, 0)$ and $(1.316, 0.368)$. Determine a point in this feasible region that will
   (a) minimize $z = x + y$
   (b) minimize $z = 10x + 3y$
   (c) minimize $z = 5x + 3y$
   (d) minimize $z = 2x + 22y$

4. Maximize: $z = x + 2y$
   subject to: $2x + y \leq 4$
   $$x \geq 0$$
   $$y \geq 0$$

5. Solve Problem 4 if instead the objective function is

$$z = 10x + 3y$$

6. Solve Problem 4 if instead the objective function is to be minimized.

7. Maximize: $z = x + y$
   subject to: $5x + 8y \leq 40$
   $$7x + 4y \leq 28$$
   $$x \geq 0$$
   $$y \geq 0$$

8. Solve Problem 7 if instead the objective function is

$$z = 3x + 7y$$

9. Solve Problem 7 if instead the objective function is

$$z = 4x$$

10. Minimize: $z = 9x + 9y$
subject to: $x + 3y \geq 6$
$\qquad\qquad x \leq 4$
$\qquad\qquad y \leq 2$

11. Solve Problem 10 if instead the objective function is to be maximized.

12. Minimize: $z = 50x + 20y$
subject to: $2x + \ y \geq 20$
$\qquad\qquad 5x + \ y \geq 25$
$\qquad\qquad 5x + 2y \leq 50$

13. Solve Problem 12 if instead the objective function is

$$z = 20x + 50y$$

14. Solve Problem 12 if instead we seek to maximize the objective function

$$z = 20x + 50y$$

15. Maximize: $z = 3x + 2y$
subject to: $6x + \ y \leq 24{,}000$
$\qquad\qquad 2x + 5y \leq 47{,}200$
$\qquad\qquad\quad x \geq 0$
$\qquad\qquad\quad y \geq 0$

16. Solve Problem 15 if instead the objective function is

$$z = 20x + 7y$$

17. Minimize: $z = 8000x + 7000y$
subject to: $4x + \ \ 3y \geq 160$
$\qquad\qquad 6x + 10y \geq 350$
$\qquad\qquad\quad x \geq \ \ 0$
$\qquad\qquad\quad y \geq \ \ 0$
$\qquad\qquad\quad x \leq 31$
$\qquad\qquad\quad y \leq 31$

Are any of the constraints superfluous?

18. Solve Problem 17 if instead the objective function is to be maximized.

19. Maximize: $z = 30x + 12y$
subject to: $7x + 9y \leq 630$
$\qquad\qquad\quad x \geq 100$
$\qquad\qquad\quad y \geq \ \ 0$

20. Maximize: $z = 8x + 15y$
subject to: $0.4x + 0.8y \leq 2240$
$\qquad\qquad 0.6x + 0.4y \leq 1440$
$\qquad\qquad\qquad x \geq \ \ 0$
$\qquad\qquad\qquad y \geq \ \ 0$

21. Solve Problem 20 if instead the objective function is

$$z = 7x + 15y$$

22. Solve Problem 20 if instead the objective function is

$$z = 15x + 7y$$

23. Minimize $\ \ z = 160x + 210y$
subject to: $6x + 5y \geq 300$
$\qquad\qquad 3x + 4y \leq 120$
$\qquad\qquad\quad x \geq \ \ 0$
$\qquad\qquad\quad y \geq \ \ 0$

24. Solve Problem 23 if instead the objective function is to be maximized.

25. Maximize: $z = 40x + 50y$
subject to: $9x + 12y \leq 252$
$5x + 10y \leq 160$
$x \geq 0$
$y \geq 0$

26. Maximize: $z = x$
subject to: $10x + 4y \leq 1800$
$4x + 5y \leq 1400$
$x \geq 0$
$y \geq 100$

27. Minimize: $z = 30x + 20y$
subject to: $10x + 2y \geq 15$
$4x + y \geq 10$
$x \leq 8$
$y \leq 12$

28. Minimize: $z = 120x + 75y$
subject to: $-3x + y \leq 0$
$x - y \leq 0$
$x \geq 1$
$x \leq 8$

29. Maximize: $z = 30x + 23y$
subject to: $-3x + y \leq 0$
$x - y \leq 0$
$7x + 3y \leq 80$

30. Maximize: $z = 40x + 60y$
subject to: $30x + 50y \leq 80{,}000$
$10x + 5y \leq 14{,}000$
$4x + 4y \leq 7{,}200$
$x \geq 0$
$y \geq 0$

31. Maximize: $z = 270x + 180y$
subject to: $x + \frac{1}{2}y \leq 90$
$5x + 3y \leq 225$
$2x + \frac{1}{2}y \leq 76$
$x \geq 0$
$y \geq 0$
$x \leq 40$
$y \leq 60$
Are any of these constraints superfluous?

32. Minimize: $z = x + \frac{1}{4}y$
subject to: $0.60x + 0.05y \geq 100$
$0.40x + 0.10y \geq 100$
$0.08y \geq 20$
$x \geq 0$
$y \geq 0$
$x \leq 1200$
$y \leq 2000$

33. Solve Problem 32 if instead the objective function is to be maximized.

34. Maximize: $z = 2x + 3y$
    subject to:
    $$12x + 14y \leq 14{,}952$$
    $$x + 4y \leq 1{,}892$$
    $$0.03x + 0.12y \leq 72$$
    $$x \geq 0$$
    $$y \geq 0$$

CREATING MODELS

In Problems 35 through 53, model each process as a linear programming problem and solve using graphical techniques.

35. A wood cabinet manufacturer produces cabinets for television consoles and lamp bases, both of which must be assembled and decorated. Every television cabinet requires 3 hours to assemble and 5 hours to decorate, and each lamp base requires 2 hours to assemble and 1 hour to decorate. The profit on each cabinet and each lamp base is $7 and $4, respectively, and the manufacturer can sell as many units as can be produced. The manufacturer has 14,000 hours available each week for assembling these products and another 14,000 hours per week for decorating. How many of each product should the manufacturer produce weekly to maximize profit?

36. Continental Motors manufactures taxis and airport limousines at its two plants. Plant A produces 75 taxis and 8 limousines daily, while plant B produces 50 taxis and 4 limousines each day. To fulfill its contractual obligations for April, the company must manufacture 3,000 taxis and 300 limousines. The daily operating costs for plants A and B are $250,000 and $150,000, respectively. How many days should each plant be operated during April to fulfill all contractual obligations at minimum cost?

37. The Atlas Pipe Company manufactures two types of metal joints, each of which must be molded and threaded. Type I joints require 60 seconds on the molding machine and 2 seconds on the threading machine, while Type II joints require 10 seconds on the molding machine and 5 seconds on the threading machine. Each day the molding machines are operable for 240,000 seconds and the threading machines for 47,200 seconds. The profits on each Type I and Type II joint are 3¢ and 2¢, respectively. How many joints of each type should be produced to maximize the daily profit?

38. Eagle Mining Enterprises owns two mines in West Virginia. The ore from each mine is separated into two grades before it is shipped. Mine I produces 4 tons of high-grade ore and 6 tons of low-grade ore daily with an operating cost of $8,000. Mine II produces 3 tons of high-grade ore and 10 tons of low-grade ore daily at a cost of $7,000 per day. The mining company has contracts that require it to deliver 160 tons of high-grade ore and 350 tons of low-grade ore each month. Determine the number of days each mine should operate to meet the company's contractual obligations at a minimum cost.

ROBINSON RUN PORTAL
MINE NO. 95

39. Solve Problem 38 if, in addition, mine I produces 4 tons of medium-grade ore each day and mine II produces 5 tons of medium-grade ore each day.

40. Solve Problem 38 if, in addition, mine I produces 4 tons of medium-grade ore each day, mine II produces 5 tons of medium-grade ore each day, and the company has contracts to deliver 208 tons of medium-grade ore each month.

41. The Heaven Rest Mattress Company produces a regular mattress labeled Sleeper and an extra-firm mattress labeled Johnny Firm. Each Sleeper requires 0.4 hours for basic assembling (binding the springs) and 0.6 hours for finishing (adding the padding and covering the unit with material). Each Johnny Firm requires 0.8 hours for assembling

and 0.4 hours for finishing. The current labor force provides 2,240 hours and 1,440 hours per week, respectively, for assembling and finishing mattresses. The profit on the *Sleeper* is $15 per mattress; each *Johnny firm* returns a profit of $30. How many mattresses of each type should Heaven Rest produce weekly to maximize its profit?

42. The King Sleep Corporation also produces two lines of mattresses, regular and extra-firm. Each regular mattress requires 30 springs, 10 pounds of padding, and 4 yards of material, while each extra-firm mattress requires 50 springs, 5 pounds of padding, and 4 yards of material. The wholesale price is $40 for each regular mattress and $60 for each extra-firm mattress. The maximum weekly inventory King Sleep can stock is 80,000 springs, 14,000 pounds of padding, and 7,200 yards of material. How many mattresses should King Sleep produce weekly to maximize its income, and how much inventory does such a production schedule consume?

43. A truck traveling from Davenport, Iowa, to Fort Meyers, Colorado, is in demand from two manufacturers who need to ship goods between those cities. Crates from the first manufacturer measure 60 cubic feet, weigh 100 pounds, and cost $80 each to ship. Crates from the second manufacturer measure 40 cubic feet, weigh 120 pounds, and cost $95 each to ship. The truck can carry no more than 12,000 pounds and has room for at most 8,000 cubic feet of cargo. How many crates of each type should the trucker accept if the objective is to maximize revenue? (Note: Shipping costs for each manufacturer are revenues to the trucker.)

44. A doctor recommends that a patient's daily diet be supplemented with at least 500 milligrams of vitamin C, 10,000 USP units of vitamin A, and 1,000 USP units of vitamin D. For this purpose, the patient has identified two granular vitamin supplements that are stirred into liquid. Each teaspoon of Supplement I contains 200 milligrams of vitamin C, 1,000 USP units of vitamin A, and 100 USP units of vitamin D and costs 5¢. Each teaspoon of Supplement II contains 50 milligrams of vitamin C, 2,500 USP units of vitamin A, and 300 USP units of vitamin D and costs 8¢. How many teaspoons of each supplement should the patient use each day if the objective is to meet the doctor's recommendations at minimum cost?

45. Organizers of a conference have reserved $63,000 for travel grants for conference participants. Regional participants live 1,000 miles or less from the conference site and receive a grant of $500; all others are considered national participants and receive $800. To provide a national flavor to the conference, there should be at least twice as many national participants as regional participants. How many participants of each type should the organizers invite if they want to maximize total attendance?

46. A savings and loan association has $10 million available for home loans and car loans. Its policy is to have at least five times as much money invested in home loans as in car loans. Applications on file from prospective borrowers total $6 million for home loans and $8 million for car loans. Currently, the savings and loan earns 9% on home loans and 15% on car loans. How much should be allocated to each type of loan to maximize earnings?

47. A manufacturer has 14,952 feet of rope, 1,892 feet of lumber, and 72 gallons of varnish with which to produce outdoor swings with wood seats and rope sides and tree-house ladders with wood steps and rope sides. Each swing requires 12 feet of rope, 1 foot of lumber, and 0.03 gallons of varnish and returns a profit of $2. Each ladder requires 14 feet of rope, 4 feet of lumber, and 0.12 gallons of varnish and returns a profit of $3. How many units of each product should the manufacturer produce to maximize profit?

48. A caterer has in stock 1,200 gallons of a premixed fruit juice that contains 60% orange juice and 40% grapefruit juice and 2,000 gallons of a premixed fruit drink containing 5% orange juice, 10% grapefruit juice, 8% cranberry juice, and 77% filler (sugar, water and flavorings). The fruit juice costs $1 per gallon, while the fruit drink costs

25¢ per gallon. From this stock, the caterer is to produce 1,000 gallons of a fruit punch for a corporation picnic. This drink must contain at least 10% orange juice, 10% grapefruit juice, and 2% cranberry juice. How much of each stock item should be used to produce the fruit punch at the lowest possible cost?

49. The manager of a supermarket meat department finds that she has 160 pounds of round steak, 600 pounds of chuck steak, and 300 pounds of pork in stock on Saturday morning. From experience, she knows that she can sell half these quantities as straight cuts at a high profit. The remaining meat will be ground and combined into hamburger meat and picnic patties for which there is a large weekend demand that exceeds existing supplies. Each pound of hamburger meat must contain at least 20% ground round and 60% ground chuck. Each pound of picnic patties must contain at least 30% ground pork and 50% ground chuck. The remainder of each product can consist of an inexpensive nonmeat filler that the store has in large supply. The profit on each pound of hamburger meat and each pound of picnic patties is 40¢. How many pounds of each product should be made if the objective is to maximize total profit on these items?

50. The inventory of a deodorant company contains 900,000 ounces of an active ingredient that is used to produce two different products, regular and super. Each carton of regular deodorant uses 30 oz of the active ingredient and costs $2 to produce, while each carton of super deodorant uses 40 oz of the active ingredient and costs $2.50 to produce. To justify start-up costs for any production run, the company must produce at least 30,000 cartons of deodorant. In addition, because of market demand, the company always produces at least three times as much super as regular. How many cartons of each type of deodorant should the company produce to minimize costs?

51. A local kennel has 50 cages available for housing dogs and cats. The kennel charges $80 per week to house a dog and $50 per week to house a cat. Each dog requires 2 hours of employee time per week to groom and feed, while each cat requires 1 hour of time. The kennel has a maximum of 80 hours of employee time for grooming and feeding animals. How many dogs and cats should the kennel accept if it wants to maximize income?

52. A local rent-a-car company needs to purchase at least 25 new cars, which can be either compacts or midsize cars. Compacts cost $10,000 each, midsize cars cost $15,000 each, and the company has $300,000 reserved for purchases. From experience, the company estimates that its yearly operating costs will be $2000 for each compact and $2300 for each midsize car. How many cars of each type should the company buy if it wants to minimize total operating costs?

53. A refining company always produces more gasoline than fuel oil. To meet the fuel oil demand of its distributors, the refinery must produce at least 60 million gallons of fuel oil; it can sell an additional 70 million gallons to other refineries. To meet the gasoline demand at its own stations, the refinery must produce at least 50 million gallons of gasoline, and it can sell an additional 60 million gallons to other stations. Each gallon of fuel oil sells for $1.05, and each gallon of gasoline sells for 90¢. How many gallons of fuel oil and how many gallons of gasoline should the refinery produce to maximize profit?

In Problems 54 through 60, model each process as a linear program but do not solve.

54. Model Problem 38 if the company also has a third mine that costs $6,000 per day to operate and produces two tons of high-grade ore and eight tons of low-grade ore daily.

55. Model Problem 42 if the King Sleep Corporation also produces an economy mattress that requires 20 springs, 14 pounds of padding, and 3 yards of material and sells for $35 per mattress wholesale.

56. Model Problem 44 if the patient is also willing to use a third supplement, each teaspoon of which provides 100 milligrams of vitamin C, 2,000 USP units of vitamin A, and 150 USP units of vitamin D and costs 6¢.

57. A wood cabinet manufacturer produces cabinets for television consoles, frames for grandfather clocks, and lamp bases, all of which must be assembled, decorated, and crated. Each television console requires 3 hours to assemble, 5 hours to decorate, and 0.2 hours to crate and returns a profit of $7. Each clock frame requires 10 hours to assemble, 8 hours to decorate, and 0.6 hours to crate and returns a profit of $22. Each lamp base requires 2 hours to assemble, 1 hour to decorate, and 0.1 hour to crate and returns a profit of $4. The manufacturer has 30,000, 40,000, and 1,600 hours available weekly for assembling, decorating, and crating, respectively. How many units of each product should it produce to maximize profit?

58. A can of dog food is guaranteed to contain at least 10 units of protein, 20 units of mineral matter, and 6 units of fat. The product is composed of a blend of four different ingredients. Ingredient I contains 10 units of protein, 2 units of mineral matter, and $\frac{1}{2}$ unit of fat per ounce. Ingredient II contains 1 unit of protein, 40 units of mineral matter, and 3 units of fat per ounce. Ingredient III contains 1 unit of protein, 1 unit of mineral matter, and 6 units of fat per ounce. Ingredient IV contains 5 units of protein, 10 units of mineral matter, and 3 units of fat per ounce. These four ingredients cost 2¢, 2¢, 1¢, and 3¢ per ounce, respectively. How many ounces of each ingredient should be used to minimize the cost of the dog food yet still meet the guaranteed composition?

59. An investor has identified three attractive stocks and will divide $10,000 among the three. The first stock is low-risk and returns 4% per year, the second is medium-risk and returns 6% per year, and the last is high-risk and returns 9% per year. At least $6,000 must be invested in low- and medium-risk stocks, and the amount invested in the high-risk stock must be no more than $2,000 more than the amount invested in the low-risk stock. How much stock of each type should the investor purchase to maximize total return?

60. A publisher of cookbooks, garden books, and self-help books will limit the list of new titles to at most 24 next year. To guarantee diversity, the policy of the publisher is to bring out at least four new titles in each group each year. Furthermore, the number of new titles in cookbooks must be at least as great as all other new titles combined. On average, each new cookbook title generates $8,500 in profits, each new garden book title generates $8,000 in profits, and each new self-help book generates $9,000 in profits. How many new titles in each category should the publisher produce to maximize profits?

EXPLORING IN TEAMS

61. Show that the linear programming problem

$$\text{Maximize:} \quad z = 12x + 15y$$
$$\text{subject to:} \quad 4x + y \le 24$$
$$4x + 5y \le 40$$
$$x \ge 0$$
$$y \ge 0$$

has more than one solution. Locate three other points on the line segment joining the two corner points that maximize the objective. Show that each of these three points yields the same value of the objective as the two corner points. What do you conclude? What opportunities present themselves?

62. Show that the linear programming problem

$$\text{Minimize:} \quad z = 14x + 10y$$

$$\text{subject to:} \quad 7x + 5y \geq 35$$
$$x \leq 4$$
$$y \leq 6$$

has more than one solution. Locate three other points on the line segment joining the two corner points that minimize the objective. Show that each of these three points yields the same value of the objective as the two corner points. What do you conclude? What opportunities present themselves?

EXPLORING WITH TECHNOLOGY

63. Replace each of the inequalities in the constraint equation for Example 1 with equalities, and then use a graphing calculator to plot all the resulting equations on the same coordinate system. Adjust the viewing window so that it contains the entire feasible region. Use the cursor and the zoom feature on the calculator to locate all the corner points identified in Table 3.5.

64. Repeat Problem 63 for the linear programming model in Example 3. Use the cursor and the zoom feature on the calculator to locate all the corner points identified in Table 3.8.

65. Use a graphing calculator to assist in solving Problems 4, 7, 10, and 12.

REVIEWING MATERIAL

66. (Section 1.3) A high school guidance department believes that the number of graduates from that school who will go on to obtain college degrees in four years is directly proportional to the number of graduates who completed high school on time. Four years ago, 240 students graduated from the high school on time, and of those, 156 graduated from college four years later. This year's graduating class has 280 students who completed high school on time. How many of them are expected to graduate from college four years later?

67. (Section 2.2) Use elementary row operations to transform the augmented matrix

$$\begin{bmatrix} 2 & 10 & -10 & | & 80 \\ 3 & 20 & 0 & | & 80 \\ 5 & -5 & 20 & | & 80 \end{bmatrix}$$

into upper triangular form.

RECOMMENDING ACTION

68. Respond by memo to the following request:

MEMORANDUM

To: J. Doe Reader

From: Superintendent's Office

Date: Today

Subject: **Curriculum in High School Mathematics**

A few parents visited me recently arguing that our curriculum in mathematics is antiquated. They believe that most of the topics in our courses deal with equations, while most of the world deals with inequalities. Are they correct? Should we be putting more emphasis on inequalities and less on equations? I will be meeting with our math supervisor later this month and I would like to have your thoughts on this matter before then. If you disagree, I need to know why. If you agree, then I need some suggestions on what material should be deleted to make room for new material on inequalities.

3.4 INITIALIZING THE SIMPLEX METHOD

As long as linear programming problems involve just two variables and a few constraints, they can be solved efficiently with graphical methods. Having more variables requires correspondingly more coordinates, with one axis for each variable, and graphical procedures become impractical. In fact, it is not uncommon for a linear programming model to have hundreds of important factors. In these situations nongraphical methods are required, and one of the very best is the *simplex method*.

A *decision variable* is a factor in a process that is under the direct control of someone managing the process, someone who has the authority to assign values to the variable. These variables may represent the number of different items to manufacture, the number of items to ship, or the quantities of ingredients to blend into a product. In each case, their values must be specified by a decision maker. It is customary to designate n decision variables as $x_1, x_2, x_3, \ldots, x_n$, using the letter x as a generic variable label and distinguishing different variables by different subscripts. Two decision variables are represented as x_1 and x_2; 500 decision variables are represented as $x_1, x_2, x_3, \ldots, x_{499}, x_{500}$.

The goal of a linear programming problem is to choose the decision variables so that all constraints are satisfied and an objective function is optimized. The objective function in a linear programming model with n decision variables has the general form

$$\text{Optimize: } z = c_1 x_1 + c_2 x_2 + c_3 x_3 + \ldots + c_n x_n$$

where $c_1, c_2, \ldots, c_n$ are known numbers. Here, the variable z is *not* a decision variable because the decision maker has only indirect control over its value. The decision maker does have direct control over the assignment of values to the decision variables, $x_1, x_2, x_3, \ldots, x_n$, whereupon z is determined by the equation defining it. An example of an objective function with four decision variables, x_1, x_2, x_3, x_4, is $z = 5x_1 + 8x_2 - 4x_3 + x_4$, with $c_1 = 5$, $c_2 = 8$, $c_3 = -4$, and $c_4 = 1$.

A linear programming model is in standard form if it satisfies four conditions:

STANDARD FORM FOR A LINEAR PROGRAMMING PROBLEM

1. All decision variables are nonnegative.

2. The objective function is to be maximized.

3. All constraints on the decision variables (other than nonnegativity conditions) have the form

$$a_1x_1 + a_2x_2 + a_3x_3 + \ldots + a_nx_n \leq b$$

where $a_1, a_2, a_3, \ldots, a_n$, and b are known numbers. That is, all constraints are of the $\leq$ type.

4. The right side of each constraint in condition 3, designated by the letter b, is also nonnegative.

Two examples of linear programming models in standard form are

$$\text{Maximize:} \quad z = 7x_1 + 22x_2$$
$$\text{subject to:} \quad 3x_1 + 10x_2 \leq 33{,}000$$
$$5x_1 + 8x_2 \leq 42{,}000 \tag{8}$$
$$\text{assuming:} \quad x_1 \text{ and } x_2 \text{ are nonnegative}$$

and

$$\text{Maximize:} \quad z = 4x_1 + 10x_2 + 6x_3$$
$$\text{subject to:} \quad 2x_1 + 4x_2 + x_3 \leq 12$$
$$6x_1 + 2x_2 + x_3 \leq 26 \tag{9}$$
$$5x_1 + x_2 + 2x_3 \leq 80$$
$$\text{assuming:} \quad x_1, x_2, \text{ and } x_3 \text{ are nonnegative}$$

The basic simplex method is a sequence of matrix-based operations applied to linear programming problems in standard form. An augmented matrix containing all the essential information is created, and the simplex method is applied to that matrix. In this section, we show how to construct such an augmented matrix. This step is called the initialization process. In the next section, we give the basic steps of the simplex algorithm, and then finally in Section 3.6 we enhance the basic algorithm so that it can be applied to all linear programming problems, not just those in standard form.

Augmented matrices hold linear equations (see Chapter 2). Thus, the first step in the initialization process is to convert all linear inequalities of the $\leq$ type to linear equations. This is done by adding a nonnegative quantity called a *slack variable* to the left side of each inequality. The inequality

A slack variable is a nonnegative quantity added to the left side of a $\leq$ inequality to make the left side equal the right side.

$$3x_1 + 10x_2 \leq 33{,}000 \tag{10}$$

is converted to an equality by adding an "appropriate amount" to the left side of this inequality. The "appropriate amount" here is the difference between the right side of the inequality, 33,000, and the left side, $3x_1 + 10x_2$. If we denote this amount by s_1, then we can rewrite inequality (10) as

$$3x_1 + 10x_2 + s_1 = 33,000 \qquad (11)$$

Similarly, the inequality

$$5x_1 + 8x_2 \le 42,000 \qquad (12)$$

is converted to an equality by adding an "appropriate amount" to the left side of the inequality. If we denote the "appropriate amount" by the slack variable s_2, we have

$$5x_1 + 8x_2 + s_2 = 42,000 \qquad (13)$$

Slack variables, such as s_1 and s_2 in Equations (11) and (13), represent unused capacity or slack in a process. In particular, Inequalities (10) and (12) model constraints developed in Example 2 of Section 3.1, where we dealt with the assembly and decoration of television cabinets and grandfather clock frames; x_1 and x_2 have replaced x and y, respectively, to be consistent with our new notation. The left side of (10) is the amount of assembly time used for the production of both items, and the inequality indicates that the time used to assemble television cabinets and grandfather clock frames cannot exceed the 33,000 hours of assembly time available at the manufacturing plant.

Inequality (12) states that the amount of decorating time used for the production of television cabinets and clock frames cannot exceed the 42,000 hours of decorating time available. In each case, the left side of the inequality is the time actually used, and the right side is the time available. The difference between time available and time used is unutilized capacity. Thus, s_1 in Equation (11) measures the time available for assembly operations but not used, while s_2 in Equation (13) measures the time available for decorating operations but not used. The slack variables represent, respectively, the amount of time assemblers and decorators will be idle and therefore available for other projects.

Slack variables are incorporated into the objective function by adding them to the objective function with coefficients of zero. For the cabinet–clock frame problem modeled by (8), the goal is to maximize $z = 7x_1 + 22x_2$, where z denotes profit. This is identical to writing

$$\text{Maximize: } z = 7x_1 + 22x_2 + 0s_1 + 0s_2 \qquad (14)$$

because $0s_1 + 0s_2 = 0$. Idle assembly time and idle decorating time add nothing to profits.

It follows from Equations (11), (13), and (14) that the linear programming problem

$$\begin{aligned} \text{Maximize:} \quad & z = 7x_1 + 22x_2 \\ \text{subject to:} \quad & 3x_1 + 10x_2 \le 33,000 \\ & 5x_1 + 8x_2 \le 42,000 \\ \text{assuming:} \quad & x_1 \text{ and } x_2 \text{ are nonnegative} \end{aligned} \qquad \text{(8 repeated)}$$

is equivalent to the linear programming problem

$$\begin{aligned} \text{Maximize:} \quad & z = 7x_1 + 22x_2 + 0s_1 + 0s_2 \\ \text{subject to:} \quad & 3x_1 + 10x_2 + s_1 \qquad = 33,000 \\ & 5x_1 + 8x_2 \qquad + s_2 = 42,000 \\ \text{assuming:} \quad & x_1, x_2, s_1, \text{ and } s_2 \text{ are all nonnegative} \end{aligned} \qquad (15)$$

EXAMPLE 1

Convert the linear programming problem

$$\text{Maximize:} \quad z = 4x_1 + 10x_2 + 6x_3$$

$$\text{subject to:} \quad 2x_1 + 4x_2 + x_3 \leq 12$$
$$6x_1 + 2x_2 + x_3 \leq 26 \qquad \text{(9 repeated)}$$
$$5x_1 + x_2 + 2x_3 \leq 80$$

$$\text{assuming:} \quad x_1, x_2, \text{ and } x_3 \text{ are nonnegative}$$

into an equivalent problem with slack variables.

Solution

We let the slack variables s_1, s_2, and s_3 denote the quantities that must be added to the left sides of the three inequalities so that each left side will equal its right side. Then adding these slack variables to the objective function with coefficients of zero, we obtain

$$\text{Maximize:} \quad z = 4x_1 + 10x_2 + 6x_3 + 0s_1 + 0s_2 + 0s_3$$

$$\text{subject to:} \quad 2x_1 + 4x_2 + x_3 + s_1 \qquad\qquad = 12$$
$$6x_1 + 2x_2 + x_3 \qquad + s_2 \qquad = 26 \qquad \text{(16)}$$
$$5x_1 + x_2 + 2x_3 \qquad\qquad + s_3 = 80$$

$$\text{assuming:} \quad x_1, x_2, x_3, s_1, s_2, \text{ and } s_3 \text{ are all nonnegative}$$

An objective function is written with all variables on the left side of the equation, the coefficient of z set to one, and the right side of the equation set to zero.

The second step in the initialization process is to rewrite the objective function in the same form as the constraints. This will allow us to insert the objective function into the augmented matrix that holds the constraints. Recall from Chapter 2 that an augmented matrix holds linear equations in which the variables are on the left side of the equations and the constant terms are on the right side. We can transform an objective function into this format by shifting all variables to the left side of the equation, with the coefficient of z set to one, leaving a right side of zero. In particular, the objective function

$$z = 7x_1 + 22x_2 + 0s_1 + 0s_2$$

is rewritten as

$$-7x_1 - 22x_2 + 0s_1 + 0s_2 + z = 0$$

The objective function

$$z = 4x_1 + 10x_2 + 6x_3 + 0s_1 + 0s_2 + 0s_3$$

becomes

$$-4x_1 - 10x_2 - 6x_3 + 0s_1 + 0s_2 + 0s_3 + z = 0$$

The third step in the initialization process is to construct an augmented matrix to hold the constraint equations and the objective function. We begin with an augmented matrix for just the constraint equations created by the first step in the initialization process. Then we attach to this augmented matrix one new row and one new column. The new row is attached to the bottom of the augmented matrix, and it holds the coefficients of the objective function, in the form specified by the second step in the initialization process. The new column is placed just before the last column in the augmented matrix, to accommodate the z term.

The augmented matrix for the linear programming model

$$\text{Maximize } z: \quad -7x_1 - 22x_2 + 0s_1 + 0s_2 + z = 0$$

$$\text{subject to:} \quad 3x_1 + 10x_2 + s_1 \qquad\quad = 33{,}000$$
$$5x_1 + 8x_2 \qquad\quad + s_2 = 42{,}000$$

$$\text{assuming:} \quad x_1, x_2, x_3, s_1, \text{ and } s_2 \text{ are all nonnegative}$$

is

$$
\begin{array}{ccccc}
x_1 & x_1 & s_1 & s_2 & z \\
\end{array}
$$
$$
\left[
\begin{array}{ccccc|c}
3 & 10 & 1 & 0 & 0 & 33{,}000 \\
5 & 8 & 0 & 1 & 0 & 42{,}000 \\
\hline
-7 & -22 & 0 & 0 & 1 & 0
\end{array}
\right]
\tag{17}
$$

where the dashed line separates the coefficients of the constraint equations from those of the objective. Because z does not appear explicitly in either of the two constraint equations, its coefficient is zero in each of those equations. Similarly, if a slack variable does not appear explicitly in a particular constraint equation, then its coefficient in that equation is also zero. The augmented matrix for the linear programming model

$$\text{Maximize } z: \quad -4x_1 - 10x_2 - 6x_3 + 0s_1 + 0s_2 + 0s_3 + z = 0$$

$$\text{subject to:} \quad 2x_1 + 4x_2 + x_3 + s_1 \qquad\qquad = 12$$
$$6x_1 + 2x_2 + x_3 \qquad + s_2 \qquad = 26$$
$$5x_1 + x_2 + 2x_3 \qquad\qquad + s_3 = 80$$

$$\text{assuming:} \quad x_1, x_2, x_3, s_1, s_2, \text{ and } s_3 \text{ are all nonnegative}$$

is

$$
\begin{array}{ccccccc}
x_1 & x_2 & x_3 & s_1 & s_2 & s_3 & z \\
\end{array}
$$
$$
\left[
\begin{array}{ccccccc|c}
2 & 4 & 1 & 1 & 0 & 0 & 0 & 12 \\
6 & 2 & 1 & 0 & 1 & 0 & 0 & 26 \\
5 & 1 & 2 & 0 & 0 & 1 & 0 & 80 \\
\hline
-4 & -10 & -6 & 0 & 0 & 0 & 1 & 0
\end{array}
\right]
\tag{18}
$$

> *An initial basic feasible solution is obtained by setting all decision variables equal to zero and setting each slack variable equal to the right side of the equation in which that slack variable appears.*

By adding a slack variable to the left side of each $\leq$ inequality in a linear programming problem in standard form, we create a system of equality constraints, each with a different slack variable. If we set all decision variables to zero, then we easily generate one feasible solution to the constraint equalities by setting each slack variable equal to the right side of the equation in which the slack variable appears. This solution, which sets all production to zero, is known as an *initial basic feasible solution*. Workers become inactive, inventory is not used, machines are turned off. All available resources become idle or slack. Such a do-nothing policy may not be profitable, but it is always a feasible solution to a linear programming problem in standard form.

EXAMPLE 2

Find an initial basic feasible solution to the constraint equations

$$3x_1 + 10x_2 + s_1 \qquad\quad = 33{,}000$$
$$5x_1 + 8x_2 \qquad + s_2 = 42{,}000$$

from linear programming model (*15*).

Solution

We set $x_1 = x_2 = 0$, whereupon $s_1 = 33{,}000$ and $s_2 = 42{,}000$. This is the initial basic feasible solution. Recall that these constraint equations model, respectively, the assembly time and decorating time in the production of television cabinets and clock frames. By setting the decision variables equal to zero, we are stipulating that no television cabinets or clock frames will be produced. As a result, the slack variables must equal the times available, indicating that all available time is slack or idle.

EXAMPLE 3

Find an initial basic feasible solution to the constraint equations

$$2x_1 + 4x_2 + x_3 + s_1 \qquad\qquad = 12$$
$$6x_1 + 2x_2 + x_3 \qquad + s_2 \qquad = 26$$
$$5x_1 + x_2 + 2x_3 \qquad\qquad + s_3 = 80$$

from linear programming model (*16*).

Solution

We set $x_1 = x_2 = x_3 = 0$ whereupon $s_1 = 12$, $s_2 = 26$, and $s_3 = 80$.

The simplex method, which will be described in the next section, is a procedure for improving solutions. The method checks a current solution to determine whether a better solution is available. If so, the simplex method will find a better (i.e., more optimal) solution. The simplex method is then repeated with the new solution as the current solution. This cycle of checking current solutions and improving on them continues until no improvements are possible. To begin, however, the simplex method requires a starting solution, and this is the role of the initial basic feasible solution. It is an easily obtained starting point for the simplex method, a place to begin, not to end.

The variables used to produce a solution to a system of constraint equations, one variable per equation, are called *basic variables*; all other variables are set to zero and are called *nonbasic variables*. In Examples 2 and 3, the decision variables are the nonbasic variables in the initial basic feasible solution because they were set to zero; the slack variables are the basic variables in those solutions because their values were chosen to satisfy each of the equations.

To easily identify the basic variables, we list them in front of the augmented matrix for a linear programming problem. The result is known as a *simplex tableau* to emphasize its unique design.

A simplex tableau is a matrix associated with a linear programming model in standard form.

We showed that the linear programming problem

$$\text{Maximize:} \quad z = 7x_1 + 22x_2$$
$$\text{subject to:} \quad 3x_1 + 10x_2 \le 33{,}000$$
$$5x_1 + 8x_2 \le 42{,}000$$

(8 repeated)

$$\text{assuming:} \quad x_1 \text{ and } x_2 \text{ are nonnegative}$$

could be rewritten as

$$\text{Maximize } z: \quad -7x_1 - 22x_2 + 0s_1 + 0s_2 + z = 0$$
$$\text{subject to:} \quad 3x_1 + 10x_2 + s_1 \qquad = 33{,}000$$
$$5x_1 + 8x_2 \qquad + s_2 = 42{,}000$$

$$\text{assuming:} \quad x_1, x_2, x_3, s_1, \text{ and } s_2 \text{ are all nonnegative}$$

The simplex tableau associated with this linear programming problem is

$$
\begin{array}{c}
\begin{array}{ccccc} x_1 & x_2 & s_1 & s_2 & z \end{array} \\
\begin{array}{c} s_1 \\ s_2 \\ \\ \end{array}
\left[
\begin{array}{ccccc|c}
3 & 10 & 1 & 0 & 0 & 33{,}000 \\
5 & 8 & 0 & 1 & 0 & 42{,}000 \\
\hline
-7 & -22 & 0 & 0 & 1 & 0
\end{array}
\right]
\end{array}
\tag{19}
$$

Note that the initial basic feasible solution is read immediately from a simplex tableau by assigning to each basic variable in front of the augmented matrix the number in the last column and corresponding row.

The initial basic feasible solution for Tableau (*19*) is $s_1 = 33{,}000$ and $s_2 = 42{,}000$ with x_1 and x_2 set to zero. The objective function for this linear programming problem is

$$ z = 7x_1 + 22x_2 $$

or

$$ z = 7x_1 + 22x_2 + 0s_1 + 0s_2 $$

The value of the objective function for the initial basic feasible solution is

$$ z = 7(0) + 22(0) + 0(33{,}000) + 0(42{,}000) = 0 $$

which appears as the number in the lower-right corner of the simplex tableau.

We showed that the linear programming problem

$$
\begin{aligned}
\text{Maximize:} \quad & z = 4x_1 + 10x_2 + 6x_3 \\
\text{subject to:} \quad & 2x_1 + 4x_2 + x_3 \le 12 \\
& 6x_1 + 2x_2 + x_3 \le 26 \\
& 5x_1 + x_2 + 2x_3 \le 80
\end{aligned}
\tag{9 repeated}
$$

$$ \text{assuming:} \quad x_1, x_2, \text{ and } x_3 \text{ are nonnegative} $$

can be rewritten as

$$
\begin{aligned}
\text{Maximize } z: \quad & -4x_1 - 10x_2 - 6x_3 + 0s_1 + 0s_2 + 0s_3 + z = 0 \\
\text{subject to:} \quad & 2x_1 + 4x_2 + x_3 + s_1 = 12 \\
& 6x_1 + 2x_2 + x_3 + s_2 = 26 \\
& 5x_1 + x_2 + 2x_3 + s_3 + = 80
\end{aligned}
$$

$$ \text{assuming:} \quad x_1, x_2, x_3, s_1, s_2, \text{ and } s_3 \text{ are all nonnegative} $$

The simplex tableau associated with this linear programming problem is

$$
\begin{array}{c}
\begin{array}{ccccccc} x_1 & x_2 & x_3 & s_1 & s_2 & s_3 & z \end{array} \\
\begin{array}{c} s_1 \\ s_2 \\ s_3 \\ \\ \end{array}
\left[
\begin{array}{ccccccc|c}
2 & 4 & 1 & 1 & 0 & 0 & 0 & 12 \\
6 & 2 & 1 & 0 & 1 & 0 & 0 & 26 \\
5 & 1 & 2 & 0 & 0 & 1 & 0 & 80 \\
\hline
-4 & -10 & -6 & 0 & 0 & 0 & 1 & 0
\end{array}
\right]
\end{array}
\tag{20}
$$

Here the basic variables are s_1, s_2, and s_3; the initial basic feasible solution is $s_1 = 12$, $s_2 = 26$, and $s_3 = 80$, with x_1, x_2, and x_3 all set to zero. The corresponding objective for this do-nothing policy is $z = 0$, which again is the number appearing in the lower-right corner of the simplex tableau.

We summarize the initialization process for the simplex method in the following algorithm:

INITIALIZATION PROCESS

STEP 1 Add a slack varaible to the left side of each $\leq$ inequality, thereby converting the inequality to an equality. Add to the objective function each slack variable with a zero coefficient.

STEP 2 Write the objective function so that all variables appear on the left side of the equation, with the coefficient of z set to one, and the right side of the equation set to zero.

STEP 3 Construct an augmented matrix for the constraint equations created by Step 1. Above each column write the variable associated with that column. Add one additional column, just before the last column, to accommodate the z term; add one additional row for the objective function in the form specified by Step 2. Write each slack variable in front of the augmented matrix, on the same line as the constraint equation in which it appears.

STEP 4 Generate an initial basic feasible solution by setting all decision variables equal to zero and then setting each slack variable equal to the right side of the equation in which the slack variable appears.

EXAMPLE 4

Create a simplex tableau for the following linear programming model and then determine the initial basic feasible solution.

$$\text{Maximize:} \quad z = 2x_1 + 4x_2 + x_3 + x_4$$

$$\text{subject to:} \quad x_1 + 3x_2 \qquad\quad + 4x_4 \leq 4$$

$$2x_1 + x_2 \qquad\qquad\qquad \leq 3$$

$$x_2 + 4x_3 + x_4 \leq 3$$

$$\text{assuming:} \quad x_1, x_2, x_3, \text{ and } x_4 \text{ are nonnegative}$$

Solution

This linear programming model is in standard form, so we can apply the initialization process to it. First, we add a slack variable, s_1, s_2, or s_3, to each of the three $\leq$ inequalities, transforming each into an equality:

$$x_1 + 3x_2 \qquad + 4x_4 + s_1 \qquad\qquad = 4$$

$$2x_1 + x_2 \qquad\qquad\qquad + s_2 \qquad = 3 \qquad\qquad (21)$$

$$x_2 + 4x_3 + x_4 \qquad\qquad + s_3 = 3$$

These slack variables, with zero coefficients, are also added to the objective. The second step is to rewrite the objective so that its right side is zero. Thus, the objective becomes

$$\text{Maximize } z: \; -2x_1 - 4x_2 - x_3 - x_4 + 0s_1 + 0s_2 + 0s_3 + z = 0$$

The corresponding simplex tableau is

$$
\begin{array}{c}
\\
s_1 \\
s_2 \\
s_3 \\
\\
\end{array}
\begin{array}{c}
\begin{array}{ccccccccc}
x_1 & x_2 & x_3 & x_4 & s_1 & s_2 & s_3 & z \\
\end{array} \\
\left[\begin{array}{cccccccc|c}
1 & 3 & 0 & 4 & 1 & 0 & 0 & 0 & 4 \\
2 & 1 & 0 & 0 & 0 & 1 & 0 & 0 & 3 \\
0 & 1 & 4 & 1 & 0 & 0 & 1 & 0 & 3 \\
\hline
-2 & -4 & -1 & -1 & 0 & 0 & 0 & 1 & 0 \\
\end{array}\right]
\end{array}
$$

To generate an initial basic feasible solution, we set the decision variables x_1, x_2, x_3, and x_4 all to zero and then solve System (21) for the slack variables, obtaining $s_1 = 4$, $s_2 = 3$, and $s_3 = 3$. These slack variables are the initial basic variables. We achieve the same result by setting each of the variables that appears in front of the augmented matrix, here, s_1, s_2, and s_3, equal to the number that appears in the corresponding row and last column of the augmented matrix and then setting all other variables equal to zero. The corresponding value of the objective function for this initial basic feasible solution is the number in the lower-right corner of the simplex tableau, namely, $z = 0$.

IMPROVING SKILLS

In Problems 1 through 15, construct a simplex tableau for each linear programming problem in standard form.

1. Maximize: $z = 3x_1 + 2x_2$
     subject to: $60x_1 + 10x_2 \le 240{,}000$
               $2x_1 + 5x_2 \le 47{,}200$
     assuming: x_1 and x_2 are nonnegative

2. Maximize: $z = 15x_1 + 30x_2$
     subject to: $0.4x_1 + 0.8x_2 \le 2240$
               $0.6x_1 + 0.4x_2 \le 1440$
     assuming: x_1 and x_2 are nonnegative

3. Maximize: $z = 2x_1 + 7x_2$
     subject to: $x_1 - 3x_2 \le 3$
               $2x_1 + 4x_2 \le 18$
     assuming: x_1 and x_2 are nonnegative

4. Maximize: $z = 180x_1$
     subject to: $50x_1 + 27x_2 \le 170$
               $8x_1 + 25x_2 \le 80$
     assuming: x_1 and x_2 are nonnegative

5. Maximize: $z = 2x_1 - x_2$
     subject to: $x_1 + 2x_2 \le 2$
               $-3x_1 + 4x_2 \le 7$
     assuming: x_1 and x_2 are nonnegative

6. Maximize: $z = 35x_1 + 12x_2$
     subject to: $8x_1 + 11x_2 \le 9{,}500$
               $7x_1 + 12x_2 \le 8{,}000$
               $9x_1 + 9x_2 \le 9{,}200$
     assuming: x_1 and x_2 are nonnegative

7. Maximize: $z = 15x_1 + 10s_2 + 14x_3$

   subject to: $6x_1 + 5x_2 + 3x_3 \leq 26$

   $4x_1 + 2x_2 + 5x_3 \leq 8$

   assuming: $x_1, x_2,$ and x_3 are nonnegative

8. Maximize: $z = 3x_1 - 4x_2 + 5x_3$

   subject to: $x_1 + 2x_2 + 2x_3 \leq 180$

   $3x_1 + 4x_2 + 4x_3 \leq 230$

   assuming: $x_1, x_2,$ and x_3 are nonnegative

9. Maximize: $z = 5x_1 + 6x_2 + 7x_3 + 8x_4$

   subject to: $3x_1 - 4x_1 + 5x_3 - 6x_4 \leq 90$

   $7x_1 + 7x_2 + 8x_3 - 9x_4 \leq 135$

   assuming: $x_1, x_2, x_3,$ and x_4 are nonnegative

10. Maximize: $z = 15x_1 + 10x_2 + 14x_3$

    subject to: $30x_1 + 50x_2 + 40x_3 \leq 80{,}000$

    $10x_1 + 5x_2 + 7x_3 \leq 14{,}000$

    $4x_1 + 4x_2 + 6x_3 \leq 7{,}200$

    assuming: x_1, x_2, x_3 are nonnegative

11. Maximize: $z = x_3$

    subject to: $0.3x_1 + 0.4x_2 + 0.3x_3 \leq 9{,}000$

    $0.2x_1 + 0.8x_2 \leq 5{,}500$

    $0.9x_1 + 0.1x_3 \leq 7{,}300$

    assuming: $x_1, x_2,$ and x_3 are nonnegative

12. Maximize: $z = 15x_1 + 14x_3$

    subject to: $30x_1 + 50x_2 + 40x_3 \leq 80{,}000$

    $10x_1 + 5x_2 + 7x_3 \leq 14{,}000$

    $4x_1 + 4x_2 + 6x_3 \leq 7{,}200$

    $14x_1 + 7x_2 + 10x_3 \leq 21{,}000$

    assuming: $x_1, x_2,$ and x_3 are nonnegative

13. Maximize: $z = 9x_1 + 9x_2 + x_3$

    subject to: $4x_1 + 4x_2 + 5x_3 \leq 0.03$

    $2x_1 + 3x_2 + 8x_3 \leq 0.07$

    $x_1 + x_2 + 3x_3 \leq 0.01$

    $2x_1 + x_2 + x_3 \leq 0.08$

    $4x_1 + 3x_2 + 5x_3 \leq 0.12$

    assuming: $x_1, x_2,$ and x_3 are nonnegative

14. Maximize: $z = 2x_1 + 3x_2 + 4x_3 + 6x_4$

    subject to: $x_1 + x_2 + x_3 + x_4 \leq 15$

    $7x_1 + 5x_2 + 3x_3 + 2x_4 \leq 120$

    $3x_1 + 5x_2 + 10x_3 + 15x_4 \leq 100$

    assuming: $x_1, x_2, x_3,$ and x_4 are nonnegative

15. Maximize: $z = x_2 + 2x_4$

    subject to: $2x_1 + 3x_2 + 3x_3 + 4x_4 \leq 20$

    $3x_1 + 4x_2 + 4x_3 + 5x_4 \leq 35$

    $5x_1 + 5x_2 + 4x_3 + 3x_4 \leq 35$

    $3x_1 + 3x_2 + 2x_3 + x_4 \leq 20$

    assuming: $x_1, x_2, x_3,$ and x_4 are nonnegative

In Problems 16 through 25, explain why the linear programming models are not in standard form.

16. Minimize: $z = 2x_1 + 7x_2$
    subject to: $x_1 - 3x_2 \leq 3$
    $2x_1 + 4x_2 \leq 18$
    assuming: x_1 and x_2 are nonnegative

17. Maximize: $z = 2x_1 + 7x_2$
    subject to: $x_1 - 3x_2 = -3$
    $2x_1 + 4x_2 = 18$
    assuming: x_1 and x_2 are nonnegative

18. Maximize: $z = 2x_1 + 7x_2$
    subject to: $x_1 - 3x_2 \leq -3$
    $2x_1 + 4x_2 \leq 18$
    assuming: x_1 and x_2 are nonnegative

19. Maximize: $z = 2x_1 + 7x_2$
    subject to: $x_1 - 3x_2 \leq 3$
    $2x_1 + 4x_2 \leq 18$
    $x_1 \geq 5$
    $x_2 \leq 3$

20. Maximize: $z = 3x_1 - 4x_2 + 5x_3$
    subject to: $x_1 + 2x_2 + 2x_3 \geq 180$
    $3x_1 + 4x_2 + 4x_3 \geq 230$
    assuming: x_1, x_2, and x_3 are nonnegative

21. Minimize: $z = 3x_1 - 4x_2 - 5x_3$
    subject to: $x_1 + 2x_2 + 2x_3 \leq 180$
    $3x_1 + 4x_2 + 4x_3 \leq 230$
    assuming: x_1, x_2, and x_3 are nonnegative

22. Minimize: $z = 5x_1 + 6x_2 + 7x_3 + 8x_4$
    subject to: $3x_1 - 4x_2 + 5x_3 - 6x_4 \leq 90$
    $7x_1 + 7x_2 + 8x_3 - 9x_4 \leq 135$
    assuming: x_1, x_2, x_3, and x_4 are nonnegative

23. Maximize: $z = 5x_1 + 6x_2 + 7x_3 + 8x_4$
    subject to: $3x_1 - 4x_2 + 5x_3 - 6x_4 \leq -90$
    $7x_1 + 7x_2 + 8x_3 - 9x_4 \leq 135$
    assuming: x_1, x_2, and x_3, and x_4 are nonnegative

24. Maximize: $z = x_2 + 2x_4$
    subject to: $2x_1 + 3x_2 + 3x_3 + 4x_4 \leq 20$
    $3x_1 + 4x_2 + 4x_3 + 5x_4 \leq 35$
    $5x_1 + 5x_2 + 4x_3 + 3x_4 \leq 35$
    $3x_1 + 3x_2 + 2x_3 + x_4 \leq 20$
    assuming: x_1, x_2, x_3, and x_4 are real

25. Maximize: $z = x_3$
    subject to: $0.3x_1 + 0.4x_2 + 0.3x_3 \geq 9{,}000$
    $0.2x_1 + 0.8x_2 \geq 5{,}500$
    assuming: x_1, x_2, and x_3 are nonnegative

Transform the linear programs in Problems 26 through 29 into standard form.

26. Maximize: $z = 2x_1 + 7x_2$
    subject to: $x_1 - 3x_2 \geq -3$
    $2x_1 + 4x_2 \leq 18$
    assuming: x_1 and x_2 are nonnegative

27. Maximize: $z = 3x_1 - 4x_2 + 5x_3$
    subject to: $x_1 - 2x_2 + 2x_3 \geq 0$
    $3x_1 + 4x_2 - 4x_3 \geq 0$
    assuming: $x_1, x_2,$ and x_3 are nonnegative

28. Maximize: $z = 5x_1 + 6x_2 + 7x_3 + 8x_4$
    subject to: $3x_1 - 4x_2 + 5x_3 - 6x_4 \leq 90$
    $7x_1 + 7x_2 + 8x_3 - 9x_4 \leq 135$
    $x_3 - 2x_4 \geq 0$
    assuming: $x_1, x_2, x_3,$ and x_4 are nonnegative

29. Maximize: $z = 5x_1 + 6x_2 + 7x_3 + 8x_4$
    subject to: $3x_1 - 4x_2 + 5x_3 - 6x_4 \geq -90$
    $7x_1 + 7x_2 - 8x_3 - 9x_4 \geq 0$
    assuming: $x_1, x_2, x_3,$ and x_4 are nonnegative

CREATING MODELS

In Problems 30 through 35, model each of the processes as a linear programming problem in standard form but do not solve.

30. Organizers of a conference have reserved $63,000 for travel grants for conference participants, who are categorized as either regional, national, or international. Regional participants live 1,000 miles or less from the conference site and receive a grant of $500; national participants receive $800, and international participants receive $900. To ensure a spectrum of participants, the organizers invite at least twice as many national participants as regional participants. How many participants of each type should the organizers invite if they want to maximize total attendance?

31. Model Problem 30 if, in addition, the number of international participants must be at least as many as the number of regional participants.

32. A savings and loan association has $10 million available for home loans and car loans. Its policy it to have at least five times as much money invested in home loans as in car loans. Applications on file from prospective borrowers total $6 million for home loans and $8 million for car loans. Currently, the savings and loan earns 9% on home loans and 15% on car loans. How much should be allocated to each type of loan to maximize earnings?

33. An investor has identified three attractive funds and will divide $10,000 among the three. The first is low-risk and returns 4% per year, the second is medium-risk and returns 6% per year, and the last is high-risk and returns 9% per year. The investor's strategy requires that the amounts allocated to low- and medium-risk stocks always exceed the amount invested in high-risk stocks. How much money should the investor put into each fund to maximize his total return?

34. Model Problem 33 if instead the investor's strategy requires that at least $6,000 must be invested in low- and medium-risk stocks.

35. A publisher of cookbooks, garden books, and self-help books will limit the list of new titles to at most 24 next year. To meet market demand, the number of new cookbooks must be at least as great as all other new titles combined. On average, each new cookbook title generates $8,500 in profits, each new garden book title generates $8,000 in profits, and each new self-help book generates $9,000 in profits. How many new titles in each category should the publisher produce to maximize profits?

EXPLORING IN TEAMS

36. Find a second initial basic feasible solution to the constraint equations given in Example 2 by using Gaussian elimination to solve for x_1 and x_2. What are the values of the slack variables in such a solution?

37. Give a condition on the equations to explain why the initial basic solution found in Example 2 is easier to identify than the one found in Problem 36.

38. Use Gaussian elimination to solve for x_1, x_2, and x_3 in the constraint equations of Example 3. Can you use the result as an initial basic feasible solution to those constraint equations?

39. Give conditions on the constraint equations in Example 3 to explain why the initial basic solution found in Example 3 is feasible.

REVIEWING MATERIAL

40. (Section 1.6) Data show that the number of telephone calls N received daily by a local answering service is given by the equation $N = 5t + 100$, where t denotes time (measured in weeks). Graph this equation with t on the horizontal axis. The current staff can handle a maximum of $C = 500$ calls per day. Graph this equation on a C-t coordinate system with t on the horizontal axis and the scales on both axes identical to those of the first graph. Superimpose one graph on the other and locate the point of intersection. What is the physical significance of this point?

41. (Section 2.5) Find the inverse of

$$\mathbf{A} = \begin{bmatrix} -1 & -5 & 7 \\ 2 & 8 & -11 \\ 1 & 2 & -3 \end{bmatrix}$$

RECOMMENDING ACTION

42. Respond by memo to the following request:

> ### MEMORANDUM
>
> To: J. Doe Reader
>
> From: Director of In-House Education
>
> Date: Today
>
> Subject: **Seminars**
>
> Last week, I paid for a seminar on solving linear programming problems by graphical means. Our company needs to solve linear programming problems, and if a seminar gets our employees up to speed, that is fine. But now you want me to hire other consultants for a new seminar on something called the simplex method. Why? Isn't one solution procedure enough? Surely, this additional expense is not warranted, is it?

3.5 THE BASIC SIMPLEX ALGORITHM

The basic simplex algorithm is a sequence of steps for solving linear programming problems in standard form. The steps are applied to simplex tableaus, so before using the algorithm, one must construct such a tableau, following the initialization process described in Section 3.4. The basic algorithm begins with an initial basic feasible solution and iteratively locates better solutions to the constraints until it has a solution that maximizes the objective function.

We will develop the basic simplex algorithm by analyzing the following linear programming problem in standard form:

$$\text{Maximize:} \quad z = 4x_1 + 10x_2 + 6x_3$$

$$\text{subject to:} \quad 2x_1 + 4x_2 + x_3 \le 12$$
$$6x_1 + 2x_2 + x_3 \le 26 \qquad \text{(9 repeated)}$$
$$5x_1 + x_2 + 2x_3 \le 80$$

$$\text{assuming:} \quad x_1, x_2, \text{ and } x_3 \text{ are nonnegative}$$

Using the initialization process described in Section 3.4, we showed that this model can be rewritten as

$$\text{Maximize:} \quad z = 4x_1 + 10x_2 + 6x_3 + 0s_1 + 0s_2 + 0s_3$$

$$\text{subject to:} \quad 2x_1 + 4x_2 + x_3 + s_1 \qquad\qquad = 12$$
$$6x_1 + 2x_2 + x_3 \qquad + s_2 \qquad = 26 \qquad \text{(16 repeated)}$$
$$5x_1 + x_2 + 2x_3 \qquad\qquad + s_3 = 80$$

$$\text{assuming:} \quad x_1, x_2, x_3, s_1, s_2, \text{ and } s_3 \text{ are all nonnegative}$$

with the corresponding initial simplex tableau

$$
\begin{array}{c}
\begin{array}{ccccccc}
x_1 & x_2 & x_3 & s_1 & s_2 & s_3 & z
\end{array} \\
\begin{array}{c}
s_1 \\ s_2 \\ s_3 \\ \\
\end{array}
\left[
\begin{array}{ccccccc|c}
2 & 4 & 1 & 1 & 0 & 0 & 0 & 12 \\
6 & 2 & 1 & 0 & 1 & 0 & 0 & 26 \\
5 & 1 & 2 & 0 & 0 & 1 & 0 & 80 \\
\hline
-4 & -10 & -6 & 0 & 0 & 0 & 1 & 0
\end{array}
\right]
\end{array}
\qquad \text{(20 repeated)}
$$

An initial basic feasible solution is $x_1 = x_2 = x_3 = 0$, $s_1 = 12$, $s_2 = 26$, $s_3 = 80$, with $z = 0$ as the corresponding value of the objective function.

With one solution in hand, we ask whether we can do better? Can we improve upon this solution and obtain a different solution with a larger z value? If so, we see that one of the variables currently set to zero must become nonzero, because if we leave $x_1 = x_2 = x_3 = 0$, then all the other variables will remain the same and the solution will be unchanged. But which x variable should we allow to become nonzero? Consider the objective function in (16):

$$z = 4x_1 + 10x_2 + 6x_3 + 0s_1 + 0s_2 + 0s_3$$

If we allow x_1 to become nonzero, then for every positive unit change in x_1, the objective function increases by 4. If we allow x_2 to become nonzero, then for every positive unit change in x_2, the objective function increases by 10. Similarly, if we allow x_3 to become nonzero, then for every positive unit change in x_3, the objective function increases by 6. Notice that we get the largest increase in the objective function by letting x_2 become nonzero. Since our aim is to maximize the objective function we will let x_2 become nonzero.

The coefficients in the objective function appear in the last row of the simplex tableau (20) with opposite signs. The largest positive coefficient in the objective function is the most negative coefficient in the last row of the simplex tableau. Thus, we direct our attention to the variable associated with the most negative coefficient in the last row of the simplex tableau. This is the logic for the first step of the basic simplex algorithm:

> STEP 1 Locate the most negative number in the last row of the simplex tableau, ignoring the last column, and designate the column in which this number appears as the *work column*. If more than one equally negative number exists, arbitrarily choose one.

The elements in the last row of simplex tableau

$$
\begin{array}{c}
\quad\; x_1 \quad x_2 \quad x_3 \quad s_1 \quad s_2 \quad s_3 \quad z \\
\begin{array}{c} s_1 \\ s_2 \\ s_3 \\ \\ \end{array}
\left[
\begin{array}{ccccccc|c}
2 & 4 & 1 & 1 & 0 & 0 & 0 & 12 \\
6 & 2 & 1 & 0 & 1 & 0 & 0 & 26 \\
5 & 1 & 2 & 0 & 0 & 1 & 0 & 80 \\
\hline
-4 & -10 & -6 & 0 & 0 & 0 & 1 & 0
\end{array}
\right]
\end{array}
\qquad \text{(20 repeated)}
$$

are -4, -10, -6, 1 and 0. Of these, the first three are negative, with -10 being the most negative. Since -10 appears in the second column, the x_2 column becomes the work column.

The variables x_2 will be increased from its current value of zero. Each time we increase x_2 by 1 unit, we increase the objective function by 10 units. The more we increase x_2, the more we increase z. This leads to the next question: How much should we increase x_2? The answer is, as *much as possible without violating the constraints*. The constraints for this problem are given in System (9) as

$$2x_1 + 4x_2 + x_3 \leq 12$$

$$6x_1 + 2x_2 + x_3 \leq 26$$

$$5x_1 + x_2 + 2x_3 \leq 80$$

We see that x_2 cannot exceed $12/4 = 3$ if the first constraint is to be honored, x_2 cannot exceed $26/2 = 13$ if the second constraint is to be honored, and x_2 cannot exceed 80 if the third constraint is to be honored. To honor all three constraints, the most severe limitation on x_2 must hold, that is, x_2 cannot exceed 3. This is the logic for the second step of the basic simplex algorithm:

> STEP 2 Form ratios by dividing each positive element of the work column into the corresponding element in the last column of the same row. Designate the element in the work column that yields the smallest ratio as the pivot. If more than one element yields the smallest ratio, arbitrarily choose one to be the pivot.

Step 2 directs us to form ratios only with *positive* elements in the work column, because zeros or negative coefficients would place no restrictions on the value of the corresponding variable. If our first constraint had been

$$2x_1 - 4x_2 + x_3 \le 12$$

where the coefficient of x_2 has been changed from 4 to -4, then x_2 could become as large as we want and still not violate the constraint.

For the simplex tableau

$$
\begin{array}{c}
 \\
s_1 \\
s_2 \\
s_3 \\

\end{array}
\begin{array}{cccccccc}
x_1 & x_2 & x_3 & s_1 & s_2 & s_3 & z & \\
\left[\begin{array}{ccccccc|c}
2 & 4 & 1 & 1 & 0 & 0 & 0 & 12 \\
6 & 2 & 1 & 0 & 1 & 0 & 0 & 26 \\
5 & 1 & 2 & 0 & 0 & 1 & 0 & 80 \\
\hline
-4 & -10 & -6 & 0 & 0 & 0 & 1 & 0
\end{array} \right]
\end{array}
\qquad \text{(20 repeated)}
$$

the work column, from Step 1, is the x_2 column, and the ratios of interest are $12/4 = 3$, $26/2 = 13$, and $80/1 = 80$. Since $12/4$ is the smallest ratio, the element 4 becomes the pivot. Note that we did not consider the ratio $0/-10$ because -10 is not a positive element of the work column.

Thus the binding constraint for x_2 is the first constraint, which defines the first slack variable, s_1. Recall that we introduced the first slack variable to convert the first constraint into the equation

$$2x_1 + 4x_2 + x_3 + s_1 = 12$$

If x_2 is to be 3, the largest possible value that does not violate the first constraint, then s_1 must become zero and both x_1 and x_3 must remain zero for the first constraint equation to remain true. The original constraints of System (*16*) are

$$
\begin{aligned}
2x_1 + 4x_2 + x_3 + s_1 &= 12 \\
6x_1 + 2x_2 + x_3 + s_2 &= 26 \\
5x_1 + x_2 + 2x_3 + s_3 &= 80
\end{aligned}
$$

Setting $s_1 = x_1 = x_3 = 0$, we reduce this system to

$$
\begin{aligned}
4x_2 &= 12 \\
2x_2 + s_2 &= 26 \\
x_2 + s_3 &= 80
\end{aligned}
$$

To obtain our new solution, we must solve this system of three equations for the three variables x_2, s_2, and s_3. We do so using Gauss-Jordan reduction (see Section 2.2), which is the third step of the basic simplex algorithm.

> **STEP 3** Use the second elementary row operation to convert the pivot to 1, if it is not already 1, and then use the pivot with the third elementary row operation to transform all other elements in the work column to 0.

As we did in Section 2.1, we shall use a caret ($\hat{}$) to mark the pivot and we shall circle the next element to be transformed to zero. Applying Step 3 of the basic simplex algorithm to Tableau (*20*), we find

$$
\begin{array}{c}
\begin{array}{ccccccc}
x_1 & x_2 & x_3 & s_1 & s_2 & s_3 & z
\end{array}\\[2pt]
\begin{array}{c}
s_1\\ s_2\\ s_3\\ \\
\end{array}
\left[
\begin{array}{ccccccc|c}
2 & 4 & 1 & 1 & 0 & 0 & 0 & 12\\
6 & 2 & 1 & 0 & 1 & 0 & 0 & 26\\
5 & 1 & 2 & 0 & 0 & 1 & 0 & 80\\
\hline
-4 & -10 & -6 & 0 & 0 & 0 & 1 & 0
\end{array}
\right]
\end{array}
$$

$$
\begin{array}{c}
\begin{array}{ccccccc}
x_1 & x_2 & x_3 & s_1 & s_2 & s_3 & z
\end{array}\\[2pt]
\begin{array}{c}
\rightarrow s_1\\ s_2\\ s_3\\ \\
\end{array}
\left[
\begin{array}{ccccccc|c}
\tfrac12 & \hat{1} & \tfrac14 & \tfrac14 & 0 & 0 & 0 & 3\\
6 & (2) & 1 & 0 & 1 & 0 & 0 & 26\\
5 & 1 & 2 & 0 & 0 & 1 & 0 & 80\\
\hline
-4 & -10 & -6 & 0 & 0 & 0 & 1 & 0
\end{array}
\right]
\quad \tfrac14 R_1 \rightarrow R_1
\end{array}
$$

$$
\begin{array}{c}
\begin{array}{ccccccc}
x_1 & x_2 & x_3 & s_1 & s_2 & s_3 & z
\end{array}\\[2pt]
\begin{array}{c}
s_1\\ \rightarrow s_2\\ s_3\\ \\
\end{array}
\left[
\begin{array}{ccccccc|c}
\tfrac12 & \hat{1} & \tfrac14 & \tfrac14 & 0 & 0 & 0 & 3\\
5 & 0 & \tfrac12 & -\tfrac12 & 1 & 0 & 0 & 20\\
5 & ① & 2 & 0 & 0 & 1 & 0 & 80\\
\hline
-4 & -10 & -6 & 0 & 0 & 0 & 1 & 0
\end{array}
\right]
\quad R_2 + (-2)R_1 \rightarrow R_2
\end{array}
$$

$$
\begin{array}{c}
\begin{array}{ccccccc}
x_1 & x_2 & x_3 & s_1 & s_2 & s_3 & z
\end{array}\\[2pt]
\begin{array}{c}
s_1\\ \rightarrow s_2\\ s_3\\ \\
\end{array}
\left[
\begin{array}{ccccccc|c}
\tfrac12 & \hat{1} & \tfrac14 & \tfrac14 & 0 & 0 & 0 & 3\\
5 & 0 & \tfrac12 & -\tfrac12 & 1 & 0 & 0 & 20\\
\tfrac92 & 0 & \tfrac74 & -\tfrac14 & 0 & 1 & 0 & 77\\
\hline
-4 & (-10) & -6 & 0 & 0 & 0 & 1 & 0
\end{array}
\right]
\quad R_3 + (-1)R_1 \rightarrow R_3
\end{array}
$$

$$
\begin{array}{c}
\begin{array}{ccccccc}
x_1 & x_2 & x_3 & s_1 & s_2 & s_3 & z
\end{array}\\[2pt]
\begin{array}{c}
s_1\\ \rightarrow s_2\\ s_3\\ \\
\end{array}
\left[
\begin{array}{ccccccc|c}
\tfrac12 & \hat{1} & \tfrac14 & \tfrac14 & 0 & 0 & 0 & 3\\
5 & 0 & \tfrac12 & -\tfrac12 & 1 & 0 & 0 & 20\\
\tfrac92 & 0 & \tfrac74 & -\tfrac14 & 0 & 1 & 0 & 77\\
\hline
1 & 0 & -\tfrac72 & \tfrac52 & 0 & 0 & 1 & 30
\end{array}
\right]
\quad R_4 + (10)R_1 \rightarrow R_4
\end{array}
$$

$$(22)$$

A reduced column has a single element equal to one with all other elements equal to zero.

Note that after Step 3 is completed, the elements in the work column are all zeros except for a single one. We call any column having this form a *reduced column*. Tableau (22) has four reduced columns, the columns associated with x_2, s_2, s_3, and z.

The purpose of Step 3 was to create a new solution with $x_2 = 3$, the largest possible value this variable can assume without violating a constraint. As result, s_1 had to become zero. That is, s_1 changed from a basic variable to a nonbasic variable, and x_2 changed from a nonbasic variable to a basic variable. Step 4 of the basic simplex algorithm just records these changes.

> **STEP 4** Replace the basic variable appearing in the same row as the pivot and in front of the augmented matrix with the variable appearing in the same column as the pivot and above the augmented matrix.

For Tableau (22), the pivot used in Step 3 was in the first row and second column. The basic variables in the first row (the same row as the pivot) and in front of the augmented matrix is s_1; the variable in the second column (the same column as the pivot) and above the augmented matrix is x_2. Therefore, we replace s_1 with x_2 as a new basic variable, and Tableau (22) becomes

$$
\begin{array}{c}
\begin{array}{ccccccc} x_1 & x_2 & x_3 & s_1 & s_2 & s_3 & z \end{array} \\
\begin{array}{c} x_2 \\ s_2 \\ s_3 \\ \end{array}
\left[
\begin{array}{ccccccc|c}
\frac{1}{2} & 1 & \frac{1}{4} & \frac{1}{4} & 0 & 0 & 0 & 3 \\
5 & 0 & \frac{1}{2} & -\frac{1}{2} & 1 & 0 & 0 & 20 \\
\frac{9}{2} & 0 & \frac{7}{4} & -\frac{1}{4} & 0 & 1 & 0 & 77 \\
\hline
1 & 0 & -\frac{7}{2} & \frac{5}{2} & 0 & 0 & 1 & 30
\end{array}
\right]
\end{array}
\qquad (23)
$$

We can read the new solution from Tableau (23) in the same manner that we read the initial solution from the initial simplex tableau. We set each basic variable appearing in front of the augmented matrix equal to its corresponding element in the same row and last column in the augmented matrix and set all other variables equal to zero. The value of the objective function is the number appearing in the last row and last column of the simplex tableau. From Tableau (23), we have $x_2 = 3$, $s_2 = 20$, $s_3 = 77$, with $x_1 = x_3 = s_1 = 0$, and $z = 30$.

In Step 3 we applied the third elementary row operation to all the rows in the simplex tableau, even the last, which holds the objective function. By using the pivot in the first row and second column to place a zero in the fourth row and second column, we effectively removed x_2 from the equation defining the objective function. Looking back at our original linear programming problem

Maximize: $z = 4x_1 + 10x_2 + 6x_2 + 0s_1 + 0s_2 + 0s_3$

subject to: $2x_1 + 4x_2 + x_3 + s_1 \qquad\qquad = 12$
$\qquad\qquad 6x_1 + 2x_2 + x_3 \qquad + s_2 \qquad = 26$ (16 repeated)
$\qquad\qquad 5x_1 + x_2 + 2x_3 \qquad\qquad + s_3 = 80$

assuming: $x_1, x_2, x_3, s_1, s_2,$ and s_3 are all nonnegative

we can solve the first constraint equation for x_2 and obtain

$$
x_2 = 3 - \frac{1}{2}x_1 - \frac{1}{4}x_3 - \frac{1}{4}s_1
$$

If we then substitute this equation for x_2 into the objective function, we get

$$
z = 4x_1 + 10\left(3 - \frac{1}{2}x_1 - \frac{1}{4}x_3 - \frac{1}{4}s_1\right) + 6x_3
$$
$$
= 30 - x_1 + \frac{7}{2}x_3 - \frac{5}{2}s_1
$$

Finally, rewriting this last equation with all the variables on the left side of the equation and the constant term on the right side, we generate the equation represented by the last row of Tableau (23). Furthermore, by setting $x_1 = x_3 = s_1 = 0$, the objective function reduces to a value of $z = 30$, which is better than the value $z = 0$ associated with the initial basic feasible solution. We have identified a better solution!

We will now repeat our entire analysis, using our current solution and starting again with the question: Can we improve upon this solution and obtain a different solution with a larger z value? However, rather than working with the original linear programming problem, we will work instead with the equivalent problem defined by our latest simplex tableau. We worked hard to obtain our latest tableau, and it is foolish not to use the results. Our current objective function is

$$
z = 30 - x_1 + \frac{7}{2}x_3 - \frac{5}{2}s_1
$$

and our current solution has $x_1 = x_3 = s_1 = 0$ with $z = 30$. Can we increase z? We can not increase it by increasing either x_1 or s_1, because either increase would decrease z. We can only increase z by increasing x_3. The coefficients of x_1 and s_1 are both negative, while the coefficient of x_3 is positive. If we rewrite the objective function with all the variables on the left side, as it appears in a simplex tableau, we have

$$x_1 - \frac{7}{2}x_3 + \frac{5}{2}s_1 + z = 30$$

In this form, we see that we can increase z only if this equation has a variable with a negative coefficient. Consequently, we have Step 5 of the basic simplex algorithm:

> **STEP 5** Repeat Steps 1 through 4 until there are no negative numbers in the last row of the current simplex tableau, ignoring the element in the last row and last column.

Our current simplex tableau is

$$
\begin{array}{c}
\\ x_2 \\ s_2 \\ s_3 \\ \\
\end{array}
\begin{bmatrix}
x_1 & x_2 & x_3 & s_1 & s_2 & s_3 & z & \\
\frac{1}{2} & 1 & \frac{1}{4} & \frac{1}{4} & 0 & 0 & 0 & 3 \\
5 & 0 & \frac{1}{2} & -\frac{1}{2} & 1 & 0 & 0 & 20 \\
\frac{9}{2} & 0 & \frac{7}{4} & -\frac{1}{4} & 0 & 1 & 0 & 77 \\
\hline
1 & 0 & -\frac{7}{2} & \frac{5}{2} & 0 & 0 & 1 & 30
\end{bmatrix}
$$

(23 repeated)

and we apply Steps 1 through 5 to it:

Step 1 The only negative entry in the last row is $-\frac{7}{2}$, so the x_3 column becomes the new work column.

Step 2 We compare the ratios $3/(1/4) = 12$, $20/(1/2) = 40$, and $77/(7/4) = 44$. The ratio $30/(-7/2)$ is not considered because $-7/2$ is not positive. The smallest ratio is 12, so the $\frac{1}{4}$ element in the first row of the work column becomes the new pivot.

Step 3 Using the second elementary row operation on Tableau (23) to transform the pivot to 1 and then using the third elementary row operation to transform all other elements in the work column to 0, we obtain

$$
\begin{array}{c}
\\ x_2 \\ s_2 \\ s_3 \\ \\
\end{array}
\begin{bmatrix}
x_1 & x_2 & x_3 & s_1 & s_2 & s_3 & z & \\
\frac{1}{2} & 1 & \frac{1}{4} & \frac{1}{4} & 0 & 0 & 0 & 3 \\
5 & 0 & \frac{1}{2} & -\frac{1}{2} & 1 & 0 & 0 & 20 \\
\frac{9}{2} & 0 & \frac{7}{4} & -\frac{1}{4} & 0 & 1 & 0 & 77 \\
\hline
1 & 0 & -\frac{7}{2} & \frac{5}{2} & 0 & 0 & 1 & 30
\end{bmatrix}
$$

$$
\begin{array}{c}
\\ x_2 \\ \rightarrow \ s_2 \\ s_3 \\ \\
\end{array}
\begin{bmatrix}
x_1 & x_2 & x_3 & s_1 & s_2 & s_3 & z & \\
2 & 4 & \hat{1} & 1 & 0 & 0 & 0 & 12 \\
5 & 0 & \tfrac{1}{2} & -\frac{1}{2} & 1 & 0 & 0 & 20 \\
\frac{9}{2} & 0 & \frac{7}{4} & -\frac{1}{4} & 0 & 1 & 0 & 77 \\
\hline
1 & 0 & -\frac{7}{2} & \frac{5}{2} & 0 & 0 & 1 & 30
\end{bmatrix}
\quad 4R_1 \rightarrow R_1
$$

$$
\begin{array}{c}
\begin{array}{ccccccc} x_1 & x_2 & x_3 & s_1 & s_2 & s_3 & z \end{array}\\
\begin{array}{c} x_2 \\ \rightarrow s_2 \\ s_3 \\ {} \end{array}
\left[
\begin{array}{ccccccc|c}
2 & 4 & \hat{1} & 1 & 0 & 0 & 0 & 12 \\
4 & -2 & 0 & -1 & 1 & 0 & 0 & 14 \\
\frac{9}{2} & 0 & \tfrac{7}{4} & -\frac{1}{4} & 0 & 1 & 0 & 77 \\
\hline
1 & 0 & -\frac{7}{2} & \frac{5}{2} & 0 & 0 & 1 & 30
\end{array}
\right]
\begin{array}{l} {} \\ R_2 + (-\tfrac{1}{2})R_1 \rightarrow R_2 \\ {} \\ {} \end{array}
\end{array}
$$

$$
\begin{array}{c}
\begin{array}{ccccccc} x_1 & x_2 & x_3 & s_1 & s_2 & s_3 & z \end{array}\\
\begin{array}{c} x_2 \\ \rightarrow s_2 \\ s_3 \\ {} \end{array}
\left[
\begin{array}{ccccccc|c}
2 & 4 & \hat{1} & 1 & 0 & 0 & 0 & 12 \\
4 & -2 & 0 & -1 & 1 & 0 & 0 & 14 \\
1 & -7 & 0 & -2 & 0 & 1 & 0 & 56 \\
\hline
1 & 0 & -\tfrac{7}{2} & \frac{5}{2} & 0 & 0 & 1 & 30
\end{array}
\right]
\begin{array}{l} {} \\ R_3 + (-\tfrac{7}{4})R_1 \rightarrow R_3 \\ {} \\ {} \end{array}
\end{array}
$$

$$
\begin{array}{c}
\begin{array}{ccccccc} x_1 & x_2 & x_3 & s_1 & s_2 & s_3 & z \end{array}\\
\begin{array}{c} x_2 \\ \rightarrow s_2 \\ s_3 \\ {} \end{array}
\left[
\begin{array}{ccccccc|c}
2 & 4 & \hat{1} & 1 & 0 & 0 & 0 & 12 \\
4 & -2 & 0 & -1 & 1 & 0 & 0 & 14 \\
1 & -7 & 0 & -2 & 0 & 1 & 0 & 56 \\
\hline
8 & 14 & 0 & 6 & 0 & 0 & 1 & 72
\end{array}
\right]
\begin{array}{l} {} \\ R_4 + (\tfrac{7}{2})R_1 \rightarrow R_4 \\ {} \\ {} \end{array}
\end{array}
$$

Step 4 The pivot used to generate Tableau (*24*) is in the first row and third column. The basic variable in the first row (the same row as the pivot) and in front of the augmented matrix is x_2; the variable in the third column (the same column as the pivot) and above the augmented matrix is x_3. Therefore, x_3 replaces x_2 as a new basic variable and Tableau (*24*) becomes

$$
\begin{array}{c}
\begin{array}{ccccccc} x_1 & x_2 & x_3 & s_1 & s_2 & s_3 & z \end{array}\\
\begin{array}{c} x_3 \\ s_2 \\ s_3 \\ {} \end{array}
\left[
\begin{array}{ccccccc|c}
2 & 4 & 1 & 1 & 0 & 0 & 0 & 12 \\
4 & -2 & 0 & -1 & 1 & 0 & 0 & 14 \\
1 & -7 & 0 & -2 & 0 & 1 & 0 & 56 \\
\hline
8 & 14 & 0 & 6 & 0 & 0 & 1 & 72
\end{array}
\right]
\end{array}
\qquad (25)
$$

Our second iteration of the basic simplex algorithm is complete. The linear programming model associated with Tableau (*25*) has the constraint equations

$$
\begin{aligned}
2x_1 + 4x_2 + x_3 + s_1 &= 12 \\
4x_1 - 2x_2 \quad\; - s_1 + s_2 &= 14 \\
x_1 - 7x_2 \quad -2s_1 \quad\; + s_3 &= 56
\end{aligned}
$$

If we set $x_1 = x_2 = s_1 = 0$, then these constraints reduce to $x_3 = 12$, $s_2 = 14$, and $s_3 = 56$. This is precisely the solution we obtain from Tableau (*25*) by setting each basic variable appearing in front of the augmented matrix equal to the corresponding element in the same row and last column of the matrix, with all other variables set to zero. The last row of Tableau (*25*) holds the objective function in the form

$$8x_1 + 14x_2 + 0x_3 + 6s_1 + 0s_2 + 0s_3 + z = 72$$

or, more simply,

$$z = 72 - 8x_1 - 14x_2 - 6s_1$$

With $x_1 = x_2 = s_1 = 0$, the objective function attains the value $z = 72$, which is the entry in the last row and last column of Tableau (*25*). It is also clear that any attempt to increase

x_1, x_2, or s_1 from zero will reduce z. Thus, we see that we cannot improve the value of the objective function beyond 72 (this is equivalent to noting that there are no negative entries in the last row of Tableau (25)). Step 5 is complete, and the last solution is the optimal solution.

The optimal solution from Tableau (25) is $x_3 = 12$, $s_2 = 14$, $s_3 = 56$ with $x_1 = x_2 = s_1 = 0$. This solution satisfies all the constraints and maximizes the objective function. To check this, let us return to the original linear programming problem:

$$\text{Maximize:} \quad z = 4x_1 + 10x_2 + 6x_3$$

$$\text{subject to:} \quad 2x_1 + 4x_2 + x_3 \le 12$$
$$6x_1 + 2x_2 + x_3 \le 26 \qquad \text{(9 repeated)}$$
$$5x_1 + x_2 + 2x_3 \le 80$$

$$\text{assuming:} \quad x_1, x_2, \text{ and } x_3 \text{ are nonnegative}$$

Substituting values from the optimal solution into the first constraint, we have $2(0) + 4(0) + 12 = 12$. The first constraint is satisfied, and there is no slack between the left and right sides of the inequality, which is why $s_1 = 0$ in the optimal solution. Substituting values from the optimal solution into the second constraint, we have $6(0) + 2(0) + 12 = 12$, which is 14 units less than the right side of 26. The second constraint is satisfied, but there are 14 units of slack between the left and right sides of the inequality, which is why $s_2 = 14$ in the optimal solution. Substituting values from the optimal solution into the third constraint, we have $5(0) + 1(0) + 2(12) = 24$, which is 56 units less than the right side of 80. The third constraint is satisfied, but there are 56 units of slack between the left and right sides of the inequality, which is why $s_3 = 56$ in the optimal solution.

Finally, evaluating the objective function with values from the optimal solution, we have

$$z = 4(0) + 10(0) + 6(12) = 72$$

We can use Steps 1 through 5 to solve all linear programming problems in standard form.

EXAMPLE 1

A wood cabinet manufacturer produces cabinets for television consoles and frames for grandfather clocks, both of which must be assembled and decorated. Every television cabinet requires 3 hours to assemble and 5 hours to decorate. Each grandfather clock frame requires 10 hours to assemble and 8 hours to decorate. The profits on each cabinet and clock frame are $7 and $22, respectively, and the manufacturer can sell all the units it produces. Each week the manufacturer has 33,000 hours available for assembling these products and 42,000 hours available for decorating. How many of each product should the manufacturer produce weekly to maximize profits?

Solution

Our goal is to maximize the profit from the production of television cabinets and grandfather clock frames. We set

$$x_1 = \text{the number of television cabinets produced}$$

$$x_2 = \text{the number of grandfather clock frames produced}$$

Profit z is modeled by the equation $z = 7x_1 + 22x_2$. The constraints were modeled in Example 2 of Section 3.1. Combining those constraints with this objective function, we have the linear programming model

$$\text{Maximize:} \quad z = 7x_1 + 22x_2$$

$$\text{subject to:} \quad 3x_1 + 10x_2 \leq 33{,}000$$
$$5x_1 + 8x_2 \leq 42{,}000$$

$$\text{assuming:} \quad x_1 \text{ and } x_2 \text{ are nonnegative}$$

This model involves only two decision variables, x_1 and x_2, so it can be solved by graphical methods, as was done in Example 1 of Section 3.3. The simplex method, however, is applicable to all linear programming problems in standard form and can therefore be used to solve this linear programming problem too.

The simplex tableau associated with this linear programming model in standard form was determined in Section 3.4 to be

$$
\begin{array}{c}
\\
s_1 \\
s_2 \\
\\
\end{array}
\begin{array}{c}
\begin{array}{ccccc}
x_1 & x_2 & s_1 & s_2 & z
\end{array} \\
\left[
\begin{array}{ccccc|c}
3 & 10 & 1 & 0 & 0 & 33{,}000 \\
5 & 8 & 0 & 1 & 0 & 42{,}000 \\
\hline
-7 & -22 & 0 & 0 & 1 & 0
\end{array}
\right]
\end{array}
\qquad \text{(19 repeated)}
$$

An initial basic feasible solution is $s_1 = 33{,}000$, $s_2 = 42{,}000$, $x_1 = x_2 = 0$ with $z = 0$, and we use the basic simplex algorithm on Tableau (19) to improve upon this solution.

Step 1 The last row of Tableau (19), ignoring the last column, contains two negative numbers with the most negative, -22, located in the second column, so the x_2 column becomes the work column.

Step 2 We compare the ratios $33{,}000/10 = 3{,}300$ and $42{,}000/8 = 5{,}250$. The smallest is $3{,}300$, so the element 10 in the first row of the work column becomes the pivot.

Step 3 Using the second elementary row operation to transform the pivot 10 to 1 and then using the pivot with the third elementary row operation to transform all other elements in the work column to zero, we generate

$$
\begin{array}{c}
\\
s_1 \\
s_2 \\
\\
\end{array}
\begin{array}{c}
\begin{array}{ccccc}
x_1 & x_2 & s_1 & s_2 & z
\end{array} \\
\left[
\begin{array}{ccccc|c}
0.3 & \hat{1} & 0.1 & 0 & 0 & 3{,}300 \\
2.6 & 0 & -0.8 & 1 & 0 & 15{,}600 \\
\hline
-0.4 & 0 & 2.2 & 0 & 1 & 72{,}600
\end{array}
\right]
\end{array}
$$

Step 4 The pivot was in the first row and second column, hence the basic variable in the first row (the same row as the pivot) and in front of the augmented matrix (namely, s_1) is replaced by the variable in the second column (the same column as the pivot) and above the augmented matrix (namely, x_2). The new simplex tableau becomes

$$
\begin{array}{c}
\\
x_2 \\
s_2 \\
\\
\end{array}
\begin{array}{c}
\begin{array}{ccccc}
x_1 & x_2 & s_1 & s_2 & z
\end{array} \\
\left[
\begin{array}{ccccc|c}
0.3 & 1 & 0.1 & 0 & 0 & 3{,}300 \\
2.6 & 0 & -0.8 & 1 & 0 & 15{,}600 \\
\hline
-0.4 & 0 & 2.2 & 0 & 1 & 72{,}600
\end{array}
\right]
\end{array}
\qquad \text{(26)}
$$

Step 5 The last row of Tableau (26) contains a negative entry that is not in the last column, so we *repeat* Steps 1 through 4 on Tableau (26). Before doing so, we note that the new solution is $x_2 = 3{,}300$, $s_2 = 15{,}600$, $x_1 = s_1 = 0$ with $z = 72{,}600$, which is clearly better than the initial solution.

Step 1 The last row of Tableau (26), ignoring the last column, contains one negative number, -0.4, so it is the most negative by default. This element appears in the first column, hence the x_1 column becomes the work column.

Step 2 We compare the ratios $3{,}300/0.3 = 11{,}000$ and $15{,}600/2.6 = 6{,}000$. The smallest ratio is 6,000, so the element 2.6 in the second row of the work column becomes the pivot.

Step 3 Using the second elementary row operation to transform the pivot 2.6 to 1 and then using the pivot with the third elementary row operation to transform all other elements in the work column to 0, we obtain

$$
\begin{array}{c}
\\
x_2 \\
s_2 \\
\\
\end{array}
\begin{array}{c}
x_1 \quad x_2 \quad\quad s_1 \quad\quad s_2 \quad z \\
\left[
\begin{array}{ccccc|c}
0 & 1 & 0.19 & -0.12 & 0 & 1{,}500 \\
\hat{1} & 0 & -0.31 & 0.38 & 0 & 6{,}000 \\
\hline
0 & 0 & 2.08 & 0.15 & 1 & 75{,}000
\end{array}
\right]
\end{array}
$$

Step 4 The pivot was in the second row and first column, hence the basic variable in the second row (the same row as the pivot) and in front of the augmented matrix (namely, s_2) is replaced by the variable in the first column (the same column as the pivot) and above the augmented matrix, namely x_1. The new simplex tableau becomes

$$
\begin{array}{c}
\\
x_2 \\
x_1 \\
\\
\end{array}
\begin{array}{c}
x_1 \quad x_2 \quad\quad s_1 \quad\quad s_2 \quad z \\
\left[
\begin{array}{ccccc|c}
0 & 1 & 0.19 & -0.12 & 0 & 1{,}500 \\
1 & 0 & -0.31 & 0.38 & 0 & 6{,}000 \\
\hline
0 & 0 & 2.08 & 0.15 & 1 & 75{,}000
\end{array}
\right]
\end{array}
\tag{27}
$$

Step 5 The last row of Tableau (27) contains no negative entries, so the basic simplex algorithm is complete.

The optimal solution is $x_2 = 1{,}500$, $x_1 = 6{,}000$, $s_1 = s_2 = 0$ with $z = 75{,}000$. That is, the best production schedule for the wood cabinet manufacturer in terms of maximizing profit is to produce 6,000 television cabinets and 1,500 grandfather clock frames for a weekly profit of $75,000.

The model in Example 1 involves just two decision variables and can be solved either by the basic simplex algorithm or graphically. The power of the basic simplex algorithm is its utility in solving linear programming models containing more than two decision variables, such as the linear programming problem at the beginning of this section. One advantage of applying the basic simplex algorithm to a linear programming problem with two decision variables is that we can analyze the method graphically. For example, the constraints for the linear programming model in Example 1 are

$$3x_1 + 10x_2 \le 33{,}000$$

$$5x_1 + 8x_2 \le 42{,}000$$

which define the feasible region shaded in Figure 3.18. The initial basic feasible solution has $x_1 = x_2 = 0$. This is point A in Figure 3.18. After one iteration of the basic simplex algorithm, we identified a second solution with $x_1 = 0$ and $x_2 = 3{,}300$. This is point B in Figure 3.18. After the second iteration, we located the optimal solution with $x_1 = 6{,}000$

FIGURE 3.18

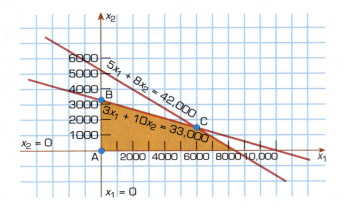

and $x_2 = 1,500$. This is point C in Figure 3.18. Observe that the solution after each iteration of the basic simplex algorithm is a corner point! This is no coincidence. We know from our work in Section 3.3 that the optimal solution for a linear programming problem must lie on a corner point of the feasible region defined by the constraints. The basic simplex algorithm begins at the origin, which is always a corner point for a linear programming problem with nonnegative variables, and then sequentially probes other corner points for better values of the objective function.

The bulk of the work in the basic simplex algorithm revolves around the elementary row operations in Step 3. Fortunately, graphing calculators have menus for performing these operations, and this technology should be used whenever possible. Simpler still are the host of computer software packages available for automating the entire algorithm. Professionals routinely use such packages when they need to solve a linear programming model of a real-world system.

One modification in Step 2 of the basic simplex algorithm is required if it is to apply to all linear programming problems in standard form. In Step 2, ratios are formed by dividing each positive element of the work column into the corresponding element in the last column of the same row. It can happen, however, that a work column has no positive elements. In such a case, the linear programming problem does not have a solution and the algorithm is terminated.

> **STEP 2 (ADDITION)** If a work column has no positive elements, then the linear program has no solution.

EXAMPLE 2

Use the basic simplex algorithm to solve

$$\text{Maximize:} \quad z = 2x_1 + 7x_2$$

$$\text{subject to:} \quad -x_1 + x_2 \leq 3$$

$$\text{assuming:} \quad x_1 \text{ and } x_2 \text{ are nonnegative}$$

Solution

Following the initialization process described in Secton 3.4, we add a slack variable s_1 to the right side of the constraint inequality, converting it to the equality

$$-x_1 + x_2 + s_1 = 3$$

The objective function is then rewritten as

$$-2x_1 - 7x_2 + 0s_1 + z = 0$$

and the initial simplex tableau becomes

$$
\begin{array}{c}
\quad\;\; x_1 \quad x_2 \quad s_1 \quad z \\
s_1 \begin{bmatrix} -1 & 1 & 1 & 0 & 3 \\ \hline -2 & -7 & 0 & 1 & 0 \end{bmatrix}
\end{array}
\tag{28}
$$

Step 1   The last row of Table (28), ignoring the last column, contains two negative numbers with the most negative, -7, located in the second column, so the x_2-column becomes the work column.

Step 2   There is only one positive entry in the work column, the element 1, so it becomes the pivot by default.

Step 3   Using elementary row operations to convert the work column into a reduced column, we obtain

$$
\begin{array}{c}
\quad\;\; x_1 \quad x_2 \quad s_1 \quad z \\
s_1 \begin{bmatrix} -1 & \hat{1} & 1 & 0 & 3 \\ \hline -9 & 0 & 7 & 1 & 21 \end{bmatrix}
\end{array}
$$

Step 4   The pivot was in the first row and second column, hence the basic variable in the first row (the same row as the pivot) and in front of the augmented matrix (namely, s_1) is replaced by the variable in the second column (the same column as the pivot) and above the augmented matrix (namely, x_2). After one iteration of the basic simplex algorithm, the simplex tableau becomes

$$
\begin{array}{c}
\quad\;\; x_1 \quad x_2 \quad s_1 \quad z \\
x_2 \begin{bmatrix} -1 & 1 & 1 & 0 & 3 \\ \hline -9 & 0 & 7 & 1 & 21 \end{bmatrix}
\end{array}
\tag{29}
$$

Step 5   The last row of Tableau (29) contains a negative entry that is not in the last column, so we repeat Steps 1 through 4 on Tableau (29).

Step 1   The last row of Tableau (29), ignoring the last column, contains one negative number, -9, so it is the most negative by default. This element appears in the first column, hence the x_1 column becomes the work column.

Step 2   There are no positive numbers in this work column, so no ratios can be formed and this step cannot be executed. Consequently, the given linear programming problem has no solution.

    The conclusion to Example 2 is not surprising once we analyze the problem graphically. The feasible region is shaded in Figure 3.19, and it is unbounded! The dashed lines in Figure 3.19 are graphs of the objective function $z = 2x_1 + 7x_2$ when $z = 30, 40, 50$, and 60. Since each of these lines contains points in the feasible region, each value of the objective function is realizable. Furthermore, every larger value of z is also obtainable. The line corresponding to any value of z greater than 60 has points in the feasible region; the larger the value of z, the further the line will be from the origin. For each value of z, there exist points x_1 and x_2 in the feasible region that yield an even greater value of the objective function.

    In Section 3.3, we noted that an unbounded feasible region indicates the existence of additional hidden constraints. Consequently, the linear programming problem in Example 2 is not a complete model for a real-world process. A complete model requires additional constraints to bound the feasible region. When modeling a real-world process, practitioners sometimes fail to identify all relevant constraints, resulting in an unbounded feasible region. This omission is obvious when dealing with only two decision variables, because we can graph the feasible region. Most models, however, involve many decision variables, making a graph impractical. At such times, the addition to Step 2 is our indicator that other constraints, not yet modeled, are also operative.

With the modification to Step 2, we have the complete basic simplex algorithm. A sixth step is included to formalize the procedure used to identify a solution from a simplex tableau.

THE BASIC SIMPLEX ALGORITHM

STEP 1 Locate the most negative number in the last row of the simplex tableau, ignoring the last column, and designate the column in which this number appears as the work column. If more than one equally negative number exists, arbitrarily choose one.

STEP 2 Form ratios by dividing each positive element of the work column into the corresponding element in the last column of the same row. If no element in the work column is positive, stop: the problem has no solution; otherwise, designate the element in the work column that yields the smallest ratio as the pivot. If more than one element yields the smallest ratio, arbitrarily choose one to be the pivot.

STEP 3 Use the second elementary row operation to convert the pivot to one if it is not one already, and then use the third elementary row operation to transform all other elements in the work column to zero.

STEP 4 Replace the basic variable appearing in the same row as the pivot and in front of the augmented matrix portion of the simplex tableau with the variable appearing in the same column as the pivot and above the augmented matrix portion of the simplex tableau.

STEP 5 Repeat Steps 1 through 4 until there are no negative numbers in the last row of the current simplex tableau, ignoring the element in the last row and last column.

STEP 6 An optimal solution is obtained by setting each basic variable appearing in front of the augmented matrix equal to its corresponding element in the last column of the same row of the augmented matrix, and then setting all other variables equal to zero. The value of the objective function is the number appearing in the last row and last column of the simplex tableau.

Finally, we note that the basic simplex algorithm does *not* produce all the solutions of a linear programming problem in standard form when more than one solution exists (see Problem 53). The goal in linear programming is to produce a solution that maximizes an objective function, subject to constraints, and the basic simplex algorithm does this. Other solutions that have the same value for the objective function may exist, but there is no solution that has a better value of the objective function than the one found by the basic simplex algorithm.

FIGURE 3.19

IMPROVING SKILLS

In Problems 1 through 10, apply one interation of the simplex method to the given tableaus; that is, apply Steps 1 through 4 once and then stop. Determine whether a second iteration is necessary. If it is, identify the pivot for that iteration; if not, determine the optimal solution.

1.

	x_1	x_2	s_1	s_2	z	
s_1	1	2	1	0	0	10
s_2	3	4	0	1	0	11
	-5	-6	0	0	1	0

2.

	x_1	x_2	s_1	s_2	z	
s_1	3	-3	1	0	0	6
s_2	2	5	0	1	0	10
	-8	-5	0	0	1	0

3.

	x_1	x_2	s_1	s_2	z	
s_1	1	2	1	0	0	1000
s_2	3	4	0	1	0	3600
	20	-35	0	0	1	0

4.

	x_1	x_2	s_1	s_2	z	
s_1	1	-2	1	0	0	1000
s_2	2	4	0	1	0	3600
	20	-35	0	0	1	0

5.

	x_1	x_2	s_1	s_2	z	
s_1	1	-2	1	0	0	1000
s_2	3	0	0	1	0	3600
	20	-35	0	0	1	0

6.

	x_1	x_2	s_1	s_2	z	
s_1	1	0	1	0	0	39,000
s_2	2	5	0	1	0	73,000
	0	-4	0	0	1	0

7.

	x_1	x_2	s_1	s_2	s_3	z	
s_1	11	12	1	0	0	0	460
s_2	25	22	0	1	0	0	385
s_3	7	15	0	0	1	0	520
	-31	-28	0	0	0	1	0

8.

	x_1	x_2	s_1	s_2	s_3	z	
s_1	35	-18	1	0	0	0	1
s_2	0	11	0	1	0	0	2
s_3	17	9	0	0	1	0	1
	-0.01	-0.02	0	0	0	1	0

9.

	x_1	x_2	s_1	s_2	s_3	s_4	z	
s_1	2	3	1	0	0	0	0	20.4
s_2	5	2	0	1	0	0	0	18.3
s_3	3	5	0	0	1	0	0	35.7
s_4	3	4	0	0	0	1	0	29.6
	-1	-1	0	0	0	0	1	0

10.

	x_1	x_2	s_1	s_2	s_3	s_4	s_5	z	
s_1	1	1	1	0	0	0	0	0	4
s_1	1	-1	0	1	0	0	0	0	1
s_3	1	0	0	0	1	0	0	0	2
s_4	0	1	0	0	0	1	0	0	2
s_5	2	3	0	0	0	0	1	0	7
	-30	-35	0	0	0	0	0	1	0

In Problems 11 through 34, use the basic simplex algorithm to solve the given linear programming models.

11. Maximize: $z = x_1 + 2x_2$
subject to: $2x_1 + x_2 \le 4$
assuming: x_1 and x_2 are nonnegative

12. Maximize: $z = 10x_1 + 3x_1$
subject to: $2x_1 + x_2 \le 4$
assuming: x_1 and x_2 are nonnegative

13. Maximize: $z = x_1 + x_2$
subject to: $5x_1 + 8x_2 \le 40$
$7x_1 + 4x_2 \le 28$
assuming: x_1 and x_2 are nonnegative

14. Maximize: $z = 4x_1$
subject to: $5x_1 + 8x_2 \le 40$
$7x_1 + 4x_2 \le 28$
assuming: x_1 and x_2 are nonnegative

15. Maximize: $z = 3x_1 + 2x_2$
subject to: $6x_1 + x_2 \le 24{,}000$
$2x_1 + 5x_2 \le 47{,}200$
assuming: x_1 and x_2 are nonnegative

16. Maximize: $z = 8x_1 + 15x_2$
subject to: $0.4x_1 + 0.8x_2 \le 2{,}240$
$0.6x_1 + 0.4x_2 \le 1{,}440$
assuming: x_1 and x_2 are nonnegative

17. Maximize: $z = 40x_1 + 50x_2$
subject to: $9x_1 + 12x_2 \le 252$
$5x_1 + 10x_2 \le 160$
assuming: x_1 and x_2 are nonnegative

18. Maximize: $z = 24x_2$
subject to: $10x_1 + 4x_2 \le 1{,}800$
$4x_1 + 5x_2 \le 1{,}400$
assuming: x_1 and x_2 are nonnegative

19. Maximize: $z = 3x_1 + 8x_2$
subject to: $60x_1 + 10x_2 \le 240{,}000$
$2x_1 + 5x_2 \le 47{,}200$
assuming: x_1 and x_2 are nonnegative

20. Maximize: $z = 15x_1 + 30x_2$
subject to: $0.4x_1 + 0.8x_2 \le 2{,}240$
$0.6x_1 + 0.4x_2 \le 1{,}440$
assuming: x_1 and x_2 are nonnegative

21. Maximize: $z = 2x_1 + 7x_2$
subject to: $x_1 - 3x_2 \le 3$
$2x_1 + 4x_2 \le 18$
assuming: x_1 and x_2 are nonnegative

22. Maximize: $z = 180x_1$
subject to: $50x_1 + 27x_2 \leq 170$
$8x_1 + 25x_2 \leq 80$
assuming: x_1 and x_2 are nonnegative

23. Maximize: $z = 2x_1 - x_2$
subject to: $x_1 + 2x_2 \leq 2$
$-3x_1 + 4x_2 \leq 7$
assuming: x_1 and x_2 are nonnegative

24. Maximize: $z = 40x_1 + 60x_2$
subject to: $30x_1 + 50x_2 \leq 80,000$
$10x_1 + 5x_2 \leq 14,000$
$4x_1 + 4x_2 \leq 7,200$
assuming: x_1 and x_2 are nonnegative

25. Maximize: $z = 35x_1 + 12x_2$
subject to: $8x_1 + 11x_2 \leq 9,500$
$7x_1 + 12x_2 \leq 8,000$
$9x_1 + 9x_2 \leq 9,200$
assuming: x_1 and x_2 are nonnegative

26. Maximize: $z = 15x_1 + 10x_2 + 14x_3$
subject to: $6x_1 + 5x_2 + 3x_3 \leq 26$
$4x_1 + 2x_2 + 5x_3 \leq 8$
assuming: x_1, x_2, and x_3 are nonnegative

27. Maximize: $z = 3x_1 - 4x_2 + 5x_3$
subject to: $x_1 + 2x_2 + 2x_3 \leq 180$
$3x_1 + 4x_2 + 4x_3 \leq 230$
assuming: x_1, x_2, and x_3 are nonnegative

28. Maximize: $z = 5x_1 + 6x_2 + 7x_3 + 8x_4$
subject to: $3x_1 + 4x_2 + 5x_3 - 6x_4 \leq 90$
$7x_1 + 7x_2 + 8x_3 - 9x_4 \leq 135$
assuming: x_1, x_2, x_3, and x_4 are nonnegative

29. Maximize: $z = 15x_1 + 10x_2 + 14x_3$
subject to: $30x_1 + 50x_2 + 40x_3 \leq 80,000$
$10x_1 + 5x_2 + 7x_3 \leq 14,000$
$4x_1 + 4x_2 + 6x_3 \leq 7,200$
assuming: x_1, x_2, and x_3 are nonnegative

30. Maximize: $z = x_3$
subject to: $0.3x_1 + 0.4x_2 + 0.3x_3 \leq 9,000$
$0.2x_1 + 0.8x_2 \leq 5,500$
$0.9x_1 + 0.1x_3 \leq 7,300$
assuming: x_1, x_2, and x_3 are nonnegative

31. Maximize: $z = 15x_1 + 14x_3$
subject to: $30x_1 + 50x_2 + 40x_3 \leq 80,000$
$10x_1 + 5x_2 + 7x_3 \leq 14,000$
$4x_1 + 4x_2 + 6x_3 \leq 7,200$
$14x_1 + 7x_2 + 10x_3 \leq 21,000$
assuming: x_1, x_2, and x_3 are nonnegative

32. Maximize: $z = 9x_1 + 9x_2 + x_3$
subject to: $4x_1 + 4x_2 + 5x_3 \leq 0.03$
$2x_1 + 3x_2 + 8x_3 \leq 0.07$
$x_1 + x_2 + 3x_3 \leq 0.01$
$2x_1 + x_2 + x_3 \leq 0.08$
$4x_1 + 3x_2 + 5x_3 \leq 0.12$
assuming: x_1, x_2, and x_3 are nonnegative

33. Maximize: $z = 2x_1 + 3x_2 + 4x_3 + 6x_4$
subject to: $x_1 + x_2 + x_3 + x_4 \leq 15$
$7x_1 + 5x_2 + 3x_3 + 2x_4 \leq 120$
$3x_1 + 5x_2 + 10x_3 + 15x_4 \leq 100$
assuming: x_1, x_2, x_3, and x_4 are nonnegative

34. Maximize: $z = x_2 + 2x_4$
subject to: $2x_1 + 3x_2 + 3x_3 + 4x_4 \leq 20$
$3x_1 + 4x_2 + 4x_3 + 5x_4 \leq 35$
$5x_1 + 5x_2 + 4x_3 + 3x_4 \leq 35$
$3x_1 + 3x_2 + 2x_3 + x_4 \leq 20$
assuming: x_1, x_2, x_3, and x_4 are nonnegative

CREATING MODELS

Use the basic simplex algorithm to solve the problems described in Problems 35 through 52.

35. The King Sleep Corporations produces two lines of mattresses, regular and extra-firm. Each regular mattress requires 30 springs, 10 pounds of padding, and 4 yards of material, while each extra-firm mattress requires 50 springs, 5 pounds of padding, and 4 yards of material. The wholesale price is $40 for each regular mattress and $60 for each extra-firm mattress. The maximum weekly inventory King Sleep can stock is 80,000 springs, 14,000 pounds of padding, and 7,200 yards of material. How many mattresses should King Sleep produce weekly to maximize its income? How much inventory does this production schedule consume?

36. Solve Problem 35 if the King Sleep Corporation also produces an economy mattress that requires 20 springs, 14 pounds of padding, and 3 yards of material and sells for $35 per mattress at wholesale.

37. The Atlas Pipe Company manufactures two types of metal joints, each of which must be molded and threaded. Type I joints require 60 seconds on the molding machine and 2 seconds on the threading machine, while Type II joints require 10 seconds on the molding machine and 5 seconds on the threading machine. Each day the molding machines are operable for 240,000 seconds and the threading machines for 47,200 seconds. The profit on each Type I and Type II joint is 3¢ and 2¢, respectively. How many joints of each type should be produced to maximize the daily profit?

38. Solve Problem 37 if, in addition, the company also manufactures a third type of joint that requires 20 seconds per unit on the molding machine and 6 seconds per unit on the threading machine and generates a unit profit of 3¢.

39. A truck traveling from Davenport, Iowa, to Fort Meyers, Colorado, is in demand from two manufacturers who need to ship goods between those cities. Crates from the first manufacturer measures 60 cubic feet, weigh 100 pounds, and cost $80 per crate to ship. Crates from the second manufacturer measure 40 cubic feet, weigh 120 pounds, and cost $95 per crate to ship. The truck can carry no more than 12,000 pounds and has room for at most 8,000 cubic feet of cargo. How many crates of each type should the trucker accept if the objective is to maximize revenue? (Note: Shipping costs for each manufacturer are revenues to the trucker.)

40. Solve Problem 39 if, in addition, two other manufacturers also bid for space on the truck. Crates from one of these manufacturers measure 60 cubic feet, weigh 50 pounds, and cost $70 each to ship; crates from the other manufacturer measure 80 cubic feet, weight 40 pounds, and cost $90 each to ship.

41. A wood cabinet manufacturer produces cabinets for television consoles, frames for grandfather clocks, and lamp bases, all of which must be assembled, decorated, and crated. Each television console requires 3 hours to assemble, 5 hours to decorate, and 0.1 hour to crate and returns a profit of $7. Each clock frame requires 10 hours to assemble, 8 hours to decorate, and 0.6 hours to crate and returns a profit of $22. Each lamp base requires 1 hour to assemble, 1 hour to decorate, and 0.1 hour to crate and returns a profit of $4. The manufacturer has 30,000, 40,000, and 120 hours available weekly for assembling, decorating, and crating, respectively. How many units of each product should it produce to maximize profit?

42. Solve Problem 41 if the employer hires additional packers to increase the available packing time to 1,200 hours.

43. Solve Problem 41 if the employer hires additional packers to increase the available packing time to 3,000 hours per week.

44. A local kennel has 50 cages available for housing dogs and cats. The kennel charges $80 per week to house a dog and $50 per week to house a cat. Each dog receives 2 hours of petting time per week, and each cat receives 1 hour of petting time per week. The kennel has a maximum of 80 hours of employee time available for petting animals. How many dogs and cats should the kennel accept if it wants to maximize income?

45. Solve Problem 44 if, in addition, the kennel offers dog owners an economy plan that costs $25 per week and provides no petting time for the dog.

46. A small company makes steak knives and pocket knives, both of which are in high demand. Each type of knife requires one unit of steel. Steak knives take 4 hours of labor to produce and sell for $9, while pocket knives require 10 hours of labor and sell for $22. How many knives of each type should the company produce to maximize income if it has 30 units of steel in stock and 180 hours of available labor?

47. Solve Problem 46 if, in addition, the company also makes chef's knives, each of which sells for $50 and requires 2.5 units of steel and 5 hours of labor.

48. Organizers of a conference have reserved $63,000 for travel grants for conference participants. Regional participants live 1,000 miles or less from the conference site and receive a grant of $500; all other participants are national participants and receive $800. To provide a national flavor, there will be at least twice as many national participants as regional participants. How many participants of each type should the organizers invite if they want to maximize total attendance?

49. Solve Problem 48 if, in addition, international participants to the conference receive $900 grants, and the organizers want the number of international participants to be at least as many as the number of regional participants.

50. A savings and loan association has $10 million available for home loans and car loans. Its policy is to invest at least five times as much money in home loans as in car loans. Current applications from prospective borrowers total $6 million for home loans and $8 million for car loans. The savings and loan earns 9% on home loans and 15% on car loans. How much should it allocate to each type of loan to maximize its earnings?

51. An investor has identified three attractive funds and will divide $10,000 among them. The first is low-risk and returns 4% per year, the second is medium-risk and returns 6% per year, and the last is high-risk and returns 9% per year. The investor's strategy requires that the amounts allocated to low- and medium-risk funds always exceed the amount invested in high-risk funds and that no more than $4,000 be invested in high-risk funds. How much of each fund should the investor purchase to maximize total return?

52. A publisher of cookbooks, garden books, and self-help books will limit the list of new titles to at most 24 next year. To meet market demand, the number of new titles in cookbooks must be at least as great as all other new titles combined. On average, each new cookbook title generates $8,500 in profits, each new garden book generates $8,000 in profits, and each new self-help book generates $9,000 in profits. How many new titles in each category should the publisher produce to maximize profits?

EXPLORING IN TEAMS

53. Use the basic simplex algorithm to show that one solution to the linear programming problem

$$\text{Maximize:} \quad z = 3x_1 + 3x_2$$
$$\text{subject to:} \quad x_1 + x_2 \le 5$$
$$2x_1 + x_2 \le 8$$
$$\text{assuming:} \quad x_1 \text{ and } x_2 \text{ are nonnegative}$$

is $x_1 = 0$, $x_2 = 5$, $s_1 = 0$, $s_2 = 3$ with $z = 15$. Solve this problem graphically, and show that the corner point $x_1 = 3$ and $x_2 = 2$ also yields $z = 15$ and is an equally valid solution. What can you say about the points on the line segment between (3, 2) and (0, 5)?

54. Return to Tableau (20) at the beginning of this section and apply Step 3 of the basic simplex algorithm with the element 2 in the second row and second column as the pivot. What goes wrong?

55. Return to Tableau (20) and apply Step 3 of the basic simplex algorithm with the element 1 in the third row and second column as the pivot. What goes wrong?

56. Return to Tableau (26) in Example 1 and apply Step 3 of the basic simplex algorithm with the element 0.3 in the first row and first column as the pivot. What goes wrong?

57. Use the results of Problems 54 through 56 to determine the effect of using the wrong element in the work column as the pivot.

58. Solve Example 1, beginning with the first column as the work column, in violation of Step 1 of the basic simplex algorithm. Complete the algorithm, applying each subsequent step correctly. What do you conclude?

59. Return to Tableau (20) at the beginning of this section and take the first column as the initial work column, in violation of Step 1 of the basic simplex algorithm. Complete the algorithm, applying each subsequent step correctly. What do you conclude?

60. Return to Tableau (20) and take the third column as the initial work column, in violation of Step 1 of the basic simplex algorithm. Complete the algorithm, applying each subsequent step correctly. What do you conclude?

61. Use the results of Problems 58 through 60 to make a conjecture about using the wrong column as the work column.

EXPLORING WITH TECHNOLOGY

62. Many microcomputer centers have easy-to-use software package for solving linear programming problems. Experiment with such a package by using it to solve the linear programming problems defined by System (9) at the beginning of this section and Examples 1 and 2.

63. Use a linear programming software package to solve Problems 11 through 34.

REVIEWING MATERIAL

64. (Section 1.4) Graph the equation

$$y = \frac{1}{x^2}$$

65. (Section 2.3) The following table summarizes data collected from police reports and surveying records relating the width of roads (measured in feet) to the number of accidents on those roads (in a year). Find the equation of the least-squares straight line that fits this data and then use the result to estimate the number of accidents expected each year on a road that measures 45 feet in width.

x (road width)	25	30	50	60	70
y (number of accidents)	102	92	47	35	8

RECOMMENDING ACTION

66. Respond by memo to the following request:

MEMORANDUM

To: J. Doe Reader

From: Accounting

Date: Today

Subject: **Slack Variables**

I have received a proposed production schedule that someone in your department recommended to our manufacturing unit. It is based on a simplex algorithm analysis. I am concerned because this solution includes a positive slack variable. Is it possible for an optimal solution to include a positive slack variable? It seems to me that our best production schedule would use all our resources and require all slack variables to be zero.

3.6 THE ENHANCED SIMPLEX ALGORITHM

In the previous two sections we initialized and solved linear programming models in standard form using the basic simplex algorithm. Many linear programming models, however, are not in standard form, and to solve them we must either change the model or enhance the basic algorithm.

A linear programming model is not in standard form if an inequality of the $\leq$ type has a negative number on the right side. An example of such an inequality is

$$x_1 - 2x_2 - 3x_3 \leq -10$$

We handle inequalities of this type by applying a *preprocessing* step to the initial simplex tableau before applying the basic simplex algorithm. The initialization process remains as described in Section 3.4: a slack variable is added to the left side of each $\leq$ inequality, thereby converting each inequality into an equality, and a simplex tableau is created. Now, however, the last column of the simplex tableau contains negative entries, so we add the following step.

The enhanced simplex algorithm is a combination of the preprocessing step, repeated as often as necessary to remove unwanted negative elements in the last column of a simplex tableau, followed by the basic simplex algorithm.

PREPROCESSING STEP:

Identify the most negative entry in the last column of the simplex tableau, ignoring the last row. The row containing that entry is designated as the work row. Designate the most negative element in the work row, ignoring the last column, as the pivot. If choices exist for either the work row or the pivot, then those choices may be made arbitrarily. If the pivot is not one initially, use the second elementary row operation to convert the pivot to one. Then use the third elementary row operation to transform all other elements in the column containing the pivot to zero.

The preprocessing step is repeated as often as necessary until there are no negative entries remaining in the last column (ignoring the last row). After each application of the preprocessing step, we replace the basic variable appearing in the same row as the pivot and in front of the augmented matrix by the variable appearing in the same column as the pivot and above the augmented matrix, just as we did in the basic simplex algorithm.

If the preprocessing step succeeds in removing all negative elements from the last column of a simplex tableau (still ignoring the last row), then the basic simplex algorithm is applied to the resulting tableau. The combination of the preprocessing step with the basic simplex algorithm is the *enhanced simplex algorithm.*

The preprocessing step cannot be implemented when a work row contains only non-negative elements except in the last column, because then there is no negative element to serve as a pivot. In such a situation, the linear programming model has no solution and the enhanced simplex algorithm is terminated.

EXAMPLE 1

Use the enhanced simplex algorithm to solve

$$\text{Maximize:} \quad z = -x_1 + 2x_2 - x_3$$

$$\text{subject to:} \quad x_1 + x_2 \quad\quad \leq 10$$

$$-x_2 - x_3 \leq -5$$

$$\text{assuming:} \quad x_1, x_2, \text{ and } x_3 \text{ are nonnegative}$$

Solution

This linear programming model is in standard form except for the negative right side of the second inequality, so we apply the initialization process described in Section 3.4. We add slack variables s_1 and s_2 to the left sides of the two $\leq$ inequalities, changing each into an equality. Thus, the constraints become

$$x_1 + x_2 \quad\quad + s_1 \quad\quad = 10$$

$$-x_2 - x_3 \quad\quad + s_2 = -5$$

These slack variables, with zero coefficients, are also added to the objective function, which we rewrite as

$$x_1 - 2x_2 + x_3 + 0s_1 + 0s_2 + z = 0$$

and then construct the initial simplex tableau

$$
\begin{array}{c}
\quad\quad x_1 \quad x_2 \quad x_3 \quad s_1 \quad s_2 \quad z \\
\begin{array}{c} s_1 \\ s_2 \\ \\ \end{array}
\left[
\begin{array}{cccccc|c}
1 & 1 & 0 & 1 & 0 & 0 & 10 \\
0 & -1 & -1 & 0 & 1 & 0 & -5 \\
\hline
1 & -2 & 1 & 0 & 0 & 1 & 0
\end{array}
\right]
\end{array}
$$

The preprocessing step is designed to generate an initial solution that is feasible.

If we attempt to read the current solution from this simplex tableau by setting each basic variable in front of the tableau equal to the entry in the same row and last column, we obtain $s_1 = 10$ and $s_2 = -5$, *which is not feasible*, because slack variables are presumed to be nonnegative. The preprocessing step is designed to generate an initial solution that is feasible.

There is one negative element in the last column of the initial simplex tableau, namely, -5, and since that entry appears in the second row, the second row becomes the work row. The most negative element in the work row, ignoring the last column, is -1, and it appears twice, so we have a choice for the pivot. We arbitrarily choose the -1 element in the x_2-column as the pivot. Thus, x_2, the variable in the same column as the pivot, will replace s_2, the variable in the same row as the pivot and in front of the augmented matrix, as a basic variable. Using the second elementary row operation to transform the pivot to one and then the third elementary row operation to transform all other elements in the column containing the pivot to zero, we obtain

$$
\begin{array}{c c}
 & \begin{array}{c c c c c c} x_1 & x_2 & x_3 & s_1 & s_2 & z \end{array} \\
\begin{array}{c} s_1 \\ x_2 \\ \end{array} &
\left[\begin{array}{c c c c c c | c}
1 & 0 & -1 & 1 & 1 & 0 & 5 \\
0 & 1 & 1 & 0 & -1 & 0 & 5 \\
\hline
1 & 0 & 3 & 0 & -2 & 1 & 10
\end{array} \right]
\end{array}
$$

There are no negative elements in the last column of this simplex tableau, so the preprocessing step is complete. The current solution is $s_1 = 5$, $x_2 = 5$ with $s_2 = x_1 = 0$ and $z = 10$. This solution is feasible, but it is not optimal, because the last row, ignoring the last column, has a negative entry. If we apply the basic simplex algorithm to this tableau, we generate after one iteration the simplex tableau

$$
\begin{array}{c c}
 & \begin{array}{c c c c c c} x_1 & x_2 & x_3 & s_1 & s_2 & z \end{array} \\
\begin{array}{c} s_2 \\ x_2 \\ \end{array} &
\left[\begin{array}{c c c c c c | c}
1 & 0 & -1 & 1 & 1 & 0 & 5 \\
1 & 1 & 0 & 1 & 0 & 0 & 10 \\
\hline
3 & 0 & 1 & 2 & 0 & 1 & 20
\end{array} \right]
\end{array}
$$

The new solution is $s_2 = 5$, $x_2 = 10$ with $s_1 = x_1 = 0$ and $z = 20$, which is optimal.

EXAMPLE 2

Use the enhanced simplex algorithm to solve

$$\text{Maximize:} \quad z = x_1 + x_2$$

$$\text{subject to:} \quad \begin{aligned} x_1 + x_2 &\le 125 \\ -4x_1 - 5x_2 &\le -500 \\ -2x_1 - x_2 &\le -200 \end{aligned}$$

$$\text{assuming:} \quad x_1 \text{ and } x_2 \text{ are nonnegative}$$

Solution

This linear programming model is in standard form except for the negative right sides of the second and third inequalities, so we apply the initialization process described in Section 3.4. We add slack variables s_1, s_2, and s_3 to the left sides of the three $\le$ inequalities, changing each into an equality. Thus, the constraints become

$$
\begin{aligned}
x_1 + x_2 + s_1 &= 125 \\
-4x_1 - 5x_2 + s_2 &= -500 \\
-2x_1 - x_2 + s_3 &= -200
\end{aligned}
$$

These slack variables, with zero coefficients, are also added to the objective function, which we rewrite as

$$-x_1 - x_2 + 0s_1 + 0s_2 + 0s_3 + z = 0$$

We then construct the simplex tableau

$$
\begin{array}{c}
\begin{array}{cccccc}
x_1 & x_2 & s_1 & s_2 & s_3 & z
\end{array} \\
\begin{array}{c}
s_1 \\ s_2 \\ s_3 \\ \\
\end{array}
\left[
\begin{array}{cccccc|c}
1 & 1 & 1 & 0 & 0 & 0 & 125 \\
-4 & -5 & 0 & 1 & 0 & 0 & -500 \\
-2 & -1 & 0 & 0 & 1 & 0 & -200 \\
\hline
-1 & -1 & 0 & 0 & 0 & 1 & 0
\end{array}
\right]
\end{array}
$$

The most negative element in the last column of this simplex tableau, ignoring the last row, is -500 which appears in the second row, hence the second row becomes the work row. The most negative element in the work row, ignoring the last column, is -5, and it becomes the pivot. Thus x_2, the variable in the same column as the pivot, will replace s_2, the variable in the same row as the pivot and in front of the augmented matrix, as a basic variable. Using the second elementary row operation to transform the pivot to one and then the third elementary row operation to transform all other elements in the column containing the pivot to zero, we obtain

$$
\begin{array}{c}
\begin{array}{cccccc}
x_1 & x_2 & s_1 & s_2 & s_3 & z
\end{array} \\
\begin{array}{c}
s_1 \\ x_2 \\ s_3 \\ \\
\end{array}
\left[
\begin{array}{cccccc|c}
\frac{1}{5} & 0 & 1 & \frac{1}{5} & 0 & 0 & 25 \\
\frac{4}{5} & 1 & 0 & -\frac{1}{5} & 0 & 0 & 100 \\
-\frac{6}{5} & 0 & 0 & -\frac{1}{5} & 1 & 0 & -100 \\
\hline
-\frac{1}{5} & 0 & 0 & -\frac{1}{5} & 0 & 1 & 100
\end{array}
\right]
\end{array}
$$

There is still a negative entry in the last column of this simplex tableau, so the pre-processing step is repeated. The only negative entry in the last column of the simplex tableau, ignoring the last row, is -100, which appears in the third row, so the third row becomes the new work row. The most negative element in the work row, ignoring the last column, is $-\frac{6}{5}$, which becomes the pivot. Thus, x_1, the variable in the same column as the pivot will replace s_3, the variable in the same row as the pivot and in front of the augmented matrix, as a basic variable. Using the second elementary row operation to transform the pivot to one and the third elementary row operation to transform all other elements in the column containing the pivot to zero, we obtain

$$
\begin{array}{c}
\begin{array}{cccccc}
x_1 & x_2 & s_1 & s_2 & s_3 & z
\end{array} \\
\begin{array}{c}
s_1 \\ x_2 \\ x_1 \\ \\
\end{array}
\left[
\begin{array}{cccccc|c}
0 & 0 & 1 & \frac{1}{6} & \frac{1}{6} & 0 & 8\frac{1}{3} \\
0 & 1 & 0 & -\frac{1}{3} & \frac{2}{3} & 0 & 33\frac{1}{3} \\
1 & 0 & 0 & \frac{1}{6} & -\frac{5}{6} & 0 & 83\frac{1}{3} \\
\hline
0 & 0 & 0 & -\frac{1}{6} & -\frac{1}{6} & 1 & 116\frac{2}{3}
\end{array}
\right]
\end{array}
$$

There are no negative elements in the last column of this simplex tableau, so the pre-processing step is complete. The current solution is $s_1 = 8\frac{1}{3}$, $x_2 = 33\frac{1}{3}$, $x_1 = 83\frac{1}{3}$ with $s_2 = x_3 = 0$ and $z = 116\frac{2}{3}$. This solution is feasible but not optimal, because the last row, ignoring the last column, has a negative entry.

We now apply the basic simplex algorithm to this tableau, and we have a choice for our first work column. If we select the s_2 column as the work column, then we find as the optimal solution $x_1 = 75$, $x_2 = 50$, $s_2 = 50$, $s_1 = s_3 = 0$ with $z = 125$. Alternatively, if we select the s_3 column as the work column, we also have a choice for the pivot. Arbitrarily using the element in the first row and the s_3 column as the pivot, we find as the optimal solution $x_1 = 125$, $x_2 = 0$, $s_3 = 50$, $s_1 = s_2 = 0$ with $z = 125$. Thus, we have identified two solutions to this problem, each with $z = 125$. As we noted at the end of Section 3.5, the simplex method may not produce all the solutions when more than one exist. The simplex method, however, always locates one solution that optimizes the objective function when such a solution exists.

A linear programming model is *not* in standard form if it contains an inequality of the $\geq$ type. Recall that the sense of an inequality is reversed by multiplying the inequality by -1. Consequently, we can write

$$x_1 \geq 5 \qquad \text{as} \qquad -x_1 \leq -5$$

and

$$x_1 + 8x_2 + 3x_3 - x_4 \geq 100$$

as

$$-x_1 - 8x_2 - 3x_3 + x_4 \leq -100$$

An inequality of the $\geq$ type is changed to one of the $\leq$ type by multiplying the inequality by -1.

An inequality of the $\geq$ type is changed to one of the $\leq$ type by multiplying the inequality by -1. The new inequalities may have negative right sides; if so, the preprocessing step is activated.

Furthermore, a linear programming model is *not* in standard form if a constraint is modeled by an equality. Note that the equality $a = b$ is equivalent to the *pair* of inequalities $a \leq b$ and $a \geq b$. Thus we can convert an equality into a pair of inequalities by replacing the equality first with a $\leq$ inequality and also with a $\geq$ inequality. For example, we replace

$$x_1 - 2x_2 + 3x_3 = 80$$

A constraint modeled by an equality of the form $a = b$ is replaced by the two inequalities $a \leq b$ and $a \geq b$.

by the pair of inequalities

$$x_1 - 2x_2 + 3x_3 \leq 80$$

and

$$x_1 - 2x_2 + 3x_3 \geq 80$$

This last inequality is then multiplied by -1 to transform it into

$$-x_1 + 2x_2 - 3x_3 \leq -80$$

which is an inequality of the $\leq$ type.

EXAMPLE 3

Construct an initial simplex tableau for the linear programming model

$$\text{Maximize:} \quad z = x_1 + 5x_2 + 4x_3$$

$$\text{subject to:} \quad x_1 + x_2 + x_3 = 200$$

$$x_1 \geq 30$$

$$\text{assuming:} \quad x_1, x_2, \text{ and } x_3 \text{ are nonnegative}$$

Solution
We replace the equality

$$x_1 + x_2 + x_3 = 200$$

by the pair of inequalities

$$x_1 + x_2 + x_3 \leq 200$$

and

$$x_1 + x_2 + x_3 \geq 200$$

and then multiply the second of these two inequalities by -1 to obtain

$$-x_1 - x_2 - x_3 \leq -200$$

In a similar fashion, we multiply the inequality $x_1 \geq 30$ by -1 to obtain

$$-x_1 \leq -30$$

The model becomes

$$\text{Maximize:} \quad z = x_1 + 5x_2 + 4x_3$$

$$\text{subject to:} \quad x_1 + x_2 + x_3 \leq 200$$
$$-x_1 - x_2 - x_3 \leq -200$$
$$-x_1 \leq -30$$

$$\text{assuming:} \quad x_1, x_2, \text{ and } x_3 \text{ are nonnegative}$$

Now all constraints are modeled by inequalities of the $\leq$ type, and we can apply the initialization procedure described in Section 3.4.

We add the slack variables s_1, s_2, and s_3 to the left sides of the three inequalities, changing each into an equality. These slack variables, with zero coefficients, are also added to the objective function, and the model is reformulated as

$$\text{Maximize } z: \quad -x_1 - 5x_2 - 4x_3 + 0s_1 + 0s_2 + 0s_3 + z = 0$$

$$\text{subject to:} \quad x_1 + x_2 + x_3 + s_1 = 200$$
$$-x_1 - x_2 - x_3 + s_2 = -200$$
$$-x_1 + s_3 = -30$$

$$\text{assuming:} \quad \text{all variables are nonnegative}$$

The corresponding simplex tableau is

$$
\begin{array}{c}
\\
s_1 \\
s_2 \\
s_3 \\
\\
\end{array}
\begin{array}{c}
\begin{array}{ccccccc}
x_1 & x_2 & x_3 & s_1 & s_2 & s_3 & z
\end{array} \\
\left[
\begin{array}{ccccccc|c}
1 & 1 & 1 & 1 & 0 & 0 & 0 & 200 \\
-1 & -1 & -1 & 0 & 1 & 0 & 0 & -200 \\
-1 & 0 & 0 & 0 & 0 & 1 & 0 & -30 \\
\hline
-1 & -5 & -4 & 0 & 0 & 0 & 1 & 0
\end{array}
\right]
\end{array}
\qquad (30)
$$

EXAMPLE 4

Solve the linear programming model given in Example 3.

Solution

The initial simplex tableau for this sytem is Tableau (*30*). Since the last column (ignoring the last row) contains negative entries, we apply the preprocessing step prior to activating the basic simplex algorithm.

The most negative number in the last column of Tableau (*30*) is -200, which appears in the second row; thus row two becomes the work row. The most negative element in the work row, ignoring the last column, is -1, which appears in each of the first three columns, so we have a choice for the pivot. We arbitrarily choose the -1 element in the x_1 column as the pivot. Thus, x_1, the variable in the same column as the pivot, will replace s_2, the variable in the same row as the pivot and in front of the augmented matrix, as a basic variable. Using the second elementary row operation to transform the pivot to one and then the third elementary row operation to transform all other elements in the column containing the pivot to zero, we obtain

$$
\begin{array}{c c}
 & \begin{array}{c c c c c c c} x_1 & x_2 & x_3 & s_1 & s_2 & s_3 & z \end{array} \\
\begin{array}{c} s_1 \\ x_1 \\ s_3 \\ {} \end{array} &
\left[
\begin{array}{c c c c c c c|c}
0 & 0 & 0 & 1 & 1 & 0 & 0 & 0 \\
1 & 1 & 1 & 0 & -1 & 0 & 0 & 200 \\
0 & 1 & 1 & 0 & -1 & 1 & 0 & 170 \\
\hline
0 & -4 & -3 & 0 & -1 & 0 & 1 & 200
\end{array}
\right]
\end{array}
$$

There are no negative elements in the last column of this simplex tableau, so the preprocessing step is complete. The current solution is $s_1 = 0$, $x_1 = 200$, $s_3 = 170$ with $x_2 = x_3 = s_2 = 0$ and $z = 200$. This solution is feasible, but it is not optimal because the last row (ignoring the last column) has negative entries. Applying the basic simplex algorithm to this tableau, we find after two iterations that the optimal solution is $x_1 = 30$, $x_2 = 170$, $x_3 = s_1 = s_2 = s_3 = 0$ with $z = 880$.

Minimize an objective function by maximizing its negative.

A linear programming model is not in standard form if its objective function must be minimized. We reformulate a minimization problem as a maximization problem by noting that the minimum of z occurs at the same points where $-z$ is maximized. In particular, consider the linear programming model

$$\text{Minimize:} \quad z = 8x_1 + 5x_2$$

$$
\text{subject to:} \quad
\begin{aligned}
x_1 + x_2 &\le 2 \\
-2x_1 - x_2 &\le 1 \\
-x_2 &\le 3
\end{aligned}
$$

This linear programming model has just two decision variables and can be solved graphically. The feasible region is shaded in Figure 3.20, it has corner points at $(-3, 5)$, $(5, -3)$ and $(1, -3)$. If we evaluate z and $-z$ at each corner point, we generate Table 3.12.

The minimun value of the objective function is $z = -7$ at the corner point $x_1 = 1$ and $x_2 = -3$; the maximum value of the objective function after it is multiplied by -1 is 7, again at $x_1 = 1$ and $x_2 = -3$. The minimum value of z and the maximum value of $-z$ occur at the same corner point and differ only by a sign.

In general, we minimize z by maximizing $Z = -z$. The values of the decision variables

FIGURE 3.20

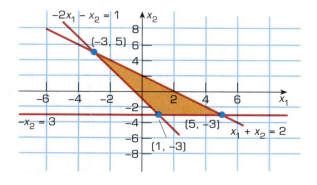

TABLE 3.12

CORNER POINTS		OBJECTIVE FUNCTION		
x_1	x_2	$z = 8x_1 + 5x_2$		$-z = -8x_1 - 5x_2$
-3	5	$z = 8(-3) + 5(5) =$	1	$-z = -8(-3) - 5(5) = -1$
5	-3	$z = 8(5) + 5(-3) =$	25	$-z = -8(5) - 5(-3) = -25$
1	-3	$z = 8(1) + 5(-3) =$	-7	$-z = -8(1) - 5(-3) = 7$

that maximize $-z$ are the same values that minimize z, with the optimal value of z being the negative of the optimal value of $-z$. In particular, we reformulate the objective function

$$\text{Minimize: } z = 20x_1 - 10x_2 + 30x_3$$

as

$$\text{Maximize: } Z = -20x_1 + 10x_2 - 30x_3$$

We reformulate the objective function

$$\text{Minimize: } z = x_1 + x_2 + 2x_3 + 3x_4$$

as

$$\text{Maximize: } Z = -x_1 - x_2 - 2x_3 - 3x_4$$

EXAMPLE 5

Construct an initial simplex tableau for the linear programming model

$$\text{Minimize: } \quad z = 20x_1 - 10x_2 + 30x_3$$

$$\text{subject to:} \quad 5x_1 + 10x_2 + 15x_3 \geq 150$$

$$-2x_1 - 2x_2 + 6x_3 \geq 12$$

$$\text{assuming:} \quad x_1, x_2, \text{ and } x_3 \text{ are nonnegative}$$

Solution

Instead of minimizing z, we maximize $Z = -z$. We also multiply both inequalities by -1 to convert them into inequalities of the $\leq$ type. The linear programming model is reformulated as

$$\text{Maximize: } \quad Z = -20x_1 + 10x_2 - 30x_3$$

$$\text{subject to:} \quad -5x_1 - 10x_2 - 15x_3 \leq -150$$

$$2x_1 + 2x_2 - 6x_3 \leq -12$$

$$\text{assuming:} \quad x_1, x_2, \text{ and } x_3 \text{ are nonnegative}$$

Now the problem involves a maximization with constraints modeled by inequalities of the $\leq$ type.

Using the initialization procedure described in Section 3.4, we add the slack variables s_1 and s_2 to the left sides of the two inequalities, changing them into equalities. These slack variables, with zero coefficients, are also added to the new objective function, which is then rearranged with all variables on one side of the equality sign and the coefficient of Z set to one. The linear programming model becomes

$$\text{Maximize } Z: \quad 20x_1 - 10x_2 + 30x_3 + 0s_1 + 0s_2 + Z = 0$$

$$\text{subject to:} \quad -5x_1 - 10x_2 - 15x_3 + s_1 \qquad\quad = -150$$

$$2x_1 + 2x_2 - 6x_3 \qquad\quad + s_2 = -12$$

$$\text{assuming:} \quad x_1, x_2, \text{ and } x_3 \text{ are nonnegative}$$

The corresponding simplex tableau is

$$
\begin{array}{c}
\\
s_1 \\
s_2 \\
\\
\end{array}
\begin{array}{cccccc}
x_1 & x_2 & x_3 & s_1 & s_2 & Z \\
\left[\begin{array}{cccccc|c}
-5 & -10 & -15 & 1 & 0 & 0 & -150 \\
2 & 2 & -6 & 0 & 1 & 0 & -12 \\
\hline
20 & -10 & 30 & 0 & 0 & 1 & 0
\end{array}\right]
\end{array}
\qquad (31)
$$

EXAMPLE 6

Solve the linear programming model given in Example 5.

Solution

The initial simplex tableau for this system is Tableau (31). Since the last column (ignoring the last row) contains negative entries, we apply the preprocessing step prior to activating the basic simplex algorithm.

The most negative number in the last column of Tableau (31) is -150, which appears in the first row; thus the first row becomes the work row. The most negative element in the work row (ignoring the last column) is -15, which becomes the pivot. Thus, x_3, the variable in the same column as the pivot, will replace s_1, the variable in the same row as the pivot and in front of the augmented matrix, as a basic variable. Using elementary row operations to transform the pivot element to one and all other elements in the column containing the pivot to zero, we obtain

$$
\begin{array}{c}
\\
x_3 \\
s_2 \\
\\
\end{array}
\begin{array}{cccccc}
s_1 & x_2 & x_3 & s_1 & s_2 & Z \\
\left[\begin{array}{cccccc|c}
\frac{1}{3} & \frac{2}{3} & 1 & -\frac{1}{15} & 0 & 0 & 10 \\
4 & 6 & 0 & -\frac{2}{5} & 1 & 0 & 48 \\
\hline
10 & -30 & 0 & 2 & 0 & 1 & -300
\end{array}\right]
\end{array}
$$

There are no negative elements in the last column of this simplex tableau (ignoring the last row) so the preprocessing step is complete. The current solution is $x_3 = 10$, $s_2 = 48$ with $x_1 = x_2 = s_1 = 0$ and $Z = -300$. This solution serves as our initial basic feasible solution, and with it we can apply the basic simplex algorithm. The last row, ignoring the last column, contains a negative element, namely, -30 in column 2, so the x_2 column becomes the new work column. The new pivot is 6. Using elementary row operations, we generate the new simplex tableau

$$
\begin{array}{c}
\\
x_3 \\
x_2 \\
\\
\end{array}
\begin{array}{cccccc}
x_1 & x_2 & x_3 & s_1 & s_2 & Z \\
\left[\begin{array}{cccccc|c}
-\frac{1}{9} & 0 & 1 & -\frac{1}{45} & -\frac{1}{9} & 0 & \frac{14}{3} \\
\frac{2}{3} & 1 & 0 & -\frac{1}{15} & \frac{1}{6} & 0 & 8 \\
\hline
30 & 0 & 0 & 0 & 5 & 1 & -60
\end{array}\right]
\end{array}
$$

The new solution, $x_3 = 14/3$, $x_2 = 8$ with $x_1 = s_1 = s_2 = 0$ is optimal, because there are no negative entries in the last row (ignoring the last column). The corresponding value of the objective function is $Z = -60$, or, in terms of the original objective, $z = -Z = 60$.

We close by recalling a comment we made in the last section. Most of the work in the enhanced simplex algorithm involves the elementary row operations. Graphing calculators have menus for performing these operations and should be used whenever possible. In addition, there are a host of computer software packages that automate the entire algorithm and are even simpler to use. Professionals routinely use these packages to solve linear programming models of real-world systems.

IMPROVING SKILLS

In Problems 1 through 30, find (*a*) the initial simplex tableau for each linear programming model, (*b*) the result of applying the preprocessing step *once* to the initial simplex tableau, and then (*c*) the optimal solution.

1. Maximize: $z = 3x_1 + 2x_2$
   subject to: $60x_1 + 10x_2 \geq 240{,}000$
   $\qquad\qquad 2x_1 + 5x_2 \leq 47{,}200$
   assuming: x_1 and x_2 are nonnegative

2. Minimize: $z = 3x_1 + 2x_2$
   subject to: $60x_1 + 10x_2 \geq 240{,}000$
   $\qquad\qquad 2x_1 + 5x_2 \leq 47{,}200$
   assuming: x_1 and x_2 are nonnegative

3. Minimize: $z = 3x_1 + 2x_2$
   subject to: $60x_1 + 10x_2 \geq 240{,}000$
   $\qquad\qquad 2x_1 + 5x_2 \geq 47{,}200$
   assuming: x_1 and x_2 are nonnegative

4. Minimize: $z = 3x_1 + 2x_2$
   subject to: $\quad 60x_1 + 10x_2 \geq 240{,}000$
   $\qquad\qquad -2x_1 - 5x_2 \geq -47{,}200$
   assuming: x_1 and x_2 are nonnegative

5. Minimize: $z = 15x_1 + 30x_2$
   subject to: $0.4x_1 + 0.8x_2 \geq 2{,}240$
   $\qquad\qquad 0.6x_1 + 0.4x_2 \geq 1{,}440$
   assuming: x_1 and x_2 are nonnegative

6. Minimize: $z = 15x_1 + 30x_2$
   subject to: $0.4x_1 + 0.8x_2 = 2{,}240$
   $\qquad\qquad 0.6x_1 + 0.4x_2 \geq 1{,}440$
   assuming: x_1 and x_2 are nonnegative

7. Minimize: $z = 15x_1 + 30x_2$
   subject to: $0.4x_1 + 0.8x_2 = 2{,}240$
   $\qquad\qquad 0.6x_1 + 0.4x_2 \leq 1{,}440$
   assuming: x_1 and x_2 are nonnegative

8. Minimize: $z = 15x_1 + 30x_2$
   subject to: $0.4x_1 + 0.8x_2 \leq 2{,}240$
   $\qquad\qquad 0.6x_1 + 0.4x_2 \leq 1{,}440$
   assuming: x_1 and x_2 are nonnegative

9. Maximize: $z = 2x_1 + 7x_2$
   subject to: $x_1 - 3x_2 = 3$
   $2x_1 + 4x_2 \leq 18$
   assuming: x_1 and x_2 are nonnegative

10. Maximize: $z = 2x_1 + 7x_2$
    subject to: $x_1 - 3x_2 = 3$
    $2x_1 + 4x_2 = 18$
    assuming: x_1 and x_2 are nonnegative

11. Maximize: $z = 2x_1 + 7x_2$
    subject to: $x_1 - 3x_2 = 3$
    $2x_1 + 4x_2 \geq 18$
    assuming: x_1 and x_2 are nonnegative

12. Maximize: $z = 2x_1 + 7x_2$
    subject to: $x_1 - 3x_2 = -3$
    $2x_1 + 4x_2 \geq 18$
    assuming: x_1 and x_2 are nonnegative

13. Minimize: $z = 180x_1$
    subject to: $50x_1 + 27x_2 \geq 170$
    $8x_1 + 25x_2 \geq 80$
    assuming: x_1 and x_2 are nonnegative

14. Minimize: $z = 180x_1$
    subject to: $50x_1 + 27x_2 = 170$
    $8x_1 + 25x_2 \geq 80$
    assuming: x_1 and x_2 are nonnegative

15. Maximize: $z = 2x_1 - x_1$
    subject to: $x_1 - 2x_2 \geq -2$
    $-3x_1 + 4x_1 \geq -7$
    assuming: x_1 and x_2 are nonnegative

16. Maximize: $z = 35x_1 + 12x_2$
    subject to: $8x_1 + 11x_2 = 9{,}500$
    $7x_1 + 12x_2 \geq 8{,}000$
    $9x_1 + 9x_2 \leq 9{,}200$
    assuming: x_1 and x_2 are nonnegative

17. Maximize: $z = 15x_1 + 10x_2 + 14x_3$
    subject to: $6x_1 + 5x_2 + 3x_3 = 26$
    $4x_1 - 2x_2 - 5x_3 = -8$
    assuming: x_1, x_2, and x_3 are nonnegative

18. Minimize: $z = 15x_1 + 10x_2 + 14x_3$
    subject to: $6x_1 + 5x_2 + 3x_3 = 26$
    $4x_1 + 2x_2 + 5x_3 = 8$
    assuming: x_1, x_2, and x_3 are nonnegative

19. Minimize: $z = 15x_1 + 10x_2 + 14x_3$
    subject to: $6x_1 + 5x_2 + 3x_3 \geq 26$
    $4x_1 + 2x_2 + 5x_3 \geq 8$
    assuming: x_1, x_2, and x_3 are nonnegative

20. Maximize: $z = 3x_1 - 4x_2 + 5x_3$

subject to: $x_1 + 2x_2 + 2x_3 \le 180$
$-3x_1 + 4x_2 - 4x_3 \le -230$

assuming: $x_1, x_2,$ and x_3 are nonnegative

21. Maximize: $z = 5x_1 + 6x_2 + 7x_3 + 8x_4$

subject to: $3x_1 - 4x_2 + 5x_3 - 6x_4 \le 90$
$7x_1 + 7x_2 + 8x_3 - 9x_4 \le -135$

assuming: $x_1, x_2, x_3,$ and x_4 are nonnegative

22. Minimize: $z = 5x_1 + 6x_2 + 7x_3 + 8x_4$

subject to: $3x_1 - 4x_2 + 5x_3 - 6x_4 \le 90$
$7x_1 + 7x_2 + 8x_3 - 9x_4 \ge -135$

assuming: $x_1, x_2, x_3,$ and x_4 are nonnegative

23. Minimize: $z = 5x_1 + 6x_2 + 7x_3 + 8x_4$

subject to: $3x_1 - 4x_2 + 5x_3 - 6x_4 \ge 90$
$7x_1 + 7x_2 + 8x_3 - 9x_4 \ge 135$

assuming: $x_1, x_2, x_3,$ and x_4 are nonnegative

24. Maximize: $z = 15x_1 + 10x_2 + 14x_3$

subject to: $30x_1 + 50x_2 + 40x_3 = 80{,}000$
$10x_1 + 5x_2 + 7x_3 \le 14{,}000$
$4x_1 + 4x_2 + 6x_3 \ge 7{,}200$

assuming: $x_1, x_2,$ and x_3 are nonnegative

25. Minimize: $z = x_3$

subject to: $0.3x_1 + 0.4x_2 + 0.3x_3 = 9{,}000$
$0.2x_1 - 0.8x_2 \ge -5{,}500$
$0.9x_1 + 0.1x_3 \le 7{,}300$

assuming: $x_1, x_2,$ and x_3 are nonnegative

26. Minimize: $z = 15x_1 + 14x_3$

subject to: $30x_1 + 50x_2 + 40x_3 \le 80{,}000$
$10x_1 + 5x_2 + 7x_3 \ge 14{,}000$
$4x_1 - 4x_2 - 6x_3 \ge -7{,}200$
$14x_1 + 7x_2 + 10x_3 = 21{,}000$

assuming: $x_1, x_2,$ and x_3 are nonnegative

27. Minimize: $z = 9x_1 + 9x_2 + 1x_3$

subject to: $4x_1 + 4x_2 + 5x_3 \ge 0.03$
$2x_1 - 3x_2 - 8x_3 \ge -0.07$
$x_1 + x_2 + 3x_3 \ge 0.01$
$2x_1 + x_2 + x_3 = 0.08$
$4x_1 + 3x_2 + 5x_3 \le 0.12$

assuming: $x_1, x_2,$ and x_3 are nonnegative

28. Maximize: $z = 2x_1 + 3x_2 + 4x_3 + 6x_4$

subject to: $x_1 + x_2 + x_3 + x_4 = 15$
$7x_1 + 5x_2 + 3x_3 + 2x_4 = 120$
$3x_1 + 5x_2 + 10x_3 + 15x_4 \le 100$

assuming: $x_1, x_2, x_3,$ and x_4 are nonnegative

29. Maximize: $z = x_2 + 2x_4$
subject to:
$$2x_1 + 3x_1 + 3x_3 + 4x_4 = 20$$
$$3x_1 + 4x_2 + 4x_3 + 5x_4 = 35$$
$$5x_1 + 5x_2 + 4x_3 + 3x_4 \leq 35$$
$$3x_1 + 3x_2 + 2x_3 + x_4 \leq 20$$
assuming: $x_1, x_2, x_3,$ and x_4 are nonnegative

30. Minimize: $z = x_2 + 2x_4$
subject to:
$$2x_1 + 3x_2 + 3x_3 + 4x_4 = 20$$
$$3x_1 + 4x_2 - 4x_3 - 5x_4 = -35$$
$$5x_1 + 5x_2 + 4x_3 + 3x_4 \geq 35$$
$$3x_1 + 3x_2 + 2x_3 + x_4 \leq 20$$
assuming: $x_1, x_2, x_3,$ and x_4 are nonnegative

CREATING MODELS

In Problems 31 through 44, use the enhanced simplex algorithm to solve each problem.

31. Eagle Mining Enterprises owns two mines in West Virginia. The ore from each mine is separated into two grades before it is shipped. Mine I produces 4 tons of high-grade ore and 6 tons of low-grade ore daily with an operating cost of $8,000. Mine II produces 3 tons of high-grade ore and 10 tons of low-grade ore daily at a cost of $7,000 per day. The mining company has contracts that require it to deliver 160 tons of high-grade ore and 350 tons of low-grade ore each month. Determine the number of days each mine should operate to meet the company's contractual obligations at a minimum cost.

32. Solve Problem 31 if, in addition, mine I produces 4 tons of medium-grade ore each day, mine II produces 5 tons of medium-grade ore each day, and the company has contracts to deliver 216 tons of medium-grade ore each month.

33. Solve Problem 31 if, in addition, the company also has a third mine that costs $6,000 per day to operate and produces 2 tons of high-grade ore and 8 tons of low-grade ore daily.

34. A doctor recommends that a patient supplement his daily diet with at least 500 milligrams of vitamin C, 10,000 USP units of vitamin A, and 1,000 USP units of vitamin D. The patient has identified two granular vitamin supplements that are stirred into liquid. Each teaspoon of Supplement I contains 200 milligrams of vitamin C, 1,000 USP units of vitamin A, and 100 USP units of vitamin D and costs 5¢. Each teaspoon of Supplement II contains 50 milligrams of vitamin C, 2,500 USP units of vitamin A, and 300 USP units of vitamin D and costs 8¢. How many teaspoons of each supplement should the patient use each day if the goal is to meet the doctor's recommendations at minimum cost?

35. Solve Problem 34 if, in addition, the patient is also willing to use a third supplement, each teaspoon of which provides 100 milligrams of vitamin C, 2,000 USP units of vitamin A, and 150 USP units of vitamin D and costs 6¢ per teaspoon.

36. A caterer has in stock 1,200 gallons of a premixed fruit juice that contains 60% orange juice and 40% grapefruit juice and 2,000 gallons of a premixed fruit drink containing 5% orange juice, 10% grapefruit juice, 8% cranberry juice, and 77% filler (sugar, water, and flavorings). The fruit juice costs $1 per gallon, while the fruit drink costs

25¢ per gallon. From this stock, the caterer is to fulfill a contract for 1,000 gallons of fruit punch for a corporation picnic. The punch must contain at least 10% orange juice, 10% grapefruit juice, and 2% cranberry juice. How much of each stock item should be used to make the punch at the lowest possible cost?

37. The manager of a supermarket meat department finds that she has 160 pounds of round steak, 600 pounds of chuck steak, and 300 pounds of pork in stock on Saturday morning. From experience, she knows that she can sell half these quantities as straight cuts at a high profit. The remaining meat will be ground and combined into hamburger meat and picnic patties for which there is a large weekend demand that exceeds existing supplies. Each pound of hamburger meat must contain at least 20% ground round and 60% ground chuck. Each pound of picnic patties must contain at least 30% ground pork and 50% ground chuck. The remainder of each product can consist of an inexpensive nonmeat filler that the store has in large supply. The profit on each pound of hamburger meat and each pound of picnic patties is 40¢. How many pounds of each product should be made if the objective is to maximize total profit on these items?

38. A can of dog food, which is guaranteed to contain at least 10 units of protein, 20 units of mineral matter, and 6 units of fat, is composed of a blend of four different ingredients. Ingredient I contain 10 units of protein, 2 units of mineral matter, and $\frac{1}{2}$ unit of fat per ounce. Ingredient II contains 1 unit of protein, 40 units of mineral matter, and 3 units of fat per ounce. Ingredient III contains 1 unit of protein, 1 unit of mineral matter, and 6 units of fat per ounce. Ingredient IV contains 5 units of protein, 10 units of mineral matter, and 3 units of fat per ounce. The cost of each ingredient is 2¢, 2¢, 1¢, and 3¢ per ounce, respectively. How many ounces of each ingredient should be used to minimize the cost of the dog food yet still meet the guaranteed composition?

39. A local rent-a-car company needs to purchase at least 25 new cars, which can be either compacts of midsize cars. Compacts cost $10,000 each, midsize cars cost $15,000 each, and the company has $300,000 reserved for purchases. From experience, the company estimates that yearly operating costs will be $2,000 for each compact and $2,300 for each midsize car. How many cars of each type should the company buy if it wants to minimize total operating costs?

40. An investor has identified three attractive stocks and will divide $10,000 in available funds among the three. The first is low-risk and returns 4% per year, the second is medium-risk and returns 6% per year, and the last is high-risk and returns 9% per year. At least $6,000 must be invested in low- and medium-risk stocks, and the amount invested in high-risk stocks must be no more than $2,000 more than the amount invested in low-risk stocks. How much stock of each type should the investor purchase to maximize total return?

41. A publisher of cookbooks, garden books, and self-help books will limit the list of new titles to at most 24 next year. To guarantee diversity, the policy of the publisher is to bring out at least four new titles in each group each year. Furthermore, the number of new cookbook titles must be at least as great as all other new titles combined. On average, each new cookbook title generates $8,500 in profits, each new garden book title generates $8,000 in profits, and each new self-help book generates $9,000 in profits. How many new titles in each category should the publisher produce to maximize profits?

42. A deodorant company's inventory contains 900,000 ounces of active ingredients that are used to produce two different products, regular and super. Each carton of regular deodorant uses 30 oz of the active ingredient and costs $2 to produce, while each carton of super deodorant uses 40 oz of the active ingredient and costs $2.50 to produce. To justify start-up costs for any production run, the company must produce

at least 30,000 cartons of deodorant. Furthermore, the company always produces at least three times as much super as regular because of market demand. How many cartons of each type should the company produce to minimize its costs?

43. A car rental company allows customers to return cars to any of its offices, so there is sometimes an uneven distribution of cars at the various locations. Currently, location A has a surplus of 15 cars, location B has a surplus of 22 cars, location C has a shortage of 18 cars, and location D has a shortage of 19 cars. Transportation costs for shipping cars between these locations are summarized in the following table:

	LOCATION C	LOCATION D
LOCATION A	$13	$21
LOCATION B	$18	$25

Determine a shipping plan that will minimize the total cost of transporting cars from where they are in surplus to where they are needed. *Hint*: Let

$$x_1 = \text{number of cars shipped from A to C}$$

$$x_2 = \text{number of cars shipped from A to D}$$

$$x_3 = \text{number of cars shipped from B to C}$$

$$x_4 = \text{number of cars shipped from B to D}$$

and write separate constraint inequalities to model the shipments of cars out of location A, the shipments of cars out of location B, the shipments of cars into location C, and the shipments of cars into location D.

44. A distributor with warehouses in New York and Atlanta must supply stores in New Orleans, Omaha, and Cleveland with chocolate for their candy counters. The distributor has 200 pounds of chocolate stored at its New York facility and 350 pounds at its Atlanta facility. The store in New Orleans has ordered 100 pounds of chocolate, the store in Omaha, 150 pounds, and the store in Cleveland, 125 pounds. Delivery costs (in cents per pound) from each warehouse to each store are summarized in the following table.

	TO NEW ORLEANS	TO OMAHA	TO CLEVELAND
FROM NEW YORK	25	20	10
FROM ATLANTA	10	15	30

Determine a shipping plan that will satisfy all store needs at minimum cost. *Hint*: Let

$$x_1 = \text{number of pounds shipped from New York to New Orleans}$$

$$x_2 = \text{number of pounds shipped from New York to Omaha}$$

$$x_3 = \text{number of pounds shipped from New York to Cleveland}$$

$$x_4 = \text{number of pounds shipped from Atlanta to New Orleans}$$

$$x_5 = \text{number of pounds shipped from Atlanta to Omaha}$$

$$x_6 = \text{number of pounds shipped from Atlanta to Cleveland}$$

and write separate constraint inequalities to model the shipments out of each warehouse and the shipments into each store.

EXPLORING IN TEAMS

45. Write the number -10 as the difference of two *nonnegative* numbers. How many ways can this be done? Do the same for -80.

46. Any variable y_1 that is unrestricted as to its sign can be expressed as the difference of two *nonnegative* variables by writing $y_1 = y_2 - y_3$ and restricting both y_2 and y_3 to be nonnegative. Use this observation to transform the following linear programming model into one in standard form.

$$\text{Maximize:} \quad z = 3x_1 + 2x_2$$
$$\text{subject to:} \quad 60x_1 + 10x_2 \le 240{,}000$$
$$2x_1 + 5x_2 \le 47{,}200$$
$$\text{assuming:} \quad x_1 \text{ and } x_2 \text{ are unrestricted}$$

47. Use the technique described in Problem 45 to transform the following linear programming model into one in standard form.

$$\text{Maximize:} \quad z = 3x_1 - 4x_2 + 5x_3$$
$$\text{subject to:} \quad x_1 + 2x_2 + 2x_3 \le 180$$
$$3x_1 + 4x_2 + 4x_3 \le 230$$
$$\text{assuming:} \quad x_1, x_2, \text{ and } x_3 \text{ are real numbers}$$

EXPLORING WITH TECHNOLOGY

48. Many microcomputer centers have software packages for solving linear programming problems. Most packages do not require a linear programming model to be in standard form and, therefore, do not use the preprocessing step. Instead, most linear programming software packages replace the preprocessing step with a more complex but efficient algorithm. However, such packages do require that the right side of each constraint be nonnegative, so any constraint that does not satisfy this condition is first multiplied by -1. Experiment with such a package by using it to solve Examples 1, 2, 4, and 6.

49. Use a linear programming software package to solve Problems 1 through 30.

REVIEWING MATERIAL

50. (Section 1.6) A manufacturer of wooden chair frames knows that the firm has a fixed cost of $118,800 per month and that labor and material costs for each chair frame are $45. Determine the break-even point if each chair frame sells for $100.

51. (Section 2.4) Calculate the product

$$\begin{bmatrix} 0 & 2 \\ -1 & 3 \\ -2 & 2 \end{bmatrix} \begin{bmatrix} 1 & -4 \\ 2 & -3 \end{bmatrix}$$

RECOMMENDING ACTION

52. Respond by memo to the following request:

> ### MEMORANDUM
>
> To: J. Doe Reader
>
> From: Manufacturing
>
> Date: Today
>
> Subject: **Noninteger Solutions**
>
> Some of our employees will be on vacation next week. When I used your linear programming model to schedule next week's production of grandfather clock frames and television cabinets with the reduced labor force (those not on vacation), I got noninteger answers. This makes no sense to me. I cannot schedule half a grandfather clock frame for assembly and decorating. Production schedules must be in integers. Is your model still useful? If so, how do I use it to obtain integer-valued production schedules?

CHAPTER 3   KEYS

KEY WORDS

basic simplex algorithm (p. 224)

basic variable (p. 204)

bounded region (p. 179)

corner point (p. 177)

decision variable (p. 199)

enhanced simplex algorithm (p. 231)

feasible region (p. 176)

hidden conditions (p. 164)

initial basic feasible solution (p. 203)

linear inequality (p. 172)

linear objective function (p. 182)

nonbasic variable (p. 204)

pivot (p. 213)

preprocessing step (p. 231)

reduced column (p. 215)

simplex tableau (p. 204)

slack variable (p. 200)

standard form (p. 200)

strict inequality (p. 163)

superfluous constraint (p. 191)

work column (p. 213)

work row (p. 231)

KEY CONCEPTS

3.1 Modeling with Inequalities

■ The statement ''X cannot exceed Y'' is modeled as $X \leq Y$; the statement ''X is at least as great as Y'' is modeled as $X \geq Y$.

- The production, shipment, or consumption of important components in a process is limited by the availability of resources, and these limitations are often modeled by inequalities.

- Hidden constraints are constraints that are not explicitly stated, and they often include nonnegativity conditions on important components.

- Adding or subtracting a number from both sides of an inequality does not change the sense of the inequality.

- Multiplying or dividing both sides of an inequality by a negative number changes the sense of the inequality; multiplying or dividing by a positive number does not change the sense of the inequality.

3.2 Graphing Linear Inequalities

- Every straight line of the form $ax + by = c$ divides the plane into two regions, one on each side of the line. The points in one region satisfy the inequality $ax + by < c$, while the points in the other region satisfy the inequality $ax + by > c$.

- A solution to a system of linear inequality constraints is a point whose coordinates simultaneously satisfy all the inequalities in the system. The set of all such solutions is the feasible region for the system.

- A corner point of a feasible region is a point on the boundary of the feasible region where two lines bounding the region intersect.

3.3 Optimizing with Graphs

- A linear programming problem involves optimizing a linear objective function subject to constraints modeled by linear equalities or linear inequalities.

- The maximum and minimum of a linear programming model occur at corner points of the feasible region when the feasible region is nonempty and bounded.

3.4 Initializing the Simplex Method

- A linear programming model is in standard form if (1) all decision variables are nonnegative, (2) the objective function is to be maximized, (3) all the constraints are of the $\leq$ type, and (4) the right side of each constraint is nonnegative.

- A slack variable is a nonnegative quantity that is added to the left side of a $\leq$ inequality to transform the inequality into an equality.

3.5 The Basic Simplex Algorithm

- The basic simplex algorithm is a matrix procedure that is applied directly to simplex tableaus to solve linear programming models in standard form.

- The basic simplex algorithm identifies one optimal solution, even when many equally optimal solutions exist.

3.6 The Enhanced Simplex Algorithm

- The preprocessing step identifies an initial feasible solution for a linear programming model that meets all the criteria for standard form, except that some or all of the right sides of the inequalities are negative.

- An inequality of the $\geq$ type is transformed into an inequality of the $\leq$ type by multiplying the inequality by -1.

- A constraint modeled by an equality of the form $a = b$ is replaced by the two inequalities $a \leq b$ and $a \geq b$. The second inequality is then multiplied by -1 and rewritten as $-a \leq -b$ to transform it into an inequality of $\leq$ type.

- A minimum problem is converted to a maximum problem by changing all the signs in the objective function.

- The enhanced simplex algorithm consists of the preprocessing step followed by the basic simplex algorithm. This matrix procedure is applied directly to the simplex tableau associated with a linear programming model in nonstandard form.

KEY PROCEDURES

- To determine the feasible region that satisfies the single inequality $ax + by \leq c$ or $ax + by \geq c$:

   Step 1   Graph the straight line defined by $ax + by = c$.

   Step 2   Pick a point on one side of the line as a test point.

   Step 3   Determine by direct substitution whether the coordinates of the test point satisfy the given inequality. If the coordinates of the test point satisfy the given inequality, then the feasible region is the set of all points on the side of the line containing the test point; if not, the feasible region is the other side of the line.

- To determine the feasible region for a system of linear inequalities in two variables, locate the feasible regions for each inequality in the system, one at a time, and then identify the region common to all of them. If there is no region common to all the constraints, then that system has no solution.

- To solve a linear programming problem in two variables graphically:

   Step 1   Find the feasible region that satisfies the constraints. If there is no feasible region, stop: the problem has no solution; otherwise, continue.

   Step 2   Determine whether the feasible region is bounded. If not, stop: this method is not applicable; otherwise, continue.

   Step 3   Locate the coordinates of all corner points.

   Step 4   Substitute the coordinates of each corner point into the objective function and determine the corner point(s) that optimizes the objective function.

- A simplex tableau can be constructed as follows for linear programming models meeting all the criteria of standard form, with the exception that the right sides of inequality constraints of the $\leq$ type can be negative.

   Step 1   Add a slack variable to the left side of each $\leq$ inequality, thereby converting the inequaity to an equality. Add to the objective function each slack variable with a zero coefficient.

   Step 2   Write the objective function so that all the variables appear on the left side of the equation, with the coefficient of z set to one, and the right side of the equation equal to zero.

Step 3 Construct an augmented matrix for the constraint equations created by Step 1. Above each column write the variable associated with that column. Add to this matrix one additional column, just before the last column, to accommodate the z term; add one additional row for the objective function in the form specified by Step 2. Write each slack variable on the augmented matrix, on the line that represents the constraint equation in which it appears.

- The enhanced simplex algorithm is applied to simplex tableaus:

Preprocessing step (1) Identify the most negative entry in the last column of the simplex tableau, ignoring the element in the last row and last column, and designate the row in which this number appears as the work row. If more than one equally negative number exists, arbitrarily choose one.

Preprocessing step (2) Identify the most negative element in the work row, ignoring the last column, and designate that element as the pivot. If more than one equally negative number exists, arbitrarily choose one. If no negative element exists to serve as a pivot, stop: the linear programming model has no solution. Otherwise, continue.

Preprocessing step (3) Use the second elementary row operation to convert the pivot to one, if it is not already one, and then use the third elementary row operation to convert all other elements in the column containing the pivot to zero.

Preprocessing step (4) Replace the basic variable appearing in the same row as the pivot and in front of the augmented matrix portion of the simplex tableau with the variable appearing in the same column as the pivot and above the augmented matrix portion of the simplex tableau.

Preprocessing step (5) Repeat preprocessing steps (1) through (4) until there are no negative numbers in the last column of the current simplex tableau, ignoring the element in the last row and last column; otherwise, continue with the basic simplex algorithm.

Step 1 Locate the most negative number in the last row of the simplex tableau, ignoring the last column, and designate the column in which this number appears as the work column. If more than one equally negative number exists, arbitrarily choose one.

Step 2 Form ratios by dividing each positive element of the work column into the corresponding element in the same row and last column. If no element in the work column is positive, stop: the problem has no solution; otherwise, designate the element in the work column that yields the smallest ratio as the pivot. If more than one element yields the smallest ratio, arbitrarily choose one to be the pivot.

Step 3 Use the second elementary row operation to convert the pivot to one, if it is not already one, and then use the third elementary row operation to convert all other elements in the work column to zero.

Step 4　Replace the basic variable appearing in the same row as the pivot and in front of the augmented matrix portion of the simplex tableau with the variable appearing in the same column as the pivot and above the augmented matrix portion of the simplex tableau.

Step 5　Repeat Steps 1 through 4 until there are no negative numbers in the last row of the current simplex tableau, ignoring the element in the last row and last column.

Step 6　An optimal solution is obtained by setting each basic variable appearing in front of the augmented matrix equal to its corresponding element in the same row and last column of the augmented matrix and then setting all other variables equal to zero. The value of the objective function is the number appearing in the last row and last column of the simplex tableau.

TESTING YOURSELF

Use the following problems to test yourself on the material in Chapter 3. The odd problems comprise one test, the even problems a second test.

1. Graph the feasible region for the following set of constraints.

$$5x + 2y \leq 1{,}000$$
$$x + 2y \leq 800$$
$$x \geq 0$$
$$y \geq 100$$

2. Model but do not solve the following problem: A company manufactures two styles of tents, a standard model and deluxe model. It requires 2 hours of labor to make a standard tent and 3 hours of labor to make a deluxe tent. Materials cost $3 for each standard tent and $4 for each deluxe tent. The profit on each standard model is $5, and the profit on each deluxe model is $7. The company has 4,000 labor hours available each week and a materials budget of $9,600 per week. How many of each model tent should the company produce weekly to maximize its profits?

3. The shaded region in Figure *A* represents a feasible region for a set of constraints. Determine which point in this feasible region will maximize $z = 10x + 23y$.

4. Apply one iteration of the simplex method to the following tableau, and then stop.

$$
\begin{array}{c}
\begin{array}{cccccc}
x_1 & x_2 & s_1 & s_2 & z & \\
\end{array} \\
\begin{array}{c} s_1 \\ x_1 \\ {} \\ {} \end{array}
\left[
\begin{array}{ccccc|c}
0 & 5 & 1 & 5 & 0 & 10 \\
1 & 2 & 0 & 3 & 0 & 12 \\
\hline
0 & -4 & 0 & -3 & 1 & 4
\end{array}
\right]
\end{array}
$$

What is the current solution for the resulting tableau? Is this solution optimal?

FIGURE A

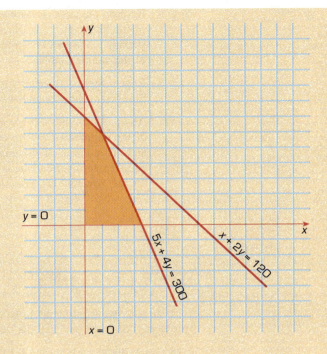

5. Construct a simplex tableau for the following linear program and stop.

Maximize: $z = 5x_1 + 6x_2 + 7x_3 + 8x_4$

subject to: $3x_1 - 4x_2 + 5x_3 - 6x_4 \leq 90$

$7x_1 + 7x_2 + 8x_3 - 9x_4 \leq 135$

assuming: $x_1, x_2,$ and $x_3,$ and x_4 are nonnegative

6. Graph the feasible region for the following set of constraints:

$$2x - y \leq 12$$

$$x + y \geq 0$$

$$x \leq 5$$

$$y \leq 2$$

7. Construct a simplex tableau for the following linear program and stop.

Minimize: $z = 2x_1 - 3x_2 - x_3$

subject to: $x_1 - 2x_2 + x_3 = 12$

$3x_1 + x_2 - 2x_3 \leq -5$

assuming: $x_1, x_2,$ and x_3 are nonnegative

8. The shaded region in Figure B is the feasible region for a set of constraints. Determine which point in this feasible region will minimize $z = 100x - 150y$.

FIGURE B

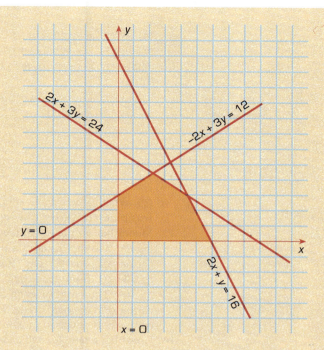

9. Apply one iteration of the simplex tableau method to the following tableau and then stop.

$$\begin{array}{c} \\ x_1 \\ s_2 \\ \\ \end{array} \begin{array}{c} \begin{array}{cccccc} x_1 & x_2 & x_3 & s_1 & s_2 & z \end{array} \\ \left[\begin{array}{cccccc|c} 1 & 2 & 1 & -1 & 0 & 0 & 100 \\ 0 & 4 & 4 & 2 & 1 & 0 & 300 \\ \hline 0 & -5 & -3 & 5 & 0 & 1 & 400 \end{array}\right] \end{array}$$

What is the current solution for the resulting tableau? Is this solution optimal?

10. Model but do not solve the following problem: A company needs to purchase a new fleet of cars and has narrowed the choice to two models from the same dealer. The SE model costs $16,000, with an expected annual operating cost of $2,600. The LE model costs $20,000, with an expected annual operating cost of $2,300. The company needs at least nine cars and has $160,000 available for their purchase. At least two of the new cars must be LE models, and the company president does not want any more than eight new SE models. How many cars of each type should the company purchase to minimize expected operating costs?

11. Apply one iteration of the preprocessing step to the following tableau and stop.

$$\begin{array}{c} \\ s_1 \\ s_2 \\ \\ \end{array} \begin{array}{c} \begin{array}{ccccc} x_1 & x_2 & s_1 & s_2 & z \end{array} \\ \left[\begin{array}{ccccc|c} 1 & -2 & 1 & 0 & 0 & -1{,}000 \\ 3 & 0 & 0 & 1 & 0 & 3{,}600 \\ \hline 20 & -35 & 0 & 0 & 1 & 0 \end{array}\right] \end{array}$$

(*a*) What is the current solution for the resulting tableau?   (*b*) Is this solution feasible?   (*c*) What can you say about the optimal solution?

12. Apply one iteration of the preprocessing step to the following tableau and then stop.

$$
\begin{array}{c}
\\
s_1 \\
s_2 \\
s_3 \\
\\
\end{array}
\begin{array}{c}
\begin{array}{ccccccc}
x_1 & x_2 & x_3 & s_1 & s_2 & s_3 & z
\end{array} \\
\left[
\begin{array}{ccccccc|c}
1 & -2 & 1 & 1 & 0 & 0 & 0 & -10 \\
-1 & 1 & 3 & 0 & 1 & 0 & 0 & -20 \\
2 & 2 & -1 & 0 & 0 & 1 & 0 & -5 \\
\hline
10 & 25 & 15 & 0 & 0 & 0 & 1 & 0
\end{array}
\right]
\end{array}
$$

(*a*) What is the current solution for the resulting tableau?   (*b*) Is this solution feasible?   (*c*) What can you say about the optimal solution?

13. Model but do not solve the following problem: A soft drink company makes regular and diet soda in a plant whose production is limited to at most 6,000 cartons of soda per day. It cost $1.10 to produce each carton of regular soda and $1.20 to produce each carton of diet soda. The daily operating budget is $5,400. The profit on each carton of regular soda is 20¢, and the profit on each carton of diet soda is 18¢. How many cartons of each type of soda should the company produce to maximize its total profits?

14. Construct a simplex tableau for the following linear program and stop.

$$\text{Mazimize:} \quad z = 3x_1 + 2x_2 + 5x_3$$

$$
\begin{aligned}
\text{subject to:} \quad 4x_1 - 4x_2 + 8x_3 &\leq 1{,}225 \\
x_1 + 2x_2 \qquad x_3 &\leq 850 \\
3x_1 \qquad\qquad 2x_3 &\leq 1{,}100
\end{aligned}
$$

$$\text{assuming:} \quad x_1,\ x_2,\ \text{and}\ x_3\ \text{are nonnegative}$$

15. Model but do not solve the following problem: A publishing company has two plants, one located in Maryland and the other in Utah, each equipped to produce copies of the same book in both paperback and hardcover formats. The Maryland plant produces 3,000 paperback books and 4,000 hardback books per day, while the Utah plant produces 4,000 paperback books and 3,000 hardcover books each day. An order is received for 30,000 paperbacks and 24,000 hardbacks. Find the number of days each plant should operate so that the combined number of days in operation is a minimum. Labor contracts stipulate that the Maryland plant may never operate fewer days than the Utah plant.

16. Construct a simplex tableau for the following linear program and stop.

$$\text{Minimize:} \quad z = 6x_1 + 3x_2 - 5x_3$$

$$
\begin{aligned}
\text{subject to:} \quad 2x_1 + 2x_2 - x_3 &\geq 100 \\
x_1 - 3x_2 + x_3 &\leq 500 \\
3x_1 + 2x_2 - x_3 &= 350
\end{aligned}
$$

$$\text{assuming:} \quad x_1,\ x_2,\ \text{and}\ x_3\ \text{are nonnegative}$$

4

FINANCIAL MODELS

4.1 INTEREST

Individuals, organizations, businesses, and countries exchange their goods and services for the products of others. Bartering was one of the earliest mechanisms for establishing trade—a farmer and a weaver might exchange one bushel of corn for one wool scarf—but bartering soon gave way to currency, first gold coin and more recently scrip (paper money), as the primary unit of trade. Scrip itself has little intrinsic worth; the real value of money is its acceptance as a recognized unit of trade. With money as a medium, a bushel of corn worth $10 and a wool scarf worth $13 can be traded fairly, generally through a succession of wholesalers and distributors.

Money is saved, borrowed, and lent. Money is saved to buy consumer goods such as television sets, money is borrowed to finance home purchases and college educations, and money is lent by depositing it in savings accounts so that banks can lend it to others. Each dollar, pound, mark, shilling, yen, ruble, or peso that is lent or borrowed exacts a charge or cost called *interest*.

The amount of money lent or borrowed is called the *principal*, usually denoted by P, and the duration of the loan is its *maturity*. In the simplest type of interest computation, the interest payment is directly proportional to the product of the principal and maturity. The constant of proportionality is the *interest rate*, r. If we let I denote total interest, t the duration of the loan, and write r as a decimal in terms of the same unit of time as t, then

$$I = Prt \qquad (1)$$

> *Simple interest is the principal times the interest rate times the duration of the loan.*

Equation (*1*) is the *simple interest formula*. It is used extensively by governments and corporations to calculate interest payments on bonds and notes. When a person buys a bond, he or she lends money to a corporation or government entity. The purchaser of the bond is the lender, the corporation or government entity that issues the bond is the borrower, and the lender expects to receive interest payments from the borrower.

EXAMPLE 1

How much interest will a person earn by purchasing a $1,000 savings bond for five years at 6% per year using simple interest calculations?

Solution

Here the principal is $P = \$1,000$, the interest rate is $r = 0.06$ (6% in decimal form) per year, and the duration of the loan is $t = 5$ years. Using the simple interest formula, we calculate

$$I = (\$1,000)(0.06)(5) = \$300$$

With simple interest, interest payments are identical from one time period to the next, because interest is always calculated on the *original* investment. Each year, the account described in Example 1 will generate $60 of interest (6% of $1,000). Over a five-year period, the total accumulated interest is $300 (see Figure 4.1).

FIGURE 4.1

Six Percent Simple Interest on $1,000

EXAMPLE 2

How much interest will a person pay if he borrows $4,200 for half a year at $1\frac{1}{2}\%$ interest per month, using simple interest calculations?

Solution

The principal is $P = \$4,200$, and the interest rate is $r = 0.015$ ($1\frac{1}{2}\%$ in decimal form) per month. Since the interest rate is given in terms of months, we measure the duration of the loan in the same time unit. Thus, the maturity of the loan is $t = 6$ months. Using the simple interest formula, we have

$$I = (\$4,200)(0.015)(6) = \$378$$

MODEL 4.1 **Lump Sum Investment with Simple Interest Payments**

Important Factors

P = initial principal

A = value of investment at maturity

I = total interest payments

r = interest rate

t = duration of the investment

Objective

Find I and A.

Constraints

$I = Prt$

$A = P + I = P(1 + rt)$

A borrower must repay both principal and interest at the conclusion of a loan. Also, an investor expects to receive the initial investment plus interest at the conclusion of an investment. If we let A denote the future value of an investment at maturity, then

$$A = P + I = P + Prt = P(1 + rt) \tag{2}$$

The investor in Example 1 will receive $A = P + I = \$1,000 + \$300 = \$1,300$ after five years. Similarly, the borrower in Example 2 must pay the investor who lent the money $A = P + I = \$4,200 + \$378 = \$4,578$ at the end of six months. The complete financial model for a lump sum investment with simple interest payments is shown in Model 4.1.

Most interest payments, from common savings accounts in banks to unpaid balances on credit cards, involve compound interest computations. The defining property of *compound interest* is that current interest payments are made on both the principal and all previous interest payments. In effect, interest is paid on interest.

EXAMPLE 3

How much interest will a person earn by depositing $1,000 in a savings account for five years at 6% per year compounded annually?

Solution

After one year, the principal earns 6%, or $1,000(0.06) = $60. The new principal is $1,060 (the initial deposit plus the $60 interest payment), and the second year's interest payment will be calculated on this new amount. Interest for the second year will be 6% of $1,060 or $1,060(0.06) = $63.60, so the balance in the savings account after two years will total $1,123.60. Interest payments for the third year will be calculated on this new amount. The results of all interest computations through five years are shown in Table 4.1. The sum of all the interest payments (the amounts on the even-numbered lines in Table 4.1) is $338.23.

Compare this amount with the total interest payments of $300 calculated in Example 1 using the simple interest formula.

With compound interest, current interest payments are made on both the principal and all previous interest payments.

The yearly interest payments for the account described in Example 3 are illustrated in Figure 4.2. The interest payment each year is greater than that of the previous year because interest is calculated on the *sum* of the original investment *and* all interest payments previously distributed. Compare Figure 4.2 with the analogous simple interest payments shown in Figure 4.1.

FIGURE 4.2

Six Percent Interest Compounded Annually on $1,000

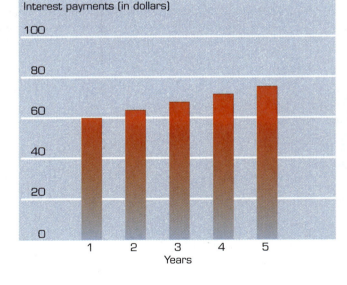

The interest rate per period, denoted by i, *is the nominal annual interest rate divided by the number of conversion periods in a year.*

Interest rates are generally quoted on an annual basis but are compounded over shorter intervals of time. This annual rate is *the nominal annual interest rate*, which we denote by r. The time between successive interest computations is called the *conversion period* or just the *period*. The *interest rate per period* i is the nominal annual interest rate r divided by the number of conversion periods in a year N, that is,

$$i = \frac{r}{N} \tag{3}$$

If interest is compounded quarterly, then $N = 4$ (there are four quarters in a year) and $i = r/4$. For interest compounded semiannually, $i = r/2$; for interest compounded monthly, $i = r/12$; and for interest compounded weekly, $i = r/52$. If no conversion period is stipulated, conversion periods are assumed to be annual and $r = i$.

TABLE 4.1 **Six Percent Interest Compounded Annually on $1,000**

	PRINCIPAL	INTEREST
1. Original deposit	$1,000.00	
2. Interest for first year (6% of line 1)		$60.00
3. Principal for second year (line 1 plus line 2)	$1,060.00	
4. Interest for second year (6% of line 3)		$63.60
5. Principal for third year (line 3 plus line 4)	$1,123.60	
6. Interest for third year (6% of line 5)		$67.42
7. Principal for fourth year (line 5 plus line 6)	$1,191.02	
8. Interest for fourth year (6% of line 7)		$71.46
9. Principal for fifth year (line 7 plus line 8)	$1,262.48	
10. Interest for fifth year (6% of line 9)		$75.75
11. Final principal (line 9 plus line 10)	$1,338.23	
Total interest payments		$338.23

Table 4.1 is a model of a bank savings account in which interest is compounded annually at the rate of 6% per year. Here, $i = r = 0.06$. We used this model to determine how much money would be in the account after five years, without having to wait five years to observe an actual account. We can extend this model to find the principal at the end of any year, for example, the 25th year, although this would require some effort. Alternatively, we can develop a more efficient model.

For notational simplicity, we let P_1 denote the principal after the first period, P_2 the principal after the second period, P_3 the principal after the third period, and so on. The original deposit is denoted by P_0. It follows from line 3 of Table 4.1 that the principal after the first period (in this case, one year) is

$$P_1 = P_0 + 0.06P_0 = P_0(1 + 0.06) \tag{4}$$

It follows from line 5 of Table 4.1 that the principal after the second period is

$$P_2 = P_1 + 0.06P_1 = P_1(1 + 0.06)$$

If we substitute Equation (4) into this last equation, we obtain

$$P_2 = [P_0(1 + 0.06)](1 + 0.06) = P_0(1 + 0.06)^2 \tag{5}$$

It follows from line 7 of Table 4.1 that the principal after the third period is

$$P_3 = P_2 + 0.06P_2 = P_2(1 + 0.06)$$

If we substitute Equation (5) into this last equation, we obtain

$$P_3 = [P_0(1 + 0.06)^2](1 + 0.06) = P_0(1 + 0.06)^3 \tag{6}$$

It follows from line 9 of Table 4.1 that

$$P_4 = P_3 + 0.06P_3 = P_3(1 + 0.06)$$

If we substitute Equation (6) into this last equation, we obtain

$$P_4 = [P_0(1 + 0.06)^3](1 + 0.06) = P_0(1 + 0.06)^4 \qquad (7)$$

Equations (4) through (7) form a pattern, and if we recognize that pattern we can use it to write a formula for the principal after any number of time periods. In particular, $P_5 = P_0(1 + 0.06)^5$, $P_6 = P_0(1 + 0.06)^6$, and in general,

$$P_n = P_0(1 + 0.06)^n \qquad (8)$$

for any number of time periods n (in this example, n is in years).

Table 4.1 and Equation (8) were derived using an interest rate of 6% compounded annually. Thus, $r = i = 0.06$. If the interest rate is different or if interest is compounded over a shorter period than a year, then Equation (8) generalizes to

$$P_n = P_0(1 + i)^n = P_0\left(1 + \frac{r}{N}\right)^n \qquad (9)$$

Again, N is the number of conversion periods in a year, $i = r/N$ is the interest rate per period, and n is the number of conversion periods in the life of the investment or loan. Equation (9) is the compound interest formula, and it is one of the most powerful mathematical models ever developed.

EXAMPLE 4

One thousand dollars is invested in a savings account that pays 6% interest compounded annually. Determine the balance (*a*) after 5 years and (*b*) after 25 years.

Solution

The initial amount of the investment is $P_0 = \$1,000$. Interest is compounded annually, so one conversion period is one year and time will be measured in units of a year. Consequently, the interest rate per period is $i = r = 0.06$ per year.

(*a*) Five years is five time periods, so $n = 5$. It follows from the compound interest formula that

$$P_5 = \$1,000\,(1 + 0.06)^5 \approx \$1,000(1.3382256) = \$1,338.23$$

rounded to the nearest penny. Compare this calculation to those in Table 4.1. Which method is more efficient?

(*b*) Twenty-five years is 25 time periods, so $n = 25$. Now

$$P_{25} = \$1,000(1 + 0.06)^{25} \approx \$1,000(4.2918707) = \$4,291.87$$

rounded to the nearest penny. Certainly it was easier to obtain P_{25} this way than to continue Table 4.1 through another 20 years.

Raising a number to any power is accomplished most efficiently by using a calculator with an exponent key. An exponent key is generally labeled either y^x or x^y (although other labels are sometimes used). The keystrokes needed (for many calculators) to evaluate $(1 + 0.06)^5$ from part (*a*) of Example 4 are:

The result is 1.3382256, rounded to seven decimal places.

EXAMPLE 5

One thousand dollars is invested in a savings account that pays 6% interest compounded quarterly. Determine the balance after 25 years.

Solution

As in Example 4, the initial amount of the investment is $P_0 = \$1,000$, but now interest is compounded quarterly, so one conversion period is one-fourth year. There are four quarters to a year, so the interest rate per period is $i = r/4 = 0.06/4 = 0.015$ per quarter. Twenty-five years is equivalent to $25(4) = 100$ quarters, hence $n = 100$ and we seek P_{100}. It follows from the compound interest formula that

$$P_{100} = \$1,000(1 + 0.015)^{100} \approx \$1,000(4.4320456) = \$4,432.05$$

rounded to the nearest penny. Compare this number to the result in part (b) of Example 4.

EXAMPLE 6

How much must a borrower repay at the end of four years for a loan of $4,500 at an interest rate of 8% interest compounded monthly?

Solution

The initial amount of the investment is $P_0 = \$4,500$. Interest is compounded monthly, so one conversion period is one month, and time will be measured in units of a month. There are 12 months to a year, hence the interest rate per period is $i = r/12 = 0.08/12 \approx 0.006667$ per month. Four years is equivalent to $4(12) = 48$ months, so $n = 48$ and we seek P_{48}. It follows from the compound interest formula that

$$P_{48} = \$4,500(1 + 0.08/12)^{48}$$

Using a calculator with the keystrokes

[4500] [×] [( ] [1] [+] [0.08] [÷] [12] [ )] [y^x] [48] [=]

we obtain $P_{48} = \$6,190.50$, rounded to the nearest penny. Observe that we let the calculator approximate 0.08/12 rather than input the estimate 0.006667, both to save keystrokes and, more importantly, to allow the calculator to store as many digits as it can, thereby minimizing roundoff errors in the calculations.

Table 4.2 illustrates the power of compounding for $1,000 invested for five years at a 6% interest rate (see Examples 1, 4, and 5). Observe that the more conversion periods, the greater the value of the investment at maturity.

The compound interest formula yields the full value of an investment at maturity. To calculate total interest payments, we simply subtract the initial investment or loan from its final value. The savings account in Example 5 will grow to $4432.05 after 25 years. The account began with a deposit of $1,000, so the difference $4432.05 - \$1,000 = \$3432.05$ is the sum total of all interest payments into the account. In Example 6, the borrower must repay $6,190.50 after four years to the investor who lent the money. The initial loan was $4,500, and the difference $6,190.50 - \$4,500 = \$1,690.50$ is interest. The complete financial model for a lump-sum investment with compound interest payments is shown in Model 4.2.

If we invest $100 at 8% compounded quarterly, our balance after one year or four quarters will be

$$P_4 = 100(1 + 0.08/4)^4 = \$108.24$$

MODEL 4.2 Lump Sum Investment with Compound Interest Payments

Important Factors

P_0 = initial principal

N = number of conversion periods per year

n = number of conversion periods until maturity

P_n = value of investment at maturity

r = annual nominal interest rate

i = interest rate conversion period

I = total interest payments

Objective

Find i, I, and P_n.

Constraints

$i = r/N$

$P_n = P_0(1 + r/N)^n$

$I = P_n - P_0$

TABLE 4.2 **The Interest Earned on $1,000 Invested for 5 Years at 6% a Year**

TYPE OF COMPOUNDING	MATURITY	FORMULA FOR VALUE AT MATURITY	VALUE AT MATURITY
None (simple interest)	5 years	$A = \$1,000[1 + 0.06(5)]$	$1,300.00
Annual	5 periods	$P_5 = \$1,000(1 + 0.06)^5$	$1,338.23
Semiannual	$2 \times 5 = 10$ periods	$P_{10} = \$1,000\left(1 + \dfrac{0.06}{2}\right)^{10}$	$1,343.92
Quarterly	$4 \times 5 = 20$ periods	$P_{20} = \$1,000\left(1 + \dfrac{0.06}{4}\right)^{20}$	$1,346.86
Monthly	$12 \times 5 = 60$ periods	$P_{60} = \$1,000\left(1 + \dfrac{0.06}{12}\right)^{60}$	$1,348.85
Weekly	$52 \times 5 = 260$ periods	$P_{260} = \$1,000\left(1 + \dfrac{0.06}{52}\right)^{260}$	$1,349.63
Daily	$365 \times 5 = 1825$ periods	$P_{1825} = \$1,000\left(1 + \dfrac{0.06}{365}\right)^{1825}$	$1,349.83

Effective interest rate is the simple annual interest rate that generates the same amount after one year as the nominal annual interest rate compounded over its stated conversion period.

FIGURE 4.3

of which $8.24 is the result of compound interest. To generate the same amount with interest compounded annually, we would have to receive a simple annual interest rate 8.24%. The *effective interest rate*, denoted by E, is the simple annual interest rate that generates the same amount *after one year* as the nominal annual interest rate compounded over its stated conversion period. Thus, 8.24% is the effective interest rate for 8% annual interest compounded quarterly. In many applications, the effective interest rate is known as the *annual percentage rate* or *APR* for short.

With a nominal interest rate r compounded N times a year, an initial investment of P_0 dollars will grow to $P_0(1 + r/N)^N$ after one year ($n = N$). With an interest rate E compounded annually, the amount after one year will be $P_0(1 + E/1)^1$. If we want these two amounts to be equal, then

$$P_0\left(1 + \frac{r}{N}\right)^N = P_0(1 + E)$$

Dividing both sides of this equation by P_0 and then solving for E, we obtain

$$E = \left(1 + \frac{r}{N}\right)^N - 1 \tag{10}$$

Equation (10) gives the effective interest rate in decimal form.

EXAMPLE 7

Verify the bank advertisement shown in Figure 4.3 if the nominal annual rate of 15.32% is compounded quarterly.

Solution
Using (*10*), we have

$$E = (1 + 0.1532/4)^4 - 1 \approx 1.162228 - 1 = 0.162228 \approx 16.22\%$$

One important application of the compound interest formula (9) is to determine the effects of constant percentage increases on prices (*i.e.*, inflation). If the price of a particular item increases 4% each year, then next year's price will be 4% greater than this year's price, and the price in two years will be 4% greater than next year's price. In the compound interest formula, the price is treated as the principal and the annual percentage change in the price is equivalent to the interest rate (which is negative when prices are decreasing). Each year, the new price is calculated as a percentage change from the most recent price—this is the defining characteristic of compounding.

EXAMPLE 8

An employee begins work at a salary of $18,500 per year and receives a 6% salary increase each year. What will be the salary of this employee be after seven years?

Solution

Salary is the price an employer pays for the services of an employee. Each year, the new salary is 6% greater than the previous year's salary, so salary increases one year become part of the base salary for calculating increases the next year.

The initial price of the employee's service is $P_0 = \$18,500$. Increases are compounded annually, so one conversion period is one year. The percentage change per year, in decimal form, is $i = 0.06$ per year. Seven years is seven time periods, hence $n = 7$. It follows from the compound interest formula that

$$P_7 = \$18,500(1 + 0.06)^7 = \$27,817.16$$

rounded to the nearest penny. After seven years, or at the beginning of the eighth year, the employee will be earning an annual salary of $27,817.16.

The financial models for a lump sum investment with simple interest (see Model 4.1) and a lump sum investment with compound interest (see Model 4.2) were developed under the assumption that interest rates *remain constant*. Neither formula is correct if interest rates change before an investment matures. If interest rates change over the life of an investment, then the entire investment period must be separated into shorter intervals of time over which interest rates remain constant. Then the appropriate formula is applied to each time interval.

EXAMPLE 9

A couple deposits $5,000 in a savings account paying 5% annual interest compounded quarterly. After $3\frac{1}{2}$ years, the bank raises its interest rate to 5.4%, compounded quarterly. How much will the couple have in their account after five years?

Solution

For the first $3\frac{1}{2}$ years, the interest rate remains constant at 5%, and for the last $1\frac{1}{2}$ years, the interest rate remains constant at 5.4%. Thus, we separate the five-year period into two time intervals, one interval of $3\frac{1}{2}$ years and a second interval of $1\frac{1}{2}$ years. After $3\frac{1}{2}$ years or 14 quarters at 5% interest a year, compounded quarterly, the initial investment of $5,000 grows to

$$P_{14} = 5,000\left(1 + \frac{0.05}{4}\right)^{14} \approx \$5,949.77$$

This amount becomes the principal for the second time interval of $1\frac{1}{2}$ years. After $1\frac{1}{2}$ years or six quarters at 5.4% interest compounded quarterly, $5,949.77 will grow to

$$P_6 = 5,949.77\left(1 + \frac{0.054}{4}\right)^6 \approx \$6,448.26$$

IMPROVING SKILLS

In Problems 1 through 10, calculate (*a*) the amount of interest and (*b*) the total value of the investment (or loan) at maturity with simple interest computations.

1. $1,000 at 6% a year for 4 years
2. $1,000 at 6% a year for 10 years
3. $1,000 at 6% a year for 9 months
4. $1,000 at $7\frac{1}{2}$% a year for 4 years
5. $25,000 at 6% a year for 4 years
6. $25,000 at $7\frac{1}{2}$% a year for 4 years
7. $25,000 at $7\frac{1}{2}$% a year for $4\frac{1}{2}$ years
8. $3,200 at $10\frac{1}{4}$% a year for one-half year
9. $6,825 at 4% a year for 20 years
10. $5,370 at 3.75% a year for 9.5 years
11. An initial investment at 8% a year simple interest is worth $14,000 after five years. What was the initial principal?
12. An initial investment at $6\frac{1}{2}$% a year simple interest is worth $1,402.50 after ten years. What was the initial principal?

In Problems 13 through 29, calculate (*a*) the total value of the investment (or loan) at maturity and (*b*) the total interest accrued.

13. $1,000 for 10 years at 6% a year compounded semiannually
14. $1,000 for 10 years at 6% a year compounded quarterly
15. $1,000 for 10 years at 6% a year compounded monthly
16. $5,000 for 7 years at 8% a year compounded semiannually
17. $5,000 for 7 years at 8% a year compounded quarterly
18. $5,000 for 7 years at 8% a year compounded weekly
19. $3,400 for 3 years at $5\frac{1}{2}$% a year compounded quarterly
20. $3,400 for 3 years at $5\frac{1}{2}$% a year compounded monthly
21. $3,400 for 3 years at $5\frac{1}{2}$% a year compounded weekly
22. $4,750 for $2\frac{1}{2}$ years at $10\frac{1}{4}$% a year compounded monthly
23. $4,750 for $2\frac{1}{2}$ years at $10\frac{1}{4}$% a year compounded weekly
24. $20,000 for 2 years at $9\frac{1}{4}$% a year compounded semiannually
25. $20,000 for 2 years at $9\frac{1}{4}$% a year compounded daily
26. $680 for $1\frac{1}{2}$ years at $11\frac{3}{4}$% a year compounded semiannually
27. $680 for $1\frac{1}{2}$ years at $11\frac{3}{4}$% a year compounded quarterly
28. $2,225 for 3 years at 7.5% a year compounded annually
29. $2,225 for 3 years at 7.5% a year compounded daily
30. Reproduce Table 4.2 for a $10,000 investment for 8 years at $7\frac{1}{4}$%.
31. Reproduce Table 4.2 for a $7,500 investment for 12 years at 10%.
32. Many banks use an *approximate year* containing twelve months of 30 days each with 360 days to a year for some interest computations. Use an approximate year to calculate the value of:

(**a**) $2,225 after earning 7.5% a year compounded daily for 3 years
(**b**) $775 after earning 4% a year compounded daily for 10 years
(**c**) $15,000 after earning 6% a year compounded daily for $3\frac{1}{2}$ years
(**d**) $1,400 after earning $5\frac{1}{2}$% a year compounded daily for $4\frac{1}{4}$ years

33. Determine the effective interest rate for a 12% interest-bearing account compounded (*a*) quarterly, (*b*) monthly, (*c*) weekly, and (*d*) daily.
34. Determine the effective interest rate for a 6.75% interest-bearing account compounded (*a*) quarterly, (*b*) monthly, (*c*) weekly, and (*d*) daily.

CREATING MODELS

35. Margaret purchases a $10,000 30-year U.S. Treasury bond that pays 4% simple interest each year. Interest payments are sent to the holder of the bond every half year. How much can Margaret expect to receive from the U.S. Treasury every half year?
36. How much total interest will Margaret receive over the life of the bond (see Problem 35).
37. Margaret (Problem 35) describes her investment to a cousin, who decides to make a similar investment a month later. At that time, however, interest rates have fallen to $3\frac{7}{8}$%. How much will the cousin receive in interest payments every half year?
38. Matt purchases a $5,000 corporate bond that pays $7\frac{1}{2}$% simple interest each year. He sells this bond four months later. How much interest should he receive?
39. Kiesha purchases a $5,000 corporate bond that pays $6\frac{1}{4}$% simple interest each year. She sells this bond five months later. How much interest should she receive?
40. A retired couple deposits $300,000 into an account that pays an interest rate of 5% a year compounded quarterly. They plan to live off the interest payments, so each quarter they withdraw all the interest from the account. How much can they expect to receive from the account every three months?
41. A mining company deposits $20,000 in a savings account to buy a new dump truck in four years. How much money will the company have at that time if the interest rate paid on the account is 8% a year compounded quarterly?
42. A wood cabinet manufacturer deposits $7,000 in a savings account to buy new tools in two years. How much money will the company have at that time if the interest rate paid on the account is $6\frac{3}{4}$% a year compounded monthly?
43. A mattress manufacturer deposits $2,100 in a savings account to buy a new computer in three years. How much money will the company have at that time if the interest rate paid on the account is $7\frac{1}{4}$% a year compounded daily? (Use a year with 365 days.)
44. The Murphys place $5,000 in an account for their newborn grandchild. How much will the child have for college in 18 years if the account pays $8\frac{1}{4}$% interest per year compounded monthly?
45. An investor places $5,000 in an account paying $6\frac{1}{4}$% interest a year compounded quarterly. After three years, he takes all the money from that account and opens a new account at another bank paying 7% interest per year compounded semiannually. How much will he have after five years (three years in the original account and two years in the second account)?
46. Juan borrows $1,000 from his aunt and agrees to repay it after two years at 5% simple interest per year. After two years, Juan requests an extension for another two years; the new interest rate will be $6\frac{1}{2}$% simple interest per year on the principal of the original loan and on all the interest payments for the first two years. How much should Juan expect to pay his aunt four years from the date of the original loan?

47. An entrepreneur borrows $20,000 from an investor and must pay it back in $2\frac{1}{2}$ years at 8% interest per year compounded quarterly. At maturity, the entrepreneur requests an extension for another two years at an interest rate of 8% compounded monthly. How much must the entrepreneur pay the investor at the end of the extension?

48. A contest winner is offered $10,000 now or $11,000 in one year. Which option is more attractive if current interest rates are 9% a year compounded daily?

49. A contest winner is offered $10,000 now or $12,000 in four years. Which option is more attractive if current interest rates are 5% a year compounded quarterly?

50. Aki can borrow money at one bank at $10\frac{3}{4}$% per year compounded daily (use 365 days to a year) or at another bank at 11% per year compounded semiannually. Which is the better deal?

51. A company can invest its money at 6% per year compounded monthly or 6.10% compounded semiannually. Which is the more attractive rate?

52. Tamar placed $15,000 in an Individual Retirement Account. How much will the account be worth in 25 years if it grows at the rate of $5\frac{1}{2}$% a year with no additional deposits or withdrawals?

53. The Yoons have $10,000 to invest and have two opportunities. One pays 5% a year interest compounded quarterly and the other pays 3% a year compounded semiannually. Which opportunity is more attractive?

54. At graduation, Susan accepts a job at a starting salary of $22,000 per year. How much will she be earning at the end of six years if she receives a 5% raise each year?

55. At graduation, Scott accepts a job at a starting salary of $22,000 per year. Raises are given every six months. How much will he be earning at the end of six years if he receives a 2% raise each half year?

56. The Atlas Pipe Company has a total debt of $900,000, which will grow at the rate of 9% a year. How much will the debt be after seven years?

57. The King Sleep Corporation has a total debt of $250,000, which will grow at the rate of $4\frac{1}{2}$% a year. How much will the debt be after five years?

58. The Anderson's average montly oil bill to heat their home is $80 per month. How much will it be in ten years, if oil prices increase at the rate of $\frac{1}{2}$% a month?

59. The Njies buy a house valued at $240,000. How much will the house be worth in 20 years if the house increases in value by 4% a year?

EXPLORING IN TEAMS

60. Occasionally, banks will make *discounted* loans. Total interest I for the duration of a loan is computed on the face value of a loan P, and the amount $P - I$ is lent. At maturity, the borrower repays the full principal P, having effectively repaid the interest at the beginning of the loan. How much does a borrower receive at the beginning of a discounted loan if the duration of the loan is two years and it has a face value of $1,000 and a 5% annual interest rate with quarterly compounding?

61. How much must a borrower repay at maturity for a discounted loan over nine months having a face value of $800 at 8% annual interest compounded monthly?

62. With a discounted loan, the borrower pays total interest I for borrowing an initial principal of $P - I$. Using the data provided in Problem 60, determine the real interest rate charged by the bank; that is, the nominal annual interest rate, compounded quarterly, that generates total interest payments of I on the dollars actually borrowed.

63. Using the data provided in Problem 61, determine the real interest rate charged by the bank; that is, the nominal annual interest rate, compounded monthly, that generates total interest payments of I on the dollars actually borrowed.

EXPLORING WITH TECHNOLOGY

64. Use a graphing calculator to graph Equation (2) with $P = \$1,000$, $r = 0.06$, y replacing A, and x replacing t. Display the graph over a 30-year period.

65. On the same coordinate system used in Problem 64, graph Equation (8) with $P_0 = \$1,000$, y replacing P_n, and x replacing n. Use the results to compare graphically the difference between calculations with simple interest and compound interest.

66. On the same coordinate system, graph Equation (8) with five different values of P_0, letting y replace P_n and x replace n. What can you say about the shapes of the various curves?

67. On the same coordinate system, graph the equation $y = 1,000\,(1 + i)^x$, for five different values of the annual interest rate i. What is the financial significance of this equation? What can you say about the shapes of the curves?

68. Use a graphing calculator to estimate the annual interest rate a person needs to receive on a \$1,000 investment to have it grow to \$3,000 in 10 years. *Hint*: Estimate the solution to the equation

$$3{,}000 = 1{,}000\,(1 + i)^{10}$$

by graphing the equations $y = 3000$ and $y = 1000\,(1 + i)^x$ on the same coordinate system. Experiment with different values of i, attempting to find one such that the two curves intersect at $x = 10$.

REVIEWING MATERIAL

69. (Section 1.3) A railroad yard is constructed to scale using *HO*-gauge track (with a scale of 1 inch to 87 inches). The dimensions of the model yard are 10 ft by 15 ft. (*a*) What are the dimensions of the real yard? (*b*) Are the areas of the two yards, the model yard and the real yard, directly proportional to each other? If so, what is the constant of proportionality?

70. (Section 2.5) Use matrix inversion to solve the following system for x, y, and z.

$$x + 2y + z = 1$$
$$2x + 5y - z = 9$$
$$2x + 3y + 2z = 1$$

71. (Section 3.2) Graph the feasible region defined by the system

$$x + 2y \geq 8$$
$$2x - 5y \leq 10$$

RECOMMENDING ACTION

72. Respond by memo to the following request:

> ### MEMORANDUM
>
> To: J. Doe Reader
>
> From: Advertising
>
> Date: Today
>
> Subject: **Highest Possible Effective Interest**
>
> Customers are attracted to banks with high interest rates, so we are always looking for ways to maximize our effective interest without changing our nominal annual rate of 5%. I see that a bank in New Jersey is advertising a very attractive effective interest by basing interest calculations on a 360-day year, 30 days to a month, but paying interest daily for the actual days money is deposited, for 365 days per calendar year. Thus, they claim to offer a 5.20% effective annual yield on a 5% nominal annual interest rate. Are their computations correct? Do they offer a higher rate than we do, using traditional effective yield calculations? Can we do better still with some other scheme?

4.2 PRESENT VALUE

A dollar today is worth more than that same dollar due any time in the future.

One hundred dollars today is worth more than $100 due in one year. At a 6% interest rate compounded annually, $100 today can be invested in a savings account that will grow to $106 in one year; effectively, one hundred dollars today is worth six dollars more than $100 due in one year. The value of money depends both on the amount under discussion *and* when that amount is paid. The same amount of money has many different values, depending on when that amount becomes available.

In the last section we developed the following compound interest formula for a lump sum investment with an interest rate r, compounded N times a year:

$$P_n = P_0\left(1 + \frac{r}{N}\right)^n$$

(9 repeated)

The present value of an investment is its current worth today.

Here P_0 represents the current amount of an investment, while P_n is the future value of that investment n conversion periods later. The current value of an investment, that is, the value of an investment today, is called its *present value* and is denoted simply as *PV*. The value of an investment at some time in the future is called the *future value* of the investment and is denoted by *FV*. For example, $100 deposited today in an account earning 6% interest a year will be worth $106 in one year, so $PV = \$100$ is the present value and $FV = \$106$ is the future value after one year. Using present value and future value notation, Equation (9) takes the form

$$FV = PV\left(1 + \frac{r}{N}\right)^n$$

(11)

If we solve Equation (*11*) for *PV*, we obtain

$$PV = \frac{FV}{\left(1 + \dfrac{r}{N}\right)^n} = FV\left(1 + \frac{r}{N}\right)^{-n} \tag{12}$$

Equation (*12*) is a financial model (see Model 4.3) for the present value of a single payment due at a specific date in the future. This model establishes the amount of money *PV* that must be invested now to grow to *FV* dollars after *n* conversion periods at the prevailing interest rate at the time of deposit.

EXAMPLE 1

MODEL 4.3 Present Value of a Single Payment Due in the Future with Compound Interest

Important Factors

PV = present value

FV = future value

n = number of conversion periods from present time until the future payment is due

N = number of conversion periods per year

r = annual nominal interest rate

Objective

Find *PV*.

Constraints

$PV = FV(1 + r/N)^{-n}$

A couple wants to start a college fund for their newborn child by depositing a lump sum of money in a savings account and allowing the account to accrue interest at 7% a year, compounded quarterly. How much must the couple deposit if they want the account to be worth $100,000 in 18 years?

Solution

The future value of this investment must be *FV* = $100,000 after 18(4) = 72 quarters. Using Equation (*12*), we find that the present value of this investment is

$$PV = \$100,000(1 + 0.07/4)^{-72}$$

Using a calculator with the keystrokes

we obtain *PV* = $28,676.22, rounded to the nearest penny. The couple should invest $28,676.22 now to have $100,000 in 18 years, assuming interest continues to be paid at the rate of 7% a year, compounded quarterly.

EXAMPLE 2

Solve Example 1 if the interest rate is instead 10% a year, compounded monthly.

Solution

Now *FV* = $100,000 *after* 18(12) = 216 months. Using Equation (*12*), we find that *PV* = $100,000(1 + 0.10/12)^{-216} = $16,653.64, rounded to the nearest penny.

To compare competing investment opportunities that mature at different times, compare their present values.

Investors sometimes must choose the most attractive of several competing opportunities. The decision is easy when all opportunities mature at the same date, because then the opportunity that pays the most at maturity is the most attractive. Clearly, an investment that pays $1,100 after two years is more attractive than an investment that pays $1,000 after two years. The decision is more difficult when different opportunities mature at different times. An investment that pays $1,000 after one year may or may not be more attractive than an investment that pays $1,100 after three years.

Competing investments can be compared only when the value of each is known *at the same instant of time*. The most convenient time is generally the present, when the choice between opportunities must be made. The opportunity with the greatest present value is the most profitable.

EXAMPLE 3

Mr. and Mrs. Kingsley plan to sell their bakery and retire. Two of their employees want to buy the bakery, but neither has any immediate cash to make the purchase. However, each expects to make money from operating the business. Employee A offers $50,000 payable at the end of eight years; Employee B offer $43,500 payable at the end of five years. Which offer is better at the prevailing interest rate of 5% a year?

Solution

Since no compounding period is stipulated, compounding is assumed to be annual. Thus, $N = 1$, and $i = r = 0.05$. Employee A's offer has a future value of $50,000 in eight years (conversion periods). Using

$$PV = FV \left(1 + \frac{r}{N}\right)^{-n}$$ (12 repeated)

we find the present value of Employee A's offer to be

$$PV_\text{A} = \$50{,}000(1 + 0.05)^{-8} \approx \$33{,}841.97$$

Employee B's offer has a future value of $43,500 after five years (conversion periods). Using (12), we calculate its present value as

$$PV_\text{B} = \$43{,}500(1 + 0.05)^{-5} \approx \$34{,}083.39$$

Since Employee B's offer has the greater present value, it is the better offer.

The present value of each offer in Example 3 represents the equivalent cash settlement

FIGURE 4.4
Five Percent Interest Compounded Annually

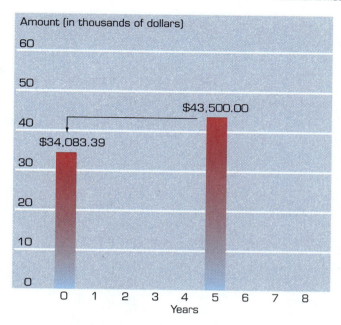

now. The $50,000 due in eight years is worth $33,841.97 to the Kingsleys today at an interest rate at 5% a year. If the Kingsleys deposit $33,841.97 today into a savings account paying 5% interest a year, compounded annually, the account will grow to $50,000 in eight years (see Figure 4.4).

Similarly, $43,500 due in five years is worth $34,083.39 to the Kingsleys today at an interest rate at 5% a year. At 5% interest a year, compounded annually, a lump sum investment of $34,083.39 will grow to $43,500 in five years and to $50,356.69 in eight years (see Figure 4.5).

In Example 3, we compared the present values of two competing opportunities. That is, we compared the worth of two opportunities in terms of their dollar values today. The current time (today) is the preferred time for comparing competing opportunities, although any other instant of time, such as eight years into the future, will do equally well. For example, Employee *A*'s offer of $50,000 is worth exactly $50,000 after eight years. Employee *B*'s offer of $43,500 is due in five years, and it could then invested at the prevailing interest rate of 5% a year for an additional three years; after eight years, it would be worth $50,356.69. The value of *B*'s offer after eight years is greater than *A*'s offer when they are compared at the same time, so again we see that *B*'s offer is more attractive.

Some investment opportunities involve multiple payments, with different amounts due at different times. To calculate the present value of an investment with multiple payments, we use Equation (12) to determine the present value of each individual payment and then sum the results.

> *To calculate the present value of an investment with multiple payments, determine the present values of the individual payments and then sum the results.*

EXAMPLE 4

Before The Kingsleys can decide between the two offers in Example 3, they receive a third offer from an outside buyer who is willing to pay $10,000 immediately plus another $29,200 in four years. How does this offer compare with the other two at the same interest rate?

Solution

This third offer comes with two payments $10,000 now and another payment of $29,200, in four years. The present value of the $10,000 down payment, payable immediately, is

$$PV_1 = \$10,000$$

The present value of $29,200 due in four years at 5% interest, compounded annually (the same interest rate given in Example 3), is

$$PV_2 = \$29{,}200(1 + 0.05)^{-4} \approx \$24{,}022.91$$

The present value of the entire offer is

$$PV = PV_1 + PV_2 = \$10{,}000 + 24{,}022.91 = \$34{,}022.91$$

which is better than the offer from Employee A but not as good as the offer from Employee B.

The value of the third offer described in Example 4 is shown as the stacked column at year 0 in Figure 4.6. The green columns represent the value of the same payment at different times, $24,022.91 at year 0 and $29,200 at year 4. When $24,022.91 is added to the down payment of $10,000, shown by the red column at year 0, the result is the present value of the entire offer.

FIGURE 4.6

Five Percent Interest Compounded Annually

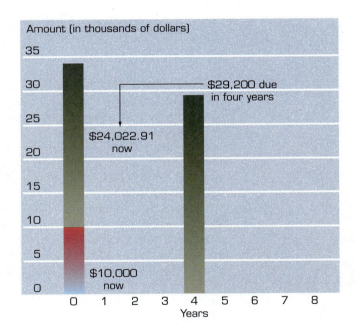

EXAMPLE 5

A friend wants to borrow $800 from you and offers to pay you $300 in two years, another $300 in three years, and a final payment of $400 in five years. If you do not lend the money, you can purchase a bank certificate of deposit that earns $7\frac{1}{2}\%$ interest, compounded quarterly, for five years. Which is the better financial opportunity, assuming your friend will not default on the loan?

Solution

With $r = 0.075$ and $N = 4$ quarters per year, the present value of the first payment of $300 due in two years (8 quarters) is

$$PV_1 = \$300\,(1 + 0.075/4)^{-8} \approx \$258.57$$

The present value of the second payment of $300 due in three years (12 quarters) is

$$PV_2 = \$300(1 + 0.075/4)^{-12} \approx \$240.05$$

The present value of the final payment of $400 due in five years (20 quarters) is

$$PV_3 = \$400(1 + 0.075/4)^{-20} \approx \$275.87$$

The present value of the entire offer, shown graphically by the stacked column at year 0 in Figure 4.7, is

$$PV = PV_1 + PV_2 + PV_3$$
$$= \$258.57 + 240.05 + 275.87 = \$774.49$$

The present value of your $800 available today for deposit in a bank is $800, its current worth. Since your friend's offer has a lower present value, it is less attractive based solely on financial considerations. Lending $800 to your friend is reasonable only if the friendship is worth at least $25.51 to you.

FIGURE 4.7

The Present Value of an Offer with Multiple Payments at 7.5% Interest Compounded Quarterly

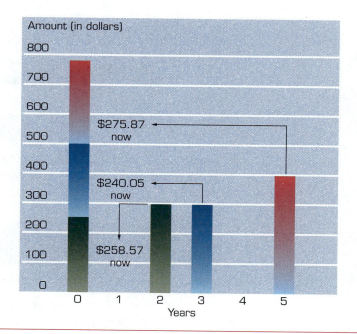

Sometimes we are interested in the interest rate needed to convert a fixed amount of money now into a larger sum at a specific time in the future. For example, we may have $3,000 now and we will need $5,000 in four years. Solving

$$FV = PV\left(1 + \frac{r}{N}\right)^n \qquad \text{(11 repeated)}$$

for r, we obtain

$$\frac{FV}{PV} = \left(1 + \frac{r}{N}\right)^n$$

$$\left(\frac{FV}{PV}\right)^{1/n} = 1 + \frac{r}{N}$$

$$\frac{r}{N} = \left(\frac{FV}{PV}\right)^{1/n} - 1$$

or

$$r = N\left[\left(\frac{FV}{PV}\right)^{1/n} - 1\right] \tag{13}$$

EXAMPLE 6

Determine the interest rate (compounded quarterly) needed to convert $3,000 now into $5,000 in four years.

Solution

There are $N = 4$ conversion periods per year with quarterly compounding. Here $PV = $3,000$ and $FV = $5,000$ after four years or $n = 16$ conversion periods. It follows from Equation (13) that

$$r = 4\left[\left(\frac{5000}{3000}\right)^{1/16} - 1\right]$$

Using a calculator with the keystrokes

we obtain $r \approx 0.1297669$. Thus $3,000 will grow to $5,000 over a four-year period if the money earns 12.98% (rounded to two decimal places) annual interest, compounded quarterly.

IMPROVING SKILLS

In Problems 1 through 20, find the present values of the given payments.

1. $12,000 due in 4 years with an interest rate of 8% a year, compounded quarterly
2. $12,000 due in $4\frac{1}{2}$ years with an interest rate of 8% a year, compounded quarterly
3. $12,000 due in 4 years with an interest rate of 10% a year, compounded quarterly
4. $1,500 due in 3 years with an interest rate of 7% a year, compounded monthly
5. $1,500 due in 4 years with an interest rate of 7% a year, compounded monthly
6. $1,500 due in 4 years with an interest rate of 5.3% a year, compounded monthly
7. $900 due in 2 years with an interest rate of 12% a year, compounded annually
8. $4,200 due in $4\frac{1}{2}$ years with an interest rate of 6.2% a year, compounded quarterly
9. $25,000 due in 10 years with an interest rate of $7\frac{1}{4}$% a year, compounded semiannually.
10. $12,500 due in 7 years with an interest rate of 6% a year, compounded annually
11. $1,000 due now and a second payment of $1,000 due in 3 years with an interest rate of 8% a year, compounded quarterly
12. $1,000 due now and a second payment of $2,000 due in 4 years with an interest rate of 6% a year, compounded quarterly
13. $1,000 due in 2 years and a second payment of $2,000 due in 3 years with an interest rate of 6% a year, compounded quarterly
14. $600 due in 2 years and a second payment of $850 due in 3 years with an interest rate of $9\frac{1}{4}$% of a year, compounded monthly

15. $10,000 due in $3\frac{1}{2}$ years and a second payment of $10,000 due in 5 years with an interest rate of 5% a year, compounded semiannually.
16. $2,000 due in 3 years, $3,000 due in 5 years, and a third payment of $4,000 due in 6 years with an interest rate of 5% a year, compounded annually
17. $600 due now, $700 due in $2\frac{1}{2}$ years, and a third payment of $1,000 due in 5 years with an interest rate of 6% a year, compounded quarterly
18. $400 due in 1 year, $400 due in 2 years, and a third payment of $1,200 due in 3 years with an interest rate of 5.1% a year, compounded monthly
19. $200 due at the end of each month for 6 consecutive months with an interest rate of 6% a year, compounded monthly
20. $200 due at the end of each quarter for 6 consecutive quarters with an interest rate of 6% a year, compounded monthly

In Problems 21 through 29, determine which of the given options has the higher present value.

21. Option 1: $1,000 due now
    Option 2: $1,100 due in 2 years
    $r = 4\%$ compounded quarterly.
22. Option 1: $1,000 due now
    Option 2: $1,100 due in 2 years
    $r = 6\%$ compounded quarterly
23. Option 1: $1,000 due in 1 year
    Option 2: $2,000 due in 10 years
    $r = 7\frac{1}{2}\%$ compounded quarterly
24. Option 1: $1,000 due in 1 year
    Option 2: $2,000 due in 10 years
    $r = 8\%$ compounded quarterly
25. Option 1: $5,000 due now
    Option 2: $3,000 due in 1 year and a second payment of $3,000 due in 2 years
    $r = 5\%$ compounded monthly
26. Option 1: $3,000 due in 2 years
    Option 2: $1,500 due in 1 year and a second payment of $2,000 due in 5 years
    $r = 8\%$ compounded semiannually
27. Option 1: $1,200 due in 3 years
    Option 2: $400 due now and a second payment of $800 due in 5 years
    $r = 6\%$ compounded semiannually
28. Option 1: $900 due now and a second payment of $900 due in $4\frac{1}{2}$ years
    Option 2: $900 due in 1 year and a second payment of $900 due in $3\frac{1}{4}$ years
    $r = 7.3\%$ compounded quarterly
29. Option 1: $1,400 due in 1 year and a second payment of $1,600 due in 3 years
    Option 2: $1,600 due in 2 years and a second payment of $1,400 due in $3\frac{1}{2}$ years
    $r = 12\%$ compounded monthly
30. Determine the annual interest rate required to convert $10,000 to $15,000 in five years.
31. Determine the annual interest rate, compounded monthly, required to convert $10,000 to $15,000 in five years.
32. Determine the annual interest rate required to convert $10,000 to $14,000 in five years.
33. Determine the annual interest rate, compounded semiannually, required to convert $1,000 to $1,350 in three years.
34. Determine the annual interest rate, compounded monthly, required to convert $1,000 to $1,350 in three years.

35. Determine the annual interest rate, compounded monthly, required to convert $1,000 to $1,350 in $3\frac{1}{2}$ years.

36. Determine the annual interest rate required to double an investment in five years.

37. Determine the annual interest rate, compounded quarterly, required to double an investment in five years.

38. Determine the annual interest rate required to double an investment in seven years.

39. Determine the annual interest rate required to triple an investment in five years.

CREATING MODELS

40. With the birth of their child, the Boswells decide to deposit a sum of money in government bonds paying 4% interest a year, compounded annually. Their objective is to accumulate $38,000 in 18 years for college tuition for their child. Determine the amount of money they should invest now to meet their objective.

41. With the birth of their daughter, the Tucks decide to deposit a sum of money in a time savings account that pays 6% annual interest, compounded semiannually. Their objective is to accumulate $20,000 in 20 years to cover their daughter's wedding expenses. How much should they invest now to meet their objective?

42. A trucking company decides to place a lump sum of money into a savings account paying 10% annual interest, compounded monthly. How much should they invest if they will need $42,500 in $4\frac{1}{2}$ years to purchase a new dump truck?

43. The Atlas Pipe Company will invest a lump sum of money into a savings account paying 8.35% yearly interest, compounded semiannually. How much should they invest if they will need $82,000 in eight years to replace an existing piece of machinery?

44. The King Sleep Corporation will invest a lump sum of money into a savings account paying $4\frac{1}{2}$% a year, compounded quarterly. How much should they invest if they want to purchase a new computer system costing $69,500 in $3\frac{1}{2}$ years?

45. Ms. Hernandez's financial advisor recommends that she invest $20,000 in a new housing development for an anticipated return of $28,000 in six years. The advisor claims that this is a better investment than keeping her money in the bank at 6% interest a year, compounded monthly. Is the advisor correct?

46. Dr. Dhaliwal has $10,000 to invest. She can put it into a friend's business, with an expected return of $14,000 in three years, or she can deposit the money in a bank paying 8% interest a year, compounded quarterly. Which opportunity is more profitable?

47. Mr. Jones has two buyers for his business. Buyer A will pay $10,000 immediately and another $25,000 in seven years. Buyer B will pay $7,000 immediately and another $26,000 in five years. Which offer is more attractive with an interest rate of 7% a year, compounded annually?

48. Ms. Jarvis has three buyers for her business. Buyer A will pay $20,000 now and another $5,000 at the end of four years. Buyer B will pay $15,000 now and another $10,000 at the end of three years. Buyer C will pay $10,000 now and another $18,000 at the end of six years. Which is the best offer with an interest rate of 8% a year, compounded annually?

49. Solve the previous problem with an interest rate of 4% a year, compounded semiannually.

50. Mr. Lind has three opportunities for investing the same amount of money. The first will return $8,000 in four years, the second will return $7,000 in two years, and the third will return $10,000 in seven years. Which opportunity is the most attractive with an interest rate of $4\frac{1}{4}\%$, compounded quarterly?

51. Solve the previous problem with an interest rate of $10\frac{1}{2}\%$, compounded annually.

52. Mr. Yoon has two land ventures available to him, each requiring an investment of $7,000. Venture A will return $15,000 in six years, while venture B will return $16,500 in eight years. Which opportunity is more attractive at current interest rates of 6.15%, compounded monthly?

53. The owner of a local dress shop offers a buyer two options: pay $20,000 immediately or pay nothing now and $37,000 at the end of 10 years. Which option is more expensive at an interest rate of 6% a year, compounded quarterly?

EXPLORING IN TEAMS

54. Solve Equation (*11*) for n and show that

$$n = \frac{\log(FV/PV)}{\log(1 + r/N)}$$

55. Use the result from Problem 54 to determine the number of quarters needed to convert $1,000 to $1,500 at 6% interest a year, compounded quarterly.

56. Use the result from Problem 54 to determine the number of months needed to convert $1,000 to $1,500 at 8% interest a year, compounded monthly.

57. Determine the number of quarters needed to convert $800 to $950 at 5.75% interest a year, compounded quarterly.

58. Determine the time it will take for an initial investment of $6,000 to triple if it earns 8% interest a year, compounded quarterly.

59. Determine the time it will take for an investment to double if earns $7\frac{1}{4}\%$ interest a year, compounded annually.

REVIEWING MATERIAL

60. (Section 1.5) Find the equation of a straight line that passes through the point (10, 25) and is perpendicular to the graph of the equation $y = 10x + 35$.

61. (Section 2.4) Find values for a and b that satisfy the equation

$$\begin{bmatrix} -3 & 4 \\ 3 & -2 \end{bmatrix} \begin{bmatrix} -2 \\ 7 \end{bmatrix} + 2 \begin{bmatrix} a \\ b \end{bmatrix} = \begin{bmatrix} 0 \\ -1 \end{bmatrix}$$

62. (Section 3.3) Use graphical methods to solve the linear programming model

$$\text{Maximize:} \quad z = 3x + 4y$$

$$\text{subject to:} \quad 5x + 3y \le 15$$
$$3x + 4y \le 12$$

$$\text{assuming:} \quad x \text{ and } y \text{ are nonnegative}$$

63. Respond by memo to the following request:

> MEMORANDUM
>
> To: J. Doe Reader
>
> From: The Kingsley Bakery
>
> Date: Today
>
> Subject: **What About Risk?**
>
> We have three offers to buy our bakery, each structured with lump sum payments due in the future. The best offer is clear under the assumption that each potential buyer will pay as promised. There is risk, however, involved with all future payments. A buyer may delay a future payment beyond the due date or be unable to pay part of a future payment, even with the best of intentions. Should risk be factored into our decision, and if so, how?

4.3 ANNUITIES

A savings plan with periodic payments is an effective strategy for accumulating funds. Individuals make weekly deposits into Christmas club accounts at local banks to meet anticipated Christmas expenses. Employees save for retirement by depositing a portion of each weekly, biweekly, or monthly salary check into a pension plan. Corporations finance the purchase of new equipment by depositingfixed sums of money each quarter into capital equipment savings accounts. The common thread of all installment savings plans is a stream of payments into an interest-bearing account.

EXAMPLE 1

Samir decides to save for a used car by depositing $600 every six months into a savings account that pays 8% interest a year, compounded semiannually. How much will he have after $2\frac{1}{2}$ years if deposits are made at the end of each six-month period?

Solution

The first payment occurs at the end of the first conversion period, so after $2\frac{1}{2}$ years the first payment will have earned interest for four conversion periods. Thus, its future value at that time will be:

$$\text{future value of first payment in } 2\tfrac{1}{2} \text{ years} = \$600\left(1 + \frac{0.08}{2}\right)^4 \approx \$701.92$$

The second payment occurs at the end of the first year or second conversion period, so after $2\frac{1}{2}$ years the second payment will have earned interest for three conversion periods. Thus,

$$\text{future value of second payment in } 2\tfrac{1}{2} \text{ years} = \$600\left(1 + \frac{0.08}{2}\right)^3 \approx \$674.92$$

Similarly,

$$\text{future value of third payment in } 2\tfrac{1}{2} \text{ years} = \$600\left(1 + \frac{0.08}{2}\right)^2 = \$648.96$$

$$\text{future value of fourth payment in } 2\tfrac{1}{2} \text{ years} = \$600\left(1 + \frac{0.08}{2}\right)^1 = \$624.00$$

The fifth and last payment, made at the end of the $2\frac{1}{2}$ year period, draws no interest but does contribute to the final sum. Thus,

$$\text{future value of fifth payment in } 2\tfrac{1}{2} = \$600\left(1 + \frac{0.08}{2}\right)^0 = \$600.00$$

Summing the individual future values, we obtain

$$\$701.92 + 674.92 + 648.96 + 624.00 + 600.00 = \$3,249.80$$

as the amount available to Samir after $2\frac{1}{2}$ years. The growth of each payment and the total amount in the savings plan at the end of the fifth semiannual conversion period are shown in Figure 4.8. A time schematic for this investment strategy is shown in Figure 4.9.

FIGURE 4.8

A Five-Payment Annuity at 8% Interest Compounded Semiannually

FIGURE 4.9

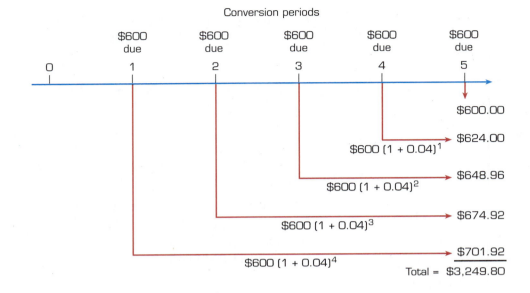

Conversion periods

An annuity is a stream of equal payments made over equal intervals of time.

An annuity is ordinary or due depending on whether payments are made at the end or at the beginning of each payment period, respectively.

An annuity is simple if the conversion period at which interest is determined coincides with the rent period.

Installment loans, mortgage payments, life insurance premiums, social security deductions, and Christmas clubs are all examples of an *annuity*, a stream of equal payments made over equal intervals of time. Annuities are used to accumulate capital, as in a Christmas Club, or to reduce debt, as in home mortgages. The payments are called *regular payments* or *rents* and are denoted by R; the time between payments is called a *rent interval* or *payment interval*. In Example 1, the regular payment is $600 and the payment interval is a half year.

An annuity is classified as an *ordinary annuity* if payments are made at the end of each rent period or as an *annuity due* if payments are made at the beginning of each rent period. Example 1 describes an ordinary annuity. Most automobile loan repayment plans are also ordinary annuities because the first monthly payment is not required until 30 days after the loan has been consummated. In contrast, most life insurance payment plans as well as car insurance payment plans are annuities due because the first premium is due immediately upon approval of the insurance application.

An annuity, whether ordinary or due, is *simple* if the conversion period at which interest is determined coincides with the rent period of the annuity. Example 1 describes a simple annuity because payments into the annuity and interest computations are both made semiannually and they are made *at the same time*. It is relatively easy to formulate mathematical models for simple annuities, so for the rest of this chapter we will consider simple annuities only.

In Example 1, payments of $600 were made semiannually into a savings plan to buy a used car. It is more likely that the individual would save $100 a month or $25 a week instead of $600 every half-year. Saving $100 at the end of every month for $2\frac{1}{2}$ years involves 30 payments. Saving $25 at the end of each week for $2\frac{1}{2}$ years involves 130 payments. We could model each of these annuities as we did in Example 1, but the sheer number of payments makes this approach impractical. It would be much more useful to have a formula for these annuities.

Figure 4.10 is a time schematic for an ordinary annuity that deposits R dollars at the end of each conversion period for n consecutive periods. The value of the entire annuity

FIGURE 4.10

Conversion periods

at the time of the last deposit is the sum of the future values of each payment at that time. In particular, the first payment of R dollars is made at the end of the first conversion period, draws interest for $n - 1$ conversion period, and has a future value of

$$R(1 + r/N)^{n-1} = R(1 + i)^{n-1}$$

at the end of the annuity. Here, as always, r is the nominal annual interest rate and N is the number of conversion periods in one year. The quotient

$$i = \frac{r}{N} \qquad \text{(3 repeated)}$$

is the interest rate per conversion period. The second payment of R dollars is made at the end of the second conversion period, draws interest for $n - 2$ conversion periods, and has a future value of $R(1 + i)^{n-2}$ at the end of the annuity. The next to the last payment draws interest for one conversion period and has a future value of $R(1 + i)^1$ at the end of the annuity. The last payment of R dollars is made at the end of the annuity; it draws no interest, and has a future value of $R(1 + i)^0 = R$ at the end of the annuity. The future value of the annuity is

$$FV = R + R(1 + i)^1 + R(1 + i)^2 + \ldots \qquad (14)$$
$$+ R(1 + i)^{n-3} + R(1 + i)^{n-2} + R(1 + i)^{n-1}$$

Equation (14) is a geometric series that can be summed by a clever strategy. If we multiply the equation by the factor $(1 + i)$, we have

$$(1 + i)FV = R(1 + i) + R(1 + i)^2 + R(1 + i)^3 + \ldots \qquad (15)$$
$$+ R(1 + i)^{n-2} + R(1 + i)^{n-1} + R(1 + i)^n$$

Observe that the right sides of Equations (14) and (15) are nearly identical. Except for the R term on the right side of Equation (14) and the term $R(1 + i)^n$ on the right side of

Equation (*15*), all other terms appear in both equations. If we subtract Equation (*14*) from Equation (*15*), most terms will cancel, leaving

$$(1 + i)FV - FV = R(1 + i)^n - R$$

which can be rewritten as

$$FV[(1 + i) - 1] = R[(1 + i)^n - 1]$$

or

$$FV = R\left[\frac{(1 + i)^n - 1}{i}\right] \tag{16}$$

The terms in the bracket are often denoted simply as $s_{\bar{n}|i}$ (read "*s sub n at i*") on many financial calculators.

EXAMPLE 2

Solve Example 1 using Equation (*16*).

Solution

$R = \$600$ and $i = 0.08/2 = 0.04$. The annuity is for $2\frac{1}{2}$ years, or five semiannual conversion periods, so $n = 5$. Equation (*16*) becomes

$$FV = 600\left[\frac{(1 + 0.04)^5 - 1}{0.04}\right]$$

Using a calculator, we find that $(1 + 0.04)^5 = 1.216653$, rounded to six decimal places. Therefore,

$$FV \approx 600\left[\frac{1.216653 - 1}{0.04}\right] \approx \$3,249.79$$

Note how much easier it is to solve the problem this way than the way we did it in Example 1.

Caution: Although we show $(1 + 0.04)^5$ as 1.216653 for presentation purposes, we did not round $(1 + 0.04)^5$ to six decimal places in the actual calculation. We let a calculator evaluate $(1 + 0.04)^5$ to as many decimal places as the calculator could store and then we used the resulting number to determine FV. The penny difference between the answers in Examples 1 and 2 is due to roundoff.

MODEL 4.4 Future Value of a Simple Ordinary Annuity

Important Factors

R = regular payments or rents

N = number of conversion periods per year

n = number of conversion periods in the annuity

FV = value of the annuity at maturity

r = annual nominal interest rate

i = interest rate per conversion period

Objective

Find FV.

Constraints

$i = r/N$

$FV = R\left[\dfrac{(1 + i)^n - 1}{i}\right]$

A complete financial model for the future value of a simple ordinary annuity is given in Model 4.4. This model is applicable only when the annuity is simple (the payment dates and the conversion dates for calculating interest coincide) and the annuity is ordinary (all payments are equal and occur at the end of each conversion period).

EXAMPLE 3

Alexis decides to save money to buy a used car by depositing \$25 at the end of each week into a savings account that earns 8% interest a year, compounded weekly. How much will he have after $2\frac{1}{2}$ years?

Solution

The annuity is for $2\frac{1}{2}$ years or 130 weeks, thus $n = 130$. The weekly payment is $R = \$25$, and the interest rate per week is $i = 0.08/52$. Equation (16) becomes

$$FV = 25\left[\dfrac{\left(1 + \dfrac{0.08}{52}\right)^{130} - 1}{\dfrac{0.08}{52}}\right]$$

The keystrokes needed (for many calculators) to evaluate $(1 + 0.08/52)^{130}$ are:

The result is 1.221215, rounded to six decimal places. We show this intermediate result for presentation purposes only, because such rounding is not recommended. Computational accuracy is improved by allowing the calculator to store as many decimal places as possible and then rounding only the final answer to the nearest penny. Doing so, we find that

$$FV = 25\left[\dfrac{1.221215 - 1}{\dfrac{0.08}{52}}\right] \approx \$3,594.74$$

For an annuity due, each regular payment is made at the *beginning* of the period. The first payment is deposited immediately, and the last payment is deposited one conversion period prior to the end of the annuity. Thus each payment earns interest for one additional period when compared to payments in an ordinary annuity. A time schematic for an annuity due is shown in Figure 4.11.

FIGURE 4.11

The future value of such an annuity is

$$
\begin{aligned}
FV &= R(1 + i)^1 + R(1 + i)^2 + R(1 + i)^3 + \ldots \\
&\quad + R(1 + i)^{n-2} + R(1 + i)^{n-1} + R(1 + i)^n \\
&= (1 + i)[R + R(1 + i)^1 + R(1 + i)^2 + \ldots \\
&\quad + R(1 + i)^{n-3} + R(1 + i)^{n-1} + R(1 + i)^{n-1}]
\end{aligned}
$$

The terms in the square brackets are just the right side of Equation (*14*), which we already summed neatly as a geometric series. Thus, the future value of an annuity due is

$$
\begin{aligned}
FV &= R(1 + i)\left[\frac{(1 + i)^n - 1}{i}\right] \\
&= R(1 + i)s_{\overline{n}|i}
\end{aligned}
\tag{17}
$$

EXAMPLE 4

The Board of Directors of Eagle Mining instructs the company president to create an equipment replacement fund by making regular payments of $25,000. The first payment is to be made immediately, and future payments are to be made every quarter thereafter. How much will be in the account after five years if interest is earned at 7.85% a year, compounded quarterly?

MODEL 4.5 Future Value of a Simple Annuity Due

Important Factors

R = regular payments or rents

N = number of conversion periods per year

n = number of conversion periods in the annuity

FV = value of the annuity at maturity

r = annual nominal interest rate

i = interest rate per conversion period

Objective

Find FV.

Constraints

$i = r/N$

$FV = R(1 + i)\left[\dfrac{(1 + i)^n - 1}{i}\right]$

Solution

This is an annuity due, because the first payment is due immediately. The annuity is for five years or 20 quarters, thus $n = 20$. The regular payment is $R = \$25,000$, and the interest rate per quarter is $i = 0.0785/4$. Equation (*17*) becomes

$$
\begin{aligned}
FV &= 25,000(1 + 0.0785/4)\left[\frac{(1 + 0.0785/4)^{20} - 1}{0.0785/4}\right] \\
&\approx \$617,047.68
\end{aligned}
$$

or $617,000, rounded to the nearest one hundred dollars.

Model 4.5 is a financial model for an annuity due. The model is applicable only when equal payments are made at the beginning of each period and when the date of each payment coincides with the date on which interest is determined.

Often the future value of an annuity is known, and the problem is to determine the amount of the regular payments. An individual may create an annuity to accumulate $80,000 in 18 years to pay for a college education, or $2,000 in 26 weeks to pay for a vacation, or $8,000 in 3 years to replace a roof. When the annuity is created to pay off a corporate or public debt, it is often called a *sinking fund*. In all of these savings plans, the regular payment R is determined directly from either Equation (*16*) when the annuity is an ordinary annuity or Equation (*17*) when the annuity is an annuity due.

EXAMPLE 5

The Board of Directors in Example 4 wants to establish a sinking fund to total one million dollars in five years. How much must the company deposit quarterly, beginning immediately, to reach their goal in five years at an interest rate of 7.85% a year, compounded quarterly?

Solution

Because the first payment will be made immediately, this sinking fund is an annuity due. As in Example 4, $n = 20$ and $i = 0.0785/4$. The future value of the annuity is specified as $FV = \$1,000,000$, and we are to find the regular payment R. Equation (*17*) becomes

$$1,000,000 = R(1 + 0.0785/4)\left[\frac{(1 + 0.0785/4)^{20} - 1}{0.0785/4}\right]$$

A sinking fund is an annuity created to pay off a corporate or public debt.

Here

$$s_{\overline{20}|0.0785/4} = \left[\frac{(1 + 0.0785/4)^{20} - 1}{0.0785/4}\right] = 24.206848$$

rounded to six decimal places. Therefore,

$$1,000,000 = R(1 + 0.0785/4)[24.206848]$$

$$1,000,000 = 24.681907R$$

or $R = 1,000,000/24.681907 \approx \$40,515.51$. The company should deposit $\$40,515.51$ into a savings account each quarter to accumulate one million dollars in five years.

Again, we emphasize that we rounded $s_{\overline{20}|0.0785/4}$ for presentation purposes only. In our computations, we stored the calculated value for $s_{\overline{20}|0.0785/4}$ in our calculator's memory, allowing the calculator to retain as many digits as possible, and then used the number stored in memory for all future calculations, rounding only at the very end when we solved for R.

IMPROVING SKILLS

In Problems 1 through 15, use the information provided to find the future value of (*a*) an ordinary annuity and (*b*) an annuity due.

1. $100 deposited quarterly over 10 years at an interest rate of 6% a year, compounded quarterly
2. $100 deposited quarterly over 20 years at an interest rate of 6% a year, compounded quarterly
3. $100 deposited quarterly over 20 years at an interest rate of 8% a year, compounded quarterly

4. $100 deposited semiannually over 20 years at an interest rate of 8% a year, compounded semiannually

5. $65 deposited weekly over two years at an interest rate of 4.35% a year, compounded weekly

6. $25 deposited weekly over five years at an interest rate of $7\frac{1}{4}$% a year, compounded weekly

7. $2,300 deposited quarterly over $4\frac{1}{2}$ years at an interest rate of 9% a year, compounded quarterly

8. $2,300 deposited quarterly over six years at an interest rate of 7% a year, compounded quarterly

9. $450 deposited semiannually over 18 years at an interest rate of $6\frac{1}{2}$% a year, compounded semiannually

10. $450 deposited semiannualy over 18 years at an interest rate of $3\frac{1}{2}$% a year, compounded semiannually

11. $450 deposited semiannually over 18 years at an interest rate of $9\frac{1}{2}$% a year, compounded semiannually

12. $90 deposited monthly over 15 years at an interest rate of 5.4% a year, compounded monthly

13. $20 deposited weekly over 15 years at an interest rate of 5.4% a year, compounded weekly

14. $25 deposited weekly over 15 years at an interest rate of 5.4% a year, compounded weekly

15. $75 deposited quarterly over 30 years at an interest rate of 7% a year, compounded quarterly

In Problems 16 through 25, determine the regular payments for the given sinking funds.

16. A total of $6,000 after three years at an interest rate of 6% a year, compounded monthly, with payments made at the end of each month

17. A total of $6,000 after three years at an interest rate of 10% a year, compounded monthly, with payments made at the end of each month

18. A total of $8,000 after three years at an interest rate of 6% a year, compounded monthly, with payments made at the end of each month

19. A total of $8,000 after $5\frac{1}{4}$ years at an interest rate of 6% a year, compounded monthly, with payments made at the end of each month

20. A total of $8,000 after $5\frac{1}{4}$ years at an interest rate of 6% a year, compounded monthly, with payments made at the beginning of each month

21. A total of $75,000 after four years at an interest rate of 6% a year, compounded quarterly, with payments made at the beginning of each quarter

22. A total of $500,000 after five years at an interest rate of 9% a year, with payments made at the beginning of each year

23. A total of $500,000 after five years at an interest rate of 9% a year, compounded quarterly, with payments made at the end of each quarter

24. A total of $10,000 after three years at an interest rate of 7.8% a year, compounded weekly, with payments made at the beginning of each week

25. A total of $10,000 after three years at an interest rate of 5% a year, compounded weekly, with payments made at the beginning of each week

26. A total of $1,500 after two years at an interest rate of 4% a year, compounded weekly, with payments made at the beginning of each week

CREATING MODELS

27. At the end of each week Mr. Neulander deposits $10 in a Christmas club account that earns 5% interest a year, compounded weekly. How much will be in the account after 52 weeks?

28. At the beginning of each week Ms. Donofrio deposits $18 in a Christmas club account that earns 4% interest a year, compounded weekly. How much will be in the account after 52 weeks?

29. The Mbakus plan on depositing $100 at the end of each month into a college fund for their newborn child. How much will be in the fund after 17 years if the fund earns 7% interest a year, compounded monthly?

30. The O'Tooles want to create a fund for their child's wedding by depositing $20 at the end of each week into an interest-bearing account earning $6\frac{1}{4}$% a year, compounded weekly. How much will they have after seven years?

31. The Bells want to save for the down payment for a house. At the end of each month they will deposit one of their salaries into a savings account that earns 4% a year interest, compounded monthly. How much will they have after five years, if they save $1,000 a month?

32. Solve Problem 31 if the salary check is $180 a week, interest is compounded weekly, and payments are made into the account at the end of each week.

33. Stephen decides to save for the down payment on a new car by depositing $20 every month into a savings account that earns 8% interest, compounded monthly. He will make his first deposit on January 1, 1997 and his last deposit on December 1, 1999. How much will he have accumulated by January 1, 2000?

34. The Johnsons buy municipal bonds that yield $300 in interest at the end of every half year. All interest payments are deposited into a savings account earning 7.75% interest per year, compounded semiannually. How much will be in this account after 10 years?

35. In 1970, Ms. Patel decided to invest $500 at the end of each year in a no-load mutual fund. Determine the value of her holdings at the end of 30 years if the fund's market value has increased by 8% a year.

36. With the receipt of his first salary check, Mr. Aziz decides to open a retirement fund by depositing $25 into the fund at the end of each week. How much will he have after 30 years if the fund appreciates an average of 10% a year, compounded weekly?

37. A couple wants to establish an annuity for their newborn child to pay for a Bar Mitzvah in 13 years. How much should they deposit at the beginning of each month into an account earning 5% interest a year, compounded monthly, if they estimate the Bar Mitzvah will cost $7,500?

38. The Nguyens want to establish an annuity for their newborn child to pay for a college education. How much should they deposit at the end of each month into an account earning $5\frac{1}{4}$% interest a year, compounded monthly, if they want $80,000 in 18 years?

39. The Browns want to establish an annuity for their newborn child to pay for a wedding. How much should they deposit at the beginning of each quarter in an account earning 8% interest a year, compounded quarterly, if they want $20,000 in 20 years?

40. Joan wants to establish an annuity for her dream vacation—an around-the-world cruise. How much should she deposit at the end of each week in an account

earning 4.75% interest a year, compounded weekly, if she wants $21,000 after three years?

41. Sean wants to establish an annuity for a Caribbean vacation. How much should he deposit at the end of each week in an account earning 7% interest a year, compounded weekly, if he wants $2,000 after 26 weeks?

42. The Vitellis want to have $30,000 for a down payment on a new house in $5\frac{1}{2}$ years. How much must they deposit at the end of each month into an annuity for this purpose if the fund earns 9% a year interest, compounded monthly?

43. Yummy Nuts will need additional warehouse space in eight years. They expect that adequate space will cost 3 million dollars, and they want to create a sinking fund to meet this need. How much should they deposit at the end of each quarter if money earns 8% a year, compounded quarterly?

44. A mattress company plans to replace an expensive piece of machinery in six years by creating a sinking fund for this purpose. How much should the company deposit in the fund at the beginning of each six-month period to reach a goal of $120,000 with an interest rate of 9.25% a year, compounded semiannually?

45. A mining company wants to create a sinking fund to replace a dump truck in two years at a cost of $50,000. How much should the company deposit into the fund at the beginning of each quarter if interest is paid at the rate of 6.8% a year, compounded quarterly?

46. The Board of Directors of a condominium votes to create a sinking fund to finance a swimming pool for the building. They want to have $200,000 in three years. How much should they deposit into the sinking fund at the end of each month if interest rates are 12% a year, compounded monthly?

47. A town must repay the face value of a $750,000 bond in 20 years, and the town council has decided to create a sinking fund to meet this obligation. Deposits will be made quarterly as taxes are received. How much must be deposited into the fund at the end of each quarter if the fund has an interest rate of 8.15% a year, compounded quarterly?

48. A couple wants to create an annuity for their retirement in 20 years by making weekly deposits. Interest rates are expected to average 6% a year, compounded weekly, over this period. How much should the couple deposit at the end of each week to reach their goal of $500,000 at retirement?

49. The Lees have been told by their roofer that they will need to replace their roof in a few years. They decide to establish an annuity to meet this expense. How much must they deposit at the beginning of each month to accumulate $8,000 in three years if interest rates are $8\frac{1}{4}$% a year, compounded monthly?

EXPLORING IN TEAMS

50. Discuss how you would model the future value of an annuity in which deposits are made monthly but interest is compounded quarterly and funds earn interest only when they have been on deposit for the entire quarter.

51. Use the model created in Problem 50 to determine the value in five years of an ordinary annuity in which monthly deposits of $75 are made in an account that earns interest at the rate of 5% a year, compounded quarterly, but only on funds that have been deposited for the entire quarter.

52. Discuss how you would model the future value of an annuity in which deposits are made yearly but interest is compounded quarterly.

53. Use the model created in Problem 52 to determine the value in 20 years of an ordinary annuity in which annual deposits of $2,500 are made in an account that earns interest at the rate of 5% a year, compounded quarterly.

EXPLORING WITH TECHNOLOGY

54. All electronic spreadsheets and financial calculators, in contrast to most scientific calculators, have built-in features for calculating some or all of the following quantities: $s_{\overline{n}|i}$, the future value FV of an annuity, and the regular payment PMT for a sinking fund. The keystrokes differ from one instrument to another, so it is essential to consult a user manual for the available functions and precise keystrokes associated with each machine. All, however, require a user to input in some fashion the interest rate per period, the number of periods, and either the regular payment R or the future value of the annuity FV. A few commands then produce the quantity of interest. Use a financial calculator or electronic spreadsheet to reproduce the results in Examples 1 through 5.

55. Use a financial calculator or electronic spreadsheet to solve Problems 1 through 26.

REVIEWING MATERIAL

56. (Section 1.4) Construct a qualitative graphical model that relates weekly attendance at a hit Broadway musical (vertical axis) to the number of weeks the musical has been running on Broadway (horizontal axis).

57. (Section 2.3) The following table, provided by the U.S. Commerce Department, lists the U.S. trade deficit with Taiwan from 1988 through 1992. Find the equation of the least squares straight line that fits this data, and then use the results to estimate the trade deficit in 1993, assuming past trends continue.

Year	1988	1989	1990	1991	1992
Trade deficit (in billions of dollars)	14.1	13.0	11.2	9.8	9.4

58. (Section 3.5) Use the basic simplex algorithm to solve the linear program:

$$\text{Maximize:} \quad z = 10x_1 + 25x_2$$
$$\text{subject to:} \quad 2x_1 + 5x_2 \leq 20{,}000$$
$$4x_1 + 3x_2 \leq 16{,}000$$
$$\text{assuming:} \quad x_1 \text{ and } x_2 \text{ are nonnegative}$$

RECOMMENDING ACTION

59. Respond by memo to the following request:

> ### MEMORANDUM
>
> To: J. Doe Reader
>
> From: Pensions and Benefits
>
> Date: Today
>
> Subject: **New Products**
>
> We propose creating two new pension plans for our clients. Each plan must grow to $500,000 after 30 years with a nominal annual interest rate of 6%. We'll call one plan Fast Start. A client deposits the same amount at the end of each month for 10 years and then makes no more deposits. For the remaining 20 years a Fast Start account sits and accumulates interest. We'll call the other plan Deferred Start. A Deferred Start client makes no deposits for 10 years and then makes monthly deposits for the next 20 years. What will the regular payment be for each of these new pension plans? To whom will these plans appeal?

4.4 PRESENT VALUE OF AN ANNUITY

In Example 3 of Section 4.3 we saw that weekly deposits of $25 grow to $3,594.74 after $2\frac{1}{2}$ years at an interest rate of 8% a year, compounded weekly. We now ask what single lump sum deposit must be made at the beginning of the annuity to equal the value of the annuity at maturity, that is, $3,594.74 after $2\frac{1}{2}$ years. The present value PV of a fixed amount FV due n conversion periods into the future is given by the formula

$$ PV = FV\left(1 + \frac{r}{N}\right)^{-n} \qquad \text{(12 repeated)} $$

The present value of an annuity is the single amount that must be deposited at the beginning of an annuity to yield the same value as the annuity at maturity.

Here r is the nominal annual interest rate and N is the number of conversion periods in a year. For future value $FV = \$3,594.74$ and an interest rate of 8% a year, compounded weekly, we have

$$ PV = \$3,594.74\left(1 + \frac{0.08}{52}\right)^{-130} \approx \$2,943.58 $$

The *present value of an annuity* is the single amount that must be deposited at the beginning of the annuity to yield the same value as the annuity at maturity. In our example, a simple ordinary annuity of $25 a week for $2\frac{1}{2}$ years at an interest rate of 8% a year, compounded weekly, has a present value of $2,943.58. This means that a lump-sum of $2,943.58 deposited now will grow after $2\frac{1}{2}$ years to $3,594.74, the future value of the annuity at that time.

A financial model for determining the future value FV of a simple ordinary annuity with periodic payments R for n consecutive conversion periods is

$$FV = R\left[\frac{(1 + i)^n - 1}{i}\right] \qquad \text{(16 repeated)}$$

As always, the interest rate per conversion period i is related to r and N by the equation

$$i = \frac{r}{N} \qquad \text{(3 repeated)}$$

Substituting Equation (*3*) into Equation (*12*), we obtain

$$PV = FV(1 + i)^{-n} \qquad \text{(18)}$$

as the present value PV of a lump sum investment.

Equation (*16*) determines the future value of an ordinary annuity after n conversion periods. Equation (*18*) determines the present value of any fixed amount due n conversion periods in the future. Therefore, if we substitute Equation (*16*) into Equation (*18*), we obtain

$$PV = R\left[\frac{(1 + i)^n - 1}{i}\right](1 + i)^{-n}$$

or

$$PV = R\left[\frac{1 - (1 + i)^{-n}}{i}\right] \qquad \text{(19)}$$

as the present value of a simple ordinary annuity. The terms in the bracket are often denoted simply as $a_{\overline{n}|i}$ (read "*a* sub *n* at *i*") on many financial calculators. Model 4.6 gives a complete financial model for the present value of a simple ordinary annuity.

MODEL 4.6 Present Value of a Simple Ordinary Annuity

Important Factors

R = periodic payments or rents

N = number of conversion periods per year

n = number of conversion periods in the annuity

PV = present value of the annuity

r = annual nominal interest rate

i = interest rate per conversion period

Objective

Find PV.

Constraints

$$i = \frac{r}{N}$$

$$PV = R\left[\frac{1 - (1 + i)^{-n}}{i}\right]$$

EXAMPLE 1

Determine the present value of an annuity that pays $200 at the end of each quarter for the next five years with an interest rate of 6% a year, compounded quarterly.

Solution

The interest rate per quarter is $i = 0.06/4 = 0.015$. There are 20 quarters in a five-year period, hence $n = 20$. With $R = \$200$, Equation (*19*) becomes

$$PV = 200\left[\frac{1 - (1 + 0.015)^{-20}}{0.015}\right]$$

The keystrokes needed (in many calculators) to evaluate

$$a_{\overline{20}|0.015} = \frac{[1 - (1 + 0.015)^{-20}]}{0.015} \quad \text{are:}$$

The result is 17.168639, rounded to six decimal places. Therefore, $PV = 200(17.168639) \approx \$3,433.73$. That is, $3,433.73 deposited today will grow to the same value as the annuity after five years.

Caution: Although we show $a_{\overline{20}|0.015}$ as 17.168639 for presentation purposes, we did not round to six decimal places in the actual calculation. We let a calculator evaluate $a_{\overline{20}|0.015}$ to as many decimal places as it could store and then used the resulting number to determine PV.

> *The present value of an annuity is its current worth.*

An annuity is a stream of equal payments over equal intervals of time. If these payments are made into an interest-bearing account, we have a savings plans like those described in the previous section. Another type of annuity is a set of payments to a person or organization for goods or services previously received. For example, one person may buy a business from someone else and agree to pay for the business on an installment plan. Similarly, a corporation may borrow money to expand from a bank and then repay the bank on an installment plan. These annuities are repayments of debt.

Sometimes a borrower wants to retire a loan in one lump sum, or a lender might want to sell a loan obligation to a third party. In these situations, the parties must know the current worth of the annuity, that is, its present value.

EXAMPLE 2

A bakery employee purchased the business from the previous owner by agreeing to pay the previous owner $800 at the end of each month for ten years. Business has been so good that the employee wants to retire the loan after three years by paying the previous owner one lump sum. How much should this payment be with an interest rate of 10% a year, compounded monthly?

Solution

After three years of loan repayments, there are still seven years left to the annuity, so *at present* this annuity has a duration $7(12) = 84$ months. The interest rate per conversion period is $i = r/N = 0.1/12$, and $R = \$800$. Using Equation (*19*), we have

$$
PV = 800 \left[\frac{1 - \left(\dfrac{1 + 0.1}{12} \right)^{-84}}{\dfrac{0.1}{12}} \right] \approx 800(60.236667) \approx \$48{,}189.33
$$

To retire the loan fairly after three years, the employee should pay the previous owner $48,189.33

EXAMPLE 3

A bank holds a home mortgage that calls for monthly payments of $550. The bank wants to sell the mortgage, which still has $17\frac{1}{2}$ years to run, to another bank for cash. What is the fair market value of this mortgage with an interest rate of $8\frac{1}{4}\%$ a year, compounded monthly?

Solution

There are $17\frac{1}{2}$ years left to the mortage, so *at present* this annuity has a duration of $17.5(12) = 210$ months. The interest rate per conversion period is $i = r/N = 0.0825/12$, and $R = \$550$. Using Equation (*19*), we have

$$
PV = 550 \left[\frac{1 - \left(1 + \dfrac{0.0825}{12} \right)^{-210}}{\dfrac{0.0825}{12}} \right] \approx \$61{,}023.21
$$

The current value of the mortgage from now until maturity is $61,023.21

> *The amount borrowed on an installment plan with equal monthly payments is the present value of that annuity at its beginning.*

Home mortgages and car loans are repaid in equal monthly installments as ordinary annuities. The amount borrowed and the duration of the loan are known, for example, $200,000 for 30 years to buy a new house or $8,000 for 4 years to buy a new car. What needs to be determined is the amount of the monthly payment. If we realize that the amount borrowed is just the present value of an annuity, then we can use Equation (*19*) to calculate the associated periodic payments R.

EXAMPLE 4

Kia borrows $8,000 from a bank to buy a pickup truck. How much will the monthly payments be if the loan is for four years at an interest rate of 9.8%, compounded monthly?

Solution

Kia's repayment plan over four years (48 months) is an ordinary annuity with a present value $PV = \$8,000$. The interest rate per month is $i = r/N = 0.098/12$. Using

$$PV = R\left[\frac{1 - (1 + i)^{-n}}{i}\right]$$ (19 repeated)

we have

$$8{,}000 = R\left[\frac{1 - \left(1 + \dfrac{0.098}{12}\right)^{-48}}{\dfrac{0.098}{12}}\right]$$

$$8{,}000 \approx R(39.577874)$$

or

$$R \approx \frac{8{,}000}{39.577874} = \$202.13$$

rounded to the nearest penny.

Home mortgages are examples of loans for which interest is paid only on the unpaid balance of the loan. Part of each installment payment is used to pay the interest for that period, and part is used to reduce the amount of the loan still outstanding. As the outstanding balance is reduced, so too are the monthly interest charges. Since the monthly payments are all equal, a larger portion of each successive payment is credited against the outstanding balance. Loans that are repaid in this manner are said to be *amortized*; the procedure itself is called *amortization*.

An *amortization schedule* is a table that displays the number of payments in the duration of a loan, the amount of each (equal) payment, the portion of each payment that goes towards interest charges, the portion of each payment that is credited against the outstanding balance of the loan, and the outstanding balance after each payment is made. The periodic payment R is determined from Equation (*19*) with PV as the amount of the loan. Since interest is calculated only on the outstanding portion of the loan, the interest charged at any time is the interest rate per conversion period i (expressed as a decimal) multiplied by the outstanding balance at that time. That is,

$$[\textit{interest charged}] = i \cdot [\textit{outstanding balance}] \tag{20}$$

EXAMPLE 5

A debt of $1,000 is to be amortized over one year with monthly payments. Determine the amount of the installment payments, and construct an amortization schedule for the repayment of the debt at an interest rate of 5% a year compounded monthly.

Solution

The duration of this annuity is 12 months with a present value $PV = \$1,000$. The interest rate per month is $i = r/N = 0.05/12 = 0.00416667$. Using Equation (*19*), we have

$$1,000 = R\left[\frac{1 - \left(1 + \dfrac{0.05}{12}\right)^{-12}}{\dfrac{0.05}{12}}\right]$$

$$1,000 \approx R(11.681222)$$

or

$$R \approx 1,000/11.681222 = \$85.61$$

> *An amortization schedule is a table that displays the number of payments for repaying a loan, the amount of each (equal) payment, the portion of each payment that goes towards interest charges, the portion of each payment that is credited against the outstanding balance of the loan, and the outstanding balance after each payment is made.*

rounded to the nearest penny.

During the first period, the outstanding balance is $1,000, the full amount of the loan, so the interest charge is

$$i \cdot (\textit{outstanding balance}) \approx 0.00416667(1,000) \approx \$4.17$$

Since each installment payment must cover the interest charges for that period, we are left with $85.61 − 4.17 = \$81.44$ to credit against the debt. The outstanding balance after the first payment is $1,000 − 81.44 = \$918.56$.

During the second period, the outstanding balance is $918.56, and the interest charge is

$$i \cdot (\textit{outstanding balance}) \approx 0.00416667(918.56) \approx \$3.83$$

This leaves $85.61 − 3.83 = \$81.78$ to credit against the debt, reducing the outstanding balance to $918.56 − 81.78 = \$836.78$ Continuing in this manner, we complete Table 4.3

After payment number 11, the outstanding balance is $85.23, so the interest charge for the last month is

$$i \cdot (\textit{outstanding balance}) \approx 0.00416667(85.23) \approx \$0.36$$

Payment 12 will be the sum of the outstanding balance plus the interest charge, $85.23 + \$0.36 = \$85.59$, which is two cents less than the other payments. In general, the last

TABLE 4.3 Amortization Schedule for a $1,000 Loan over 12 Months at 5% Interest

PAYMENT NUMBER	PAYMENT AMOUNT	PAYMENT ON INTEREST	PAYMENT ON DEBT	OUTSTANDING DEBT
0				$1,000.00
1	$85.61	$4.17	$81.44	918.56
2	85.61	3.83	81.78	836.78
3	85.61	3.49	82.12	754.66
4	85.61	3.14	82.47	672.19
5	85.61	2.80	82.81	589.38
6	85.61	2.46	83.15	506.23
7	85.61	2.11	83.50	422.73
8	85.61	1.76	83.85	338.88
9	85.61	1.41	84.20	254.68
10	85.61	1.06	84.55	170.13
11	85.61	0.71	84.90	85.23
12	85.59	0.36	85.23	0.00

payment of an amortized loan varies slightly from the other payments because all dollar figures in the calculations, in particular, the monthly payment, are rounded to the nearest penny. In this example, the monthly payment to four decimal places is $85.6075. By rounding to $85.61 each period, each payment is a fraction of a penny too much, and over the course of one year the result is a two-cent overpayment. The last payment is reduced to reflect this overpayment.

We see from Table 4.3 that the amount of interest paid each period decreases, from $4.17 at the first payment to $0.36 at the last payment. There is less interest charged at the end of an amortized loan than at the beginning because the outstanding balance is lower. At the beginning of the loan, the borrower has $1,000 of the lender's money. During the last month of the loan, the borrower has only $85.23 of the lender's money, having paid all the rest back through previous installment payments. Clearly the interest charged on $85.23 must be less than the interest charged on $1,000. The difference between the amounts of interest paid at the beginning and end of an amortized loan can be striking for a long-term loan. Table 4.4 lists the first four and last four monthly payments for a $36,000 mortgage over 30 years at $7\frac{1}{2}\%$ per year compounded monthly.

It is enlightening to see how much interest is paid for the $36,000 home mortgage whose amortization schedule is displayed in Table 4.4. The borrower pays the bank 359 payments of $251.72 plus a last payment of $247.98 for a total of

$$359(\$251.72) + \$247.98 = \$90,615.46$$

Of this amount, $36,000 goes to repay the money borrowed, while the rest,

$$\$90,615.46 - \$36,000 = \$54,615.46$$

is the payment of interest. Over 60% of all the money paid to the bank goes for interest!

TABLE 4.4   Amortization Schedule for a $36,000 Loan Over 30 Years at $7\frac{1}{2}$% Interest

PAYMENT NUMBER	PAYMENT AMOUNT	PAYMENT ON INTEREST	PAYMENT ON DEBT	OUTSTANDING DEBT
0				$36,000.00
1	$251.72	$225.00	$26.72	35,973.28
2	251.72	224.83	26.89	35,946.39
3	251.72	224.66	27.06	35,919.33
4	251.72	224.50	27.22	25,892.11
⋮	⋮	⋮	⋮	⋮
357	251.72	6.17	245.55	742.15
358	251.72	4.64	247.08	495.07
359	251.72	3.09	248.63	246.44
360	247.98	1.54	246.44	0.00

IMPROVING SKILLS

In Problems 1 through 15, use the information provided to find the present value of an ordinary annuity due.

1. $100 paid quartery over 10 years at an interest rate of 6% a year, compounded quarterly
2. $100 paid quarterly over 20 years at an interest rate of 6% a year, compounded quarterly
3. $100 paid quarterly over 20 years at an interest rate of 8% a year, compounded quarterly
4. $100 paid semiannually over 20 years at an interest rate of 8% a year, compounded semiannually
5. $65 paid weekly over 2 years at an interest rate of 4.35% a year, compounded weekly
6. $25 paid weekly over 5 years at an interest rate of $7\frac{1}{4}$% a year, compounded weekly
7. $2,300 paid quarterly over $4\frac{1}{2}$ years at an interest rate of 9% a year, compounded quarterly
8. $2,300 paid quarterly over 6 years at an interest rate of 7% a year, compounded quarterly
9. $450 paid semiannually over 18 years at an interest rate of $6\frac{1}{2}$% a year compounded semiannually
10. $450 paid semiannually over 18 years at an interest rate of $3\frac{1}{2}$% a year, compounded semiannually
11. $450 paid semiannually over 18 years at an interest rate of $9\frac{1}{2}$% a year, compounded semiannually
12. $90 paid monthly over 15 years at an interest rate of 5.4% a year, compounded monthly
13. $20 paid weekly over 15 years at an interest rate of 5.4% a year, compounded weekly
14. $25 paid weekly over 15 years at an interest rate of 5.4% a year, compounded weekly
15. $75 paid quarterly over 30 years at an interest rate of 7% a year, compounded quarterly

In Problems 16 through 25, determine the monthly payments for the following amortized loans.

16. $6,000 for 3 years at 6% a year, compounded monthly
17. $6,000 for 3 years at 10% a year, compounded monthly
18. $8,000 for 3 years at 6% a year, compounded monthly
19. $8,000 for 4 years at 6% a year, compounded monthly
20. $20,000 for 10 years at $7\frac{1}{4}$% a year, compounded monthly
21. $75,000 for 15 years at 8.3% a year, compounded monthly
22. $125,000 for 20 years at 9% a year, compounded monthly
23. $125,000 for 30 years at 9% a year, compounded monthly
24. $125,000 for 30 years at 7% a year, compounded monthly
25. $200,000 for 25 years at 6.7% a year, compounded monthly
26. Verify the first four rows of the amortization schedule shown in Table 4.4.
27. Construct an amortization schedule for the monthly payments of a six-month $1,000 loan at 12% a year, compounded monthly.

In Problems 28 through 34, construct the first three lines of an amortization schedule for the monthly payments of the given loans, and then estimate how much a borrower pays in interest charges over the life of the loan by assuming that the last payment is identical to all other payments.

28. $1,000 over 6 months at 12% a year, compounded monthly
29. $4,000 over 3 years at 10% a year, compounded monthly
30. $9,500 over 4 years at 8% a year, compounded monthly
31. $30,000 over 10 years at 8% a year, compounded monthly
32. $50,000 over 15 years at $5\frac{1}{2}$% a year, compounded monthly
33. $150,000 over 25 years at 7% a year, compounded monthly
34. $240,000 over 30 years at 9.7% a year, compounded monthly

CREATING MODELS

35. Ms. Tilson agrees to sell some land to a friend and also to act as the mortgage holder. The mortgage is for $14,000 over five years at 7% a year, compounded monthly. After two years the friend wants to retire the remaining debt with a lump sum payment. How much should this payment be?

36. Mr. Chin sells his business for a down payment of $25,000 and monthly payments of $500 for the next eight years. After four years, the buyer wants to retire the remaining debt with a lump sum payment. How much should this payment be if interest rates are 9% a year, compounded monthly?

37. Sahir borrows $4,500 from her aunt to buy a used car and agrees to pay the loan back over three years in monthly installments at 8% a year, compounded monthly. Twenty months later, Sahir wins a lottery and wants to pay off the debt to her aunt in one lump sum. How much should this single payment be?

38. A bank holds a mortgage that calls for monthly payments of $600. The bank wants to sell the mortgage, which still has 12 years to run, to another bank for cash. What is the fair market value of this mortgage if it has an interest rate of $8\frac{1}{4}$% a year, compounded monthly?

39. A bank holds a 30-year, $200,000 mortgage on a house at $8\frac{1}{4}$%, compounded monthly. What is the fair market value of this mortgage after 13 years?

40. A bank holds a 10-year, $50,000 mortgage on a house at $9\frac{1}{2}$%? What is the fair market value of this mortgage after 40 months?

41. Chris wants to buy a $20,000 new car. He has $6,000 for a down payment and will finance the rest with a new car loan for four years at 10%. Determine his monthly payments.

42. Emily wants to buy a $35,000 new car. She has $4,000 for a down payment and the trade-in value of her current car will bring another $8,200. She will finance the rest with a new car loan for five years at 12%. Determine her monthly payments.

43. Leo wants to borrow money from a bank to buy a car. Leo's budget limits his future car payments to $300 a month. How much can he afford to borrow if he takes out a four-year loan at an interest rate of 9.5% a year, compounded monthly?

44. The Gordons want to borrow money from a bank to buy a house. They can afford monthly mortgage payments of $800. How much can they borrow with a 20-year mortgage at 9% and still remain within their budget?

45. A couple with an excellent credit rating wants to buy a house. They have $80,000 for a down payment, and they can afford monthly mortgage payments of $1,000. How expensive a house can they purchase and still remain within their budget if they plan to get a 30-year mortgage at 8%?

46. The Newmans borrow $100,000 to buy a house by taking a 30-year mortgage at 12% from a bank. How much will the couple pay to the bank over the life of the mortgage, and how much of this is interest?

47. A philanthropist wants to establish a fund that will provide his favorite charity with $5,000 at the end of each year for the next 20 years. What lump sum (endowment) must he place in the fund now to provide the desired payment stream if interest rates are 8% a year?

48. Ms. Johnson is offered two investment opportunities. The first will return $500 at the end of each year for the next 20 years, while the second will return $1,000 at the end of each year for the next 7 years. Which is the more attractive opportunity at an annual interest rate of 6%?

49. Tom can invest $20,000 now and receive $2,300 at the end of each quarter for the next three years, or he can invest $18,000 now and receive $1,500 at the end of each quarter for the next four years. Which opportunity is the more attractive at an annual interest rate of 8%, compounded quarterly?

50. Winners of a million-dollar state lottery are paid $50,000 per year for 20 years. The first payment is made immediately, and subsequent payments are made at the end of each year for the next 19 years. How much must the lottery agency place in an interest-bearing account earning 6% a year to cover the payments to each winner?

EXPLORING IN TEAMS

Let P_k denote the outstanding balance of an amortized loan *after* the kth payment has been made. Thus, P_0 is the outstanding balance after zero payments have been made, that is, it is the amount of the loan. Also, let I_k represent the interest portion of the kth payment.

51. Show that $I_k = i \cdot P_{k-1}$.

52. Show that $P_k = P_{k-1} - (R - i \cdot P_{k-1}) = (1 + i)P_{k-1} - R$.

53. By evaluating the results of Problem 52 for successive integer values of k, beginning with $k = 1$, verify that

$$P_1 = (1 + i)P_0 - R$$

$$P_2 = (1 + i)P_1 - R = (1 + i)^2 P_0 - R[1 + (1 + i)]$$

$$P_3 = (1 + i)P_2 - R = (1 + i)^3 P_0 - R[1 + (1 + i) + (1 + i)^2]$$

and, in general, that

$$P_k = (1 + i)^k P_0 - R[1 + (1 + i) + (1 + i)^2 + \ldots + (1 + i)^{k-1}]$$

54. Compare the result from Problem 53, particularly the terms in the square bracket, with Equation (*14*) in Section 4.3, and show that

$$P_k = (1 + i)^k P_0 - R\left[\frac{(1 + i)^k - 1}{i}\right]$$

55. Use the result from Problem 54 to verify that P_{357} (the balance due after the 357*th* payment) in the amortization schedule given by Table 4.4 is $742.15.

56. Use the result from Problem 54 to determine P_{180}, the outstanding balance after the 180*th* payment (halfway through the loan), for the loan in Table 4.4.

57. Find the present value of a 10-year loan at 10% a year compounded monthly with monthly payments of $800. Then use the result from Problem 54 to determine P_{36}, the outstanding balance after the 36*th* payment. Compare P_{36} with the result from Example 2.

58. Find the present value of a 30-year mortgage at $8\frac{1}{4}\%$ a year, compounded monthly, with monthly payments of $550. Then use the result from Problem 54 to determine P_{150}, the outstanding balance after 150 months. Compare P_{150} with the result from Example 3.

EXPLORING WITH TECHNOLOGY

59. Electronic spreadsheets and financial calculators, in contrast to most scientific calculators, have built-in features for calculating some or all of the following quantities: $a_{\overline{n}|i}$, the present value *PV* of an annuity, and the installment payment *PMT*. The keystrokes differ from one instrument to another, so it is essential to consult a user manual for the available functions and precise keystrokes associated with each machine. All, however, require a user to input in some fashion the interest rate per period, the number of periods, and either the amount of the installment payment or the present value of the annuity *PV*. A few commands then produce the quantity of interest. Use a financial calculator or electronic spreadsheet to reproduce the results in Examples 1 through 5.

60. Use a financial calculator or electronic spreadsheet to solve Problems 1 through 26.

61. Use an electronic spreadsheet to construct a complete amortization schedule (all 360 lines) for the loan described in Table 4.4.

REVIEWING MATERIAL

62. (Section 1.4) A company expects that the cumulative profit *P* (in millions of dollars) of a new product will be related to time *t* (in years) by the equation

$$P = -0.014t^3 + 0.26t^2 - 0.128t - 1.5$$

for the first 15 years of the product's life. Graph this equation from $t = 0$ (when the product is introduced) until $t = 15$, with P on the vertical axis and t on the horizontal axis. Use your graph to estimate when cumulative profits will first reach $6\frac{1}{2}$ million dollars.

63. (Section 2.2) Use Gaussian elimination to solve the system

$$x + 3y + 2z = 10$$

$$4x + 9y - 7z = 28$$

$$3x + 6y - 9z = 12$$

64. (Sections 3.3 and 3.6) Use the enhanced simplex algorithm to solve the following linear program, and then compare your solution with one found graphically.

$$\text{Minimize:} \quad z = 10x_1 + 25x_2$$

$$\text{subject to:} \quad 2x_1 + 5x_2 \geq 20{,}000$$

$$4x_1 + 3x_2 \geq 24{,}000$$

$$x_1 \leq 8{,}000$$

$$x_2 \leq 7{,}000$$

$$\text{assuming:} \quad x_1 \text{ and } x_2 \text{ are nonnegative}$$

RECOMMENDING ACTION

65. Respond by memo to the following request:

MEMORANDUM

To: J. Doe Reader

From: J.J. Johnstone

Date: Today

Subject: **Home Mortgage**

My bank has approved my home mortgage for $120,000 over 25 years at 8% a year, compounded monthly. I have $20,000 in cash that I can use to increase my down payment and reduce the size of my mortgage to $100,000. Alternatively, I can deposit this $20,000 in a bank account earning 5% a year, compounded daily, and let it accrue interest. What should I do?

Chapter 4 Keys

KEY WORDS

amortization (p. 293)
amortization schedule (p. 293)
APR (p. 262)
annuity (p. 280)
annuity due (p. 280)
compound interest (p. 257)
conversion period (p. 258)
effective interest rate (p. 262)
future value of an annuity (p. 282, 284)
future value of a lump sum (p. 268)
FV (p. 268)
i (p. 258)
interest (p. 256)
interest rate per period (p. 258)
maturity (p. 256)

N (p. 258)
n (p. 260)
nominal annual interest rate (p. 258)
ordinary annuity (p. 280)
payment interval (p. 280)
present value of an annuity (p. 290)
present value of a lump sum (p. 269)
principal (p. 256)
PV (p. 268)
R (p. 280)
r (p. 258)
regular payment (p. 280)
simple annuity (p. 280)
simple interest (p. 256)
sinking fund (p. 284)

KEY CONCEPTS

4.1 Interest

- The simplest type of interest computation involves an interest payment that is directly proportional to the product of the principal and the duration of the loan. The constant of proportionality is the interest rate.

- With simple interest, interest payments are identical from one time period to the next because interest is always calculated on the *original* investment.

- With compound interest, interest payments are made on the current balance of an account, which is generally different from the original balance.

- With compound interest, interest payments on a lump sum deposit increase from one conversion period to the next.

4.2 Present Value

- A dollar today is worth more than a dollar due at any time in the future.

- Different investment opportunities can be compared only when the value of each is known at the same instant of time. To compare competing opportunities that mature at different times, compare either their present values or the value of each at some particular time after all opportunities have matured.

- To calculate the present value of an investment with multiple payments, determine the present values of the individual payments and then sum the results.

4.3 Annuities

- An annuity is ordinary or due depending on whether payments are made at the end or at the beginning of each payment period, respectively.

- An annuity is simple if the conversion period at which interest is paid coincides with the payment period.

4.4 Present Value of an Annuity

- The present value of an annuity is the amount that must be deposited at the beginning of an annuity to yield the same value as the annuity at maturity.

- The fair market value of an annuity is its present value.

- The amount borrowed on an installment plan with equal monthly payments is the present value of that annuity.

KEY FORMULAS

Financial modeling of investment opportunities can be confusing because of the number of formulas (models) involved. The task of selecting an appropriate model is simplified by asking two questions:

Question 1   Does the investment opportunity involve a set of equal payments over equal periods of time?

Question 2   Do we know or want to know the full value of the investment opportunity now or at some time in the future?

If the answer to Question 1 is yes, then the investment opportunity is an annuity, and the annuity formulas are applicable. If the payments are made at the beginning of each period, the annuity is due; if payments are made at the end of each period, the annuity is ordinary.

(1) *If we know or want to know the value of the investment now*, then the present value formula for an annuity is applicable. The present value of a simple ordinary annuity is

$$PV = R\left[\frac{1 - (1 + i)^{-n}}{i}\right]$$

Use this formula to solve for *PV* when the regular payment *R* is known, or use it to solve for *R* when the present value of the annuity (usually a loan) is known. If the annuity is an amortized loan, then the interest charged per period is

$$I = i \cdot (outstanding\ balance)$$

(2) *If we know or want to know the value of the investment in the future*, then a future value formula for an annuity is applicable. The future value of a simple ordinary annuity is

$$FV = R\left[\frac{(1 + i)^n - 1}{i}\right]$$

The future value of a simple annuity due is

$$FV = R(1 + i)\left[\frac{(1 + i)^n - 1}{i}\right]$$

If the answer to Question 1 is no, then the investment opportunity is *not* an annuity, and all deposits are treated as individual lump sums.

(1) *If we know or want to know the value of the investment in the future*, then the future value of a lump sum due t time units in the future under simple interest is

$$A = P(1 + rt)$$

with the total interest payments given by

$$I = Prt$$

The future value of a lump sum due n conversion periods in the future under compound interest is

$$P_n = P_0\left(1 + \frac{r}{N}\right)^n$$

(2) *If we know or want to know the value of the investment now*, then the present value of a lump sum due n conversion periods in the future is

$$PV = FV\left(1 + \frac{r}{N}\right)^{-n}$$

■ The annual percentage rate or effective interest rate E associated with a nominal interest rate r compounded N times a year is

$$E = \left(1 + \frac{r}{N}\right)^N - 1$$

TESTING YOURSELF

Use the following problems to test yourself on the material in Chapter 4. The odd problems comprise one test, the even problems a second test.

1. Eli deposits \$6,500 in an account earning $5\frac{1}{4}\%$ interest a year, compounded quarterly. How much will be in this account after $10\frac{1}{2}$ years?

2. A three-year business venture will pay \$2,000 after one year, another \$2,000 after two years, and an additional \$5,000 after three years. What is the value of this venture today with an interest rate of 6.5% a year, compounded semiannually?

3. The Swensons decide to deposit a lump sum in an account earning $8\frac{3}{4}\%$ a year, compounded monthly. How much must they deposit initially, if they want the account to be worth \$50,000 in 18 years when their newborn child is ready to attend college?

4. (*a*) Ms. Osei is paid every two weeks, and she saves a portion of each pay check for her dream vacation. How much will she have after three years if she deposits \$40 at the end of each two-week period into an account that pays interest at the rate of 5% a year, compounded biweekly?
   (*b*) How much should she deposit each week if she wants her account to total \$5,000 after three years?

5. What is the effective annual interest rate of 8.67% compounded daily?

6. Construct the first two complete lines of an amortization table for equal monthly payments on a \$50,000 loan over 20 years at an interest rate of $8\frac{1}{4}\%$ a year, compounded monthly.

7. A seller receives two offers from different buyers. One buyer will pay the seller $5,000 now and another $5,000 in 3 years. The second buyer will pay the seller $3,000 after 1 year and another $7,000 after $2\frac{1}{2}$ years. Which is the better offer for the seller, if the current interest rate is 6.2%, compounded quarterly?

8. Mr. Lin wants to borrow money from a bank to buy a business, and he proposes to repay the loan in equal monthly installments over a 10-year period. He plans to make the loan payments with profits from running the business. After a detailed analysis of his projected income and expenses, Mr. Lin decides that he can afford to pay the bank $850 at the end of each month. How much can he borrow at an interest rate of 13% per year, compounded monthly?

9. (a) Mr. Jones decides to create an IRA (Individual Retirement Account) at his bank by depositing $80 into a mutual fund at the *beginning* of each month. How much will he have after 30 years if the account is expected to grow at a rate of 8.5% a year, compounded monthly?

   (b) How much should Mr. Jones deposit each month if he wants his IRA to be worth $500,000 after 30 years?

10. A newly married couple decides to deposit a lump sum in an account earning $7\frac{3}{4}$% a year, compounded daily. How much must they deposit if they want the account to be worth $15,000 after 25 years, when they plan to use the money for a gala 25*th* anniversary party and vacation.

11. Ms. Smith wants to borrow money to buy a car. Her bank will lend her the money for four years, providing she pays the loan back in equal monthly installments. Ms. Smith determines that the most she can afford to pay on such a loan is $65 a month. How much should she borrow at an interest rate of 9% per year, compounded monthly?

12. A deposit of $4,000 is made in an account earning $3\frac{1}{4}$% interest a year, compounded weekly. How much will be in this account after $8\frac{1}{2}$ years?

13. Construct the first two complete lines of an amortization table for equal monthly payments on a $2,000 loan over three years at an interest rate of 6% a year, compounded monthly.

14. What is the effective annual interest rate of 6% compounded monthly?

5

PRESENTATION
MODELS

5.1 SETS AND VENN DIAGRAMS

Today we live in what many people call the information age, with access to increasingly large quantities of information in computer data bases, reports, books, periodicals, news-print, and videos. Our problem is to use that information effectively to improve our lives. We make decisions daily: when to buy a car, where to go on vacation, whom to hire, where to open a new office, which college to attend, and on and on and on. The better our decisions, the better our lives become, both in the workplace and in the home.

Prior to making any intelligent decision, we collect information, we determine what pieces of information are relevant to our needs, we present that information meaningfully (either to ourselves or to others), we absorb the implications of the information as best we can, and we then act on our understanding of that information. In this chapter, we focus on ways to present information so that it can be absorbed readily.

One useful technique for synthesizing hundreds of pieces of information, such as survey results, into manageable form is to collect the information into sets. A *set* is a collection of objects. Examples are the set of all clocks in the world, the set of all males in the United States, the set of all males in New Jersey, the set of all elephants in the San Diego Zoo, and the set of all real numbers between 2.05 and 7.13. Any collection, group, or class of well-defined objects is a set.

> *A set is a collection of objects called elements.*

The objects in a set are called *elements*. A simple way to specify a set containing just a few elements is to list all the elements in a row, separating the individual elements with commas and enclosing the entire list in braces. Thus, the set of all integers between 2 and 9, inclusive, is {2, 3, 4, 5, 6, 7, 8, 9}, while the set of the first five presidents of the United States is {George Washington, John Adams, Thomas Jefferson, James Madison, James Monroe}.

We shall use uppercase, italicized letters to represent sets and lowercase, italicized letters to denote elements in a set. The symbol $\in$ is read "is a member of" or "belongs to." Thus $a \in A$ is read "the element a is a member of the set A" or more simply, "a belongs to A." The symbol $\notin$ is read "*is not a member of*." An element is either in a set or it is not in a set, so a particular element is never listed twice. The set of letters in the word *meet* is $A = \{m, e, t\}$; $m \in A$, $e \in A$, but $r \notin A$. Also irrelevant is the order in which elements are listed in a set; only the elements themselves matter. The set $\{m, e, t\}$ is the same as the set $\{t, m, e\}$, because both sets contain the same elements. Two sets are *equal* if and only if they contain the same elements. Consequently,

> *Two sets are equal if and only if they contain the same elements.*

$$\{\text{John, Jim, Mary, Alice}\} = \{\text{Mary, Jim, Alice, John}\} = \{\text{Alice, Mary, John, Jim}\}.$$

A common method for specifying a set, especially sets with many elements, is the *set-builder notation*

$$S = \{s \mid s \text{ has property } P\}$$

which specifies a property shared by all members of the set. The mathematical symbol $\mid$ is read "such that." An element is in a set if and only if that element satisfies the stated property for the set. The set

$$B = \{b \mid b \text{ is an odd integer with } 3 \le b \le 15\}$$

is the set {3, 5, 7, 9, 11, 13, 15}. The set

$$S = \{s \mid s \text{ is a person sworn in as president of the United States}\}$$

is the set of all people who have served as president of the United States. Note how much simpler it is to give the set in set-builder notation than to actually list all the presidents.

The set

$$E = \{e \mid e \text{ is a person elected president of the United States}\}$$

contains fewer elements than S because people have been sworn in as president without having been elected to that position, in particular, Andrew Johnson and Gerald Ford, vice-presidents who succeeded elected presidents in midterm.

A set A is a *subset* of a set B, written $A \subseteq B$, if every element in A is also an element of B. A set of A is a *proper subset* of B, written $A \subset B$, if A is a subset of B and if there is at least one element of B that is *not* in A. We write $A \not\subseteq B$ when A is not a subset of B. For the sets S and E defined in the previous paragraph, $E \subset S$. Every person elected president of the United States has been sworn in as president, so E is a subset of S; E is a proper subset of S because S also contains Gerald Ford and Andrew Johnson who are not members of E. For the sets

> *A set A is a subset of a set B if every element in A is also an element of B.*

$$A = \{1, 2\}, \qquad B = \{1, 2, 3, 4\} \qquad C = \{1, 2, 4\}, \qquad \text{and} \qquad D = \{2, 5\}$$

we have $A \subset B$, because every element in A is also in B, yet $4 \in B$ and $4 \notin A$. Also, $C \subset B$, because each element in C is also in B, yet $3 \in B$ and $3 \notin C$. Furthermore, $A \subset C$. In contrast, $D \not\subseteq B$ because D contains an element (namely, 5) that is not an element of B.

> *The empty set ϕ is the set containing no elements.*

It follows from our definition of a subset that $P \subseteq P$ for any set P. A set containing no elements is called the *empty* (or *null* or *void*) *set*, denoted by ϕ. Two examples are

$$\phi = \{d \mid d \text{ is a dog that was born on the moon}\}$$

and

$$\phi = \{m \mid m \text{ is a pregnant male}\}$$

The empty set is a subset of every set; that is, $\phi \subseteq P$, for every set P. If P contains at least one element, that is, P itself is not empty, then $\phi \subset P$.

> *A universal set U is a set that contains all conceivable elements of interest in a particular situation.*

We define a *universal set U* to be the set of all elements of interest in a particular situation. If we are analyzing a survey of adult males, we could take the universal set to be all the adult males who responded. If we need to work in a larger context, we could expand the universal set to include all males or all people. The universal set must be large enough to include all elements of interest, thereby guaranteeing that all sets of interest are subsets of the universal set.

> *The complement of A, denoted as A', is the set of all elements in the universal set U that are not in A.*

Venn diagrams are used to depict sets graphically. A rectangle is drawn to represent a universal set U, and circles within this rectangle represent subsets of U. Elements of the universal set are portrayed by points within the rectangle, and elements of subsets of U are portrayed by points within circles. Figure 5.1 shows a single set A embedded in a universal set U, while Figure 5.2 shows two subsets A and B of a universal set U. The shaded region in Figure 5.1 depicts elements of the universal set U that are *not* in A. Such elements form the *complement of A*, which is denoted as A'. In set-builder notation,

$$A' = \{x \mid x \in U \text{ and } x \notin A\} \tag{1}$$

FIGURE 5.1

Figures 5.3 and 5.4 are other configurations for diagramming two sets. In Figure 5.3, the set B is embedded in A, indicating that B is a proper subset of A. In Figure 5.2, the circles representing A and B are separated, indicating that the two sets have no elements in common. If two sets have some but not all elements in common, their Venn diagram looks like Figure 5.4.

The *intersection* of two sets A and B, written $A \cap B$, is the set of all elements that are

FIGURE 5.2

FIGURE 5.3

FIGURE 5.4

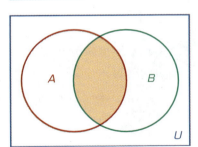

The intersection of two sets is the set of all elements that belong to both sets.

in A and *also* in B. In set-builder notation,

$$A \cap B = \{x \mid x \in A \text{ and also } x \in B\} \tag{2}$$

The shaded region in Figure 5.4 represents this intersection.

EXAMPLE 1

A new-car dealer groups people visting the dealership as follows:

$$F = \{p \mid p \text{ is a person who ordered four-wheel drive}\}$$
$$P = \{p \mid p \text{ is a person who ordered an interior package}\}$$
$$L = \{p \mid p \text{ is a person who ordered a larger engine}\}$$
$$V = \{p \mid p \text{ is a person who did not place an order}\}$$

Describe (*a*) $F \cap P$, (*b*) $P \cap L$, (*c*) $F \cap P \cap L$, and (*d*) $L \cap V$.

Solution

(*a*) $F \cap P$ is the set of all people who ordered both four-wheel drive and an interior package.
(*b*) $P \cap L$ is the set of people who ordered both an interior package and a larger engine.
(*c*) $F \cap P \cap L$ is the set of people who ordered four-wheel drive, an interior package, *and* a larger engine.
(*d*) $L \cap V = \phi$

Two sets are disjoint if their intersection is empty; a group of sets is mutually disjoint if every pair of sets in the group is disjoint.

Two sets are *disjoint* if their intersection is empty; that is, if they have no points in common. Figure 5.2 is a Venn diagram for two disjoint sets. A group of sets is *mutually disjoint* if every pair of sets in the group is disjoint. If a group of sets is mutually disjoint, then an element of one set is *not* an element of any other set in the group.

The usual Venn diagram for subsets A, B, and C of a universal set U is shown in Figure 5.5. The shaded region in Figure 5.5 depicts the subset $A \cap B \cap C$. The structure of Figure 5.5 can be rearranged appropriately when there is additional information about the sets. For example, Figure 5.6 is the Venn diagram when B is a proper subset of A and when A and C are disjoint.

FIGURE 5.5

FIGURE 5.6

FIGURE 5.7

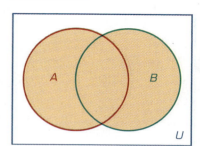

> *The union of two sets A and B is the set of all elements that are either in A or in B or in both sets.*

The *union* of two sets A and B, denoted by $A \cup B$, is the set of all elements that are either in A or in B or in both sets. That is,

$$A \cup B = \{x \mid x \in A \text{ or } x \in B\} \qquad (3)$$

The shaded region in Figure 5.7 represents this union.

EXAMPLE 2

A marketing organization groups its target audience of people as follows:

$$F = \{p \mid p \text{ is female}\}$$
$$S = \{p \mid p \text{ is a senior citizen}\}$$
$$W = \{p \mid p \text{ is wealthy}\}$$

Describe (a) $F \cup W$,   (b) $F' \cup S$,   (c) $V \cup S \cup W$, and (d) $F \cup F'$.

Solution
We take the universal set to be all people in the target audience.

(a) $F \cup W$ is the set of all individuals in the target audience who are either female or wealthy or both. It includes, among others, wealthy men (because they are wealthy) and nonwealthy females (because they are female).

(b) F' is the complement of F, hence F' is the set of all members of the target audience who are not female, i.e., who are male. Thus, $F' \cup S$ is the set of all individuals in the target audience who are either male or senior citizens or both. $F' \cup S$ includes, among others, young males (because they belong to F') and senior citizen females (because they belong to S).

(c) $F \cup S \cup W$ is the set of all individuals in the target audience who are either female or senior citizens or wealthy. Any member of the target audience who possesses one, two, or three of these characteristics is a member of this set.

(d) $F \cup F' = U$, because the set of all females combined with the set of all males constitutes the entire target audience.

FIGURE 5.8

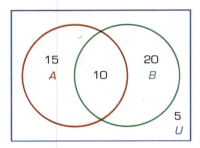

The *cardinal number* of a set A, denoted by $n(A)$, is the number of elements in A. Often the cardinal numbers of sets are placed on Venn diagrams in the regions reserved to those sets. These cardinal numbers can then be used to infer other information about the composition of a universal set under study. For example, the cardinal numbers in Figure 5.8 indicate that there are 10 elements common to both A and B, because we have placed the number 10 in the region associated with $A \cap B$. There are 15 elements in the section of A that is separated from B, so there are 15 elements in A that are not in B. In total, $n(A) = 10 + 15 = 25$ elements. There are 30 elements in B, the 10 elements belonging to $A \cap B$ and the 20 elements that are just in B. There are 5 elements that are in the universal set but are neither in A nor in B. Here $n(U) = 15 + 10 + 20 + 5 = 50$ elements.

EXAMPLE 3

The cardinal number of a set is the number of elements in that set.

The cardinal numbers in Figure 5.9(a), in units of millions, are taken from 1991 U.S. data of people in the United States who are 18 years old or older, excluding members of the armed forces who live on-post by themselves. Let M represent the set of all males, B the set of all blacks, and R the set of all married people in this data set. Determine (a) the number of people in the United States in 1991 who are 18 years old or older, excluding members of the armed forces who live on-post by themselves, (b) the number of blacks, (c) the number of black males, (d) the number of married blacks, (e) the number of married black males, and (f) the number of females.

FIGURE 5.9a

FIGURE 5.9b

FIGURE 5.9c

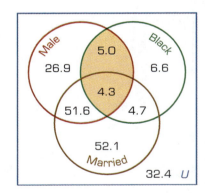

Solution

(a) The universal set of U is the set of all people in the United States in 1991 who were 18 years old or older, excluding members of the armed forced who live on-post by themselves.

$$n(U) = 26.9 + 5.0 + 4.3 + 51.6 + 6.6 + 4.7 + 52.1 + 32.4$$

$$= 183.6 \text{ million people}$$

(b) The number of blacks is the sum of the cardinal numbers in the circle representing blacks, which is indicated by the shaded region in Figure 5.9(b). Thus,

$$n(B) = 5.0 + 6.6 + 4.7 + 4.3 = 20.6 \text{ million people.}$$

This number includes all blacks, both males and females.

FIGURE 5.9d

FIGURE 5.9e

FIGURE 5.9f

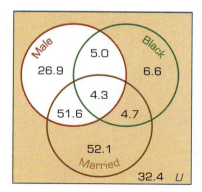

(c) The number of black males is the sum of the cardinal numbers in the regions common to both blacks and males, indicated by the shaded region in Figure 5.9(c). Here,

$$n(B \cap M) = 5.0 + 4.3 = 9.3 \text{ million people.}$$

(d) The number of married blacks is the sum of the cardinal numbers in the regions common to both blacks and married people, indicated by the shaded region in Figure 5.9(d). Here,

$$n(B \cap R) = 4.3 + 4.7 = 9.0 \text{ million people.}$$

Of these, 4.3 million are black, married, and male. The other 4.7 million are black, married, and not male, which means they are black, married, and female.

(e) Married black males are represented by the shaded region in Figure 5.9(e). Here

$$n(B \cap M \cap R) = 4.3 \text{ million people.}$$

(f) Females are the complement of males, and they are represented graphically by the shaded region in Figure 5.9 (f). Thus,

$$n(M') = 6.6 + 4.7 + 52.1 + 32.4 = 95.8 \text{ million people.}$$

EXAMPLE 4

Referring to the sets defined in Example 3, describe and count the elements in (a) $M \cup B$, (b) R', (c) $M \cap B'$, and (d) $B \cap R'$.

Solution

(a) $M \cup B = \{p \mid p$ is married *or* Black$\}$; it contains $5.0 + 6.6 + 4.3 + 4.7 + 51.6 + 52.1 = 124.3$ million people.

(b) $R' = \{p \mid p$ is not married$\}$; it contains $26.9 + 5.0 + 6.6 + 32.4 = 70.9$ million people.

(c) $M \cap B' = \{p \mid p$ is married *and* not Black$\}$; it contains $51.6 + 52.1 = 103.7$ million people.

(d) $B \cap R' = \{p \mid p$ is Black *and* not married$\}$; it contains $5.0 + 6.6 = 11.6$ million people.

EXAMPLE 5

A survey of 200 high school students yields the following information:

41 students are seniors

164 students are taking a mathematics course

105 students are female

24 students are seniors taking a mathematics course

82 students are females taking a mathematics course

22 students are females and seniors

10 students are female seniors taking a mathematics course

How many students in the survey are (*a*) male seniors not taking a mathematics course, (*b*) nonseniors taking a mathematics course, (*c*) females who are not seniors, and (*d*) males or seniors?

Solution

It is convenient to construct a Venn diagram with cardinal numbers for each distinct region written on the diagram, similar to Figure 5.9(*a*). The counts in each category can then be read directly from the completed Venn diagram. Our universal set is the set of all 200 students surveyed, and the other sets of interest are

$$S = \{p \mid p \text{ is a senior}\}$$

$$M = \{p \mid p \text{ is taking a mathematics course}\}$$

$$F = \{p \mid p \text{ is a female}\}$$

Figure 5.10(*a*) is the Venn diagram for these sets, without the cardinal numbers. We determine the cardinal number for each region by starting with the innermost region, $S \cap M \cap F$, and then working out towards the boundary of U.

We are given that $n(S \cap M \cap F) = 10$, so we write it in the innermost section of the diagram, see Figure 5.10(*b*).

The cardinal number of $S \cap M$, the region that is shaded in Figure 5.10(*b*), is 24. Since $n(S \cap M \cap F) = 10$, it follows that the remainder of this region contains $24 - 10 = 14$ people, as shown in Figure 5.10(*c*).

The cardinal number of $M \cap F$, the region that is shaded in Figure 5.10(*c*), is 82. Since $n(S \cap M \cap F) = 10$, it follows that the remainder of this region contains $82 - 10 = 72$ people, as shown in Figure 5.10(*d*).

The cardinal number of $F \cap S$, the region that is shaded in Figure 5.10(*d*), is 22. Since

> *Cardinal numbers for the distinct regions in a Venn diagram are calculated sequentially, starting with the innermost region for which information is available and then working out towards the boundary of the universal set.*

FIGURE 5.10a

FIGURE 5.10b

FIGURE 5.10c

FIGURE 5.10d

FIGURE 5.10e

FIGURE 5.10f

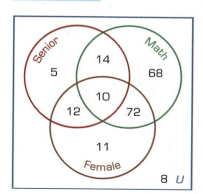

$n(S \cap M \cap F) = 10$, it follows that the remainder of this region contains $22 - 10 = 12$ people, as shown in Figure 5.10(e).

The survey included 41 seniors. We accounted for 36 of them in Figure 5.10(e), so there must be 5 seniors who are neither female nor taking a mathematics course. The survey included 164 students who are taking a mathematics course. We accounted for 96 of them in Figure 5.10(e), so there must be 68 students taking a mathematics course who are neither female nor seniors. The survey included 105 students who are female. We accounted for 94 of them in Figure 5.10(e), so there must be 11 females who are neither seniors nor taking mathematics courses. These numbers are shown in Figure 5.10(f).

The total in the three circles is 192 students, so there are 8 other students in the survey who are in none of the three categories. Using Figure 5.10(f), we determine

(a) $n(F' \cap S \cap M') = 5$

(b) $n(S' \cap M) = 68 + 72 = 140$

(c) $n(F \cap S') = 72 + 11 = 83$

(d) $n(F' \cup S) = 5 + 14 + 10 + 12 + 68 + 8 = 117$

IMPROVING SKILLS

In Problems 1 through 15, determine whether the given statements are true or false.

1. $2 \in \{1, 2, 3\}$

2. $3 \in \{1, 2, 3\}$

3. $4 \in \{1, 2, 3\}$

4. $\{2, 3\} \in \{1, 2, 3\}$

5. $\{2, 3\} \subset \{1, 2, 3\}$

6. $\{2, 3\} \subseteq \{1, 2, 3\}$

7. $\{1, 2, 3\} \subseteq \{1, 2\}$

8. $\phi \subset \{1, 2\}$

9. $\phi \subseteq \{1, 2, 3\}$

10. $\{1, 2\} \subset \phi$

11. $\phi \subset \phi$

12. $2 \notin \phi$

13. $3 \notin \{1, 2\}$

14. $2 \notin \{1, 2, 3\}$

15. $2 \in A \cap A'$

16. Let $A = \{a, b, c\}$. Determine universal sets such that

    (**a**) $A' = \{x, y, z\}$ (**b**) $A' = \{e\}$

    (**c**) $A' = \{r, s, t, u, v\}$ (**d**) $A' = \phi$

17. List all possible subsets of $A = \{1, 2, 3\}$.

18. An author defines two sets A and B to be equal if and only if $A \subseteq B$ and $B \subseteq A$. Is this definition consistent with ours?

19. Let $U = \{1, 2, 3, 4, 5, 6\}$, $A = \{1, 2, 5, 6\}$, and $B = \{2, 3, 5, 6\}$.

    List the elements of

    (**a**) $A \cup B$ (**b**) $A \cap B$ (**c**) A' (**d**) B'

    (**e**) $A' \cap B$ (**f**) $A' \cap B'$ (**g**) $A' \cup B'$ (**h**) $U \cap B$

20. Let $U = \{\text{red, white, blue, green, black, orange}\}$, $A = \{\text{red, blue, green, black}\}$, and $B = \{\text{red, white, black, orange}\}$.

    List the elements of

    (**a**) $A \cup B$ (**b**) $A \cap B$ (**c**) A' (**d**) B'

    (**e**) $A' \cap B$ (**f**) $A' \cap B'$ (**g**) $A' \cup B'$ (**h**) $U \cup B$

21. Let $A = \{\text{Buffalo, Nashville, Tampa}\}$, $B = \{\text{Richmond, Cleveland, Nashville, Tampa}\}$, and $C = \{\text{Buffalo, Cleveland}\}$.

    List the elements of

    (**a**) $A \cap B$ (**b**) $A \cap B \cap C$ (**c**) $B \cap C$

    (**d**) $A \cup B$ (**e**) $A \cup C$ (**f**) $(B \cup C) \cap A$

22. A company categorizes its employees into the following sets:

$$A = \{p \mid p \text{ is a male employee}\}$$

$$B = \{p \mid p \text{ is a female employee}\}$$

$$C = \{p \mid p \text{ is a minority employee}\}$$

$$D = \{p \mid p \text{ is a married employee}\}$$

    Describe in words each of the following sets if the universal set is the set of all employees of the company.

    (**a**) A' (**b**) $A \cup B$ (**c**) $C \cap D$ (**d**) $A \cap C$

    (**e**) $A \cup D$ (**f**) $A \cap D$ (**g**) $A \cap C \cap D$ (**h**) D'

    (**i**) $D' \cap C'$ (**j**) $D' \cap C$ (**k**) $B \cap A$

23. Let U be the set of all people living in the world today, and let

$$A = \{p \mid p \text{ is a person who smokes}\}$$

$$B = \{p \mid p \text{ is a male}\}$$

$$C = \{p \mid p \text{ is a person over 40}\}$$

    Write the following sets in terms of A, B, and C.

    (**a**) $\{p \mid p \text{ is a female}\}$

    (**b**) $\{p \mid p \text{ is a male who smokes}\}$

    (**c**) $\{p \mid p \text{ is a smoking person over 40}\}$

    (**d**) $\{p \mid p \text{ is a male over 40}\}$

    (**e**) $\{p \mid p \text{ is a female over 40}\}$

    (**f**) $\{p \mid p \text{ is a nonsmoking male}\}$

    (**g**) $\{p \mid p \text{ is a person 40 or younger}\}$

    (**h**) $\{p \mid p \text{ is a smoker 40 or younger}\}$

24. Let U be the set of all merchandise in a store, and let

$$D = \{x \mid x \text{ is defective}\}$$

$$S = \{x \mid x \text{ is on sale}\}$$

$$A = \{x \mid x \text{ is advertised}\}$$

Write the following sets in terms of D, S, and A.

(**a**) $\{x \mid x \text{ is defective and on sale}\}$
(**b**) $\{x \mid x \text{ is advertised and on sale}\}$
(**c**) $\{x \mid x \text{ is advertised but not on sale}\}$
(**d**) $\{x \mid x \text{ is unadvertised}\}$
(**e**) $\{x \mid x \text{ is advertised or on sale}\}$
(**f**) $\{x \mid x \text{ is advertised or defective}\}$
(**g**) $\{x \mid x \text{ is advertised or not defective}\}$
(**h**) $\{x \mid x \text{ is defective, advertised, and on sale}\}$
(**i**) $\{x \mid x \text{ is defective, advertised, but not on sale}\}$

CREATING MODELS

25. Figure A is a model of the U.S. population (in millions of people) in 1991, as reported by the U.S. Bureau of Labor Statistics. The universal set includes all people 18 years old or older, excluding members of the armed forces who live on-post by themselves. Use this model to determine the number of people in the universal set who are (*a*) Hispanic, (*b*) married, (*c*) not married, (*d*) not Hispanic, (*e*) married and Hispanic, and (*f*) married but not Hispanic.

26. Figure B is a model of the U.S. population (in millions of people) in 1989, as reported by the U.S. Bureau of Labor Statistics. The set 65+ includes all people 65 years old or older, while the universal set includes all people 18 years or older. Use this model to determine the number of people in the universal set who are (*a*) male, (*b*) less than 65 years old, (*c*) male of age 65 years of more, (*d*) female of age 65 years or more, (*e*) female less than 65 years old, and (*f*) male less than 65 years old.

27. Figure C is a model of the labor force of the United States (in millions of people) in 1991, as reported by the U.S. Bureau of Labor Statistics. The universal set includes

FIGURE A

FIGURE B

FIGURE C

FIGURE D

only the noninstitutionalized civilian population who are 25 years old or older and in the labor force. Use this model to determine the number of people in the universal set who are (*a*) male, (*b*) nonwhite females, (*c*) nonwhite males, (*d*) college-educated females, (*e*) college-educated males, (*f*) college-educated white males, (*g*) not college educated, and (*h*) nonwhite females who are not college educated.

28. Figure *D* is a model of new Ph.D.s in mathematics from July 1, 1992–June 30, 1993, as reported in the *Notices of the American Mathematical Society*. The universal set omits 19 individuals whose citizenship or racial/ethnic grouping is unknown. The Asian set includes Pacific Islanders. Use this model to determine the number of people in the universal set who are (*a*) U.S. citizens, (*b*) not U.S. citizens, (*c*) Asian females, (*d*) Asian males, (*e*) Asian males with U.S. citizenship, (*f*) Asian females without U.S citizenship, (*g*) non-Asian females who are U.S. citizens, and (*h*) males who are not U.S. citizens.

29. A survey of 100 people revealed that 52 were female, 23 were senior citizens, and 15 were female senior citizens. How many were neither female nor senior citizens?

30. At a high school sports banquet, 32 varsity letters were awarded in swimming and 44 varsity letters were awarded in track. How many athletes received at least one letter if 7 athletes receive letters in both swimming and track?

31. A survey of 200 people revealed the following information: 68 were married, 95 owned cars, 83 had jobs, 46 were married with jobs and owned a car, 53 were married and had jobs, 70 owned cars and had jobs, and 53 were married and had cars. Construct a Venn diagram for this population, and then use the diagram to determine how many people in the population (*a*) were married with a car but no job and (*b*) had a job but no car?

32. A survey of 600 people revealed the following information:

> 380 subscribe to a daily newspaper
>
> 158 subscribe to a weekly magazine
>
> 306 subscribe to a monthly magazine
>
> 70 subscribe to a daily newspaper and a weekly magazine
>
> 81 subscribe to a weekly and a monthly magazine
>
> 175 subscribe to a monthly magazine and a daily newspaper
>
> 58 subscribe to all three types of publications

Construct a Venn diagram for this population, and then use the diagram to determine how many people in the population (*a*) subscribe only to a daily newspaper, (*b*) subscribe only to magazines of either type, and (*c*) subscribe to at least one type of publication?

33. A survey of 1,150 college students revealed the following information:

> 837 liked country music
>
> 278 liked just country music
>
> 620 liked rock music
>
> 224 liked just rock music
>
> 375 liked country and rock music
>
> 318 liked the blues
>
> 65 liked all three types of music

Construct a Venn diagram for this population, and then use the diagram to determine how many people in the population (*a*) like only blues music, (*b*) like country and blues, (*c*) like country and blues but not rock, and (*d*) do not like any of the three types of music.

34. A survey of 164 employees of a medium-size business revealed the following information:

   63 owned General Motors vehicles

   12 owned Ford and General Motors vehicles but no Chrysler vehicles

   8 owned Ford and Chrysler vehicles

   24 owned Chrysler vehicles

   45 owned just General Motors vehicles

   6 owned Chrysler and General Motors vehicles but no Ford vehicles

   52 did not own vehicles from Ford, General Motors or Chrysler

   Construct a Venn diagram for this population, and then use the diagram to determine how many people in the population (*a*) owned Ford vehicles, (*b*) owned only Ford vehicles, (*c*) owned General Motors or Chrysler vehicles, and (*d*) owned General Motors vehicles but no Chrysler vehicles.

35. A survey of 175 people revealed the following information about where they purchase their new clothes:

   137 use department stores

   45 use department stores only

   79 use department stores and catalogs

   117 use catalogs

   61 use catalogs and department stores but not boutiques

   13 use boutiques but not catalogs

   38 use catalogs only

   Construct a Venn diagram for this population, and then use the diagram to determine how many people in the population (*a*) use boutiques, (*b*) use boutiques and catalogs but not department stores, (*c*) use boutiques or catalogs, and (*d*) use none of the three.

36. Thirty businesses were asked which overnight express mail services they used. When the responses were tallied, they showed that

   16 used Federal Express (FedEx)

   14 used United Parcel Service (UPS)

   11 used the U.S. Postal Service (USPS)

   2 used all three services

   6 used only FedEx

   5 used only UPS

   4 used UPS and USPS but not FedEx

Construct a Venn diagram for this population, and then use the diagram to determine how many companies in the population (*a*) used none of the three services, (*b*) used UPS or USPS, (*c*) used UPS or USPS but not FedEx, and (*d*) used only USPS.

EXPLORING IN TEAMS

37. The *difference* of two sets A and B is defined as

$$A - B = \{x \mid x \in A \text{ and } x \notin B\}$$

If $A = \{1, 2, 3, 4\}$, $B = \{1, 2, 5\}$, $C = \{3, 4, 5, 6\}$ and $U = \{1, 2, 3, 4, 5, 6\}$, list the elements in
(*a*) $A - B$ (*b*) $B - A$ (*c*) $A - C$
(*d*) $C - A$ (*e*) $A - C'$ (*f*) $C - A'$

38. If $A = \{1, 3, 5, 7, 9\}$, $B = \{2, 4, 6\}$, $C = \{3, 4, 5, 6\}$ and $U = \{1, 2, 3, 4, 5, 6, 7, 8, 9\}$, list the elements in
(*a*) $A - B$ (*b*) $B - A$ (*c*) $A - C$
(*d*) $C - A$ (*e*) $A - C'$ (*f*) $C - A'$

39. Using the sets defined in Problem 23, describe the following sets in words
(*a*) $A - B$ (*b*) $B - A$ (*c*) $A - C$
(*d*) $C - A$ (*e*) $A - C'$ (*f*) $C - A'$

40. An author defines the complement of a set A as $U - A$. Is this definition consistent with the one given in Equation (1)?

REVIEWING MATERIAL

41. (Section 1.3) The actual distance between city A and city B is $187\frac{1}{2}$ miles. The distance between these two cities on a road map is $\frac{3}{4}$ inch. What is the actual distance between cities C and D if the distance between these two cities on the same map is $2\frac{1}{2}$ inches?

42. (Section 2.2) Use Gaussian elimination to solve the system:

$$2x + 3y - 4z = 24$$
$$7x + 8y + 6z = 40$$
$$3x + 2y + 14z = 12$$

43. (Section 3.5) Use the simplex method to solve:

Maximize: $z = 2x_1 + 3x_3$

subject to: $2x_1 + 2x_2 + 5x_3 \le 1{,}200$
              $4x_1 + 3x_2 + 2x_3 \le 800$

assuming: x_1, x_2, and x_3 are nonnegative

44. (Section 4.2) Determine the value of an investment opportunity in two years if the investment pays \$2,500 in three years and another \$3,000 in five years and current interest rates are 4.5% a year, compounded quarterly.

RECOMMENDING ACTION

45. Respond by memo to the following request:

> MEMORANDUM
>
> To: J. Doe Reader
> From: Doris
> Date: Today
> Subject: **Sets**
>
> I would like to simplify things, if possible: Suppose I have a bunch of sets whose intersection is empty. Can I conclude that the sets are mutually disjoint? Conversely, suppose I know that a group of sets are mutually disjoint. Can I conclude that the intersection of all the sets is empty?

5.2 PIE CHARTS AND BAR GRAPHS

A pie chart is a circle that is divided into sectors, with each sector representing one category of the population.

A Venn diagram is a useful model for presenting information when the underlying population is separated into a few sets with nonempty intersections. In different situations, however, other graphical models may display information more clearly.

A group of sets is *exhaustive* if their union is the universal set. The age groupings 0–10, 11–20, and 21–30 are *not* exhaustive for the population of the United States, because the union of these three sets does not include anyone over 30 years old. If we add a fourth category for people over 30, then we would have an exhaustive group of sets. A mutually disjoint, exhaustive group of subsets has the property that every element of the universal set is an element of one and only one of the subsets.

Pie charts are popular graphical models for presenting information about populations that are grouped into a finite number of mutually disjoint, exhaustive subsets. A *pie chart* is a circle (or pie) that is divided into sectors, with each sector representing a subset of the population. The area of a sector is directly proportional to the number of elements in the subset represented by that sector, that is,

$$\frac{\text{area of a sector representing a subset}}{\text{number of elements in the subset}} = k \tag{4}$$

where k is a constant of proportionality. If one subset includes 50% of a population, that subset is represented by 50% of a pie; if a second subset includes 10% of a population, then that subset is represented by 10% of a pie.

Since Equation (4) is true for any sector, it is true when the sector is the entire circle. The entire circle represents the entire population, so Equation (4) specializes to

$$\frac{\text{area of the circle}}{\text{number of elements in the universal set}} = k \tag{5}$$

Combining Equations (*4*) and (*5*), we have

$$\frac{\text{area of a sector representing a subset}}{\text{number of elements in the subset}} = \frac{\text{area of the circle}}{\text{number of elements in the universal set}}$$

which can be rewritten as

$$\frac{\text{area of a sector representing a subset}}{\text{area of the circle}} = \frac{\text{number of elements in a subset}}{\text{number of elements in the universal set}} \qquad (6)$$

The angle of a sector is directly proportional to the area of the sector. If we denote the angle of a sector by θ, then

$$\frac{\theta}{\text{area of sector}} = m \qquad (7)$$

where m is a constant of proportionality. Since Equation (*7*) is true for any sector, it is true when the sector is the entire circle. In this case, the angle of the sector is 360°, and Equation (*7*) becomes

$$\frac{360°}{\text{area of the circle}} = m \qquad (8)$$

Equations (*7*) and (*8*) can be combined into

$$\frac{\theta}{360°} = \frac{\text{area of sector}}{\text{area of the circle}} \qquad (9)$$

Combining Equations (*6*) and (*9*), we can write

$$\frac{\theta}{360°} = \frac{\text{number of elements in a subset}}{\text{number of elements in the universal set}}$$

or

$$\theta = \frac{\text{number of elements in a subset}}{\text{number of elements in the universal set}} \times 360° \qquad (10)$$

as the angle of the pie chart sector that represents a particular subset.

EXAMPLE 1

The *New York Times* reported retail bulk sales (not single servings) of frozen dairy desserts for the year ending June 30, 1990, to be 3.272 billion dollars. The largest market share was for ice cream purchases, which totaled 2.160 billion dollars. Ice milk accounted for 366 million dollars in sales, while frozen yogurt accounted for 474 million dollars in sales. The remaining sales were attributed to sherbet and other miscellaneous frozen dairy desserts. Construct a pie chart for this information.

Solution

The universal set or population is dollars spent on frozen dairy desserts for the year ending June 30, 1990, and it totals 3.272 billion dollars. Ice cream accounts for 2.160 billion dollars, or approximately 66% of the total. It follows from Equation (*10*) that the sector of the pie chart (for popular frozen dairy desserts) representing ice cream will have an angle of

$$\frac{2.160}{3.272} \times 360° \approx 238°$$

Ice milk sales totaled 366 million or 0.366 billion dollars. The pie chart sector representing ice milk will have an angle of

$$\frac{0.366}{3.272} \times 360° \approx 40°$$

Similarly, the pie chart sector representing frozen yogurt will have an angle of

$$\frac{0.474}{3.272} \times 360° \approx 52°$$

Sherbet and other sales totaled $3.272 - 2.160 - 0.366 - 0.474 = 0.272$ billion dollars, so the pie chart sector for this subset will have an angle of

$$\frac{0.272}{3.272} \times 360° \approx 30°$$

The pie chart is shown in Figure 5.11.

FIGURE 5.11
Retail Bulk Sales of Frozen Dairy Desserts

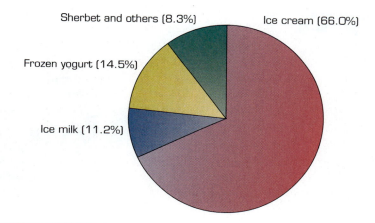

Sherbet and others (8.3%) Ice cream (66.0%)
Frozen yogurt (14.5%)
Ice milk (11.2%)

Figure 5.11 is a graphical representation of how people spent their money for certain desserts, so it is a model of that process. Observe how each sector of the pie chart is shaded differently to make the chart easier to read.

When the number of elements in a subset is reported as a percent of the number of elements in the universal set, Equation (*10*) simplifies to the product of the percent, written as a decimal, times 360°.

EXAMPLE 2

Construct a pie chart for the information in Table 5.1, which was provided by the U.S. Bureau of Census and lists the age distribution for males who married for the first time in 1987.

TABLE 5.1 Age Distribution of Males at First Marriage (1987)

Age	under 20	20–24	25–29	30–34	35–44	45 and older
Percentage	6.9	40.2	33.2	13.0	5.5	1.2

Solution

The population is all males in the United States who married for the first time in 1987, and this becomes our universal set. Males under 20 years old account for 6.9% of the total, so the pie chart sector representing this age group will have an angle of

$$0.069 \times 360° \approx 24.8°$$

The pie chart sector representing males between the ages of 20 and 24, inclusive, will have an angle of

$$0.402 \times 360° \approx 144.7°$$

Continuing in this manner, we generate the pie chart shown in Figure 5.12. We gave the pie chart a three-dimensional look to enhance its appearance, and we detached one sector for emphasis.

FIGURE 5.12

Age Distribution of Males at First Marriage (1987)

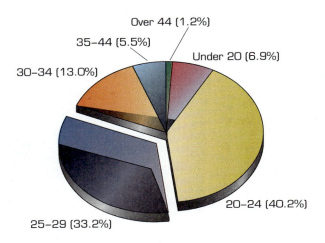

The impact of a pie chart can be accentuated by incorporating the chart into a picture, as was done by *USA Today* for the pie chart in Figure 5.13. (There is a typographical error in Figure 5.13: the category "more than 7" should be "7 or more" to ensure that the sets are exhaustive.)

A *stacked column*, which is constructed from bricks laid one on top of the other, is an alternative graphical model that is equivalent to a pie chart. Each brick represents one subset of the universal set, and the height of each brick is directly proportional to the number of elements in the subset represented by the brick. Figure 5.14 is a stacked column model for the information given in Example 1. Ice cream sales accounted for 66.0% of total sales of frozen dairy desserts, so the brick representing ice cream is 66.0% of the height of the entire column.

Figure 5.15 is a stacked column model for the information given in Example 2. Men between the ages of 25 and 29, inclusive, accounted for 33.2% of all men who married for the first time in 1987, so the brick representing this subset is 33.2% of the height of the entire column.

The *New York Times* created an interesting variation of a stacked column (see Figure 5.16) by using ice cream scoops rather than bricks to represent top-selling flavors. This model would be even more effective had the size of the scoops been proportional to the market share of the flavors the scoops represented—the chocolate scoop should be larger

A stacked column is a column of bricks in which each brick represents one subset of the universal set, and the height of each brick is directly proportional to the number of elements in that subset.

FIGURE 5.13
Reprinted, by permission, from _USA TODAY_, copyright 1993.

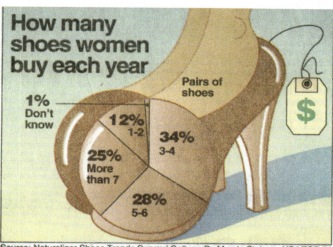

FIGURE 5.14
Retail Bulk Sales of Frozen Dairy Desserts

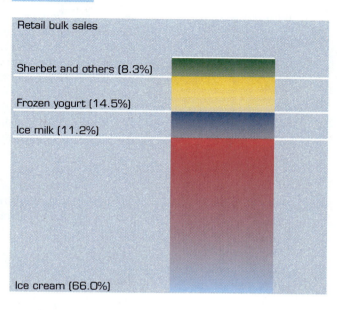

FIGURE 5.15
Age Distribution of Males at First Marriage (1987)

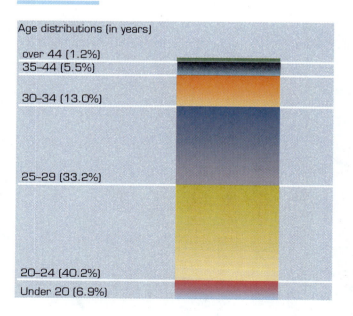

FIGURE 5.16
Reprinted, by permission, from the _New York Times_, copyright 1992/93 by The New York Times Company.

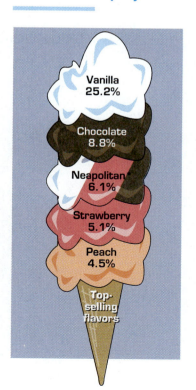

> *A bar graph is a sequence of bars in which each bar represents one subset of a population and its height reflects the number of elements in that subset.*

than the strawberry scoop, and the vanilla scoop should be more than five times larger than the peach scoop.

Pie charts and stacked columns become cluttered and lose their effectiveness as visual models when they depict too many subsets. Most pie charts are limited to eight or fewer sectors, and the fewer sectors, the clearer the chart. Furthermore, pie charts and stacked columns model only groups of subsets that are exhaustive. Other visual models are more useful if a population is separated into many mutually disjoint subsets, if information is available for only some subsets of a population, or if one is interested in population changes with respect to time. In these situations a bar graph, which we used repeatedly in Chapter 4 to display the accumulation of money over time, is often an appropriate model.

A *bar graph* is a sequence of columns or bars in which each bar represents one subset of a population or universal set. The width of each bar is the same, but the height reflects the number of elements in the subset represented by that bar.

EXAMPLE 3

Construct a bar graph for the information in Table 5.2 (provided by the National Highway Traffic Safety Administration) which lists the number of deaths in car and truck accidents per 100 million miles traveled.

TABLE 5.2 Car and Truck Fatalities

Year	1982	1983	1984	1985	1986	1987	1988	1989	1990	1991
Deaths per 100 million miles	2.8	2.6	2.6	2.5	2.5	2.4	2.3	2.2	2.1	1.9

Solution

Information is given for a ten-year period, so we use one bar for each year. The vertical axis is scaled between 0 and 3 (deaths per 100 million miles traveled), and the height of

FIGURE 5.17
Car and Truck Fatalities

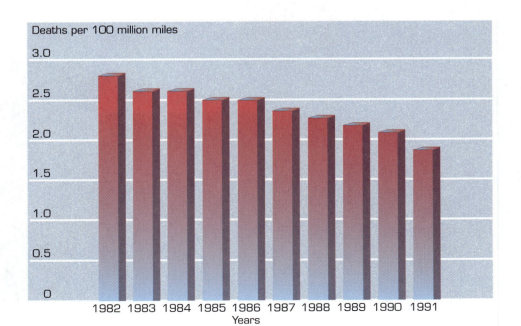

each bar reflects the number of deaths that particular year. The complete bar graph is shown in Figure 5.17. Here we gave each bar a three-dimensional look to make the graph more dramatic.

In Figure 5.18, the *New York Times* varied the traditional form of a bar graph by replacing each bar by a column of bullion bricks, a particularly appropriate symbol for the information being modeled.

FIGURE 5.18

Reprinted, by permission, from the *New York Times,* copyright 1992/93 by The New York Times Company.

Monthly sales of precious metal mutual fund shares, in millions of dollars

Source: Investment Company Institute

Bar graphs are often rotated 90°, placing the subsets on the vertical axis and their heights, which are now their lengths, along the horizontal axis. This presentation is particularly useful when the labels for the various subsets are lengthy.

EXAMPLE 4

The American Cancer Society estimates that in 1993, 182,000 women were diagnosed with breast cancer, 75,000 women were diagnosed with rectal cancer, 70,000 with lung cancer, 44,500 with uterine cancer, 22,400 with lymphoma, 22,000 with ovarian cancer, 14,200 with pancreatic cancer, and 12,600 with leukemia. Represent this information using a bar graph.

Solution

The population is the incidence of diagnosed cancers for women in 1993. It is conceivable that a particular woman might have two different types of cancer simultaneously and therefore might fall into two different categories, but the universal set here is not women but rather diagnosed incidence. Each diagnosed incidence will fall into one and only one category. A woman with two different cancers will be diagnosed twice, once for each cancer, and each diagnosis will fall into a different category. Thus, the subsets are mutually disjoint. Because the labels for each cancer are relatively lengthy, we rotate the bar graph, generating Figure 5.19. We chose two-dimensional bars here because they provide a starker appearance.

FIGURE 5.19

Diagnosed Cancers in Women (1993)

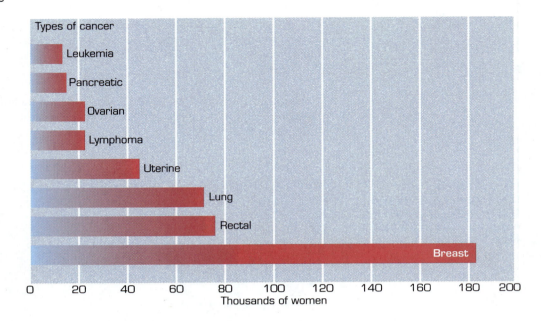

The bar graph in Figure 5.19 illustrates a significant difference between bar graphs and pie charts. The union of the subsets represented by the sectors of a pie chart is the universal set, and each element in the universal set must appear in one of the subsets. Not so with bar graphs. Some elements of the universal set may not appear in any subset represented by the bar graph. Figure 5.19, for example, includes some but not all types of cancer.

EXAMPLE 5

Construct a bar graph for the information in Table 5.3 (provided by the U.S. Bureau of Census) which lists the age distributions of females who married for the first time in 1987, and then compare it to a similar chart for the information in Table 5.1 for males. Note that the percents do not add up to 100 due to rounding.

TABLE 5.3 **Age Distribution of Females at First Marriage (1987)**

Age	under 20	20–24	25–29	30–34	35–44	45 and older
Percentage	18.1	44.2	25.2	8.4	3.5	0.8

Solution

Instead of creating two separate bar graphs for the information in Tables 5.1 and 5.3, we place both bar graphs on the same display. To distinguish information on males from information on females, we use different shadings for each gender and identify the shadings in the chart. Comparable age groupings for each gender are placed next to each other, with different age groups separated from each other by blank space. The result is Figure 5.20. Merged bar graphs are particularly useful for comparing similar information from two or more populations.

FIGURE 5.20
Distribution of Ages at First Marriage (1987)

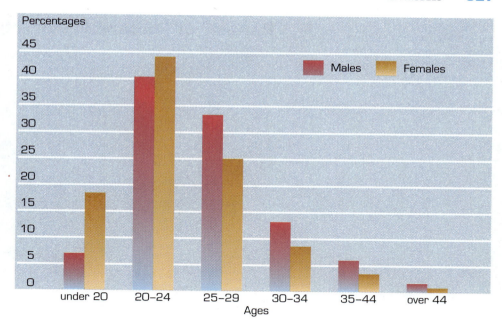

If the labels in a bar graph are too large to fit neatly under the horizontal axis, we rotate the bar graph. Figure 5.21 shows a rotated chart created by the magazine *Security Sales* (CCTV is an acronym for Closed Circuit TeleVision).

The purpose of pie charts, stacked columns, and bar graphs is to provide a concise and

FIGURE 5.21
Reprinted, by permission, from *Security Sales*, October 1993.

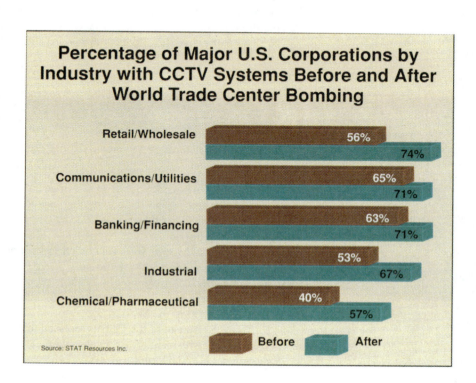

FIGURE 5.22
Reprinted, by permission, from *Newsweek,* October 18, 1993, copyright 1993, Newsweek, Inc.

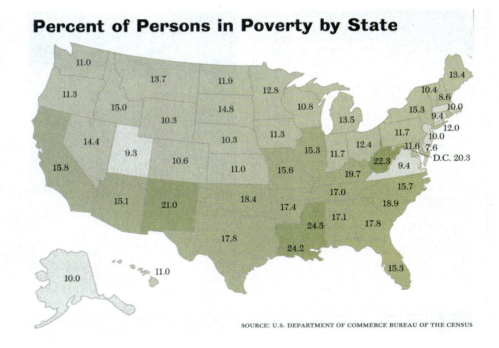

Percent of Persons in Poverty by State

SOURCE: U.S. DEPARTMENT OF COMMERCE BUREAU OF THE CENSUS

readable format for large amounts of information. If other visual models are more appropriate to this purpose, then they should be used. Figure 5.22 is a graphical model of poverty levels by state. Rather than simply listing all 50 states with corresponding percentages of people living in poverty, *Newsweek* placed the percentages on a map and used shading to highlight the differences between states. The darker the shading, the higher the percentage. Note how effective this model is in making the information accessible to readers.

CREATING MODELS

For Problems 1 through 16, create (*a*) pie charts and (*b*) stacked columns for the information provided.

1. *USA TODAY* reported that 36,421 women's bowling leagues, 28,372 men's bowling leagues, and 73,553 mixed bowling leagues existed in 1993.
2. *The General Social Survey* included 1,033 people in 1989 and found that 30% smoked and 70% did not.
3. The *General Social Survey* included 984 people in 1988 and found that 13% belonged to a labor union and 87% did not.
4. The National Association of Realtors reported that in 1988, 673,000 new one-family homes were sold in the Northeast, 929,000 in the Midwest, 1,350,000 in the South, and 642,000 in the West.
5. The U.S. Bureau of the Census reported that in 1989, 20,200 new mobile homes were placed for residential use in the Northeast, 39,100 in the Midwest, 114,900 in the South, and 30,600 in the West.

6. *USA TODAY.* reported in 1994 that the average cost of a formal wedding in the United States had risen to $17,470. The four largest costs were $7,600 for the reception and music, $2,760 for the wedding rings, $2,460 for gowns, tuxes, and attendants' dresses, and $1,570 for photography and videography.

7. The National Science Foundation reported that 400 Nobel prizes were awarded between 1901 and 1989. Of these, 156 went to citizens of the United States, 69 to citizens of the United Kingdom, 57 to citizens of West Germany, 22 to citizens of France, and 10 to citizens of the former Soviet Union. No other country received more than 4.

8. In late January 1994, United Airlines employed 80,123 people. Of these, 8,172 were pilots, 17,376 were flight attendants, 27,942 were managers and nonunion workers, and 26,633 were mechanics and other members of the machinists' union.

9. The U.S. Energy Information Administration reported that the sources of the energy consumed in the United States in 1990 were coal (23.4%), petroleum (41.3%), natural gas (23.8%), and other sources (11.5%).

10. *TV TECHNOLOGY* magazine reported that its international edition received almost 12,000 sales leads during the first six months of 1993. Of these 20% came from Europe, 31% from Latin America, 39% from Asia and the Pacific Rim combined, and the rest from Canada and Africa.

11. The *New York Times* reported that in 1992, the gross revenues from various types of legal gambling in the United States were (in billions of dollars) 11.5 from lotteries, 10.1 from casinos, 3.7 from parimutuels, 1.1 from bingo, 1.5 from Indian gaming, and 2.0 from others. The revenues listed for casinos and bingo do not include revenue from Indian gaming (casinos and bingo on Indian reservations).

12. Information reported in *An American Profile: Opinions and Behavior, 1972–1989* based on a survey of 1533 16 year olds in 1989 showed that 63% were raised as Protestants, 29% as Catholics, 4% with no religion, 2% as Jews, and 2% other.

13. According to the book *The Diversity of Life*, there are 751,000 known insect species in the world, while the other forms of living organisms include 281,000 species of other animals (including mammals), 4,800 different bacteria, 69,000 fungi, 26,900 algae, 1,000 viruses, 248,400 species of higher plants, and 30,800 different protozoa.

14. The Motor and Equipment Manufacturers Association estimated that 116.6 billion dollars were spent in 1993 on car repairs and services, excluding gas, oil, tires, and body parts. Of these, $30 billion were spent at service stations, $25.4 billion at new car dealerships, $17.3 billion at general repair shops, $15 billion at specialty shops, $9.2 billion at tire stores, and $19.7 billion at other facilities.

15. The National Highway Traffic Safety Administration reported that 27.8% of all people killed in passenger car crashes in 1992 were wearing seat belts, 61.9% were not wearing seat belts, and in the remaining cases it was not possible to determine whether a seat belt had been worn.

16. The National Highway Traffic Safety Administration reported that in 1992, 58.8% of all people who survived a passenger car crash in which at least one person was killed were wearing seat belts, 21.1% were not wearing seat belts, and in the remaining cases it was not possible to determine whether a seat belt had been worn.

17. Combine the information in Problems 15 and 16 into a single pie chart if 75% of all people survive car crashes involving fatalities.

In Problems 18 through 35, create bar graphs for the information provided.

18. The U.S. Commerce Department reported that the U.S. trade deficit with China (in billions of dollars) was 3.4 in 1987, 4.2 in 1988, 6.2 in 1989, 10.4 in 1990. 12.7 in 1991, and 18.3 in 1992.

19. The U.S. Department of the Interior reports that the number of pairs of adult bald

eagles occupying nesting areas in the 48 contiguous states between 1981 and 1993 were as follows:

Year	1981	1984	1986	1988	1990	1992	1993
Pairs of nesting eagles	1,188	1,757	1,875	2,475	3,020	3,747	4,016

20. The nine states with the most feature film starts in 1993 were California with 434; New York, 53; Florida, 16; Texas, 14; Arizona, 12; Nevada, 9; Illinois, 9; Utah, 8; and New Jersey, 7.

21. The *New York Times* reported that the ten best-attended state fairs in North America in 1993 were (with attendance in millions of people): State Fair of Texas, 3.15; State Fair of Oklahoma, 1.79; Canadian National Exposition, 1.78; New Mexico State Fair, 1.68; Minnesota State Fair, 1.60; Houston Livestock Show, 1.57; Western Washington Fair, 1.42; Los Angeles County Fair, 1.40; Calgary Stampede, 1.13; Del Mar Fair (California), 1.11.

22. *An American Profile: Opinions and Behavior, 1972–1989* reports that in 1988, 3% of the surveyed population watched, on a daily average, no television, 19% watched one hour, 25% watched two hours, 19% watched three hours, 14% watched four hours, 14% watched five hours, and 6% watched six or more hours.

23. The *Miami Herald* reported that the Florida Power and Light Company employed 14,608 people in 1987, 15,018 in 1988, 15,124 in 1989, 15,497 in 1990, 14,510 in 1991, 14,047 in 1992, and an estimated 12,700 in 1993.

24. *Omaha Magazine* reported that the cost of living index for the fourth quarter of 1991 was 220.3 in New York City, 132.4 in Boston, 117.8 in Seattle, 102.5 in Denver, 99.6 in Boise, 95.0 in Little Rock, 93.5 in Louisville, and 89.5 in Omaha.

25. The following table presents data on the yearly number of new immigrants to the United States (in thousands), as reported in The *Official Guide to the American Marketplace*:

Year	1980	1982	1984	1986	1988	1990
New immigrants	531	594	544	602	643	656

26. In 1993, 31,695 job discrimination claims based on race were filed with the Equal Employment Opportunity Commission. In addition, 23,919 claims were based on sex discrimination, 19,884 on age discrimination, 15,274 on disability discrimination, 7,454 on national origin discrimination, 1,449 on religious discrimination, and 1,334 claims were based on unequal pay.

27. The National Sporting Goods Association estimates revenues (in millions of dollars) from 1993 sales of athletic shoes to be 357 from hiking shoes, 92 from soccer shoes, 189 from golf shoes, 586 from running shoes, 47 from football shoes, 465 from basketball shoes, 601 from tennis shoes, 368 from aerobic shoes, 1,124 from sneakers, and 76 from baseball shoes.

28. The World Health Organization estimated that from 1990 to 1992, the average number of cigarettes smoked per year by persons 15 years old or older was 2,590 in Australia, 2,540 in Canada, 3,590 in Greece, 3,260 in Hungary, 3,240 in Japan, 3,620 in Poland, 3,010 in South Korea, 2,670 in Spain, 2,910 in Switzerland, and 2,670 in the United States.

29. *Florida Journal* reported that the numbers of Europeans visiting Southwest Florida between 1988 and 1993 were as follows:

Year	1988	1989	1990	1991	1992	1993
Visitors	144,700	198,400	265,700	297,800	404,800	441,500

30. The number of people who sought help for heroin problems in U.S. emergency rooms was 38,063 in 1988, 41,656 in 1989, 33,884 in 1990, 35,898 in 1991, and 48,004 in 1992.

31. In 1993, U.S. airlines reported the following operating costs to fly each available passenger seat one mile: American, 8.82¢; America West, 7.08¢; Continental, 8.15¢; Delta, 9.52¢; Northwest, 9.32¢; Southwest, 7.21¢; TWA, 9.37¢; United, 9.35¢; and USAir, 11.35¢.

32. Use the information in Problem 5.

33. Use the information in Problem 8.

34. Use the information in Problem 11.

35. Use the information in Problem 14.

In Problems 36 through 41, create merged bar graphs for the two given sets of information.

36. Compare the information provided in Table 5.1 (Example 2) with the following information on the age distribution of males who remarried in 1987.

Age	under 20	20–24	25–29	30–34	35–44	45 and older
Percentage	0.1	4.2	15.2	20.1	32.8	29.6

37. Compare the information in Problem 18 with U.S. trade deficits with Taiwan, which were (in billions of dollars) 19.0 in 1987, 14.1 in 1988, 13.0 in 1989, 11.2 in 1990, 9.8 in 1991, and 9.4 in 1992.

38. Compare the information in Problem 37 with U.S. trade deficits with Hong Kong, which were (in billions of dollars) 6.5 in 1987, 5.1 in 1988, 3.4 in 1989, 2.8 in 1990, 1.1 in 1991, and 0.7 in 1992.

39. Compare the information in Problems 18, 37, and 38.

40. Use the following data on the number of vacuum cleaners and freezers per 1,000 people sold in the years 1986–1988, as reported in *The Economist Book of Vital World Statistics*.

Country	Denmark	France	Greece	Ireland	Netherlands	Norway
Vacuum cleaners	37.04	48.68	13.99	56.50	25.07	80.95
Freezers	15.59	13.07	34.97	14.12	12.87	16.67

41. Use the following data (in units of kilos per person) on the consumption of cereal and potatoes in the period 1987–1988, as reported in *The Economist Book of Vital World Statistics*.

Country	Mexico	Spain	Turkey	India	Argentina	Indonesia	United States
Cereal	17.5	95.7	192.2	18.0	50.0	3.3	67.3
Potatoes	12.0	102.6	67.6	0.9	63.0	1.1	29.5

EXPLORING IN TEAMS

42. Many of the government statistics cited in this section are found in the book *Statistical Abstracts of the United States*, which is available in most libraries. Locate a recent copy of this book and use information in it to write a short report on some aspect of American life. Include pie charts and bar graphs in your report.

EXPLORING WITH TECHNOLOGY

43. Pie charts and bar graphs can be constructed very easily using electronic spreadsheets or graphical software packages, which are available in many microcomputer centers. Use one of these software packages to reproduce Figures 5.11, 5.12, 5.17, 5.19, and 5.20.

44. Use a graphical software package or electronic spreadsheet to solve Problems 1 through 41.

REVIEWING MATERIAL

45. (Section 1.4) Graph the equation

$$y = \frac{x^3}{2} - 2$$

46. (Section 2.3) Determine the equation of the least-squares straight line that fits the information given in Example 3. Use this result to estimate the number of deaths in cars and truck accidents per 100 million miles traveled in 1995, assuming the trend continues.

47. (Section 3.3) Use graphical methods to solve the following linear program:

$$\text{Maximize:} \quad z = 2x_1 + 3x_2$$

$$\text{subject to:} \quad 8x_1 + 3x_2 \le 2,400$$

$$4x_1 + 7x_2 \le 2,800$$

$$x_1 \ge 0$$

$$x_2 \ge 0$$

48. (Section 4.1) Determine the annual effective interest rate for 7.2% a year compounded daily. (Use 365 days to a year.)

RECOMMENDING ACTION

49. Respond by memo to the following request:

> **MEMORANDUM**
>
> To: J. Doe Reader
>
> From: The Publisher
>
> Date: Today
>
> Subject: **Standards**
>
> I was horrified to see in yesterday's paper a bar graph on cancer deaths that used coffins to represent deaths. I though the graphics were in bad taste and detracted from the point we were trying to make in the article. We need some standards here. I encourage the use of interesting icons to jazz up bar graphs, but I insist that this usage be in good taste. Please prepare a memo that establishes standards on the use of icons in bar graphs.

5.3 STANDARD MEASURES FOR NUMERICAL DATA

Data are recorded observations of a process.

The mean of n data points is the sum of the data divided by n.

Data are recorded observations of a process. These observations may be numerical or textual (for example, names). We collect data to better understand a process of interest, and we often display data using the presentation models described in the previous section.

A convenient way to characterize a large amount of numerical data is to identify a typical element in the process. We do this routinely when we report the average salary of baseball players, the life expectancy of a typical American male, or the income of a typical American family. A *measure of central tendency* is a single number that reasonably represents all the numerical data. The most common measure of central tendency is the *mean* or *average* value. If we denote a data list as $x_1, x_2, x_3, \ldots, x_n$, then the mean $\bar{x}$ is

$$\bar{x} = \frac{x_1 + x_2 + x_3 + \ldots + x_n}{n} \tag{11}$$

EXAMPLE 1

The final examination scores for a senior seminar are 93, 81, 69, 86, 97, 50, 86, 77, and 72. What is the mean examination grade?

Solution

$$\bar{x} = \frac{93 + 81 + 69 + 86 + 97 + 50 + 86 + 77 + 72}{9} = 79$$

Even though the mean represents a *typical* element in the process, the mean itself may not be an actual element. None of the test scores in Example 1 equals the mean test score of 79.

A company employs 100 people, and each employee receives the same salary, $15,000 a year, except for the president, who receives an annual salary of 7 million dollars. What is the mean annual salary of all company workers?

Solution

There are 100 employees, including the president. The president receives 7 million dollars a year, and the other 99 employees each receive $15,000 a year. The mean of these salaries is

$$\bar{x} = \frac{\$7,000,000 + 99(\$15,000)}{100} = \$84,850$$

The median of n data points is a number such that half the data points have values greater than or equal to that number and the other half have values less than or equal to that number.

Example 2 illustrates a case in which the mean is a deceptive representative of the data. Most people would jump at the chance to work for a company boasting an average annual salary of $84,850, but probably not the company described in Example 2. The mean is a skewed indicator of data if one or more pieces of data are significantly larger than all the others. In such situations, other measures of central tendency are preferred.

The *median* of *n* data points is a number such that half the data points have values greater than or equal to the median and the other half have values less than or equal to the median. To determine the median of data, we first list the data in either ascending or descending order. If *n* is odd, that is, if there is an odd number of data points, then the median is the middle number in the list; if *n* is even, then the median is the mean of the two numbers in the middle of the list.

EXAMPLE 3

Determine the medians for the two sets of data given in Examples 1 and 2.

Solution

Arranging the data in Example 1 in descending order, we obtain the list

97, 93, 86, 86, 81, 77, 72, 69, 50.

The middle number in this list of nine data points is the fifth number, 81, which is the median. There are four data points with values less than 81 and another four data points with values greater than 81.

Arranging the data in Example 2 in ascending order, we would obtain a list of ninety-nine values of $15,000 followed by $7,000,000. This list contains 100 entries. The 50th and the 51st entries are the two middle numbers, and both are $15,000. The median is

$$\frac{\$15,000 + \$15,000}{2} = \$15,000$$

The mode of a list of data is the number that appears most frequently in the list.

The *mode* is a less common measure of central tendency than the mean and the median. The *mode* is the number that appears most frequently in a list of data. If two or more values appear equally often and more frequently than all other numbers in the data list, then each is a mode. If no data value repeats, then there is no mode. A list of data may have no mode, one mode, or many modes.

EXAMPLE 4

Determine the modes for the two sets of data given in Examples 1 and 2.

Solution

The mode for the data in Example 1 is 86, because that is the only score that occurs more than once. The mode for the data in Example 2 is $15,000, because that value is repeated 99 times, while the only other value, $7,000,000, appears just once.

EXAMPLE 5

Determine the mode for each data list:
(*a*) {5, 2, 3, 8, 8, 3, 4, 5, 5, 1, 3}
(*b*) {5, 2, 3, 18, 8, 12, 4, 15, 25, 1, 33}

Solution

(*a*) The numbers 3 and 5 appear most frequently (three times) and equally often, so both numbers are modes. When a set of data has two modes, as is the case here, the set is said to be *bimodal*.
(*b*) No number appears more than once, so there is no mode for this data list.

A moving average is a sequence of means used to smooth data recorded over equal intervals of time.

Data are often recorded over equal intervals of time, for example, every hour or every day. In particular, some U.S. population data is reported annually. If there are significant fluctuations in the data from one time interval to the next, a *moving average*, which essentially smoothes the data, is used to model the underlying process. To compute an n-time moving average, where n is a specified positive integer, we calculate a mean at each recorded time using the observation at that time and the $n - 1$ preceding observations. No means are calculated for times when there are not enough preceding reporting intervals to calculate a mean. Three-day moving averages ($n = 3$) involve means for successive three-day periods, using the data for each day and for the two preceding days; four-month moving averages ($n = 4$) involve means for successive four-month periods, using the data for each month and for the three preceding months. The result is a sequence of means that "move with the data."

EXAMPLE 6

Calculate three-month moving averages for the sales data in Table 5.4.

Solution

A bar graph of the data is displayed in Figure 5.23. Three-month moving averages ($n = 3$) involve means for successive three-month periods, using the data for each month and for the two preceding months. For the data in Table 5.4, we begin by calculating the

TABLE 5.4 **Monthly Sales (Millions of Dollars)**

January	28	July	29
February	32	August	31
March	29	September	27
April	33	October	29
May	32	November	31
June	27	December	33

FIGURE 5.23
Monthly Sales

three-month moving average for March, because March is the first month for which we have data for that month and for the two preceding months. The means, rounded to one decimal, are:

$$\text{March:} \quad \frac{29 + 32 + 28}{3} \approx 29.7$$

$$\text{April:} \quad \frac{33 + 29 + 32}{3} \approx 31.3$$

$$\text{May:} \quad \frac{32 + 33 + 29}{3} \approx 31.3$$

$$\text{June:} \quad \frac{27 + 32 + 33}{3} \approx 30.7$$

Continuing in this manner, we obtain three-month means of 29.3 for July, 29 for August,

FIGURE 5.24
Three-Month Moving Averages for Sales

September, October, and November, and 31 for December. Figure 5.24 is a bar graph of these means.

EXAMPLE 7

Calculate five-year moving averages for the data in Table 5.5, which lists the annual rainfall in Baltimore, Maryland, from 1951 through 1965, as reported by U.S. Weather Service.

TABLE 5.5 Annual Rainfall (Inches) Baltimore, Maryland

1951	46.9	1959	35.8
1952	55.9	1960	43.9
1953	49.3	1961	40.0
1954	30.5	1962	38.1
1955	47.9	1963	34.1
1956	37.8	1964	37.2
1957	37.7	1965	30.8
1958	50.4		

Solution

The data are displayed in Figure 5.25. Not surprisingly, there are significant fluctuations in total rainfall from one year to the next, so long-range trends may be difficult to discern. Five-year moving averages ($n = 5$) involve means for successive five-year periods, using the data for each year and for the four preceding years. For the data in Table 5.4, we begin by calculating a five-year moving average for 1955, because that is the first

FIGURE 5.25
Rainfall in Baltimore

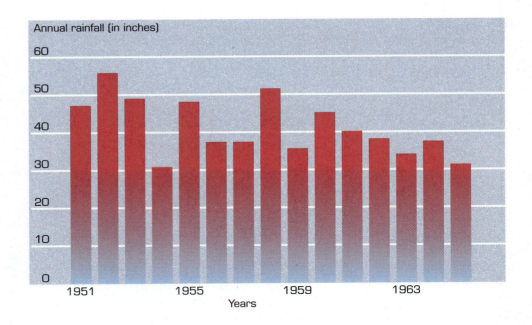

year for which we have data for that year and also for the four preceding years. The means are:

$$1955: \quad \frac{47.9 + 30.5 + 49.3 + 55.9 + 46.9}{5} = 46.1$$

$$1956: \quad \frac{37.8 + 47.9 + 30.5 + 49.3 + 55.9}{5} = 44.28$$

FIGURE 5.26
Five-Year Moving Averages for Rainfall in Baltimore

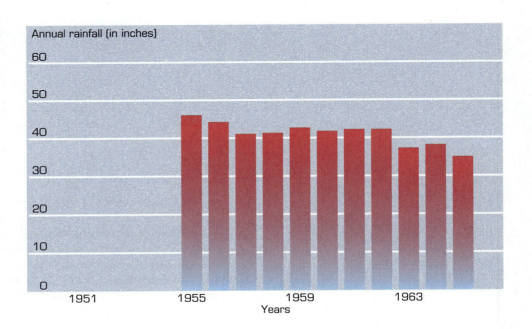

$$1957: \quad \frac{37.7 + 37.8 + 47.9 + 30.5 + 49.3}{5} = 40.64$$

$$1958: \quad \frac{50.4 + 37.7 + 37.8 + 47.9 + 30.5}{5} = 40.86$$

Continuing in this manner, we obtain five-year means of 41.92 for 1959, 41.12 for 1960, 41.56 for 1961, 41.64 for 1962, 38.38 for 1963, 38.66 for 1964, and 36.04 for 1965. Figure 5.26 is a bar graph of these means. Note how much smoother the five-year averages are than the original data in Figure 5.25. It seems reasonable to conclude that yearly rainfalls in Baltimore decreased, as a general trend, from 1951 through 1965.

The arithmetic for calculating consecutive moving averages can be simplified if we note that each mean differs from its predecessor by the inclusion of one new data point and the deletion of one old data point. In Example 7, the 1956 mean differed from the 1955 mean by the inclusion of the 1956 rainfall and the deletion of the 1951 rainfall. Similarly, the 1957 mean differed from the 1956 mean by the inclusion of the 1957 rainfall and the deletion of the 1952 rainfall. In general, each new mean can be calculated from the previous mean by adding to the previous mean the difference between the new data point and the oldest data point in the previous mean, divided by n. In Example 7 (with $n = 5$),

$$1956 \text{ mean} = 1955 \text{ mean} + \left(\frac{37.8 - 46.9}{5} \right)$$
$$= 46.1 + (-1.82)$$
$$= 44.28$$

Similarly,

$$1957 \text{ mean} = 1956 \text{ mean} + \left(\frac{37.7 - 55.9}{5} \right)$$
$$= 44.28 + (-3.64)$$
$$= 40.64$$

Of the three measures of central tendency—the mean, median, and mode—the most commonly used measure is the mean. The mean is a particularly good measure for summarizing gross amounts, such as salaries, number of children, test scores, and sales. The mean can be distorted as a reliable measure of central tendency by a few data points that are significantly larger or smaller than the rest, in which case other measures are preferred. The median is useful as a benchmark for positioning data. Individual data points can be ranked as being either below the median or above it. The mode is a measure of popularity, identifying the data point that appears most frequently. Modes are particularly important in determining consumer preferences.

Measures of central tendency characterize data by a *typical* observation. Missing in such a characterization is some indication about the variation between the data and this typical element. It is one thing to know that an average starting salary is $30,000; it is something else again to also know that there is *no* variation in starting salaries, or that there is a $15,000 variation around this average. A number that quantifies the variation in data is called a *measure of dispersion*.

The simplest measure of dispersion is the *range*, which is the difference between the largest number and the smallest number in a list of data. The range is often used in tandem with the median to describe data.

The range is the difference between the largest number and the smallest number in a data list.

EXAMPLE 8

Determine the range of the final examination scores reported in Example 1 (93, 81, 69, 86, 97, 50, 86, 77, and 72).

Solution

The highest score is 97 and the lowest score is 50, so the range is $97 - 50 = 47$. There is a 47-point spread in the data.

One seemingly natural way to measure dispersion using the mean is to calculate the *deviation* (*i.e.*, the difference) between the mean and each data point and sum all the deviations. If we do this for the examination data in Example 8, which has a mean $\bar{x} = 79$ (see Example 1), we generate Table 5.6. Notice that the sum of the deviations is zero, because some deviations are positive while others are negative, and each type cancels out the other. We want to reserve zero as a measure of dispersion that indicates that there is no variation whatsoever in the data, so we do not want to use a measure that allows sign cancellations of different nonzero deviations. One way to avoid this situation is to square each deviation, thus guaranteeing that each quantity to be summed is nonnegative. The *variance* is the mean of the squares of the deviations. If we list the data as $x_1, x_2, x_3, \ldots, x_n$ with *mean $\bar{x}$*, then

> *The variance is the mean of the squares of the deviations of each data point from the mean.*

$$\text{variance} = \frac{(x_1 - \bar{x})^2 + (x_2 - \bar{x})^2 + (x_3 - \bar{x})^2 + \ldots + (x_n - \bar{x})^2}{n} \tag{12}$$

Equation (*12*) is also known as the *variance for a population*. This formula is applicable only when we have observations for *all* members of a population. A slightly different formula, known as the variance for a sample, is used when we have observations for only some of the members of a population. Sampling and the variance for a sample will be discussed in Section 9.3.

TABLE 5.6

DATA	MEAN	DEVIATION FROM THE MEAN
93	79	$93 - 79 = 14$
81	79	$81 - 79 = 2$
69	79	$69 - 79 = -10$
86	79	$86 - 79 = 7$
97	79	$97 - 79 = 18$
50	79	$50 - 79 = -29$
86	79	$86 - 79 = 7$
77	79	$77 - 79 = -2$
72	79	$72 - 79 = -7$
		Sum of deviations $= 0$

EXAMPLE 9

Determine the variance for the data displayed in Table 5.6.

Solution

The data are the final examination scores described in Example 1. We presume that these are *all* the scores and that we have an observation for every student in the senior seminar. Squaring each of the deviations listed in the last column of Table 5.6 and then computing the mean of those squares, we obtain

$$\frac{14^2 + 2^2 + (-10)^2 + 7^2 + 18^2 + (-29)^2 + 7^2 + (-2)^2 + (-7)^2}{9}$$

$$= \frac{1,616}{9}$$

$$= 179.5556$$

rounded to four decimal places, as the variance.

> *The standard deviation is the square root of the variance.*

The *standard deviation* is the square root of the variance, and it is the measure of dispersion most commonly associated with the mean.

EXAMPLE 10

Determine the standard deviation for the data displayed in Table 5.6.

Solution

Using the results of Example 9, we have

$$\text{standard deviation} = \sqrt{\text{variance}} = \sqrt{179.5556} = 13.40$$

rounded to two decimal places.

EXAMPLE 11

Find the range and standard deviation of the annual rainfall data for Baltimore displayed in Table 5.5.

Solution

The data set is reproduced in the first two columns of Table 5.7. We see that the largest amount of rainfall was 55.9 inches in 1952 and the smallest amount was 30.5 inches in 1954. The range of the data is $55.9 - 30.5 = 25.4$ inches. To calculate the standard deviation, we need the mean. Summing the data entries in column 2 of Table 7, we obtain 616.3 total inches of rainfall over the period 1951–1965. There are 15 years in this period, so the mean is

$$\bar{x} = \frac{616.3}{15} = 41.0867 \text{ inches}$$

rounded to four decimal places. This mean fills column 3 of Table 5.7. Subtracting the mean from each annual rainfall gives the deviations listed in column 4. Squaring each deviation, we obtain the entries in column 5, again rounded to four decimal places. The sum of the deviations squared is 803.8973. With 15 data entries, we calculate

$$\text{variance} \approx \frac{803.8973}{15} \approx 53.593 \text{ in}^2$$

TABLE 5.7

YEAR	RAINFALL	MEAN	DEVIATION	DEVIATION SQUARED
1951	46.9	41.0867	5.8133	33.7948
1952	55.9	41.0867	14.8133	219.4348
1953	49.3	41.0867	8.2133	67.4588
1954	30.5	41.0867	−10.5867	112.0775
1955	47.9	41.0867	6.8133	46.4215
1956	37.8	41.0867	−3.2867	10.8022
1957	37.7	41.0867	−3.3867	11.4695
1958	50.4	41.0867	9.3133	86.7382
1959	35.8	41.0867	−5.2867	27.9488
1960	43.9	41.0867	2.8133	7.9148
1961	40.0	41.0867	−1.0867	1.1808
1962	38.1	41.0867	−2.9867	8.9202
1963	34.1	41.0867	−6.9867	48.8135
1964	37.2	41.0867	−3.8867	15.1062
1985	30.8	41.0867	−10.2867	105.8155
	sum = 616.3			sum = 803.8973

and

$$\text{standard deviation} = \sqrt{\text{variance}} \approx \sqrt{53.593} = 7.32 \text{ inches}$$

rounded to two decimal places.

A number y lies within k standard deviations of a mean $\bar{x}$ if

$$\bar{x} - k(\textit{standard deviation}) \leq y \leq \bar{x} + k(\textit{standard deviation})$$

An important result, known as *Chebyshev's Theorem* states that the fraction of data points lying within k standard deviations of the mean is at least as large as $1 - \dfrac{1}{k^2}$. For $k = 2$,

$$1 - \frac{1}{2^2} = 1 - \frac{1}{4} = \frac{3}{4}$$

Thus at least $\frac{3}{4}$ or 75% of the data lies within two standard deviations of the mean. For $k = 3$,

$$1 - \frac{1}{3^2} = 1 - \frac{1}{9} = \frac{8}{9}$$

This means that at least 8/9 or approximately 89% of the data lies within three standard deviations of the mean.

> *Chebyshev's Theorem:*
> *The fraction of data points lying within* k *standard deviations of the mean is at least as large as* $1 - \dfrac{1}{k^2}$.

EXAMPLE 12

Verify Chebyshev's Theorem for the rainfall data listed in Table 5.7, using $k = 2$ and $k = 3$.

Solution

It follows from Example 11 that the mean is $\bar{x} = \dfrac{616.3}{15} = 41.09$ in, rounded to two decimal places, and the standard deviation is 7.32 in. Two standard deviations is $2(7.32) = 14.64$ inches, so two standard deviations from the mean is the interval

$$\text{from } 41.09 - 14.64 \quad \text{to} \quad 41.09 + 14.64$$

or

$$\text{from } 26.45 \quad \text{to} \quad 55.73$$

According to Chebyshev's Theorem, *at least* 75% of the data points must fall within this interval. In fact, we see that 14 of the 15 data points (all but the data point for 1952) are within this interval.

Three standard deviations is $3(7.32) = 21.96$ inches, and three standard deviations from the mean is the interval

$$\text{from } 41.09 - 21.96 \quad \text{to} \quad 41.09 + 21.96$$

or

$$\text{from } 19.13 \quad \text{to} \quad 63.05$$

According to Chebyshev's Theorem, *at least* 89% of the data points must fall within this interval. In fact, all the data are in this interval.

Standard measures for data, both measures of central tendency and measures of dispersion, use just a few numbers to characterize a process. Such indicators are limited, however, and provide no indication about trends. Review again Example 3 of Section 5.2, which deals with traffic deaths in the United States. All of the standard measures described in this section can be calculated for that data, but none would be as revealing as the bar graph in Figure 5.17. Figure 5.17 displays ten pieces of information, one for each year, so we expect it to reveal more than one or two pieces of information. The more information we have about a process, the more we understand the process and the better our decisions should be for directing that process.

IMPROVING SKILLS

In Problems 1 through 14, calculate the (*a*) mode, (*b*) median, (*c*) range, (*d*) mean, and (*e*) standard deviation for the given lists of numbers.

1. 8, 8, 9, 9, 10, 10, 11, 12, 12, 12, 13
2. 8, 8, 9, 9, 10, 10, 11, 12, 12, 12, 13, 13
3. 8, 8, 9, 9, 9, 10, 11, 12, 12, 12, 13
4. 8, 8, 9, 9, 9, 10, 11, 12, 12, 12
5. 8, 8, 9, 9, 9, 10, 11, 12, 12

6. 1, 2, 3, 4, 5, 6, 7, 8, 9
7. 1, 3, 5, 7, 9
8. 2, 4, 6, 8, 10
9. 2, 4, 6, 8, 10, 12
10. 2, 2, 2, 2, 2, 2, 2, 2, 2
11. 2, 2, 2, 2, 2, 2, 2, 2, 900
12. 1, 1, 1, 2, 2, 2
13. 1
14. 1, 1, 1
15. A list of data has a mean of 72 and a standard deviation of 6. Use Chebyshev's theorem to determine the fraction of data points that lie in the following intervals:
    (**a**) [60, 84]   (**b**) [54, 90]   (**c**) [48, 96]   (**d**) [42, 102]
    (**e**) less than 60 or more than 84   (**f**) less than 54 or more than 90
16. A list of data has a mean of 235 and a standard deviation of 13. Use Chebyshev's theorem to determine the fraction of data points that lie in the following intervals:
    (**a**) [209, 261]   (**b**) [196, 274]   (**c**) [183, 287]   (**d**) [170, 300]
    (**e**) less than 196 or more than 274   (**f**) less than 170 or more than 300

CREATING MODELS

17. The annual rainfall in inches in San Francisco between 1959 and 1970 is listed below:

1959	12.5	1962	20.0	1965	19.9	1968	18.0
1960	17.8	1963	18.8	1966	16.5	1969	27.0
1961	14.6	1964	17.7	1967	24.3	1970	24.3

Find the mean and the standard deviation for this data. Construct a bar graph for this data, and then make a separate graph showing the four-year moving averages over this period.

18. The annual domestic fish catch (in millions of pounds) from 1980 through 1987 is as follows:

1980	6,482	1982	6,367	1984	6,438	1986	6,031
1981	5,977	1983	6,439	1985	6,258	1987	6,896

Find the mean and the standard deviation for this data. Construct a bar graph for this data, and then make separate graph of the three-year moving averages over this period.

19. The U.S. Department of Agriculture estimates that the annual corn harvest (in millions of metric tons) in the United States from 1980 through 1992 is as follows:

1980/1	168.8	1984/5	194.9	1988/9	125.2
1981/2	208.3	1985/6	225.2	1989/90	191.2
1982/3	212.3	1986/7	209.6	1990/1	201.6
1983/4	106.8	1987/8	179.4	1991/2	190.2

Find the mean and standard deviation for this data. Construct a bar graph for this data, and then make a separate graph showing the five-year moving averages over this period.

20. The total sales (in thousands of units) of new, privately owned, one-family houses in the United States from 1971 through 1990 is as follows:

1971	656	1976	646	1981	436	1986	750
1972	718	1977	819	1982	412	1987	671
1973	634	1978	817	1983	623	1988	676
1974	519	1979	709	1984	639	1989	650
1975	549	1980	545	1985	688	1990	534

Find the mean and standard deviation for this data. Construct a bar graph for this data, and then make a separate graph of the four-year moving averages over this period.

21. The U.S. Bureau of Labor Statistics reports the total number of work stoppages between 1970 and 1980 (excluding stoppages of less than one day in duration) for corporations having 1,000 or more employees as follows:

1970	381	1973	317	1976	231	1979	235
1971	298	1974	424	1977	298	1980	187
1972	250	1975	235	1978	219		

Find the mean and standard deviation for this data. Construct a bar graph for this data, and then make a separate graph of the three-year moving averages over this period.

22. The Crop Reporting Board of the U.S. Department of Agriculture estimates that the average price farmers received for hogs (in cents per pound) over an 18-month period from July 1990 through December 1991 was as follows

July	1990	60.8	Jan.	1991	50.0	July	1991	54.2
Aug.	1990	55.9	Feb.	1991	52.2	Aug.	1991	51.2
Sept.	1990	54.3	Mar.	1991	51.5	Sept.	1991	46.4
Oct.	1990	56.8	Apr.	1991	50.9	Oct.	1991	43.6
Nov.	1990	50.2	May	1991	54.1	Nov.	1991	38.0
Dec.	1990	47.8	June	1991	54.7	Dec.	1991	38.6

Find the mean and standard deviation for this data. Construct a bar graph for this data, and then make a separate graph showing the four-month moving averages over this period.

23. The average spot price for unleaded gasoline on the Chicago Mercantile Exchange (in cents per gallon) over an 18-month period from July 1991 through December 1992 is as follow

July	1991	65.23	Jan	1992	53.04	July	1992	59.94
Aug.	1991	69.90	Feb.	1992	55.14	Aug.	1992	62.06
Sept.	1991	62.17	Mar.	1992	54.10	Sept.	1992	61.63
Oct.	1991	64.41	Apr.	1992	59.75	Oct.	1992	60.81
Nov.	1991	65.10	May	1992	63.48	Nov.	1992	58.22
Dec.	1991	55.55	June	1992	64.51	Dec.	1992	53.24

Find the mean and standard deviation for this data. Construct a bar graph for this data, and then make a separate graph of the six-month moving averages over this period.

EXPLORING IN TEAMS

24. *Exponential moving averages* are calculated from the last computed mean and the next data point in sequence by the formula

$$\text{new mean} = 0.9*(\text{old mean}) + 0.1*(\text{current data point})$$

The advantages of exponential moving averages over standard moving averages are (*i*) the most current data point is weighted more heavily than other data points in computing each successive mean, and (*ii*) all of the previous data points are used in calculating each new mean. The sequence of exponential moving averages is initialized by calculating the mean of the first 10 data points in the normal fashion. Then exponential means are calculated beginning with the 11*th* data point.

Calculate exponential moving averages for the rainfall data displayed in Table 5.5 if it is known that the mean annual rainfall for the 10-year period from 1941 to 1950 is 43.0 inches.

25. Calculate exponential moving averages for the data in Problem 22 if it is known that the mean price per pound of hog was 50.53¢ for the preceding ten-month period.

26. Calculate exponential moving averages for the data in Problem 20 for the period 1981 to 1990.

EXPLORING WITH TECHNOLOGY

27. Many calculators have statistical capabilities for inputting data and then calculating the mean, variance, and standard deviation of that data. Using such a calculator, reproduce the results of Examples 1, 9, 10, and 11.

28. Using a calculator with statistical capabilities, find the mean and standard deviation for the data in Problems 1 through 14 and Problems 17 through 23.

REVIEWING MATERIAL

29. (Section 1.5) Find the equation of a straight line that passes through the point $(2, -10)$ and is parallel to the line defined by the equation $5x + 4y = 2$.

30. (Section 2.4) Find $\mathbf{A}^2$ and $\mathbf{A}^3$ when

$$\mathbf{A} = \begin{bmatrix} 1 & 0 & 1 \\ 1 & 1 & 0 \\ 0 & 1 & 2 \end{bmatrix}$$

31. (Sections 3.1 and 3.3) A seamstress finishes skirts and trousers for a clothing manufacturer. Skirts take two minutes to finish and earn the seamstress 50¢ each; trousers take five minutes to finish and earn the seamstress $1.00 each. She must finish at least four trousers and six skirts each hour. How many skirts and trousers should the seamstress finish each hour to maximize her hourly income?

32. (Section 4.3) A couple saves $25 a week in a simple annuity due for four years at a

5.3% annual interest rate, compounded weekly. How much will be in the account at the end of the annuity?

RECOMMENDING ACTION

33. Respond by memo to the following request:

> ## MEMORANDUM
>
> To: J. Doe Reader
>
> From: Office of the Executive Vice President
>
> Date: Today
>
> Subject: **Averages**
>
> I enjoyed your seminar last week on moving averages and the average as defined by the mean for data taken over time. A very nice job, indeed! I regret to admit, however, that I remain unclear as to how these two concepts relate to each other, if at all. Do me a favor and plot the mean on the same graph with the moving average for any available data and see if you can uncover any relationships. I would be grateful for any insights you provide.

5.4 TREE DIAGRAMS

A process is an action or decision that can result in one or more possible outcomes.

A sample space for a process is the set of all possible outcomes for that process.

Generally, when a person or organization takes an action or makes a decision, several outcomes are possible. The decision-maker anticipates a particular outcome, but other less desirable possibilities may occur. A person may decide to take a new job, expecting better pay or better working conditions, but the new job may not meet the worker's expectations. A corporation may fund new research in anticipation of a more profitable product; the research may succeed, but it may also fail, take too long, or cost too much.

Any action or decision that can result in one or more possible outcomes is a *process*. A decision to initiate research is a process. Hiring a new employee from a list of applicants is a process. Flipping a coin is a process.

The set of all possible outcomes for a particular process is a *sample space* for that process. Sample spaces are denoted by U, the same notation we used in Section 5.1 for a universal set. Since a sample space for a process includes all possible outcomes, it is the universal set for the process. The cardinal number of a sample space, $n(U)$, is the number of different outcomes in the sample space.

Defining a sample space for a process can be subjective, because a possible outcome to one person may not be an outcome to a second person. The sample space for the outcomes of flipping a coin is an example. One person may list the sample space as $U_1 = \{$head, tail, side, coin lost$\}$, with $n(U_1) = 4$. A second person may not consider a coin landing on its side a realistic possibility (most people do not) and may take $U_2 = \{$head, tail, coin lost$\}$ as the appropriate sample space, with $n(U_2) = 3$. A third person may not consider losing a coin a realistic outcome and may take $U_3 = \{\text{head, tail}\}$ as the standard sample space with $n(U_3) = 2$.

EXAMPLE 1

Determine a sample space for the process of selecting a high school student and identifying (*a*) the student's gender, (*b*) the student's grade level, and (*c*) the student's grade level *and* gender.

Solution

(*a*) $U = \{\text{male, female}\}$, with $n(U) = 2$

(*b*) $U = \{\text{freshman, sophomore, junior, senior}\}$, with $n(U) = 4$

(*c*) $U = \{\text{freshman-female, freshman-male, sophomore-female, sophomore-male, junior-female, junior-male, senior-female, senior-male}\}$, with $n(U) = 8$

If follows from Example 1 that the sample space changes as the process changes. If a process admits only a few possible outcomes, then a sample space is constructed easily by identifying elements as they come to mind, as we did in Example 1, and then counting the number of outcomes directly. We need a more systematic approach, however, to avoid overlooking possibilities when a sample space contains many outcomes.

Consider a corporation that must close two of its four plants, which are located in Buffalo, Milwaukee, Cleveland, and Nashville; one plant will be closed each year over a two-year period. We can develop a systematic procedure for listing all possible outcomes by placing ourselves in the position of the decision-maker. To start the process, we must decide which plant will close first. We have a choice, because we can close any one of the four plants. Therefore, we begin graphically with Figure 5.27.

Figure 5.27 is an example of a one-level tree. It is one level because only one decision is illustrated: which plant to close first. Care must be taken in reading the diagram. Each branch from "Start" represents one *possible* outcome. In Figure 5.27, we have four branches, so there are four possible ways to schedule the first closing. We can close the first plant in Buffalo (B), Milwaukee (M), Cleveland (C), or Nashville (N).

Once we decide which plant will be closed first, we then must decide which plant will be closed second. Again we have choices, but now the choices depend on which plant was closed first. If we decide to close Buffalo first, we cannot schedule Buffalo for the second closing, too. If Buffalo is closed first, we have only three options for the second closing: Milwaukee, Cleveland, or Nashville. These options are shown in Figure 5.28.

If, instead, we decide to close the Nashville plant first, we again have three choices for the second closing, but we have three different choices—Buffalo, Milwaukee, or Cleveland. Adding these choices to Figure 5.28, we obtain Figure 5.29.

We complete the tree by adding the possible second choices when the Milwaukee plant is closed first and when the Cleveland plant is closed first. The result is the two-level tree shown in Figure 5.30, which is a model of the closing process.

Each complete path in Figure 5.30, beginning at "Start" and moving from left to right through the nodes and branches without any backtracking, is one possible outcome to the process of closing two plants, one each year. The top path is Buffalo first and Milwaukee second. Directly under the top path is the path Buffalo first and Cleveland second, which represents a second closing option. Using this tree, we can easily list all 12 possible schedules by starting with the top path and moving sequentially to the bottom path:

$$U = \{\text{BM, BC, BN, MB, MC, MN, CB, CM, CN, NB, NM, NC}\}$$

FIGURE 5.27

First closing

FIGURE 5.28

FIGURE 5.29

FIGURE 5.30

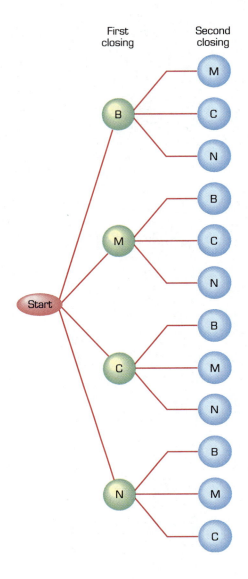

First
closing

Second
closing

EXAMPLE 2

A used-car dealer wants to display two Fords, one red and one white, and one white Buick next to each other in a showroom. Construct a tree diagram that lists all possible arrangements of the three cars.

Solution

We put ourselves in the place of the dealer. We must decide which car to position first. We have a choice of any of the three cars, the red Ford (RF), the white Ford (WF), or the white Buick (WB). The first level of the tree is shown in Figure 5.31.

The next decision focuses on the car to be positioned second, and our choices depend on the car positioned first. If the first car was the red Ford, then the second car can be either the white Ford or the white Buick. If the first car was the white Ford, then the second car can be either the red Ford or the white Buick. If the first car was the white Buick, then

FIGURE 5.31

FIGURE 5.32

FIGURE 5.33

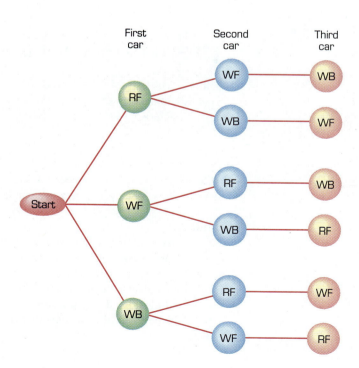

the second car can be either the red Ford or the white Ford. Figure 5.32 shows this two-level tree.

The next decision is which car to position third. After two cars have been positioned, only one other car remains to be placed, even though its type will depend on which cars were positioned previously. The complete tree is shown in Figure 5.33.

Using this tree as a model of the process, we can list all six possible arrangements, starting from the top path and moving sequentially to the bottom path:

$$U = \{\text{RF-WF-WB, RF-WB-WF, WF-RF-WB, WF-WB-RF, WB-RF-WF, WB-WF-RF}\}$$

A tree diagram is a useful model for presenting a sample space with a small number of outcomes, but it becomes less attractive as the list of possible outcomes becomes large. Indeed, any listing mechanism is unwieldy when the list includes over 1,000 possible outcomes. If we are interested in just the *number* of possible outcomes rather than the outcomes themselves, then we may be able to count the outcomes without first listing them.

Figure 5.30 is a model for all possible closings of two plants, one each year, from company plants in Buffalo, Milwuakee, Cleveland, and Nashville. It is a two-level tree, with the first level representing the first closing and the second level representing the second closing. Note that there are four ways to reach the first level (that is, four different possibilities) and that each first-level entry has three second-level entries associated with it. This is important—although each first-level entry has a different group of second-level entries associated with it, the total *number* of second-level entries associated with each first-level entry on the tree is the same, three. The total number of different paths is four (the number of first-level entries) times three (the number of second-level entries associated with each first-level entry) or 12. This example illustrates the following general rule:

THE FUNDAMENTAL THEOREM OF COUNTING

If the first level of a tree has r different entries and if every first-level entry has exactly s second-level entries associated with it, then the two-level tree contains $r \times s$ different paths.

EXAMPLE 3

A chain with seven stores needs to close two stores, one each month over the next two months. How many different ways can this be done?

Solution

Without constructing a tree, we can visualize it as having two levels: the first represents the decision to close the first store and the second represents the decision to close the second store. There are seven entries on the first level, because there are seven possibilities for the first closing. Regardless of which store is closed first, there are six possibilities for the second closing. It follows from the fundamental theorem of counting that there are $7(6) = 42$ paths on the two-level tree, with each path representing a different possible outcome.

EXAMPLE 4

An airline has ten different flights from Boston to Chicago daily and seven flights each day from Chicago to Boston. How many different round-trip schedules can it offer Bostonians who plan to leave and return on different days?

Solution

If we constructed a tree of the round-trip itinerary, it would contain two levels: the first represents a Boston-to-Chicago flight, and the second represents a Chicago-to-Boston flight. There would be ten entries on the first level, and each of them would be associated with the same seven second-level entries. Consequently, there are $10(7) = 70$ different round-trip schedules.

The fundamental theorem of counting is easily generalized to trees with three or more levels. In particular, if the first level of a three-level tree contains r different entries, and if each first-level entry has s associated second-level entries, and if each second-level entry has t associated third-level entries, then the tree contains $r \times s \times t$ paths.

Figure 5.33 is an example of a three-level tree. There are three entries on the first level, each first-level entry has two associated second-level entries, and each second-level entry has one associated third-level entry. The tree contains $3 \times 2 \times 1 = 6$ different paths.

EXAMPLE 5

How many different combinations of 3-letter initials can be made using the 26 letters of the English alphabet?

Solution

We visualize but do not construct a tree for the process of combining initials. It is a three-level tree. The first level represents the first initial, and there are 26 different possibilities. The second level represents the second initial. Each first initial is associated with 26 possible second initials, because the first and second initials can be the same. The third level represents the third initial, and each second initial is associated with 26 different third initials. Thus there are $26(26)(26) = 17,576$ different combinations.

EXAMPLE 6

A chain with seven stores needs to close five stores, one each month over the next five months. How many different ways can this be done?

Solution

A tree for this process has five levels, one for each successive closing. There are seven possibilities for the first closing. Associated with each first closing are six possible second closings. For each second closing, there are five possible third closings. Continuing, we find $7(6)(5)(4)(3) = 2,520$ different closing schedules.

When modeling with trees or totaling possibilities with the fundamental theorem of counting, *order always matters*. Different arrangements of the same distinguishable objects are different outcomes. In Figure 5.30, our model of 12 different closing schedules, a

Different arrangements of the same distinguishable objects are different outcomes when modeling with trees or counting with the fundamental theorem of counting.

schedule of Buffalo first and Nashville second is different from a schedule of Nashville first and Buffalo second. Although the same two cities are involved in both outcomes, the order in which they are chosen is crucial, especially to the employees. The same is true with the initials in Example 5. The initials JFK are different from the initials FKJ, which are the same three distinguishable letters arranged differently.

In some processes, order does not matter. For example, consider the process of making a sandwich from two of four lunch meats—salami, ham, baloney, and roast beef. A sandwich of salami and ham is not different from a sandwich of ham and salami, the same two lunch meats rearranged. The order of the ingredients does not matter, only the ingredients themselves. Trees and the fundamental theorem of counting are *not* appropriate to model or count such processes.

The fundamental theorem of counting is *not* applicable when entries at one level are associated with different numbers of entries at a higher level, even when order does matter. Furthermore, the fundamental theorem of counting gives no information about the individual elements in a sample space. We know from Example 6 that there are 2,520 different ways to close five stores, one a month, from a pool of seven stores. If we actually need the schedules, then we must list them.

EXAMPLE 7

Associated Caterers, a business specializing in box lunches, receives an order for 240 lunches, each containing an entree, a side dish, and a dessert. Each lunch must cost under $2, and the customer wants as much variety among the lunches as possible. Associated Caterers has two types of entrees, fried chicken ($1.50) and ham sandwiches (90¢); two different side dishes, potato salad (20¢) and deluxe salad (55¢); and three types of desserts, chocolate bars (15¢), cake (65¢), and apples (25¢). Two box lunches are considered different if any item is different. How many different lunches can the caterer prepare?

Solution

Box lunches can be modeled by a three-level tree. The first level represents the entree, and there are two choices, either chicken (CH) or a ham sandwich (HS). These options are displayed in Figure 5.34.

Next we add the side dish. If the entree is fried chicken, the side dish cannot be the deluxe salad (DS) because the cost of fried chicken plus the deluxe salad is greater than the $2 limit. Only the potato salad (PS) is possible. If the entree is a ham sandwich, both side dishes are acceptable. Adding these possibilities to Figure 5.34, we obtain Figure 5.35.

Finally, we add dessert to the menu. If the box lunch already contains fried chicken and potato salad, then cake (C) is not a possibility because the total cost would exceed $2. The other two desserts are acceptable. If the box lunch contains a ham sandwich and the deluxe salad, then again the cake is too expensive, and the only possibilities are the chocolate bar (CB) or the apple (A). Any of the three desserts is affordable with an entree of a ham sandwich and a side dish of potato salad. Figure 5.36 is a model of all possible box lunches.

Each path in Figure 5.36 represents a different affordable box lunch. There are seven paths in the model, so seven different box lunches can be prepared ranging from fried chicken–potato salad–chocolate bar (the top path) to a ham sandwich–deluxe salad–apple (the bottom path). The fundamental theorem of counting cannot be used to determine the total number of different box lunches because the conditions for applying the theorem are not satisfied.

FIGURE 5.34

Entrée

FIGURE 5.35

FIGURE 5.36

CREATING MODELS

In Problems 1 through 15, construct a tree for each process and then use the tree to list a sample space.

1. A luxury car manufacturer offers three body colors, silver, black, and green, and three interior colors, black, brown, and red. What are the possible color combinations available to a prospective buyer?

2. Frozen dinners consist of one entree, either chicken, salisbury steak, or ham steak, one vegetable, either peas or corn, and one dessert, either apple pie or brownie. How many different frozen dinners can a customer purchase? The dinners are considered different if any item is different.

3. A builder will construct three houses on a block, one next to the other. Each house will be either a ranch, a colonial, or a split-level. How many ways can the three houses be constructed if identical models are not to be placed directly next to one another?

4. Messages are relayed to passing trains by hoisting two flags up a flagpole, one under the other. There are three types of flags, red, white, and striped, and each different combination signifies a different message. A red flag under a white flag is a different message from a white flag under a red flag. How many different messages are possible?

5. Using the flags described in the previous problem, determine how many different three-flag messages are possible if two flags of the same color are not allowed next to each other (two red flags separated by a white flag is an acceptable message, but two red flags followed by a white flag is not).

6. A local department store has two vacancies, one in the jewelry department and one in the housewares department. There are four applicants for these positions, Marita and Nancy, who are female, and Douglas and Jose, who are male. How many different ways can the two positions be filled if *at least one* position must go to a female?

7. A well-known fruit drink is a combination of orange, grape, and apple ingredients. The manufacturer can use either real fruit juice or artificial flavoring for each ingredient, but *at least one* of the ingredients must be real juice, because the product will be advertised as a "real fruit drink." How many different combinations of artificial and real ingredients can be used in the production of the drink?

8. How many different three-letter combinations can be made from the four letters *a*, *b*, *c*, and *d* if no letter can be used twice in the same arrangement?

9. Solve Problem 8 for two-letter combinations if the same letter can be used twice.

10. A drugstore chain with five stores in Tampa decides to close one store every Saturday and *another* store every Sunday. The store chosen for Saturday closings will close every Saturday, and the store chosen for Sunday closings will close every Sunday. How many different ways can the choice of closings be made?

11. A bank gives free gifts to all new depositors. Those who depoist $50 can choose one gift: either a set of bridge cards or a lighter. Those who deposit $100 can choose either a radio, a silver-plated baby spoon, or a stuffed animal. Those who deposit $500 can choose a toaster or a camera. A $1,000 new depositor can choose one item from *each* of the three categories. How many different sets of gifts can the bank offer to new depositors of $1,000?

12. A condominium developer offers four different types of lights to buyers, and buyers can select the same or different types of lights for the foyer and living room. How many different combinations of lights are available to each buyer?

13. Solve Example 7 if the chicken costs $1.25 and the potato salad costs 40¢.

14. A Los Angeles distributor imports merchandise from Italy by way of New York. The merchandise is shipped from Italy to New York by either plane or ship, and from there to Los Angeles by plane, train, or truck. How many different ways can merchandise be shipped from Italy to Los Angeles?

15. A survey categorizes respondents as (*i*) male or female, (*ii*) white, black, or other, and (*iii*) under 20 years old, between 20 and 65, or over 65 years old. Into how many different subsets has the population been divided?

16. A department store advertises a watch sale, offering a watch body and a band for $65. The store has 15 different bodies and 20 different bands from which to choose. How many different sets of one body and one band can the store assemble?

17. A furniture store advertises ensembles, furniture for three different rooms, for $5,000. Buyers can choose from seven different living-room sets, five different dining-room sets, and ten different bedroom sets. How many different ensembles can the store advertise?

18. How many different license plates are available if each plate has six places, the first three reserved for letters and the last three for numbers?

19. An executive needs to fly from Boston to Chicago on Monday, from Chicago to San Francisco on Wednesday, and return to Boston from San Francisco on Friday. There are 15 different flights from Boston to Chicago, 21 different flights from Chicago to San Francisco, and 5 different flights from San Francisco to Boston, all available on a daily basis. From how many different itineraries can the executive choose?

20. An automobile manufacturer offers six different models, each with a choice of ten different exterior colors and seven different interior colors. How many different cars does it offer?

21. A buyer of a particular car model is offered the following options: automatic or stan-

dard transmission; with or without air-conditioning; one of seven different tires; AM radio, AM-FM radio, or no radio; and none, one, or two air bags. How many different ways can a buyer equip a car?

22. A restaurant offers a fixed-price dinner for $24.95. The menu lists 7 appetizers, 8 entrees, 4 vegetables, 3 types of potatoes, 8 desserts, and 12 wines, and a diner may choose one item from each category. How many different complete dinners does the restaurant offer?

23. Solve Problem 22 if diners may elect not to choose a selection from one or more categories.

24. A motel has 31 rooms and six reservations. How many different ways can the six parties be assigned rooms?

25. A rent-a-car agency has 14 available cars and four customers in the office. How many different ways can the four customers be assigned cars?

26. Messages are relayed to passing trains by displaying two single-colored flags, one under the other. Each combination of colors represents a different message. How many messages are possible from a set of (*a*) six flags, each of a different color, (*b*) eight flags, each of a different color, (*c*) eight flags, two each of four different colors, if two flags of the same color cannot be placed under one another?

27. A baseball team has three pitchers, five outfielders, six infielders, and two catchers. Each outfielder can play each of the three outfield positions, and each infielder can play each of the four infield positions. How many different defensive lineups can be assembled for this team if a defensive line-up involves assigning a particular player to each of the nine different positions? Two defensive lineups are different if any one assignment is different.

28. Solve Problem 27 if pitchers can also play in the outfield.

29. A telephone repair depot has nine crews and five requests for repairs. How many different ways can repairs be scheduled if no crew is assigned more than one repair?

30. A computer is given the names of five females and eight males. How many different ways can the computer match each of the five females with a male?

31. A personal identification number (PIN) for an automatic teller machine (ATM) contains four digits. How many different PIN numbers are there?

32. A woman packs six blouses, two vests, and four skirts for a business trip. The garments are all blue-gray coloring and can be matched with one another. How many different outfits does the woman have?

EXPLORING IN TEAMS

33. A tree is a special type of graph. In general, a graph is a set of circles, called *nodes*, and a set of lines or curves, called *arcs*, that connect pairs of nodes. For convenience, we number the nodes using consecutive positive integers, beginning with 1. Arrowheads are placed on each arc to indicate directions of allowable movement between nodes. It may be convenient to think of nodes as cities and arcs as roads between cities. Traffic flow can be one-way or two-way, as indicated by the arrowheads.

An *adjacency matrix* $\mathbf{M} = [m_{ij}]$ associated with a particular graph is defined as

$$m_{ij} = \text{number of distinct arcs from node } i \text{ to node } j$$

Construct an adjacency matrix $\mathbf{M}$ for the graph displayed in Figure *A*. Then calculate

M^2 and verify that the i-j element of M^2 is the number of different paths containing two arcs that connect node i to node j.

34. Construct an adjacency matrix M for the graph displayed in Figure B. Then calculate M^2 and verify that the i-j element of M^2 is the number of different paths containing two arcs that connect node i to node j.

35. Calculate M^3 for the matrix M identified in Problem 34, and verify that the i-j element of M^3 is the number of different paths containing three arcs that connect node i to node j.

FIGURE A

FIGURE B

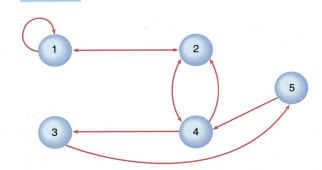

REVIEWING
MATERIAL

36. (Section 1.6) Use graphical methods to solve the following problem. A rancher owned 120 cattle. Adult female cattle have a market value of $1,500 each, and adult male cattle have a market value of $1,000 each. How many cattle of each type did the rancher own if it is known that the herd contained twice as many adult female cattle as adult male cattle and that the entire herd was sold for $160,000?

37. (Section 2.5) Use elementary row operations to show that the following matrix does not have an inverse:

$$A = \begin{bmatrix} 1 & 2 & 3 \\ 2 & -1 & 1 \\ 4 & 3 & 7 \end{bmatrix}$$

38. (Section 3.6) Use the enhanced simplex method to solve:

$$\text{Minimize:} \quad z = 2x_1 + 2x_2 + 3x_3$$

$$\text{subject to:} \quad x_1 + 2x_2 + 2x_3 \geq 100$$

$$\text{assuming:} \quad x_1, x_2, \text{ and } x_3 \text{ are nonnegative}$$

39. (Section 4.4) Construct the first three lines of an amortization schedule for repaying a $6,000 loan in equal monthly installments over a four-year period at a nominal annual interest rate of 9.6%. Assume payments are made at the end of each month.

RECOMMENDING ACTION

40. Respond by memo to the following request:

MEMORANDUM

To: J. Doe Reader

From: Vice President for Planning

Date: Today

Subject: **Downsizing**

As you know, we planned on downsizing our operation in Raleigh by closing two of our five stores there, one this year and one next year. We still must decide which of the stores to close and then move the inventory from those stores to the three remaining stores. Yesterday, the Board of Directors changed its mind and voted to close both stores at the same time. Does this decision change the number of possible closing schedules available to us? If so, by how many?

CHAPTER 5   KEYS

KEY WORDS

bar graph (p. 324)
bimodal (p. 335)
cardinal number of a set (p. 310)
complement of a set (p. 307)
data (p. 333)
deviation from the mean (p. 340)
disjoint sets (p. 308)
element (p. 306)
empty set (p. 307)
exhaustive group of sets (p. 319)
intersection of sets (p. 308)
mean (p. 333)
measure of central tendency (p. 333)
measure of dispersion (p. 339)
median (p. 334)
merged bar graph (p. 326)

mode (p. 335)
moving averages (p. 335)
mutually disjoint sets (p. 308)
pie chart (p. 319)
process (p. 347)
proper subset (p. 307)
range (p. 339)
sample space (p. 347)
set (p. 306)
stacked column (p. 322)
standard deviation (p. 341)
subset (p. 307)
tree (p. 348)
union of sets (p. 309)
universal set (p. 307)
variance (p. 340)

KEY CONCEPTS

5.1 Sets and Venn Diagrams

- The symbol $\in$ is read "is a member of" or "belongs to"; the symbol $\notin$ is read "is not a member of."

- The symbol $|$ is read "such that."

- Two sets are equal if and only if they contain the same elements.

- Set-builder notation defines a set by specifying a property shared by all members of the set.

- ϕ denotes the empty set, U denotes the universal set, and A' denotes the complement of A.

5.2 Pie Charts and Bar Graphs

- Pie charts and stacked columns model only groups of subsets that are mutually exhaustive.

- Bar graphs are often rotated 90°, especially when the labels for the various subsets are lengthy.

- Merged bar graphs are particularly useful for comparing similar information from two or more populations.

5.3 Standard Measures for Numerical Data

- The mean of a list of data need not be a data point.

- The mean is a deceptive indicator for data when one or more pieces of data are significantly larger or smaller than all the others.

- The median is useful as a benchmark for positioning data. Individual data points are ranked as being either below the median or above it.

- The mode is a measure of popularity; it identifies the data point that appears most frequently. Such measures are particularly important in determining consumer preferences.

- The fraction of data points lying within k standard deviations of the mean is at least as large as $1 - \dfrac{1}{k^2}$.

5.4 Tree Diagrams

- Defining a sample space for a process can be subjective, because a possible outcome to one person may not be an outcome to a second person.

- Each complete path in a tree, beginning at the "start" and moving through each level of the tree without any backtracking, is one possible outcome to the underlying process.

- If the first level of a tree has r different entries and if every first-level entry has exactly *s* second-level entries associated with it, then the two-level tree contains $r \times s$ different paths.

- When modeling with trees or totaling possibilities with the fundamental theorem of counting, *order always matters*. Different arrangements of the same distinguishable objects are different outcomes.

KEY FORMULAS

- The angle of a sector in a pie chart representing a particular subset is

$$\theta = \frac{\text{number of elements in a subset}}{\text{number of elements in the universal set}} \times 360°$$

- The mean $\bar{x}$ of the set of data $\{x_1, x_2, x_3, \ldots, x_n\}$ is

$$\bar{x} = \frac{x_1 + x_2 + x_3 + \ldots + x_n}{n}$$

- The variance is

$$\frac{(x_1 - \bar{x})^2 + (x_2 - \bar{x})^2 + (x_3 - \bar{x})^2 + \ldots + (x_n - \bar{x})^2}{n}$$

- The standard deviation is

$$\sqrt{\frac{(x_1 - \bar{x})^2 + (x_2 - \bar{x})^2 + (x_3 - \bar{x})^2 + \ldots + (x_n - \bar{x})^2}{n}}$$

KEY PROCEDURES

- Sets can be depicted graphically by a Venn diagram. A rectangle is drawn to represent a universal set U, and circles within this rectangle represent subsets of U. Elements of the universal set are portrayed by points within the rectangle, and elements of subsets are portrayed by points within circles.

- Cardinal numbers for the distinct regions in a Venn diagram are calculated sequentially, starting with the innermost region for which information is available and then working out towards the boundary of the universal set.

- To determine the median of numerical data, first list the data in either ascending or descending order. If there are an odd number of data points, then the median is the middle number in the list; if there are an even number of data points, then the median is the mean of the two numbers in the middle of the list.

- To computer an *n*-time moving average, where *n* is a specified positive integer, we calculate a mean at each recorded time using the observation at that time and the $n - 1$ preceding observations. No means are calculated for times where there are not enough preceding reporting intervals to calculate a mean.

TESTING YOURSELF

Use the following problems to test yourself on the material in Chapter 5. The odd problems comprise one test, and the even problems a second test.

1. Let

$$U = \{A, B, C, D, E, F, G\}$$

$$A = \{A, C, D, E\}$$

$$B = \{A, B, D, F, G\}$$

$$C = \{A, C, E, F\}$$

Find

(**a**) A' (**b**) $C \cap B$ (**c**) $A \cup C$ (**d**) a subset of B

(**e**) $\{s \mid s \in U$ and s is a letter in the word BABBLES$\}$

2. Construct a bar graph for the following information on the consumption of eggs in 1987–1988. The consumption is given in units of kilos per person, as reported in The *Economist Book of Vital World Statistics*.

Country	Mexico	Japan	Canada	Poland	Brazil	Philippines	United States
Eggs	5.7	15.0	11.5	10.3	6.5	4.2	14.5

3. Let U be the set of all people living in the world today, and let

$$A = \{p \mid p \text{ is a person who is female}\}$$

$$B = \{p \mid p \text{ is a person who is retired}\}$$

$$C = \{p \mid p \text{ is a person who has blue eyes}\}$$

Write the following sets in terms of A, B, and C.

(**a**) $\{p \mid p$ is a retired female$\}$

(**b**) $\{p \mid p$ is male$\}$

(**c**) $\{p \mid p$ is a retired male$\}$

(**d**) $\{p \mid p$ is either retired or has blue eyes$\}$

(**e**) $\{p \mid p$ is a blue-eyed female$\}$

(**f**) $\{p \mid p$ is not female and not retired$\}$

4. (**a**) A truck manufacturer offers three engines (220 horsepower, 240 horsepower, and 260 horsepower), three interior colors (gray, tan, and black), and either standard or automatic shift. Construct a tree diagram and then count the number of different trucks a customer can order.

(**b**) How many different cars can a customer order if the manufacturer offers four different engines, nine different interior colors, and three different shifting mechanisms?

5. A survey asked 250 people where they vacationed in the last three years and received the following responses.

40 had been to a beach

125 had been to a lake

89 had been to the mountains

2 had been to both a lake and a beach

4 had been to a beach and to the mountains

13 had been to a lake and to the mountains

2 had been to all three types of locations

Construct a Venn diagram for this population, and then use the diagram to determine how many people in the survey (a) had been to only a beach, (b) had been to either a lake or a mountain, (c) had *not* been to a lake, and (d) had not been to any of the three locations.

6. Let

$$U = \{0, 1, 2, 3, 4, 5, 6, 7, 8, 9\}$$

$$A = \{1, 2, 3, 4, 5\}$$

$$B = \{0, 2, 4, 6, 8\}$$

$$C = \{0, 1, 2, 7, 8, 9\}$$

Find

(**a**) A' (**b**) $A \cap B$ (**c**) $B \cup C$ (**d**) $B \cup B'$
(**e**) $\{s \mid s \in U$ and   (**f**) a nonempty set that is disjoint from C $s > 6\}$

7. Construct a pie chart for the following information: On Sunday, February 20, 1994, *CBS* broadcast three hours of prime time Winter Olympics coverage from Norway. In this period, 85 minutes were devoted to Olympic events, 43 minutes were devoted to features and interviews, and 52 minutes were devoted to commercials.

8. (**a**) Monthly sales (in millions of dollars) for a corporation last year were 1.2 in January, 1.3 in February, 1.4 in March, 1.4 in April, 1.7 in May, 1.5 in June, 1.2 in July, 1.3 in August, 1.6 in September, 1.5 in October, 1.5 in November, and 1.8 in December. Calculate three-month moving averages for these sales figures.

(**b**) Find the mean, mode, median, range, and standard deviation for the sales figures given in part (a).

9. Use the following data to construct a bar graph showing average hourly earnings by profession (these data were reported by the Bureau of Labor Statistics for 1993): business services, $10.11; construction, $14.35; insurance and real estate, $11.32; manufacturing, $11.76; mining, $14.60; retail trade, $7.29; social services, $7.86; and wholesale trade, $11.71.

10. Six hundred people were asked what type of motor vehicles they owned. After the results were tallied, it was discovered that 179 people owned a truck, 350 people owned a car, 238 people owned only a car, 20 people owned a car and a truck, 24 people owned only a van, 11 people owned a van and a truck, and 8 people owned all three types of vehicles. Construct a Venn diagram for this population, and then use the diagram to determine how many people in the survey (a) owned a van, (b) owned a van but not a car, (c) owned only a truck, and (d) did *not* own a truck.

11. Determine the mean, mode, median, range, and standard deviation for the following data:

1, 1, 7, 2, 4, 3, 1, 2, 5, 4, 2

12. Construct a merged bar graph combining the information in Problem 2 with the following information on the consumption of poultry (in units of kilos per person) in 1987–1988 as reported in *The Economist Book of Vital World Statistics*.

Country	Mexico	Japan	Canada	Poland	Brazil	Philippines	United States
Poultry	8.0	13.3	28.2	8.1	12.1	5.0	37.1

13. (*a*) Frozen dinners consist of one entree, either fish sticks, turkey, roast beef, or ham; one vegetable, either carrots or peas; and one dessert, either apple pie or cherries jubilee. Construct a tree diagram and then count the number of different frozen dinners available to a customer. Frozen dinners are considered different if any item is different.

    (*b*) How many different frozen dinners can a company offer its customers if has eight different entrees, ten different vegetables, and five different desserts? Each dinner must contain one item from each category.

14. Create a pie chart using the following information from the Internal Revenue Service on the sources of Federal income in 1992: 35% came from personal income taxes, 7% from corporate income taxes, 30% from retirement taxes, including social security and medicare taxes, 7% came from other taxes such excise, gift, and customs taxes, and 21% came from borrowing.

6

PROBABILITY MODELS

6.1 EQUALLY LIKELY OUTCOMES

A sample space with equally likely outcomes for a process is a sample space in which each outcome has the same chance of happening.

Very few things in life are guaranteed. Uncertainties abound. The Weather Service may predict a sunny day tomorrow, but there is always a chance the day will be cloudy or rainy. An airline may schedule an arrival for 9:51, but there is always the possibility that the plane will arrive at a different time. A corporation may project a profit of $6 million next year, but it is very likely that the actual profit will differ from this forecast.

Most decisions, actions, or projections have several possible outcomes, not all of which are desirable. Decision makers must account for all possible outcomes and the likelihood of each outcome actually occurring. They need to recognize what is possible and what is probable. It is one thing to know that a suggested course of action can either succeed or fail. It is quite another thing to know that there is a 90% chance of failure and a 10% chance of success.

At the end of Chapter 5, we developed methods for listing and counting sample spaces—possible outcomes for a process of interest. In this chapter, we turn our attention to determining the probability or likelihood of each outcome occurring.

The simplest type of probability modeling occurs when a sample space U for a process is chosen so that all outcomes have the same chance of happening. We call such sets *sample spaces with equally likely outcomes*. For example, consider the process of throwing an ordinary die (die is the singular of dice) and announcing the number that is on top. If the die is fair, then a sample space for this process is $U = \{1, 2, 3, 4, 5, 6\}$. Each outcome has the same chance of happening, so U is a sample space with equally likely outcomes.

Often a process can be modeled with several different sample spaces. Probability models are easier to analyze when a sample space contains equally likely outcomes, so we prefer such a sample space when it is possible to define one.

Consider a barrel that contains four red balls, numbered from 1 through 4, and one white ball (see Figure 6.1). The process of closing your eyes and picking a ball from the barrel can be modeled by the sample space $U_1 = \{\text{red ball, white ball}\}$. One selects either a red ball or a white ball. This sample space does *not* contain equally likely outcomes. Since there are four red balls and only one white ball in the barrel, there is a greater chance of picking a red ball then there is of picking a white ball. A sample space of equally likely outcomes is

$$U_2 = \{\text{red 1, red 2, red 3, red 4, white}\}$$

where each red ball in the barrel is described by both its color and its number. Now each outcome has the same likelihood of occurring.

All American coins have the head of a famous figure on one side. For example, Abraham Lincoln's face appears on the U.S. penny, and George Washington's face appears on the U.S. quarter. If the face side lands up when a coin is flipped, the result of the flip is said to be "heads"; if the face side lands down, the result is called "tails."

FIGURE 6.1

EXAMPLE 1

Determine a sample space with equally likely outcomes for the process of flipping a coin twice and describing the result in terms of heads or tails.

Solution

A sample space is

$$U_1 = \{\text{two heads, two tails, one of each}\}$$

One either flips heads twice, flips tails twice, or flips one head and one tail. U_1 is *not* a sample space with equally likely outcomes, because the outcome "one of each" is twice as likely to occur as the outcome "two heads." "One of each" can happen in two ways:

a head first and a tail second or a tail first and a head second; "two heads" can occur only one way, by flipping a head first and a head second.

A second sample for the same process is

$$U_2 = \{HH, HT, TH, TT\}$$

where the first and second letter of each outcome represent the results of the first and second toss, respectively, with H denoting a head and T a tail. U is a sample space with equally likely outcomes because each outcome has the same chance of happening.

EXAMPLE 2

Determine a sample space with equally likely outcomes for the process of throwing two dice summing the faces that turn up.

Solution

When two dice are thrown, any integer sum between 2 and 12 can occur, so we are tempted to take

$$U_1 = \{2, 3, 4, 5, 6, 7, 8, 9, 10, 11, 12\}$$

as the sample space. However, the outcomes in this sample space are *not* equally likely. The outcome 7, which can be rolled as $6 + 1$, $5 + 2$, or $4 + 3$, is more likely to occur than the outcome 2, which can be rolled only as $1 + 1$.

To construct a sample space with equally likely outcomes for this process, we *imagine* that the two dice are distinguishable (perhaps one is red and the other is green). We then list the outcomes in sum format with the first number representing the face of the first (red) die and the second number representing the face of the second (green) die (see Figure 6.2). Here each of the 36 different outcomes has the same chance of occurring.

FIGURE 6.2

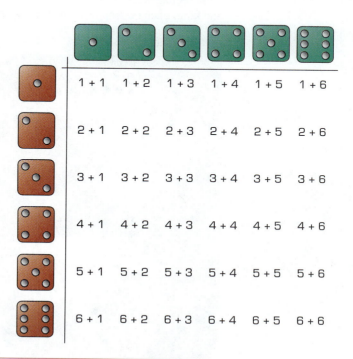

More often than not, we are interested in the likelihood that a set of outcomes will occur, rather than the likelihood of only one outcome occurring. We define an *event*,

> *An event is a subset of a sample space.*

denoted by E, to be a subset of a sample space, so an event is a set of outcomes. Two different subsets of a sample space represent two different events. In general, we are interested in determining the likelihood that a particular event will occur. We do this by first constructing a sample space for the underlying process, choosing a sample space with equally likely outcomes, if one is available, and then selecting a subset of the sample space to define the event of interest.

EXAMPLE 3

A coin is flipped twice. Find the subset that defines the event that each toss has a different face.

Solution

A sample space with equally likely outcomes was found in Example 1 as $U = \{HH, HT, TH, TT\}$. Then $E = \{HT, TH\}$ contains all the elements in this sample space with the property that each toss has a different face, and E is the event of interest.

EXAMPLE 4

Two dice are rolled. Find the subset for the event that the sum of the dice is 7.

Solution

A sample space with equally likely outcomes is displayed in Figure 6.2. For this sample space,

$$E = \{1 + 6, 2 + 5, 3 + 4, 4 + 3, 5 + 2, 6 + 1\}$$

is the subset of all outcomes in the sample space that sum to 7.

A regular deck of 52 cards contains 13 different ranks: one through ten, jack, queen, and king. Each rank appears four times, once in each of the four suits: clubs (♣), diamonds

A regular deck of 52 cards

($\blacklozenge$), hearts ($\heartsuit$), and spades ($\spadesuit$). A one is generally called an ace. Jacks, queens, and kings are called face cards. Diamonds and hearts are colored red, and clubs and spades are colored black. Individual cards are labeled by two characters, the first denoting rank and the second suit. Thus, 3$\heartsuit$ or 3H is the three of hearts, Q$\clubsuit$ or QC is the queen of clubs, and A$\blacklozenge$ or AD is the ace of diamonds.

EXAMPLE 5

One card is selected from a well-shuffled regular deck of cards spread face down on a table. Take the sample space to be the set consisting of each one of the 52 cards. Describe in words the events defined by the following subsets:

(*a*) {3$\clubsuit$, 3$\blacklozenge$, 3$\heartsuit$, 3$\spadesuit$ }

(*b*) {3$\clubsuit$, 4$\clubsuit$, 3$\blacklozenge$, 4$\blacklozenge$, 3$\heartsuit$, 4$\heartsuit$, 3$\spadesuit$, 4$\spadesuit$ }

(*c*) {A$\blacklozenge$, 2$\blacklozenge$, 3$\blacklozenge$, 4$\blacklozenge$, 5$\blacklozenge$, 6$\blacklozenge$, 7$\blacklozenge$, 8$\blacklozenge$, 9$\blacklozenge$, 10$\blacklozenge$, J$\blacklozenge$, Q$\blacklozenge$, K$\blacklozenge$, A$\heartsuit$, 2$\heartsuit$, 3$\heartsuit$, 4$\heartsuit$, 5$\heartsuit$, 6$\heartsuit$, 7$\heartsuit$, 8$\heartsuit$, 9$\heartsuit$, 10$\heartsuit$, J$\heartsuit$, Q$\heartsuit$, K$\heartsuit$}

Solution

(*a*) Selecting a 3.   (*b*) Selecting a 3 or a 4.   (*c*) Selecting a red card.

> *If* U *is a sample space with a finite number of equally likely outcomes, then the probability that an event* E *will happen is the cardinal number of* E *divided by the cardinal number of* U.

An event is a subset of a sample space, regardless of whether the outcomes in the sample space are equally likely to occur. If, however, the sample space consists of a finite number of equally likely outcomes, then we have an easy way to determine the probability or likelihood that a particular event will occur. Denote the probability that an event E will happen as $P(E)$. If each of the finite number of outcomes in U is equally likely to occur, then

$$P(E) = \frac{n(E)}{n(U)} \tag{1}$$

The probability that an event will happen is $n(E)$, the cardinal number of the event E, divided by $n(U)$, the cardinal number of the sample space U.

EXAMPLE 6

A coin is flipped twice. Find the probability that each toss has a different face.

Solution

We saw in Example 1 that a sample space with equally likely outcomes is $U = \{HH, HT, TH, TT\}$, so $n(U) = 4$. In Example 3 we determined that the event that each toss has a different face is the subset $E = \{HT, TH\}$, so $n(E) = 2$. Using Equation (*1*), we have

$$P(E) = \frac{n(E)}{n(U)} = \frac{2}{4} = 0.5$$

The probability that each toss has a different face is $\frac{1}{2}$ or 0.5.

EXAMPLE 7

A coin is flipped twice. Find the probability that both tosses are heads.

Solution

As in Example 6, the sample space with equally likely outcomes is $U = \{HH, HT, TH, TT\}$, with $n(U) = 4$. The event that both tosses are heads is described by the subset $E = \{HH\}$, so $n(E) = 1$. Using Equation (*1*), we have

$$P(E) = \frac{n(E)}{n(U)} = \frac{1}{4} = 0.25$$

EXAMPLE 8

Two dice are thrown. Determine the probability that the sum of the faces that turn up is 7.

Solution

A sample space U with equally likely outcomes is shown in Figure 6.2; for this sample space, $n(U) = 36$. In Example 4 we saw that the subset of all outcomes that sum to 7 is $E = \{1 + 6, 2 + 5, 3 + 4, 4 + 3, 5 + 1, 6 + 1\}$, so $n(E) = 6$. Using Equation (1) we have

$$P(E) = \frac{n(E)}{n(U)} = \frac{6}{36} = \frac{1}{6}$$

or 0.1667 rounded to four decimal places.

A process is random if nothing is done by anyone involved with the process to favor one outcome over another.

A process is *random* if nothing is done by anyone involved with the process to favor one outcome over another. Shuffling a regular deck of cards, fanning them out *face down*, and then selecting one card is a random process, because there is no way of favoring the selection of one card over another. The process is not random if the cards are fanned face up, so that the selector can see the cards before choosing, or if the deck is marked. Selecting one ball from the barrel shown in Figure 6.1 is a random process if the balls are well mixed and if the selection is made without looking at the balls. The process would not be random if the selector looked at the balls as the selection was made, and it would not be random if the selector knew in advance where each ball was placed.

EXAMPLE 9

Pick one card randomly from a regular deck. Determine the probability that the card is a heart.

Solution

The process is to pick a card, while the event of interest is that the card is a heart. We can solve this problem two different ways, depending how we initially model the sample space.

Take U to be the set of all 52 cards in the deck. Since each card is as likely to be picked as any other card, the outcomes are equally likely and $n(U) = 52$. With this sample space, the event E is the set of all hearts, ranging from the ace of hearts through the king of hearts, that is,

$$E = \{A♥, 2♥, 3♥, 4♥, 5♥, 6♥, 7♥, 8♥, 9♥, 10♥, J♥, Q♥, K♥\}$$

This subset of U contains 13 elements so $n(E) = 13$. It now follows from Equation (1) that

$$P(E) = \frac{n(E)}{n(U)} = \frac{13}{52} = 0.25$$

Alternatively, we can model the sample space as $U_1 = \{\text{club, diamond, heart, spade}\}$, which also contains equally likely outcomes because there are the same number of cards (13) of each suit in a regular deck. Here $n(U_1) = 4$. The subset of U_1 that defines the event of interest is $E_1 = \{\text{heart}\}$, with $n(E_1) = 1$. Using Equation (1), we determine

$$P(E_1) = \frac{n(E_1)}{n(U_1)} = \frac{1}{4} = 0.25$$

Using either model, we find that the probability of picking a heart is $\frac{1}{4}$ or 0.25.

To construct sample spaces with equally likely outcomes, we may create distinctions that are not apparent.

A sample space for a process is a model of that process. Often we can associate a number of sample spaces (models) with the same process; some of these sample spaces will have equally likely outcomes and some will not. Some models may even be conceptualizations of outcomes. Figure 6.1 shows a barrel containing four red balls and one white ball. If the balls are not numbered, then the four red balls are indistinguishable from each other. Even so, we can *think* of the red balls as being numbered and distinguishable, and in doing so we create an abstract model of the process in which the sample space consists of equally likely outcomes. We have created a conceptualization of the balls that adequately represents the process of interest.

We can apply similar reasoning to the process of throwing two dice and summing the faces that turn up. Generally, the two dice in a pair are indistinguishable from one another—in reality, we cannot tell which is the first and which is the second. Nonetheless, we can replace, at least in our minds, one of the dice with another of a different color. This will not alter the process, but it will change the model and lead to the sample space with equally likely outcomes displayed in Figure 6.2.

Sample spaces with equally likely outcomes are extremely useful models of random processes because Equation (*1*)

$$P(E) = \frac{n(E)}{n(U)}$$

(1 repeated)

can be applied to such models to calculate probabilities. To construct such sample spaces, we may create artificial distinctions, such as thinking of a pair of dice being of different colors.

IMPROVING SKILLS

1. Two identical bags each contain the numbers 1 through 4 written on separate but identical slips of paper. One number is picked from each bag and their sum is found. Determine which of the following sample spaces are equally likely.

   (*a*) {odd, even}

   (*b*) {2, 3, 4, 5, 6, 7, 8}

   (*c*) {1 + 1, 1 + 2, 1 + 3, 1 + 4, 2 + 1, 2 + 2, 2 + 3, 2 + 4, 3 + 1, 3 + 2, 3 + 3, 3 + 4, 4 + 1, 4 + 2, 4 + 3, 4 + 4}

   (*d*) {1 + 1, 1 + 2, 1 + 3, 1 + 4, 2 + 2, 2 + 3, 2 + 4, 3 + 3, 3 + 4, 4 + 4} (*Note*: In this sample space the outcomes are displayed by listing the smaller number first.)

   (*e*) {less than 5, 5, greater than 5}

2. Four cards, a king of clubs, a king of diamonds, a king of hearts, and a king of spades, are shuffled and placed face down on a table. One card is then turned over. Determine which of the following sample spaces for this process contain equally likely outcomes.

   (*a*) {red king, black king}

   (*b*) {king of clubs, not the king of clubs}

   (*c*) {king, not a king}

   (*d*) {K♣, K♦, K♥, K♠}

3. Nine plastic tiles, each with a different integer between 1 and 9 written on one side, are placed face down on a table. The tiles are identical in all respects except for their face values. A person is asked to mix the tiles and randomly pick one. Determine which of the following sample spaces for this process contain equally likely outcomes.

   (*a*) {odd number, even number}

   (*b*) {less than 5, 5, greater than 5}

   (*c*) {1, 2, 3, 4, 5, 6, 7, 8, 9}

   (*d*) {9, 8, 7, 6, 5, 4, 3, 2, 1}

4. Determine which of the following sample spaces contain equally likely outcomes for the process of describing the genders of three children in a family.

   (*a*) {no girls, at least one girl}

   (*b*) {all girls, all boys, some of each sex}

   (*c*) {all girls, 2 girls and 1 boy, 1 girl and 2 boys, 3 boys}

   (*d*) {ggg, ggb, gbg, gbb, bgg, bgb, bbg, bbb}, where g denotes a girl, b denotes a boy, and each listings specifies the order of birth of each child from oldest to youngest

5. Consider the sample space in Figure 6.2 for the process of throwing two dice and summing the faces that turn up. Describe in words the events defined by the following subsets:

   (*a*) $A = \{1 + 1\}$

   (*b*) $B = \{1 + 1, 6 + 6\}$

   (*c*) $C = \{1 + 1, 1 + 2, 2 + 1, 2 + 2\}$

   (*d*) $D = \{1 + 1, 2 + 2, 3 + 3, 4 + 4, 5 + 5, 6 + 6\}$

   (*e*) $B \cup C$

   (*f*) D'

6. Consider the sample space consisting of each one of the 52 cards in a regular deck for the process of randomly selecting one card from the deck. Describe in words the events defined by the following subsets.

   (*a*) $A = \{2\heartsuit, 2\clubsuit, 2\spadesuit, 2\diamondsuit\}$

   (*b*) $B = \{A\spadesuit, 2\spadesuit, 3\spadesuit, 4\spadesuit, 5\spadesuit, 6\spadesuit, 7\spadesuit, 8\spadesuit, 9\spadesuit, 10\spadesuit, J\spadesuit, Q\spadesuit, K\spadesuit\}$

   (*c*) $C = \{J\clubsuit, J\diamondsuit, J\heartsuit, J\spadesuit, Q\clubsuit, Q\diamondsuit, Q\heartsuit, Q\spadesuit, K\clubsuit, K\diamondsuit, K\heartsuit, K\spadesuit\}$

   (*d*) $A \cap B$

   (*e*) B'

   (*f*) $B \cap C$

7. Consider the sample space

$$U = \{ggg, ggb, gbg, gbb, bgg, bgb, bbg, bbb\}$$

for the genders of three children in a family, where g denotes a girl, b denotes a boy and each listing specifies the oldest first and the youngest last. Describe in words the events defined by the following subsets.

   (*a*) $A = \{bbb\}$

   (*b*) A'

   (*c*) $B = \{ggb, gbg, bgg\}$

   (*d*) $C = \{gbb, bgb, bbg\}$

   (*e*) $B \cup C$

8. Consider the process in which a die is thrown and then a coin is flipped.

   (*a*) Determine a sample space of equally likely outcomes for this process.

   (*b*) Find the subset of this sample space that defines the event of obtaining an odd number on the die and a head on the coin.

   (*c*) Find the subset of this sample space that defines the event of obtaining an odd number on the die.

   (*d*) Find the subset of this sample space that defines the event of obtaining a head on the coin.

9. A coin is flipped three times in succession, and the face that turns up each time is recorded.

   (*a*) Determine a sample space of equally likely outcomes for this process.

(*b*) Find the subset of this sample space that defines the event of obtaining exactly two heads.

(*c*) Find the subset of this sample space that defines the event of obtaining at least two tails.

(*d*) Determine the probability of flipping a coin three times and obtaining exactly two heads.

(*e*) Determine the probability of flipping a coin three times and obtaining at least two tails.

10. A bag contains two spoons and three forks, all of equal size and weight. A magnet is placed in the bag and is withdrawn after it attaches itself to one of the utensils.
(*a*) Determine a sample space of equally likely outcomes for this process.
(*b*) Find the subset of this sample space that defines the event of drawing a spoon.
(*c*) Determine the probability of drawing a spoon from the bag.

11. A man has three green ties, two red ties, and two blue ties. His valet who is colorblind, packs two of these ties in a suitcase.
(*a*) Determine a sample space of equally likely outcomes for the process of packing two ties in the suitcase.
(*b*) Find the subset of this sample space that defines the event that the two packed ties are of different colors.
(*c*) Determine the probability that the two ties in the suitcase are different colors.

12. The probability of selecting a woman from a group of 80 people is 0.3. How many women are in the group?

13. The probability of selecting a college-educated person from a group of 620 people is 0.05. How many college-educated people are in the group?

14. The probability of selecting a left-handed person from a group of 500 people is 0.15. How many left-handed people are in a group of 500?

15. A person is selected randomly from a jury list of 40 names. The probability that this person is a white male is 0.2. How many white males are on the jury list?

16. A barrel contains five red balls along with balls of other colors. The probability of randomly selecting one red ball from the barrel is known to be 0.125. How many nonred balls are in the barrel?

CREATING MODELS

17. One die is thrown and the number that turns up is recorded.
(*a*) What is the probability that the number is less than 5?
(*b*) What is the probability that the number is odd?

18. Two dice are thrown and the sum of the faces that turn up is computed. What is the probability that this sum is less than 5?

19. Two dice are thrown. What is the probability that identical faces that turn up on the two dice?

20. One card is selected blindly from a regular deck of cards.
(*a*) What is the probability that the card is red?
(*b*) What is the probability that the card is a face card?
(*c*) What is the probability that the card is a 5?
(*d*) What is the probability that the card is a red 5?

21. Four hundred tickets are sold for a raffle, and Ms. Lin buys 8 of those tickets. One ticket stub is drawn and the holder of that ticket wins a new car. What is the probability that Ms. Lin will win the car?

22. One ball is selected blindly from a barrel containing two red balls, three white balls, and four green balls.
    (**a**) What is the probability that the ball selected is red?
    (**b**) What is the probability that the ball selected is not green?
23. A jury list of 100 names contains names of 55 men and 45 women. Forty-two of the men and 31 of the women have college degrees, and the others do not. A person is selected randomly from this jury list.
    (**a**) What is the probability that this person is female?
    (**b**) What is the probability that this person has a college degree?
    (**c**) What is the probability that this person is female and has a college degree?

The data in Problems 24 through 27 are based on information reported in *An American Profile: Opinions and Behavior, 1972–1989* for surveys taken in 1989.

24. The following table summarizes the responses to the question, "Do you have any guns or revolvers in your home or garage?

	YES	NO	TOTALS
Male	245	201	446
Female	228	356	584
Totals	473	557	1,030

What is the probability that a person in the survey (*a*) is male, (*b*) has a gun, (*c*) is female and does not have a gun?

25. The following table summarizes the responses to the question, "Do you personally know anyone, living or dead, who has been infected with AIDS?"

	YES	NO	TOTALS
Male	66	594	660
Female	70	805	875
Totals	136	1,399	1,535

What is the probability that a person in the survey (*a*) is female, (*b*) knows someone who has been infected with AIDS, (*c*) is male and does not know anyone infected with AIDS?

26. The following table summarizes the responses to the question, "Do you ever have occasion to use alcoholic beverages?"

	UA	TA	TOTALS
White	617	264	881
Black	57	53	110
Totals	674	317	991

White and black refer to race, *UA* denotes a user of alcohol on occasion, and *TA* denotes a total abstainer from alcohol. What is the probability that a person in the survey is (*a*) black, (*b*) a total abstainer, (*c*) white and drinks alcohol on occasion?

27. The following table summarizes the responses to the question, "During the last year, did someone burglarize your home or apartment?"

	YES	NO	TOTALS
White	45	845	890
Black	10	92	102
Totals	55	937	992

What is the probability that a person in the survey (*a*) is white, (*b*) has been burglarized, (*c*) is black and has not been burglarized?

28. The editor of a corporation's annual report is informed by mail that a photographer has taken a picture of two lawyers in the firm's Tucson office for the report. The editor knows that the Tucson office employs five lawyers, four females and one male. What is the probability that the male lawyer is in the picture?

29. A man has three suits, each of a different color, and three ties, one matched to each suit. Without looking, the man quickly grabs one suit and one tie. What is the probability that the suit and tie match?

30. Mr. Jones purchases two cans of chicken soup and three cans of vegetable soup at a bargain price. The cans have no labels and look identical. On the way home, the cans become thoroughly mixed. Once home, Mr. Jones opens two cans. Determine the probability that the contents of the two cans are the same.

31. A couple's wedding anniversary is in April, but a friend cannot remember the date. What is the probability that the wedding anniversary will fall on a weekend (either Saturday or Sunday) if it is known that April 1 is a Friday.

In Problems 32 through 40, use the fundamental theorem of counting to determine the cardinal numbers of interest.

32. Keyless entry to a new car consists of pushing three buttons in sequence from a menu of five buttons. The correct sequence may require one button to be pressed more than once. A thief attempts to steal the car by arbitrarily pressing a three-button sequence. What is the probability that the thief will gain entry on his or her first try?

33. Assume that a license plate has five places, the first two reserved for numbers and the last three for letters. What is the probability of getting a license plate in which the numbers match and the letters are different?

34. An automobile manufacturer offers ten exterior colors and seven interior colors. Each interior color is available as an exterior color, but there are three exterior colors that are not available for the interior. What is the probability that a car will have different interior and exterior colors?

35. A restaurant offers a fixed-price dinner for $24.95. A diner may choose one item for each course. There are seven appetizers, three of which are fat-free, eight entrees, four of which are fat-free, and eight desserts, two of which are fat-free. A customer who does not read English orders dinner at random. What is the probability that the diner will order a totally fat-free dinner?

36. A motel has 31 rooms, numbered 1 through 31, and six reservations. What is the probability that all of the six reservations are assigned to the first 15 rooms?

37. A rent-a-car agency has 14 available cars (6 Fords and 8 Dodges) and four customers in the office. What is the probability that each of the four customers is assigned a Ford?

38. Messages are relayed to passing trains by displaying two single-colored flags, one under the other. Messages are formed by selecting flags from a box of eight flags (two each of four different colors). Each combination of colors represents a different message, and all possible messages are equally likely to occur. What is the probability of seeing a message containing two flags of the same color?

39. A personal identification number (PIN) for an automatic teller machine (ATM) contains four digits. What is the probability that someone's PIN begins with a zero and ends with an even number?

40. An airport taxi service has a fleet of 20 Pontiacs, 15 Chryslers, and 10 Fords. A customer calls for a taxi. What is the probability that the customer will ride in (*a*) a Ford, (*b*) in either a Ford or a Pontiac?

EXPLORING IN TEAMS

41. Figure *A* is a pie chart showing the age distribution of males who married for the first time in 1987, based on data provided by the U.S. Bureau of the Census. Using this information, determine the probability that a man who married for the first time in 1987 is (*a*) between the ages of 25 and 29, inclusive, (*b*) 30 years old or older.

42. The following table shows the age distribution of females who married for the first time in 1987, based on data provided by the U.S. Bureau of the Census.

Age	under 20	20–24	25–29	30–34	35–44	45 and older
Percentage	18.1	44.2	25.2	8.4	3.5	0.8

Use this information to determine the probability that a female who married for the first time in 1987 is (*a*) between the ages of 25 and 29, inclusive (*b*) 30 years old or older.

43. Figure *B* is a Venn diagram model of the U.S. population (in millions of people) in 1991, as reported by the U.S. Bureau of Labor Statistics. The universal set includes all people 18 years or older, excluding members of the Armed Forces who live on-post by themselves. Determine the probability that a randomly selected person from the universal set is (*a*) married, (*b*) Hispanic, or (*c*) neither married nor Hispanic.

44. Throw a pair of dice and record the difference between the face value of the two dice by subtracting the smaller value from the larger value. If the two dice show the same face value, then the difference is 0. Repeat this process many times—well over 200 times, if possible—and construct a pie chart from your results. Use your results to estimate the probability that the difference is 2.

FIGURE A

Age distribution of males at first marriage (1987)

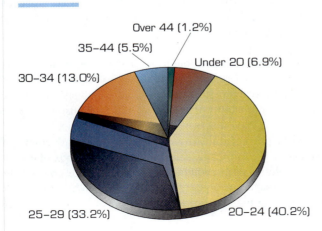

Over 44 (1.2%)
35–44 (5.5%)
30–34 (13.0%)
Under 20 (6.9%)
25–29 (33.2%)
20–24 (40.2%)

FIGURE B

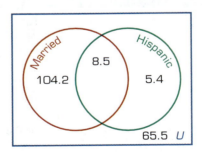

Married 104.2
8.5
Hispanic 5.4
65.5 *U*

REVIEWING MATERIAL

45. (Section 1.4) Figure C is a window of a graph relating the profits of a company (vertical axis) to time (horizontal axis). Using this graph, estimate the time when (*a*) the company first lost money, (*b*) the profits first reached $60 million, (*c*) company profits were $20 million.

FIGURE C

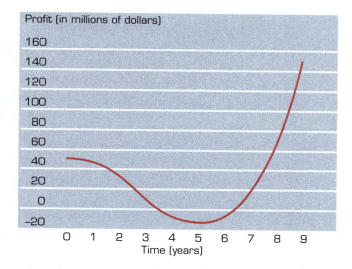

46. (Section 2.3) The following table, provided by the U.S. Patent and Trademark Office, lists the number of patent applications (in thousands) filed with the U.S. government each year.

Year	1982	1984	1986	1988	1990
Patent applications filed (thousands)	118.4	120.6	133.0	151.9	176.7

Find the equation of the least-squares straight line that fits this data, and then use the results to estimate the number of patents filed in (*a*) 1985, (*b*) 1989.

47. (Section 4.3) Ms. Cortes decides to create an IRA (Individual Retirement Account) by depositing $80 into a mutual fund at the end of each month. How much will she have after 35 years if the account grows 6.5% a year, compounded monthly?

48. (Section 5.1) Let

$$U = \{0, 1, 2, 3, 4, 5, 6\}$$

$$A = \{1, 2, 3\}$$

$$B = \{0, 2, 4, 6\}$$

$$C = \{0, 1, 2\}$$

Find (*a*) A' (*b*) $A \cap B$ (*c*) $B \cup C$ (*d*) $\{x \mid s \in U \text{ and } s > 6\}$

RECOMMENDING ACTION

49. Respond by memo to the following request:

> MEMORANDUM
>
> To: J. Doe Reader
>
> From: The Carnival Committee
>
> Date: Today
>
> Subject: **A New Game**
>
> We are now planning for next year's carnival. To keep the carnival fresh and exciting, we always need new games, games people have never before played. Can you create one for us?
>
> We need a game that can be played by a single person, a game that will appeal to a wide audience, and one in which the likelihood of losing is a bit greater than the likelihood of winning. We want the game to produce many winners, but because this carnival is our major fund raiser for the year, we need participants, as a whole, to lose more often than they win.

6.2 THE LAWS OF PROBABILITY

Most probability problems involve sample spaces in which the outcomes are not all equally likely to occur and for which the formula

$$P(E) = \frac{n(E)}{n(U)} \qquad \text{(1 repeated)}$$

is *not* applicable. In these cases we must develop other formulas and to do so, we must first understand the rules and properties we want probabilities to possess.

All of us have an intuitive feeling for probabilities, but we have not yet formally defined a probability. We have not answered the question, What is a probability? We will use our intuition and experiences with sample spaces of equally likely outcomes to develop a set of rules that are applicable to all probability problems, regardless of whether their outcomes are equally likely to occur.

Probabilities are numbers associated with events.

We first observe that probabilities are numbers. In Section 6.1, we found that the probability of flipping a coin twice and obtaining one head and one tail is $\frac{1}{2}$ (Example 6), that the probability of throwing two dice and having their sum be 7 is $\frac{1}{6}$ (Example 8), and that the probability of randomly selecting a heart from a regular deck of cards is $\frac{1}{4}$ (Example 9). Probabilities are numbers associated with sets of outcomes called events. Each event has one and only one probability associated with it, and that probability is a measure of the chance or likelihood that the event will actually occur.

Observe that all the probabilities determined in Section 6.1 were numbers between 0 and 1, inclusive. This can be understood by looking at Equation (*1*): the number of elements in a subset of a sample space is always less than or equal to the number of elements in the entire sample space. Thus, the numerator in Equation (*1*) is always less than or equal to the denominator, so the quotient is always between 0 and 1, inclusive. The first law of probability is thus:

THE RANGE LAW OF PROBABILITY

A probability is a number in the range between 0 and 1, inclusive.

A sample space U is the set of *all* possible outcomes. Once a process is initiated, the result must be an element of U. If it is not, we do not have a valid sample space. Accordingly, there is a 100% chance that some outcome in the sample space will occur when a process is run. The decimal equivalent of 100% is 1, so we have

THE CERTAINTY LAW OF PROBABILITY

$$P(U) = 1 \qquad\qquad [2]$$

A process is an action or decision that results in one or more possible outcomes. Once a process is initiated, something must happen and that something must be in the sample space, because a sample space is a set of all possible outcomes. There is no chance of running a process and having no outcome occur. The event of no outcome occurring is the empty set ϕ. Thus, we have

THE NULL LAW OF PROBABILITY

$$P(\phi) = 0 \qquad\qquad [3]$$

A probability is a number assigned to an event, and that number is restricted to the range between 0 and 1, inclusive. An event is a subset of a sample space U, so a probability is a number assigned to a subset of a sample set. Each subset has a probability assigned to it, although the number can be different for different subsets. The probability assigned to the entire sample space U must be 1, and the probability assigned to the subset ϕ, the empty set, must be 0.

The next law of probability is not as obvious as the previous three. Figure 6.3 is a Venn diagram showing two events E and F, which are subsets of a sample space U.

According to this diagram, there are b elements in $E \cap F$, a elements in E but not in F, and c elements in F but not in E. In terms of cardinal numbers, we have

$$n(E \cap F) = b$$
$$n(E) = a + b$$
$$n(F) = b + c$$
$$n(E \cup F) = a + b + c$$

FIGURE 6.3

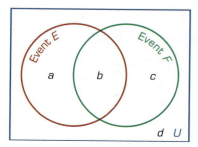

Consequently,

$$n(E \cup F) = a + b + c$$
$$= (a + b) + (b + c) - b$$
$$= n(E) + n(F) - n(E \cap F)$$

Dividing both sides of this last equation by the cardinal number of the sample space, $n(U)$, we have

$$\frac{n(E \cup F)}{n(U)} = \frac{n(E)}{n(U)} + \frac{n(F)}{n(U)} - \frac{n(E \cap F)}{n(U)} \qquad (4)$$

If U is a sample space of equally likely outcomes, then we have

$$P(E) = \frac{n(E)}{n(U)} \qquad \text{(1 repeated)}$$

and by extension, we also have

$$P(F) = \frac{n(F)}{n(U)}$$

$$P(E \cup F) = \frac{n(E \cup F)}{n(U)}$$

$$P(E \cap F) = \frac{n(E \cap F)}{n(U)}$$

Substituting these equalities into Equation (4), we obtain

THE INCLUSION-EXCLUSION LAW OF PROBABILITY

$$P(E \cup F) = P(E) + P(F) - P(E \cap F) \qquad (5)$$

EXAMPLE 1

Table 6.1 is based on information reported in *An American Profile: Opinions and Behavior, 1972–1989*. The table summarizes the responses of 979 people who were asked in 1988 how many hours they watched television on an average day. Verify Equation (5) for the process of randomly selecting one person from this sampled population when E is the event of selecting a male and F is the event of selecting someone who watches three hours of television or more on an average day.

TABLE 6.1 Survey on Time Spent Watching Television in 1988

	MALE	FEMALE	TOTALS
Less than three hours	226	234	460
Three hours or more	209	310	519
Totals	435	544	979

Solution

Take U to be all the 979 people in the sampled population. Then U is a sample space of equally likely outcomes. With respect to this sample space, E is the subset containing all males.

$$n(E) = 226 + 209 = 435$$

and

$$P(E) = 435/979 = .444, \text{ rounded to three decimal places}$$

F is the subset containing all people sampled who watch three hours or more of television on an average day.

$$n(F) = 209 + 310 = 519$$

and

$$P(F) = 519/979 \approx .530$$

$E \cup F$ is the subset of people sampled who are either male or who watch three hours or more of television on an average day.

$$n(E \cup F) = 226 + 209 + 310 = 745$$

and

$$P(E \cup F) = 745/979 \approx .761$$

$E \cap F$ is the subset of people sampled who are male *and* watch three hours or more of television on an average day.

$$n(E \cap F) = 209$$

and

$$P(E \cap F) = 209/979 \approx .213$$

We have

$$P(E) + P(F) - P(E \cap F) = \frac{435}{979} + \frac{519}{979} - \frac{209}{979}$$

$$= \frac{745}{979}$$

$$= P(E \cup F)$$

This verifies Equation (5).

Before continuing, let us reiterate the relationship between sets and events. An event is a subset of a set U, called the sample space. Even though we often describe an event using a phrase, that phrase just defines a property p that every element in the event must satisfy. The event E is

$$E = \{u \mid u \in U \text{ and } u \text{ satisfies the property } p\}$$

Furthermore, $E \cup F$ is the subset containing all elements of U that are either in E or in F; $E \cap F$ is the subset containing all elements of U that are in both E and F, simultaneously; and E' is the subset containing all elements of U that are not in E.

If E and F are disjoint, then $E \cap F = \phi$, and $P(E \cap F) = P(\phi) = 0$, as a result of the Null Law of Probability. In this case, the Inclusion-Exclusion Law of Probability, Equation (5), simplifies to

THE EXCLUSIVITY LAW OF PROBABILITY

$$\text{If } E \cap F = \phi, \text{ then } P(E \cup F) = P(E) + P(F). \tag{6}$$

EXAMPLE 2

Verify Equation (6) for the process of randomly picking one card from a regular deck when E is the event of picking an ace and F is the event of picking a face card (jack, queen, or king).

Solution

Take U to be the set of all 52 cards in the deck, distinguished by rank and suit. Then U is a sample space of equally likely outcomes. With respect to this sample space,

$$E = \{A\clubsuit, A\diamondsuit, A\heartsuit, A\spadesuit\}, \text{ with } n(E) = 4 \text{ and } P(E) = 4/52$$

$$F = \{J\clubsuit, J\diamondsuit, J\heartsuit, J\spadesuit, Q\clubsuit, Q\diamondsuit, Q\heartsuit, Q\spadesuit, K\clubsuit, K\diamondsuit, K\heartsuit, K\spadesuit\},$$
$$\text{with } n(F) = 12 \text{ and } P(F) = 12/52$$

$$E \cup F = \{A\clubsuit, A\diamondsuit, A\heartsuit, A\spadesuit, J\clubsuit, J\diamondsuit, J\heartsuit, J\spadesuit, Q\clubsuit, Q\diamondsuit, Q\heartsuit, Q\spadesuit, K\clubsuit, K\diamondsuit,$$
$$K\heartsuit, K\spadesuit\} \text{ is the event of picking either an ace or}$$
$$\text{a face card, with } n(E \cup F) = 16 \text{ and } P(E \cup F) = 16/52$$

$E \cap F$ is the event of picking one card that is both an ace *and* a face card. There is no such card in a regular deck, so the set $E \cap F$ is empty, and

$$P(E \cap F) = P(\phi) = 0$$

For these events, we have

$$P(E \cup F) = \frac{16}{52} = \frac{12}{52} + \frac{4}{52} = P(E) + P(F)$$

which verifies Equation (6).

EXAMPLE 3

The probability that a particular investment will make money (event E) is known to be .42, while the probability that it will lose money (event F) is .35. What is the probability that the investment will either make money or lose money?

Solution

A sample space for the process of making such an investment is

$$U = \{\text{make money, lose money, break even}\}$$

The two events $E = \{\text{make money}\}$ and $F = \{\text{lose money}\}$ are disjoint, so $P(E \cap F) = 0$. The event $E \cup F$ is the event of making money *or* losing money. It follows from the Exclusivity Law of Probability, Equation (6) that

$$P(E \cup F) = P(E) + P(F) = .42 + .35 = .77$$

A group of sets is mutually exclusive if the intersection of every pair of sets in the group is the empty set.

The Exclusivity Law of Probability can be generalized to more than two sets, providing the sets are mutually disjoint (see Section 5.1). When dealing with probabilities, the term *mutually exclusive* is often used instead of the term *mutually disjoint*. Mutually exclusive sets (or events, in probability) refer to a group of sets having the property that the intersection of every pair of sets in the group is the empty set.

THE EXCLUSIVITY LAW OF PROBABILITY (GENERALIZED)

If $E_1, E_2, \ldots, E_n$ are mutually exclusive (that is, each pair of sets is disjoint), then

$$P(E_1 \cup E_2 \cup \ldots \cup E_n) = P(E_1) + P(E_2) + \ldots + P(E_n) \qquad [7]$$

A probability distribution is a rule that assigns a unique number, called a probability, to each subset of a sample space, in accordance with the laws of probability.

We are now ready to define a probability distribution. Let a process have a finite number of possible outcomes, which constitute a sample space U. A *probability distribution* is a rule of correspondence between subsets of U and real numbers. This rule assigns a unique number, called the *probability of event E occurring* and denoted by $P(E)$, to each subset E of U, in accordance with the laws of probability developed in this section. Any rule that satisfies all these laws defines a probability distribution. Equation (*1*) is one way to assign probabilities. Other ways involve the concept of a simple event.

A sample space is *simple* if it has the property that one and only one of its elements can occur when the underlying process is run. In Example 1 of Section 6.1 we listed two different sample spaces for the process of flipping a coin twice and describing the results in terms of heads or tails:

$$U_1 = \{\text{two heads, two tails, one of each}\}$$

and

$$U_2 = \{\text{HH, HT, TH, TT}\}$$

U_2 is a sample space of equally likely outcomes, while U_1 is not, but both are simple sample spaces. Each time the process is run, that is, each time a coin is flipped twice, one and only one of the outcomes in each sample space can occur. In contrast, the sample space

$$U_3 = \{\text{at least one head, no heads, at least one tail, no tails}\}$$

is not simple. The outcomes *at least one head* and *at least one tail* both occur whenever the two flips show one head and one tail.

A simple event is a set containing a single element of a simple sample space.

A *simple event* is a set containing a single element of a simple sample space. There are as many simple events associated with a sample space as there are elements in that sample space. For example, the simple sample space $U = \{\text{red, green, blue}\}$ has as its simple events $E_1 = \{\text{red}\}$, $E_2 = \{\text{green}\}$, and $E_3 = \{\text{blue}\}$. Simple events are mutually exclusive. Since no two simple events occur at the same time, the intersection of any two simple events is empty.

Consider a process modeled by the simple sample space

$$U = \{s_1, s_2, s_3, \ldots, s_n\}$$

with corresponding simple events $E_1 = \{s_1\}$, $E_2 = \{s_2\}$, $E_3 = \{s_3\}$, $\ldots$, $E_n = \{s_n\}$. Clearly,

$$U = E_1 \cup E_2 \cup E_3 \cup \ldots \cup E_n$$

and

$$P(U) = P(E_1 \cup E_2 \cup E_3 \cup \ldots \cup E_n)$$

It follows from the Generalized Exclusivity Law of Probability, Equation (7), that

$$P(U) = P(E_1) + P(E_2) + P(E_3) + \ldots + P(E_n)$$

and from the Certainty Law of Probability, Equation (2), that $P(U) = 1$. Therefore,

$$1 = P(E_1) + P(E_2) + P(E_3) + \ldots + P(E_n)$$

That is, the sum of the probabilities assigned to each simple event in a simple sample space must be 1.

A probability distribution can be created for a simple sample space by assigning to each simple event a nonnegative number between zero and one, while ensuring that the sum of all the assigned numbers is 1. In particular, if we assign

$$P(E_1) = 0.1$$
$$P(E_2) = 0.3 \qquad (8)$$
$$P(E_3) = 0.6$$

> *The sum of the probabilities assigned to each simple event in a simple sample space must be 1.*

for the simple events $E_1 = \{red\}$, $E_2 = \{green\}$, and $E_3 = \{blue\}$ of the simple sample space $U = \{red, green, blue\}$, we have a probability distribution. A different assignment of numbers will also define a probability distribution, as long as each number in the assignment is between 0 and 1 in value and their sum is 1. The assignment

$$P(E_1) = 0.35$$
$$P(E_2) = 0.50 \qquad (9)$$
$$P(E_3) = 0.15$$

defines another probability distribution for the simple sample space $U = \{red, green, blue\}$.

Each sample space and each assignment of probabilities to events in a sample space is a model of a process. If the probabilities reflect reality, then the model is a good one; otherwise the model does not adequately represent the process under study. Consider the process of choosing one ball at random from a barrel of balls of identical size. If the barrel contains 100 balls, with 10 red balls, 30 green balls, and 60 blue balls, then the probabilities defined in (8) describe a good mathematical model for the process. If, however, the barrel contains 35 red balls, 50 green balls, and 15 blue balls, then the probabilities defined in (9) describe the appropriate model for the process.

EXAMPLE 4

A die is weighted (loaded) so that the side that comes up when the die is rolled is directly proportional to the number of dots on the face. Determine the probability of each face of the die coming up when the die is rolled.

Solution

A sample space for the process of rolling one die is

$$U = \{1, 2, 3, 4, 5, 6\}$$

where each outcome denotes the number of dots on the side of the die that comes up. The simple events are $\{1\}$, $\{2\}$, $\{3\}$, $\{4\}$, $\{5\}$, and $\{6\}$. We are told that the probability of a simple event occurring is directly proportional to the number of dots associated with that event, so we know (see Section 1.3) that

$$\frac{P(\{1\})}{1} = k$$

$$\frac{P(\{2\})}{2} = k$$

$$\frac{P(\{3\})}{3} = k$$

$$\frac{P(\{4\})}{4} = k$$

$$\frac{P(\{5\})}{5} = k$$

$$\frac{P(\{6\})}{6} = k$$

where k is a constant of proportionality. Consequently,

$$P(\{1\}) = 1k$$
$$P(\{2\}) = 2k$$
$$P(\{3\}) = 3k$$
$$P(\{4\}) = 4k$$
$$P(\{5\}) = 5k$$
$$P(\{6\}) = 6k$$

The sum of the probabilities must equal one, so

$$1k + 2k + 3k + 4k + 5k + 6k = 1$$

Therefore, $21k = 1$, $k = \frac{1}{21}$, and

$$P(\{1\}) = 1/21 \approx .048, \; P(\{2\}) = 2/21 \approx .095, \; P(\{3\}) = 3/21 \approx .143,$$
$$P(\{4\}) = 4/21 \approx .190, \; P(\{5\}) = 5/21 \approx .238, \; P(\{6\}) = 6/21 \approx .286$$

Probabilities associated with simple events can be represented graphically by a pie chart. Each simple event is associated with a different sector of the pie chart, with the fractional

FIGURE 6.4

Probabilities for a loaded die

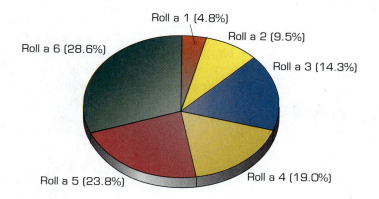

Roll a 1 (4.8%)
Roll a 2 (9.5%)
Roll a 6 (28.6%)
Roll a 3 (14.3%)
Roll a 5 (23.8%)
Roll a 4 (19.0%)

FIGURE 6.5
Probabilities for a fair die

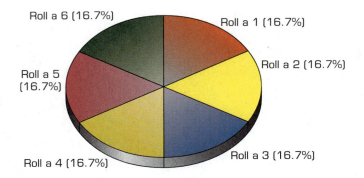

Roll a 6 (16.7%)   Roll a 1 (16.7%)

Roll a 2 (16.7%)

Roll a 5 (16.7%)

Roll a 3 (16.7%)

Roll a 4 (16.7%)

area of each sector equal to the probability that the associated simple event will occur. If, in particular, a simple event has a probability of .25 of occurring, then the sector representing that event will occupy .25 or 25% of the area of the pie chart. A pie chart for the probabilities calculated in Example 4 is shown in Figure 6.4. For comparison, Figure 6.5 is a pie chart for a fair die, for which each simple event is equally likely to occur and has a probability of $1/6 \approx .167$.

EXAMPLE 5

A die is weighted as described in Example 4. Determine the probability of rolling such a die and having an even number of dots turn up.

Solution

$U = \{1, 2, 3, 4, 5, 6\}$ is a sample space for the process of rolling one die. The event E of rolling an even number of dots is described by the subset $E = \{2, 4, 6\}$. Using the probabilities found in Example 4 and the Generalized Exclusivity Law of Probability, Equation (7), we have

$$P(E) = P(\{2\} \cup \{4\} \cup \{6\})$$
$$= P(\{2\}) + P(\{4\}) + P(\{6\})$$
$$= 2/21 + 4/21 + 6/21$$
$$= 12/21 = .5714$$

rounded to four decimals. Note that Equation (7) is valid here because the sets $\{2\}$, $\{4\}$, and $\{6\}$ are mutually exclusive.

EXAMPLE 6

A football coach responds to a reporter's question about winning Saturday's game by saying, "We are five times more likely to win than we are to lose." When the reporter asks about a tie, the coach replies, "The chance of a tie is one in a hundred." What is the probability of the team winning?

Solution

A sample space is $U = \{$win, loss, tie$\}$. Mathematically modeling both of the coach's replies, we have

$$P(\{\text{win}\}) = 5 \cdot P(\{\text{loss}\}) \qquad \text{and} \qquad P(\{\text{tie}\}) = 1/100$$

Now,

$$1 = P(\{\text{win}\}) + P(\{\text{loss}\}) + P(\{\text{tie}\})$$
$$= 5 \cdot P(\{\text{loss}\}) + P(\{\text{loss}\}) + \frac{1}{100}$$

Solving this equation for $P(\{\text{loss}\})$, we obtain

$$6 \cdot P(\{\text{loss}\}) = \frac{99}{100}$$

$$P(\{\text{loss}\}) = \frac{99}{600} = .165$$

Therefore,

$$P(\{\text{win}\}) = 5 \cdot P(\{\text{loss}\}) = 5\left(\frac{99}{600}\right) = 0.825$$

Figure 6.6 is a pie chart for each of these simple events and the probabilities associated with them.

FIGURE 6.6

Coach predicts outcome

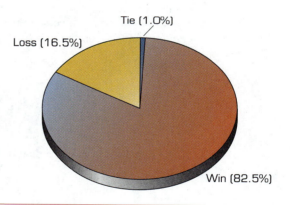

Tie (1.0%)

Loss (16.5%)

Win (82.5%)

If a process has been run a number of times in the past, then we use historical data to calculate the fraction of times a particular event occurs and let this fraction be the probability of the event occurring.

One effective way to assign probabilities to simple events is to use historical data when they are available. If a process has been run a number of times in the past, or if a process can be run repeatedly in the present, so that information about outcomes and their frequency of occurring exists, then we can use that data to calculate the fraction of times a particular event occurs. This fraction becomes the probability of the event. Such probabilities are called *empirical probabilities*. For example, if we flip a coin 800 times and obtain 450 heads and 350 tails, then we would assign the fraction $\frac{450}{800} = .5625$ as the probability of obtaining a head, and the fraction $\frac{350}{800} = .4375$ as the probability of obtaining a tail.

EXAMPLE 7

Between 1972 and 1993, the *General Social Survey* asked people how often they went to a bar or tavern. The results of those who responded are as follows:

Almost everyday	274
Once or twice a week	1,488
Several times a month	1,250
About once a month	1,629
Several times a year	1,950
About once a year	1,789
Never	9,045

Assuming that this sample is representative of the entire U.S. population, determine the empirical probabilities that a person will go to a bar or tavern (*a*) almost every day, (*b*) once or twice a week, (*c*) several times a month, and (*d*) more than once a month.

Solution

The sample consists of 17,425 respondents. There are seven outcomes listed, and they are mutually exclusive, so there are seven simple events: E_1: a person goes to a bar or tavern almost everyday; E_2: a person goes once or twice a week; E_3: a person goes several times a month; and so on through E_7: a person never goes to a tavern.

(*a*) $P(E_1) = \dfrac{274}{17,425} \approx .016$

(*b*) $P(E_2) = \dfrac{1,488}{17,425} \approx .085$

(*c*) $P(E_3) = \dfrac{1,250}{17,425} \approx .072$

(*d*) P(a person goes more than once a month) $= P(E_1 \cup E_2 \cup E_3)$
$= P(E_1) + P(E_2) + P(E_3)$ (because the events are mutually exclusive)
$\approx .016 + .085 + .072 = .173.$

EXAMPLE 8

A department store survey shows that of all women entering the store on the day of the survey, 22% made a purchase in the jewelry department, 17% made a purchase in the cosmetics department, and 8% made purchases in both departments. Assume that the survey accurately reflects what happens in the store everyday. What is the probability that a woman leaving the store made a purchase in at least one of the two departments?

Solution

Let E be the event that a woman makes a purchase in the jewelry department and F the event that a woman makes a purchase in the cosmetics department. Using the survey as historical data, we have the empirical probabilities $P(E) = .22$ and $P(F) = .17$, as well as $P(E \cap F) = 0.08$. The event of making a purchase in at least one of the two departments is $E \cup F$, and it follows from Equation (5) that

$$P(E \cup F) = P(E) + P(F) - P(E \cap F)$$
$$= .22 + .17 - .08$$
$$= .31$$

A subset E is always disjoint from its complement E', because an element is either in a set or it is not in the set. Thus, for any event E in a sample space U,

$$E \cap E' = \phi \qquad \text{and} \qquad E \cup E' = U$$

It follows from the laws of probability that

$$1 = P(U) = P(E \cup E') = P(E) + P(E')$$

or

THE COMPLEMENT LAW OF PROBABILITY

$$P(E') = 1 - P(E) \qquad\qquad [10]$$

The probability that an event does not happen is 1 minus the probability that the event happens.

EXAMPLE 9

Using the results of Example 7, determine the probability that a person goes to a bar or tavern once a month or less.

Solution

Let E be the event of a person going to a bar or tavern more than once a month. Then E' is the event of a person going once a month or less. We have from Example 7 that $P(E) = .173$. It follows from Equation (10) that $P(E') = 1 - P(E) = 1 - .173 = .827$.

EXAMPLE 10

The birthday paradox. Determine the probability that at least 2 people in a group of 23 have the same birthday (day and month). Assume that a year has 365 days.

Solution

Let E be the event that at least 2 people have the same birthday. We can calculate $P(E)$ directly, or we can first determine $P(E')$ and then use the formula $P(E') = 1 - P(E)$ to find $P(E)$. If one way seems easier than the other, we will choose the easier approach.

The event that at least 2 people have the same birthday can happen many ways—2 people can have the same birthday, 3 people can have the same birthday, and so on. On the other hand, E' is the event of all people having different birthdays, which is much more definitive. It appears that it might be easier to calculate $P(E')$, so we will take that approach.

We let U be all possible birthday arrangements for 23 people. To count U, we note that the first person in the group can have a birthday on any one of 365 possible days. Regardless of the birthday of this first person, the second person can also have a birthday on any one of 365 possible days. The same is true for the third person, the fourth person, and every other person in the group. Using the fundamental theorem of counting (see Section 5.4), we have

$$n(U) = \underbrace{(365)(365)(365) \cdots (365)}_{23 \text{ times}} = (365)^{23}$$

To count E', the event that no two people have the same birthday, we note that the first person in the group can have a birthday on any one of the 365 days. Once this birthday is chosen, however, the second person can have a birthday on any one of the 364 *remaining* days (any day except the birthday of the first person). Once this second birthday is chosen, the third person can have a birthday on any one of the 363 remaining days (any day except the birthdays of the first two people). Continuing this analysis for each of the 23 people in the group and using the fundamental theorem of counting, we obtain

$$n(E') = (365)(364)(363) \ldots (343)$$

Since U is a sample space of equally likely outcomes, we have that

$$P(E') = \frac{n(E')}{n(U)} = \frac{(365)(364)(363) \ldots (343)}{(365)(365)(365) \ldots (365)} \approx .493$$

From Equation (*10*) we know that $P(E') = 1 - P(E)$, so

$$P(E) = 1 - P(E')$$
$$\approx 1 - .493$$
$$= .507$$

The probability that 2 people in a group of 23 have the same birthday is .507, rounded to three decimal places. Therefore there is a better than 50% chance of having a match. Does this seem surprising?

If $P(E') \neq 0$, *then the quotient* $P(E)/P(E')$, *normally written as a ratio of integers, is the odds in favor of the event* E.

If $P(E') \neq 0$, then the quotient $P(E)/P(E')$ is the *odds in favor of the event E*. Analogously, if $P(E) \neq 0$, then the quotient $P(E')/P(E)$ is the odds in favor of the event E', which is also *the odds against E*. Odds are written as a fraction a/b, with both a and b as integers, and read as "odds of a to b." For example, if $P(E) = .4$, so that $P(E') = 1 - .4 = .6$, then the odds in favor of E are

$$\frac{P(E)}{P(E')} = \frac{.4}{.6} = \frac{4}{6} = \frac{2}{3}$$

or 2 to 3.

EXAMPLE 11

Determine the odds for and against 2 or more people in a group of 23 having the same birthday.

Solution

Let E be the event that at least 2 people in a group of 23 have the same birthday. It follows from Example 10 that $P(E) = .507$ and $P(E') = .493$. The odds for E are

$$\frac{P(E)}{P(E')} = \frac{.507}{.493} = \frac{507}{493}$$

or 507 to 493. The odds against E are

$$\frac{P(E')}{P(E)} = \frac{.493}{.507} = \frac{493}{507}$$

or 493 to 507.

If we know the odds in favor of an event E are a to b, then

$$\frac{P(E)}{P(E')} = \frac{a}{b}$$

Substituting the right side of Equation (10) $(P(E') = 1 - P(E))$ into this equation and solving for $P(E)$, we obtain

$$\frac{P(E)}{1 - P(E)} = \frac{a}{b}$$

$$bP(E) = a[1 - P(E)]$$

$$bP(E) = a - aP(E)$$

$$aP(E) + bP(E) = a$$

$$(a + b)P(E) = a$$

or

$$P(E) = \frac{a}{a + b}$$

EXAMPLE 12

The odds in favor of rain tomorrow are 3 to 5. What is the probability of (a) rain, and (b) no rain tomorrow?

Solution

Let E be the event that it rains tomorrow. We are given that

$$\frac{P(E)}{P(E')} = \frac{3}{5}$$

so $a = 3$ and $b = 5$. Then

(a) $P(E) = \dfrac{3}{3 + 5} = \dfrac{3}{8} = .375$, and

(b) $P(E') = 1 - P(E) = 1 - \dfrac{3}{8} = \dfrac{5}{8} = .625$

IMPROVING SKILLS

In Problems 1 through 6, a set of numbers is assigned to the three simple events of the sample space $U = \{A, B, C\}$. Determine whether the assignments represent a probability distribution.

1. $P(A) = .25$, $P(B) = .10$, $P(C) = .65$
2. $P(A) = 0$, $P(B) = 0$, $P(C) = 1$
3. $P(A) = 1$, $P(B) = 1$, $P(C) = 0$
4. $P(A) = .25$, $P(B) = .20$, $P(C) = .55$
5. $P(A) = .25$, $P(B) = .15$, $P(C) = .50$
6. $P(A) = .25$, $P(B) = -.10$, $P(C) = .85$

In Problems 7 through 15, a set of numbers is assigned to the two simple events of the sample space $U = \{\text{true, false}\}$. Determine whether the assignments represent a probability distribution.

7. $P(\text{true}) = .25,\quad P(\text{false}) = .75$
8. $P(\text{true}) = 1,\quad P(\text{false}) = 0$
9. $P(\text{true}) = 0,\quad P(\text{false}) = 1$
10. $P(\text{true}) = 1,\quad P(\text{false}) = 1$
11. $P(\text{true}) = 0,\quad P(\text{false}) = 0$
12. $P(\text{true}) = .3,\quad P(\text{false}) = .4$
13. $P(\text{true}) = .3,\quad P(\text{false}) = .8$
14. $P(\text{true}) = .3,\quad P(\text{false}) = .7$
15. $P(\text{true}) = 1,\quad P(\text{false}) = 2$
16. Let E be the event that a person gets a raise, and let F be the event that a person gets a promotion. Describe in words the events defined by
    (a) E' (b) $E \cup F$ (c) $E \cap F$
    (d) $E' \cap F$ (e) $E \cup F'$ (f) $E' \cap F'$
17. Let E be the event that a customer is female, and let F be the event that a customer makes a purchase. Describe the following probabilities in words:
    (a) $P(E')$ (b) $P(F')$ (c) $P(E \cap F)$
    (d) $P(E \cup F)$ (e) $P(E \cup F')$ (f) $P(E \cap F')$
18. E and F are disjoint events with $P(E) = .45$ and $P(F) = .40$. Determine
    (a) $P(E')$ (b) $P(F')$ (c) $P(E \cap F)$
    (d) $P(E \cup F)$ (e) $P(E \cup E')$
19. E, F, and G are mutually exclusive events with $P(E) = .25$, $P(F) = .35$, and $P(G) = .15$. Determine
    (a) $P(E')$ (b) $P(F')$ (c) $P(E \cap F)$
    (d) $P(E \cup F)$ (e) $P(E \cup F \cup G)$ (f) $P(E \cup G)$
20. A shoe company commissions a survey on its advertising effectiveness. The results are as follows: The probability of a person hearing the advertisement is .23; the probability of a person not hearing the advertisement but being told about it is .37; and the probability of a person not hearing the advertisement and not being told about it is .63. What is wrong with the survey?
21. A department head wants to fund a new project. He claims that the project can result in only one of three outcomes and that the probability of each outcome occurring is .20. What is wrong with his analysis?
22. A coin is flipped 650 times and comes up heads 372 times. (a) What is the probability that this particular coin will come up heads when flipped? (b) What is the probability that this coin will come up tails?
23. A die is rolled repeatedly and the total number of times (frequency) each face occurs is recorded. The results are as follows:

Face value	1	2	3	4	5	6
Frequency	237	168	180	103	165	147

Use this data to determine empirical probabilities for each side of this die to turn up on any single roll. Use these results to determine the probability that (a) the die comes up even, (b) the die shows a number greater than 4, and (c) the die does not come up 1.

24. Between 1972 and 1993, the *General Social Survey* asked people in which month they were born. The results for those who responded are as follows:

January	1,922	May	1,784	September	2,031
February	1,806	June	1,909	October	1,934
March	2,028	July	1,933	November	1,900
April	1,837	August	2,080	December	1,918

Assuming that this sample is representative of the entire U.S. population, use this data to determine empirical probabilities of being born in each month of the year. Then determine the probability that (a) of being born in the first quarter of a year, (b) of being born in the last three quarters of a year, (c) of being born in either August or September, and (d) not being born in January.

25. The following table summarizes responses to the question, "Do you smoke?" in a survey taken in 1989, based on information reported in *An American Profile: Opinions and Behavior, 1972–1989*:

AGES	YES	NO
18–23	30	63
24–29	47	95
30–35	46	99
36–41	44	89
42–53	73	130
over 53	68	248

Assuming that this sample is representative of the entire U.S. population, use this data to determine empirical probabilities that a person in each age group will be a smoker.

26. People in a survey taken between 1985 and 1989 were asked whether they approved of a married person having sexual relations with someone other than the marriage partner. The following table, based on information reported in *A Statistical Handbook on the American Family*, summarize their responses:

	GENERALLY WRONG	GENERALLY ACCEPTABLE
Male 18–24	241	33
Male 25–34	466	73
Male 35–44	405	74
Male 45–64	545	67
Male over 64	324	37
Female 18–24	281	30
Female 25–34	614	77
Female 35–44	501	76
Female 45–64	716	56
Female over 64	612	27

Assuming that this sample is representative of the entire U.S. population, use this data to determine the probability that (a) a male feels that the practice is generally wrong, (b) a female feels that the practice is generally wrong, and (c) a person of either gender under the age of 35 feels that the practice is wrong.

27. Determine the odds in favor of flipping a fair coin and obtaining a head.
28. Determine the odds in favor of rolling one die and having a 3 appear.
29. Determine the odds in favor of rolling two dice and having their face values sum to 7.

30. Determine the odds against picking one card from a regular deck and having that card be an ace.

In Problems 31 through 38, the odds in favor of an event happening are given. Determine the probability of that event occurring.

31. 3 to 2   **32.** 2 to 3   **33.** 8 to 9

34. 8 to 5   **35.** 20 to 30.   **36.** 100 to 1

37. 28 to 13   **38.** 6 to 5

39. Show that if the odds in favor of an event occurring are a to b, then the probability that the event will *not* happen is $b/(a + b)$.

40. E and F are known to be mutually exclusive events. Must E' and F' also be mutually exclusive?

41. E and F are known to be mutually exclusive events. Must E and $(E \cup F)'$ also be mutually exclusive?

CREATING MODELS

42. A political party is running two candidates for two different positions in state government. The probability that candidate A will win is .74, and the probability that candidate B will win is .55. The probability that both will win is .38. What is the probability that at least one candidate will win?

43. A government spokesperson claims that there is a 25% chance that the rate of inflation will decrease and a 37% chance that it will remain constant. What is the probability that the rate will increase?

44. Four candidates are running for mayor. Three of the candidates commission pollsters to determine their chances of winning. Candidate A is given a 15% chance of winning, candidate B is given a 37% chance of winning, and candidate C is given a 29% chance of winning. Determing the probability that candidate D will win.

45. The probability of having a college-educated woman on a jury list is .59. The probability of having a woman who is not college-educated on the jury list is .20. One person from the jury list is selected at random. Determine the probability that (a) this person is a woman, and (b) this person is a man.

46. Referring to the jury list described in Problem 45, suppose that it is also known that the probability of having a college-educated male on the list is .10. Determine the probability that a randomly selected person from this jury list is a male who is not college-educated.

47. A bridge toll plaza has two booths. The probability that booth 1 has a line of waiting cars at any point in time is .87. The probability for booth 2 is .62. The probability that both booths have a line at the same time is .55. Determine the probability that there is a line for at least one booth.

48. A survey taken in a particular city showed that 43% of all wives and 51% of all husbands had a college degree. The probability that both spouses had a college degree was .37. Determine the probability that at least one spouse in a randomly chosen couple in this town has a college degree.

49. From experience, an automobile dealership knows that 14% of its new cars will require mechanical repairs and 9% of its new cars will require electrical repairs within the first year of operation. In addition, 4% of all new cars sold will require both types of repairs during the first year. Determine the probability that the car just sold to Dr. Lin will require mechanical repairs, electrical repairs, or both during its first year of operation.

50. A coin is weighted so that a head comes up twice as often as a tail. Determine the probability of flipping this coin and obtaining a tail.

51. Solve Problem 50 if a tail comes up three times as often as a head.

52. An investment is three times as likely to make money as it is to lose money, but the probability of the investment breaking even is .5. What is the probability that (*a*) the investment will make money, and (*b*) the investment will not make money?

53. A coach claims that the chances of his team winning the next game are twice as likely as the chances of losing, and the chances of the game ending in a tie are only half the chances of losing. What is the probability that the team will win the next game if the coach's claims are correct?

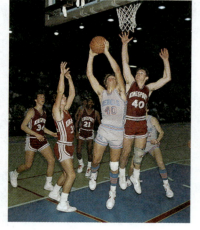

54. A barrel contains many balls, each numbered 1, 2, 3, 4, or 5. The probability of a number appearing when a ball is selected from the barrel at random is directly proportional to the number on the ball, that is, a 5 is five times more likely to be picked than a 1, and a 4 is twice as likely to be picked as a 2. Determine the probability of selecting one ball at random and having that ball (*a*) be a 1, (*b*) be an even number.

55. A die is weighted so that the numbers 1 through 5 are all equally likely to appear when the die is thrown but the number 6 is twice as likely to appear as any other number. What is the probability of rolling such a die and (*a*) obtaining a 1, (*b*) obtaining a 6?

EXPLORING IN TEAMS

56. It is known that $P(E) = .35$, $P(F) = .42$, and $P(E \cap F) = .18$. Use a Venn diagram to evaluate

(*a*) $P(E \cup F)$ (*b*) $P(E' \cap F)$ (*c*) $P(E' \cap F')$

57. It is known that $P(E) = .5$, $P(F) = .3$, $P(G) = .1$, $F \cap G = \phi$, $P(E \cap F) = .2$, and $P(E \cap G) = .1$. Use a Venn diagram to evaluate

(*a*) $P(E \cap F \cap G)$ (*b*) $P(E' \cap F)$ (*c*) $P(E \cup F)$

REVIEWING MATERIAL

58. (Section 2.2) Use Gaussian elimination to solve the system

$$x + 2y + z = 10$$
$$2x + 7y - 3x = 50$$
$$2y - z = 20$$

59. (Section 3.5) Apply the simplex method to the following linear program:

$$\text{Maximize:} \quad z = 5x_1 + 6x_2 + 7x_3 + 8x_4$$

$$\text{subject to:} \quad 3x_1 + 4x_2 + 5x_3 - 6x_4 \leq 90$$

$$7x_1 + 7x_2 + 8x_3 - 9x_4 \leq 135$$

$$\text{assuming:} \quad x_1, x_2, x_3, \text{ and } x_4 \text{ are nonnegative}$$

60. (Section 4.4) A person receives two offers for her business. One offer will pay $400 at the end of each month for five years. The second offer will pay $300 at the end of each month for seven years. Which is the better offer if current interest rates are $4\frac{1}{2}\%$ a year compounded monthly?

61. (Section 5.2) According to figures supplied by the U.S. Department of Agriculture, U.S. agricultural exports to Mexico in 1991 totaled $2.884 billion, while imports from Mexico that year totaled $2.536 billion. These imports consisted of $338 million in coffee, $371 million in fruits and nuts, $894 million in vegetables, $409 million in livestock, $39 million in citrus, and $485 in all other types of agricultural products. Construct a pie chart and a bar graph to present this information on imports.

RECOMMENDING ACTION

62. Respond by memo to the following request:

> ### MEMORANDUM
>
> To: J. Doe Reader
>
> From: Kemal
>
> Date: Today
>
> Subject: **Simple Events**
>
> I am having difficulty with the relationship between simple events and equally likely outcomes, and I am hoping that you can help me. Are simple events necessarily equally likely outcomes for a given process? And what about the converse: Are equally likely outcomes for a process necessarily simple events?

6.3 CONDITIONAL PROBABILITY AND INDEPENDENT EVENTS

Probabilities measure the likelihoods that the various outcomes of a particular process will occur. Probability distributions are numbers, determined through experimentation or past experience, that model reality as we understand it. As new information becomes available, however, our perception of reality may change, and so too may the probabilities we associate with various events in our models. We may, for example, determine that the probability of randomly picking one card from a regular deck and having it be a heart is .25 (Example 9 of Section 6.1), but we may want to revise our analysis if we are told that the card we chose is red.

In almost all the personal and professional situations that affect our daily lives, the conditions surrounding us change. Some changes, such as the birth of a child, occur infrequently. Other changes, such as changes in economic conditions, can happen rapidly and without forewarning. "Nothing endures but change," Heraclitus wrote, to which we may add the anonymous Latin saying translated as, "Times change, and we change with them." As we become aware of changes, we update our previous estimates and forecasts and revise probabilities for success and failure.

Conditional probabilities are probabilities determined with the availability of additional information. In effect, the probabilities associated with various events are affected or *conditioned* by the new information. The probability that a randomly selected card from a regular deck is a heart, *knowing* that the selected card is red, is a conditional probability. The probability will be determined using the additional information that the color of the selected card is red. The probability that a company's stock will decrease in value *knowing* that the company's president will announce a large loss at tomorrow's press conference is a conditional probability. The probability is affected by the additional information that a loss will be disclosed.

> *The conditional probability* P(E|F) *is the probability that event* E *will occur, knowing that event* F *is true.*

We create the notation $E|F$, read "E given F," to denote the event E when the event F is known to be true. We use the phrase "F is known to be *true*" to mean that either F has occurred or we know with certainty that F will occur. Then $P(E|F)$ is the probability that event E will occur, knowing that event F is true. Consider the process of randomly selecting one card from a regular deck. If E denotes the event of choosing a heart and F denotes the event of choosing a red card, then $P(E|F)$ is the probability of selecting a heart, knowing (or *given*) that the selected card is red.

EXAMPLE 1

A person makes a reservation at a motel with 31 single rooms; some rooms have computers, some have waterbeds, and some have both, as listed in Table 6.2. Determine (*a*) the probability of getting a single room with a computer, and (*b*) the conditional probability of getting a single room with a computer if the reservation guarantees a waterbed.

TABLE 6.2   Motel Room Features

	REGULAR BED	WATERBED	TOTALS
Computer	2	8	10
No computer	18	3	21
Totals	20	11	31

Solution

(*a*) There are 31 single rooms, and with no other information available to us, we assume that each is equally likely to be assigned. Let E be the event of getting a room with a computer. Since 10 of the 31 rooms have computers, $P(E) = \frac{10}{31} \approx .323$. It follows that $P(E') = 1 - P(E) = \frac{21}{31} \approx .677$ is the probability of not getting a room with a computer. (These probabilities are shown graphically in Figure 6.7.)

(*b*) Let F be the event of getting a room with a waterbed. If the reservation guarantees a waterbed, then the sample space of possible room assignments is reduced to the 11 rooms with waterbeds. Of these, 8 have computers, so $P(E|F) = \frac{8}{11} \approx .727$. Now $P(E'|F) = 1 - P(E|F) = \frac{3}{11} \approx .273$ is the probability of not getting a room with a computer. (These probabilities are displayed in Figure 6.8.) Observe how much the probabilities change with the additional knowledge that event F will occur.

FIGURE 6.7

All rooms

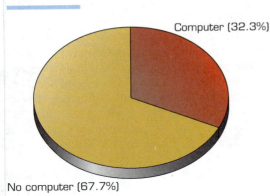

Computer (32.3%)

No computer (67.7%)

FIGURE 6.8

Rooms with a waterbed

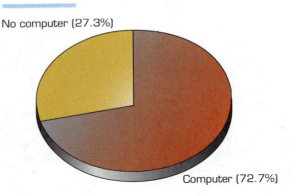

No computer (27.3%)

Computer (72.7%)

One approach to determining conditional probabilities is to modify the sample space each time new information is introduced, thereby revising what is possible and what is not. We did this in Example 1, when we pruned the sample space from 31 rooms to just the 11 rooms with waterbeds. If a sample space can be modified easily to accommodate new information, then doing so is an efficient way to calculate conditional probabilities.

In Example 1, U is the set of all 31 possible room assignments, E is the event of getting a room with a computer, and F is the event of getting a room with a waterbed. The cardinal numbers of E and U are, respectively, $n(E) = 10$ and $n(U) = 31$, which led us to the computation

$$P(E) = \frac{n(E)}{n(U)} = \frac{10}{31}$$

The cardinal number of F is $n(F) = 11$, and these 11 rooms became the revised sample space for calculating the conditional probability in part (*b*). The event of getting a room with both a computer and a waterbed is the event $E \cap F$, with cardinal number $n(E \cap F) = 8$. The conditional probability $P(E|F)$ was calculated as

$$P(E|F) = \frac{n(E \cap F)}{n(F)} \tag{11}$$

$$= \frac{8}{11}$$

FIGURE 6.9

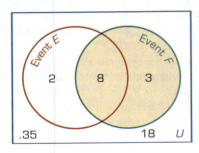

To determine the conditional probability $P(E|F)$ by modifying the sample space, we prune the sample space to just F, as has been done in the Venn diagram in Figure 6.13, where E and F refer to the events in Example 1. The event F is shaded in Figure 6.9. Knowing that event F occurs means that we restrict ourselves to this shaded region, which becomes our new sample space. Restricted in this way, E can occur only as a subset of F, that is, E is restricted to $E \cap F$.

A very interesting formula develops when we divide both the numerator and the denominator on the right side of Equation (*11*) by $n(U)$, the cardinal number of the *original* sample space. We get

$$P(E|F) = \frac{n(E \cap F)/n(U)}{n(F)/n(U)} \tag{12}$$

If U is a sample space of equally likely outcomes, we recognize that

$$\frac{n(E \cap F)}{n(U)} = P(E \cap F) \quad \text{and} \quad \frac{n(F)}{n(U)} = P(F)$$

Thus, Equation (*12*) becomes

$$P(E|F) = \frac{P(E \cap F)}{P(F)}$$

where all the probabilities on the right side are calculated with respect to the original sample space, assuming no additional information! We now use this result to define conditional probabilities for all sample spaces, with or without equally likely outcomes.

CONDITIONAL PROBABILITY

If E and F are events in a sample space U, with $P(F) > 0$, then

$$P(E|F) = \frac{P(E \cap F)}{P(F)} \qquad (13)$$

If we need to calculate conditional probabilities due to new information, then one approach, as we have seen, is to modify the sample space. Sometimes, however, it is difficult to change a sample space (see Example 2). Other times, so much work has gone into establishing a probability distribution that it is not feasible to begin anew with a revised sample space. In these situations, it is essential to have a formula for calculating conditional probabilities from the probability distribution associated with the original sample space, and Equation (*13*) is such a formula!

EXAMPLE 2

A recent survey of new car buyers indicates that 40% have two or more children and currently own a van, 44% currently own a van, and 61% have two or more children. A person walks into a car showroom and claims to own a van. What is the probability that this person has two or more children?

FIGURE 6.10

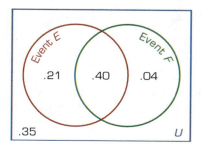

Solution

Let E be the event of having two or more children, and let F be the event of currently owning a van. As a result of the survey, we have $P(E) = .61$, $P(F) = .44$, and $P(E \cap F) = .40$ (see Figure 6.10). We are interested in $P(E|F)$, the probability that a customer has two or more children, *knowing that* the customer currently owns a van. Using Equation (*13*), we find

$$P(E|F) = \frac{P(E \cap F)}{P(F)} = \frac{.40}{.44} \approx .909$$

Initially, the probability of a person having two or more children was $P(E) = .61$. This probability increases to .909 when we know the person owns a van.

EXAMPLE 3

Using the information given in Example 2, determine the probability that a customer currently owns a van if she enters the showroom with two children who call her Mom.

Solution

The events E and F remain as before, but now we are interested in $P(F|E)$. It follows from Equation (13), with the roles of E and F reversed, that

$$P(F|E) = \frac{P(F \cap E)}{P(E)} = \frac{.40}{.61} \approx .656$$

Note that $F \cap E = E \cap F$.

EXAMPLE 4

One card is selected randomly from a regular deck of 52 cards. Determine the probability that the card is a heart if it is known that the card is red.

Solution

Let H be the event of selecting a heart, and let R be the event of selecting a red card.

(a) Each of the 52 cards in the deck are equally likely to be selected. There are 13 hearts in a deck and 26 red cards, so $P(H) = \frac{13}{52} = .25$ and $P(R) = \frac{26}{52} = .5$. The event $H \cap R$ is the event of being a heart *and* a red card, which is the same as being a heart. Thus, $P(H \cap R) = P(H) = .25$, and it follows from

$$P(E|F) = \frac{P(E \cap F)}{P(F)} \qquad \text{(13 repeated)}$$

with H replacing E and R replacing F, that

$$P(H|R) = \frac{P(H \cap R)}{P(R)} = \frac{.25}{.5} = .5$$

(b) This problem can also be solved by modifying the sample space. If we know the card is red, then the sample space is reduced to the 13 cards that are hearts and the 13 cards that are diamonds, for a total of 26 possible outcomes, all of which are equally likely to occur. Only 13 of these 26 red cards are hearts, so the probability of selecting a heart is $\frac{13}{26} = .5$, as before.

We can generate another important formula from the conditional probability equation

$$P(E|F) = \frac{P(E \cap F)}{P(F)} \qquad \text{(13 repeated)}$$

by multiplying both sides of this equation by $P(F)$. We obtain

---

THE PRODUCT RULE FOR PROBABILITIES

$$P(E \cap F) = P(F) \cdot P(E|F) \qquad (14)$$

The probability that both events E *and* F occur is the probability that F occurs times the probability that E occurs *knowing* that F is true. Often, E and F refer to sequential outcomes, with F occurring first and E second. In this context, $E|F$ may be interpreted as the second event occurring, knowing that the first event occurs.

EXAMPLE 5

Determine the probability of receiving two kings when two cards are dealt from a regular deck.

Solution

Let F be the event that the first card dealt is a king (*i.e.*, the first event in a sequence of two outcomes), and let E be the event that the second card dealt is a king (*i.e.*, the second event in the sequence). We seek $P(E \cap F)$. To determine $P(F)$, we reason that there are 52 cards in the deck, all equally likely to be dealt, and 4 of them are kings. Thus, $P(F) = \frac{4}{52}$. To determine $P(E|F)$, the probability that the second card is a king *knowing* that the first card dealt is a king, we reason as follows: Once a king has been dealt (event F), there are only 51 cards left, all equally likely to be dealt, and 3 of them are kings. Therefore, $P(E|F) = \frac{3}{51}$. Using Equation (*14*), we calculate

$$P(E \cap F) = P(F) \cdot P(E|F) = \left(\frac{4}{52}\right)\left(\frac{3}{51}\right) \approx .005$$

EXAMPLE 6

The probability that the Lincoln Hamburger Company will report a profit this year is 0.63. If the company reports a profit, there is a 70% chance the company will hire new employees next year; if the company does not report a profit, the chance of the company hiring new employees if only 15%. What is the probability that the company will report a profit this year *and* hire new employees next year?

Solution

Let M (denoting the making of money) be the event that the company reports a profit (the first outcome in the sequence of interest), and let H be the event of hiring new employees (the second outcome in the sequence of interest). We seek $P(H \cap M)$, and we are given

$$P(M) = .63$$
$$P(H|M) = .7$$
$$P(H|M') = .15$$

It now follows from

$$P(E \cap F) = P(F) \cdot P(E|F) \qquad \text{(14 repeated)}$$

with M replacing F and H replacing E, that

$$P(H \cap M) = P(M) \cdot P(H|M) = .63(.7) = .441$$

If outcome A occurs before outcome B in a tree diagram, then the conditional probability P(B|A) is placed on the line connecting node A to node B.

The events and the probabilities associated with each event in Example 6 can be illustrated compactly with a tree diagram (see Section 5.4). Probabilities are placed on the lines connecting two circles or nodes. The probability $P(H|M)$ is placed on the line connecting node H to node M when outcome M precedes outcome H. Figure 6.11 is a tree diagram for the events and probabilities described in Example 6.

We have, in fact, all the information required to place probabilities on the lines connecting every pair of nodes in Figure 6.11. With $P(M) = .63$, we conclude that $P(M') = 1 - .63 = .37$; if the company has a .63 probability of making money, then it also has a .37 probability of *not* making money. Furthermore, with $P(H|M) = .7$, we have $P(H'|M) = $

FIGURE 6.11

FIGURE 6.11

FIGURE 6.12

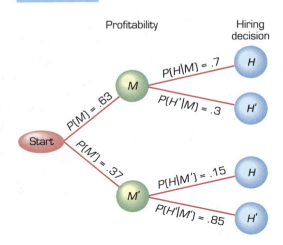

The probability of a path is the product of all the probabilities on the path.

$1 - .7 = .3$; once the company reports a profit, the probability of hiring is .7 and the probability of *not* hiring is .3. We therefore place .3 on the line joining node H' to node M. Similarly, $P(H|M') = .15$ implies that $P(H'|M') = 1 - .15 = .85$, so .85 is placed on the line joining node H' to node M'. The result is Figure 6.12.

A *path* is a sequential set of connecting nodes and lines that begins with the starting node and never backtracks. The product rule provides a heuristic procedure for determining probabilities of paths: The probability of a path is the product of all the probabilities on the path. For example, the bottom path in Figure 6.12 represents the events not reporting a profit *and* not hiring; the probability of this occurring is the product $P(H' \cap M') = P(M') \cdot P(H'|M') = .37(.85) = .3145$.

If a number of events are mutually exclusive, then the probability of their union is the sum of their individual probabilities (see Section 6.2). Trees are generally constructed from mutually exclusive events. That is, each time we branch from a node, we branch to a group of mutually exclusive events. In Figure 6.12 the first branching was to M and M', to making money and *not* making money, which are disjoint events. The next branchings were to H and H', to hiring and *not* hiring, which are also disjoint events. The event associated with one path in a tree is, therefore, generally disjoint from events associated with other paths in the tree. This leads to another heuristic for tree diagrams, based on the Exclusivity Law of Probability: The probability that at least one of a number of events will occur, when the individual events are mutually exclusive and depicted by distinct paths on a tree diagram, is the sum of the probabilities associated with each of those paths.

To calculate the probability of an event on a tree diagram, locate all nodes on the tree corresponding to this event. For each such node, determine the probability of the path leading to this node, beginning at start, and then sum the results.

In Figure 6.12, the path associated with the event of making money and hiring, the top path, has probability $.63(.7) = .441$. The path associated with the event of not making money and hiring has probability $.37(.15) = .0555$. Thus, the probability of at least one of these two events, making money and hiring *or* not making money and hiring, is .441 $+ .0555 = .4965$. This is the probability of the company hiring new employees next year.

We now have our final and, perhaps, most important heuristic dealing with tree diagrams: To calculate the probability of an event, locate all nodes on a tree diagram corresponding to this event. For each such node, determine the probability of the path leading to that node, beginning at Start, *and then sum the results.*

EXAMPLE 7

Figure 6.13 is a tree diagram for the following process: An evening-dress buyer must place orders with a dress manufacturer nine months before the dresses are needed, so the buyer must forecast fashion. The buyer must first determine the length of dresses that will be in fashion, either knee length (event K), above the knee (event AK), or below the knee (event BK), and then the color that will be in fashion, black (event B) or red (event R). Using Figure 6.13, determine the probability that (*a*) knee length is in fashion, (*b*) black is in fashion if it is known that knee length is in fashion, (*c*) red is in fashion if it is known that above the knee is in fashion, (*d*) below the knee and red are in fashion, and (*e*) black is in fashion.

FIGURE 6.13

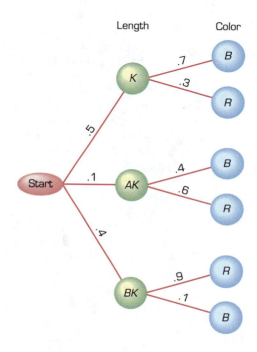

Solution
(*a*) $P(K) = .5$, taken directly from the tree diagram
(*b*) $P(B|K) = .7$, taken directly from the tree diagram
(*c*) $P(R|AK) = .6$, taken directly from the tree diagram
(*d*) $P(R \cap BK) = P(BK) \cdot P(R|BK) = .4(.1) = .04$, which is the probability associated with the bottom path in the tree diagram
(*e*) There are three nodes on the diagram corresponding to black: (*i*) black when knee length is in fashion, (*ii*) black when above the knee is in fashion, and (*iii*) black when below the knee is in fashion. The probabilities of the paths that end with black, beginning at *Start* and ending with B, are, going from left to right and top to bottom .5(.7) = .35, .1(.4) = .04, and .4(.9) = .36. Consequently, $P(B) = .35 + .04 + .36 = .75$.

Although the probability of an event can change as additional information becomes available, the original probability should not be affected by new information that is unrelated or irrelevant to the event of interest. As an example of irrelevant additional information, consider again the process of throwing one die and the event E that the face that turns up is a 3. Assuming the die is fair, $P(E) = \frac{1}{6}$. Suppose we are now told the color of the die. Clearly, this new information does not change our previous calculation, because the color of the die has no bearing on the outcome. $P(E) = \frac{1}{6}$ if the die is red, and $P(E) = \frac{1}{6}$ if the die is white.

Let F be the event that the die is red and let E remain the event that the face turns up is a 3. We have

$$P(E|F) = P(E) \qquad (15)$$

The probability that E occurs is the same *whether or not* we know F. We say two events E and F are *independent* whenever Equation (*15*) is true.

Usually, it is clear from the context of the process whether two events are independent. The die process just described is one example. When independence is in question, however, one must calculate $P(E)$ and $P(E|F)$. If both quantities are equal, then E and F are independent; if Equation (*15*) is *not* true, then the events are *not* independent.

> *Two events* E *and* F *are independent whenever* $P(E|F) = P(E)$.

EXAMPLE 8

Consider all families with two children, and pick one of these families at random. Let E be the event that the family has children of both genders, and let F be the event that the family has at most one girl. Are E and F independent?

Solution

A sample space is $U = \{bb, bg, gb, gg\}$, where the first letter indicates the gender of the older child and the second letter indicates the gender of the younger child. If we assume there is an equal chance for a child to be of either gender, then U is a sample space of equally likely outcomes. Here $E = \{bg, gb\}$, $F = \{bb, bg, gb\}$, and $E \cap F = \{bg, gb\} \cap \{bb, bg, gb\} = \{bg, gb\}$. Therefore,

$$P(E) = \frac{2}{4} = .5$$

$$P(F) = \frac{3}{4} = .75$$

$$P(E \cap F) = \frac{2}{4} = .5$$

from which it follows that

$$P(E|F) = \frac{P(E \cap F)}{P(F)} = \frac{.5}{.75} \approx .6667$$

Since $P(E)$ does *not* equal $P(E|F)$, the events E and F are *not* independent.

EXAMPLE 9

Solve Example 8 for families with three children.

Solution

A sample space is

$$U = \{bbb, bbg, bgb, bgg, gbb, gbg, ggb, ggg\}$$

where the first, second, and third letters designate, respectively, the gender of the oldest, middle, and youngest child. Now,

$$E = \{bbg, bgb, bgg, gbb, gbg, ggb\}$$

$$F = \{bbb, bbg, bgb, gbb\}$$

and

$$E \cap F = \{bbg, bgb, gbb\}$$

with

$$P(E) = \frac{6}{8} = .75$$

$$P(F) = \frac{4}{8} = .5$$

$$P(E \cap F) = \frac{3}{8} = .375$$

Now

$$P(E|F) = \frac{P(E \cap F)}{P(F)} = \frac{.375}{.5} = .75$$

which does equal $P(E)$, so the events E and F are independent.

Examples 8 and 9 illustrate the difficulty of determining the independence of certain events. The same two events, in words, can be independent with respect to one sample space and not independent with respect to a second sample space. Fortunately, such situations occur infrequently. It is generally clear from context whether two events are independent.

The product rule

$$P(E \cap F) = P(F) \cdot P(E|F) \qquad \text{(14 repeated)}$$

is always true for any two events E and F. If, in addition, the events are independent, then

$$P(E|F) = P(E) \qquad \text{(15 repeated)}$$

Substituting Equation (*15*) into Equation (*14*) and rearranging, we obtain

$$P(E \cap F) = P(E) \cdot P(F) \qquad (16)$$

Equation (*16*) states that for independent events, the probability that both events will occur is the product of the probabilities of each event occurring.

> *If two events are independent, then the probability of both events occurring is the product of the probabilities of each event occurring.*

EXAMPLE 10

An appliance store knows from experience that the probability of a new washing machine needing repair in its first year of operation (event E) is .13 and the probability of a new dryer needing repair in its first year of operation (event F) is .09. A customer buys a washer and a dryer. What is the probability that both machines will need repairs within a year?

Solution

The washer and dryer are different machines, and a breakdown in one should not affect the performance of the other. Thus, we assume that E and F are independent. It follows from Equation (16) that

$$P(E \cap F) = P(E) \cdot P(F) = .13(.09) = .0117$$

This is the probability that both machines will fail during their first year of operation.

The equation

$$P(E \cap F) = P(E) \cdot P(F) \qquad \text{(16 repeated)}$$

is applicable only when E and F are independent. If it is unclear whether two events are independent, as was the case in Examples 8 and 9, then it is wise not to use Equation (16).

IMPROVING SKILLS

In Problems 1 through 10, determine the conditional probabilities $P(E|F)$ and $P(F|E)$ from the given information.

1. $P(E) = .4$, $P(F) = .5$, $P(E \cap F) = .1$
2. $P(E) = .8$, $P(F) = .2$, $P(E \cap F) = .16$
3. $P(E) = .2$, $P(F) = .3$, $P(E \cap F) = .1$
4. $P(E) = .78$, $P(F) = .5$, $P(E \cap F) = .39$
5. $P(E) = .42$, $P(F) = .63$, $P(E \cap F) = .27$
6. $P(E) = .83$, $P(F) = .48$, $P(E \cap F) = .44$
7. $P(E) = .2$, $P(F) = .3$, $P(E \cup F) = .4$ (*Hint*: Use Equation (4) from Section 6.2 to calculate $P(E \cap F)$.)
8. $P(E) = .5$, $P(F) = .6$, $P(E \cup F) = .8$
9. $P(E) = .25$, $P(F) = .4$, $P(E \cup F) = .55$
10. $P(E) = .57$, $P(F) = .53$, $P(E \cup F) = .58$
11. Determine whether the events E and F are independent in Problems 1 through 10.

The data in Problems 12 through 15 are based on information reported in *An American Profile: Opinions and Behavior, 1972–1989* for surveys taken in 1989.

12. The following table summarizes the responses to the question, "Do you have any guns or revolvers in your home or garage?"

	YES	NO	TOTALS
Male	245	201	446
Female	228	356	584
Totals	473	557	1,030

Let M denote the event of being male, F the event of being female, HG the event of having a gun, and NG the event of not having a gun. Using the set of all respondents as a sample space, describe the following quantities in words and then calculate them:

(*a*) $P(M)$ (*b*) $P(M|HG)$ (*c*) $P(M|NG)$
(*d*) $P(NG|M)$ (*e*) $P(F|HG)$ (*f*) $P(F|NG)$

13. Survey participants were also asked, "Do you know personally anyone, living or dead, who has been infected with AIDS?" The following table summarizes their responses.

	YES	NO	TOTALS
Male	66	594	660
Female	70	805	875
Totals	136	1,399	1,535

Let M denote the event of being male, F the event of being female, KA the event of knowing someone with AIDS, DK the event of not knowing someone with AIDS. Using the set of all respondents as a sample space, describe the following quantities in words and then calculate them:

(a) $P(KA)$ (b) $P(KA|F)$ (c) $P(KA|M)$

(d) $P(M|KA)$ (e) $P(F|DK)$ (f) $P(DK|M)$

14. Participants' responses to the question, "Do you ever have occasion to use alcoholic beverages?" are summarized in the following table:

	UA	TA	TOTALS
White	617	264	881
Black	57	53	110
Totals	674	317	991

White (W) and black (B) refer to race, UA denotes a user of alcohol on occasion, while TA denotes a total abstainer from alcohol. Using the set of all respondents as a sample space, describe the following quantities in words and then calculate them:

(a) $P(W)$ (b) $P(W|TA)$ (c) $P(B|TA)$

(d) $P(TA|B)$ (e) $P(UA|W)$

15. The following table summarizes the responses to the question, "During the last year, did someone burglarize your home or apartment?"

	YES	NO	TOTALS
White	45	845	890
Black	10	92	102
Totals	55	937	992

White (W) and black (B) refer to race, WB denotes someone who was burglarized, while NB denotes someone who was not burglarized. Using the set of all respondents as a sample space, describe the following quantities in words and then calculate them:

(a) $P(B)$ (b) $P(B|NB)$ (c) $P(B|WB)$

(d) $P(WB|B)$ (e) $P(W|WB)$ (f) $P(NB|W)$

16. Use the tree diagram shown in Figure A to determine the following probabilities:

(a) $P(A)$ (b) $P(B)$ (c) $P(C|A)$

(d) $P(C|B)$ (e) $P(C \cap A)$ (f) $P(D \cap B)$

(g) $P(C)$ (h) $P(D)$

17. Use the tree diagram shown in Figure B to determine the following probabilities:

(a) $P(A)$ (b) $P(B)$ (c) $P(G|A)$

(d) $P(S|A)$ (e) $P(S \cap A)$ (f) $P(G \cap B)$

(g) $P(G)$ (h) $P(S)$

FIGURE A

FIGURE B

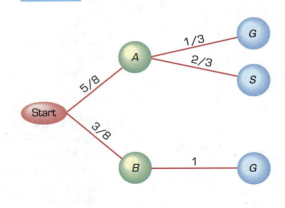

18. Use the tree diagram shown in Figure C to determine the following probabilities:

(**a**) $P(C)$ (**b**) $P(D)$ (**c**) $P(S|B)$

(**d**) $P(S|D)$ (**e**) $P(R|C)$ (**f**) $P(T|D)$

(**g**) $P(S \cap B)$ (**h**) $P(T \cap C)$ (**i**) $P(R)$

(**j**) $P(T)$

FIGURE C

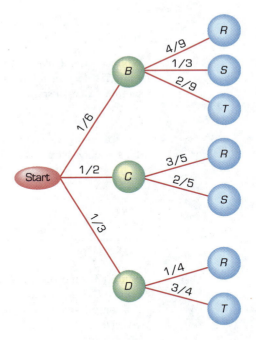

In Problems 19 through 23, use the given information to find $P(E \cap F)$ if E and F are independent events.

19. $P(E) = P(F) = \frac{1}{2}$

20. $P(E) = P(F) = .1$

21. $P(E) = .8$, $P(F) = .9$

22. $P(E) = .6$, $P(F) = .8$

23. $P(E) = .2$, $P(F) = .25$

24. Find $P(E)$ if $P(E \cap F) = .3$, $P(F) = .4$, and E and F are independent events.

25. Find $P(E)$ if $P(E \cup F) = .8$, $P(F) = .4$, and E and F are independent events.

CREATING MODELS

The data in Problems 26 through 30 are based on information reported in *An American Profile: Opinions and Behavior, 1972–1989* for surveys taken in 1989. Model each of the following processes with a tree diagram and then use those models to supply the requested information.

26. People were surveyed as to whether they ever had occasion to use alcoholic beverages such as liquor, wine, or beer during the year or whether they were total abstainers. Of those surveyed, 85% were white (by race) and the remainder were nonwhite. Thirty percent of all whites surveyed declared themselves total abstainers, while 52% of nonwhites declared themselves total abstainers. Assuming that this survey is representative of all adults in the United States, what is the probability that an adult, chosen at random, abstains totally from alcoholic beverages?

27. People were surveyed as to whether they were members of school fraternities or sororities. Of those surveyed, 57% were female and 43% were male. Five percent of all females answered yes to the survey, while 7% of all males answered yes. Assuming that this survey is representative of all adults in the United States, what is the probability that an adult, chosen at random, is a member of a school fraternity or sorority?

28. Another survey asked people whether they were members of labor unions. Of those surveyed, 57% were female and 43% were male. Fourteen percent of all respondents answered yes to the survey, but 21% of the males answered yes. Assuming that this survey is representative of all adults in the United States, what is the probability that a female, chosen at random, belongs to a labor union?

29. People were also asked whether someone broke into or somehow illegally gained entry into their homes or apartments during the last year. Of those surveyed, 10% were black (by race) and 90% were nonblack. Six percent of all respondents answered yes to the survey, but 10% of blacks answered yes. Assuming that this survey is representative of all adults in the United States, what is the probability that a nonblack, chosen at random, answered no to the survey?

30. Survey participants were asked whether they currently smoke. Of those surveyed, 23% were under the age of 30. Thirty percent of all respondents answered yes to the survey, but 33% of all people under the age of 30 answered yes. Assuming that this survey is representative of all adults in the United States, what is the probability that a person 30 years old or older, chosen at random, answered yes to the survey?

31. During a test program for a new drug, the following data were gathered: The drug was effective on 73% of those tested and not effective on the remaining 27%; 45% of all those tested experienced aftereffects; for 5% of all those tested, the drug was not effective and aftereffects were experienced. Determine the probability that (*a*) the drug was not effective on a person who experienced aftereffects, and (*b*) a person would experience aftereffects if the drug was not effective.

32. In a small town, test marketing for a new type of bleach resulted in the following data: 10% of the households received a free sample of the bleach, 3% of all household shoppers bought the bleach at least once, and 25% of the people who received a free sample later bought the bleach. What is the probability that a household shopper received a free sample and later bought the bleach?

33. A health service knows that 52% of all adults in one area of the country smoke. Furthermore, 11% of the deaths in that area are due to cancer. A smoker has a 20% chance of dying of cancer. (*a*) What is the probability that a person from this area

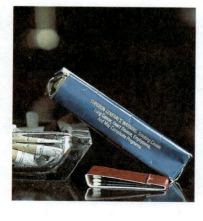

selected at random will be a smoker and will die from cancer? (*b*) What is the probability that a nonsmoker from this area will die from cancer?

34. A survey commissioned by a meat packer resulted in the following data: The probability that a person eats meat for dinner at least three times a week is .72. The probability that a person eats fish for dinner at least once a week is .45. The probability that a person does both is .27. Determine the probability that (*a*) a person who eats meat at least three times a week will have fish at least once a week, and (*b*) a person who eats fish at least once a week will have meat at least three times a week.

35. An insurance company knows that 19% of all policy holders will have at least one accident in a year, and 11% of all policy holders will have at least two accidents in a year. What is the probability that a person who has had one accident will have a second in the same year?

36. Determine the probability of rolling two dice whose faces total 4. What is the probability if we know that the two dice will not have the same face value?

37. Determine the probability of rolling two dice whose faces total 5. What is the probability if we know that the face value of one of the dice is a 1?

38. Solve Example 8 if the probability of having a male child is .52 and the probability of having a female child is .48. *Hint*: Construct a tree diagram of the process.

39. A fair coin is tossed three times. Let E be the event that at least one toss is a head, and let F be the event that the three tosses result in both a head and a tail occurring at least once. Are these events independent?

40. A barrel contains three red balls and four green balls. Two balls are selected at random, one after the other. Determine the probability that both balls will be red if (*a*) the first ball is returned to the barrel before the second ball is selected, and (*b*) the first ball is *not* returned to the barrel.

41. Consider the barrel of balls described in the previous problem. Determine the probability of selecting two balls of different colors if it is known that the first ball selected is green and that ball is *not* returned to the barrel.

42. A die is thrown and then a coin is flipped. What is the probability that the die will show 6 and the coin will be tails?

43. A vaccine is effective on 97% of all people inoculated. Last week a doctor inoculated two people. What is the probability that the vaccine will be effective on both?

44. The probability of a licensed driver having at least one accident within a year is .21. Two licensed drivers are selected at random. What is the probability that neither of them will have an accident within a year?

45. An aptitude test given to all applicants for jobs at the Chubby Cat Food Corporation is 90% effective. That is, the test accurately predicts the aptitude of 90% of those tested. Determine the probability that the test will correctly predict the aptitudes of the next two people who take it. The next three people?

46. The American Citrus Corporation employs 30 lawyers in its legal department. The probability that a lawyer selected at random in this company will be a woman is .2. The company randomly picks two of its lawyers to handle a new case. What is the probability that both lawyers are women?

47. A dress manufacturer knows that 3% of all the dresses it produces are defective. What is the probability that the first two dresses checked by the quality control department will be perfect?

48. A plumbing manufacturer guarantees that the probability of a new faucet working properly when correctly installed is .99. Determine the probability that (*a*) the next two faucets correctly installed by a plumber will work properly, and (*b*) the next two faucets installed will work properly if the plumber knows that the last faucet installed worked properly?

EXPLORING IN TEAMS

49. Bernoulli trials are repeated runs of the same process. The process is assumed to have two outcomes, one called a *success* and the other called a *failure*, with the outcome of each run independent from all other runs. Note that a sample space can always be represented by just two elements, either a particular event happens (a success) or it does not happen (a failure). Construct a tree diagram, with appropriate probabilities on each branch, for the Bernoulli trial of throwing a die three times in succession, when a success is rolling a 6 and a failure is not rolling a 6. Use this tree diagram to determine the probability of rolling a 6 three times in succession. What is the probability of not rolling a single 6 on three successive attempts?

50. A barrel contains four red balls, two green balls, and two yellow balls. A ball is selected at random and then replaced. The process is repeated four times. Construct a tree diagram, with appropriate probabilities, for this Bernoulli trial when a success is picking a green ball and a failure is not picking a green ball. Use this tree diagram to determine the probability of selecting a green ball every time. What is the probability of never selecting a green ball in four successive attempts?

51. Let p denote the probability of a success in one run of a Bernoulli trial, and assume that the trial consists of n identical runs of the same process. Use the tree diagrams from Problems 49 and 50 to convince yourself that the probability of obtaining n successes is p^n and the probability of obtaining n failures is $(1 - p)^n$.

52. A vaccine is known to produce unpleasant aftereffects on 1% of the population. One day a doctor inoculates five patients with this vaccine. Use the results of Problem 51 to determine the probability that none of the patients will experience unpleasant aftereffects.

53. Five people take a lie detector test that is accurate 80% of the time. What is the probability that the test will be accurate for all five people? For none of the five people?

REVIEWING MATERIAL

54. (Section 1.3) The depreciation of a building is proportional to its age. How much will the building depreciate after 35 months if it depreciates $121,000 after 30 months?

55. (Section 3.3) A small company makes steak knives and pocketknives, both of which are in high demand. Each type of knife requires one unit of steel. Steak knives take four hours of labor to produce and sell for $9, while pocket knives require nine hours of labor and sell for $20. Use graphical methods to determine the number of knives of each type the company should produce to maximize income, if the company has 25 units of steel in stock and 180 hours of available labor.

56. (Section 4.2) A person receives two offers for her business. One offer will pay $2,000 now, another $3,000 at the end of 2 years, and a final payment of $4,000 at the end of 5 years. The second offer will pay $4,000 now and $5,000 at the end of 6 years. Which is the better offer if current interest rates are $4\frac{1}{2}\%$ a year compounded monthly?

57. (Section 5.3) The U.S. Department of Agriculture estimates that the amount of corn planted in the U.S. from 1984 through 1993 (in millions of acres) was as follows

1984–5	80.5	1987–8	65.7	1990–1	74.2
1985–6	83.4	1988–9	67.6	1992–2	75.9
1986–7	76.7	1989–90	72.3	1992–3	79.3

Find the mean, median, mode, range, and standard deviation for this data.

RECOMMENDING ACTION

58. Respond by memo to the following request:

> ### MEMORANDUM
>
> To: J. Doe Reader
>
> From: Rheza
>
> Date: Today
>
> Subject: **Winning**
>
> I was watching a simple carnival game last week, and I think I can beat it. Players bet $1 on the outcome of a coin flip, either heads or tails, and a player can enter the game at any flip. If a player guesses correctly, the player wins $1; otherwise the player loses $1. Each time a player wins, the carnival takes a nickel of the winnings as its charge for running the game.
>
> Frequently, the coin flipper would throw two successive heads. If I wait until this occurs to bet, then I believe a good bet is tails on the next flip. After all, the probability of flipping three consecutive heads is $\frac{1}{8}$, so there is a $\frac{7}{8}$ chance that I would win my bet. Don't you agree?

6.4 THE BAYESEAN METHOD

Figure 6.12 shows the probabilities of a company making money (event M) or not making money (event M') and *then* either hiring (event H) or not hiring (event H') new employees. It follows immediately from Figure 6.12 that

FIGURE 6.12

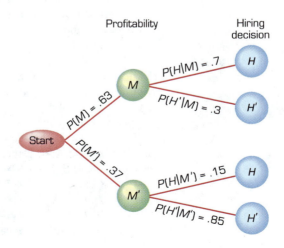

$$P(M) = .63$$

$$P(H|M) = .7$$

$$P(H|M') = .15$$

$$P(M') = .37$$

$$P(H'|M) = .3$$

$$P(H'|M') = .85$$

$$P(H \cap M) = P(M) \cdot P(H|M) = .63(.7) = .441$$

$$P(H' \cap M) = P(M) \cdot P(H'|M) = .63(.3) = .189$$

$$P(H \cap M') = P(M') \cdot P(H|M') = .37(.15) = .0555$$

$$P(H' \cap M') = P(M') \cdot P(H'|M') = .37(.85) = .3145$$

$$P(H) = .63(.7) + .37(.15) = .441 + .0555 = .4965$$

$$P(H') = .63(.3) + .37(.85) = .189 + .3145 = .5035$$

These equations follow from the material developed in Section 6.3. If you are uncomfortable with any of these calculations, you should review the previous section.

Refer again to Figure 6.12. Suppose we know that the company made money. With this information, can we determine the probability that the company will hire a new employee? This probability is $P(H|M)$, and we can read this conditional probability directly from the tree diagram as .7. Now, instead, suppose we know that the company hired a new employee. With this information, can we determine the probability that the company made money? That is, can we calculate $P(M|H)$? We cannot read this probability directly from the tree diagram, but we can calculate it from the information in Figure 6.12.

In Section 6.3 we developed the formula

$$P(E|F) = \frac{P(E \cap F)}{P(F)} \qquad \text{(13 repeated)}$$

for calculating conditional probabilities. Replacing E with M and F with H in this formula, we obtain

$$P(M|H) = \frac{P(M \cap H)}{P(H)}$$

Using the probabilities listed above, we have $P(M \cap H) = P(H \cap M) = .441$ and $P(H) = .4965$, so $P(M|H) = .441/.4965 \approx .888$.

The Bayesean method is a procedure for calculating the conditional probability $P(F|E)$ from $P(E|F)$. We have a conditional probability of the form $P(E|F)$, the probability that E will occur given that F is true, and we want to use this information to calculate a related conditional probability, namely $P(F|E)$, the probability that F will occur given that E is true. The letters E and F represent events and, as always, these letters can be replaced by any other two labels, such as M and H, when other labels are more suitable to a particular model. In our example, we knew the probability that a company would hire a new employee, event H, when they made money, event M, and we wanted the related probability that the company made money, event M, once we knew that they hired a new employee, event H.

Using the conditional probability formula, we wrote

$$P(M|H) = \frac{P(M \cap H)}{P(H)}$$

The Bayesean method is a procedure for calculating the conditional probability P(F|E) from the conditional probability P(E|F).

Reviewing Figure 6.12, we see that $P(M \cap H)$ is the probability of making money *and* hiring a new employee. This probability is the probability of the path that passes through node M and ends at node H. $P(H)$ is the probability of hiring a new employee, and it is the sum of the probabilities of all paths that end at node H.

EXAMPLE 1

Figure 6.13 shows the tree diagram for the department store buyer who must determine nine months in advance the length of dresses that will be in fashion—knee length (event K), above the knee (event AK), or below the knee (event BK)—and the color that will be in fashion—black (event B) or red (event R). What is the probability that the fashion length will be at the knee if it is known that red will be the color in fashion?

FIGURE 6.13

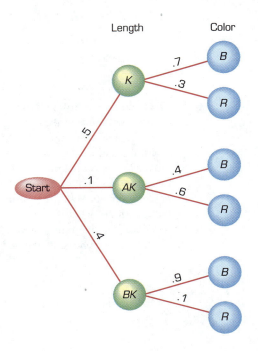

Solution

We seek $P(K|R)$. Observe that we can read $P(R|K) = .3$ directly from the tree diagram. We know $P(R|K)$, and we want $P(K|R)$. Using conditional probability, we have

$$P(K|R) = \frac{P(K \cap R)}{P(R)}$$

Reviewing Figure 6.13, we see that $P(K \cap R)$ is the probability that knee length dresses will be in fashion *and* that red will be the color in fashion. This probability is the probability of the path in Figure 6.13 that passes through node K and ends at node R. Therefore,

$$P(K \cap R) = .5(.3) = .15$$

Furthermore, $P(R)$ is the probability that red will be the color in fashion, and it is the sum of the probabilities of all paths in Figure 6.13 that end at node R:

$$P(R) = .5(.3) + .1(.6) + .4(.1) = .25$$

We can now calculate

$$P(K|R) = \frac{.15}{.25} = .6$$

Originally, before we had any additional information, the probability that knee length would be the length in fashion was $P(K) = .5$. Now, with the additional information that the fashion color is red, the probability that knee length will be the length in fashion increases to $P(K|R) = .6$.

Calculate P(F|E) *from a tree diagram by dividing the probability of the path that passes through node F and ends at node E by the sum of the probabilities of all paths that end at node E.*

In the more general problem, we know the conditional probability $P(E|F)$, from a tree diagram, and we want the related conditional probability $P(F|E)$. Since

$$P(F|E) = \frac{P(F \cap E)}{P(E)}$$

we have the following heuristic when dealing with tree diagrams: To calculate $P(F|E)$, divide the probability of the path that passes through node F and ends at node E by the sum of the probabilities of all paths that end at node E.

EXAMPLE 2

In 1989 the *General Social Survey* asked a large sample of 16 year olds to identify the religion in which they were raised. Of the survey respondents, 63% were raised as Protestants (event *PR*), 29% as Catholics (event *C*), 2% as Jews (event *J*), 2% in other religions (event *O*), and 4% with no religion (event *N*). Females (event *F*) were 59% of the surveyed Protestants, 53% of the surveyed Catholics, 58% of the surveyed Jews, 60% of those surveyed who were raised in other religions, and 59% of those who were raised with no religion; the other survey respondents were males (event *M*). A male is selected at random from this surveyed population. Determine the probability that this male was raised Catholic.

Solution

All of the information we are given is presented compactly on the tree diagram in Figure 6.14. We seek $P(C|M)$, the probability that a person was raised Catholic *knowing* the person is male. The probability of the path that passes through node C and ends at node M is

$$P(C \cap M) = .29(.47) = .1363$$

The sum of the probabilities of *all* paths that end at node M (there are five such paths) is

$$P(M) = .63(.41) + .29(.47) + .02(.42) + .02(.40) + .04(.41) = .4274$$

Thus,

$$P(C|M) = \frac{P(C \cap M)}{P(M)} = \frac{.1363}{.4274} = .32, \text{ rounded to two decimal places}$$

FIGURE 6.14

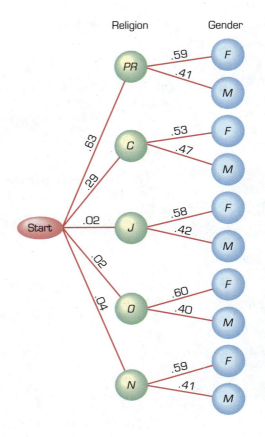

If we start with the product rule

$$P(E \cap F) = P(F) \cdot P(E|F) \qquad (14 \text{ repeated})$$

we can generate a formula for calculating the conditional probability $P(F|E)$ from the conditional probability $P(E|F)$. By reversing the roles of E and F, Equation (14) becomes

$$P(F \cap E) = P(E) \cdot P(F|E) \qquad (17)$$

But $E \cap F = F \cap E$, so $P(E \cap F) = P(F \cap E)$, and Equations (14) and (17) together imply that

$$P(E) \cdot P(F|E) = P(F) \cdot P(E|F)$$

or

BAYES' FORMULA

If E and F are events in a sample space U, with $P(E) \neq 0$, then

$$P(F|E) = \frac{P(F) \cdot P(E|F)}{P(E)} \qquad (18)$$

Bayes' Formula provides a way to calculate $P(F|E)$ directly from $P(E|F)$. This formula is, however, nothing more than a formal statement of the heuristic developed earlier. The numerator on the right side of Equation (*18*) is the product of $P(F)$ and the conditional probability $P(E|F)$. On a tree diagram, this is simply the probability of the path leading to node F times the probability of the path from node F to node E. In other words, the numerator in Bayes' Formula is the probability of the path that passes through node F and ends at node E.

EXAMPLE 3

Use Bayes' Formula and the information in Figure 6.12 to determine the probability that the company made money, *knowing* that the company hired a new employee.

Solution

We seek $P(M|H)$, where M denotes the event of the company making money and H denotes the event of the company hiring a new employee. We can read $P(H|M)$ directly from Figure 6.12, so Bayes' Formula is promising. With M replacing F and H replacing E, Equation (*18*) becomes

$$P(M|H) = \frac{P(M) \cdot P(H|M)}{P(H)}$$

Using the probabilities calculated in the beginning of this section, we have $P(M) = .63$, $P(H|M) = .7$, $P(H) = .4965$, and

$$P(M|H) = \frac{.63(.7)}{.4965} \approx .888$$

EXAMPLE 4

Use Bayes' Formula and the information in Figure 6.13 to determine the probability that the fashion length will be above the knee if it is known that black will be the color in fashion.

Solution

We seek $P(AK|B)$. We can read $P(B|AK) = .4$ directly from Figure 6.13, so Bayes' Formula is promising. Letting AK and B replace F and E, respectively, in Equation (18), we have

$$P(AK|B) = \frac{P(AK) \cdot P(B|AK)}{P(B)}$$

$P(AK) = .1$ also comes directly from Figure 6.13. We calculate $P(B)$ as the sum of the probabilities of all paths leading to a black node:

$$P(B) = .5(.7) + .1(.4) + .4(.9) = .75$$

Therefore,

$$P(AK|B) = \frac{.1(.4)}{.75} \approx .053$$

Equation (*18*) is the theoretical underpinning for the Bayesean method. We could use it to generate other, more specific formulas (see Problem 44), but we will not do so here. Such formulas are intimidating and are generally useful only to experienced modelers in probability theory.

IMPROVING SKILLS

Use the information in Figure *A* to calculate the conditional probabilities specified in Problems 1 through 4.

1. $P(A|C)$ **2.** $P(B|C)$ **3.** $P(A|D)$ **4.** $P(B|D)$

FIGURE A

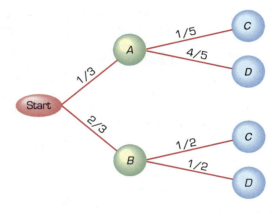

Use the information in Figure *B* to calculate the conditional probabilities specified in Problems 5 through 8.

5. $P(A|G)$ **6.** $P(A|S)$ **7.** $P(B|G)$ **8.** $P(B|S)$

FIGURE B

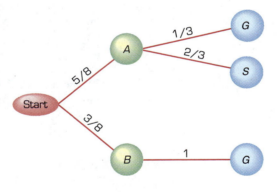

Use the information in Figure *C* to calculate the conditional probabilities specified in Problems 9 through 16.

9. $P(B|R)$ **10.** $P(B|S)$ **11.** $P(D|R)$ **12.** $P(D|S)$
13. $P(D|T)$ **14.** $P(C|R)$ **15.** $P(C|T)$ **16.** $P(C|S)$

Use the information in Figure 6.14 (Example 2) to calculate the conditional probabilities specified in Problems 17 through 24.

17. $P(PR|M)$ **18.** $P(PR|F)$ **19.** $P(C|F)$ **20.** $P(J|F)$
21. $P(J|M)$ **22.** $P(O|F)$ **23.** $P(N|M)$ **24.** $P(N|M')$

FIGURE C

CREATING MODELS

25. A test program for a new drug generated the following data: The drug was effective for 73% of those tested and not effective for the rest. Of those for whom the drug was effective, 40% experienced aftereffects and 60% did not. Of those for whom the drug was not effective, 5% experienced aftereffects and the rest did not. Determine the probability that the drug was effective on a person who did *not* experience aftereffects.

26. In a small town, test marketing for a new bleach yielded the following data: 10% of the households received a free sample of the bleach, and the rest did not; 25% of the people who received a free sample later bought the bleach, while only 5% of the people who did not receive a free sample later bought the bleach. What is the probability that a household shopper spotted buying the bleach in a supermarket actually received a free sample earlier?

27. National statistics indicate that 32% of all adults smoke. Assume that in one region, 14% of those who smoke contract lung cancer at some point in their lives, while only 5% of those who do not smoke contract lung cancer. What is the probability that a randomly selected person from this region smoked, if the person has lung cancer?

28. A survey commissioned by a meat packer produced the following data: The probability that a person eats meat for dinner at least three times a week is .72. Of this group, 45% also eat fish for dinner at least once a week. The probability that a person does *not* eat meat for dinner at least three times a week is .28, and of this group, 92% eat fish for dinner at least once a week. What is the probability that a particular person eats meat at least three times a week if this person does *not* eat fish?

29. Barrel 1 contains three red balls and four green balls, while barrel two contains five red balls and one green ball. A die is thrown, and then a ball is chosen. If the face of the thrown die comes up 1, the ball is chosen from barrel 1; otherwise, the ball is chosen from barrel 2. What is the probability that the die came up 1 if the ball selected is green?

30. Consider the process described in the previous problem, but now there are six barrels, each containing six balls. The barrels are numbered 1 to 6, and the number of each barrel indicates how many red balls it contains. For example, barrel 2 has two red balls and the rest are green balls. The die is thrown, and a ball is picked from the barrel having the same number as the die face that turns up. That is, if the die comes up 4, then a ball is selected from barrel 4. What is the probability that the die came up 1 if the ball selected is green?

31. The Chubby Cat Food Corporation wants to check the reliability of an aptitude test it gives to all applicants for jobs. It will do so by hiring everyone who applies, regardless of how they do on the test, and then monitoring whether each person is still employed one year later. The test is given to 124 people, and 65 of these people pass (according to past practices) while 59 do not. Nevertheless, all 124 are hired. Of the 65 people who passed, 52 are still working at the company a year later, and 13 are not. Of the 59 who did not pass the aptitude test, 24 are still working at the company a year later, and 25 are not. What is the probability that a person from the test group of 124 people who still works at the company a year later (*a*) actually passed the aptitude test, and (*b*) did *not* pass the test?

32. A dress manufacturer produces garments at two plants. Plant *A* is larger than Plant *B* and produces 60% of all dresses for the manufacturer. It is known that 3% of all the dresses produced at Plant *A* are defective, while only 2% of all dresses produced at Plant *B* are defective. A quality control inspector chooses one dress at random and finds it defective. What is the probability that this dress was made at Plant *A*?

33. Eight hundred women, half of them pregnant and half of them not pregnant, agree to test a new home screening pregnancy test for a consumer research magazine. The test correctly identifies 312 of the pregnant women as pregnant and identifies the other 88 pregnant women as not pregnant. The test correctly identifies 377 of the nonpregnant women as not pregnant, but it identifies the other 23 nonpregnant women as pregnant. Based on this test, what is the probability that (*a*) a woman is actually pregnant when the screening test indicates that she is, and (*b*) a woman is pregnant when the screening test indicates that she is not pregnant?

34. The police department in a large city is considering purchasing a new lie-detector test, so they subject it to a reliability test. Forty-eight people are asked to take the test; half are instructed to tell the truth and the other half are instructed to lie. The test indicates lying in 23 of the people who were lying and in 5 of the people who were not lying. Based on this test, what is the probability that (*a*) a person is lying if the test indicates lying, and (*b*) a person is not lying if the test indicated not lying?

35. Voting records in a particular town show that 53% of the voters are registered Republicans, 18% are registered Democrats, and the rest are independent. In the last presidential election, exit polls indicated that the Republican candidate received 88% of all votes cast by registered Republican voters, 23% of all votes cast by registered Democratic voters, and only 11% of all votes cast by independents. All other votes went to the Democratic candidate. What is the probability that a person who voted for the Republican candidate is an independent?

36. A carpet store uses four different installers to lay the carpet the store sells to customers. The store prefers to use Installer *A*, and it gives 65% of its work to that installer. Installer *B* gets 10% of the store's work, Installer *C* gets 18%, and Installer *D* gets 7%. From past experience, the store receives complaints from 1% of the customers whose carpeting was installed by Installer *A*, from 7% of the customers served by *B*, and from 3% of the customers served by either *C* or *D*. The last customer for whom

carpet was laid calls in with a complaint. What is the probability that Installer B laid this carpet?

37. Residents in a particular town have access to five commercial VHS television channels. During *each* hour of prime-time weekday broadcasts, truck commercials air for three minutes on channel 2, two minutes on channel 4, $1\frac{1}{2}$ minutes on channel 5, and one minute each on channels 8 and 10. During one of these hours, a person enters a room with a television on and sees a truck commercial being broadcast. Assuming that each channel is equally likely to be on at any given moment, what is the probability that the television is currently on channel 2?

EXPLORING IN TEAMS

38. Determine whether the following equation is true as it pertains to the events defined in Example 2

$$P(M) = P(M|PR){\cdot}P(PR) + P(M|C){\cdot}P(C) + P(M|J){\cdot}P(J)$$
$$+ P(M|O){\cdot}P(O) + P(M|N){\cdot}P(N)$$

39. Let $A_1, A_2, \ldots, A_N$ denote a finite set of events that are mutually exclusive (see Section 6.2) and have the property that $U = A_1 \cup A_2 \cup \ldots \cup A_N$. That is, each element in the sample space is a member of one and only one of the sets $A_1, A_2 \ldots, A_N$. Identify two different partitions of this type for the process described in Example 2.
40. Identify a partition of the type described in the previous problem for the process described in Example 1.
41. Let $A_1, A_2, \ldots, A_N$ have the properties described in Problem 39, and let B be some other event. Construct a tree diagram in which the first stage contains the events $A_1, A_2, \ldots, A_N$ and the second stage, obtained from branching each node at the first stage, contains the event B and perhaps some other events.
42. Use the tree diagram constructed in Problem 41 to develop a mathematical equation for $P(B)$. *Hint*: This equation will be similar to the one in Problem 38.
43. Use the result from Problem 42 to write a formula for $P(A_1|B)$ in terms of the conditional probabilities $P(B|A_1), P(B|A_2), \ldots, P(B|A_N)$.
44. Generalize the result of Problem 43 to an equation for $P(A_k|B)$ for any value of $k = 1, 2, \ldots, N$. This equation is an expanded version of Bayes' Formula.

REVIEWING MATERIAL

45. (Sections 2.4 and 2.5) Find $\mathbf{x}$ if

$$\mathbf{Ax} + \mathbf{b} = \mathbf{Cx}$$

when

$$\mathbf{A} = \begin{bmatrix} 1 & 2 \\ 3 & 4 \end{bmatrix}, \qquad \mathbf{C} = \begin{bmatrix} -1 & 4 \\ 3 & -2 \end{bmatrix}, \qquad \text{and} \qquad \mathbf{b} = \begin{bmatrix} 10 \\ -2 \end{bmatrix}$$

46. (Section 3.3) Model but do not solve: The Old Town Butcher Shoppe makes its ground meat packages from only select cuts of beef and pork. The beef is 72% meat and 28% fat and costs the shop $1.05 per pound; the pork is 80% meat and 20% fat and costs

the shop 1.20¢ per pound. How much beef and how much pork should the shop use in *each* pound of ground meat if it wants to minimize cost and yet still guarantee that the ground meat contains at least 75% meat?

47. (Section 4.1) A couple invests $5,000 in a mutual fund that grows at the rate of 6% a year, compounded monthly, for the first 10 years and at the rate of 8% a year, compounded monthly, thereafter. How much will their account be worth after 15 years, if no funds are withdrawn in that period?

48. (Section 5.4) (*a*) How many different seven-digit telephone numbers can be assigned if the first digit cannot be either a 0 or a 1? (*b*) Area codes increase this number by making each telephone exchange a ten-digit number. How many different telephone numbers (with area codes) exist, if the first digit of an area code also cannot be 0 or 1?

RECOMMENDING ACTION

49. Respond by memo to the following request:

> ### MEMORANDUM
>
> To: J. Doe Reader
>
> From: Graphics Department
>
> Date: Today
>
> Subject: **Altering Your Diagram**
>
> Last week, I displayed your tree diagram (Figure 6.13) at the Board of Trustees meeting. Recall that you placed the dress length on the first level and the color on the second level. The Board wants me to redraw the graph so that the color appears on the first level and the length on the second level. What would such a tree diagram look like? What would the numbers be?

6.5 COMBINATORICS

In Section 6.1, we developed the formula

$$P(E) = \frac{n(E)}{n(U)} \qquad \text{(1 repeated)}$$

for calculating probabilities in sample spaces with equally likely outcomes. To use this equation, we must count the elements in a subset E and then divide by the number of elements in the entire sample space U. Clearly, our ability to use Equation (*1*) depends on our ability to count $n(E)$ and $n(U)$.

So far, we have relied on tree diagrams to help us list and then count the outcomes in a sample space. Tree diagrams have served us well, but for more complicated processes, we require more sophisticated counting techniques.

A permutation of a set of elements is one arrangement of those elements.

Many useful counting procedures stem directly from the fundamental theorem of counting (see Section 5.4). A *permutation* of a set of elements is a particular listing or arrangement of all the elements in that set. For example, three permutations of the set {1, 2, 3, 4, 5} are 54321, 12453, and 35142. Two permutations of the set {red, green, blue, yellow} are the arrangements red-blue-green-yellow and green-yellow-blue-red. All possible permutations of the set {a, b, c} are

$$abc \quad acb \quad bac \quad bca \quad cab \quad cba \tag{19}$$

The fundamental theorem of counting is ideal for counting the total number of permutations of any set of elements. The set {a, b, c} contains three elements. In any arrangement of these three elements, we have three choices for the first entry, either a, b, or c. For each choice, there are two possible choices for the second entry—either of the two letters not chosen first—and then there is only one choice for the last entry. Using the fundamental theorem of counting, we conclude that there are $3(2)(1) = 6$ different arrangements of the set {a, b, c}, and they are listed in (*19*).

EXAMPLE 1

Determine the number of permutations that can be found from a set of five objects.

Solution

A permutation is a listing of all five objects, and each different arrangement is counted as a different permutation. In a five-element set, there are five choices for the first entry in an arrangement. For each possible first entry, there are four possible second entries—any of the objects not listed first. For each second entry there are three possible third entries, for each third entry there are two possible fourth entries, and for each fourth entry there is only one remaining object available as a last entry. It follows from the fundamental theorem of counting that a set of five objects may be arranged in $5(4)(3)(2)(1) = 120$ different permutations.

A pattern is developing, and it involves the product of successive integers. For any positive integer n, n factorial, denoted $n!$, is the product

$$n! = n(n - 1)(n - 2) \ldots (1) \tag{20}$$

n factorial is the product of successively decreasing positive integers, beginning with n and ending with 1. In particular,

$$
\begin{aligned}
1! &= 1 &&\text{(one factorial)} \\
2! &= 2(1) = 2 &&\text{(two factorial)} \\
3! &= 3(2)(1) = 6 &&\text{(three factorial)} \\
4! &= 4(3)(2)(1) = 24 &&\text{(four factorial)} \\
5! &= 5(4)(3)(2)(1) = 120 &&\text{(five factorial)} \\
6! &= 6(5)(4)(3)(2)(1) = 720 &&\text{(six factorial)}
\end{aligned}
$$

Factorials quickly become large. Eight factorial (8!) is 40,320. Eleven factorial (11!) is nearly 40 million and 13! is over 6 billion. It is convenient to define

A set of n *elements can be arranged in* n! *different permutations.*

$$0! = 1 \tag{21}$$

It is now a direct result of the fundamental theorem of counting that a set of n elements has a total of $n!$ different permutations.

EXAMPLE 2

A seven-store chain is ordered by bankruptcy court to go out of business by closing one store every month for the next seven months. How many different ways can the closings be scheduled?

Solution

If we label the stores A through G, respectively, then each closing schedule is a permutation of those seven letters. For example, the permutation $BGAFECD$ represents closing B first, then G, then A, . . . , until finally store D is closed last. There are as many different closing schedules as there are permutations of the seven letters A through G, exactly $7! = 7(6)(5)(4)(3)(2)(1) = 5{,}040$.

At times, we do not need to list all the elements of a particular set, but only those in a subset. Perhaps we need to close only four of the seven stores in Example 2, in which case we need all possible arrangements of four letters from the set of seven. For example, $FDAB$ represents closing F first, D second, A third, and B last. There are seven choices for the first closing, any one of the seven stores in the chain. For each possible first closing, there are six possible second closings—any store not closed in the first month. For each second closing there are five choices for the third closing, and for each third closing there are four choices for the fourth closing. It follows from the fundamental theorem of counting that there are a total of

$$7(6)(5)(4) = 840$$

different schedules for closing four stores out of seven, one each month. We can write the product $7(6)(5)(4)$ compactly in terms of factorials, as follows:

$$7(6)(5)(4) = 7(6)(5)(4)\left[\frac{3(2)(1)}{3(2)(1)}\right] = \frac{7(6)(5)(4)(3)(2)(1)}{3(2)(1)} = \frac{7!}{3!}$$

P(n, r) is the total number of ways to list r objects in order from a set of n objects, with n ≥ r.

A permutation of r objects from a set of n objects ($n \geq r$) is a listing of any r objects in the set. The total number of such listings is denoted by $P(n, r)$, and it follows from the fundamental theorem of counting that

$$P(n, r) = n(n - 1)(n - 2) \ldots (n - r + 1) \tag{22}$$

The right side of Equation (22) can be written in terms of factorials as

$$n(n - 1)(n - 2) \ldots (n - r + 1)$$
$$= \frac{n(n - 1)(n - 2) \ldots (n - r + 1)(n - r)(n - r - 1) \ldots (3)(2)(1)}{(n - r)(n - r - 1) \ldots (3)(2)(1)}$$
$$= \frac{n!}{(n - r)!}$$

Thus, we also have

$$P(n, r) = \frac{n!}{(n - r)!} \tag{23}$$

When closing four out of seven stores, one each month, $n = 7$, $r = 4$, and

$$P(7, 4) = \frac{7!}{(7 - 4)!} = \frac{7!}{3!} = \frac{5{,}040}{6} = 840$$

Many calculators have a key, often labeled *nPr*, for evaluating the quotient in Equation (23), providing the result is not so large it challenges the calculator's ability to store the number. Alternatively, almost all calculators have a factorial key, which allows us to evaluate the numerator and denominator of Equation (23) separately, and then divide. As a last resort, we can calculate $P(n, r)$ directly using elementary multiplication and Equation (22).

EXAMPLE 3

Judges at an ice-skating competition with 22 contestants must choose a winner and a runner-up. How many different ways can the competition end?

Solution

Each listing of 2 people from the 22-person field, with the winner listed first and the runner-up listed second, is a possible outcome. There are as many different endings to the contest as there are permutations of $r = 2$ people from a set of $n = 22$ contestants, namely

$$P(22, 2) = \frac{22!}{(22 - 2)!} = \frac{22!}{20!} = 22(21) = 462$$

In all our examples so far, a different arrangement of the same distinguishable objects is a different outcome. In Example 3, the outcome McNultey-Pria (McNultey is the winner and Pria is the runner-up) is a different outcome than the outcome Pria-McNultey (Pria is the winner and McNultey is the runner-up). In Example 2, *BGAFECD* is a different closing schedule from *GBAFECD*. There are processes, however, in which different arrangements of the same distinguishable objects are *not* different outcomes. Sometimes two outcomes are different *only* when they involve different elements.

> *With permutations, a different arrangement of the same distinguishable objects is a different outcome.*

A police patrol unit consists of two police officers working together. Suppose that a desk sergeant must dispatch one unit to respond to a call for assistance, and the sergeant has four police officers available for duty: Ali, Carl, Jamila, and Monica. How many different ways can the desk sergeant form a responding patrol unit? If we construct a tree listing all the possible ways the sergeant can form the patrol, we obtain Figure 6.15, where each officer is identified by his or her first initial.

Starting from the top branch and moving sequentially to the bottom branch, we have

$$U = \{AC, AJ, AM, CA, CJ, CM, JA, JC, JM, MA, MC, MJ\}$$

Apparently, there are 12 different patrol units that can be formed from the four officers. But this sample space differentiates between different orderings of the same people, which constitute the same unit. The patrol unit of Ali and Carl (*AC*) is the same unit as Carl and Ali (*CA*). The team of Carl and Monica (*CM*) is the same team as Monica and Carl (*MC*). In Figure 6.15, each distinguishable team is listed two times, where $2 = 2(1) = 2!$ is the number of different permutations of the same two people. There are really only $12/2 = 6$ different patrol units of two people that can be formed from four officers when order does *not* matter.

As a second example of a process in which order does not matter, we consider the creation of a hero sandwich. A store stocks four types of ingredients: salami, ham, American Cheese, and provolone, and a regular hero contains three of these ingredients, chosen

FIGURE 6.15

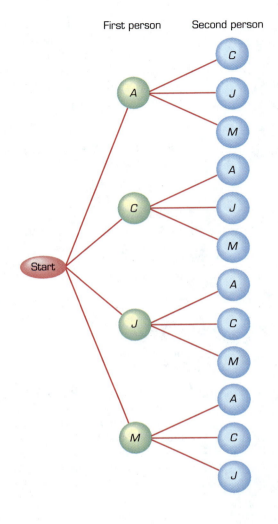

First person   Second person

by the customer. Figure 6.16 is a tree diagram of all possible permutations of three ingredients on a sandwich.

It appears from Figure 6.16 that the store offers 24 different sandwiches, one corresponding to each branch, but this is not so. The order in which ingredients are placed in a sandwich is irrelevant, only the ingredients themselves matter. Starting from the top branch, the first, third, seventh, ninth, thirteenth, and fifteenth branch all contain ham (H), American cheese (A), and salami (S) and all represent the same sandwich. In fact, if we analyze Figure 6.16 closely, we see that each sandwich is repeated six times, where $6 = 3(2)(1) = 3!$ is the number of different ways the same three ingredients can be arranged among themselves. There are only $\frac{24}{6} = 4$ different sandwiches of three ingredients that can be created from a menu of four ingredients when order does not matter.

The previous two examples suggest a procedure for counting sample spaces or events when order does *not* matter. First, count the number of possibilities assuming that order does matter, and then divide that result by the number of ways the objects in each outcome can be arranged among themselves.

To count the number of ways a police patrol unit of $r = 2$ officers can be formed from

FIGURE 6.16

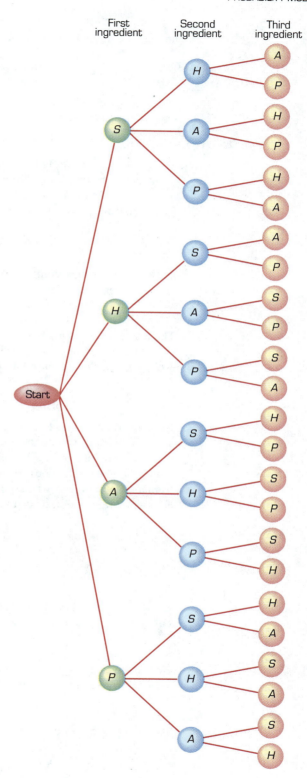

a set of $n = 4$ officers, when order does *not* matter, we first count the number of ways the unit can be formed assuming order matters:

$$P(4, 2) = \frac{4!}{(4 - 2)!} = \frac{4!}{2!} = \frac{24}{2} = 12$$

and then divide this number by the number of ways two people can be arranged among themselves, namely,

$$2! = 2(1) = 2$$

The result is $\frac{12}{2} = 6$ different patrol units.

To count the number of different hero sandwiches that can be made using $r = 3$ out of $n = 4$ ingredients, when order does *not* matter, we first count the number of ways a sandwich can be made assuming order matters:

$$P(4, 3) = \frac{4!}{(4 - 3)!} = \frac{4!}{1!} = \frac{24}{1} = 24$$

and then divide this number by the number of ways three ingredients can be arranged among themselves, which is

$$3! = 3(2)(1) = 6$$

The result is $\frac{24}{6} = 4$ different types of sandwiches.

Two common notations for the total number of ways to select r objects from a set of n objects when order does not matter are

$$C(n, r) \qquad \text{and} \qquad \binom{n}{r}$$

Both notations are read "a combination of n objects taken r at a time" or "n choose r," and the choice of which notation to adopt is one of personal preference. In this book, we shall use both interchangeably. To count the total number of ways of listing r objects from a set of n when order does not matter, we first count the number of ways to list r objects from a set of n objects when order does matter, namely $P(n, r)$, and then we divide by the number of ways r objects can be arranged among themselves, namely $r!$. The result is

$$C(n, r) = \binom{n}{r} = \frac{P(n, r)}{r!} = \frac{n!}{(n - r)!r!} \tag{24}$$

Many calculators have a key, often labeled nCr, for evaluating the quotient in (24), providing the result is not so large it challenges the calculator's ability to store the number.

EXAMPLE 4

How many ways can a committee of 4 members be selected from a club of 80 people?

Solution

The order in which committee members are selected does not matter, only the people selected for the committee are important. Thus, there are

$$C(80, 4) = \binom{80}{4} = \frac{80!}{(80 - 4)!4!} = \frac{80!}{76!4!} = 1,581,580$$

different possible committees.

EXAMPLE 5

Following a complimentary newspaper review, a bakery receives requests from 20 restaurants for daily orders of baked goods. How many ways can the bakery choose its new customers if it has the capacity to service only three new accounts?

Solution

The bakery will select $r = 3$ restaurants from a pool of $n = 20$ requests. Because order does not matter, only the name of the chosen restaurants, the number is

$$C(20, 3) = \binom{20}{3} = \frac{20!}{(20-3)!3!} = \frac{20!}{17!3!} = 1,140$$

EXAMPLE 6

How many different lottery tickets exist if a ticket contains 6 different numbers selected from the integers 1 through 46?

Solution

The order in which numbers are selected does not matter, only the numbers themselves; two tickets are different only when they differ by at least one number. There are as many different tickets as there are ways to choose $r = 6$ numbers from a set of $n = 46$ numbers, namely

$$C(46, 6) = \binom{46}{6} = \frac{46!}{(46-6)!6!} = \frac{46!}{40!6!} = 9,366,819$$

If a player wanted to purchase tickets to cover all possible outcomes, that player would need over nine million tickets.

EXAMPLE 7

A standard poker hand consists of 5 cards dealt from a standard deck of 52 cards. How many different poker hands exist?

Solution

In all card games, the order in which cards are dealt or held in a hand does not matter—only the cards themselves matter; two hands are differently only when they differ by at least one card. Thus, there are as many different poker hands as there are ways to deal $r = 5$ cards from a deck of $n = 52$ cards when order does not matter. The answer is

$$C(52, 5) = \binom{52}{5} = \frac{52!}{(52-5)!5!} = \frac{52!}{47!5!} = 2,598,960$$

No counting procedure—not the fundamental theorem of counting, nor Equation (*23*), the formula for permutations, nor Equation (*24*), the formula for combinations—is applicable to all counting problems. Sometimes, however, we can combine procedures to count the total number of outcomes in a particular process.

EXAMPLE 8

A rent-a-car agency receives an order from a corporation for 2 luxury cars, 5 standard cars, and 7 compact cars. How many ways can the agency fill this order if it currently has 10 luxury cars, 12 standard cars, and 11 compact cars available?

Solution

We first consider the luxury cars. The agency must select $r = 2$ from its stock of $n = 10$ luxury cards. Both cars go to the same corporation, so the order of selection does not matter. Therefore, there are $C(10, 2)$ ways to fill the luxury car portion of the order. Next we consider the standard cars. The agency must select $r = 5$ from its stock of $n = 12$ standard cars, and again the order of selection does not matter. Consequently, there are $C(12, 5)$ ways to fill the standard car portion of the order. Finally we consider the compact cars. The agency must select $r = 7$ from its stock of $n = 11$ compact cars, and the order of selection does not matter. Therefore, there are $C(11, 7)$ ways to fill the compact car portion of the order.

To determine the number of different ways to fill the entire order, we use the fundamental theorem of counting. There are $C(10, 2)$ ways to select the luxury cars. For each choice, there are $C(12, 5)$ ways to select the standard cars. For each one of these possibilities, there are $C(11, 7)$ ways to select the compact cars. Thus, there are

$$C(10, 2) \cdot C(12, 5) \cdot C(11, 7) = 45(792)(330) = 11,761,200$$

different ways to fill the order.

EXAMPLE 9

How many different ways can a committee of 6 people be formed from a group of 12 women and 10 men if the committee must have an equal number of men and women?

Solution

This problem is of the same type as Example 8, but now we are dealing with different types of people rather than different types of cars. We need to select $r = 3$ women from a set of $n = 12$ women, and order does not matter. Two committees are different only if they differ by at least one member. There are $C(12, 3)$ ways to choose the women. Similarly, we need to select $r = 3$ men from a pool of $n = 10$ men, and there are $C(10, 3)$ ways to do this when order does not matter.

To determine the number of different ways to form the entire committee, we use the fundamental theorem of counting. There are $C(12, 3)$ ways to select the female members. For each one of these possibilities, there are $C(10, 3)$ ways to select the male members. Thus, there are

$$C(12, 3) \cdot C(10, 3) = 220(120) = 26,400$$

different ways to constitute the committee.

If one part of a process is subject to more restrictions than other parts, model the most constrained part first.

In Examples 8 and 9, we broke each problem into sequential parts and modeled each part separately. If one part of a process is subject to more restrictions than the other parts, it is often advantageous to model the most constrained part of the process first.

EXAMPLE 10

A rent-a-car agency receives an order from one corporation for 6 cars and another order from a second corporation for 5 standard cars. How many ways can the orders be filled if the agency currently has 10 luxury cars, 12 standard cars, and 11 compact cars available for renting?

Solution

The order from the second corporation is more restrictive because it specifies the type of car, so we handle it first. The agency must select $r = 5$ standard cars from its stock of

$n = 12$ standard cars. Order does not matter, because all 5 cars are going to the same customer, hence there are $C(12, 5)$ ways to fill the requirements of the second corporation. We now handle the needs of the first corporation. The agency has 28 cars unassigned—10 luxury cars, 11 compact cars, and the $12 - 5 = 7$ standard cars not already allocated to the second corporation. Any of these cars can be reserved for the first corporation, and there are $C(28, 6)$ ways to do so.

There are $C(12, 5)$ ways to fill the second corporation's order for 5 standard cars, and then for each one of those ways there are $C(28, 6)$ ways to fill the first corporation's order. The total number of ways to fill both orders is

$$C(12, 5) \cdot C(28, 6) = 792(376,740) = 298,378,080$$

EXAMPLE 11

How many ways can a five-card (poker) hand be dealt from a standard deck of cards and contain exactly two kings?

Solution

Order does not matter in a hand of cards. The hands of interest must contain two kings and three other cards that are not kings. There are no restrictions on the cards that are not kings; these three cards may or may not match in rank. The only restrictions are on the kings, so we will consider the kings first. There are $n = 4$ kings in a deck, and we need $r = 2$ of them in the hand. There are $C(4, 2)$ ways to have two different kings in the same hand when order does not matter. Next we consider the other three cards. They cannot be kings, so they must come from the other $n = 48$ cards in the deck. We need $r = 3$ of these cards for the hand, hence there are $C(48, 3)$ ways to choose them. The product

$$C(4, 2) \cdot C(48, 3) = 6(17,296) = 103,766$$

is the number of different poker hands that contain exactly two kings.

EXAMPLE 12

How many different poker hands contain a full house (a full house is three cards of one rank and two cards of another rank)?

Solution

There are 13 choices for the rank of the matching three cards. It can be three aces, three kings, three queens, . . . , all the way down to three deuces. Once we choose the rank of the triple, then we have $n = 4$ cards with that rank in the deck, and from them we must choose $r = 3$ cards. Order does not matter, so there are $C(4, 3)$ ways to choose the three cards. Since there are 13 ways to choose the rank of the triple and then for each rank $C(4, 3)$ ways to choose the triple, it follows from the fundamental theorem of counting that there are $13 \cdot C(4, 3) = 13(4) = 52$ different ways to deal a triple.

Next we consider the pair. It can be any rank except the rank of the triple, so there are 12 choices for the rank of the pair. Once we choose the rank of the pair, then we have $n = 4$ cards with that rank in the deck, and from them we must choose $r = 2$ cards. Order does not matter, so there are $C(4, 2)$ ways to do that. Since there are 12 ways to choose the rank of the pair and then for each rank $C(4, 2)$ ways to choose the pair, it follows from the fundamental theorem of counting that there are $12 \cdot C(4, 2) = 12(6) = 72$ different ways to deal a pair after a triple is dealt.

There are 52 ways to deal a triple, and for each triple there are 72 ways to deal a pair, so it follows from the fundamental theorem of counting that there are 52(72) = 3,744 different poker hands that contain a full house.

EXAMPLE 13

What is the probability of being dealt five cards from a standard deck and getting a full house?

Solution

We found in Example 7 that 2,598,960 different poker hands can be dealt from a standard deck of cards. Each of these hands is equally likely to occur. We found in Example 12 that 3,744 of these hands contain a full house, so it follows that the probability of getting a full house is

$$\frac{3,744}{2,598,960} \approx .0014406$$

Probabilities associated with other poker hands are listed in Table 6.3.

TABLE 6.3   Poker Hands

TYPE OF POKER HAND	DESCRIPTION OF HAND	NUMBER OF HANDS OF THIS TYPE	PROBABILITY OF BEING DEALT A HAND OF THIS TYPE
Royal flush	Ten-jack-queen-king-ace of the same suit	4	.0000015
Straight flush	Five cards of consecutive rank and the same suit, but not a royal flush	36	.0000139
Four of a kind	Four cards that match in rank and one other card	624	.0002401
Full house	Three cards that match in one rank and two cards that match in another rank (see Example 12)	3,744	.0014406
Flush	Five cards that match in suit but are not consecutive in rank	5,108	.0019654
Straight	Five cards that are consecutive in rank but not all of the same suit	10,200	.0039246
Three of a kind	Three cards of the same rank and two other cards of different ranks	54,912	.0211285
Two pair	Two cards of one rank, two cards of a second rank, and one card of a third rank	123,552	.0475390
One pair	two cards of one rank and three other cards of different ranks	1,098,240	.4225690
Bust	None of the above	1,302,540	.5011774

IMPROVING SKILLS

In Problems 1 through 24, evaluate the given quantities.

1. 9!

2. 4!4!

3. 5!3!

4. $\dfrac{8!}{4!}$

5. $\dfrac{7!}{5!}$

6. $\dfrac{0!}{3!}$

7. $\dfrac{3!4!}{5!}$

8. $\dfrac{6!}{3!3!}$

9. $P(10, 3)$

10. $P(15, 2)$

11. $C(10, 3)$

12. $C(15, 2)$

13. $P(8, 5)$

14. $P(12, 6)$

15. $\dbinom{8}{5}$

16. $\dbinom{12}{6}$

17. $P(20, 1)$

18. $P(7, 7)$

19. $C(20, 1)$

20. $C(7, 7)$

21. $P(80, 3)$

22. $P(9, 3)$

23. $\dbinom{80}{3}$

24. $\dbinom{9}{3}$

CREATING MODELS

25. An organization of 50 members meets to elect 3 officers—president, vice-president, and secretary—from its ranks. In how many different ways can this executive slate be formed?

26. A developer who plans to build 7 houses on a particular block has a group of 42 different house designs. In how many different ways can the block be built if no two houses will be of the same design?

27. A hotel has 141 rooms and 5 reservations from different parties. In how many different ways can the parties be assigned rooms?

28. In how many ways can the hotel in the previous problem reserve rooms if all 5 reservations are made by the same person?

29. A rent-a-car company has 14 available cars on its lot and 3 customers in the office. In how many different ways can the customers be assigned cars?

30. A telephone repair depot has seven crews and five requests for service. In how many ways can repairs be scheduled if no crew is assigned more than one repair?

31. A dating service has the names of six females and eight males who seek matches. In how many ways can the service match each female with one male?

32. The Army has 9 newly promoted captains and 17 companies without captains. In how many ways can the captains be assigned to companies if no company is assigned more than one captain?

33. Seventeen colonels will be promoted to generals. In how many ways can the promotions be awarded if 31 colonels are eligible for promotion?

34. Eight horses are entered in a race that will post a winner, a runner-up (place position), and third place (show position). In how many ways can the race end if there are no ties?

35. In how many ways can two horses tie for the win position in an eight-horse race?

36. Nine players are picked for a team in a baseball game. How many different batting orders can the manager create from this team?

37. An ice cream store stocks 30 flavors and offers a rainbow banana split that contains 3 scoops of ice cream, each of a different flavor. How many different rainbow splits can the store advertise?

38. A department store chain has options to lease 20 different locations for new stores in different parts of the country. The management decides to build 3 new stores next year. In how many different ways can sites for these new stores be selected?

39. A candy manufacturer produces Halloween surprise bags by filling bags with 5 different surprises. How many different surprise bags can the company create if it stocks 14 different types of surprises?

40. The maintenance department of an industrial plant receives calls from 17 offices for new neon bulbs. In how many ways can the maintenance department respond if it has only 8 bulbs in stock?

41. In some civil cases, eight people form a jury. At the end of presentations, two of these jurors are chosen by lot to be alternates for the other six jurors, who deliberate and arrive at a verdict. In how many ways can the alternates be selected?

42. In how many ways can a person be dealt five cards from a standard deck and receive all hearts?

43. In how many ways can a person be dealt five cards from a standard deck and receive all red cards, either hearts or diamonds?

44. In how many ways can a person be dealt 13 cards (a bridge hand) from a regular deck and receive no aces?

45. In how many ways can a person be dealt 13 cards from a regular deck and receive all four aces and nine other cards?

46. A Chinese dinner for six consists of one selection from group A, three selections from group B, and two selection from group C. There are 8, 20, and 13 entrees listed in groups A, B, and C, respectively. How many different dinners for six does the restaurant offer?

47. The residence office of a university has three dormitory rooms unassigned, Room 103, which is a triple, and Rooms 102 and 202, which are doubles. Ten male students have room requests on file. In how many ways can assignments be made to the available rooms?

48. A 9-person bargaining committee and a 3-person grievance committee are to be formed from the membership of a local union. No member may serve on both committees. In how many different ways can the committees be formed if 25 people belong to the union?

49. The head of the legal department of a large corporation plans to send two lawyers to Arizona and three lawyers to Boston on the same day to conclude stock merger plans. In how many different ways can the department head make these assignments from a pool of ten lawyers?

50. A manager of a new store must hire 5 sales people from 20 applicants, 2 cashiers from 5 applicants, and 3 stock people from 11 applicants. Determine the number of different ways the manager can fill the openings.

51. The manager in the previous problem decides that none of the jobs require experience and that each job can be filled by any of the applicants. Now how many ways can the manager fill the openings?

52. A football team plays nine different opponents in a season. How many ways can the schedule end with six wins, two losses, and one tie?

53. A man owns three suits, ten ties, and ten shirts. How many ways can he select a traveling wardrobe of two suits, four ties, and six shirts?

54. A jai alai game involves eight teams, numbered 1 through 8, in competition. A quinella wager is a bet on the two teams that will place first and second, regardless of order. A 5-6 quinella wager is the same as 6-5 quinella wager, and both win if team 5 finishes first and team 6 finishes second, or vice versa. How many different quinella wagers are possible?

55. An exact wager in jai alai is a bet on the two teams that will place first and second, respectively. A 5-6 exacta wager wins only if team 5 wins and team 6 comes in second. How many different exacta wagers are possible in a game involving eight teams?

56. What is the probability of winning an exacta wager in the previous problem if the selection of teams is made randomly?

57. What is the probability of winning the quinella wager described in Problem 54 if the selection of teams is made randomly?

58. A combination lock is a sequence of three numbers, where each number is restricted to the range from 0 to 39, inclusive. What is the probability of buying such a lock and having the combination not include any number less than 20?

59. A motel has 60 rooms, 20 on the first floor and 40 on the second floor. An organization reserves 10 rooms. What is the probability that all 10 rooms will be on the second floor?

60. A teacher will choose 6 of his 20 students to present homework problems from the previous day. What is the probability that neither Raj nor Jane will be selected?

61. A company leader must select 3 men for a platoon assignment. What is the probability of Paul being chosen if there are 14 men in the company?

62. What is the probability that a random seven-digit number, not starting with zero, has no repeated numbers?

63. A box of 25 light bulbs contains 5 bulbs that are defective. Two bulbs are selected randomly from the box. What is the probability that both bulbs are defective?

64. Assume that four bulbs are selected randomly from the box described in the previous problem. What is the probability that two of these bulbs will be defective and two will not?

65. Determine the probability of receiving four cards that match in rank in a poker hand (five cards dealt from a standard deck).

66. Determine the probability of receiving no face cards and no aces in a poker hand.

67. Determine the probability of receiving only queens and jacks in a poker hand.

68. Determine the probability of receiving no aces in a bridge hand (13 cards from a standard deck).

69. Determine the probability of receiving all four aces in a bridge hand.

70. Determine the probability of receiving exactly seven cards of the same suit in a bridge hand.

71. A barrel contains seven red balls and three white balls, all identical except for color. Two balls are selected blindly from the barrel. What is the probability that both balls are red?

72. What is the probability that the two balls drawn in the previous problem are different in color?

73. A daily double wager is a bet on the winners of two horse races. One day, the first race in the daily double involves eight horses and the second race involves ten horses. What is the probability of winning the daily double if a bettor chooses the winners randomly?

EXPLORING IN TEAMS

74. The number of distinguishable six-letter arrangements that can be created from the letters in the word *banana* is

$$\frac{6!}{3!2!} = 60$$

Develop a rationale for this formula.

75. Use the rationale developed in the previous problem to calculate the number of different eleven-letter words that can be created from the letters in *Mississippi*.

76. Verify the entry in Table 6.3 for the probability of receiving three of a kind in a poker hand.

77. Verify the entry in Table 6.3 for the probability of receiving one pair in a poker hand.

78. Verify the entry in Table 6.3 for the probability of receiving two pairs in a poker hand.

EXPLORING WITH TECHNOLOGY

Often the first step in proving a mathematical formula is believing the formula is true. Using a calculator with the capability of evaluating $C(n, r)$, arbitrarily select values for n and r, with $n \geq r$, and determine which of the following formulas you believe to be true.

79. $\dbinom{n}{r} = \dbinom{n}{n-r}$

80. $\dbinom{n+r-1}{r} = \dbinom{n+r-1}{n-1}$

81. $\dbinom{n}{r} + \dbinom{n}{r-1} = \dbinom{n+1}{r}$

82. $\dbinom{n}{r} + \dbinom{n-1}{r} = \dbinom{n}{r-1}$

83. $\dbinom{n}{0} + \dbinom{n}{1} + \dbinom{n}{2} + \ldots + \dbinom{n}{n-1} + \dbinom{n}{n} = 2^n$

REVIEWING MATERIAL

84. (Section 1.4) A woman leaves her home at 8:00 a.m. and drives to a business meeting two hundred miles away at a constant speed of 50 miles per hour. The meeting takes exactly two hours, whereupon the woman drives back to her home at a constant speed of 40 miles per hour. Draw a graph that shows her distance from home at various times of the day. Construct a window with time (from 8:00 a.m. to 5:00 p.m.) on the horizontal axis and distance from home (from 0 to 200 miles) on the vertical axis.

85. (Section 1.5) Find the equation of the straight line that passes through the two points (5, 890) and (10, 650).

86. (Section 2.2) Use Gaussian elimination to solve the following system:

$$2x + 5y - 3z = 80$$
$$3x - 2y + 4z = 50$$
$$5x + 3y + z = 70$$

87. (Section 4.3) A company estimates that it will need $1.5 million in five years to refurbish its plant and wants to establish a sinking fund to meet this need. How much must the company deposit at the end of each quarter to reach its goal at an interest rate of 4.82% a year, compounded quarterly?

RECOMMENDING ACTION

88. Respond by memo to the following request:

MEMORANDUM

To: J. Doe Reader

From: Conservation and Wildlife Commission

Date: Today

Subject: **Generating Funds**

The Governor has cut our funds by $8 million, so we must generate this amount another way. One idea, which has the Governor's support, is to offer vanity license plates at $50 each. Each plate will include a picture of a woodpecker on a tree, so the number of identifying characters on a plate would be limited to a maximum of 4.

Originally, we hoped to use only numbers on each license plate, but this will not create enough plates to meet our financial goal. I can accept also the letters CW (for conservation and wildlife) in tandem, thus adding plates 8CW5, 44CW, and the like. But will this be enough? What do you recommend?

6.6 BERNOULLI TRIALS

Some processes can be modeled as multiple repetitions of a simpler process. The simpler the process, the easier it is to analyze. Therefore, it is to our advantage to decompose complex processes into simpler ones whenever we can.

Flipping five coins can be modeled as flipping one coin five times in succession. Rolling two dice can be modeled as rolling a single die twice. The process of testing ten bottles of beer for purity is the tenfold repetition of testing one bottle of beer.

Bernoulli trials are a set of identical repetitions of a process having just two outcomes, with the outcome of each repetition being independent of the others and the probability of a success remaining the same for each trial.

We call each repetition of the same process a *trial* or *run*. *Bernoulli trials* are a set of repeated runs of the same process when the following three conditions are satisfied:

1. The process has just two outcomes, one designated a *success* and the other a *failure*.

2. The outcome of each trial is independent of the outcome of every other trial.

3. The probability of a success (or a failure) is the same for each trial.

The process of flipping the same coin five times is a set of Bernoulli trials. The only possible outcomes of each coin flip are a head, which we could call a success, and a tail, which we could call a failure. The probability of each outcome is $\frac{1}{2}$, and the outcome of one flip of the coin is independent of any other flip.

Any process can be modeled with just two outcomes: a particular event happens (a success) or it does not happen (a failure). A manager, for example, may evaluate the outcome of each decision as either good (a success) or bad (a failure). An investor may classify the outcome of each day's activity on the stock market as either profitable (a success) or not profitable (a failure). In Example 6 of Section 6.5, we showed that there are over nine million outcomes for a state lottery in which six integers between 1 and 46 are chosen in each trial. A player, however, is likely to view the outcome of each lottery as either having the winning number (a success) or not having it (a failure).

Generally, with Bernoulli trials, we want to determine the probability of obtaining a specified number of successes, say r, when the same process is repeated n times ($n \geq r$). For example, we may be interested in the probability of flipping a coin five times ($n = 5$) and obtaining exactly two heads ($r = 2$), or we may be interested in the probability of testing ten bottles of beer for purity ($n = 10$) and finding that exactly seven bottles passed ($r = 7$).

We begin analyzing the general Bernoulli trials problem by considering a special case: determining the probability that two of the next three calls to a fire department will be legitimate (not false alarms) if it is known that the probability of any call being legitimate is .8.

The process of receiving three calls can be decomposed into the simpler process of receiving one call, repeated three times. We define a legitimate call as a success S and a false alarm as a failure F, and note that the outcomes of different calls are independent. A call has a probability of .8 of being legitimate and .2 of being false, regardless of the legitimacy of any previous call. Thus we can create a set of Bernoulli trials to model the process of receiving three calls.

Figure 6.17 is a tree diagram for the sample space of outcomes from three successive calls. We see that the second, third, and fifth paths from the top satisfy the event of receiving two successes and one failure. It follows directly from this tree diagram and from our work in Section 6.3 that the probability of receiving two successes and one failure in three calls is

$$.8(.8)(.2) + .8(.2)(.8) + .2(.8)(.8) = 3(.8)^2(.2) = .384$$

The second, third, and fifth branches in Figure 6.17 have identical probabilities associated with them (namely, $(.8)^2(.2)$) because each branch contains two successes and one failure. There are three branches that satisfy the event of interest, so the probabilities of the three branches are added together (or, equivalently, the probability of any of the branches is multiplied by 3) to obtain the final answer.

FIGURE 6.17
Fire calls

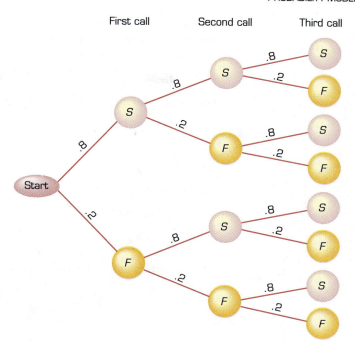

In Figure 6.17, the three branches that have two successes and one failure correspond to the outcomes

$$SSF \quad SFS \quad FSS$$

These outcomes represent all possible ways of filling three positions with two Ss and one F. That is, we have $n = 3$ positions to fill, and $r = 2$ of them must be filled with Ss. The ordering of the Ss in their positions does not matter, because different arrangements of the same two Ss among themselves generate the same outcome. Thus, there are $\binom{3}{2}$ ways to place two Ss in three available positions. Once the Ss are placed, there is only one place to put the single F—in the slot not filled by the Ss.

To calculate the probability of receiving two successes and one failure in the next three calls *without* constructing Figure 6.17, we reason as follows: There are as many branches to the tree having two successes out of three trials as there are ways of choosing $r = 2$ positions out of a set of $n = 3$ positions when order does not matter, namely, $\binom{3}{2}$. Each way contains two Ss, each S having probability .8, and one F, with probability .2. Because each call is independent of the others, the probability of a branch with two Ss and one F is $(.8)^2(.2)$. We have $\binom{3}{2}$ branches of interest, and each branch has a probability $(.8)^2(.2)$. The sum of these probabilities is

$$\binom{3}{2}(.8)^2(.2) = 3(.8)^2(.2) = .384$$

Now let us generalize to the probability of obtaining exactly r successes in n Bernoulli trials by visualizing but not drawing a corresponding tree diagram. We assume that the probability of an individual success is known, and we denote it by p. A failure is the complement of a success, so the probability of a failure is $1 - p$ (see Section 6.2). The probability of a path in a tree diagram with exactly r successes and $n - r$ failures is the product of r factors of the probability p and $n - r$ factors of the probability $1 - p$, that is,

$$p^r(1 - p)^{n-r}$$

There are as many branches having r successes out of n trials as there are ways of choosing r positions out of a set of n positions when order does not matter, namely, $\binom{n}{r}$. If we sum the probabilities with each branch that has exactly r successes and $n - r$ failures, we find that

> The probability of obtaining exactly r successes in n bernoulli trials is
>
> $$\binom{n}{r}p^r(1 - p)^{n-r} \tag{25}$$

where p denotes the probability of an individual success.

EXAMPLE 1

Determine the probability of flipping a fair coin $n = 5$ times and obtaining exactly $r = 2$ heads.

Solution
Define the outcome of flipping a head as a success. It has probability $p = .5$ of occurring with a fair coin. Using Equation (25), we calculate the probability of interest as

$$\binom{5}{2}(.5)^2(1 - .5)^{5-2} = 10(.5)^2(.5)^3 = .3125$$

EXAMPLE 2

A distributor of toys knows from experience that 5% of all Halloween masks in each carton of masks arrive damaged from the manufacturer. The distributor receives 20 cartons of masks and picks one mask from each carton for inspection. Determine the probability that exactly three of these masks are damaged.

Solution
Since 5% of every carton arrive damaged, the probability of randomly selecting one damaged mask from a carton is .05. If we let a success S be the event of picking a damaged mask, then the probability of a success is $p = .05$.

Selecting one mask from each of 20 cartons is a repetition of the same process 20 times. The results of one pick have no bearing on the results of any other pick, so the outcomes of each selection are independent. We have, therefore, a set of Bernoulli trials with $n =$

20, $r = 3$, and $p = .05$. Using Equation (25), we calculate the probability of picking exactly three damaged masks as

$$\binom{20}{3}(.05)^3(1 - .05)^{20-3} = 1,140(.05)^3(.95)^{17} \approx .06$$

EXAMPLE 3

Over a set distance, an archer hits the bull's-eye of a target 80% of the time. What is the probability that at least eight of the next ten arrows shot by the archer over this distance will be bull's-eyes?

Solution

Hitting the bull's-eye is a success and missing it is a failure. Each arrow shot at the target is a trial, and the probability of any shot being a success is $p = .8$ The probability of *at least* eight successes in ten trials means either exactly eight successes (event E, for eight) or exactly nine successes (event N, for nine) or exactly 10 successes (event T for ten). We will calculate each of these probabilities separately using Equation (25). For $r = 8$ successes in $n = 10$ trials, the probability is

$$P(E) = \binom{10}{8}(.8)^8(1 - .8)^{10-8} = 45(.8)^8(.2)^2 \approx .3020$$

For $r = 9$ successes in $n = 10$ trials, the probability is

$$P(N) = \binom{10}{9}(.8)^9(1 - .8)^{10-9} = 10(.8)^9(.2)^1 \approx .2684$$

For $r = 10$ successes in $n = 10$ trials, the probability is

$$P(T) = \binom{10}{10}(.8)^{10}(1 - .8)^{10-10} = 1(.8)^{10}(.2)^0 \approx .1074$$

The probability of at least eight successes is $P(E \cup N \cup T)$. But since the events E, N, and T are mutually exclusive, we have from the Exclusivity Law of Probability that

$$P(E \cup N \cup T) = P(E) + P(N) + P(T)$$
$$\approx .3020 + .2684 + .1074$$
$$= .6778$$

There is nearly a 68% chance that the archer will hit the bull's-eye eight or more times in ten tries when the probability of each bull's-eye is .8.

EXAMPLE 4

A hero sandwich is made from three ingredients selected by the customer from a menu listing 4 ingredients: ham, salami, provolone, and American cheese. Determine the probability that none of the next five customers will order provolone for their hero.

Solution

Each hero order is a repetition of the same process. We assume that the order of one customer has no bearing on the order of another customer (no peer pressure) and that each order is an independent event. We also assume that each of the four ingredients is equally likely to be selected.

Let S denote the event of a customer ordering a hero with provolone. It is not obvious how to determine p, the probability of a success, because there are many ways an order can include provolone and, therefore, be a success. In contrast, it is straightforward to calculate the probability of a failure. A failure occurs when a customer orders a sandwich without provolone. If each ingredient is equally likely to be selected, then each of the four ingredients is equally likely to be omitted. Since only one ingredient is omitted from each hero, the probability that the missing ingredient will be provolone is $\frac{1}{4}$. The probability of a failure is $\frac{1}{4}$ and, consequently, the probability of a success is $p = 1 - \frac{1}{4} = \frac{3}{4}$.

We now have a set of Bernoulli trials with $n = 5$ repetitions. Using Equation (25), we calculate the probability of obtaining $r = 0$ successes as

$$\binom{5}{0}(\tfrac{3}{4})^0(1 - \tfrac{3}{4})^{5-0} = 1(1)(\tfrac{1}{4})^5 \approx .001$$

IMPROVING SKILLS

In Problems 1 through 10, use the given information to determine the probability of obtaining exactly r successes in n Bernoulli trials, with the probability of a single success denoted by p.

1. $n = 15, r = 3, p = .3$
2. $n = 15, r = 4, p = .3$
3. $n = 14, r = 3, p = .3$
4. $n = 14, r = 3, p = .4$
5. $n = 8, r = 5, p = \frac{1}{3}$
6. $n = 100, r = 2, p = \frac{7}{8}$
7. $n = 12, r = 12, p = \frac{1}{4}$
8. $n = 30, r = 0, p = \frac{2}{3}$
9. $n = 9, r = 5, p = 1$
10. $n = 15, r = 20, p = .45$

CREATING MODELS

11. A coin is flipped three times. Determine the probability that two heads will turn up.
12. A coin is flipped ten times. Determine the probability that five heads will turn up.
13. A coin is flipped five times. Determine the probability that heads will turn up no more than one time.
14. Four dice are thrown. Determine the probability of obtaining one 6.
15. Four dice are thrown. Determine the probability that the face that appears on each die is greater than 3.

16. A card is picked from a standard deck and then replaced. The deck is shuffled and the process repeated three more times for a total of four trials. What is the probability that exactly one of the four cards is a spade?

17. What is the probability that exactly one of the four cards selected in the previous problem is a face card?

18. What is the probability that exactly two of the four cards selected in Problem 16 are kings?

19. A barrel contains four red balls, two green balls, and two yellow balls. A ball is selected at random and then replaced. This process is repeated four more times for a total of five trials. What is the probability that exactly two of the five balls picked will *not* be green?

20. What is the probability that at least one of the balls picked in the previous problems is green?

21. A student takes a true-false test with ten questions and guesses on every answer. What is the probability that the student will pass the test (that is, answer seven or more questions correctly)?

22. A quiz has ten multiple choice questions with five answers to each question. What is the probability that a student will pass this test (answer seven or more questions correctly) if he guesses on every answer?

23. A plumbing manufacturer guarantees the probability of a new faucet working properly when correctly installed is .99. Determine the probability that one of the next 10 faucets correctly installed by a plumber will *not* work properly.

24. The probability that a drug will cure a particular infection is .9. The drug is given to ten patients with the disease. What is the probability that the drug will be effective on nine of these patients?

25. The probability that a particular eye operation will be successful is .85. During one week, eight patients undergo this operation. (*a*) What is the probability that the operation will be successful on all eight patients? (*b*) on at least six patients?

26. The probability of a licensed driver having an accident within a year is .21. Four licensed drivers are selected at random. Determine the probability that only one of these drivers will be involved in an accident within the year.

27. A manufacturer of baseballs knows from experience that under ordinary circumstances, 1% of all baseballs produced are defective. A quality-control engineer randomly inspects 50 baseballs from all those produced over the last month. Determine the probability that exactly 5 of these balls are defective.

28. A candy distributor knows from experience that 2% of the candy bars in every shipment arrive damaged. He receives 20 shipments and picks one candy bar from each. Determine the probability that exactly two of these candy bars are damaged.

29. A manufacturer of ball bearings knows from experience that 95% of the bearings produced meet a prescribed tolerance. A quality-control engineer randomly samples ten bearings from the production line. What is the probability that at least nine of these bearings will be within the prescribed tolerance?

30. From experience, oil explorers expect to find oil in 1 out of every 30 tries. If a company sinks 6 wells in the next year, what is the probability of finding oil in at least one of them? *Hint*: First find the probability of not finding oil in any of the wells.

31. A quarterback completes 62% of his passes. What is the probability that the quarterback will complete all of his next five passes?

32. A basketball player hits 35% of her three-point attempts. What is the probability the player will hit three of her next seven attempts?

33. A baseball player has a lifetime batting average of .282, that is, he gets 282 hits for every 1000 official at bats. What is the probability that such a player will get no hits in his next four official at bats?

34. Determine the probability that a baseball player with a .312 lifetime batting average will get four hits in his next four official at bats?

35. Tests show that the probability of an enemy missile penetrating air defenses is 1 out of 20. Three missiles are fired. What is the probability of at least one getting through?

36. Assume that the probability of a marriage ending in divorce within 15 years in .6. Five couples, all friends, get married the same year. (*a*) What is the probability that none of these couples will be divorced after 15 years? (*b*) that all will be divorced?

EXPLORING IN TEAMS

37. The probability of a missile penetrating enemy air defenses is 1 out of 20. How many missiles must be fired if we want the probability of at least one missile getting through to be at least .95?

38. Oil explorers expect to find oil in 1 out of every 30 tries. How many wells must a company drill if it wants to be at least 90% certain of striking oil at least once?

REVIEWING MATERIAL

39. (Section 1.6) Use graphical methods to estimate the solution to the following system of equations:

$$50x + 80y = 40,000$$

$$70x + 40y = 28,000$$

40. (Section 2.5) Find the inverse of

$$\mathbf{A} = \begin{bmatrix} 1 & -2 & 0 \\ -3 & 6 & -1 \\ 0 & 1 & -1 \end{bmatrix}$$

41. (Section 3.6) Use the enhanced simplex method to

$$\text{Minimize:} \quad z = 2x_1 + x_2 + 3x_3$$

$$\text{subject to:} \quad x_1 + 3x_2 + x_3 \geq 100$$

$$x_1 + 8x_2 \quad\quad \leq 150$$

with: all variables nonnegative

42. (Section 5.3) The heights of the girls in a sixth grade class are (to the nearest inch):

49, 50, 50, 51, 51, 51, 53, 53, 53, 53, 54, 55, 57, 59, 60, 61, 61, 62, 65

Determine the mean, mode, median, range, and standard deviation of these heights.

RECOMMENDING ACTION

43. Respond by memo to the following request:

MEMORANDUM

To: J. Doe Reader

From: The Commissioner

Date: Today

Subject: **Ted Williams**

Ted Williams is the last player in the major leagues to average at least four hits in every ten at-bats. Going into the last day of the 1941 season, a scheduled double-header against Philadelphia, Williams' average was .39955, which rounded to .400. Even so, Ted decided to play.

Philadelphia pitchers told Williams that they would pitch to him, and not walk him as they often did. Consequently, Ted Williams could expect eight at–bats. What was the probability that Ted would finish the day with an average of .400 or better? How likely was it for Ted to get six hits in eight at–bats that day, which is what happened?

6.7 EXPECTED VALUE

Life is a series of gambles. Whenever a person or organization undertakes a plan of action whose outcome is uncertain, the decision maker gambles on the future. World leaders gamble when they take their countries to war. Corporations gamble when they merge. Individuals gamble when they change jobs, buy a house, choose a college, or even eat at a new restaurant. People gamble because they must, because the outcomes of their decisions are rarely known with certainty.

Sometimes we cannot imagine all the possible outcomes of a particular decision. When scientists cross two genes for the first time or when surgeons operate on a brain, all of the ramifications are not known in advance. Indeed, this uncertainty fuels science fiction from *Frankenstein* to the *Twilight Zone*. Often, however, we do know in advance the set of all possible outcomes for a particular course of action—we then have a sample space for the process of interest (see Section 5.4).

Since gamble we must, we try to limit our exposure to loss or failure by determining both the set of all possible outcomes and also the likelihood of each outcome occurring. We do this by assigning probabilities to outcomes whenever we can (see Sections 6.1 and 6.2). A person might be willing to experiment with a new medicine if the probability of a cure is .9, but the same person might reconsider if the probability of death is .9.

Although probabilities are a key factor in many decision processes, they are not, in general, the decisive factor. If they were, people would not play state lotteries, undergo risky operations, or volunteer to fly into space. People agree to participate in risky ventures because the payoff from a successful outcome is large enough to compensate for the risk

To model risk, we need to balance the probability of success with the payoff realized from a success.

of a failure. Thus, if we are to model risk, we must find measures that balance the probability of success with the gain to be realized from a success.

Raffles are popular games for raising money. In one school raffle, for example, 500 tickets are sold at $3 each for prizes donated by local merchants. The first prize is a television set valued at $460; each of the three second prizes is a $50 gift certificate at a local florist; each of the ten third prizes is a pair of tickets to the local movie house, worth $14 per pair. What is the value of this raffle to a player who buys a single ticket?

One player receives a television set for first prize and realizes a net gain of $457, the value of the prize less the cost of the raffle ticket. Three players receive $50 in gift certificates and realize net gains of $50 − $3 = $47. The ten players who win third prize realize net gains of $14 − $3 = $11, and each of the remaining 486 players leaves with a net loss of $3, the price of a raffle ticket. Using plus signs to denote gains and negative signs to denote losses, the average gain to all ticket holders is

$$\frac{1(\$457) + 3(\$47) + 10(\$11) + 486(-\$3)}{500} = -\$1.50 \qquad (26)$$

Of course, no one actually loses $1.50. Most people lose $3, and a few lucky players show a profit. Nonetheless, the *average* (see Section 5.3) loss to all players is $1.50. Equation (*1*) is the expected value of the game to each ticket buyer.

We can rewrite the left side of Equation (*1*) as

$$\$457\left(\frac{1}{500}\right) + \$47\left(\frac{3}{500}\right) + \$11\left(\frac{10}{500}\right) + (-\$3)\left(\frac{486}{500}\right) \qquad (27)$$

If we recognize that $\frac{1}{500}$ is the probability of winning first prize, $\frac{3}{500}$ is the probability of winning a second prize, $\frac{10}{100}$ is the probability of winning a third prize, and $\frac{486}{500}$ is the probability of losing, then (*2*) is a sum of products, where each product is the value of an outcome times the probability of that outcome occurring.

More generally, if a process has N possible *numerical* outcomes, $d_1, d_2, \ldots, d_N$, with corresponding probabilities, $p_1, p_2, \ldots, p_N$, then the *expected value* of the process is

$$E = d_1 p_1 + d_2 p_2 + \cdots + d_N p_N \qquad (28)$$

Expected values are calculated by multiplying the numerical outcomes in the sample space by their probability of occurring and then summing the results.

EXAMPLE 1

A coin is flipped. If the coin turns up heads, you win $1, and if the coin turns up tails, you lose $1. What is your expected payoff?

Solution
This simple game involves the flipping of a coin and the subsequent exchange of money. We take a sample space to be $S = \{\$1, -\$1\}$, which lists the two possible cash outcomes

for the process, with $d_1 = \$1$ and $d_2 = -\$1$. Assuming that the coin is a fair one, there is an equal chance of flipping heads or tails, which means that there is an equal chance of winning or losing one dollar. Consequently, $p_1 = p_2 = \frac{1}{2}$ and

$$E = d_1 p_1 + d_2 p_2 = \$1(\tfrac{1}{2}) + (-\$1)(\tfrac{1}{2}) = \$0$$

The expected value of this game is zero, even though each time you play you either win $1 or lose $1. On the average, however, over many plays of the game, you can expect to break even.

A game is fair if its expected value is zero.

A game is a *fair game* if the expected value of the game is zero. Example 1 describes a fair game. Most games are not fair. If we multiply an expected value by the number of times a particular process (game) is repeated (played), we have a good *estimate* of what a person can expect after *many* repetitions of the game.

EXAMPLE 2

A person selects a three-digit number, pays $1 to the operator of the numbers game, and receives a receipt stipulating the selected number. The same number may be selected by many players. The next day, a winning three-digit number, picked at random, is posted. If a person selected the winning number, he or she wins $600; otherwise, there is no payoff to the player. What is the expected value of this game (*a*) to a player and (*b*) to the operator of the game with respect to each player?

Solution

There are 1,000 three-digit numbers, ranging from 000 to 999. If the winning number is picked at random, then there is an equal probability of $\frac{1}{1000} = .001$ that each three-digit number will be the winning number. Consequently, $p_1 = .001$ is the probability that a selected number is the winning number, and $p_2 = .999$ is the probability that a selected number is not the winning number.

(*a*) There are two monetary outcomes of the game to a player: either the player wins $599 (the $600 payoff less the entrance fee of $1) or the players loses $1. Thus, $d_1 = \$599$ and $d_2 = -\$1$. The expected value to the player is

$$E = d_1 p_1 + d_2 p_2 = \$599(.001) + (-\$1)(.999) = -\$0.40$$

(*b*) There are two monetary outcomes of the game to the game's operator with respect to each player: either the operator loses $599 or the operator wins $1. Thus, $d_1 = -\$599$ and $d_2 = \$1$. The expected value to the operator is

$$E = d_1 p_1 + d_2 p_2 = -\$599(.001) + \$1(.999) = \$0.40$$

Example 2 is the *numbers game* that was played illegally from prohibition through the 1960s. Originally, the entrance fee was a dime with a payoff of $60, but with inflation the entrance fee and payoffs gradually rose to those in the example. Often, the operator of the game was an organized crime syndicate known as the mob, and it follows from Example 2 that the mob won, on average, 40¢ from each player each day. In large cities, 100,000 people might play daily, and from 100,000 players the mob could expect a total daily profit of 100,000($0.40) = $40,000.

Although the mob might *expect* a profit of $40,000 from 100,000 players, it rarely won exactly that amount. Some days the mob would win more, some days less. On a day when many people selected the winning number, the mob might experience a cash-flow problem and have to pay out more money than it received. Today, many state lotteries that go by names such as *Pick 3* are based on the payoff scheme described in Example 2, with one difference: state lotteries often vary the payoffs to guarantee themselves a specific profit and to guarantee they never have a bad day or a cash-flow problem. (See Problem 63.)

Expected value is a useful measure for describing processes when the outcomes are monetary payoffs. The measure, however, is applicable to any process with numerical outcomes.

EXAMPLE 3

A person has four keys and only one of them can open a particular lock. Unfortunately, he does not know which key is the correct one, so he tries them one at a time. How many keys can he expect to try before opening the lock?

Solution

The numerical outcomes in this process are the number of keys that must be tried: $d_1 = 1$, $d_2 = 2$, $d_3 = 3$, and $d_4 = 4$. Since each key is equally likely to be the correct one, each outcome is equally likely to occur. The chance that the first key tried will open the lock is the same as the chance that the last key tried will do the job. Therefore, $p_1 = p_2 = p_3 = p_4 = \frac{1}{4}$, as illustrated in Figure 6.18. In this tree diagram c signifies that the *correct* key was used and nc denotes *not correct*. The expected value is

FIGURE 6.18

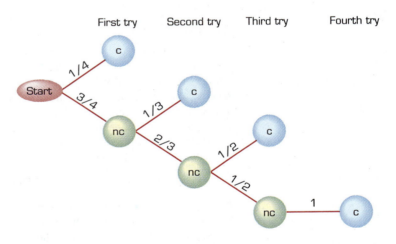

$$E = 1\left(\frac{1}{4}\right) + 2\left(\frac{3}{4} \cdot \frac{1}{3}\right) + 3\left(\frac{3}{4} \cdot \frac{2}{3} \cdot \frac{1}{2}\right) + 4\left(\frac{3}{4} \cdot \frac{2}{3} \cdot \frac{1}{2} \cdot 1\right)$$

$$= 1\left(\frac{1}{4}\right) + 2\left(\frac{1}{4}\right) + 3\left(\frac{1}{4}\right) + 4\left(\frac{1}{4}\right)$$

$$= 2.5$$

Obviously, one cannot try $2\frac{1}{2}$ keys. Each time the process is repeated, the number of actual keys tried will be an integer between 1 and 4. Sometimes the lock will be opened on the first try. Other times two tries will be needed, still other times three or four tries will be needed. The expected value is an average number of tries. If 100 people use the four keys to open the lock, we expect them to try a total of $100(2\frac{1}{2}) = 250$ keys.

EXAMPLE 4

In 1994, Publishers Clearing House mailed personal identification numbers to approximately 40 million people. If a person returned his or her number, at the cost of a first-class stamp, and it matched the winning number, that person won $10 million. What was the expected value of the game to each player?

Solution

A person either wins $10 million less 29¢ (the cost of a first-class stamp in 1994) or loses 29¢. Thus, $d_1 = \$9,999,999.71$ and $d_2 = -\$0.29$. Assuming that 40 million numbers were mailed and that each stood an equal chance of being the winning number, we have

$$p_1 = \frac{1}{40,000,000} \quad \text{and} \quad p_2 = \frac{39,999,999}{40,000,000}$$

Therefore, the expected value of this game to a player is

$$E = \$9,999,999.71\left(\frac{1}{40,000,000}\right) + (-\$0.29)\left(\frac{39,999,999}{40,000,000}\right) = -\$0.04$$

Expected value is of limited significance in the context of Example 4, because a player *cannot* play this game very often. If a player *could* play the game millions of times, then on average he or she would lose 4¢ each time. But in this case a player is limited to one or two plays each year, depending on the number of times Publishers Clearing House runs a contest, and with each play, a player will lose 29¢ with high probability or win $10 million with very low probability.

Expected value is most significant when a process is repeated many times, as is the case with casino games such as roulette. A roulette wheel is divided into thirty-eight slots numbered 1 through 36, 0, and 00 (double zero). Half the numbers from 1 to 36 are colored red, the other half black. Zero and double zero are colored green. A roulette game starts by whirling a small white ball around the outside of the wheel, and a game ends when the ball falls into one of the numbered slots on the wheel. The number of the slot holding the ball becomes the winning number.

EXAMPLE 5

In one variation of roulette, a player bets $5 on a number of his or her choice. If the player's number matches the winning number, the player retrieves the $5 bet and receives an additional $175 from the casino; if the player's number is not the winning number, the casino keeps the bet. What is the expected value of this game to a player?

Solution

There are two monetary outcomes: either the player wins $175 or the player loses $5. Thus, $d_1 = \$175$ and $d_2 = -\$5$. There are 38 different numbers, and assuming the wheel is fair, each number has an equal chance of being the winning number. Thus, $p_1 = \frac{1}{38}$ is the probability that a player's selected number will win, and $p_2 = \frac{37}{38}$ is the probability that the player's selected number will lose. Therefore,

$$E = \$175\left(\frac{1}{38}\right) + (-\$5)\left(\frac{37}{38}\right) \approx -\$0.2632$$

A player can expect to lose, on average, approximately 26¢ every time he or she plays this version of roulette.

EXAMPLE 6

In a second variation of roulette, a player bets $5 on either red or black. If the player's color matches the color of the winning number, the player retrieves his or her $5 bet and receives an additional $5 from the casino; if the player's color does not match the color of the winning number, the casino keeps the player's bet. What is the expected value of this game to a player?

Solution
Eighteen of the numbers have the same color as that bet by the player, and twenty numbers have a different color. If a player bets red, there are 18 red numbers that match and 18 black numbers and 2 green numbers (0 and 00) that do not match. There are two monetary outcomes: either the player wins $5 or the player loses $5. Thus, $d_1 = \$5$ and $d_2 = -\$5$. There are 38 different numbers, and assuming the wheel is fair, each number has an equal chance of being the winning number. Thus, $p_1 = \frac{18}{38}$ is the probability that a player's color will win, and $p_2 = \frac{20}{38}$ is the probability that it will lose. Here,

$$E = \$5\left(\frac{18}{38}\right) + (-5)\left(\frac{20}{38}\right) \approx -\$0.2632$$

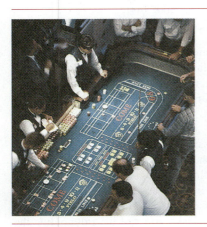

Another popular casino game is craps, which is played by throwing or rolling two dice onto a table and summing the number of dots that appear on the two dice. The number of throws needed to complete a standard game of craps varies. If the first roll results in either 7 or 11, the player wins and the game is over; if the first roll results in either a 2, 3, or 12, the player loses and the game is over; if, however, the first roll results in either a 4, 5, 6, 8, 9, or 10, then that number becomes the *player's number* or *point* and the game continues with additional throws of the dice. All successive throws of the dice are ignored until either a 7 appears or the player's number appears for the second time. If the player's number appears before a 7 is thrown, then the player wins and the game is over; if a 7 appears first, then the player loses and the game is over.

EXAMPLE 7

Determine the probability of a player winning and the probability of a player losing in a standard game of craps.

Solution
The first roll either ends the game or establishes the player's number. If successive rolls are needed, they are taken and ignored until either the player's number or a 7 appears. Consequently, only two rolls matter: the first and the last. Everything in-between is irrelevant to the outcome of the game. Figure 6.19 is a tree diagram for craps; it has two stages, one for the first roll and one for the last roll. Each path ends with a win or a loss, which are designated by W and L, respectively.

If we refer to the equally-likely sample space in Figure 6.2 of Section 6.1, we see that there are 36 different possible outcomes on the first roll. Six of these outcomes sum to 7:

$$1 + 6 \qquad 2 + 5 \qquad 3 + 4 \qquad 4 + 3 \qquad 5 + 2 \qquad 6 + 1$$

so the probability of rolling a 7 on the first throw is $\frac{6}{36}$. Similarly, five of the outcomes sum to 6:

$$1 + 5 \qquad 2 + 4 \qquad 3 + 3 \qquad 4 + 2 \qquad 5 + 1$$

so the probability of rolling a 6 on the first throw is $\frac{5}{36}$. The probabilities of rolling other sums are determined analogously, and these probabilities appear on the branches leading to the first-stage outcomes in Figure 6.19.

FIGURE 6.19
A Tree Diagram for Craps

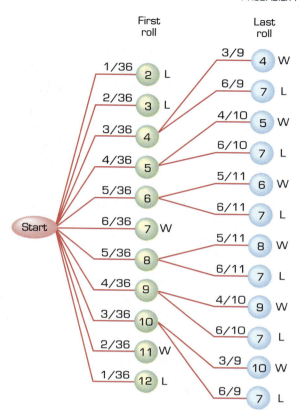

If a player's number is 4 after the first roll, then the game continues through successive rolls until either a 4 or a 7 appears. These are the only two sums that matter; every other outcome is irrelevant and has no bearing on the game. We modify the sample space to include only the outcomes that end the game, that is, the six outcomes that sum to 7 along with the three outcomes that sum to 4:

$$1 + 3 \qquad 2 + 2 \qquad 3 + 1$$

Thus, there are nine ways for the game to end when the player's number is 4, and each way is equally likely to occur. Six of the ways sum to 7 and end in a loss for the player, and three of the ways sum to 4 and end in a win for the player. Consequently, if the player's number is 4, the player will win with probability $\frac{3}{9}$ and lose with probability $\frac{6}{9}$.

Similarly, if the player's number is 5 after the first roll, the game continues through successive rolls until either a 5 or a 7 appears. Again, we modify the sample space to include only the outcomes that matter, and we thus have the six outcomes that sum to 7 along with the four outcomes that sum to 5:

$$1 + 4 \qquad 2 + 3 \qquad 3 + 2 \qquad 4 + 1$$

Now there are ten equally likely ways for the game to end. Six of the ways sum to 7 and end in a loss for the player, and four of the ways sum to 5 and end in a win for the player. Consequently, if the player's number is 5, he or she will win with probability $\frac{4}{10}$ and lose with probability $\frac{6}{10}$.

The probabilities of winning or losing associated with player's numbers of 6, 8, 9, and 10 are calculated analogously. They appear on the corresponding branches leading to the second stage outcomes in Figure 6.19. It follows from Section 6.2 that the probability of a player winning is the sum of all the probabilities of all the branches in Figure 6.19 that end in a win. Thus,

$$P\{\text{player's win}\} = \frac{3}{36}\left(\frac{3}{9}\right) + \frac{4}{36}\left(\frac{4}{10}\right) + \frac{5}{36}\left(\frac{5}{11}\right) + \frac{6}{36} + \frac{5}{36}\left(\frac{5}{11}\right) + \frac{4}{36}\left(\frac{4}{10}\right)$$
$$+ \frac{3}{36}\left(\frac{3}{9}\right) + \frac{2}{36}$$
$$= \frac{244}{495}$$
$$\approx .49292929$$

A loss is the complement of a win, so

$$P\{\text{player's loss}\} = 1 - \frac{244}{495} = \frac{251}{495} \approx .50707070$$

EXAMPLE 8

In a standard game of craps, a player bets $5. If the player wins, the player retrieves his or her $5 bet and receives an additional $5 from the casino; if the player loses, the casino keeps the player's bet. What is the expected value of this game to a player?

Solution
There are two monetary outcomes: either the player wins $5 or the player loses $5. Thus, $d_1 = \$5$ and $d_1 = -\$5$. Using the results of Example 7, we have $p_1 = \frac{244}{495}$ as the probability of a player winning and $p_2 = \frac{251}{495}$ as the probability of a player losing. Therefore,

$$E = \$5\left(\frac{244}{495}\right) + (-\$5)\left(\frac{251}{495}\right) \approx -\$0.07$$

On average, a player can expect to lose approximately 7¢ every time he or she plays standard craps. Of course, on each play, the player will actually win or lose $5. However, if a person plays standard craps repeatedly, say five hundred time, then that player expects his or her *net* winnings to be approximately $500(-\$0.07) = -\$35$.

IMPROVING SKILLS

A complete set of outcomes is specified in Problems 1 through 18, along with corresponding probability for each outcome. Determine the expected value for each set of outcomes.

1.
d_i	1	2	3	4
p_i	.25	.25	.25	.25

2.
d_i	1	2	3	4
p_i	.1	.2	.3	.4

3.
d_i	1	2	3	4
p_i	.4	.3	.2	.1

4.
d_i	10	12	22	35
p_i	.4	.3	.2	.1

5.
d_i	1	2	3	4	5
p_i	.1	.3	.3	.2	.1

6.
d_i	1	2	3	4	5
p_i	.45	.25	.15	.10	.05

7.

d_i	-5	-2	0	2	5
p_i	.45	.25	.15	.10	.05

8.

d_i	1	2	3	4	5	6
p_i	.2	.2	.1	.1	.2	.2

9.

d_i	1	2	3	4	5	6
p_i	.1	.2	.3	.2	.1	.1

10.

d_i	-50	-10	-5	20	25	40
p_i	.2	.2	.1	.1	.2	.2

11.

FIGURE A

12.

FIGURE B

13.   **FIGURE C**

14.   **FIGURE D**

15.   **FIGURE E**

16.   **FIGURE F**

17.   **FIGURE G**

18.

FIGURE H

−80 (27.1%)   50 (24.9%)

100 (15.7%)

−25 (32.3%)

In Problems 19 through 26, processes are repeated and n_i, the number of times each outcome occurs, is reported. Using empirical probabilities based on the data, determine the expected value for each process.

19.

d_i	−1	1	2	5
n_i	240	140	18	2

20.

d_i	−1	2	5	10
n_i	850	40	5	5

21.

d_i	1	2	3	4	5
n_i	85	49	15	61	90

22.

d_i	−5	−1	0	1	2	10
n_i	65	433	328	144	28	2

23.

FIGURE I

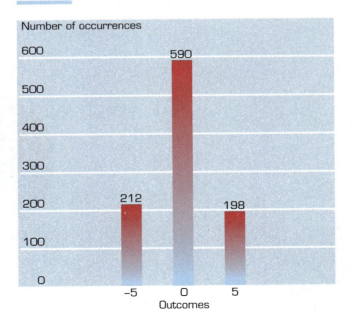

Number of occurrences

600	590
500	
400	
300	
200	212 198
100	
0	

−5   0   5

Outcomes

24. **FIGURE J**

25. **FIGURE K**

26.

FIGURE L

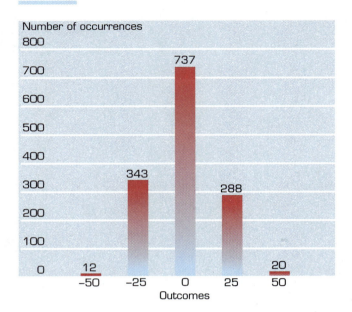

Number of occurrences

In Problems 27 through 34, assume that the data refers to a game and the outcomes are in terms of dollars. What entrance fee should you pay for playing the game once if the game is to be a fair one?

27. Use the data in Problem 1.
28. Use the data in Problem 2.
29. Use the data in Problem 3.
30. Use the data in Problem 6.
31. Use the data in Problem 8.
32. Use the data in Problem 11.
33. Use the data in Problem 13.
34. Use the data in Problem 14.

CREATING MODELS

35. A person works some days at home and some days at the office. During 50% of the weeks, she goes to the office four days and stays home one day; during 30% of the weeks, she goes to the office three days and stays home two days; for the other 20% of the weeks, she goes to the office two days a week and stays home three days. What is the expected number of days this person works at home each week?

36. On those days when the person in Problem 35 goes to the office, her daily transportation costs are $4.00. What are this person's average weekly transportation expenses?

37. A person buys a parcel of land in New Mexico for $5,000 with the intent of selling the land at the end of the year for $15,000 and realizing a profit of $10,000. The profit depends on the construction of a new dam in the region, but the dam still requires approval. If the dam is not approved, the land is worthless. What is the expected value of this investment opportunity if the probability of the dam being approved is $\frac{1}{4}$?

38. In a certain animal species, an adult female gives birth to either 0, 1, 2, or 3 healthy newborns each year with probabilities .11, .60, .25, and .04, respectively. What is the expected number of healthy newborns each year for each adult female?

39. A new car dealership sells between zero and ten cars each business day. The dealership was open 300 days last year and tallied the following sales figures:

Daily sales	0	1	2	3	4	5	6	7	8	9	10
Number of days	10	15	36	82	67	50	25	8	4	2	1

Determine the expected number of cars sold daily if this data represents typical sales for the dealership.

40. A manufacturer is considering a proposal to replace its existing machinery at a cost of $50,000. If the new machinery is an improvement, the manufacturer will realize an additional profit of $15,000 per year. If the new machinery is no better than existing machinery, the manufacturer gains nothing, effectively losing the $50,000 cost of the new machinery. If the new machinery is worse than current equipment, the manufacturer can return it and reinstate the old equipment at a net cost of $75,000. The probabilities associated with the various outcomes are .75, .20, and .05, respectively. What is the expected value of approving the replacement proposal over a 10-year period?

41. In a variation of the numbers game, a person selects a four-digit number and pays $1 to the operator of the game. If the person selected the winning number, picked at random at the end of the day, that person wins $5,000; otherwise there is no payoff to a player. What is the expected value of this game to the player?

42. A person flips a coin twice. What is the expected number of times that heads appears?

43. A person flips a coin three times. What is the expected number of times that tails appears?

44. A person pays $2 to play the following game. He flips a coin three times and receives $1 each time a head appears. What is the expected value of this game to the player?

45. Every day a person puts a $1 bill into a vending machine to purchase a cup of coffee that cost 80¢. He receives coffee and the correct change 85% of the time. At those times, he gets full value for the one-dollar gamble of entrusting a dollar to a machine. However, 10% of the time, he receives coffee and no change, and 5% of the time, he receives the correct change but no coffee. What is the expected value of this daily venture to the individual?

46. A barrel contains five red balls, two white balls, and one green ball. A person selects one ball at random. If the ball is red, she loses $1; if it is white, she wins $1; and if it is green, she wins $3. What is the expected value of this game to the player?

47. Three balls are selected, one at a time, from the barrel described in the previous problem. Each time a ball is selected, its color is recorded and the ball is returned to the barrel before the next selection is made. Determine the expected number of red balls selected.

48. One card is selected from each of two different standard decks. Determine the expected number of face cards selected.

49. Solve the previous problem if three decks are used.

50. Two cards are drawn in succession without replacement from a standard deck of cards (that is, the first card remains out of the deck while the second card is chosen). What is the expected number of red cards, either diamonds or hearts, selected?

51. Two cards are drawn in succession without replacement (see previous problem) from a standard deck of cards. What is the expected number of hearts?

52. A person pays a $1 entrance fee to play a game based on the process described in the previous problem. In return the person receives $3 for every heart selected. What is the expected value of this game to the person?

53. In a variation of roulette, a player bets $5 on even. If the winning number is even, the player retrieves his or her $5 bet and receives an additional $5 from the casino; if the winning number is odd, 0, or 00, the casino keeps the player's bet. What is the expected value of this game to the player?

54. In another variation of roulette, a player bets $5 on a set of three numbers. If the winning number matches one of the player's three numbers, the player retrieves his or her $5 bet and receives an additional $55 from the casino; otherwise, the casino keeps the player's bet. What is the expected value of this game to the player?

55. In still another variation of roulette, a player bets $5 on a set of four numbers. If the winning number matches one of the player's four numbers, the player retrieves his or her $5 bet and receives an additional $40 from the casino; otherwise, the casino keeps the player's bet. What is the expected value of this game to the player?

56. In another variation of roulette, a player bets $5 on a set of twelve numbers. If the winning number matches one of the player's twelve numbers, the player retrieves his or her $5 bet and receives an additional $10 from the casino; otherwise, the casino keeps the player's bet. What is the expected value of this game to the player?

57. What is the expected number of dots that appear face up when a single die is rolled once? $1\left(\frac{1}{6}\right) + 2\left(\frac{1}{6}\right) + 3\left(\frac{1}{6}\right) + 4\left(\frac{1}{6}\right) + 5\left(\frac{1}{6}\right) + 6\left(\frac{1}{6}\right)$ 1.5

58. What is the expected sum when two dice are rolled and the number of dots that appear face up on each die are summed?

59. In a variation of craps, a player bets $5 on *any craps*, which are the numbers 2, 3, and 12. The dice are rolled once. If one of these numbers appears, the player retrieves his or her $5 bet and receives an additional $35; if any other number appears, the player loses $5. What is the expected value of this game?

60. In another variation of craps, a player bets $5 on the *field*, which is the set of numbers containing 2, 3, 4, 9, 10, 11, and 12. The dice are rolled once. If a 5, 6, 7, or 8 appears, the player loses $5; if a field number appears, the player retrieves his or her $5 bet and receives either an additional $10 if the roll was a 2 or 12 or an additional $5 if the roll was 3, 4, 9, 10, or 11. What is the expected value of this game?

61. In another variation of craps, a player bets $5 on 4 as the player's number. The dice are rolled as many times as needed until either a 7 or a 4 appears. If a 7 appears first, the player loses $5; if 4 appears first, the player retrieves his or her $5 and receives an additional $9 from the casino. What is the expected value of this game?

62. In still another variation of craps, a player bets $5 on 9 as the player's number. The dice are rolled as many times as needed until either a 7 or a 9 appears. If a 7 appears first, the player loses $5; if 9 appears first, the player retrieves his or her $5 and receives an additional $7 from the casino. What is the expected value of this game?

EXPLORING IN TEAMS

63. Most state lotteries are *pari-mutuels*. A percentage of all monies bet by individuals is taken by the state to support state initiatives, such as aid for education. The remaining money is reserved for prizes and is divided equally by all winners. Consider a numbers game like the one described in Example 2. Assume that one million people played,

each betting $1, and the state's take is 50% of all money bet. How many winners would you expect? What is the payoff to each winner if the expected number of winners is the actual number of winners?

64. What is the probability that the expected number of winners in the previous problem is the actual number of winners?

65. Solve Problem 63 if only 700,000 people play the game.

66. Solve Problem 63 if the state's take is 55% of all money bet.

67. What is the expected value of the game described in Problem 63 to a player who bets $1.

EXPLORING WITH TECHNOLOGY

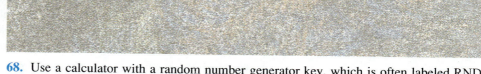

68. Use a calculator with a random number generator key, which is often labeled RND. Each time this key is pressed, a different number should appear on the calculator's display. Press this key many times, over 200 if possible, and each time sum the first three digits that appear. For example, if the display shows .938552, then the first three digits are 9, 3, and 8 and have a sum of $9 + 3 + 8 = 20$. The only sums that can occur are the integers between 0 and 27, inclusive. Calculate the empirical probability of each sum occurring, and then use those probabilities to estimate the expected value of the sum.

REVIEWING MATERIAL

69. (Section 1.6) Estimate a solution for the following set of simultaneous equations by graphing each equation on the same coordinate system.

$$y = -5x + 3$$
$$y = 7x - 4$$

70. (Section 2.3) According to the New Jersey Board of Regulatory Commissioners, the average monthly rate for basic cable television service in New Jersey was $10.02 in 1986, $12.10 in 1987, $13.43 in 1988, $15.32 in 1990, $18.96 in 1991, and $20.88 in 1992. Estimate the cost of service in 1995 based on this trend.

71. (Section 3.5) Use the simplex method to solve the linear programming problem:

$$\text{Maximize:} \quad z = 2x_1 + 3x_2 + 2x_3$$
$$\text{subject to:} \quad 5x_1 + 2x_2 + 4x_3 \leq 800$$
$$x_1 + 4x_2 + 3x_3 \leq 1,000$$
$$\text{with:} \quad \text{all variables nonnegative.}$$

72. (Section 6.4) A person plays one game of craps, as described in Example 7, and wins. What is the probability that the person won on his or her first roll of the dice?

RECOMMENDING ACTION

73. Respond by memo to the following request:

> ### MEMORANDUM
>
> To: J. Doe Reader
>
> From: Promotions Department
>
> Date: Today
>
> Subject: **The Casino's 25th Anniversary**
>
> As part of our 25th anniversary, the casino will offer special promotions during the last week in May. One idea is to replace the roulette wheel with the one we had when we first opened. You recall that the old roulette wheel had only 37 slots, O through 36, *without* the number OO.
>
> Obviously the payoffs to the casino are diminished with the old wheel, and Tom is worried about the loss of revenue. We believe that the lure of playing the old wheel will bring 1,000 new players to roulette during the week and that the extra traffic will more than compensate for the smaller take per customer. Furthermore, with a more favorable wheel, we expect the average wager per player to increase by 10%.
>
> I need your analysis of the situation for my meeting with Tom next week. Historically, we average 5,800 players at roulette during the last week in May, with each player betting on average a total of $100.

CHAPTER 6   KEYS

KEY WORDS

KEY CONCEPTS

6.1 Equally Likely Outcomes

- A sample space with equally likely outcomes for a process is a sample space in which each outcome has the same chance of happening.

- A sample space for a process is a model of that process. Often, we can associate a number of sample spaces (models) to the same process; some of these sample spaces will have equally likely outcomes and some will not.

- To construct a sample space with equally likely outcomes, we sometimes create abstract distinctions that are not apparent in actuality, such as thinking of a pair of dice being of different colors.

- If a sample space for a process has a finite number of equally likely outcomes, then the probability that an event will happen when the process is run is the cardinal number of the event divided by the cardinal number of the sample space.

6.2 The Laws of Probability

- Probabilities are numbers ranging between 0 and 1, inclusive, associated with events.

- A probability distribution is a rule of correspondence between subsets of a sample space and real numbers; it assigns a unique number to each subset in accordance with the Range Law, the Certainty Law, the Null Law, and the Generalized Exclusivity Law of probability.

- Probabilities of simple events can be represented graphically by a pie chart in which each simple event corresponds to a different sector of the pie chart and the fractional area of each sector equals the probability that the corresponding simple event will occur.

- Odds are a ratio of probabilities, algebraically rewritten as a ratio of integers.

6.3 Conditional Probability and Independent Events

- One approach to determining conditional probabilities is to modify the sample space each time new information is introduced, thereby revising what is possible and what is not.

- A second approach to determining conditional probabilities is to develop a formula that depends only on probabilities associated with the original sample space.

- If outcome A occurs before outcome B in a tree diagram (that is, if A is a node on the first stage of the tree diagram and B is a node on the second stage, with B attached to A by a branch), then the conditional probability $P(B|A)$ is placed on the line connecting the two nodes.

6.4 The Bayesean Method

- The Bayesean method is a procedure for calculating the conditional probability $P(F|E)$ from $P(E|F)$.

6.5 Combinatorics

- A set of n elements can be arranged in $n!$ different permutations.

- With permutations, a different arrangement of the same distinguishable objects is a different outcome.

- To count combinations, that is, the number of elements in a sample space when order does not matter, first count the number of possibilities when order does matter and then divide the result by the number of ways objects in each outcome can be arranged among themselves.

- If one part of a process is subject to more restrictions than other parts, model the most constrained part first.

6.6 Bernoulli Trials

- Bernoulli trials are a set of identical repetitions of a process that has just two outcomes, with the outcome of each repetition being independent of the others.

6.7 Expected Value

- Whenever a person or organization undertakes a plan of action for which the outcome is not certain, the decision maker gambles on the future.

- Expected value balances the probability of success with the payoff realized from a success.

- An expected value represents an average return over many repetitions of the same decision for the same process.

KEY FORMULAS

- If E is an event in a sample space U of equally likely outcomes, then
$$P(E) = \frac{n(E)}{n(U)}$$

- *The Certainty Law of Probability* If U denotes the sample space of a process, then
$$P(U) = 1$$

- *The Null Law of Probability*
$$P(\phi) = 0$$

- *The Inclusion-Exclusion Law of Probability*
$$P(E \cup F) = P(E) + P(F) - P(E \cap F)$$

- *The Exclusivity Law of Probability (Generalized)* If $E_1, E_2, \ldots, E_n$ are mutually exclusive, then
$$P(E_1 \cup E_2 \cup \ldots \cup E_n) = P(E_1) + P(E_2) + \ldots + P(E_n)$$

- *The Complement Law of Probability*
$$P(E') = 1 - P(E)$$

- If $P(E') \neq 0$ and $P(E) \neq 0$, then the odds in favor of the event E and the odds against the event E are, respectively,

$$\frac{P(E)}{P(E')} \quad \text{and} \quad \frac{P(E')}{P(E)}$$

algebraically simplified to a ratio of integers.

- *Conditional Probability* If $P(F) > 0$, then

$$P(E|F) = \frac{P(E \cap F)}{P(F)}$$

- *Product Rule for Probabilities*

$$P(E \cap F) = P(F) \cdot P(E|F)$$

- Two events E and F are independent when $P(E|F) = P(E)$.

- If two events E and F are independent, then $P(E \cap F) = P(E) \cdot P(F)$.

- *Bayes' Formula*

$$P(F|E) = \frac{P(F) \cdot P(E|F)}{P(E)}$$

$$n! = n(n - 1)(n - 2) \ldots (1)$$

$$0! = 1$$

$$P(n, r) = n(n - 1)(n - 2) \ldots (n - r + 1) = \frac{n!}{(n - r)!}$$

$$C(n, r) = \binom{n}{r} = \frac{P(n, r)}{r!} = \frac{n!}{(n - r)!r!}$$

- The probability of obtaining exactly r successes in n Bernoulli trials with probability p of an individual success equals

$$\binom{n}{r}p^r(1 - p)^{n-r}$$

- If a process has N possible numerical outcomes, $d_1, d_2, \ldots, d_N$, with corresponding probabilities $p_1, p_2, \ldots, p_N$, then the expected value of the process is

$$E = d_1p_1 + d_2p_2 + \ldots + d_Np_N$$

KEY PROCEDURES

- If a process has been run a number of times in the past, then we can use historical data to calculate the fraction of times a particular simple event occurs and let this fraction be the probability of the event occurring.

- The probability of a path in a tree diagram is the product of all probabilities on the path.

- To calculate the probability of an event on a tree diagram, locate all the nodes on the tree that correspond to this event. For each such node, determine the probability of the path leading to it (beginning at the start), and then sum the results.

- To calculate $P(F|E)$ when F is a node on the first stage of a tree diagram and E is a node on the second stage, divide the probability of the path that passes through node F and ends at node E by the sum of the probabilities of all paths that end at node E.

TESTING YOURSELF

Use the following problems to test yourself on the material in Chapter 6. The odd problems comprise one test, the even problems a second test.

1. A barrel contains four balls, of which three are green and one is red. The process of interest is to select two balls at random, leaving two balls in the barrel, and note their colors.
   (a) Construct a sample space of equally likely outcomes for this process.
   (b) Determine the probability of getting two balls of different colors.
   (c) Determine $P(E|F)$ when E is the event that the second ball is green and F is the event that the first ball is red.

2. A barrel contains 50 red balls and 25 green balls. What is the probability that two balls selected at random from the barrel are both red?

3. Determine $P(A|B)$ and $P(B|A)$ if it is known that $P(A) = .35$, $P(B) = .42$, and $P(A \cup B) = .57$.

4. Two nonstandard decks of cards are placed on a table. The first deck contains only red cards, all 26 hearts and diamonds, while the second deck contains only face cards, the 12 jacks, queens, and kings from the four suits. A coin is flipped. If the coin comes up heads, a card is selected randomly from the first deck; if the coin comes up tails, a card is selected from the second deck. Determine the probability of
   (a) not selecting a queen
   (b) selecting a queen knowing that the coin came up heads
   (c) having the coin come up heads knowing that a queen was selected

5. Let W represent the event that a particular candidate for elected state office will win, and let U represent the event that the candidate will have the support of the state's largest union. Suppose that $P(W) = .6$, $P(U) = .5$, and $P(W|U) = .9$.
   (a) Determine whether W and U are independent events.
   (b) Find $P(U|W)$.

6. What is the probability that all three of a person's initials are different? An example of such a set is JFK.

7. The odds in favor of a candidate winning election to office are 8 to 9. What is the probability that this candidate will win?

8. A photographer wants to line up three horses in a row and take their picture. How many ways can this be done if the photographer has seven horses available for the picture?

9. A person pays $2 to enter a card game and then selects one card at random from a standard deck of 52 cards. If he selects an ace, he receives $10 (effectively winning $8 because of the $2 entrance fee); if he selects a face card, he receives $5; otherwise he receives nothing and loses the $2 entrance fee. What is the expected value of this game to a player?

10. You have an interview with a recruiter on the second floor of a hotel, but you do not remember the room number. There are ten rooms on the second floor, and you decide to knock on doors one at a time until you arrive at the correct room. How many doors do you expect to try?

11. The following data were gathered during a test program for a new medication: The new medication was effective on 80% of those given the medication, and of those patients, only 5% experienced dizziness. In contrast, 20% of all

patients given the medication experienced dizziness. Determine the probability that

(*a*) the medication was effective on a patient who experiences dizziness, and

(*b*) a person given the medication would find it effective and also *not* experience dizziness.

12. The sample space $\{A, B, C, D\}$ is simple with

$$P(\{A\}) = 0.3, \qquad P(\{B\}) = 0.1, \qquad \text{and} \qquad P(\{C\}) = 0.4$$

Determine

(*a*) $P(\{D\})$ (*b*) $P(\{A\} \cup \{C\})$

(*c*) $P(\{B\} \cap \{C\})$ (*d*) $P(\{A\}')$

13. The following table summarizes the responses to a survey in which company employees were asked, "Do you drive to work in your own car?"

	YES	NO
Male	235	132
Female	274	73

(*a*) What is the probability that an employee drives to work?

(*b*) What is the probability that a male employee drives to work?

(*c*) What is the probability that an employee does *not* drive to work if it is known that the employee is female?

14. The probability that a drug will cure a particular infection is .95. The drug is given to eight patients with the disease. What is the probability that the drug will be effective on at least seven of these patients?

15. In February, a computer company produced 5,000 computers, 2,800 at its Ohio plant and the rest at its Nevada plant. From experience, the manufacturer expects to have 2% of all computers made in Ohio and 1% of all computers made in Nevada returned for repairs during the 90-day warranty period. What is the probability that the first returned computer from the February production came from the Ohio plant?

16. In seven-card stud a person is dealt a hand of 7 cards from a standard deck of 52 cards. What is the probability of a person receiving 5 spades and 2 cards that are not spades?

17. A survey taken in a particular company showed that 53% of all employees were married and 62% of all employees had children. In addition, the probability of an employee being married and having children was .47. What is the probability of an employee being married or having children?

18. A high school club has three freshman members, two sophomore members, six junior members, and one senior member. One member is selected at random to act as the club's representative to the student government.

(*a*) Construct a sample space of equally likely outcomes for this process.

(*b*) Determine the probability that the representative is an upperclassman (that is, either a junior or a senior).

(*c*) Determine $P(E|F)$ when E is the event that the representative is a junior and F is the event that the representative is *not* a freshman.

19. A city has 25 high school basketball teams, and each week a local newspaper lists the best five teams, ranked from 1 to 5. How many different ranking lists are possible?

20. A stock is four times as likely to rise in price as it is to fall in price, although the probability of the stock price remaining constant is .6. What are the odds of the stock rising in price?

21. A committee of five people is to be formed from a group of seven Republicans, six Democrats, and three independents. What is the probability that this committee will have exactly two Republicans, two Democrats, and one Independent?

22. Use the tree diagram in Figure A to determine
    (a) $P(E|A)$, (b) $P(F|A)$, (c) $P(F|B)$, (d) $P(B|F)$, (e) $P(A|E)$

FIGURE A

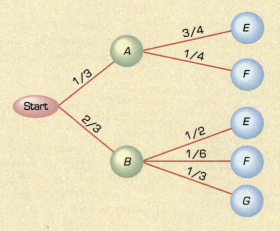

23. On June 9, 1994, the same three-digit number won in both New York's and Connecticut's state lottery games. What is the probability of this event, if each state selects a random three-digit number between 000 and 999, inclusive, each day?

24. A person lives in Nebraska and has a mother who lives in Colorado. According to national statistics, the probability of the person's mother dying next year is .035 and the probability of the person's cat dying next year is .46. What is the probability of both the person's mother and cat dying next year?

25. A barrel contains ten red balls, eight green balls, and three yellow balls. A player blindly picks two balls from the barrel in succession without replacement (that is, the first ball remains outside the barrel while the second ball is picked). If the two balls are of the same color, the player wins $2; otherwise the player loses $1. What is the expected value of this game to a player?

7

MARKOV CHAIN
MODELS

7.1 MARKOV CHAINS

The focal point of most models is a *system*, a set of interrelated parts working towards a common goal. A watch is a set of parts that includes hands, a face, a band, gears, and a casing, all working together to keep time. Disconnected and spread on a table, these same parts no longer form a system, because they no longer work together towards a common goal. The United States—a group of people with laws, customs, and institutions designed to make life more comfortable for its citizens—is an example of a much more complicated system.

In Section 5.4, we defined a process as an action or decision that results in one of a set of possible outcomes. In general, most processes are part of a larger system. Flipping a coin is a process whose outcome is either a head or a tail. This process is embedded in a larger system that contains, at a minimum, the coin and the flipper and perhaps also contains other individuals affected by the outcome, such as two football teams awaiting the outcome of a coin flip at the beginning of a game.

> *A system is a set of interrelated parts working towards a common goal.*

EXAMPLE 1

Congress is to vote on a new tax bill. Model this event as a process and describe a system that includes the process.

Solution

The vote on the tax bill can be modeled as a process with four possible outcomes: the bill may become law, the bill may become law in an amended form, the bill may not become law, or the bill may be tabled for future consideration. As always, there are other ways to model this process. One could list outcomes according to the actions taken separately by the Senate, the House of Representatives, and the President. In this model, one possible outcome is that the Senate passes the bill and the House does not, while a second is that the Senate passes the bill, the House passes the bill, and the President vetoes the bill.

The process is embedded in a system that includes the members of Congress, Congressional committees, the President, staff members, lobbyists, and both citizens and corporations affected by the bill. If the effects of the bill have an international impact, then the system could be expanded to include other countries affected by the bill.

> *A state of a system is one of a group of disjoint sets into which objects of the system can be placed.*

Sometimes objects in a system can be grouped into mutually disjoint categories or sets called the *states* of the system. A harvest can be classified as either poor, average, or good, and these classifications become the three states of harvests in that system. A patient in a hospital can be classified as either ambulatory, bedridden, discharged, or deceased, which then become the four states of patients in that system.

Many systems are evaluated periodically over equally spaced time intervals. Census counts of the U.S. population are traditionally done every ten years. Harvests are often determined yearly. The conditions of hospital patients are generally updated daily. These evaluation intervals, whether they are decades, years, or days, are called *periods*.

> *A period is a time interval between successive determinations of the status of objects in a system.*

EXAMPLE 2

Every five years, the town of Hopewell surveys its residents to determine whether people are living in single-family houses, multiple-unit dwellings, or trailers. Describe the states and the periods of this system.

Solution

Each period is five years, because a new evaluation is done every five years. There are three possible states for a resident's living accommodations: single-family house, multiple-unit dwelling, or trailer.

EXAMPLE 3

Every year, *Fortune* magazine ranks American corporations by total sales and then prints the names of the top 500. Advertising agencies then refer to a company as a *Fortune* 100 company, if the company's ranking is between 1 and 100, inclusive, a *Fortune* 200 company, if the company's ranking is between 101 and 200, inclusive, a *Fortune* 300 company, if the company's ranking is between 201 and 300, inclusive, a *Fortune* 400 company, if the company's ranking is between 301 and 400, inclusive, or a *Fortune* 500 company, if the company's ranking is between 401 and 500, inclusive. Describe the states and the periods of this system.

Solution

Each period is one year, because *Fortune* magazine prints a new ranking annually. There are six possible states for a company: *Fortune* 100, *Fortune* 200, *Fortune* 300, *Fortune* 400, *Fortune* 500, and not on the *Fortune* list.

MARKOV CHAINS

A *Markov Chain* is a set of objects, a finite set of states, and a set of consecutive time periods such that:

1. At any given time, each object is in one and only one state, although different objects can be in the same state.

2. The probability that an object moves from one state to another or remains in the same state over a single time period depends only on the beginning and ending states.

Markov chains are represented graphically by *transition diagrams* similar to the one in Figure 7.1. Each state is designated by a unique integer and represented by a node consisting of that integer enclosed in a circle. Possible *transitions* or movements between

FIGURE 7.1

states are represented by arrows, with the arrowheads indicating the direction of allowable movement. The probability associated with each transition, called a *transitional probability*, is placed on the arrow representing the transition.

In Figure 7.1, there are two states, state 1 and state 2, denoted by the two circled numbers. The arrow with .6 on it, which starts at node 1 and ends at node 2, indicates that the probability is .6 for an object to move to state 2 from state 1 over one time period. The arrow with .3 on it, which starts at node 2 and ends at node 1, indicates that the probability is .3 for an object to move to state 1 from state 2 over one time period. The arrow with .4 on it, which starts and ends at node 1, indicates that the probability is .4 for an object in state 1 to remain in state 1 over one time period. Finally, the arrow with .7 on it, which starts and ends at node 2, indicates that the probability is .7 for an object in state 2 to remain in state 2 over one time period.

EXAMPLE 4

Describe the possible transitions and their associated probabilities for the transition diagram in Figure 7.2.

FIGURE 7.2

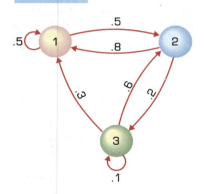

Solution

As Figure 7.2 shows, there are three states in this system, 1, 2, and 3. If an object is in state 1, then the probability that the object will move to state 2 in one time period is .5 and the probability that it will remain in state 1 over one time period is also .5. If an object is in state 2, there is a probability of .8 that it will move to state 1 in one time period and a probability of .2 that it will move to state 3 in one time period. If an object is in state 3, there is a probability of .3 that it will move to state 1 in one time period, a probability of .6 that it will move to state 2 in one time period, and a probability of .1 that it will remain in state 3 over one time period. Note that there are no arrows from node 1 to node 3 or from node 2 back to itself—thus an object cannot move from state 1 to state 3 over one time period, and an object may not remain in state 2 over one time period.

EXAMPLE 5

An airline with a 5:30 P.M. commuter flight between Los Angeles and San Francisco does not want departure delays on that flight two days in a row. If the flight leaves late one day, the airline makes a concerted effort to have the plane leave on time the next day and succeeds 95% of the time. If the flight leaves on time one day, then less effort is made the next day, and the next day's flight leaves on time 80% of the time. Model this system as a Markov chain.

Solution

The objects of interest are the 5:30 commuter flights each business day, and these objects can be in only one of two states: on time, which we denote as state 1, and late, which we denote as state 2. A time period is one business day. If a plane is late one day, then 95% of the time that flight is *not* late the next day, which means that 5% of the time the flight is late the next day. A transition from state 2 to state 1 indicates that the flight left late one day and then left on time one business day later—the probability of this occurring is .95.

A transition from state 2 to state 2 indicates that the flight left late one day and also left late one business day later—the probability of this occurring is .05. If a flight is on time one day, then 80% of the time it is not late the next day, but 20% of the time it is. A transition from state 1 to state 2 indicates that the flight left on time one day and then left late one business day later—the probability of this occurring is .2. A transition from state 1 to state 1 indicates that the flight left on time two business days in a row, and the probability of this occurring is .8. A transition diagram for this Markov chain is shown in Figure 7.3.

FIGURE 7.3

The sum of all probabilities of transitions from a given state over one time period is 1.

In the transition diagrams in Figure 7.1 through 7.3, the sum of the probabilities on the arrows leaving each node is 1. This must always be the case. Each object in the system must be in one and only one state at any point in time. One period later, each object must again be in one and only one state, although the state may have changed. Thus, an object in one state during one period *must* either remain in that state or move to another state one period later.

If a Markov chain contains N possible states, then we denote the probability of moving from state i to state j over one time period by p_{ij}. Thus, p_{12} is the probability of moving from state 1 to state 2 over one time period, while p_{52} is the probability of moving from state 5 to state 2 in one time period. For the transition diagram in Figure 7.3, we have two states, and the transition probabilities are

$$p_{11} = .8 \qquad p_{12} = .2$$
$$p_{21} = .95 \qquad p_{22} = .05$$

These probabilities are summarized neatly in the *transition matrix*

The elements of a transition matrix are transition probabilities; the rows of the matrix represent the state from which an object is coming and the columns represent the state to which an object is going.

$$\mathbf{P} = \begin{array}{c} \\ \text{from state 1} \\ \text{from state 2} \end{array} \begin{array}{cc} \overset{to}{\text{state 1}} & \overset{to}{\text{state 2}} \\ \left[\begin{array}{cc} .8 & .2 \\ .95 & .05 \end{array}\right] \end{array} \qquad (1)$$

The rows in a transition matrix indicate the state an object is *coming from* and the columns indicate the state an object is *going to*. The sum of the elements in each row must be 1.

EXAMPLE 6

Construct a transition matrix for the transition diagram shown in Figure 7.2.

Solution

$$\mathbf{P} = \begin{array}{c} \\ \text{from state 1} \\ \text{from state 2} \\ \text{from state 3} \end{array} \begin{array}{ccc} \overset{to}{\text{state 1}} & \overset{to}{\text{state 2}} & \overset{to}{\text{state 3}} \\ \left[\begin{array}{ccc} .5 & .5 & 0 \\ .8 & 0 & .2 \\ .3 & .6 & .1 \end{array}\right] \end{array}$$

There are no arrows from node 1 to node 3 or from node 2 back to itself, hence no movement in these directions is allowed over one time period and the corresponding transitional probabilities, p_{13} and p_{22}, respectively, are zero.

EXAMPLE 7

The mayor of a large city was elected on a platform of improving public housing. Consequently, the mayor wants quarterly reports from the housing authority on the condition of all residential apartments under its jurisdiction. The housing authority rates each apartment as either good, serviceable, or decayed. Over a three-month period, 2% of all good apartments deteriorate to serviceable, another $\frac{1}{2}$% deteriorate to decayed, and the others remain in good condition. Over the same time period, 4% of all serviceable apartments deteriorate to decayed while the other 96% remain serviceable. Housing authority funds are spent to upgrade decayed units, but budget constraints limit the housing authority to upgrading 1% of all decayed apartments to good condition every quarter, leaving the remaining units in decayed condition. Model this system with both a transition diagram and transition matrix.

Solution

FIGURE 7.4

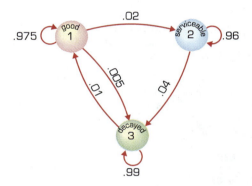

The objects of interest are the apartments under the jurisdiction of the city housing authority, and these units can be in only one of three states: good, serviceable, or decayed, which we denote as states 1, 2, and 3, respectively. A time period is one quarter of a year or three months. Figure 7.4 is a transition diagram for this Markov chain. The corresponding transition matrix is

$$
\mathbf{P} = \begin{array}{c} \\ \\ \textit{from good} \\ \textit{from serviceable} \\ \textit{from decayed} \end{array}
\begin{array}{ccc} \textit{to} & \textit{to} & \textit{to} \\ \textit{good} & \textit{serviceable} & \textit{decayed} \\ \left[\begin{array}{ccc} .975 & .02 & .005 \\ 0 & .96 & .04 \\ .01 & 0 & .99 \end{array}\right] \end{array} \tag{2}
$$

Transitional probabilities apply to movements over one time period, although often we are interested in movements over multiple time periods. In Example 7, the transitional probabilities apply to a time period of three months. The mayor, however, may also want information over one year ot even over four years, the time period between elections. In Example 7, note that one year is equivalent to four quarters or 4 time periods, while four years corresponds to 16 time periods.

$p_{ij}^{(n)}$ denotes the transitional probability of moving from state i to state j over n time periods.

We use superscripts to denote probabilities over multiple time periods. In general, $p_{ij}^{(n)}$ denotes the transitional probability of moving from state i to state j over n time periods. For example, $p_{12}^{(3)}$ is the probability of moving from state 1 to state 2 over 3 time periods, while $p_{32}^{(16)}$ is the transitional probability of moving from state 3 to state 2 over 16 time periods.

Transitional probabilities over multiple periods are obtained directly from the transitional probabilities over one period. To see how, we return to Example 5, which dealt with commuter flights between Los Angeles and San Francisco, and calculate transitional probabilities over two time periods. There are only two states to this system, state 1, representing an on-time flight, and state 2, representing a late flight. The tree diagram in Figure 7.5 is a simple graphical portrait showing the different possibilities over two time periods for a flight that begins in state 1.

FIGURE 7.5
Starting in state 1.

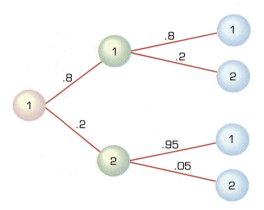

There are two paths that lead from state 1 to state 2 over *two* time periods: one path that starts in state 1 and remains in state 1 over the first time period and then moves to state 2 over the second time period, and a second path that starts in state 1 and moves to state 2 over the first time period and then remains in state 2 over the second time period. It follows from Section 6.3 that the probability associated with the first path is .8(.2), while the probability associated with the second path, the bottom path in Figure 7.5, is .2(.05). The transitional probability of starting in state 1 and ending in state 2 *two* time periods later is the sum of the probabilities of all paths that begin at state 1 and end at state 2:

$$p_{12}^{(2)} = .8(.2) + .2(.05) = .17 \tag{3}$$

To determine the transitional probability of remaining in state 1 over two time periods, we sum the probabilities of all paths in Figure 7.5 that begin in state 1 and end in state 1 two time periods later. There are two such paths: one path that starts in state 1 and remains in state 1 over each time period (this is the top path in Figure 7.5, and it has probability .8(.8)) and a second path that starts in state 1, moves to state 2 after one time period, and then returns to state 1 after the second time period (this path has probability .2(.95)). Thus,

$$p_{11}^{(2)} = .8(.8) + .2(.95) = .83 \tag{4}$$

The tree diagram shown in Figure 7.6 lists the different possibilities over two time periods for a flight that begins in state 2. There are two paths that lead to state 1 from state 2 over *two* time periods: one path that starts in state 2, moves to state 1 over the first time period, and then remains in state 1 over the second time period (this is the top path in Figure 7.6, and it has probability .95(.8)) and a second path that starts in state 2, remains

FIGURE 7.6
Starting in state 2.

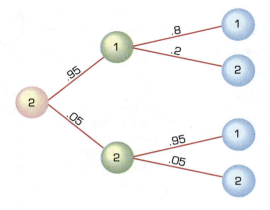

in state 2 over the first time period, and then moves to state 1 over the second time period with probability .05(.95). Thus,

$$p_{21}^{(2)} = .95(.8) + .05(.95) = .8075 \tag{5}$$

To determine the transitional probability of remaining in state 2 over two time periods, we sum the probabilities of all paths that begin in state 2 and end in state 2 two time periods later. There are two such paths: one path that starts in state 2, moves to state 1 after one time period, and then returns to state 2 after the second time period (its probability is .95(.2)) and a second path that starts in state 2 and remains in state 2 over each time period, the bottom path in Figure 7.6, which has probability .05(.05). Thus,

$$p_{22}^{(2)} = .95(.2) + .05(.05) = .1925 \tag{6}$$

The sums in Equations (3) through (6) have a familiar format! Recall how matrices are multiplied (see Section 2.4). The transitional probabilities over two time periods are just the elements of the matrix product **PP**, when

$$
\mathbf{P} = \begin{array}{c} \\ \textit{from state 1} \\ \textit{from state 2} \end{array}
\begin{array}{cc} \textit{to} & \textit{to} \\ \textit{state 1} & \textit{state 2} \end{array}
\begin{bmatrix} .8 & .2 \\ .95 & .05 \end{bmatrix}
\qquad \text{(1 repeated)}
$$

Here

$$
\begin{aligned}
\mathbf{P} &= \begin{bmatrix} .8 & .2 \\ .95 & .05 \end{bmatrix}\begin{bmatrix} .8 & .2 \\ .95 & .05 \end{bmatrix} \\
&= \begin{bmatrix} .8(.8) + .2(.95) & .8(.2) + .2(.05) \\ .95(.8) + .05(.95) & .95(.2) + .05(.05) \end{bmatrix} \tag{7}
\end{aligned}
$$

$$
= \begin{array}{c} \\ \textit{from state 1} \\ \textit{from state 2} \end{array}
\begin{array}{cc} \textit{to} & \textit{to} \\ \textit{state 1} & \textit{state 2} \end{array}
\begin{bmatrix} .83 & .17 \\ .8075 & .1925 \end{bmatrix}
$$

*If **P** is a transition matrix for a Markov chain, then for any positive integer* n, *the sum of the elements in each row of **P**^{n} is 1.*

It is common to write the product **PP** as **P**^{2}. We see from the Matrix (7) that the i-j element of **P**^{2} is $p_{ij}^{(2)}$. Furthermore,

$$p_{11}^{(2)} + p_{12}^{(2)} = .83 + .17 = 1$$

and

$$p_{21}^{(2)} + p_{22}^{(2)} = .8075 + .1925 = 1$$

That is, the sum of the elements in each row of $\mathbf{P}^2$ is 1, just as the sum of the elements in each row of $\mathbf{P}$ is 1.

The transitional probability of moving from state i to state j over n time periods is the i-j element of $\mathbf{P}^n$, the nth power of the transition matrix $\mathbf{P}$ for the Markov chain.

In general, we write the product $\mathbf{PPP} = \mathbf{PP}^2$ as $\mathbf{P}^3$, and, for any positive integer n, we set

$$\mathbf{P}^n = \underbrace{\mathbf{PP} \cdots \mathbf{P}}_{n \text{ times}} = \mathbf{PP}^{n-1} = \mathbf{P}^{n-1}\mathbf{P} \tag{8}$$

The analysis used to generate Equations (3) through (6) can be extended to three or more time periods. Doing so, we would find that

$$\mathbf{P}^n = [p_{ij}^{(n)}] \tag{9}$$

Also, for any positive integer n, the sum of the elements in each row of $\mathbf{P}^n$ is 1, reflecting the fact that the regardless of where an object starts, that object must end in one of the states of the system after n time periods.

EXAMPLE 8

Calculate $\mathbf{P}^3$ for the transition matrix

$$\mathbf{P} = \begin{bmatrix} .1 & .9 \\ .4 & .6 \end{bmatrix}$$

and describe the significance of each element in the resulting matrix.

Solution

$$\mathbf{P}^3 = \mathbf{PPP} = \begin{bmatrix} .1 & .9 \\ .4 & .6 \end{bmatrix}\begin{bmatrix} .1 & .9 \\ .4 & .6 \end{bmatrix}\begin{bmatrix} .1 & .9 \\ .4 & .6 \end{bmatrix}$$

$$= \begin{bmatrix} .37 & .63 \\ .28 & .72 \end{bmatrix}\begin{bmatrix} .1 & .9 \\ .4 & .6 \end{bmatrix}$$

$$= \begin{bmatrix} .289 & .711 \\ .316 & .684 \end{bmatrix}$$

Here $p_{11}^{(3)} = .289$ is the probability starting in state 1 and ending in state 1 after three time periods; $p_{12}^{(3)} = .711$ is the probability of moving from state 1 to state 2 after three time periods; $p_{21}^{(3)} = .316$ is the probability of moving from state 2 to state 1 after three time periods; $p_{22}^{(3)} = .684$ is the probability starting in state 2 and ending in state 2 after three time periods. Observe that the sum of the elements in each row of each matrix is 1.

EXAMPLE 9

Using the information provided in Example 7, determine the probability that a decayed apartment will be in good condition after one year.

Solution

One time period is three months, so one year is four time periods. The transitional probability we seek is embedded in $\mathbf{P}^4$, where

$$\mathbf{P} = \begin{array}{c} \textit{from good} \\ \textit{from serviceable} \\ \textit{from decayed} \end{array} \begin{array}{ccc} \textit{to} & \textit{to} & \textit{to} \\ \textit{good} & \textit{serviceable} & \textit{decayed} \\ \begin{bmatrix} .975 & .02 & .005 \\ 0 & .96 & .04 \\ .01 & 0 & .99 \end{bmatrix} \end{array} \quad \text{(2 repeated)}$$

Now

$$\mathbf{P}^2 = \mathbf{PP} = \begin{bmatrix} .975 & .02 & .005 \\ 0 & .96 & .04 \\ .01 & 0 & .99 \end{bmatrix} \begin{bmatrix} .975 & .02 & .005 \\ 0 & .96 & .04 \\ .01 & 0 & .99 \end{bmatrix}$$

$$= \begin{bmatrix} .950675 & .0387 & .010625 \\ .0004 & .9216 & .078 \\ .01965 & .0002 & .98015 \end{bmatrix}$$

and

$$\mathbf{P}^4 = \mathbf{PPPP} = \mathbf{P}^2\mathbf{P}^2$$

$$= \begin{bmatrix} .950675 & .0387 & .010625 \\ .0004 & .9216 & .078 \\ .01965 & .0002 & .98015 \end{bmatrix} \begin{bmatrix} .950675 & .0387 & .010625 \\ .0004 & .9216 & .078 \\ .01965 & .0002 & .98015 \end{bmatrix}$$

$$\approx \begin{array}{c} \textit{from good} \\ \textit{from serviceable} \\ \textit{from decayed} \end{array} \begin{array}{ccc} \textit{to} & \textit{to} & \textit{to} \\ \textit{good} & \textit{serviceable} & \textit{decayed} \\ \begin{bmatrix} .9040 & .0725 & .0235 \\ .0023 & .8494 & .1483 \\ .0379 & .0011 & .9609 \end{bmatrix} \end{array}$$

where the elements in this last matrix are rounded to four decimal places for presentation purposes. Roundoff error is responsible for the sum of the third row not being 1. The probability of moving from decayed (state 3) to good (state 1) over four time periods is $p_{31}^{(4)} = .0379$.

IMPROVING SKILLS

1. Explain why each of the following matrices *cannot* be a transition matrix:

   (a) $\begin{bmatrix} .1 & .7 \\ .9 & .1 \end{bmatrix}$ (b) $\begin{bmatrix} 1 & 0 \\ 2 & 1 \end{bmatrix}$ (c) $\begin{bmatrix} 1 & 0 \\ 2 & -1 \end{bmatrix}$

   (d) $\begin{bmatrix} 1 & 0 \\ 0 & 2 \end{bmatrix}$ (e) $\begin{bmatrix} .1 & .3 & .6 \\ .7 & .5 & -.2 \\ .1 & .4 & .5 \end{bmatrix}$ (f) $\begin{bmatrix} .1 & .5 & .4 \\ .2 & .5 & .2 \\ .7 & .1 & .3 \end{bmatrix}$

2. Using the transition matrix

$$\mathbf{P} = \begin{bmatrix} .1 & .9 \\ .7 & .3 \end{bmatrix}$$

determine

(*a*) the number of states,

(*b*) the probability of starting in state 1 and ending in state 2 after one time period,

(*c*) the probability of starting in state 1 and ending in state 1 after one time period,

(*d*) the probability of starting in state 2 and ending in state 1 after one time period.

3. Using the transition matrix

$$\mathbf{P} = \begin{bmatrix} 0 & 1 \\ .4 & .6 \end{bmatrix}$$

determine

(*a*) the number of states,

(*b*) the probability of starting in state 1 and ending in state 2 after one time period,

(*c*) the probability of starting in state 2 and ending in state 2 after one time period,

(*d*) the probability of starting in state 2 and ending in state 1 after one time period.

4. Using the transition matrix

$$\mathbf{P} = \begin{bmatrix} .1 & .7 & .2 \\ .2 & .8 & 0 \\ 0 & .6 & .4 \end{bmatrix}$$

determine

(*a*) the number of states,

(*b*) the probability of starting in state 1 and ending in state 3 after one time period,

(*c*) the probability of starting in state 1 and ending in state 2 after one time period,

(*d*) the probability of starting in state 3 and ending in state 3 after one time period,

(*e*) the probability of starting in state 2 and ending in state 3 after one time period.

5. Using the transition matrix

$$\mathbf{P} = \begin{bmatrix} .1 & .5 & .2 & .2 \\ .2 & .2 & 0 & .6 \\ .4 & .4 & .1 & .1 \\ 0 & 0 & 0 & 1 \end{bmatrix}$$

determine

(*a*) the number of states,

(*b*) the probability of starting in state 4 and ending in state 1 after one time period,

(*c*) the probability of starting in state 1 and ending in state 3 after one time period,

(*d*) the probability of starting in state 2 and ending in state 3 after one time period,

(*e*) the probability of starting in state 3 and ending in state 3 after one time period.

6. Calculate $\mathbf{P}^2$ for the transition matrix defined in Problem 2, and then determine

(*a*) the probability of starting in state 1 and ending in state 1 after two time periods.

(*b*) the probability of starting in state 1 and ending in state 2 after two time periods.

(*c*) the probability of starting in state 2 and ending in state 2 after two time periods,

(*d*) the probability of starting in state 2 and ending in state 1 after two time periods.

7. Solve Problem 6 for the transition matrix defined in Problem 3.

8. Calculate $\mathbf{P}^3$ for the transition matrix defined in Problem 2, and then determine

(*a*) the probability of starting in state 1 and ending in state 1 after three time periods,

(*b*) the probability of starting in state 1 and ending in state 2 after three time periods,

(*c*) the probability of starting in state 2 and ending in state 2 after three time periods,

(*d*) the probability of starting in state 2 and ending in state 1 after three time periods.

9. Calculate P^2 for the transition matrix defined in Problem 4, and then determine
   (a) the probability of starting in state 1 and ending in state 3 after two time periods,
   (b) the probability of starting in state 1 and ending in state 2 after two time periods,
   (c) the probability of starting in state 3 and ending in state 3 after two time periods,
   (d) the probability of starting in state 2 and ending in state 3 after two time periods.
10. Calculate P^4 for the transition matrix defined in Problem 4, and then determine
    (a) the probability of starting in state 1 and ending in state 3 after four time periods,
    (b) the probability of starting in state 1 and ending in state 2 after four time periods,
    (c) the probability of starting in state 3 and ending in state 1 after four time periods,
    (d) the probability of starting in state 2 and ending in state 3 after four time periods.
11. Calculate P^2 for the transition matrix defined in Problem 5, and then determine
    (a) the probability of starting in state 4 and ending in state 1 after two time periods,
    (b) the probability of starting in state 1 and ending in state 3 after two time periods,
    (c) the probability of starting in state 2 and ending in state 2 after two time periods,
    (d) the probability of starting in state 4 and ending in state 3 after two time periods.
12. Construct a transition matrix for Figure A.

FIGURE A

13. Determine the probability of starting in state 1 and ending in state 1 after 2 time periods for the Markov chain represented by Figure A.
14. Construct a transition matrix for Figure B.

FIGURE B

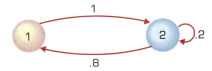

15. Determine the probability of starting in state 1 and ending in state 2 after three time periods for the Markov chain represented by Figure B.
16. Construct a transition matrix for Figure C.

FIGURE C

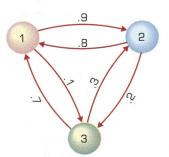

17. Determine the probability of starting in state 3 and ending in state 1 after two time periods for the Markov chain represented by Figure C.

18. Construct a transition matrix for Figure D.

FIGURE D

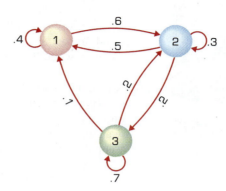

19. Determine the probability of starting in state 2 and ending in state 1 after two time periods for the Markov chain represented by Figure D.

20. Construct a transition matrix for Figure E.

FIGURE E

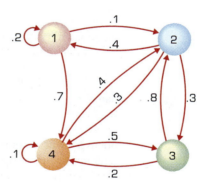

21. Construct a transition diagram from the transition matrix given in Problem 2.

22. Construct a transition diagram from the transition matrix given in Problem 3.

23. Construct a transition diagram from the transition matrix given in Problem 4.

24. Construct a transition diagram from the transition matrix given in Problem 5.

CREATING MODELS

In Problems 25 through 34, construct a transition matrix for each of the given systems so that they can be modeled as Markov chains under the assumption that existing trends will continue without change into the future.

25. A town mayor is elected every four years. Voting records show that if a Democrat is mayor one year, then the probabilities are .45 that a Democrat will be mayor and .55

that a Republican will be mayor four years later. If a Republican is mayor one year, then the probabilities are .4 that a Democrat will be mayor and .6 that a Republican will be mayor four years later.

26. The ABC Company, a maker of a well-known laundry bleach, commissioned a survey and found that 85% of its customers remained loyal to the ABC brand over the past year, while the other 15% switched to a competitor's brand over the same time period. In contrast, 22% of all competitors' customers switched to ABC's bleach over one year, while the remaining 78% of other customers did not switch to the ABC bleach.

27. Farmers in a particular region of the country describe their summers as either dry or wet. Historical data indicates that there is a 20% chance that a dry summer will be followed by another dry summer and a 35% chance that a wet summer will be followed by another wet summer.

28. A population can be divided into people who are overweight and those who are not. Over a five-year period, the probability is .95 that an overweight person will remain overweight and the probability is .22 that a person who was not overweight will become overweight.

29. A state lottery commission separates adult residents into two categories, those who play the lottery regularly and those who do not. Over a two-year period, the probability is .98 that a regular player will continue to play regularly; the probability is .04 that over the same time period a person who does not play regularly will become a regular player.

30. A town zones its land as either residential, commercial, or public. Over the last 10 years, the probability that a parcel of residential land remained residential is .84, while the probability of that same parcel becoming commercial is .16. The probability that a parcel of commercial land remained commercial is .7, the probability that it became residential is .25, and the probability that it became public land is .05. The probability that a parcel of public land remained public is .9, while the probability that the same parcel became commercial or residential is .05 for each.

31. Residents in a major urban area are classified as either regular users, occasional users, or nonusers of a new rapid transit system. It is estimated that each month, 10% of the nonusers will become regular users and 15% will become occasional users. The remaining nonusers will remain nonusers. It is also estimated that 85% of all regular users will remain regular users each month, 5% will become nonusers, and the other 10% will become occasional users. All occasional users remain occasional users from month to month.

32. An insurance company classifies customers as high risk, medium risk, and low risk. Each year, 10% of the high-risk customers are reclassified as low risk, with the remaining 90% continuing as high risk. Each year, 40% of all medium-risk customers are reclassified as low risk, another 35% are classified as high risk, and the remaining 25% continue as medium risk. Each year, 15% of all low-risk customers are reclassified as high risk, 40% are reclassified as medium risk, and the rest remain low risk.

33. A publishing firm has four new books that it will promote through radio advertising. Each day, one book will be selected by the editors for promotion. Book A is thought to be the most likely to succeed, so it will be promoted the most, although no book will be promoted two days in a row. If book A is not promoted one day, then it is twice as likely to be promoted the next day as either of the other two books not promoted that day. If book A is promoted one day, then there is an equal chance that either of the other three books will be promoted the next day.

34. Figure F shows a maze with four compartments. The time it takes the mouse to move from one compartment to another is considered one time period. Water is placed in compartment 3, so once the mouse is in that compartment, the mouse stays there. If

FIGURE F

the mouse is in compartment 2 or 4, the mouse always moves next to compartment 3. If the mouse is in compartment 1, the mouse moves with equal probability to either compartment 2 or 4.

EXPLORING IN TEAMS

35. If a matrix **A** has an inverse, then we define

$$\mathbf{A}^{-n} = (\mathbf{A}^{-1})^n$$

for any positive integer n. We use the right side of this equation, which is well defined, to determine the left side. Find $\mathbf{A}^{-2}$ for

$$\mathbf{A} = \begin{bmatrix} 1 & 2 \\ 3 & 4 \end{bmatrix}$$

36. Use the definition in Problem 35 to calculate $\mathbf{A}^{-3}$ for the matrix in that problem.

37. Use the definition in Problem 35 to calculate $\mathbf{B}^{-2}$ for

$$\mathbf{B} = \begin{bmatrix} 5 & 3 \\ 3 & 2 \end{bmatrix}$$

38. Use the results of Problem 35 to calculate $\mathbf{B}^{-4}$ for the matrix in Problem 37.

39. Use the definition in Problem 35 to calculate $\mathbf{P}^{-2}$ for

$$\mathbf{P} = \begin{bmatrix} .1 & 0 & 0 \\ .7 & .6 & 0 \\ .2 & .4 & 1 \end{bmatrix}$$

EXPLORING WITH TECHNOLOGY

Use a graphing calculator or electronic spreadsheet to calculate

$$\mathbf{P}^2, \mathbf{P}^4, \mathbf{P}^8, \mathbf{P}^{16}, \mathbf{P}^{32}, \mathbf{P}^{64}, \mathbf{P}^{128}, \mathbf{P}^{256}, \mathbf{P}^{512}, \text{ and } \mathbf{P}^{1024}$$

for the matrices given in Problems 40 through 43. What observations can you make about the elements in these matrices?

40. $\mathbf{P} = \begin{bmatrix} .8 & .2 \\ .95 & .05 \end{bmatrix}$ (*1 repeated*)

41. $\mathbf{P} = \begin{bmatrix} .7 & .3 \\ .4 & .6 \end{bmatrix}$

42. $\mathbf{P} = \begin{bmatrix} .5 & .5 & 0 \\ .8 & 0 & .2 \\ .3 & .6 & .1 \end{bmatrix}$ (*from Example 6*)

43. $\mathbf{P} = \begin{bmatrix} .975 & .02 & .005 \\ 0 & .96 & .04 \\ .01 & 0 & .99 \end{bmatrix}$ (*2 repeated*)

REVIEWING MATERIAL

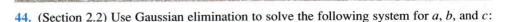

44. (Section 2.2) Use Gaussian elimination to solve the following system for a, b, and c:

$$
\begin{aligned}
a + \quad b + \quad c &= 1 \\
.025a \quad\quad\quad - .01c &= 0 \\
-.02a + .04b \quad\quad &= 0 \\
-.005a - .04b + .01c &= 0
\end{aligned}
$$

45. (Section 4.1) Determine the effective interest of an account paying 10.3% annual interest compounded daily.

46. (Section 5.2) The Treasury Department reports that the federal government spent just under $1.4 trillion in fiscal 1993. Of this total, the Department of Defense spent $308 billion, Social Security payments accounted for $298 billion, interest on public debt cost $293 billion, social services payments (other than social security) totaled $283 billion, the Department of Agriculture spent $63 billion, and all other sources accounted for $150 billion. Construct a pie chart of these expenditures.

47. (Section 6.1) Alia, who lives in Tampa, and Kareem, who lives in Cleveland, each call their mother in Houston once a week. What is the probability that the two children call on the same day if there is no pattern to when either child calls?

RECOMMENDING ACTION

48. Respond by memo to the following request:

MEMORANDUM

To: J. Doe Reader

From: Graphics Department

Date: Today

Subject: **Modeling Population Growth**

I am creating the graphics for the President's talk next month on population growth. I though I would model the population as a Markov chain with *Children, Mature Adults,* and *Senior Citizens* as three of the major states. Clearly, *Deceased* must also be a state, because there are transitions from each of the major states into *Deceased,* but what about newborns? Is *Newborns* also a state? If so, what are the transitions into it and from what state do they come? After all, if there are no transitions into *New Born,* only transitions out of it, then we must deplete the state eventually, which seems absurd. Do you see my problem?

7.2 DISTRIBUTION VECTORS

The components of a distribution vector are the probabilities of a randomly chosen object being in each state.

The transition matrix for a two-state Markov chain contains four probabilities for transitions over a single time period: the probability of moving from state 1 to state 2, the probability of moving from state 2 to state 1, the probability of remaining in state 1, and the probability of remaining in state 2. A transition matrix does not, however, stipulate the likelihood of an object actually being in a particular state. It is one thing to know the probability of moving from state 1 to state 2 is .9; it is quite another thing to know that no object is actually in state 1.

A *distribution vector* for a Markov chain is a row vector (see Section 2.4) that lists the probabilities of a randomly chosen object being in each state. The first component is the probability of being in state 1, the second component is the probability of being in state 2, and so on. A distribution vector has one component for each state in the Markov chain. A distribution vector **d** for an N-state Markov chain has the form

$$\mathbf{d} = [p_1 \quad p_2 \quad p_3 \ldots p_N] \tag{10}$$

where p_1 is the probability of being in state 1, p_2 is the probability of being in state 2, p_3 is the probability of being in state 3, and so on.

EXAMPLE 1

The housing authority of a large city rates residential apartments under its jurisdiction as either good, serviceable, or decayed. Currently, 20% of all apartments are rated good, 47% are rated serviceable, and the rest are decayed. Construct a distribution vector for this system.

Solution

The objects in this system are apartments under the jurisdiction of the city housing authority, and each of these units can be in only one of three states: good, serviceable, or decayed, which we denote as states 1, 2, and 3, respectively. Currently, 20 of every 100 apartments are in good condition and 47 out of every 100 apartments are in serviceable condition, leaving 33 out of every 100 apartments in decayed condition. The probability of a randomly selected apartment being in good condition (state 1) is $p_1 = \frac{20}{100} = .20$; the probability of a randomly selected apartment being in serviceable condition (state 2) is $p_2 = \frac{33}{100} = .47$; and the probability of a randomly selected apartment being decayed (state 3) is $p_3 = \frac{13}{300} = .33$. The distribution vector for this system is

$$\mathbf{d} = [p_1 \quad p_2 \quad p_3] = [.20 \quad .47 \quad .33]$$

EXAMPLE 2

An airline with a reputation for undependable service hires a new management team to repair its image. The new team decides first to improve the on-time performance of the company's premier route, the 5:30 P.M. commuter flight between Los Angeles and San Francisco. Currently, 60% of those flights leave late; the other 40% of the time the flights leave on time. Under new management, if the flight leaves late one day, the airline makes a concerted effort to have the plane leave on time the next day and succeeds 95% of the time. If the flight leaves on time one day, then less effort is made the next day, and the next day's flight leaves on time 80% of the time. Determine a distribution vector for this flight as the new management team takes control.

Solution

The objects of interest are the 5:30 commuter flights each business day, and these objects can be in only one of two states: on time, which we denote as state 1, or late, which we denote as state 2. As the new management team takes control, this flight has been leaving late 60% of the time. That is, 60 out of every 100 flights leave late, and 40 out of every 100 flights leave on time. The probability of the 5:30 flight leaving on time (state 1) on a randomly chosen business day is $p_1 = \frac{40}{100} = .4$; the probability of the 5:30 flight leaving late (state 2) on a randomly chosen business day is $p_1 = \frac{60}{100} = .6$. The distribution vector for this Markov chain under old management (or just as new management assumes control) is

$$\mathbf{d} = [p_1 \quad p_2] = [.4 \quad .6] \tag{11}$$

Let us consider what happens after the new management team in Example 2 has been in charge for a full business day, one period. When this management team starts, the most recent 5:30 flight will be the one the team inherited from the previous administration. From the initial distribution vector (*11*), the probability is .4 that that flight was in state 1 (on time) and .6 that it was in state 2 (late). Over the next business day, steps are taken to improve on-time performance, and the probabilities of moving from one state to another state over one period are specified by the transition matrix for this Markov chain. In Section 7.1, we found this transition matrix to be

$$\mathbf{P} = \begin{array}{c} \\ \textit{from state 1} \\ \textit{from state 2} \end{array} \begin{array}{cc} \textit{to} & \textit{to} \\ \textit{state 1} & \textit{state 2} \\ \begin{bmatrix} .8 & .2 \\ .95 & .05 \end{bmatrix} \end{array} \qquad \text{(1 repeated)}$$

The tree diagram in Figure 7.7 shows all the possible states after one time period.

FIGURE 7.7

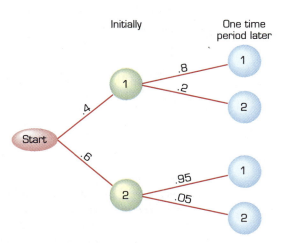

To calculate the probability of being in state 1 after one time period, we determine the probability of each path in Figure 7.7 that terminates in state 1 and we sum the results (see Section 6.3). Thus,

$$P\{\text{state 1 after one period}\} = .4(.8) + .6(.95) = .89 \tag{12}$$

Similarly, the probability of being in state 2 after one time period is

$$P\{\text{state 2 after one period}\} = .4(.2) + .6(.05) = .11 \tag{13}$$

Note the marked improvement in performance after just one day under new management! The probability of having an on-time flight has improved from .4 to .89. The two probabilities from Equations (12) and (13) are precisely the matrix product **dP**, with **P** defined by Matrix (*1*) and **d** defined by Vector (*11*).

$$\mathbf{dP} = [.4 \quad .6]\begin{bmatrix} .8 & .2 \\ .95 & .05 \end{bmatrix}$$

$$= [.4(.8) + .6(.95) \quad .4(.2) + .6(.05)]$$

$$= [.89 \quad .11]$$

If we let $\mathbf{d}_1$ be the distribution vector after one time period and $\mathbf{d}_0$ be the distribution vector at the beginning of the process, often called the *initial distribution vector*, then we can write

$$\mathbf{d}_1 = \mathbf{d}_0\mathbf{P} \tag{14}$$

Equation (*14*) is noteworthy. It implies that the distribution vector for a Markov chain after one period can be obtained by multiplying the initial distribution vector by the transition matrix for the Markov chain. This is, in fact, always the case, and the relationship can be extended to distribution vectors over two or more consecutive time periods.

Let us determine the probability of the flight leaving on-time after the new management team in Example 2 has been in place for two business days. We know from Equation (*14*) that the probability of being in state 1 after one day is .89 and the probability of being in state 2 is .11. The transitional probabilities of moving from one state to another state over *one business day*, in this case from day one to day two, are specified in the transition matrix (*1*). Figure 7.8 is a tree diagram for the process from day 1 to day 2.

> *The distribution vector for a Markov chain after one time period is the product of the initial distribution vector and the transition matrix for the process.*

FIGURE 7.8

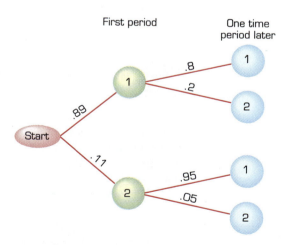

The probability of being in state 1 after day 2, or two time periods, is

$$P\{\text{state 1 after two periods}\} = .89(.8) + .11(.95) = .8165 \tag{15}$$

Similarly, the probability of being in state 2 after day 2 is

$$P\{\text{state 2 after two periods}\} = .89(.2) + .11(.05) = .1835 \tag{16}$$

> *The distribution vector of a Markov chain at any time is the product of the distribution vector one time period earlier and the transition matrix for the process.*

These two probabilities are the components of the matrix product $\mathbf{d}_1\mathbf{P}$:

$$[.89 \quad .11]\begin{bmatrix} .8 & .2 \\ .95 & .05 \end{bmatrix} = [.8165 \quad .1835]$$

If we let $\mathbf{d}_2$ designate the distribution vector after two time periods, then we can write

$$\mathbf{d}_2 = \mathbf{d}_1\mathbf{P} \tag{17}$$

Equations (14) and (17) begin to form a pattern, namely

$$\mathbf{d}_j = \mathbf{d}_{j-1}\mathbf{P} \tag{18}$$

for any positive integer j. When $j = 1$, Equation (18) reduces to Equation (14); when $j = 2$, Equation (18) reduces to Equation (17); and when $j = 3$, Equation (18) becomes $\mathbf{d}_3 = \mathbf{d}_2\mathbf{P}$. The distribution vector of a Markov chain at any time is the product of the distribution vector one time period earlier and the transition matrix for the process.

EXAMPLE 3

Find the probability of an on-time departure for the 5:30 commuter flight described in Example 2 for each day of the new management team's first week.

Solution
We have from Equations (12), (13), (15), and (16) that

$$\mathbf{d}_1 = [.89 \quad .11] \quad \text{and} \quad \mathbf{d}_2 = [.8165 \quad .1835]$$

It now follows from Equation (18), first with $j = 3$, then with $j = 4$, and finally with $j = 5$ that

$$\mathbf{d}_3 = \mathbf{d}_2\mathbf{P} = [.8165 \quad .1835]\begin{bmatrix} .8 & .2 \\ .95 & .05 \end{bmatrix} \approx [.8275 \quad .1725]$$

$$\mathbf{d}_4 = \mathbf{d}_3\mathbf{P} \approx [.8275 \quad .1725]\begin{bmatrix} .8 & .2 \\ .95 & .05 \end{bmatrix} \approx [.8259 \quad .1741]$$

and

$$\mathbf{d}_5 = \mathbf{d}_4\mathbf{P} \approx [.8259 \quad .1741]\begin{bmatrix} .8 & .2 \\ .95 & .05 \end{bmatrix} \approx [.8261 \quad .1739]$$

Here all computations are carried through fifteen significant digits and rounded to four decimal places for presentation purposes. The first component of each distribution vector corresponds to state 1 (on time), so we have the following probabilities of an on-time flight for the five business days in the first week: .89 for Monday, .8165 for Tuesday, .8275 for Wednesday, .8259 for Thursday, and .8261 for Friday.

EXAMPLE 4

The mayor of a large city was elected on a platform of improving public housing. The housing authority rates public residential apartments as either good, serviceable, or decayed, and when the mayor took office 20% of all apartments were rated good, 47% were rated serviceable, and the rest were decayed. Upon taking office, the mayor directed that the housing authority allocate all refurbishing funds to the upgrade of decayed units. The mayor's policy is to upgrade 1% of all decayed apartments to good condition every three months, leaving the remaining units in decayed condition. As a result, 2% of all good apartments deteriorate to serviceable each quarter, another $\frac{1}{2}$% deteriorate to decayed, and the others remain in good condition. Also, 4% of all serviceable apartments deteriorate to decayed each quarter, while the other 96% remain serviceable. Find the distribution of apartments one year into the mayor's term.

Solution

One time period is one quarter, or three months. One year is equivalent to four quarters, so we seek $\mathbf{d}_4$, the distribution vector after four time periods. The initial distribution vector as the mayor takes office is

$$\mathbf{d}_0 = [.20 \quad .47 \quad .33]$$

The transition matrix for this Markov chain was found in Example 7 of Section 7.1 to be

$$
\mathbf{P} = \begin{array}{c} \\ \textit{from good} \\ \textit{from serviceable} \\ \textit{from decayed} \end{array}
\begin{array}{ccc} \textit{to} & \textit{to} & \textit{to} \\ \textit{good} & \textit{serviceable} & \textit{decayed} \\ \left[\begin{array}{ccc} .975 & .02 & .005 \\ 0 & .96 & .04 \\ .01 & 0 & .99 \end{array}\right] \end{array}
$$

It now follows from repeated use of (18) that

$$\mathbf{d}_1 = \mathbf{d}_0\mathbf{P} = [.20 \quad .47 \quad .33]\begin{bmatrix} .975 & .02 & .005 \\ 0 & .96 & .04 \\ .01 & 0 & .99 \end{bmatrix}$$

$$= [.1983 \quad .4552 \quad .3465]$$

$$\mathbf{d}_2 = \mathbf{d}_1\mathbf{P} = [.1983 \quad .4552 \quad .3465]\begin{bmatrix} .975 & .02 & .005 \\ 0 & .96 & .04 \\ .01 & 0 & .99 \end{bmatrix}$$

$$\approx [.1968 \quad .4410 \quad .3622]$$

$$\mathbf{d}_3 = \mathbf{d}_2\mathbf{P} \approx [.1968 \quad .4410 \quad .3622]\begin{bmatrix} .975 & .02 & .005 \\ 0 & .96 & .04 \\ .01 & 0 & .99 \end{bmatrix}$$

$$\approx [.1955 \quad .4273 \quad .3722]$$

$$\mathbf{d}_4 = \mathbf{d}_3\mathbf{P} \approx [.1955 \quad .4273 \quad .3772]\begin{bmatrix} .975 & .02 & .005 \\ 0 & .96 & .04 \\ .01 & 0 & .99 \end{bmatrix}$$

$$\approx [.1944 \quad .4141 \quad .3915]$$

Here all computations are carried through fifteen significant digits and rounded to four decimal places for presentation purposes. Reading the components of $\mathbf{d}_4$, we conclude that after four quarters, approximately 19% of all apartments will be in good condition, 41% will be in serviceable condition, and 39% will be in decayed condition. This is not good news for the mayor. The overall situation has deteriorated after one year.

Substituting successive integer values of j into the formula

$$\mathbf{d}_j = \mathbf{d}_{j-1}\mathbf{P}$$

(18 repeated)

we have

$$\mathbf{d}_1 = \mathbf{d}_0\mathbf{P} \tag{19}$$

$$\mathbf{d}_2 = \mathbf{d}_1\mathbf{P} \tag{20}$$

$$\mathbf{d}_3 = \mathbf{d}_2\mathbf{P} \tag{21}$$

$$\mathbf{d}_4 = \mathbf{d}_3\mathbf{P} \tag{22}$$

and so on. Substituting Equation (*19*) into Equation (*20*), we obtain

$$\mathbf{d}_2 = (\mathbf{d}_0\mathbf{P})\mathbf{P} = \mathbf{d}_0\mathbf{P}^2 \tag{23}$$

Substituting Equation (*23*) into Equation (*21*), we obtain

$$\mathbf{d}_3 = (\mathbf{d}_0\mathbf{P}^2)\mathbf{P} = \mathbf{d}_0\mathbf{P}^3 \tag{24}$$

Substituting Equation (*24*) into Equation (*22*), we obtain

$$\mathbf{d}_4 = (\mathbf{d}_0\mathbf{P}^3)\mathbf{P} = \mathbf{d}_0\mathbf{P}^4 \tag{25}$$

Equations (*19*) and (*23*) through (*25*) form a pattern, namely

$$\mathbf{d}_j = \mathbf{d}_0\mathbf{P}^j \tag{26}$$

The distribution vector $\mathbf{d}_j$ for the states in a Markov chain after j periods is related to the initial distribution vector $\mathbf{d}_0$ by the formula $\mathbf{d}_j = \mathbf{d}_0\mathbf{P}^j$.

for any positive integer value of *j*. That is, the distribution vector after *j* periods in a Markov chain is related to the initial distribution vector by Equation (*26*). Thus, we can use either

$$\mathbf{d}_j = \mathbf{d}_{j-1}\mathbf{P} \tag{18 repeated}$$

or Equation (*26*) to calculate distribution vectors *j* periods into the future.

Table 7.1 lists the distribution vectors for each of the first ten time periods for the Markov chain described in Example 2, which dealt with commuter flights between Los Angeles and San Francisco. The first five lines come directly from the distribution vectors found in Example 3; the last five lines are obtained in a similar fashion. All calculations are carried through 15 significant figures and rounded to 6 decimal places for presentation purposes. Observe how the distribution vectors stabilize over time. After day 5, the first three decimal places of the probability of an on-time performance remain .826. After day 6, the first five decimal places have stabilized to .82608. After day 8, the probability of an on-time department has stabilized to .826087, and the probability of departing late has stabilized to .173913.

We compiled Table 7.1 under the assumption that we did not know the precise state of

TABLE 7.1

DAY	ON TIME	LATE
initially	.4	.6
after day 1	.89	.11
after day 2	.8165	.1835
after day 3	.827525	.172475
after day 4	.825871	.174129
after day 5	.826119	.173881
after day 6	.826082	.173918
after day 7	.826088	.173912
after day 8	.826087	.173913
after day 9	.826087	.173913
after day 10	.826087	.173913

TABLE 7.2

DAY	ON TIME	LATE
initially	1	0
after day 1	.8	.2
after day 2	.83	.17
after day 3	.8255	.1745
after day 4	.826175	.173825
after day 5	.826074	.173926
after day 6	.826089	.173911
after day 7	.826087	.173913
after day 8	.826087	.173913
after day 9	.826087	.173913
after day 10	.826087	.173913

the system when the new management team took control. Under the old management team, the 5:30 commuter flight left on time with probability .4 and left late with probability .6, so we took as our distribution vector

$$\mathbf{d} = [p_1 \quad p_2] = [.4 \quad .6] \qquad \text{(11 repeated)}$$

However, once the new management team takes control, it knows what happened the day before. If the 5:30 commuter flight left on-time on the last day under the old management, then the initial distribution vector for the new management team would be

$$\mathbf{d}_0 = [1 \quad 0]$$

and with this initial distribution vector, Table 7.1 would become Table 7.2.

On the other hand, if the 5:30 commuter flight left late on the last day under old management, then the initial distribution vector for the new management team would be

$$\mathbf{d}_0 = [0 \quad 1]$$

With this initial distribution vector, Table 7.1 would become Table 7.3.

Note that in each case, the distribution vectors in Table 7.1, 7.2, and 7.3 stabilize to the same row vector, [.826087   .173913].

TABLE 7.3

DAY	ON TIME	LATE
initially	0	1
after day 1	.95	.05
after day 2	.8075	.1925
after day 3	.828875	.171125
after day 4	.825669	.174331
after day 5	.826150	.173850
after day 6	.826078	.173922
after day 7	.826088	.173912
after day 8	.826087	.173913
after day 9	.826087	.173913
after day 10	.826087	.173913

EXAMPLE 5

Determine whether the distribution vectors in Example 4 stabilize over time.

Solution

We found the first four distribution vectors in Example 4. Continuing into the future with the aid of a computer, we generate Table 7.4. Table 7.4 lists selected distribution vectors over a 68-year time frame (recall that four time periods make one year). The distribution vectors do stabilize, but slowly. If the mayor's directives remain city policy for the next 68 years, then eventually 25% of all city apartments will be in good condition, 12.5% will be in serviceable condition, and 62.5% will be decayed.

TABLE 7.4

TIME	GOOD	SERVICEABLE	DECAYED
initially	.20	.47	.33
after period 4	.1944	.4141	.3915
after period 8	.1915	.3662	.4422
after period 16	.1915	.2908	.5176
after period 32	.2029	.1985	.5985
after period 64	.2295	.1338	.6367
after period 128	.2482	.1233	.6285
after period 192	.2500	.1248	.6253
after period 224	.2500	.1249	.6251
after period 256	.2500	.1250	.6250
after period 272	.2500	.1250	.6250

If we replicate the computations in Table 7.4 with any other initial distribution vector, we would find that the result is always the same. That is, regardless of the state of the system initially, over time it will stabilize with 25% of all city apartments in good condition, 12.5% in serviceable condition, and 62.5% in decayed condition. This will occur whether we begin with all brand new apartments in good condition, or all apartments in decayed condition, or any other initial distribution.

The Markov chains for Example 2 (the 5:30 commuter flight) and Example 4 (city apartment conditions) share the property that their distribution vectors stabilize over time and that the long range trends do not depend on the initial distribution vector of the system. Unfortunately, not *all* Markov chains share this property. Among those that do are Markov chains defined by *regular* transition matrices. A transition matrix is *regular* if some power of that matrix contains all positive elements.

A transition matrix is regular if some power of that matrix contains all positive elements.

EXAMPLE 6

Determine whether the transition matrix

$$\mathbf{P} = \begin{bmatrix} .8 & .2 \\ .95 & .05 \end{bmatrix}$$

is regular.

Solution

This is the transition matrix from Example 2. The matrix has only positive elements, so its *first* power is a matrix containing all positive elements, and **P** is regular.

EXAMPLE 7

Determine whether the transition matrix

$$
\mathbf{P} = \begin{array}{c} \\ \text{from good} \\ \text{from serviceable} \\ \text{from decayed} \end{array}
\begin{array}{ccc} \overset{to}{good} & \overset{to}{serviceable} & \overset{to}{decayed} \end{array}
\begin{bmatrix} .975 & .02 & .005 \\ 0 & .96 & .04 \\ .01 & 0 & .99 \end{bmatrix} \quad \text{(2 repeated)}
$$

is regular.

Solution

This is the transition matrix from Example 4. The matrix has two elements that are *not* positive (two of its elements are zero). Nonetheless, the second power of this matrix,

$$
\mathbf{P}^2 = \begin{bmatrix} .975 & .02 & .005 \\ 0 & .96 & .04 \\ .01 & 0 & .99 \end{bmatrix} \begin{bmatrix} .975 & .02 & .005 \\ 0 & .96 & .04 \\ .01 & 0 & .99 \end{bmatrix}
$$

$$
= \begin{bmatrix} .950675 & .0387 & .010625 \\ .0004 & .9216 & .078 \\ .01965 & .0002 & .98015 \end{bmatrix}
$$

contains only positive elements, so **P** is regular.

We say that a Markov chain has a *stable distribution vector*, denoted by **s**, if successive distribution vectors $\mathbf{d}_j = \mathbf{d}_0 \mathbf{P}^j$ get closer and closer to **s** as the number of periods j gets larger and larger, for *every* choice of $\mathbf{d}_0$. That is, for each choice of $\mathbf{d}_0$, there is some value of j (which may be large), say $j = n$, such that $\mathbf{d}_n$ and all successive distribution vectors equal **s** to whatever degree of accuracy we are carrying.

*A vector **s** is a stable distribution vector for a Markov chain if for each choice of $\mathbf{d}_0$, the initial distribution vector, there is some value of j, say j = n, such that $\mathbf{d}_n$ and all successive distribution vectors equal **s** to within a specified degree of accuracy.*

The stable distribution vector for the Markov chain associated with Example 2 (the 5:30 commuter flight) is $\mathbf{s} = [.826087 \quad .173913]$, rounded to six decimal places. In Table 7.1, where we started with the initial distribution vector $\mathbf{d}_0 = [.4 \quad .6]$, we found that

$$[.826087 \quad .173913] = \mathbf{d}_8 = \mathbf{d}_9 = \mathbf{d}_{10} = \dots$$

rounded to six decimal places. In Table 7.2, where we started with the initial distribution vectors [1 0], we found that

$$[.826087 \quad .173913] = \mathbf{d}_7 = \mathbf{d}_8 = \mathbf{d}_9 = \mathbf{d}_{10} = \dots$$

rounded to six decimal places.

The calculations required to complete Tables 7.1 and 7.2 are lengthy and tiresome. It would be nice to have a simpler procedure for finding stable distribution vectors, and indeed one exists for Markov chains with regular transition matrices. The following development of a simple formula for locating stable distribution vectors illustrates the usefulness of mathematical models and the power of mathematics to provide simplifications that save time and effort.

If **s** is a stable distribution vector for a Markov chain with transition matrix **P**, then for some (perhaps very large) integer n and some specific number of decimals,

$$\mathbf{d}_n = \mathbf{d}_{n+1} = \mathbf{d}_{n+2} = \ldots = \mathbf{s} \qquad (27)$$

It follows from

$$\mathbf{d}_j = \mathbf{d}_0 \mathbf{P}^j \qquad \text{(26 repeated)}$$

(with $j = n$) that

$$\mathbf{d}_n = \mathbf{d}_0 \mathbf{P}^n \qquad (28)$$

and (with $j = n + 1$) that

$$\mathbf{d}_{n+1} = \mathbf{d}_0 \mathbf{P}^{n+1} \qquad (29)$$

Combining Equations (27), (28), and (29), we can write

$$\mathbf{s} = \mathbf{d}_n = \mathbf{d}_0 \mathbf{P}^n \qquad \text{and} \qquad \mathbf{s} = \mathbf{d}_{n+1} = \mathbf{d}_0 \mathbf{P}^{n+1}$$

But $\mathbf{P}^{n+1} = \mathbf{P}^n \mathbf{P}$, so

$$\mathbf{s} = \mathbf{d}_{n+1} = \mathbf{d}_0 \mathbf{P}^{n+1} = \mathbf{d}_0[\mathbf{P}^n \mathbf{P}] = [\mathbf{d}_0 \mathbf{P}^n]\mathbf{P} = \mathbf{d}_n \mathbf{P} = \mathbf{s}\mathbf{P}$$

From the extreme left and right sides of this equation, we have the equality

$$\mathbf{s} = \mathbf{s}\mathbf{P} \qquad (30)$$

where **s** is the stable distribution vector for the transition matrix **P**.

Equation (30) is the simplified formula we seek for calculating a stable distribution vector. Rather than calculating successive distribution vectors until stabilization occurs, we just find a distribution vector that satisfies Equation (30). Equation (30) is equivalent to a set of linear equations. To this set, *we add the requirement that the sum of the components of* **s** *be 1*, a condition of all distribution vectors, and then we solve the resulting set of equations by Gaussian elimination (see Section 2.2).

EXAMPLE 8

Use Equation (30) to locate the stable distribution vector for the Markov chain described in Example 2 (the 5:30 commuter flight).

Solution

The transition matrix for this Markov chain is

$$\mathbf{P} = \begin{bmatrix} .8 & .2 \\ .95 & .05 \end{bmatrix} \qquad \text{(1 repeated)}$$

and we know from Example 6 that it is regular. This Markov chain has two states, hence its distribution vectors are two-dimensional and **s** has the form

$$\mathbf{s} = [a \quad b]$$

where the values of a and b must be determined. Equation (30) becomes

$$[a \quad b] = \mathbf{s} = \mathbf{s}\mathbf{P} = [a \quad b]\begin{bmatrix} .8 & .2 \\ .95 & .05 \end{bmatrix} = [(.8a + .95b) \quad (.2a + .05b)]$$

which is equivalent to the system of equations

$$a = .8a + .95b$$

$$b = .2a + .05b$$

Bringing all variables to the left side of each equation, we have

$$.2a - .95b = 0$$

$$-.2a + .95b = 0$$

To this set of equations, we add the requirement that $a + b = 1$, obtaining the new system

$$a + \quad b = 1$$

$$.2a - .95b = 0$$

$$-.2a + .95b = 0$$

Solving this system of equation by Gaussian elimination, we find $b = \frac{4}{23}$ and $a = \frac{19}{23}$. Thus, the stable distribution vector is

$$\mathbf{s} = [a \quad b] = \begin{bmatrix} \dfrac{19}{23} & \dfrac{4}{23} \end{bmatrix} \approx [.826087 \quad .173913]$$

Compare this result with the entries in Tables 7.1, 7.2, and 7.3.

EXAMPLE 9

Use Equation (30) to locate the stable distribution vector for the Markov chain in Example 4 (the condition of public housing apartments).

Solution

The transition matrix for this Markov chain is

$$
\mathbf{P} = \begin{array}{c} \\ \textit{from good} \\ \textit{from serviceable} \\ \textit{from decayed} \end{array}
\begin{array}{ccc} \textit{to} & \textit{to} & \textit{to} \\ \textit{good} & \textit{serviceable} & \textit{decayed} \end{array}
\begin{bmatrix} .975 & .02 & .005 \\ 0 & .96 & .04 \\ .01 & 0 & .99 \end{bmatrix} \quad \text{(2 repeated)}
$$

and we know from Example 7 that it is regular. This Markov chain has three states, hence distribution vectors are three-dimensional and $\mathbf{s}$ has the form

$$\mathbf{s} = [a \quad b \quad c]$$

where the values of a, b, and c must be determined. Equation (30) becomes

$$[a \quad b \quad c] = \mathbf{s} = \mathbf{s}\mathbf{P} = [a \quad b \quad c] \begin{bmatrix} .975 & .02 & .005 \\ 0 & .96 & .04 \\ .01 & 0 & .99 \end{bmatrix}$$

$$= [(.975a + .01c) \quad (.02a + .96b) \quad (.005a + .04b + .99c)]$$

which is equivalent to the system of equations

$$a = .975a \qquad\qquad + .01c$$

$$b = \quad .02a + .96b$$

$$c = .005a + .04b + .99c$$

Bringing all variables to the left side of each equation, we have

$$
\begin{aligned}
.025a \qquad\quad - .01c &= 0 \\
-.02a + .04b \qquad\quad &= 0 \\
-.005a - .04b + .01c &= 0
\end{aligned}
$$

To this set of equations, we add the requirement that $a + b + c = 1$, obtaining the new system

$$
\begin{aligned}
a + \quad b + \quad c &= 1 \\
.025a \qquad\quad - .01c &= 0 \\
-.02a + .04b \qquad\quad &= 0 \\
-.005a - .04b + .01c &= 0
\end{aligned}
$$

Solving this set of four equations in three unknowns by Gaussian elimination, we find that $a = \frac{1}{4}$, $b = \frac{1}{8}$, and $c = \frac{5}{8}$. The stable distribution vector is

$$
\mathbf{s} = [a \quad b \quad c] = \left[\frac{1}{4} \quad \frac{1}{8} \quad \frac{5}{8}\right] = [.25 \quad .125 \quad .625]
$$

Note how much quicker it is to locate the stable distribution this way, compared with constructing Table 7.4 (Example 5).

A Markov chain with a regular transition matrix has one and only one stable distribution vector, and this stable distribution vector **s** gives the long-range trends for the underlying Markov process. Regardless how the process begins, successive distribution vectors must move towards and stabilize at **s**. If a transition matrix is not regular, then its distribution vectors may not stabilize. We will consider some nonregular transition matrices in the next section. So far, we can summarize our results as follows:

STABLE DISTRIBUTION VECTORS AND REGULAR TRANSITION MATRICES:

1. A transition matrix **P** is regular if some power of **P** contains only positive elements.

2. A regular transition matrix has one and only one stable distribution vector.

3. The unique stable distribution vector **s** for a Markov chain with a regular transition matrix **P** is found by solving the system $\mathbf{s} = \mathbf{sP}$ with the additional equation obtained by setting the sum of the components of **s** equal to 1.

IMPROVING SKILLS

1. Determine which of the following vectors *cannot* be distribution vectors.
   (*a*) [.1 .7]   (*b*) [.2 .8]   (*c*) [.4 .7]
   (*d*) [1 0]   (*e*) [2 −1]   (*f*) [0 0]

   (*g*) $\left[\dfrac{1}{2} \quad 0 \quad \dfrac{1}{2}\right]$

   (*h*) $\left[\dfrac{1}{3} \quad \dfrac{1}{3} \quad \dfrac{1}{3}\right]$

   (*i*) [1 1 1]

   (*j*) $\left[\dfrac{22}{61} \quad \dfrac{28}{61} \quad \dfrac{13}{61}\right]$

   (*k*) $\left[\dfrac{8}{22} \quad \dfrac{15}{22} \quad -\dfrac{1}{22}\right]$

   (*l*) $\left[\dfrac{17}{35} \quad \dfrac{8}{35} \quad \dfrac{2}{7}\right]$

In Problems 2 through 8, construct an initial distribution vector for the given Markov chains.

2. Initially, a two-state Markov chain has 53% of the objects in state 1 and 47% of the objects in state 2.
3. Initially, the probability of an object being in state 1 of a two-state Markov chain is .35.
4. Initially, a three-state Markov chain has 28% of the objects in state 1 and 31% of the objects in state 3.
5. Initially, the probability of an object being in state 1 of a three-state Markov chain is .235; the probability for state 2 is .487, and the probability for state 3 is .278.
6. Initially, all of the objects in a three-state Markov chain are in state 3.
7. Initially, the objects in a four-state Markov chain are equally distributed between the states.
8. Initially, half the objects in a five-state Markov chain are in state 2 and the rest are in state 5.

In Problems 9 through 26, a transition matrix and an initial distribution vector are given for particular Markov chains. For each system, find distribution vectors after one, two, and three time periods.

9. $\mathbf{P} = \begin{bmatrix} .1 & .9 \\ .7 & .3 \end{bmatrix}$, $\mathbf{d}_0 = [1 \quad 0]$

10. $\mathbf{P} = \begin{bmatrix} .1 & .9 \\ .7 & .3 \end{bmatrix}$, $\mathbf{d}_0 = \left[\dfrac{1}{2} \quad \dfrac{1}{2}\right]$

11. $\mathbf{P} = \begin{bmatrix} .1 & .9 \\ .7 & .3 \end{bmatrix}$, $\mathbf{d}_0 = [.6 \quad .4]$

12. $\mathbf{P} = \begin{bmatrix} .1 & .9 \\ .7 & .3 \end{bmatrix}$, $\mathbf{d}_0 = [.4375 \quad .5625]$

13. $\mathbf{P} = \begin{bmatrix} 0 & 1 \\ .4 & .6 \end{bmatrix}$, $\mathbf{d}_0 = [1 \quad 0]$

14. $\mathbf{P} = \begin{bmatrix} 0 & 1 \\ .4 & .6 \end{bmatrix}$, $\mathbf{d}_0 = \left[\dfrac{1}{4} \quad \dfrac{3}{4}\right]$

15. $\mathbf{P} = \begin{bmatrix} 0 & 1 \\ .4 & .6 \end{bmatrix}$, $\mathbf{d}_0 = \begin{bmatrix} \frac{3}{4} & \frac{1}{4} \end{bmatrix}$

16. $\mathbf{P} = \begin{bmatrix} .5 & .5 \\ .1 & .9 \end{bmatrix}$, $\mathbf{d}_0 = [.2 \quad .8]$

17. $\mathbf{P} = \begin{bmatrix} .5 & .5 \\ .1 & .9 \end{bmatrix}$, $\mathbf{d}_0 = [.7 \quad .3]$

18. $\mathbf{P} = \begin{bmatrix} \frac{1}{3} & \frac{2}{3} \\ \frac{2}{7} & \frac{5}{7} \end{bmatrix}$, $\mathbf{d}_0 = [.5 \quad .5]$

19. $\mathbf{P} = \begin{bmatrix} \frac{1}{3} & \frac{2}{3} \\ \frac{2}{7} & \frac{5}{7} \end{bmatrix}$, $\mathbf{d}_0 = [0 \quad 1]$

20. $\mathbf{P} = \begin{bmatrix} .1 & .7 & .2 \\ .2 & .8 & 0 \\ 0 & .6 & .4 \end{bmatrix}$, $\mathbf{d}_0 = [1 \quad 0 \quad 0]$

21. $\mathbf{P} = \begin{bmatrix} .1 & .7 & .2 \\ .2 & .8 & 0 \\ 0 & .6 & .4 \end{bmatrix}$, $\mathbf{d}_0 = [.5 \quad .5 \quad 0]$

22. $\mathbf{P} = \begin{bmatrix} .1 & .7 & .2 \\ .2 & .8 & 0 \\ 0 & .6 & .4 \end{bmatrix}$, $\mathbf{d}_0 = [0 \quad .5 \quad .5]$

23. $\mathbf{P} = \begin{bmatrix} .1 & 0 & .9 \\ 0 & .1 & .9 \\ .1 & 0 & .9 \end{bmatrix}$, $\mathbf{d}_0 = [.6 \quad .3 \quad .1]$

24. $\mathbf{P} = \begin{bmatrix} .1 & 0 & .9 \\ 0 & .1 & .9 \\ .1 & 0 & .9 \end{bmatrix}$, $\mathbf{d}_0 = [.2 \quad .2 \quad .6]$

25. $\mathbf{P} = \begin{bmatrix} .1 & .2 & .7 \\ .1 & .3 & .6 \\ .1 & .4 & .5 \end{bmatrix}$, $\mathbf{d}_0 = [0 \quad 1 \quad 0]$

26. $\mathbf{P} = \begin{bmatrix} .1 & .5 & .2 & .2 \\ .2 & .2 & 0 & .6 \\ .3 & 0 & 0 & .7 \\ 0 & .5 & .5 & 0 \end{bmatrix}$, $\mathbf{d}_0 = [0 \quad 0 \quad 1 \quad 0]$

In Problems 27 through 36, determine whether each transition matrix is regular.

27. $\mathbf{P} = \begin{bmatrix} .1 & .9 \\ 1 & 0 \end{bmatrix}$

28. $\mathbf{P} = \begin{bmatrix} 1 & 0 \\ 0 & 1 \end{bmatrix}$

29. $\mathbf{P} = \begin{bmatrix} .1 & .9 \\ 0 & 1 \end{bmatrix}$

30. The matrix in Problem 9
31. The matrix in Problem 13
32. The matrix in Problem 20
33. The matrix in Problem 23

34. $\mathbf{P} = \begin{bmatrix} .1 & .7 & .2 \\ 0 & .8 & .2 \\ 0 & .9 & 1 \end{bmatrix}$

35. The matrix in Problem 26

36. $\mathbf{P} = \begin{bmatrix} .1 & .9 & 0 & 0 \\ .1 & .8 & .1 & 0 \\ 0 & .1 & .7 & .2 \\ 0 & 0 & .6 & .4 \end{bmatrix}$

In Problems 37 through 47, find stabilized distribution vectors for the Markov chains described by the given transition matrices.

37. The matrix in Problem 9
38. The matrix in Problem 13
39. The matrix in Problem 16
40. The matrix in Problem 18
41. The matrix in Problem 27
42. The matrix in Problem 20
43. The matrix in Problem 25

44. $\mathbf{P} = \begin{bmatrix} \frac{1}{2} & \frac{1}{2} & 0 \\ \frac{1}{2} & \frac{1}{4} & \frac{1}{4} \\ \frac{1}{8} & \frac{3}{8} & \frac{1}{2} \end{bmatrix}$

45. $\mathbf{P} = \begin{bmatrix} \frac{1}{2} & \frac{1}{4} & \frac{1}{4} \\ \frac{1}{2} & \frac{1}{4} & \frac{1}{4} \\ \frac{1}{2} & \frac{1}{4} & \frac{1}{4} \end{bmatrix}$

46. The matrix in Problem 26
47. The matrix in Problem 36

CREATING MODELS

48. A town mayor is elected every four years. Voting records show that if a Democrat is mayor one year, then the probabilities are .45 that a Democrat will be mayor and .55 that a Republican will be mayor four years later. If a Republican is mayor one year, then the probabilities are .4 that a Democrat will be mayor and .6 that a Republican will be mayor four years later. The current mayor is a Democrat. Determine the probability of having a Democratic mayor (*a*) after two more elections and (*b*) over the long run.

49. The ABC Company, a maker of well-known laundry bleach, currently controls 55% of the bleach market. Data from a recently commissioned survey indicates that 85% of its customers remain loyal to the ABC brand each year, while the other 15% switch to a competitor's brand over the same time period. In contrast, 22% of all competitors' customers switch to ABC's bleach each year, while the remaining 78% of other customers do not switch to the ABC bleach. Assuming that these trends continue, determine ABC's share of the market (*a*) next year and (*b*) over the long run.

50. Farmers in a particular region of the country describe their summers as either dry or wet. Historical data indicates that there is a 20% chance that a dry summer will be followed by another dry summer and a 35% chance that a wet summer will be followed by another wet summer. Determine the probabilities for wet summers over the next five years if the most recent summer was dry.

51. A population can be divided into people who are overweight and those who are not. Over a five-year period, the probability is .95 that an overweight person will remain overweight, and the probability is .22 that a person who was not overweight will become overweight. Currently, 40% of the population is overweight. What will the distribution be (*a*) in five years and (*b*) in ten years?

52. A state lottery commission separates adult residents into two categories, those who play the lottery regularly and those who do not. Currently, 15% of the population play regularly. Over a two-year period, the probability is .98 that a regular player will continue to play regularly; the probability is .04 that over the same time period a person who does not play regularly will become a regular player. Determine the percent of the population that will play regularly (*a*) after two years and (*b*) over the long run.

53. A town zones its land as either residential, commercial, or public. Over the last 10 years, the probability that a parcel of residential land remained residential is .84, while the probability that a residential parcel became commercial is .16. The probability that a parcel of commercial land remained commercial is .7, while the probability that it became residential is .25, and the probability that it became public land is .05. The probability that a parcel of public land remained public is .9, while the probability that it became commercial or residential is .05 for each. Assuming that these trends continue, determine the long-range distribution of the town's land.

54. Residents in a major urban area are classified as either regular users, occasional users, or nonusers of a new rapid transit system. Currently, 5% of the population are regular users and another 5% are occasional users. It is estimated that each month, 10% of the nonusers will become regular users and 15% will become occasional users. The remaining nonusers will remain nonusers. It is also estimated that 85% of all regular users will remain regular users each month, 5% will become nonusers, and the other 10% will become occasional users. All occasional users remain occasional users from month to month. Determine the distribution of riders over each of the next three months. Is this transition matrix regular?

55. An insurance company classifies customers as high risk, medium risk, and low risk. Currently 15% of their customers are high risk, 25% are medium risk, and the remaining customers are low risk. Each year, 10% of the high risk customers are reclassified as low risk with the remaining 90% continuing as high risk. Each year, 40% of all medium-risk customers are reclassified as low risk, another 35% are reclassified as high risk, and the remaining 25% continue as medium risk. Each year, 15% of all low-risk customers are reclassified as high risk, 40% are reclassified as medium risk, with the rest remaining low risk. Determine the distribution of customers (*a*) after two years and (*b*) over the long run.

56. A publishing firm has four new books it will promote through radio advertising. Each day, one book will be selected by the editors for promotion. Book A is thought to be the most likely to succeed, so it will be promoted the most, although no book will be promoted two days in a row. If book A is not promoted one day, then it is twice as likely to be promoted the next day as either of the other two books not promoted that day. If book A is promoted one day, then there is an equal chance for any of the other three books to be promoted the next day. The company begins the campaign by promoting book A. What is the probability of promoting book A on each of the next four days?

EXPLORING IN TEAMS

57. Consider the transition matrix for the Markov chain described in Problem 54. Show that the matrix is not regular, yet the distribution vectors still stabilize over time.

58. Consider the transition matrix and initial distribution vector

$$\mathbf{P} = \begin{bmatrix} 0 & 1 \\ 1 & 0 \end{bmatrix} \qquad \mathbf{d}_0 = [1 \quad 0]$$

Show that the matrix is not regular. Then calculate distribution vectors for each of the next eight periods and use your results to show that the distribution vectors *never* stabilize.

59. Consider the transition matrix and initial distribution vector

$$\mathbf{P} = \begin{bmatrix} 0 & 1 & 0 \\ 0 & 0 & 1 \\ 1 & 0 & 0 \end{bmatrix} \qquad \mathbf{d}_0 = [1 \quad 0 \quad 0]$$

Show that the matrix is not regular. Then calculate distribution vectors for each of the next eight periods and use your results to show that the distribution vectors *never* stabilize.

60. Use the results of Problems 57 through 59 to develop a hypothesis about transition matrices whose distribution vectors do not stabilize over time.

EXPLORING WITH TECHNOLOGY

61. Use an electronic spreadsheet or a graphing calculator to reproduce the results in Table 7.2.

In Problems 62 through 65, use a graphing calculator or electronic spreadsheet to calculate

$$\mathbf{d}^2, \mathbf{d}^4, \mathbf{d}^8, \mathbf{d}^{16}, \mathbf{d}^{32}, \mathbf{d}^{64}, \mathbf{d}^{128}, \mathbf{d}^{256}, \mathbf{d}^{512}, \text{ and } \mathbf{d}^{1024}$$

for the given transition matrices. For each problem, start with any initial distribution vector of your choice. Then solve each problem again starting with a different distribution vector of your choice. Finally, compare your stabilized distribution vectors with the results obtained in Problems 40 through 43 of Section 7.1. What conclusions do you draw?

62. $\mathbf{P} = \begin{bmatrix} .8 & .2 \\ .95 & .05 \end{bmatrix}$ (*1 repeated*)

63. $\mathbf{P} = \begin{bmatrix} .7 & .3 \\ .4 & .6 \end{bmatrix}$

64. $\mathbf{P} = \begin{bmatrix} .5 & .5 & 0 \\ .8 & 0 & .2 \\ .3 & .6 & .1 \end{bmatrix}$ (*from Example 6*)

65. $\mathbf{P} = \begin{bmatrix} .975 & .02 & .005 \\ 0 & .96 & .04 \\ .01 & 0 & .99 \end{bmatrix}$ (*2 repeated*)

REVIEWING MATERIAL

66. (Sections 3.1 and 3.5) A couple decides to invest $20,000 of their savings in three ventures: a stock fund, a real estate fund, and a bank deposit. The stock fund pays 1% interest a year, is expected to grow 12% a year, and has a risk factor of 10. The real estate fund pays 8% a year, is expected to grow 5% a year, and has a risk factor of 5. Bank deposits pay 3% a year, do not grow at all, and have a risk factor of 0. The couple needs their savings to generate at least $800 a year in interest, they want their money to grow at least as fast as inflation, which is expected to average 4% a year, and they want the average risk factor of all their money be no greater than 4. How much should they invest in each venture if they want to maximize growth?

67. (Sections 4.1 and 4.3) A couple decides to create a college fund for their new baby. The baby received $4,000 at birth in gifts from friends and relatives, and they deposit the entire amount into the fund. In addition, they will add $100 to the fund at the end of each month. How much money will be in the fund after 18 years if the fund grows at the rate of 8% a year, compounded monthly?

68. (Section 5.3) The *Record* reported that juvenile arrests in New Jersey from 1988 through 1992 are

| 1988 | 94,586 | 1989 | 89,634 | 1990 | 90,062 |
| 1991 | 89,659 | 1992 | 88,682 | | |

Find the mean, median, mode, range, and standard deviation for this data.

69. (Section 6.3) In a particular high school, it was found that 44% of all female students were absent at least five days during the school year, as were 58% of all males. The student body was 52% female. Determine the probability that a randomly selected student is male *and* was absent at least five days during the school year.

RECOMMENDING ACTION

70. Respond by memo to the following request:

MEMORANDUM

To: J. Doe Reader

From: Mayor Washington

Date: Today

Subject: **Long-Term Trends**

I have your analysis regarding the condition of city apartments over the long run, but who cares? You show stabilization occurring after some fifty years!! This city is very much different today than it was fifty years ago, and the same can be expected fifty years from now. There seems very little, if any, justification for analyzing city trends fifty years into the future. Am I wrong, or are you just wasting the taxpayers' money?

ABSORBING STATES

A state in a Markov chain is an absorbing state if the state cannot be left after it is entered.

We saw in the last section that distribution vectors stabilize over time for Markov chains modeled by regular transition matrices. Such Markov chains, however, are not the only ones whose distribution vectors stabilize.

A state in a Markov chain is an *absorbing state* if an object must remain in that state once the state is entered. That is, an object in the system cannot leave an absorbing state after entering that state. A Markov chain is an *absorbing Markov chain* if it contains at least one absorbing state *and* if it is possible for an object in any nonabsorbing state to enter an absorbing state after a finite number of time periods.

Figure 7.9 is a transition diagram for an absorbing Markov chain with three states. State 2 is an absorbing state because once an object enters state 2, it remains there forever. Furthermore, it is *possible* for an object in state 1 to enter state 2 after one time period (although the probability of this occurring is only .2), and it is *possible* for an object in

FIGURE 7.9

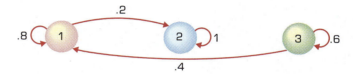

state 3 to enter the absorbing state after two time periods by moving first to state 1 and then to state 2. Thus, it is possible for an object in *any* nonabsorbing state to enter an absorbing state after a finite number of time periods. The transition matrix for Figure 7.9 is

$$\mathbf{P} = \begin{bmatrix} .8 & .2 & 0 \\ 0 & 1 & 0 \\ .4 & 0 & .6 \end{bmatrix} \tag{31}$$

Figure 7.10, in contrast, is *not* a model of an absorbing Markov chain even though it has state 1 as an absorbing state. An object in either state 2 or state 3, which are nonab-

FIGURE 7.10

sorbing states, will always be in one of those two states because it is not possible for an object in state 2 or state 3 to move into state 1. In an absorbing Markov chain, it must be possible for an object in a nonabsorbing state to move to an absorbing state over a finite number of time periods.

EXAMPLE 1

Model the following system as a Markov chain: The neurology department of a hospital classifies patients under its care as either bedridden or ambulatory (i.e., able to leave bed on their own). Data for the last ten years reveals that over a one-day time period 20% of all ambulatory patients are discharged from the hospital, 75% remain ambulatory, and the other 5% are remanded to complete bed rest. In contrast, 10% of all bedridden patients become ambulatory over a 24-hour period, 80% remain bedridden, and the other 10% die. Currently, the hospital has 80 patients under the care of neurologists, and 30 of them are ambulatory.

Solution

We make two assumptions: First, we are interested in tracking only existing patients while they are in the neurology department. Thus, once a patient is discharged, that patient is presumed to remain discharged. If a patient is subsequently readmitted, that patient then becomes a new record for this data set. Second, we are interested in a patient only while the patient is under hospital care. All patient tracking ends when a patient leaves the hospital. Consequently, a discharged patient who later dies outside the hospital is still classified as a discharged patient, not a deceased patient.

FIGURE 7.11

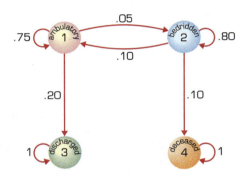

A time period is one day, and the states of interest are ambulatory, bedridden, discharged, and deceased, which we denote as states 1 through 4, respectively. Figure 7.11 is a transition diagram for this system. The corresponding transition matrix is

$$
\mathbf{P} = \begin{array}{c} \\ \textit{ambulatory} \\ \textit{bedridden} \\ \textit{discharged} \\ \textit{deceased} \end{array}
\begin{array}{cccc} \textit{ambulatory} & \textit{bedridden} & \textit{discharged} & \textit{deceased} \end{array}
\begin{bmatrix} .75 & .05 & .20 & 0 \\ .10 & .80 & 0 & .10 \\ 0 & 0 & 1 & 0 \\ 0 & 0 & 0 & 1 \end{bmatrix} \quad (32)
$$

States 3 and 4 (discharged and deceased) are absorbing states. Once a patient dies, the patient remains dead for all future time periods; once a patient is discharged, the patient remains discharged (for the purpose of patient tracking) for all future time periods.

Initially, 30 of the 80 patients are ambulatory (state 1) and 50 of the 80 are bedridden (state 2). Thus, the initial distribution vector is

$$
\mathbf{d}_0 = \begin{bmatrix} \dfrac{30}{80} & \dfrac{50}{80} & 0 & 0 \end{bmatrix} = [.375 \quad .625 \quad 0 \quad 0]
$$

The absorbing state in

$$
\mathbf{P} = \begin{bmatrix} .8 & .2 & 0 \\ 0 & 1 & 0 \\ .4 & 0 & .6 \end{bmatrix} \quad (31 \text{ repeated})
$$

is state 2. Note that the second row in that matrix contains all zeros except for $p_{22} = 1$. The absorbing states in transition matrix (32) are states 3 and 4. In that matrix, the third and fourth rows contain all zeros except for $p_{33} = p_{44} = 1$. A transition matrix for an absorbing Markov chain having state i as an absorbing state contains all zeros in its ith row except for the entry in the ith row and ith column, which is 1. All powers of this matrix will exhibit this characteristic. Thus, all powers of a transition matrix for an absorbing Markov chain have at least one row that is mostly zero. Consequently, transition matrices for absorbing Markov chains are *not* regular.

In an absorbing Markov chain, an object can never leave an absorbing state after such a state is entered, and each object in an nonabsorbing state can reach an absorbing state over a finite number of time periods. It is not surprising, therefore, to find that all objects in an absorbing Markov chain are captured eventually by the absorbing states.

If an absorbing Markov chain has only one absorbing state, then all objects in the system eventually gravitate to that absorbing state. The transition diagram in Figure 7.9, for example, depicts an absorbing Markov chain in which state 2 is the only absorbing state. Consequently, over time, all objects in this Markov chain are captured in state 2, and all distribution vectors gravitate to

$$\mathbf{s} = [0 \quad 1 \quad 0]$$

Distribution vectors for the system defined by Figure 7.9 stabilize to $\mathbf{s}$ regardless of how the objects in the system are initially distributed. Consequently, $\mathbf{s}$ is the stable distribution vector (see Section 7.2) for this Markov chain.

If an absorbing Markov chain contains more than one absorbing state, then eventually all objects in the system gravitate to the absorbing states and are captured by them. However, *the final distribution of objects in each absorbing state depends on the initial distribution of objects in the nonabsorbing states.* To see how this final distribution depends on the initial distribution, it is convenient to index the nonabsorbing states with lower state numbers than the absorbing states. We did this in Example 1 when we indexed the nonabsorbing states (bedridden and ambulatory) as states 1 and 2, respectively, and the absorbing states (discharged and deceased) as states 3 and 4, respectively. If an absorbing Markov chain has N states, of which k are absorbing states, then by indexing the k absorbing states last, we can generate a transition matrix of the partitioned form

$$\mathbf{P} = \left[\begin{array}{c|c} \mathbf{Q} & \mathbf{R} \\ \hline \mathbf{O} & \mathbf{I_k} \end{array} \right] \tag{33}$$

where $\mathbf{I_k}$ is a $k \times k$ identity matrix having one row and one column for each absorbing state and $\mathbf{O}$ is a matrix of all zero elements. In particular, the transition matrix

$$\mathbf{P} = \begin{array}{c} \\ \textit{ambulatory} \\ \textit{bedridden} \\ \textit{discharged} \\ \textit{deceased} \end{array} \begin{array}{cccc} \textit{ambulatory} & \textit{bedridden} & \textit{discharged} & \textit{deceased} \\ \left[\begin{array}{cccc} .75 & .05 & .20 & 0 \\ .10 & .80 & 0 & .10 \\ 0 & 0 & 1 & 0 \\ 0 & 0 & 0 & 1 \end{array} \right] \end{array} \tag{32 repeated}$$

for the Markov chain described in Example 1 has

$$\mathbf{Q} = \begin{bmatrix} .75 & .05 \\ .10 & .80 \end{bmatrix} \qquad \mathbf{R} = \begin{bmatrix} .20 & 0 \\ 0 & .10 \end{bmatrix}$$

$$\mathbf{O} = \begin{bmatrix} 0 & 0 \\ 0 & 0 \end{bmatrix} \qquad \mathbf{I_2} = \begin{bmatrix} 1 & 0 \\ 0 & 1 \end{bmatrix} \tag{34}$$

This Markov chain has two absorbing states, so the identity matrix in the lower right partition of Matrix (33) will have two rows and two columns.

We define a *fundamental matrix* **F** for an absorbing Markov chain with transition matrix

$$\mathbf{P} = \left[\begin{array}{c|c} \mathbf{Q} & \mathbf{R} \\ \hline \mathbf{O} & \mathbf{I_k} \end{array}\right] \qquad \text{(33 repeated)}$$

to be the inverse (see Section 2.5) of $\mathbf{I} - \mathbf{Q}$; that is,

$$\mathbf{F} = (\mathbf{I} - \mathbf{Q})^{-1} \qquad (35)$$

where **I** is the identity matrix of the same order as **Q**.

EXAMPLE 2

Find the fundamental matrix associated with the matrices in (34).

Solution

Since **Q** is a *2 × 2* matrix, we take **I** to be the *2 × 2* identity matrix. Then

$$\mathbf{I} - \mathbf{Q} = \begin{bmatrix} 1 & 0 \\ 0 & 1 \end{bmatrix} - \begin{bmatrix} .75 & .05 \\ .10 & .80 \end{bmatrix} = \begin{bmatrix} .25 & -.05 \\ -.10 & .20 \end{bmatrix}$$

and

$$\mathbf{F} = (\mathbf{I} - \mathbf{Q})^{-1} = \begin{bmatrix} .25 & -.05 \\ -.10 & .20 \end{bmatrix}^{-1} = \begin{bmatrix} \frac{40}{9} & \frac{10}{9} \\ \frac{20}{9} & \frac{50}{9} \end{bmatrix}$$

An *initial absorption vector* **a** for an absorbing Markov chain is a row matrix that has the same number of components as there are nonabsorbing states and where each component is equal to the initial fraction of objects in the corresponding nonabsorbing state. An initial absorption vector is derived from an initial distribution vector by retaining only the components that correspond to nonabsorbing states. For example, we found the initial distribution vector for Example 1 to be

$$\mathbf{d}_0 = \begin{bmatrix} \dfrac{30}{80} & \dfrac{50}{80} & 0 & 0 \end{bmatrix} = [.375 \quad .625 \quad 0 \quad 0]$$

The nonabsorbing states are states 1 and 2, so the initial absorption vector is

$$\mathbf{a} = [.375 \quad .625] \qquad (36)$$

The components of an initial absorption vector need *not* sum to 1, although they can, as in (36). Consequently, an initial absorption vector may not be a distribution vector. If some objects are initially in an absorbing state, then those objects will *not* be represented in an initial absorption vector. An initial absorption vector lists only the proportion of objects in the system that are not yet in an absorbing state. These are the main objects of interest, however, because once an object is in an absorbing state, it must remain there forever.

We know that each object in an nonabsorbing state must gravitate eventually to an absorbing state, but we know little about the relative abilities of the different absorbing states to attract such objects. Thus, we want a procedure for calculating the proportion of objects that began in nonabsorbing states and eventually land in each absorbing state. Such a stabilization algorithm exists. Although its development is beyond the scope of this book, the algorithm itself is straightforward.

STABILIZATION IN ABSORBING MARKOV CHAINS:

Let **a** denote an initial absorption vector for an absorbing Markov chain with a transition matrix

$$\mathbf{P} = \left[\begin{array}{c|c}\mathbf{Q} & \mathbf{R} \\ \hline \mathbf{O} & \mathbf{I_k}\end{array}\right]$$

and a fundamental matrix

$$\mathbf{F} = (\mathbf{I} - \mathbf{Q})^{-1}$$

The matrix product **aFR** is the proportion of all objects in the system that begin in nonabsorbing states and eventually land in each of the absorbing states, listed by ascending state numbers.

Using the matrices **R**, as defined in (*34*), **F**, as defined in Example 2, and **a**, as defined in Equation (*36*), we have

$$\mathbf{aFR} = [.375 \quad .625]\begin{bmatrix}\frac{40}{9} & \frac{10}{9} \\ \frac{20}{9} & \frac{50}{9}\end{bmatrix}\begin{bmatrix}.20 & 0 \\ 0 & .10\end{bmatrix} \approx [.6111 \quad .3889]$$

where for presentation purposes the components are rounded to four decimal places. It follows that approximately 61% of all objects in the system begin in nonabsorbing states and will end up in the first absorbing state, state 3; approximately 39% of all objects in the system begin in nonabsorbing states and will end up in the second absorbing state, state 4. In Example 1, state 3 corresponds to discharged patients and state 4 corresponds to deceased patients. Of the 80 patients initially in the neurology department, approximately 61% (49 patients) will be discharged eventually, while the other 39% (31 patients) will die in the hospital, if historical trends prevail.

EXAMPLE 3

A manufacturer subjects all products to quality-control testing before shipment. A new product passes the tests with probability .7 and fails with probability .3. New products that pass the quality-control tests are shipped, while those that that fail are returned to manufacturing for reprocessing and then retesting. A reprocessed product passes the quality-control tests with probability .9 and fails with probability .1. Reprocessed products that pass the quality-control tests are shipped, while those that fail are classified as recycled and returned to manufacturing for a second reprocessing. After being reprocessed a second time, a recycled product passes the quality-control test with probability .6 and fails with probability .4. Recycled products that pass the quality-control test are shipped, while those that fail are destroyed. Current inventory lists 417 new products, 55 reprocessed products, and 28 recycled products. How many of these items will be shipped eventually?

Solution

The objects in this system are manufactured products, and they can be in one of five states: new, shipped, reprocessed, recycled, or destroyed. A destroyed item remains destroyed forever and a shipped item is forever shipped, so these two states are absorbing states and will be listed last. We label new items as state 1, reprocessed items as state 2, recycled items as state 3, shipped items as state 4, and destroyed items as state 5. Figure 7.12 is a transition diagram for this system.

FIGURE 7.12

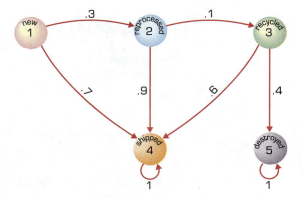

The corresponding transition matrix is

$$
\mathbf{P} = \begin{array}{c} \\ new \\ reprocessed \\ recycled \\ shipped \\ destroyed \end{array}
\begin{array}{ccccc} new & reprocessed & recycled & shipped & destroyed \end{array}
\left[\begin{array}{ccc|cc}
0 & .3 & 0 & .7 & 0 \\
0 & 0 & .1 & .9 & 0 \\
0 & 0 & 0 & .6 & .4 \\
\hline
0 & 0 & 0 & 1 & 0 \\
0 & 0 & 0 & 0 & 1
\end{array}\right]
$$

with

$$
\mathbf{Q} = \begin{bmatrix} 0 & .3 & 0 \\ 0 & 0 & .1 \\ 0 & 0 & 0 \end{bmatrix}
\quad \text{and} \quad
\mathbf{R} = \begin{bmatrix} .7 & 0 \\ .9 & 0 \\ .6 & .4 \end{bmatrix}
$$

Here

$$
\mathbf{F} = (\mathbf{I} - \mathbf{Q})^{-1} = \left(\begin{bmatrix} 1 & 0 & 0 \\ 0 & 1 & 0 \\ 0 & 0 & 1 \end{bmatrix} - \begin{bmatrix} 0 & .3 & 0 \\ 0 & 0 & .1 \\ 0 & 0 & 0 \end{bmatrix} \right)^{-1}
$$

$$
= \begin{bmatrix} 1 & -.3 & 0 \\ 0 & 1 & -.1 \\ 0 & 0 & 1 \end{bmatrix}^{-1}
$$

$$
= \begin{bmatrix} 1 & .3 & .03 \\ 0 & 1 & .1 \\ 0 & 0 & 1 \end{bmatrix}
$$

The nonabsorbing states are states 1, 2, and 3, and the initial fractions of items in these states are $\frac{417}{500} = .834$, $\frac{55}{500} = .11$, and $\frac{28}{500} = .056$, respectively. The initial absorption vector is $\mathbf{a} = [.834 \quad .11 \quad .056]$, whereupon

$$
\mathbf{aFR} = [.834 \quad .11 \quad .056] \begin{bmatrix} 1 & .3 & .03 \\ 0 & 1 & .1 \\ 0 & 0 & 1 \end{bmatrix} \begin{bmatrix} .7 & 0 \\ .9 & 0 \\ .6 & .4 \end{bmatrix} = [.963192 \quad .036808]
$$

Thus, .963192 of all items in the system begin in nonabsorbing states and will end up in the first absorption state, shipped. There are 500 items in the system, so over the long run .963192(500) ≈ 482 will be shipped, and the remaining 18 will eventually be destroyed.

EXAMPLE 4

Solve Example 3 if the current inventory lists 300 new products, 100 reprocessed products, 75 recycled products, and 25 destroyed products.

Solution

The states and transition diagram remain as defined in Example 4, so **F** and **R** remain unchanged. The fractions of items in the *nonabsorbing* states are, initially, $\frac{300}{500} = .6$ in state 1, $\frac{100}{500} = .2$ in state 2, and $\frac{75}{500} = .15$ in state 3. The initial absorption vector is

$$\mathbf{a} = [.6 \quad .2 \quad .15]$$

Note that the sum of the components in **a** is not 1 because initially 25 of the 500 objects (5% of the objects in the system) are in an absorbing state. Now

$$\mathbf{aFR} = [.6 \quad .2 \quad .15]\begin{bmatrix} 1 & .3 & .03 \\ 0 & 1 & .1 \\ 0 & 0 & 1 \end{bmatrix}\begin{bmatrix} .7 & 0 \\ .9 & 0 \\ .6 & .4 \end{bmatrix} = [.8748 \quad .0752]$$

Thus, .8748 of all items in the system begin in nonabsorbing states and will end up in the first absorption state, state 4 (shipped). There are 500 items in the system, so .8748(500) ≈ 437 will be shipped. In addition, .0752(500) ≈ 38 items that begin in nonabsorbing states will end up in the second absorption state, destroyed. When these 38 items are added to the current inventory of 25 destroyed products, we have 38 + 25 = 63 items that are or will be destroyed from current inventory.

Examples 3 and 4 highlight an important difference between absorbing Markov chains and Markov chains with regular transition matrices. Distribution vectors for a Markov chain with a regular transition matrix stabilize over time to the same vector **s**, regardless of the initial distribution of objects in the various states. In contrast, distribution vectors in an absorbing Markov chain also stabilize over time, but the vector they stabilize to may depend on the initial distribution vector. With 500 products initially in inventory in Example 3, approximately 482 eventually shipped and 18 were destroyed eventually. When we altered the initial distribution vector in Example 4, we found that the long-range trend changed to approximately 437 items shipped and 63 destroyed.

If an absorbing Markov chain has only one absorbing state, then we know that all objects must gravitate over time to that single absorbing state. If we denote this absorbing state as state N, then all initial distribution vectors stabilize over time to the vector having a 1 as its Nth component and 0s everywhere else. Such a vector is a stable distribution as defined in Section 7.2. It is only *when an absorbing Markov chain has more than one absorbing state*, as is the case in Examples 3 and 4, that *the long-run trends depend on the initial distribution*.

Example 3 also illustrates how the concept of a time period is generalized to extend the applicability of Markov chain models. In Sections 7.1 and 7.2, successive time periods involved equal time intervals, such as a business day, a quarter of a year, a year, or a decade. In contrast, a time period in Example 3 is the time between successive inspections of a product, and this time period is not a constant. We no longer require successive time periods to be equal; time periods may vary in length when appropriate.

Another troublesome issue with time periods in Example 3 is that different products are inspected at different times, so the time periods for one product do not coincide with time periods for other products. For modeling purposes, however, we *assume* that all products are inspected at the same time. We model the process as if all first inspections take place at the same time, as do all second inspections of reprocessed prod-

If an absorbing Markov chain has more than one absorbing state, then the long-run trends depend on the initial distribution.

Time periods may vary in length when appropriate.

ucts and all third inspections of recycled products. With this simplification, we create a mathematical model with different timing patterns from the real system but with the same long-range characteristics. This simplification is acceptable in Example 3 because the objective of the model is the long-range trend, not inventory levels at particular moments in time.

Markov chains are useful models because they adequately represent many systems and because we know so much about Markov chains. We can use our knowledge of Markov chains to analyze systems that can be modeled by Markov chains. In constructing models, we sometimes make distinctions or assumptions that are not apparent in actuality. We do so when those assumptions do not alter the fundamental characteristics of a real system and when those assumptions provide a means (that is, a model) to understand adequately those characteristics of a system that do interest us. We did this before, in Section 6.1, when we assumed that two dice had different colors, and we did it again in Example 3 by assuming that all inspections take place at the same time.

An interesting application of time periods of varying length occurs in the field of genetics and deals with the passing of hereditary traits from parents to children through the genes. In this case the appropriate time interval is a generation. Person *A* is born, matures, mates, and then becomes a parent to Person *B*. The time between successive births is one time period. Now Person *B* is born, matures, mates, and then becomes a parent to Person *C*, and again the time between successive births is one time period.

In Mendel's theory of pairing, each parent imparts one gene from each pair of genes to each offspring. The child receives one gene from each parent, and these two genes then form the genetic pair of the offspring. In the simplest case, genes can be of two types, dominant or recessive, which are denoted by *A* and *a*, respectively. Consequently, a gene pair can be *AA*, *Aa*, *aA*, or *aa*. The pairs *Aa* and *aA* are often the same genetically, and an individual with either pairing has a hybrid genetic makeup or *genotype*. An individual with an *AA* pairing has a dominant genotype, and a person with an *aa* pairing has a recessive genotype. A dominant parent can impart only a dominant gene to an offspring, and a recessive parent can impart only a recessive gene to an offspring. A hybrid parent with an *aA* pairing can impart either a dominant or a recessive gene to an offspring, and we assume that these two possibilities occur with equal probabilities of .5.

> *When modeling the genetic history of a family, a time interval is often defined as a generation, the time between successive births in direct lineage.*

EXAMPLE 5

A person or unknown genotype mates with a person of recessive genotype and gives birth to a child. The offspring then mates with a another person of recessive genotype, and this mating pattern continues through the generations: offspring always mate with persons of recessive genotype. What conclusions can be made about the genotype of the offspring over the long run?

Solution

We take a time period to be one generation, and since the genotype of an offspring is completely determined by the genotype of the parents one generation (time period) earlier, we attempt to model the dynamics of this system with a Markov chain. We take the states in the system to be dominant, recessive, and hybrid. The genotype of one parent is always recessive. The other parent's genotype is unknown, and the genotypes become the objects of interest in the system.

If the parent of unknown genotype is recessive, then a mating with a recessive partner must result in a recessive offspring, as illustrated in Table 7.5. The top row lists the two genes of the parent of unknown genotype under the assumption that that parent is recessive; the left column lists the two *aa* genes of the recessive parent. The child inherits one gene

TABLE 7.5 Unknown Genotype Is Recessive. Known Genotype Is Recessive.

	a	a
a	aa	aa
a	aa	aa

TABLE 7.6 Unknown Genotype Is Dominant. Known Genotype Is Recessive.

	A	A
a	aA	aA
a	aA	aA

from each parent, and the result is always the recessive pair *aa*. If this pattern continues, that is, if each offspring mates with a person of recessive genotype, then the result is always recessive and recessive becomes an absorbing state.

In contrast, if the parent of unknown genotype is dominant, then a mating with a recessive partner must result in a hybrid offspring with probability 1, as illustrated in Table 7.6. Again, the top row lists the two genes of the parent of unknown genotype, now under the assumption that that parent is dominant; the left column lists the two *aa* genes of the recessive parent. The genotype of the offspring is always the hybrid pair *aA*.

Finally, if the parent of unknown genotype is hybrid, then a mating with a recessive partner results in an offspring with either a recessive genotype or a hybrid genotype, as illustrated in Table 7.7. Of the four possible genetic results for the offspring, two are recessive and two are hybrid, and these pairings occur with equal probabilities of .5.

FIGURE 7.13
Genotype of offspring

TABLE 7.7 Unknown Genotype Is Hybrid. Known Genotype Is Recessive.

	a	A
a	aa	aA
a	aa	aA

A transition diagram for this system is shown in Figure 7.13. A transition matrix for the system, following our convention of listing absorbing states last, is

$$\mathbf{P} = \begin{array}{c} \text{dominant} \\ \text{hybrid} \\ \text{recessive} \end{array} \begin{array}{ccc} \text{dominant} & \text{hybrid} & \text{recessive} \\ \left[\begin{array}{ccc} 0 & 1 & 0 \\ 0 & .5 & .5 \\ 0 & 0 & 1 \end{array}\right] \end{array}$$

This model is an absorbing Markov chain with a single absorbing state. This absorbing state is recessive, so eventually all offspring must enter and remain in this state. Over the long run, all offspring must be of recessive genotype.

IMPROVING SKILLS

In Problems 1 through 13, determine whether the given matrices are transition matrices for absorbing Markov chains. For those that are, identify the absorbing states. *Warning:* The convention of listing absorbing states last is *not* necessarily observed in these problems.

1. $\begin{bmatrix} .1 & .9 \\ 0 & 1 \end{bmatrix}$ 2. $\begin{bmatrix} 1 & 0 \\ .4 & .4 \end{bmatrix}$ 3. $\begin{bmatrix} 1 & 0 \\ 0 & 1 \end{bmatrix}$

4. $\begin{bmatrix} 0 & 1 \\ 1 & 0 \end{bmatrix}$ 5. $\begin{bmatrix} .3 & 0 & .7 \\ 0 & 1 & 0 \\ 0 & 0 & 1 \end{bmatrix}$ 6. $\begin{bmatrix} .3 & .7 & 0 \\ .6 & .4 & 0 \\ 0 & 0 & 1 \end{bmatrix}$

7. $\begin{bmatrix} .1 & .2 & .7 \\ .2 & .8 & 0 \\ 1 & 0 & 0 \end{bmatrix}$ 8. $\begin{bmatrix} 1 & 0 & 0 \\ .2 & .8 & 0 \\ .1 & .2 & .7 \end{bmatrix}$ 9. $\begin{bmatrix} .1 & .9 & 0 & 0 \\ .2 & .8 & 0 & 0 \\ 0 & 0 & 1 & 0 \\ 0 & 0 & 0 & 1 \end{bmatrix}$

10. $\begin{bmatrix} .1 & .9 & 0 & 0 \\ 0 & .2 & .8 & 0 \\ 0 & 0 & 1 & 0 \\ 0 & 0 & 0 & 1 \end{bmatrix}$ **11.** $\begin{bmatrix} .1 & .9 & 0 & 0 \\ 0 & 0 & 1 & 0 \\ .2 & .8 & 0 & 0 \\ 0 & 0 & 0 & 1 \end{bmatrix}$

12. $\begin{bmatrix} .1 & .9 & 0 & 0 \\ 0 & 1 & 0 & 0 \\ 0 & 0 & .5 & .5 \\ 0 & 0 & 0 & 1 \end{bmatrix}$ **13.** $\begin{bmatrix} .1 & .9 & 0 & 0 \\ 0 & 1 & 0 & 0 \\ 0 & 0 & 0 & 1 \\ 0 & 0 & .5 & .5 \end{bmatrix}$

In Problems 14 through 26, transition matrices for absorbing Markov chains are given. Determine the fundamental matrix **F** and the corresponding matrix **R** for each.

14. $\begin{bmatrix} .3 & .1 & .6 \\ .2 & .6 & .2 \\ 0 & 0 & 1 \end{bmatrix}$ **15.** $\begin{bmatrix} .6 & .1 & .3 \\ .2 & .6 & .2 \\ 0 & 0 & 1 \end{bmatrix}$ **16.** $\begin{bmatrix} .1 & .2 & .7 \\ .2 & .3 & .5 \\ 0 & 0 & 1 \end{bmatrix}$

17. $\begin{bmatrix} .1 & .2 & .7 \\ 0 & .5 & .5 \\ 0 & 0 & 1 \end{bmatrix}$ **18.** $\begin{bmatrix} .1 & .2 & .7 \\ .5 & 0 & .5 \\ 0 & 0 & 1 \end{bmatrix}$ **19.** $\begin{bmatrix} .1 & .2 & .7 \\ .5 & .5 & 0 \\ 0 & 0 & 1 \end{bmatrix}$

20. $\begin{bmatrix} .3 & 0 & .7 \\ 1 & 0 & 0 \\ 0 & 0 & 1 \end{bmatrix}$ **21.** $\begin{bmatrix} .4 & .6 & 0 \\ .9 & 0 & .1 \\ 0 & 0 & 1 \end{bmatrix}$ **22.** $\begin{bmatrix} .25 & .25 & .25 & .25 \\ .2 & .3 & 0 & .5 \\ 0 & 0 & 1 & 0 \\ 0 & 0 & 0 & 1 \end{bmatrix}$

23. $\begin{bmatrix} 0 & .2 & .3 & .5 \\ .2 & .3 & 0 & .5 \\ 0 & 0 & 1 & 0 \\ 0 & 0 & 0 & 1 \end{bmatrix}$ **24.** $\begin{bmatrix} .1 & .9 & 0 & 0 \\ 0 & 0 & 1 & 0 \\ .2 & .5 & 0 & .3 \\ 0 & 0 & 0 & 1 \end{bmatrix}$

25. $\begin{bmatrix} .25 & .25 & .25 & .25 \\ .25 & .25 & .25 & .25 \\ .25 & .25 & .25 & .25 \\ 0 & 0 & 0 & 1 \end{bmatrix}$ **26.** $\begin{bmatrix} .25 & .25 & .25 & .25 \\ 0 & 1 & 0 & 0 \\ 0 & 0 & 1 & 0 \\ 0 & 0 & 0 & 1 \end{bmatrix}$

CREATING MODELS

27. Determine the long-range distribution of patients in the neurology department of the hospital described in Example 1 if initially 50 patients are ambulatory and 30 are bedridden.

28. Solve Problem 27 if the 80 patients are split evenly between ambulatory and bedridden.

29. Determine the genotype over the long run of the offspring in Example 5 if a person of unknown genotype always mates with a person of dominant genotype.

30. Determine the genotype over the long run of the offspring in Example 5 if a person of unknown genotype always mates with a person of hybrid genotype.

31. A town zones its land as either residential, commercial, or public. Over the last 10 years, the probability that a parcel of residential land remained residential is .84, while the probability that it became commercial is .16. The probability that a parcel of commercial land remained commercial is .7, the probability that it became residential is .25, and the probability that it became public is .05. All public land remained public over the same period. Assuming that these trends continue into the future, what will be the eventual distribution of land in town? Is this trend likely to continue indefinitely?

32. Residents in a major urban area are classified as either regular users, occasional users, or nonusers of a new rapid transit system. It is estimated that each month, 10% of the nonusers will become regular users and 15% will become occasional users. The remaining nonusers will remain nonusers. It is also estimated that each month 85% of all regular users will remain regular users, 5% will become nonusers, and the other 10% will become occasional users. All occasional users remain occasional users from month to month. What will be the eventual distribution of users if, in the beginning, 2% of the population are regular users, and 1% are occasional users?

33. A manufacturer subjects all products to quality-control testing before shipment. A new product passes the tests with probability .7 and fails with probability .3. New products that pass the quality-control tests are shipped, while those that fail are returned to manufacturing for reprocessing and then retesting. A reprocessed product passes the quality-control tests with probability .9 and fails with probability .1. Reprocessed products that pass the quality-control tests are shipped, while those that fail are destroyed. Current inventory lists 200 new products and 50 reprocessed products. How many of these items will ship eventually?

34. Solve Problem 33 for an inventory of 150 new products and 100 reprocessed products.

35. Solve Problem 33 if all 250 items in inventory are new products.

36. A training program for account executives at a brokerage firm involves a half year of classroom training followed by a half year of on-the-job training. From past experience, the company expects 60% of all trainees to leave the company before completing classroom training, while the remaining 40% will move to the next stage. Seventy percent of all those who begin on-the-job training complete it and become account executives for the firm; the other 30% leave the firm prior to completing this stage of training. Assuming these trends continues, what is expected to happen to the current crop of 82 trainees, 50 of whom are in classroom training and 32 of whom are in on-the-job training?

37. A two-year community college expects from past experience that 60% of all freshmen will become sophomores after one year, 10% will remain freshmen, and the other 30% will leave the school. Over the same time period, 70% of all sophomores will graduate, 25% will remain sophomores, and the other 5% will leave the school without graduating. Currently, the school has an enrollment of 2,450 freshmen and 1,680 sophomores. What is expected to happen to these students over the long run?

38. Solve the previous problem if the current enrollment is 1,680 freshmen and 2,450 sophomores.

39. A player has $2, and the player's opponent has $1. A game is played by flipping a fair coin. If the coin turns up heads, the player wins $1 from the opponent; if the coin turns up tails, the opponent wins $1 from the player. The game continues through succeeding flips of the coin until one player has no money. What is the eventual outcome of this game?

40. Solve the previous problem if both players begin with $2.

41. Another application of varying time periods is found with Markov chains that model the spread of information in one of two possible forms, perhaps *go* or *no go*. Information is passed from person to person, and with each passage the informer accurately tells what he or she heard with probability p and tells it inaccurately with probability $1 - p$. Model this process as a Markov chain. What is a time period? Over the long run, what fraction of the population will hear the information correctly?

42. A candidate for a job is interviewed, and the head of the personnel department informs a friend in the company that the candidate will be hired. That friend tells someone else in the company, who then passes the information to still another person, and this dissemination continues indefinitely. At each telling, however, the probability is .95 that the information will be passed correctly and .05 that it will not. Over the long run, what fraction of the company will think the candidate is being hired?

43. Use a graphing calculator or electronic spreadsheet to calculate successively

$$\mathbf{P}^2, \ \mathbf{P}^4, \ \mathbf{P}^8, \ \text{and} \ \mathbf{P}^{16}$$

for the transition matrix

	new	reprocessed	recycled	shipped	destroyed
new	0	.3	0	.7	0
reprocessed	0	0	.1	.9	0
recycled	0	0	0	.6	.4
shipped	0	0	0	1	0
destroyed	0	0	0	0	1

$\mathbf{P} =$

from Example 3. Compare your results with the matrix product $\mathbf{FR}$ from that example. What do you conclude?

44. Use a graphing calculator or electronic spreadsheet to calculate successively

$$\mathbf{P}^2, \ \mathbf{P}^4, \ \mathbf{P}^8, \ \mathbf{P}^{16}, \mathbf{P}^{32}, \ \text{and} \ \mathbf{P}^{64}$$

for the transition matrix

	ambulatory	bedridden	discharged	deceased
ambulatory	.75	.05	.20	0
bedridden	.10	.80	0	.10
discharged	0	0	1	0
deceased	0	0	0	1

$\mathbf{P} =$

from Example 1. Compare your results with the matrix product **FR** obtained from **P** (see Equation (*34*) and Example 2). What do you conclude?

45. Use your conclusions from Problems 43 and 44 to develop a conjecture about the form of P^n as n gets larger and larger when **P** is a transition matrix for an absorbing Markov chain.

REVIEWING MATERIAL

46. (Section 1.4) Construct a qualitative graphical model that relates the distance a business traveler is from her local airport (vertical axis) to time (horizontal axis), starting the moment the traveler departs on a homebound flight from another city. The flight is nonstop, but because of air traffic, the flight must circle the local airport for some time prior to landing.

47. (Section 4.4) How much must a person invest now in one lump sum to have the same amount of money in 25 years as a second person who invests $100 at the end of each month for 25 years, at an average annual interest rate of 9% a year, compounded monthly?

48. (Section 5.4) Construct a tree diagram for the holdings of the player in Problem 39 over four flips of the coin.

49. (Section 6.4) A person listens to one of three radio stations while driving. Station 1, which plays only country music, is his favorite, and he listens to it 50% of the time. The format at station 2 is half country music and half talk radio, and he listens to that station 40% of the time. Station 3 plays country music only 10% of the time, so he listens to that station only 10% of the time. A passenger gets into this person's car and hears a country music song playing on the radio. What is the probability that the radio is tuned to station 1?

RECOMMENDING ACTION

50. Respond by memo to the following request:

> ### MEMORANDUM
>
> To: J. Doe Reader
>
> From: President's Office
>
> Date: Today
>
> Subject: **Life Cycle of Aluminum Cans**
>
> I am scheduled to address the Save-A-Can Foundation next month, and I am thinking of presenting the life cycle of the aluminum cans we use for our soft drink products as a Markov Chain. Can I do this? Would discarded aluminum cans be an absorbing state? What about recycling?

CHAPTER 7 KEYS

KEY WORDS

absorbing Markov chain (p. 485)

absorbing state (p. 485)

distribution vector (p. 467)

fundamental matrix (p. 488)

initial absorption vector (p. 488)

Markov chain (p. 453)

period (p. 452)

regular transition matrix (p. 474)

stable distribution vector (p. 475)

state (p. 452)

system (p. 452)

transitional probabilities (p. 454)

transition diagrams (p. 453)

transition matrix (p. 455)

KEY CONCEPTS

7.1 Markov Chains

- A Markov Chain is a set of objects, a finite set of states, and a set of consecutive time periods such that:

  1. At any given time, each object must be in one and only one state, although different objects can be in the same state.
  2. The probability that an object moves from one state to another or remains in the same state over a single time period depends only on the beginning and ending states.

- The sum of all probabilities of transitions from a given state over one time period is 1.

- A Markov chain can be represented graphically by a transition diagram in which each state is designated by a unique integer enclosed in a circle. Movements between states are represented by arrows, with the arrowheads indicating the direction of movement. The probability associated with each move is placed on the arrow representing the transition.

- A Markov chain can be represented by a transition matrix whose elements are transitional probabilities. The rows indicate the states an object is coming from, and the columns indicate the states an object is going to.

- The elements in each row of a transition matrix sum to 1.

- The probability of moving from state i to state j over n time periods is the i-j element of $\mathbf{P}^n$, the nth power of the transition matrix for the Markov chain.

7.2 Distribution Vectors

- The sum of the components of a distribution vector is 1.

- The distribution vector for a Markov chain after one period is the product of the initial distribution vector and the transition matrix for the Markov chain.

- A regular transition matrix has a unique stable distribution vector.

7.3 Absorbing States

- Transition matrices for absorbing Markov chains are not regular.

- All objects in an absorbing Markov chain are captured eventually by the absorbing states. If there is only one absorbing state, then all objects in the system eventually gravitate to that state.

- If an absorbing Markov chain has more than one absorbing state, then the fraction of objects that gravitates to each absorbing state over the long run depends on the initial distribution vector.

- By convention, we label the nonabsorbing states with smaller positive integers than the absorbing states.

- Time periods for a Markov chain may vary in length when appropriate.

KEY FORMULAS

- If $\mathbf{P}$ is a transition matrix for a Markov chain, $\mathbf{d}_0$ is the initial distribution vector, and $\mathbf{d}_{n-1}$ is the distribution vector after $n-1$ periods, then the distribution vector after n periods is

$$\mathbf{d}_n = \mathbf{d}_0 \mathbf{P}^n$$

or

$$\mathbf{d}_n = \mathbf{d}_{n-1} \mathbf{P}$$

- If $\mathbf{P}$ is a regular transition matrix for a Markov chain, then its stable distribution vector $\mathbf{s}$ is found by solving the system

$$\mathbf{s} = \mathbf{s}\mathbf{P}$$

with an additional equation obtained by setting the sum of the components of $\mathbf{s}$ equal to 1.

- If an absorbing Markov chain has N states, of which k are absorbing, then by indexing the k absorbing states last, we generate a transition matrix in the partitioned form

$$\mathbf{P} = \left[\begin{array}{c|c} \mathbf{Q} & \mathbf{R} \\ \hline \mathbf{O} & \mathbf{I_k} \end{array} \right]$$

where $\mathbf{I_k}$ is a $k \times k$ identity matrix with one row and one column for each absorbing state and $\mathbf{O}$ is a matrix of all zero elements.

- The fundamental matrix for an absorbing Markov chain with a transition matrix in the partitioned form given above is

$$\mathbf{F} = (\mathbf{I} - \mathbf{Q})^{-1}$$

where $\mathbf{I}$ is an identity matrix of the same order as $\mathbf{Q}$.

- Let $\mathbf{a}$ be an initial absorption vector for an absorbing Markov chain with a transition matrix in the partitioned form given above, and let $\mathbf{F}$ be the associated fundamental matrix. The matrix product $\mathbf{aFR}$ is the fraction of objects in the system that begin in nonabsorbing states and eventually land in each absorbing state, listed by ascending state numbers.

TESTING YOURSELF

Use the following problems to test yourself on the material in Chapter 7. The odd problems comprise one test, the even problems a second test.

1. The planning commission of a large metropolitan city studied the commuting habits of residents and found that 98% of all commuters who began a year using mass transportation continued to do so two years later. The other 2% no longer used mass transportation and commuted to work in private cars. The commission also found that 95% of all commuters who began a year driving to work in private cars continued to do so two years later. The other 5% switched to mass transportation. Model this system as a Markov chain under the assumption that existing trends will continue unchanged into the future. Describe the objects, states, and periods in this system, and construct a transition matrix for the system.

2. Consider a system modeled by the transition matrix

$$\mathbf{P} = \begin{bmatrix} 1 & 0 & 0 \\ 0 & 0 & 1 \\ 1 & 0 & 0 \end{bmatrix}$$

Find $\mathbf{P}^2$, $\mathbf{P}^3$, and $\mathbf{P}^4$. What conclusions can you make about where objects will be at different points in time?

3. Construct a transition diagram from the transition matrix

$$\mathbf{P} = \begin{bmatrix} .4 & 0 & .3 & .3 \\ .2 & .7 & 0 & .1 \\ 0 & 0 & 1 & 0 \\ .5 & .5 & 0 & 0 \end{bmatrix}$$

4. Find the stable distribution vector for a Markov chain model with the transition matrix

$$\mathbf{P} = \begin{bmatrix} .6 & .4 \\ .3 & .7 \end{bmatrix}$$

5. Consider a system modeled by the transition matrix

$$\mathbf{P} = \begin{bmatrix} .5 & .5 \\ .3 & .7 \end{bmatrix}$$

Determine the probability of moving from state 1 to state 2
(*a*) after two periods
(*b*) over the long run.

6. Construct a transition matrix from the following transition diagram:

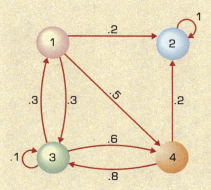

7. Find the stable distribution vector for a Markov chain model with transition matrix

$$\mathbf{P} = \begin{bmatrix} .2 & .8 & 0 \\ 0 & .3 & .7 \\ 0 & 1 & 0 \end{bmatrix}$$

8. A university has three undergraduate colleges: liberal arts, business, and engineering. A recent study by the registrar's office revealed that each year 15% of all liberal arts majors transfer to the business college, 1% of all liberal arts majors transfer to engineering, 10% of all engineering majors transfer to liberal arts, 20% of all engineering majors transfer to business, 30% of all business majors transfer to liberal arts, and all other students remain in their current major. Model this system as a Markov chain under the assumption that existing trends will continue unchanged into the future. Describe the objects, states, and periods in this system, and construct a transition matrix for the system.

9. Find the stable distribution vector for a Markov chain model with the transition matrix

$$\mathbf{P} = \begin{bmatrix} .1 & .9 & 0 \\ 0 & 1 & 0 \\ .3 & 0 & .7 \end{bmatrix}$$

10. A system with four states, two of which are absorbing, is modeled by a Markov chain with the transition matrix

$$\mathbf{P} = \begin{bmatrix} 1 & 0 & 0 & 0 \\ .1 & 0 & .4 & .5 \\ .3 & .2 & .2 & .3 \\ 0 & 0 & 0 & 1 \end{bmatrix}$$

Initially, the same number of objects are in each state. Determine the fraction of objects that will be attracted to state 1 over the long run.

PROBABILITY
DISTRIBUTION MODELS

RANDOM NUMBER GENERATORS

We noted in Section 6.2 that an effective way to assign probabilities to events is through historical data. By replicating the same process many times, we observe the frequency of outcomes that actually occur and then use this information to calculate empirical probabilities. For example, to determine the probability that a light bulb will operate 1000 hours without failing, we could test them. If we keep 1600 light bulbs on for 1000 hours each and find that only 21 failed, then the fraction $\frac{21}{1600}$ is an empirical probability that any lightbulb will fail in the first 1000 hours of use.

To generate historical data on a process, we can run the process itself repeatedly or, alternatively, we can build a model of the process and then run the model. This latter approach is called simulation. A popular type of simulation is a *Monte Carlo simulation*. At its core is a *random number generator* that produces real numbers that are equally likely to have any value between 0 and 1 (including 0 but not including 1). Calculators often have a key labeled RAND, RDM, or RANDOM for this purpose. Many spreadsheets such as Quattro Pro, Lotus 1-2-3, and Excel include random number generators under the function name RAND. In each case, a single keystroke on a calculator or a simple call to the resident software produces a number between 0 and 1. The important feature of a random number generator is that any number in the interval [0, 1) is equally likely to occur.

Throughout this chapter we adopt the convention of using a bracket next to a number to indicate that the number is included in an interval and a parenthesis next to a number to indicate that the number is not included in the interval. The interval [0, 1) includes zero but not one.

If each number in the interval [0, 1) is as likely to occur as any other number, then half the numbers will fall in the interval [0, .5) and the other half in the interval [.5, 1). Thus, the probability of getting a random number r between 0 and .5 is $\frac{1}{2}$, and the probability of getting a random number r in [.5, 1) is also $\frac{1}{2}$. Similarly, one-tenth of the numbers will fall in the interval [0, .1), another tenth in the interval [.1, .2), another tenth in the interval [.2, .3), and so on. Thus, the probability of getting a random number r between 0 and .1 is $\frac{1}{10}$, the probability of getting a random number r between .1 and .2 is $\frac{1}{10}$, and the probability of getting a random number r between .9 and 1 is $\frac{1}{10}$. In general, the probability of getting a random number in the interval [c, d) when $0 \le c < d \le 1$ is $d - c$. That is

$$P(c \le r < d) = d - c \tag{1}$$

> *A random number generator is a mechanism for creating real numbers that are equally likely to be any value between 0 and 1, including 0 but not including 1.*

EXAMPLE 1

What is the probability of obtaining from a random number generator a number (*a*) between $\frac{1}{3}$ and $\frac{2}{3}$, (*b*) between .75 and .84, and (*c*) between .15 and .446?

Solution

(*a*)
$$P\left(\frac{1}{3} \le r < \frac{2}{3}\right) = \frac{2}{3} - \frac{1}{3} = \frac{1}{3}$$

(*b*) $P(.75 \le r < .84) = .84 - .75 = .09$

(*c*) $P(.15 \le r < .446) = .446 - .15 = .296$

Equation (*1*) provides the rationale for using random number generators to replicate other probability distributions. Suppose, for example, we want to replicate a coin flip.

Rather than actually flipping a coin many times, we can model the process with random numbers, assigning to each random number the result of one coin flip. If a random number is less than .5, it can represent a head; otherwise, it represents a tail. We know from Equation (*1*) that

$$P(0 \leq r < .5) = .5 - 0 = .5$$

and

$$P(.5 \leq r < 1) = 1 - .5 = .5$$

Therefore, this model of a coin flip will give us a head with probability $\frac{1}{2}$ and a tail with probability $\frac{1}{2}$. Table 8.1 shows the results of ten different coin flips obtained using this model and a random number generator. For presentation purposes, each random number r is rounded to three decimal places.

TABLE 8.1 Coin Toss Simulation

TOSS	1	2	3	4	5	6	7	8	9	10
r	.024	.737	.945	.100	.282	.515	.253	.610	.621	.627
Coin	H	T	T	H	H	T	H	T	T	T

EXAMPLE 2

Use a random number generator to model the process of throwing two dice and identifying the faces that turn up.

Solution

We model each die with a single random number, hence two random numbers are required to model two dice. Each die has six faces, numbered 1 through 6, and each face has the same probability of appearing, namely, $\frac{1}{6}$.

To model the roll of one die, we use a random number generator to obtain a random number r. If $0 \leq r < \frac{1}{6}$, which according to Equation (*1*) occurs with probability

$$P\left(0 \leq r < \frac{1}{6}\right) = \frac{1}{6} - 0 = \frac{1}{6}$$

we model the face of the die as 1. If $\frac{1}{6} \leq r < \frac{2}{6}$, which according to Equation (*1*) occurs with probability

$$P\left(\frac{1}{6} \leq r < \frac{2}{6}\right) = \frac{2}{6} - \frac{1}{6} = \frac{1}{6}$$

we model the face of the die as 2. Similarly, if r is between $\frac{2}{6}$ and $\frac{3}{6}$, we model the face of the die as 3; if it is between $\frac{3}{6}$ and $\frac{4}{6}$, we model the face of the die as 4; between $\frac{4}{6}$ and $\frac{5}{6}$, we model the face of the die as 5; and between $\frac{5}{6}$ and 1, we model the face of the die as 6.

Table 8.2 shows the results of using this model and a random number generator for ten rolls of two dice. For each roll, we generated two random numbers r_1 and r_2, one for each of the two dice. For presentation purposes, all random numbers are rounded to four decimal places.

TABLE 8.2   **Two Dice Simulation**

ROLL	r_1	r_2	FIRST DIE	SECOND DIE
1	.5348	.8112	4	5
2	.9804	.1222	6	1
3	.5953	.4024	4	3
4	.4058	.3935	3	3
5	.4321	.6926	3	5
6	.9734	.7033	6	5
7	.5812	.9196	4	6
8	.2624	.4464	2	3
9	.8329	.9739	5	6
10	.2282	.0758	2	1

EXAMPLE 3

In a two-finger game, Player A wants to reveal one finger $\frac{3}{4}$ of the time and two fingers $\frac{1}{4}$ of the time. Player B wants to reveal one finger $\frac{2}{5}$ of the time and two fingers $\frac{3}{5}$ of the time. How can both players implement their strategies without being predictable?

Solution

Prior to each move, Player A uses a random number generator to obtain a number r_A. If the number is less than .75, which will occur with probability $\frac{3}{4}$, Player A shows one finger, otherwise Player A shows two fingers. Player B also produces a number r_B with a random number generator prior to each move (Player B can use the same calculator or software package that Player A uses or a different one). If r_B is less than .4, which will occur with probability $\frac{2}{5}$, Player B shows one finger, otherwise Player B shows two fingers. Table 8.3 shows the result of using this model for ten plays of the game. For presentation purposes, all random numbers are rounded to three decimal places.

TABLE 8.3   **Two-Finger Game Simulation**

GAME	r_A	r_B	PLAYER A REVEALS	PLAYER B REVEALS
1	.619	.088	1	1
2	.116	.830	1	2
3	.223	.118	1	1
4	.043	.955	1	2
5	.204	.559	1	2
6	.048	.649	1	2
7	.963	.192	2	1
8	.270	.340	1	1
9	.635	.936	1	2
10	.808	.966	2	2

EXAMPLE 4

Describe how to simulate the demand for gasoline if demand follows the probability distribution listed in Table 8.4.

TABLE 8.4

Demand (gallons)	2,000	3,000	4,000	5,000	6,000
Probability	.09	.16	.37	.24	.14

Solution

We use a random number generator and partition the interval [0, 1) into the five subintervals

$$[0, .09), \text{ which has length } .09$$

$$[.09, .09 + .16) = [.09, .25), \text{ which has length } .16$$

$$[.25, .25 + .37) = [.25, .62), \text{ which has length } .37$$

$$[.62, .62 + .24) = [.62, .86), \text{ which has length } .24$$

$$[.86, .86 + .14) = [.86, 1), \text{ which has length } .14$$

Each time we generate a random number r, we identify the subinterval that contains r. If $0 \leq r < .09$, which according to Equation (*1*) occurs with probability .09, then we model the demand at 2,000 gallons. If $.09 \leq r < .25$, which according to Equation (*1*) occurs with probability .16, then we model the demand at 3,000 gallons. If $.25 \leq r < .62$, we model the demand at 4,000 gallons; if $.62 \leq r < .86$, we model the demand at 5,000 gallons; and if $.86 \leq r < 1$, which according to Equation (*1*) occurs with probability .14, then we model the demand at 6,000 gallons.

The power of simulation is its ability to reproduce the outcomes of games quickly.

Monte Carlo simulations use random number generators to run a process or play a game many times in a short period. Through these replications we generate an historical record on the outcomes which we can use to estimate cumulative measures of performance. For example, let us consider a simple game in which a person tosses a coin and wins a dollar when the coin turns up heads and loses a dollar when the coin turns up tails. If the probability of both outcomes is .5, then the game is a fair one and on average the person should expect to break even. But what percentage of the time will the person actually be even? This is not a question we can answer easily with the methods developed in Chapter 6, but if we play the game sufficiently often, we can use empirical probabilities to estimate the answer, and Monte Carlo simulations allow us to do just that.

Table 8.1 listed the results of simulating ten tosses of a coin. The first toss is a head, and the player wins $1. The player's total winnings at this point are $1, and the player is ahead, not even. The second toss is a tail, and the player loses $1. The player's total winnings are now $0—the player has won a dollar and then lost a dollar—and the player is even for the first time since play began. The third toss is also a tail, and the player loses $1. Total winnings are now −$1, and the player is behind. After three tosses, the player has been even only once. The fourth toss is a head, and the player wins $1. Total winnings are now $0, and the player is even for the second time in the game. Continuing, we generate Table 8.5. After ten coin tosses, the player is behind by $2 and has broken even four times.

TABLE 8.1 (REPEATED) **Coin Toss Simulation**

TOSS	1	2	3	4	5	6	7	8	9	10
r	.024	.737	.945	.100	.282	.515	.253	.610	.621	.627
Coin	H	T	T	H	H	T	H	T	T	T

Of course, Table 8.5 represents only one set of ten coin tosses, or one *run* of the simulation. If we repeat the experiment, and simulate other runs of ten coin tosses, as we did in Tables 8.6 and 8.7, we are likely to get different outcomes. In Table 8.6, the player broke even after ten tosses but was only even twice during the game, once after the second toss and again after the tenth toss. For most of this game, the player was ahead with positive total winnings. In Table 8.7, the player never broke even. He was behind after the first toss and remained behind throughout all ten tosses.

Tables 8.5 through 8.7 are taken directly from a spreadsheet that was programmed to simulate the coin-toss game. Once the program is coded, a single keystroke produces each simulation, and we can generate as many simulations as we want by simply hitting the recalculating key as many times as we choose. Doing so twenty times, we generate the summarized information in Table 8.8.

We are interested in the total number of times a player breaks even while flipping a coin ten times. The entry for run 1 in Table 8.8 is taken from the last line in Table 8.5, the entry for run 2 is taken from the last line of Table 8.6, and the entry for run 3 is taken from the last line in Table 8.7. The average (see Section 5.3) number of times a player breaks even during ten coin tosses, based on 20 simulation runs, is

$$\frac{3(0) + 1(1) + 11(2) + 3(3) + 1(4) + 1(5)}{20} = 2.05$$

Because 2.05 is an average for a specific number of runs, it is only an *estimate* of the answer we seek. If we did the simulation with another group of 20 runs, we would probably get a slightly different answer. We expect, however, that the average will remain fairly stable when we average over a large enough set of runs.

TABLE 8.5 **Coin Toss Game Simulation: Run #1**

TOSS	RANDOM NUMBER	COIN FACE	PAYOFF	TOTAL WINNINGS	TIMES EVEN
1	0.024390	H	1	1	0
2	0.736792	T	−1	0	1
3	0.945015	T	−1	−1	1
4	0.100406	H	1	0	2
5	0.282385	H	1	1	2
6	0.514565	T	−1	0	3
7	0.253310	H	1	1	3
8	0.610121	T	−1	0	4
9	0.621430	T	−1	−1	4
10	0.627292	T	−1	−2	4

TABLE 8.6   Coin Toss Game Simulation: Run #2

TOSS	RANDOM NUMBER	COIN FACE	PAYOFF	TOTAL WINNINGS	TIMES EVEN
1	0.010103	H	1	1	0
2	0.992527	T	−1	0	1
3	0.382885	H	1	1	1
4	0.140333	H	1	2	1
5	0.695844	T	−1	1	1
6	0.393349	H	1	2	1
7	0.603286	T	−1	1	1
8	0.069956	H	1	2	1
9	0.806370	T	−1	1	1
10	0.597877	T	−1	0	2

TABLE 8.7   Coin Toss Game Simulation: Run #3

TOSS	RANDOM NUMBER	COIN FACE	PAYOFF	TOTAL WINNINGS	TIMES EVEN
1	0.894385	T	−1	−1	0
2	0.628761	T	−1	−2	0
3	0.982003	T	−1	−3	0
4	0.694529	T	−1	−4	0
5	0.353550	H	1	−3	0
6	0.334270	H	1	−2	0
7	0.413341	H	1	−1	0
8	0.616334	T	−1	−2	0
9	0.612518	T	−1	−3	0
10	0.718490	T	−1	−4	0

TABLE 8.8   Coin Toss Game Simulations

RUN #	TIMES EVEN	RUN #	TIMES EVEN	RUN #	TIMES EVEN
1	4	8	2	15	2
2	2	9	2	16	2
3	0	10	5	17	3
4	1	11	0	18	2
5	2	12	0	19	2
6	3	13	2	20	3
7	2	14	2		

The procedure for using simulation to estimate a measure of performance is as follows:

ESTIMATING PERFORMANCE MEASURES WITH MONTE CARLO SIMULATION:

Step 1 Using a random number generator, create a model to simulate the game.

Step 2 Make a number of different runs using the model, thereby playing the game many times. After each run, record the value of the performance measure.

Step 3 Calculate the mean of all recorded performance measures, and use the mean as an estimate of that performance measure.

EXAMPLE 5

A gasoline station on Cape Cod begins the summer fully stocked with 6,000 gallons of regular gasoline and then receives a delivery at the beginning of each week. Demand for regular gasoline during the ten-week summer period depends on the number of tourists that vacation on the Cape, and the number fluctuates according to weather conditions in the Northeast. Based on past experience, the station believes that the weekly demand pattern is given by Table 8.4.

TABLE 8.4 (REPEATED)

Demand (gallons)	2,000	3,000	4,000	5,000	6,000
Probability	.09	.16	.37	.24	.14

Because of distance, the station must order and accept the same amount of gasoline from its supplier each week. If that amount is too small, the station depletes its inventory and loses profit at the rate of 11¢ a gallon. If the amount of gasoline delivered is too much, then whatever cannot fit into the tank at the station, which can hold a maximum of 6,000 gallons, is returned to the supplier and incurs a restocking fee of 15¢ a gallon. This year, the station owner is thinking of contracting for 4,500 gallons each week. What losses, if any, can be expected?

Solution

The performance measure of interest here is lost revenue, either lost sales when the gasoline inventory is too low or the restocking fee when inventory is too high. We model the ten-week summer period using a random number generator to set the demand for each week, as described in Example 4. One run of the simulation is summarized in Table 8.9.

We begin with a stock of 6,000 gallons. The first random number is between 0.09 and 0.25, indicating a demand of 3,000 gallons for the first week. The station can meet this

TABLE 8.9 Gasoline Station Simulation: Run #1

WEEK	DELIVERY	RETURNED DELIVERY	RESTOCKING FEE	STOCK AT BEGINNING	RANDOM NUMBER	DEMAND	STOCK AT END	SHORTFALL	LOST PROFIT
1				6000	0.09937684	3000	3000	0	$0
2	4500	1500	$225	6000	0.41834836	4000	2000	0	$0
3	4500	500	$75	6000	0.93351055	6000	0	0	$0
4	4500	0	$0	4500	0.34586173	4000	500	0	$0
5	4500	0	$0	5000	0.26153855	4000	1000	0	$0
6	4500	0	$0	5500	0.96942239	6000	0	500	$55
7	4500	0	$0	4500	0.95261264	6000	0	1500	$165
8	4500	0	$0	4500	0.57224534	4000	500	0	$0
9	4500	0	$0	5000	0.11023702	3000	2000	0	$0
10	4500	500	$75	6000	0.45053981	4000	2000	0	$0
			$375						$220

Total lost revenue = $595

demand, so there is no shortfall and there are no lost sales. At the end of the first week, the station has an inventory of 6,000 − 3,000 = 3,000 gallons. At the beginning of week 2, a shipment of 4,500 gallons arrives, but the station has room in its tank for only 3,000 more gallons. The station fills its 6,000-gallon tank and returns 1,500 gallons, incurring a restocking fee of $0.15(1500) = $225.

The second random number is between 0.25 and 0.62, indicating a demand of 4,000 gallons for the second week. The station can meet this demand, so there is no shortfall and there are no lost sales. At the end of the second week, the station has an inventory of 6,000 − 4,000 = 2,000 gallons. At the beginning of week 3, a shipment of 4,500 gallons arrives, but the station has room to accept only 4,000 gallons. The station fills its 6,000-gallon tank and returns 500 gallons, incurring a restocking fee of $0.15(500) = $75.

The third random number is between 0.86 and 1, representing a demand of 6,000 gallons for the third week. The station can meet this demand, so there is no shortfall and there are no lost sales. At the end of the third week, the station has an inventory of 0 gallons. At

TABLE 8.10 Gasoline Station Simulations

RUN	REVENUE LOST (DOLLARS)	RUN	REVENUE LOST (DOLLARS)	RUN	REVENUE LOST (DOLLARS)
1	$595	11	$540	21	$1,210
2	$405	12	$675	22	$1,650
3	$555	13	$1,255	23	$760
4	$525	14	$1,120	24	$390
5	$675	15	$505	25	$425
6	$935	16	$745	26	$515
7	$975	17	$1,155	27	$855
8	$1,425	18	$975	28	$595
9	$595	19	$805	29	$525
10	$0	20	$445	30	$385

the beginning of week 4, a shipment of 4,500 gallons arrives and the station accepts it all, avoiding a restocking fee.

Skipping to week 6 in Table 8.9, we find that the station has 1,000 gallons in stock at the end of week 5 and can accept the entire 4,500-gallon delivery for week 6, bringing the stock to 5,500 gallons. The sixth random number is between 0.86 and 1, so the demand for the sixth week is 6,000 gallons, which the station cannot satisfy. The station will experience a shortfall of 500 gallons, the difference between the demand for 6,000 gallons and the inventory of 5,500 gallons, and lost profits of $0.11(500) = $55.

Table 8.9 is one simulation of the station's expenses over a single summer time period. If we run the simulation again, we generate a different set of results. Table 8.10 summarizes the results of 30 runs, with the first run taken from Table 8.9. The mean revenue lost over these 30 runs is $740.50. Equally interesting, perhaps, are the median revenue loss of $635 and the maximum revenue loss of $1,650.

IMPROVING SKILLS

Problems 1 through 8 refer to a three-finger game in which one of the players wants to reveal one finger with probability p_1, two fingers with probability p_2, and three fingers with probability p_3. In each case, describe how to use a random number generator to implement the strategy.

1. $p_1 = .2$, $p_2 = .5$, $p_3 = .3$
2. $p_1 = .13$, $p_2 = .34$, $p_3 = .53$
3. $p_1 = .65$, $p_2 = .15$, $p_3 = .2$
4. $p_1 = \frac{1}{3}$ $p_2 = \frac{1}{2}$ $p_3 = \frac{1}{6}$
5. $p_1 = \frac{3}{7}$, $p_2 = \frac{2}{7}$, $p_3 = \frac{2}{7}$
6. $p_1 = 0$, $p_2 = \frac{1}{2}$, $p_3 = \frac{1}{2}$
7. $p_1 = .05$, $p_2 = 0$, $p_3 = .95$
8. $p_1 = .65$, $p_2 = .35$, $p_3 = 0$
9. Use the ten random numbers listed under the r_1 heading in Table 8.2 to create another simulation of the coin toss game shown in Table 8.5.
10. Use the ten random numbers listed under the r_2 heading in Table 8.2 to create another simulation of the coin toss game shown in Table 8.5.
11. Use the ten random numbers listed in Table 8.5 to create another simulation of the gasoline station simulation shown in Table 8.9.
12. Use the ten random numbers listed in Table 8.6 to create another simulation of the gasoline station simulation shown in Table 8.9.
13. Revise Table 8.9, using the same random numbers, if weekly deliveries are 4,000 gallons.
14. Revise Table 8.9, using the same random numbers, if weekly deliveries are 5,000 gallons.

CREATING MODELS

Simulate each of the processes described in Problems 15 through 27 using random numbers selected sequentially from the following list. Stop after generating ten lines of output.

.92630	.78240	.19267	.95457	.53497	.23894	.37708	.79862
.79445	.78735	.71549	.44843	.26104	.67318	.00701	.34986
.59654	.71966	.27386	.50004	.05358	.94031	.29281	.18544
.31524	.49587	.76612	.39789	.13537	.48086	.59483	.60680
.06348	.76938	.90379	.51392	.55887	.71015	.09209	.79157

15. A person repeatedly flips a coin, winning $2 each time the coin matches the previous flip and losing $1 each time the coin differs from the previous flip. The first flip establishes the game and involves no money. How much can the person expect to win?

16. A person plays roulette, always betting on red. The player wins $1 when a red number appears and loses $1 otherwise. How many times will the player break even? Recall that a roulette wheel has 18 red numbers, 18 black numbers, and 2 green numbers.

17. The roulette player in the previous problem decides never to bet on two reds in a row, so he bets red only when the previous number was not red and declines to wager when the previous number was red. How much can this player expect to win?

18. A person invests $1,000 in a bank account and plans to leave it there for five years. The account accrues interest, compounded semiannually, but the interest rate can vary. Currently, the nominal interest rate is 4% a year, but in future periods the probability that it will remain 4% is .4. Future rates may be $4\frac{1}{4}$% with probability .2, $4\frac{1}{2}$% with probability .1, $4\frac{3}{4}$% with probability .1, and $3\frac{3}{4}$% with probability .2. How much can the depositor expect to have in the account after five years?

19. A corporation creates a sinking fund by depositing $50,000 every quarter into an account that accrues interest, compounded quarterly. The nominal interest rate on the account is currently 6% a year, but it may vary in future quarters. The probability of the interest rate remaining 6% is .2. Future rates may be 7% with probability .3, 8% with probability .2, 5% with probability .2, and 4% with probability .1. How much can the corporation expect in the account after $2\frac{1}{2}$ years?

20. A child's allowance is set at five dollars a week, payable on Saturday, at the beginning of the summer. The child's main chore is to mow the family lawn every Friday. To motivate the child, the parents agree to increase the child's allowance every week by one dollar when the lawn is mowed on time and to decrease the child's allowance every week by two dollars when the lawn is not mowed on time, although the allowance is never decreased to less than zero. What will the child's allowance be at the end of summer, ten weeks later, if each week there is an 80% chance that the child mows the lawn on time?

21. Each week, a real estate broker sells zero houses with probability .60, one house with probability .25, two houses with probability .10, and three houses with probability .05. The commission on a sold house is $5,000 with probability .6, $6,000 with probability .2, $7,000 with probability .1, and $8,000 with probability .1. How much commission can the broker expect over the next ten weeks?

22. The time between trains arriving in New York City from a Long Island railroad station during the morning commuter hours is 15 minutes with probability .4, 16 minutes with probability .2, 17 minutes with probability .1, 14 minutes with probability .2, and 13 minutes with probability .1. What will the arrival pattern of trains be between 6:30 a.m. and 8:30 a.m., assuming the first train comes in at 6:42 a.m.?

23. Customers begin arriving at a bank 15 minutes before it opens. The time between customer arrivals in this quarter-hour period is one minute with probability .3, two minutes with probability .2, three minutes with probability .4, and four minutes with probability .1. How many customers are in line when that bank opens?

24. Three costs are associated with each item that a business sells: a production cost, a shipping cost, and an administrative cost. The production cost per item is either $22, $35, or $47 with probabilities .3, .5, and .2, respectively. The shipping cost per item is either $2, $4, or $7 with probabilities .1, .5, and .4, respectively. The administrative cost per item is either $11 or $28 with probabilities $\frac{2}{3}$ and $\frac{1}{3}$, respectively. What is the average cost of each item sold?

25. A slot machine has three windows, and the picture of either an apple, an orange, or a watermelon appears in each window. Each fruit is equally likely to appear in each window. A player wins $10 if the same fruit appears in all three windows and loses $1 otherwise. What are her expected winnings after 10 plays of this slot machine?

26. Suppose the slot machine in the previous problem is fixed so that the probabilities of getting an apple in window 1, an orange in window 2, and a watermelon in window 3 are all .4, and the probability of getting either of the other two fruits in each window is .3. What are the player's expected winnings after ten plays of this slot machine?

27. In one dental office, the hygienist first cleans a patient's teeth and then the patient is examined by the dentist. The hygienist completes a cleaning in 20 minutes 80% of the time, but the other 20% of the time a cleaning takes 30 minutes. The dentist completes the examination in 10 minutes for 65% of all patients and needs 15 minutes for the remaining 35% of patients. However, the dentist tries to complete an examination quickly when the hygienist has used a full 30 minutes; in these cases, the examination can be reduced by one minute 50% of the time, by two minutes 10% of the time, and not at all the remaining times. How much time does a patient spend with the two practitioners?

EXPLORING IN TEAMS

28. The gasoline station in Example 5 can contract for any amount of gasoline, providing the same amount is delivered each week. The station can contract for 4,500 gallons, as in Example 5, or 3,280 gallons, or any other amount. Ideally, management wants to order the amount that will maximize profit, which is 11¢ a gallon for each gallon sold less restocking fees. Use simulation to recommend a contractual amount to the station's manager.

29. In the game of Guava, each player places $1 on the center of the table, called the pot, and then rolls a single die. If neither player rolls a 6 or if both players roll a 6, the game continues with another roll and the pot is increased by $1 from each player. If only one player rolls a 6, that player wins all the money in the pot. What is the expected amount won each time in this two-person game?

EXPLORING WITH TECHNOLOGY

30. Using an electronic spreadsheet, create your own simulation of the coin-toss game described in Tables 8.5 through 8.7. Using your model, run 20 simulations of the game, with each run involving 10 coin flips. Use your results to construct a table similar to Table 8.8, and calculate the average number of times a player breaks even.

31. Using an electronic spreadsheet, construct your own simulation of the gas station problem described in Example 5. Using your model, run 30 simulations for the pro-

cess, each covering a 10-week period. Use your results to construct a table similar to Table 8.9, and calculate the mean summer revenue loss.

32. Create spreadsheet models for the processes described in Problems 15 through 27 and estimate an answer for each.

REVIEWING MATERIAL

33. (Section 1.4) Construct a qualitative graphical model that relates the number of students in a high school (vertical axis) who are absent because of the flu to time (horizontal axis), covering the period of a school year. On the same coordinate system, also construct a qualitative graphical model that relates the number of students in the same high school (vertical axis) who recover from the flu.

34. (Section 3.3) Use graphical methods to solve the linear program:

$$\text{Minimize:} \quad z = x + 10y$$

$$\text{subject to:} \quad 2x + 5y \geq 40{,}000$$

$$3x - 4y \leq 36{,}000$$

$$y \leq 14{,}000$$

35. (Section 6.3) The probability of a company reporting a profit next quarter is .65. If the company reports a profit, there is a 85% chance that the company will declare a dividend; if the company does not report a profit, there is only a 40% chance that the company will declare a dividend. What is the probability that the company will report a profit next quarter and *not* declare a dividend?

RECOMMENDING ACTION

36. Respond by memo to the following request:

MEMORANDUM

To: J. Doe Reader

From: Housing Department

Date: Today

Subject: **Assigning Available Apartments**

We have 118 applications for 22 apartments in our newest low-cost housing facility, and we need a fair way to assign apartments to applicants. We have decided not to assign apartments on a first-come, first-served basis because the mayor feels strongly that all applicants who meet the filing deadline should have the same opportunity for an apartment.

Can we use a random number generator to match apartments with applicants? If so, how? Please be specific as to which applicants would get the apartments and which would not.

UNIFORM DISTRIBUTIONS

> *A sample space is uniformly distributed between the numbers $a < b$ when the outcomes are equally likely to be any value in the interval [a, b).*

Random number generators produce numbers that are equally likely to be any real number between 0 and 1, including 0 but not including 1. These outcomes are a special case of a sample space that is *uniformly distributed*. In general, a sample space is uniformly distributed on the interval $[a, b)$ if outcomes are equally likely to be any real number between a and b, including a but not including b. For example, a sample space is uniformly distributed on $[3, 7)$ if outcomes are equally likely to be *any* real number between 3 and 7, including 3 but not including 7. A sample space is uniformly distributed on $[-10, 950)$ if outcomes are equally likely to be *any* real number between -10 and 950, including -10 but not including 950. Random number generators produce outcomes that are uniformly distributed on $[0, 1)$.

If each number in the interval $[a, b)$ is as likely to occur as any other number, then when we divide this interval into halves, we expect half the outcomes to fall in the first subinterval and the other half to fall in the second subinterval. The midpoint of the interval

> *The median and the mean of a sample space uniformly distributed on [a, b) occur at the midpoint $\frac{a + b}{2}$.*

$[a, b)$ is $\frac{a + b}{2}$. Thus the probability of getting an outcome in the interval $\left[a, \frac{a + b}{2}\right)$ is $\frac{1}{2}$, the same as the probability of getting an outcome in the interval $\left[\frac{a + b}{2}, b\right)$. It follows that the median (see Section 5.3) of a sample space uniformly distributed on $[a, b)$ occurs at the midpoint $\frac{a + b}{2}$. Furthermore, an outcome is as likely to be a little above the midpoint as it is to be a little below the midpoint, so the mean of the sample space also occurs at the midpoint.

EXAMPLE 1

The time between bus arrivals at a particular location is equally likely to be any value between 12 and 18 minutes, including 12 minutes but excluding 18 minutes. What does this mean to riders who want to board a bus at that stop?

Solution

Bus arrivals are uniformly distributed on the interval $[12, 18)$, with both numbers in units of minutes. The mean and the median time between arrivals is $\frac{(12 + 18)}{2} = 15$ minutes, but buses can arrive up to 3 minutes later or up to 3 minutes earlier, and each possibility in this range is as likely to occur as any other.

EXAMPLE 2

Weekly demand for gasoline at a service station varies between 2,000 and 6,000 gallons, uniformly distributed. What does this mean?

Solution

Weekly demand is never less than 2,000 gallons and never more than 6,000 gallons, but every real number in this range is as likely to be the actual demand as any other number. The mean and median demands are $\frac{(2,000 + 6,000)}{2} = 4,000$ gallons.

A random number generator uniformly distributed on [0, 1) can be used to create random numbers that are uniformly distributed on any finite interval.

With our knowledge of straight lines (see Section 1.5), we can use a random number generator uniformly distributed on [0, 1) to create random numbers that are uniformly distributed on any interval [a, b). The approach we will use is shown schematically in Figure 8.1.

We limit our interest to the interval [0, 1) on the horizontal axis and to the interval [a, b) on the vertical axis. Points between 0 and 1 on the horizontal axis are associated with random numbers uniformly distributed on [0, 1) and are denoted by r. Points between a and b on the vertical axis are associated with random numbers that are uniformly distributed on [a, b) and are denoted by R. The endpoints of the line segment in Figure 8.1 are $(0, a)$ and $(1, b)$. The slope of the straight line passing through these two endpoints is

$$m = \frac{b - a}{1 - 0} = b - a$$

and the equation of the line is

$$R = (b - a)r + a \tag{2}$$

To generate numbers R that are uniformly distributed on [a,b), we first use a random number generator to produce numbers r that are uniformly distributed on [0, 1). Then we substitute each r into Equation (2) and calculate R. This procedure is illustrated graphically by the dotted lines in Figure 8.1. The random number r_1 is obtained from a random number generator that produces uniformly distributed random numbers on the interval [0, 1). r_1 is then substituted into the right side of Equation (2) to produce R_1, which is an element of a set of random numbers uniformly distributed on the interval [a, b). An identical procedure is used to calculate R_2 from r_2 and to calculate any subsequent random numbers that may be required on the interval [a, b).

FIGURE 8.1

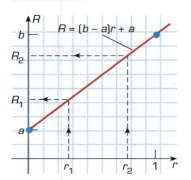

EXAMPLE 3

Generate ten numbers from a sample space that is uniformly distributed on [12, 18].

Solution
Here $a = 12$, $b = 18$, and Equation (2) reduces to

$$R = (18 - 12)r + 12 = 6r + 12 \tag{3}$$

Using a random number generator, we obtain the ten numbers shown in the r columns of Table 8.11. Substituting each of these numbers in turn into Equation (3), we calculate the numbers in the R columns of Table 8.11, where they are rounded to one decimal place. The entries in the R columns model outcomes from a sample space that is uniformly distributed on [12, 18].

TABLE 8.11

r	$R = 6r + 12$	r	$R = 6r + 12$
0.819499	16.9	0.390961	14.3
0.059492	12.4	0.812566	16.9
0.649792	15.9	0.976262	17.9
0.826910	17.0	0.813032	16.9
0.252296	13.5	0.166790	13.0

EXAMPLE 4

Demand for regular gasoline at a station on Cape Cod during the ten-week summer period is uniformly distributed between 2,000 and 6,000 gallons per week. Use a random number generator to simulate the demand over a ten-week period.

Solution

Here $a = 2,000$, $b = 6,000$, and Equation (2) reduces to

$$R = 4,000r + 2,000 \qquad (4)$$

Using a random number generator, we obtain the ten numbers listed in the r columns of Table 8.12. Substituting each of these numbers in turn into Equation (4), we calculate the numbers in the R columns of Table 8.12, where they are rounded to two decimal places. These entries model weekly gasoline demands from a sample space that is uniformly distributed on [2,000 6,000).

TABLE 8.12

r	$R = 4,000r + 2,000$	r	$R = 4,000r + 2,000$
.819499	5,278.00	.806528	5,226.11
.216748	2,866.99	.967748	5,870.99
.900191	5,600.76	.562969	4,251.88
.042029	2,168.12	.997746	5,990.98
.184579	2,738.32	.848450	5,393.80

In Example 5 of Section 8.1, we simulated the cost to a service station of contracting for 4,500 gallons of gasoline each week over the ten-week summer period. Recall that in this example the station had a 6,000-gallon storage tank, and the station had to pay its supplier a restocking fee of 15¢ a gallon for every gallon the station could not accept. The station also lost 11¢ a gallon in profits if it ran out of gasoline and could not meet customer

TABLE 8.13 Gasoline Station Simulation

WEEK	DELIVERY	RETURNED DELIVERY	RESTOCKING FEE	STOCK AT BEGINNING	RANDOM NUMBER	DEMAND	STOCK AT END	SHORTFALL	LOST PROFIT
1				6000.00	0.81949906	5278.00	722.00	0.00	$0
2	4500	0.00	$0	5222.00	0.21674841	2866.99	2355.01	0.00	$0
3	4500	855.01	$128	6000.00	0.90019104	5600.76	399.24	0.00	$0
4	4500	0.00	$0	4899.24	0.04202913	2168.12	2731.12	0.00	$0
5	4500	1231.12	$185	6000.00	0.18457914	2738.32	3261.68	0.00	$0
6	4500	1761.68	$264	6000.00	0.80652781	5226.11	773.89	0.00	$0
7	4500	0.00	$0	5273.89	0.96774768	5870.99	0.00	597.10	$66
8	4500	0.00	$0	4500.00	0.56296871	4251.87	248.13	0.00	$0
9	4500	0.00	$0	4748.13	0.99774604	5990.98	0.00	1242.86	$137
10	4500	0.00	$0	4500.00	0.84845039	5393.80	0.00	893.80	$98
			$577						$301

Total lost revenue = $878

demand. However, the demand pattern in that example was unrealistic. Weekly demands were hypothesized to be between 2,000 and 6,000 gallons but were restricted to 1,000-gallon units. A demand of 4,000 gallons was possible, but not 3,890.7 gallons; a demand of 5,000 gallons was possible, but not 5,371 gallons. If we presume that the demand for gasoline is uniformly distributed between 2,000 and 6,000 gallons, then a weekly demand of 5,371 gallons is as likely to occur as a demand for 5,000 gallons or 3,890.7 gallons.

The entries in column 2 and 4 of Table 8.12 represent one set of weekly demands over a ten-week period when those demands are uniformly distributed on [2,000, 6,000). These demands were used to generate Table 8.13, which is one simulation of the lost revenues to the station over a summer.

EXAMPLE 5

During the morning commuter hours from approximately 6:15 to 8:30 a.m., the times between bus arrivals at a particular stop are uniformly distributed between 12 and 18 minutes. Each bus carries between 36 and 48 passengers when it arrives at this stop, and all passenger loads in this range, including 36 and 48, are equally likely to occur. No passengers disembark at this stop. Riders who want to catch a bus at this stop arrive at the rate of approximately one passenger per minute and expect to board the next bus that comes. The buses, however, have a maximum capacity of 60 passengers, and drivers will not allow more than this number on any bus. Once a bus is full, riders who are denied entrance must wait for the next bus. The bus company would like to know how many passengers must wait for a second bus each morning.

Solution

We assume that passengers begin arriving at the stop as early as 6:00 a.m. There are approximately ten buses in the time period of interest, hence we simulate ten bus arrivals. One set of interarrival times was calculated in Example 3, and it becomes the third column of Table 8.14.

Passengers arrive at the stop at the rate of approximately one per minute, so there will be as many passengers arriving between buses as there are minutes between buses. If the time between buses is 16 minutes, then approximately 16 passengers will arrive in that time period. The interarrival times are not restricted to integers, so we round these times

TABLE 8.14 Bus Stop Simulation: Run #1

BUS NUMBER	RANDON NUMBER	INTERARRIVAL TIME	PEOPLE WAITING	RANDOM NUMBER	PEOPLE ON BUS	PEOPLE REJECTED
1	0.819499	16.917	17	0.9080	47	4
2	0.059492	12.357	16	0.2809	39	0
3	0.649792	15.899	16	0.7986	46	2
4	0.826910	16.961	19	0.7463	45	4
5	0.252296	13.514	18	0.1079	37	0
6	0.390961	14.346	14	0.4752	42	0
7	0.812566	16.875	17	0.2558	39	0
8	0.976262	17.858	18	0.2389	39	0
9	0.813032	16.878	17	0.3341	40	0
10	0.166790	13.001	13	0.0435	36	0

Total = 10

to the nearest minute to estimate the number of arriving passengers. If we add to this figure the number of riders denied entrance to the previous bus, we have the fourth column in Table 8.14, which lists the total number of people waiting for each bus as it arrives.

There are 13 possible passenger loads for an arriving bus—any integer between 36 and 48, inclusive—and each is equally likely to occur. Therefore, we simulate passenger load using the techniques developed in Section 8.1. We use a random number generator to obtain a random number r. If r is between 0 and $\frac{1}{13}$, we assign the bus a passenger load of 36; if r is between $\frac{1}{13}$ and $\frac{2}{13}$, we assign the bus a load of 37; if r is between $\frac{2}{13}$ and $\frac{3}{13}$, we assign the bus a load of 38, and so on. Doing so, we generate columns 5 and 6 in Table 8.14.

Using Table 8.14, we see that the first bus arrives approximately 16.9 minutes after 6:00 a.m., which means that 17 people are waiting for it. The bus arrives carrying 47 people, so there is room for only 13 additional riders and 4 people must wait for the next bus.

The second bus arrives approximately 12 minutes later, which means that 12 new people have arrived at the stop. They line up behind the 4 people denied entrance to the first bus, bringing the total waiting to $12 + 4 = 16$ people. The second bus arrives with 39 passengers on board. This bus has room for 21 passengers, so it accepts all 16 people waiting at the stop.

The third bus arrives approximately 16 minutes later, which means that 16 people have arrived and are waiting for it. The bus arrives carrying 46 people, leaving room for only 14 additional riders. Therefore, 2 people must wait for the next bus.

The fourth bus arrives approximately 17 minutes later, which means that 17 new people have arrived at the stop. They line up behind the 2 people denied entrance to the preceding bus, bringing the line to $17 + 2 = 19$ people. The fourth bus arrives carrying 45 passengers, leaving room for only 15 additional riders, so 4 people must wait for the next bus.

Continuing in this manner, we complete Table 8.14, which represents one run of the morning rush hour. On that particular morning, a total of 10 people were denied entrance to the first bus they wanted to board. Repeating the analysis for 39 additional runs, we complete Table 8.15, which contains simulated results for 40 days (8 weeks).

The mean number of people denied entrance to a bus in a single morning is 10, based on the 40 measurements in Table 8.15. The median is 5 and the mode is 0. The largest number is 58, and on 5 of the 40 simulated days (12.5% of the time), the total number of people denied entrance to a bus exceeded 20 people.

TABLE 8.15 **Bus Stop Simulations**

RUN	TOTAL REJECTED	RUN	TOTAL REJECTED	RUN	TOTAL REJECTED	RUN	TOTAL REJECTED
1	10	11	23	21	26	31	8
2	14	12	6	22	3	32	12
3	15	13	34	23	19	33	2
4	5	14	0	24	3	34	0
5	19	15	1	25	0	35	2
6	0	16	7	26	18	36	3
7	7	17	6	27	5	37	5
8	1	18	1	28	58	38	39
9	5	19	17	29	3	39	18
10	0	20	0	30	3	40	2

Monte Carlo simulations are used to estimate area.

Monte Carlo simulations are also useful for estimating area, providing the region of interest can be enclosed in a rectangle of known dimensions. We illustrate the procedure with the unit circle (a circle of radius 1 centered at the origin) as our region of interest, as shown in Figure 8.2. The interior of the unit circle is is an ideal test region because we know its area from elementary geometry. The area of any circle with radius r is πr^2, hence the area of the unit circle is $\pi(1)^2 = \pi \approx 3.14$ square units.

The unit circle can be surrounded by many rectangles. One such rectangle is drawn in Figure 8.2, but any larger rectangle would do equally as well.

FIGURE 8.2

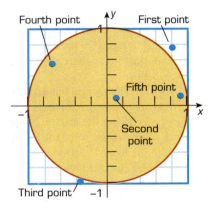

The Monte Carlo approach for calculating area is to randomly produce points within the rectangle. Five such points are shown by the bullets in Figure 8.2. We can think of these bullets as darts thrown blindly at the rectangle. If we throw enough darts, then the fraction of darts that land within the region of interest (the circle) is a good estimate for the fraction of the rectangle's area that is occupied by the region of interest. In mathematical terms, the number of randomly chosen points that fall in a region is directly proportional (see Section 1.3) to the area of the region. That is,

$$\frac{\text{number of points in the region}}{\text{area of the region}} = k \tag{5}$$

where k is a constant of proportionality. Equation (5) is true for any region, so it is also true for the rectangle that encloses the region of interest. Thus,

$$\frac{\text{number of points in the rectangle}}{\text{area of the rectangle}} = k \tag{6}$$

Eliminating k from Equations (5) and (6), we obtain

$$\frac{\text{number of points in the region}}{\text{area of the region}} = \frac{\text{number of points in the rectangle}}{\text{area of the rectangle}}$$

which can be rewritten as

$$\text{area of the region} = \left(\frac{\text{number of points in the region}}{\text{number of points in the rectangle}}\right)\text{area of the rectangle} \tag{7}$$

To estimate the area of a bounded region, multiply the fraction of randomly generated points within the region by the area of a surrounding rectangle.

Multiplying the fraction of points within a region of interest by the area of a surrounding rectangle is an estimate of the area of the region of interest. If, for example, one-fourth of the points fall within the region of interest, then the region should fill one-fourth of the area of the rectangle. Areas of rectangles are, of course, simple to calculate.

EXAMPLE 6

Use Monte Carlo simulations to estimate the area of the unit circle.

Solution

We first enclose the unit circle within a rectangle—the square illustrated in Figure 8.2, centered at the origin with sides of length 2, will do nicely. We then randomly produce points that fall within this square. The x- and y-coordinates of such points are restricted to the range between −1 and 1. Thus, we generate one set of numbers that are uniformly distributed between −1 and 1 to simulate the x-coordinates, and we generate another set of numbers uniformly distributed between −1 and 1 to simulate the y-coordinates. The results are shown partially in Table 8.16.

To find the x-coordinate of our first point, we use a random number generator and we obtain 0.902897 as a random number between 0 and 1. To transform this random number into a number that falls between −1 and 1, we use Equation (2) with $a = -1$ and $b = 1$. The random number

$$R = [1 - (-1)](0.902897) + (-1) \approx 0.8058$$

becomes a simulated x-coordinate. Our random number generator then produces 0.889503 as a second random number between 0 and 1, and we transform it into

$$R = [1 - (-1)](0.889503) + (-1) \approx 0.7790$$

for our simulated y-coordinate. We now have both coordinates for our first point: (0.8058, 0.7790). Plotting this point on Figure 8.2, we see that it falls outside the unit circle, so we do not count it as a point within the unit circle.

We generated 1,000 trial points using this process; only the first five and last five are listed in Table 8.16. The first five points are plotted in Figure 8.2. Each trial point falls within the square selected to surround the unit circle. The last column in Table 8.16 is a running total of the number of points that also fall within the unit circle; in this simulation it ends at 773. The area of a square with sides of length 2 is $(2)^2 = 4$, so it follows from Equation (7) that an estimate for the area of the unit circle is

$$\text{area of unit circle} = \left(\frac{773}{1000}\right)4 = 3.092$$

TABLE 8.16 **Unit Circle Simulation: Run #1**

TRIAL POINT	RANDOM NUMBER	x-VALUE	RANDOM NUMBER	y-VALUE	ACCEPT POINT	TOTAL ACCEPTED
1	0.902897	0.8058	0.889503	0.7790	no	0
2	0.547893	0.0958	0.572459	0.1449	yes	1
3	0.318614	−0.3628	0.016878	−0.9662	no	1
4	0.166238	−0.6675	0.791147	0.5823	yes	2
5	0.963814	0.9276	0.572970	0.1459	yes	3
⋮						
996	0.718970	0.4379	0.117311	−0.7654	yes	770
997	0.000925	−0.9982	0.821018	0.6420	no	770
998	0.277748	−0.4445	0.569400	0.1388	yes	771
999	0.891242	0.7825	0.383952	−0.2321	yes	772
1000	0.730081	0.4602	0.520522	0.0410	yes	773

If we repeat this simulation nine more times, with 1,000 trial points in each simulation, we obtain the estimates shown in Table 8.17.

The mean of these ten values is 3.139, which becomes our final estimate of the area of the unit circle. Compare this estimate to $\pi \approx 3.142$.

TABLE 8.17 Unit Circle Simulations

RUN	AREA	RUN	AREA
1	3.092	6	3.152
2	3.196	7	3.124
3	3.180	8	3.136
4	3.156	9	3.096
5	3.188	10	3.068

Monte Carlo simulations yield good estimates for area only after many trial points. Five, 10, or even 100 trials points are rarely sufficient—thousands are generally required. Fortunately, the necessary computations can be completed quickly with a spreadsheet or other comparable computer software.

It may appear from Figure 8.2 that we determined which points fell within the unit circle by plotting the points. Indeed, we did plot the first five points to illustrate the Monte Carlo method graphically, but that is not how the actual decision was made. Graphing is too labor intensive to be efficient.

The equation of the unit circle is

$$x^2 + y^2 = 1 \qquad (8)$$

A point (x, y) is inside the unit circle when $x^2 + y^2 < 1$ and is outside the unit circle when $x^2 + y^2 > 1$. To see why this is so, we apply the same logic that we used to graph feasible regions in Chapter 3. The equation defines the curve. Replacing the equality sign with an inequality sign results in a mathematical relationship that is satisfied by points on one side of the curve. Consequently, to determine where a trial point (x, y) falls in relation to the unit circle, we calculate $x^2 + y^2$ for the point. If the result is less than 1, the point is in the interior of the unit circle; if $x^2 + y^2 > 1$, the point is exterior to the unit circle. The actual computations used to complete the next-to-last column for the first five points in Table 8.16 were as follows:

POINT			RELATIONSHIP		ACCEPT
x	y	$x^2 + y^2$	TO 1	INTERIOR/EXTERIOR	POINT
0.8058	0.7790	1.256	greater than	exterior	no
0.0958	0.1449	0.030	less than	interior	yes
−0.3628	−0.9662	1.065	greater than	exterior	no
−0.6675	0.5823	0.785	less than	interior	yes
0.9276	0.1459	0.882	less than	interior	yes

EXAMPLE 7

Use a Monte Carlo simulation to estimate the area under the graph of the parabola $y = 8x - x^2$ and above the horizontal x-axis.

Solution

The graph of this equation falls partly above the x-axis and partly below. The part above the horizontal x-axis is shown in Figure 8.3. The yellow region is the area of interest. Figure 8.3 shows a window from 0 to 8 on the x-axis and from 0 to 20 on the vertical axis and its perimeter forms a rectangle around the area of interest.

FIGURE 8.3
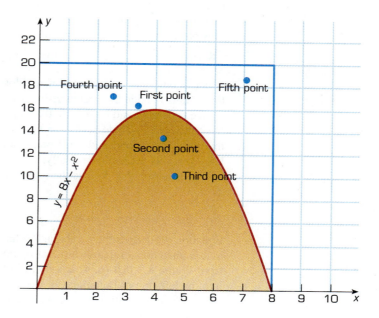

We randomly produce points that fall within this rectangular window. Such points have x-coordinates between 0 and 8 and y-coordinates between 0 and 20. Thus, we generate one set of numbers that are uniformly distributed between 0 and 8 to simulate the x-coordinates and we generate another set of numbers uniformly distributed between 0 and 20 to simulate the y-coordinates. The results are shown partially in Table 8.18.

TABLE 8.18 Parabola Simulation: Run #1

TRIAL POINT	RANDOM NUMBER	x-VALUE	RANDOM NUMBER	y-VALUE	ACCEPT POINT	TOTAL ACCEPTED
1	0.428865	3.43	0.787726	15.75	no	0
2	0.543218	4.35	0.698990	13.98	yes	1
3	0.579941	4.64	0.501601	10.03	yes	2
4	0.353370	2.83	0.855209	17.10	no	2
5	0.964425	7.72	0.920150	18.40	no	2
⋮						
996	0.688085	5.50	0.622532	12.45	yes	543
997	0.073555	0.59	0.684543	13.69	no	543
998	0.798773	6.39	0.000192	0.00	yes	544
999	0.341596	2.73	0.864566	17.29	no	544
1000	0.155317	1.24	0.325843	6.52	yes	545

To obtain our first x-coordinate, we use a random number generator and we obtain 0.428865 as a random number between 0 and 1. To transform this random number into a number that falls between 0 and 8, we use Equation (2) with $a = 0$ and $b = 8$. The random number

$$R = [8 - 0](0.428865) + 0 \approx 3.43$$

is a simulated x-coordinate. Our random number generator then produces 0.787726 as a second random number between 0 and 1. To transform this random number into one that falls between 0 and 20 we use Equation (2) with $a = 0$ and $b = 20$. The result is

$$R = [20 - 0](0.787726) + 0 \approx 15.75$$

which is our simulated y-coordinate. Plotting the point (3.43, 15.75) in Figure 8.3, we see that it falls outside the region of interest.

We generated and tested 1,000 trial points using this process; only the first five and last five are listed in Table 8.18. The first five points are plotted in Figure 8.3, where we see that only the second and third points are in the region of interest. The actual decision, however, was made algebraically.

The equation that defines the region of interest is $y = 8x - x^2$, which can be rewritten as

$$y - 8x + x^2 = 0 \tag{9}$$

Equation (9) defines the parabola. Replacing the equality sign by an inequality sign results in a mathematical relationship that is satisfied by points on one side of the curve. We see from Figure 8.3 that (6, 4) is a point in the region of interest. Substituting this test point into the left side of Equation (9), we obtain $4 - 8(6) + (6)^2 = -8$, which is less than the right side of Equation (9). Consequently, points in the region of interest satisfy the relationship

$$y - 8x + x^2 < 0 \tag{10}$$

To test whether a trial point (x, y) falls in the shaded region of Figure 8.3, we calculate $y - 8x + x^2$ for the point. If the result is less than zero, the point is in our region of interest, otherwise it is not. For the first five trial points in Table 8.18, we have

| POINT | | | RELATIONSHIP | IN REGION OF | ACCEPT |
x	y	$y - 8x + x^2$	TO 0	INTEREST?	POINT
3.43	15.75	0.075	greater than	no	no
4.35	13.98	−1.898	less than	yes	yes
4.64	10.03	−5.560	less than	yes	yes
2.83	17.10	2.469	greater than	no	no
7.72	18.40	16.234	greater than	no	no

Each of the 1,000 trial points used in Table 8.18 is inside the rectangular window shown in Figure 8.3. The last column in Table 8.18 is a running total of the number of trial points that fall within our region of interest. The total is 545 in this simulation. The area of a rectangle with sides of lengths 8 and 20 is $8(20) = 160$, so it follows from Equation (7) that an estimate for the area of the shaded region in Figure 8.3 is

$$\text{area of region} = \left(\frac{545}{1000}\right)160 = 87.2$$

Running nine more simulations, each with 1,000 trial points, we obtain the estimates shown in Table 8.19. The mean of these 10 values if 85.10, which we take as our best estimate of the area of the region under the graph of $y = 8x - x^2$ and above the horizontal x-axis.

TABLE 8.19 **Parabola Simulations**

RUN	AREA	RUN	AREA
1	87.20	6	84.80
2	83.20	7	85.44
3	86.56	8	83.04
4	87.36	9	88.00
5	83.52	10	81.92

IMPROVING SKILLS

In Problems 1 through 8, use the following ten random numbers to generate a set of 10 random numbers that are uniformly distributed on the given intervals.

.53479	.81115	.98036	.12217	.59526
.40238	.40577	.39351	.03172	.69255

1. [0, 5) **2.** [0, 8) **3.** [3, 8)
4. [−5, 0) **5.** [−6, 3) **6.** [100, 550)
7. [2,500, 4,900) **8.** [−6,000, 3,500)

9. Use the ten random numbers listed in Table 8.11 to create another run of the gasoline station simulation shown in Table 8.13.

10. Replace the ten random numbers in the second column of Table 8.14 with the ten random numbers listed in Table 8.12, and create another run of the bus stop simulation. Do not alter the random numbers in column 5.

11. Replace the ten random numbers in the fifth column of Table 8.14 with the ten random numbers listed in Table 8.12, and create another run of the bus stop simulation. Do not alter the random numbers in column 2.

In Problems 12 through 20, use Monte Carlo simulations to estimate the areas of the given regions. Stop after generating ten lines of output. Use the first ten random numbers in the following list to simulate x-coordinates and the last ten random numbers to simulate y-coordinates.

.00149	.84745	.63222	.50533	.50159
.60433	.04822	.49577	.89049	.16162
.53250	.73200	.84066	.59620	.61009
.38542	.05758	.06178	.80193	.26466

12. The region under the straight line $y = 2x + 1$ and above the x-axis between $x = 0$ and $x = 2$.

13. The region under the straight line $y = 2x + 1$ and above the x-axis between $x = 1$ and $x = 4$.

14. The region under the straight line $y = 2x + 1$ and above the x-axis between $x = -1$ and $x = 3$.

15. The region under the straight line $y = 5 - x$ and above the x-axis between $x = 0$ and $x = 3$.

16. The region under the straight line $y = 5 - x$ and above the x-axis between $x = -3$ and $x = -1$.

17. The region above the curve $y = x^2$ and below the line $y = 4$.

18. The region below the curve $y = 5 - x^2$ and above the line $y = 1$.

19. The region below the curve $y = 5 - x^2$ and above the line $y = -1$.

20. The region below the curve $y = x^3$ and above the x-axis between $x = 0$ and $x = 1$.

CREATING MODELS

Simulate each of the processes described in Problems 21 through 33 using random numbers selected sequentially from the following list. Stop after generating ten lines of output.

.15025	.20237	.63386	.71122	.06620	.07415	.94982	.32324
.95610	.08030	.81469	.91066	.88857	.56583	.01224	.28097
.09026	.40378	.05731	.55128	.74298	.49196	.31669	.42605
.81431	.99955	.52462	.67667	.97332	.69808	.21240	.65952
.21431	.59335	.58627	.94822	.65484	.09641	.41018	.85100

21. A person invests $1,000 in a bank account and plans to leave it there for five years. The account accrues interest compounded semiannually, but the interest rate can vary. Currently, the nominal interest rate is 4% a year, but in future periods it is equally likely to be any percentage between $3\frac{1}{2}$% and 5%. How much can the depositor expect to have in the account after five years?

22. A corporation creates a sinking fund by depositing $50,000 every quarter into an account that accrues interest compounded quarterly. The nominal interest rate on the account is currently 6%, but in future periods it is equally likely to be any percentage between 5% and 10%. How much can the corporation expect to have in the account after $2\frac{1}{2}$ years?

23. Each week, a real estate broker sells zero houses with probability .60, one house with probability .25, two houses with probability .10, and three houses with probability .05. The commission on a sale is uniformly distributed between $4,000 and $10,000. How much commission can the broker expect over the next ten weeks?

24. During the morning commuter hours, the time between trains arriving in New York

City from a Long Island railroad station is uniformly distributed between 13 and 20 minutes. What will the arrival pattern of trains be between 6:30 a.m. and 8:30 a.m., assuming the first train comes in at 6:42 a.m.?

25. Customers begin arriving at a bank 15 minutes before it opens. The time between customer arrivals in this quarter-hour period is uniformly distributed between 30 and 90 seconds. How many customers will be in line when the bank opens?

26. A machine produces bearings with diameters that are uniformly distributed between 2.2 and 2.65 centimeters. Simulate the diameters of the next ten bearings from this machine.

27. Balls produced by a toy company have diameters that are uniformly distributed between 11.8 and 12.1 inches. Cubical packaging boxes have sides that are uniformly distributed between 11.95 and 12.05 inches. A ball will fit into a box if the ball's diameter is less than the length of the box's side. Machines scan each ball and box to ensure that a ball will fit. If not, the ball is routed back to inventory and the machine attempts to package a new ball. How many balls are routed back to inventory?

28. In one dental office, the hygienist first cleans a patient's teeth and then the dentist examines the patient. Cleaning time is uniformly distributed between 20 and 30 minutes. Normally, a dental examination is uniformly distributed between 10 and 15 minutes. However, the dentist tries to complete an examination quickly when the hygienist needs more than 25 minutes for cleaning. In these cases, the examination is shortened by 0 to 2 minutes, uniformly distributed. Simulate the time spent with both practitioners for each of the next ten patients.

29. Prior to opening a new store, a site must be selected, contracts signed, and a building constructed. It takes a firm between 100 and 130 days, uniformly distributed, to complete the site selection process, between 30 and 50 days, uniformly distributed, to complete contract negotiations once a site is selected, and between 250 and 310 days, uniformly distributed, to complete construction once contracts are signed. How long will it take to open the next five stores?

30. The times between arrivals of automobiles at an automated car wash are uniformly distributed between one and five minutes. The washing area accommodates one car at a time, and the washing process takes exactly two minutes to complete. If a car arrives while another car is being washed, the arriving car must wait in line. Simulate the activities at the car wash during the first hour of its operation each morning. How long does the waiting line become?

31. A barber runs a one-person shop, and because of the barber's reputation, arriving customers are willing to wait for a haircut if the barber is occupied. The times between customer arrivals are uniformly distributed between 15 and 25 minutes. A haircut takes between 18 and 22 minutes, uniformly distributed. How long does the waiting line become over an eight-hour day?

32. An assembly line delivers a new television to a quality-control inspector every 20 ± 5 minutes, uniformly distributed. Inspection takes 11 ± 4 minutes, uniformly distributed. What percentage of the time is the inspector idle over an eight-hour day?

33. Vehicles begin arriving at a highway gas station 15 minutes before it opens and then wait for gasoline if they can. The station has room for a maximum of 5 cars, and because the station is on a highway, cars that cannot be accommodated cannot wait and their business is lost for the day. If interarrival times are uniformly distributed between 2 and 3 minutes, how many cars would wait for the service station to open but cannot because of inadequate waiting space?

EXPLORING IN TEAMS

34. A simple way to randomly generate *integers* between *a* and *b* (inclusive) that are equally likely to occur is to produce numbers that are uniformly distributed on the interval

$$[a - 0.5, b + 0.5)$$

and then round to the nearest integer. Use this process with the random numbers in Table 8.2 in Section 8.1, and show that the results are the same for each die. Here *a* = 1 and *b* = 6.

35. Repeat the procedure described in the previous problem using the random numbers in column 5 of Table 8.14 to reproduce column 6. Here *a* = 36 and *b* = 48.

36. Discuss why the procedure described in Problem 34 would be incorrect if we used the interval [*a, b*) instead.

EXPLORING WITH TECHNOLOGY

37. Using an electronic spreadsheet, construct your own simulation of the gasoline station problem described in Table 8.13. Then use your model to run 30 simulations of the process, each covering a ten-week period. With your results, estimate the mean summer revenue loss.

38. Use an electronic spreadsheet to construct your own simulation of the bus problem described in Example 5. Then use your model to run 30 simulations of the process and construct a table similar to Table 8.15. What is your estimate of the number of people denied entrance to a bus in a single morning?

39. Use an electronic spreadsheet to create your own simulation and estimate the answer for the problem in Example 6.

40. Create simulation models on an electronic spreadsheet to estimate the areas specified in Problems 12 through 20.

41. Create simulation models on an electronic spreadsheet for the processes described in Problems 21 through 33, and estimate an answer for each.

REVIEWING MATERIAL

42. (Section 1.5) Find the equation of the straight line passing through the two points (50, 20,000) and (100, 8000). Does this line include (200, 2000)?

43. (Section 4.4) Construct the first three lines of an amortization table to repay a $2,000 loan over 24 months at 10% interest a year, compounded monthly.

44. (Section 5.1) A survey of 300 high school seniors revealed the following information:

289 had passing grades

105 played a varsity sport

160 had a library card

158 had a library card and passing grades

96 had passing grades and played a varsity sport

2 played a varsity sport and had a library card

2 had a library card, had passing grades and played a varsity sport

Using this information, determine the number of students in the survey who (*a*) did not have passing grades, (*b*) did not have a library card, (*c*) had a library card but did not play a varsity sport, and (*d*) played a varsity sport and did not have passing grades.

RECOMMENDING ACTION

45. Respond by memo to the following request:

> ### MEMORANDUM
>
> To: J. Doe Reader
>
> From: Lottery Commissioner
>
> Date: Today
>
> Subject: **Machine Selections**
>
> As the new Commissioner, I want to understand fully how our state lottery systems work. Perhaps, you can explain one aspect to me. To play our million-dollar lottery, a player must select six different numbers from 1 to 48. Many players do not actually pick the numbers themselves but rather let the computer choose the numbers for them. My question is: How does the computer make the selections?

 # SAMPLING

In the previous section, we simulated processes with uniformly distributed outcomes. Left unanswered was the question of how we determine that outcomes are uniformly distributed—the answer is by analyzing the process through sampling, and we will show how to do this in this section.

First, however, we note that many of the sample spaces in Section 8.2 contained infinitely many possible outcomes and are therefore structurally different from any sample spaces previously considered in this book. In Section 5.1, we defined the cardinal number of a set to be the number of elements in that set. But from Chapter 5 through Section 8.1, the only sets we considered were sample spaces with a finite number of outcomes.

All the tree diagrams in Chapter 6 had a finite number of branches, and all the Markov chains in Chapter 7 had a finite number of states. In contrast, each interarrival time in Example 1 of Section 8.2 can be any of the infinitely many real numbers between 12 and 18 minutes.

TABLE 8.20

LINE NUMBER	OUTCOME
1	AAA
2	AAB
3	AAC
4	AAD
5	AAE
⋮	⋮
17,572	ZZV
17,573	ZZW
17,574	ZZX
17,575	ZZY
17,576	ZZZ

As an example of a sample space with a finite number of outcomes, consider the 17,576 different sets of 3 initials that can be formed from the 26 letters in the English alphabet (see Example 5 of Section 5.4). The fundamental theorem of counting allows us to calculate this total without actually listing each possibility, but the fact remains that we can list all the possible outcomes in this sample space if we must, as demonstrated in Table 8.20.

Table 8.20 establishes a one-to-one correspondence between the first 17,576 positive integers and the outcomes in the sample space. Each outcome is listed once, and different integers are associated with different outcomes. We say that a process is *discrete* if the outcomes can be put into a one-to-one correspondence with a subset of the positive integers. One subset of the positive integers is the set of positive integers itself, so discrete does not mean finite. A discrete process may allow a finite number of outcomes, as in Table 8.20, or infinitely many outcomes, as long as we can establish a one-to-one relationship between each outcome in the process and a positive integer.

EXAMPLE 1

A process is discrete if the set of all possible outcomes can be put into a one-to-one correspondence with a subset of the positive integers.

A person buys 100 shares of stock, holds it for 40 years, and then sells the stock and deposits the proceeds in a bank account. The item of interest here is the value of the account after the process is completed. Is this process discrete?

Solution

Bank accounts are rounded to the nearest penny, so the person's account can be any figure in dollars and cents from $0.00 on up. After 40 years, the stock may be worthless or it may be worth a fortune. Theoretically, there are infinitely many values for the stock, including $0.00, $2,321.76 and $4,677,322.10, for example, but all of the possibilities can be placed into a one-to-one correspondence with the positive integers. The beginning of such a list is shown in Table 8.21. $0.00 appears on line number 1, $2,321.76 would appear on line number 232,177 and $4,677,322.10 would appear on line number 467,732,211. This process is discrete.

TABLE 8.21

LINE NUMBER	OUTCOME
1	$0.00
2	$0.01
3	$0.02
4	$0.03
5	$0.04
⋮	⋮

A process is *continuous* if the outcomes can assume any value in some interval on the real number line. Table 8.13 in Section 8.2 deals with a demand process for gasoline that is uniformly distributed on [2,000, 6,000). Theoretically, the demand can be any real number between 2,000 and 6,000, including 2,010.5, 3,504.37 and 5,667.889 gallons, hence this process is continuous. Other examples of continuous processes are measuring people's heights weighing sausage, or estimating the time it takes to complete a task.

A process is continuous if outcomes can assume any value in some interval on the real number line.

Many continuous processes are transformed into discrete processes by rounding. It is a common practice to round people's heights to the nearest quarter inch. If we also agree that no person is taller than 10 feet or 120 inches (an assumption akin to the hypothesis that a coin never lands on its edge), then the process of measuring heights becomes discrete with outcomes restricted to the sample space

$$\{0.00'', 0.25'', 0.50'', \ldots, 119.50'', 119.75'', 120,00''\}$$

which contains 481 elements. Most delicatessen scales round weights to the nearest thousandth of a pound and are calibrated to a maximum of 100 pounds. Weighing sausage on such a scale is therefore a discrete process. We distinguish between determining the true weight of the sausage, a continuous process, and its measurement on the scale, a discrete process.

Taking data on a continuous process involves measuring outcomes from a sample and then collecting those measurements into a small number of subsets for analysis. Both tactics have the effect of modeling a continuous process with a discrete process. Measurements are inherently a rounding process; collecting outcomes into a few subsets reduces the number of outcomes considerably. The collected data can then be presented in a bar graph.

A histogram is a bar graph of a continuous process and has no gaps between adjacent bars.

A bar graph is a sequence of bars in which one bar represents each subset of the sample space (see Section 5.2). The heights of the bars are directly proportional to the number of elements in each subset. The widths of the bars are generally equal. When dealing with continuous processes, however, we do not leave space between adjacent bars. Neighboring bars abut one another. The resulting bar graph model of a continuous process is often called a *histogram*.

The two major decisions a modeler must make in creating a histogram are the number of categories to use and the choice of endpoints for each category. The following rules of thumb are generally followed:

ESTABLISHING CATEGORIES FOR HISTOGRAMS

1. Use between 5 and 20 categories.

2. Try to have at least five data points in each category.

3. Choose endpoints so that no data point falls into two categories.

4. Choose endpoints so that each data point is in some category.

5. Try to choose endpoints so that the widths of bars are the same.

6. Use endpoints that are sensible.

The number of categories is influenced by the amount of available data and the need for sensible endpoints. If each category is to contain at least 5 data points, then a sample of 50 data points should not be partitioned into more than 10 categories and probably should have fewer because some subintervals will generally contain more data than others. If the data falls between the values of 0 and 7, then the dictum for sensible endpoints would suggest either 7 subintervals of length 1 or 14 subintervals of length $\frac{1}{2}$. Lengths such as $\frac{7}{15}$, for example, are generally not recommended.

No data point can belong to more than one category, and this requirement must also be satisfied by the endpoints, especially endpoints that define two adjacent categories. One category may go from 200 to 300 and the next from 300 to 400, but it must be clear to which category 300 belongs. It cannot belong to both. In general, we shall define categories to include their left endpoints but not their right endpoints. Thus, the interval [200, 300) includes 200 but not 300, while the adjacent interval [300, 400) includes 300 but not 400.

Although we try to choose intervals that are of equal widths and contain at least five points, neither condition is a requirement. If a natural partition of data has one or two categories with fewer than 5 points, then so be it. Also, we will accept intervals of unequal widths when they are meaningful to a particular process. For example, in sampling test grades, we may take widths to be [0, 59), [60, 69), [70, 79), [80, 89), and [90, 100] when those categories represent the different letter grades awarded to students.

EXAMPLE 2

Construct a histogram for the following set of interarrival times, originally recorded in minutes and seconds and then converted to the following decimal equivalents in minutes.

17.66	17.89	15.18	15.51	12.92	14.69	17.70	12.05
17.17	12.20	12.06	13.82	14.97	13.63	13.48	17.07
16.11	13.73	15.37	13.85	14.18	15.87	15.19	17.70
12.63	12.10	14.55	16.72	14.59	16.65	14.77	14.88
16.93	12.54	13.39	15.32	17.60	16.84	16.56	14.79

TABLE 8.22 Frequency Table of Interarrival Times

CATEGORY	NUMBER OF DATA POINTS (FREQUENCY)
[12, 13)	7
[13, 14)	6
[14, 15)	8
[15, 16)	6
[16, 17)	6
[17, 18)	7

Solution

This data set contains 40 points with values between 12.05 and 17.89. These two values are not sensible endpoints, so we use instead the interval [12, 18) as our sample space. We can partition this sample space into 6 subintervals of length 1 or 12 subintervals of length $\frac{1}{2}$. With 12 subintervals, many categories would contain fewer than 5 data points, because we have only 40 data points in our sample, so we opt instead for subintervals of length 1. Counting the number of data points in each category, we construct Table 8.22. Figure 8.4 is a histogram of this information. For clarity, we labeled the midpoint of each category. Alternatively, we could label the endpoints of each category.

FIGURE 8.4

Histogram of interarrival times

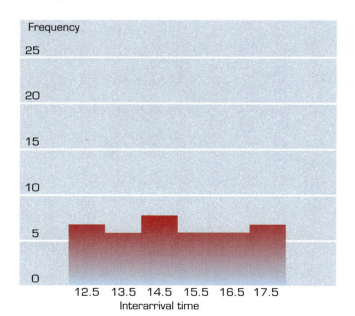

The number of data points that fall into each partitioned category is its *frequency* or frequency of occurrence. A table that lists each category with its frequency, such as Table 8.22, is a *frequency table*.

EXAMPLE 3

A medical research team is conducting a test on males at a large state university. Height is postulated to be an important characteristic in this test, so 50 males are selected at random and their heights are measured. The following data shows their heights rounded to the nearest quarter of an inch. Construct a histogram of this sampled data.

5'11"	5'9½"	5'10¼"	5'7¾"	5'8"	5'7½"	5'9½"	5'8¾"	5'9¼"	6'1½"
5'9"	5'9¾"	5'11¾"	5'10¼"	5'8¾"	5'10"	5'7"	5'10¾"	5'9½"	5'10"
5'11¾"	5'3¼"	6'2¼"	5'5¼"	5'6¾"	5'5½"	5'9½"	5'9"	6'1¼"	5'8½"
6'0"	5'10¼"	6'1¼"	5'11½"	5'8¼"	5'10"	5'10¼"	5'5¼"	5'10"	5'11¼"
5'8¾"	5'6½"	5'7¼"	5'10"	5'4¼"	5'9½"	5'10¾"	5'8¾"	5'7"	6'3¼"

Solution

This data set contains 50 points with values between $5'3\frac{1}{4}''$ and $6'3\frac{1}{4}''$. These two values are not the most pleasing endpoints, so we consider instead the interval $[5'3'', 6'4'')$ as our sample space. If we partition this sample space into subintervals of length 1, we will have 13 subintervals, with most containing fewer than 5 points. If we partition into subintervals

TABLE 8.23 **Frequency Table of Heights**

CATEGORY	NUMBER OF DATA (FREQUENCY)
[5'3", 5'5")	2
[5'5", 5'7")	5
[5'7", 5',9")	12
[5'9", 5'11")	20
[5'11", 6'1")	6
[6'1", 6'3")	4
[6'3", 6'5")	1

FIGURE 8.5
Histogram of heights

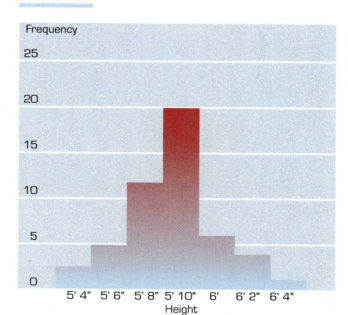

of length 2, we will have 7 subintervals with some containing fewer than 5 points. Scanning the data, we see that most are clustered around 5′10″ and only a few are near the boundaries of our sample space. It is unlikely, therefore, that many data points will belong to the outlying subintervals in any reasonable partition, so we opt for subintervals of length 2. Counting the number of data points in each category, we construct a frequency table (Table 8.23). Figure 8.5 is a histogram of this information with the midpoint of each category suitably labeled.

Observe that by combining the first two categories and the last two categories in Table 8.23, we can create a group of partitions that all contain 5 or more points. The subinterval [5′3″, 5′7″) would have 7 points and the subinterval [6′1″, 6′5″) would have 5 points. Of course, the partitions are no longer of equal width. By demanding partitions with sensible endpoints, we either create subintervals of equal width with some having fewer than 5 points or subintervals of unequal widths with all having at least 5 points. In Example 3, we opted for subintervals with equal widths.

The frequency table shown in Table 8.24 uses a different set of partitions but meets the same criteria as Table 8.23. Its corresponding histogram is shown in Figure 8.6. Figures 8.5 and 8.6 are alternative graphical models and both are acceptable histograms for the same data.

TABLE 8.24 Frequency Table of Heights

CATEGORY	NUMBER OF DATA POINTS (FREQUENCY)
[5′2″, 5′4″)	1
[5′4″, 5′6″)	4
[5′6″, 5′8″)	7
[5′8″, 5′10″)	16
[5′10″, 6′0″)	16
[6′0″, 6′2″)	4
[6′2″, 6′4″)	2

FIGURE 8.6

Histogram of heights

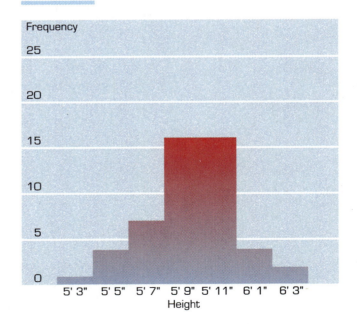

A frequency polygon can be drawn from a histogram by connecting the midpoints of the tops of successive bars with line segments.

Each category in a histogram can be represented by a point whose horizontal coordinate is the midpoint of the subinterval that defines the category and whose vertical coordinate is equal to the frequency of the category. If these points are connected by line segments, then the resulting curve is a *frequency polygon*.

EXAMPLE 4

Construct a frequency polygon for the data on interarrival times listed in Example 2.

Solution

That data is summarized in Table 8.22, which forms the first and third columns of Table 8.25. The points associated with the last two columns in Table 8.25 are plotted in Figure 8.7 and connected by line segments. Figure 8.7 is a frequency polygon for the given data.

TABLE 8.25   Frequency Table of Interarrival Times

CATEGORY	MIDPOINT	FREQUENCY
[12, 13)	12.5	7
[13, 14)	13.5	6
[14, 15)	14.5	8
[15, 16)	15.5	6
[16, 17)	16.5	6
[17, 18)	17.5	7

FIGURE 8.7

Frequency polygon of interarrival times

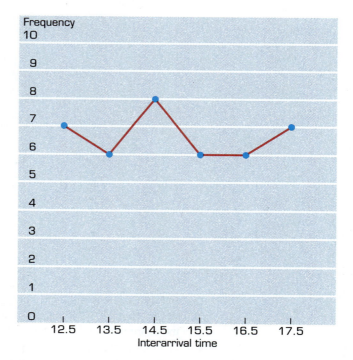

EXAMPLE 5

Construct a frequency polygon for the data in Table 8.23 of Example 3, which dealt with heights of males at a large state university.

Solution

The data in Table 8.23 become the first and third columns of Table 8.26. The coordinates of interest are the midpoints and frequencies in the last two columns of Table 8.26. Figure 8.8 is the corresponding frequency polygon.

TABLE 8.26   **Frequency Table of Heights**

CATEGORY	MIDPOINT	NUMBER OF DATA (FREQUENCY)
[5′3″, 5′5″)	5′4″	2
[5′5″, 5′7″)	5′6″	5
[5′7″, 5′9″)	5′8″	12
[5′9″, 5′11″)	5′10″	20
[5′11″, 6′1″)	6′0″	6
[6′1″, 6′3″)	6′2″	4
[6′3″, 6′5″)	6′4″	1

FIGURE 8.8

Frequency polygon of heights

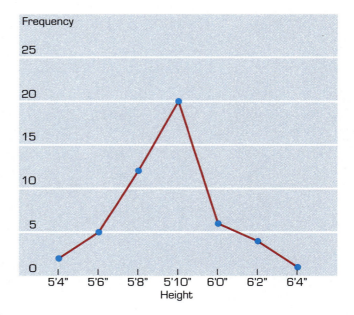

A slightly different frequency polygon can be created for Example 5 if we use the categories in Table 8.24, which are repeated in the first and third columns of Table 8.27. Again the coordinates of interest are the midpoints and frequencies in the last two columns. Figure 8.9 is the corresponding frequency polygon.

We see from Figures 8.8 and 8.9 that the same data can have different frequency polygons that reflect different partitions of the same sample space. Still, Figures 8.8 and 8.9 have similar shapes—both are somewhat bell-shaped. In contrast, the frequency polygon in Figure 8.7 is not bell-shaped.

Although a data set can be represented by many frequency polygons, depending on how we partition the sample space, we *expect* that all frequency polygons for the same data will exhibit similar characteristics because they represent the same sample. Therefore we *expect* all frequency polygons for the same data to have similar shapes! If one frequency polygon is close to a straight line, then all frequency polygons for the same data should

All frequency polygons for the same data should have similar shapes.

TABLE 8.27   Frequency Table of Heights

CATEGORY	MIDPOINT	FREQUENCY
[5'2", 5'4")	5'3"	1
[5'4", 5'6")	5'5"	4
[5'6", 5'8")	5'7"	7
[5'8", 5'10")	5'9"	16
[5'10", 6'0")	5'11"	16
[6'0", 6'2")	6'1"	4
[6'2", 6'4")	6'3"	2

FIGURE 8.9

Frequency polygon of heights

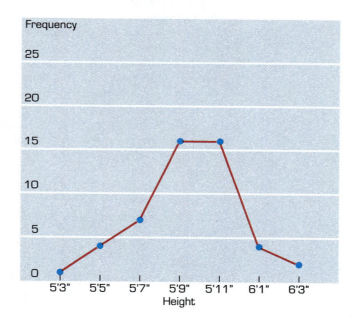

be close to the same straight line. If one frequency polygon approximates a bell-shaped curve, then all frequency polygons for the same data should approximate the same bell-shaped curve.

We also expect different samples from the same population to exhibit similar characteristics. Example 3 listed the heights of 50 males selected at random from students at a large state university. If we chose another sample of 50 males, we would get a different set of data. The histogram for this new sample likely would be different from those in either Figure 8.5 or 8.6, and the corresponding frequency polygon for this new sample likely would be different from those in either Figure 8.8 or 8.9. However, if both samples adequately represent the same population, that is, if both samples are good models, then their frequency polygons should exhibit the same characteristics.

Frequency polygons for different samples of the same population should have similar shapes, and frequency polygons for the same sample partitioned differently should have

Frequency polygons for samples from populations that are uniformly distributed approximate a horizontal line.

similar shapes. All of these frequency polygons should be approximations of the curve that defines the entire population. If a population is uniformly distributed, this curve is a horizontal line.

If a population is uniformly distributed, then each value in its sample space is equally likely to occur. If we partition the sample space into subintervals of equal length, we expect the same number of outcomes to appear in each subinterval, and we expect the corresponding histogram to contain bars of equal height. Thus, the midpoints of the tops of these bars should all have the same height. When we connect these midpoints with line segments, we expect to get a horizontal line.

We do not expect a straight line for any particular sample, however. If we flip a coin 100 times, we expect 50 heads and 50 tails, but this is a theoretical expectation based on our knowledge that a head has the same probability of occurring as a tail. Any particular set of 100 throws may end with 53 heads and 47 tails, or 48 heads and 52 tails, or some other combination close to 50-50. Similarly, we do not expect the frequency polygon of any *sample* from a uniformly distributed population to be a straight line, but we expect such a frequency polygon to be close to a horizontal line.

This is how we will determine whether a population is uniformly distributed: We sample the population, taking care that the sample size is large enough to provide a good model for the population. How large a sample to take is a statistical question that we shall not address in this book, but the number is often governed by the resources available—either time, money, or personnel—for obtaining the data. We construct a frequency polygon from the data, and if the frequency polygon is reasonably close to a horizontal line, then we postulate that the population is uniformly distributed. Reasonably close is also a statistical concept beyond the scope of this book. Here we shall be content with the "eyeball" approach—feeling comfortable that a curve is reasonably close to another curve from visual inspection.

EXAMPLE 6

Determine whether the data in Example 2 comes from a population that is uniformly distributed.

Solution

A frequency polygon for this data is drawn in Figure 8.7. The curve is reasonably flat, so we hypothesize that the underlying population is uniformly distributed. In Figure 8.7, two points are at 7, three are at 6, and one is at 8, which are all reasonably close to one another, since all plotted points must have vertical coordinates that are integer valued.

EXAMPLE 7

Determine whether the data in Example 3 comes from a population that is uniformly distributed.

Solution

Two frequency polygons for this data are shown in Figures 8.8 and 8.9. Neither is reasonably flat, so we hypothesize that the underlying population is *not* uniformly distributed. Both curves are bell-shaped and approximate normal distributions, which are the subject of Section 8.4.

The histograms and frequency polygons constructed for the data in Example 3 (heights of males at a university) provide a good overview of the *distribution* of heights. But, if our need for information about male heights is met by knowing the average height, then these presentation models are, in a sense, overkill. Similarly, the graphical models created for the data in Example 2 (interarrival times) may be more complex than necessary, especially if our need for information is met by knowing just the median interarrival time.

As we stated in Section 1.1, a model is adequate if it meets the needs of the people using the model. Furthermore, the simpler the model, the better, as long as it is adequate. *If* data can be summarized adequately by just one or two numbers, then that is the type of model we should use.

We know from our work in Section 5.3 that the various measures of central tendency, such as the mean, the median, and the mode, are simple models of sets of numerical data.

EXAMPLE 8

Determine the mean, median, and mode for the interarrival times given in Example 2.

Solution

For convenience, we first sort the data in ascending order and obtain the following list:

12.05	12.06	12.10	12.20	12.54	12.63	12.92	13.39
13.48	13.63	13.73	13.82	13.85	14.18	14.55	14.59
14.69	14.77	14.79	14.88	14.97	15.18	15.19	15.32
15.37	15.51	15.87	16.11	16.56	16.65	16.72	16.84
16.93	17.07	17.17	17.60	17.66	17.70	17.70	17.89

There are 40 data points, so the median is the mean of the 20*th* and 21*st* points in this sorted list:

$$\frac{14.88 + 14.97}{2} = 14.925$$

The sum of all 40 data values is 600.86, so the mean of this sample is

$$\bar{x} = \frac{600.86}{40} \approx 15.02$$

The only sampled time that appears more than once is 17.70, so it is the mode. Here the mode occurs only twice in 40 samples and thus is not a useful measure of central tendency. The median and mean are much more significant.

We can also calculate measures of dispersion for sampled data, but the formulas for variance and standard deviation must be modified from those in Section 5.3. These modifications are needed because we no longer have data from an entire population, as we did in Section 5.3, but rather data from a sample.

To calculate variance in a population, we summed the deviations (*i.e.*, the differences) between the mean and each data point. With sample data, however, we do not know the mean of the population. We know only the mean of the sample, which we use as an *estimate* of the mean of the population. To distinguish the two means, we use $\bar{x}$ to denote the mean of a sample and the Greek letter μ (mu) to denote the mean of an entire population. By using $\bar{x}$ as an estimate of μ in calculating the variance of a population, we lose one degree of freedom in statistical terminology. To compensate, we divide the sum of the deviations by $n - 1$ rather than n, where n is the number of data points in the sample. If we list the data as $x_1, x_2, x_3, \ldots, x_n$ with mean $\bar{x}$, then the variance of the population *as estimated by the sample* is

$$variance = \frac{(x_1 - \bar{x})^2 + (x_2 - \bar{x})^2 + (x_3 - \bar{x})^2 + \ldots + (x_n - \bar{x})^2}{n - 1} \tag{11}$$

Compare this to the formula developed in Section 5.3 for variance, namely,

$$variance = \frac{(x_1 - \mu)^2 + (x_2 - \mu)^2 + (x_3 - \mu)^2 + \ldots + (x_n - \mu)^2}{n} \qquad (12)$$

where now we use μ to represent the mean of the entire population.

The formulas for variance and standard deviation are modified when data comes from a sample rather than from the entire population.

Equation (*11*) is commonly referred to as the variance of a sample, while Equation (*12*) is the variance of a populaton. When we have data from *all* members of a population, we use Equation (*12*) to calculate variance; when we have data from a sample of a population, we use Equation (*11*) to estimate the variance of the population.

Standard deviation is the square root of the variance. Thus, we also have two standard deviations: the standard deviation of a sample, when the variance is given by Equation (*11*), and the standard deviation of a population, when the variance is given by Equation (*12*). Many calculators and spreadsheets have both standard deviations programmed into them. If only one is provided, however, then the user must check the appropriate documentation to determine which standard deviation is available on a given machine.

EXAMPLE 9

The following list of nine test grades was chosen at random from a class of 200.

93, 81, 69, 86, 97, 50, 86, 77, 72

Use this sample to estimate the mean and standard deviation of the entire class on that test.

Solution

The mean of the sample is

$$\bar{x} = \frac{93 + 81 + 69 + 86 + 97 + 50 + 86 + 77 + 72}{9} = 79$$

and we use this as an estimate of the mean for the class. The deviations for each data point are listed in Table 8.28. Equation (*11*) becomes

TABLE 8.28

DATA	MEAN	DEVIATION FROM THE MEAN
93	79	93 − 79 = 14
81	79	81 − 79 = 2
69	79	69 − 79 = −10
86	79	86 − 79 = 7
97	79	97 − 79 = 18
50	79	50 − 79 = −29
86	79	86 − 79 = 7
77	79	77 − 79 = −2
72	79	72 − 79 = −7

Therefore, an estimate of the standard deviation of the test grades for the class of 200 is $\sqrt{202} \approx 14.21$.

EXAMPLE 10

Estimate the mean and standard deviation of a population from the sampled values 1, 3, 8, 6, 6.

Solution

The mean of the sample is

$$\bar{x} = \frac{1 + 3 + 8 + 6 + 6}{5} = 4.8$$

which becomes an estimate of the mean of the population. Equation (*11*) becomes

$$variance = \frac{(1 - 4.8)^2 + (3 - 4.8)^2 + (8 - 4.8)^2 + (6 - 4.8)^2 + (6 - 4.8)^2}{5 - 1}$$

$$= \frac{30.8}{4}$$

$$= 7.7$$

An estimate for the standard deviation of the population is $\sqrt{7.7} \approx 2.77$.

IMPROVING SKILLS

In Problems 1 through 10, estimate the mean and standard deviation of a population from the given *samples*.

1. 8, 8, 9, 9, 10, 10, 11, 12, 12, 12
2. 8, 8, 9, 10, 10, 11, 11, 12, 12, 13
3. 8, 9, 9, 9, 10, 11, 12, 12, 12, 13
4. 8, 8, 9, 9, 9, 10, 10, 11, 12, 13
5. 8, 8, 8, 9, 9, 10, 11, 12, 12, 13
6. 1, 2, 3, 4, 5, 6, 7, 8, 9
7. 1, 3, 5, 7, 9
8. 2, 4, 6, 8, 10
9. 2, 4, 6, 8, 10, 12
10. 1, 1, 1, 2, 2, 2

In Problems 11 through 19, determine whether the given frequency polygon suggests an underlying population that is uniformly distributed.

11.

FIGURE A

12.

FIGURE B

13.

FIGURE C

14.

FIGURE D

15.

16.

17.

FIGURE G

18.

FIGURE H

19.

FIGURE I

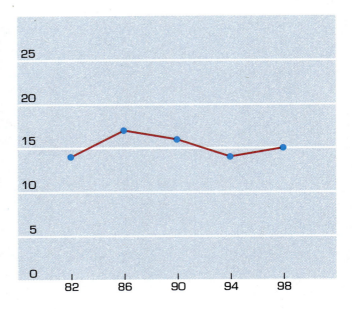

20. Determine the categories associated with the frequency polygon in Problem 11.
21. Determine the categories associated with the frequency polygon in Problem 14.
22. Determine the categories associated with the frequency polygon in Problem 17.
23. Determine the categories associated with the frequency polygon in Problem 19.

CREATING MODELS

In Problems 24 through 31, use the sampled data to model the sample space with a histogram and frequency polygon and then decide whether the sample space is uniformly distributed.

24. The following data shows the weights of a random sample of females at a large state university, rounded to the nearest pound.

120	122	123	142	145	144	109	139	139	134
142	121	142	142	143	125	136	145	148	129
124	152	148	133	124	118	132	136	129	124
130	122	132	126	124	142	128	123	143	133
142	136	125	116	139	143	135	131	127	

25. Service times for customers at the delicatessen counter of a local supermarket were recorded one morning, with the following results (the results are shown in minutes, rounded to two decimal places):

2.62	2.37	3.05	4.32	3.33	2.14	2.29	3.80	4.45	2.29
3.38	2.99	3.96	3.78	3.62	2.80	2.17	3.64	4.06	4.12
3.89	2.14	2.86	3.70	3.09	2.48	3.80	2.66	2.62	2.98
2.16	2.38	3.88	4.24	3.19	2.96	4.01	8.89	3.30	4.13
3.09	3.72	2.18	3.73	3.81	4.39	3.89	2.61		

26. The times required by mechanics at a service center to complete an oil change and minor lubrication were recorded one day. The results are shown below in minutes, rounded to two decimal places.

10.92	8.11	6.52	8.77	10.67	4.79	9.48	8.11	11.41	10.10
9.31	8.28	8.17	5.42	6.27	8.31	9.18	5.03	6.68	9.70
9.97	8.05	4.08	4.14	8.52	8.59	10.67	10.03	6.42	8.46
8.18	8.38	7.43	5.21	8.69	6.37	10.26	6.10	9.08	8.45
7.78	9.17	8.01	7.85	6.03					

27. The following data (in seconds) shows the amount of time customers at a local bank spent with a teller.

6	147	40	106	42	44	139	41	256	153
188	33	79	197	13	79	170	58	28	139
65	98	212	92	24	177	24	92	56	232
130	100	181	31	267	91	135	31	77	25
171	40	43	139	20	266	61	59	193	

28. A random sample of family incomes in a midwestern town resulted in the following data, rounded to the nearest hundred dollars:

$35,800	$38,200	$49,000	$44,700	$35,300	$45,500
$36,200	$46,300	$49,300	$31,200	$32,200	$36,000
$47,500	$36,000	$42,100	$37,700	$47,700	$41,500
$43,300	$37,200	$36,100	$47,500	$32,700	$30,500
$44,900	$49,900	$34,500	$30,500	$49,700	$38,200
$47,700	$34,600	$46,800	$32,500	$33,000	$32,900
$31,600	$41,600	$41,500	$34,700	$34,600	$40,100
$41,100	$36,700	$40,200	$46,100	$44,200	$30,300

29. A random sample of family incomes in a southern town resulted in the following data, rounded to the nearest hundred dollars:

$61,300	$61,800	$59,500	$65,200	$56,100	$46,000
$71,800	$67,500	$57,400	$57,900	$61,000	$61,800
$54,500	$43,500	$52,300	$59,000	$56,900	$58,000
$56,600	$58,300	$57,800	$63,700	$52,100	$53,400
$59,500	$55,200	$49,600	$67,000	$50,100	$53,500
$54,200	$52,200	$57,200	$63,600	$46,500	$48,900
$68,000	$51,500	$56,000	$58,000	$62,400	$46,400
$58,800	$53,800	$51,300	$61,800	$59,300	$47,300

30. The distance traveled each day by a postal delivery van on a local route was recorded over an eight-week period with the following results, rounded to the nearest tenth of a mile:

6.3	7.2	7.3	6.9	7.2	6.2	7.4	6.5	6.0	7.2
6.0	6.2	6.7	7.0	6.1	6.8	7.1	6.5	6.8	6.7
6.4	6.5	6.4	7.3	6.9	6.0	6.3	6.7	6.3	7.1
7.2	6.4	6.2	6.2	7.2	6.1	6.9	7.5	6.2	7.3
6.0	6.9	6.7	7.2	6.7	7.3	6.4	6.9		

31. The times between arrivals at the drive-through windows of a fast-food restaurant were recorded one afternoon. The results are listed in minutes, rounded to two decimal places.

1.57	1.12	7.70	0.31	3.85	0.44	4.43	1.32	6.51	0.62
4.27	2.76	9.49	0.20	0.24	1.92	0.45	13.79	9.34	2.90
0.62	5.20	1.97	1.22	2.87	1.06	0.08	0.56	1.02	0.26
1.97	2.59	3.10	8.29	4.06	2.20	5.24	4.99	1.29	0.31
4.88	13.24	0.36	0.60	2.08	0.69	3.90	0.10		

EXPLORING IN TEAMS

32. Randomly select 100 or more individuals in an organization (either a university or place of work) and ask each for the *total* amount of money he or she is carrying in his or her wallet, pocket, purse, and so forth. Construct a frequency polygon for your sample, and determine whether the outcomes are uniformly distributed. Based on your sample, what do you estimate to be the average amount of money people in that organization carry, and what is the standard deviation?

33. Locate any table of values from a book in your personal library or a public library. Record the last digit for each entry. For example, if an entry in your table is 54.338, record the digit 8. If the table is large, work with a sample from the table, but try to record with at least 100 tabular entries. Construct a frequency polygon for your recorded digits and determine whether the outcomes are uniformly distributed.

EXPLORING WITH TECHNOLOGY

34. Many calculators have statistical capabilities for inputting data from a sample and then calculating the mean, variance, and standard deviation of that sample. Using such a calculator, reproduce the results of Examples 9 and 10.

35. Using a calculator with statistical capabilities, find the mean and standard deviation for the sample data in Problems 24 through 31.

REVIEWING MATERIAL

36. (Section 2.4) Determine whether

$$A = \begin{bmatrix} 1 & 2 & -1 \\ 1 & -1 & 2 \\ -1 & 2 & -1 \end{bmatrix} \quad \text{and} \quad B = \begin{bmatrix} 4 & -2 & 4 \\ -2 & 7 & -5 \\ 2 & -6 & 6 \end{bmatrix}$$

commute under the operation of multiplication.

37. (Section 4.3) How much must a person deposit at the end of each week in an account earning 4.5% a year, compounded weekly, if the goal is to have $12,000 after five years?

38. (Section 6.5) A committee with one chairperson and four members must be selected from a faculty of 80 members. How many different ways can such a committee be formed?

RECOMMENDING ACTION

39. Respond by memo to the following request:

> ### MEMORANDUM
>
> To: J. Doe Reader
>
> From: Bureau of the Census
>
> Date: Today
>
> Subject: **Taking the Census**
>
> Currently we take the census every ten years by counting each and every person in the United States. Most counts are done by mail, with individual households responding to questionnaires sent to them. When responses are not returned, we follow with in-person visits and interviews.
>
> Some members of Congress are pressing us to drop this exhaustive counting procedure and, instead, rely on sampling to obtain our statistics. What do you think?

NORMAL DISTRIBUTIONS

In the previous section, we saw that frequency polygons can be graphical models for sampled data and that they can give clues about the underlying process. In particular, if a frequency polygon was reasonably close to a horizontal line, we inferred that the sample space was uniformly distributed.

As we saw, however, some frequency polygons suggest shapes other than straight lines. In particular, some are bell-shaped. The family of *normal curves* is a well-known family of bell-shaped curves that model many sets of sampled data very well. This family of curves includes the three curves displayed in Figure 8.10.

FIGURE 8.10
Three normal curves

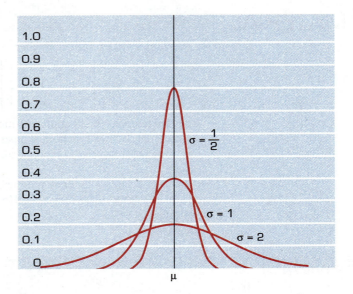

Just as every straight line is completely determined from two pieces of information— either two distinct points or one point and a slope—each normal curve is also completely determined by two values, a mean μ and a standard deviation σ.

> *A normal curve is symmetric around a vertical line that intersects the horizontal axis at the mean μ.*

A normal curve is symmetric around a vertical line that intersects the horizontal axis at the mean μ. The larger the standard deviation σ, the shorter and fatter the normal curve; the smaller the standard deviation, the taller and narrower the normal curve. The three normal curves in Figure 8.10 correspond to standard deviations of $\frac{1}{2}$, 1, and 2.

Normal curves model processes in which outcomes cluster around a mean and become more sparse in intervals farther from the mean. Many processes share this trait, including scores on College Board tests, people's weights, and measurements of machine-made parts. When a sample space can be modeled by a normal curve, we say that the outcomes of the underlying continuous process are *normally distributed*.

> *A normal curve is completely determined by its mean μ and its standard deviation σ.*

In truth, we rarely know whether a sample space is normally distributed, because we rarely know all the elements in the sample space. Populations are too large and there are too many outcomes to measure. Instead, we sample the population. We use the mean $\bar{x}$ of a sample as an estimate of the mean μ of the population, and we use the standard deviation s of a sample as an estimate of the standard deviation σ of the population. If a normal curve with these estimates for μ and σ models the sample reasonably well, then we *deduce* that the underlying sample space is normally distributed.

Regardless of its shape, every normal curve shares the property that 68.27% (rounded to two decimal places) of the area between the curve and the horizontal axis falls within one standard deviation on either side of the mean, as illustrated by the shaded region in Figure 8.11. Furthermore 95.45% (rounded to two decimal places) of the area between a normal curve and the horizontal axis falls within two standard deviations on either side of the mean, 99.73% of the area falls within three standard deviations, and 99.99% of the area falls within four standard deviations. Thus if outcomes are normally distributed, more than two-thirds of the outcomes will fall within one standard deviation on either side of the mean, and over 95% of all outcomes will fall within two standard deviations on either side of the mean.

FIGURE 8.11

A normal curve

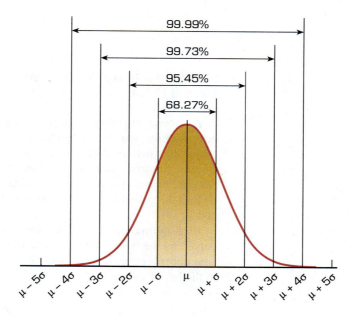

In Example 3 of Section 8.3, we analyzed a sample of heights of 50 randomly selected males from a large state university. The mean of that sample is $5'9\frac{1}{2}''$ with a standard deviation of $2\frac{1}{2}''$, both rounded to the nearest quarter inch. *If* male heights at this university are normally distributed around $5'9\frac{1}{2}''$ with a standard deviation of $2\frac{1}{2}''$, then we expect 68.27% of all males at the university to have heights between $5'9\frac{1}{2}'' - 2\frac{1}{2}'' = 5'7''$ and $5'9\frac{1}{2}'' + 2\frac{1}{2}'' = 6'0''$ and over 99% of all males to have heights within three standard deviations of this mean or between $5'9\frac{1}{2}'' - 3(2\frac{1}{2}'') = 5'2''$ and $5'9\frac{1}{2}'' + 3(2\frac{1}{2}'') = 6'5''$.

The shaded area in Figure 8.12 is the area between a normal curve and the horizontal axis, bounded on the left by a vertical line that intersects the horizontal axis at the mean μ and on the right by a line that intersects the horizontal axis at a value $x > \mu$. An important feature of normal curves is that this area depends only on the number of standard deviations that x is from μ. If x is greater than μ, then

$$z = \frac{x - \mu}{\sigma}$$

(13)

FIGURE 8.12

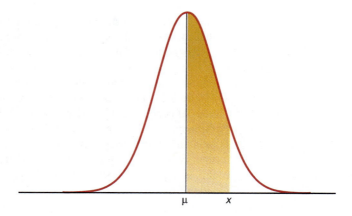

is the number of standard deviations that x is from μ. The area of the shaded region in Figure 8.12 is listed in Table 8.29 for all values of z (rounded to two decimal places) between 0 and 4.

A second important feature of normal curves is that the entire area between any normal curve and the horizontal axis is 1. Consequently, the area in Figure 8.12 is the probability that an outcome falls in the interval $[\mu, x]$, which includes both μ and x. Because the area under a point is zero, the shaded area in Figure 8.12 is also the probability that an outcome falls in the interval $[\mu, x)$, which includes μ but does not include x, as well as the probability that an outcome falls in the interval (μ, x), which excludes both μ and x.

Each normal curve is symmetric around its mean, so half the area between a normal curve and the horizontal axis falls to the right of the mean and half falls to the left. The area under a normal curve is 1, so the area to the right of the mean is 0.5, as is the area to the left of the mean.

FIGURE 8.13
Normal curve associated with Table 8.29

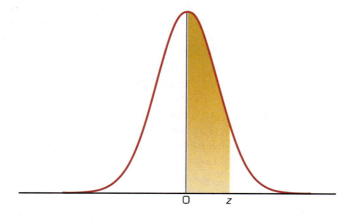

TABLE 8.29 Each Entry in This Table Is the Area Under the Normal Curve and Above the Horizontal Axis, Between z = 0 and a Positive Value of z. Areas for Negative Values of z Are Obtained by Symmetry.

z	.00	.01	.02	.03	.04	.05	.06	.07	.08	.09
0.0	.00000	.00399	.00798	.01197	.01595	.01994	.02392	.02790	.03188	.03586
0.1	.03983	.04380	.04776	.05172	.05567	.05962	.06356	.06749	.07142	.07535
0.2	.07926	.08317	.08706	.09095	.09483	.09871	.10257	.10642	.11026	.11409
0.3	.11791	.12172	.12552	.12930	.13307	.13683	.14058	.14431	.14803	.15173
0.4	.15542	.15910	.16276	.16640	.17003	.17364	.17724	.18082	.18439	.18793
0.5	.19146	.19497	.19847	.20194	.20540	.20884	.21226	.21566	.21904	.22240
0.6	.22575	.22907	.23237	.23565	.23891	.24215	.24537	.24857	.25175	.25490
0.7	.25804	.26115	.26424	.26730	.27035	.27337	.27637	.27935	.28230	.28524
0.8	.28814	.29103	.29389	.29673	.29955	.30234	.30511	.30785	.31057	.31327
0.9	.31594	.31859	.32121	.32381	.32639	.32894	.33147	.33398	.33646	.33891
1.0	.34134	.34375	.34614	.34849	.35083	.35314	.35543	.35769	.35993	.36214
1.1	.36433	.36650	.36864	.37076	.37286	.37493	.37698	.37900	.38100	.38298
1.2	.38493	.38686	.38877	.39065	.39251	.39435	.39617	.39796	.39973	.40147
1.3	.40320	.40490	.40658	.40824	.40988	.41149	.41309	.41466	.41621	.41774
1.4	.41924	.42073	.42220	.42364	.42507	.42647	.42785	.42922	.43056	.43189
1.5	.43319	.43448	.43574	.43699	.43822	.43943	.44062	.44179	.44295	.44408
1.6	.44520	.44630	.44738	.44845	.44950	.45053	.45154	.45254	.45352	.45449
1.7	.45543	.45637	.45728	.45818	.45907	.45994	.46080	.46164	.46246	.46327
1.8	.46407	.46485	.46562	.46638	.46712	.46784	.46856	.46926	.46995	.47062
1.9	.47128	.47193	.47257	.47320	.47381	.47441	.47500	.47558	.47615	.47670
2.0	.47725	.47778	.47831	.47882	.47932	.47982	.48030	.48077	.48124	.48169
2.1	.48214	.48257	.48300	.48341	.48382	.48422	.48461	.48500	.48537	.48574
2.2	.48610	.48645	.48679	.48713	.48745	.48778	.48809	.48840	.48870	.48899
2.3	.48928	.48956	.48983	.49010	.49036	.49061	.49086	.49111	.49134	.49158
2.4	.49180	.49202	.49224	.49245	.49266	.49286	.49305	.49324	.49343	.49361
2.5	.49379	.49396	.49413	.49430	.49446	.49461	.49477	.49492	.49506	.49520
2.6	.49534	.49547	.49560	.49573	.49585	.49598	.49609	.49620	.49632	.49643
2.7	.49653	.49664	.49674	.49683	.49693	.49702	.49711	.49720	.49728	.49736
2.8	.49744	.49752	.49760	.49767	.49774	.49781	.49788	.49795	.49801	.49807
2.9	.49813	.49819	.49825	.49831	.49836	.49841	.49846	.49851	.49856	.49861
3.0	.49865	.49869	.49874	.49878	.49882	.49886	.49889	.49893	.49896	.49900
3.1	.49903	.49906	.49910	.49913	.49916	.49918	.49921	.49924	.49926	.49929
3.2	.49931	.49934	.49936	.49938	.49940	.49942	.49944	.49946	.49948	.49950
3.3	.49952	.49953	.49955	.49957	.49958	.49960	.49961	.49962	.49964	.49965
3.4	.49966	.49968	.49969	.49970	.49971	.49972	.49973	.49974	.49975	.49976
3.5	.49977	.49978	.49978	.49979	.49980	.49981	.49981	.49982	.49983	.49983
3.6	.49984	.49985	.49985	.49986	.49986	.49987	.49987	.49988	.49988	.49989
3.7	.49989	.49990	.49990	.49990	.49991	.49991	.49992	.49992	.49992	.49992
3.8	.49993	.49993	.49993	.49994	.49994	.49994	.49994	.49995	.49995	.49995
3.9	.49995	.49995	.49996	.49996	.49996	.49996	.49996	.49996	.49997	.49997
4.0	.49997	.49997	.49997	.49997	.49997	.49997	.49998	.49998	.49998	.49998

EXAMPLE 1

A sample space of blood pressures is normally distributed with a mean of 120 and a standard deviation of 10. What is the probability that a person chosen at random from this population will have a blood pressure between 120 and 132?

The probability that an outcome in a normally distributed sample space with mean μ and standard deviation σ falls between two numbers a and b, with b > a, is the same as the area under the normal curve with mean μ and standard deviation σ, above the horizontal axis, and between a and b.

Solution

Using Equation (*13*) with $\mu = 120$ and $\sigma = 10$, we find that $x = 132$ is

$$z = \frac{132 - 120}{10} = 1.2$$

standard deviations from the mean. Going to Table 8.29, we move down the z column until we find 1.2. The value to the right of $z = 1.2$ is .38493. It follows that 38.493% of the area between the horizontal axis and the normal curve with $\mu = 120$ and $\sigma = 10$ falls in the interval [120, 132]. It also follows that .38493 is the probability that a randomly selected blood pressure will be between 120 and 132.

The values in Table 8.29 list probabilities of outcomes in intervals bounded on the left by the mean μ. The table is restricted to intervals $[\mu, x]$ for values of $x > \mu$. Often, however, we are interested in other intervals that are not bounded on the left by μ. If $[a, b]$ is the interval of interest, with $b > a > \mu$, then we write

$$[a, b] = [\mu, b] - [\mu, a]$$

and use Table 8.29 with Equation (*13*) to find the probabilities associated with both intervals on the right side of this last equality.

EXAMPLE 2

The diameters of machine-made ball bearings are uniformly distributed with a mean of 325 millimeters and a standard deviation of 6 millimeters. What is the probability that a randomly selected ball bearing will have a diameter between 330 and 338 millimeters?

Solution

We require the area of the shaded region in Figure 8.14, over the interval [330, 338]. We calculate this area by first finding the area over the interval [325, 338] and then *subtracting* the area over the interval [325, 330].

Using Equation (*13*) with $\mu = 325$ and $\sigma = 6$, we find that $x = 338$ is

$$z = \frac{338 - 325}{6} \approx 2.17$$

standard deviations from the mean. Going to Table 8.29, we move down the z column until we find 2.1 and then move 8 columns to the right until we are under the column headed by .07, which corresponds to the second decimal place of z. The tabular value is .48500, which represents the area under the normal curve with $\mu = 325$ and $\sigma = 6$ and above the horizontal x-axis between $x = 325$ and $x = 338$.

For $x = 330$, the corresponding z is

$$z = \frac{330 - 325}{6} \approx 0.83$$

FIGURE 8.14
Distribution of ball bearings

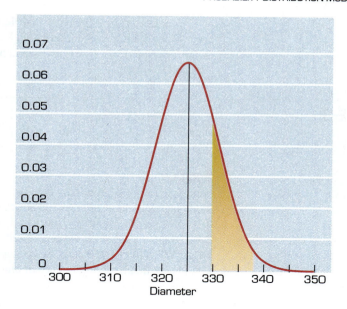

Going to Table 8.29, we move down the *z* column until we find 0.8 and then right to the column headed by .03. The tabular value is .29673, which represents the area between 325 and 330. Subtracting this area from the first area we found, we have .48500 − .29673 = .18827 as the probability that a ball bearing has a diameter between 330 and 338 millimeters.

Because a normal curve with mean μ is symmetric around a vertical line that intersects the horizontal axis at μ, the area between a normal curve and the horizontal axis over an interval extending *h* units to the *left* of the mean is the same as the area between the same normal curve and the horizontal axis over an interval extending *h* units to the right of the mean. This relationship is shown graphically, in Figure 8.15, where the shaded region to the left of the mean has the same area as the shaded region to the right of the mean.

FIGURE 8.15

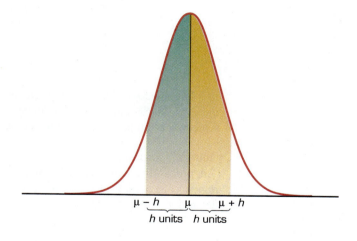

Consequently, in a normally distributed sample space with mean μ, the probability of an outcome falling in the interval $[\mu - h, \mu]$ is the same as the probability of its falling in the interval $[\mu, \mu + h]$, and we can determine this latter probability using Equation (13) and Table 8.29.

EXAMPLE 3

A sample space of blood pressures is normally distributed with a mean of 120 and a standard deviation of 10. What is the probability that a person chosen at random from this population will have a blood pressure between 108 and 120?

Solution

Note that 108 is 12 units less than the mean of 120. The probability of a normally distributed outcome being between 108 and the mean 120 is the area of the green region in Figure 8.16. Because of symmetry, this area is the same as the area of the yellow region in Figure 8.16. The yellow area is the probability of having blood pressure between 120 and 120 + 12 = 132, which we found in Example 1 to be .38493. Therefore, .38493 is also the probability that a randomly chosen blood pressure will fall between 108 and 120.

FIGURE 8.16
Distribution of blood pressures

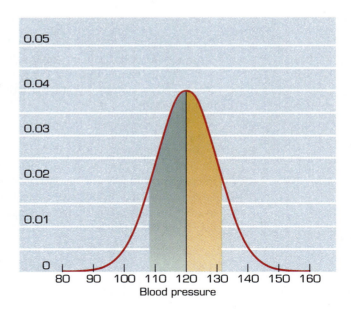

EXAMPLE 4

It is hypothesized that the body temperatures of humans are normally distributed with a mean of 98.6°F and a standard deviation of 0.2°F. What is the probability of a randomly selected person having a body temperature between 98.25°F and 99.05°F?

Solution

The probability is the same as the area of the shaded region in Figure 8.17. Part of this region is to the left of the mean (98.6°) and part is to the right of this mean, so we will determine each area separately.

FIGURE 8.17
Distribution of body temperatures

The region to the right of the mean is between 98.6° and 99.05°. Using Equation (*13*) with $\mu = 98.6$ and $\sigma = 0.2$, we find that $x = 99.05$ is

$$z = \frac{99.05 - 98.6}{0.2} = 2.25$$

standard deviations from the mean. Going to Table 8.29, we move down the z column until we find 2.2 and then move six columns to the right until we are under the column headed by .05, which corresponds to the second decimal place of z. The tabular value .48778 is the area under the normal curve with $\mu = 98.6°$ and $\sigma = 0.2°$, above the horizontal x-axis, and between $x = 98.6°$ and $x = 99.05°$. It is also the probability that a randomly selected individual in the population has a body temperature between 98.6° and 99.05°.

The region to the left of the mean is between 98.25° and 98.6°, an interval of 98.6 − 98.25 = 0.35 degrees. It has the same area as the region to the right of mean between 98.6° and 98.6° + 0.35° = 98.95°. To find this area, we first calculate

$$z = \frac{98.95 - 98.6}{0.2} = 1.75$$

Going to Table 8.29, we move down the z column until we find 1.7 and then move six columns to the right until we are under the column headed by .05, which corresponds to the second decimal place of z. The tabular value .45994 is the probability that a randomly selected individual in the population has a body temperature between 98.6° and 98.95°, and it is also the probability of having a body temperature between 98.25° and 98.6°.

The probability of a temperature falling in the interval [98.25°, 98.6°] is .45994, and that the probability of a temperature falling in the interval [98.6°, 99.05°] is .48778. Therefore, the probability of a temperature falling in the interval [98.25°, 99.05°] is .45994 + .48778 = .94772

EXAMPLE 5

A sample space of blood pressures is normally distributed with a mean of 120 and a standard deviation of 10. What is the probability that a person chosen at random from this population will have a blood pressure greater than 132?

Solution

The probability we seek is the same as the area of the shaded region in Figure 8.18. The entire area between the normal curve and the horizontal axis is 1, with half falling to the right of the mean $\mu = 120$ and half falling to the left. Therefore, the area to the right of the mean $\mu = 120$ is .5. From the results of Example 1, we know that the area between 120 and 132 is .38493. Consequently, the probability of having a blood pressure greater than 132 is

$$.5 - .38493 = .11507$$

FIGURE 8.18
Distribution of blood pressures

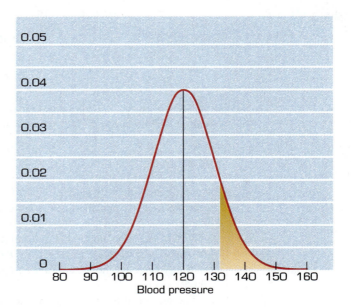

An interesting application of normal distribution involves Bernoulli trials. Recall from Section 6.6 that a Bernoulli trial is a process with two outcomes, one called a success and the other called a failure. We let p denote the probability of a success, hence the probability of a failure is $1 - p$. If we repeat a Bernoulli trial n times, then we have a set of n Bernoulli trials, and the probability of obtaining exactly r successes is

$$C(n, r)p^r(1 - p)^{n-r}$$

where $C(n, r)$ is a combination of n objects taken r at a time, defined in Section 6.5 as

$$C(n, r) = \frac{n!}{r!(n - r)!}$$

A simple set of Bernoulli trials occurs when we flip a fair coin ten times. Let's take a head to be a success and a tail to be failure; the probability of a success is $p = \frac{1}{2}$, the same

TABLE 8.30 **Flip a Fair Coin Ten Times**

NUMBER OF HEADS	PROBABILITY
r	$C(10,r)(\frac{1}{2})^r(\frac{1}{2})^{10-r}$
0	.00098
1	.00977
2	.04395
3	.11719
4	.20508
5	.24609
6	.20508
7	.11719
8	.04395
9	.00977
10	.00098

as the probability of a failure. With ten tosses, we must obtain an integer number of successes between 0 and 10, inclusive. The probabilities associated with these eleven outcomes are listed in Table 8.30, rounded to five decimal places.

Figure 8.19 is a bar graph of the data in Table 8.30. We use a bar graph rather than a histogram, leaving spaces between the columns, because Bernoulli trials represent a discrete process. Histograms are reserved for continuous processes for which the outcomes can be any real number in an interval. The Bernoulli trials listed in Table 8.30 allow for only 11 possible outcomes, a finite number.

The shape of the bar graph in Figure 8.19 is strikingly familiar! Indeed, if we connect the midpoints of the tops of adjacent columns, we would create a curve that *appears* to follow a normal distribution. To see if it really does follow a normal distribution, we must graph a normal curve, and to do that we need a mean and a standard deviation.

The probability of a success in a single Bernoulli trial is represented by p. If we repeat the process n times, then we expect np successes, so the mean number of successes is $\mu = np$. It is shown in more advanced statistics books that the standard deviation for a set of n Bernoulli trials is $\sigma = \sqrt{np(1-p)}$. With $n = 10$ and $p = \frac{1}{2}$, we have

$$\mu = 10\left(\frac{1}{2}\right) = 5 \quad \text{and} \quad \sigma = \sqrt{10\left(\frac{1}{2}\right)\left(1 - \frac{1}{2}\right)} = \sqrt{2.5} \approx 1.58$$

Figure 8.20 is a graph of a normal curve with mean 5 and standard deviation 1.58. The points in Table 8.30 are superimposed on this curve as bullets. We see that some of these points to do not quite fall on the curve; still, the normal curve seems to be a very good approximation to the actual curve that connects these points.

In general, the probabilities associated with a set of Bernoulli trials closely follow a normal curve when the probability of a success is near $p = \frac{1}{2}$ or when n, the total number of trials, is large. Thus, we can use normal curves to *estimate* the probabilities associated with Bernoulli trials. When we do so, however, we are using information about a continuous process, modeled by a normal distribution, to obtain information about a discrete process defined by a set of Bernoulli trials. Some adjustments, therefore, must be made.

FIGURE 8.19

Possible Outcomes of Flipping a Coin Ten Times

FIGURE 8.20

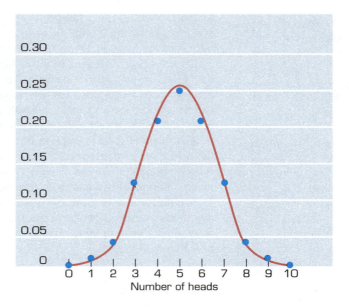

When we estimate probabilities about Bernoulli trials from a normal curve, we must associate with each integer of interest in a set of Bernoulli trials the interval of all real numbers that round to that integer. If we are interested in exactly five successes in a set of Bernoulli trials, then we use information from the *interval* [4.5, 5.5) on a normal curve, because each real number in this interval rounds to the integer 5. If we are interested in the probability of obtaining between 240 and 260 successes inclusive in a set of Bernoulli trials, then we use information from the interval [239.5, 260.5) on a normal curve, because each number in this interval rounds to an integer between 240 and 260, inclusive.

To appreciate how good such estimates can be, let us calculate the probability of flipping a coin ten times and getting either 5, 6, or 7 heads. This probability, rounded to five decimal places, is obtained easily from Table 8.30 as

$$P(5 \text{ heads}) + P(6 \text{ heads}) + P(7 \text{ heads}) = .24609 + .20508 + .11719 = .56836$$

To estimate this probability from a normal curve with mean $\mu = 5$ and standard deviation $\sigma = \sqrt{2.5}$, we need the area of the region under the normal curve in Figure 8.20 and above the horizontal axis between $x = 4.5$ and $x = 7.5$. We use the interval [4.5, 7.5) because every real number in this interval rounds to either 5, 6, or 7.

The region to the right of the mean is between $x = 5$ and $x = 7.5$. Using Equation (*13*) with $\mu = 5$ and $\sigma = \sqrt{2.5}$, we find that $x = 7.5$ is

$$z = \frac{7.5 - 5}{\sqrt{2.5}} \approx 1.58$$

standard deviations from the mean. Going to Table 8.29, we locate the area associated with this z-value to be .44295.

The region to the left of the mean is between 4.5 and 5, an interval of length 0.5. It has the same area as the region to the right of the mean between 5 and 5.5. To find this area, we first calculate

$$z = \frac{5.5 - 5}{\sqrt{2.5}} \approx 0.32$$

Going to Table 8.29, we find the corresponding area to be .12552.

The probability of an outcome falling in the interval [5, 7.5) is .44295, and the probability of its falling in the interval [4.5, 5) is .12552. Therefore, the probability of an outcome falling in the interval [4.5, 7.5) is

$$.44295 + 1.2552 = .56847$$

This is our *estimate* for the probability of obtaining exactly 5, 6, or 7 heads when flipping a coin ten times, based on a normal curve. Note how close this estimate is to the real answer, .56836.

Using a normal curve to approximate probabilities for Bernoulli trials is most appealing when *n* is large.

EXAMPLE 6

A game of roulette is played 1,000 times, and each time the color of the winning number is recorded. What is the probability of obtaining between 480 and 515 red numbers, inclusive?

Solution

We view this process as a set of 1,000 Bernoulli trials, with a success defined as a red number. To calculate the probability of interest precisely, we would need to determine the individual probabilities associated with 36 distinct outcomes—every outcome between 480 red numbers and 515 red numbers, inclusive—and then sum the results. Such an approach would be quite laborious. Instead, we will *estimate* the probability of interest using a normal curve. This estimate should be a good one because the number of trials, 1000, is large.

A roulette wheel has 38 numbers—18 red, 18 black, and 2 green—so the probability of a single success, that is, a single red number, is $\frac{18}{38}$. For such a process,

$$\mu = np = 1000\left(\frac{18}{38}\right) \approx 473.68$$

and

$$\sigma = \sqrt{np(1 - p)} = \sqrt{1000\left(\frac{18}{38}\right)\left(1 - \frac{18}{38}\right)} \approx 15.79$$

A normal curve with this mean and standard deviation is drawn in Figure 8.21. We seek the area of the shaded region in Figure 8.21, the region under the normal curve, above the horizontal axis, and between $x = 479.5$ and 515.5. (All numbers in the interval [479.5, 515.5) round to integers between 480 and 515, inclusive, which are the outcomes of interest).

Here, $x = 479.5$ is

$$z = \frac{479.5 - 473.68}{15.79} \approx 0.37$$

standard deviations from the mean. Using Table 8.29, we find the corresponding tabular area to be .14431. Similarly, $x = 515.5$ is

$$z = \frac{515.5 - 473.68}{15.79} \approx 2.65$$

FIGURE 8.21

Playing roulette 1000 times

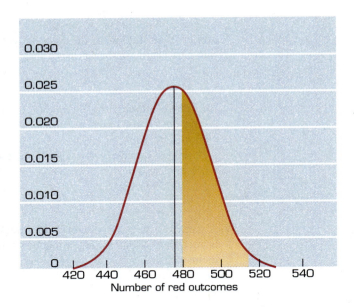

standard deviations from the mean, with a corresponding tabular area of .49598. The area of the shaded region in Figure 8.21 is thus

$$.49598 - .14431 = .35167$$

The probability of obtaining between 480 and 515 red numbers in 1000 roulette plays is approximately .35167.

We are interested in simulating outcomes from a normal distribution with mean μ and standard deviation σ. To do so, we must generate random numbers that come from a normal distribution with mean μ and standard deviation σ. A number of procedures exist for this purpose, and although their derivations are well beyond the scope of this book, we can use such procedures in our simulations.

One well-known procedure for generating a normal distribution with mean μ and standard deviation σ is as follows: Use a random number generator to produce *two* numbers r_1 and r_2 that are uniformly distributed between 0 and 1. Then calculate the expression

$$n = \mu + \sigma \cdot \cos(2\pi r_1)\sqrt{-2 \cdot \ln(r_2)} \tag{14}$$

Some of the functions in this expression may be unfamiliar, but they are readily available as keys on most calculators and as library functions in many computer programs and electronic spreadsheets. *cos* is the trigonometric function, cosine; *ln* refers to the natural logarithm, a function that is similar to but distinct from the common logarithm *log*; and π is the real number 3.14159. . . . If a calculator is used, it must be set to radian mode rather than degree mode before using the cosine function key. The resulting number n in Equation (*14*) is an outcome from a normally distributed sample space with mean μ and standard deviation σ.

EXAMPLE 7

Use a random number generator to produce temperatures that are normally distributed with a mean of 98.6°F and a standard deviation of 0.2°F.

Solution

Here $\mu = 98.6°$ and $\sigma = 0.2°$. Accessing a random number generator, we obtain

$$r_1 = 0.312175 \quad \text{and} \quad r_2 = 0.516666$$

as two numbers from a population that is uniformly distributed on $[0, 1]$. Then, sequentially calculating the terms in Equation (*14*), we obtain

$$2\pi r_1 = 2(3.14159)(0.312175) \approx 1.96145$$

$$cos(2\pi r_1) = cos(1.96145) \approx -0.380796$$

$$ln(r_2) = ln(0.516666) \approx -0.660359$$

$$\sqrt{-2 \cdot \ln(r_2)} = \sqrt{-2(-0.660359)} \approx 1.149225$$

and finally

$$n = \mu + \sigma \cdot cos(2\pi r_1)\sqrt{-2 \cdot \ln(r_2)}$$
$$= 98.6 + 0.2(-0.380796)(1.149225)$$
$$\approx 98.51$$

This value of n becomes the first temperature entry in Table 8.31.

To obtain a second entry, we repeat the entire procedure with a new set of random numbers. We use a random number generator to generate two more numbers from a population that is uniformly distributed on $[0, 1]$, and we get

$$r_1 = 0.511149 \quad \text{and} \quad r_2 = 0.529704$$

Then, sequentially calculating the terms in Equation (*14*), we obtain

$$2\pi r_1 = 2(3.14159)(0.511149) \approx 3.21164$$

$$cos(2\pi r_1) = cos(3.21164) \approx -0.997547$$

TABLE 8.31   Simulating Normally Distributed Temperatures

TRIAL	RANDOM NUMBER	RANDOM NUMBER	TEMPERATURE (degrees)
1	0.312175	0.516666	98.51
2	0.511149	0.529704	98.38
3	0.814643	0.341134	98.72
4	0.630753	0.846835	98.52
5	0.946331	0.964791	98.65
6	0.502784	0.124030	98.19
7	0.444183	0.513289	98.38
8	0.363832	0.213480	98.37
9	0.939675	0.023328	99.11
10	0.174745	0.326635	98.74

and finally

$$n = \mu + \sigma \cdot \cos(2\pi r_1)\sqrt{-2 \cdot \ln(r_2)}$$
$$= 98.6 + 0.2(-0.997547)(1.127330)$$
$$\approx 98.38.$$

Continuing in this manner, we complete Table 8.31.

EXAMPLE 8

Use a random number generator to produce diameters of machine-made ball bearings that are uniformly distributed with a mean of 325 millimeters and a standard deviation of 6 millimeters.

Solution

Here $\mu = 325$ millimeters and $\sigma = 6$ millimeters. Accessing a random number generator, we obtain

$$r_1 = 0.209729 \quad \text{and} \quad r_2 = 0.517216$$

as two numbers from a population that is uniformly distributed on [0, 1). Then, sequentially calculating the terms in Equation (*14*), we obtain

$$2\pi r_1 = 2(3.14159)(0.209729) \approx 1.31777$$
$$cos(2\pi r_1) = cos(1.31777) \approx 0.250339$$
$$ln(r_2) = ln(0.517216) \approx -0.659295$$
$$\sqrt{-2 \cdot \ln(r_2)} = \sqrt{-2(-0.659295)} \approx 1.14830$$

and finally

$$n = \mu + \sigma \cdot \cos(2\pi r_1)\sqrt{-2 \cdot \ln(r_2)}$$
$$= 325 + 6(0.250339)(1.14830) \approx 326.72$$

which becomes the first diameter entry in Table 8.32. Continuing in this manner, we complete Table 8.32.

TABLE 8.32   Simulating Normally Distributed Diameters

TRIAL	RANDOM NUMBER	RANDOM NUMBER	DIAMETER (mm)
1	0.209729	0.517216	326.72
2	0.672490	0.251967	320.34
3	0.983139	0.957622	326.76
4	0.364483	0.146088	317.25
5	0.987746	0.496411	332.08

EXAMPLE 9

In a manufacturing process, a robotic arm attempts to place a ball bearing into a grooved slot. The fit is good when the diameter of the ball bearing is less than that of the grooved slot, otherwise the fit is bad. If the fit is bad, the robotic arm returns both the ball bearing and the grooved slot to inventory to be used again later.

Ball bearings come from a supplier who stipulates that their diameters are normally distributed with a mean of 325 millimeters and a standard deviation of 6 millimeters. The grooved slot is provided by a different supplier, who stipulates that their diameters are normally distributed with a mean of 330 millimeters and a standard deviation of 10 millimeters. What percentage of the time can one expect a good fit between randomly selected ball bearings and grooved slots?

Solution

We solve this problem by simulation. We simulate 1,000 ball bearings and 1,000 grooved slots, each time comparing whether the ball bearing and the grooved slot are well matched. The diameters of the ball bearings are simulated as in Example 8. In fact, the diameters in Table 8.32 are the first five ball bearing diameters in Table 8.33. The diameters of the grooved slots are simulated similarly, using Equation (14) with $\mu = 330$ and $\sigma = 10$ and are shown in the slot diameter column in Table 8.33.

TABLE 8.33 Robotic Arm Simulation: Run #1

TRIAL	BEARING DIAMETER	SLOT DIAMETER	GOOD FIT	TOTAL NUMBER OF SUCCESSES
1	326.72	334.54	yes	1
2	320.34	346.74	yes	2
3	326.76	320.14	no	2
4	317.25	350.33	yes	3
5	332.08	311.60	no	3
⋮				
996	322.16	338.31	yes	663
997	329.37	325.97	no	663
998	332.53	319.08	no	663
999	316.56	330.70	yes	664
1,000	319.77	340.94	yes	665

The first ball bearing in Table 8.33 has a diameter of 326.72 millimeters, and the first grooved slot has a diameter of 334.54 millimeters. This is a good fit, and we now have one successful match. The second ball bearing in Table 8.33 has a diameter of 320.34 millimeters, and the second grooved slot has a diameter of 346.74 millimeters. This is also a good fit, and we have a second successful match. The third ball bearing in Table 8.33 has a diameter of 326.76 millimeters and the third grooved slot has a diameter of 320.14 millimeters. This is not a good fit, so we still have just two successful matches. Continuing this way, we complete the simulation shown in Table 8.33, where only the first and last five trials are exhibited.

In this simulation, 665 of 1,000 attempts to match a ball bearing with a grooved slot ended successfully in good fits. We run this simulation 19 more times, each with 1,000 trials, and generate the entries in Table 8.34.

TABLE 8.34 **Robotic Arm Simulations**

RUN	TOTAL NUMBER OF SUCCESSES	RUN	TOTAL NUMBER OF SUCCESSES
1	665	11	672
2	671	12	651
3	680	13	657
4	634	14	662
5	687	15	651
6	653	16	635
7	654	17	650
8	702	18	686
9	685	19	644
10	645	20	666

The mean of the 20 values for total number of successful fits is 662.5. Based on these simulations, we expect about 662.5 good fits out of every 1,000 attempts. We expect a good fit about 66.25% of the time.

IMPROVING SKILLS

In Problems 1 through 6, determine the probability that a randomly selected outcome will fall in the specified interval if the sample space is normally distributed with a mean of 22 and a standard deviation of 3.

1. [22, 24]   **2.** [22, 26]   **3.** [19, 22]
4. [23, 25]   **5.** [19, 21]   **6.** [21, 25]

In Problems 7 through 12, determine the probability that a randomly selected outcome will fall in the specified interval if the sample space is normally distributed with a mean of 200 and a standard deviation of 15.

7. [200, 225]   **8.** [180, 200]   **9.** [210, 230]
10. [205, 208]   **11.** [190, 210]   **12.** [173, 195]

In Problems 13 through 18, determine the probability that a randomly selected outcome will fall in the specified interval if the sample space is normally distributed with a mean of −50 and a standard deviation of 8.

13. [−60, −50]   **14.** [−62, −54]   **15.** [−54, −48]
16. [−48, −45]   **17.** [−45, −40]   **18.** [−40, −39]

19. A sample space is normally distributed with a mean of 150 and a standard deviation of 22. Find the probability that a randomly selected outcome will be (*a*) greater than 170, (*b*) greater than 200, (*c*) less than 100.

20. A sample space is normally distributed with a mean of 30 and a standard deviation of 0.3. Find the probability that a randomly selected outcome will be (*a*) greater than 31, (*b*) greater than 29.5, (*c*) less than 29.

21. A sample space is normally distributed with a mean of -10 and a standard deviation of 1. Find the probability that a randomly selected outcome will be (*a*) greater than -8, (*b*) less than -10, (*c*) less than -9.
22. Use the random numbers in Table 8.31 to generate a sample from a normally distributed sample space with mean 100 and standard deviation 10.
23. Use the random numbers in Table 8.32 to generate a sample from a normally distributed sample space with mean 8 and standard deviation 0.3.
24. Use the random numbers in Table 8.32 to generate a sample from a normally distributed sample space with mean -20 and standard deviation 4.

CREATING MODELS

25. A standard IQ test is prepared so that scores will be normally distributed with a mean of 100 and a standard deviation of 15. What is the probability that a person will score between 100 and 130?
26. Scholastic Aptitude Tests are scaled so that scores are approximately normally distributed with a mean of 500 and a standard deviation of 100. What is the probability that a person will score 700 or higher?
27. Salaries for technicians at a particular company are normally distributed with a mean of $30,000 and a standard deviation of $5,500. (*a*) What percentage of all technicians make between $25,000 and $35,000? (*b*) What percentage make more than $40,000?

28. The body weights of males at a large state university are normally distributed with a mean of 170 lb and a standard deviation of 20 lb. What is the probability that the next male student who enters the infirmary (*a*) will weigh between 160 and 180 lbs? (*b*) will weigh less than 140 lb?
29. The life span of a particular model muffler is normally distributed with a mean of 40,000 miles and a standard deviation of 2,500 miles. What is the probability that such a muffler will wear out before 35,000 miles?
30. A teacher grades "on a curve" if the final grade distribution is normally distributed around the mean grade for a class. Students within one standard deviation of the mean receive a *C*, and students with grades at least two standard deviations above the mean receive an *A*. In this system, what percentage of students will receive a final grade of *C* and what percentage will receive a final grade of *A*?
31. The average life span of a certain type of 100-watt light bulb is normally distributed with a mean of 750 hours and a standard deviation of 20 hours. A manufacturing company operates around the clock and, as a matter of policy, replaces all bulbs with new ones every 720 hours. What percentage of the 100-watt bulbs are still functioning when they are replaced?
32. The company in the previous problem reevaluates its replacement policy for light bulbs. For efficiency, all bulbs will continue to be replaced at the same time, but the company wants to ensure that at least 70% of all 100-watt bulbs are producing light at all times. How often should the light bulbs be replaced?
33. Annual rainfall in a region is normally distributed with a mean of 40″ and a standard deviation of 5″. Flood conditions occur, on average, once every 100 years. What amount of rainfall triggers a flood condition?
34. The breaking strength of a certain type of rope is normally distributed with a mean of 150 lb and a standard deviation of 5 lb. A customer buys this rope and uses it to lift a 154-lb weight. What is the probability that the rope will break?

35. The life span of a particular model tire is normally distributed with a mean of 60,000 miles and a standard deviation of 3,000 miles. The manufacturer warranties all tires for 50,000 miles. What percentage of all tires sold will be returned by customers for refunds on their warranties?

36. A fair coin is flipped 500 times. Estimate the probability of getting between 240 and 260 heads, inclusive.

37. A game of roulette is played 600 times. Estimate the probability of of getting 40 or more green numbers.

38. A pair of dice is thrown 500 times, and the numbers on the dice are summed each time. Estimate the probability of obtaining a sum of seven 100 or more times.

39. A card is picked from a standard deck and then replaced. The deck is shuffled, and the process repeated for a total of 800 trials. Estimate the probability of getting between 180 and 220 spades, inclusive.

40. Estimate the probability of getting at least 200 face cards from the process described in the previous problem.

41. A student takes a true-false test with 200 questions and guesses on every answer. Estimate the probability that he will pass the test (that is answer, 70% or more of the questions correctly)?

42. A test has 225 multiple-choice questions with five answers to each question. What is the probability that a student will pass this test (answering 158 or more questions correctly) if she guesses on every answer?

43. A plumbing manufacturer guarantees the probability of a new faucet working properly when correctly installed is .99. Estimate the probability that at most 10 of the next 1,000 faucets correctly installed by plumbers will *not* work properly.

44. The probability that a drug will cure a particular infection is .9. The drug is given to 1,200 patients with the disease. Estimate the probability that the drug will be effective on at least 920 of these patients.

45. The probability that a particular eye operation will be successful is .85. During one year, approximately 800 patients undergo this operation at one hospital. Estimate the probability that the operation will be successful on at least 700 of those patients.

46. From experience, a manufacturer of baseballs knows that under ordinary circumstances 3% of all baseballs produced are defective. The manufacturer delivers 6,000 baseballs to a city for their summer leagues. Estimate the probability that between 150 and 200, inclusive, of these balls are defective.

47. Assume that the probability of a marriage ending in divorce within 15 years is .6. A study follows 240 couples, all married the same year. Estimate the probability that 150 or more of these couples will be divorced after 15 years.

EXPLORING IN TEAMS

48. Sample the heights of either males *or* females by collecting the heights of 100 or more randomly selected individuals of the same gender. Take your samples either at a school where you study or at a corporation where you work. Construct a frequency polygon for your sample, and determine whether the outcomes are normally distributed.

49. The area under a normal curve and above the horizontal axis is always 1, but the area under most bell-shaped frequency polygons is generally much greater than 1. However, the construction of histograms and frequency polygons can be modified to accommodate this area requirement. First, a histogram is constructed as in Section 8.3 and its area A is determined. Determining the area is straightforward because histo-

grams are collections of rectangles. Each frequency is then divided by A, resulting in a set of relative frequencies. A modified histogram is constructed using the relative frequencies as the heights of the bars. A frequency polygon is then drawn from the modified histogram. Construct a modified histogram and corresponding frequency polygon for Figure 8.6 in Section 8.3.

50. Use the procedure described in the previous problem to construct a modified histogram and corresponding frequency polygon for Figure 8.4 in Section 8.3.

51. Determine the area of a modified histogram.

52. Frequencies are the number of outcomes in each histogram category. What is the significance of a relative frequency?

53. How does the shape of a histogram constructed with frequencies compare with the shape of a histogram constructed with relative frequencies, if the two histograms are based on the same data and use the same categories?

EXPLORING WITH TECHNOLOGY

Use a spreadsheet to simulate each of the processes described in Problems 54 through 64.

54. A machine fills 33-ounce bottles with a toxic liquid. The amount of liquid dispensed is normally distributed with a mean of 32 ounces and a standard deviation of 0.4 ounces. If more liquid than a bottle can hold is dispensed, the excess runs into a holding container and is treated as hazardous waste. How much liquid becomes hazardous waste?

55. Customers begin arriving at a bank 15 minutes before it opens. The time between customer arrivals in this quarter-hour period is normally distributed with a mean of 3 minutes and a standard deviation of 30 seconds. How many customers are in line when the bank opens?

56. A business has three costs associated with each item it sells: a production cost, a shipping cost, and an administrative cost. The production cost per item is normally distributed with a mean of $35 and a standard deviation of $4. The shipping cost is fixed at $4 per item. The administrative cost per item is normally distributed with a mean of $18 and a standard deviation of $3. What is the average cost of each item sold?

57. What is the average cost of each item sold in the previous problem if the shipping costs are uniformly distributed between $2 and $7?

58. A person invests $1,000 in a bank account and plans to leave it there for five years. The account accrues interest compounded semiannually, but the interest rate can vary. Currently, the nominal interest rate is 4%, but in future periods it is expected to be normally distributed with a mean of 4% and a standard deviation of 0.75%. How much can the depositor expect in the account after five years?

59. A corporation creates a sinking fund by depositing $50,000 every quarter into an account that accrues interest compounded quarterly. The nominal annual interest rate on the account is currently 6%, but in future periods it is expected to be normally distributed with a mean of 6.5% and a standard deviation of 1.2%. How much can the corporation expect in the account after $2\frac{1}{2}$ years?

60. Balls produced by a toy company have a diameter that is normally distributed with a mean of 11.9 inches and a standard deviation of 0.1 inches. Cubical packaging boxes

have sides that are normally distributed with a mean of 12.1 inches and a standard deviation of 0.05 inches. A ball will fit into a box if the ball's diameter is less than the length of a box's side. Machines scan each ball and box to ensure that a ball will fit. If not, the ball and box are routed back to inventory and the machine attempts to package a new ball in a new box. How many balls will be routed back to inventory?

61. How many balls will be routed back to inventory in the previous problem if boxes are never rejected but remain to be filled with the next ball from inventory?

62. Before a new store is opened, a site must be selected, contracts signed, and a building constructed. The time required for site selection and completion of contract negotiations once a site is selected is normally distributed with a mean of 100 days and a standard deviation of 20 days. The time required to complete construction once contracts are signed is normally distributed with a mean of 200 days and a standard deviation of 30 days. How long will it take to open each of the next five stores?

63. On an assembly line, the time between deliveries of new television sets to a quality control inspector is normally distributed with a mean of 20 minutes and a standard deviation of 5 minutes. Inspection time is normally distributed with a mean of 11 minutes and a standard deviation of 3 minutes. What percentage of the time is the inspector idle?

64. Vehicles begin arriving at a highway gas station 15 minutes before it opens and then wait for gasoline if they can. The station has room for a maximum of four cars, and because the station is on a highway, cars that cannot be accommodated cannot wait and their business is lost for the day. How many cars would wait for the service station to open but cannot because of inadequate waiting space if interarrival times are normally distributed with a mean of 3 minutes and a standard deviation of 45 seconds?

REVIEWING MATERIAL

65. (Section 2.2) Use Gaussian elimination to solve the following system of equations:

$$2x - 3y + 4z = 40$$

$$3x + 2y - 2z = 24$$

$$y + 5z = 63$$

66. (Section 5.2) In 1993, the U.S. imported 2.384 billion pounds of unroasted coffee: 446.6 million pounds came from Brazil, 391.2 from Colombia, 389.8 from Mexico, 240 from Guatemala, 168.6 from El Salvador, and 141.4 from Thailand. No other source accounted for more than 100 million pounds. Construct a pie chart from this information.

67. (Section 6.2) If one card is chosen from a standard deck, what are the odds of having that card be a heart?

68. (Section 6.4) Two hundred people with the same disease are screened for that disease using a new test, and the test is positive for 170 of these patients. The test is also given to another 150 people who do not have the disease, and the test is positive for 25 of them. Based on this data, what is the probability that a person who takes the test actually has the disease when the test is positive?

RECOMMENDING ACTION

69. Respond by memo to the following request:

MEMORANDUM

To: J. Doe Reader

From: Superintendent's Office

Date: Today

Subject: **Grading**

Many of our teachers grade on a tradition scale: 90 to 100 is an *A*, 80 to 89 a *B*, 70 to 79 a *C*, 60 to 69 a *D*, and below 60 an *F*. The student government wants a system based on a curve. The most natural, of course, is the normal curve.

With this system, the average grade for the class becomes the class mean. A *C* is a grade within one standard deviation of the mean, an *A* is a grade that is two or more standard deviations above the mean, and an *F* is a grade that is two or more standard deviations below the mean. A *B* is between an *A* and *C*, while a *D* is between an *F* and *C*. Do you think students really understand the ramifications of such a system? Will it make them content? What do you recommend?

8.5 EXPONENTIAL DISTRIBUTIONS

Frequency polygons come in many shapes. Some are reasonably close to horizontal lines, from which we infer that the underlying sample spaces are uniformly distributed (see Section 8.2). Some frequency polygons resemble bell-shaped curves, and we infer that their underlying sample spaces are normally distributed (see Section 8.4). In this section, we consider yet another common shape for frequency polygons.

Table 8.35 lists data for the time between arrivals at the drive-in window of a bank over the same 90-minute period, 11:30 a.m. to 1:00 p.m., for five different Mondays. During these time periods, 197 customers arrived, ranging from a low of 29 on the third Monday to a high of 49 on the fifth Monday.

Grouping the time between arrivals into categories of one-minute duration, we create Table 8.36. In this grouping, we have 12 categories of equal width with 5 categories containing less than five observations. One alternative for avoiding categories with few observations is to combine the last four categories into a single category over the interval 8 to 12 minutes and to join the intervals between 5 and 6 minutes and 6 and 7 minutes. We would thus obtain categories of unequal widths, each with five or more observations. A second alternative is to group the data into intervals of four-minutes duration, defining the categories as [0, 4), [4, 8), [8, 12]. Although such a partition yields categories of equal widths with each containing five or more points, the partition results in only three distinct categories. We opt instead for the categories shown in Table 8.36.

Using the first and third columns of Table 8.36, we construct the histogram displayed in Figure 8.22. Connecting the midpoints of the tops of the bars in this histogram with

TABLE 8.35

	INTERARRIVAL TIMES				
Customer	First Monday	Second Monday	Third Monday	Fourth Monday	Fifth Monday
1	03:05	00:01	00:10	00:07	00:05
2	00:39	01:07	00:15	00:54	01:31
3	01:09	07:56	11:01	00:04	00:19
4	01:43	01:01	00:34	01:54	01:10
5	01:13	00:08	05:02	02:56	02:18
6	00:06	01:52	01:43	00:14	02:08
7	01:02	02:01	05:14	00:26	02:34
8	00:43	00:13	01:03	01:13	01:38
9	04:12	00:42	00:31	01:01	00:42
10	02:23	00:22	03:07	00:29	00:02
11	01:29	05:02	04:54	04:23	05:59
12	02:20	00:29	00:01	01:27	01:21
13	04:49	00:47	00:10	00:08	00:20
14	03:27	00:26	05:13	06:13	00:25
15	00:54	01:32	02:42	00:41	02:56
16	00:02	07:06	00:58	00:16	02:15
17	02:06	01:09	00:21	01:52	01:22
18	08:53	04:01	00:34	07:56	00:03
19	00:42	00:06	04:45	02:12	04:37
20	00:59	04:57	01:32	01:11	00:41
21	02:14	00:39	07:14	08:39	03:17
22	01:23	02:29	00:13	01:21	01:27
23	01:14	01:17	00:26	04:12	00:07
24	05:42	00:45	07:41	00:38	01:46
25	02:15	00:07	02:11	00:23	01:13
26	03:22	00:03	00:26	00:01	00:14
27	01:02	00:50	00:04	00:02	00:16
28	02:46	01:08	10:30	00:31	01:48
29	06:07	11:52	03:40	01:08	07:06
30	01:41	02:35		00:55	01:12
31	00:32	00:14		00:48	01:30
32	00:57	01:41		03:44	07:25
33	00:26	00:59		01:39	01:27
34	00:16	03:14		03:09	04:12
35	00:42	01:38		00:50	00:38
36	02:19	00:24		00:53	00:18
37	01:33	00:30		00:18	00:06
38	11:39	03:00		04:11	00:47
39		01:22			00:45
40		00:03			01:08
41		04:37			00:55
42		00:41			00:53
43		02:17			04:00
44					01:39
45					00:57
46					00:50
47					03:04
48					00:18
49					04:11

TABLE 8.36   **Frequency Table of Interarrival Times**

CATEGORY	MIDPOINT	FREQUENCY
[0, 1)	0.5	85
[1, 2)	1.5	46
[2, 3)	2.5	20
[3, 4)	3.5	11
[4, 5)	4.5	14
[5, 6)	5.5	7
[6, 7)	6.5	1
[7, 8)	7.5	7
[8, 9)	8.5	2
[9, 10)	9.5	0
[10, 11)	10.5	1
[11, 12)	11.5	3

FIGURE 8.22

Histogram of interarrival times

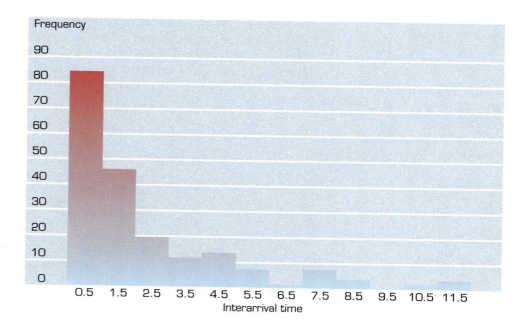

straight line segments, or simply using the second and third columns of Table 8.36, we graph the frequency polygon shown in Figure 8.23. Clearly, this frequency polygon does not approximate either a horizontal line or a bell-shaped curve.

FIGURE 8.23

Frequency polygon of interarrival times

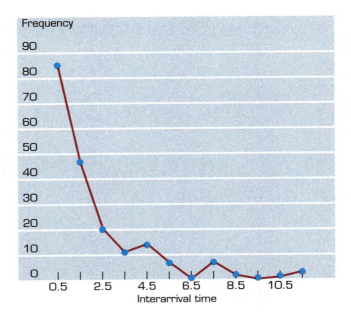

Figure 8.23 is a nonnegative curve located entirely within the first quadrant, beginning at its maximum and decreasing to zero as we move in the positive direction along the horizontal axis. Such behavior is modeled well by a family of curves known as *exponential distribution curves*; two examples are shown in Figure 8.24.

FIGURE 8.24

Two exponential curves

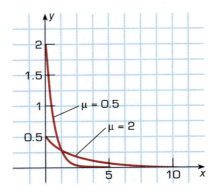

Exponential distribution curves are completely determined by specifying a nonnegative mean μ, and all have the feature that the entire area in the first quadrant between the exponential distribution curve and the horizontal x-axis is 1. In an x-y coordinate system, an exponential distribution curve attains its maximum at $x = 0$, and this maximum always equals the reciprocal of mean μ, that is at $x = 0$, $y = \dfrac{1}{\mu}$. The smaller the mean μ, the quicker the descent to zero as x increases in the positive direction; the larger the mean, the more leisurely the descent.

When a sample space can be modeled by an exponential distribution curve, we say that the outcomes of the underlying continuous process are *exponentially distributed*. In such a process, outcomes are always nonnegative, cluster close to zero, and become sparser in intervals further from the origin. Exponentially distributed processes are characterized by numerous outcomes of very short duration and a few outcomes of long duration. In Table 8.35, 48 out of 197 observations are less than 30 seconds and 85 are less than one minute, compared with 14 observations greater than 6 minutes and only 6 observations greater than 8 minutes.

Since the entire area in the first quadrant between the exponential distribution curve and the horizontal x-axis is 1, the area of the shaded region in Figure 8.25 is also the probability that an outcome will fall in the interval $[a, b]$, when both a and b are nonnegative.

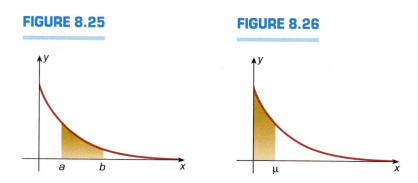

FIGURE 8.25

FIGURE 8.26

An exponential distribution curve has the property that 63.21% (rounded to two decimal places) of the area in the first quadrant between the curve and the horizontal axis falls between $x = 0$ and $x = \mu$, as illustrated by the shaded region in Figure 8.26. Consequently, the probability that an outcome will be less than or equal to the mean of a sample space that is exponentially distributed is .6321 (rounded to four decimal places), and the mean of such a sample space does *not* equal its median. In an exponentially distributed sample space, an outcome is nearly twice as likely to be less than the mean as it is to be greater than the mean. In contrast, the probability of an outcome being less than the mean in either a uniformly distributed or a normally distributed sample space is .5, the same as the probability of being greater than the mean.

The area of the shaded region in Figure 8.25 can be determined using techniques from calculus. We will do so in Section 12.6, but for now we simply use the results. If we denote a random outcome from an exponentially distributed sample space by X, then the probability that X falls in the intervals $[a, b]$ is given by the formula

$$P(a \leq X \leq b) = e^{-a/\mu} - e^{-b/\mu} \tag{15}$$

where e is the real number 2.7182818284590 ... , and μ is the mean of the underlying sample space. Many calculators have an e^x key, which reduces to a single keystroke the process of raising e to any power of x. In Equation (15), we raise e to the powers $-a/\mu$ and $-b/\mu$. To perform these operations on calculators with an e^x key, first calculate either $-a/\mu$ or $-b/\mu$ and then press the e^x key. On some calculators and in many computer programs, e is designated by the alternative notation *exp*, which stands for the word *exponential*.

EXAMPLE 1

Calculate $e^{-x/2}$ for (*a*) $x = 0.5$ and (*b*) $x = 3$.

Solution
We use a calculator with an e^x key.
(*a*) For $x = 0.5$, we have

$$e^{-0.5/2} = e^{-0.25} \approx 0.7788$$

On many calculators, the keystrokes for this computation are:

(*b*) For $x = 3$, we have

$$e^{-3/2} = e^{-1.5} \approx 0.2231$$

The keystrokes for this computation on many calculators are:

EXAMPLE 2

The interarrival time of cars at the drive-in window of a fast-food restaurant is exponentially distributed with a mean of 2 minutes. Determine the probability that the next car will arrive (*a*) between 0.5 and 3 minutes after the previous arrival and (*b*) 3 minutes or less after the previous arrival?

Solution
(*a*) Here $a = 0.5$ and $b = 3$. Using Equation (*15*) with $\mu = 2$ and the results of Example 1, we have

$$
\begin{aligned}
P(0.5 \le X \le 3) &= e^{-0.5/2} - e^{-3/2} \\
&= e^{-0.25} - e^{-1.5} \\
&\approx 0.7788 - 0.2231 \\
&= .5557
\end{aligned}
$$

(*b*) Now $a = 0$ and $b = 3$, and Equation (*15*) becomes

$$
\begin{aligned}
P(0 \le X \le 3) &= e^{-0/2} - e^{-3/2} \\
&= e^0 - e^{-1.5} \\
&\approx 1 - 0.2231 \\
&= .7769
\end{aligned}
$$

EXAMPLE 3

Calculate the probabilities in Example 2 if instead the mean interarrival time is 3 minutes.

Solution

(a) Here $a = 0.5$, $b = 3$, and $\mu = 3$. Using Equation (15), we have

$$P(0.5 \leq X \leq 3) = e^{-0.5/3} - e^{-3/3}$$
$$= e^{-0.5/3} - e^{-1}$$
$$\approx 0.8465 - 0.3679$$
$$= .4786$$

(b) Now $a = 0$ and $b = 3$, and Equation (15) becomes

$$P(0 \leq X \leq 3) = e^{-0/3} - e^{-3/3}$$
$$= e^0 - e^{-1}$$
$$\approx 1 - 0.3679$$
$$= .6321$$

EXAMPLE 4

The time between successive incoming calls to a telephone answering service is exponentially distributed with a mean of 40 seconds. What is the probability that the time between two consecutive calls will be greater than 30 seconds?

FIGURE 8.27

Solution

The probability we seek is the same as the area of the shaded region in Figure 8.27. Recall that the area in the first quadrant between an exponential distribution curve and the horizontal axis is 1. Therefore, we use Equation (15) to calculate the area between 0 and 30 seconds, and then we subtract that number from 1 to obtain the area of interest.

With $a = 0$, $b = 30$, and $\mu = 40$, Equation (15) becomes

$$P(0 \leq X \leq 30) = e^{-0/40} - e^{-30/40}$$
$$= e^0 - e^{-0.75}$$
$$\approx 1 - 0.4724$$
$$= .5276$$

Therefore,

$$P(X > 30) = 1 - P(0 \leq X \leq 30)$$
$$= 1 - .5276$$
$$= .4724$$

We are interested in simulating outcomes from an exponential distribution with mean μ. To do this, we must generate random numbers from such a population. A popular procedure for this purpose is based on calculus: First, use a random number generator to produce numbers r that are uniformly distributed between 0 and 1. Then use the following formula to transform these numbers into numbers s that are exponentially distributed with mean μ:

$$s = -\mu \cdot \ln(r) \qquad (16)$$

As before, ln refers to the natural logarithm. Most calculators have an $\ln(x)$ key, and the natural logarithm is also available as a library function in almost all computer software packages designed for numerical computations.

EXAMPLE 5

Use a random number generator to produce a set of ten interarrival times from a sample space that is exponentially distributed and has a mean of 40 seconds.

Solution

Here $\mu = 40$. Accessing a random number generator, we obtain $r = 0.650582$ as a random number from a population that is uniformly distributed on $[0, 1)$. Using Equation (*16*), we convert this number to

$$s = -\mu \cdot \ln(r)$$
$$= -40\ln(0.650582)$$
$$\approx -40(-0.429888)$$
$$\approx 17$$

rounded to the nearest second. On many calculators the keystrokes needed for this computation are:

This value of s becomes the first interarrival time in Table 8.37.

TABLE 8.37   Simulating Exponentially Distributed Times with Mean 40

TRIAL	r	$\ln(r)$	$s = -40 \cdot \ln(r)$
1	0.650582	−0.429888	17
2	0.213081	−1.546083	62
3	0.069642	−2.664387	107
4	0.679353	−0.386614	15
5	0.534584	−0.626266	25
6	0.185438	−1.685035	67
7	0.962169	−0.038565	2
8	0.744978	−0.294401	12
9	0.842627	−0.171231	7
10	0.793525	−0.231270	9

To create a second entry, we use a random number generator to obtain $r = 0.213081$ as another random number from a population that is uniformly distributed on $[0, 1)$. Using Equation (*16*), we convert this number to

$$s = -\mu \cdot \ln(r)$$
$$= -40\ln(0.213081)$$
$$\approx -40(-1.546083)$$
$$\approx 62$$

rounded to the nearest second. Continuing in this manner, we complete Table 8.37.

EXAMPLE 6

Use a random number generator to produce a set of six interarrival times from a sample space that is exponentially distributed with a mean of 5 minutes.

Solution

Here $\mu = 5$. Accessing a random number generator, we obtain $r = 0.293296$ as a random number from a population that is uniformly distributed on [0, 1). Using Equation (16), we convert this number to

$$
\begin{aligned}
s &= -\mu \cdot \ln(r) \\
&= -5\ln(0.293296) \\
&\approx -5(-1.226573) \\
&\approx 6.13
\end{aligned}
$$

in minutes, rounded to the second decimal place. Continuing in this manner, we complete Table 8.38.

TABLE 8.38   Simulating Exponentially Distributed Times with Mean 5

TRIAL	r	$\ln(r)$	$s = -5 \cdot \ln(r)$
1	0.293296	−1.226573	6.13
2	0.966885	−0.033676	0.17
3	0.459493	−0.777632	3.89
4	0.776727	−0.252666	1.26
5	0.646329	−0.436447	2.18
6	0.080450	−2.520119	12.60

EXAMPLE 7

The returns desk at a department store is staffed by a single employee, who takes customers on a first-come, first-served basis. If the employee is busy with a customer when a new customer arrives, the arriving customer waits in line. The time between customer arrivals at the returns desk is exponentially distributed with a mean of 5 minutes. The time required for the employee to completely service a customer is normally distributed with a mean of 4 minutes and a standard deviation of 30 seconds. Simulate the activity at the returns desk for the first five customers on a particular day.

Solution

Activity at the returns desk is governed by two events, the arrival of customers and the time required to service each customer. We are interested in simulating the activity for five customers, so we require five interarrival times and five service times. The interarrival times must be drawn from a sample space that is exponentially distributed with a mean of 5 minutes. We simulated interarrival times from this sample space in Example 6, where the first five times were

Interarrival times: 6.13,   0.17,   3.89,   1.26,   2.18

all expressed in units of minutes. The service times must be drawn from a sample space that is normally distributed with a mean of 4 minutes and a standard deviation of 30 seconds or 0.5 minutes. Following the procedures described in Section 8.4, we obtain

Service times: 2.57, 4.37, 4.23, 2.84, 2.26

all in units of minutes.

We begin the simulation at time 0.00 on a simulated clock. To facilitate our analysis, we will keep track of the next scheduled arrival time, the status of the employee as busy or idle, the time when a busy employee will complete service for the current customer, and the status of the line. Customers will be labeled with consecutive integers in the order they arrive.

At time 0:00 there are no customers at the return desk and the employee is idle. The first interarrival time is 6.13, so customer 1 does not arrive until $0:00 + 6.13 = 6.13$ minutes on our simulated clock. Initially, the status of the system is:

Status Update 1

CURRENT TIME	NEXT ARRIVAL	SERVER STATUS	NEXT SERVICE COMPLETION	CUSTOMERS IN LINE
0.00	6.13	idle	—	none

At time 6.13, customer 1 arrives and is immediately served. This customer will require 2.57 minutes of service, which will be completed at time $6.13 + 2.57 = 8.70$. The second interarrival time is 0.17 minutes after the first arrival, at $6.13 + 0.17 = 6.30$ on our simulated clock. Thus, we update the status of this system to:

Status Update 2

CURRENT TIME	NEXT ARRIVAL	SERVER STATUS	NEXT SERVICE COMPLETION	CUSTOMERS IN LINE
6.13	6.30	busy	8.70	none

Customer 2 is scheduled to arrive at 6.30, and customer 1 is scheduled to complete business and depart the returns desk at 8.70. Since 6.30 occurs before 8.70, we process the next customer arrival as the next transaction. At time 6.30, customer 2 arrives, finds the employee busy with customer 1, and waits in line for service. The third interarrival time is 3.89 minutes after the second arrival, at $6.30 + 3.89 = 10.19$ on our simulated clock. The status of this system becomes:

Status Update 3

CURRENT TIME	NEXT ARRIVAL	SERVER STATUS	NEXT SERVICE COMPLETION	CUSTOMERS IN LINE
6.30	10.19	busy	8.70	#2

The next customer is scheduled to arrive at 10.19 and the customer being served is scheduled to complete business and depart the returns desk at 8.70. Since 8.70 occurs before 10.19, we process the completion of service for the first customer as the next transaction. At time 8.70, customer 1 is finished at the returns desk and departs. Customer

2 is in line waiting for service and is immediately helped by the employee. The second service time is 4.37 minutes, so customer 2 will complete service at $8.70 + 4.37 = 13.07$ on our simulated clock. The status of this system is now:

Status Update 4

CURRENT TIME	NEXT ARRIVAL	SERVER STATUS	NEXT SERVICE COMPLETION	CUSTOMERS IN LINE
8.70	10.19	busy	13.07	none

The next arrival is scheduled at 10.19, and the second customer is scheduled to complete business and depart at 13.07. Since 10.19 occurs before 13.07, we process the next customer arrival as the next transaction. At time 10.19, customer 3 arrives, finds the server busy with customer 2, and waits in line for service. The fourth interarrival time is 1.26 minutes after the third arrival, at $10.19 + 1.26 = 11.45$ on our simulated clock. The status of the system becomes:

Status Update 5

CURRENT TIME	NEXT ARRIVAL	SERVER STATUS	NEXT SERVICE COMPLETION	CUSTOMERS IN LINE
10.19	11.45	busy	13.07	#3

The next arrival is scheduled at 11.45, and the customer currently being served is scheduled to complete business and depart at 13.07, so we process the next customer arrival as the next transaction. At time 11.45, customer 4 arrives, finds the server busy with customer 2, and waits in line for service behind cusomter 3. The fifth interarrival time is 2.18 minutes after the fourth arrival, at $11.45 + 2.18 = 13.63$ on our simulated clock. The status of the system is now:

Status Update 6

CURRENT TIME	NEXT ARRIVAL	SERVER STATUS	NEXT SERVICE COMPLETION	CUSTOMERS IN LINE
11.45	13.63	busy	13.07	#3, #4

The next customer is scheduled to arrive at 13.63, and the customer currently being served (customer 2) is scheduled to complete business and depart the return desk at 13.07, so we process the completion of service for customer 2 as the next transaction. At time 13.07, customer 2 is finished at the returns desk and departs. Customer 3 is in line waiting for service and is immediately helped by the employee. The third service time is 4.23 minutes, so customer 3 will complete service at $13.07 + 4.23 = 17.30$ on our simulated clock. Now the status of this system is:

Status Update 7

CURRENT TIME	NEXT ARRIVAL	SERVER STATUS	NEXT SERVICE COMPLETION	CUSTOMERS IN LINE
13.07	13.63	busy	17.30	#4

At time 13.63, customer 5 arrives and joins the line behind customer 4. To schedule the next customer arrival, we require another random number from a sample space that is exponentially distributed with a mean of 5 minutes. Using the last entry in Table 8.38, we take 12.60 minutes as the next interarrival time. The actual arrival time will occur at 13.63 + 12.60 = 26.23, and the status of the system becomes:

Status Update 8

CURRENT TIME	NEXT ARRIVAL	SERVER STATUS	NEXT SERVICE COMPLETION	CUSTOMERS IN LINE
13.63	26.23	busy	17.30	#4, #5

Customers 4 and 5 will require 2.84 and 2.26 minutes of service, respectively. Consequently, the next three status updates are

Status Update 9

CURRENT TIME	NEXT ARRIVAL	SERVER STATUS	NEXT SERVICE COMPLETION	CUSTOMERS IN LINE
17.30	26.23	busy	20.14	#5

Status Update 10

CURRENT TIME	NEXT ARRIVAL	SERVER STATUS	NEXT SERVICE COMPLETION	CUSTOMERS IN LINES
20.14	26.23	busy	22.40	none

Status Update 11

CURRENT TIME	NEXT ARRIVAL	SERVER STATUS	NEXT SERVICE COMPLETION	CUSTOMERS IN LINE
22.40	26.23	idle	—	none

The first five customers are now completely processed and the simulation is done. Status updates 1 through 11 provide a concise record of the events in the simulation.

IMPROVING SKILLS

In Problems 1 through 6, determine the probability that a randomly selected outcome will fall in the specified interval if the sample space is exponentially distributed with a mean of 22.

1. [22, 24]   **2.** [20, 22]   **3.** [19, 22]
4. [16, 22]   **5.** [10, 18]   **6.** [0, 11]

In Problems 7 through 12, determine the probability that a randomly selected outcome will fall in the specified interval if the sample space is exponentially distributed with a mean of 200.

7. [200, 225]   **8.** [175, 200]   **9.** [100, 200]
10. [100, 300]   **11.** [200, 250]   **12.** [150,270]

In Problems 13 through 18, determine the probability that a randomly selected outcome will fall in the specified interval if the sample space is exponentially distributed with a mean of 50.

13. [60, 70]   **14.** [0, 70]   **15.** [54, 68]
16. [48, 55]   **17.** [0, 20]   **18.** [10, 45]

19. A sample space is exponentially distributed with a mean of 5. Find the probability that a randomly selected outcome will be (*a*) greater than 8, (*b*) greater than 4, (*c*) less than or equal to 7.

20. A sample space is exponentially distributed with a mean of 30. Find the probability that a randomly selected outcome will be (*a*) greater than 31, (*b*) greater than 20, (*c*) greater than 40.

21. A sample space is exponentially distributed with a mean of 10. Find the probability that a randomly selected outcome will be (*a*) greater than 3, (*b*) greater than 10, (*c*) less than −1.

22. Use the random numbers *r* in Table 8.37 to generate a sample from an exponentially distributed sample space with mean 100.

23. Use the random numbers in Table 8.38 to generate a sample from an exponentially distributed sample space with mean 8.

24. Use the random numbers in Table 8.38 to generate a sample from an exponentially distributed sample space with mean 0.35.

CREATING MODELS

25. Salaries for employees at a company are exponentially distributed with a mean of $45,000. (*a*) What percentage of all employees make between $25,000 and $35,000? (*b*) What percentage make more than $60,000?

26. The sale of coffee beans at a coffee house is exponentially distributed with a mean of 2 lb. What is the probability that the next sale (*a*) will be for a pound or less? (*b*) will be between 1 and 3 lb?

27. Rainfall in a region is exponentially distributed with an annual mean of 250 inches. What is the probability that the rainfall next year (*a*) will exceed 300 inches? (*b*) will be 100 inches or less?

28. Using the random numbers *r* in Table 8.37, simulate the annual rainfall for the next ten years in the region described in Problem 27.

29. Using the random numbers *r* in Table 8.37, simulate the next ten coffee sales at the store described in Problem 26.

30. Create another run of the simulation in Example 7 (activity at the returns desk at a department store), using the same interarrival times but the following service times: 3.32, 3.48, 3.36, 2.70, 2.78.

31. Create another run of the simulation in Example 7, using the same service times but the following interarrival times: 7.58, 0.75, 1.49, 0.15, 5.28, 18.11.

EXPLORING IN TEAMS

32. The area between an exponential distribution curve and the horizontal axis in the first quadrant is always 1, but the area under most frequency polygons is generally much greater than 1. However, the constructions of histograms and frequency polygons can be modified as we saw in Problem 49 of Section 8.4. Use this procedure to construct a modified histogram and frequency polygon for the data in Table 8.35.

33. The equation that models an exponential distribution is

$$y = \frac{1}{\mu}\, e^{-x/\mu}, \text{ for } x \geq 0$$

Graph this equation when $\mu = 2$.

34. Graph the equation in the previous problem when $\mu = 0.5$.

35. The mean of the data in Table 8.35, rounded to one decimal place, is $\bar{x} = 2.1$ minutes. Using this value as an estimate for μ, graph the equation in Problem 33 and then superimpose the graph on the modified frequency polygon constructed in Problem 32. How good is the fit?

EXPLORING WITH TECHNOLOGY

Use a spreadsheet to simulate each of the processes described in Problems 36 through 40.

36. A barber shop is staffed by a single barber who takes customers on a first-come, first-served basis. If the barber is busy with a customer when a new customer arrives, the arriving customer waits in line. The time between customer arrivals is exponentially distributed with a mean of 22 minutes. The time required for the barber to completely serve a customer is uniformly distributed between 15 and 20 minutes. Simulate the activity at the barber shop through the first two hours of operation.

37. The interarrival times of television sets at a quality-control station are exponentially distributed with a mean of 10 minutes. The station is staffed by a single engineer, who inspects each set and sends it on to the next station on the assembly line. Inspection time is uniformly distributed between 8 and 12 minutes. Sets are inspected in the order they arrive, and if the inspector is busy when a set arrives, then the arriving set sits at the quality-control station until it can be inspected. Simulate the activity at this inspection station through the first five sets.

38. Expand the simulation in the previous problem to include the routing of sets from the inspector, either to shipping or to an adjustment center, if 80% of all sets pass inspection and are cleared for shipping and 20% fail inspection and require further adjustments.

39. The times between the arrivals of shoppers at the custom meat counter at a supermarket are exponentially distributed with a mean of 8 minutes. The meat counter is staffed by a single butcher who takes customers on a first-come, first-served basis. The time required for the butcher to completely serve a customer is uniformly distributed between 5 and 7 minutes. Simulate the activity at the meat counter through the first hour of operation if one customer is waiting for service when the butcher opens.

40. Cars arrive at a car wash on average every 3 minutes, although the actual distribution of times is exponential. Cars are washed automatically in a facility that holds one car at a time. A complete car wash takes exactly 90 seconds, and cars are washed on a first-come, first-served basis. Simulate the activity at the car wash through the first fifteen minutes of operation if two cars are waiting for service when the operation opens.

REVIEWING MATERIAL

41. (Section 2.5) Use matrix inversion to solve the following system of equations:

$$2x - 3y + 4z = 40$$
$$3x + 2y - 2z = 24$$
$$y + 5z = 63$$

42. (Section 4.2) A person will receive $350,000 from a trust fund in four years but wants to borrow against the fund now. How much should she be able to borrow if current interest rates are 12% a year, compounded quarterly?

43. (Section 5.4) A traveler packs three pairs of slacks, one gray, one tan, and one navy blue, two blazers, one gray and one navy blue, and three shirts, each of a different color. Construct a tree diagram listing all possible ensembles consisting of one pair of slacks, one blazer, and one shirt if the color of the blazer must always be different from the color of the slacks.

44. (Section 6.6) A card is picked from a standard deck and then replaced. The deck is shuffled and the process is repeated four more times. What is the probability that at least two of the five cards picked from the deck are hearts?

RECOMMENDING ACTION

45. Respond by memo to the following request:

MEMORANDUM

To: J. Doe Reader

From: Operations Research Department

Date: Today

Subject: **Distributions**

We are having some heated discussions in-house on when to use the different probability distributions. I think that some of our problems stem from a basic lack of understanding of the relevance of each distribution. Can you give us examples of some real-life situations in which we would expect a normal distribution, an exponential distribution, and a uniform distribution, and explain why?

CHAPTER 8 KEYS

KEY WORDS

continuous process (p. 548)
discrete process (p. 548)
exponential distribution (p. 592)
frequency (p. 551)
frequency polygon (p. 552)

frequency table (p. 551)
histogram (p. 549)
normal distribution (p. 568)
random number generator (p. 521)
uniform distribution (p. 533)

KEY CONCEPTS

8.1 Random Number Generators

■ A random number generator is a mechanism for creating real numbers that are equally likely to be any value between 0 and 1, including 0 but excluding 1.

8.2 Uniform Distribution

■ A random number generator that produces numbers uniformly distributed on [0, 1) can be used to create numbers that are uniformly distributed on any interval.

■ The median and the mean of a sample space uniformly distributed on [a, b) occur at the midpoint $\dfrac{(a + b)}{2}$.

■ Simulation is useful for estimating areas of bounded regions.

8.3 Sampling

■ A process is continuous if outcomes can assume any value in some interval on the real number line.

■ Taking data on a continuous process involves measuring outcomes from a sample (and rounding to some acceptable tolerance); analyzing data often begins by collecting those measurements into a small number of subsets. Both tactics have the effect of modeling a continuous process as a discrete process.

■ Frequency polygons approximate horizontal straight lines when the samples are from sample spaces that are uniformly distributed.

■ The mean μ and standard deviation of σ of a sample space from a continuous process are estimated from the mean $\bar{x}$ and the standard deviation s of the sample.

8.4 Normal Distributions

- A normal curve is completely determined by two values, a mean μ and a standard deviation σ.

- A normal curve is symmetric around a vertical line that intersects the horizontal axis at the mean μ. The larger the standard deviation σ, the shorter and fatter the normal curve.

- Every normal curve shares the property that 68.27% of the area between the curve and the horizontal axis falls within one standard deviation on either side of the mean.

- The area between a normal curve and the horizontal axis is 1.

- In a normally distributed sample space with mean μ and standard deviation σ, the probability that an outcome falls in the interval $[a, b]$ is the same as the area under the normal curve with mean μ and standard deviation σ, above the horizontal axis, and between a and b.

8.5 Exponential Distributions

- Exponential distribution curves model processes in which outcomes are always non-negative, cluster close to zero, and become sparser in intervals further from the origin.

- Exponential distribution curves are often good models for the distribution of times between customer arrivals at a facility.

KEY FORMULAS

- When a random number generator is used, the probability of getting a number r that falls in the interval $[c, d)$ when $0 \le c < d \le 1$ is
$$P(c \le r < d) = d - c$$

- To generate numbers R that are uniformly distributed on $[a, b)$, use a random number generator to produce numbers r that are uniformly distributed on $[0, 1)$ and then substitute r into the equation
$$R = (b - a)r + a$$

- If $\bar{x}$ is the mean of numerical data $x_1, x_2, x_3, \ldots, x_n$ from a sample, then the variance of the population as estimated by the sample is
$$\text{variance} = \frac{(x_1 - \bar{x})^2 + (x_2 - \bar{x})^2 + (x_3 - \bar{x})^2 + \ldots + (x_n - \bar{x})^2}{n - 1}$$

- If a population is normally distributed with mean μ and standard deviation σ, then the number of standard deviations that an outcome $x > \mu$ is from the mean is
$$z = \frac{x - \mu}{\sigma}$$

- To generate numbers n from a population that is normally distributed with mean μ and standard deviation σ, use a random number generator to produce two numbers r_1 and r_2 that are uniformly distributed on $[0, 1)$ and then evaluate the equation
$$n = \mu + \sigma \cdot \cos(2\pi r_1)\sqrt{-2 \cdot \ln(r_2)}$$

■ The probability that a numerical outcome X falls in the interval $[a, b]$ of a sample space that is exponentially distributed with mean μ is

$$P(a \le X \le b) = e^{-a/\mu} - e^{-b/\mu}$$

■ To generate numbers s from a population that is exponentially distributed with mean μ, use a random number generator to produce numbers r that are uniformly distributed on $[0, 1)$ and then evaluate the equation

$$s = -\mu \cdot \ln(r)$$

KEY PROCEDURES

■ To estimate performance measures using Monte Carlo simulation:

Step 1 Create a model of the game using a random number generator.

Step 2 Make a number of different runs of the model, thereby playing the game many times. After each run, record the value of the performance measure.

Step 3 Calculate the mean of all recorded performance measures and use the mean as an estimate of that performance measure.

■ To estimate the area of a bounded region:

Step 1 Enclose the region within a rectangle.

Step 2 Use a random number generator to produce points inside this rectangle.

Step 3 Determine what fraction of the points generated in Step 2 fall within the region of interest.

Step 4 Multiply the fraction obtained in Step 3 by the area of the rectangle used in Step 1.

■ Histograms are constructed according to the following four criteria:

Criterion 1 Use between 5 and 20 categories.

Criterion 2 Choose endpoints so that no data point falls into two categories.

Criterion 3 Choose endpoints so that each data point is in some category.

Criterion 4 Use endpoints that are sensible.

When possible, the following two criteria are also observed:

Criterion 5 Try to choose endpoints so that the widths of the bars are the same.

Criterion 6 Try to have at least five data points in each category.

TESTING YOURSELF

Use the following problems to test yourself on the material in Chapter 8. The odd problems comprise one test, the even problems a second test. In each problem that requires random numbers, use the following list of random numbers, indicating clearly how each random number was used.

.771	.035	.189	.992	.501
.530	.627	.100	.726	.334

1. A fair coin is flipped and the face value, either heads or tails, announced. Simulate ten consecutive flips of this coin.

2. Use random numbers to create a list of five numbers from an exponential distribution with a mean of 12.5.

3. Weekly washer and dryer sales at an appliance store appear to follow the following probability distributions:

SALES (NUMBER OF UNITS)	0	1	2	3
PROBABILITIES OF WASHER SALES	.20	.35	.30	.15
PROBABILITIES OF DRYER SALES	.55	.20	.15	.10

Describe how to use a random number generator to simulate weekly sales of both types of appliances, using these probability distributions. Simulate the sale of washers and dryers at the store for a five-week period.

4. Estimate the mean and standard deviation of a population from the following sample:

$$0, \ 0, \ 0, \ 1, \ 1, \ 2, \ 2, \ 2, \ 4, \ 8$$

5. A store has room to stock at most four computers. Each week, the store sells between one and four computers, with each outcome equally likely. If the store has fewer than two computers in stock at the end of the week, it orders enough new computers from its supplier to bring the stock to four by the beginning of the next week. If the store ends a week with two or more computers in stock, it does not order any new ones for the following week. If a customer wants a computer and the store has none in stock, then that sale is lost. Simulate computer sales at the store for ten weeks and estimate the average number of sales lost each week. Assume that the store begins the 10-week period with three computers in stock.

6. Create a histogram and a frequency polygon from the following data, and then use the results to identify a reasonably likely distribution of the underlying sample space.

MIDPOINT	10	30	50	70	90
FREQUENCY	8	32	114	41	5

7. Use random numbers to create a list of five numbers that are uniformly distributed between -3.5 and 4.5.

8. A barrel contains 3 red balls, 7 green balls, and 15 white balls. One ball is selected from the barrel at random and then returned to the barrel before the next ball is chosen. Simulate the results of this game for ten consecutive plays, each time identifying the color of the selected ball.

9. Create the first five lines of an output table for a Monte Carlo simulation for estimating the area under the curve $y = 5x + 2$, above the x-axis, and between $x = 0$ and $x = 2$.

10. Test scores are scaled so that they are normally distributed with a mean of 60 and a standard deviation of 10. What is the probability that the next test graded will have a score greater than 70?

11. Estimate the mean and standard deviation of a population from the following sample:

$$8, \quad 9, \quad 11, \quad 12, \quad 12, \quad 14, \quad 17, \quad 17$$

12. Use random numbers to create a list of ten numbers that are uniformly distributed between 200 and 500.

13. Create a histogram and a frequency polygon from the following data, and then use the results to identify a reasonably likely distribution of the underlying sample space.

CATEGORY	[0, 5)	[5, 10)	[10, 15)	[15, 20)	[20, 25)	[25, 30)	[30, 35)
FREQUENCY	38	23	16	10	7	5	1

14. A player places a bet and then chooses a card at random from a standard deck of 52 cards. If the card is an ace, the player retrieves the wager and also receives 10 times the wager for winning; if the card is not an ace, the player loses the wager. Robin bets \$1 whenever her losses are more than or equal to her winnings and \$2 when her winnings are more than her losses. Simulate Robin's play for the first five games.

15. A population is known to be normally distributed with a mean of 32 and a standard deviation of 2. What is the probability a number drawn randomly from this population will be between 29 and 36?

16. Create the first five lines of an output table for a Monte Carlo simulation for estimating the area above the curve $y = x^2 - 1$ and under the x-axis.

17. Use random numbers to create a list of ten numbers that are exponentially distributed with a mean of 5.

18. Each evening, the staff of a local coffee shop prepares a special house blend so that the store can start each morning with 100 pounds of coffee ready for sale. The daily demand for this blend is uniformly distributed between 50 and 120 pounds. The staff is too busy during the day to prepare new quantities of its house blend, so sales are lost whenever demand exceeds supply. Simulate coffee sales of the house blend for ten days, and estimate the average amount of lost sales (in pounds) each day.

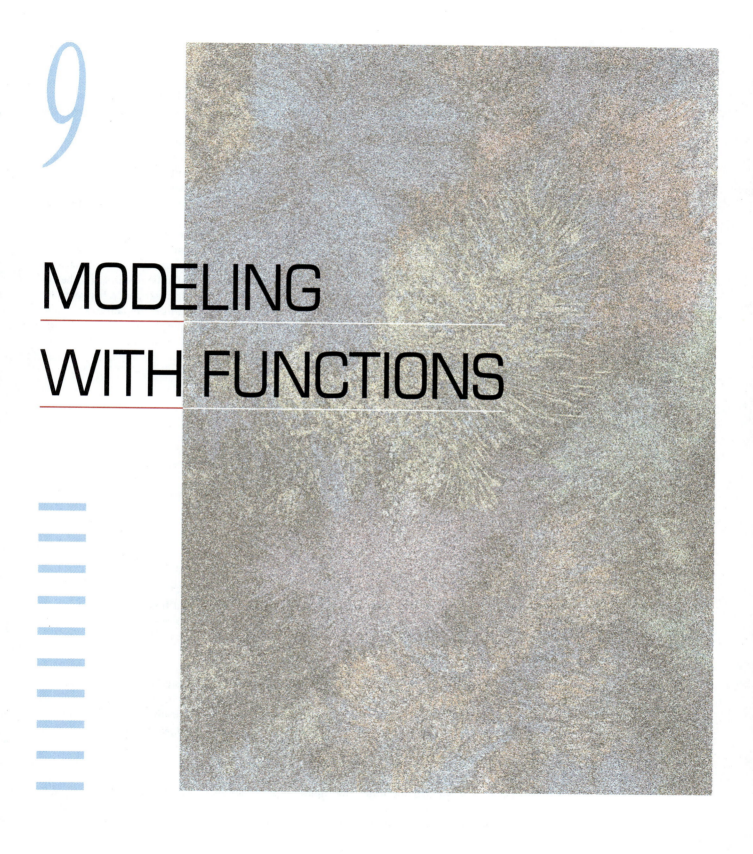

9

MODELING
WITH FUNCTIONS

9.1 FUNCTIONS

Many processes can be modeled as input–output relationships. In the production of automobiles, we mix the inputs of paint, steel, plastic, fabric, rubber, capital, and labor in an assembly process to create the output of a car. To produce a house as an output, contractors combine concrete, wood, glass, metal, capital, and labor in a mix called the construction process. Each of these input–output relationships can be represented schematically by Figure 9.1. The black box in Figure 9.1 symbolizes a mechanism or a production process for converting three inputs (for example, wood, paint, and labor) into two outputs (say, chairs and tables). In the simplest case, a process creates a single output from a single input.

A function is a rule of correspondence that assigns to each element in a domain exactly one element in the codomain.

Input–output relationships lie at the core of many everyday processes, and if mathematics is to model such processes successfully, it must account for those relationships. Ideally there is a rule of correspondence to describe how a particular process changes an input to an output. In general, we call the set of allowable inputs the *domain* and the set of allowable outputs the *codomain*. A *function* is a rule of correspondence between the domain and the codomain, a rule that assigns to each element in the domain exactly one element in the codomain. *Note*: Each input results in only one output, but different inputs may lead to different outputs or to the same output. A function, therefore, has three components: a domain, a codomain, and a rule of correspondence.

FIGURE 9.1
A Production Input–Output Process

EXAMPLE 1

Determine whether the following input–output relationship is a function: The domain is the set of all people in the world, the codomain is the set of all real numbers (in units of pounds), and the rule of correspondence is "Assign to each person his or her weight."

Solution

This is a function. The mechanism for converting inputs (in this case people) into outputs (in this case weights) is a scale, depicted by the black-box in Figure 9.1. The scale can weigh any person in the domain, and each input (person) results in a unique output (weight). Different people may have the same weight or different people may have different weights, but each person has only one weight.

EXAMPLE 2

Determine whether the following input–output relationship is a function: The domain is the set of all cars in the world, the codomain is the set of all colors, and the rule of correspondence is "Assign to each car its colors."

Solution

This is not a function. Although we have a well-defined domain (all cars) and a well-defined codomain (all colors), the rule of correspondence would assign two colors to a car with a red body and white roof, which violates the requirement that each input be matched with exactly one output.

The range of a function is a subset of the codomain; it contains only those outputs actually used.

It is useful to visualize the rule of correspondence for a function as a robot or a machine, as in Figure 9.2. The robot is programmed to accept inputs from the domain, transform each input according to a specific rule of correspondence, and emit an output that belongs to the codomain.

In Example 1, we defined the codomain to be all real numbers (in pounds). Thus -20 pounds is in the codomain, even though no person has such a weight. A function need not assign every element in the codomain as an output. The only requirement is that each output be in the codomain. The set of values in the codomain that are actually assigned to at least one input in the domain is called the *range* of the function.

FIGURE 9.2

A Rule of Correspondence Associated with a Function as Represented by a Robot

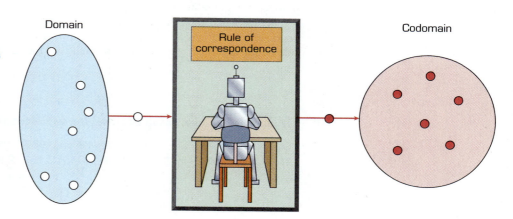

EXAMPLE 3

Determine whether either of Figures 9.3(a) or (b) depicts a function. In both cases, the domain is the set $\{1, 2, 3\}$ and the codomain is the set $\{A, B, C, D\}$. The rule of correspondence in each is defined by the arrows.

FIGURE 9.3

Examples of Rules of Correspondence

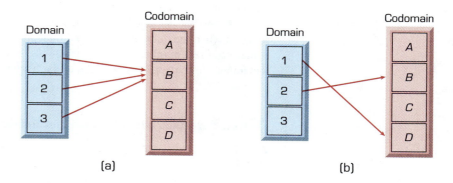

(a)   (b)

Solution

Figure 9.3(a) depicts a function. Here all elements in the domain are assigned to the same element in the codomain, namely, B. A function must pair every element in the domain with exactly one element in the codomain, but not necessarily a different element. The range of this function is $\{B\}$.

Figure 9.3(b) does *not* depict a function because the element 3 in the domain is not assigned to any element in the codomain. A function must match each and every element in the domain with some element in the codomain.

The functions of greatest interest are those where both the domain and the codomain are subsets of the real numbers. These numbers can have any units, for example, prices in dollars, time in hours, sales in cartons, or weights in pounds. The rule of correspondence can be depicted by arrows (as in Example 3), by a table, a graph, an equation, or a verbal description.

EXAMPLE 4

Table 9.1 reproduces data from Example 6 in Section 2.3. It deals with the estimated number of maltreated children reported in the U.S. Determine whether the table represents a function when the domain is the set of years between 1985 and 1992, inclusive, the codomain is the set of real numbers (in units of millions of maltreated children), and the rule of correspondence is to assign to each year the number of maltreated children for that year.

TABLE 9.1   Maltreated Children in the United States

YEAR	1985	1986	1987	1988	1989	1990	1991	1992
ESTIMATED NUMBER (millions)	1.919	2.086	2.157	2.265	2.435	2.557	2.723	2.936

Solution

This relationship is a function because each number in the domain has assigned to it a unique number in the codomain. The range of this function is the set of eight numbers

$$\{1.919, 2.086, 2.157, 2.265, 2.435, 2.557, 2.723, 2.936\}.$$

EXAMPLE 5

Table 9.2 reproduces the data from Problem 14 in Section 2.3. The table lists both the number of weeks eight clients of a well-known weight-loss center have been on its diet program and the weight lost by each client. Determine whether the table represents a function when the domain is the set of integers $\{2, 3, 4\}$ (in units of weeks), the codomain is the set of real numbers (in units of pounds), and the rule of correspondence is to assign the amount of a client's weight loss to the length of time on the program.

TABLE 9.2   Diet Results from Eight Individuals

WEEKS ON DIET	2	2	2	3	3	4	4	4
POUNDS LOST	7	7	8	10	15	15	17	18

Solution
This relationship is not a function because each number in the domain has more than one number assigned to it in the codomain. The input of 2 weeks is assigned both 7 pounds and 8 pounds as outputs; the input of 3 weeks is assigned both 10 pounds and 15 pounds; and the input of 4 weeks is assigned three different outputs. A function has the property that each input is assigned only one output in the codomain.

Curves in the plane can also model functions. In this case, the domain is always placed on the horizontal axis and the codomain on the vertical axis. If the rule of correspondence is a function, then the rule assigns to each input x a single number y having the property that the ordered pair (x, y) is a point on the curve.

EXAMPLE 6

Figure 9.4 is a graph of the population growth in the U.S. from 1790 through 1990. Does this curve represent a function?

FIGURE 9.4
U.S. Population Figures

Solution
We take the domain to be the set of all real numbers in the closed interval [1790, 1990] and the codomain to be the set of all nonnegative real numbers. The rule of correspondence is indicated by the arrows. For each point x in the domain, we go up a vertical line to the curve and then across a horizontal line to a point y on the vertical axis. In particular, when $x = 1870$ (an input), the corresponding y (the output) is approximately 40 (million); when $x = 1950$, the corresponding y is approximately 152 (million). Each point in the domain is matched with one and only one point in the codomain, so the graph does model a function.

EXAMPLE 7

Figure 9.5 is a graph of the equation $x^2 + y^2 = 25$, first considered in Example 3 of Section 1.4. Determine whether this curve represents a function.

FIGURE 9.5

The Graph of the Equation
$x^2 + y^2 = 25$

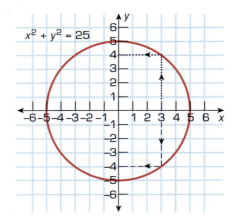

Solution

This curve does *not* model a function because the circle has two *y*-values (outputs) for some of the *x*-values (inputs), such as $x = 3$. If we follow the path of dotted lines upward in Figure 9.5, we find that $y = 4$ is a corresponding output; if we follow the path of dashed lines downward, we find that $y = -4$ is also a corresponding output for $x = 3$. Both points $(3, 4)$ and $(3, -4)$ lie on the circle.

> *The natural domain of a function is the largest set of real numbers having corresponding output values under the rule of correspondence.*

If the domain in Example 7 includes real numbers greater than 5 or less than -5, then *not every* input value is matched with an output value and the graph fails as a function for still another reason. For example, there is no *y* assigned to $x = 6$ because there is no value of *y* for which the point $(6, y)$ lies on the circle. If Figure 9.5 is to have any chance of being a function, we must restrict the domain to the interval $[-5, 5]$. Consequently, we define the *natural domain* of a function to be the largest set of real numbers such that each input value has a corresponding output value under the rule of correspondence. Unless stipulations are made to the contrary, we assume that the domain of a function is its natural domain.

For Figure 9.5, the domain must be a subset of $[-5, 5]$. Even then, we saw in Example 7 that the curve does not represent a function. Therefore not all graphs represent functions. To do so, a graph must pass the *vertical line test*.

VERTICAL LINE TEST

Any vertical line that crosses the horizontal axis at a value in the domain of a function must intersect the graph of that function at one and only one point.

The graph in Figure 9.4 passes this test because each vertical line through the horizontal axis in the interval $[1790, 1990]$ intersects the graph at one and only one point. The graph in Figure 9.5, however, does *not* pass the vertical line test because the vertical line through $x = 3$ intersects the circle at two points.

When we use a function denoted by the letter *f* to model an input–output relationship, we often write

$$\text{output} = f(\text{input}) \tag{1}$$

to capture that relationship. The rule f operates on an input to produce an output. Other letters, such as g and h, are often used instead of f.

It is common to use the letters x and y as generic labels for the input and output, respectively, although other labels are used when they are more appropriate to a particular model. Equation (1) then takes the form

$$y = f(x) \qquad (2)$$

This equation states that an input x is transformed into an output y according to the rule of correspondence represented by f. The notation $f(x)$, read "f of x," signifies that a rule f operates on an input x.

We use the notation $y = f(x)$ most often when the rule of correspondence is described by an algebraic formula. For example, we use the formula $f(x) = x^2$ as shorthand for the rule "Assign to each number its square." We use the formula $g(x) = 2x - 5$ as shorthand for the rule "Multiply each input by 2 and then subtract 5 from the result."

If we have a rule of correspondence defined by a formula $f(x)$, then we find the element in the codomain associated with a particular value of x by substituting that value for x into the formula. Thus $f(3)$ is the effect of applying the rule of correspondence to the input 3, while $f(5)$ is the effect of applying the rule of correspondence to the input 5.

EXAMPLE 8

Determine the natural domain for the function $f(x) = 2/(x^2 - 1)$ and then find $f(3)$, $f(5)$, and $f(-2)$.

Solution

The domain is not specified, so we presume it is the largest set of real numbers with corresponding outputs under the given formula. We cannot evaluate the formula for $f(x)$ at either $x = 1$ or $x = -1$ because that would require division by zero, which is undefined. Thus, neither $x = 1$ nor $x = -1$ has an output associated with it under this relationship. Consequently, the natural domain is the set of all real numbers excluding ± 1. For the inputs $x = 3$, $x = 5$, and $x = -2$, we have the following outputs:

$$f(3) = \frac{2}{(3)^2 - 1} = \frac{1}{4}$$

$$f(5) = \frac{2}{(5)^2 - 1} = \frac{1}{12}$$

$$f(-2) = \frac{2}{(-2)^2 - 1} = \frac{2}{3}$$

The labels chosen to represent the inputs of a function and those selected for the outputs can be any letter or acronym that makes sense to the modeler. Such labels are called *variables*. In Chapter 3 we called the input variables *decision variables,* because they are under the direct control of the decision-maker. Another common name is *independent variables,* which emphasizes the fact that the values of these variables can be chosen independently of one another by the decision-maker. The output variable for the function is then called the *dependent variable,* because its value depends completely on the choices made for the inputs and on the rule of correspondence.

The function $f(x) = 2/(x^2 - 1)$ considered in Example 8 is a function of the single variable x. The value of the input x can be chosen by the decision-maker, so x is a decision or independent variable. We can choose x to be 3, or 5, or -2, or any real number other than ± 1. Once chosen, however, the value of the output y is completely determined by the rule of correspondence, so it is a dependent variable.

We will use letters other than x for the independent variable and other than y for the dependent variable whenever it is appropriate to do so. For example, the equation

$$D = f(p)$$

signifies a function in which the output of D (demand) is matched with input p (price) according to the rule of correspondence f.

We use functions routinely every day. We understand that the value of a car depends on its age (the older the car the less its value). If we denote the car's age by a and its value by V, then the relationship between age and value can be represented by a function $V = g(a)$, with a as the independent variable and V as the dependent variable. If instead we prefer to use y to denote the age of the car and D for its dollar value, then the relationship takes the form $D = g(y)$. We also know that the number of hamburgers consumed at a family picnic depends on the number of people attending (the more people the more hamburgers). If we denote the number of people attending the picnic by n and the pounds of hamburger consumed by p, then the relationship can be represented by a function $p = h(n)$, with n as the independent variable and p as the dependent variable.

IMPROVING SKILLS

In Problems 1 through 10, the rules of correspondence are illustrated by arrows. Determine if the given relationships are functions and, for those that are, identify their ranges.

1.

2.

3.

4.

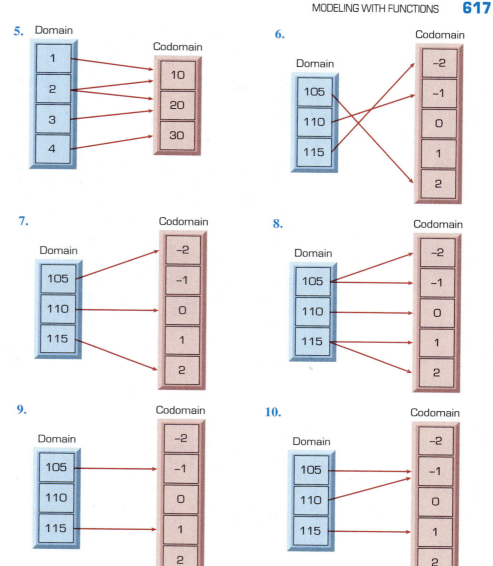

In Problems 11 through 18, determine if the given tables represent functions and, for each that does, identify the range. In each table, the rule of correspondence is to assign to each element in the top row (the domain) the element directly below it in the bottom row (the codomain).

11.

x	1	2	3	4	5	6
y	2	2	2	2	2	2

12.

x	2	2	2	2	2	2
y	1	2	3	4	5	6

13.

x	1	2	3	4	5	6
y	2	2	3	3	4	4

14.

x	2	2	3	3	4	4
y	1	2	3	4	5	6

15.

x	1	2	3	4	5	6
y	6	5	4	3	2	1

16.

x	1	2	3	4	5	6
y	10	11	−1	−2	0	3

17.

x	10	11	−1	−2	0	3
y	1	2	3	4	5	6

18.

x	1	2	3	4	5	6
y	1	2	3	4	5	6

In Problems 19 through 30, determine whether each curve is a function and, if so, identify its natural domain. The rule of correspondence assigns to each *x*-value on the horizontal axis (the domain) all *y*-values on the vertical axis (the codomain) such that (*x*, *y*) lies on the curve.

19.

20.

21.

22.

23.

24.

25.

26.

27.

28.

29.

30.

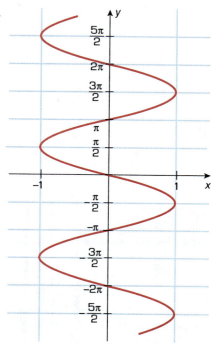

In Problems 31 through 40, determine the natural domain for each function of x.

31. $y = x - 18$ **32.** $y = 1/(x - 18)$

33. $y = (x - 1)(x - 2)$ **34.** $y = 1/(x - 1)(x - 2)$

35. $y = x^2 - 1$ **36.** $y = 1/(x^2 - 1)$

37. $y = 2/(x^2 - 5x + 6)$ **38.** $y = (2x + 1)/(x^2 + 1)$

39. $y = (2x - 4)/(x^2 - 6)$ **40.** $y = 1/x$

41. Given the function $f(x) = 2x^2 + 3$, find the following:
     (a) $f(0)$ **(b)** $f(1)$ **(c)** $f(-1)$ **(d)** $f(0.5)$

42. Given the function $g(x) = 2x^2 - x$, find the following:
     (a) $g(0)$ **(b)** $g(1)$ **(c)** $g(-1)$ **(d)** $g(0.5)$

43. Given the function $h(x) = -3x + 3$, find the following:
     (a) $h(2)$ **(b)** $h(-2)$ **(c)** $h(2a)$ **(d)** $h(a + b)$

44. Given the function $p(x) = x^2$, find the following:
     (a) $p(2)$ **(b)** $p(-2)$ **(c)** $p(2a)$ **(d)** $p(a^2)$

45. Given the function $f(x) = (x + 2)/(2x - 4)$, find the following:
     (a) $f(1)$ **(b)** $f(-4)$ **(c)** $f(2a)$ **(d)** $f(a + b)$

46. Given the function $g(x) = (x^2 + x)/(x - 4)$, find the following:
     (a) $g(1)$ **(b)** $g(-4)$ **(c)** $g(3a)$ **(d)** $g(a^2)$

CREATING MODELS

In Problems 47 through 70, model, if possible, one factor in each process as a function of the other factors in the process. In each situation, identify the inputs (there may be more than one) and the output. Where possible, specify the natural domain.

47. Revenues to a movie theater come from the sale of $8 tickets.

48. Revenues to a movie theater come from the sale of tickets: $8 for adults and $5 for senior citizens.

49. Revenues to a movie theater come from the sale of $8 tickets and popcorn, $2.50/bag.

50. Revenues to a movie theater come from the sale of tickets and popcorn: tickets are $8 for adults and $5 for senior citizens, and popcorn is $2.50/bag.

51. A certificate of deposit (CD) earns 6% a year, compounded quarterly. The interest is directly proportional to the amount in the account at the beginning of the quarter.

52. Jarronn has up to $1,500 to invest in the CD in Exercise 51, which earns 6% a year, compounded quarterly. The interest paid is directly proportional to the amount in the account at the beginning of the quarter.

53. The amount of interest earned by $1,000 in a year depends solely on the annual interest rate.

54. The elongation of a spring mounted on the ceiling is directly proportional to the weight attached to the spring.

55. The perimeter of a square depends on the length of a side.

56. The perimeter of a square with area greater than 16 in.^{2} depends on the length of a side.

57. The perimeter of a rectangle depends on its height and width.

58. The perimeter of a rectangle with no side greater than 4 in. depends on the lengths of its sides.

59. The area of a square depends on the length of a side.

60. The area of a square with no side less than 4 in. depends on the length of any side.

61. The area of a rectangle with a side of 4 in. depends on the length of the other side.

62. The area of a rectangle with no side greater than 4 in. depends on its height and width.

63. The area of a circle depends on its radius.

64. The area of any circle with radius between 2 in. and 11 in. depends on its radius.

65. The area of a circle depends on its diameter.

66. The area of a triangle 5 in. high depends on the length of its base.

67. The area of a triangle with a 4 in. base depends on its height.

68. The area of a triangle depends on both its height and the length of its base.

69. The area of a triangle standing no higher than 5 in. depends on both its height and the length of its base.

70. The area of a triangle having no side longer than 5 in. depends on both its height and the length of its base.

EXPLORING IN TEAMS

71. A function $f(x)$ is *even* if $f(-x) = f(x)$ for every value of x in its domain. Determine which of the following functions are even.

(**a**) $f(x) = x$ (**b**) $f(x) = x^2$ (**c**) $f(x) = 2$

(**d**) $f(x) = x + x^2$ (**e**) $f(x) = 2 + x^2$ (**f**) $f(x) = 1/x$

72. What can you say about evaluating an even function at the values ± 3, ± 5, ± 8, and ± 10 if they're all in the domain of the function?

73. Using the results of the previous problem, discuss the characteristics of the graph of an even function.

74. What can you say about the sum of two even functions? What about their product?

EXPLORING WITH TECHNOLOGY

75. Use a graphing calculator to graph the function $f(x) = x^2$. Then superimpose the graphs of $g(x) = f(x) + 1 = x^2 + 1$, $h(x) = f(x) + 3 = x^2 + 3$, and $j(x) = f(x) + 5 = x^2 + 5$ on the same coordinate system. Now make a conjecture as to what happens to the graph of a function when you add a positive number.

76. Graph on the same coordinate system: $f(x) = 2/(x^2 + 1)$, $f(x) + 1$, $f(x) + 3$, and $f(x) + 5$. Use the results to test the conjecture you developed in Exercise 75, and reformulate the conjecture if you feel it is necessary.

77. Graph on the same coordinate system: $f(x) = x^3$, $f(x) - 1$, $f(x) - 3$, and $f(x) - 5$. Based on your results, make a conjecture about the graph of a function when you subtract a positive number.

78. Graph on the same coordinate system: $f(x) = 2/(x^2 + 1)$, $f(x) - 1$, $f(x) - 3$, and $f(x) - 5$. Use the results to test the conjecture you developed in Exercise 77, and reformulate the conjecture if you feel it is necessary.

REVIEWING MATERIAL

79. (Section 1.4) Construct a qualitative graphical model for River City High School relating the number of students absent due to flu (vertical axis) to the passage of time during the school year (horizontal axis). On the same coordinate system, also construct a qualitative graphical model that relates the number of students who recover from the flu (vertical axis) over the same time period.

80. (Section 3.3) Use graphical methods to solve the following linear program:

$$\text{Minimize:} \quad z = x + 10y$$

$$\text{Subject to:} \quad 2x + 5y \geq 40,000$$

$$3x - 4y \leq 36,000$$

$$y \leq 14,000$$

81. (Section 6.3) The probability of a company's reporting a profit next quarter is 0.65. If there is a profit, there is an 85% chance that the company will declare a dividend; if the company does not report a profit, there is only a 40% chance of a dividend. What is the probability that the company will report a profit next quarter and *not* declare a dividend?

82. (Section 6.7) Joe is a realtor with City Realty Company. In a given week, there is a 60% chance that he won't sell any houses; a 25% chance that he'll sell one house; a 10% chance that he'll sell two houses; and a 5% chance he'll sell three. If he does sell a house, the probability that the profit will be $5,000 is 3/5 and that it will be $6,000 is 1/5. There is a 10% possibility that a sale will yield a $7,000 profit and the same possibility of an $8,000 profit. What is the expected number of houses Joe will sell in a week? What is his expected profit on a house? Using these two results, what is Joe's expected weekly profit?

RECOMMENDING ACTION

83. Respond by memo to the following request:

> MEMORANDUM
>
> To: J. Doe Reader
>
> From: Armand
>
> Date: Today
>
> Subject: **Functions**
>
> I make rocking chairs using wood, paint, and labor (in hours), so it seems to me I have a process with one output and three inputs. Does your definition of a function apply to my process? If not, can we expand your definition of a function so that it does?

9.2 POLYNOMIAL, RATIONAL, AND POWER FUNCTIONS

> *A polynomial function of degree n has the form*
> $$f(x) = a_nx^n + a_{n-1}x^{n-1} + \cdots + a_2x^2 + a_1x + a_0$$
> *where n is a nonnegative integer and the coefficients*
> $$a_n, a_{n-1}, \cdots, a_2, a_1, a_0$$
> *are known real numbers with $a_n \neq 0$.*

Functions model input–output relationships, which are the core of many real-world processes. Consequently, functions are essential tools for modelers and decision-makers alike. In this section, as we list and describe some of the most useful modeling functions, you will see that we have been using functions as models all along; we simply did not identify them as such.

A function f is a *polynomial function* if it has the form

$$f(x) = a_nx^n + a_{n-1}x^{n-1} + \cdots + a_2x^2 + a_1x + a_0 \tag{3}$$

where $a_n, a_{n-1}, \ldots, a_2, a_1, a_0$ are all known real numbers. The lead coefficient a_n cannot be zero, but any of the other coefficients can be. The constant term a_0 in Equation (3) is the coefficient of $x^0 = 1$. The powers on the input variable x are all nonnegative integers, and the highest power with a nonzero coefficient is called the *degree* of the polynomial function.

EXAMPLE 1

Determine which of the following functions are polynomial functions. For those that are, state their degree.

(a) $f(x) = 3x^2 - 2x$

(b) $f(x) = 0.5x^5 - 2.25x^3 + \sqrt{5}$

(c) $f(x) = \sqrt{x}$

(d) $f(x) = \dfrac{1}{3}$

(e) $f(x) = \dfrac{1}{x}$

Solution

(a) This is a polynomial function of degree 2 with $a_2 = 3$, $a_1 = -2$, and $a_0 = 0$.

(b) This is a polynomial function of degree 5 with $a_5 = 0.5$, $a_4 = 0$, $a_3 = -2.25$, $a_2 = a_1 = 0$, and $a_0 = \sqrt{5}$.

(c) This is *not* a polynomial function because $\sqrt{x} = x^{1/2}$ has the input variable x raised to the power of $\frac{1}{2}$, which is not an integer.

(d) This is a polynomial function of degree zero with $a_0 = \frac{1}{3}$.

(e) This is *not* a polynomial function because $\dfrac{1}{x} = x^{-1}$ has the input variable x raised to -1, which is a *negative* integer.

The graph of a polynomial function of degree zero is a horizontal line.

Zero-degree polynomial functions take the form $f(x) = a_0$. We generally use the letter c to denote a constant, so zero-degree polynomial functions are often written

$$f(x) = c \tag{4}$$

This function says that every input is assigned the same nonzero output c. We know from our work in Section 1.5 that the graph of an equation of the form $y = c$ is a horizontal straight line, so the graph of any zero-degree polynomial function is a horizontal straight line. Two such graphs are shown in Figure 9.6.

The x-axis in a Cartesian coordinate system is the graph of the *zero function*, $f(x) = 0$, which assigns an output of zero to each input value. This function can be written in the form $f(x) = a_n x^n + a_{n-1} x^{n-1} + \cdots + a_2 x^2 + a_1 x + a_0$, where all of the coefficients are zero. Thus the zero function is not a polynomial function, because the condition $a_n \neq 0$ is violated. Furthermore, the zero function does not have a degree, because every power of the input variable x is multiplied by a coefficient of zero.

First-degree polynomials are functions of the form $f(x) = a_1 x + a_0$, with $a_1 \neq 0$. The constants a_1 and a_0 are often replaced by m and b, respectively, and the function is written

$$f(x) = mx + b \qquad (m \neq 0) \tag{5}$$

or

$$y = mx + b \qquad (m \neq 0)$$

FIGURE 9.6

Graphs of Two Polynomial Functions of Zero Degree

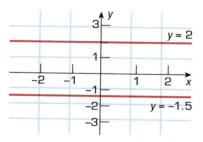

This last equation is the slope–intercept equation of a straight line, which we developed in Section 1.5. The graph of the equation is a straight line with slope m; it intercepts the y-axis at the point $(0, b)$. Thus the graph of a first-degree polynomial function is a straight line with a nonzero slope; the line is neither horizontal nor vertical. The graph of the line $y = -100x + 1500$ is shown in Figure 9.7. The line intercepts the y-axis at $(0, 1500)$ and has slope $m = -100$.

A horizontal line has slope zero, and such a line is the graph of a polynomial function of degree zero or of the zero function. A straight line with nonzero slope, that is, a straight line that is neither horizontal nor vertical, is the graph of a polynomial function of degree 1.

The graph of a polynomial function of degree 1 is a straight line with nonzero slope.

FIGURE 9.7
The Graph of the First-Degree Polynomial
$y = -100x + 1500$

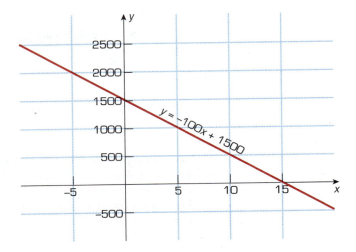

A linear function is a polynomial function of either degree 0 or degree 1, or the zero function.

A vertical line, such as the one illustrated in Figure 9.8, has undefined slope. Also it will not pass the vertical line test (there is a vertical line that intersects it more than once!), so it cannot be the graph of a function. (See Section 9.1.) The vertical line through the point (1.5, 0) intersects (in fact, coincides with) the line $x = 1.5$ at every point on the graph, not just a single point, as required by the vertical line test.

Functions whose graphs are nonvertical straight lines—the zero function, polynomial functions of degree zero, and polynomial functions of degree 1—are called *linear functions*. We used straight lines as models extensively in Chapters 1 through 3, so we already know that linear functions are extremely useful models.

Second-degree polynomial functions, often called *quadratic functions*, are of the form

$$f(x) = a_2 x^2 + a_1 x + a_0$$

A quadratic function is a polynomial function of degree 2; its graph is a parabola.

with $a_2 \neq 0$. If we replace the constants a_2, a_1, and a_0 by a, b, and c, respectively, we can write a quadratic function simply as

$$f(x) = ax^2 + bx + c \qquad (a \neq 0) \tag{6}$$

This is the graph of a parabola. (It has a shape similar to the nose cone of a rocket.) Two different quadratic functions are graphed in Figures 9.9 and 9.10.

FIGURE 9.8
The Vertical Line $x = 1.5$

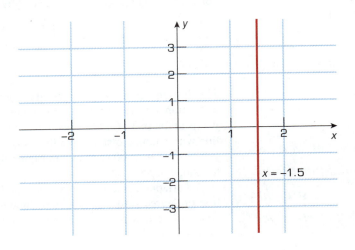

FIGURE 9.9
The Graph of the Second-Degree Polynomial
$y = x^2 - 4x + 9$

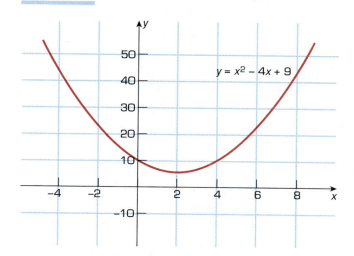

FIGURE 9.10
The Graph of the Second-Degree Polynomial
$y = -0.1x^2 - 2x + 20$

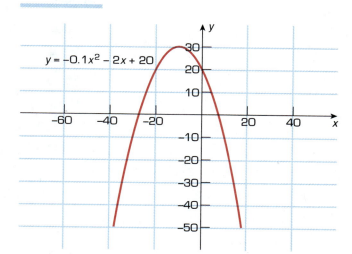

EXAMPLE 2

From past experience, a washing machine manufacturer knows that the relationship between its monthly profit P and the number N of machines that it produces is expressed by the equation

$$P = -22N^2 + 210N - 100$$

where P is in thousands of dollars and N is in thousands of machines. Graph the equation and explain the practical significance of the curve.

Solution

In this production process, the output is P, the monthly profit, which depends on N, the number of washing machines produced. The decision variable under the direct control of the manufacturer is the number of washing machines produced. The relationship between production and profit,

$$P = f(N) = -22N^2 + 210N - 100$$

is a quadratic function.

If no washing machines are produced, $N = 0$ and the corresponding profit is

$$P = -22(0)^2 + 210(0) - 100 = -100 \text{ thousand dollars}$$

that is, the manufacturer experiences a loss of $100,000. This loss reflects fixed costs such as rent, insurance, and leasing payments on equipment. Evaluating $f(N)$ for other input values, we obtain Table 9.3. These points are plotted in Figure 9.11 with a partial graph

TABLE 9.3 Input–Output Relationships for Example 2

N	0	1	2	3	4	5	6	7	8	9	10
P	−100	88	232	332	388	400	368	292	172	8	−200

FIGURE 9.11
The Graph of
$P = -22N^2 + 210N - 100$

of the function. We restrict the window in Figure 9.11 to nonnegative inputs because negative production has no practical significance. The natural domain here is $\{N \mid N \geq 0\}$.

The production of 1,000 washing machines ($N = 1$) results in a profit of $88,000. Profits continue to increase with higher production levels until approximately $N = 5$, when 5,000 machines are produced. At that point, the corresponding profit of $400,000 is near its maximum. Increasing production beyond 5,000 machines appears to reduce profits. One explanation is that the marketplace can absorb about 5,000 washing machines monthly from this manufacturer. Additional machines increase costs but add nothing to income, since they can't be sold. Thus profits are reduced.

A rational function is the quotient of two polynomial functions.

The quotient of two polynomial functions is called a *rational function*. It has the general form

$$f(x) = \frac{p(x)}{q(x)} \tag{7}$$

where both $p(x)$ and $q(x)$ are polynomial functions.

EXAMPLE 3

Determine which of the following functions are rational functions:

(a) $f(x) = \dfrac{1}{x}$

(b) $f(x) = \dfrac{0.5x^5 - 2.25x^3 + \sqrt{5}}{x^2 - 1}$

(c) $f(x) = \dfrac{x^2 - 1}{x^3 + 5}$

(d) $f(x) = \dfrac{\sqrt{x} + 3}{2x}$

Solution

(a) This is a rational function because both the numerator and denominator are linear functions. The numerator $p(x) = 1$ is a polynomial function of degree zero, while the denominator $q(x) = x$ is a polynomial function of degree 1.

(b) This is a rational function: the numerator

$$p(x) = 0.5x^5 - 2.25x^3 + \sqrt{5}$$

is a polynomial function of degree 5, while the denominator

$$q(x) = x^2 - 1$$

is a polynomial function of degree 2.

(c) This is a rational function because the numerator and denominator are polynomial functions of degrees 2 and 3, respectively.

(d) This is not a rational function. Although the denominator is a linear function, the numerator is not a polynomial function because of the $x^{1/2}$-term, which raises the input to a noninteger power.

> *The natural domain of a rational function is the set of all real numbers for which the function's denominator is nonzero.*

A rational function is a quotient and therefore not defined when its denominator is zero. Consequently, the natural domain of the rational function $f(x) = p(x)/q(x)$ is the set of real numbers for which $q(x) \neq 0$. The denominator of the rational function

$$f(x) = \frac{1}{x}$$

is zero when $x = 0$, therefore the natural domain for this function is the set of all nonzero real numbers. The denominator of the rational function

$$f(x) = \frac{0.5x^5 - 2.25x^3 + \sqrt{5}}{x^2 - 1}$$

is zero when $x^2 - 1 = (x - 1)(x + 1)$ is zero, which occurs when $x = \pm 1$, hence the natural domain for this function is the set of all real numbers except 1 and -1. The denominator of the rational function

$$f(x) = \frac{x^3 - 1}{x^2 + 5}$$

is never zero for any real-valued input, so the natural domain is the set of all real numbers.

A class of functions that is related to both polynomial functions and rational functions is that of the *power functions*, functions of the form

> *A power function is a function of the form $f(x) = x^p$, where p is a known real number.*

$$f(x) = x^p \tag{8}$$

where p is a known real number. If p is a nonnegative integer, $p = n$, then Equation (8) reduces to $f(x) = x^n$, a polynomial function of degree n. If p is a negative integer, $p = -n$, then Equation (8) reduces to $f(x) = x^{-n} = 1/x^n$, a rational function. When p is not an integer, however, power functions give us a whole new set of input–output relationships.

We can use a calculator to evaluate the output of a power function for a specific input. To evaluate $f(x) = x^{2.1}$ when $x = 3.4$, many graphing calculators use the following sequence of keystrokes:

The result is 13.064889, rounded to six decimal places. When $x = 0.24$, the following sequence of keystrokes is needed to evaluate $f(x) = x^{-5.38}$.

The result is 2160.05, rounded to two decimal places. Graphing calculators generally have a menu operation for evaluating a function at a point; such a menu is recommended when evaluating the same function at many different points.

Some calculators will not take powers of negative numbers. In this situation, for the power function $f(x) = x^p$ with noninteger exponents ($p \neq$ integer), we take the natural domain to be the set of all nonnegative real numbers when p is positive and the set of all positive real numbers when p is negative. When p is an integer, power functions are polynomial or rational functions and procedures for determining their domains have been established previously.

Figure 9.12 illustrates the graphs of some power functions for input values from the interval [0, 5]. Note that all of these power functions pass through the points (0, 0) and (1, 0). Furthermore, when we restrict our attention to the behavior of these functions for inputs greater than 1, we see that the higher the exponent, the larger the output.

FIGURE 9.12
Some Graphs of Power Functions on the Interval [0, 5]

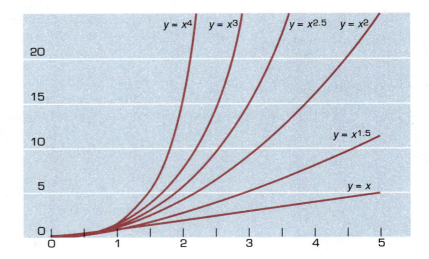

Now if we zoom in on Figure 9.12 and magnify the graph for just the input values in the interval [0, 1], we obtain Figure 9.13. We see that on the interval [0, 1], the higher the exponent, the *smaller* the output.

FIGURE 9.13
Some Graphs of Power Functions on the Interval [0, 1]

IMPROVING SKILLS

In Problems 1 through 18, determine which relationships are polynomial functions and, for those that are, state their degrees.

1. $f(x) = 2.1x^2$
2. $f(x) = 2x^{2.1}$
3. $f(x) = 3x^2 - 2.1$
4. $f(x) = 2x^3 - 2.1$
5. $f(x) = \pi$
6. $f(x) = \pi^2 - 3$
7. $f(x) = \pi^{2.1}$
8. $f(x) = x^\pi$
9. $f(x) = 1 - x^5$
10. $f(x) = x^3 - 3x^4$
11. $f(x) = \dfrac{3x^4 + 2x^2 - 0.5}{2 - x}$
12. $f(x) = \dfrac{3x^4 + 2x^2 - 0.5}{2 - \sqrt{x}}$
13. $f(x) = \dfrac{3x^4 + 2x^2 - 0.5}{\sqrt{2} - x}$
14. $f(x) = \dfrac{3x^4 + 2x^2 - 0.5}{\sqrt{2} - \sqrt{3}}$
15. $f(x) = \dfrac{3x^4 + 2x^2 - 0.5}{\sqrt{2 - x}}$
16. $f(x) = \dfrac{2.5x^3 + x^2 - x}{2 - x^2}$
17. $f(x) = \dfrac{x^2 + 2x - 3}{x^2 + 3x + 5}$
18. $f(x) = \dfrac{x^4 + x^2 + 4}{x^6 - 3x^7 - 5x^8}$

19. Determine which of the non-polynomial functions in Problems 1 through 18 are rational functions.
20. Determine which of the functions in Problems 1 through 18 are power functions.
21. What is the degree of the product of two polynomial functions?
22. Show by example that it is possible to add together two polynomial functions of degree 2 and obtain a polynomial function of degree 1.
23. Prove that every polynomial function must intersect the y-axis.

CREATING MODELS

24. The relationship between F, the temperature measured in degrees Fahrenheit, and C, the temperature measured in degrees Celsius, is given by the formula

$$F = \frac{9}{5}C + 32$$

(a) Is F a polynomial function of the variable C? If so, what is its degree?
(b) What is the Fahrenheit equivalent of 22° Celsius?
(c) What is the Celsius equivalent of 90° Fahrenheit?
(d) Solve for C in terms of F. What type of relationship do you get?

25. A publisher of a current economics textbook determines that the manufacturing costs per book are $10 and that the fixed costs of the business are $20,000. Show that total cost can be modeled by a linear function.

26. A manufacturer of electric can openers has a daily fixed cost of $21,000 and knows that the variable cost of producing one can opener is $5.50. The can openers sell for $9.00 each. Show that total daily cost and the total daily revenue can be modeled by linear functions.

27. A bakery currently has a fixed cost of $120,000 per month and a variable production cost of approximately 23¢ per loaf of bread. Loaves sell at a wholesale price of 55¢/loaf. Show that total monthly cost, the total monthly revenue, and the total

monthly profit all can be modeled by linear functions, assuming that all loaves can be sold.

28. Starting from rest, a ball dropped from an airplane will fall a distance d feet after t seconds according to the formula $d = 16t^2$, assuming no air resistance. Using this quadratic model, find: **(a)** the distance the ball falls in 5 seconds; **(b)** the distance the ball falls in 10 seconds; and **(c)** the time it takes for the ball to hit the ground if the plane is at an altitude of 5,000 feet when the ball is released.

29. Starting with a velocity of 800 ft/sec, a projectile fired vertically from the ground for t seconds will be at a height of h feet as determined by the formula $h = 800t - 16t^2$, until it returns to the ground. Using this quadratic model, find **(a)** the height of the projectile after 2 seconds; **(b)** the height of the projectile after 5 seconds; and **(c)** the time it takes for the projectile to return to the ground.

30. Show that the area of a square can be modeled as a polynomial function of its perimeter. What is the degree of this function?

31. Show that the volume of a cube can be modeled as a polynomial function of the length of a side. What is the degree of this function?

32. Show that the area of a circle can be modeled as a polynomial function of its diameter. What is the degree of this function?

33. Show that the volume of a sphere can be modeled as a polynomial function of its radius. What is the degree of this function?

EXPLORING IN TEAMS

34. Use the procedure described in Problem 26 of Section 2.3 to show that the least-squares quadratic fit to the data listed in the following table is $y = -22x^2 + 210x - 100$. (This is Table 9.3 reproduced here in terms of x and y.)

x	0	1	2	3	4	5	6	7	8	9	10
y	−100	88	232	332	388	400	368	292	172	8	−200

35. The following data, provided by the Bureau of the Census, shows the population of the United States from 1790 to 1990.

YEAR	POPULATION millions	YEAR	POPULATION (millions)
1790	3.9	1900	76.0
1800	5.3	1910	92.0
1810	7.2	1920	105.7
1820	9.6	1930	122.8
1830	12.9	1940	131.7
1840	17.1	1950	151.3
1850	23.2	1960	179.3
1860	31.4	1970	203.3
1870	39.8	1980	226.5
1880	50.2	1990	248.7
1890	62.9		

Use the procedure described in Problem 26 of Section 2.3 to model this data with a least-squares quadratic fit. To simplify calculations, call 1890 Year zero, 1900 Year 1; 1880 Year −1; 1910 Year 2, and so on. Graph the resulting equation and the data on the same coordinate system. Would you recommend using this model?

EXPLORING WITH TECHNOLOGY

FIGURE A

Width = w

Length = l

36. Use a graphing calculator to graph the equation $y = -22x^2 + 210x - 100$, initially establishing a window from 0 to 12 on the horizontal axis and -500 to 500 on the vertical axis. (This equation is from Example 2, written in terms of x and y.) Use the trace feature of your calculator to estimate, to the nearest integer, the value of x that yields the maximum value of y.

37. Use a graphing calculator to graph the equation $y = 800x - 16x^2$, initially establishing a window from 0 to 50 on the horizontal axis and 0 to 12,000 on the vertical axis. (This equation is from Exercise 29 in terms of x and y.) Use the trace and zoom features of your calculator to estimate the value of x that yields the maximum value of y.

38. A farmer has 200 feet of fence with which to enclose a rectangular plot of land. To save on fencing, the farmer will use an existing wall as one side of the plot (see Figure A). Show that the area A of the enclosed plot can be modeled by the formula

$$A = 200l - 2l^2$$

where l is the length of fence perpendicular to the wall on one side of the plot. Rewrite this equation as $y = 200x - 2x^2$ and then use a graphing calculator to estimate the plot dimensions that yield maximum enclosed area.

39. A 20-foot piece of rope is cut into two pieces. Use a graphing calculator to determine the length of each piece if the sum of the squares of each length is minimum.

40. Use a graphing calculator to graph each of the following quadratic functions and then hypothesize a relationship between the orientation of a parabola and the sign of the coefficient of x^2.

(**a**) $f(x) = 2x^2 - 1$ (**b**) $f(x) = 2x^2 + 1$
(**c**) $f(x) = -2x^2 - 1$ (**d**) $f(x) = -2x^2 + 1$
(**e**) $f(x) = 0.5x^2 + x + 2$ (**f**) $f(x) = -0.5x^2 + x + 2$

41. Use a graphing calculator to graph about thirty different fourth-degree polynomials of the form

$$y = a_4x^4 + a_3x^3 + a_2x^2 + a_1x + a_0$$

for specific choices of the coefficients a_4, a_3, a_2, a_1, a_0. Include in your choices zero values for some of the coefficients and both positive and negative values for the coefficient a_4. Examples are $y = 3x^4 - 9x^3 - 2x^2 + 1$, where $a_1 = 0$ and $a_4 = 3 > 0$; also $y = -x^4 + 5x^2 - 1$, where $a_3 = a_1 = 0$ and $a_4 = -1 < 0$. Use your graphs to make an educated guess about the number of peaks and pits on such a graph. What is the maximum number of times a polynomial function of degree 4 can cross the x-axis? Can you extend these conjectures to polynomial functions of other degrees?

42. Use a graphing calculator to graph several different rational functions, each with the property that its numerator and denominator have the same degree. Two examples are

$$y = \frac{6x^2 - 3}{2x^2 + 2x + 5} \quad \text{and} \quad y = \frac{7x^4 + 2x^3 - 5x + 2}{x^4 + 4x^3 - 6x^2}$$

For each of your functions, use the graph to make an educated guess as to some values of the output for large input values. Compare your guess to the ratio of the coefficient of the highest power of x in the numerator to that in the denominator. In the first of the preceding examples, both the numerator and the denominator are of degree 2 and the ratio of the coefficients of x^2 is 3. In the second example, both numerator and denominator have degree 4 and the ratio of the coefficients of x^4 is 7.

REVIEWING MATERIAL

43. (Section 1.5) Find the equation of the straight line passing through the two points (50, 20,000) and (100, 8000). Does this line include the point (200, 2000)?

44. (Section 4.4) Construct the first three lines of an amortization table to repay a $2,000 loan over 24 months at an annual rate of 10% compounded monthly.

45. (Section 5.1) A survey of 300 graduating high school seniors reveals the following information about their senior years:

289 had passing grades

105 played a varsity sport

160 had a library card

158 had a library card and passing grades

96 had passing grades and played a varsity sport

2 played a varsity sport and had a library card

2 had a library card, passing grades, and played a varsity sport

Using this information, determine the number of students in the survey who **(a)** did not have passing grades; **(b)** did not have a library card; **(c)** had a library card but did not play a varsity sport; and **(d)** played a varsity sport but did not have passing grades.

RECOMMENDING ACTION

46. Respond by memo to the following request.

MEMORANDUM

To: J. Doe Reader

From: Alison

Date: Today

Subject: **A Little Help, Please!**

While reviewing my class notes, I see that I wrote the following:

> With large values (both positive and negative) of the input variable, the behavior of a polynomial function of degree n is determined almost completely by the single term $a_n x^n$. All other terms are negligible and may be ignored. Furthermore, the most important feature of the polynomial function in such a case is the sign of a_n.

I must have copied this comment from the board, but I have no idea what it means. Can you help? I think this will be on the next test.

9.3 EXPONENTIAL FUNCTIONS

> *An exponential function has the form $f(x) = a(b^x)$, where a is a nonzero real number and b is a positive real number, $b \neq 1$.*

The graphs of functions introduced in Section 9.2 include straight lines and parabolas. These functions are some of the simplest yet most valuable functions available to modelers. By themselves, however, they aren't sufficient for modeling all real-world phenomena. Many input–output processes follow other patterns, so it's advantageous to expand the collection of familiar functions. The more functions at our disposal, the more successful we will be as modelers.

An *exponential function* is a function of the form

$$f(x) = a(b^x) \tag{9}$$

where a denotes a nonzero real number and b denotes a positive real number not equal to 1. The number b is called the *base*. The distinguishing feature of an exponential function, and the rationale for its name, is that the input variable x is the exponent.

EXAMPLE 1

Determine which of the following are exponential functions.

(a) $f(x) = 5^x$ (b) $f(x) = 2.1(5^x)$ (c) $f(x) = (\frac{1}{3})^x$
(d) $f(x) = -2(\frac{1}{3})^x$ (e) $f(x) = (-2)^x$ (f) $f(x) = x^5$

Solution

(a) This is an exponential function with base $b = 5$ and constant $a = 1$.
(b) This is an exponential function with base $b = 5$ and constant $a = 2.1$.
(c) This is an exponential function with base $b = 1/3$ and constant $a = 1$.
(d) This is an exponential function with base $b = 1/3$ and constant $a = -2$.
(e) This is not an exponential function because the base $b = -2$ is not positive.
(f) This is not an exponential function but rather a polynomial function of degree 5 (as well as a power function).

We don't allow the base b to be 1 in our definition of an exponential function because if $b = 1$, then

$$f(x) = a(b^x) = a(1^x) = a(1) = a$$

which is a polynomial function of degree zero.

With the aid of a graphing calculator, it is a simple matter to evaluate the output of an exponential function corresponding to a specific input value. Keystrokes commonly used to evaluate $f(x) = 5.4(2.35)^x$ at $x = 1.5$ are

The result is 19.453402 rounded to six decimal places. Evaluating $y = 5.4(2.35)^x$ at other input values of x generates Table 9.4, with all output values rounded to two decimal places. These points are plotted in Figure 9.14.

TABLE 9.4 **Some Evaluations of $y = 5.4(2.35)^x$**

x	−2	−1.5	−1	−0.5	0	0.5	1	1.5	2
y	0.98	1.50	2.30	3.52	5.4	8.28	12.69	19.45	29.82

FIGURE 9.14
The Graph of the Exponential Function $y = 5.4(2.35)^x$

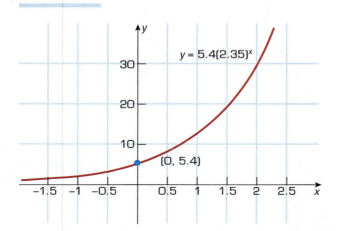

FIGURE 9.15
The Graph of the Exponential Function $y = 2(1/3)^x$

EXAMPLE 2

Graph the function $y = 2(1/3)^x$.

Solution

One possibility is to go to the graph menu on a graphing calculator immediately, but instead we first plot a few points to get a feeling for how the function behaves with different inputs. The keystrokes needed to evaluate the function $f(x) = 2(\frac{1}{3})^x$ when $x = -2$ are

The result is 18. Continuing in this manner, we generate Table 9.5 with all output values rounded to two decimal places. These points and a graph of the given function are plotted in Figure 9.15.

TABLE 9.5 **Some Evaluations of $y = 2(1/3)^x$**

x	-2	-1.5	-1	-0.5	0	0.5	1.0	1.5	2.0
y	18	10.39	6	3.46	2	1.15	0.67	0.38	0.22

Figures 9.14 and 9.15 are representative of the graphs of exponential functions of the form $f(x) = a(b^x)$ with $a > 0$. Such graphs never intersect the x-axis but approach it from above in one direction and increase steeply in the other. These graphs always intersect the y-axis at $(0, a)$. Exponential functions with $a > 0$ have one of two general shapes shown in Figures 9.14 and 9.15, depending on whether the base b is greater than 1 or less than 1.

Figure 9.14 is the graph of $y = 5.4(2.35)^x$. Here the base is $b = 2.35 > 1$ and the coefficient is $a = 5.4$. We say this function *increases exponentially*. If we restrict our attention to the first quadrant, the output becomes very large very quickly as the values of the input increase. And the greater the base the quicker the increase, as demonstrated in Figure 9.16.

The graph of the exponential function $y = a(b)^x$ always intersects the y-axis at $(0, a)$.

FIGURE 9.16
Some Graphs of Exponential Functions with Different Bases

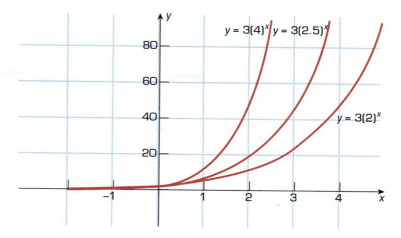

Figure 9.15 is the graph of $y = 2(\frac{1}{3})^x$. Here the base is $b = \frac{1}{3} < 1$ and the coefficient is $a = 2 > 0$. This function *decreases exponentially*. Now if we restrict our attention to the first quadrant, the output drops to near zero very quickly as the values of the input increase. This time, the *smaller* the base, the quicker the *decrease*.

When $a < 0$, the curves stay below the x-axis, and they look like reflections of Figures 9.14 and 9.15. Their general shapes are illustrated in Figures 9.17 and 9.18. As with $a > 0$, the graphs never reach the x-axis but here they approach it from below in one direction and decrease dramatically in the other. These graphs also intersect the y-axis at $(0, a)$, but now a is negative.

FIGURE 9.17
The Graph of the Exponential Function $a(b^x)$ with $a < 0$ and $b > 1$

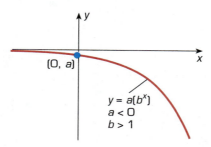

FIGURE 9.18
The Graph of the Exponential Function $a(b^x)$ with $a < 0$ and $b < 1$

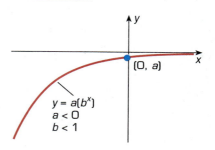

EXAMPLE 3

Two strands of bacteria are placed in a large petri dish and allowed to replicate. Every day at the same time, the number of strands in the culture are counted; based on the tallies, the number of strands appears to double each day. Model the growth of this bacterial culture with a function, then use the model to estimate the number of strands in the culture after $6\frac{1}{2}$ days.

Solution

On day 1, the beginning of the experiment, there are two strands of bacteria. If the number of strands doubles each day, then on day 2 there will be $4 = 2^2$ strands; on day 3 there will be $2(4) = 8 = 2^3$ strands; on day 4 there will be $2(8) = 16 = 2^4$ strands; and so on. Results for the first seven days are compiled in Table 9.6.

In this input–output relationship, the length of time of the culturing process is determined by the lab personnel, so time is the independent or decision variable. Each day they will measure the size of the population; this number is the output of the process. If we designate time by t and the number of strands by N, then Table 9.6 can be summarized by the relationship $N = 2^t$.

TABLE 9.6   **The Population of a Bacterial Culture**

DAYS	1	2	3	4	5	6	7
STRANDS OF BACTERIA	2	2^2	2^3	2^4	2^5	2^6	2^7

In Table 9.6, time is restricted to integer values—day 1, day 2, etc.—for the convenience of the experimenters. (They count only once a day.) There are $2^6 = 64$ strands at the beginning of day 6 and $2^7 = 128$ strands at the beginning of day 7. We know, however, that 64 strands don't suddenly appear at the beginning of day 6. In fact, the replications take place over the course of a day, so it seems reasonable for the input variable t to be real-valued rather than integer-valued. Also, because the experiment begins at $t = 1$, there is no data generated before that time. Thus the natural domain is all real numbers t greater than or equal to 1, and the input–output process is modeled by the exponential function

$$N(t) = 2^t, \quad t \geq 1$$

The base here is 2.

After $6\frac{1}{2}$ days, $t = 6.5$ and $f(6.5) = 2^{6.5} \approx 91$. So based on this model, the experimenters should anticipate approximately 91 strands of bacteria in the culture after $6\frac{1}{2}$ days.

Exponential functions follow all the familiar rules of exponents (see Appendix A). We summarize these rules in terms of the exponential functions b^x and c^x, when b and c are positive real numbers not equal to 1.

Properties of Exponential Functions

1. $b^x b^y = b^{x+y}$

2. $(b^x)^y = b^{xy}$

3. $\dfrac{b^x}{b^y} = b^{x-y}$

4. $\left(\dfrac{1}{b}\right)^x = \dfrac{1}{b^x} = b^{-x}$

5. $(bc)^x = b^x c^x$

6. $\left(\dfrac{b}{c}\right)^x = \dfrac{b^x}{c^x}$

Just as polynomial functions are characterized by their degree, exponential functions are characterized by their base. All polynomial functions are useful for modeling, but linear functions—those of degree 0 and 1—are the most important. Similarly, for exponential functions, some bases are more valuable than others. The base $b = 10$ is the fundamental base of our decimal system. Also important in this age of computers is the base $b = 2$. The most useful base for modeling, however, is the irrational number

$$e = 2.7182\ 8182\ 8459 \cdots$$

known as *Euler's number*. The exponential function

$$f(x) = e^x \tag{10}$$

that uses Euler's number as its base is called the *natural exponential function*. Its graph is shown in Figure 9.19.

The natural exponential function is $f(x) = e^x$, where Euler's number, e, is the base.

FIGURE 9.19

The Graph of the Natural Exponential Function $y = e^x$

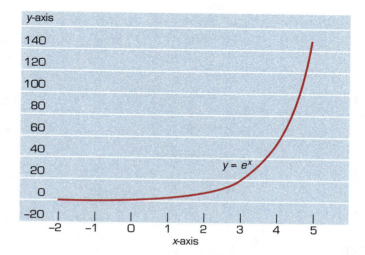

The Euler number e is not the first irrational number we've encountered as central to a particular branch of mathematics. The number $\pi = 3.141592 \cdots$ is well known as essential in geometry.

As an example of how e can arise in modeling, let's consider the growth of $1 in an interest-bearing account paying an annual rate of 100%. Although this is surely an exorbitant interest rate (we'll consider more common rates shortly), it has the pleasant feature that its decimal equivalent is $r = 1$. If this interest rate is compounded over N periods, then by the compound interest formula (see Section 4.1), $1 will grow to a future value of

$$FV = \$1\left(1 + \frac{1}{N}\right)^N \tag{11}$$

after N conversion periods, or one year. With the aid of a calculator, we evaluate Equation (11) for common values of N and obtain Table 9.7, with all computations rounded to six decimal places.

We know from Section 4.1 that the value of an account increases as the number of conversion periods in a year increases and, indeed, the values in the last column of Table 9.7 are increasing: they are approaching Euler's number e.

If we repeat this exercise with a different nominal annual interest rate r (expressed as a decimal), instead of the decimal equivalent of 100%, the future value becomes

$$FV = \$1\left(1 + \frac{r}{N}\right)^N \tag{12}$$

TABLE 9.7 $1 Invested for One Year at an Annual Rate of 100%

TYPE OF COMPOUNDING	FORMULA FOR VALUE AT MATURITY	VALUE AT MATURITY
Annually	$\$1\left(1 + \dfrac{1}{1}\right)^{1}$	$2.00
Semiannually	$\$1\left(1 + \dfrac{1}{2}\right)^{2}$	$2.25
Quarterly	$\$1\left(1 + \dfrac{1}{4}\right)^{4}$	$2.441406
Monthly	$\$1\left(1 + \dfrac{1}{12}\right)^{12}$	$2.613035
Weekly	$\$1\left(1 + \dfrac{1}{52}\right)^{52}$	$2.692597
Daily	$\$1\left(1 + \dfrac{1}{365}\right)^{365}$	$2.714567
Hourly	$\$1\left(1 + \dfrac{1}{8760}\right)^{8760}$	$2.718127
Every minute	$\$1\left(1 + \dfrac{1}{525,600}\right)^{525,600}$	$2.718279
Every second	$\$1\left(1 + \dfrac{1}{31,536,000}\right)^{31,536,000}$	$2.718282

One dollar grows to e^r dollars after one year at an interest rate of r%, compounded continuously.

As we evaluate this quantity with increasing values of N, the results get closer and closer to e^r. (See Exercises 25 and 26.) Decreasing the conversion period from quarters to years to days to minutes to seconds and so on, ad infinitum, we eventually reach what bankers call *compounding continuously*. One dollar will grow to e^r dollars after one year when it is placed in an account that earns interest at the rate r, compounded continuously.

Evaluating e to a power is accomplished easily with a calculator that has an e^x key. To evaluate $e^{1.5}$, the sequence of keystrokes on many graphing calculators is

The result is 4.481689, rounded to six decimal places. The keystrokes for $e^{-0.7}$ are

The result is 0.496585, rounded to six decimal places.

If PV (present value) dollars are deposited instead of $1, Equation (12) becomes

$$FV = PV\left(1 + \frac{r}{N}\right)^{N}$$

and as the number of conversion periods N increases, the value of the account after one year approaches PVe^r. That is, after one year, PV dollars grow to PVe^r dollars when the interest rate is compounded continuously.

We can take this analysis one step further and determine the future value (FV) of an initial deposit of PV dollars after t years at a nominal annual interest rate r, compounded continuously. We begin with regular compounding over N conversion periods in a year.

In t years, the initial deposit draws interest for Nt conversion periods, and the deposit will grow to a future value of

$$FV = PV\left(1 + \frac{r}{N}\right)^{Nt}$$

It follows from the properties of exponents that this last equation can be rewritten

$$FV = PV\left[\left(1 + \frac{r}{N}\right)^{N}\right]^{t}$$

> *An initial deposit of PV dollars earning an annual interest rate r, compounded continuously, will grow after t years to*
> $$FV = PVe^{rt}.$$

As N increases, the quantity inside the square brackets increases to e^r and the amount in the account grows to $PV[e^r]^t = PVe^{rt}$. That is, in t years, an initial deposit of PV dollars invested in an account earning a nominal annual interest rate r (expressed as a decimal), compounded continuously, will grow to

$$FV = PVe^{rt} \tag{13}$$

EXAMPLE 4

Determine the future value of an $800 deposit that draws interest for 25 years in an account earning 6% a year, compounded continuously.

Solution

Using Equation (13) with $PV = \$800$, $r = 0.06$, and $t = 25$, we find that the future value of this lump-sum deposit is

$$FV = \$800e^{(0.06)25} = \$800e^{1.5} = \$3,585.35$$

rounded to the nearest penny.

Under continuous compounding at a nominal annual interest rate r, one dollar grows to e^r dollars after one year. Consequently, the interest payments over this time frame come to

$$E_c = e^r - 1 \tag{14}$$

The quantity E_c is called the *continuous conversion rate,* expressed as a decimal. It is equal to the effective interest rate (see Section 4.1), in decimal form, for an interest rate of r (in decimal form), compounded continuously.

EXAMPLE 5

Determine the effective interest rate offered by a bank with a nominal annual rate of 7% compounded continuously.

Solution

With $r = 0.07$, Equation (14) becomes

$$E_c = e^{0.07} - 1 \approx 1.072508 - 1 = 0.072508$$

which can be rounded to 0.0725 or 7.25%.

IMPROVING SKILLS

In Problems 1 through 12, determine whether each function is (or can be written as) an exponential function. If so, identify the base.

1. $f(x) = 7^x$
2. $f(x) = x^7$
3. $f(x) = x^x$
4. $f(x) = 2(7^x)$
5. $f(x) = 7(2^x)$
6. $f(x) = 7^x 2^x$
7. $f(x) = 7^{2x}$
8. $f(x) = 2^{7x}$
9. $f(x) = \pi^x$
10. $f(x) = e^{\pi x}$
11. $f(x) = \pi e^x$
12. $f(x) = e^\pi$

In Problems 13 through 20, use the properties of exponents to simplify each expression into a single exponential function.

13. $2^x 3^{2x}$
14. $(\frac{1}{3})^x 3^{2x}$
15. $2^x 3^x / 5^x$
16. $2^{(x-3)}$
17. $2^{2(x-3)}$
18. e^{2x}
19. $2^{(2x+1)}$
20. $e^{2x} 3^{3x}$

21. Use a calculator to evaluate the exponential function $f(x) = 3(1.2^x)$ for the following input values:
   (**a**) $x = 5$ (**b**) $x = 2.3$ (**c**) $x = \pi$
   (**d**) $x = -3$ (**e**) $x = -2.3$ (**f**) $x = 1/3$

22. Use a calculator to evaluate the natural exponential function $f(x) = e^x$ for the input values specified in the previous problem.

23. Use a calculator to evaluate the exponential function $f(x) = (-1.1)(3.05^x)$ for the following input values:
   (**a**) $x = 3.1$ (**b**) $x = 1.3$ (**c**) $x = -1.3$
   (**d**) $x = 0.3$ (**e**) $x = -0.3$ (**f**) $x = 0.05$

24. Use a calculator to evaluate the natural exponential function $f(x) = e^x$ for the input values specified in the previous problem.

25. Reevaluate Table 9.7 using the formula $\$1[1 + (r/n)]^N$ with $r = 0.09$ and show that the entries in the last column approach $e^{0.09}$.

26. Reevaluate Table 9.7 using the formula $\$1[1 + (r/n)]^N$ with $r = 0.04$ and show that the entries in the last column approach $e^{0.04}$.

27. Determine the effective interest rate for a 4% interest-bearing account compounded continuously.

28. Determine the effective interest rate for a 5.5% interest-bearing account compounded continuously.

29. Determine the effective interest rate for a 8.2% interest-bearing account compounded continuously.

30. Determine the effective interest rate for a 12% interest-bearing account compounded continuously.

CREATING MODELS

31. A mining company deposits $20,000 in a savings account for the purpose of buying a new dump truck after four years. If the interest rate is 8%/year compounded continuously, how much money will the company have after four years?

32. A wood-cabinet manufacturer deposits $7,000 in a savings account for new tools. How much money will there be in two years if the interest rate is $6\frac{3}{4}\%$/year compounded continuously?

33. A mattress manufacturer deposits $2,100 in a savings account to buy a new computer in three years. How much money will there be at that time if the interest on the account is $7\frac{1}{4}\%$/year, compounded continuously?

34. A retired couple deposits $300,000 into an account paying 5%/year, compounded continuously; they plan to live off the interest payments, withdrawn quarterly. How much can they expect to withdraw at the end of each quarter?

35. A $5,000 account is opened for a newborn baby by her grandparents. How much will she have for college in 18 years if the account pays $8\frac{1}{4}\%$ annual interest, compounded continuously?

36. An investor places $5,000 in an account paying $6\frac{1}{4}\%$ interest a year, compounded quarterly. Three years later, she takes all the money from that account and opens a new account at another bank paying 7%/year, compounded continuously. How much will she have in another two years?

37. An entrepreneur borrows $20,000 from an investor, to be paid back after $2\frac{1}{2}$ years at 8% annual interest, compounded continuously. However, after $2\frac{1}{2}$ years he requests an extension for another two years at a rate of 7%, again compounded continuously. How much must the entrepreneur pay the investor $4\frac{1}{2}$ years later, at the end of the extension?

Some populations tend to double their size over constant time intervals, and they're often modeled by the exponential function $N = N_0 2^{t/c}$, where t denotes time in some specified time unit, N is the size of the population at time t, the initial population at time $t = 0$ is denoted N_0, and c is the amount of time required for the population to double. Use this model to solve Exercises 38 through 42.

38. Fifteen strands of a bacteria are cultured for 30 hours. Estimate the size of the population then if it is known that this species of bacteria doubles every eight hours.

39. If ten strands of a bacteria are cultured for two days and this bacteria doubles every ten hours, how many strands will be in the culture at the end of that time period?

40. The bacteria *Escherichia coli*, better known as *E. coli*, lives in the intestines of many mammals. Laboratory experiments indicate that it doubles every 24 minutes. If researchers introduce 1,000 strands of this bacteria into the intestine of a host cow, how many strands will be present in the cow after 18 minutes? (You may recall the contamination by *E. coli* of some fast-food hamburgers a few years back, resulting in several deaths.)

41. The population of a country doubles approximately every 20 years. How many inhabitants will it have in ten years if the current population is 25 million?

42. The present population of a country is 12.5 million people and it doubles about every 18 years. How many inhabitants will there be in 50 years?

EXPLORING IN TEAMS

43. Consider an exponential function of the form $P = P_0 b^t$, where P denotes the number of inhabitants in a population and t represents time. What does the quantity P_0 represent? Show that the base b represents the factor by which the population grows each year.

44. Complete the following table for two types of population growth, one linear and the second exponential. Use the result of the preceding problem to complete the third column.

t	LINEAR GROWTH	EXPONENTIAL GROWTH
0	100	100
1	300	300
2		
3		
4		

45. Complete this table for two types of population growth, one linear, one exponential.

t	LINEAR GROWTH	EXPONENTIAL GROWTH
0	100	100
1		
2	225	225
3		
4		

46. Assume that a population modeled by the exponential function in Problem 43 increases at the rate of 5%/year. Complete the following table:

t (years)	TOTAL POPULATION	GROWTH IN A YEAR
0		
1		
2		
3		
4		

47. Show that if a population grows at an annual rate of r%/year, the growth rate can be modeled by an exponential function with base $b = 1 + r$, with r expressed as a decimal.

48. The population of a certain country increases at the rate of 3.4%/year. If there were 1.2 million people in the country in 1980, estimate the population in 2000. (Hint: Use Problem 47.)

EXPLORING WITH TECHNOLOGY

49. Use a graphing calculator to graph the functions $y_1 = x^2$ and $y_2 = 2^x$ on the same coordinate system. Which function grows faster as x becomes large?

50. Use a graphing calculator to graph the functions $y_1 = x^3$ and $y_2 = 3^x$ on the same coordinate system. Which function grows faster as x becomes large?

51. Use a graphing calculator to sketch the graph of the function $f(x) = x^2 - 3$. On the same coordinate system, graph $f(x - 1)$, $f(x - 2)$, and $f(x - 5)$. Based on your

results, make a conjecture as to what happens to the graph of a function when the input variable x is replaced by $x - c$, for a nonzero constant c.

52. On the same coordinate system, graph $f(x) = e^{x/2}$, $g(x) = e^{(x-1)/2}$, $h(x) = e^{(x-2)/2}$, and $j(x) = e^{(x-5)/2}$. Try out the conjecture you developed in Exercise 51 and see if it works for the function $f(x) = e^{x/2}$. Reformulate the conjecture if you feel it's necessary.

53. On the same coordinate system, graph $f(x) = x^3$, $f(x + 1)$, $f(x + 3)$, and $f(x + 5)$. Based on your results, make a conjecture on the graph of a function when the input variable x is replaced by $x + c$, where c is a nonzero constant.

54. On the same coordinate system, graph $f(x) = e^{2x}$, $g(x) = e^{2(x+1)}$, $h(x) = e^{2(x+2)}$, and $j(x) = e^{2(x+5)/2}$. Use the results to test the conjecture you developed in Exercise 53 and reformulate your conjecture, if necessary.

REVIEWING MATERIAL

55. (Section 2.4) Determine whether the matrices

$$A = \begin{bmatrix} 1 & 2 & -1 \\ 1 & -1 & 2 \\ -1 & 2 & -1 \end{bmatrix} \quad \text{and} \quad B = \begin{bmatrix} 4 & -2 & 4 \\ -2 & 7 & -5 \\ 2 & -6 & 6 \end{bmatrix}$$

commute under the operation of multiplication.

56. (Section 4.3) How much must a person deposit at the end of each week in an account earning 4.5%/year, compounded weekly, if the goal is to have $12,000 in 5 years?

57. (Section 6.5) A committee with one chairperson and four members must be selected from a faculty of 80. How many different ways can such a committee be formed?

RECOMMENDING ACTION

58. Respond by memo to the following request.

MEMORANDUM

To: J. Doe Reader

From: Advertising Department

Date: Today

Subject: **Coffee Temperature**

I was at a seminar last week and heard a consultant say that if a hot cup of coffee is placed in a cool room, the temperature of the coffee will decrease exponentially until it reaches the temperature of the room. I think we can use this fact in our advertising copy, but I'm not sure I understand it. Does the consultant's remark make sense to you? If so, could you please explain it to me? Also, please provide me with a suitable graph for the process, one I can use in advertising copy.

THE NATURAL LOGARITHM

Sometimes we want to reverse an input–output process, either to undo an unwanted result or, more often, so that we can reproduce a desirable result. To do this, we must identify the input that caused the output to occur. For example, the president of a company might review sales data and observe that sales during one week in March were significantly higher than those in other weeks. Of course, the president would want to know the reason. Did the company offer different financing? Was advertising different? Were rebates offered? What input caused this highly desirable output?

When we model input–output relationships with functions, we apply a given rule of correspondence to an input, producing a corresponding output. Now we want to identify the input that generates a known output. In mathematics, reversing a process like this is called *finding an inverse*.

Let's "reverse the process" for the input–output operation described by the natural exponential function $y = e^x$. Suppose we want to produce the output $y = 10$. Can we do it? That is, can we identify an input value for x that will produce the output value $e^x = 10$?

Figure 9.20 is a graph of $y = e^x$ on the interval $[-2, 5]$. Following the procedures developed in Section 1.4, we can estimate the input x corresponding to the output $y = 10$ by first locating $y = 10$ on the vertical axis and then following a horizontal line to the

FIGURE 9.20

The Graph of the Natural Exponential Function $y = e^x$ on the Interval $[-2, 5]$

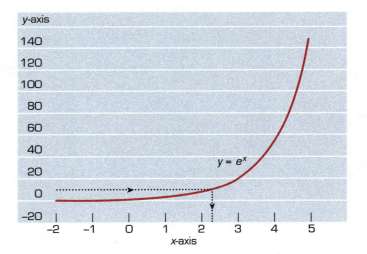

right until it intersects the curve. From the point of intersection, we follow a vertical line downward until it intersects the x-axis. This path is illustrated by the dotted lines in Figure 9.20. We see that the input that corresponds to $y = 10$ is a value of x between 2 and 2.5, close to 2.3.

To improve on our estimate, we zoom in on a smaller window, between 0 and 2.5 on the x-axis and between 0 and 20 on the y-axis; see Figure 9.21. Now using the trace feature, we find that the input $x = 2.302$ yields the output $y \approx 9.99$. Similarly, the input $x = 2.305$ yields the output $y \approx 10.02$. Thus the input corresponding to $y = 10$ is a value of x between 2.302 and 2.305. By zooming in on an even smaller window around $x = 2.30$ and $y = 10$, we can continue to improve our estimate. Eventually we will obtain $x = 2.302585$, rounded to six decimal places, as the input that yields the output $y = 10$ for the natural exponential function.

We can use this same approach to estimate the input of the natural exponential function for any output. Unfortunately, the work we do finding the inverse of one output is usually of little value in finding the inverse of another. None of our work in locating the inverse of $y = 10$ in Figures 9.20 and 9.21 is of any value for identifying inputs for x that produce the outputs $e^x = 50$, $e^x = 100$, or $e^x = 0.1$. We need a more efficient procedure for finding inverses for many different inputs.

The inverse of the natural exponential function is readily available on many calculators as the keystroke labeled "ln," for *natural logarithm*. To solve the equation $10 = e^x$, the sequence of keystrokes on many graphing calculators is

The inverse of the natural exponential function is the natural logarithm function.

The result is 2.302585, rounded to six decimal places. To solve the equation $50 = e^x$, the keystrokes are

The result is 3.912023, rounded to six decimal places.

The natural logarithm function is defined to be the inverse of the natural exponential function. In other words,

$$\ln(b) = a \quad \text{if and only if} \quad e^a = b \tag{15}$$

We have $\ln(50) \approx 3.912023$ because $e^{3.912023} \approx 50$. We have $\ln(10) \approx 2.302585$ because $e^{2.302585} \approx 10$. In general, for a given input b, the natural logarithm "searches" for an exponent a that is a solution to the equation $e^a = b$. That exponent a is then the output. For example, we know that $e^0 = 1$, so we have $\ln(1) = 0$. Furthermore, we know that $e^1 = e$, hence $\ln(e) = 1$.

Readers familiar with common logarithms should not confuse them with natural logarithms. The common logarithm function is the inverse of the exponential function $y = 10^x$; it is available on many calculators and computer software packages under the menu label "log." The natural logarithm function is the inverse of the exponential function $y = e^x$; it is available on many calculators and computer software packages under the menu label "ln" (or LN). Of the two, the natural logarithm function is the more useful for modeling and analysis.

The natural domain of the natural logarithm function is the set of all positive real numbers.

The output of the natural exponential function is always positive because Euler's number $e = 2.718 \cdots$ raised to any power—positive, negative, or zero—is a positive number. Hence *the natural domain of the natural logarithm function $y = \ln x$ is the set of all positive real numbers.*

Let's use the relationships in Equations (15) to draw the graph of the natural logarithm function. Table 9.8 lists some evaluations of the natural exponential function $b = e^a$, with all outputs rounded to two decimal places. In this case a is the input and b is the output.

TABLE 9.8 Some Evaluations of $b = e^a$

a	−2	−1.5	−1	−0.5	0	0.5	1	1.5	2	2.5	3
b	0.14	0.22	0.37	0.61	1	1.65	2.72	4.48	7.39	12.18	20.09

If we reverse the two rows of Table 9.8, as shown in Table 9.9, then it follows from Equations (15) that Table 9.9 is a set of functional evaluations for the function $a = \ln b$. The roles of a and b are reversed when dealing with the inverse function. Now b is the input and a is the output.

TABLE 9.9 Some Evaluations of $a = \ln b$

b	0.14	0.22	0.37	0.61	1	1.65	2.72	4.48	7.39	12.18	20.09
a	−2	−1.5	−1	−0.5	0	0.5	1	1.5	2	1.5	3

To generate a graph of the function $a = \ln b$, we plot the points in Table 9.9, putting b-values on the horizontal axis and a-values on the vertical axis; then we connect the points with a curve. The result is the graph in Figure 9.22. If we relabel b and a, respectively, as x and y, then Figure 9.22 is the graph of the function $y = \ln x$.

FIGURE 9.22
The Graph of the Natural Logarithm $y = \ln x$

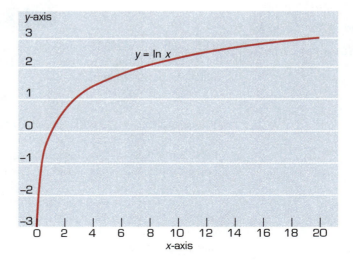

The natural logarithm function and the natural exponential function are said to be *inverses* of one another because each reverses the effect of the other. If we start with the input 0.5, then $e^{0.5} \approx 1.648721$ and $\ln(1.648721) \approx 0.5$. Whatever the natural exponential function does to the input 0.5, the natural logarithm function reverse that result and restores it to the original value. Similarly, $\ln(0.5) \approx -0.693147$ and $e^{-0.693147} \approx 0.5$. Whatever the natural logarithm function does to the input 0.5, the natural exponential function reverses that result and restores it to the original value. Therefore, for any positive real number x,

$$e^{\ln(x)} = x \tag{16}$$

and for *any* real number x,

$$\ln(e^x) = x \tag{17}$$

It follows from Equation (16) that

$$e^{\ln(6)} = 6$$

$$e^{\ln(14.32)} = 14.32$$

$$e^{\ln(c+d)} = c + d \qquad (c + d > 0)$$

and

$$e^{\ln(p^2)} = p^2$$

However, $e^{\ln(-3)}$ is *not* defined because the natural logarithm function is not defined for negative inputs.

It follows from Equation (17) that

$$\ln(e^6) = 6$$

$$\ln(e^{-2.33}) = -2.33$$

$$\ln(e^{c+d}) = c + d$$

and

$$\ln(e^{2p^3}) = 2p^3$$

EXAMPLE 1

Simplify the following expressions:

(**a**) $\ln(e^{-2}e^5)$ (**b**) $\ln\left(\dfrac{e^{2p}}{e^3}\right)$ (**c**) $\ln(e^{3p})^2$

Solution

(**a**) $\ln(e^{-2}e^5) = \ln(e^{-2+5}) = \ln(e^3) = 3$

(**b**) $\ln\left(\dfrac{e^{2p}}{e^3}\right) = \ln(e^{2p-3}) = 2p - 3$

(**c**) $\ln(e^{3p})^2 = \ln(e^{6p}) = 6p$

Because the natural exponential function and the natural logarithm function are so closely related, it's not surprising that many properties of the natural logarithm function follow directly from those of exponentials. In particular, we know that for any two real numbers a and b,

$$e^a e^b = e^{a+b}$$

If we set $u = e^a$ and $v = e^b$, then

$$a = \ln(u) \qquad \text{and} \qquad b = \ln(v)$$

and

$$\ln(uv) = \ln(e^a e^b) = \ln(e^{a+b}) = a + b = \ln(u) + \ln(v)$$

Also,

$$\ln\left(\frac{u}{v}\right) = \ln\left(\frac{e^a}{e^b}\right) = \ln(e^{a-b}) = a - b = \ln(u) - \ln(v)$$

Furthermore, if p is any real number, then

$$\ln(u^p) = \ln(e^a)^p = \ln(e^{ap}) = ap = pa = p\ln(u)$$

We now summarize the basic properties of the natural logarithm function.

Properties of the Natural Logarithm

1. $\ln(1) = 0$ **2.** $\ln(e) = 1$

3. $\ln(uv) = \ln(u) + \ln(v)$ **4.** $\ln\left(\dfrac{u}{v}\right) = \ln(u) - \ln(v)$

5. $\ln(u^p) = p\ln(u)$

Property 5 is particularly useful for analyzing models with exponentials, especially when we want to solve for an exponent.

EXAMPLE 2

Find t if $2.3 = 5(1.06)^{-4t}$.

Solution
We are given that

$$2.3 = 5(1.06)^{-4t}$$

so

$$(1.06)^{-4t} = \frac{2.3}{5} = 0.46$$

Taking the natural logarithm of both sides of this last equation, we have

$$\ln[(1.06)^{-4t}] = \ln(0.46)$$

Now using Property 5, with $p = -4t$ and $u = 1.06$, we simplify:

$$-4t\ln(1.06) = \ln(0.46)$$

$$t = \frac{\ln(0.46)}{-4\ln(1.06)} \approx 3.3317$$

EXAMPLE 3

A bank pays interest at the rate of 6% a year, compounded quarterly. How long will it take for a lump-sum deposit to double in value?

Solution

From Sections 4.1 and 4.2, we know that if PV dollars are deposited in an account at 6% a year, compounded quarterly, then after n periods (quarters), the future value of the account will be

$$FV = PV\left(1 + \frac{0.06}{4}\right)^n = PV(1.015)^n$$

The account doubles in value when $FV = 2PV$, or when

$$2PV = PV(1.015)^n$$

Dividing both sides of this equation by the common factor PV, we obtain

$$2 = (1.015)^n$$

Now we want to solve for n:

$$\ln(2) = \ln[(1.015)^n]$$
$$= n\ln(1.015)$$

$$n = \frac{\ln(2)}{\ln(1.015)} \approx 46.56$$

Since interest is paid quarterly, the deposit must remain at the bank for 47 quarters, or $11\frac{3}{4}$ years, to double in value.

Natural logarithms are useful for analyzing substances that decay or grow at rates directly proportional to existing amounts. For example, some radioactive materials decay at rates that are directly proportional to the amount present: the more radioactive material there is, the faster it decays; the less there is, the slower it decays. In contrast, some populations *grow* at rates that are directly proportional to their size: the larger the population, the more births; the smaller the population, the less births.

We let $N(t)$ denote the amount of a substance under analysis at time t. Then the quantity present at time $t = 0$ is $N(0)$, which we write simply as N_0. If the decay or growth in such a substance is directly proportional to the quantity present, then a good model for $N(t)$ is

$$N(t) = N_0 e^{\pm kt} \tag{18}$$

where k is a positive constant of proportionality called the *continuous growth rate*. Expressed as a percentage, k is the rate of growth over one time period. We use the plus sign to model a growth process and the minus sign to model decay. The quantity $N(t)$ increases exponentially over time when k is positive and it decreases exponentially over time when k is negative. Properties of the natural logarithm function are used to solve Equation (18) for either k or t.

EXAMPLE 4

A bacterial culture is known to grow at a rate directly proportional to the amount present. Initially there are 200 strands of this bacteria in a culture and, after one day, that number increases to 500 strands. Find (*a*) an expression for the approximate number of strands of bacteria present in the culture at any time t and (**b**) the approximate number of strands in the culture after 3 days.

Solution

(**a**) We seek $N(t)$, the amount of bacteria in the culture at any time t (measured in days), knowing that $N_0 = 200$ and $N(1) = 500$. This population of bacteria grows at a rate directly proportional to the amount present in the culture, so

$$N(t) = N_0 e^{kt} \tag{19}$$

is a good model of the population size. Here we use the plus sign in Equation (18) because the substance under study is increasing over time. We know that $N(0) = N_0 = 200$, but we still need to determine k, the constant proportionality. Setting $t = 1$ in Equation (19), we have

$$N(1) = N_0 e^{k(1)}$$

$$500 = 200 e^k$$

$$e^k = \frac{500}{200} = 2.5$$

$$\ln(e^k) = \ln(2.5)$$

But $\ln(e^k) = k$, so $k = \ln(2.5) \approx 0.9163$, and our model is

$$N(t) \approx 200 e^{0.9163t}$$

The population grows at the continuous rate of 91.63% a day.

(**b**) If we substitute $t = 3$ into this last equation, our estimate for the number of strands of bacteria in the culture after 3 days is

$$N(3) \approx 200 e^{0.9163(3)} = 200 e^{2.7489} \approx 200(15.625) \approx 3125$$

The half-life of a substance is the time required for the substance to decrease to one-half of its original amount.

If a quantity decreases in value over time, then its *half-life* is the time required for the quantity to decrease to one-half of its original amount. Half-life is an important concept used to determine, for example, the speed of decomposition of pesticides and carbon 14.

EXAMPLE 5

Years ago, the insecticide DDT was sprayed on crops. Some of the spray eventually seeped into the soil, where it remains today. Over time, all DDT decomposes into harmless substances, but the degradation process is slow, with a decay rate that is directly proportional to the amount of DDT present. Develop a decomposition model for DDT in soil if laboratory tests show that the half-life of DDT in soil is approximately 10.5 years.

Solution

Let $N(t)$ denote the amount of DDT active at time t. Since the rate of change of $N(t)$ is directly proportional to the amount present, a general decay model is

$$N(t) = N_0 e^{-kt} \tag{20}$$

Here we use the minus sign in Equation (18) because the DDT is *decreasing* over time. A half-life of 10.5 years implies that $N(10.5) = \frac{1}{2}N_0$, so substituting $t = 10.5$ into Equation (20), we have

$$N(10.5) = N_0 e^{-10.5k}$$

$$\tfrac{1}{2}N_0 = N_0 e^{-10.5k}$$

$$\tfrac{1}{2} = e^{-10.5k}$$

$$e^{-10.5k} = 0.5$$

Now we take logarithms of both sides of this last equation and simplify:

$$\ln(e^{-10.5k}) = \ln(0.5)$$

$$-10.5k = \ln(0.5)$$

$$k = \frac{\ln(0.5)}{-10.5} \approx 0.06601$$

Substituting this value of k into Equation (20), we obtain

$$N(t) = N_0 e^{-0.06601t}$$

as a model for the amount of DDT present in the soil at any time t after an initial amount N_0 is introduced. DDT decomposes at a continuous rate of approximately 6.6% a year.

EXAMPLE 6

Carbon 14, a radioactive form of carbon present in all living organisms (including humans), can be used for carbon-dating fossils. The amount of carbon 14 present in any live mammal, for example, remains constant throughout the life of that organism. Once a mammal dies, however, the amount of carbon 14 decreases with a half-life of 5,730 years. The rate of decay is directly proportional to the amount of carbon 14 still present in a fossil. Estimate the age of bones that retain 80% of the carbon 14 present in live mammals.

Solution

Let $N(t)$ denote the amount of carbon 14 in the fossil at time t (years). Since the rate of change of $N(t)$ is directly proportional to the amount present, a general decay model is

$$N(t) = N_0 e^{-kt} \qquad (21)$$

A half-life of 5,730 years implies that $N(5730) = \frac{1}{2}N_0$, so, substituting $t = 5730$ into Equation (21), we have

$$N(5730) = N_0 e^{-5730k}$$

$$\tfrac{1}{2}N_0 = N_0 e^{-5730k}$$

$$\tfrac{1}{2} = e^{-5730k}$$

$$e^{-5730k} = 0.5$$

Now, taking logarithms of both sides of this equation and simplifying, we find that

$$\ln(e^{-5730k}) = \ln(0.5)$$

$$-5730k = \ln(0.5)$$

$$k = \frac{\ln(0.5)}{-5730} \approx 0.0001210$$

Thus, carbon 14 decays at the continuous rate of 0.0121% a year. Substituting this value of k into Equation (21), we obtain

$$N(t) = N_0 e^{-0.0001210t}$$

as a model for the amount of carbon 14 in the bones t years after the mammal died.

If the bones retain 80% or 0.8 of the carbon 14 they had up until the time t when the animal died, $(t = 0)$, then $N(t) = 0.8N_0$. Our decay model for carbon 14 now becomes

$$0.8N_0 = N_0 e^{-0.0001210t}$$

$$0.8 = e^{-0.0001210t}$$

Taking logarithms of both sides of this equation and solving for t, we find that

$$\ln(0.8) = \ln(e^{-0.0001210t}) = -0.0001210t$$

Hence

$$t = \frac{\ln(0.8)}{-0.0001210} \approx 1844$$

The bones are from a mammal that died approximately 1,844 years ago.

IMPROVING SKILLS

1. Use a calculator to evaluate the following logarithms:
   - (a) $\ln(2.5)$
   - (b) $\ln(0.32)$
   - (c) $\ln(100)$
   - (d) $\ln(45)$
   - (e) $\ln(-2)$
   - (f) $\ln(3,235)$

2. Use a calculator to evaluate the following quantities:
   - (a) $\ln(e^{2.3})$
   - (b) $\ln(e^{0.4})$
   - (c) $\ln(e^{-4.1})$
   - (d) $e^{\ln(3.82)}$
   - (e) $e^{\ln(0.04)}$
   - (f) $e^{\ln(5.25)}$

3. Evaluate the following expressions *without* using a calculator:
   - (a) $\ln(e^{6.15})$
   - (b) $\ln(e^{-2.04})$
   - (c) $\ln(e^{-18})$
   - (d) $e^{\ln(8.28)}$
   - (e) $e^{\ln(0.14)}$
   - (f) $e^{\ln(5.25)}$

In Problems 4 through 19, use the properties of logarithms to prove each equality.

4. $\ln(2) = -\ln(\frac{1}{2})$
5. $\ln(0.25) = -\ln(4)$
6. $\ln(16) = 4\ln(2)$
7. $\ln(16) = -2\ln(\frac{1}{4})$
8. $\ln(4) = \frac{1}{2}\ln(16)$
9. $\ln(64) = 1.5\ln(16)$
10. $\ln(64) = -6\ln(0.5)$
11. $\ln(6) = \ln(2) + \ln(3)$
12. $\ln(10) = \ln(5) + \ln(2)$
13. $\ln(18) = 2\ln(3) + \ln(2)$
14. $\ln(30) = \ln(60) - \ln(2)$
15. $\ln(9.5) = \ln(19) - \ln(2)$

16. $\ln(3) = \ln(12) - 2\ln(2)$

17. $\ln(72) = 3\ln(2) + 2\ln(3)$

18. $\ln(20) = \ln(80) - 2\ln(2)$

19. $\ln(9) = \ln(81) - 2\ln(3)$

In Problems 20 through 29, solve each equation for x.

20. $e^{2x} = 3$

21. $e^{3x} = 2$

22. $e^{-2x} = 3$

23. $2e^{-3x} = 1$

24. $e^{2x+1} = 8$

25. $e^{2x-3} = 0.2$

26. $e^{2x+1} = 0$

27. $4e^{3x-1} = 16$

28. $e^{2x+1} = 2^x$

29. $e^{3x-1} = 3^{2x}$

CREATING MODELS

30. A person places \$5,000 in a bank account paying 6% a year, compounded monthly. How long will it take this deposit to grow to \$7,000?

31. A person places \$4,000 in a mutual fund that has been growing at the rate of 9% a year. How long will it take for this investment to grow to \$15,000 if the fund's past performance continues?

32. A bank pays interest at the rate of 7%/year, compounded quarterly. How long will it take for a lump-sum deposit to double in value?

33. A bank pays interest at the rate of 7%/year, compounded quarterly. How long will it take for a lump-sum deposit to triple in value?

34. A bank pays interest at the rate of 6%/year, compounded daily, using 360 days to a year. How long will it take for a lump-sum deposit to double in value?

35. A bank compounds interest quarterly. What nominal annual interest rate must a depositor receive from the bank to have an initial deposit double in value in five years?

36. A couple applies to a bank for a \$100,000 mortgage to buy a house. This bank charges 9% a year, compounded monthly, on all new mortgages. How long will it take the couple to pay off the mortgage if they make payments of \$1,100/month?

37. How long will it take to pay off the mortgage if the couple pays \$1,200/month?

38. A bacterial culture is known to grow at a rate directly proportional to the amount present. Initially, there are 30 strands of this bacteria in the culture and, after one day, there are 45 strands.

(*a*) Find an expression for the approximate number of strands of bacteria present in the culture at any time t.

(*b*) Determine the approximate number of strands in the culture after five days.

(*c*) Calculate the continuous daily growth rate.

39. A bacterial culture is known to grow at a rate directly proportional to the amount present. Initially, there are 100 strands of this bacteria in the culture and, after eight hours, there are 125 strands. Find (*a*) an expression for the approximate number of strands present at any time t, and (*b*) the approximate number of strands in the culture after 50 hours.

40. A bacterial culture is known to grow at a rate directly proportional to the amount present. Initially, there are 45 strands of this bacteria in the culture and, after six hours, that number is 60 strands. Find (*a*) an expression for the approximate number of strands present at any time t, and (*b*) the approximate number of strands after three days.

41. A bacterial culture grows at a rate directly proportional to the amount present and doubles after two days. How many strands of the bacteria were cultured initially if there are 2,000 strands after ten days?

42. The population of Africa was 280 million people in 1960 and 363 million people in 1970.

 (**a**) Estimate the population in 1980 and again in 1985 under the assumption that population growth is directly proportional to the size of the population.

 (**b**) What is the continuous annual growth rate of this population?

43. Statistical data suggest that monthly wages for Brazilian workers increased during the early 1980s at a rate directly proportional to the current wage. In 1984, the average monthly wage was 424 cruzeiro; by 1986, the average monthly wage grew to 3,507 cruzeiro. Construct a model for the average monthly wage as a function of time; then use the model to estimate the average monthly wage in 1985, in 1987, and again in 1988. How much confidence do you have in this model once you find out that the average monthly wage in Brazil during 1987 was 10,949 cruzeiro?

44. An economist believes that the price of a one-pound loaf of white bread increases each year at a rate directly proportional to its current price. In 1955, a one-pound loaf of white bread cost, on average, 19.3¢ in Los Angeles. By 1960, that price had risen to 23.4¢.

 (**a**) If this economist is correct, what will a one-pound loaf of white bread cost in Los Angeles in the year 2000?

 (**b**) How long did it take for the 1955 price of bread to double?

 (**c**) What is the continuous annual growth rate for a loaf of bread?

45. The economist from the previous problem also believes that the price of a quart of milk increases in direct proportion to its current price, at a continuous rate of 3.5% a year. In 1960, a quart of milk cost, on average, 28.10¢ in Chicago.

 (**a**) If this economist is correct, what will a quart of milk cost in Chicago in the year 2000?

 (**b**) How long did it take for the 1960 price of milk to double?

46. United Nations data for Egypt show that in 1975, the number of students enrolled in school preceding the first level was 41,948. By 1980, this number increased to 74,921. Model the data by assuming that the growth rate is directly proportional to total enrollment.

 (**a**) How many students were enrolled in 1977?

 (**b**) How many students were enrolled in 1985?

 (**c**) What is the continuous annual growth rate?

 (**d**) What do you think of this model if U.N. statistics show an actual enrollment of 128,272 in 1985?

47. Use the results of Example 5 to estimate the percentage of DDT still active in the soil 100 years after initially being absorbed.

48. A skull is found at an archeological dig and laboratory tests reveal that this skull retains 70% of the carbon 14 found in living organisms. Estimate the age of this skull. *Hint*: See Example 6.

49. What percent of the amount of carbon 14 in a living organism will be in a 5,000-year-old fossil? *Hint*: See Example 6.

50. Cobalt 60 has a half-life of 5.27 years and decays at a rate directly proportional to the amount present. A laboratory receives a shipment of 100 grams for a project, but the project is canceled right after the shipment arrives. If the laboratory stores the cobalt 60, how much will remain after two years?

51. Radium 226 is known to decay at a rate proportional to the amount present. Initially, there are 20 grams present in a sample and, after one year, the amount of radium in this same sample measures 19.9913 grams. What is the half-life of radium 226?

52. The decay of cobalt 57 is directly proportional to the amount present with a continuous radiation rate of 93.34%/year.
    (a) What is the half-life of cobalt 57?
    (b) How long will it take for 50 grams of this substance to decay to 10 grams?
53. One of the main contaminants from the nuclear accident at Chernobyl is strontium 90, which has a half-life of 29 years. Scientists believe it will take 100 years before the level of strontium 90 in Chernobyl will be low enough to be safe for humans. What fraction of the original strontium 90 contamination at Chernobyl must dissipate before the area is considered safe?

EXPLORING IN TEAMS

54. Use the identity $a(b^x) = e^{\ln[a(b^x)]}$ along with the properties of the natural logarithm function to show that $a(b^x) = ce^{mx}$, where c is the constant $e^{\ln(a)}$ and m is the constant $\ln(b)$.
55. Use the results of the previous problem to show that each of the following exponential functions can be written in terms of the natural exponential function.
    (a) 2^x (b) 3^x (c) $2(3^x)$
    (d) $3(7^x)$ (e) 7^{2x} (f) 3^{x+2}

EXPLORING WITH TECHNOLOGY

56. Use a graphing calculator to graph the function $y = \ln(x)/x^n$ for various values of n. In particular, include $n = 0.1, 0.3, 0.5, 2, 3,$ and 4. Determine the effect on each graph as x becomes large.
57. Use the results of the previous problem to develop a conjecture about the growth rate of $\ln(x)$ and that of x^n, for any positive integer of n. Which of these functions grows more quickly as x becomes large?

REVIEWING MATERIAL

58. (Section 2.3) A study by the Consortium on Financing Higher Education reported that the average annual tuition at 31 of the nation's most prestigious private universities was $5,126 in 1980, $9,389 in 1985, $14,009 in 1990, and $19,110 in 1995. Find the equation of the least-squares straight line that fits this data and then use the result to estimate average tuition charges in the year 2000.
59. (Section 5.2) In 1993, the U.S. imported 2.384 billion pounds of unroasted coffee. In units of a million pounds, 446.6 came from Brazil, 391.2 from Colombia, 389.8 from Mexico, 240 from Guatemala, 168.6 from El Salvador, and 141.4 from Thailand. No other source accounted for more than 100 million pounds. Construct a pie chart from this information.

60. (Section 5.4) A traveler packs three pairs of slacks (one each of gray, tan, and navy blue), two blazers (one gray, one navy blue), and three shirts (each a different color). Construct a tree diagram showing all possible ensembles consisting of a pair of slacks, a blazer, and a shirt if the color of the blazer must be different from that of the slacks.

61. (Section 6.4) Two hundred people with the same disease are screened with a new test, which comes back positive for 170 of these patients. When the test is given to another 150 people who do *not* have the disease, it comes back positive for 25 of them. Based on this data, what is the probability that a person who takes the test actually has the disease when the test shows positive?

RECOMMENDING ACTION

62. Respond by memo to the following request:

> MEMORANDUM
>
> To: J. Doe Reader
>
> From: Director of Children's Services
>
> Date: Today
>
> Subject: **Schoolchildren in Egypt**
>
> I have in front of me a model for the number of Egyptian students enrolled in school preceding the first level. (See Problem 46.) This model was created under the assumption that the growth rate is directly proportional to total enrollment. Does this assumption make sense? Why should the rate at which new students enroll have anything to do with the number of students already in the school? If this model is reasonable, can I assume that it will continue to be useful for planning purposes in the foreseeable future?

9.5 TRIGONOMETRIC FUNCTIONS

Some processes, especially those with time as the independent variable, have output patterns that are cyclical, that is, repeat with regularity. For example, the Earth retraces its orbit around the sun every year, while our moon retraces its orbit around the earth every 27.32 days.

Table 9.10 lists the average daily high temperatures for each month in Billings, Montana, based on historical data as reported by the Automobile Association of America. Figure 9.23 is a graph of this data.

We can alter the appearance of this graph somewhat by changing either the scales on the axes or the width between the tick marks. If we narrow to one-fifth the spacing between tick marks on the horizontal axis, we generate Figure 9.24.

Figure 9.25 consists of five juxtaposed copies of Figure 9.24. Because the graph repeats every 12 months, it is a useful graphical model for forecasting the average daily high temperatures over the next five years.

TABLE 9.10 **Average Daily High Temperatures for Billings, Montana**

MONTH	TEMPERATURE (°F)	MONTH	TEMPERATURE (°F)
January	33	July	89
February	36	August	86
March	43	September	75
April	57	October	63
May	68	November	46
June	76	December	39

FIGURE 9.23

Average Daily High Temperatures for Billings, Montana, for One Year

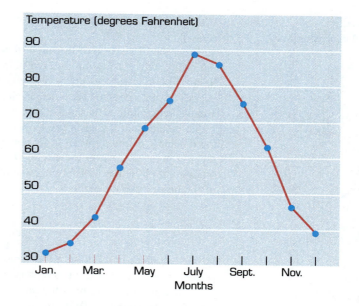

FIGURE 9.24

Average Daily High Temperatures for Billings, Montana, for One Year

FIGURE 9.25

Average Daily High Temperatures for Billings, Montana, for Five Years

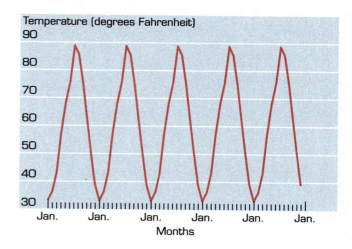

The unit circle in a Cartesian coordinate system is a circle centered at the origin with a radius of one unit.

FIGURE 9.26
An Angle θ with the Unit Circle

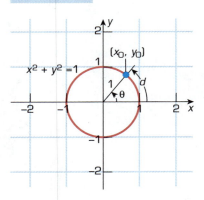

A radian is a dimensionless number measuring the angle swept out by an arc of length 1, starting at the point (1, 0) on the unit circle and moving in a counterclockwise direction.

To model real-world cyclical processes, we need functions that repeat themselves with regularity. A useful class of functions with this property are the *trigonometric functions,* which originated from the study of angles and triangles. There is nothing inherently cyclical about either angles or triangles, though, so we'll take another approach. We will develop trigonometric functions by considering points on the circumference of the *unit circle,* which is centered at the origin of the Cartesian plane and has a radius of one unit. The unit may be inches, feet, miles, millimeters, or any other unit of length that is appropriate for a particular model.

The equation for the unit circle in an *xy*-coordinate system is

$$x^2 + y^2 = 1$$

which is pictured in Figure 9.26. Note that as we move around the circle, we retrace the same path every 360°.

We want a measure of the angle θ between the positive *x*-axis and a ray beginning at the origin and passing through the point (x_0, y_0) on the unit circle, as pictured in Figure 9.26. Although degrees are a common measure for angles, we do not multiply degrees because square-degrees have no physical significance. For modeling purposes, we prefer a measure for angles that makes sense under multiplication.

To create such a measure, we begin at the point (1, 0) on the unit circle and measure the distance we must travel in the counterclockwise direction along the circumference to reach a point (x_0, y_0) on the circle. Denote this arc length by d, for distance. Each point on the unit circle has a different value of d associated with it. If we divide d by the radius of the unit circle, we obtain

$$\frac{d \text{ units of length}}{1 \text{ unit of length}} = d$$

a dimensionless number with the same value as the arc length. This measure is called a *radian*; it is a number used to represent an angle.

A measure of 1 radian is the angle corresponding to an arc length one unit long, beginning at the point (1, 0) on the unit circle and moving in a counterclockwise direction. Similarly, a measure of 2 radians is the angle corresponding to an arc length two units long, beginning at the point (1, 0) on the unit circle and moving in a counterclockwise direction. *A negative angle measures length in the clockwise direction.*

We know from geometry that the circumference C of the unit circle with radius $r = 1$ is $C = 2\pi(1) = 2\pi$ units. Therefore if we begin at (1, 0) and make one full revolution around the unit circle, we trace out an angle of 2π radians. Also, one full rotation around a circle corresponds to 360°, so we have

$$2\pi \text{ radians} = 360°$$

It then follows that

$$\pi \text{ radians} = 180°, \qquad \frac{\pi}{2} \text{ radians} = 90°$$

and

$$1 \text{ radian} = \frac{360°}{2\pi} \approx 57.3°$$

These conversions are included in Table 9.11 with other commonly used angles.

TABLE 9.11 **Degree/Radian Conversions of Commonly Used Angles**

DEGREES	RADIANS	DEGREES	RADIANS
0°	0	360°	2π
30°	$\pi/6$	−30°	$-\pi/6$
45°	$\pi/4$	−45°	$-\pi/4$
60°	$\pi/3$	−90°	$-\pi/2$
90°	$\pi/2$	−180°	$-\pi$
120°	$2\pi/3$	−360°	-2π
135°	$3\pi/4$	1°	$\pi/180$
150°	$5\pi/6$	$180°/\pi \approx 57.3°$	1
180°	π	$360°/\pi \approx 114.6°$	2
270°	$3\pi/2$	$540°/\pi \approx 171.9°$	3

EXAMPLE 1

Graph the angles having radian measures of −2 and 1/2, respectively.

Solution

An angle of radian measure −2 sweeps out an arc length of 2 units on the unit circle, beginning at (1, 0) and moving in a clockwise (*negative*) direction. An angle of radian measure 1/2 sweeps out an arc length of 1/2 unit on the unit circle, beginning at (1, 0) and moving in a counterclockwise (*positive*) direction. Both are illustrated in Figure 9.27.

FIGURE 9.27
Angles of −2 and 1/2
Radians

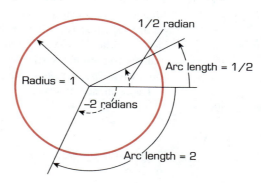

EXAMPLE 2

Determine the radian measure for an angle of 37°.

Solution

We know that π radians = 180°, so

$$1° = \frac{\pi}{180} \text{ radians}$$

Then

$$37° = 37\left(\frac{\pi}{180}\right) \text{ radians} \approx 0.6458 \text{ radians}$$

EXAMPLE 3

Determine the degree measure for an angle of 37 radians.

Solution

We know that π radians $= 180°$, so

$$1 \text{ radian} = \left(\frac{180}{\pi}\right)°$$

so that

$$37 \text{ radians} = 37\left(\frac{180}{\pi}\right)° \approx 2{,}119.9°$$

> *If (x, y) is the terminal point of an arc that begins at (1, 0) on the unit circle and sweeps out an arc length of |θ| units, then cos θ = x and sin θ = y.*

FIGURE 9.28
An Angle of θ Radians

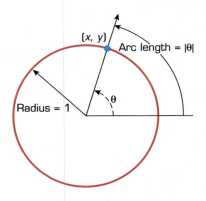

> *Set the mode of a calculator to radian measure before using the sine and cosine keys.*

An angle of θ radians sweeps out an arc length of $|\theta|$ units on the unit circle, beginning at (1, 0) and moving in a counterclockwise direction when θ is positive and in the clockwise direction when θ is negative. In Figure 9.28, we have denoted the terminal point of this arc length by (x, y). This point is uniquely determined by the angle θ. We define the *cosine* of the angle θ, denoted by "cos θ," to be x and the *sine* of the angle θ, denoted by "sin θ," to be y. That is,

$$\cos \theta = x \quad \text{and} \quad \sin \theta = y \tag{22}$$

Both the cosine and the sine are functions, having the set of all real numbers (radians) as their natural domains. To find the output of either function at θ, we draw a unit circle and sweep out an arc length of $|\theta|$ units, beginning at the point (1, 0). When θ is positive we go in a counterclockwise direction and when θ is negative we go in a clockwise direction. We locate the terminal point of this arc length and then, noting the quadrant, we use Equation (22). Easier, by far, is to use a calculator with "sin" and "cos" keys. Most calculators have different modes available for measuring angles, allowing them to evaluate trigonometric functions with inputs measured in either degrees or radians. In this book, *we always set the mode of the calculator to radians prior to any evaluations.*

The sequence of keystrokes on most graphing calculators for evaluating sin 2 is

The result is 0.909297, rounded to six decimal places. The keystrokes needed to evaluate cos 2 are

The result is -0.416147, rounded to six decimal places. Similarly, to evaluate cos 37 (with the mode of the calculator set to radians!), the keystroke sequence is usually

We obtain cos 37 ≈ 0.765414.

The domains of the sine and cosine functions, the set of real numbers, are in fact radians, measurements of angles. When we evaluate cos 37, we are evaluating the cosine of 37 radian (see Example 3). Similarly, when we evaluate sin 2, we are evaluating the sine of 2 radians.

Some useful evaluations are listed in Table 9.12. All entries with three decimal places have been rounded.

TABLE 9.12 **Selected Evaluations of the Sine and Cosine Functions**

θ (radians)	0	$\pi/6$	$\pi/4$	$\pi/3$	$\pi/2$	$2\pi/3$	$3\pi/4$	$5\pi/6$	π	$5\pi/4$	$3\pi/2$	$7\pi/4$	2π
$\cos\theta$	1	0.866	0.707	0.5	0	−0.5	−0.707	−0.866	1	−0.707	0	0.707	1
$\sin\theta$	0	0.5	0.707	0.866	1	0.866	0.707	0.5	0	−0.707	−1	−0.707	0

FIGURE 9.29
Angles $\pi/6 \pm 2n\pi$ Radians

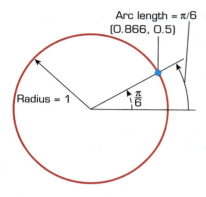

From Table 9.12 we see that $\cos(\pi/6) \approx 0.866$ and $\sin(\pi/6) = 0.5$. If we begin at the point $(1, 0)$ and sweep out an arc of length $\pi/6$ in the counterclockwise direction on the unit circle, the terminal point is the point $(0.866, 0.5)$, with coordinates rounded to three decimal places, as shown in Figure 9.29. Now suppose we continue in a counterclockwise direction and sweep out an additional arc of length 2π radians $(= 360°)$. We have generated an angle of

$$\frac{\pi}{6} + 2\pi = \frac{13\pi}{6}$$

radians, so the terminal point will again be $(0.866, 0.5)$. Indeed, if we add or subtract any integral multiple of 2π, which is equivalent to making one or more complete revolutions around the unit circle, we generate different angles but each has the same sine and cosine. This is true at any point on the unit circle. In other words, for any integer n and angle θ,

$$\sin(\theta + 2n\pi) = \sin\theta \quad \text{and} \quad \cos(\theta + 2n\pi) = \cos\theta \qquad (23)$$

sin θ and cos θ are periodic functions with period 2π.

sin θ and cos θ both have amplitude 1.

The sine function and the cosine function repeat every 2π radians; we say they are "periodic."

A function $f(x)$ is *periodic* if there exists a nonzero number p such that

$$f(x + p) = f(x) \qquad (24)$$

for every input value x in the domain of $f(x)$. The smallest positive value of p that makes Equation (24) true is called the *period* of the function. Comparing Equations (23) and (24), we conclude that both the sine and cosine functions are periodic. They repeat every $2n\pi$ radians, for every integer n. Of these, the smallest positive value is 2π, so the period of both the sine and cosine functions is 2π. If we graph the sine function using values from Table 9.12 and the fact that it has period 2π, we obtain Figure 9.30. The graph of the cosine function is shown in Figure 9.31. Note the cyclical behavior of both graphs.

FIGURE 9.30
The Function sin θ Has Period 2π and Amplitude 1

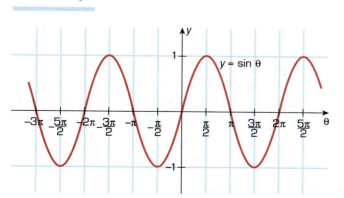

FIGURE 9.31
The Function cos θ Has Period 2π and Amplitude 1

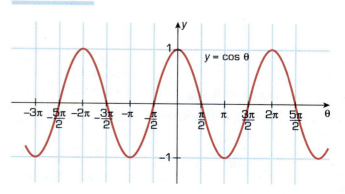

The *amplitude* of a periodic function is half the distance between its maximum output and its minimum output. For both the sine and cosine functions, the maximum output is 1 and the minimum output is -1. Therefore the amplitude of both the sine and cosine functions is $\frac{1}{2}[1 - (-1)] = \frac{1}{2}(2) = 1$.

If we want to use the sine and cosine functions to model cyclical behavior, we need to generate graphs of these functions with various amplitudes. The simplest way is to multiply either the sine function or the cosine function by a real, nonzero number A. The results,

$$A \sin \theta \quad \text{and} \quad A \cos \theta$$

are periodic functions with period 2π and amplitude $|A|$, where $|A| = A$ if $A > 0$ and $|A| = -A$ if $A < 0$.

> *$A \sin \theta$ and $A \cos \theta$ have period 2π and amplitude $|A|$.*

EXAMPLE 4

Graph the functions $2 \sin \theta$ and $5 \sin \theta$.

Solution

These functions are graphed in Figure 9.32 along with the graph of $\sin \theta$, for reference. The function $2 \sin \theta$ has amplitude 2 while the function $5 \sin \theta$ has amplitude 5.

FIGURE 9.32
Sine Functions with Various Amplitudes

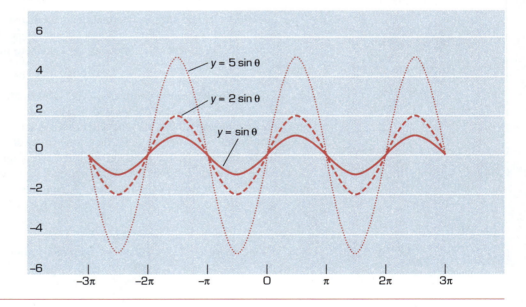

Cyclical processes are those that repeat, but not necessarily with a period of 2π. Some processes may have a period of 1 year, while others may have a period of 5 hours. If we want to model cyclical processes with sines and cosines, we need to modify these functions to allow for periods other than 2π. The simplest way is to multiply the input θ by a positive number b. The results,

$$A \sin(b\theta) \quad \text{and} \quad A \cos(b\theta)$$

are periodic functions with period $2\pi/b$ and amplitude $|A|$.

> *The functions $A \sin(b\theta)$ and $A \cos(b\theta)$ both have period $2\pi/b$ and amplitude $|A|$.*

EXAMPLE 5

Graph the function $y = 4\sin(\theta/2)$.

Solution

This function has the form $y = A\sin(b\theta)$ with $A = 4$ and $b = 1/2$. Its amplitude is 4, so the output will range between -4 and 4. Its period is $2\pi/b = 2\pi/\frac{1}{2} = 4\pi$. The graph of this function is displayed in Figure 9.33.

FIGURE 9.33

A Sine Function with Amplitude 4 and Period 4π

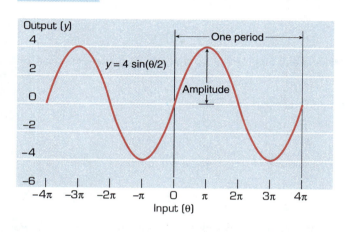

FIGURE 9.34

A Cosine Function with Amplitude 7 and Period 1

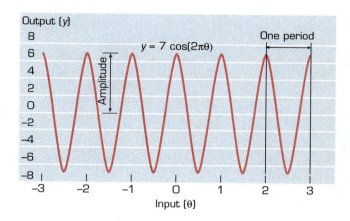

EXAMPLE 6

Graph the function $f(\theta) = 7\cos(2\pi\theta)$.

Solution

This function has the form $y = A\cos(b\theta)$ with $A = 7$ and $b = 2\pi$. Its amplitude is 7, so the output will range between -7 and 7. Its period is $2\pi/b = 2\pi/2\pi = 1$. The graph of this function is displayed in Figure 9.34.

We vary the amplitudes and the periods of the functions $y = A\sin(b\theta)$ and $y = A\cos(b\theta)$ by varying our choices for the constants A and b. In both functions, the amplitude A establishes the spread in the output values. For example, if $A = 7$, as in Figure 9.34, the output is always between -7 and 7; if $A = 4$, as in Figure 9.33, the output is always between -4 and 4. Unfortunately, most real-world processes are not so symmetrical. Cyclical outputs may range between 20 and 50, for example, or between -10 and 100, or between 0 and 5000.

If we add a *positive* constant k to a function $f(x)$, we *increase* each output by k units. The effect is to raise the graph of $f(x)$ by k units in the vertical direction, as demonstrated in Figure 9.35. Similarly, if we add a *negative* constant k to a function $f(x)$, we *reduce* each output by k units. The effect is to lower the graph of $f(x)$ by k units in the vertical direction. Consequently, the functions

$$k + A\sin(b\theta) \quad \text{and} \quad k + A\cos(b\theta)$$

are periodic functions with period $2\pi/b$, amplitude A, and output values between $k - A$ and $k + A$.

The functions $k + A\sin(b\theta)$ and $k + A\cos(b\theta)$ have period $2\pi/b$ and output values between $k - A$ and $k + A$.

FIGURE 9.35
The Graphical Effect of Adding a Constant to a Function

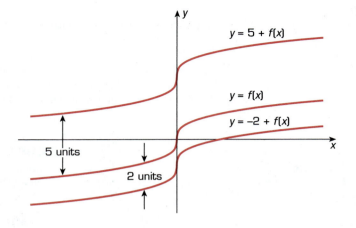

EXAMPLE 7

Create a sine function with period 5 and outputs that assume all values between 2 and 10.

Solution

The sine function $y = k + A\sin(b\theta)$ has period 5 when

$$\frac{2\pi}{b} = 5 \quad \text{or} \quad b = \frac{2\pi}{5}$$

The outputs range between $k - A$ and $k + A$, so we require that

$$k - A = 2$$
$$k + A = 10$$

The solution to this system of equations is $k = 6$ and $A = 4$. Therefore the function we seek is

$$y = 6 + 4\sin(2\pi\theta/5)$$

which is plotted in Figure 9.36.

FIGURE 9.36
The Graph of
$$y = 6 + 4\sin\left(\frac{2\pi\theta}{5}\right)$$

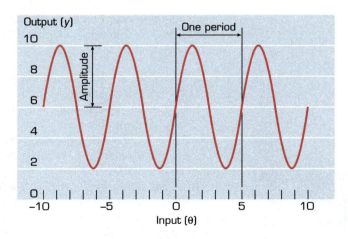

A process is sinusoidal if its graph duplicates a sine or cosine curve.

An input–output process that can be modeled by either sines or cosines is called *sinusoidal*. Such processes are cyclical and have graphs that duplicate sine or cosine curves. Although we developed sines and cosines from angles associated with the unit circle, we're not limited just to that scenario. Now that the sine and cosine functions are available to us, we can use them to model any sinusoidal process, independent of angles. In particular, if the graph of an input–output relationship duplicates the graph of a particular sine or cosine function, then we can use that function as a model for the corresponding input–output relationship.

Throughout this section we have been using θ to designate the input variable, but we can use any label we want to designate an input variable. Since many sinusoidal processes are cyclical with respect to time, the label t is often the appropriate input designator.

EXAMPLE 8

The difference between high tide and low tide at the Bay of Fundy in Canada is approximately 50 feet under normal conditions. Assume that on a particular day, high tide occurs at midnight, and that the tides are periodic every $12\frac{1}{2}$ hours. Use a cosine function to model the rise in the water level over time.

Solution

We take the water level at low tide as our reference point for height and code it as zero height. Then high tide has a height of 50 feet. On the day previously mentioned, high tide occurred at midnight; we take this as our reference point for time and code it as time $t = 0$. We seek a relationship that gives the height of a tide at any time t, so t is our label for the input variable. Since the maximum height occurred at $t = 0$, the cosine function $y = k + A\cos(bt)$, which is maximum at $t = 0$, is a reasonable function to use as a model.

The tides repeat themselves every 12.5 hours, so that is the period. The function $k + A\cos(bt)$ has period 12.5 when

$$\frac{2\pi}{b} = 12.5 \quad \text{or} \quad b = \frac{2\pi}{12.5} = 0.16\pi$$

This function has outputs that range between $k - A$ and $k + A$, so we require that

$$k - A = 0$$

$$k + A = 50$$

The solution to this system of equations is $k = A = 25$. The function we seek is $y = 25 + 25\cos(0.16\pi t)$, which is plotted in Figure 9.37.

FIGURE 9.37

The Graph of
$y = 25 + 25 \cos(0.16\pi t)$

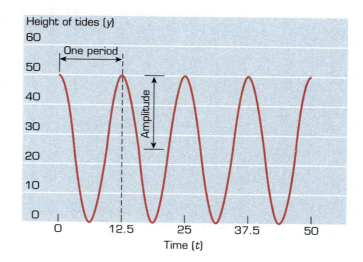

EXAMPLE 9

Figure 9.25 is a graph of the daily high temperature in Billings, Montana, for five years. Model that weather pattern using a sine or cosine function.

Solution

The graph in Figure 9.25 appears to be sinusoidal, with maximum temperatures recorded each July. At input $t = 0$ the output of a sine function is midway between its maximum and its minimum, while at $t = 0$ the output of a cosine function is at its maximum. If we measure time in months, with July as our zero point, then $t = 0$ corresponds to July and $t = \pm 6$ corresponds to January. With this coding, maximum temperatures occur at $t = 0$, so we use the cosine function $y = k + A \cos(bt)$ as our model.

The curve in Figure 9.25 repeats every 12 months, which is its period. The function $k + A \cos(bt)$ has period 12 when

$$\frac{2\pi}{b} = 12 \quad \text{or} \quad b = \frac{2\pi}{12} = \frac{\pi}{b}$$

This function has outputs that range between $k - A$ and $k + A$, and the temperatures in Figure 9.25 range between 33° and 89°. Therefore we want

$$k - A = 33$$
$$k + A = 89$$

The solution to this system of equations is $k = 61$ and $A = 28$, so the function we seek as our model is

$$y = 61 + 28 \cos(\pi t/6)$$

This is plotted in Figure 9.38 as a dotted line, while the solid line is the curve from Figure 9.25.

FIGURE 9.38

Average Daily High Temperatures for Billings, Montana, for Five Years

IMPROVING SKILLS

1. Transform the following angles from degrees to radians.
   (**a**) 22° (**b**) 80° (**c**) 167° (**d**) 410°
   (**e**) −14° (**f**) −138° (**g**) −675° (**h**) $\pi°$

2. Transform the following angles from radians to degrees.
   (**a**) 0.4 (**b**) 1.3 (**c**) $19\pi/12$ (**d**) 6.95
   (**e**) −1.31 (**f**) −5 (**g**) −8.11 (**h**) e

In Problems 3 through 10, sketch the angle and then determine from your sketch (without doing any calculating) whether the sine and cosine of the angle is positive or negative.

3. $\pi/12$ 4. $13\pi/12$ 5. $-13\pi/12$ 6. $-23\pi/12$
7. $25\pi/12$ 8. 3   9. 4   10. −4

In Problems 11 through 22, use a calculator to evaluate the following functions, rounding to four decimal places.

11. sin 12   12. cos 12   13. sin 1   14. cos 53
15. sin 2.8   16. cos 3.5   17. $\sin(\pi/9)$ 18. $\cos(2\pi/7)$
19. $\sin(-5)$ 20. $\cos(-5)$ 21. $\sin(-5\pi/17)$ 22. $\cos(-2\pi/7)$

In Problems 23 through 28, determine values of the constants A and b so that the function $y = A\sin(b\theta)$ has the prescribed period and amplitude.

23. Period $= \pi$ and amplitude $= 2$ 24. Period $= \pi$ and amplitude $= 1/2$
25. Period $= \pi/2$ and amplitude $= 1/2$ 26. Period $= 1$ and amplitude $= 5$
27. Period $= 2$ and amplitude $= 5$ 28. Period $= 7$ and amplitude $= 3$

In Problems 29 through 46, sketch the graphs of the given functions and then check your answers using a graphing calculator.

29. $y = 3\sin\theta$ 30. $y = 3\cos\theta$ 31. $y = 3 + 3\sin\theta$
32. $y = 5 + \cos\theta$ 33. $y = \sin 2\theta$ 34. $y = 5\cos 2\theta$
35. $y = 2 + \sin 2\theta$ 36. $y = -2 + 5\cos 2\theta$ 37. $y = 7\sin(\theta/4)$
38. $y = 2\cos(\theta/2)$ 39. $y = -2 + 7\sin(\theta/4)$ 40. $y = 2 + 2\cos(\theta/2)$
41. $y = \sin\pi\theta$ 42. $y = 5\cos 2\pi\theta$ 43. $y = 100 + \sin\pi\theta$
44. $y = -10 + \cos 2\pi\theta$ 45. $y = 3\sin(\pi\theta/4)$ 46. $y = 2\cos(\pi\theta/3)$

FIGURE A

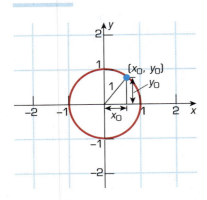

In Problems 47 through 52, determine values of the constants k, A, and b so that the function $y = k + A \cos(b\theta)$ has the prescribed period and range.

47. Period $= \pi$ and range $[0, 2]$
48. Period $= \pi$ and range $[2, 3]$
49. Period $= \pi/2$ and range $[5, 10]$
50. Period $= 1$ and range $[0, 100]$
51. Period $= 2$ and range $[-2, 5]$
52. Period $= 7$ and range $[-5, 5]$
53. Using Figure A and the Pythagorean Theorem, show that

$$\sin^2 \theta + \cos^2 \theta = 1$$

where $\sin^2 \theta$ is shorthand notation for $(\sin \theta)^2$ and $\cos^2 \theta$ is shorthand for $(\cos \theta)^2$.

54. Use the unit circle to derive a formula relating $\sin \theta$ to $\sin(-\theta)$.
55. Use the unit circle to derive a formula relating $\cos \theta$ to $\cos(-\theta)$.

CREATING MODELS

56. The average daily low temperatures in Billings, Montana, range from 13°F in January to 58°F in July. Use a cosine function to construct a model from this data that will forecast the low temperatures in Billings. The independent variable will be time, measured in months.
57. The average daily high temperatures in Bakersfield, California, range from 100°F in July to 57°F in January. Construct a model from this data using a cosine function that will forecast the high temperatures in Bakersfield, using time, in months, as the independent variable.
58. Under normal conditions, the lower portion of the Hackensack River in New Jersey is a tidal estuary, so the height of the river is determined by the tides. At low tide the river is essentially a mud flat, with a 6-inch-deep trickle of water in the middle of the river. At high tide, the depth of the river is 6 feet. Assume that high tide occurs at midnight on one particular day and assume that the tides have a period of $12\frac{1}{2}$ hours. Model the rise in water level over time with a cosine function.
59. Solve Problem 58 if high tide occurs at 9:00 A.M.
60. An animal population on an island varies sinusoidally from a low of 500 every January to a high of 2,000 every July. Model the population of this species as a function of time with either a sine or cosine function.
61. The Earth rotates around the sun once a year, in a nearly circular orbit at a distance of 150 million kilometers from the sun. Place a coordinate system on the solar system that has Earth at an angle of zero each January 1. Construct a graphical model for the rotation of Earth around the sun as a function of time, measured in years.

EXPLORING IN TEAMS

62. Four other trigonometric functions—the tangent (tan), cotangent (ctn), secant (sec), and cosecant (csc)—are related to the sine and cosine functions by these formulas:

$$\tan \theta = \frac{\sin \theta}{\cos \theta} \qquad \text{ctn } \theta = \frac{\cos \theta}{\sin \theta}$$

$$\sec \theta = \frac{1}{\cos \theta} \qquad \csc \theta = \frac{1}{\sin \theta}$$

Determine the domains of each of these functions.

63. Use the graphs of $\sin \theta$ and $\cos \theta$ to graph each of the functions defined in the previous problem and determine the periods of each of these functions.

64. Use the results of Problems 54 and 55 to determine a relationship between $\tan \theta$ and $\tan(-\theta)$.

65. Use the result of Problem 53 to show that, for any value of θ in the domain of both the tangent and secant functions,

$$1 + \tan^2 \theta = \sec^2 \theta$$

EXPLORING WITH TECHNOLOGY

66. Use a graphing calculator to display the graph of the function $y = \sin x$ in a window with $-10 \le x \le 10$ and $-2 \le y \le 2$. Use the zoom-in feature of your calculator to magnify the region around $(0, 0)$. Continue to zoom in on the origin until the shape of the curve does not change. What type of curve do you get? Repeat this procedure by zooming in on other points of the sine curve. What do you conclude?

67. Use a graphing calculator to display the graph of the function $y = e^x$ in a window with $-3 \le x \le 3$ and $0 \le y \le 20$. Choose a point on this graph and use the zoom-in feature of your calculator to magnify the region around that point. Continue to zoom in on this point until the shape of the curve does not change. What type of curve do you get? Repeat this procedure by zooming in on other points on the graph of the natural exponential function. What do you conclude?

68. Repeat Problem 70 with the function $y = x^4 - 5x^2 + 4$ in a window with $-3 \le x \le 3$ and $-5 \le y \le 10$.

69. Repeat Problem 70 with other functions in this chapter. Make a conjecture based on your explorations.

70. The graphs of the functions in Problems 69 and 71 have peaks and pits. Zoom in on a peak and a pit until the shape of the curve does not change. What do you conclude? Does your conclusion agree with the conjecture you made in the previous problem? Do you want to expand on that conjecture?

71. (Section 2.5) Use matrix inversion to solve the following system of equations:

$$2x - 3y + 4z = 40$$
$$3x + 2y - 2z = 24$$
$$y + 5z = 63$$

72. (Section 4.2) A depositor purchases a four-year certificate of deposit (CD) that earns 8% a year, compounded quarterly. At maturity, the CD is worth $80,000. What was the initial deposit?

73. (Section 6.6) A card is picked from a standard deck and then replaced. The deck is shuffled and the process is repeated four more times. What is the probability that at least two of the cards picked from the deck are hearts?

74. Respond by memo to the following request:

MEMORANDUM

To: J. Doe Reader

From: Director of Planning

Date: Today

Subject: **Business Patterns**

Our business is cyclical. Each year, we generate our largest monthly revenues in December (the Christmas season) and our smallest monthly revenues in June. Because of rising prices (due to inflation), our monthly revenues increase from one year to the next. December revenues this year are greater than those from last December and will be less than the revenues next December. This is true for all months, even though the number of products we sell in a given month is constant from year to year. Is our revenue stream sinusoidal? Can we model our business with a sine or cosine function?

CHAPTER 9   KEYS

KEY WORDS

amplitude (p. 662)

codomain (p. 610)

compounding continuously (p. 638)

continuous conversion rate (p. 639)

continuous growth rate (p. 649)

cosine function (p. 660)

degree of a polynomial function (p. 622)

domain (p. 610)

Euler's number e (p. 637)

exponential function (p. 633)

first-degree polynomial function (p. 623)

function (p. 610)

half-life (p. 650)

linear function (p. 624)

natural domain (p. 614)

natural exponential function (p. 637)

natural logarithm function (p. 645)

period (p. 661)

periodic function (p. 661)

polynomial function (p. 622)

power function (p. 627)

quadratic function (p. 624)

radian (p. 658)

range (p. 611)

rational function (p. 626)

sine function (p. 660)

sinusoidal (p. 664)

unit circle (p. 658)

zero function (p. 623)

KEY CONCEPTS

9.1 Functions

- The functions of greatest interest are those in which both the domain and the codomain are subsets of the real numbers.

- Functions are usually specified by words (rules), equations, graphs, or tables.

- When a graph depicts a function, the domain is always placed on the horizontal axis and the codomain on the vertical axis. The rule of correspondence assigns a number y to each input x such that the ordered pair (x, y) is a point on the curve.

- A graph represents a function if and only if the graph passes the vertical line test: any vertical line that crosses the horizontal axis at a value in the domain must intersect the graph at one and only one point.

9.2 Polynomial, Rational, and Power Functions

- The graph of a polynomial function of degree zero is a horizontal line.

- The graph of a polynomial function of degree 1 is a straight line with a nonzero slope.

- The graph of a quadratic function is a parabola.

- Unless stipulations are made to the contrary, the domain of a rational function is the set of all real numbers at which the function's denominator is nonzero.

9.3 Exponential Functions

■ The graph of an exponential function of the form $y = a(b^x)$ approaches the horizontal axis from above when $a > 0$; it increases steeply when $b > 1$ and decreases steeply when $0 < b < 1$. When $a < 0$, the exponential function $y = a(b^x)$ approaches the x-axis from below; it decreases steeply when $b > 1$ and increases steeply when $0 < b < 1$.

■ The amount of \$1 grows to e^r dollars after 1 year in an account with an interest rate of $r\%$, compounded continuously.

9.4 The Natural Logarithm

■ The natural logarithm function ln is the inverse of the natural exponential function. That is, $\ln(b) = a$, for any positive real number b, if and only if $e^a = b$.

■ The domain of the natural logarithm function is the set of positive real numbers; its range is the set of all real numbers.

9.5 Trigonometric Functions

■ A radian is a dimensionless number that represents the angle beginning at the point $(1, 0)$ on the unit circle, and sweeping out an arc of length 1, moving in the counter-clockwise direction.

■ Always set the mode of a calculator to radian measure before using the sine or cosine keys.

■ The functions $A \sin \theta$ and $A \cos \theta$ have period 2π and amplitude $|A|$.

■ The functions $A \sin(b\theta)$ and $A \cos(b\theta)$ have period $2\pi/b$ and amplitude $|A|$.

■ The functions $k + A \sin(b\theta)$ and $k + A \cos(b\theta)$ have period $2\pi/b$ and output values between $k - A$ and $k + A$.

KEY FORMULAS

■ A function f is a *polynomial function* of degree n if it has the form
$$f(x) = a_n x^n + a_{n-1} x^{n-1} + \cdots + a_2 x^2 + a_2 x + a_0$$

where $a_n, a_{n-1}, \cdots, a_2, a_1, a_0$ are all known numbers, with a_n not equal to zero.

■ A *rational function* is the quotient of two polynomial functions. It has the general form
$$f(x) = \frac{p(x)}{q(x)}$$

where both $p(x)$ and $q(x)$ are polynomial functions and x is any real number such that $q(x) \neq 0$.

■ A *power function* is a function of the form
$$f(x) = x^p,$$

where p is a known real number.

■ An *exponential function* is a function of the form
$$f(x) = a(b^x)$$

where a is a nonzero real number and b is a positive real number not equal to 1.

- If b and c denote positive real numbers not equal to 1, then the following properties of exponential functions are true:

1. $b^x b^y = b^{x+y}$

2. $(b^x)^y = b^{xy}$

3. $\dfrac{b^x}{b^y} = b^{x-y}$

4. $\left(\dfrac{1}{b}\right)^x = \dfrac{1}{b^x} = b^{-x}$

5. $(bc)^x = b^x c^x$

6. $\left(\dfrac{b}{c}\right)^x = \dfrac{b^x}{c^x}$

- The natural exponential function is
$$f(x) = e^x$$
with Euler's number as its base.

- After t years, a deposit of PV dollars in an account with a nominal annual interest rate r (expressed as a decimal), compounded continuously, will grow to
$$FV = PVe^{rt}$$

- If u and v denote positive real numbers, then the following properties of the natural logarithm function are true:

1. $\ln(1) = 0$

2. $\ln(e) = 1$

3. $\ln(uv) = \ln(u) + \ln(v)$

4. $\ln\left(\dfrac{u}{v}\right) = \ln(u) - \ln(v)$

5. $\ln(u^b) = b \ln(u)$

- 2π radians $= 360°$

- If (x, y) is the terminal point of an arc that begins at $(1, 0)$ on the unit circle and that defines an angle θ (in radians), then
$$\cos \theta = x \quad \text{and} \quad \sin \theta = y$$

TESTING YOURSELF

Use the following problems to test yourself on the material in Chapter 9. The odd problems comprise one test, the even problems a second test.

1. Determine whether either of the following relationships are functions and explain why. The rules of correspondence are described by arrows.

2. Construct sine functions that have the following characteristics:
   (*a*) Period π and amplitude 1.5
   (*b*) Period 2 and amplitude π
3. Given the function $y = f(x) = (1/x^2) - 1$, which is defined on all real numbers except $x = 0$, find: (*a*) $f(1)$; (*b*) $f(1/2)$; (*c*) $f(-2)$; (*d*) $f(z^3)$.
4. A bacterial culture beginning with 100 strands grows at a continuous rate of 30%/day. Find: (*a*) an expression for the approximate number of strands in the culture at time t (measured in days); and (*b*) the approximate number of strands in the culture after one week.
5. Use a calculator to evaluate the following expressions:
   (*a*) $f(-2.1)$ when $f(x) = e^x$ (*b*) $g(24)$ when $g(x) = \ln x$
   (*c*) $h(5)$ when $h(x) = x^{-2.1}$ (*d*) $\sin 5$
   (*e*) $-2\cos(1/4)$ (*f*) $f(0.05)$ when $f(x) = 5^x$
6. Find x if $\ln 2x - 3\ln 2 = 5$.
7. Sketch the graph of the following functions:
   (*a*) $f(x) = 3\sin(2x)$ (*b*) $f(x) = \ln x$
   Is either graph periodic? If so, what is the period?
8. Determine if either of the following two graphs represents functions:

   (*a*)

   (*b*)

9. A person deposits \$5,250 in a savings account at 5%/year, compounded continuously. How much will the account be worth after $3\frac{1}{2}$ years?
10. Transform $234°$ into radians.
11. Find x if $3e^{2x+2} = 12$.
12. Carbon 14 has a half-life of 5,730 years, and its rate of decay in a fossil is directly proportional to the amount still present. How old is a fossil that still retains 60% of its original amount of carbon 14?
13. A person places \$2,000 in a bank account at 6.4%/year, compounded quarterly. How long will it take for this deposit to grow to \$3,000?
14. Sketch the graph of each of the following functions:
    (*a*) $f(x) = e^{-x}$ (*b*) $f(x) = 2 + \cos x$
15. An angle measures -1.4 radians. How many degrees is this angle?
16. Classify each of the following functions as either a polynomial function, a rational function, a power function, or an exponential function:
    (*a*) $f(x) = 1 + 2x - 3x^7$ (*b*) $f(x) = x^{3.4}$ (*c*) $f(x) = e^2 + 3$
    (*d*) $f(x) = x^e$ (*e*) $f(x) = x^5 - 3x^4$ (*f*) $f(x) = e^x$
    (*g*) $f(x) = \dfrac{3x^2 - 1}{2 + 3x}$ (*h*) $f(x) = \dfrac{3x^6 + 5x^3 - 2.1}{7}$
17. Determine the amplitude and period of the following functions:
    (*a*) $2\sin 3\theta$ (*b*) $5 + 7\cos(\pi\theta/3)$
    Sketch the graph of each function.
18. Model, if possible, one factor in the following process as function of the others: A piece of machinery that costs \$15,000 depreciates at the rate of \$200/month for the first two years of its life. Identify the inputs and the output.

10

MODELING CHANGE

10.1 RATES OF CHANGE

We use functions to model input–output processes and we evaluate functions to find outputs associated with various inputs. The ability to calculate the outputs for different inputs is essential for decision-making. For example, the amount of jet fuel needed for a flight depends on the cargo weight. Nighttime temperatures will determine the number of beds to prepare for a homeless shelter. The number of police needed to maintain order at a rally depends on the expected size of the crowd.

Although evaluating functions is essential in many processes, such information may not be sufficient. Often managers need to know how quickly a process is changing. It's one thing for a city health agency to expect 1,000 new cases of a certain disease in the next month, but it is something else again for the agency to suspect that the number of new cases is doubling each month. It is one thing for a car dealer to expect 1,000 sales next year, and it is quite another thing for the dealer to know that sales have been declining at the rate of 50 sales a year!

Dealing with change is basic to the financial well-being of most enterprises. The first step in managing change is to understand it, which is best done by measuring it.

One of the most familiar measures of change is average speed, defined by the formula

$$\text{average speed} = \frac{\text{change in distance}}{\text{change in time}}$$

EXAMPLE 1

A woman drove from her house to a destination 400 miles away. Table 10.1 is a log of her progress beginning at 1:00 P.M., when she started her trip, and continuing every hour thereafter until she reached her destination. Determine the average speed for (*a*) the entire trip, (*b*) the last four hours of her trip, and (*c*) the time period from 6:00 P.M. to 7:00 P.M., when she took a dinner break.

TABLE 10.1 **Trip Log**

TIME (P.M.)	DISTANCE TRAVELED (miles)	TIME (P.M.)	DISTANCE TRAVELED (miles)
1:00	0	6:00	280
2:00	50	7:00	290
3:00	115	8:00	350
4:00	175	9:00	400
5:00	230		

Solution

(a) average speed for trip $= \dfrac{\text{change in distance}}{\text{change in time}}$

$\qquad\qquad = \dfrac{400 - 0}{9 - 1}$

$\qquad\qquad = \dfrac{400 \text{ miles}}{8 \text{ hours}}$

$\qquad\qquad = 50 \text{ mph}$

(b) average speed for last 4 hours $= \dfrac{\text{change in distance}}{\text{change in time}}$

$\qquad\qquad = \dfrac{400 - 230}{9 - 5}$

$\qquad\qquad = \dfrac{170}{4} = 42.5 \text{ mph}$

(c) average speed from 6:00 P.M. to 7:00 P.M. $= \dfrac{\text{change in distance}}{\text{change in time}}$

$\qquad\qquad = \dfrac{290 - 280}{7 - 6}$

$\qquad\qquad = \dfrac{10}{1} = 10 \text{ mph}$

Average speed may vary as we vary the time interval over which we measure the change. In Example 1, the average speed over the 8-hour time interval (1:00 P.M. to 9:00 P.M.) was 50 mph. Over that time period, the distance from home changed, on average, by 50 miles for *each* hour traveled. In contrast, the average speed over the last 4-hour time interval (1:00 P.M. to 5:00 P.M.) was 42.5 mph. Over this time period, the distance from home changed, on average, by 42.5 miles for each hour traveled.

If we recognize distance d as a function of elapsed time t, then we can write $d = f(t)$. In Example 1, the function is modeled by the entries in Table 10.1. If we set $d_1 = f(t_1)$ and $d_2 = f(t_2)$, the average speed over the time interval $t_1 \le t \le t_2$, usually denoted $[t_1, t_2]$, is

$$\frac{d_2 - d_1}{t_2 - t_1} = \frac{f(t_2) - f(t_1)}{t_2 - t_1} \tag{1}$$

In part (a) of Example 1, the time interval of interest is $[1, 9]$, where for $t_1 = 1$,

$$d_1 = f(t_1) = f(1) = 0$$

and for $t_2 = 9$,

$$d_2 = f(t_2) = f(9) = 400$$

In part (c) of Example 1, the time interval of interest is $[6, 7]$. Here $t_1 = 6$ and $t_2 = 7$, so

$$d_1 = f(t_1) = f(6) = 280$$

$$d_2 = f(t_2) = f(7) = 290$$

Equation (1) defines the *average rate of change* of the function $d = f(t)$ over the interval $[t_1, t_2]$. If we change the variables and write $y = f(x)$, with x denoting the input and y the output, then the average rate of change of $f(x)$ over the interval $[x_1, x_2]$ is $\dfrac{f(x_2) - f(x_1)}{x_2 - x_1}$

The average rate of change of the function $y = f(x)$ over the input interval $[x_1, x_2]$ is the quotient

$$\frac{f(x_2) - f(x_1)}{x_2 - x_1}$$

EXAMPLE 2

Determine the average rate of change of the function $y = x^2 - 2x + 2$ over the intervals (*a*) $[-1, 2]$ and (*b*) $[3, 5]$.

Solution

(*a*) Here $y = f(x) = x^2 - 2x + 2$. For the interval $[-1, 2]$ we have $x_1 = -1$ and $x_2 = 2$, so

$$f(x_1) = (-1)^2 - 2(-1) + 2 = 5$$

$$f(x_2) = (2)^2 - 2(2) + 2 = 2$$

Therefore the average rate of change of the function $f(x) = x^2 - 2x + 2$ over the interval $[-1, 2]$ is

$$\frac{f(x_2) - f(x_1)}{x_2 - x_1} = \frac{2 - 5}{2 - (-1)} = -1$$

This tells us that over the interval $[-1, 2]$, the value of the output variable y changes, on average, by -1 unit for each unit of change in the input variable x; that is, the output *decreases* on this interval as the input increases.

(*b*) For the interval $[3, 5]$, the endpoints are $x_1 = 3$ and $x_2 = 5$, which gives us

$$f(x_1) = (3)^2 - 2(3) + 2 = 5$$

$$f(x_2) = (5)^2 - 2(5) + 2 = 17$$

hence the average rate of change of the function over this interval is

$$\frac{f(x_2) - f(x_1)}{x_2 - x_1} = \frac{17 - 5}{5 - 3} = 6$$

Over the interval $[3, 5]$, the value of the output variable y changes, on average, by 6 units for each unit change in the input variable x; that is, the output *increases* on this interval as the input increases.

EXAMPLE 3

FIGURE 10.1

A Graph of Function
$y = f(x)$ over the Interval
[1, 4]

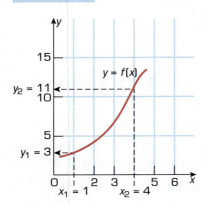

Determine the average rate of change between the endpoints $x_1 = 1$ and $x_2 = 4$ of the function illustrated in Figure 10.1.

Solution

At the input values $x_1 = 1$ and $x_2 = 4$, the corresponding output values are $y_1 = f(x_1) = 3$ and $y_2 = f(x_2) = 11$. The average rate of change of the function f over the interval $[1, 4]$ is

$$\frac{f(x_2) - f(x_1)}{x_2 - x_1} = \frac{11 - 3}{4 - 1} = \frac{8}{3}$$

Over the interval $[1, 4]$, the value of the output variable y increases, on average, by $2\frac{2}{3}$ units for each unit of change in the input variable x.

If we set $y_1 = f(x_1)$ and $y_2 = f(x_2)$, then

$$\frac{f(x_2) - f(x_1)}{x_2 - x_1} = \frac{y_2 - y_1}{x_2 - x_1} = m \qquad (2)$$

The average rate of change of the function $y = f(x)$ over the interval $[x_1, x_2]$, is the slope of the line through the two points $(x_1, f(x_1))$ and $(x_2, f(x_2))$.

which is the formula we used in Chapter 1 to calculate the slope m of the straight line passing through the two points (x_1, y_1) and (x_2, y_2). This leads to an interesting geometric interpretation of the average rate of change, as shown in Figure 10.2. According to Equation (2), the average rate of change of the function $y = f(x)$ over the input interval $[x_1, x_2]$ is the slope m of the line passing through the two points (x_1, y_1) and (x_2, y_2), where $y_1 = f(x_1)$ and $y_2 = f(x_2)$.

Now returning to Example 3, we can obtain the average rate of change of the function in Figure 10.1 over the interval $[1, 4]$ by calculating the slope of the straight line passing through the two points $(1, 3)$ and $(4, 11)$, as shown in Figure 10.3.

In Example 1, we determined that a woman's average speed over an 8-hour trip was 50 mph, well under the posted speed limits on most highways. Even so, we can't say whether she got a speeding ticket during her trip, because her average speed may be significantly different from the actual speed on her speedometer at 2:30 P.M. or 5:45 P.M.

Average rates of change are useful for many decision-making purposes; however, sometimes events change so rapidly that averages—weekly, daily, or even hourly—are not current enough. Sometimes decision-makers need to know how quickly a process is changing *right now*. Rather than an average rate of change, which measures change over an interval, a decision-maker might require an *instantaneous* rate of change, which measures change at a specific instant of time.

To approximate the instantaneous rate of change at a particular input value x_1, we compute the average rate of change over smaller and smaller intervals, where each interval has x_1 as one of its endpoints. This approach is illustrated in Figure 10.4, which is a graph of the same function $f(x)$ as in Figure 10.3. We want to approximate the instantaneous rate of change of $f(x)$ at the input value $x_1 = 1$, so we need to find the average rate of change of $f(x)$ over smaller and smaller intervals, each with $x_1 = 1$ as an endpoint.

FIGURE 10.2

The Interpretation of the Average Rate of Change as Slope

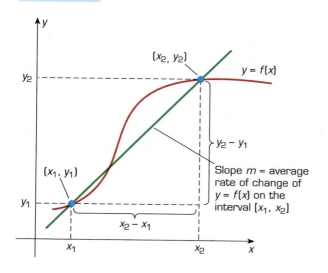

FIGURE 10.3

The Average Rate of Change of $y = f(x)$ in Example 3 is Equal to the Slope of the Line between $(1, f(1))$ and $(4, f(4))$

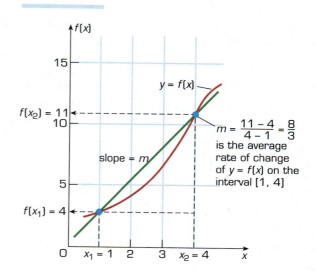

An average rate of change can be represented graphically by the slope of a specific line, and three such lines are displayed in Figure 10.4, one between the points A and D over the interval $[1, 4]$, a second line between the points A and C over the interval $[1, 3.2]$, and a third line between the points A and B over the interval $[1, 2.4]$. The slope of each of these lines is the *average* rate of change of the function over the corresponding interval on the horizontal axis. As these intervals shrink, the slopes of the corresponding lines approach that of the green line in Figure 10.4. We define the slope of this line to be the instantaneous rate of change of the function $f(x)$ at $x_1 = 1$.

FIGURE 10.4

The Instantaneous Rate of Change of $y = f(x)$ at $x_1 = 1$ shown as the Limit of Average Rates of Change

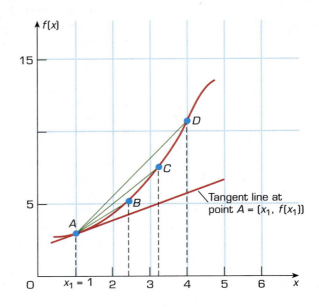

As we saw in Equation (2), the slope of the line through the two points (x_1, y_1) and (x_2, y_2) on the graph of $y = f(x)$, where $y_1 = f(x_1)$ and $y_2 = f(x_2)$, is

$$\frac{f(x_2) - f(x_1)}{x_2 - x_1} \tag{3}$$

The instantaneous rate of change of $f(x)$ at $x = x_1$ is

$$\lim_{x_2 \to x_1} \frac{f(x_2) - f(x_1)}{x_2 - x_1}.$$

We shrink the interval $[x_1, x_2]$ by letting the distance between x_2 and x_1 approach zero, keeping x_1 fixed and x_1 and x_2 separate, so that we always have two distinct points (x_1, y_1) and (x_2, y_2) through which to draw a line. We use the notation $x_2 \to x_1$, read "x_2 approaches x_1," for the process of forcing the difference between x_2 and x_1 to become arbitrarily small without becoming zero. For each choice of x_2, we use the quotient (3) to calculate the slope of the corresponding line. If these slopes approach a number L, called the *limit* of the quantity in (3), then this number is the instantaneous rate of change of the function $f(x)$ at $x = x_1$. The line with slope L passing through the point (x_1, y_1) is called *the tangent line to the graph of $y = f(x)$ at $x = x_1$*. A convenient way to picture a tangent line is to think of an automobile traveling along a curved road at night. At any point on the road, the line generated by the headlights is part of the tangent line.

The notation $\lim_{x_2 \to x_1} g(x)$ is read "the limit of $g(x)$ as $x_2 \to x_1$" and, when the difference between $g(x)$ and L approaches zero as x_2 approaches x_1, we write

$$\lim_{x_2 \to x_1} g(x) = L$$

The instantaneous rate of change of the function $y = f(x)$ at x_1 is the slope of the tangent line to the graph of $f(x)$ at the point (x_1, y_1) where $y_1 = f(x_1)$.

With this notation, we write the slope of the tangent line to the curve $y = f(x)$ at $x = x_1$ as

$$\lim_{x_2 \to x_1} \frac{f(x_2) - f(x_1)}{x_2 - x_1} \tag{4}$$

when this limit exists.

A numerical procedure for calculating (4) begins by selecting some values for x_2 that approach x_1. For each such x_2, we calculate the average rate of change of the function $f(x)$ over the interval $[x_1, x_2]$. If these rates of change approach a single number L as x_2 approaches x_1, then L is in fact the instantaneous rate of change of the function $f(x)$ at $x = x_1$, as well as the slope of the tangent line for $f(x)$ at x_1.

In choosing values for x_2 that approach x_1, we must select some points that are greater than x_1 and some that are less. When x_2 is greater than x_1, we say that x_2 is approaching x_1 *from the right*; when x_2 is less than x_1, we say that x_2 is approaching x_1 *from the left*. For the limit in (4) to exist, it must approach the same number L regardless of the direction in which x_2 approaches x_1.

EXAMPLE 4

Determine the instantaneous rate of change of the function $f(x) = x^3 + 1$ at $x_1 = 2$.

Solution

We calculate the average rates of change over smaller and smaller intervals, each with $x_1 = 2$ as an endpoint. The results are tabulated in Table 10.2, with all calculations rounded to six significant figures. These average rates of change approach 12, so the instantaneous rate of change of $f(x) = x^3 + 1$ at $x_1 = 2$ is 12.

TABLE 10.2   Average Rates of Change for $f(x) = x^3 + 1$

x_1	x_2	$f(x_1)$	$f(x_2)$	$\dfrac{f(x_2) - f(x_1)}{x_2 - x_1}$
2	2.5	9	16.625	15.25
2	2.2	9	11.648	13.24
2	2.1	9	10.261	12.61
2	2.05	9	9.61513	12.3025
2	2.01	9	9.12060	12.0601
2	2.001	9	9.01201	12.0060
2	2.0001	9	9.00120	12.0006
2	2.00001	9	9.00012	12.0001
2	1.5	9	4.375	9.25
2	1.8	9	6.832	10.84
2	1.9	9	7.859	11.41
2	1.95	9	8.41488	11.7025
2	1.99	9	8.88060	11.9401
2	1.999	9	8.98801	11.9940
2	1.9999	9	8.99880	11.9994
2	1.99999	9	8.99988	11.9999

Figure 10.5 is a graphical view of the limiting process in Example 4. On the horizontal axis we placed values of x_2 from the second column in Table 10.2, and on the vertical axis we placed the corresponding values for the average rate of change, taken from the last column. Notice that as x_2 approaches 2 from the right, passing successively through 2.2, 2.1, 2.05, and 2.01, the corresponding average rates of change decrease to 12. As x_2 approaches 2 from the left, passing successively through 1.8, 1.9, 1.95, and 1.99, the corresponding average rates of change *increase* to 12. Regardless of direction, the closer x_2 gets to 2, as illustrated by the dashed arrows, the closer the average rate of change over the interval $[x_2, 2]$ gets to 12, as illustrated by the solid arrows.

FIGURE 10.5

The Limiting Process in Example 4

EXAMPLE 5

Determine the instantaneous rate of change of the function $f(x) = x^3 + 1$ at $x_1 = -1.3$.

Solution

We calculate the average rates of change over smaller and smaller intervals, each with $x_1 = -1.3$ as an endpoint. The results are tabulated in Table 10.3, with all calculations rounded to seven significant figures. These average rates of change approach 5.07, so the instantaneous rate of change of $f(x) = x^3 + 1$ at $x_1 = -1.3$ is 5.07.

Figure 10.6 is a graphical view of the limiting process in Example 5, using some of the corresponding values in the second and last columns of Table 10.3. We see that as x_2 approaches -1.3, illustrated by the dashed arrows, the average rate of change approaches a number close to 5, as illustrated by the solid arrows. It's not clear from the graph whether that number is 5.07 or 5.05 or 5.063; the numerical approach used to generate Table 10.3 provides a more precise answer.

We conclude from Examples 4 and 5 that the instantaneous rate of change of a function may differ at different values of the input variable. The same conclusion is apparent from inspecting the two tangent lines associated with Examples 4 and 5. These lines are shown

TABLE 10.3   **Average Rates of Change for $f(x) = x^3 + 1$**

x_1	x_2	$f(x_1)$	$f(x_2)$	$\dfrac{f(x_2) - f(x_1)}{x_2 - x_1}$
−1.3	−1.5	−1.197	−2.375	5.89
−1.3	−1.4	−1.197	−1.744	5.47
−1.3	−1.35	−1.197	−1.460375	5.2675
−1.3	−1.31	−1.197	−1.248091	5.1091
−1.3	−1.305	−1.197	−1.222448	5.089525
−1.3	−1.301	−1.197	−1.202074	5.073901
−1.3	−1.3001	−1.197	−1.197507	5.070390
−1.3	−1.30001	−1.197	−1.197051	5.070039
−1.3	−1.300001	−1.197	−1.197005	5.070004
−1.3	−1.1	−1.197	−0.331	4.33
−1.3	−1.2	−1.197	−0.728	4.69
−1.3	−1.25	−1.197	−0.953125	4.8775
−1.3	−1.29	−1.197	−1.146689	5.0311
−1.3	−1.295	−1.197	−1.171747	5.050525
−1.3	−1.299	−1.197	−1.191934	5.066101
−1.3	−1.2999	−1.197	−1.196493	5.069610
−1.3	−1.29999	−1.197	−1.196949	5.069961
−1.3	−1.299999	−1.197	−1.196995	5.069996

FIGURE 10.6

The Limiting Process in Example 5

in Figure 10.7: one passes through the point $(2, f(2)) = (2, 9)$ and the second passes through the point $(-1.3, f(-1.3)) = (-1.3, -1.197)$. The slope of each tangent line is different.

Many graphing calculators have menus for quickly duplicating Figure 10.7. Once the curve for $y = f(x)$ is displayed in a window, a tangent line to any point on the curve is created by placing the cursor on that point and then accessing the keystroke designator for

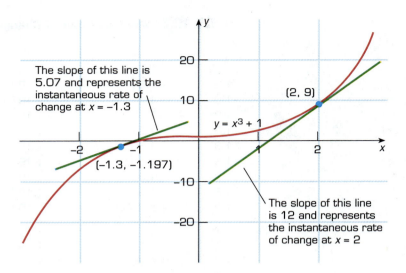

The slope of this line is 5.07 and represents the instantaneous rate of change at $x = -1.3$

$(2, 9)$

$y = x^3 + 1$

$(-1.3, -1.197)$

The slope of this line is 12 and represents the instantaneous rate of change at $x = 2$

a tangent line, often labeled tanln. The tangent line will then superimpose on the graph of the function, and the slope of this line will appear on the calculator screen. Using such technology when available is more efficient and is therefore preferable to constructing tables.

EXAMPLE 6

Find the equations of the tangent lines shown in Figure 10.7.

Solution

(*a*) We have from Example 4 that the instantaneous rate of change of the function $f(x) = x^3 + 1$ at $x_1 = 2$ is 12, which is also the slope of the tangent line to the function at the point (2, 9). Using the point–slope equation developed in Section 1.5, we obtain

$$y - 9 = 12(x - 2)$$

$$y = 12x - 15$$

as the equation of the tangent line to $f(x)$ at (2, 9).

(*b*) As a result of Example 5, we know that 5.07 is a reasonable estimate (to two decimal places) of the slope of the tangent line to the function $f(x) = x^3 + 1$ at $(-1.3, -1.197)$. The point–slope form of the equation of this tangent line is

$$y - (-1.197) = 5.07[x - (-1.3)]$$

or

$$y = 5.07x + 5.394$$

We find the instantaneous rate of change for a function $f(x)$ at the input value x_1 by calculating the average rate of change $[f(x_2) - f(x_1)]/(x_2 - x_1)$ between x_1 and x_2 for a sequence of input values x_2 that approach x_1. We did this in Table 10.2 with $x_1 = 2$ and again in Table 10.3 with $x_1 = -1.3$. The average rate of change is a quotient: its denom-

An instantaneous rate of change for a function has the units of output per one unit of input.

inator is a measure of the change in input, $x_2 - x_1$, while its numerator is a measure of the corresponding change in output, $f(x_2) - f(x_1)$. The unit of measure associated with this quotient is the output units per one input unit, for both the average rate of change and the instantaneous rate of change.

In Example 1, Table 10.1, the function $d = f(t)$ has hours as the input unit and miles as the output unit. An average or instantaneous rate of change for this function will therefore be in units of miles per hour, a common unit of measure for speed.

EXAMPLE 7

A function f that measures fuel efficiency for automobiles gives the distance traveled (m miles) as the output; it is a function of the amount of gasoline used (g gallons), that is, $m = f(g)$. What are the units associated with the instantaneous rate of change of f with respect to gasoline consumption?

Solution

The output m is measured in units of miles and the input g is measured in units of gallons, so the instantaneous rate of change of output with respect to input has units of miles per gallon.

EXAMPLE 8

The graph in Figure 10.8 shows the cost of producing a textbook as a function of the number of copies produced. What are the units associated with the instantaneous rate of change of this function with respect to the size of a printing?

FIGURE 10.8
Textbook Production Costs

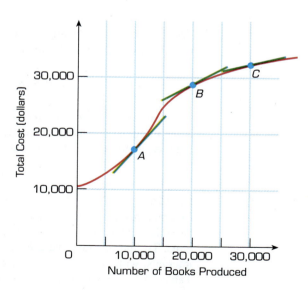

Solution

The input variable is the number of books printed and the output variable is the total production cost measured in dollars. The instantaneous rate of change is in units of dollars per book.

In business, the instantaneous rate of change of cost per unit of input is called *marginal cost* while the instantaneous rate of change of revenue (i.e., income) per unit of input is called *marginal revenue*. Each is a measure of how quickly the output variable—cost or revenue—changes as the input variable is changed. Marginal cost provides an answer to the question, "How much will my cost change if I alter the current input?" Marginal revenue answers the question, "How much will my revenue change if I alter the current input?"

The curve shown in Figure 10.8 is the graph of a cost function for a book publisher. The input variable is the number of books to be produced (a decision made by management), and the output variable is the cost associated with each production schedule. There are three tangent lines to the curve, at points A, B, and C, corresponding to production runs of 10,000 books, 20,000 books, and 30,000 books, respectively. The slope of each tangent line, in units of dollars per book, is a marginal cost. The slope of the tangent line at A is greater than that of the tangent line at B, which is in turn greater than the slope of the tangent line at C. Consequently, the marginal cost at a production run of 10,000 books is greater than the marginal cost at a production run of 20,000 books, which is greater than the marginal cost at a production run of 30,000 books. It costs more in dollars/book to increase production from a base of 10,000 books than it does from a base of 20,000 books. Similarly, the cost of increasing production from 20,000 to 20,001 is greater than the cost of increasing production from 30,000 to 30,001 books.

We'll have much more to say about marginal revenue and marginal cost in Section 10.6; we're simply introducing the terms here to suggest the relevance of instantaneous rates of change to commerce and economics.

EXAMPLE 9

The Undercover Publishing Company's revenue R is related to their advertising expenditures a by the model

$$R(a) = 800 - 800e^{-0.1a}$$

with both variables, a and $R(a)$, measured in thousand-dollar units. What is the marginal revenue at the current advertising budget of \$45,000; is it advisable to increase the current advertising budget?

Solution

Marginal revenue is the instantaneous rate of change of revenue with respect to the input variable, here advertising expenditures, at an input level of \$45,000, or 45 thousand-dollar units. To find this rate, we calculate the average rate of change over smaller and smaller intervals, with each interval having $a_1 = 45$ as an endpoint. The results are tabulated in Table 10.4, with all calculations rounded to six decimal places.

These average rates of change seem to approach 0.88872, which is both the marginal revenue at $a_1 = 45$ thousand dollars and the slope of the tangent line displayed in Figure 10.9. Sales are increasing at the rate of 0.88872 units (of \$1,000) or \$888.72 for every increase of 1 unit of input or \$1,000 of advertising. Clearly, it is not profitable for Undercover to invest \$1,000 for a return of approximately \$888.72, so the current advertising budget should not be increased.

TABLE 10.4 **Average Rates of Change for $R(a) = 800 - 800e^{-0.1a}$**

a_1	a_2	$R(a_1)$	$R(a_2)$	$\dfrac{R(a_2) - R(a_1)}{a_2 - a_1}$
45	46	791.112803	791.958531	0.845729
45	45.5	791.112803	791.546236	0.866867
45	45.1	791.112803	791.201232	0.884291
45	45.05	791.112803	791.157128	0.886502
45	45.01	791.112803	791.121686	0.888276
45	45.001	791.112803	791.113691	0.888675
45	45.0001	791.112803	791.112892	0.888715
45	45.00001	791.112803	791.112812	0.888719
45	44	791.112803	790.178128	0.934675
45	44.5	791.112803	790.657146	0.911313
45	44.9	791.112803	791.023485	0.893178
45	44.95	791.112803	791.068256	0.890945
45	44.99	791.112803	791.103911	0.889164
45	44.999	791.112803	791.111914	0.888764
45	44.9999	791.112803	791.112714	0.888724
45	44.99999	791.112803	791.112794	0.888720

FIGURE 10.9
Marginal Revenue as a Slope

IMPROVING SKILLS

In Problems 1 through 10, find the average rate of change for each function over the given interval.

1. $f(x) = x^2 - 4x + 5$; [1, 8]
2. $f(x) = x^2 - 4x + 5$; [−2, 1]
3. $f(x) = 1/x$; [1, 3]
4. $f(x) = 1/x$; [1, 4]
5. $f(x) = 1/x$; [−4, −1]
6. $f(x) = e^x$; [0, 2]
7. $f(x) = \sin x$; [0, 1]
8. $f(x) = \cos x$; [0 2π]
9. $f(x) = \ln x$; [1, π]
10. $f(x) = x^{0.2}$; [2.5, 3.6]

In Problems 11 through 14, use the tangent line that appears in each figure to estimate the instantaneous rate of change of each function at the x-component of the point of tangency.

11.

12.

13.

14.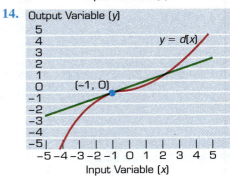

15. Complete the following table and use the results to determine the instantaneous rate of change of the function $f(x) = x^3 + 1$ at $x_1 = 1$.

x_1	x_2	$f(x_1)$	$f(x_2)$	$\dfrac{f(x_2) - f(x_1)}{x_2 - x_1}$
1	1.5	2	4.375	4.75
1	0.5	2	1.125	1.75
1	1.2	2	2.728	3.64
1	0.8	2	1.512	2.44
1	1.1	2	2.331	3.31
1	0.9	2	1.729	
1	1.05	2	2.157625	
1	0.95	2	1.857375	
1	1.01	2	2.030301	
1	0.99	2		
1	1.001	2		
1	0.999	2		
1	1.0001	2		
1	0.9999	2		
1	1.00001	2		
1	0.99999	2		

16. Complete the accompanying table and use the results to determine the instantaneous rate of change of the function $f(x) = 1/(x - 1)$ at $x_1 = 2$.

x_1	x_2	$f(x_1)$	$f(x_2)$	$\dfrac{f(x_2) - f(x_1)}{x_2 - x_1}$
2	2.2	1	0.833333	−0.833333
2	1.8	1	1.25	−1.25
2	2.1	1	0.909091	−0.909091
2	1.9	1	1.111111	−1.111111
2	2.05	1	0.952381	−0.952381
2	1.95	1	1.052632	
2	2.01	1	0.990099	
2	1.99	1	1.010101	
2	2.005	1	0.995025	
2	1.995	1		
2	2.001	1		
2	1.999	1		
2	2.0001	1		
2	1.9999	1		
2	2.00001	1		
2	1.99999	1		

In Problems 17 through 26, find the instantaneous rate of change of each function at the given point x_1. If you have a graphing calculator with a tangent line menu, use it; if not, construct tables similar to Tables 10.2 and 10.3.

17. $f(x) = x^2 - 4x + 5$; $x_1 = 1$
18. $f(x) = x^2 - 4x + 5$; $x_1 = 8$
19. $f(x) = 1/x$; $x_1 = 0.5$
20. $f(x) = 2/(x^2 + 1)$; $x_1 = 0$
21. $f(x) = \sqrt{1 + x^2}$; $x_1 = 1.2$
22. $f(x) = e^x$; $x_1 = 2$
23. $f(x) = \sin x$; $x_1 = 0$
24. $f(x) = \cos x$; $x_1 = 0$
25. $f(x) = \ln x$; $x_1 = 0.8$
26. $f(x) = x^{0.2}$; $x_1 = 5$

In Problems 27 through 31, find the equation of the tangent line to each function at the point $(x_1, f(x_1))$.

27. $f(x) = x^2 - 4x + 5$; $x_1 = 1$
28. $f(x) = 1/x$; $x_1 = 0.5$
29. $f(x) = 2/(x^2 + 1)$; $x_1 = 0$
30. $f(x) = \sqrt{1 + x^2}$; $x_1 = 1.2$
31. $f(x) = e^x$; $x_1 = 2$

32. Using the graph in Figure 10.4, arrange the following numbers in ascending order:
    (a) The average rate of change over the interval [1, 4]
    (b) The average rate of change over the interval [1, 3]
    (c) The average rate of change over the interval [1, 2]
    (d) The instantaneous rate of change at $x = 1$
33. Using the graph in Figure 10.7, arrange the following numbers in ascending order:
    (a) The instantaneous rate of change at $x = -1.3$
    (b) The instantaneous rate of change at $x = 2$
    (c) The average rate of change over the interval $[-1.3, 2]$
34. Using the graph in Figure 10.8, arrange the following numbers in ascending order:
    (a) The instantaneous rate of change at $x = 10,000$
    (b) The instantaneous rate of change at $x = 20,000$
    (c) The instantaneous rate of change at $x = 30,000$

CREATING
MODELS

35. The following table lists data provided by the Bureau of the Census on the population of the U.S. Use this information to determine the average rate of change of the population over the intervals (**a**) [1800, 1890] and (**b**) [1900, 1990].

YEAR	POPULATION (millions)	YEAR	POPULATION (millions)
1800	5.3	1900	76.0
1830	12.9	1930	122.8
1860	31.4	1960	179.3
1890	62.9	1990	248.7

36. A model for population growth in the U.S. for the years 1800–1990 is

$$y = 6.027403 + 0.023731x + 0.006641x^2$$

where x denotes the year ($x = 0$ is 1800, $x = 100$ is 1900, and $x = 190$ is 1990), and y denotes the corresponding population (in millions). Use this model to estimate the average rate of change in the U.S. population between 1970 and 1985.

37. Use the model in Problem 36 to estimate the instantaneous rate of change in the U.S. population in 1970.

38. The accompanying table records the balance in an account over a 200-year period. Use this information to determine the average rate of change in the account over the intervals (**a**) [0, 50]; (**b**) [50, 100]; (**c**) [100, 150]; (**d**) [150, 200].

YEAR	BALANCE	YEAR	BALANCE
0	$1,000.00		
10	$1,628.89	120	$348,911.99
30	$4,321.94	140	$925,767.37
50	$11,467.40	150	$1,507,977.50
70	$30,426.43	160	$2,456,336.44
90	$80,730.37	180	$6,517,391.84
100	$131,501.26	200	$17,292,580.82

39. The account in Problem 38 earns interest at the rate of 5% a year, compounded annually. A model for this growth is given by the formula

$$y = 1000(1.05)^x$$

where x denotes the year and y the balance in the account that year. Use this model to estimate the average rate of change in the account for year 130 to year 155.

40. Use the model in Problem 39 to estimate the instantaneous rate of change in the account at year 110.

41. The following table lists data on the number (in millions) of maltreated children in the United States, as reported in the March 1994 issue of the *Journal of Interpersonal*

Violence. Use this data to estimate the average rate of change of this population over the intervals (*a*) [1985, 1992] and (*b*) [1990, 1992].

YEAR	ESTIMATED NUMBER OF CHILDREN REPORTED FOR MALTREATMENT (in millions)
1985	1.919
1986	2.086
1987	2.157
1988	2.265
1989	2.435
1990	2.557
1991	2.723
1992	2.936

42. The data in the table in Problem 41 was modeled in Example 6 of Section 2.3; rounding the coefficients of that model to three significant figures, we get

$$y = 0.139x - 274$$

where x denotes the year and y denotes the number of children (millions) maltreated that year. Use this model to estimate the instantaneous rate of growth of this population in 1991.

43. Sketch a graphical model for each of the following situations. Take the vertical axis to be distance traveled and the horizontal axis to be time.
(*a*) A car travels at a constant speed.
(*b*) The speed of a car is always increasing.
(*c*) The speed of a car is always decreasing.

44. One morning Mary left home to drive to school, but halfway there, she realized that she had forgotten her physics project. She returned home, retrieved the project, and again set out for school. Mary's car has cruise control, so she almost always travels at a constant speed. However, once she remembered her project, she sped up to make up for lost time. Sketch a graphical model of Mary's commute that morning.

45. A simple model for the distance a bungee jumper travels in free fall is $d = 16t^2$, where the distance d is measured in feet and time t in seconds. Find the speed of the jumper after 2 seconds in free fall.

46. A company's sales are known to be related to advertising expenditures by the model $S = 150{,}000 + 6000x - 50x^2$, where x denotes the monthly advertising expenditures in thousands of dollars.
(*a*) Find marginal sales at an advertising budget of $45,000.
(*b*) Based on your results from part (*a*), would you recommend that the advertising budget be increased?

47. A firm's total sales, in millions of dollars, is modeled by the equation $y = 3x + \frac{1}{2}x^2$, where x denotes the number of years the firm has been in operation.
(*a*) Determine the firm's average growth rate in sales over its first seven years in business.
(*b*) Determine the firm's marginal sales after its seventh year in business.
(*c*) What will the firm's total sales be at the end of the tenth year if sales continue to follow the given equation?
(*d*) What will the firm's total sales be at the end of the tenth year if the growth rate in sales after the seventh year doesn't change from what it was at the end of that year?

EXPLORING IN TEAMS

FIGURE A

FIGURE B

48. Consider the function

$$f(x) = \begin{cases} x^2 + 1 & \text{when } x \le 1 \\ 2x & \text{when } x > 1 \end{cases}$$

which is graphed in Figure A. What can you say about the instantaneous rate of change of this function at $x_1 = 1$?

49. Consider the function

$$f(x) = \begin{cases} 2 + 5x & \text{when } x \le 3 \\ 23 - 2x & \text{when } x > 3 \end{cases}$$

which is graphed in Figure B. What can you say about the instantaneous rate of change of this function at $x_1 = 3$?

50. Consider the function

$$f(x) = \begin{cases} 7 & \text{when } x \le 5 \\ 12 & \text{when } x > 5 \end{cases}$$

which is shown below. What can you say about the instantaneous rate of change of this function at $x_1 = 5$?

51. Relate your findings in each of the three previous problems to tangent lines.

EXPLORING WITH TECHNOLOGY

52. In Example 9, we found that when the Undercover Publishing Co. has an advertising budget of $45,000, their marginal revenue is $888.72; in other words, an additional expenditure of $1,000 in advertising will return approximately $888.72 in additional sales, resulting in a revenue loss of $1,000 − $888.72 = $111.28.

   (a) Show that the marginal revenue for an advertising budget of $20,000 is approximately $10,827, that is, increasing advertising expenditures by $1,000 will return approximately $10,827 in additional sales, resulting in a revenue *gain* of $9,827.

   (b) Show that the marginal revenue for an advertising budget of $75,000 is approximately $44.25, resulting in a revenue *loss* of $955.75.

(*c*) Use a graphing calculator to estimate the marginal revenue for various advertising budgets between \$20,000 ($x = 20$) and \$75,000 ($x = 75$). Based on your findings, recommend an advertising budget for the Undercover Company.

REVIEWING MATERIAL

53. (Section 2.2) Use Gaussian elimination to solve the following system of equations:

$$2x - 3y + 4z = 40$$
$$3x + 2y - 2z = 24$$
$$y + 5z = 63$$

54. (Section 6.2) What are the odds in favor of randomly choosing one card from a regular deck and having that card be a heart?

55. (Section 7.2) Brooks City is divided into three zones. Repair crews for the Brooks City Utility Company report in to the central dispatching office when they complete a service call to receive their next assignment. Records kept by the utility company show that 50% of all crews completing a repair in Zone 1 are given a subsequent assignment in Zone 1, while 40% are sent to Zone 2 and 10% to Zone 3. Forty percent of all crews that finish a service call in Zone 2 are sent to Zone 1 for their next repair, 30% remain in Zone 2, and 30% are sent to Zone 3. For crews that complete a service call in Zone 3, twenty percent are routed next to Zone 1, while 30% go on to Zone 2 and 50% remain in Zone 3. At the beginning of a day, 40% of all repair crews are sent to Zone 1, whereas 40% are sent to Zone 2 and 20% to Zone 3. What is the distribution of repair crews in different zones after each crew has completed two service calls?

56. (Sections 9.1–9.5) Determine which of the following functions *cannot* have Figure C as its graph.

FIGURE C

   (*a*) x^2 (*b*) x^3 (*c*) e^x (*d*) $\ln x$ (*e*) $2 \sin 3x$

RECOMMENDING ACTION

57. Respond by memo to the following request.

MEMORANDUM

To: J. Doe Reader

From: Office of the President

Date: Today

Subject: **Advertising Budget**

Thank you for your analysis [see Example 9] of Marketing's proposal to increase our advertising expenditures beyond \$45,000 a year. Clearly I will not authorize such an increase. I was wondering, however, if your analysis provides insight into a reduction of our current budget. Based on your work, would you recommend a *decrease* in our current advertising budget of \$45,000?

10.2 LIMITS AND CONTINUITY

In Section 10.1, we introduced a limiting operation to define the instantaneous rate of change of a function at a specific point. This was our first encounter with limits but it will not be our last, because limits are one of the fundamental concepts in calculus.

The notation $x \to x_1$ indicates that the variable x is getting closer and closer to the value x_1 without ever reaching it. Similarly, the notation $x \to a$, read "x approaches a," indicates that the variable x is getting closer and closer to the value a without equaling it. As an example, think of starting at a point $x = 3$ and moving toward $x_1 = 2$, each time covering half of the remaining distance. After the first move we are at $x = 2\frac{1}{2}$; after the second move we are at $x = 2\frac{1}{4}$; after 3 moves we are at $x = 2\frac{1}{8}$; etc. Our locations form the sequence

$$2\tfrac{1}{2},\ 2\tfrac{1}{4},\ 2\tfrac{1}{8},\ 2\tfrac{1}{16},\ 2\tfrac{1}{32},\ 2\tfrac{1}{64}, \cdots \tag{5}$$

This process is shown schematically in Figure 10.10. With each move, we get closer to $x = 2$, but we never actually reach it because each step takes us only halfway to our destination. Sequence (5) is one way to approach 2. There are many other ways to do this.

FIGURE 10.10

Approaching $x_1 = 2$ from the Right

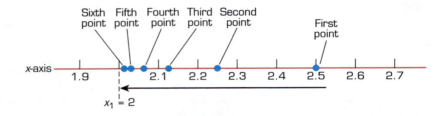

Sequence (5) approaches $x_1 = 2$ from the right on the number line in Figure 10.10. On the other hand, starting at $x = 1$, the sequence

$$1\tfrac{1}{2},\ 1\tfrac{3}{4},\ 1\tfrac{7}{8},\ 1\tfrac{15}{16},\ 1\tfrac{31}{32},\ 1\tfrac{63}{64}, \cdots \tag{6}$$

is diagrammed in Figure 10.11, approaching $x_1 = 2$ from the left. To distinguish between the two approaches, we adopt the notation $x \to a^+$ to indicate that x approaches a from the right (x is always greater than a), and $x \to a^-$ to indicate that x approaches a from the left (x is always less than a). Sequence (5) is one way to have $x \to 2^+$, while sequence (6) is a way to have $x \to 2^-$.

FIGURE 10.11

Approaching $x_1 = 2$ from the Left

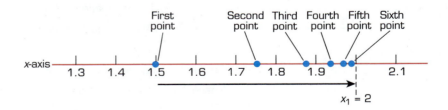

In general, we're interested in what happens to the output of a function as the input variable approaches a particular value. As an example, consider the function $y = x^2 + 1$ as $x \to 2^+$. We want to determine what happens to the output of this function as the input variable gets closer and closer to $x_1 = 2$ from the right.

Figure 10.12 is a graph of the function $y = x^2 + 1$. Since we're interested in the behavior of this function near $x_1 = 2$, we zoom in on the highlight region with $1.5 \leq x \leq 2.5$ and $3 \leq y \leq 7.5$. The result is the graph in Figure 10.13. One way for $x \to 2^+$ is illustrated by the disks on the x-axis of Figure 10.13. We begin with $x = 2.3$ and move toward $x_1 = 2$ from the right. The corresponding y-values, illustrated by the disks on the y-axis, approach 5. Therefore we write

$$\lim_{x \to 2^+} (x^2 + 1) = 5$$

read "As x approaches 2 from the right, the limit of the function $y = x^2 + 1$ is 5."

FIGURE 10.12

The Graph of the Function
$y = x^2 + 1$

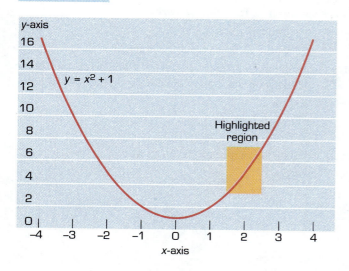

FIGURE 10.13

Approaching $x_1 = 2$ from
the Right for the Function
$y = x^2 + 1$

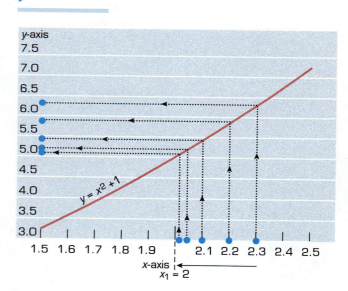

We can obtain the same answer numerically by creating Table 10.5, with all entries in the second row rounded to four decimal places. The first row is a sequence of input values that approach $x_1 = 2$ from the right. They include all points marked by disks on the x-axis in Figure 10.13. (Any other sequence will do, as long as it approaches 2 from the right.)

TABLE 10.5 $\lim_{x \to 2^+} (x^2 + 1)$

x	2.5	2.3	2.2	2.1	2.05	2.02	2.01	2.001	2.0001
$x^2 + 1$	7.25	6.29	5.84	5.41	5.2025	5.0804	5.0401	5.0040	5.0004

EXAMPLE 1

Determine $\lim\limits_{x \to 2^-} (x^2 + 1)$.

Solution

We have to determine what happens to the output of the function $y = x^2 + 1$ as the input variable gets closer and closer to $x_1 = 2$ from the left. We take a numerical approach to this problem by constructing Table 10.6, with all entries in the last row rounded to four decimal places.

TABLE 10.6 $\lim\limits_{x \to 2^-} (x^2 + 1)$

x	1.5	1.7	1.8	1.9	1.95	1.98	1.99	1.995	1.999
$x^2 + 1$	3.25	3.89	4.24	4.61	4.8025	4.9204	4.9601	4.9800	4.9960

The first row is a sequence of input values that approach $x_1 = 2$ from the left. As $x \to 2^-$, the corresponding output values approach 5. Therefore we write

$$\lim\limits_{x \to 2^-} (x^2 + 1) = 5$$

Alternatively, we can solve this problem graphically. Figure 10.14 displays some of the points in Table 10.6. As $x \to 2^-$ (the disks on the x-axis) the corresponding y-values (the disks on the y-axis) approach 5:

$$\lim\limits_{x \to 2^-} (x^2 + 1) = 5$$

This indicates that the limiting value of the output of the function $y = x^2 + 1$ is 5 as x approaches $x_1 = 2$ from the left.

FIGURE 10.14

Approaching $x_1 = 2$ from the Left for the Function $y = x^2 + 1$

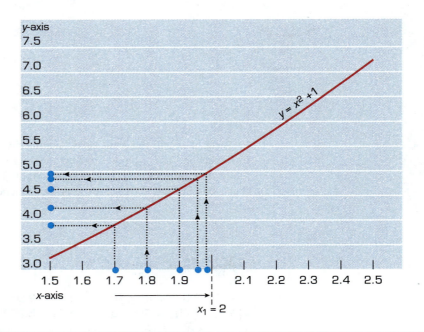

We solved Example 1 using both numerical and graphical approaches to evaluate limits. If a graphing calculator is available, then a graphical approach is usually preferred. Most graphing calculators have a zoom feature, which focuses on a small part of a graph around a point of interest, and a trace feature, which follows the graph of a function as the input is changed, displaying the corresponding outputs on the screen. By observing how these outputs change, we see how the function behaves as the input approaches a fixed value from either the right or the left.

The trace *and* zoom *features on a graphing calculator are useful for determining limits.*

EXAMPLE 2

Determine

$$\lim_{x \to 1^+} \frac{x^2 + 3x - 4}{x^2 - 1}$$

Solution

Since the denominator of this function, $x^2 - 1$, is zero when $x = \pm 1$, the natural domain of $f(x) = (x^2 + 3x - 4)/(x^2 - 1)$ is the set of all real numbers except ± 1. In particular, $f(x)$ is not defined at $x = 1$. Figure 10.15 is a graph of this function in a small window centered at $x_1 = 1$. The circle above x_1 on the x-axis indicates that the graph has a ''hole'' there: there is no output associated with $x_1 = 1$ because it is not in the domain. We see from Figure 10.15 that as x approaches 1 from the right, the corresponding outputs approach 2.5. Thus we write

$$\lim_{x \to 1^+} \frac{x^2 + 3x - 4}{x^2 - 1} = 2.5$$

By similar reasoning,

$$\lim_{x \to 1^-} \frac{x^2 + 3x - 4}{x^2 - 1} = 2.5$$

FIGURE 10.15

Approaching $x_1 = 1$ from the Right for the Function $y = (x^2 + 3x - 4)/(x^2 - 1)$

EXAMPLE 3

Determine

$$\lim_{x \to 1^+} \frac{x^2 + 3x + 4}{x^2 - 1}$$

Solution

This function differs from the one in Example 2 by a single sign in the numerator. The denominator is the same, so once again the natural domain is the set of all real numbers excluding ± 1. Figure 10.16 is a graph of this function in a small window centered at

FIGURE 10.16

Approaching $x_1 = 1$ from the Right for the Function $y = (x^2 + 3x + 4)/(x^2 - 1)$

$x_1 = 1$. Now, as x approaches 1 from the right, the corresponding outputs grow increasingly large. These outputs do not approach any fixed value, so for $x \to 1^+$, the limit of the function $y = (x^2 + 3x + 4)/(x^2 - 1)$ does not exist. It is common notation in this case to write

$$\lim_{x \to 1^+} \frac{x^2 + 3x + 4}{x^2 - 1} = \infty$$

to indicate that the outputs are becoming larger than any number.

If we prefer a numerical approach, we obtain the same result from Table 10.7:

TABLE 10.7 $f(x) = \lim\limits_{x \to 1^+} \dfrac{x^2 + 3x + 4}{x^2 - 1}$

x	1.5	1.3	1.2	1.1	1.05	1.01	1.001	1.0001	1.00001
$f(x)$	8.6	13.9	20.6	40.5	80.5	400.5	4,001	40,001	400,001

We also see from Figure 10.16 that the outputs are assuming increasingly large negative values as x approaches 1 from the left. Common notation for this is

$$\lim_{x \to 1^-} \frac{x^2 + 3x + 4}{x^2 - 1} = -\infty$$

The limit of this function as $x \to 1^-$ also does not exist.

Although many functions can be modeled by a single equation, there are others that cannot. Sometimes a function must be modeled by a *piecewise-defined* formula, consisting of different equations defined on non-overlapping intervals. For example, the function graphed in Figure 10.17 is modeled mathematically by the formula

$$f(x) = \begin{cases} 5 - x^2 & x < 3 \\ 4 - x & x \geq 3 \end{cases} \tag{7}$$

FIGURE 10.17

A Graph of the Function
$$y = \begin{cases} 5 - x^2, & x < 3 \\ 4 - x, & x \geq 3 \end{cases}$$

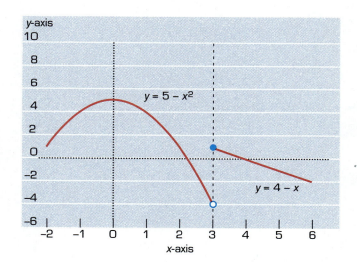

A disk on a graph indicates that a point is part of the graph, while a circle indicates that the point is not included.

When x is less than 3, we use the formula $5 - x^2$ to evaluate the outputs; when x is greater than or equal to 3, we use the formula $4 - x$ to evaluate the outputs. For example,

$$f(1) = 5 - (1)^2 = 4, \qquad f(2.5) = 5 - (2.5)^2 = -1.25$$

$$f(3) = 4 - 3 = 1, \qquad f(3.95) = 4 - 3.95 = 0.05$$

To eliminate uncertainty about a functional value associated with a particular element of the domain, we use a disk to indicate that a point is included in the graph and a circle to indicate that it's not. See Figure 10.17.

EXAMPLE 4

Determine $\lim_{x \to 3^+} f(x)$ and $\lim_{x \to 3^-} f(x)$ for the function modeled by Equation (7).

Solution

We are interested in the behavior of the function near $x = 3$, so we restrict our attention to a small domain centered at 3. Figures 10.18 and 10.19 are magnified portions of Figure

10.17 both with $2 \leq x \leq 4$ and $-4 \leq y \leq 4$. Using Figure 10.18, we see that as x approaches 3 from the right, the corresponding output values approach 1. Therefore

$$\lim_{x \to 3^+} f(x) = 1$$

We see from Figure 10.19 that as x approaches 3 from the left, the corresponding output values approach -4. Therefore,

$$\lim_{x \to 3^-} f(x) = -4$$

FIGURE 10.18

Approaching 3 from the Right for the Function $y = 4 - x$

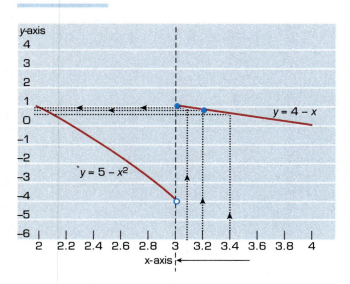

FIGURE 10.19

Approaching 3 from the Left for the Function $y = 5 - x^2$

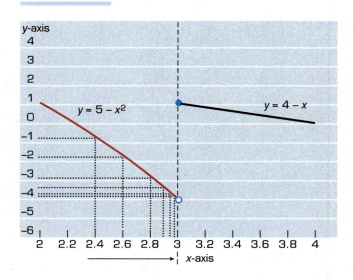

We used Tables 10.5 and 10.6 to show that

$$\lim_{x \to 2^+} (x^2 + 1) = 5 = \lim_{x \to 2^-} (x^2 + 1)$$

Here the limit of the function $f(x) = x^2 + 1$ is the same when x approaches 2 from the right and when it approaches 2 from the left. In Example 2, we showed that

$$\lim_{x \to 1^+} \frac{x^2 + 3x - 4}{x^2 - 1} = 2.5 = \lim_{x \to 1^-} \frac{x^2 + 3x - 4}{x^2 - 1}$$

Again, the limiting value is the same when we approach $x = 1$ from both the right and the left. We want to acknowledge this situation when it occurs, so we introduce the notion of *a limit at a point*. We say that a function $f(x)$ has a *limit L* at $x = a$ when

$$\lim_{x \to a^+} f(x) = L = \lim_{x \to a^-} f(x)$$

and we write

$$\lim_{x \to a} f(x) = L \tag{8}$$

Determine whether the function graphed in Figure 10.20 has a limit as x approaches -1.

FIGURE 10.20
The Graph of the Function
Modeled by Equation (9)

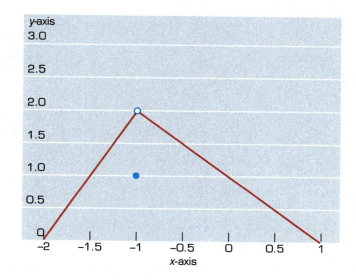

Solution
The function in Figure 10.20 is modeled by the formula

$$f(x) = \begin{cases} 2x + 4 & x < -1 \\ 1 & x = -1 \\ 1 - x & x > -1 \end{cases} \tag{9}$$

At $x = -1$ the output is 1, which is shown graphically by the disk in Figure 10.20; for any input values less than -1, we evaluate the corresponding output with the expression $2x + 4$; and for any input value greater than -1, we use the expression $1 - x$ to determine the output.

We see from Figure 10.21 that

$$\lim_{x \to -1^+} f(x) = 2$$

and Figure 10.22 shows us that

$$\lim_{x \to -1^-} f(x) = 2$$

These limits are equal, hence

$$\lim_{x \to -1} f(x) = 2$$

FIGURE 10.21

Approaching −1 from the
Right for the Function in
Equation (9)

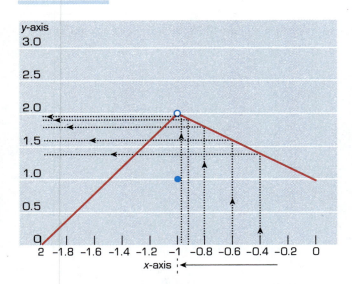

FIGURE 10.22

Approaching −1 from the
Left for the Function in
Equation (9)

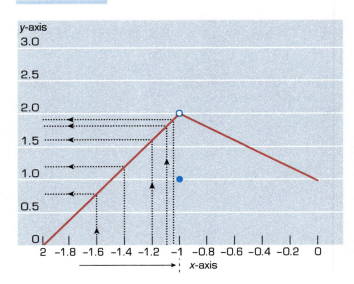

In summary, a function has a limit L at an input $x = a$ if it has a limit at a from the right and a limit at a from the left and these limits both equal L. Intuitively, this definition is equivalent to the following two conditions:

1. For all values of x sufficiently close but not equal to a, the value of $f(x)$ is approximately equal to L.

2. The value of $f(x)$ can be made as close to L as we need by taking x sufficiently close to a, as long as $x \neq a$.

If either the limit from the left or the limit from the right does not exist at a particular input value, then the function does not have a limit there.

EXAMPLE 6

Determine $\lim_{x \to 1} f(x)$ for the function

$$f(x) = \begin{cases} \dfrac{x^2 + 3x + 4}{x^2 - 1} & x \neq 1 \\ 5 & x = 1 \end{cases}$$

Solution

For this function $f(x)$ $(x \neq 1)$, we showed in Example 3 that $\lim_{x \to 1^+} f(x)$ does not exist. Therefore we conclude that $\lim_{x \to 1} f(x)$ does not exist.

EXAMPLE 7

Determine $\lim_{x \to 3} f(x)$ for the function defined by Equation (7):

$$f(x) = \begin{cases} 5 - x^2 & x < 3 \\ 4 - x & x \geq 3 \end{cases}$$

Solution

We have from Example 4 that

$$\lim_{x \to 3^+} f(x) = 1 \qquad \text{and} \qquad \lim_{x \to 3^-} f(x) = -4$$

The limit of this function as x approaches 3 from the right does *not* equal the limit of the function as x approaches 3 from the left; therefore $\lim_{x \to 3} f(x)$ does not exist.

Limits, when they exist, have the following properties.

PROPERTIES OF LIMITS

If $\lim_{x \to a} f(x) = L$ and $\lim_{x \to a} g(x) = M$, then the following rules hold:

SUM RULE

$$\lim_{x \to a} [f(x) + g(x)] = \lim_{x \to a} f(x) + \lim_{x \to a} g(x) = L + M$$

DIFFERENCE RULE

$$\lim_{x \to a} [f(x) - g(x)] = \lim_{x \to a} f(x) - \lim_{x \to a} g(x) = L - M$$

PRODUCT RULE

$$\lim_{x \to a} [f(x)g(x)] = \left[\lim_{x \to a} f(x)\right]\left[\lim_{x \to a} g(x)\right] = LM$$

QUOTIENT RULE

If $M \neq 0$, then

$$\lim_{x \to a} \frac{f(x)}{g(x)} = \frac{\lim_{x \to a} f(x)}{\lim_{x \to a} g(x)} = \frac{L}{M}$$

When we determine the limit of a function at an input value, we approach that value from both the right and the left. Even though we get closer and closer to the input value of interest, we never actually evaluate the function at that value. In Example 7, we were interested in the limit of a function as x approached 3. Consequently, we determined what happens to the function as x got very near to *3* but we ignored $f(3)$, the value of the function at $x = 3$. Often, however, we do want to compare the limit of a function at a point to the value of the function at that point, when it is defined there.

A function $f(x)$ is continuous at $x = a$ if the function has a limit at a and this limit equals $f(a)$.

We say that a function $f(x)$ is *continuous at $x = a$* if the function has a limit at a and if this limit equals the value of the function at a. To be continuous at $x = a$, the function f must satisfy the condition

$$f(a) = \lim_{x \to a^+} f(x) = \lim_{x \to a^-} f(x) \tag{10}$$

The function $f(x)$ is *not* continuous at a if its limit does not exist at a, or if the function is not defined at a, or if any of the equalities in Equation (10) are not true. We say that a function is *discontinuous* at an input value when it is not continuous there.

EXAMPLE 8

Determine whether the function $f(x) = x^2 + 1$ is continuous at $x = 2$.

Solution

We used Tables 10.5 and 10.6 earlier in this section to show that

$$\lim_{x \to 2^+} (x^2 + 1) = 5 = \lim_{x \to 2^-} (x^2 + 1)$$

Furthermore, $f(2) = (2)^2 + 1 = 5$, so $f(2) = \lim_{x \to 2} (x^2 + 1)$ and f is continuous at $x = 2$.

EXAMPLE 9

Determine whether the function

$$f(x) = \begin{cases} 2x + 4 & x < -1 \\ 1 & x = -1 \\ 1 - x & x > -1 \end{cases}$$

is continuous at $x = -1$.

Solution

From Example 5, we know that

$$\lim_{x \to -1} f(x) = 2$$

However, $f(-1) = 1$, so

$$f(-1) \neq \lim_{x \to -1} f(x)$$

and the function is discontinuous at $x = -1$.

EXAMPLE 10

Determine whether the function

$$f(x) = \begin{cases} 5 - x^2 & x < 3 \\ 4 - x & x \geq 3 \end{cases}$$

is continuous at $x = 3$.

Solution

Here $f(3) = 4 - 3 = 1$. However, in Example 7 we showed that $\lim_{x \to 3} f(x)$ does not exist, so f is discontinuous at $x = 3$.

If a function is continuous at $x = a$ then its graph must show a seamless transition of functional values as it moves through the point $(a, f(a))$.

If a function has a limit at a, then as x approaches a, the values of $f(x)$ approach a number L. If the function is also continuous at a, then $f(a) = L$ and there is a seamless transition of functional values as x moves from near a onto a itself. If a function is not continuous at a, then that transition is not seamless, that is, in moving along the graph of $f(x)$ through the point $(a, f(a))$, there will be a hole or a ledge at a. This gives us a simple heuristic test for continuity at a point: *A function is continuous at all points where its graph has no holes or ledges.*

Consider the graph for the function $f(x) = x^2 + 1$, shown in Figure 10.12. This graph is seamless, no holes or ledges, so the function is continuous at every point on the x-axis. In contrast, the graph in Figure 10.17 has a ledge at $x = 3$, so the associated function is not continuous there. This graph is seamless at every other point, however, so the function graphed in Figure 10.17 is continuous at every point on the x-axis except $x = 3$. The graph in Figure 10.20 has a hole at $x = -1$, so its associated function is not continuous there;

however, it's seamless at every other point, so the graphed function is continuous at every point on the x-axis except $x = -1$.

EXAMPLE 11

Determine where the function graphed in Figure 10.23 is continuous.

Solution

The graph in Figure 10.23 has a hole at $x = -1$, so $f(x)$ is not continuous there. At $x = -1$ the limit of the function is -2, but $f(-1) = 1$. Figure 10.23 also has a ledge at $x = 1$, so $f(x)$ is not continuous there, either. In this case, $\lim_{x \to 1^-} f(x) = 1$ and $\lim_{x \to 1^+} f(x) = f(1) = 2$, so $\lim_{x \to 1} f(x)$ does not exist. The graph is seamless everywhere else, hence $f(x)$ is continuous at all points on the x-axis except $x = \pm 1$.

FIGURE 10.23

An Example of a Function That is Discontinuous at Two Points

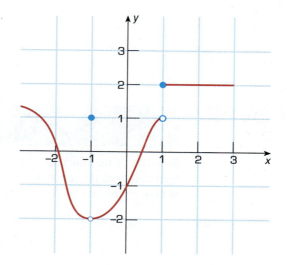

EXAMPLE 12

In 1996, first-class postage rates in the United States for letters weighing up to 6 ounces were 32¢ for the first ounce and 23¢ for each additional ounce or part thereof. Create a function that models these postal rates for first-class mail weighing between 0.5 and 3.5 ounces, inclusive; then determine where on this domain the function is continuous.

Solution

We let x denote the weight of a first-class letter. A mathematical model for first-class postage for letters weighing six ounces or less is

$$f(x) = \begin{cases} 32 & 0 \le x \le 1 \\ 55 & 1 < x \le 2 \\ 78 & 2 < x \le 3 \\ 101 & 3 < x \le 4 \\ 124 & 4 < x \le 5 \\ 147 & 5 < x \le 6 \end{cases}$$

The graph of this function is shown in Figure 10.24. If we restrict the domain to $0.5 \le x \le 3.5$, then the region of interest shrinks to the highlight box in Figure 10.24. Within this box, the function is continuous everywhere except at $x = 1, 2,$ and 3.

FIGURE 10.24

A Graph of a Model for First-Class Postage

IMPROVING SKILLS

For each of the functions defined in Problems 1 through 20, determine limits from the right and from the left at $x = 1$, 2, and 3. Then decide whether the functions have limits at those points and if they are continuous there.

1.

W100FA

2.

3.

4.

5.

6.

7.

8.

9.

10.

11.

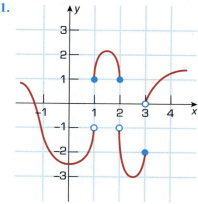

12.
$$f(x) = \begin{cases} x & x < 1 \\ 1 & 1 \le x < 2 \\ x & x \ge 2 \end{cases}$$

13.
$$f(x) = \begin{cases} x & x < 1 \\ 1 & 1 \le x < 2 \\ x + 1 & x \ge 2 \end{cases}$$

14.
$$f(x) = \begin{cases} x & x < 1 \\ 2 & 1 \le x < 2 \\ x + 1 & x \ge 2 \end{cases}$$

15.
$$f(x) = \begin{cases} x & x < 1 \\ 3 & 1 \le x < 2 \\ x + 1 & x \ge 2 \end{cases}$$

16.
$$f(x) = \begin{cases} x^3 & x \le 2 \\ 6 + x & 2 < x < 3 \\ 2 + x & x \ge 3 \end{cases}$$

17.
$$f(x) = \begin{cases} x^3 & x < 2 \\ 6 & x = 2 \\ 6 + x & 2 < x < 3 \\ 2 + x & x \ge 3 \end{cases}$$

18.
$$f(x) = \begin{cases} x^3 & x < 2 \\ 8 & x = 2 \\ 6 + x & 2 < x < 3 \\ x^2 & x \ge 3 \end{cases}$$

19.
$$f(x) = \begin{cases} 2x - 1 & x \le 1 \\ 1 & 1 < x \le 3 \\ 4 - x & x > 3 \end{cases}$$

20.
$$f(x) = \begin{cases} 2 & x < 1 \\ 1 & 1 \le x < 2 \\ 0 & 2 \le x < 3 \\ 1 & x \ge 3 \end{cases}$$

In Problems 21 through 40, determine whether the indicated limits exist.

21. $\lim\limits_{x \to 1}(3x + 2)$

22. $\lim\limits_{x \to 2}(1 - x^2)$

23. $\lim\limits_{x \to 1}\left(\dfrac{1}{x + 1}\right)$

24. $\lim\limits_{x \to 1}\left(\dfrac{1}{x - 1}\right)$

25. $\lim\limits_{x \to 0}\left(\dfrac{1}{x + 1}\right)$

26. $\lim\limits_{x \to 0}\left(\dfrac{1}{x^2}\right)$

27. $\lim\limits_{x \to 3} \sqrt{x + 1}$

28. $\lim\limits_{x \to 1}\left(\dfrac{x^2 - 1}{x - 1}\right)$

29. $\lim\limits_{x \to -1}\left(\dfrac{x^2 + 3x + 2}{x + 1}\right)$

30. $\lim\limits_{x \to -1}\left(\dfrac{x^2 + 3x + 2}{x + 2}\right)$

31. $\lim\limits_{x \to -1}\left(\dfrac{x^2 + 3x + 1}{x + 1}\right)$

32. $\lim\limits_{x \to 0}\left(\dfrac{x + 4}{3x + 8}\right)$

33. $\lim\limits_{x \to 0}\left(\dfrac{1}{x}\right)$

34. $\lim\limits_{x \to 0}\left(\dfrac{e^x - 1}{2x}\right)$

35. $\lim\limits_{x \to 0}\left(\dfrac{e^x - 1}{x + 2}\right)$

36. $\lim\limits_{x \to 0}\left(\dfrac{x + 2}{e^x - 1}\right)$

37. $\lim\limits_{x \to 0}(\ln x)$

38. $\lim\limits_{x \to 1}(\ln x)$

39. $\lim\limits_{x \to 1}\left(\dfrac{\ln x}{x - 1}\right)$

40. $\lim\limits_{x \to 1}\left(\dfrac{\ln x}{x + 1}\right)$

CREATING MODELS

In problems 41 through 49, create functions that model each process, determine where those functions are *not* continuous, and then answer the questions.

41. During an advertising campaign, a telephone company offers to charge its customers 10¢ for each minute or part thereof for calls within the continental United States. How much does this company charge its customers for calls of $4\frac{1}{2}$ minutes or less?

42. The Dickensons deposit $1,000 into a savings account that earns 5% interest a year, compounded annually. If they make no withdrawals or other deposits, how much will they have in the account at any time over the next three years?

43. How much will the Dickensons have in their account if interest is compounded semiannually?

44. How much will the Dickensons have in their account if interest is compounded continuously?

45. How much will a salesman earn in a day if his daily salary is $50 plus 2% of that day's sales?

46. Solve Problem 45 if the employer caps all salaries at a maximum of $400/day.

47. The All-Day Parking Garage in New York City charges $12 for the first hour, $6 for the second hour, and $3 for each additional hour or part thereof, subject to a maximum of $24 per day. How much is a customer charged for parking a car for a day?

48. The 1994 Federal Income Tax 1040 booklet contains the following table as Schedule X for single taxpayers.

IF THE AMOUNT ON FORM 1040, LINE 37, IS:		ENTER ON FORM 1040, LINE 38		OF THE AMOUNT OVER —
Over —	**But not over —**			
$0	$22,750	....	15%	$0
22,750	55,100	$3,412.50 +	28%	22,750
55,100	115,000	12,470 +	31%	55,100
115,000	250,000	31,039.50 +	36%	115,000
250,000	....	79,639.50 + 39.6%		250,000

How much income tax must be listed on line 38 by a filer who uses this rate schedule in the 1040 form?

49. The Roasted Bean Coffee Distributors has the following rate schedule for each order: $4/pound for orders up to 50 pounds; $200 for the first 50 pounds of the order plus $3.80/pound for any additional coffee between 50 and 100 pounds; $390 for the first 100 pounds ordered plus $3.50/pound for coffee in excess of 100 pounds. How much do they charge for each order?

EXPLORING IN TEAMS

50. Consider the function

$$f(x) = \left(1 + \frac{1}{x}\right)^x$$

(a) Suggest a reason for concluding that the value of this function approaches 1 as x gets larger and larger. Use the fact that 1 raised to any power is 1.

(b) Suggest a reason for concluding that the value of this function grows without bound as x gets larger and larger. Use the fact that any number greater than 1 becomes very large when raised to a large positive power.

(c) Complete the following table and use your results to predict the behavior of $f(x)$ as x becomes large.

x	1	2	10	100	1,000	10,000	100,000	1,000,000
$f(x)$								

51. Use a calculator to evaluate the function in Problem 50 for $x = 10^{50}$ (the number 1 followed by 50 zeros). How do you explain the result? Does it change your answer to Problem 50?

52. Use a calculator to determine

$$\lim_{x \to 0}\left(\frac{x}{\sin x}\right) \quad \text{and} \quad \lim_{x \to 0}\left(\frac{x}{\cos x}\right)$$

Then, using these two results and the properties of limits, determine the following:

(a) $\lim_{x \to 0}\left(\dfrac{x}{\sin x} + \dfrac{x}{\cos x}\right)$

(b) $\lim_{x \to 0}\left(\dfrac{x^2}{\sin x \cos x}\right)$

(c) $\lim_{x \to 0}\left(\dfrac{3x}{\sin x} - \dfrac{4x}{\cos x}\right)$

(d) $\lim_{x \to 0}\left(\dfrac{x^2}{\sin x}\right)$

(e) $\lim_{x \to 0}\left(\dfrac{\sin x}{x}\right)$

(f) $\lim_{x \to 0}(\tan x)$

REVIEWING MATERIAL

53. (Section 1.2) If A is directly proportional to B with a constant of proportionality of 3, and if B is directly proportional to C with a constant of proportionality of 5, is C directly proportional to A? If so, what is the constant of proportionality?

54. (Section 2.5) Find the inverse of the matrix

$$\mathbf{A} = \begin{bmatrix} 2 & 0 & 3 \\ 0 & 2 & 6 \\ -1 & 1 & 0 \end{bmatrix}$$

55. (Section 9.5) Using the sine function, create a function that is periodic with amplitude 6 and period 10 and has outputs that assume all values between -2 and 10, inclusive.

RECOMMENDING ACTION

56. Respond by memo to the following request.

MEMORANDUM

To: J. Doe Reader

From: Station Manager

Date: Today

Subject: **Gasoline Inventory**

Every Monday morning we receive a delivery of regular gasoline that fills our tanks to their maximum holding capacity of 6,000 gallons. Last week we took 3,775.6 gallons; the week before it was 5,192.8 gallons. During the week, we deplete our inventory through sales to individual customers. I understand that our inventory can be modeled as a function, with time as the input variable and gasoline in stock as the output variable. What would the graph of such a function look like? Is this function continuous?

THE DERIVATIVE

In Section 10.1 we developed the formula

$$\lim_{x_2 \to x_1} \frac{f(x_2) - f(x_1)}{x_2 - x_1}$$

to calculate the instantaneous rate of change of the function $f(x)$ at $x = x_1$. When this limit exists, a common notation for this rate of change is $f'(x_1)$ (read "f prime of x):

$$f'(x_1) = \lim_{x_2 \to x_1} \frac{f(x_2) - f(x_1)}{x_2 - x_1} \tag{11}$$

Once $f(x)$ is specified, we can determine the limit in Equation (11) by using either a numerical or a graphical approach, as described in Section 10.2. For polynomial functions, a more efficient procedure can be developed algebraically. In doing so, it is convenient, because it simplifies the algebra, to change notation by setting $h = x_2 - x_1$, where h represents the directed distance between the two values $x = x_2$ and $x = x_1$. The quantity h is positive when x_2 is greater than x_1 and negative when x_2 is less than x_1. Solving the equation $h = x_2 - x_1$ for x_2, we find that $x_2 = x_1 + h$. Then letting x_2 approach x_1 is equivalent to having h approach zero, so Equation (11) may be rewritten

$$f'(x_1) = \lim_{h \to 0} \frac{f(x_1 + h) - f(x_1)}{h} \tag{12}$$

Equations (11) and (12) are two different but equivalent ways of expressing the same rate of change, as illustrated in Figure 10.25.

FIGURE 10.25

Two Different Notations for Rate of Change

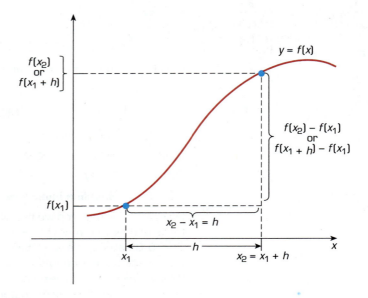

To evaluate Equation (12) when $f(x)$ is a polynomial function, we first cancel any factors of h in the numerator with the h in the denominator, and then let h approach zero. The function $f(x_1 + h)$ is evaluated most easily by replacing x in $f(x)$ with $x_1 + h$. For example, if

$$f(x) = 2x^5 + 3x \quad \text{then} \quad f(x_1 + h) = 2(x_1 + h)^5 + 3(x_1 + h)$$

and if

$$f(x) = x^2 - 8x + 5 \quad \text{then} \quad f(x_1 + h) = (x_1 + h)^2 - 8(x_1 + h) + 5$$

EXAMPLE 1

Find the instantaneous rate of change of $f(x) = 5x^2 - 3x + 4$ at $x_1 = 1$.

Solution

Here $f(x)$ is a second-degree polynomial with

$$f(x_1) = f(1) = 5(1)^2 - 3(1) + 4 = 6$$

and

$$f(x_1 + h) = f(1 + h) = 5(1 + h)^2 - 3(1 + h) + 4$$

Using Equation (12) with $x_1 = 1$, we obtain

$$f'(1) = \lim_{h \to 0} \frac{f(1 + h) - f(1)}{h}$$

$$= \lim_{h \to 0} \frac{[5(1 + h)^2 - 3(1 + h) + 4] - 6}{h}$$

$$= \lim_{h \to 0} \frac{5(1 + 2h + h^2) - 3 - 3h + 4 - 6}{h}$$

$$= \lim_{h \to 0} \frac{5h^2 + 7h}{h} = \lim_{h \to 0} \frac{h(5h + 7)}{h}$$

$$= \lim_{h \to 0} (5h + 7)$$

As h approaches zero, $5h$ also approaches zero, and $5h + 7$ approaches 7. Therefore $f'(1) = 7$.

EXAMPLE 2

Find the instantaneous rate of change of $f(x) = x^3 + 1$ at $x_1 = 2$.

Solution

Here $f(x)$ is the third-degree polynomial with

$$f(x_1) = f(2) = (2)^3 + 1 = 9$$

and

$$f(x_1 + h) = f(2 + h) = (2 + h)^3 + 1$$

Using Equation (12) with $x_1 = 2$, we obtain

$$f'(2) = \lim_{h \to 0} \frac{f(2 + h) - f(2)}{h}$$

$$= \lim_{h \to 0} \frac{[(2 + h)^3 + 1] - 9}{h}$$

$$= \lim_{h \to 0} \frac{(8 + 12h + 6h^2 + h^3) + 1 - 9}{h}$$

$$= \lim_{h \to 0} \frac{h^3 + 6h^2 + 12h}{h}$$

$$= \lim_{h \to 0} \frac{h(h^2 + 6h + 12)}{h}$$

$$= \lim_{h \to 0} (h^2 + 6h + 12) = 12$$

Notice how much faster this approach is compared to the numerical procedure we used in Example 4 of Section 10.1.

Equation (12),

$$f'(x_1) = \lim_{h \to 0} \frac{f(x_1 + h) - f(x_1)}{h}$$

defines the instantaneous rate of change of the function $y = f(x)$ at $x = x_1$. Sometimes, as in Examples 4 and 5 in Section 10.1, we need the instantaneous rate of change of a function at many input values. We can proceed in two different ways: we can evaluate Equation (12) at each input value of interest, or we can evaluate Equation (12) with $x_1 = x$ and then substitute the required input values into the resulting expression. This second approach is often quicker and the one preferred by practitioners. Replacing x_1 with x in Equation (12), we obtain

$$f'(x) = \lim_{h \to 0} \frac{f(x + h) - f(x)}{h} \tag{13}$$

as the instantaneous rate of change of the function $f(x)$ at an arbitrary input value x. If we require the instantaneous rate of change of $f(x)$ at a particular input value x_1, we simply evaluate the final formula for $f'(x)$ at $x = x_1$.

EXAMPLE 3

Find $f'(x)$ for the function $f(x) = x^2$, then use the result to determine the instantaneous rate of change of the function at $x = -1$ and at $x = 2$.

Solution

Here $f(x + h) = (x + h)^2 = x^2 + 2xh + h^2$, and so

$$f'(x) = \lim_{h \to 0} \frac{f(x + h) - f(x)}{h}$$

$$= \lim_{h \to 0} \frac{[x^2 + 2xh + h^2] - x^2}{h}$$

$$= \lim_{h \to 0} \frac{2xh + h^2}{h}$$

$$= \lim_{h \to 0} \frac{h(2x + h)}{h}$$

$$= \lim_{h \to 0}(2x + h) = 2x$$

Thus the instantaneous rate of change of $f(x) = x^2$ at any input value x is $2x$. When $x = -1$, we get $f'(-1) = 2(-1) = -2$; when $x = 2$, the result is $f'(2) = 2(2) = 4$.

Figure 10.26 displays a window from the graph of the function $f(x) = x^2$ with two tangent lines: one at $(-1, 1)$ and the other at $(2, 4)$. We know from Example 3 that the slopes of these tangent lines are -2 and 4, respectively. Moreover, we now know that the slope of any tangent line to the curve at a point (x, y) is twice the value of its x-coordinate. This is quite a bit of information for very little work!

FIGURE 10.26
Two Tangent Lines to the
Curve $y = x^2$

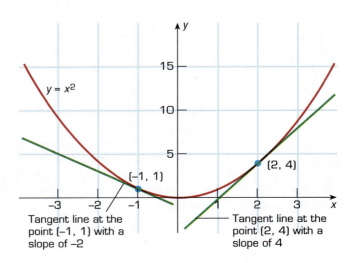

Tangent line at the point $(-1, 1)$ with a slope of -2

Tangent line at the point $(2, 4)$ with a slope of 4

EXAMPLE 4

Find $f'(x)$ for the function $f(x) = x$, then use the result to determine the instantaneous rate of change of the function at $x = 1.3$ and at $x = 2.5$.

Solution

Here $f(x + h) = (x + h)$, which gives us

$$f'(x) = \lim_{h \to 0} \frac{f(x + h) - f(x)}{h}$$

$$= \lim_{h \to 0} \frac{[x + h] - x}{h}$$

$$= \lim_{h \to 0} \frac{h}{h} = \lim_{h \to 0} 1$$

The number 1 does not depend on h and therefore is not influenced by anything h does. Specifically, as h approaches zero, the number 1 does not change. Thus the instantaneous rate of change of $f(x) = x$ is always 1 at any input value. For example, if $x = 1.3$ then $f'(1.3) = 1$, and if $x = 2.5$ then $f'(2.5) = 1$.

FIGURE 10.27

The Line $y = x$ and Its Tangent Lines

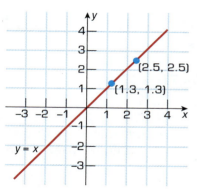

$y = x$

The derivative of the function $y = f(x)$ at the input value x is $f'(x) =$
$$\lim_{h \to 0} \frac{f(x + h) - f(x)}{h}.$$

A derivative represents an instantaneous rate of change and also the slope of a tangent line.

The results of Example 4 show clearly in a graph of the function. Since the function $f(x) = x$ is linear, its graph is a straight line, as shown in Figure 10.27. The slope of a straight line is a constant, in this case 1, and a tangent line at any point on the line is the line itself. Thus, every tangent line to the graph of $y = x$ has the same equation ($y = x$) and the same slope, 1.

The right side of Equation (13) is so fundamental in modeling that we give it a special name, the *derivative* of $y = f(x)$ at the input value x. The process of finding a derivative is called *differentiation*. Common notations for the derivative of $y = f(x)$ besides $f'(x)$ are y' (read "y prime"), $\frac{dy}{dx}$ (read "dee y dee x"), and $\frac{d}{dx}[f(x)]$ (read "dee $f(x)$ dee x"). The choice of notation is one of personal preference or convenience; we will use them interchangeably in this text. In each case we are differentiating the function $f(x)$ with respect to the independent variable x.

Equation (13),

$$f'(x) = \lim_{h \to 0} \frac{f(x + h) - f(x)}{h}$$

is the derivative of the function $f(x)$ at the input value x. It is also the instantaneous rate of change of $f(x)$ at the point x, as well as the slope of the tangent line to the graph of $f(x)$ at the point $(x, f(x))$. Thus, a physical interpretation of the derivative is an instantaneous rate of change, and a graphical interpretation of the derivative is the slope of a tangent line.

In Example 3, we found that the derivative of $f(x) = x^2$ is $f'(x) = 2x$. Using the notation $y = 2x$, we can write $y' = \frac{dy}{dx} = 2x$. In Example 4, we found that the derivative of $f(x) = x$ is $f'(x) = 1$, or $y' = \frac{dy}{dx} = 1$. By evaluating Equation (13) for other power functions, we will see that the derivative of x^3 is $3x^2$, the derivative of x^4 is $4x^3$, and the derivative of x^5 is $5x^4$. A pattern is emerging, and this pattern is the basis for the *power rule of differentiation*, extended to both integer and noninteger powers.

THE POWER RULE OF DIFFERENTIATION

If n is a real number, then the derivative of the power function $y = f(x) = x^n$ is

$$\frac{dy}{dx} = f'(x) = nx^{n-1}$$

EXAMPLE 5

Find the derivatives of the following power functions:

(a) $f(x) = x^7$ (b) $f(x) = x^1$
(c) $f(x) = x^{-8}$ (d) $f(x) = x^{5.2}$
(e) $f(x) = x^{-5.2}$ (f) $f(x) = \sqrt{x}$

Solution
(a) $n = 7$, so $f'(x) = 7x^{7-1} = 7x^6$.
(b) $n = 1$, so $f'(x) = 1x^{1-1} = 1x^0 = 1$.
(c) $n = -8$, so $f'(x) = -8x^{-8-1} = -8x^{-9}$.
(d) $n = 5.2$, so $f'(x) = 5.2x^{5.2-1} = 5.2x^{4.2}$.
(e) $n = -5.2$, so $f'(x) = -5.2x^{-5.2-1} = -5.2x^{-6.2}$.
(f) $\sqrt{x} = x^{1/2}$ is a power function with $n = 1/2$. Thus

$$f'(x) = \frac{1}{2} x^{\frac{1}{2}-1} = \frac{1}{2} x^{-1/2} = \frac{1}{2x^{1/2}} = \frac{1}{2\sqrt{x}}$$

A constant function has the form $f(x) = c$, for some fixed real number c: each input has the same output, c. In particular, $f(x) = f(x + h) = c$, and the derivative of this function reduces to

$$f'(x) = \lim_{h \to 0} \frac{f(x + h) - f(x)}{h} = \lim_{h \to 0} \frac{c - c}{h} = \lim_{h \to 0} 0 = 0$$

As h approaches zero, the number 0 is unaffected; it's still zero! Here is what we have proven:

FIGURE 10.28
The Horizontal Line
$y = f(x) = c$

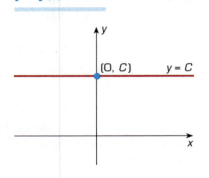

THE CONSTANT RULE OF DIFFERENTIATION

If $y = f(x) = c$, a constant, then $\frac{dy}{dx} = f'(x) = 0$.

The constant rule of differentiation is straightforward when viewed graphically. Figure 10.28 shows the graph of the equation $y = f(x) = c$. (We've pictured c as positive, but it can be zero or negative.) It is a line parallel to the horizontal axis, and a tangent line at any point (x, c) on the line is the line itself. The derivative of $f(x) = c$ at an input value x is the slope of the tangent line at (x, c). Since the slope of a horizontal line is zero, the derivative of $y = c$ is zero.

EXAMPLE 6

Find the derivatives of (**a**) $f(x) = 10$; (**b**) $f(x) = -19$; (**c**) $f(x) = \pi$.

Solution

Each of these three functions has the form $f(x) = c$, with $c = 10, -19$, and π, respectively, so the derivative of each of these functions is zero.

The previous two rules of differentiation are limited to power functions and constant functions. These rules become even more useful when combined with other rules dealing with the derivatives of sums, differences, and constant multiples of functions.

If a given function $f(x)$ can be expressed as a constant c times another function $u(x)$, that is, if $f(x) = cu(x)$, then $f(x + h) = cu(x + h)$ and, using the product rule of limits from Section 10.2, we find that

$$f'(x) = \lim_{h \to 0} \frac{f(x + h) - f(x)}{h}$$

$$= \lim_{h \to 0} \frac{cu(x + h) - cu(x)}{h}$$

$$= \lim_{h \to 0} \frac{c[u(x + h) - u(x)]}{h}$$

$$= \lim_{h \to 0} c\left(\frac{u(x + h) - u(x)}{h}\right)$$

$$= \left[\lim_{h \to 0}(c)\right]\left[\lim_{h \to 0} \frac{u(x + h) - u(x)}{h}\right]$$

Now, the constant c does not depend on h, thus $\lim_{h \to 0}(c) = c$. Then from Equation (13), with $u(x)$ replacing $f(x)$,

$$u'(x) = \lim_{h \to 0} \frac{u(x + h) - u(x)}{h}$$

Putting all of this together, we have proven that

$$f'(x) = cu'(x)$$

CONSTANT MULTIPLE RULE OF DIFFERENTIATION

If $y = cu(x)$, where c is a real number and the function $u(x)$ has a known derivative $u'(x)$, then

$$\frac{d}{dx}[cu(x)] = c\frac{d}{dx}[u(x)] = cu'(x)$$

EXAMPLE 7

Find the derivatives of the following functions:

(**a**) $y = 5x^7$ (**b**) $y = 3x^{21}$ (**c**) $y = 2.7x^{-8}$ (**d**) $y = -\frac{1}{2}x^{-5.2}$

Solution

(**a**) The derivative of $f(x) = x^7$ is $f'(x) = 7x^6$, so the derivative of $y = 5x^7$ is $y' = 5(7x^6) = 35x^6$.

(b) The derivative of $f(x) = x^{21}$ is $f'(x) = 21x^{20}$, so the derivative of $y = 3x^{21}$ is $y' = 3(21x^{20}) = 63x^{20}$.

(c) The derivative of $f(x) = x^{-8}$ is $f'(x) = -8x^{-9}$, so the derivative of $y = 2.7x^{-8}$ is $y' = 2.7(-8x^{-9}) = -21.6x^{-9}$.

(d) The derivative of $f(x) = x^{-5.2}$ is $f'(x) = -5.2x^{-6.2}$, so the derivative of $y = -\frac{1}{2}x^{-5.2}$ is $y' = -\frac{1}{2}(-5.2x^{-6.2}) = 2.6x^{-6.2}$.

If a function $f(x)$ can be expressed as the sum of two other functions $u(x)$ and $v(x)$ with known derivatives, $f(x) = u(x) + v(x)$, then, using the sum rule of limits from Section 10.2, we have the following:

$$f'(x) = \lim_{h \to 0} \frac{f(x + h) - f(x)}{h}$$

$$= \lim_{h \to 0} \frac{[u(x + h) + v(x + h)] - [u(x) + v(x)]}{h}$$

$$= \lim_{h \to 0} \frac{[u(x + h) - u(x)] + [v(x + h) - v(x)]}{h}$$

$$= \lim_{h \to 0} \left[\frac{u(x + h) - u(x)}{h} + \frac{v(x + h) - v(x)}{h} \right]$$

$$= u'(x) + v'(x)$$

We have proven the sum rule of differentiation; the same analysis applies for subtraction.

SUM AND DIFFERENCE RULES OF DIFFERENTIATION

If $u(x)$ and $v(x)$ are two functions with known derivatives, $u'(x)$ and $v'(x)$, respectively, then the derivative of the sum $y = u(x) + v(x)$ is the sum of the derivatives:

$$y' = u'(x) + v'(x)$$

while the derivative of the difference $y = u(x) - v(x)$ is the difference of the derivatives:

$$y' = u'(x) - v'(x)$$

EXAMPLE 8

Find the derivatives of the following functions:

(a) $y = x^5 + x^4$ **(b)** $y = x^2 + 1$ **(c)** $y = x^{-2.1} - x^{2.1}$ **(d)** $y = x^{-3} - \frac{1}{2}$

Solution

(a) The derivative of x^5 is $5x^4$, while the derivative of x^4 is $4x^3$, so the derivative of $y = x^5 + x^4$ is $y' = 5x^4 + 4x^3$.

(b) The derivative of x^2 is $2x$, while the derivative of the constant function 1 is 0, so the derivative of $y = x^4 + 1$ is $y' = 2x + 0 = 2x$.

(c) The derivative of $x^{-2.1}$ is $-2.1x^{-3.1}$, while the derivative of $x^{2.1}$ is $2.1x^{1.1}$, so the derivative of $y = x^{-2.1} - x^{2.1}$ is $y' = -2.1x^{-3.1} - 2.1x^{1.1}$.

(d) The derivative of x^{-3} is $-3x^{-4}$, while the derivative of the constant function $\frac{1}{2}$ is 0, so the derivative of $y = x^{-3} - \frac{1}{2}$ is $y' = -3x^{-4} - 0 = -3x^{-4}$.

Even though the sum and the difference rules of differentiation are stipulated for the sum or difference of two functions, the rules are applicable to the sums and differences of more than two functions.

EXAMPLE 9

Find the derivative of $y = -\frac{1}{2}x^6 + 5x^3 + 7x - 2$.

Solution

$$\frac{dy}{dx} = \frac{d}{dx}\left(-\frac{1}{2}x^6 + 5x^3 + 7x - 2\right)$$

$$= \frac{d}{dx}\left(-\frac{1}{2}x^6\right) + \frac{d}{dx}(5x^3) + \frac{d}{dx}(7x) - \frac{d}{dx}(2)$$

$$= -\frac{1}{2}\frac{d}{dx}(x^6) + 5\frac{d}{dx}(x^3) + 7\frac{d}{dx}(x) - \frac{d}{dx}(2)$$

$$= -\frac{1}{2}(6x^5) + 5(3x^2) + 7(1) - 0$$

$$= -3x^5 + 15x^2 + 7$$

We now have an easy way of finding the instantaneous rate of change of a function $f(x)$ at a particular input value $x = x_1$; it is $f'(x_1)$. We first find the derivative $f'(x)$ at an arbitrary input value x and then evaluate the result at x_1. Common notations for the derivative of the function $y = f(x)$ evaluated at $x = x_1$ are

$$y'(x_1), \quad f'(x_1), \quad \left.\frac{dy}{dx}\right|_{x=x_1} \quad \text{and} \quad \left.\frac{d}{dx}[f(x)]\right|_{x=x_1}$$

EXAMPLE 10

Find $y'(3)$ and $y'(-2)$ for the function $y = x^5 + x^4$.

Solution

Using the result of part (a) of Example 8, we have $y' = 5x^4 + 4x^3$. Evaluating this derivative at the specified input values, we obtain

$$y'(3) = \left.\frac{dy}{dx}\right|_{x=3} = 5(3)^4 + 4(3)^3 = 5(81) + 4(27) = 513$$

$$y'(-2) = \left.\frac{dy}{dx}\right|_{x=-2} = 5(-2)^4 + 4(-2)^3 = 5(16) + 4(-8) = 48$$

The notation $y'(3)$ means the "derivative of $y = f(x)$ evaluated at $x = 3$." The derivative is taken first and then evaluated. To calculate the derivative at $x = 3$, we substitute $x = 3$ into the expression for the derivative $f'(x)$. The function $f(x)$ is *not* evaluated first and then differentiated.

We presented all of the rules for differentiation in terms of a function $y = f(x)$. If other labels are used instead of x for the input variable and/or y for the output variable, then we simply replace x and/or y with the new labels in each rule.

EXAMPLE 11

FIGURE 10.29

A Spherical Tumor of Radius r

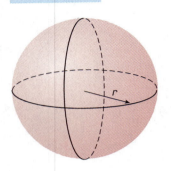

Colon tumors have an approximately spherical shape, as shown in Figure 10.29. As a tumor grows, its radius increases and its volume expands. Find the instantaneous rate of the change of the volume of the tumor with respect to its radius when it is 2 cm long.

Solution

The volume V of a sphere is related to its radius r by the formula $V(r) = \frac{4}{3}\pi r^3$. Here the output V is a function of the input r. Differentiating this function with respect to the input variable, we obtain

$$\frac{dV}{dr} = \frac{d}{dr}\left(\frac{4}{3}\pi r^3\right) = \frac{4}{3}\pi \frac{d}{dr}(r^3) = \frac{4}{3}\pi(3r^2) = 4\pi r^2$$

Evaluating this derivative at $r = 2$, we obtain

$$\left.\frac{dV}{dr}\right|_{r=2} = 4\pi(2)^2 = 16\pi$$

At the moment $r = 2$, the volume is growing at a rate of $16\pi \approx 50.27$ cm^3 for each centimeter increase in r.

EXAMPLE 12

From past experience, a washing machine manufacturing company knows that its monthly profit P (in thousands of dollars) is related to the number of machines N (in thousands of units) by the equation $P(N) = -22N^2 + 210N - 100$. What is the marginal profit when $N = \$4{,}000$?

Solution

Marginal profit is the instantaneous rate of change of profit with respect to the input variable. Here the input variable is N, the output variable is P, and the required instantaneous rate of change is $P'(4)$. Using our rules of differentiation, we have

$$P'(N) = \frac{dP}{dN} = \frac{d}{dN}(-22N^2 + 210N - 100)$$

$$= \frac{d}{dN}(-22N^2) + \frac{d}{dN}(210N) - \frac{d}{dN}(100)$$

$$= -22\frac{d}{dN}(N^2) + 210\frac{d}{dN}(N) - \frac{d}{dN}(100)$$

$$= -22(2N^1) + 210(N^0) - 0$$

$$= -44N + 210$$

Evaluating this derivative at $N = 4$, we obtain a marginal profit of

$$\left.\frac{dP}{dN}\right|_{N=4} = -44(4) + 210 = 34$$

in output units per input unit, or thousands of dollars per thousands of machines. Therefore at a production level of $N = 4$ thousand machines, the marginal profit will be $34,000 if another thousand machines are produced.

IMPROVING SKILLS

In Problems 1–44, find the derivatives for each of the given functions.

1. $f(x) = x^6$
2. $f(x) = 3x^6$
3. $f(x) = 5x^6$
4. $f(x) = -5.4x^6$
5. $y = x^{-8}$
6. $y = 3x^{-8}$
7. $y = 5x^{-8}$
8. $y = -5.4x^{-8}$
9. $f(x) = x^{0.2}$
10. $y = 4x^{0.2}$
11. $f(x) = x^{-0.2}$
12. $y = -2x^{-0.2}$
13. $f(x) = x^6 + x^5$
14. $y = 2x^6 + 3x^5$
15. $y = x^7 - x^{12}$
16. $y = -0.4x^7 - 0.3x^{12}$
17. $y = 2x^{6.1} - 3x^{-6.1}$
18. $f(x) = 3x^{0.7} + 4x^{-1.2}$
19. $y = x^\pi - 1$
20. $f(x) = x^3 - \pi^3$
21. $y = e^\pi - 1$
22. $f(x) = 5^{2.7}$
23. $y = x^5 - 7x^3 + 4x + 2$
24. $f(x) = x^5 - 7x^2 + 3$
25. $y = x^7 + 6x^3 - 4x^2$
26. $f(x) = 9x^2 + 3x + 4$
27. $f(x) = 2x^8 + 6x^5 - 3x^3 + 9x$
28. $f(x) = 2.1x^5 - 7x^{-2} + 3$
29. $f(x) = -3 - 4x^{-4} - 6x^3 - 4x$
30. $f(x) = x^{1/3} - x^{-1/3} + 3$
31. $f(x) = \dfrac{x^5}{5} - \dfrac{x^3}{3} - \dfrac{x^2}{2} + 2$
32. $f(x) = \dfrac{1}{\sqrt{x}}$
33. $y = (x^2 + 3)x^5$
34. $y = (x + 3)(x - 9)$
35. $y = t^4 - 3t$
36. $y = 2t^2 - 3t + 4$
37. $y = 2N^{4.2} + 3N$
38. $y = N^4 - 4N^{-4}$
39. $y = \theta^4 - 2\theta^2$
40. $z = 3t^5 - 2t^4 + t^3$
41. $z = 2r^2 - 3$
42. $r = z^8 - z^4 + 4$
43. $s = 1/t$
44. $t = 2N^7 + 7N^2 - N$

45. Find the instantaneous rate of change of the function $f(x) = x^7 - 3x^4$ at (**a**) $x = 1$ and (**b**) $x = 2$.

46. Find the instantaneous rate of change of the function $f(x) = 2x^3 + 3x^2 - 10$ at (**a**) $x = 2$ and (**b**) $x = -2$.

47. Find $\dfrac{dy}{dx}\bigg|_{x=-1}$ when $y = x^{10} - x^7$.

48. Find $\dfrac{dy}{dt}\bigg|_{t=2}$ when $y = 3t^3 - 5t^2 + 2t - 8$.

49. Find the slope of the tangent line to the graph of $y = x^4 - 2x^2 + 5$ at $x = 3$.
50. Find the slope of the tangent line to the graph of $y = 1 - 2\sqrt{x}$ at $x = 5$.
51. Find the equation of the tangent line to the graph of $y = \sqrt{x}$ at $x = 4$.
52. Find the equation of the tangent line to the graph of $y = x^3 - 2x^2$ at $x = -2$.

53. Consider the function graphed below. Determine the values of the input variable x at which the derivative is (**a**) positive, (**b**) negative, and (**c**) zero.

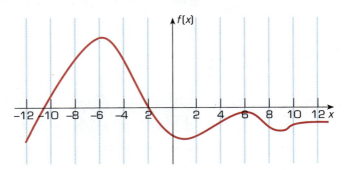

In Problems 54–59, use Equation (13) to find the derivatives of the functions.

54. $f(x) = 3x^2 - 1$ **55.** $f(x) = 3x^2 - 2x$ **56.** $f(x) = x^3$
57. $f(x) = x^4$ **58.** $f(x) = 7x + 5$ **59.** $f(x) = 1/x$

CREATING MODELS

60. A model for population growth in the United States from 1800 through 1990 is

$$y = 6.027403 + 0.023731x + 0.006641x^2$$

where x denotes the year ($x = 0$ is 1800, $x = 100$ is 1900, and $x = 190$ is 1990), and y denotes the corresponding population (in millions). Use this model to calculate the instantaneous rate of change in the U.S. population in (**a**) 1850, (**b**) 1950, and (**c**) 1990.

61. The relationship between temperature measured in degrees Fahrenheit, F, and in degrees Celsius, C, is given by the formula

$$F = \frac{9}{5}C + 32$$

Find the instantaneous rate of change of temperature measured in Fahrenheit degrees with respect to each change of 1°C.

62. Solve the equation in Problem 61 for C in terms of F, thereby making F the input variable and C the output variable. Find the instantaneous rate of change of temperature measured in degrees Celsius with respect to each change of 1°F.

63. A simple model for the distance a bungee jumper travels in free fall is $d = 16t^2$, where d is measured in feet and t in seconds. Find the speed of the jumper after 2 seconds in free fall.

64. Figure A models the distance traveled as a function of time for a driver's 2-mile trip through the Holland Tunnel from New Jersey to New York City. The equation of this curve is

$$D = -0.05t^3 + 0.25t^2 + 0.3t$$

where d is the distance traveled in miles and t is the time through the tunnel in minutes (from the start of the trip).

(**a**) Determine the driver's average speed for the complete 2-mile trip.

FIGURE A

(**b**) Trace Figure A on a separate sheet of paper and draw a line such that the slope represents the average speed determined in part (**a**).

(**c**) Find the speed on the driver's speedometer one minute into the tunnel.

65. A firm's total sales, in millions of dollars, is given by the equation $y = 3x + \frac{1}{2}x^2$, where x denotes the number of years the firm has been in operation. What is the firm's marginal sales after its seventh year in business?

66. The cost C in dollars to produce n handcrafted chess sets is modeled by the equation $C = 500 + 4n - n^2 + 2n^3$. Find the marginal cost when (**a**) $n = 2$, (**b**) $n = 5$, and (**c**) $n = 10$.

67. A manufacturer of electric can openers has a daily fixed cost of $21,000 and knows that the variable cost of producing each can opener is $5.50.

(**a**) Find an expression relating total cost C to N, the number of can openers produced.

(**b**) What is the instantaneous rate of change of C with respect to N?

(**c**) What is the commercial significance of this instantaneous rate of change?

68. A company's sales S are known to be related to advertising expenditures E by the equation

$$S = 150,000 + 6,000E - 50E^2$$

where S is measured in dollars and E is measured in thousands of dollars.

(**a**) Determine whether an increase in advertising is recommended if the present advertising budget is $45,000.

(**b**) Would the situation be different if the present advertising budget were $60,000?

69. A bakery currently has a fixed cost of $120,000 per month and estimates its variable cost at 23¢ per loaf of bread. Loaves are sold at wholesale for 55¢/loaf. Find (**a**) the marginal cost and (**b**) the marginal revenue for this bakery.

70. A projectile is fired vertically from the ground with an initial velocity of 800 ft/sec. After t seconds, in the absence of air resistance, the projectile will attain a height of h feet,

$$h = 800t - 16t^2$$

until the projectile returns to the ground.

(**a**) What is $h'(t)$ when the projectile reaches its apex?

(**b**) How high will the projectile rise before gravity forces it back to earth?

71. Find the instantaneous rate of change of the area of a circle with respect to its radius.

72. Find the instantaneous rate of change of the area of a square with respect to the length of a side.

73. Find the instantaneous rate of change of the area of a square with respect to its perimeter.

EXPLORING IN TEAMS

74. A company believes that the selling price of its product must depend on the amount it produces if they expect to sell all items. The more items the company makes in a month, the lower the unit price must be. Sketch a graph of a possible price–production function for this company. What can you say about the derivative of this function? Do you think that the company's belief is justified?

75. A company introduces a new product to the marketplace with the expectation that the monthly demand for this product will increase rapidly over the first two years, stabilize for a period of time, and finally wane as the product becomes obsolete. Sketch a graph of a demand function for this product over time. What can you say about the derivative of this function? Do you think that the company's expectations for the life cycle of its product is justified?

76. A company must budget for the development of a new product, knowing that the more money budgeted, the quicker the development process. Sketch a graph of a function that models this relationship between money and time. What is the independent variable? What is the dependent variable? What can you say about the derivative of this function?

77. A U.S. airplane takes off at sea level from an East Cost airport. It rises to its cruising altitude and eventually lands at a West Coast airport that is also at sea level. Sketch a graph of the altitude of the plane as a function of time. What can you say about the derivative of this function? What are the units associated with this derivative and what do they signify about the aircraft?

78. Consider the airplane described in Problem 77 with $x(t)$ denoting the distance traveled from the East Coast as a function of time, with $t = 0$ the time the aircraft became airborne. Sketch a graph of this function. What can you say about the derivative of this function?

EXPLORING WITH TECHNOLOGY

79. Many graphing calculators have menus for evaluating a derivative at a point. Use such a calculator to solve Problems 45 through 50.

REVIEWING MATERIAL

80. (Section 5.3) Find the mode, median, mean, and standard derivative for the following data:

$$2, 2, 2, 4, 4, 5, 5, 7, 7, 7, 7, 8, 9, 9$$

81. (Section 6.3) A population is equally divided between females and males, and 8% of the male population is color blind. Find the probability that a randomly selected person from this population is a color-blind male.

82. (Sections 9.2–9.5) From past experience, a company believes that the increase in price P of its product results in a decrease in demand D for that product. Based on this information, which of the following equations *cannot* be a model for this relationship?

(*a*) $D = 5P - 1000$ (*b*) $D = 3000/P$ (*c*) $D = P^2 - 2000$

(*d*) $D = e^P$ (*e*) $D = e^{-P}$ (*f*) $D = \ln P$

RECOMMENDING ACTION

83. Respond by memo to the following request:

> MEMORANDUM
>
> To: J. Doe Reader
>
> From: Vice President for Marketing
>
> Date: Today
>
> Subject: **Advertising Budget**
>
> We know that our profit depends on our advertising expenditures, and we have a good model for this relationship consisting of a function that relates these two variables. I have a recommendation on my desk to base future advertising budgets on our marginal profit. The recommendation calls for increasing the budget as long as our marginal profit is positive. Do you agree?

10.4 MORE RULES OF DIFFERENTIATION

If all functions were polynomials or power functions, the rules of differentiation presented so far would be sufficient. We know, however, that other functions are important for modeling, including exponentials, logarithms, and trigonometric functions, and none of these functions can be differentiated with our current collection of rules. Let us start with a rule for differentiating the natural exponential function.

THE NATURAL EXPONENTIAL RULE OF DIFFERENTIATION

The derivative of the function $f(x) = e^x$ is

$$\frac{d(e^x)}{dx} = e^x$$

The natural exponential function has the remarkable property of not changing under differentiation. In other words, the derivative of e^x is itself. The slope of the tangent line to the graph of $y = e^x$ at any point (x, e^x) is the same as the value of the function e^x at that point, as illustrated in Figure 10.30.

We can combine the natural exponential rule of differentiation with the other rules to differentiate constant multiples of e^x as well as functions where e^x is part of a sum.

FIGURE 10.30

Some Tangent Lines to the Graph of the Natural Exponential Function

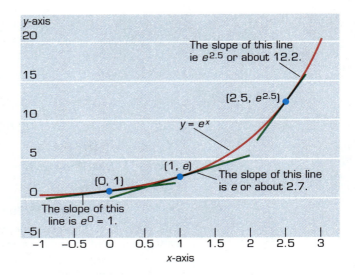

EXAMPLE 1

Find the derivative of the function $y = 10e^x$.

Solution

The derivative of e^x is itself, so it follows from the constant multiple rule of differentiation that

$$\frac{dy}{dx} = \frac{d}{dx}(10e^x) = 10\frac{d}{dx}(e^x) = 10e^x$$

EXAMPLE 2

Find the derivative of $y = 3x^5 + 5x^2 - 4x + 5 - 8e^x$, evaluated at $x = -1$.

Solution

First we differentiate:

$$\frac{dy}{dx} = \frac{d}{dx}(3x^5 + 5x^2 - 4x + 5 - 8e^x) = 15x^4 + 10x - 4 + 0 - 8e^x$$

Therefore,

$$\left.\frac{dy}{dx}\right|_{x=-1} = 15(-1)^4 + 10(-1) - 4 - 8e^{-1} \approx -1.943036$$

Now let us consider the derivative of the natural logarithm function.

THE NATURAL LOGARITHMIC RULE OF DIFFERENTIATION

The derivative of the function $f(x) = \ln x$ is

$$\frac{d(\ln x)}{dx} = \frac{1}{x}$$

The natural logarithm function has the interesting property that the slope of the tangent line to the graph of $y = \ln x$ at any point $(x, \ln x)$ is the reciprocal of x, as illustrated in Figure 10.31.

FIGURE 10.31

Two Tangent Lines to the Graph of the Natural Logarithm Function

---

EXAMPLE 3

Find the derivative of $y = -4 \ln x$.

Solution

$$\frac{dy}{dx} = \frac{d}{dx}(-4 \ln x) = -4\frac{d}{dx}(\ln x) = -4\left(\frac{1}{x}\right) = -\frac{4}{x}$$

EXAMPLE 4

Find the derivative of $y = \frac{1}{2}x^2 - \ln x + e^2$, evaluated at $x = 2$.

Solution

Because $e^2 \approx 7.389$ is a constant, its derivative is zero:

$$\frac{dy}{dx} = \frac{d}{dx}\left(\frac{1}{2}x^2 - \ln x + e^2\right) = x - \frac{1}{x} + 0 = x - \frac{1}{x}$$

Therefore

$$\left.\frac{dy}{dx}\right|_{x=2} = 2 - \frac{1}{2} = 1.5$$

The derivatives of the sine and cosine functions are closely related to each other.

THE SINE RULE OF DIFFERENTIATION

The derivative of the function $f(x) = \sin x$ is

$$\frac{d(\sin x)}{dx} = \cos x$$

EXAMPLE 5

Find the derivative of the function $y = 2.1 \sin x$.

Solution

$$\frac{dy}{dx} = \frac{d}{dx}(2.1 \sin x) = 2.1 \frac{d}{dx}(\sin x) = 2.1 \cos x$$

THE COSINE RULE OF DIFFERENTIATION

The derivative of the function $f(x) = \cos x$ is

$$\frac{d(\cos x)}{dx} = -\sin x$$

EXAMPLE 6

Evaluate the derivative of $y = 3 \sin x - 4 \cos x + 5$ at $x = \pi$.

Solution

We must use several different rules of differentiation here:

$$\frac{dy}{dx} = \frac{d}{dx}(3 \sin x - 4 \cos x + 5) = 3(\cos x) - 4(-\sin x) + 0 = 3 \cos x + 4 \cos x$$

Now, evaluating at $x = \pi$, we get

$$\left.\frac{dy}{dx}\right|_{x=\pi} = 3 \cos \pi + 4 \sin \pi = 3(-1) + 4(0) = -3$$

In Section 10.3, we developed the sum and difference rules for differentiation. These two rules, along with the constant multiple rule, allowed us to differentiate combinations of functions, which significantly increased our differentiation skills. Now we have differentiation rules for the exponential, logarithmic, and trigonometric functions, and we can link these rules with the combination rules for differentiation. This allows us to differentiate such functions as $2 \sin x$, $-3 \sin x$, $\frac{1}{2}\cos x$, $e^x + \sin x$, $2e^x - \ln x$, and $2e^x - 3 \sin x$.

The sum and difference of two functions are natural arithmetic combinations. So too are the product and quotient of two functions, but the differentiation rules for these combinations are not as straightforward.

THE PRODUCT RULE OF DIFFERENTIATION

If $u(x)$ and $v(x)$ are two functions with known derivatives $u'(x)$ and $v'(x)$, respectively, then the derivative of the product $y = u(x) \cdot v(x)$ is

$$y' = u(x) \cdot v'(x) + u'(x) \cdot v(x)$$

Note: The derivative of a product is *not* equal to the product of the individual derivatives.

EXAMPLE 7

Find the derivatives of the following functions: (*a*) $y = x^2 e^x$; (*b*) $y = x \ln x$; (*c*) $y = \sin x \cos x$.

Solution

(*a*) In this case $x^2 e^x$ is the product of x^2 with e^x. Therefore

$$\frac{d}{dx}(x^2 e^x) = x^2 \cdot \left[\frac{d}{dx}(e^x)\right] + \left[\frac{d}{dx}(x^2)\right] \cdot e^x$$

$$= x^2 e^x + 2x e^x$$

(*b*) Here $x \ln x$ is the product of x with $\ln x$. Therefore

$$\frac{d}{dx}(x \ln x) = x \cdot \left[\frac{d}{dx}(\ln x)\right] + \left[\frac{d}{dx}(x)\right] \cdot \ln x$$

$$= x \cdot \frac{1}{x} + 1 \cdot \ln x$$

$$= 1 + \ln x$$

(*c*) The product $\sin x \cos x$ has factors $\sin x$ and $\cos x$; therefore

$$\frac{d}{dx}(\sin x \cos x) = \sin x \cdot \left[\frac{d}{dx}(\cos x)\right] + \left[\frac{d}{dx}(\sin x)\right] \cdot \cos x$$

$$= (\sin x)(-\sin x) + (\cos x)(\cos x)$$

$$= -(\sin x)^2 + (\cos x)^2$$

This last quantity is often written more simply as $-\sin^2 x + \cos^2 x$.

For consistency, we'll continue stating the rules of differentiation in terms of a function $y = f(x)$. However, if other labels—θ for an angle, t for time, T for temperature, S for sales—are more appropriate for a particular model, then we'll replace x and/or y with those labels.

EXAMPLE 8

A major soft drink producer test-marketed a new high-energy cola in one region of the country for 6 months and found that daily sales S (in thousands of cans) at time t (in months) were given by the model

$$S(t) = 200 + 10(t^{1.5} - 5)\ln t \quad (1 \le t \le 6) \qquad (14)$$

The company plans to introduce the cola to a second region and expects a similar growth pattern. What growth rate can it expect after $1\frac{1}{2}$ months and after 3 months?

Solution

Figure 10.32 displays a window of the graph of Equation (14). We want the instantaneous rate of change of sales with respect to time, $S'(t)$. Since the formula for sales is the sum of two terms, we apply the sum rule of differentiation, noting that the derivative of 200 is zero. Thus,

$$\frac{dS}{dt} = \frac{d}{dt}(200) + \frac{d}{dt}[10(t^{1.5} - 5)\ln t]$$

$$= \frac{d}{dt}[10(t^{1.5} - 5)\ln t]$$

FIGURE 10.32

A Graphical Representation of Cola Sales

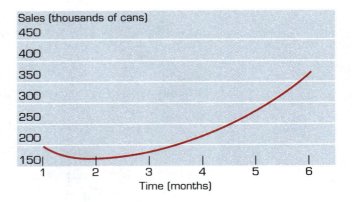

The term yet to be differentiated is a product, so now we use the product rule of differentiation:

$$\frac{d}{dt}[10(t^{1.5} - 5)\ln t] = \frac{d}{dt}(10t^{1.5} - 50)\ln t$$

$$= (10t^{1.5} - 50) \cdot \frac{d}{dt}(\ln t) + \left[\frac{d}{dt}(10t^{1.5} - 50)\right] \cdot \ln t$$

$$= (10t^{1.5} - 50) \cdot \frac{1}{t} + (15t^{0.5} - 0) \cdot \ln t$$

$$= 10t^{0.5} - \frac{50}{t} + 15t^{0.5}\ln t$$

$$= (15 \ln t + 10)t^{0.5} - \frac{50}{t}$$

Therefore the growth rate of sales at the end of 1.5 months is

$$\frac{dS}{dt}\bigg|_{t=1.5} = (15 \ln 1.5 + 10)(1.5)^{0.5} - \frac{50}{1.5} \approx -13.6$$

At the end of a month and a half, sales are declining at the rate of 13,600 cans/month. However, the growth rate of sales at the end of 3 months is

$$\frac{dS}{dt}\bigg|_{t=3} = (15 \ln 3 + 10)(3)^{0.5} - \frac{50}{3} \approx 29.2$$

so sales are increasing at the rate of 29,200 cans/month.

THE QUOTIENT RULE OF DIFFERENTIATION

If $u(x)$ and $v(x)$ are two functions with known derivatives $u'(x)$ and $v'(x)$, respectively, then the derivative of the quotient $y(x) = u(x)/v(x)$ is

$$y'(x) = \frac{u'(x) \cdot v(x) - u(x) \cdot v'(x)}{[v(x)]^2}$$

The product and quotient rules of differentiation are not as straightforward as the rules for sums and differences, so some care is required when applying them.

EXAMPLE 9

Find the derivatives of the following functions: (*a*) $y = (\ln x)/x^5$; (*b*) $y = e^x/(x^2 + 4)$; (*c*) $y = (\sin x)/(\cos x)$.

Solution

(*a*) $\dfrac{d}{dx}\left(\dfrac{\ln x}{x^5}\right) = \dfrac{\left[\dfrac{d}{dx}(\ln x)\right] \cdot x^5 - (\ln x)\dfrac{d}{dx}(x^5)}{[x^5]^2}$

$= \dfrac{\dfrac{1}{x} \cdot x^5 - (\ln x)\, 5x^4}{x^{10}}$

$= \dfrac{x^4 - 5x^4 \ln x}{x^{10}} = \dfrac{1 - 5\ln x}{x^6}$

(*b*) $\dfrac{d}{dx}\left(\dfrac{e^x}{x^2 + 4}\right) = \dfrac{\left[\dfrac{d}{dx}(e^x)\right](x^2 + 4) - e^x\left[\dfrac{d}{dx}(x^2 + 4)\right]}{(x^2 + 4)^2}$

$= \dfrac{e^x(x^2 + 4) - e^x(2x)}{(x^2 + 4)^2}$

$= \dfrac{(x^2 - 2x + 4)e^x}{(x^2 + 4)^2}$

(*c*) $\dfrac{d}{dx}\left(\dfrac{\sin x}{\cos x}\right) = \dfrac{\left[\dfrac{d}{dx}(\sin x)\right](\cos x) - (\sin x)\dfrac{d}{dx}(\cos x)}{(\cos x)^2}$

$= \dfrac{(\cos x)(\cos x) - (\sin x)(-\sin x)}{\cos^2 x}$

$= \dfrac{\cos^2 x + \sin^2 x}{\cos^2 x}$

There are several trigonometric identities that will be helpful here:

$$\cos^2 x + \sin^2 x = 1 \qquad \frac{\sin x}{\cos x} = \tan x \qquad \frac{1}{\cos x} = \sec x$$

With them, we can rewrite part (*c*) of Example 9 as

$$\frac{d}{dx}(\tan x) = \frac{d}{dx}\left(\frac{\sin x}{\cos x}\right) = \frac{\cos^2 x + \sin^2 x}{\cos^2 x}$$

$$= \frac{1}{\cos^2 x} = \left(\frac{1}{\cos x}\right)^2 = \sec^2 x$$

which is the rule for differentiating the tangent function.

EXAMPLE 10

The Cuddly Company, which manufactures teddy bears, believes that the number n of bears sold depends on the price p (in dollars) of each bear according to the formula

$$n = \frac{200{,}000}{1 + 0.01p^2} \qquad (15)$$

Determine the marginal revenue when $p = \$20$. What is advisable regarding a proposed price increase?

Solution

Figure 10.33 is a window of the graph for Equation (15): it shows us that the demand decreases as the price increases. Total revenue is $R = np$, the number of bears sold times the price charged per bear. Therefore

$$R(p) = \left(\frac{200{,}000}{1 + 0.01p^2}\right)p = \frac{200{,}000p}{1 + 0.01p^2}$$

FIGURE 10.33

A Portion of the Graph of Teddy Bear Sales

Using the quotient rule, we get a formula for the change in revenue:

$$\frac{dR}{dp} = \frac{d}{dp}\left(\frac{200{,}000p}{1 + 0.01p^2}\right)$$

$$= \frac{\left[\dfrac{d}{dp}(200{,}000p)\right](1 + 0.01p^2) - 200{,}000p\left[\dfrac{d}{dp}(1 + 0.01p^2)\right]}{(1 + 0.01p^2)^2}$$

$$= \frac{200{,}000(1 + 0.01p^2) - 200{,}000p(0 + 0.02p)}{(1 + 0.01p^2)^2}$$

$$= \frac{200{,}000(1 - 0.01p^2)}{(1 + 0.01p^2)^2}$$

Evaluating this derivative at $p = 20$, we obtain the corresponding marginal revenue:

$$\left.\frac{dR}{dp}\right|_{p=20} = \frac{200{,}000(1 - 0.01(20)^2)}{[1 + 0.01(20)^2]^2} = -\$24{,}000$$

At a price of $20 per bear, marginal revenue is negative, indicating that an increase in price will decrease total income. Not only is a price increase not advisable, the Cuddly Co. should *cut* their prices!

IMPROVING SKILLS

Find the derivatives of each of the functions given in Problems 1 through 50.

1. $f(x) = xe^x$
2. $f(x) = 2xe^x$
3. $f(x) = x^4 e^x$
4. $f(t) = t^{3.75} e^t$
5. $f(x) = (x^3 + x)e^x$
6. $f(t) = (2t^2 + 3t - 14)e^t$
7. $f(z) = (z^{0.8} - z^{-0.8})e^z$
8. $f(x) = e^x \sin x$
9. $y = e^x \cos x$
10. $y = e^x \ln x$
11. $y = x^2 \ln x$
12. $f(x) = 5x^3 \ln x$
13. $y = t^{-2} \ln t$
14. $y = (s^2 + 1)\ln s$
15. $N = (x^3 - 5x^2)\ln x$
16. $y = \sin x \ln x$
17. $f(\theta) = -\frac{1}{4}\cos \theta \ln \theta$
18. $z = t \cos t$
19. $P = x \sin x$
20. $y = \theta^2 \sin \theta$
21. $x = q^{-6} \sin q$
22. $f(t) = (t^2 + 3t - 1)\cos t$
23. $N = \pi + \sin x$
24. $f(x) = xe^2 + 2e^x - \pi^3$
25. $y = \sqrt{x}e^x$
26. $f(x) = \sqrt{x} \ln x$

27. $y = \dfrac{3x + 5}{x^2 - 1}$
28. $f(x) = \dfrac{5}{1 - x^2}$

29. $y = \dfrac{x^2 + 2x}{3x + 4}$
30. $f(x) = \dfrac{2x - 3}{3x - 2}$

31. $f(x) = \dfrac{x^2 + 2x - 3}{x^2 - 5x + 1}$
32. $z = \dfrac{2x^2 + 5x}{x^2 + 7x - 2}$

33. $f(t) = \dfrac{-7t}{t^2 + 8}$
34. $z = \dfrac{t^4 + t^2 + 1}{t^3 - t}$

35. $z = \dfrac{\sqrt{x} + 1}{\sqrt{x} - 1}$
36. $f(t) = \dfrac{1}{\sqrt{t} - t}$

37. $f(x) = \dfrac{e^x + 5}{e^x - 3}$
38. $f(x) = \dfrac{e^x}{\sin x}$

39. $f(x) = \dfrac{x}{\ln x}$
40. $f(x) = \dfrac{5}{\ln x}$

41. $f(t) = \dfrac{\ln t}{t^3}$
42. $f(\theta) = \dfrac{\ln \theta}{\theta^2 - 3\theta + 4}$

43. $y = \dfrac{\cos \theta}{\sin \theta}$
44. $f(x) = \dfrac{xe^x}{\sin x}$

45. $y = \dfrac{x \ln x}{e^x + 1}$
46. $f(x) = \dfrac{(x^2 + x)\ln x}{x^2 - 1}$

47. $f\theta) = \dfrac{\theta \sin \theta}{1 + 2\theta}$

48. $z = \dfrac{(x + 4)e^x}{xe^x + 4}$

49. $f(t) = \dfrac{(t^3 - 1)\ln t}{t^2 e^t}$

50. $f(t) = \dfrac{(3 - e^\pi)\sin t}{(t^2 + t + 2)\ln 2}$

51. Find the instantaneous rate of change of the function $f(x) = x \cos x$ at (**a**) $x = 1$; (**b**) $x = \pi$.

52. Find the instantaneous rate of change of the function $f(x) = (2x - 1)/(x + 3)$ at (**a**) $x = 2$; (**b**) $x = -2$.

53. Find $y'(-1)$ when $y = (x - 2)e^x$.

54. Find $\left.\dfrac{dy}{dt}\right|_{t=2}$ when $y = \dfrac{e^t - 5}{t + 2}$.

55. Find the slope of the tangent line to the graph of $y = (x^4 + 5)\ln x$ at $x = 3$.

56. Find the slope of the tangent line to the graph of $y = \sqrt{x}/e^x$ at $x = 1.3$.

57. Find the equation of the tangent line to the graph of $y = (2x - 3)/(1 - x^2)$ at $x = 4$.

58. Find the equation of the tangent line to the graph of $y = (x^2 + 5)\sin x$ at $x = -2$.

CREATING MODELS

59. Hobbystop Toys always has one employee on the staff to assemble dollhouses for customers. The more experienced the employee, the quicker the assembly process, but the work is so repetitive, the dollhouse assemblers leave as soon as other opportunities become available. From experience, the owner of Hobbystop knows that the number n of dollhouses assembled each day is directly related to the number d of days that the assembler has been on the job; this relation is modeled by the function

$$n(d) = \frac{5d}{d + 4}$$

What is an employee's rate of improvement after (**a**) one day on the job? (**b**) ten days on the job?

60. Psychological studies show that an individual's interest in accomplishing a specific task increases until mastery is attained: then interest wanes once the task is routine. Test results from a sample of individuals learning the same task suggest that interest I (measured on a scale from 0 to 1) is related to learning time t (in hours) by the formula

$$\frac{1 + \ln\left(t + \dfrac{1}{e}\right)}{1.5\left(t + \dfrac{1}{e}\right)^2} \qquad (t > 0)$$

How does an individual's interest change with respect to time after (**a**) $\frac{1}{2}$ hour? (**b**) 3 hours?

61. The population size P of a species of newt is being studied in a controlled experiment. The population size changes with time $t \geq 0$ according to the function $P = f(t)$. What can you say about this newt population if you know that $f'(t) \geq 0$ for all values of $t \geq 0$?

62. The population P of white rhinoceroses has been declining ever since 1950. Assume that P is related to time by the function $P = g(t)$. What can you say about the sign of $g'(t)$ during this time period?

63. The population E of adult bald eagles in the 48 contiguous states declined from 1950 through 1973, when the Endangered Species Act became law. From that year to the present, the population has increased yearly. What can you say about the sign of E' from 1950 to the present?

64. The only animals on an island are rabbits and foxes. The rabbits eat vegetation and the foxes eat rabbits. Biologists studying this island use the equation

$$R(t) = \sqrt{t}(100 + 50 \sin t)$$

as their model of the rabbit population, where R denotes the number of rabbits on the island and t denotes the time in years from the start of the study. Determine the rate of change in rabbit population after (**a**) one year; (**b**) four years.

65. A model for the fox population in Problem 64 is

$$F(t) = (10 + 2 \cos t)\ln t$$

where F denotes the number of foxes on the island and t again denotes the number of years since the start of the study. Determine the rate of change in fox population after (**a**) one year; (**b**) four years.

66. The demand d for a product is related to its price p by a function $d = f(p)$. Suppose that $f(15) = 3000$ and $f'(15) = -100$.
    (**a**) What does this information tell you about the demand for the product at a price of \$15?
    (**b**) What can you deduce about the demand for the product at another price?

67. The demand d for a product is related to its price p by a function $d = f(p)$. Suppose that $f(15) = 3000$ and $f'(15) = -100$, and suppose that revenue is $R = pd$. Determine the marginal revenue at a price of \$15.

68. Solve Problem 67 if $f'(15) = -400$.

69. Solve Problem 67 if $f(15) = 1250$.

70. A company believes that the number n of items it sells depends on their unit price p (in dollars) according to the formula

$$n = \frac{60,000}{4 + p}$$

Determine the marginal revenue at a unit price of \$3. What recommendation would you make to management regarding a price increase for this item?

71. What recommendation would you make to the management in Problem 70 if the formula relating n to p is changed to

$$n = \frac{60,000}{4 + p^2}$$

EXPLORING IN TEAMS

72. Generalize the product rule of differentiation to the product of *three* functions,

$$y = f(x) = u(x) \cdot v(x) \cdot w(x)$$

Hint: Set $g(x) = v(x) \cdot w(x)$ and then start by differentiating $f(x) = u(x) \cdot g(x)$.

73. Generalize the product rule of differentiation to the product of *four* functions,

$$y = f(x) = u(x) \cdot v(x) \cdot w(x) \cdot z(x)$$

Do you see a pattern developing? Can you use this pattern to take the derivative of the product of *five* functions?

74. Use the results of Problems 72 and 73 to calculate the derivatives of the following products:
    (a) $(x - 1)(x - 2)$
    (b) $(x - 1)(x - 2)(x - 3)$
    (c) $(x - 1)(x - 2)(x - 3)(x - 4)$
    (d) $(x - 1)(x - 2)(x - 3)(x - 4) \cdots (x - n)$

75. Use the properties of exponents along with other rules of differentiation to find the derivatives of e^{2x} and e^{3x}. Using your results, can you guess the derivatives of e^{4x} and e^{5x}?

76. Use the properties of exponents along with other rules of differentiation to find the derivative of e^{-x}.

77. Show that the constant rule of differentiation is an immediate consequence of the product rule of differentiation.

EXPLORING WITH TECHNOLOGY

78. Many graphing calculators have menus for evaluating a derivative at a point. Use such a calculator to solves Problems 51 through 56.

79. Use a graphing calculator to evaluate the derivative of the function for total revenue used in Example 10, namely

$$R(p) = \frac{200,000p}{1 + 0.01p^2}$$

at different values of p between 1 and 20. Based on your results, what selling price for teddy bears do you recommend to the owner of Hobbystop Toys?

80. Graph the function $R(p)$ in Problem 79. At what input value is the output the largest? How does this compare to your answer to Problem 79? What is $R'(p)$ at that point?

81. Graph the function

$$y = \frac{\ln x}{x} \quad (x > 0)$$

and determine the value of x that generates the largest value of y. What is dy/dx at this point? Does this surprise you?

REVIEWING MATERIAL

82. (Section 4.1) Shirley invests $3,000 in a 3-year certificate of deposit that pays 5.3% a year compounded monthly. At the end of three years, she reinvests the full proceeds into a new 2-year CD that earns 6.7% a year compounded quarterly. How much will the initial $3,000 deposit be worth at the end of the 5-year period?

83. (Section 7.3) Victims of a particular disease are classified as either cured, infected, or dead. Once cured, a person is immune from the disease and can never be infected again. Each year, 80% of all people infected with this disease are cured, 5% die from the disease, and 15% remain infected. Currently, a city has 1,000 people infected with the disease. What will happen to these people over the long run?

84. (Sections 8.2–8.5) Match each of the graphs in Figures *A–D* with one of the following functions:

(*a*) x^3 (*b*) x^4 (*c*) $\ln x$ (*d*) e^x

(*e*) $\sin(\pi x/2)$ (*f*) $\sin(2\pi x)$ (*g*) $\cos(2\pi x)$ (*h*) $\cos(x/2\pi)$

FIGURE A

FIGURE B

FIGURE C

FIGURE D

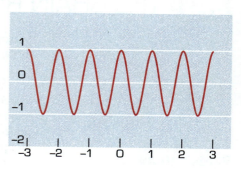

RECOMMENDING ACTION

FIGURE E

Items sold (×1000)   s = items sold; d = dollars

$$s = 15 - \frac{14}{1+d}$$

Advertising dollars (× $100,000)

85. Respond by memo to the following request:

MEMORANDUM

To:	J. Doe Reader
From:	Vice President for Marketing
Date:	Today
Subject:	**Advertising Budget**

We know that sales are influenced by advertising but only to the point where we saturate our market. Our operations research staff contends that this relationship can be modeled by Figure E, where S denotes sales and d represents our advertising dollars. The price of our product is currently $100/item, and this price is determined by the competition. Our income is derived completely from sales and our revenue is income minus dollars spent on advertising. Based on this information, how much money should we budget for advertising if our goal is to maximize revenue?

10.5 THE CHAIN RULE

Most processes considered in this chapter and in Chapter 9 can be modeled by equations of the form

$$\text{output} = f(\text{input}) \tag{16}$$

Equation (16) models a single operation with a rule of correspondence f that assigns one output to each input. We now want to consider processes involving a sequence of operations, where the output from one is the input to the next. Sequential operations are commonplace in any society that promotes job specializations.

Consider a company that sequentially converts iron ore into body fenders for automobiles, as illustrated by Figure 10.34. The ore is delivered to a smelting plant, where it is converted into sheets of rolled steel. The rolled steel is then delivered to a manufacturing plant that cuts and shapes the steel into fenders. Iron ore is the input to the smelting plant, and rolled steel is both the output from the smelting plant and the input to the fender plant. This sequential set of operations is modeled schematically in Figure 10.35.

FIGURE 10.34

FIGURE 10.35

The Sequential Operation of Transforming Iron Ore into Automobile Fenders

The progression of operations shown in Figure 10.35 can be expanded, depending on the scope of the system we're modeling. Fabrication plants take fenders as inputs and convert them into new vehicles. Dealerships take new vehicles as inputs and process them into sold vehicles. Sold vehicles depreciate and, eventually, are converted into scrap metal.

EXAMPLE 1

Construct a schematic model for a sequence of operations that transforms corn into bread.

Solution

A farmer sells ears of corn to a mill that grinds the corn into cornmeal. The cornmeal is then sold to a bakery that uses it as an ingredient in cornbread. This sequence of operations is shown schematically in Figure 10.36.

FIGURE 10.36

Transforming Corn into Cornbread

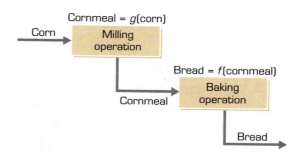

Let's denote an ear of corn by the letter c, cornmeal by the letter m, and bread by the letter b. Following our convention of letting f and g denote operations on inputs, we model the schematic diagram in Figure 10.36 with two equations:

$$b = f(m) \quad \text{and} \quad m = g(c) \tag{17}$$

We noted in Section 9.1 that the functions of greatest interest are those in which both the domain and the range are subsets of the real numbers, so we continue to confine ourselves to functions of this type. For Equations (17), the units of b, m, and c could be loaves of bread, pounds of meal, and bushels of corn, respectively. Alternatively, each variable might be measured by its monetary value, in which case all three variables would have units of dollars.

It is apparent from Equations (17) that m is both an output of one operation (the mill) and an input to a second operation (the bakery). Combining the two equations by substituting $g(c)$ into the first equation for m, we have

$$b = f(g(c)) \tag{18}$$

The composite function of f with g, denoted by $f \circ g$, is

$$f \circ g(x) = f(g(x))$$

as a direct relationship between bread and corn. The right side of Equation (18) is a combination of the functions f and g, known as the *composite function* of f with g, denoted $f \circ g$. We write

$$f \circ g(x) = f(g(x)) \tag{19}$$

EXAMPLE 2

Find $b = f \circ g(c)$ when

$$b = f(m) = \ln(2m) \quad \text{and} \quad m = g(c) = 3c + 1$$

Solution

The output of g is the input to f. If we replace m by $3c + 1$ in the formula $b = \ln(2m)$, we obtain

$$b = \ln(2m) = \ln[2(3c + 1)] = \ln(6c + 2)$$

This is equivalent to writing

$$b = f(g(c)) = f(3c + 1) = \ln(6c + 2)$$

Either way,

$$b = f \circ g(c) = \ln(6c + 2)$$

EXAMPLE 3

Find $f \circ g(x)$ when $y = f(u) = e^u$ and $u = g(x) = x^2$.

Solution

To calculate $f \circ g$, we note that the output of g becomes the input to f. If we replace u in the formula $y = f(u) = e^u$ with $g(x) = x^2$, we have

$$y = e^u = e^{x^2}$$

This is equivalent to writing

$$y = f(g(x)) = f(x^2) = e^{x^2}$$

Either way,

$$f \circ g(x) = e^{x^2}$$

We developed the equations

$$b = f(m) \quad \text{and} \quad m = g(c)$$

in Example 1 to model the process in which cornbread, b, depends on cornmeal, m, and then the process in which cornmeal, m, depends on corn, c. If we measure all of the factors in terms of their monetary value, then these equations relate the price of cornbread to the price of cornmeal and the price of cornmeal to the price of corn. The composite function $b = f \circ g(c)$ relates the price of cornbread directly to the price of corn.

If we are interested only in the relationship between the price of cornbread to the price of corn, then the equation $b = f \circ g(c)$ is sufficient. If, however, we want to know the rate of change in the price of cornbread with respect to a price-change for corn, then we need to look at the derivative db/dc. One technique for finding db/dc is to calculate the composite function $b = f \circ g(c)$ and then differentiate both sides of this equation with respect to c. An alternate approach is given by the *chain rule*.

THE CHAIN RULE OF DIFFERENTIATION

If $y = f(u)$ and $u = g(x)$ and if both $\dfrac{dy}{du}$ and $\dfrac{du}{dx}$ exist, then

$$\frac{dy}{dx} = \frac{dy}{du} \cdot \frac{du}{dx} \tag{20}$$

Equation (20) can be rewritten

$$\frac{dy}{dx} = \left[\frac{d}{du} f(u)\right]\left[\frac{d}{dx} g(x)\right]$$

EXAMPLE 4

Use the chain rule to find dy/dx when $y = e^u$ and $u = x^2$.

Solution

It follows from the natural exponential rule of differentiation that

$$\frac{dy}{du} = \frac{d}{du}(e^u) = e^u$$

and from the power rule of differentiation that

$$\frac{du}{dx} = \frac{d}{dx}(x^2) = 2x$$

Using the chain rule, we now obtain

$$\frac{dy}{dx} = \frac{dy}{du} \cdot \frac{du}{dx} = (e^u)(2x)$$

Since $u = x^2$, we can simplify this last line to

$$\frac{dy}{dx} = (e^{x^2})(2x) = 2xe^{x^2}$$

The preceding example forcefully illustrates the astounding power of the chain rule. We are given $y = e^u$ and $u = x^2$ and we are asked to find dy/dx. As we noted, one technique is to first find the composite function $f \circ g$ and then differentiate it. In Example 3, we identified this particular composite function as $y = e^{x^2}$. Unfortunately, none of the differentiation rules given in Sections 10.3 or 10.4 applies to this function. We do not know how to differentiate $y = e^{x^2}$ directly! The natural exponential rule of differentiation allows

The method of u-substitution is used to decompose a complicated function with unknown derivatives into a composite function of simpler functions with known derivatives.

us to calculate $\dfrac{d}{dx}(e^x)$ but *not* $\dfrac{d}{dx}(e^{x^2})$. Using the chain rule, however, we now know that

$$\frac{d}{dx}(e^{x^2}) = 2xe^{x^2}.$$

The technique provided by the chain rule allows us to take a complicated function with an unknown derivative and decompose it into a function made up of simpler functions with known derivatives. This is our most powerful procedure for differentiating complicated functions. The decomposition is accomplished by creating a new variable u for part of the original function. The selection of u is governed by our ability to differentiate it. We next substitute u for the chosen part of the original function, which we then write in terms of the new, single variable u. Hopefully the resulting function is differentiable, so we can use the chain rule to identify the derivative of the original function. This procedure is known as *u-substitution*.

In Example 4, we needed to differentiate e^{x^2}. We did this by setting $u = x^2$, which we know how to differentiate; furthermore, we also know how to differentiate $y = e^{x^2} = e^u$ as a function of u. Thus we were able to use the chain rule to differentiate e^{x^2}.

EXAMPLE 5

Find the derivative of $y = (1 + x^3)^{3.8}$.

Solution

None of our rules of differentiation applies to this function, so we try a *u-substitution*. If we set $u = 1 + x^3$, then

$$y = (1 + x^3)^{3.8} = u^{3.8}$$

which is a candidate for the power rule of differentiation:

$$\frac{dy}{du} = \frac{d}{du}(u^{3.8}) = 3.8u^{3.8-1} = 3.8u^{2.8}$$

We can use the sum rule of differentiation to obtain

$$\frac{du}{dx} = \frac{d}{dx}(1 + x^3) = 3x^2$$

It follows from the chain rule that

$$\frac{dy}{dx} = \frac{dy}{du} \cdot \frac{du}{dx} = [3.8u^{2.8}][3x^2] = 11.4x^2 u^{2.8}$$

Substituting $u = 1 + x^3$ into this last equation, we have

$$\frac{dy}{dx} = \frac{d}{dx}(1 + x^3)^{3.8} = 11.4x^2(1 + x^3)^{2.8}$$

In Example 5, we set $u = 1 + x^3$. Suppose, instead, we had chosen $u = x^3$, whose derivative $\dfrac{du}{dx} = \dfrac{d}{dx}(x^3) = 3x^2$ is also known. Now

$$y = (1 + x^3)^{3.8} = (1 + u)^{3.8}$$

A RULE OF THUMB:

If a function is a quantity raised to a power, choose u to be that quantity.

Here we have y as a function of u but we don't know how to calculate $\dfrac{dy}{du} = \dfrac{d}{du}(1 + u)^{3.8}$,

so we can't proceed with the chain rule, based on rules we currently have.

Often some u-substitutions will be more effective than others. One useful rule of thumb is this: If a function is a quantity raised to a power, choose u to be that quantity. In Example 5, we had the quantity $1 + x^3$ raised to the 3.8th power. Consequently we set $u = 1 + x^3$. Experience gives us directions—rules of thumb—that have been effective in the past, but we must always be willing to discard a u-substitution when it is not productive and begin anew with a different choice.

EXAMPLE 6

Find the derivative of the function $y = \ln(3x^2 - 8x + 5)$.

Solution

None of our rules of differentiation applies to this function, so let's look for a u-substitution. There are a number of obvious choices, such as $u = x^2$, or $u = 3x^2$, or $u = -8x$, or $u = -8x + 5$. Instead, we take the entire polynomial:

$$u = 3x^2 - 8x + 5$$

ANOTHER RULE OF THUMB:

If the function has an argument, take u to be that argument.

based on another rule of thumb derived from experience: If the composite function has an argument, take u to be that argument. In this example, our choice of u is the argument of the natural logarithm function. This gives us

$$y = \ln(3x^2 - 8x + 5) = \ln u$$

We know that

$$\frac{du}{dx} = \frac{d}{dx}(3x^2 - 8x + 5) = 6x - 8$$

and we have from the natural logarithm rule of differentiation that

$$\frac{dy}{du} = \frac{d}{du}(\ln u) = \frac{1}{u}$$

It now follows from the chain rule that

$$\frac{dy}{dx} = \frac{dy}{du} \cdot \frac{du}{dx} = \left(\frac{1}{u}\right)(6x - 8) = \frac{6x - 8}{u}$$

Substituting $\ln(3x^2 - 8x + 5)$ for y into the left side of this last equation and $3x^2 - 8x + 5$ for u into the right side, we can simplify the equation to

$$\frac{d}{dx}\ln(3x^2 - 8x + 5) = \frac{6x - 8}{3x^2 - 8x + 5}$$

Let's see what happens in Example 6 when we choose $u = 3x^2$, whose derivative is $y'(u) = 6x$. If we substitute this back into the equation for y, we have

$$y = \ln(3x^2 - 8x + 5) = \ln(u - 8x + 5)$$

which is a function of *both* u and x. A u-substitution requires that the resulting function be expressed in terms of the single variable u, not u and some other variable. Note that $u = 3x^3$ implies that $x^3 = u/3$, which in turn gives us $x = (u/3)^{1/3}$. Now y can be written

$$y = \ln(3x^2 - 8x + 5) = \ln[u - 8(u/3)^{1/3} + 5]$$

which is a function solely of u. Unfortunately, we do not know how to calculate dy/du, so this u-substitution is not productive. The u-substitution in the Solution of Example 6 is much better.

The u-substitution method for differentiation can be summarized as follows:

THE METHOD OF u-SUBSTITUTION FOR DIFFERENTIATING $y = f(x)$

STEP 1 Choose a part of $f(x)$ and set it equal to u. If the derivative du/dx is known, continue; if not, make another selection for u.

STEP 2 Substitute u into $y = f(x)$ and write y as a function of the single variable u. If this cannot be done, return to Step 1 and make another selection for u; otherwise, continue.

STEP 3 Determine whether the resulting function of u is simpler in form than the original function of x. If so, continue; if not, return to Step 1.

STEP 4 Calculate dy/du. If this is not possible, go to Step 6; otherwise continue.

STEP 5 Use the chain rule with the derivatives du/dx and dy/du from Steps 1 and 4 to determine dy/dx. Write the result solely in terms of the variable x and stop.

STEP 6 Return to Step 1 with either the original function or the function from Step 3, if it is simpler.

The chain rule can be used in tandem with the other rules of differentiation.

EXAMPLE 7

Find the derivative of $y = (1 + x^3)^{3.8} \ln(3x^2 - 8x + 5)$.

Solution

Here y is the product of two complicated functions, so we first apply the product rule of differentiation to obtain

$$\frac{dy}{dx} = (1 + x^3)^{3.8}\left[\frac{d}{dx}\ln(3x^2 - 8x + 5)\right] + \left[\frac{d}{dx}(1 + x^3)^{3.8}\right]\ln(3x^2 - 8x + 5)$$

Using the results of Examples 5 and 6, we find that

$$\frac{dy}{dx} = (1 + x^3)^{3.8}\left(\frac{6x - 8}{3x^2 - 8x + 5}\right) + 11.4x^2(1 + x^3)^{2.8}\ln(3x^2 - 8x + 5)$$

Although a proof of the chain rule is beyond the scope of this book, we can verify Equation (20) when $y = f(u) = u^n$ and u is a function of x. Since

$$\frac{dy}{du} = \frac{d}{du}(u^n) = nu^{n-1}$$

and

$$\frac{d}{dx}(u^n) = \frac{dy}{dx} = \frac{dy}{du} \cdot \frac{du}{dx} = nu^{n-1}\frac{du}{dx}$$

Equation (20) reduces to

$$\frac{d}{dx}(u^n) = nu^{n-1}\frac{du}{dx} \tag{21}$$

For the special case $n = 2$, the product rule of differentiation shows that

$$\frac{d}{dx}(u^2) = \frac{d}{dx}[u(x)u(x)]$$

$$= u(x)\frac{du}{dx} + \frac{du}{dx}u(x)$$

Therefore

$$\frac{d}{dx}(u^2) = 2u(x)\frac{du}{dx} \tag{22}$$

Equation (22) is Equation (21) when $n = 2$.

For the special case $n = 3$, we use the product rule of differentiation:

$$\frac{d}{dx}(u^3) = \frac{d}{dx}[u(x)u^2(x)]$$

$$= u(x)\frac{du^2}{dx} + \frac{du}{dx}u^2(x)$$

Substituting Equation (22) into the first term, we have

$$\frac{d}{dx}(u^3) = u(x)\left[2u(x)\frac{du}{dx}\right] + \frac{du}{dx}u^2(x)$$

Hence

$$\frac{d}{dx}(u^3) = 3u^2(x)\frac{du}{dx} \tag{23}$$

Equation (23) is Equation (21) when $n = 3$.

We presented the chain rule for functions of the form $y = f(u)$ and $u = g(x)$. If other labels are more appropriate to a particular model, the chain rule is revised to accommodate them. For example, if $S = f(u)$ and $u = g(\theta)$, then the composite function $f \circ g$ is a function of the variable θ and the chain rule becomes

$$\frac{dS}{d\theta} = \frac{dS}{du} \cdot \frac{du}{d\theta}$$

If, instead, $z = f(u)$ and $u = g(t)$, then the composite function $f \circ g$ is a function of t and the chain rule is written

$$\frac{dz}{dt} = \frac{dz}{du} \cdot \frac{du}{dt}$$

EXAMPLE 8

An investor deposits $800 in an account that earns interest at the rate of 6% a year, compounded continuously. How quickly is this account growing at the end of 10 years?

Solution

The future value (*FV*) of a lump-sum investment (the present value, or *PV*) earning interest under continuous compounding is given by the formula [see Equation (13) of Section 9.3]

$$FV = PVe^{rt}$$

where *r* is the nominal annual interest rate in decimal form and *t* denotes time, measured in years. Here $PV = \$800$, $r = 0.06$, and the future value of the investment at any time *t* is $FV = 800e^{0.06t}$.

The growth rate of this account after 10 years is $\dfrac{d}{dt}(FV)\Big|_{t=10}$ and to evaluate it we need $d(800e^{0.06t})/dt = 800d(e^{0.06t}) \neq dt$. The natural exponential rule of differentiation instructs us how to handle $\dfrac{d}{dt}(e^t)$ but unfortunately not $\dfrac{d}{dt}(e^{0.06t})$. Therefore we'll try a *u*-substitution to help differentiate $y = e^{0.06t}$. If we set $u = 0.06t$, then $y = e^u$ and by the chain rule,

$$\frac{dy}{dt} = \frac{dy}{du} \cdot \frac{du}{dt} = \left[\frac{d}{du}(e^u)\right]\left[\frac{d}{dt}(0.06t)\right] = (e^u)(0.06)$$

Therefore

$$\frac{d}{dt}(e^{0.06t}) = 0.06e^{0.06t}$$

It now follows that

$$\frac{d}{dt}(FV) = 800\frac{d}{dt}(e^{0.06t}) = 800(0.06e^{0.06t}) = 48e^{0.06t}$$

and

$$\frac{d}{dt}(FV)\Big|_{t=10} = 48e^{0.06t}\big|_{t=10} = 48e^{0.06(10)} \approx 87.46$$

At the end of 10 years, the account is growing at the rate of $87.46/year.

EXAMPLE 9

The average daily high temperatures each month in Billings, Montana, are listed in Table 10.8. Model this process with a function and then use the function to estimate the October 1st instantaneous rate of change in temperature highs for Montana with respect to time.

TABLE 10.8 Average Daily High Temperatures for Billings, Montana

MONTH	TEMPERATURE (°F)	MONTH	TEMPERATURE (°F)
January	33	July	89
February	36	August	86
March	43	September	75
April	57	October	63
May	68	November	46
June	76	December	39

Solution

In Example 9 of Section 9.5, we modeled this data with the periodic function

$$y = 61 + 28 \cos\left(\frac{\pi t}{6}\right)$$

where t denotes time (in months) with July 1 initialized as $t = 0$. Under this coding scheme, October 1 corresponds to $t = 3$ months, and the rate of change we seek is $y'(3)$. To evaluate this, we first need to find $y'(t)$:

$$y'(t) = \frac{dy}{dt} = \frac{d}{dt}\left[61 + 28 \cos\left(\frac{\pi t}{6}\right)\right] = 28 \frac{d}{dt}\left[\cos\left(\frac{\pi t}{6}\right)\right]$$

We do not know how to differentiate $\cos(\pi t/6)$ directly, so we try a u-substitution. Note that the cosine rule of differentiation instructs us how to handle $\frac{d}{dt}(\cos t)$ but not $\frac{d}{dt}\left[\cos\left(\frac{\pi t}{6}\right)\right]$. If we set $z = \cos(\pi t/6)$ and $u = \pi t/6$, the argument of the cosine function, then $z = \cos u$ and we can now use the chain rule:

$$\frac{dz}{dt} = \frac{dz}{du} \cdot \frac{du}{dt} = \left[\frac{d}{du}(\cos u)\right]\left[\frac{d}{dt}\left(\frac{\pi}{6}t\right)\right] = (-\sin u)(\pi/6)$$

hence

$$\frac{d}{dt}\left[\cos\left(\frac{\pi t}{6}\right)\right] = -\frac{\pi}{6} \sin \frac{\pi t}{6}$$

It now follows that

$$\frac{dy}{dt} = 28 \frac{d}{dt}\left[\cos\left(\frac{\pi t}{6}\right)\right] = 28\left(-\frac{\pi}{6} \sin \frac{\pi t}{6}\right) = -\frac{14\pi}{3} \sin \frac{\pi t}{6}$$

and

$$\frac{dy}{dt}\bigg|_{t=3} = -\frac{14\pi}{3} \sin \frac{\pi t}{6}\bigg|_{t=3}$$

$$= -\frac{14\pi}{3} \sin \frac{\pi}{2} \approx -14.7°/\text{month}$$

On October 1, the average daily high temperature is *decreasing* at the rate of 14.7°/month or nearly one-half of a degree a day.

At this point, we have restricted composite functions to be of the form $f \circ g$, that is, a composition of two functions f and g that model a sequence of two operations. Yet, some processes are a sequence of three or more operations. For example, we may have a model of the form

$$y = f(v), \quad v = g(u), \quad \text{and} \quad u = h(x)$$

where the output u from the operation h becomes the input to operation g, and the output v from operation g becomes the input to operation f. This sequence of three operations is shown schematically in Figure 10.37.

As an example, Figure 10.37 might depict the production process we discussed at the beginning of the section, in which operation h transforms iron ore into rolled steel, operation g shapes the rolled steel into automobile fenders, and operation f creates an auto-

FIGURE 10.37
A Sequential Process of Three
Operations

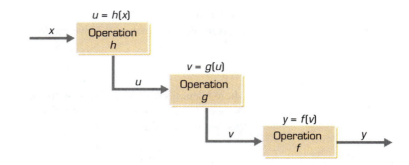

mobile body from the fenders. The production of automobiles, y, depends on the supply of ore, x, and we may need the instantaneous rate of change of y with respect to x. In other words, we may need to know how a change in the iron ore supply (caused by a new contract or a strike, for example) will affect the production of car bodies. The chain rule is again applicable, although in this case it must be generalized to three functions:

$$\frac{dy}{dx} = \frac{dy}{dv} \cdot \frac{dv}{du} \cdot \frac{du}{dx} \tag{24}$$

We can use Equation (24) to differentiate complicated functions when they are decomposed into three simpler functions with known derivatives. This idea is the rationale for Steps 3 and 6 of the u-substitution method presented earlier. If a u-substitution simplifies a complicated function $f(x)$, then we retain the substitution, even if we're not able to differentiate $y = f(u)$. We then repeat the u-substitution process, this time restricting ourselves to the simpler function $y = f(u)$. Because the letter u has been used already, we use a v-substitution instead on the second round of substitutions.

EXAMPLE 10   Find the derivative of the function $y = \sin^5[1/(1 + x)]$.

Solution
Recall that $y = \sin^5[1/(1 + x)]$ is shorthand notation for $y = (\sin[1/(1 + x)])^5$. This is a complicated function, one we cannot differentiate directly, so we need a u-substitution. Here we have a quantity $\sin[1/(1 + x)]$ raised to a power and another quantity $1/(1 + x)$ as the argument of the sine function, so we can apply either one of our two rules of thumb. We note that $1/(1 + x)$ is the simpler of the two quantities, so we start there and let

$$u = 1/(1 + x) \tag{25}$$

Then

$$\frac{du}{dx} = \frac{d}{dx}\left(\frac{1}{1 + x}\right) = \frac{-1}{(1 + x)^2}$$

from the quotient rule of differentiation. Now substituting u into the original function, we have

$$y = \left[\sin\left(\frac{1}{1 + x}\right)\right]^5 = (\sin u)^5$$

We don't know how to find dy/du directly, so we repeat the entire substitution process. We note that $(\sin u)^5$ is a quantity raised to a power, so we set $v = \sin u$, which gives us

$$\frac{dv}{du} = \frac{d}{du}(\sin u) = \cos u$$

and

$$y = (\sin u)^5 = v^5$$

which gives us

$$\frac{dy}{dv} = \frac{d}{dv}(v^5) = 5v^4$$

using the power rule of differentiation. Now from Equation (24), we determine that

$$\frac{dy}{dx} = \frac{dy}{dv} \cdot \frac{dv}{du} \cdot \frac{du}{dx} = 5v^4(\cos u)\left[\frac{-1}{(1+x)^2}\right]$$

Substituting $v = \sin u$ into this last equation, we obtain

$$\frac{dy}{dx} = 5(\sin u)^4(\cos u)\left[\frac{-1}{(1-x)^2}\right]$$

Then, using Equation (25), we have

$$\frac{dy}{dx} = 5\sin^4\left(\frac{1}{1+x}\right)\cos\left(\frac{1}{1+x}\right)\left[\frac{-1}{(1+x)^2}\right]$$

$$= \frac{-5}{(1+x)^2}\sin^4\left(\frac{1}{1+x}\right)\cos\left(\frac{1}{1+x}\right)$$

IMPROVING SKILLS

In Problems 1 through 10, find $f \circ g(x)$ for the given pairs of functions.

1. $f(t) = t^2$ and $g(x) = 2x + 1$
2. $f(t) = t^2 + 1$ and $g(x) = 2x$
3. $f(t) = t^2 + 1$ and $g(x) = 2x + 1$
4. $f(t) = t^2 + 3t + 3$ and $g(x) = x^2 - 2x - 2$
5. $f(u) = \sin u$ and $g(x) = x^2 + 2x - 5$
6. $f(u) = u^2 + 2u - 5$ and $g(x) = \sin x$
7. $f(m) = e^{2m}$ and $g(x) = \sin(x^2)$
8. $f(m) = e^{m^2}$ and $g(x) = \sin(x^2)$
9. $f(m) = e^{m^2}$ and $g(x) = \sin^2 x$
10. $f(m) = e^{2m}$ and $g(x) = \ln(x^2)$

In Problems 11 through 60, find dy/dx.

11. $y = \sqrt{1 - x^4}$
12. $y = \sqrt{x^2 - 3x + 2}$
13. $y = (x^2 + 4)^{-0.5}$
14. $y = (2x^4 + 4x + 6)^{-1.5}$
15. $y = \sqrt{1/(1 + x^4)}$
16. $y = \sqrt{x/(1 + x^4)}$

17. $y = \sqrt{\dfrac{x+1}{x-1}}$

18. $y = \sqrt{\dfrac{x^2 + 2x - 7}{x^2 + x + 5}}$

19. $y = \left(\dfrac{2}{x^2 - 5}\right)^{10}$

20. $y = \left(1 - \dfrac{1}{x}\right)^{-10}$

21. $y = e^{2x}$

22. $y = e^{3x}$

23. $y = e^{-3x}$

24. $y = e^{5x+4}$

25. $y = e^{3x^2 - 5}$

26. $y = e^{3 - x^3}$

27. $y = e^{\sin x}$

28. $y = e^{3\cos x - 4}$

29. $y = e^{x/(x-5)}$

30. $y = e^{(x-5)/x}$

31. $y = \ln 7x$

32. $y = \ln(7x - \tfrac{1}{2})$

33. $y = \ln \dfrac{1}{x}$

34. $y = \ln \dfrac{2x+3}{3x+2}$

35. $y = \sin 2x$

36. $y = \sin(x^2)$

37. $y = \sin^2 x$

38. $y = \sin(2x - 2)$

39. $y = \cos(x^2 + 2x + 3)$

40. $y = \cos(2x^2 + 3x)$

41. $y = 2x + 3\sin 2x$

42. $y = \sin 2x \cos 2x$

43. $y = xe^{2x}$

44. $y = e^{2x} \ln 3x$

45. $y = x^2 e^{2x}$

46. $y = x^2 \ln 2x$

47. $y = x^2 e^{x^2}$

48. $x^4 \ln(2x - 3)$

49. $y = \dfrac{x^2}{\ln x}$

50. $y = \dfrac{\ln(2x+3)}{x^2 - 3x}$

51. $y = \sin^3(x^2)$

52. $y = \cos^4(6x - 2)$

53. $y = (1 + e^{x/2})^5$

54. $y = (2x + 3e^{5x})^{18}$

55. $\cos^5[x/(x+3)]$

56. $y = \ln^5[x/(x+3)]$

57. $y = \sqrt{6 - \sqrt{x}}$

58. $y = \ln^{-3}(\sin 7x)$

59. $y = \sin(\ln x^7)$

60. $y = \cos^{-8}(\sin 2x)$

CREATING MODELS

61. Construct a schematic model for a sequence of operations that transforms oil in the ground into combustible energy in an automobile engine.

62. Construct a schematic model for a sequence of operations that transforms pine seedlings into pine paneling.

63. Construct a schematic model for a sequence of operations that transforms tomato seeds into tomato soup.

64. Construct a schematic model for a sequence of operations that transforms cows into hot dogs.

65. A mining company deposits $20,000 into an account at 8% a year, compounded continuously, for the purpose of buying a new dump truck. How fast is the account growing after 2 years?   *Hint:*   See Equation (13) in Section 9.3.

66. A couple places $5,000 in an account for their newborn grandchild; the account earns interest at the rate of $8\frac{1}{4}$%/year, compounded continuously. How fast is the account growing at the end of 16 years?

67. A mattress manufacturer deposits $2,100 in a savings account at $7\frac{1}{4}$% a year compounded continuously, for the purpose of buying a new computer. How fast is the account growing at 18 months?

68. A model for population growth in a monitored region is given by the equation $N(t) = 20{,}000e^{0.347t}$, where N denotes the size of the population and t is time, measured in years from the time monitoring began.

    (**a**) How many people were present at the beginning of the study?

    (**b**) How quickly was the population growing then?

    (**c**) How quickly was the population growing at 10 years?

69. A model for radioactive decay in a certain material is given by the equation $N(t) = 50e^{-0.053t}$, where N denotes the weight of a material (in mg) and t is time (in hours).

    (**a**) How quickly is the material decaying after 2 hours?

    (**b**) How quickly is the material decaying after 4 hours?

Problems 70 through 78 require the model-building skills developed in Sections 9.4 and 9.5.

70. Cobalt 60 has a half-life of 5.27 years and decays at a rate directly proportional to the amount present. A laboratory receives a shipment of 100 g of this material for a project, but the project is canceled just as the shipment arrives. As a result, the laboratory has to store the cobalt 60. At what rate will it be decaying at 2 years?

71. Radium 226 is known to decay at a rate proportional to the amount present. Initially a sample of radium has 20 g, but after 1 year the amount of radium in the sample is 19.9913 g. How quickly will the sample be decaying after 20 years?

72. The population of Africa was 280 million in 1960 and 363 million in 1970.

    (**a**) Create a model for this population under the assumption that its growth rate is directly proportional to its size.

    (**b**) Use the model to estimate population growth in 1970.

73. An economist believes that the price of white bread increases yearly at a rate directly proportional to the current price per loaf. In 1955, a one-pound loaf cost, on average, 19.3¢ in Los Angeles. By 1960, that average price had risen to 23.4¢.

    (**a**) Create a model for the price of bread.

    (**b**) Use the model to estimate the rate at which the price of bread was increasing in 1970; in 1990.

74. The average daily low temperature in Billings, Montana, has a minimum of 13°F in January and a maximum of 58°F in July.

    (**a**) Using a cosine function, construct a model from this data that will forecast the low temperature in Billings over time (in months).

    (**b**) Use this model to estimate how quickly the average daily low is changing on October 1.

75. The average daily high temperature in Bakersfield, California, has a maximum of 100°F in July and a minimum of 57°F in January.

    (**a**) Use a cosine function to construct a model from this data that will forecast the temperature highs in Bakersfield over time (in months).

    (**b**) Use this model to estimate how quickly the average daily high is changing on July 1.

76. Under normal conditions, the lower portion of the Hackensack River in New Jersey is a tidal estuary, with the depth determined by the tides. At low tide, the river is essentially a mud flat, and the depth of a small trickle of water in the middle is 6 in. At high tide, the depth of the river is 6 ft. Assume that high tide occurs on one particular day at midnight and that the tides are periodic, with a period of 12 hours and 30 minutes.

    (**a**) Model the rise in water level over time with a cosine function.

(*b*) Use this model to estimate how quickly the tides are changing at 9:00 A.M. that same day.

77. Solve Problem 76 if high tide occurs at 9:00 A.M.

78. The tortoise population on an island varies sinusoidally from a low of 500 every January to a high of 2,000 every July.

(*a*) Model this population as a function of time, using either a sine or cosine function.

(*b*) Use your model to estimate how quickly the population is changing on August 1.

EXPLORING IN TEAMS

79. Suppose that a sequence of four operations are modeled by the equations

$$y = f(w), \quad w = g(v), \quad v = p(u), \quad \text{and} \quad u = m(x)$$

(*a*) Develop a formula for *dy/dx*.

(*b*) Does it make sense to ask about *dy/du*? If so, how would you calculate it?

Find *dy/dx* for each of the functions in Problems 80–83.

80. $y = \dfrac{\sin^3 2x}{3x - 1}$

81. $y = \dfrac{\sqrt{x}\,\sin^3 2x}{3x - 1}$

82. $y = \left(\dfrac{\sqrt{x}\,\sin^3 2x}{3x - 1}\right)^{4.75}$

83. $y = \left[\left(\dfrac{\sqrt{x}\,\sin^3 2x}{3x - 1}\right)^{4.75} + x^9\right]^{-7.1}$

EXPLORING WITH TECHNOLOGY

Use a graphing calculator to answer the questions in Problems 84–87. In each problem, *S* denotes sales (in millions of dollars) and *t* denotes time (in years).

84. Company sales are modeled by the equation

$$S = 50 + 10 \sin(2\pi t)$$

(*a*) Graph this relationship and explain what the graph tells you about sales.

(*b*) When does the company enjoy its maximum sales?

(*c*) What is the rate of change of sales with respect to time when maximum sales occur?

85. Solve Problem 84 if sales are modeled by the equation

$$S = 50 + 10e^{-t/4}\sin(2\pi t)$$

86. Solve Problem 84 if sales are modeled by the equation

$$S = 50(1 - e^{-t/4})$$

87. Solve Problem 84 if sales are modeled by the equation $S = t$.

REVIEWING
MATERIAL

88. (Section 3.5) Solve the following linear programming problem:

$$\text{Maximize:} \quad z = 4x_1 + 2x_2 - 5x_3$$

$$\text{Subject to:} \quad -x_1 + 2x_2 + 5x_3 \le 100$$

$$2x_1 + 5x_2 + 4x_3 \le 200$$

$$\text{Assuming:} \quad x_1, x_2, \text{ and } x_3 \text{ are nonnegative}$$

89. (Section 4.4) A couple wants to buy a house, and they can afford monthly mortgage payments of \$950. How much can the couple borrow and still remain within their budget if they are applying for a 30-year mortgage at 9%?

90. (Section 6.5) A committee of three is to be chosen from a set of five Democrats, three Republicans, and four independents.
   (*a*) How many different ways can the committee be chosen?
   (*b*) How many ways can it be chosen if the committee must have one member from each political group?
   (*c*) What is the probability that a committee selected at random will have one member from each group?

91. (Section 9.1) Determine which of the following relationships are functions:
   (*a*) Assign to each person his or her height
   (*b*) Assign to each person his or her child
   (*c*) Assign to each person his or her parent
   (*d*) Assign to each person his or her first name
   (*e*) Assign to each first name a person who has that name

RECOMMENDING
ACTION

92. Respond by memo to the following request:

MEMORANDUM

To: J. Doe Reader

From: Congressional Staff Office

Date: Today

Subject: **The Budget**

At our last budget hearing, many senators supported decreases in the rate of growth of our federal programs. I thought this meant a budget cut, but the Senators informed me otherwise. Apparently the programs will still increase. How is this possible? I need to understand their thinking so I can present the committee's thoughts to the Office of Budget Management (OBM) next week. The people at OBM know their mathematics, so I'll look foolish unless I fully understand the Senators' position. Can you help me?

10.6 DIFFERENTIABLE FUNCTIONS

In Section 10.1 we developed the formula

$$f'(x_1) = \lim_{x_2 \to x_1} \frac{f(x_2) - f(x_1)}{x_2 - x_1} \tag{26}$$

A function is differentiable at a specific input if the function has a derivative there.

for the instantaneous rate of change of a function $f(x)$ at $x = x_1$, when this limit exists. We then used this limit to define the derivative of a function $y = f(x)$ at $x = x_1$. We say that a function $f(x)$ is *differentiable* at x_1 when the function has a derivative at x_1; that is, when the limit on the right side of Equation (26) exists. If this limit does not exist, then $f(x)$ does not have a derivative at $x = x_1$, and the function is not differentiable there.

From a graphical perspective, we know that the derivative of the function $y = f(x)$ at $x = x_1$ is the slope of the tangent line to the graph of the function at the point $(x_1, f(x_1))$, as shown in Figure 10.38. A function has a derivative at x_1 if the function has a tangent line to its graph at $(x_1, f(x_1))$ *and* if that tangent line has a slope. If a function does not have a tangent line at that point or if that tangent line is vertical (has an undefined slope), then the function is not differentiable there.

FIGURE 10.38
This Function $f(x)$ Has a Derivative at the Point $(x, f(x))$.

We can determine whether a function is differentiable at x_1 in two different ways: we can evaluate the limit in Equation (26) or we can look for a tangent line to the graph of the function. Both approaches provide insights into the types of functions that have derivatives. Once we know that a function is differentiable, we can use the rules developed in the last three sections to obtain the derivative efficiently.

If the limit in Equation (26) exists, then the numerator in the quotient

$$\frac{f(x_2) - f(x_1)}{x_2 - x_1}$$

must approach zero as $x_2 \to x_1$ (see Problem 39). That is, a necessary condition for a function $f(x)$ to be differentiable at $x = x_1$ is

$$\lim_{x_2 \to x_1} [f(x_2) - f(x_1)] = 0$$

or, equivalently,

$$\lim_{x_2 \to x_1} f(x_2) = f(x_1)$$

> *A function must be continuous at a specific input if it is to be differentiable there.*

The limit of $f(x_2)$ as $x_2 \to x_1$ must exist, and this limit must equal $f(x_1)$. In other words, $f(x)$ must be continuous at $x = x_1$.

Continuity is a necessary condition for differentiability. If a function is continuous at a specific input, then that function has the *possibility* of being differentiable there. However, if a function is *not* continuous at an input, it definitely cannot be differentiable there.

EXAMPLE 1

Determine whether the function

$$f(x) = \begin{cases} x & x < 2 \\ x^2 & x \geq 2 \end{cases}$$

is differentiable at $x_1 = 2$.

Solution

We see from the graph of this function in Figure 10.39 that the function is not continuous at $x = 2$, so it cannot be differentiable there.

FIGURE 10.39
This Function is Not Differentiable at $x_1 = 2$.

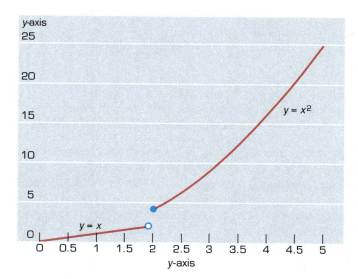

EXAMPLE 2

Identify points where the function graphed in Figure 10.40 is not differentiable.

Solution

We showed in Example 11 of Section 10.2 that this function is not continuous at $x = \pm 1$, so it cannot be differentiable at those two locations.

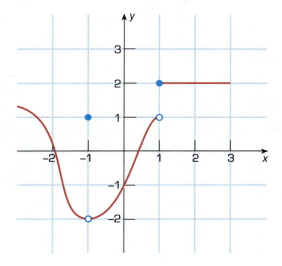

If a function is differentiable at $x = x_1$, then the function must be continuous there. However, continuity by itself is not sufficient for differentiability. To see why, let's try to evaluate the derivative of the absolute value function:

$$f(x) = |x| = \begin{cases} x & x \geq 0 \\ -x & x < 0 \end{cases}$$

which is graphed in Figure 10.41. This function is continuous everywhere, including at $x_1 = 0$, where $f(x_1) = f(0) = 0$. If we let x_2 approach 0 from the right, where $x_2 > 0$, then $f(x_2) = x_2$. Therefore the limit from the right is

$$\lim_{x_2 \to 0^+} \frac{f(x_2) - f(x_1)}{x_2 - x_1} = \lim_{x_2 \to 0^+} \frac{x_2 - 0}{x_2 - 0}$$

$$= \lim_{x_2 \to 0^+} 1 = 1$$

FIGURE 10.41

The Absolute Value Function

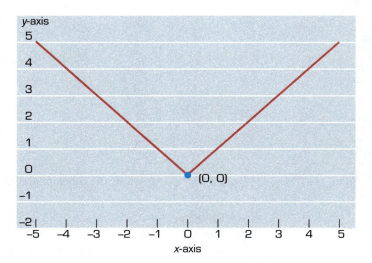

If, on the other hand, we let x_2 approach 0 from the left, where $x_2 < 0$, then $f(x_2) = -x_2$. Therefore the limit from the left is

$$\lim_{x_2 \to 0^-} \frac{f(x_2) - f(x_1)}{x_2 - x_1} = \lim_{x_2 \to 0^-} \frac{-x_2 - 0}{x_2 - 0}$$

$$= \lim_{x_2 \to 0^-} \frac{-x_2}{x_2} = \lim_{x_2 \to 0^-} (-1) = -1$$

The limit from the right does not equal the limit from the left, hence the limit

$$\lim_{x_2 \to x_1} \frac{f(x_2) - f(x_1)}{x_2 - x_1}$$

does not exist for this function at $x_1 = 0$. Therefore the absolute value function is *not* differentiable at zero.

We see from Figure 10.41 that the graph of the absolute value function has no holes and no ledges, so it is continuous. At $x = 0$, however, the graph has a corner, and at this corner the graph does not have a tangent line.

A continuous function $f(x)$ has a *corner* at x_1 if the limit from the left,

$$\lim_{x_2 \to x_1^+} \frac{f(x_2) - f(x_1)}{x_2 - x_1}$$

and the limit from the right,

$$\lim_{x_2 \to x_1^+} \frac{f(x_2) - f(x_1)}{x_2 - x_1}$$

exist but are not equal. Consequently the limit does not exist there, so $f(x)$ is not differentiable at x_1.

If a function has a corner, then the graph of the function will have an abrupt bend there. Corners, however, can be deceptive. Consider the graph of the function

$$y = f(x) = \sqrt{x^2 + 0.0001}$$

shown in Figure 10.42. The graph appears to have a corner at $x = 0$. However, if we zoom in on smaller and smaller windows around the origin, as we do in Figures 10.43 and 10.44, we notice that there is no corner on the graph. The graph changes direction quite smoothly. Indeed, if we zoom in even more, as in Figure 10.45, we see that the graph actually *flattens out* at $x = 0$. The graph has a horizontal tangent line at $x = 0$, with a slope of zero.

The zoom feature on most graphing calculators is a powerful tool for determining whether a function is differentiable at a point. As we zoom in on a point, a magnified graph often appears as a straight line. Consider, for example, the sine function displayed in Figure 10.46. This graph is certainly not a straight line. Nevertheless, if we sufficiently magnify small portions of this graph—any of the highlight boxes displayed in Figure 10.46 or a region around any other point on the graph—the result will be a straight line. Indeed, this occurs as long as the function is continuous and does not have a corner in the region being magnified. Therefore, we use the zoom feature of a graphing calculator as the basis for the following test.

FIGURE 10.42
The Graph of
$y = \sqrt{x^2 + 0.0001}$
on the Interval $[-4, 4]$

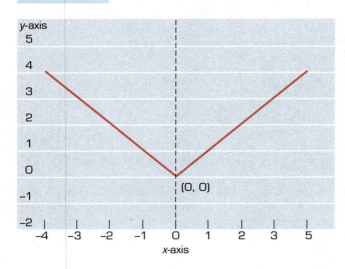

FIGURE 10.43.
The Graph of
$y = \sqrt{x^2 + 0.0001}$
on the Interval $[-0.1, 0.1]$

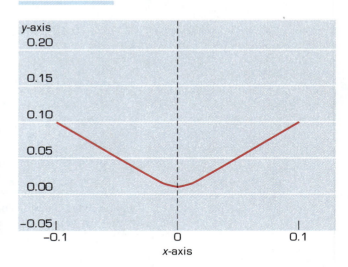

FIGURE 10.44
The Graph of
$y = \sqrt{x^2 + 0.0001}$
on the Interval $[-0.01, 0.01]$

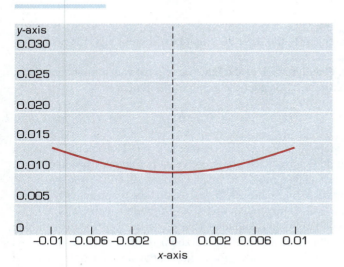

FIGURE 10.45
The Graph of
$y = \sqrt{x^2 + 0.0001}$ on
the Interval $[-0.001, 0.001]$

FIGURE 10.46
The Graph of $y = \sin x$ on the
Interval $[-10, 10]$

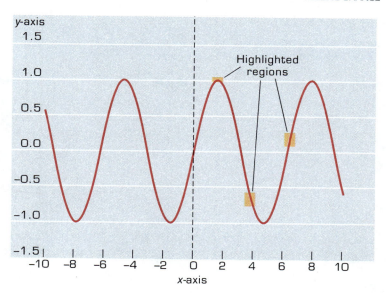

DETERMINING DIFFERENTIABILITY AT A POINT:

Use a graphing calculator to zoom in on the graph of a continuous function at x_1.
Do so many times.

1. If the resulting graphs take the shape of a single nonvertical line, then the function is differentiable at x_1 and the graphs are approaching a tangent line with slope equal to the derivative of the function at x_1.

2. If the resulting graphs continue to display a corner, then the function is not differentiable at x_1.

3. If the resulting graphs take the shape of a vertical line, then the function is not differentiable at x_1.

EXAMPLE 3

Use a graphing calculator to determine whether the function

$$f(x) = \begin{cases} x^2 & x < 1 \\ x^3 & x \geq 1 \end{cases}$$

is differentiable at $x_1 = 1$.

Solution
We see from the graph of this function in Figure 10.47 that the function is continuous at $x_1 = 1$. Furthermore, the graph does not appear to have any corners. If we magnify the graph around $(1, 1)$, we obtain Figure 10.48, which is not as smooth as we first thought. One additional magnification gives us Figure 10.49, and we now see that the graph does indeed have a corner at $(1, 1)$, so the function is *not* differentiable at $x_1 = 1$.

FIGURE 10.47

FIGURE 10.47

The Graph of the Function in
Example 3 on the Interval
[−1, 3]

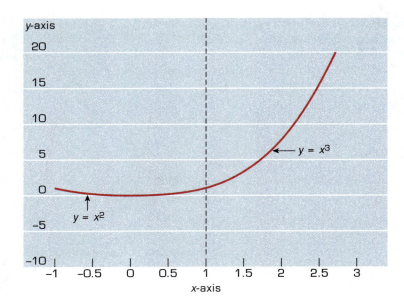

FIGURE 10.48

The Graph of the Function in
Example 3 on the Interval
[0.75, 1.25]

FIGURE 10.49

The Graph of the Function in
Example 3 on the Interval
[0.99, 1.01]

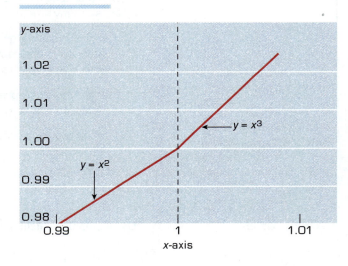

EXAMPLE 4

Use a graphing calculator to determine whether the function

$$f(x) = \begin{cases} x^2 & x > -1 \\ -2x - 1 & x \le -1 \end{cases}$$

is differentiable at $x_1 = -1$.

Solution

We see from a graph of this function in Figure 10.50 that the function is continuous at $x_1 = -1$. Magnifying a portion of this graph around $(-1, 1)$, we obtain Figure 10.51 and

FIGURE 10.50

The Graph of the Function in Example 4 on the Interval [−3, 1]

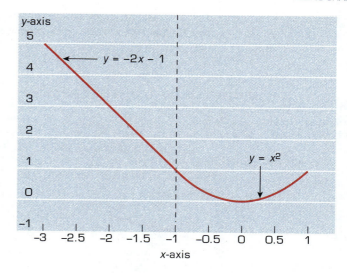

FIGURE 10.51

The Graph of the Function in Example 4 on the Interval [−1.25, −0.75]

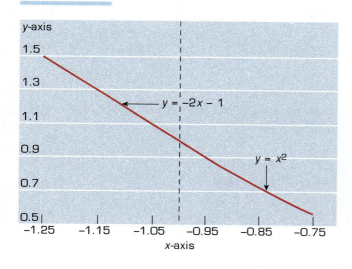

FIGURE 10.52

The Graph of the Function in Example 4 on the Interval [−1.01, −0.99]

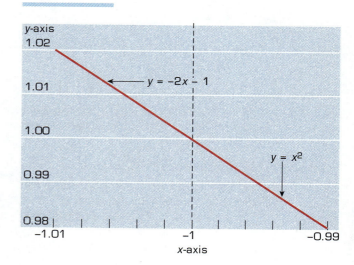

then Figure 10.52. These graphs approach a single nonvertical straight line with a slope of -2, so the function is differentiable at $x_1 = -1$ with $f'(-1) = -2$.

EXAMPLE 5

Use a graphing calculator to determine whether the function $f(x) = x^{1/3}$ is differentiable at $x_1 = 0$.

Solution

In general, we defined power functions only for nonnegative inputs. This particular power function, however, can be evaluated for all real inputs, so we extend its domain accordingly. We see from a graph of this function, Figure 10.53, that the function is continuous at $x_1 = 0$. Magnifying a portion of this graph around $(-1, 1)$, we obtain Figure 10.54 and then Figure 10.55. These graphs approach a vertical straight line, so the function is not differentiable at $x_1 = 0$.

FIGURE 10.53

The Graph of the Function $y = x^{1/3}$ on the Interval $[-5, 5]$

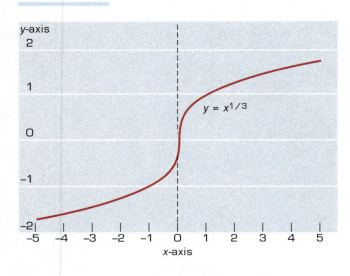

FIGURE 10.54

The Graph of the Function $y = x^{1/3}$ on the Interval $[-1.5, 1.5]$

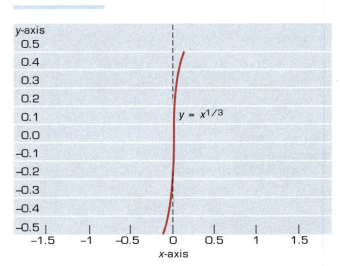

FIGURE 10.55

The Graph of the Function $y = x^{1/3}$ on the Interval $[-0.4, 0.4]$

As we magnify the graph of a function around a point where it is differentiable, the graph takes the shape of a nonvertical straight line, and at high magnifications, we find no discernible differences between the graph of the function around that point and its tangent line at the point. The two are indistinguishable. Consequently, we can use one to approximate the other.

The tangent line to the graph of $f(x)$ must pass through the point $(x_1, f(x_1))$, having slope $f'(x_1)$. Using the point–slope formula developed in Section 1.5, we find the equation of the tangent line to be

$$y - f(x_1) = f'(x_1)(x - x_1)$$

This gives us the equation of the tangent line in functional form:

$$y = f(x_1) + f'(x_1)(x - x_1) \tag{27}$$

EXAMPLE 6

Find the equation of the tangent line to the graph of $y = x^{1/3}$ at $x_1 = 2$.

Solution

Here $f(x) = x^{1/3}$, hence $f(x_1) = f(2) = 2^{1/3} \approx 1.2599$. Now, differentiating $f'(x) = \frac{1}{3}x^{-2/3}$, so

$$f'(x_1) = f'(2) = \frac{1}{3}(2)^{-2/3} \approx 0.2100$$

Substituting these results into Equation (27), we have

$$y = 1.2599 + 0.2100(x - 2)$$
$$= 0.2100x + 0.8399$$

as the equation of the tangent line to the function at $(2, 1.2599)$, with all numbers rounded to four decimal places. This tangent line is shown in Figure 10.56.

FIGURE 10.56

The Tangent Line to $y = x^{1/3}$ at $x = 2$

The tangent line in Example 6 follows the graph of the function $y = f(x) = x^{1/3}$ as long as x remains close to $x_1 = 2$. Indeed, if we use a graphing calculator to zoom in tightly enough on both curves at $(2, 1.2599)$, they will be indistinguishable. Therefore we can use the y-values of the tangent line, designated by Equation (27), to approximate the y-values on the graph of the function $f(x)$. That is,

A linear approximation to $f(x)$ near x_1 is the equation of the tangent line to the graph of $f(x)$ at x_1.

$$f(x) \approx f(x_1) + f'(x_1)(x - x_1) \tag{28}$$

These relationships are illustrated in Figure 10.57 with $x_1 = 2$ and $x = 3$. The right side of Equation (28) is called a *linear approximation to f(x) near x_1*.

Equation (28) follows directly from the definition of the derivative,

$$f'(x_1) = \lim_{x_2 \to x_1} \frac{f(x_2) - f(x_1)}{x_2 - x_1}$$

If we replace x_2 with x, this equation becomes

$$f'(x_1) = \lim_{x \to x_1} \frac{f(x) - f(x_1)}{x - x_1}$$

The quotient on the right side of this last equation approaches $f'(x_1)$ as x approaches x_1. The closer x is to x_1, the closer the quotient $[f(x) - f(x_1)]/(x - x_1)$ is to $f'(x_1)$. For any value of x reasonably close to x_1,

$$f'(x_1) \approx \frac{f(x) - f(x_1)}{x - x_1}$$

which is algebraically equivalent to Equation (28).

FIGURE 10.57

An Illustration of a Linear Approximation of $f(x) = x^{1/3}$ by a tangent line

---

EXAMPLE 7

Find the linear approximation to the function $f(x) = x^{1/3}$ at $x_1 = 2$, and then use the approximation to estimate the value of $f(2.1)$ and $f(3)$.

Solution

We approximate the function near x_1 with the equation of the tangent line to the graph of the function at x_1, obtained in Example 6:

$$f(x) \approx 0.2100x + 0.8399$$

Therefore,

$$f(2.1) \approx 0.2100(2.1) + 0.8399 = 1.2809$$

and

$$f(3) \approx 0.2100(3) + 0.8399 = 1.4699$$

Let us compare these approximations with the true values:

$$f(2.1) = (2.1)^{1/3} = 1.2806 \quad \text{and} \quad f(3) = (3)^{1/3} = 1.4422$$

each rounded to four decimal places.

In Example 7, the linear approximation was less accurate at $x = 3$ than it was at $x = 2.1$. This is to be expected: in general, the farther we are from x_1 (which is 2 in Example 7), the worse the approximation.

When x differs from x_1 by a single unit, that is, when $x = x_1 + 1$, the linear approximation

$$f(x) \approx f(x_1) + f'(x_1)(x - x_1)$$

reduces to

$$f(x_1 + 1) \approx f(x_1) + f'(x_1)$$

Rearranging terms, we get an important approximation equation:

$$f(x_1 + 1) - f(x_1) \approx f'(x_1) \tag{29}$$

This equation is the foundation for all marginal analysis. If we're analyzing a cost function, then we often replace f with the letter C, denoting cost, and rewrite Equation (29) as

$$C(x_1 + 1) - C(x_1) \approx C'(x_1) \tag{30}$$

The right side of Equation (30) is the derivative of a cost function at an input level x_1, which was defined in Section 10.1 as marginal cost. It measures how quickly cost changes as we change the input from a value of x_1. The left side of Equation (30) measures the total change in cost as we move from an input level of x_1 to an input level of $x_1 + 1$, that is, as we increase the input by a single unit. Because we have such good rules for calculating derivatives, we often use the right side of Equation (30) to approximate the left side. Marginal cost provides an *estimate* to the question, "How much will cost change if the input to the process is increased by another unit?"

Similarly, if we're analyzing a revenue function, then we often replace f with the letter R and rewrite Equation (29) as

$$R(x_1 + 1) - R(x_1) \approx R'(x_1) \tag{31}$$

The right side of Equation (31) is the derivative of a revenue function at an input level x_1, which is marginal revenue. It measures how quickly revenue changes as we change the input from a value of x_1. The left side of Equation (31) measures the total change in revenue as the input is increased by a single unit from x_1, and we often approximate this change with the right side of Equation (31). Marginal revenue provides an *estimate* to the question, "How much will revenue change if the input is increased by one unit?"

EXAMPLE 8

The Undercover Publishing Company's revenue is related to advertising expenditures by the formula

$$R = 800 - 800e^{-0.1A}$$

where A is the amount of money spent on advertising and R is revenue the company receives from book sales, both measured in thousands of dollars. Currently the company has budgeted \$20,000 to advertise the first book from a new author. What is the marginal revenue at this amount and how much will the company gain in income if it can increase its advertising budget by an additional $1,000?

Solution

Here marginal revenue is the instantaneous rate of change of revenue with respect to advertising expenditures,

$$\frac{dR}{dA} = 80e^{-0.1A}$$

Evaluating this derivative at 20 thousand-dollar units, we obtain

$$\frac{dR}{dA}\Big|_{A=20} = 80e^{-0.1(20)} \approx 10.827$$

The marginal revenue at $20,000 is 10.827 thousand-dollar units, or $10,827 of additional revenue for a $1,000 unit increase in advertising. Thus, if the advertising budget is increased from $20,000 to $21,000, the company can expect *approximately* $10,827 of additional income.

The *actual* increase in income is determined by calculating the difference between the revenues for advertising budgets of $21,000 and $20,000. Here

$$R\big|_{A=21} = 800 - 800e^{-0.1(21)} \approx 702.035 \text{ thousand-dollar units}$$

and

$$R\big|_{A=20} = 800 - 800e^{-0.1(20)} \approx 691.732 \text{ thousand-dollar units}$$

The difference is

$$R\big|_{A=21} - R\big|_{A=20} \approx \$702,035 - \$691,732 = \$10,303$$

This is the true increase in revenue from an additional $1,000 in advertising.

IMPROVING SKILLS

In Problems 1 through 6, find the equation of the tangent line to the graph of the given function at the specified value of x_1.

1. $y = x^3 + 2$; $x_1 = 3$
2. $y = 1 - 2x^2$; $x_1 = 2$
3. $y = 2x + 1$; $x_1 = 5$
4. $y = 1/(x - 1)$; $x_1 = 0$
5. $y = e^{-x}$; $x_1 = -1$
6. $y = \cos x$; $x_1 = \pi/4$
7. Find the linear approximation to $f(x) = x^2$ near $x_1 = 1$ and use it to estimate $(1.1)^2$. What is the error in this estimate?
8. Find the linear approximation to $f(x) = e^x$ near $x_1 = 0$ and use it to estimate $e^{-0.05}$. What is the error in this estimate?
9. Find the linear approximation to $f(x) = \sqrt{x}$ near $x_1 = 4$ and use it to estimate $\sqrt{4.15}$. What is the error in this estimate?
10. Find the linear approximation to $f(x) = 1/x$ near $x_1 = 1$ and use it to estimate $1/1.03$. What is the error in this estimate?

11. Find the linear approximation to $f(x) = \sin x$ near $x_1 = 0$ and use it to estimate $\sin(0.02)$. What is the error in this estimate?

12. Find the linear approximation to $f(x) = \ln x$ near $x_1 = 1$ and use it to estimate $\ln(1.07)$. What is the error in this estimate?

13. Use linear approximations to estimate **(a)** $\sqrt{1.23}$; **(b)** $\sqrt{8.95}$; **(c)** $\sqrt{36.4}$.

14. Show that $\sin x \approx x$ near $x = 0$.

15. Show that $e^x \approx 1 + x$ near $x = 0$. Use this result to show that $e^{x^2} \approx 1 + x^2$ near $x = 0$.

CREATING MODELS

16. A company's sales are known to be related to advertising expenditures by the equation

$$S(a) = 150,000 + 6000a - 50a^2$$

where a denotes the monthly advertising expenditures in thousands of dollars.

   (a) What are the marginal sales at an advertising budget of $45,000?

   (b) Use the results of part **(a)** to estimate the change in sales if advertising is increased from $45,000 to $46,000.

   (c) What is the actual change in sales from this increase in advertising?

17. A firm's total sales (in millions of dollars) is given by the equation $y(t) = 3t + \frac{1}{2}t^2$, where t denotes the number of years the firm has been in operation.

   (a) Determine the firm's marginal sales after seven years in business.

   (b) Use the results of part **(a)** to estimate the change in sales from year 7 to year 8.

   (c) What is the actual change in sales from year 7 to year 8?

18. The Can-Do Company, which manufactures electric can openers, has a daily fixed cost of $21,000 and knows that the variable cost of producing each can opener is $5.50.

   (a) Find an expression relating total cost C to the number of can openers produced, N.

   (b) Use this expression to find marginal cost.

   (c) Use the results of part **(b)** to estimate the additional cost of producing another can opener.

   (d) What is the actual cost of producing one more can opener?

   (e) How do you reconcile the results of parts **(c)** and **(d)**?

19. A company believes that the number n of items it sells depends on the unit price p (in dollars) according to the formula

$$n(p) = \frac{60,000}{4 + p}$$

   (a) Determine the marginal revenue at a unit price of $3.

   (b) Use the results of part **(a)** to estimate the change in revenue if unit price is increased an additional $1.

   (c) What is the actual change in revenue from this price increase?

20. The only animals on a remote Pacific island are rabbits and foxes. The rabbits eat vegetation and the foxes eat rabbits. Biologists studying this island use the equation

$$R(t) = \sqrt{t}(100 + 50 \sin t)$$

as their model of the rabbit population, where R denotes the number of rabbits on the island and t denotes time, measured in years from the start of the study.

(a) Determine the rate of change in the rabbit population after one year.

(b) Use the results of part (a) to estimate the change in the rabbit population from year 1 to year 2.

(c) What is the actual change in the rabbit population from year 1 to year 2?

21. A model for population growth in a monitored region is given by the equation

$$N(t) = 20{,}000e^{0.347t}$$

where N denotes the size of the population and t is time, measured in years from the time monitoring began.

(a) How quickly was the population growing as of 10 years?

(b) Use the results of part (a) to estimate the change in the population from year 10 to year 11.

(c) What is the actual change in this population from year 10 to year 11?

22. A model for radioactive decay in a certain material is given by the equation

$$N(t) = 50e^{-0.053t}$$

where N denotes the weight of the material (in mg) and t is time (in hours).

(a) How quickly is the material decaying initially?

(b) Use the results of part (a) to estimate the change in weight of the material over the first hour.

(c) What is the actual change in weight over the first hour?

EXPLORING WITH TECHNOLOGY

In Problems 23–37, use a graphing calculator and the testing procedure described in this section to determine whether the following functions are differentiable at either $x = 0$ or $x = 1$.

23. $f(x) = \begin{cases} x^2 & x \geq 0 \\ -x^2 & x < 0 \end{cases}$

24. $f(x) = \begin{cases} x^2 & x \geq 0 \\ -x^2 - 1 & x < 0 \end{cases}$

25. $f(x) = \begin{cases} x^2 & x > 0 \\ 1 & x = 0 \\ -x^2 & x < 0 \end{cases}$

26. $f(x) = \begin{cases} \sqrt{x} & x \geq 0 \\ \sqrt{-x} & x < 0 \end{cases}$

27. $f(x) = \begin{cases} x & x < 0 \\ x^2 & 0 \leq x < 1 \\ x^3 & x \geq 1 \end{cases}$

28. $f(x) = \begin{cases} x & x < 0 \\ \frac{1}{2}x^2 & 0 \leq x < 1 \\ x^3 & x \geq 1 \end{cases}$

29. $f(x) = \begin{cases} x & x < 0 \\ \frac{3}{2}x^2 & 0 \leq x < 1 \\ x^3 & x \geq 1 \end{cases}$

30. $f(x) = \begin{cases} x - \frac{1}{2} & x < 0 \\ \frac{3}{2}x^2 - \frac{1}{2} & 0 \leq x < 1 \\ x^3 & x \geq 1 \end{cases}$

31. $f(x) = \begin{cases} x & x < 0 \\ \frac{3}{2}x^2 + x & 0 \leq x < 1 \\ x^3 + x + \frac{1}{2} & x \geq 1 \end{cases}$

32. $f(x) = \begin{cases} e^x & x < 0 \\ e^{-x} & x \geq 0 \end{cases}$

33. $f(x) = \begin{cases} e^x & x < 0 \\ x + 1 & x \geq 0 \end{cases}$

34. $f(x) = \begin{cases} e^x & x < 0 \\ x + 1 & 0 \leq x < 1 \\ 3 - x & x \geq 1 \end{cases}$

35. $f(x) = \begin{cases} x^2 & x \leq 0 \\ x^3 & 0 < x < 1 \\ x^4 & x \geq 1 \end{cases}$

36. $f(x) = \begin{cases} x^2 & x \leq 0 \\ x^3 - 1 & 0 < x < 1 \\ x^4 - x & x \geq 1 \end{cases}$

37. $f(x) = \begin{cases} x^2 - 1 & x \leq 0 \\ x^3 - 1 & 0 < x < 1 \\ x^4 - x & x \geq 1 \end{cases}$

EXPLORING IN TEAMS

38. Attempt to evaluate the limit

$$f'(x_1) = \lim_{x_2 \to x_1} \frac{f(x_2) - f(x_1)}{x_2 - x_1}$$

at $x_1 = 2$ for the function $f(x)$ in Example 1; show directly that the function is not differentiable there.

39. Explain why the numerator in the quotient

$$\frac{f(x_2) - f(x_1)}{x_2 - x_1}$$

must approach zero as $x_2 \to x_1$ if the right side of Equation (11) (on the first page of Section 10.3) exists. Describe the behavior of the quotient if the numerator does *not* approach zero when $x_2 \to x_1$.

40. Use a linear approximation to estimate the values of $f(1.13)$ and $f(1.38)$ for the function defined by the following table:

x	0	1.1	1.2	1.3	1.4
$f(x)$	1	0.0953	0.1823	0.2624	0.3365

REVIEWING MATERIAL

41. (Section 3.2) Find all corner points for the region defined by the system

$$2x + 5y \geq 100$$

$$5x - 3y \leq 150$$

$$x \geq 100$$

42. (Section 5.2) Construct a visual model of the following data reported in *Business Week*. In each of the six most expensive locations in the world, the yearly cost in 1995 of retail rental space in dollars/ft^2 was: Fifth Avenue (from 49th St. to 57th St.), New York City: $500; East 57th St. (from Fifth Ave. to Madison Ave.), New York City: $440; The Ginza, Tokyo: $350; Madison Ave. (from 57th St. to 69th St.), New York City: $300; the intersection of Post St. and Stockton St., San Francisco: $300; The Champs Elysees, Paris: $280.

43. (Section 7.1) Using the transition matrix

$$\mathbf{P} = \begin{bmatrix} 0.1 & 0 & 0.9 \\ 0 & 1 & 0 \\ 0 & 0.4 & 0.6 \end{bmatrix}$$

determine the probability of (**a**) ending in state 3 after one time period if you start in state 1; (**b**) ending in state 1 after two time periods if you start in state 2; (**c**) ending in state 3 after two time periods if you start in state 1.

44. (Section 9.4) A bacterial culture is known to grow at a rate directly proportional to the present population of the bacteria. Initially there are 20 strands of this bacteria in the culture and after two days, that number increases to 100 strands. Find (**a**) a formula for the approximate number of strands of bacteria present in the culture at any time t; (**b**) the approximate number of strands in the culture at the end of 3 days; and (**c**) the amount of time it takes for the population to double in size.

RECOMMENDING ACTION

45. Respond by memo to the following request:

> ### MEMORANDUM
>
> To: J. Doe Reader
>
> From: Vice President for Finance and Budget
>
> Date: Today
>
> Subject: **Estimates from Marginal Rates**
>
> I noticed from your examples [see Examples 7 and 8] that both linear approximations overestimated the actual changes. Will this always happen? If not, is there some way to know when a linear approximation will overestimate a change and when it will underestimate?

CHAPTER 10 KEYS

KEY WORDS

KEY CONCEPTS

10.1 Rates of Change

- The average rate of change of the function $y = f(x)$ over the input interval $[x_1, x_2]$, is the slope of the line that passes through the two points (x_1, y_1) and (x_2, y_2), where $y_1 = f(x_1)$ and $y_2 = f(x_2)$.

- The instantaneous rate of change of the function $y = f(x)$ at x_1 is the slope of the tangent line to the graph of $f(x)$ at the point (x_1, y_1), where $y_1 = f(x_1)$.

- Speed is the instantaneous rate of change of distance with respect to time.

10.2 Limits and Continuity

- The trace and zoom features on a graphical calculator are useful for determining limits.

- If a function is continuous at x_1, then its graph must show a seamless transition of functional values as x moves from near x_1 onto x_1 itself. The graph of the function cannot have a hole or ledge at x_1.

10.3 The Derivative

- To find a derivative of a function at a particular input value $x = x_1$, first obtain the derivative at an arbitrary input value x and then evaluate the result at x_1.

10.5 The Chain Rule

- The chain rule is used to differentiate a complicated function with an unknown derivative by expressing that function as a composite function of simpler functions with known derivatives.

- The method of decomposing a complex function into a composite function of simpler functions is called u-substitution.

- When a function involves a quantity raised to a power, a useful rule of thumb for choosing a u-substitution is to choose u to be that quantity.

- A useful rule of thumb for choosing a u-substitution when a function involves an argument is to choose u to be that argument.

10.6 Differentiable Functions

- A function must be continuous at a point if it is to be differentiable there.

- A function is differentiable at a point if it is continuous and does not have a corner there.

- To approximate functional values of $f(x)$ near x_1, use the y-values on the tangent line to the graph at x_1.

KEY FORMULAS

- A function $f(x)$ has the limit L at $x = x_1$ if
$$\lim_{x \to x_1^+} f(x) = L = \lim_{x \to x_1^-} f(x)$$

- A function $f(x)$ is continuous at $x = x_1$ if
$$f(x_1) = \lim_{x \to x_1^+} f(x) = \lim_{x \to x_1^-} f(x)$$

- The derivative of the function $y = f(x)$ at x is
$$\lim_{h \to 0} \frac{f(x + h) - f(x)}{h}$$

- The power rule of differentiation: If n is a real number, then
$$\frac{d}{dx}(x^n) = nx^{n-1}$$

- The constant rule of differentiation: If c is a real number, then
$$\frac{d}{dx}(c) = 0$$

- The natural exponential rule of differentiation:
$$\frac{d}{dx}(e^x) = e^x$$

- The natural logarithm rule of differentiation:
$$\frac{d}{dx}(\ln x) = \frac{1}{x}$$

- The sine rule of differentiation:
$$\frac{d}{dx}(\sin x) = \cos x$$

- The cosine rule of differentiation:
$$\frac{d}{dx}(\cos x) = -\sin x$$

- The constant multiple rule of differentiation: If c is a real number and if $f(x)$ has a derivative, then
$$\frac{d}{dx}cf(x) = c\frac{d}{dx}f(x)$$

- The sum rule of differentiation: If $f(x)$ and $g(x)$ have derivatives, then

$$\frac{d}{dx}[f(x) + g(x)] = \frac{d}{dx}f(x) + \frac{d}{dx}g(x)$$

- The difference rule of differentiation: If $f(x)$ and $g(x)$ have derivatives, then

$$\frac{d}{dx}[f(x) - g(x)] = \frac{d}{dx}f(x) - \frac{d}{dx}g(x)$$

- The product rule of differentiation: If $f(x)$ and $g(x)$ have derivatives, then

$$\frac{d}{dx}[f(x) \cdot g(x)] = f(x) \cdot \frac{d}{dx}g(x) + \left[\frac{d}{dx}f(x)\right] \cdot g(x)$$

- The quotient rule of differentiation: If $f(x)$ and $g(x) \neq 0$ have derivatives, then

$$\frac{d}{dx}\left[\frac{f(x)}{g(x)}\right] = \frac{g(x) \cdot \dfrac{d}{dx}f(x) - f(x) \cdot \dfrac{d}{dx}g(x)}{[g(x)]^2}$$

- The chain rule of differentiation: If $y = f(u)$ and $u = g(x)$ are both differentiable, then

$$\frac{dy}{dx} = \frac{dy}{du} \cdot \frac{du}{dx}$$

- The linear approximation to $f(x)$ near x_1 is

$$f(x_1) + f'(x_1)(x - x_1)$$

KEY PROCEDURES

- To evaluate the limit

$$f'(x) = \lim_{h \to 0} \frac{f(x + h) - f(x)}{h}$$

analytically when $f(x)$ is a polynomial function, perform the following steps:

Step 1 Factor h from the numerator.

Step 2 Cancel h from the numerator and the denominator.

Step 3 Let h approach zero.

- The method of u-substitution for differentiating $y = f(x)$ consists of the following steps:

Step 1 Choose a part of $f(x)$ and set it equal to u. If the derivative du/dx is known, continue; it not, make another selection for u.

Step 2 Substitute u into $y = f(x)$ and write y as a function of the single variable u. If this cannot be done, return to Step 1 and make another selection for u; otherwise, continue.

Step 3 Determine whether the resulting function of u is simpler in form than the original function of x. If so, continue; if not, return to Step 1 and make another selection for u.

Step 4 Calculate dy/du. If this is not possible, go to Step 6; otherwise continue.

Step 5 Use the chain rule with the derivatives du/dx and dy/du from Steps 1 and 4 to determine dy/dx. Write the result solely in terms of the variable x and stop.

Step 6 Return to Step 1 with either the original function or the function from Step 3, if it is simpler.

■ To test graphically whether a function has a derivative at x_1, use a graphing calculator to zoom in on the graph of the function at x_1.

1. If the series of graphs eventually take the shape of a single nonvertical line, then the function is differentiable at x_1 and the graphs are approaching the tangent line to the function at x_1, with the slope of the tangent line equal to the derivative of the function at x_1.
2. If the resulting graphs eventually display a corner, then the function is not differentiable at x_1.
3. If the resulting graphs eventually take the shape of a vertical line, then the function is not differentiable at x_1.

TESTING YOURSELF

Use the following problems to test yourself on the material in Chapter 10. The odd-numbered problems comprise one test, the even-numbered problems a second test.

1. Find the average rate of change of the function $f(x) = 1/(2 - x)$ over the interval $[1, 4]$.
2. A company's sales are known to have been $2.8 million in 1990 and $3.3 million in 1993. What was the average rate of change in sales over this time period?
3. Create a table of functional values that allows you to determine whether $\lim_{x \to 1} f(x)$ exists for the function

$$f(x) = \begin{cases} 2x^2 - 1 & x < 1 \\ x^4 & x \geq 1 \end{cases}$$

4. Use the zoom feature of a graphing calculator to determine whether the function in Problem 3 is differentiable at $x = 1$. Is this function differentiable at $x = 2$?
5. Graph the function

$$f(x) = \begin{cases} 2x + 1 & x < -1 \\ x^3 & -1 \leq x < 1 \\ 3x - 2 & x \geq 1 \end{cases}$$

and then determine whether it is continuous at any of $x = -1$, $x = 0$, or $x = 1$.
6. Determine whether the function in Problem 5 is differentiable at any of $x = -1$, $x = 0$, or $x = 1$.
7. (a) Use a calculator and the definition of the derivative to estimate $f'(2)$ numerically, where $f(x) = 3x^2 + 2$.
   (b) Solve part (a) analytically.
8. Find the linear approximation to $f(x) = x^{2/3}$ near $x_1 = 1$ and use it to estimate $(1.08)^{2/3}$.

Find dy/dx for each of the functions in Problems 9–20.

9. $y = x^5 - 8$
10. $y = -5e^{2x} \ln(1.3x)$

11. $y = \dfrac{2x^5 - 5x}{1 - 3x^4}$

12. $y = 3e^{2\cos(\pi x)}$

13. $y = (x^5 - 6x^3 + x^{12})(x^{-4} - x^4)$

14. $y = [4 - \sin(\cos \pi)]/\sqrt{5}$

15. $y = (x - e^x)/(4 + \cos x)$

16. $y = \sqrt{-3x + x^3}$

17. $y = e^{-2x}\sin x$

18. $y = \ln(2x + 7)$

19. $y = \cos^4(1 - x^3)$

20. $y = 8x^7 - 7x^3 + 3$

21. A company believes that its sales, S (in millions of products), are related to advertising expenditures, E (in millions of dollars), by the equation

$$S(E) = 3 - 2e^{-E}$$

Find the instantaneous rate of change in sales with respect to advertising expenditures when the advertising budget is \$1 million. What are the units of this rate of change? Is it advisable to increase the advertising budget?

22. For a circle with a 3-inch circumference, find the instantaneous rate of change of its area with respect to its circumference.

23. Find the equation of the line that is tangent to the curve $y = (\cos 2x)/(2 + x)$ at $x = 0$.

FIGURE A

24. Complete the following table for the function $f(x)$ graphed in Figure A.

x_1	−1	0	1	2	3	4
$\displaystyle\lim_{x \to x_1^+} f(x)$						
$\displaystyle\lim_{x \to x_1^-} f(x)$						
Continuous						
Differentiable						

Insert numerical answers for the limits when they exist or write NL (for no limit) when the limits do not exist. Write either *yes* or *no* for continuity and differentiability at a specific input.

25. Show that $\ln x \approx x - 1$ near $x = 1$.

26. (*a*) Use a calculator and the definition of the derivative to estimate $f'(-1)$ numerically, where $f(x) = 2/x^2$.

    (*b*) Solve part (*a*) analytically.

27. The manufacturing cost V (in dollars) is related to x, the number of single items produced, by the formula

$$V(x) = 0.25 + 1.75e^{-0.001x}$$

What is the marginal total manufacturing cost at a production run of 1,000 items? Use this result to estimate the cost of increasing production to 1,001 items.

28. Graph the function

$$f(x) = \begin{cases} x^2 & x < -1 \\ x^3 & -1 \leq x < 1 \\ x^2 & x \geq 1 \end{cases}$$

and determine whether it is continuous or differentiable at any of $x = -1$, $x = 0$, or $x = 1$.

11

OPTIMIZATION
MODELS

11.1 CRITICAL POINTS

Real-world processes are constrained by finite resources, so all inputs to a process are bounded within a finite interval.

Optimization is the process of maximizing or minimizing a quantity; this enterprise is at the core of our daily activities. For example, we may try to maximize recreational opportunities, minimize stress, maximize satisfaction, or minimize discord. Businesses try to maximize profits and minimize costs; the government tries to maximize employment and minimize crime. All of us strive to enhance the good and diminish the bad.

The simplex method (see Chapter 3), a procedure for solving optimization problems, is limited to models with a linear objective function and linear constraints. Another optimization technique is based on differentiation; it is applicable to any process that can be modeled by a differentiable function.

An effective method for optimizing a function with existing technologies is to graph the function over its domain and then locate where the graph reaches its highest and lowest points.

We noted in Chapter 3 that all real-world processes are constrained by finite resources. The labor force is finite, demands are finite, capital is finite, and even time is essentially finite because we require solutions within specified time periods. This means that all inputs to an input–output relationship are finite and bounded within some region. If we let x denote an input, then x must be restricted to some interval $[a, b]$, where the limits a and b are defined by the process being modeled. In other words, the domain of the function is constrained to $a \leq x \leq b$. All of the points in the set $\{x \mid a < x < b\} = (a, b)$ are called *interior points*.

The maximum value of an input–output function is the largest output it can attain, while the minimum value is the smallest output it can attain. In Figure 11.1, the maximum value of the continuous function $f(x)$ occurs at the highest point on the graph and its minimum occurs at the lowest point, with inputs of f limited to the interval $[a, b]$. The minimum output of $y = y_2$ corresponds to the input value $x = x_2$, while the maximum output $y = y_5$ corresponds to two different inputs, $x = x_5$ and $x = x_7$.

One effective procedure with existing technologies for optimizing a function over a finite domain $[a, b]$ is graphing: with a "picture" in front of us, we can usually observe where the function attains its highest (maximum) and lowest (minimum) values.

FIGURE 11.1

The Graph of a Continuous Function on [a, b] with Peaks and Pits

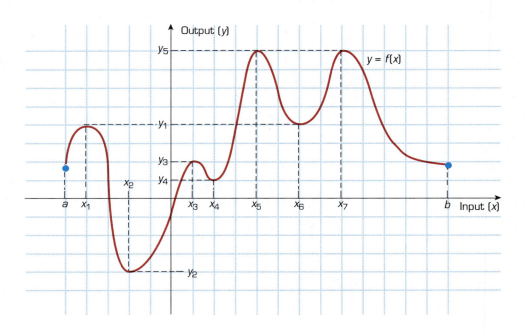

EXAMPLE 1

Find the maximum of the function $y = x \sin(\pi x)$ on the interval $[0, 6]$.

Solution

Using a graphing calculator, we generate the window shown in Figure 11.2. It appears that the function attains a maximum near $x = 4.5$, and indeed, utilizing the zoom and trace features, we find that the maximum value, $y \approx 4.51$, occurs at the input value $x \approx 4.52$. From the same graph, we can also see that the minimum value, $y \approx -5.51$, occurs at the input value $x \approx 5.52$. Here all values are rounded to two decimal places.

FIGURE 11.2

The Graph of $y = x \sin \pi x$ on the Interval [0, 6]

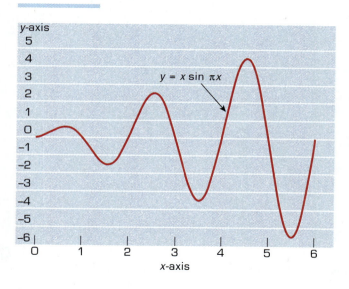

FIGURE 11.3

The Graph of the Function $f(x)$ in Example 2, on the Interval [0, 2]

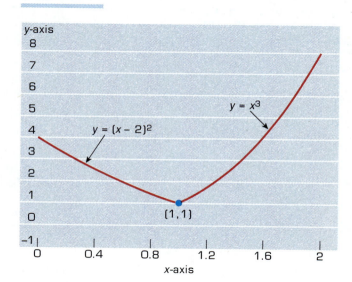

EXAMPLE 2

Find the maximum and the minimum of the function

$$y = f(x) = \begin{cases} (x - 2)^2 & x < 1 \\ x^3 & x \geq 1 \end{cases}$$

with the domain restricted to $[0, 2]$.

Solution

The graph in Figure 11.3 shows that the maximum value, $y = 8$, occurs at $x = 2$, the right endpoint of the domain, while the minimum value, $y = 1$, occurs at $x = 1$. Notice that this function is continuous everywhere on its domain and differentiable everywhere except at $x = 1$, where the graph has a corner.

EXAMPLE 3

Find the maximum and minimum values of the function

$$y = f(x) = x^3 - 21x^2 + 120x - 100$$

on the interval $[2, 12]$.

Solution

The graph in Figure 11.4 shows that the maximum value, $y = 108$, occurs at the input value $x = 4$, and the minimum value, $y = 0$, occurs at $x = 10$. This function is continuous and differentiable everywhere in its domain.

FIGURE 11.4

The Graph of
$y = x^3 - 21x^2 + 120x - 100$
on the Interval [2, 12]

In Figure 11.1, the function $f(x)$ has peaks that are not maxima on $[a, b]$. Instead they are what we call *relative* or *local maxima* because they are high points on the graph over small intervals around x_1 and x_3. If we restrict the inputs to the interval $[a, x_2]$, then *in this interval* the function $y = f(x)$ attains its maximum at the input x_1. Similarly, if we restrict the inputs to the interval $[0, x_4]$, then *in this interval* the function $y = f(x)$ attains its maximum at the input x_3. We reserve the terms *absolute* or *global maximum* for the largest value of a function over its entire domain. The global maximum for the function in Figure 11.1 is y_5.

A peak on the graph of a continuous function is called a relative or local maximum; a pit is called a relative or local minimum.

In Figure 11.1, the function $f(x)$ has a pit at the inputs x_4 and x_6. These pits are called *relative* or *local minima* because they are low points on the graph over small intervals around x_4 and x_6. If we restrict the inputs to the interval $[x_3, x_5]$, then *in this interval* the function $y = f(x)$ attains its minimum at the input x_4. Similarly, if we restrict the inputs to the interval $[x_5, x_7]$, then *in this interval* the function $y = f(x)$ attains its minimum at the input x_6. We reserve the terms *absolute* or *global minimum* for the smallest value of a function over its entire domain. The global minimum for the function in Figure 11.1 is y_2.

A function $y = f(x)$ on an interval $[a, b]$ attains its global maximum at $x = x_1$ if

$$f(x_1) \geq f(x) \tag{1}$$

for all values of x in $[a, b]$. The input x_1 is the location of a *local* maximum if Equation (1) is true for all inputs restricted to a (small) subinterval of $[a, b]$ surrounding x_1. The function $y = f(x)$ on the interval $[a, b]$ attains its global minimum at $x = x_2$ if

$$f(x_2) \leq f(x) \tag{2}$$

for all values of x in $[a, b]$. The input x_2 is the location of a local minimum if Equation (2) is true for all inputs restricted to a (small) subinterval of $[a, b]$ around x_2.

The function in Figure 11.1 is differentiable at every interior point of the domain $[a, b]$. If we zoom in on any local maximum, we obtain the window like the one shown in Figure 11.5. Five tangent lines to this curve are shown on the graph. A line with positive slope slants upward to the right while a line with negative slope slants downward to the right (see Section 1.5). We see in Figure 11.5 that the tangent lines have positive slopes to the left of the local maximum and negative slopes to the right. The slope of a tangent line to a curve $y = f(x)$ at a point x_1 is the derivative of $f(x)$ at x_1, so as we move along the curve in Figure 11.5, the derivatives at each point to the left of the local maximum are positive and to the right of the local maximum, negative. The derivative of $f(x)$ is zero at the local maximum. Consequently, at each local maximum in Figure 11.5, the tangent line to the curve is horizontal; the tangent line has a slope of zero.

FIGURE 11.5

Tangent Lines near a Local Maximum for a Differentiable Function

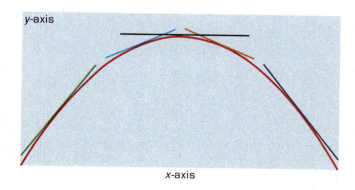

y-axis

x-axis

FIGURE 11.6

Tangent Lines near a Local Minimum for a Differentiable Function

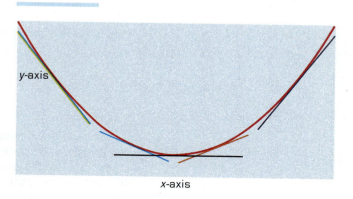

y-axis

x-axis

> *A function has a critical point at an interior point in the domain if the derivative of the function is zero or does not exist at that point.*

If we zoom in on any local minimum in Figure 11.1, we obtain a window similar to Figure 11.6. Now the derivatives of the function are negative to the left of the local minimum and positive to the right, while at the local minimum itself, the derivative of the function is zero.

From Figures 11.5 and 11.6 we see that local maxima and minima occur at input values where the derivative of the function is zero. However, Figure 11.3 shows us that a local optimum can also occur at a point where the derivative does not exist. We define a *critical point* for a function to be an interior point of the domain where the derivative of the function either is zero or does not exist. Therefore a local maximum or minimum always occurs at a critical point.

EXAMPLE 4

Find the critical points of the function

$$y = f(x) = x^3 - 21x^2 + 120x - 100$$

on the interval [2, 12].

Solution

Critical points of a function occur at points in the open interval $2 < x < 12$ where the derivative is zero or does not exist. Differentiating $y = f(x)$ with respect to x, we have

$$\frac{dy}{dx} = 3x^2 - 42x + 120$$

This derivative can be evaluated at all values of x and so exists everywhere. We want to find any values of x at which the derivative of $y = f(x)$ is zero, that is, where

$$3x^2 - 42x + 120 = 0$$

Solving this equation for x, we have

$$3(x^2 - 14x + 40) = 3(x - 4)(x - 10) = 0$$

Thus $x = 4$ and $x = 10$, which are both interior points in the domain [2, 12], are the critical points. We see from Figure 11.4 that $f(x)$ attains its maximum and minimum, respectively, at these critical points.

EXAMPLE 5

Find the critical points of the function

$$y = f(x) = \begin{cases} (x - 2)^2 & 0 \le x < 1 \\ x^3 & 1 \le x \le 2 \end{cases}$$

Solution

If there is a critical point of the function $y = f(x)$, it will occur in the open interval $0 < x < 2$ where either the derivative is zero or does not exist. We know from Example 2 and Figure 11.3 that the graph of this function has a corner at $x = 1$; consequently, $f(x)$ is not differentiable there and $x = 1$ is a critical point.

We can also see in Figure 11.3 that the tangent line to any point on the curve is never horizontal, so the slope of any tangent line is nonzero; in other words, the derivative of $f(x)$ is never zero. This same conclusion can be obtained analytically.

In the interval $0 < x < 1$, the function is defined to be $f(x) = (x - 2)^2$. Using the chain rule, we have that

$$\frac{d}{dx} f(x) = 2(x - 2)$$

which is always negative for $x < 1$. In the interval $1 < x < 2$, the function is $f(x) = x^3$ and so $f'(x) = 3x^2$, which is always positive for $x > 1$. Thus $f'(x)$ is never zero on [0, 2]. The only critical point of $f(x)$ on the interval [0, 2] is $x = 1$, where the derivative does not exist.

EXAMPLE 6

Find the critical points of the function

$$y = f(x) = (x - 20)^2(x - 30)^2$$

on the interval [15, 35].

Solution

Within the interior of the domain of the function, any critical points will occur where the first derivative is zero or does not exist. To find such points, we must differentiate $f(x)$ with respect to x. Using the product rule, we find that

$$\frac{dy}{dx} = (x - 20)^2\left[\frac{d}{dx}(x - 30)^2\right] + \left[\frac{d}{dx}(x - 20)^2\right](x - 30)^2$$

Now we can use the chain rule:

$$\frac{dy}{dx} = (x - 20)^2[2(x - 30)] + [2(x - 20)](x - 30)^2$$

Since this derivative can be evaluated at all values of x, it exists everywhere. In particular, it is zero when

$$2(x - 20)^2(x - 30) + 2(x - 20)(x - 30)^2 = 0$$
$$2(x - 20)(x - 30)[(x - 20) + (x - 30)] = 0$$
$$2(x - 20)(x - 30)(2x - 50) = 0$$

Thus $x = 20$, $x = 25$, and $x = 30$ are critical points; see Figure 11.7.

FIGURE 11.7
The Graph of
$y = (x - 20)^2 (x - 30)^2$
on the Interval [15, 35]

Figure 11.7 is a graph of the function analyzed in Example 6. The function's global minimum value, $y = 0$, occurs at both of the critical points $x = 20$ and $x = 30$, and a local maximum, $y = 625$, occurs at the critical point $x = 25$. The global maximum, $y = 5625$, is attained at $x = 15$ and $x = 35$, the endpoints of the domain. The endpoints are not critical points because they are not in the interior of the domain.

A critical point need not be the location of either a local maximum or a local minimum.

The critical points in Examples 2, 3, and 6 correspond to the locations of peaks or pits on Figures 11.3, 11.4, and 11.7, respectively. A critical point is not always the location of a local optimum. For example, for the function $y = x^3$ on the interval $[-3, 3]$, the point $x = 0$ is a critical point, because $f'(x) = 3x^2$ is zero at $x = 0$. However, we can see in Figure 11.8 that $x = 0$ is not the location of either a local maximum or a local minimum.

A function can also exhibit curious behavior at a point of discontinuity, which is therefore also a point of nondifferentiability. Although it is a critical point for the function, it need not be the location of a local maximum or minimum. (See Problems 56–58.) Fortunately, most real-world processes are modeled by continuous functions, which are generally quite well behaved.

FIGURE 11.8

A Graph of $y = x^3$ on the Interval $[-3, 3]$

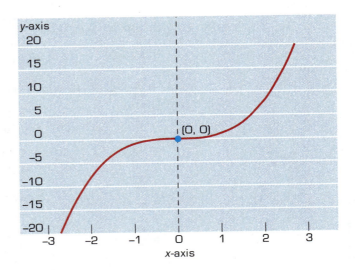

Because a function is a relationship between the inputs and outputs of a process, any limitations on the inputs can impose corresponding restrictions on the outputs. By restricting the inputs in Figure 11.7 to the interval $[15, 35]$, we limited the output values to a maximum of 5,625. It looks like we can increase the maximum value of the outputs by expanding the domain beyond $[15, 35]$. Indeed, some calculating shows that

$$f(13) = 14{,}161 > 5{,}625 = f(15)$$

and

$$f(36) = 9{,}216 > 5{,}625 = f(35)$$

A global maximum or minimum for a function can occur only at an endpoint or critical point of the domain.

However, because a domain represents the set of all possible inputs, values of the function for inputs outside the domain are not relevant. For the model in Example 6, the maximum output is 5,625 over $[15, 35]$, the domain specified as feasible.

We saw in Figure 11.7 that global optima can occur at the endpoints of a domain, and in Figure 11.4 we saw that global optima can occur at interior points of the domain. If a global optimum of a function $f(x)$ does occur at an interior point x_0, then either $f'(x_0)$ does not exist, or, if it does, it must be zero. Thus we can limit our search for global optima to two types of input values: endpoints of the domain or critical points within it.

We summarize our results in the following algorithm:

LOCATING GLOBAL OPTIMA FOR A CONTINUOUS FUNCTION $f(x)$ ON A CLOSED INTERVAL

STEP 1 Identify the endpoints of the domain.

STEP 2 Locate the critical points of $f(x)$ in the interior of the domain.

STEP 3 Evaluate $f(x)$ at each of the points found in Steps 1 and 2. The largest of these values is the global maximum and the smallest is the global minimum over the function's domain.

This algorithm allows us to search for and identify the global optima of a function without graphing. We require that the function be continuous and that the domain be a closed interval of the form $[a, b]$; however, most functions that model real-world processes are continuous with domains that are closed intervals.

EXAMPLE 7

Find the global maximum and the global minimum for the function

$$y = f(x) = x^3 - 21x^2 + 120x - 100$$

on the interval $[1, 12]$.

Solution

STEP 1 The endpoints of the domain are $x = 1$ and $x = 12$.

STEP 2 Using the results of Example 4 we find the critical points for this function to be $x = 4$ and $x = 10$.

STEP 3 We construct Table 11.1 by evaluating $y = f(x)$ at each point identified in Steps 1 and 2. The global maximum of the function over $[1, 12]$ is the largest y-value in Table 11.1, namely, $y = 108$, which occurs at the critical point $x = 4$. The global minimum is the smallest y-value, which is zero, corresponding to two input values, the endpoint $x = 1$ and the critical point $x = 10$.

TABLE 11.1 **The Optima of $y = x^3 - 21x^2 + 120x - 100$ over [1, 12]**

	INPUT x	OUTPUT $y = x^3 - 21x^2 + 120x - 100$
ENDPOINTS:	1	0
	12	44
CRITICAL POINTS:	4	108
	10	0

EXAMPLE 8

Solve Example 7 when the domain is [0, 8].

Solution

STEP 1 The endpoints of the domain are now $x = 0$ and $x = 8$.

STEP 2 Only $x = 4$ is a critical point. Since $x = 10$ is not an interior point of the new domain, it can no longer be a critical point.

STEP 3 By evaluating the function at each point identified in Steps 1 and 2, we can construct Table 11.2. The global maximum is $y = 108$ at the critical point $x = 4$. The global minimum is $y = -100$ at the endpoint $x = 0$.

TABLE 11.2   The Optima of $y = x^3 - 21x^2 + 120x - 100$ over [0, 8]

	INPUT x	OUTPUT $y = x^3 - 21x^2 + 120x - 100$
ENDPOINTS:	0 8	−100 28
CRITICAL POINTS:	4	108

We see from Examples 7 and 8 that the number of critical points and the location of global optima (here the global minimum) can change as we alter the domain.

EXAMPLE 9

Find the global maximum and the global minimum for the function $y = f(x) = (x^2)^{1/5}$ on the interval $[-3, 3]$.

Solution

STEP 1 The endpoints of the domain are $x = -3$ and $x = 3$.

STEP 2 If we rewrite the function as $y = (x^2)^{1/5} = x^{2/5}$, we can use the power rule of differentiation to obtain

$$\frac{dy}{dx} = \frac{2}{5} x^{-3/5} = \frac{2}{5(x^3)^{1/5}} \tag{3}$$

This derivative does not exist at $x = 0$ because division by zero is an undefined arithmetic operation. Also, the derivative in Equation (3) can never be zero (see Problem 64). Thus the only critical point is $x = 0$.

TABLE 11.3   The Optima of $y = x^{2/5}$ over [−3, 3]

	INPUT x	OUTPUT $y = x^{2/5}$
ENDPOINTS:	−3 3	$(-3)^{2/5} \approx 1.552$ $3^{2/5} \approx 1.552$
CRITICAL POINTS:	0	0

STEP 3 We construct Table 11.3 by evaluating $y = f(x)$ at each point identified in Steps 1 and 2. The global maximum, $y \approx 1.552$, occurs at both endpoints. The global minimum, $y = 0$, occurs at $x = 0$, where the derivative does not exist. These optima are apparent from the graph of the function displayed in Figure 11.9.

FIGURE 11.9

The Graph of $y = x^{2/5}$ on the Interval $[-3, 3]$

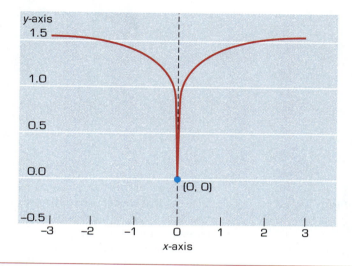

We now have two approaches for locating the global optima of a continuous function: the first is to graph it and locate the optima within a specified tolerance using the zoom and trace features of a graphing calculator; while the second approach is to use the algorithm. Both methods have their advantages.

The algorithmic approach has the advantage of not requiring a graph, but this advantage is not great with modern technology, where graphing functions on a computer or graphing calculator has become fairly routine. The algorithmic approach also has the advantage of yielding an exact solution, while the graphing technique generally provides only an approximate one. This advantage is significant but not necessarily crucial: a solution may be needed to only a particular level of precision.

A disadvantage to the algorithmic approach is that the process of setting a derivative equal to zero and solving for the independent variable can be quite challenging. Consider the function $y = (\sin x)(\ln x)$ on the interval $[1, 9]$. Its derivative is

$$\frac{dy}{dx} = \frac{1}{x} \sin x + (\cos x)(\ln x)$$

which is zero when

$$\frac{x(\cos x)(\ln x)}{\sin x} = -1$$

This is not an easy equation to solve! In contrast, we can use a graphing calculator to obtain Figure 11.10 and then use the trace and zoom features to identify the global maximum, $y \approx 2.065$, at the input $x \approx 7.915$, and the global minimum, $y \approx -1.564$, at $x \approx 4.845$. Because the graphing approach is sometimes simpler, it is now a very popular procedure for locating global optima to a specified degree of accuracy.

FIGURE 11.10
The Graph of $y = (\sin x)(\ln x)$ on the Interval [1, 9]

We noted earlier that because real-world processes are constrained by finite resources, functions that model such processes are bounded within a finite interval, including the endpoints. From a mathematical perspective, though, we can specify the domain of a continuous function to be an open interval. However, then we must modify our algorithm for locating global optima since the endpoints are no longer feasible inputs.

Instead of evaluating a continuous function $f(x)$ over (a, b) at the endpoints a and b, we can investigate the two limits

$$\lim_{x \to a^+} f(x) \quad \text{and} \quad \lim_{x \to b^-} f(x)$$

which involve only feasible inputs. If $f(x)$ grows large positively without bound as either $x \to a^+$ or $x \to b^-$, then there is no global maximum; if $f(x)$ grows large negatively without bound as either $x \to a^+$ or $x \to b^-$, then there is no global minimum. If both limits exist, we compare them to the values of $f(x)$ at each critical point in (a, b). If the largest value of $f(x)$ occurs at one of these critical points, then that value is the global maximum; if not, there is no global maximum. If the smallest value of $f(x)$ occurs at a critical point, then that value is the global minimum; if not, there is no global minimum.

Alternatively, we can graph the function on an open interval and inspect the resulting curve visually.

> *A domain should include its endpoints when a function models a real-world process.*

EXAMPLE 10

Find the global maximum and global minimum for the function $y = f(x) = 1/x$ on $(0, 2)$.

Solution
The left endpoint is zero, so we want to examine the behavior of $f(x)$ as x approaches zero from the right. What we get is

$$\lim_{x \to 0^+} \left(\frac{1}{x}\right) = \infty$$

where the symbol ∞ (infinity) denotes the fact that $1/x$ grows large beyond bound as $x \to 0^+$. The right endpoint is 2, and in this case,

$$\lim_{x \to 2^-} \left(\frac{1}{x}\right) = 0.5$$

The derivative of $y = 1/x$ is $y' = -1/x^2$, which can never be zero. Therefore the function $f(x) = 1/x$ has no critical points and no global optima.

Figure 11.11 is a graph of the function. There can be no global maximum because the y-values grow increasingly large as the x-values approach zero from the right. There is no global minimum because, as the x-values approach 2 from the left, the corresponding y-values *approach* 0.5 but never equal it. (Recall that $x = 2$ is not included in the domain.)

FIGURE 11.11

The Graph of $y = 1/x$ on the Open Interval (0, 2)

IMPROVING SKILLS

1. Locate all global and local optima for the function graphed in Figure A on the interval $[-2, 8]$.
2. Locate all global and local optima for the function graphed in Figure B on the interval $[100, 200]$.

FIGURE A

FIGURE B

Find all critical points for the functions in Problems 3 through 28, then use that information to locate the global maxima and minima on the specified intervals.

3. $y = 3x^2 - 24x - 5$; $[0, 6]$
4. $y = -x^2 + 25x - 3.5$; $[4, 16]$
5. $y = 7x^2 + 56x + 25$; $[0, 5]$
6. $y = -5x^2 + 100x$; $[0, 20]$
7. $y = 2x - 1$; $[25, 50]$
8. $y = x^3 - 12x + 7$; $[-3, 3]$
9. $y = x^3 - 12x + 7$; $(-3, 3)$
10. $y = x^3 - 12x + 7$; $[-4, 4]$
11. $y = x^3 - 12x + 7$; $[-5, 5]$
12. $y = x^3 - 12x + 7$; $(-5, 5)$
13. $y = x^3 - 12x + 7$; $[0, 3]$
14. $y = 2x^3 - 54x + 2800$; $[-3, 5]$
15. $y = 2x^3 - 54x + 2800$; $[0, 6]$
16. $y = 2x^3 - 54x + 2800$; $(0, 6)$
17. $y = x^3 - 48x + 57$; $[0, 4]$
18. $y = x^3 - 12x^2 + 36$; $[0, 9]$
19. $y = x^3 - 12x^2 + 36$; $(0, 9)$
20. $y = x^3 - 6x^2 + 9x + 6$; $[0, 3]$
21. $y = x^3 - 6x^2 + 9x + 6$; $[1, 4]$
22. $y = x^3 - 6x^2 + 9x + 6$; $[-1, 5]$
23. $y = x^3 - 9x^2 + 15x - 10$; $[-1, 7]$
24. $y = x^3 - 9x^2 + 15x - 10$; $(-1, 7)$
25. $y = x^3 - 10x^2 + 12x + 10$; $[0, 10]$
26. $y = 3x^4 - 8x^3 - 6x^2 + 24x + 1$; $[0, 4]$
27. $y = 3x^4 - 8x^3 - 6x^2 + 24x + 1$; $(0, 4)$
28. $y = x^4 - 4x^3 + 6x^2 - 4x + 1$; $[0, 4]$

$3x^2 - 24x$
$3x(x - 8)$

EXPLORING WITH TECHNOLOGY

Use a graphing calculator to graph each of the functions in Problems 29 through 55 on the given interval. Using your graphs, approximate to within three decimal places the locations of any global maxima and minima.

29. $y = 3x^4 - 8x^3 - 6x^2 + 24x + 1$; $[0, 4]$
30. $y = x^4 - 4x^3 + 6x^2 - 4x + 1$; $[0, 4]$
31. $y = x^4 - 4x^3 + 6x^2 - 4x + 1$; $(0, 4)$
32. $y = \sin x \cos x$; $[0, 8]$
33. $y = x \cos x$; $[0, 8]$
34. $y = x \cos x$; $(0, 8)$
35. $y = x^2 \cos x$; $[0, 8]$
36. $y = e^{x \cos x}$; $[-4, 0]$
37. $y = e^{-x} \sin x$; $[0, 5]$
38. $y = e^{-x} \sin x$; $[2, 4]$
39. $y = e^{-x} \sin x$; $(0, 1)$
40. $y = (x - 3)\ln x$; $[2, 5]$
41. $y = (x - 3)\ln x$; $(0, 5]$

42. $y = \begin{cases} -x^2 + 2x + 5 & 0 \le x < 2 \\ x + 4 & 2 \le x \le 3 \end{cases}$

43. $y = \begin{cases} -x^2 + 2x + 5 & 0 \le x < 2 \\ x^2 - 8x + 17 & 2 \le x \le 5 \end{cases}$

44. $y = \begin{cases} x^2 & -1 \le x < 0 \\ -x^2 & 0 \le x \le 2 \end{cases}$

45. $y = \begin{cases} x^2 & -1 < x < 0 \\ -x^2 & 0 \le x < 2 \end{cases}$

46. $y = \begin{cases} \sqrt{-x} & -1 \le x \le 0 \\ \sqrt{x} & 0 < x \le 1 \end{cases}$

47. $y = \begin{cases} \sqrt{-x} & -1 < x < 0 \\ \sqrt{x} & 0 \le x \le 1 \end{cases}$

48. $y = \begin{cases} \sqrt{-x} & -1 < x < 0 \\ \sqrt{x} & 0 \le x < 1 \end{cases}$

49. $y = \begin{cases} -x & -1 \le x < 0 \\ x^2 & 0 \le x < 1 \\ x^3 & 1 \le x \le 2 \end{cases}$

50. $y = \begin{cases} x - \frac{1}{2} & -1 \le x < 0 \\ \frac{3}{2}x^2 - \frac{1}{2} & 0 \le x < 1 \\ x^3 & 1 \le x \le 2 \end{cases}$

51. $y = \begin{cases} e^x & -2 \le x < 0 \\ e^{-x} & 0 \le x \le 2 \end{cases}$

52. $y = \begin{cases} e^x & -2 < x < 0 \\ e^{-x} & 0 \le x \le 2 \end{cases}$

53. $y = \begin{cases} e^x & -2 < x < 0 \\ e^{-x} & 0 \le x < 2 \end{cases}$

54. $y = \begin{cases} e^x & -2 \le x < 0 \\ 1 - x & 0 \le x \le 2 \end{cases}$

55. $y = \begin{cases} e^x & -2 < x < 0 \\ 1 - x & 0 \le x \le 2 \end{cases}$

EXPLORING IN TEAMS

56. Determine whether the function

$$y = f(x) = \begin{cases} 2x & 0 \le x < 1 \\ 2 - x & 1 \le x \le 2 \end{cases}$$

has a global maximum or a global minimum. Does this function have any critical points?

57. Determine whether the function

$$y = f(x) = \begin{cases} x & 0 \le x < 1 \\ 2 & x = 1 \\ 2 - x & 1 < x \le 2 \end{cases}$$

has a global maximum or a global minimum. Does this function have any critical points?

58. Determine whether the function

$$y = f(x) = \begin{cases} x & 0 \le x < 1 \\ -2 & x = 1 \\ 2 - x & 1 < x \le 2 \end{cases}$$

has a global maximum or a global minimum. Does this function have any critical points?

59. Identify all local and global optima for the function

$$f(x) = \begin{cases} 0 & x = 0 \\ e^{-x} \sin\left(\dfrac{1}{x}\right) & 0 < x \le 1 \end{cases}$$

State why the algorithm given in this section does not apply to this function.

60. Under what conditions is a critical point the location of a local optimum?

REVIEWING MATERIAL

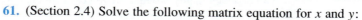

61. (Section 2.4) Solve the following matrix equation for x and y:

$$\begin{bmatrix} x & 5 \\ 4 & -3 \end{bmatrix}\begin{bmatrix} 2 \\ -1 \end{bmatrix} - 2\begin{bmatrix} 3 \\ y \end{bmatrix} = \begin{bmatrix} 2 \\ -1 \end{bmatrix}$$

62. (Section 4.3) The Board of Education of a local school district votes to create a sinking fund to finance new bleacher seats for the football field. They need to have $650,000 in four years. How much should they deposit monthly into this fund if interest rates are $5\frac{3}{4}\%$ a year, compounded monthly?

63. (Section 6.4) In 1989 the General Social Survey asked a large sample of 16-year-olds to identify the religion in which they were raised. It was found that 63% were raised Protestant and the rest were raised in either a non-Protestant religion or no religion at all. Males accounted for 41% of the surveyed Protestants and 46% of all others. Determine the probability that a female, selected at random from this surveyed population, was raised Protestant.

64. (Section 10.3) Use the definition of the derivative,

$$f'(x) = \lim_{h \to 0} \frac{f(x + h) - f(x)}{h}$$

and a calculator to create a table that shows that the function $y = f(x) = (x^2)^{1/5}$ does not have a derivative at $x = 0$.

RECOMMENDING ACTION

65. Respond by memo to the following request:

> ### MEMORANDUM
>
> To: J. Doe Reader
>
> From: Terry
>
> Date: Today
>
> Subject: **Endpoints**
>
> I'm teaching a friend's class on global optima while she's at a math conference, and I need some help. I want to be able to answer all questions from the students. I understand that we must locate critical points by analyzing derivatives at all interior points, and we must consider endpoints. But the book does not say anything about the derivatives at the endpoints. What if a student asks about them? What should I say?

11.2 DESIGN PROBLEMS

The first step in any production process is to design the final product. For example, a house is built from architectural drawings, and a newspaper is printed from page layouts. A good design can minimize waste and maximize available resources.

In Section 11.1, we developed two methods for optimizing a function over a closed interval: graphing the function (preferably using a graphing calculator or a computer software package), and then locating optima directly from the graph; or else applying the optimization algorithm. We will apply both methods to solve a variety of design problems.

EXAMPLE 1

A couple wants to leave their dog out during the day while they're at work, so they decide to build a rectangular dog yard, using their house for one side and fencing for the other three sides. Find the dimensions of the dog yard with maximum area if the couple can afford 60 feet of fencing.

Solution

If we are to design a yard with the largest possible area, there are three quantities we must use in our calculations:

l = the length of the yard (in feet)

w = the width of the yard (in feet)

A = the area of the yard (in square feet)

FIGURE 11.12

A Proposed Dog Yard

The area of a rectangle is the product of its length and width: $A = lw$. Figure 11.12 is a diagram of the proposed dog yard. Its dimensions are constrained by the fact that the length of the fence is 60 ft, that is, $l + 2w = 60$, or

$$l = 60 - 2w$$

MODEL 11.1

Important factors

w = width of the dog yard (ft)

l = length of the dog yard (ft)

A = area of the dog yard (sq. ft)

Objective

Find w and l so that A is maximized.

Constraints

$A = 60w - 2w^2$

$l = 60 - 2w$

$0 \le w \le 30$

Now we can express the area as a function relating the output A to the single input variable w:

$$A = lw = (60 - 2w)w = 60w - 2w^2$$

The width of the dog yard cannot be negative, hence $w \ge 0$. Similarly, $l \ge 0$, that is, $60 - 2w \ge 0$. Therefore w must be less than 30 ft, that is, $w \le 30$. We summarize this information in Model 11.1.

We want to maximize the function

$$A(w) = 60w - 2w^2 \qquad (4)$$

on the interval [0, 30]. The global maximum occurs either at the endpoint of this interval, $w = 30$, or at a critical point, if one exists. To find out, we need the derivative of A with respect to w:

$$\frac{dA}{dw} = 60 - 4w$$

TABLE 11.4 The Optima of $A = 60w - 2w^2$

| | INPUT | OUTPUT |
	w	A
ENDPOINT:	0	0
	30	0
CRITICAL POINT:	15	450

This derivative can be evaluated at all values of w, so the derivative exists everywhere. It is zero when $60 - 4w = 0$. Therefore, $w = 15$, which is an interior point in the domain $[0, 30]$, is the only critical point.

Evaluating Equation (4) at each prospective point for a global maximum, we generate Table 11.4. The global maximum, $A = 450$ ft^2, occurs when $w = 15$ ft, and it follows that $l = 60 - 2w = 60 - 2(15) = 30$ ft. The dimensions of the dog yard should be 15 ft wide by 30 ft long. We obtain the same answer by graphing Equation (4) on the interval $[0, 30]$, as shown in Figure 11.13, and noting that the high point on the graph appears to occur at $w = 15$.

FIGURE 11.13

The Graph of the Area Function $A(w) = 60w - 2w^2$ for $0 \leq w \leq 30$ ft

Area of dog yard (A)

(15, 450)

$A(w) = 60w - 2w^2$

Width of dog yard (w)

We solved Example 1 using differentiation. Since all of our differentiation rules apply to a function of a single input variable, it is essential that our models have this feature. We cannot use the algorithm developed in Section 11.1 unless we have models that are functions of a single input variable.

EXAMPLE 2

FIGURE 11.14

A Diagram of the Proposed Athletic Field

The Board of Education of a local school district has purchased a plot of land adjacent to the high school for a groomed athletic field that will support football, track, and soccer. The field will be a rectangular shape with two identical semicircular regions at each end. The rectangular part must be at least 120 yards long and 60 yards wide in order to meet the school's football and soccer needs, and the perimeter of the field must be exactly one-quarter mile long to meet the track requirements. Find the dimensions of the field so that the area of the rectangular portion is as large as possible.

Solution

Figure 11.14 presents a diagram of the proposed athletic field. We will standardize our measurements in feet, although yards or meters would serve equally well. Our objective is to maximize the rectangular area of the field, so we have three quantities to work with:

$$l = \text{the length of the rectangle (in feet)}$$
$$w = \text{the width of the rectangle (in feet)}$$
$$A = \text{the area of the rectangle (in square feet)}$$

Since the area of the rectangle is $A = lw$, the output A depends on two input variables, l and w. Although values for both variables are required to solve the problem, we need A to be a function of a single variable if we expect to optimize by either of the techniques described in Section 11.1.

One constraint on the dimensions of the entire field is its perimeter, which must be one-quarter mile or 1320 ft. The distance around the curved part of a half-circle of radius r is $\frac{1}{2}(2\pi r) = \pi r$, so we need to know the radius (in ft) of the semicircular portion of the field. The perimeter constraint is modeled by the equation

$$l + l + \pi r + \pi r = 2l + 2\pi r = 1320$$

or

$$l + \pi r = 660 \tag{5}$$

An inspection of Figure 11.14 shows us that the width of the field is twice the radius of each semicircular end, that is, $w = 2r$. Solving for r, we have $r = 0.5w$. Now substituting this value into Equation (5) and solving for l, we obtain

$$l = 660 - \pi r = 660 - 0.5\pi w \tag{6}$$

We can now solve for the area A of the rectangle:

$$A = lw = (660 - 0.5\pi w)w = 660w - 0.5\pi w^2$$

which models the output variable A as a function of the single input variable w.

The width of the field must be at least 60 yd = 180 ft wide, so $w \geq 180$. In addition, the length of the rectangular portion of the field must be at least 120 yd = 360 ft long, so $l \geq 360$. It follows from Equation (6) that

$$660 - 0.5\pi w = l \geq 360$$

Solving for w (see Section 3.1), we get $w \leq 600/\pi$.

The important data for this problem are displayed in Model 11.2.

We want to maximize the function

$$A(w) = 660w - 0.5\pi w^2 \tag{7}$$

on the interval $[180, 600/\pi]$. A graph of this function is shown in Figure 11.15. Using the trace and zoom features of a graphing calculator, we find that the global maximum, $A \approx 68{,}755$ ft^2, occurs at the right endpoint of the domain, $w = 600/\pi \approx 191$ ft. In other words, maximum area is achieved by making the width of the rectangular portion of the field as large as possible. It then follows that $r = w/2 \approx 95.5$ ft and, from Equation (6), that $l = 660 - 0.5\pi(600/\pi) = 360$ ft.

MODEL 11.2

Important factors

l = length of the rectangular portion (in ft)

w = width of the rectangular portion (in ft)

r = radius of the semicircular portion (in ft)

A = area of the rectangular portion (in sq. ft)

Objective

Find l, w, and r so that A is maximized.

Constraints

$$A = 660w - 0.5\pi w^2$$
$$r = 0.5w$$
$$l = 660 - 0.5\pi w$$
$$180 \leq w \leq 600/\pi$$

FIGURE 11.15

The Graph of the Area of the Field as a Function of Width

We can also solve Example 2 without graphing. The global maximum occurs either at an endpoint of the domain, $w = 180$ or $w = 600/\pi$, or at a critical point inside the domain, if such a point exists. Differentiating both sides of Equation (7) with respect to the input variable w, we obtain

$$\frac{dA}{dw} = 660 - \pi w$$

This derivative can be evaluated at every value of w, so it exists everywhere. The derivative is zero when $660 - \pi w = 0$, that is, when

$$w = 660/\pi \approx 210.08 \text{ ft}$$

Since this input is outside the domain, we discard it. Evaluating Equation (7) at each point where a global maximum may occur, we generate the entries in Table 11.5, with all areas rounded to the nearest square foot. The global maximum, $A \approx 68{,}755$, occurs when $w = 600/\pi \approx 191$ ft.

TABLE 11.5 The Optima of $A = 660w - 0.5\pi w^2$

	INPUT w	OUTPUT $A = 660w - 0.5\pi w^2$
ENDPOINTS:	180	67,906
	$600/\pi \approx 191$	68,755
CRITICAL POINTS:	None	—

In both of Examples 1 and 2, we found it helpful to draw a picture of the proposed facility prior to modeling it with mathematical equations. A diagram is a visual model and it is often a valuable intermediate step in creating a mathematical model from a verbal description of a problem. Consequently, we have the following algorithm for solving optimization problems.

SOLVING OPTIMIZATION PROBLEMS

STEP 1   Draw a diagram of the process, if possible.

STEP 2   Identify the important quantities in the process, including the item to be optimized, label them with variables, and then write an equation that relates the item to be optimized to other variables.

STEP 3   Identify all constraints involving the variables.

STEP 4   Simplify the model so that the item to be optimized is expressed as a function of a single input variable.

STEP 5   Establish a feasible domain, including endpoints, for the input variable identified in Step 4.

STEP 6   Optimize the function in Step 4 within the domain established in Step 5, using either the optimization algorithm or the graphing procedure.

EXAMPLE 3

A sheet of tin that is $20'' \times 10''$ will be shaped into an open box by cutting identical squares from each corner of the sheet [Figure 11.16(a)], then folding up and welding the sides [Figure 11.16(b)]. What size squares should be cut from each corner if the volume of the box is to be as large as possible?

FIGURE 11.16
A Future Tin Box: (a) A Diagram of the Sheet of Tin with the Proposed Cuts (b) A Diagram of the Completed Box

(a)

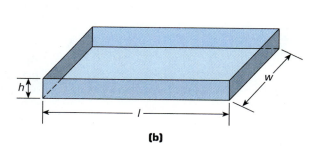

(b)

Solution

Since identical squares will be cut from each corner, the length of each cut is the side of the square [the dashed line segments in Figure 11.16(a)]. Furthermore, the length of this cut determines the height of the box. [See Figure 11.16(b).]

Our objective is to find the length of that cut so we can produce a box with the largest possible volume. The volume of a rectangle is the product of its height, width, and length, so there are four variables:

h = the height of the box (in inches)

l = the length of the box (in inches)

w = the width of the box (in inches)

V = the volume of the box (in cubic inches)

Thus the volume of a rectangular box is defined by the equation $V = hwl$.

The dimensions of the box are limited by the size of the original sheet of tin. After an h-inch square is cut from each corner, the remaining box has width $w = 10 - 2h$ and length $l = 20 - 2h$. Now, substituting these equations for w and l into the formula for volume, we obtain

$$V = hwl = h(10 - 2h)(20 - 2h) = 200h - 60h^2 + 4h^3$$

This equation relates the quantity we want to optimize, V, to the single input variable h.

We cannot make a cut of negative length, so $h \geq 0$. The shorter dimension of the original sheet of tin is 10 in., and the width of the box, $10 - 2h$, cannot be negative, so $h \leq 5$. We summarize this problem in Model 11.3.

We want to maximize the function

$$V(h) = 200h - 60h^2 + 4h^3 \tag{8}$$

on the interval [0, 5]. A graph of this function is shown in Figure 11.17. Using the trace and zoom features of a graphing calculator, we find that the global maximum, $V \approx 192.450$ in.^{3}, occurs when $h \approx 2.113$ in.

MODEL 11.3

Important factors

h = height of the box (in inches)

w = width of the box (in inches)

l = length of the box (in inches)

V = volume of the box (in cubic inches)

Objective

Find h so that V is maximized.

Constraints

$V = 200h - 60h^2 + 4h^3$

$l = 20 - 2h$

$w = 10 - 2h$

$0 \leq h \leq 5$

FIGURE 11.17

The Graph of the Volume of the Box as a Function of Height

Volume of box [V] (cubic inches)

$V = 200h - 60h^2 + 4h^3$

Height of box (h) (inches)

We can solve Example 3, also, without graphing. The global maximum of $V(h)$ occurs either at an endpoint of the domain, $h = 0$ or $h = 5$, or at a critical point inside the domain. Differentiating $V(h)$ with respect to h [see Equation (8)], we get

$$\frac{dV}{dh} = 200 - 120h + 12h^2$$

This derivative can be evaluated at every value of h, hence the derivative exists everywhere. It is zero when

$$12h^2 - 120h + 200 = 0$$

We solve for h with the quadratic formula:

$$h = \frac{-(-120) \pm \sqrt{(-120)^2 - 4(12)(200)}}{2(12)}$$

$$= \frac{120 \pm \sqrt{4800}}{24}$$

TABLE 11.6 **The Optima of**
$V = 200h - 60h^2 + 4h^3$

	INPUT	OUTPUT
	h	V
ENDPOINTS:	0	0
	5	0
CRITICAL POINT:	2.113	192.450

Thus we have two candidates for critical points: $h \approx 7.887$ and $h \approx 2.113$. The first of these points is outside the domain, so we discard it. The second, $h \approx 2.133$, is a critical point. Evaluating Equation (8) at each point where a global maximum may occur, we generate the entries in Table 11.6, with all figures rounded to three decimals. The global maximum, $V \approx 192.450$ in.^{3}, occurs when $h \approx 2.113$ in.

EXAMPLE 4

A chemical company uses cylindrical drums that hold 900 cubic centimeters of a corrosive solvent. The material used for the top and bottom of a drum costs 2¢/square centimeter while that used for the sides costs $\frac{1}{2}$¢/square centimeter. Find the dimensions of the drum that minimize the cost of materials.

Solution

FIGURE 11.18

A Cylindrical Drum of Radius r cm and Height h cm

Figure 11.18 is a diagram of the drum. Our objective is to find the dimensions that minimize the cost of manufacturing the drum, so the relevant variables are

h = the height of the drum (in centimeters)

r = the radius of the drum (in centimeters)

C = the cost of the drum (in cents)

The cost of the drum is related to its height (of the side) and its radius (of the top and bottom). The area of the top and the bottom is πr^2 cm^2, so the cost of materials to produce either is $2\pi r^2$ cents. The surface area of the side of the drum is $2\pi rh$ cm^2, so its cost is $\frac{1}{2}(2\pi rh) = \pi rh$ cents. The total cost C of the drum is the sum of the costs of the top, bottom, and side:

$$C = 2\pi r^2 + 2\pi r^2 + \pi rh = 4\pi r^2 + \pi rh \qquad (9)$$

We now have the quantity we wish to optimize, C, as a function of two input variables, r and h, so we need to eliminate one of these two inputs.

The volume of a right cylinder is πr^2h, and we require this volume to be 900 cm^3, that is, $\pi r^2h = 900$. We can solve this last equation for either r or h. Solving for h, we obtain

$$h = 900/\pi r^2$$

Substituting this into Equation (9), we have

$$C(r) = 4\pi r^2 + \pi r(900/\pi r^2) = 4\pi r^2 + (900/r) \qquad (10)$$

MODEL 11.4

Important factors

h = the height of the drum (in centimeters)

r = the radius of the drum (in centimeters)

C = the material cost of the drum (in cents)

Objective

Find r and h so that C is minimized.

Constraints

$C = 4\pi r^2 + (900/r)$

$h = 900/\pi r^2$

$0.1 \leq r \leq 53.5$

We now have modeled the cost C as a function of a single input variable, r, the radius of the drum.

To establish a domain, we first note that Equation (10) has r as a denominator. Thus r cannot be zero. Furthermore, the radius of a drum cannot be negative, so $r > 0$. This strict inequality is a lower bound for r. We could use the limiting operations discussed at the end of Section 11.1 to locate the global minimum, however, the domain of a function that models a real-world process should include its endpoints.

How small can r be? The top of a can is cut by a machine, which has limits beyond which it physically cannot cut. Let's assume that this limit is 0.1 cm, that is, that the machine that cuts the can tops is capable of cutting a circle of radius no smaller than 0.1 cm. (We shall see shortly that the actual number, which would be provided by the machine's manufacturer, is not important.) Hence, $r \geq 0.1$. Similar considerations force $h \geq 0.1$, and since $h = 900/\pi r^3$, it follows that $r \leq 53.5$ cm, rounded to one decimal place. There are many constraints that bound r from above, for example, the diameter of the drum cannot be larger than the space where it's made or the largest door out of the factory, and the drum must fit on the delivery vehicle. The situation is summarized in Model 11.4.

Figure 11.19 is a graph of the function

$$C(r) = 4\pi r^2 + (900/r)$$

FIGURE 11.19

The Graph of $C(r) = 4\pi r^2 + (900/r)$ on [0, 53.5]

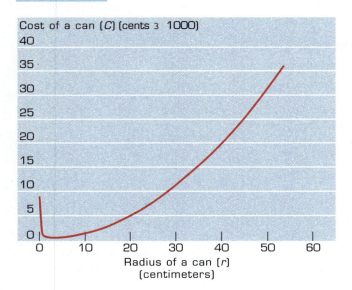

Cost of a can (C) (cents 3 1000)

Radius of a can (r) (centimeters)

FIGURE 11.20

The Graph of $C = 4\pi r^2 + (900/r)$ on [1, 5]

Cost of a can (C)

Radius of a can (r)

on the interval [0.1, 53.5]. It appears that the minimum occurs near $r = 3$. If we magnify the portion of the graph that lies between $r = 1$ and $r = 5$, we obtain Figure 11.20. Continuing to zoom in on the global minimum to obtain finer and finer resolutions, we eventually identify the global minimum as $C \approx 410¢ = \$4.10$ (rounded to the nearest penny) when $r \approx 3.296$ cm. The corresponding height of the drum is $h = 900/\pi r^2 \approx 900/\pi(3.296)^2 \approx 26.369$ cm.

From Figure 11.19 we see that we do not need to know the precise limits of the machine's capabilities as long as the lower limit is less than 3 cm. Therefore any domain that includes the interval $3 \leq r \leq 4$ contains $r \approx 3.296$ cm, the location of the global minimum. If, however, the machine cannot cut a circle with a radius that small, then $r \approx 3.296$ is no longer feasible, and the global minimum occurs at the left endpoint of whatever domain is feasible.

CREATING MODELS

FIGURE A

FIGURE B

1. A farmer has 300 ft of fence that he wants to use to enclose a rectangular plot of land. To save fencing, the farmer will use an existing wall as part of one side of the plot (see Figure A). Find the dimensions of each side of the plot if the enclosed area is to be as large as possible.

2. A rancher has 500 m of fencing for building two square, free-standing corrals. They need not be identical in size, but each corral must have sides at least 3 m long so that it is large enough to accommodate at least one animal comfortably. Find the dimensions of each corral that result in a maximum combined enclosed area.

3. A rancher has 500 m of fencing for building two free-standing corrals, one square and the other rectangular, with length twice the width. Find the dimensions of each corral that result in the maximum combined enclosed area if each corral must enclose at least 100 m².

4. A store manager wants to fence in an area outside his store for a garden center. An exterior wall of the store will serve as one side and the other three sides will be enclosed by a fence. Two sides of the enclosed area will be fenced in wood, at a cost of \$6 per linear foot, while the side parallel to the store wall will be galvanized steel fencing, at \$16 per linear foot. Determine the dimensions of the garden center that will maximize total enclosed area if the manager wants to limit the cost of fencing materials to \$500.

5. Solve Problem 4 if the manager has no limitations on the cost of fencing materials but instead needs a 1200 ft² garden shop and wants to build it at minimum cost. What are reasonable minimum dimensions for the sides if the garden center must have room for both customers and garden supplies?

6. Solve Problem 5 if zoning laws limit the sides perpendicular to the existing wall to 25 feet.

7. For the Undercover Publishing Co., a textbook page contains 56 sq. in. of text, with $1\frac{1}{2}$ in. margins at top and bottom and 1 in. margins on each side; see Figure B. Find the dimensions of a page if they have minimized the area of each page.

8. A shipping company limits the total of the height plus the girth (the dimensions around) of parcels to a maximum of 84 in. Find the dimensions of a rectangular box with a square base that meets the company's regulations and encloses the most volume.

FIGURE C

9. Find the dimensions of a cylindrical tube (see Figure *C*) that meets the regulations described in Problem 8 and encloses the most volume.

10. A storage shed is to be a closed rectangular shape with a square base and a volume of 150 ft³, with each side at least 3 ft long. The floor will be made from concrete at $8/ft², and the walls from plywood at $3/ft², while the roof will cost $5/ft². Find the dimensions of the most economical shed.

11. The Tiny Tricycle Co. needs a rectangular shipping carton with a square base, 2 ft³ of volume, and sides at least 6 in. wide. The material for the top and bottom of the box costs 20¢/ft² while the material for the sides costs 15¢/ft². Determine the dimensions of the least expensive box that meets the company's requirements.

12. A proposed batting practice cage will be built in an indoor gymnasium. The cage will be rectangular in shape with one open side and three net sides and net ceiling, as shown in Figure *D*. Determine what size cage requires the least amount of netting if the volume is to be 1200 ft³ and all sides are to be at least 10 ft long.

FIGURE D

13. The length-to-width ratio must be 1.5 for the base of a closed rectangular box. Design such a box with maximum volume if its total surface area cannot exceed 27 ft².

14. A 40-inch-long wire can be cut into two pieces and shaped into a circle and a square, or it can be left whole and shaped into a circle *or* a square. Determine how to cut the wire to maximize the total enclosed area.

15. A Norman window has the shape of a rectangle topped by a semicircle. (See Figure *E*.) Find the dimensions of a Norman window with maximum area if the outside perimeter is at most 4 m.

16. Find the dimensions of a Norman window with minimum outside perimeter if its area must be exactly 4 m².

17. Find the area of the largest rectangle that can be inscribed within a semicircle of radius r.

FIGURE F

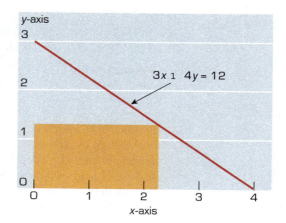

y-axis

3x 1 4y = 12

x-axis

18. A rectangle in Quadrant I is bounded by the x- and y-coordinate axes and the line $3x + 4y = 12$. (See Figure F.) Find the dimensions of the rectangle with maximum area.

19. The sides of a right triangle in Quadrant I lie on the x- and y-coordinate axes and on a line that passes through the point $(3, 4)$. Find the equation of the line that generates the triangle with minimum area.

20. Model the following process as simply as possible and explain why it cannot be solved with the methods described in Section 11.1: Tex has 500 m of fencing that will be used for building one or two free-standing square corrals, possibly with different dimensions. How many corrals should Tex build and with what dimensions to maximize the combined enclosed area of the corral(s)?

EXPLORING IN TEAMS

The notation "$x \to \infty$" signifies that x gets large beyond all bounds. Use tables to evaluate the limits, if they exist, in Problems 21 through 24.

21. $\lim\limits_{x \to \infty} \left(\dfrac{x^2 + 2x + 10}{2x^2 - 2} \right)$

22. $\lim\limits_{x \to \infty} \left(\dfrac{3x^2 + 10}{x^3 + x - 1} \right)$

23. $\lim\limits_{x \to \infty} \left(\dfrac{x^2 + 6x}{3x + 7} \right)$

24. $\lim\limits_{x \to \infty} \left(1 + \dfrac{1}{x} \right)^x$

25. Problems 21 through 23 involve rational functions, and their limits can be evaluated analytically as follows: (1) divide each numerator and denominator by the highest power of x that appears in the denominator; (2) determine the limit directly, without the aid of a calculator. Do this and compare your answers with the ones you obtained numerically.

26. Discuss how to locate the global minimum of the function

$$C(r) = 4\pi r^2 + (900/r)$$

on the interval $(0, \infty)$. How is this helpful in solving Example 4?

REVIEWING MATERIAL

FIGURE G

27. (Section 1.4) A ball is tethered to a rope hanging from a ceiling. The ball is displaced from its rest position by a positive angle θ (see Figure G) and then released. Construct a qualitative graphical model of the motion of the ball relating the displacement angle θ (vertical axis) to time (horizontal axis).

28. (Section 5.4) Construct a tree diagram to solve the following problem: The early-bird meal offered by a restaurant consists of soup, an entree, and a dessert. There are two choices for the soup (vegetable or chicken); three choices for the entree (meatloaf, chicken, or fish); and three choices for dessert (pie, ice cream, or Jell-O). How many different dinners does the restaurant offer the early birds?

29. (Section 6.1) The restaurant in Problem 28 knows from experience that it serves an equal number of each type of soup, that chicken is twice as likely to be requested as either of the other two entrees, and that ice cream is twice as likely to be chosen as Jell-O, which is twice as likely to be chosen as pie. What is the probability that a randomly selected customer will order chicken and Jell-O?

30. (Section 6.7) The restaurant in the previous two problems calculates its costs per serving as follows: vegetable soup, 10¢; chicken soup, 15¢; meatloaf, 90¢; chicken, 75¢; fish, 85¢; ice cream, 40¢; pie, 60¢; Jell-O, 30¢. What is the restaurant's expected cost per meal?

RECOMMENDING ACTION

31. Respond by memo to the following request:

> ### MEMORANDUM
>
> To: J. Doe Reader
>
> From: Superintendent's Office
>
> Date: Today
>
> Subject: **Athletic Field**
>
> I have your recommendation regarding the design of our new athletic field [Example 2], and I was wondering whether we chose the best measure for optimization. Your recommendation is based on maximizing the total area of the rectangular portion of the groomed field. What other measures might we optimize instead? Would that change the design specifications for the field?

11.3 PRODUCTION PROBLEMS

Techniques that allow businesses to minimize costs and maximize profits are vital to both retailers and wholesalers. Consequently, some of the most avid users of optimization procedures are manufacturers. These companies make daily decisions based on all facets of production, including the size of production runs, inventory levels, anticipated demands, pricing policies, and labor needs. Each decision affects a company's bottom line, and the wiser the decisions, the more successful the company.

Many production models are not linear, so the simplex method and other optimization techniques introduced in Chapter 3 do not apply. The optimization techniques developed in Section 11.1 are applicable to nonlinear models as long as those models are functions of a single input variable.

EXAMPLE 1

A manufacturer sells speakers to retailers at the wholesale price of $90 each, and the speakers are in such high demand that the company can sell all of the speakers that "come off the conveyer belt." The daily cost C of manufacturing x speakers depends on the number of speakers they produce, according to the formula $C = 5000 - 100x + x^2$. How many speakers should the manufacturer turn out each day to maximize profit if the maximum production capacity of the plant is 120 speakers a day?

Solution

Profit is total income minus total cost. Each of these three quantities is important, so we set P = daily profit, I = daily income, and C = daily cost, all in dollars. Then

$$P = I - C \tag{11}$$

The output variable P is a function of I and C. If we want to use our optimization methods from Section 11.1 to find the maximum value of P, we must express it as a function of a single input variable.

The company accumulates income by selling speakers at the wholesale price of $90 apiece. If the company produces and sells x speakers each day, their daily income is

$$I = 90x \tag{12}$$

and daily cost is specified by the formula

$$C(x) = 5000 - 100x + x^2 \tag{13}$$

Substituting Equations (12) and (13) into Equation (11), we obtain

$$P(x) = 90x - (5000 - 100x + x^2) = -5000 + 190x - x^2$$

Profit is now a function of the single input variable x.

Because x is a manufacturing quantity, $x \geq 0$. In addition, daily production is limited to 120 speakers, so $x \leq 120$. A simplification of this problem is shown in Model 11.5.

We want to maximize the function $P(x) = -5000 + 190x - x^2$ on the interval $[0, 120]$. We see from Figure 11.21 that the global maximum, $P = \$4{,}025$, occurs when $x = 95$, so the company should produce 95 speakers daily.

MODEL 11.5

Important factors

x = the number of speakers produced/sold daily

P = daily profit (in dollars)

Objective

Find x so that P is maximized.

Constraints

$P = -5000 + 190x - x^2$

$0 \leq x \leq 120$

FIGURE 11.21
The Graph of
$P(x) = -5000 + 190x - x^2$
on the Interval [0, 120]

TABLE 11.7 **The Optima of**
$P = -5000 + 190x - x^2$

	INPUT	OUTPUT
	x	P
ENDPOINTS:	0	0
	120	3,400
CRITICAL POINT:	95	4,025

Unit cost depends on the number of items produced when fixed costs are averaged over an entire production run.

The same answer is obtained by using the optimization algorithm developed in Section 11.1. The results are displayed in Table 11.7.

Unit cost is the cost of producing a single item. If we know the unit cost, we can determine *total cost* by multiplying the unit cost by the number of items produced. Production costs include both variable and fixed costs: *Variable costs* cover the cost of the materials that go into each item and often the associated labor costs, whereas *fixed costs*, sometimes called *overhead,* cover insurance, security, mortgages, lease payments on equipment, and other costs that are independent of the number of items produced. However, since fixed costs can be averaged over all items produced, the larger a production run, the smaller the overhead per item. Therefore the unit cost itself can be a function of the number of items produced.

As a simple example of this, consider a manufacturing process with a fixed cost of $10,000/month and a variable cost of 25¢/item. If the company produces no items, the total cost is just the fixed cost; otherwise the total cost is the fixed cost of $10,000 plus 25¢ times the number of items made. For a monthly production run of 2500 items, the overhead on each item is $10,000/2500 = $4, so the unit cost is $4.25. For a monthly production run of 5000 items, the overhead on each item is $10,000/5000 = $2 and the unit cost is $2.25. Notice that the unit cost associated with 5000 items is less than that associated with 2500 items.

EXAMPLE 2

A catalog company sells tennis balls nationwide. The unit cost *UC* (in dollars) for manufacturing a tennis ball is related to *n*, the number of balls produced each month, by the equation

$$UC = \frac{10,000}{n} + 0.25$$

up to the capacity of the plant, which is 100,000 balls per month. The demand *d* for the company's tennis balls depends on *UP*, the unit price (in dollars) of a ball, according to the equation

$$d = 90,000 - 40,000UP \qquad (14)$$

How many tennis balls should the company manufacture each month to maximize profit and sell the entire production run?

Solution

When all balls are sold, supply equals demand, $n = d$. We want to maximize profit, P, which is total income minus total cost:

$$P = TI - TC \tag{15}$$

Total cost is unit cost times the size of a run, so

$$TC = UC \cdot n = \left(\frac{10,000}{n} + 0.25\right)n = 10,000 + 0.25n \tag{16}$$

Total income is the unit price times the size of the run, so

$$TI = UP \cdot d = UP \cdot n$$

Solving Equation (14) for UP, we have

$$UP = 2.25 - \frac{d}{40,000} = 2.25 - \frac{n}{40,000}$$

Therefore

$$TI = UP \cdot n = \left(2.25 - \frac{n}{40,000}\right)n = 2.25n - \frac{n^2}{40,000} \tag{17}$$

Substituting Equations (16) and (17) into Equation (15), we obtain

$$P(n) = \left(2.25n - \frac{n^2}{40,000}\right) - (10,000 + 0.25n)$$

$$= -\frac{n^2}{40,000} + 2n - 10,000 \tag{18}$$

Because n is the size of a production run, $n \geq 0$. In addition, monthly production is limited to 100,000 balls, so $n \leq 100,000$. The data for this problem are organized in Model 11.6.

Equation (18) relates the output variable P to the single input variable n, so we can use the optimization algorithm developed in Section 11.1 to find maximum profit. The derivative of P with respect to n is

$$\frac{dP}{dn} = -\frac{n}{20,000} + 2$$

which exists for all values of n. Setting this derivative equal to zero and solving for n, we obtain $n = 40,000$ as the only critical point. A maximum can also occur at the endpoints, $n = 0$ and $n = 100,000$. Evaluating Equation (18) at each point in the domain where a maximum can occur, we complete Table 11.8. The global maximum is $P = \$30,000$ at a production run of 40,000 tennis balls.

The same answer can be obtained graphically. Figure 11.22 is a graph of Equation (18), which attains its peak of $P = 30,000$ at $n = 40,000$.

MODEL 11.6

Important factors

n = the number of tennis balls produced monthly

P = monthly profit (in dollars)

Objective

Find n so that P is maximized.

Constraints

$P = -\dfrac{n^2}{40,000} + 2n - 10,000$

$0 \leq n \leq 100,000$

TABLE 11.8 The Optima of
$P = -\dfrac{n^2}{40,000} + 2n - 10,000$

	INPUT n	OUTPUT P
ENDPOINTS:	0	−10,000
	100,000	−60,000
CRITICAL POINT:	40,000	30,000

FIGURE 11.22
The Graph of
$$P = -\frac{h^2}{40,000} + 2n - 10,000$$
on the Interval [0, 100]

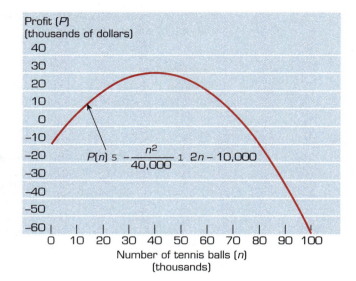

Profit (P)
(thousands of dollars)

$$P(n) = -\frac{n^2}{40,000} + 2n - 10,000$$

Number of tennis balls (n)
(thousands)

EXAMPLE 3

An artist agrees with a gallery to exhibit and sell signed prints. The gallery owner believes that they can sell 30 prints at $500 per print or even more if they reduce the unit price by $10 for each print they sell in excess of 30. For example, the owner thinks that 31 prints can be sold at a unit price of $490 while 32 prints can be sold at a unit price of $480. How many prints should the gallery accept from the artist to maximize total income?

Solution

The objective is to maximize the gallery's total income, which is the product of the selling price of each print and the number of prints sold. Three important quantities are:

I = total income (in dollars)

p = the price of each print (in dollars)

n = the number of prints sold

These quantities are related by the equation

$$I = p \cdot n \tag{19}$$

Equation (19) displays the output variable I as a function of the two input variables p and n; it is better to have I as a function of just one.

Certainly the artist will offer the gallery at least 30 prints, because this will earn $500 for each additional print sold. The real question is, how many additional prints (over 30) should the artist supply? Let us denote by x, the number of prints above 30 that are sold. It follows that $x \geq 0$. The relationship between n and x is given by the equation

$$n = 30 + x \tag{20}$$

The unit selling price is $500 if the number of prints exhibited is 30 or less, and this price is reduced by $10 per print if more than 30 are supplied. Thus

$$p = 500 - 10x \tag{21}$$

Substituting Equations (20) and (21) into Equation (19), we obtain

$$I = (500 - 10x)(30 + x) = 15,000 + 200x - 10x^2$$

MODEL 11.7

Important factors

x = the number of prints above 30

I = total income (in dollars)

Objective

Find x so that I is maximized.

Constraints

$I = 15{,}000 + 200x - 10x^2$

$x \geq 0$

$x \leq 49$

Now the output variable I is expressed as a function of the single input variable x.

The variable x denotes the number of prints in excess of 30 that the artist will supply to the gallery. Next, we need an upper bound on x. What is the most x can be? As x increases, the unit selling price decreases in accordance with Equation (21). Note that when $x = 50$, the price p is 0. The only way the gallery can dispose of $n = 30 + 50 = 80$ prints is to give them away. This is not the artist's goal so we can assume that $x \leq 49$. A complete model for this problem is shown in Model 11.7.

We want to maximize the function

$$I(x) = 15{,}000 + 200x - 10x^2$$

on the interval [0, 49]. We see from Figure 11.23 that the global maximum is $I = \$16{,}000$, which occurs when $x = 10$: the artist should supply the gallery with 40 prints.

This same answer is displayed in Table 11.9, which was constructed by using the optimization algorithm developed in Section 11.1.

FIGURE 11.23

The Graph of
$I(x) = 15{,}000 + 200x - 10x^2$
on the Interval [0, 49]

TABLE 11.9 The Optima of
$I = 15{,}000 + 200x - 10x^2$

	INPUT	OUTPUT
	x	I
ENDPOINTS:	0	15,000
	49	790
CRITICAL POINT:	10	16,000

Many companies want to minimize the cost of holding inventory while maintaining sufficient stock to meet customer demands.

In some corporations, one of the most critical costs affecting profitability is *storage cost*, the cost of maintaining inventory. Storage space is required to stock unsold goods, as well as the raw materials needed for manufacturing the goods, and that space costs money. Often optimizing techniques are used to minimize storage costs while maintaining stock so that current needs are met. The intent is to avoid overstocking, with its resulting increase in storage cost, and understocking, with its resulting loss in sales.

To apply the optimization algorithm to inventory problems, we must model the process with a function of a single input variable. We begin with the following simplifying assumptions.

(a) The demand for items in inventory is known.
(b) The demand for items in inventory is uniform.
(c) The cost of buying an item for inventory is a constant.
(d) Inventory is reordered at equal intervals of time and in equal lots.
(e) We are interested in minimizing inventory costs over a year.

A uniform demand for a given item means that its inventory, I, is being depleted as a linear function of time, t; that is, $I = mt + b$, where m and b are constants. Since inventory is being depleted by demand, m is negative; the constant b is the amount of inventory available at $t = 0$.

Although the depletion of inventory is assumed to be uniform, the inventory level is not always decreasing. Each time a new supply order arrives, there is a jump in the size of the inventory. Once this order is stocked, the inventory level will decline uniformly until the next shipment arrives. Ideally, the inventory level should decline to zero just as each new order arrives. As a direct consequence of assumptions (b) and (d), the graph of available inventory as a function of time looks like Figure 11.24.

FIGURE 11.24
An Inventory Model

When a manufacturer orders materials, the supplier often imposes handling and shipping fees. In addition, people at the manufacturing plant spend time processing the order to the supplier, which costs money in the form of wages. All of these costs are folded into a single amount called the *reordering cost.*

Let F denote the fixed cost associated with each order, including the accounting and processing time required of employees; let V denote the variable cost, including shipping charges, associated with the delivery of each item from a supplier to a manufacturer; and let x denote the number of items per order. The *individual reordering cost, IRC,* associated with an individual order of x items is

$$IRC = F + Vx$$

Although x is unknown, it follows from assumption (d) that it is the same for each order, and so the individual reordering cost is the same for every order. The *yearly reordering cost, YRC,* is the product of *IRC* and the number of orders placed in a year. If we denote by Q the quantity of items needed over a year, it follows from assumption (a) that the value of Q is known. To guarantee the arrival of items over a one-year time period, Q/x individual deliveries must be scheduled over a year.

$$
\begin{aligned}
YRC &= (IRC)(Q/x) \\
&= (F + Vx)(Q/x) \qquad\qquad (22) \\
&= \frac{FQ}{x} + VQ
\end{aligned}
$$

EXAMPLE 4

Each year a manufacturer orders a total of 25,000 drums of fiberglass resin from a supplier who imposes a $10 handling charge per order and a $2 shipping charge per drum. What is the manufacturer's yearly reordering cost based on just these charges?

Solution
Here $Q = 25,000$, $V = \$2$, and $F = \$10$. The order size is not known, so we denote it by x. Substituting these quantities into Equation (22), we determine that

$$YRC = \frac{250{,}000}{x} + 50{,}000 \qquad\qquad (23)$$

The variable x in Equation (23) appears only in the denominator, so as x increases, YRC decreases. The domain is the interval [1, 25,000] since the smallest order is for one drum and the largest is for the entire yearly supply, 25,000 drums. The quantity YRC assumes its maximum at the left endpoint $x = 1$: if the company orders only one drum at a time, then it will take 25,000 orders to meet the yearly inventory of fiberglass resin, with a $10 handling fee per order. The yearly reorder cost in this case is $YRC = \$300,000$. The quantity YRC assumes its minimum at the right endpoint, $x = 25,000$: if the company gets its entire yearly order all at once, it will incur the $10 handling fee only once, so the yearly reorder cost is $YRC = \$50,010$.

The larger an individual restocking order, the less often an order is necessary, reducing the number of times the handling charge is imposed. It is frequently cost effective to order larger quantities, however, storage costs mount with increasing order size, counteracting the benefits of large orders.

The *yearly storage cost, YSC,* is the product of the cost W of warehousing one unit of inventory for one year and the average amount of inventory stored over a year. Referring back to Figure 11.24, inventory ranges from no items just before a delivery to x items just after delivery. Depletion is uniform, so the average amount of inventory is $x/2$ and $YSC = W(x/2)$.

The *yearly inventory cost, YIC,* is the yearly restocking cost plus the yearly storage cost. Using Equation (22), we determine that

$$YIC = YRC + YSC = \frac{FQ}{x} + VQ + \frac{Wx}{2} \tag{24}$$

This equation displays the output variable—yearly inventory cost—as a function of the single input variable x, the number of items per order. The domain of this function is the interval [1, Q]. The smallest possible order is for one item and the largest possible order is for Q items, the entire year's supply. The complete inventory problem is organized in Model 11.8.

MODEL 11.8

Important factors

x = the number of items in each order

YIC = yearly inventory cost (in dollars)

Q = the quantity of items required each year

F = the fixed cost of each order (in dollars)

V = the variable cost of ordering each item (in dollars)

W = the cost of warehousing one unit for one year (in dollars)

Objective

Find x so that YIC is minimized.

Constraints

$$YIC = \frac{FQ}{x} + VQ + \frac{Wx}{2}$$

$$1 \le x \le Q$$

To determine what size order minimizes yearly inventory cost, we apply the optimization algorithm developed in Section 11.1. Taking the derivative of the function defined by Equation (24), we have

$$\frac{d}{dx}(YIC) = -\frac{FQ}{x^2} + \frac{W}{2} \tag{25}$$

This derivative does not exist at $x = 0$, but zero is outside the domain anyway. Setting Equation (25) equal to zero and solving for x, we look for critical points:

$$-\frac{FQ}{x^2} + \frac{W}{2} = 0$$

$$\frac{W}{2} = \frac{FQ}{x^2}$$

$$x^2 = \frac{2FQ}{W}$$

$$x = \pm\sqrt{\frac{2FQ}{W}}$$

> *The global minimum for the yearly inventory cost can occur only at $x = \sqrt{2FQ/W}$ or at the endpoints of the domain $x = 1$ and $x = Q$.*

Since $x = -\sqrt{2FQ/W} < 0$, it is not in the domain $[1, Q]$. So if there is a global minimum for the yearly inventory cost, it can occur only at the critical point $x = \sqrt{2FQ/W}$ or at the endpoints, $x = 1$ and $x = Q$.

EXAMPLE 5

A computer store sells computer systems that the staff assembles from component parts, including a monitor. The store estimates it will need 800 monitors during next fiscal year. Each order for monitors incurs a handling fee of $15 from the wholesale supplier, in addition to a shipping charge of $8/monitor. The store warehouses monitors at a yearly cost of $60/unit. Determine the optimum order size for monitors and the frequency of the orders to minimize the yearly inventory cost for the store.

Solution

Using Model 11.8, we have $Q = 800$, $F = 15$, $V = 8$, and $W = 60$, which we substitute into Equation (24):

$$YIC = \frac{12,000}{x} + 6400 + 30x$$

The minimum of YIC occurs either at

$$x = \sqrt{2FQ/W} = \sqrt{2(15)(800)/60} = 20$$

or at the endpoints of the domain, $x = 1$ and $x = 800$. Evaluating the formula for YIC at each of these points generates Table 11.10, from which we learn that the global minimum, $YIC = \$7,600$, occurs when $x = 20$.

Thus each order should be for 20 monitors. Eight hundred sets are needed over the year, so the appliance center will need to place 40 orders. Since orders must be placed at equal intervals of time [assumption (d)], each order must be placed approximately every nine days.

TABLE 11.10 Evaluations of the Function $YIC = \dfrac{12,000}{x} + 6400 + 30x$

	INPUT x	OUTPUT YIC
ENDPOINTS:	1	18,430
	800	30,415
CRITICAL POINT:	20	7,600

EXAMPLE 6

A buyer for Jewelry Unlimited estimates that her department will sell 200 gold necklaces during the year. Fixed costs associated with each wholesale order are $50, while variable costs are 10¢/item. Yearly warehousing costs are $12\frac{1}{2}$¢/item. Determine the optimum order size that minimizes yearly inventory cost.

Solution

Using Model 11.8, we have $Q = 200$, $F = 50$, $V = 0.10$, and $W = 0.125$, which gives us

$$YIC = \frac{10,000}{x} + 20 + 0.0625x$$

The minimum of *YIC* occurs either at the point

$$x = \sqrt{2FQ/W} = \sqrt{2(50)(200)/0.125} = 400$$

or at the endpoints of the domain, $x = 1$ and $x = 200$. Because the point $x = 400$ is outside the domain, the minimum can occur only at the endpoints. At $x = 1$, the yearly inventory cost is about $10,020.06, whereas at $x = 200$ it is $82.50. The minimum inventory cost is achieved by ordering all 200 units in one lot.

The economic order quantity is $\sqrt{2FQ/W}$

The minimum yearly storage cost is usually obtained when each order is for $x = \sqrt{2FQ/W}$ items. This value is known as the *optimum order size* or the *economic order quantity*. As Example 6 illustrates, however, the minimum is not always achieved at this point. Occasionally the global minimum occurs at an endpoint of the domain.

CREATING MODELS

1. A television manufacturer has found that the profit P (in dollars) obtained from selling x large-screen television sets per week is given by the formula

$$P = -x^2 + 300x - 5000$$

Determine how many large-screen television sets the manufacturer should produce to maximize profits if its production capacity is 400 sets/week. What is that maximum profit?

2. A manufacturing company has found that the profit P (in dollars) from selling x items a month is given by the formula

$$P = 22x - \frac{x^2}{2000} - 10,000$$

How many items should the manufacturer produce to maximize profit if production capacity is 30,000 items/month? What is the maximum profit?

3. Solve Problem 2 if $P = 225,000 \sin(\pi x/50,000)$.

4. The manufacturer in Problem 2 has determined that the total cost C of producing x items a month is given by the formula

$$C = \frac{x^2}{2000} - 7x + 10,000$$

How many items should the manufacturer produce if the objective is to minimize cost? Is this answer reasonable?

5. A refrigerator manufacturer can sell x refrigerators weekly at a unit price of $300. The total cost C (in dollars) is given by the formula

$$C = 2.5x^2 - 200x + 20,000$$

How many units should the firm produce to maximize profit if weekly production capacity is 110 refrigerators? What is the maximum profit? What is the total cost when profits are maximized?

6. Solve Problem 5 if the production capacity is 90 refrigerators/week.

7. The firm in Problem 5 can sell all units it produces at $325 each. Determine how many units should be produced to maximize profit.

8. A paperback edition of a popular travel book sells wholesale for $15. At this price, all copies can be sold up to the capacity of the printers, which is 20,000 books/week. The unit cost UC for this book is related to x, the number published per week, by the equation

$$UC = \frac{x}{5000} + 8 + \frac{12,000}{x} \quad (x \geq 1000)$$

How many books should the publisher print each week to maximize profits if they have a contract with a bookstore chain to deliver 5000 copies weekly?

9. Solve Problem 8 when $UC = 6 \ln(0.005x)$.

10. A candy company has determined that the total cost C (in dollars) of producing x boxes of chocolates is given by the equation

$$C = \tfrac{1}{3}x^3 - 10x^2 - 800x + 12,000$$

Determine the number of boxes the company should produce to minimize its total cost.

11. A book company's revenue R is related to advertising expenditures A by the formula

$$R = 800 - 800e^{-0.1A}$$

with both variables measured in thousand-dollar units. The company measures the effectiveness E of its advertising to be the revenue generated by advertising minus the cost of advertising. How much money should they spend next year on advertising their books if the objective is to maximize the effectiveness of its half-million-dollar advertising budget?

12. The owner of a new apartment building must decide on the rent for each of the 100 single-bedroom units in the complex. The owner believes she can rent all 100 apartments at $500/month, but for each additional $10 increase, another apartment will remain vacant. What rents should the owner charge to maximize her total income from the building?

13. A grower has determined that if at most 30 nut trees are planted on an acre, the yearly harvest from each tree will be 60 lb of nuts. An acre can hold a maximum of 50 trees, but each additional tree over 30 decreases the harvest of all of the trees by 2 lb/tree. How many trees should be planted per acre to maximize the total nut harvest?

14. A delivery company signs a contract with a furniture store to deliver chairs, with payments made at the end of each month. The delivery company will receive $20 for each chair, up to 100. If the furniture company orders more than 100 chairs, then for

each chair over 100, the unit price on every chair is reduced by 0.5%. The contract also specifies that the delivery company is not obligated to accept more than 200 chairs a month. What are the maximum and minimum amounts of money the delivery company can receive in a month from the furniture store under this contract?

15. A warehouse facility has a handling cost of $10 per order and a unit storage cost of $3 for one year. What is the economic order quantity if 540 items are ordered during the year?

16. In Problem 15, suppose the handling fee is $1 and the unit storage cost is $25/year. What is the economic order quantity if 31,250 items are needed during a year?

17. Solve Example 6 if the variable cost associated with each item is $1 but all other costs remain unchanged.

18. The computer store in Example 5 needs a software package for each computer system it assembles. The supplier imposes a handling charge of $9.60 on each order. The store estimates that it will need 1000 software packages over the course of a year: 800 for the computer systems it assembles and another 200 for over-the-counter sales. The cost of storing each box of software is $1.92/yr. What is the optimum order size and frequency that minimizes yearly inventory costs?

19. Construct a graphical model of the number of software packages in stock for the store in Problem 18.

20. Figure A models the inventory level of blank video tapes for a store. Determine (*a*) the economic order quantity; (*b*) the total number of orders placed each year; and (*c*) the total number of tapes sold each year.

FIGURE A

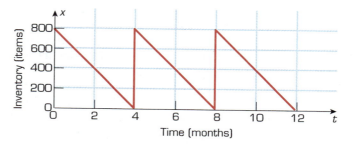

21. The manager of a franchise automobile service center installs 900 of its top-selling batteries each year. Batteries are ordered directly from the chain's main warehouse. Each order incurs a fixed cost of $9 and a variable cost of 50¢/battery. The cost of keeping a battery in inventory is $2/yr. Determine the optimum order size and frequency to minimize yearly inventory cost for this battery type.

22. A gourmet restaurant serves 450 cases of its house wine to customers over a year. Wine is ordered from a distributor that charges a variable cost of $1/case and a fixed cost of $18/order. The restaurant estimates its annual storage costs to be $2/case. Determine the optimum order size and frequency to minimize the restaurant's yearly inventory cost for its house wine.

23. A paint store sells about 1600 cans of white latex paint over a year. Paint is ordered from the manufacturer at 80¢/can for delivery and $50/order for handling. The manager estimates storage costs for a can of paint to be $4/yr. Determine the optimum order size and frequency to minimize the store's inventory cost for white latex paint.

24. The restaurant manager in Problem 22 discovers that if 500 cases of wine are used in a year, the handling cost drops to $4.80/order but the *monthly* storage cost rises to $2/case. What is the economic order quantity over six months?

EXPLORING IN TEAMS

25. A concept closely related to the economic order quantity is the *reorder point*. If available inventory falls below a specified level, which is the reorder point, an order is placed for the economic order quantity. Use the information in Figure A and the fact that it takes one month between the time an order is placed and when it's received to determine the reorder point.

26. Determine the reorder point for the inventory process in Example 5 if it takes a week to receive the monitors after they're ordered.

27. Determine the reorder point for the inventory process in Example 6 if it takes a month to get the necklaces once they've been ordered from the wholesalers.

28. Equation (22) was developed under the assumption that the individual reorder cost, $IRC = F + Vx$, is a linear function of x. Determine the economic order quantity when the individual reorder cost is quadratic: $IRC = F + Vx^2$.

EXPLORING WITH TECHNOLOGY

29. Use the same coordinate system to graph both the daily cost function and the daily income function in Example 1. Determine the break-even point, when cost equals income. Is the break-even point interesting with respect to locating global optima?

30. Repeat Problem 29 using the total cost function and total income function in Example 2.

31. Use a graphing calculator to evaluate the daily cost function and the daily income function in Example 1, where $x = 95$ is the production level that generates maximum profit. Then use a graphing calculator to evaluate the marginal cost and the marginal revenue at that production level. (See Section 10.6.) Which of these four numbers is the same? Should you expect any of these numbers to be equal?

32. Repeat Problem 31 with the total cost and total income functions in Example 2, where $n = 40,000$, the production level that generates maximum profit. Are the results surprising?

REVIEWING MATERIAL

33. (Section 1.3) Is A directly proportional to $5B$ if it is known that A is directly proportional to B? Why or why not?

34. (Section 3.5) A furniture manufacturer builds sofas, love seats, and chairs. Each sofa requires 3 hr to assemble, 6 hr to upholster, 1 hr to decorate, and returns a profit of $60. Each love seat requires 2 hr to assemble, 5 hr to upholster, no time to decorate, and returns a profit of $40. Each chair requires 3 hr to assemble, 1 hr to upholster, 3 hr to decorate, and returns a profit of $50. The manufacturer employs a labor force that can provide 240 hr of assembly time, 600 hr of upholsterers' time, and 360 hr of

decorators' time each week. How many sofas, love seats, and chairs should the manufacturer schedule for production to maximize weekly profit? Is any segment of the labor force underutilized?

35. (Section 7.2) Locate the stable distribution vector for the Markov chain defined by the transition matrix

$$
\mathbf{P} = \begin{array}{c} \\ \text{from state 1} \\ \text{from state 2} \\ \text{from state 3} \end{array}
\begin{array}{ccc}
\overset{to}{\text{state 1}} & \overset{to}{\text{state 2}} & \overset{to}{\text{state 3}} \\
\left[\begin{array}{ccc}
.6 & .3 & .1 \\
.2 & .6 & .2 \\
.3 & .5 & .2
\end{array}\right]
\end{array}
$$

36. (Sections 11.3 through 11.5) Find dy/dx when

$$ y = 3x^2 - xe^{-2x} $$

RECOMMENDING ACTION

37. Respond by memo to the following request:

MEMORANDUM

To: J. Doe Reader

From: Vice President for Planning

Date: Today

Subject: **Production of Speakers**

I have your analysis for our optimal production of speakers [Example 1]. Aren't you overdoing it? Wouldn't it be easier to minimize our cost function or, perhaps even simpler, to maximize our revenue function? It seems to me that maximizing profit is equivalent to minimizing cost or to maximizing income.

RELATED RATES

We noted in Section 10.5 that multistage processes where the output of one operation becomes the input to a subsequent one are common in modern life. Figure 11.25 is an illustration of a process with two sequential operations. First function g transforms Input 1 into Output 1, then function f transforms Output 1 into Output 2. Thus,

$$ \textit{Output 1} = g(\textit{Input 1}) \quad \text{and} \quad \textit{Output 2} = f(\textit{Output 1}) $$

The relationship between Input 1 and Output 2 is defined by the composite function

$$ \textit{Output 2} = f(g(\textit{Input 1})) = f \circ g(\textit{Input 1}) $$

FIGURE 11.25

A Process with Two Sequential Operations

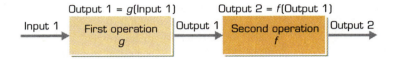

EXAMPLE 1

A company opens a new manufacturing plant in stages. In the beginning, the plant produces 1000 units a week. As the company develops, weekly production, x, increases according to the formula

$$x = g(t) = 1000 + 200t \quad (0 \le t \le 20)$$

with time t measured in weeks from the start of production. The unit cost (UC) of producing an item is related to x by the formula

$$UC = f(x) = 2 + 5e^{-0.001x}$$

Determine the relationship between unit cost and time.

Solution

The entire process is illustrated schematically in Figure 11.26. The unit cost is related to weekly production by the function $UC = f(x)$, and weekly production is related to time by the function $x = g(t)$. Thus, UC is related directly to t by the composite function

$$UC = f \circ g(t) = f(g(t)) = f(1000 + 200t)$$
$$= 2 + 5e^{-0.001(1000+200t)} = 2 + 5e^{-(1+0.2t)}$$

FIGURE 11.26

Unit Cost as a Function of Time

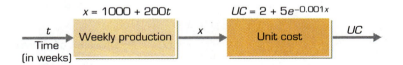

In Example 1, we were given the two functions f and g:

$$x = g(t) = 1000 + 200t$$
$$UC = f(x) = 2 + 5e^{-0.001x}$$

We can calculate the rates of change dx/dt and $d(UC)/dx$ directly:

$$\frac{dx}{dt} = \frac{d}{dt}(1000 + 200t) = 200$$

$$\frac{d}{dx}(UC) = \frac{d}{dx}(2 + 5e^{-0.001x}) = -0.005e^{-0.001x}$$

(26)

We can obtain the rate of change of unit cost with respect to time in two ways. As a result of Example 1, we know that UC is related directly to t through the composite function

$$UC = f \circ g(t) = 2 + 5e^{-(1+0.2t)}$$

Differentiating this function with respect to t, we obtain

$$\frac{d}{dt}(UC) = \frac{d}{dt}(2 + 5e^{-(1+0.2t)}) = -e^{-(1+0.2t)}$$

(27)

Alternatively, we can use the chain rule. (See Section 10.5.) Since UC is a function of x, which in turn is a function of t, the derivative of UC with respect to t is

$$\frac{d}{dt}(UC) = \frac{d}{dx}(UC) \cdot \frac{dx}{dt} \tag{28}$$

Substituting Equations (26) into Equation (27), we have

$$\frac{d}{dt}(UC) = (-0.005e^{-0.001x})200 = -e^{-0.001x} \tag{29}$$

Substituting $x = 1000 + 200t$ into Equation (29) yields Equation (27).

The chain rule is a powerful tool for calculating the rate of change of a composite function for a sequence of operations when we know the rates of change of the functions for each of the individual operations. This is particularly useful when we know just the rates of change but not the functions themselves. For example, suppose we know that a variable y is related to a variable x and that x is also related to another variable t, but we do not know the actual relationships. That is, we know there exist functions f and g such that $y = f(x)$ and $x = g(t)$, but we do not have formulas for $f(x)$ and $g(t)$. If we have information about the rates of change dy/dx and dx/dt, then we can use the chain rule to obtain information about dy/dt, namely,

$$\frac{dy}{dt} = \frac{dy}{dx} \cdot \frac{dx}{dt} \tag{30}$$

This technique is illustrated in Example 2.

> *The chain rule is used to calculate the rate of change of a composite function for a sequence of operations when the rates of change of the functions for each of the individual operations are known.*

EXAMPLE 2

Find dy/dt when $y = 2/(3x + 5)$ and $dx/dt = 15$, and evaluate dy/dt at $x = 4$.

Solution

We are given dx/dt, so we know that x is a function of t even though a precise formula has not been specified. We also know that y is a function of x, and because the functional relationship is specified, we can differentiate it:

$$\frac{dy}{dx} = \frac{d}{dx}\left(\frac{2}{3x + 5}\right) = \frac{-6}{(3x + 5)^2}$$

It now follows from Equation (30) that

$$\frac{dy}{dt} = \frac{dy}{dx} \cdot \frac{dx}{dt} = \left[\frac{-6}{(3x + 5)^2}\right]15 = \frac{-90}{(3x + 5)^2}$$

When $x = 4$,

$$\left.\frac{dy}{dt}\right|_{x=4} = \frac{-90}{[3(4) + 5]^2} = -\frac{90}{289} \approx -0.3114$$

Equation (30) is the chain rule for y as a function of x and x as a function of t. If other variables are used, then the formula is modified accordingly, to accommodate the new notation. This was done in Equation (28), where the variable UC was used instead of y.

EXAMPLE 3

A farmer's cooperative collects oranges from all of the member farms, then sells them under a single brand name. The wholesale price of oranges, p (in dollars/ton), is related to s, the supply (in tons), by the formula

$$p = 800 - \frac{500s}{3s + 200}$$

There are currently 2000 tons of oranges in stock at the cooperative and that supply is decreasing at the rate of 25 tons/day. At what rate is the wholesale price of oranges changing with respect to time?

Solution

We seek dp/dt, the rate of change of wholesale price with respect to time t (in days). We are told that the supply s of oranges is decreasing at the rate of 25 tons/day, so we know that $ds/dt = -25$. The minus sign is required because the supply is a decreasing function of time. We also know that price is a function of supply, because that relationship is specified. We have p as a function of s and s as a function of t, so p is a composite function of t. The chain rule now takes the form

$$\frac{dp}{dt} = \frac{dp}{ds} \cdot \frac{ds}{dt}$$

where

$$\frac{dp}{ds} = \frac{d}{ds}\left(800 - \frac{500s}{3s + 200}\right) = \frac{-100,000}{(3s + 200)^2}$$

Therefore

$$\frac{dp}{dt} = \frac{dp}{ds} \cdot \frac{ds}{dt} = \left[\frac{-100,000}{(3s + 200)^2}\right](-25) = \frac{2,500,000}{(3s + 200)^2}$$

When $s = 2000$,

$$\left.\frac{dp}{ds}\right|_{s=2000} = \frac{2,500,000}{[3(2000) + 200]^2} \approx 0.065$$

The price for a ton of oranges is increasing at the rate of approximately $6\frac{1}{2}$¢ a day.

EXAMPLE 4

In Example 3, at what rate is the wholesale price of oranges changing when the current supply is 1500 tons?

Solution

All of the work done in Exercise 3 remains relevant except for the last step—we simply set s to 1500 and calculate:

$$\left.\frac{dp}{ds}\right|_{s=1500} = \frac{2,500,000}{[3(1500) + 200]^2} \approx 0.113$$

Now, the rate of increase in the price of a ton of oranges is slightly more 11¢ a day.

A diagram of a process is a useful intermediate step in creating a mathematical model.

We noted in Section 11.2 that whenever possible, it is helpful to draw a diagram of a process prior to modeling it mathematically. This continues to be sound advice.

EXAMPLE 5

FIGURE 11.27

A Diagram of an Expanding Metropolitan Area

Metropolitan region

A city's metropolitan area expands outward from its center as its population grows, adding new homes and businesses. The growth is essentially concentric, with the radius increasing at the rate of $\frac{1}{2}$ mile/yr. How fast is the area of this metropolitan region expanding when its radius is 5 miles?

Solution

Figure 11.27 is a graphical model of the growth of the metropolitan region. If we let A denote the area of the circular region (in sq. miles) and t time (in years), then the rate of change we seek is described by the derivative dA/dt. From elementary geometry, $A = \pi r^2$, where r denotes the radius of the circle. Furthermore, we are given that

$$\frac{dr}{dt} = \frac{1}{2} \text{ mile/yr}$$

The radius of the metropolitan region is increasing with time, so the sign of its derivative is positive. We now have area A as a function of r and r as a function of t, so A is a composite function of t. The chain rule takes the form

$$\frac{dA}{dt} = \frac{dA}{dr} \cdot \frac{dr}{dt}$$

where

$$\frac{dA}{dr} = \frac{d}{dr}(\pi r^2) = 2\pi r$$

Therefore

$$\frac{dA}{dt} = \frac{dA}{dr} \cdot \frac{dr}{dt} = (2\pi r)\left(\frac{1}{2}\right) = \pi r$$

When $r = 5$,

$$\left.\frac{dA}{dt}\right|_{r=5} = 5\pi \approx 15.7 \text{ sq. miles/yr}$$

IMPROVING SKILLS

In Problems 1 through 10, find dy/dt by (**a**) writing y as an explicit function of t and then differentiating directly, and (**b**) using the chain rule. Show that both results are equivalent.

1. $y = 2x^2 - 3x + 1$; $x = t + 2$
2. $y = x^2 - 4$; $x = 2t + 1$
3. $y = x^2 - 2x$; $x = t - 3$
4. $y = 5x + 3$; $x = t^2 + 2$
5. $y = x^3 + 3x$; $x = 3t^2 + 6$
6. $y = \sin x$; $x = 5t^2 - 8$
7. $y = \cos x$; $x = 1/t$
8. $y = \cos 2x$; $x = 1/t$
9. $y = \ln x$; $x = 1/(2t)$
10. $y = \ln x^2$; $x = e^{-0.4t}$

In Problems 11 through 20, find dy/dt and evaluate it at the given point.

11. $y = 3x^2 - 9x + 16$; $dx/dt = 8$; $x = 10$
12. $y = x^3 - 3x$; $dx/dt = -2$; $x = 2$
13. $y = x^3 - 3x$; $dx/dt = -2$; $x = 3$
14. $y = x^3 - 3x$; $dx/dt = -1$; $x = 3$
15. $y = 2/(3x - 100)$; $dx/dt = -20$; $x = 45$
16. $y = -2x/(x^2 + 50)$; $dx/dt = 20$; $x = 10$
17. $y = \sin 3x$; $dx/dt = 5$; $x = 1$
18. $y = \cos(2x + 1)$; $dx/dt = -5$; $x = \pi$
19. $y = \ln x$; $dx/dt = 6$; $x = 0.05$
20. $y = e^{-2x}$; $dx/dt = -1$; $x = 0.2$

CREATING MODELS

21. Total weekly profit P (in dollars) is related to x, the number of items produced each week by the formula

$$P(x) = -0.001x^3 + 6x^2 - 5000x$$

Find the rate of change in profit with respect to time t (in weeks), when the production level is at 1500 items but increasing at the rate of 50 items per week.

22. The demand d for an item is related to its price p by the equation

$$d = 800 + \frac{20{,}000}{1 + p}$$

Escalating labor costs are increasing the price of each item at the rate of $2/yr. What is the rate of change in demand with respect to time when the current price is $49 for each item?

23. The demand d for an item (in thousands of units) is related to its price p by the equation

$$p^2 + d^2 = 12{,}500$$

What is the rate of change in demand with respect to time when the current price is $20 for each item but decreasing at the rate of $5 a month?   *Hint:*   Solve the equation $p^2 + d^2 = 10{,}000$ for d.

24. The supply s of an item is related to its price p by the equation

$$s = 800 + 0.1p^2$$

What is the rate of change in supply with respect to time if the current price is $49 for each item but increasing at the rate of $2/yr?

25. The supply s of an item (in thousands of units) is related to its price p by the equation

$$5s^2 - 0.02p^2 = 450$$

What is the rate of change in supply with respect to time if the current price is $50 for each item but decreasing at the rate of 50¢/yr?   *Hint:*   Solve the equation $5s^2 - 0.02p^2 = 450$ for s.

26. The number C of cases of rabies in a region is related to the number r of raccoons by the formula $C = 0.003r^{2/5}$. To reduce the incidence of rabies, the region will exterminate part of the raccoon population. At what rate will rabies decrease if 50 raccoons are destroyed each day from an existing population of 2000?

27. Solve Problem 26 one day later, when the raccoon population is 1950 animals.

28. A stone is thrown into the middle of a still pond and creates a circular ripple. How fast is the area of the circle increasing if it currently has a radius of 7 ft and is expanding at the rate of $\frac{1}{4}$ ft/sec?

29. A cylindrical vat of height 100 ft and radius 25 ft is filled at the rate of 800 ft³/min. How fast is the surface rising when the depth is 12 ft?

30. Solve Problem 29 if the vat is being filled at the rate of 700 ft³/min.

31. A spherical snowball of radius 8 cm is placed in the sun and begins to melt. Find the rate of change of the volume of the snowball with respect to time if the melting process reduces the radius of the snowball at the rate of 2 cm/hr.

32. How fast is the circumference of the snowball in Problem 31 changing with respect to time?

33. Kim inflates a spherical balloon. How quickly is the volume increasing if the current radius is 6 in. and expanding at the rate of 5 in./min?

34. Solve Problem 33 thirty seconds later.

35. Lee blows up a spherical balloon at the rate of 1 cm³/sec. What is the rate of change in radius with respect to time when the balloon has a radius of 10 cm?

36. How fast is the perimeter of a square changing if a side is increasing at the rate of 5 cm/min?

37. A pieceworker at an automobile plant earns $5.75 for each piece she produces. At what rate will her hourly salary increase if she increases her productivity by three pieces an hour?

38. A 50 ft ladder is placed against the side of a building, as shown in Figure A. The base of the ladder begins to slip away from the building at the rate of 5 ft/min. Find the rate at which the top of the ladder is sliding down the building when the bottom of the ladder is 12 ft from the base of the building.

EXPLORING IN TEAMS

39. An output y is related to an input x by the formula $xy = 1$. One way to calculate dy/dx is first to solve for y, in this case obtaining $y = 1/x$, and then to differentiate the result. A second way is to differentiate the original equation directly with respect to the independent variable x. We use the product rule on the left side and note that the right side is the constant 1 with derivative zero. Although we don't know an expression for y, because we do not solve for it, we still know that its derivative with respect to x is dy/dx. Using this approach, show that

$$\frac{d}{dx}(xy) = x \cdot \frac{d}{dx}(y) + y \cdot \frac{d}{dx}(x) = 0$$

and then conclude that

$$\frac{dy}{dx} = -\frac{y}{x}$$

40. Let $w = xy^2$, where y is known to be a function of x but is not explicitly given. Find an expression for dw/dx in terms of dy/dx.

41. Let $w = x^3y^5$, where y is known to be a function of x but is not explicitly given. Find an expression for dw/dx in terms of dy/dx.

42. Use the results of the previous three problems to find an expression for dy/dx when the output y is related to the input x by the formula $xy^2 + x^3y^5 = 1$.

REVIEWING MATERIAL

43. (Section 2.2) Solve the following system of equations for x, y, and z.

$$2x + 3y - 2z = 5$$
$$5x + 2y + 2z = 3$$
$$3x - y + 4z = 1$$

44. (Section 6.6) A manufacturer of sports equipment knows from experience that under ordinary circumstances, 1% of their production line is defective. A quality-control engineer randomly inspects 50 baseballs from all those produced over the last month. Determine the probability that three or more of these items are defective.

45. (Section 10.1) A savings account earns interest at the rate of 6% a year, compounded annually. The value V of this savings account can be modeled by the formula $V = 500(1.06)^t$, where t denotes time (in years). Find the average rate of change in the account over the first 10 years and then construct a table to estimate the instantaneous rate of change at $t = 10$.

RECOMMENDING ACTION

46. Respond by memo to the following request:

MEMORANDUM

To: J. Doe Reader

From: President's Office

Date: Today

Subject: **Bringing the Plant On-Line Faster**

I have your analysis regarding our unit costs for production [see Example 1]. At the moment, our plan is to increase production at the rate of 200 items per week. However, I can reallocate resources to bring more of the plant on-line each week, thereby increasing our production rate by an additional 10%. If I do, can I assume that our unit cost will decrease by the same amount?

11.5 SECOND DERIVATIVES AND LOCAL OPTIMA

The derivative of a function is another function, for example, the derivative of $y = 5x^7$ is the function $y' = 35x^6$; the derivative of $y = x \ln x$ is the function $y' = 1 + \ln x$; and the derivative of $y = e^{x^2}$ is the function $y' = 2xe^{x^2}$. Therefore we can use our rules of differentiation to find the derivative of the derivative. The derivative of a function is called its *first derivative,* and the derivative of the first derivative is called the *second derivative.* We denote the second derivative of the function $y = f(x)$ by any one of the following notations:

$$y'', \quad f''(x), \quad \frac{d^2y}{dx^2}, \quad \frac{d^2f(x)}{dx^2}$$

This last notation is a contraction of

$$\frac{d}{dx}\left[\frac{d}{dx}f(x)\right] = \frac{d^2f(x)}{dx^2}$$

which indicates clearly that the second derivative of a function $f(x)$ is the derivative of its first derivative.

EXAMPLE 1

Find the first and second derivatives of the function

$$y = 3x^5 + 4x^2 + 7$$

> *The second derivative of a function is the derivative of its first derivative.*

Solution

Using the rules of differentiation we presented in Chapter 10, we have

$$\frac{dy}{dx} = \frac{d}{dx}(3x^5 + 4x^2 + 7) = 15x^4 + 8x$$

and

$$\frac{d^2y}{dx^2} = \frac{d}{dx}\left(\frac{dy}{dx}\right) = \frac{d}{dx}(15x^4 + 8x) = 60x^3 + 8$$

EXAMPLE 2

Find d^2y/dx^2 for the function $y = x^2e^x$.

Solution

Here we need to use the rule of exponential differentiation and the chain rule:

$$\frac{dy}{dx} = \frac{d}{dx}(x^2e^x) = x^2e^x + 2xe^x$$

This derivative is the sum of two terms, each one a product of two simpler functions. Therefore

$$\frac{d^2y}{dx^2} = \frac{d}{dx}\left(\frac{dy}{dx}\right) = \frac{d}{dx}(x^2e^x + 2xe^x)$$

$$= \frac{d}{dx}(x^2e^x) + \frac{d}{dx}(2xe^x)$$

$$= (x^2e^x + 2xe^x) + (2xe^x + 2e^x)$$

$$= x^2e^x + 4xe^x + 2e^x$$

$$= (x^2 + 4x + 2)e^x$$

As with first derivatives, we evaluate a second derivative at a specific value of the input variable by using our rules of differentiation to find the second derivative and then evaluating that derivative at the input value of interest. All of the following notations will be used interchangeably to denote the second derivative of $y = f(x)$ evaluated at $x = x_0$:

$$y''(x_0), \quad f''(x_0), \quad \left.\frac{d^2y}{dx^2}\right|_{x=x_0}, \quad \left.\frac{d^2f(x)}{dx^2}\right|_{x=x_0}$$

EXAMPLE 3

Find $y''(-2)$ for $y = x^2e^x$.

Solution

We have from Example 2 that

$$\frac{d^2y}{dx^2} = (x^2 + 4x + 2)e^x$$

Therefore

$$y''(-2) = \left.\frac{d^2y}{dx^2}\right|_{x=-2} = (x^2 + 4x + 2)e^x \Big|_{x=-2}$$
$$= [(-2)^2 + 4(-2) + 2]e^{-2}$$
$$= -2e^{-2} \approx -0.27067$$

In Sections 11.1 through 11.3 we used information about first derivatives to locate global optima (maxima and minima). Now we will use information about first and second derivatives to locate *local* optima. Before doing so, however, it is useful to introduce the notions of *increasing* and *decreasing functions*.

A function $y = f(x)$ is *strictly increasing* on the interval $[a, b]$ if the output y gets larger as the input x gets larger, that is, for x_1, x_2 in $[a, b]$,

$$f(x_2) > f(x_1) \quad \text{when} \quad x_2 > x_1$$

The graph of a strictly increasing function climbs from lower left to upper right, as illustrated in Figure 11.28. If a strictly increasing function is differentiable throughout the interval $[a, b]$, the graph of the function has a tangent line at each point and the slopes of those tangent lines are always positive. Conversely, a function is strictly increasing at points where its first derivative is positive.

A function $y = f(x)$ is *strictly decreasing* on the interval $[a, b]$ if the output y gets smaller as the input x gets larger, that is, for x_1, x_2 in $[a, b]$,

$$f(x_2) < f(x_1) \quad \text{when} \quad x_2 > x_1$$

The graph of a strictly decreasing function drops from upper left to lower right, as illustrated in Figure 11.29; the slopes of the tangent lines are always negative. Therefore a function is strictly decreasing at points where its first derivative is negative.

> *A function is strictly increasing at points where its first derivative is positive and strictly decreasing at points where its first derivative is negative.*

FIGURE 11.28
A Strictly Increasing Function

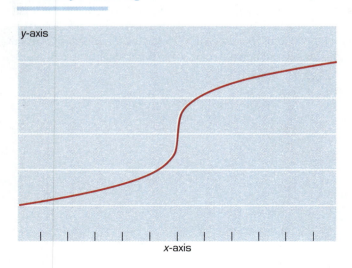

y-axis

x-axis

FIGURE 11.29
A Strictly Decreasing Function

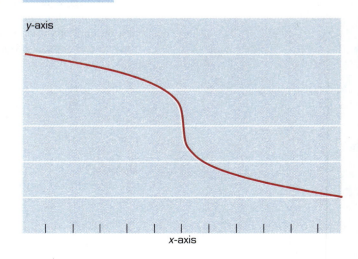

y-axis

x-axis

EXAMPLE 4

Determine where the function $y = x^4$ is strictly increasing and where it is strictly decreasing.

Solution

The derivative of $y = x^4$ is $y' = 4x^3$, which exists for all values of x. Now, x^3 is positive when x is positive, negative when x is negative, and zero when $x = 0$. Thus,

$$y' > 0 \quad \text{when} \quad x > 0$$

$$y' = 0 \quad \text{when} \quad x = 0$$

$$y' < 0 \quad \text{when} \quad x < 0$$

The function is strictly increasing when x is positive and strictly decreasing when x is negative. These regions are marked in Figure 11.30.

FIGURE 11.30

The Graph of $y = x^4$

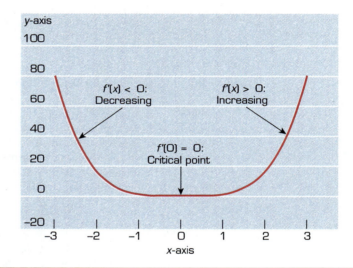

We noted in Section 11.1 that a local maximum for a differentiable function has a shape similar to Figure 11.31 while a local minimum for a differentiable function has a shape similar to Figure 11.32.

FIGURE 11.31

A Local Maximum

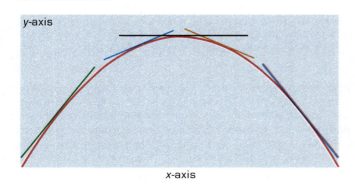

FIGURE 11.32

A Local Minimum

Near a local maximum, the function is strictly increasing to the left of the maximum and strictly decreasing to the right of the maximum. Near a local minimum, the function is strictly decreasing to the left of the minimum and strictly increasing to the right of the minimum. At each local optimum, the function has a critical point where the function is neither increasing nor decreasing. Together, these observations form the *first-derivative test*:

THE FIRST-DERIVATIVE TEST FOR LOCAL OPTIMA

STEP 1 Identify a point p in the domain of $f(x)$ where $f'(p) = 0$.

STEP 2 If $f'(x) > 0$ in some interval to the left of p and also if $f'(x) < 0$ in some interval to the right of p, then $f(x)$ has a local maximum at p.

STEP 3. If $f'(x) < 0$ in some interval to the left of p and also if $f'(x) > 0$ in some interval to the right of p, then $f(x)$ has a local minimum at p.

EXAMPLE 5

Find all local optima for the function $y = f(x) = x^4$ on $[-5, 5]$.

Solution

Here $f'(x) = 4x^3$, which exists everywhere and is zero only at $x = 0$. Thus $x = 0$ is the only critical point for this function and, therefore, the only possible location for a local optimum. Using the results of Example 4, we have that $f'(x) = y' > 0$ when $x > 0$ and $f'(x) = y' < 0$ when $x < 0$. It now follows from the first-derivative test that $x = 0$ is the location of a local minimum, as shown in the graph of the function in Figure 11.30.

We use the first derivative of a function to determine where the function is strictly increasing and where it is strictly decreasing. The second derivative has information, too: we can use it to determine where the *first derivative* is strictly increasing and where it is strictly decreasing. This combined information gives us insights into the behavior of the function itself.

If the second derivative of a function is positive at an input value x, then the first derivative (the slope) is strictly increasing there; if the second derivative is negative at that input value, the first derivative (the slope) is strictly decreasing there. If the second derivative is zero at an input x, then the slope of the graph of the function is a constant, that is, the slope stays the same.

Figure 11.33 is the graph of a function with positive slope: it is strictly increasing, rising from lower left to upper right in the window. This graph is a straight line, and we know from our work in Chapter 1 that the slope of a straight line does not change from one point to the next on the line: it is constant. Its second derivative is zero.

If a function has a positive first derivative, it is strictly increasing. If it also has a positive second derivative, then the *slope* of its graph is increasing: the slope becomes steeper as the values of the input variable increase, and the graph has the shape illustrated in Figure 11.34.

If a function has a positive first derivative and a *negative* second derivative, then the function has a *decreasing* positive slope. In other words, the slope gets closer to zero as the values of the input variable increase. The graph of a strictly increasing function with decreasing slope has the shape illustrated in Figure 11.35.

Figure 11.36 is the graph of a function with negative slope: it is strictly decreasing,

FIGURE 11.33
A Strictly Increasing Function with Constant Slope

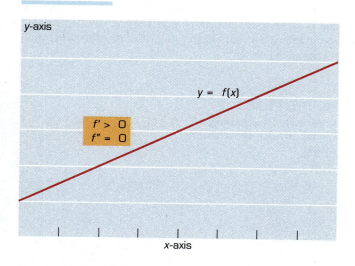

FIGURE 11.34
A Strictly Increasing Function with Increasing Slope

FIGURE 11.35
A Strictly Increasing Function with Decreasing Slope

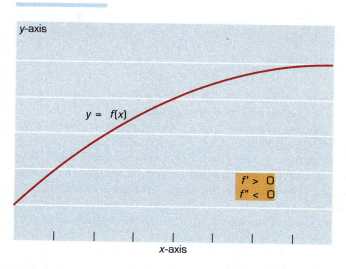

FIGURE 11.36
A Strictly Decreasing Function with Constant Slope

dropping from upper left to lower right in the window. The graph is a straight line with constant slope, with its equivalent to saying that its second derivative is zero.

If a function has a negative first derivative, it is strictly decreasing. Then if it has a positive second derivative, the *slope* of its graph is increasing, that is, the function has a negative slope that is increasing: it is getting closer to zero as the values of the input variable increase. In other words, the curve flattens out. The graph of a strictly decreasing function with increasing slope has the shape illustrated in Figure 11.37.

If a function has a negative first derivative, and a *negative* second derivative, then the function has a negative slope that is *decreasing*—the graph gets steeper—with increasing values of the input variable. The graph of a strictly decreasing function with decreasing slope has the shape illustrated in Figure 11.38.

FIGURE 11.37

A Strictly Decreasing Function with Increasing Slope

FIGURE 11.38

A Strictly Decreasing Function with Decreasing Slope

The graph of a function $f(x)$ is concave up on an interval where $f''(x) > 0$ and concave down on an interval where $f''(x) < 0$.

Figures 11.34 and 11.37 both involve functions with positive second derivatives. Figure 11.37 shows a strictly decreasing function with a positive second derivative, while Figure 11.34 is the graph of a strictly increasing function with a positive second derivative. If we paste these two figures together, we get Figure 11.39, the graph of a function with both strictly decreasing and strictly increasing pieces but with a positive second derivative throughout. It has the shape of a bowl that will hold liquid. Consequently, when the slope of a curve is increasing, that is, when $f''(x) > 0$, we say that the graph of the function $y = f(x)$ *cups up* or is *concave up*.

Figures 11.35 and 11.38 both involves functions with negative second derivatives. The function in Figure 11.35 is strictly increasing while that in Figure 11.38 is strictly decreasing. If we paste the two figures together, we get Figure 11.40, the graph of a function with both strictly increasing and strictly decreasing pieces and with a negative second derivative throughout. This figure has the shape of a bowl that will dump liquid. Consequently, when the slope of a curve is decreasing, that is, when $f''(x) < 0$, we say that the graph of the function $y = f(x)$ *cups down* or is *concave down*.

FIGURE 11.39
The Graph of a Function That Is Concave Up

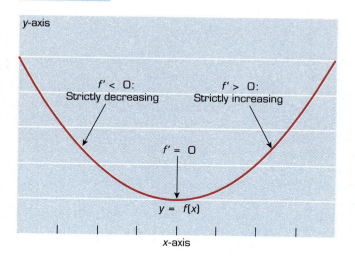

FIGURE 11.40
The Graph of a Function That Is Concave Down

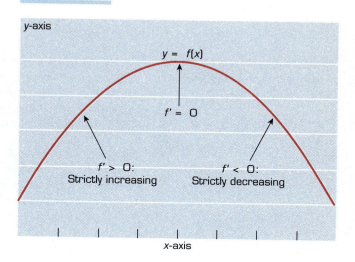

EXAMPLE 6

Figure 11.41 is the graph of a function $y = f(x)$. What does the graph indicate about the first and second derivatives of this function?

FIGURE 11.41
The Graph of a Function That Is Both Concave Down and Concave Up

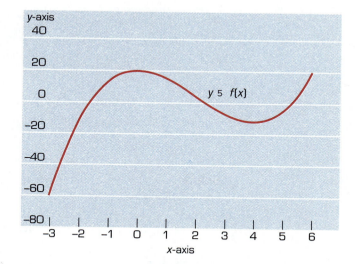

Solution

The function $y = f(x)$ is strictly increasing between $x = -3$ and $x = 0$, strictly decreasing from $x = 0$ to $x = 4$, and then strictly increasing again between $x = 4$ and $x = 6$. Thus

$$f'(x) > 0 \quad \text{when} \quad -3 < x < 0$$

$$f'(x) < 0 \quad \text{when} \quad 0 < x < 4$$

$$f'(x) > 0 \quad \text{when} \quad 4 < x < 6$$

It appears that $f'(x) = 0$ at $x = 0$ and $x = 4$. The function is certainly concave down between $x = -3$ and $x = 2$ and concave up between $x = 2$ and $x = 6$. Thus

$$f''(x) < 0 \quad \text{when} \quad -3 < x < 2$$

$$f''(x) > 0 \quad \text{when} \quad 2 < x < 6$$

The second derivative of a function changes its sign at a point of inflection and the graph of the function changes from concave down to concave up, or vice versa.

A point $(x_0, f(x_0))$ is a *point of inflection* for the graph of a continuous function $y = f(x)$ if there exists an interval centered at x_0 such that, in this interval, $f''(x)$ is positive on one side of x_0 and negative on the other. For example, the point $(2, f(2))$ is a point of inflection for the graph in Figure 11.41. If we consider the interval $[0, 4]$ centered at $x = 2$ (any smaller interval will do equally as well), we see that the second derivative is negative to the left of $x = 2$ and positive to the right of $x = 2$. At a point of inflection, a graph changes from concave down to concave up, or vice versa.

Prior to the introduction of graphing calculators and graphing software packages for computers, information about the first and second derivatives of a function was essential to the graphing process. Today, modern technologies allow us to plot hundreds of individual points quickly, and the use of this technology is the most common method for constructing a graph. However, with second derivatives we can, on occasion, locate local (or relative) optima of functions without graphing.

THE SECOND-DERIVATIVE TEST FOR LOCAL OPTIMA

If the first derivative of a function is zero at x_0 and if its second derivative is positive there, then the function has a local minimum at x_0.

If the first derivative of a function is zero at x_0 and if its second derivative is negative there, then the function has a local maximum at x_0.

EXAMPLE 7

Find the locations of all local optima for the function

$$y = f(x) = x^3 - 21x^2 + 120x - 100$$

Solution

For this function,

$$\frac{dy}{dx} = 3x^2 - 42x + 120 = 3(x - 4)(x - 10)$$

and is zero only when $x = 4$ or $x = 10$. These are the only two critical points for $f(x)$ and so the only possible locations for local optima. The second derivative of $f(x)$ is

$$\frac{d^2y}{dx^2} = \frac{d}{dx}\left(\frac{dy}{dx}\right) = \frac{d}{dx}(3x^2 - 42x + 120) = 6x - 42$$

At the critical point $x = 4$,

$$\left.\frac{d^2y}{dx^2}\right|_{x=4} = 6(4) - 42 = -18$$

is negative. It follows from the second-derivative test that $x = 4$ is the location of a local maximum.

At the critical point $x = 10$,

$$\left.\frac{d^2y}{dx^2}\right|_{x=10} = 6(10) - 42 = 18$$

is positive. It follows from the second-derivative test that $x = 10$ is the location of a local minimum.

Figure 11.42 is the graph of $f(x)$. We can see from this graph that $f(x)$ has a local maximum at $x = 4$ and a local minimum at $x = 10$. The advantage of the second-derivative test is that we did not have to graph $f(x)$ to identify its local optima and to determine that it has only two of them. This last piece of information is not apparent from Figure 11.42. There are only two local optima displayed in the window, but that does not preclude the possibility of others.

FIGURE 11.42

The Local Optima of the Function $f(x) = x^3 - 21x^2 + 120x - 100$

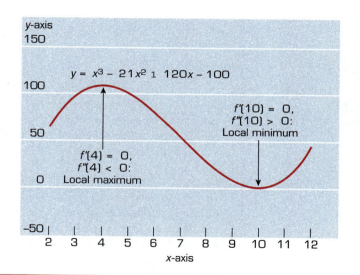

The second-derivative test is silent when $f''(x) = 0$ at a critical point, because many possibilities exist. Figures 11.30, 11.43, and 11.44 display portions of the graphs of three different functions. At $x = 0$, the function $y = x^4$ has a local minimum; the function $y = -x^4$ has a local maximum; and the function $y = x^3$ has a point of inflection, although the first and second derivatives of each of these functions is zero there.

Whenever the second-derivative test fails, we return to the first-derivative test to determine the behavior of the function around a critical point. We showed in Example 5 that $y = x^4$ has a local minimum at $x = 0$. By similar reasoning, the derivative of $y = -x^4$, which is $y' = -4x^3$, is negative when $x > 0$ and positive when $x < 0$. It follows from the first-derivative test that $y = -x^4$ has a local maximum at $x = 0$. Finally, the derivative of $y = x^3$ is $y' = 3x^2$, which is positive when $x \neq 0$. Thus, $y = x^3$ is strictly increasing for all values of x less than zero and strictly increasing for all values of x greater than zero. We conclude that $x = 0$ is the location of neither a local maximum nor a local minimum; it is a point of inflection.

The primary use of the second derivative in applications is as a measure of changing growth rates.

FIGURE 11.30

The Graph of $y = x^4$

FIGURE 11.43
The Graph of $y = -x^4$

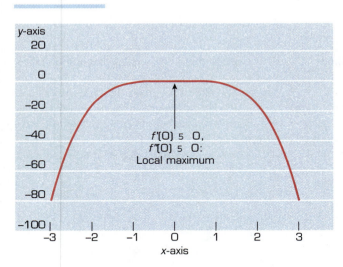

FIGURE 11.44
The Graph of $y = x^3$

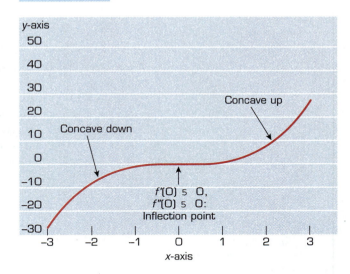

EXAMPLE 8

Some federal officials claim that the cost of health care in the United States would become manageable if the country could reduce the annual rate of growth of health care costs. What do they mean by this?

Solution

Health care costs in the United States have increased significantly during the last half of the twentieth century. If the trend continues, a graph of annual health care costs (the vertical axis) as a function of time in years (the horizontal axis) will look like Figure 11.34. Both the first and second derivatives are positive everywhere: costs are increasing and those increases are accelerating over time.

Some federal officials believe that health care costs would become manageable if, instead, the graph looked like Figure 11.35, with a positive first derivative and a negative second derivative. A positive first derivative means that actual costs increase from year to year, but a negative second derivative means that the increases become smaller as time progresses.

IMPROVING SKILLS

In Problems 1 through 20, find the second derivatives of the given functions and evaluate each at $x = 1$ and $x = 2$.

1. $y = x^5 + 4x^2 + 3x$
2. $y = x^4 + 3x^2 - 4x + 4$
3. $y = x + 3$
4. $y = 5$
5. $y = 3x^2 - 3x + 8$
6. $y = 3 - 5x^6$

7. $y = 2 - x^2$
8. $y = 1/(x + 2)$
9. $y = x - (1/x)$
10. $y = \sqrt{1 + x}$
11. $y = \sin x$
12. $y = \cos 2x$
13. $y = x \cos 2x$
14. $y = x^2 \ln x$
15. $y = \ln x^2$
16. $y = xe^x$
17. $y = e^{x^2}$
18. $y = xe^{x^2}$
19. $y = e^x \sin x$
20. $y = e^{2x} \cos 3x$

What do the graphs in Problems 21 through 30 indicate about the first and second derivatives of each function on the displayed domains?

21.

22.

23.

24.

25.

26.

27.

28.

29.

30.

In Problems 31 through 42, use information about the first and second derivatives of the given functions to determine intervals where the function is strictly increasing, strictly decreasing, concave up, and concave down.

31. $y = x^2$

32. $y = 1 - x^2$

33. $y = x^2 - 2x + 2$

34. $y = x^3 + x$

35. $y = x^3 - 6x^2 + 12x - 6$

36. $y = x^3 - 6x^2 + 9x + 6$

37. $y = e^x$

38. $y = e^{-x}$

39. $y = 1 - e^{-x}$

40. $y = \sin x \; (0 \le x \le 2\pi)$

41. $y = 1/x \; (x \ne 0)$

42. $y = 1/x^2 \; (x \ne 0)$

In Problems 43 through 60, use information about the first and second derivatives to find all local optima of the given functions. Do not graph them.

43. $y = 10 - 2x^2$

44. $y = 10 - 2x$

45. $y = x^2 - 12x$

46. $y = 12x - x^2$

47. $y = x^3 - 12x$

48. $y = x^3 + x$

49. $y = 3x^2 - x^3$

50. $y = x^3 + 2x^2 - x + 1$

51. $y = 2x^3 + 3x^2 - 12x + 5$

52. $y = x^3 - 3x^2 - 24x$

53. $y = x^3 - 3x^2 + x - 2$

54. $y = x^3 - 3x^2 - 4$

55. $y = x^4 - 2x^2$

56. $y = 3x^5 - 5x^3$

57. $y = 1/(x^2 + 1)$

58. $y = x + (1/x) \; (x > 0)$

59. $y = x/(x^2 + 1)$

60. $y = x - 2\sqrt{x} \; (x > 0)$

EXPLORING IN TEAMS

61. The function $d = f(t)$ converts an input variable t (in seconds) into an output variable d (in feet). Using the definition of the derivative (Section 10.3), determine the units associated with the first and second derivatives, $f'(t)$ and $f''(t)$.

62. The function $d = f(t)$ gives distance as a function of time. Use the results of Problem 61 to suggest a physical interpretation of both the first and second derivatives of $f(t)$.

63. The distance traveled by an object dropped from a height of 2000 ft is given by the function

$$d = 2000 - 32t + 16t^2$$

while the object remains in flight. What is the object's speed and acceleration when it hits the ground?

EXPLORING WITH TECHNOLOGY

64. Many graphing calculators have menus for evaluating a second derivative at a point. Use such a calculator to evaluate each of the functions in Problems 47 through 56 at $x = 5.2$ and at $x = -4$.

65. Use a graphing calculator to evaluate the function $f(x) = x/(x^2 + 1)$, as well as its first and second derivatives, at the five points $x = 0$, $x = \pm 1$, and $x = \pm 2$. What conclusions can you make regarding the graph of the function at these points?

66. Use a graphing calculator to evaluate the function $f(x) = xe^{-x}$ and its first and second derivatives at $x = 1$. What conclusions can you make regarding the graph of this function at $x = 1$?

REVIEWING MATERIAL

67. (Section 4.2) Tanya owes her aunt two remaining payments on a loan at 7% a year, compounded monthly. One payment of $1,000 is due in 10 months, and the final payment of $2,500 is due in 34 months. Tanya would like to retire the loan early. How much should she pay her aunt today to eliminate the debt fairly?

68. (Section 7.2) Two companies sell all of the operating systems for a particular type of personal computer. Currently Company I controls 70% of the market and Company II has the remaining 30% of the market. Surveys show that over the past two years, 15% of Company I's customers changed to the operating system of Company II, while the other 85% remained loyal to Company I. Over the same time period, 20% of Company II's customers changed to the operating system of Company I, while the other 80% remained loyal to Company II. Assuming that these trends continue into the future, what will Company I's market share be after (**a**) 2 years? (**b**) 6 years? (**c**) 10 years?

69. (Section 9.5) Determine values of the constants k, A, and b so that the function $y = k + A \sin(b\theta)$ has a period of 5 units and outputs that range inclusively between 20 and 100.

RECOMMENDING ACTION

70. Respond by memo to the following request:

> ### MEMORANDUM
>
> To: J. Doe Reader
>
> From: Robin
>
> Date: Today
>
> Subject: **Critical Points**
>
> I notice that both the first-derivative test and the second-derivative test refer to local optima at critical points where the first derivative is zero. However, critical points also include locations where the first derivative of a function does not exist. What can we say about local optima at these critical points?

CHAPTER 11 KEYS

KEY WORDS

absolute maximum (p. 780)
absolute minimum (p. 780)
concave down (p. 830)
concave up (p. 830)
critical point (p. 781)
economic order quantity (p. 813)
first derivative (p. 824)
fixed cost (p. 806)
global maximum (p. 780)
global minimum (p. 780)
interior point (p. 778)
local maximum (p. 780)

local minimum (p. 780)
overhead (p. 806)
point of inflection (p. 832)
relative maximum (p. 780)
relative minimum (p. 780)
reordering cost (p. 810)
second derivative (p. 824)
storage costs (p. 809)
strictly decreasing (p. 826)
strictly increasing (p. 826)
unit cost (p. 806)
variable cost (p. 806)

KEY CONCEPTS

11.1 Critical Points

- Optimization is the process of maximizing or minimizing a quantity.

- Since all real-world processes are constrained by finite resources, each input to an input–output relationship is bounded within some closed interval.

- A function may attain its global maximum or global minimum at two or more input values.

- A global optimum (maximum or minimum) for a continuous function can occur only at critical points within the domain or endpoints.

11.2 Design Problems

- A mathematical model must be a function of a single input variable if derivatives are to be used to analyze its behavior.

- A diagram is a valuable intermediate step in creating a mathematical model from a verbal description of a problem.

- A domain should include its endpoints when a function models a real-world process.

11.3 Production Problems

- The unit cost of an item will depend on the number of items produced when fixed costs are averaged over the entire production run.

- Optimizing techniques are used in inventory theory to minimize storage costs while ensuring that sufficient items are available to meet demand.

11.4 Related Rates

- The chain rule is used to calculate the rate of change of a composite function for a sequence of operations when the rate of change of the function for each individual operation is known.

11.5 Second Derivatives and Local Optima

- To evaluate a second derivative at a specific value of an input variable, first find the second derivative using the rules of differentiation, then evaluate that derivative at the input value of interest.

- If the first derivative of a function is positive at a specific input value, then the function is strictly increasing there; if the first derivative of the function is negative at that value, then the function is strictly decreasing there.

- If the second derivative of a function is positive in an interval, then the graph of the function is concave up over that interval; if the second derivative of the function is negative in that interval, then the graph of the function is concave down over that interval.

KEY FORMULAS

- Let F denote the fixed cost of an order, Q the quantity of items ordered each year, and W the cost of warehousing one item for one year. The economic order quantity is

$$x = \sqrt{2FQ/W}$$

- If $y = f(x)$ and $x = g(t)$ and if both functions have first derivatives, then

$$\frac{dy}{dt} = \frac{dy}{dx} \cdot \frac{dx}{dt}$$

- The second derivative of a function $f(x)$ is the derivative of its first derivative, that is,

$$\frac{d^2 f(x)}{dx^2} = \frac{d}{dx}\left[\frac{d}{dx} f(x)\right]$$

KEY PROCEDURES

- The global maximum and global minimum of a function occur where the graph of the function attains its highest and lowest points, respectively.

- To model an optimization problem:

    Step 1 Draw a diagram of the process, if possible.

    Step 2 Identify the important variables in the process, including one for the item to be optimized, then write an equation that relates that variable to some (or all) of the others.

    Step 3 Identify all constraints involving the variables.

    Step 4 Simplify the model so that the quantity to be optimized is expressed as a function of a single input variable.

    Step 5 Establish a feasible domain that includes its endpoints for the input variable identified in Step 4.

- To locate global optima for a function $f(x)$:

    Step 1 Identify the endpoints of the domain.

    Step 2 Find the critical points for $f(x)$, limiting the search to the interior of the domain.

    Step 3 Evaluate $f(x)$ at each of the points identified in Steps 1 and 2. The largest of these values is the global maximum and the smallest is the global minimum.

- The first-derivative test for finding local optima:

    Step 1 Identify a point p in the domain of $f(x)$ where $f'(p) = 0$.

    Step 2 If $f'(x) > 0$ in some interval to the left of p and also if $f'(x) < 0$ in some interval to the right of p, then $f(x)$ has a local maximum at p.

    Step 3 If $f'(x) < 0$ in some interval to the left of p and also if $f'(x) > 0$ in some interval to the right of p, then $f(x)$ has a local minimum at p.

■ The second-derivative test for finding local optima:

If the first derivative of a function is zero at a point x_0 and if its second derivative is positive there, then the function has a local minimum at x_0.

If the first derivative of a function is zero at x_0 and if its second derivative is negative there, then the function has a local maximum at x_0.

TESTING YOURSELF

Use the following problems to test yourself on the material in Chapter 11. The odd-numbered problems comprise one test, the even-numbered problems a second test.

1. Identify all critical points for the function $y = x^3 - 45x^2 + 600x - 50$ on the interval [5, 25] and then use this information to locate the global maximum and global minimum of the function on that interval.
2. Use information about the first and second derivatives of $y = x^3 - 45x^2 + 600x - 50$ to find all local maxima and local minima on the interval [0, 30].
3. Find two positive numbers whose sum is 20 and whose product is as large as possible.
4. Use information about the first and second derivatives to analyze the graph of $y = \sqrt{x}$ for $x > 0$.
5. A transit system currently charges \$1.00 per fare and is debating the amount of a fare increase. The system currently has 20,000 daily riders, but surveys suggest that each nickel added to the fare will reduce daily ridership by 500 people. What fare should be charged to maximize revenue to the system?
6. Find the second derivatives of the following functions: (**a**) $f(x) = 3x^3 + 2x^2 + 2x + 2$; (**b**) $f(x) = 2x^6 - 5x^3 + 2x^2 - 10$; (**c**) $f(x) = x\cos(2x)$.
7. The upper part of a wineglass has the shape of a cone (see Figure A) with a height of 6 cm and a diameter of 4 cm. Find how fast the height of wine is increasing from half full if the wine is being poured into it at the rate of 20 cm³/sec. *Hint:* The volume of a cone with radius r and height h is $V = \frac{1}{3}\pi r^2 h$.

FIGURE A

4 cm

6 cm

8. Sketch the graph of a function having all of the following four characteristics: (**a**) $f(2) = -5$; (**b**) $f'(x) > 0$ when $x < 2$; (**c**) $f'(x) < 0$ when $x > 2$; (**d**) $f''(x) < 0$ for all values of x.

9. The cost of placing an order is \$5 and the storage cost per unit is \$4/month. What is the economic order quantity if 1,000 items are needed during the next year?

10. Identify all critical points for the function $y = x^4 - 8x^2 + 1$ on the interval $[-1, 3]$, then use this information to find the global maximum and global minimum of the function on that interval.

11. Use information about the first and second derivatives of the function $y = x^4 - 8x^2 + 1$ to find all local maxima and minima on the interval $[-5, 5]$.

12. Consider the function graphed in Figure B. Determine the intervals where the function is (**a**) increasing, (**b**) decreasing, (**c**) concave up, and (**d**) concave down.

FIGURE B

13. Identify the locations of all global and local optima of the function displayed in Figure B, clearly distinguishing maxima from minima.

14. In calm waters, oil spilling from a ruptured tanker spreads outward in a circular fashion from the tanker. How fast is the polluted area expanding when its radius is 500 ft and increasing at the rate of 10 ft/hr?

15. Sketch the graph of a function having all of the following four characteristics: (**a**) $f(1) = 2$; (**b**) $f'(x) < 0$ for all values of x; (**c**) $f''(x) < 0$ when $x < 0$; (**d**) $f''(x) > 0$ when $x > 0$.

16. A piece of rope 20 ft. long is cut into two pieces and used to form two squares. Determine the length of both pieces if the sum of squares of the enclosed areas is to be as small as possible.

17. Use information about the first and second derivatives to determine the intervals where the function $y = x^3 - 6x^2 + 9x - 3$ is (**a**) increasing, (**b**) decreasing, (**c**) concave up, and (**d**) concave down.

18. The speed at which air exits a person's windpipe is a function of the radius of the windpipe. When a person coughs, his windpipe contracts to reduce its radius

and increase the speed of the expelled air. Let R denote the radius of a normal windpipe. The speed v at which air is expelled when the radius is reduced to r is

$$v = a(R - r)r^2 \quad (r \leq R)$$

where a is a positive constant. What value of r maximizes v?

19. Evaluate $y''(-1)$ for the functions (**a**) $y = 2x^4 - 3x^3 + 4x + 5$; (**b**) $y = x^3 e^x$.

20. A dry-cleaning chain uses 1600 drums of solvent a year. The solvent is ordered from a manufacturer who charges \$3.20/drum for delivery and \$50/order for handling. The manager calculates the storage cost for a drum of solvent at \$64/yr. Determine the optimum order size and frequency to minimize the chain's yearly inventory cost for solvent.

12

MODELING
WITH INTEGRALS

12.1 ANTIDERIVATIVES AND INDEFINITE INTEGRALS

In Chapters 10 and 11, we used derivatives of functions to enhance our understanding of real-world behavior. We learned that if a function models the position of an object, then its derivative models speed. If a function models cost, then its derivative models marginal cost. We used derivatives to determine intervals in which a function is increasing, intervals in which a function is decreasing, and the location of optima. Derivatives measure change, and the more we know about change in a system the better we understand the system itself.

Now we want to retrace our path through Chapters 10 and 11 in the reverse direction. There we began with a function and found its derivative; now we begin with a derivative and reconstruct the function. For example, we would like to pinpoint an object's position knowing its speed. We would like to produce a cost function from a marginal cost function. We want to create a model of a real-world process from knowledge about its rate of change.

When two functions $u(x)$ and $v(x)$ are related through differentiation by the equation

> *A function $v(x)$ is an antiderivative of $u(x)$ if $u(x)$ is the derivative of $v(x)$.*

$$u(x) = \frac{d}{dx} v(x)$$

we say that $u(x)$ is the derivative of $v(x)$ and that $v(x)$ is an *antiderivative* of $u(x)$. For example, $d(5x^7)/dx = 35x^6$, so $5x^7$ is an antiderivative of $35x^6$. The derivative of $x \ln x$ is $1 + \ln x$, hence $x \ln x$ is an antiderivative of the function $1 + \ln x$. Also $2xe^{x^2}$ is the derivative of e^{x^2}, so e^{x^2} is an antiderivative of $2xe^{x^2}$. These corresponding relationships are shown schematically in Figure 12.1.

We often use the notation $f'(x)$ for the derivative of the function $f(x)$. We now adopt the symbol $F(x)$ to represent an antiderivative of $f(x)$.

FIGURE 12.1

A Schematic of the Relationship between a Derivative and an Antiderivative

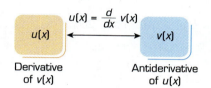

Derivative of $v(x)$ Antiderivative of $u(x)$

EXAMPLE 1

Determine whether $F(x) = 5x^3 + 3x^2 + 10x + 6$ is an antiderivative of the function $f(x) = 15x^2 + 6x + 10$.

Solution

A function $F(x)$ is an antiderivative of $f(x)$ if $F'(x) = f(x)$. Differentiating $F(x)$, we find that

$$F'(x) = \frac{d}{dx} (5x^3 + 3x^2 + 10x + 6) = 15x^2 + 6x + 10$$

which is $f(x)$. Therefore $F(x)$ is an antiderivative of $f(x)$.

EXAMPLE 2

Determine whether $G(x) = 5x^3 + 3x^2 + 10x + 37$ is also an antiderivative of the function $f(x) = 15x^2 + 6x + 10$.

Solution

Again, $G(x)$ is an antiderivative of $f(x)$ if and only if $G'(x) = f(x)$. Differentiating $F(x)$ gives us

$$G'(x) = \frac{d}{dx} (5x^3 + 3x^2 + 10x + 37) = 15x^2 + 6x + 10$$

which is $f(x)$. Therefore $G(x)$ is another antiderivative of $f(x)$.

We see from Examples 1 and 2 that a function $f(x)$ can have more than one antiderivative. In fact, since the derivative of a constant is zero, any constant added to one antiderivative is another antiderivative. Consequently, we can construct infinitely many antiderivatives of a function $f(x)$ by adding constants to a known antiderivative of $f(x)$. The converse is also true. If we have two antiderivatives of the same function, then they must differ by a real number (perhaps zero).

FIGURE 12.2

Several Antiderivatives of
$y = 15x^2 + 6x + 10$

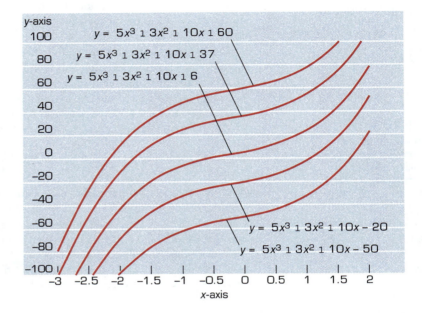

Figure 12.2 is a graph of several antiderivatives of the function

$$f(x) = 15x^2 + 6x + 10$$

including the two identified in Examples 1 and 2. Each of these antiderivatives has the property that its derivative is $f(x)$. Thus at a particular value of x, the tangent lines to all of the curves in Figure 12.2 have the same slope.

If $F(x)$ is one antiderivative of $f(x)$, then $F(x) + c$ is another antiderivative for any choice of the constant c. For any unspecified ("arbitrary") constant c, the function $F(x) + c$ is called the *indefinite integral* of $f(x)$. *The* indefinite integral of the function $f(x) = 15x^2 + 6x + 10$ is $5x^3 + 3x^2 + 10x + c$, where c denotes an arbitrary constant. Every choice of c (for instance, $c = 6$ in Example 1 and $c = 37$ in Example 2) results in a bona fide antiderivative of $f(x)$.

The standard notation for the indefinite integral of $f(x)$ is

> *If $F(x)$ is an antiderivative of $f(x)$ then so, too, is $F(x) + c$, for every choice of arbitrary constant c.*

$$\int f(x)\, dx$$

The function $f(x)$ is "sandwiched" between two symbols: the symbol $\int$, called the "integral sign," and the symbol dx, called the "differential," which reinforces the notion that the variable of interest is x. If $F(x)$ is *any* antiderivative of $f(x)$, then the indefinite integral of $f(x)$ is

$$\int f(x)\, dx = F(x) + c \tag{1}$$

FIGURE 12.3
The Relationships between a Function, Its Derivative, and Its Antiderivative

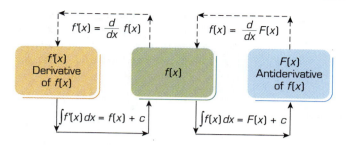

where c denotes an arbitrary real number. The relationships between a function, its derivative, and its antiderivative are shown schematically in Figure 12.3.

The process of finding an indefinite integral is called *integration,* and it is simplified by a set of rules analogous to those of differentiation, presented in Sections 10.3 and 10.4. This is not surprising, since the process of indefinite integration is the reverse of differentiation. We devote the remainder of this section to developing basic skills in integration, which we can then apply to real-world problems.

THE POWER RULE OF INTEGRATION

If n is a real number, $n \neq -1$, then the indefinite integral of the power function $f(x) = x^n$ is

$$\int x^n \, dx = \frac{x^{n+1}}{n+1} + c$$

where c is an arbitrary constant.

EXAMPLE 3

Evaluate the indefinite integral $\int x^3 \, dx$.

Solution
We seek the indefinite integral of the power function $f(x) = x^3$. Using the power rule of integration with $n = 3$, we have

$$\int x^3 \, dx = \frac{x^{3+1}}{3+1} + c = \frac{1}{4} x^4 + c$$

A convenient way to check whether $F(x)$ is the indefinite integral of $f(x)$ is to confirm that

$$\frac{d}{dx} F(x) = f(x)$$

To verify that $F(x) = \frac{1}{4}x^4 + c$ is indeed an antiderivative of $f(x) = x^3$, we differentiate $F(x)$ and check whether the result equals $f(x)$:

$$F'(x) = \frac{d}{dx}\left(\frac{1}{4} x^4 + c\right) = \frac{1}{4}(4x^3) + 0 = x^3$$

We have evaluated the indefinite integral correctly in Example 3.

For each value of c, the indefinite integral in Example 3 yields a different antiderivative of $f(x) = x^3$. A few of these antiderivatives are graphed in Figure 12.4. Each antiderivative has the property that its derivative is x^3. Thus at each value of x, the tangent lines to all of the curves in Figure 12.4 have the same slope.

FIGURE 12.4

A Few Antiderivatives of $y = x^3$

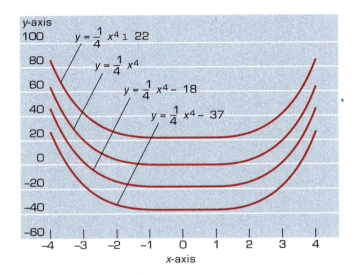

EXAMPLE 4

Evaluate $\int \sqrt{x}\, dx$.

Solution

We seek the indefinite integral of the power function $f(x) = \sqrt{x} = x^{1/2}$. Using the power rule of integration with $n = 1/2$, we have

$$\int x^{1/2}\, dx = \frac{x^{\frac{1}{2}+1}}{\frac{1}{2}+1} + c = \tfrac{2}{3} x^{3/2} + c$$

We can check the answer in Example 4 by differentiating the result and confirming that it equals $f(x)$:

$$\frac{d}{dx}\left(\frac{2}{3} x^{3/2} + c\right) = \frac{2}{3}\left(\frac{3}{2} x^{3/2-1}\right) + 0 = x^{1/2} = \sqrt{x}$$

This is indeed $f(x)$.

For each value of c, the indefinite integral in Example 4 yields a different antiderivative of $f(x) = \sqrt{x}$. A few of these antiderivatives are graphed in Figure 12.5. The domain of $\sqrt{x}$ is restricted to nonnegative inputs, so each antiderivative of $\sqrt{x}$ is similarly constrained.

EXAMPLE 5

Evaluate $\int 1\, dx$.

Solution

Since $1 = x^0$, the power rule of integration applies, with $n = 0$:

$$\int 1\, dx = \int x^0\, dx = \frac{x^{0+1}}{0+1} + c = x + c$$

We check this answer by noting that $\dfrac{d}{dx}(x + c) = 1$.

FIGURE 12.5
A Few Antiderivatives of $y = \sqrt{x}$

The power rule of integration,

$$\int x^n \, dx = \frac{x^{n+1}}{n+1} + c$$

is not applicable when $n = -1$ because we cannot divide by zero. So how can we find the antiderivative of $1/x$? The natural logarithm rule of differentiation states that $d(\ln x)/dx = 1/x$. However, the domain of $\ln x$ is restricted to nonnegative values of x, so $1/x$ is the derivative of $\ln x$ only when $x > 0$. Conversely, $\ln x$ is an antiderivative of $1/x$ when $x > 0$. That is,

$$\int x^{-1} \, dx = \int \frac{1}{x} \, dx = \ln x + c \quad (x > 0)$$

If $x < 0$ then $-x > 0$, and we have from the chain rule of differentiation that

$$\frac{d}{dx} \ln(-x) = \left(\frac{1}{-x}\right) \frac{d}{dx}(-x) = \left(\frac{1}{-x}\right)(-1) = \frac{1}{x}$$

Thus, $1/x$ is the derivative of $\ln(-x)$ when $x < 0$, and, conversely, $\ln(-x)$ is an antiderivative of $1/x$ when $x < 0$. That is,

$$\int x^{-1} \, dx = \int \frac{1}{x} \, dx = \ln(-x) + c \quad (x < 0)$$

Since

$$|x| = \begin{cases} x & \text{when } x \geq 0 \\ -x & \text{when } x < 0 \end{cases}$$

we can combine these two antiderivative rules for $1/x$.

THE INTEGRATION RULE FOR $f(x) = \dfrac{1}{x}$

The indefinite integral of the special power function $y = f(x) = x^{-1}$ is

$$\int x^{-1}\, dx = \int \frac{1}{x}\, dx = \ln|x| + c \quad (x \neq 0)$$

where c is an arbitrary constant.

The previous two rules of integration are limited to power functions. As with the rules of differentiation, the rules of integration become significantly more useful when they are combined with rules that deal with sums, differences, and constant multiples of such functions.

THE CONSTANT MULTIPLE RULE OF INTEGRATION

If $y = kf(x)$, where k is a known real number and the indefinite integral of the function $f(x)$ is known, then

$$\int kf(x)\, dx = k \int f(x)\, dx$$

EXAMPLE 6

Evaluate $\int 6x^3\, dx$.

Solution

The constant multiple rule of integration with $k = 6$ and $f(x) = x^3$ becomes

$$\int 6x^3\, dx = 6 \int x^3\, dx$$

Using the results of Example 3, we conclude that

$$6 \int x^3\, dx = 6\left(\tfrac{1}{4}x^4 + c\right) = 1.5x^4 + 6c$$

It is common to replace a constant such as $6c$ simply by c since c denotes *any* arbitrary constant, and $6c$ is also an arbitrary constant. Thus the two notations c and $6c$ both represent any real number, so we can write

$$\int 6x^3\, dx = 1.5x^4 + c$$

Check: $\dfrac{d}{dx}(1.5x^4 + c) = 6x^3$

EXAMPLE 7

Evaluate $\int -\dfrac{3}{x}\,dx$.

Solution

Combining the constant multiple rule of integration, where $k = -3$, with the integration rule for $1/x$, we obtain

$$\int -\frac{3}{x}\,dx = -3\int \frac{1}{x}\,dx = -3(\ln|x| + c) = -3\ln|x| - 3c$$

Since c represents an arbitrary constant, so too does $-3c$, and we write

$$\int -\frac{3}{x}\,dx = -3\ln|x| + c$$

Check: $\dfrac{d}{dx}(-3\ln|x| + c) = \dfrac{d}{dx}(-3\ln|x|) + \dfrac{d}{dx}(c) = -3\dfrac{d}{dx}\ln|x|$

$$= \begin{cases} -3\,\dfrac{d}{dx}\ln x & \text{when } x > 0 \\[2mm] -3\,\dfrac{d}{dx}\ln(-x) & \text{when } x < 0 \end{cases}$$

$$= \begin{cases} -3\left(\dfrac{1}{x}\right) & \text{when } x > 0 \\[2mm] -3\left(-\dfrac{1}{-x}\right) & \text{when } x < 0 \end{cases}$$

$$= -3/x \text{ when } x \neq 0$$

THE SUM AND DIFFERENCE RULES OF INTEGRATION

If $f(x)$ and $g(x)$ are two functions whose indefinite integrals are known, then the indefinite integral of their sum is

$$\int [f(x) + g(x)]\,dx = \int f(x)\,dx + \int g(x)\,dx$$

while the indefinite integral of the difference is

$$\int [f(x) - g(x)]\,dx = \int f(x)\,dx - \int g(x)\,dx$$

EXAMPLE 8

Evaluate $\int(x^3 + x^5)\,dx$.

Solution

First applying the sum rule of integration and then the power rule of integration, we obtain

$$\int (x^3 + x^5)\,dx = \int x^3\,dx + \int x^5\,dx$$
$$= (\tfrac{1}{4}x^4 + c_1) + (\tfrac{1}{6}x^6 + c_2)$$
$$= \tfrac{1}{4}x^4 + \tfrac{1}{6}x^6 + (c_1 + c_2)$$

Since c_1 and c_2 are both arbitrary constants, so too is their sum. Therefore, we can replace the quantity $c_1 + c_2$ with the single arbitrary constant c, obtaining

$$\int (x^3 + x^5)\, dx = \tfrac{1}{4}x^4 + \tfrac{1}{6}x^6 + c$$

Check: $\dfrac{d}{dx}(\tfrac{1}{4}x^4 + \tfrac{1}{6}x^6 + c) = x^3 + x^5$

EXAMPLE 9

Evaluate $\displaystyle\int \left(3x - \frac{5}{x}\right) dx$.

Solution

This operation will require the use of most of the rules of integration we've learned so far.

$$\int \left(3x - \frac{5}{x}\right) dx = \int 3x\, dx - \int \frac{5}{x}\, dx \qquad \text{(difference rule)}$$

$$= 3 \int x\, dx - 5 \int \frac{1}{x}\, dx \qquad \text{(constant multiple rule twice)}$$

$$= 3\left(\frac{1}{2}x^2 + c_1\right) - 5\int \frac{1}{x}\, dx \qquad \text{(power rule with } n = 1\text{)}$$

$$= 3(\tfrac{1}{2}x^2 + c_1) - 5(\ln|x| + c_2) \qquad \text{(integration rule for } 1/x\text{)}$$

$$= \tfrac{3}{2}x^2 - 5\ln|x| + (3c_1 - 5c_2)$$

$$= \tfrac{3}{2}x^2 - 5\ln|x| + c$$

where we have replaced the quantity $3c_1 - 5c_2$ with the single arbitrary constant c.

Check: $\dfrac{d}{dx}\left(\dfrac{3}{2}x^2 - 5\ln|x| + c\right) = 3x - \dfrac{5}{x}$

> *Combinations of arbitrary constants can be combined into a single arbitrary constant.*

The rules for evaluating the indefinite integrals of the sum or difference of two functions can be generalized in a straightforward manner to the sums and differences of more than two functions. The resulting integrations usually lead to a combination of arbitrary constants—such as $3c_1 - 5c_2$ in Example 9—which can always be replaced by a single arbitrary constant c.

EXAMPLE 10

Evaluate $\int (2x^{4.4} - 3x^{-2.3} + 12)\, dx$.

Solution

It follows from our rules of integration and Example 5 that

$$\int (2x^{4.4} - 3x^{-2.3} + 12)\, dx$$

$$= \int 2x^{4.4}\, dx - \int 3x^{-2.3}\, dx + \int 12\, dx$$

$$= 2\int x^{4.4}\, dx - 3\int x^{-2.3}\, dx + 12\int 1\, dx$$

$$= 2\left(\frac{x^{4.4+1}}{4.4+1} + c_1\right) - 3\left(\frac{x^{-2.3+1}}{-2.3+1} + c_2\right) + 12(x + c_3)$$

$$= \frac{2}{5.4} x^{5.4} + 2c_1 - \frac{3}{-1.3} x^{-1.3} - 3c_2 + 12x + 12c_3$$

$$= \frac{20}{54} x^{5.4} + \frac{30}{13} x^{-1.3} + 12x + (2c_1 - 3c_2 + 12c_3)$$

$$= \frac{10}{27} x^{5.4} + \frac{30}{13} x^{-1.3} + 12x + c$$

Check: $\dfrac{d}{dx}\left(\dfrac{10}{27} x^{5.4} + \dfrac{30}{13} x^{-1.3} + 12x + c\right) = 2x^{4.4} - 3x^{-2.3} + 12$

There is little point in carrying a lot of arbitrary constants along in an indefinite integration when we know that in the end, we will replace all of them by a single one. Therefore we don't bother with any constants until the end, when we simply add a single arbitrary constant c.

THE NATURAL EXPONENTIAL RULE OF INTEGRATION

The indefinite integral of the function $y = f(x) = e^x$ is

$$\int e^x \, dx = e^x + c$$

where c is an arbitrary constant.

EXAMPLE 11

Evaluate $\int(3e^x - \tfrac{1}{2}) \, dx$.

Solution

The natural exponential rule is very straightforward to use:

$$\int (3e^x - \tfrac{1}{2}) \, dx = \int 3e^x \, dx - \int \tfrac{1}{2} \, dx \qquad \text{(difference rule)}$$

$$= 3 \int e^x \, dx - \tfrac{1}{2} \int 1 \, dx \qquad \text{(constant multiple rule twice)}$$

$$= 3e^x - \tfrac{1}{2}x + c \qquad \text{(see Example 5)}$$

Check: $\dfrac{d}{dx}\left(3e^x - \dfrac{1}{2} x + c\right) = 3e^x - \dfrac{1}{2}$

THE SINE RULE OF INTEGRATION

The indefinite integral of the function $y = f(x) = \sin x$ is

$$\int \sin x \, dx = -\cos x + c$$

where c is an arbitrary constant.

This last rule follows from the fact that

$$\frac{d}{dx}(-\cos x) = -\frac{d}{dx}\cos x = -(-\sin x) = \sin x$$

The derivative of $-\cos x$ is $\sin x$, hence an antiderivative of $\sin x$ is $-\cos x$.

THE COSINE RULE OF INTEGRATION

The indefinite integral of the function $y = f(x) = \cos x$ is

$$\int \cos x \, dx = \sin x + c$$

where c is an arbitrary constant.

EXAMPLE 12

Evaluate $\displaystyle\int \left(2.5 \cos x - 6.3 \sin x + \frac{12}{x} - 1\right) dx$.

Solution

We combine the sine and cosine rules with several of the previous rules:

$$\int \left(2.5 \cos x - 6.3 \sin x + \frac{12}{x} - 1\right) dx$$

$$= \int 2.5 \cos x \, dx - \int 6.3 \sin x \, dx + \int \frac{12}{x} \, dx - \int 1 \, dx$$

$$= 2.5 \int \cos x \, dx - 6.3 \int \sin x \, dx + 12 \int \frac{1}{x} \, dx - \int 1 \, dx$$

$$= 2.5 \sin x - 6.3(-\cos x) + 12 \ln |x| - x + c$$

$$= 2.5 \sin x + 6.3 \cos x + 12 \ln |x| - x + c$$

Check: $\dfrac{d}{dx}(2.5 \sin x + 6.3 \cos x + 12 \ln |x| - x + c) = 2.5 \cos x - 6.3 \sin x + \dfrac{12}{x} - 1$

IMPROVING SKILLS

Evaluate the indefinite integrals in Problems 1 through 40.

1. $\int x^4 \, dx$
2. $\int 8x^4 \, dx$
3. $\int 5.5x^4 \, dx$
4. $\int x^7 \, dx$
5. $\int 2x^7 \, dx$
6. $\int (-\frac{1}{2}x^7) \, dx$
7. $\int (x^4 - x^7) \, dx$
8. $\int (3x^7 + 9x^4) \, dx$
9. $\int 2 \, dx$
10. $\int (x + 5) \, dx$
11. $\int (3x - 22) \, dx$
12. $\int (x - \sqrt{5}) \, dx$
13. $\int (\sqrt{x} - 5) \, dx$
14. $\int (\sqrt{x} - \sqrt{5}) \, dx$

15. $\int(x^2 + x + 1)\, dx$

16. $\int(3x^2 - 4x + \frac{1}{2})\, dx$

17. $\int(x^{-5} - x^{-4} + x^{-3})\, dx$

18. $\int(x^{-3} + x^{-2} - x^{-1})\, dx$

19. $\int(x^{-2} + x^2 + 2)\, dx$

20. $\int(\frac{1}{2}x^2 - \frac{1}{3}x^3 + \frac{1}{4}x^4)\, dx$

21. $\int[x - 1/x]\, dx$

22. $\int(2.3x^{3.2} + 0.6x^{5.8})\, dx$

23. $\int(2.3x^{3.2} + 0.6x^{-5.8})\, dx$

24. $\int(2.3x^{-3.2} - 0.6x^{-5.8})\, dx$

25. $\int(1/\sqrt{x})\, dx$

26. $\int[2\sqrt{x} - (0.5/\sqrt{x})]\, dx$

27. $\int(x^{1/3} + 2x^{1/4} - 3x^{-1/3})\, dx$

28. $\int(4x^{1/4} + 8x^{-1/4} - 4)\, dx$

29. $\int 3 \sin x\, dx$

30. $\int(\frac{1}{3}\sin x + 3)\, dx$

31. $\int \frac{5}{7}\cos x\, dx$

32. $\int(1 - 3\cos x)\, dx$

33. $\int(\sin x - \cos x)\, dx$

34. $\int(18 \sin x + 15 \cos x - 20)\, dx$

35. $\int(3 + 2e^x)\, dx$

36. $\int(x - e^x)\, dx$

37. $\int(e^x - 1)\, dx$

38. $\int(2 \sin x - 4e^x)\, dx$

39. $\int\left(e^x - \dfrac{1}{x}\right) dx$

40. $\int\left(e^x - \dfrac{1}{x^2} + \dfrac{1}{3}\right) dx$

EXPLORING IN TEAMS

In this section we introduced the rules of integration for elementary functions. Generally functions are more complex and cannot be integrated directly with these rules. In such cases, we try to transform complex functions into a combination of elementary functions using algebraic procedures; then we can integrate the resulting elementary functions with our present rules of integration. We did this in Example 4 when we rewrote $f(x) = \sqrt{x}$ as $f(x) = x^{1/2}$ so we could apply the power rule of integration.

Use algebra to simplify the function in each of the following indefinite integrals and then integrate.

41. $\int x\sqrt{x}\, dx$

42. $\int \sqrt{8x}\, dx$

43. $\int \dfrac{\sqrt{x}}{\sqrt[3]{x}}\, dx$

44. $\int \dfrac{x^2 + 5x - 2}{\sqrt{x}}\, dx$

45. $\int \dfrac{x - 1}{x}\, dx$

46. $\int \dfrac{x^2 + 5}{x^2}\, dx$

47. $\int(x - 1)(x + 1)\, dx$

48. $\int (x - 1)\left(\dfrac{1}{\sqrt{x}} + 1\right) dx$

49. $\int 3x^2(x^2 - 2x + 3)\, dx$

50. $\int \dfrac{(4x^4 + 1)^2}{x^4}\, dx$

51. $\int \dfrac{e^{2x} - e^x}{e^x}\, dx$

52. $\int \dfrac{\sin 2x}{\sin x}\, dx$

EXPLORING WITH TECHNOLOGY

53. Reproduce Figure 12.2 on a graphing calculator. Choose a value of x and use your calculator to draw a tangent line to each curve at that input value. Each time a tangent line is drawn, record its slope as it appears in the calculator window. After they have all been drawn, review the slopes you recorded and determine whether a pattern emerges. Now repeat the process at a different value of x. Does the same pattern emerge?

54. Use a graphing calculator to draw the graphs of the three different antiderivatives from Example 6. Repeat the process in Problem 53 with these curves. What pattern emerges? Can you produce a formula for these slopes?

55. Use a graphing calculator to draw the graphs of the three different antiderivatives from Example 7. Repeat the process in Problem 53 with these curves. What pattern emerges? Can you produce a formula for these slopes?

REVIEWING MATERIAL

56. (Section 6.3) During a test program for a new drug, the following data were gathered: The drug was effective on 82% of those tested and not effective on the remaining 18%; of those tested, 36% experienced aftereffects; for 12% of those tested, the drug was not effective but they experienced aftereffects anyway. Determine the probability that (**a**) the drug was not effective on a person who experienced aftereffects; (**b**) a person would experience aftereffects if the drug was effective.

FIGURE A

FIGURE B

57. (Section 9.1) Determine whether either of the graphs in Figures A and B are functions.
58. (Section 10.4) Find the derivative of the function $f(x) = 3e^{-2x} - 5$.
59. (Section 11.1) Find the global maximum and global minimum for

$$f(x) = 3x^4 + 4x^3 - 12x^2$$

on the interval $[-3, 3]$.

RECOMMENDING ACTION

60. Respond by memo to the following request:

> MEMORANDUM
>
> To: J. Doe Reader
>
> From: Ted
>
> Date: Today
>
> Subject: **Units**
>
> I am having a difficult time trying to determine the dimensions that go with an indefinite integral. I can't seem to find any reference to this subject. Is the unit of measure associated with the indefinite integral of $f(x)$ the same as $f(x)$ or are they different? Can you help?

12.2 APPLICATIONS OF THE INDEFINITE INTEGRAL

The indefinite integral of $f(x)$,

$$\int f(x)\, dx = F(x) + c$$

The graphs of functions from the same family of indefinite integrals is a set of parallel curves.

is an infinite set of functions, one for each choice of constant c. If $F(x)$ is one antiderivative of the function $f(x)$, then $F(x) + c$ is another antiderivative of $f(x)$, for any nonzero value of c; they differ by only a constant. Furthermore, all of these antiderivatives share the property that their derivative is $f(x)$. The graphs of the functions $F(x) + c$ comprise a set of parallel curves, and the difference in height between any two of them is the difference between their constants. Three examples are shown in Figures 12.2, 12.4, and 12.5.

FIGURE 12.4 (repeated)
A Few Antiderivatives
of $y = x^3$

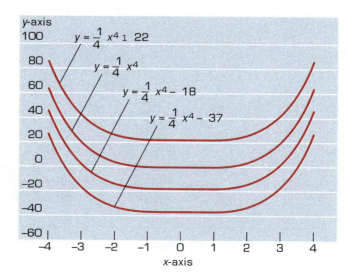

There is at most one function in an indefinite integral that yields a given output value for a particular input value.

Because parallel curves never intersect, no two curves in a family of antiderivatives pass through the same point in the plane. To isolate a curve in the family, we specify either a value of c or a point on the curve. From an input–output perspective, given a specific function, it has exactly one antiderivative that yields a required output for a specific input. For example, we saw in Example 3 of Section 12.1 that the indefinite integral of $f(x) = x^3$ is $F(x) = \frac{1}{4}x^4 + c$, and we graphed some of the antiderivatives in this family. (See Figure 12.4.) Only one of its curves passes through the point $(0, 22)$: it is the graph of the antiderivative $y = \frac{1}{4}x^4 + 22$. This is the only function in the family of antiderivatives that has an output of 22 corresponding to the input of zero.

None of the four curves we graphed in Figure 12.4, however, passes through the point $(2, 60)$, so if we want the antiderivative of $f(x) = x^3$ that passes through this point, we must draw an additional curve that does pass through $(2, 60)$. Alternatively, we can take an algebraic approach. We seek the antiderivative

$$y = F(x) = \frac{1}{4}x^4 + c \tag{2}$$

that yields an output of $y = 60$ when the input is fixed at $x = 2$. Substituting these values into Equation (2), we get $60 = \frac{1}{4}(2)^4 + c$ or $c = 56$, whereupon Equation (2) becomes

$$y = F(x) = \frac{1}{4}x^4 + 56$$

This is the antiderivative of $f(x) = x^3$ whose graph passes through the point $(2, 60)$. If we add its graph to Figure 12.4, we create Figure 12.6.

FIGURE 12.6

Graphs of $y = \frac{1}{4}x^4 + c$ for Several Values of c

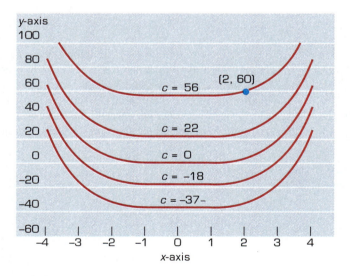

In general, given a family of functions defined by an indefinite integral, we can isolate one antiderivative $F(x)$ in one of two equivalent ways. Algebraically, we find the antiderivative $F(x)$ that has a specified output value y_0 corresponding to a particular input value x_0. Geometrically, we find the antiderivative whose graph passes through the point (x_0, y_0).

EXAMPLE 1

Find the antiderivative of the function $f(x) = 3x - (5/x)$ that has an output of 4 when the input is 1.

Solution

We saw in Example 9 of Section 12.1 that the indefinite integral of $f(x)$ is

$$F(x) = \tfrac{3}{2}x^2 - 5 \ln |x| + c \tag{3}$$

where c is an arbitrary constant. We seek a value for c to ensure that $F(x) = 4$ when $x = 1$. Substituting these values into Equation (3), we have

$$4 = F(1) = \tfrac{3}{2}(1)^2 - 5 \ln |1| + c$$
$$= \tfrac{3}{2} - 5 \cdot 0 + c$$

Therefore $c = 5/2$ and Equation (3) becomes

$$F(x) = \tfrac{3}{2}x^2 - 5 \ln |x| + \tfrac{5}{2} = \tfrac{1}{2}(3x^2 + 5) - 5 \ln |x|$$

This is the antiderivative of $f(x) = 3x - (5/x)$ whose graph passes through the point $(1, 4)$. Its graph is the solid curve in Figure 12.7; the others represent antiderivatives of $f(x)$ with different constants.

FIGURE 12.7

Graphs of $y = \tfrac{3}{2}x^2 - \ln |x| + c$ for Several Values of c

A total cost function is an antiderivative of a marginal cost function.

Fixed costs are the total cost when production is zero; they can be used to distinguish a particular cost function from the other cost functions in its family of antiderivatives.

In Chapter 10, we defined marginal cost to be the derivative of a total cost function. We modeled the total cost of a commercial process by an equation of the form $TC = f(x)$, where x denotes the number of items produced; then we found the marginal cost by differentiating this function:

$$Marginal\ cost = \frac{d}{dx}(TC) = \frac{d}{dx}f(x) = f'(x)$$

Now we want to reverse the process: starting with a specific marginal cost function, we want to find the corresponding total cost function. We begin by evaluating the indefinite integral of the marginal cost function. A particular antiderivative is then extracted from the resulting family of functions by fixing a value for the constant c.

Generally, management has a good handle on its fixed costs, which include costs associated with insurance premiums, mortgage payments, leasing payments on equipment, and all other costs that are independent of production. Fixed costs equal total cost when production is zero, and management can use this fact to distinguish a particular cost function from the others in its family of antiderivatives.

EXAMPLE 2

A company manufactures magnifying glasses for stamp collectors and other hobbyists. Fixed costs are \$5000/week and the weekly marginal cost function for the production of x magnifying glasses is

$$f(x) = 0.5 + 0.002x$$

where $f(x)$ is in units of dollars/item. What is the total cost of producing 2500 magnifying glasses a week?

Solution

Total cost is an antiderivative of marginal cost:

$$TC = \int f(x)\,dx = \int (0.5 + 0.002x)\,dx = 0.5x + 0.001x^2 + c \tag{4}$$

where c is an arbitrary constant. When $x = 0$, total cost equals the fixed cost of \$5000. Consequently,

$$5000 = TC\big|_{x=0} = 0.5(0) + 0.001(0)^2 + c$$

from which we conclude that $c = 5000$. Equation (4) becomes

$$TC = 0.5x + 0.001x^2 + 5000 \tag{5}$$

which is the total cost function for this process. Its graph is the solid curve in Figure 12.8. The dashed curves represent antiderivatives from Equation (4) that have other choices for c.

FIGURE 12.8

The Total Cost Function,
$TC = 0.001x^2 + 0.5x + c$,
for Several Values of c

We require the total cost for a production run of 2500 magnifying glasses. Setting $x = 2500$ in Equation (5), we find that

$$TC\big|_{x=2500} = 0.5(2500) + 0.001(2500)^2 + 5000 = \$12,500$$

The total cost function is given by Equation (5). We recognize the $5000-term as the fixed cost. The rest of the function, $0.5x + 0.001x^2$, depends explicitly on x, the number of items produced; it is the variable cost for the process.

Another label may be used instead of x in each rule of integration, as long as the new label replaces x throughout, including the x in the differential dx.

In Section 12.1, we developed rules for evaluating the indefinite integral of a function $f(x)$ with independent variable x. We know, however, that it is often more convenient to model input variables with other labels, such as t for time or p for price. If another label is used for the input variable x, then we replace x by that label in each of the integration rules presented in Section 12.1. For example, for the power rule of integration, we can also write

$$\int t^n \, dt = \frac{t^{n+1}}{n+1} + c, \qquad \int p^n \, dp = \frac{p^{n+1}}{n+1} + c, \qquad \int z^n \, dz = \frac{z^{n+1}}{n+1} + c$$

Similarly, for the natural exponential rule of integration, we can use labels t, u, or θ:

$$\int e^t \, dt = e^t + c, \qquad \int e^u \, du = e^u + c, \qquad \int e^\theta \, d\theta = e^\theta + c$$

Notice that the new label even replaces the x in the differential dx that appears in the indefinite integral.

The natural logarithm rule of integration states that

$$\int \frac{1}{x} \, dx = \ln |x| + c$$

When we replace the variable x with u throughout the equation, we get

$$\int \frac{1}{u} \, du = \ln |u| + c$$

Total revenue is usually zero when production is zero, and these values are used to distinguish a particular revenue function from others in its family of antiderivatives of the marginal revenue.

In contrast, we have *no* rule for evaluating $\int (1/u) \, dx$ because there is no match between the label for the variable u in the function $1/u$ and the variable x in the differential dx.

In Example 2, we determined the total cost of a process by locating a particular antiderivative of the marginal cost. Similarly, the total revenue obtained from a process is an antiderivative of the marginal revenue. Total revenue is usually zero when production is zero, because there is no income when there are no products available for sale. We use this fact to distinguish a particular revenue function from the others in its family of antiderivatives.

EXAMPLE 3

A company knows from past experience that its marginal revenue is given by the function

$$f(n) = 562.5 - 0.225n$$

where n represents the number of items offered for sale and $f(n)$ is in units of dollars per item. What is the total revenue for the production of 1200 items?

Solution

Revenue, R, is an antiderivative of marginal revenue. Here marginal revenue is given as a function of the independent variable n, so

$$R(n) = \int f(n) \, dn = \int (562.5 - 0.225n) \, dn$$

$$= 562.5 \int 1 \, dn - 0.225 \int n \, dn \qquad (6)$$

$$= 562.5n - 0.1125n^2 + c$$

where c is an arbitrary constant. When $n = 0$, total revenue (income) is zero, hence

$$0 = R(0) = 562.5(0) - 0.1125(0)^2 + c$$

from which we conclude that $c = 0$. Equation (6) becomes

$$R(n) = 562.5n - 0.1125n^2 \tag{7}$$

The graph of Equation (7) is the solid curve in Figure 12.9. The dashed curves are other antiderivatives that correspond to different choices for the constant c.

FIGURE 12.9

Graphs of the Revenue Function $R(n) = 562.5n - 0.1125n^2 + c$ for Several Values of c

We are interested in the total revenue from a production run of 1200 items, so we set $n = 1200$ in Equation (7):

$$R(1200) = 562.5(1200) - 0.1125(1200)^2 = \$513,000$$

A derivative of a function is an instantaneous rate of change. If we know a formula for the rate of change of a quantity Q with respect to time, that is, a formula for the function dQ/dt, we can use it to produce the function Q, which is an antiderivative of dQ/dt. In fact, we can obtain a whole family of antiderivatives by evaluating the indefinite integral of the function dQ/dt. We can then isolate one particular antiderivative by specifying a value of the output quantity Q at a particular value of the input variable t.

EXAMPLE 4

City planners believe that the growth rate of the population, P, will increase as a function of time (in years) according to the model

$$\frac{dP}{dt} = 200\sqrt{t}$$

Use this model to predict the population of the city in 10 years if the current population is 28,000 people.

Solution

The size P of the population is an antiderivative of the rate of change, dP/dt, with time, t, as the independent variable. We identify a family of antiderivatives by evaluating the indefinite integral of dP/dt.

$$
\begin{aligned}
P(t) = \int 200\sqrt{t}\, dt &= 200 \int t^{1/2}\, dt \\
&= 200\left(\frac{2}{3} t^{3/2} + c_1\right) \\
&= \frac{400}{3} t^{3/2} + c
\end{aligned}
\tag{8}
$$

where c is an arbitrary constant. If we measure time from the present, so that $t = 0$ represents today, then we know that $P(0) = 28{,}000$. Substituting this result into Equation (8), we find that

$$
28{,}000 = P(0) = \frac{400}{3} (0)^{3/2} + c = c
$$

Thus, $c = 28{,}000$ and Equation (8) becomes

$$
P(t) = \frac{400}{3} t^{3/2} + 28{,}000
$$

This is an expression for population as a function of time. When $t = 10$, we have

$$
P(10) = \frac{400}{3} (10)^{3/2} + 28{,}000 \approx 32{,}200
$$

City planners predict a population of approximately 32,200 people in 10 years, based on their growth model.

Velocity is the rate of change of position with respect to time. If we let $x(t)$ denote the position of an object as a function of time and $v(t)$ its velocity, then

$$
v(t) = \frac{d}{dt} x(t)
$$

Consequently, we see that position is an antiderivative of velocity.

Acceleration, denoted by $a(t)$, is the rate of change of velocity with respect to time:

$$
a(t) = \frac{d}{dt} v(t)
$$

Position is an antiderivative of velocity; velocity is an antiderivative of acceleration.

From this we see that velocity is an antiderivative of acceleration.

The laws of physics often provide us with information about the acceleration of an object. For example, we know that the acceleration of a free-falling body, in the absence of any air resistance, is equal to the gravitational constant g, which, near the surface of the earth, is approximately 32 ft/sec² in the English system and 9.8 m/sec² in the metric system. Since velocity is an antiderivative of acceleration and position is an antiderivative of velocity, we can use indefinite integration to find both the velocity and the position of an object when we know its acceleration.

EXAMPLE 5

A steel ball is released from a height of 800 feet with zero velocity. Assuming no air resistance, find an expression for the speed and position of the body at any time t while the ball remains in free fall.

FIGURE 12.10

A Schematic of the Situation When x = 0 is the Point of Impact

FIGURE 12.11

A Schematic of the Situation When x = 0 is the Point of Release

Solution

In the absence of air resistance, we know that the acceleration of the ball is the gravitational constant g. Since the ball is dropped from a height of 800 ft, we retain consistency of units by setting $g = 32$ ft/sec^2.

Measuring position requires a coordinate system, because the position of an object is its position with respect to a reference point. Since the steel ball falls vertically, we take the x-axis to be the vertical axis. Two natural points for establishing a position of $x = 0$ are the point of impact and the point of release.

If we take the point of impact to be $x = 0$, then it is natural to take the upward direction as the positive direction, as shown in Figure 12.10. The ball begins its descent at a height of 800 ft with acceleration $a(t)$ in the downward or negative direction. With respect to this coordinate system, we have $a(t) = -32$ ft/sec^2. The minus sign signifies that the acceleration is in the negative direction.

Alternatively, if we choose the release point to be $x = 0$, then it is natural to take the downward direction as the positive direction, as shown in Figure 12.11. This puts the ground at $+800$ feet. With respect to this coordinate system, we have $a(t) = 32$ ft/sec^2. The (implied) positive sign in front of the 32 signifies that acceleration is in the positive direction with respect to the new coordinate system. We can use either coordinate system, but we will use Figure 12.11 so we can work with positive numbers.

Velocity, $v(t)$, is an antiderivative of acceleration; therefore

$$v(t) = \int a(t)\, dt = \int 32\, dt = 32t + c_1 \tag{9}$$

where c_1 is an arbitrary constant. We will measure the time from the moment the steel ball is dropped. It is released with zero velocity, so $v(0) = 0$. Substituting this information into Equation (9), we find that

$$0 = v(0) = 32(0) + c_1 = c_1$$

whereupon Equation (9) becomes

$$v(t) = 32t \text{ ft/sec} \tag{10}$$

Equation (10) is an expression for the velocity of the steel ball as a function of time while it remains in free fall.

Position, $x(t)$, is an antiderivative of velocity; therefore,

$$x(t) = \int v(t)\, dt = \int 32t\, dt = 16t^2 + c_2 \tag{11}$$

where c_2 is an arbitrary constant. At time $t = 0$ we have $x(0) = 0$; substituting this information into Equation (11), we find that

$$0 = x(0) = 16(0)^2 + c_2 = c_2$$

whereupon Equation (11) becomes

$$x(t) = 16t^2 \text{ ft} \tag{12}$$

Equation (12) is an expression for the position of the steel ball as a function of time while in free fall.

The steel ball reaches the ground when $x(t) = 800$. Substituting this value into Equation (12) and then solving for t, we find that $800 = 16t^2$ and so

$$t = \sqrt{800/16} \approx 7.07 \text{ sec}$$

The steel ball falls for approximately 7.07 seconds. Thus Equations (10) and (12) are both valid for the time interval $0 \le t \le 7.07$.

IMPROVING SKILLS

Evaluate the indefinite integrals in Problems 1 through 30.

1. $\int (3u^2 - 8u + 6)\, du$
2. $\int (\tfrac{1}{2}y^4 + 8y^2 - 20)\, dy$
3. $\int (u^{0.3} + u^{-0.3})\, du$
4. $\int (2p^5 + 3p^{-5} + 3.8)\, dp$
5. $\int \sqrt{z}\, dz$
6. $\int \sqrt{3z}\, dz$
7. $\int \sqrt{3/n}\, dn$
8. $\int (5u - \sqrt{5})\, du$
9. $\int (\sqrt{5}y - \sqrt{5})\, dy$
10. $\int [3 - (2/Q)]\, dQ$
11. $\int [5.1t - (1/3.6)]\, dt$
12. $\int [2Q - (5/Q)]\, dQ$
13. $\int (1/\pi)\, dt$
14. $\int 5.3e^t\, dt$
15. $\int (2e^n - 4)\, dn$
16. $\int (u^2 - 3e^u)\, du$
17. $\int 2e\, dy$
18. $\int (\theta^{-5} + \theta^{-3} + \theta^{-1})\, d\theta$
19. $\int 2e^\theta\, d\theta$
20. $\int dP$
21. $\int 5\cos u\, du$
22. $\int (\tfrac{1}{5}\cos \theta + 109)\, d\theta$
23. $\int \tfrac{503}{75} \sin \theta\, d\theta$
24. $\int (3 - \sin t)\, dt$
25. $\int (2\sin t - 3\cos t)\, dt$
26. $\int (\tfrac{1}{7}\sin u + \tfrac{3}{7}\cos u - 5)\, du$
27. $\int (y - 8\cos y)\, dy$
28. $\int y\cos \pi\, dy$
29. $\int e^{\cos 5}\, dt$
30. $\int t^{\cos 5}\, dt$

FIGURE A

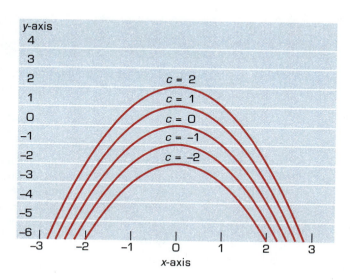

31. Figure A displays the graphs of some antiderivatives $F(x)$ of a function $f(x)$, with each curve corresponding to a different constant c. Determine values of c that satisfy the following constraints:

(a) $F(1) = 0$ (b) $F(1) = 1$ (c) $F(1) = -2$
(d) $F(2) = -3$ (e) $F(0) = -1$ (f) $F(0) = 0$

FIGURE B

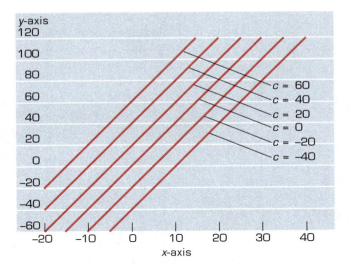

32. Figure *B* displays the graphs of some antiderivatives $R(x)$ of a function $r(x)$, with each curve corresponding to a different constant c. Determine values of c that satisfy the following constraints:

(a) $R(20) = 100$ (b) $R(10) = 100$ (c) $R(0) = 60$
(d) $R(20) = 60$ (e) $R(0) = -20$ (f) $R(10) = 0$

CREATING MODELS

33. An entrepreneurial couple produces handcrafted chess sets in their home. The marginal cost (in dollars/item) of producing these sets is modeled by the function

$$f(n) = 4 - 2n + 6n^2$$

where n represents the total number of chess sets produced each month. Find the total cost of producing five chess sets a month if the couple's monthly fixed cost is $500.

34. A manufacturer of electric can openers has a daily fixed cost of $21,000 and knows that its marginal cost is modeled by the equation $MC = \$5.50/\text{item}$. Find the total cost of producing 3200 can openers in one day.

35. The owner of a hang-glider company believes that their marginal revenue is given by the function

$$f(n) = -0.008n^3 + 0.255n^2 - 2.3n + 5.87$$

where n is the number of gliders (in thousands) offered for sale. What is the total revenue (in millions of dollars) for a production run of 12,000 gliders?

36. A publishing company's yearly marginal revenue is given by the equation $MR(t) = 3 + t$, where t denotes the number of years of operation and total revenue is measured in millions of dollars. What total revenue can the company expect over its first three years in business?

37. A car dealership believes that its monthly marginal revenue is related to its advertising expenditure E by the function $f(E) = 60,000 - 1200E$ where total revenue is measured in dollars and advertising expenditures are measured in thousands of dollars. Without any advertising, the company will generate $150,000 in sales. What total revenues can the company expect if it budgets $45,000 for advertising?

38. The population $P(t)$ of a city is declining at the rate

$$\frac{dP}{dt} = -4(t + 105)$$

where t is measured in years. Estimate the population of this city in 50 years if the current population is 250,000 people.

39. Scientists believe that the growth rate for the population of certain bacteria in a culture is modeled by the equation

$$\frac{dN}{dt} = 0.005e^t \quad (0 \le t \le 12)$$

where time is measured in hours and N is the number of strands of bacteria. Based on this model, estimate the number of strands of bacteria after 10 hr if the culture is initialized with 100 strands at $t = 0$.

40. The rate of change of the quantity Q of antique dealers in a region of the U.S. is directly proportional to time (in years). Find an equation for Q as a function of time if currently $Q = 5000$ dealers, while ten years ago, Q was 4800 dealers.

41. A study commissioned by a daily newspaper estimates that the growth rate of the number of residential subscriptions is $\dfrac{dS}{dt} = 36\sqrt{t}$, where time is measured in months.

How many subscriptions (rounded to the nearest hundred) will the company have in two years if home deliveries currently total 26,400?

42. Solve Example 5 if the ball is dropped with an initial velocity of 16 ft/sec.

43. Solve Example 5 if the ball is thrown with an initial velocity of 16 ft/sec in the upward direction.

44. Solve Example 5 using the coordinate system shown in Figure 12.10.

45. A sandbag is accidentally dropped from a hot-air balloon hovering 1200 ft above the ground. Assume that air resistance is negligible and that the only force on the sandbag is due to gravity. With what velocity will the sandbag hit the ground?

46. A ball is thrown vertically upward with an initial velocity of 64 ft/sec. Assume that air resistance is negligible and that the only force on the ball is due to gravity. How high will the ball go?

47. The ball in Problem 46 is thrown from a platform 20 ft above ground level. How long will it take for the ball to hit the ground?

EXPLORING IN TEAMS

48. A curve $y = f(x)$ passes through the point $(10, 20)$. The tangent line to this curve at the point (x, y) has a slope given by the formula $y = 8x^3 - 6x^2$. Graph the equation of the curve.

49. A curve $y = f(x)$ passes through the point $(1, -3)$. The tangent line to this curve at the point (x, y) has a slope given by the formula $1 + 2e^x$. Graph the equation of the curve.

EXPLORING WITH TECHNOLOGY

50. For each of the indefinite integrals in Problems 1, 5, 10, and 21, use a graphing calculator to graph a set of five antiderivatives on the same coordinate system. Use a different coordinate system for each problem, restricting each domain to the input interval $[1, 5]$. With each graph, *estimate* a value for the arbitrary constant c that yields a curve through the point $(2, 0)$.

REVIEWING MATERIAL

51. (Section 2.5) Find the inverse of the matrix

$$\mathbf{A} = \begin{bmatrix} 1 & 3 & 3 \\ 2 & 5 & 4 \\ 0 & 2 & -1 \end{bmatrix}$$

52. (Section 9.5) Calculate the amplitude and the period of the sinusoidal function $y = 3\cos(4\theta)$.

53. (Section 10.3) Find the derivative of $y = 2x^4 \ln x$.

54. (Section 11.2) A couple wants to leave their dog outdoors when they are at work, so they will build an enclosed, rectangular dog yard, using the side of a storage shed as part of one side of the dog yard and fencing for the rest of the perimeter (see Figure C). Find the dimensions of the dog yard with maximum area if the couple can afford 100 feet of fencing.

FIGURE C

RECOMMENDING ACTION

55. Respond by memo to the following request:

MEMORANDUM

To: J. Doe Reader

From: V.P. for Finance

Date: Today

Subject: **Profit**

I found your presentation last week illuminating. To calculate total cost from marginal cost you used the fact that total cost is fixed cost at a production run of zero. To calculate total revenue from marginal revenue, you used the fact that total revenue is zero at a production run of zero.

My department deals with profit. Often we have a model for marginal profit as a function of the number of units produced and sold, and we need to calculate total profit. What fact can we use to isolate the firm's profit from the other indefinite integrals of marginal profit?

12.3 SUBSTITUTION OF VARIABLES

Although the rules of integration listed in Section 12.1 allow us to integrate a large set of important functions, the rules by themselves are inadequate for evaluating the overwhelming majority of integrals that appear in mathematical models. The natural exponential rule of integration states that

$$\int e^x \, dx = e^x + c$$

Yet this same rule does not apply to the indefinite integrals

$$\int e^{2x} \, dx, \qquad \int e^{3x} \, dx, \qquad \int e^{4x} \, dx$$

or to any other indefinite integral of the natural exponential function having an exponent other than x. Our rules of integration allow us to integrate the polynomial functions

$$\int x \, dx = \tfrac{1}{2}x^2 + c$$

$$\int (x^2 + 1) \, dx = \tfrac{1}{3}x^3 + x + c$$

but they do not apply to the quotient

$$\int \frac{x}{x^2 + 1} \, dx$$

Sometimes a difficult indefinite integral can be rewritten as a combination of simpler indefinite integrals.

Sometimes we can use algebraic manipulations to transform a function with a difficult indefinite integral into a combination of functions with simpler indefinite integrals. Such simplifications should be performed whenever possible.

EXAMPLE 1

Evaluate $\int \sqrt{5x} \, dx$.

Solution

We use one of the properties of radicals to write $\sqrt{5x} = \sqrt{5}\sqrt{x} = \sqrt{5}x^{1/2}$. Then

$$\int \sqrt{5x} \, dx = \int \sqrt{5}x^{1/2} \, dx = \sqrt{5} \int x^{1/2} \, dx = \sqrt{5}\left(\frac{2}{3}x^{3/2} + c_1\right) = \frac{2\sqrt{5}}{3}x^{3/2} + c$$

EXAMPLE 2

Evaluate $\int 4x^3(1 + 3x^4)^2 \, dx$.

Solution

Performing each of the polynomial multiplications, we have

$$(1 + 3x^4)^2 = (1 + 3x^4)(1 + 3x^4) = 1 + 6x^4 + 9x^8$$

and therefore

$$4x^3(1 + 3x^4)^2 = 4x^3(1 + 6x^4 + 9x^8) = 4x^3 + 24x^7 + 36x^{11}$$

Then

$$\int 4x^3(1 + 3x^4)^2 \, dx = \int (4x^3 + 24x^7 + 36x^{11}) \, dx$$

$$= 4 \int x^3 \, dx + 24 \int x^7 \, dx + 36 \int x^{11} \, dx$$

$$= x^4 + 3x^8 + 3x^{12} + c$$

We have listed a few familiar indefinite integrals for convenience. These indefinite integrals have u as the independent variable instead of x, to reinforce the point that the label chosen for the independent variable is optional as long as we use it consistently.

The power rule of integration:	$\int u^n \, du = \dfrac{u^{n+1}}{n+1} + c, \quad n \neq -1$
The integration rule for $f(u) = 1/u$:	$\int u^{-1} \, du = \int \dfrac{1}{u} \, du = \ln\lvert u \rvert + c$
The natural exponential rule of integration:	$\int e^u \, du = e^u + c$
The sine rule of integration:	$\int \sin u \, du = -\cos u + c$
The cosine rule of integration:	$\int \cos u \, du = \sin u + c$

The other rules of integration developed in Section 12.1 are actually *transformation methods* that allow us to rewrite one indefinite integral in terms of others. They are:

The constant multiple rule of integration:

$$\int kf(u) \, du = k \int f(u) \, du \quad (k \text{ a known real number})$$

The sum rule of integration:

$$\int [f(u) + g(u)] \, du = \int f(u) \, du + \int g(u) \, du$$

The difference rule of integration:

$$\int [f(u) - g(u)] \, du = \int f(u) \, du - \int g(u) \, du$$

A powerful transformation method is *substitution of variables* or *u-substitution*. It allows us to rewrite an indefinite integral into another form, ideally a new indefinite integral that we recognize from our rules of integration. We describe the method by breaking it down into a series of steps.

THE METHOD OF SUBSTITUTION OF VARIABLES

STEP 1 Set $u = g(x)$, where $g(x)$ is part of the function to be integrated.

STEP 2 Calculate $du = g'(x)\,dx$. If the quantity $g'(x)\,dx$ is not or cannot easily be made to be the rest of the original indefinite integral, then return to Step 1 and try another substitution; otherwise continue.

STEP 3 Use the substitutions $u = g(x)$ and $du = g'(x)\,dx$ to write the original indefinite integral completely in terms of the new independent variable u. The resulting indefinite integral must not involve x in any way. If this cannot be done, then return to Step 1 and try another substitution; otherwise continue.

STEP 4 If the resulting indefinite integral is simpler to evaluate than the original indefinite integral, do so; if not, return to Step 1 and try another substitution.

STEP 5 After the indefinite integral is evaluated, replace u by $g(x)$ to obtain the final answer in terms of the original independent variable.

Rule of thumb:

If an integral involves a function g(x) raised to a power, then set

$$u = g(x)$$

The choice of which part of a function to set equal to the new variable u is a matter of trial and error. For example, replacing one variable with another (say, x with u, or y with x) does nothing but change the label and does not simplify the integral. Experience, however, teaches us that some substitutions are more likely to be successful than others, and these become the rules of thumb used by most modelers. One rule of thumb is: If an indefinite integral contains a function $g(x)$ raised to a power, set $u = g(x)$.

EXAMPLE 3

Evaluate $\displaystyle\int \frac{2x}{\sqrt{x^2 + 16}}\,dx$.

Solution

This indefinite integral involves the function $x^2 + 16$ raised to the $\frac{1}{2}$ power. Therefore we set $u = g(x) = x^2 + 16$ and calculate: $du = g'(x)\,dx = 2x\,dx$. With these substitutions, the original indefinite integral may be rewritten as

$$\int \frac{2x}{\sqrt{x^2 + 16}}\,dx = \int \frac{1}{(x^2 + 16)^{1/2}}\,2x\,dx$$

$$= \int \frac{1}{u^{1/2}}\,du = \int u^{-1/2}\,du$$

$$= \frac{u^{-1/2+1}}{-\frac{1}{2} + 1} + c \qquad \text{(power rule of integration with } n = -\tfrac{1}{2})$$

$$= 2\sqrt{u} + c$$

Finally, replacing u with $g(x) = x^2 + 16$, we have

$$\int \frac{2x}{\sqrt{x^2 + 16}}\, dx = 2\sqrt{x^2 + 16} + c$$

EXAMPLE 4

Evaluate $\int (3 \sin 5x)(2 + \cos 5x)^{10}\, dx$.

Solution

The quantity $2 + \cos 5x$ is raised to the tenth power, so we set

$$u = g(x) = 2 + \cos 5x$$

It then follows from the chain rule that

$$du = g'(x)\, dx = -5 \sin 5x\, dx$$

The quantity $-5 \sin 5x\, dx$ is not part of the original indefinite integral because there, $\sin 5x$ has a coefficient of 3, not -5. This, however, is easily remedied by multiplying by $1 = (-5)/(-5)$:

$$\int (3 \sin 5x)(2 + \cos 5x)^{10}\, dx = \int \left(\frac{-5}{-5}\right)(3 \sin 5x)(2 + \cos 5x)^{10}\, dx$$

$$= \int \left(\frac{3}{-5}\right)(-5 \sin 5x)(2 + \cos 5x)^{10}\, dx$$

$$= -\tfrac{3}{5} \int (2 + \cos 5x)^{10}(-5 \sin 5x)\, dx$$

Substituting u and du into this last indefinite integral, we simplify it to

$$-\tfrac{3}{5} \int u^{10}\, du = -\tfrac{3}{55} u^{11} + c$$

using the power rule of integration with $n = 10$. Now replacing u with $g(x) = 2 + \cos 5x$, we have

$$\int 3 \sin 5x(2 + \cos 5x)^{10}\, dx = -\tfrac{3}{55}(2 + \cos 5x)^{11} + c$$

Step 2 stipulates that the choice of $u = g(x)$ is useful only if the differential $du = g'(x)\, dx$ appears intact in the original indefinite integral or can be easily modified to do so. Often, as in Example 4, the quantity $g'(x)dx$ differs by just a multiplicative constant. Then multiplication by a suitable form of 1 converts the resulting indefinite integral into exactly the one we need. In general, for any constants a and b, $b \neq 0$, we can write

If a and b are constants, $b \neq 0$, then $\int af(x)\, dx = \dfrac{a}{b} \int bf(x)\, dx$

$$\int af(x)\, dx = \int a\left(\frac{b}{b}\right)f(x)\, dx$$

$$= \int \left(\frac{a}{b}\right)bf(x)\, dx = \frac{a}{b} \int bf(x)\, dx$$

thereby converting an indefinite integral with a multiplicative constant a into one with a multiplicative constant b.

EXAMPLE 5

A company manufacturers magnifying glasses for stamp collectors and other hobbyists, with fixed costs of $5000/week. The weekly marginal cost function for the production of x magnifying glasses is

$$f(x) = 4 - \frac{3.5}{1 + 0.001x}$$

where $f(x)$ is in units of dollars/item. What is the total cost for producing 2500 magnifying glasses a week?

Solution

This problem is similar to Example 2 of Section 12.2 but with a different marginal cost function. Total cost is an antiderivative of marginal cost, so

$$TC(x) = \int f(x)\, dx = \int \left(4 - \frac{3.5}{1 + 0.001x}\right) dx = \int 4\, dx - \int \frac{3.5}{1 + 0.001x}\, dx \quad (13)$$

The first term in Equation (13) is $\int 4\, dx = 4\int 1\, dx = 4x + c_1$, but the second term does not conform to any of our rules, so we will use the method of substitution of variables. We set $u = g(x) = 1 + 0.001x$; then

$$du = g'(x)\, dx = 0.001\, dx$$

The quantity $0.001\, dx$ is not part of the original indefinite integral, but this is easy to fix:

$$\int \frac{3.5}{1 + 0.001x}\, dx = \int \left(\frac{0.001}{0.001}\right) \frac{3.5}{1 + 0.001x}\, dx$$

$$= \frac{3.5}{0.001} \int \frac{0.001}{1 + 0.001x}\, dx$$

$$= 3500 \int \frac{0.001}{1 + 0.001x}\, dx$$

$$= 3500 \int \frac{du}{u} = 3500 \ln|u| + c_2$$

$$= 3500 \ln|1 + 0.001x| + c_2$$

Now we can rewrite Equation (13) as

$$TC(x) = \int 4\, dx - \int \frac{3.5}{1 + 0.001x}\, dx = 4x - 3500 \ln|1 + 0.001x| + c$$

Since $x \ge 0$ so too is $1 + 0.001x$, and we can drop the absolute value bars. When $x = 0$, total cost equals the fixed cost of $5000. Consequently,

$$5000 = 4(0) - 3500 \ln[1 + 0.001(0)] + c$$

from which we conclude that $c = 5000$, and the total cost is

$$TC(x) = 4x - 3500 \ln(1 + 0.001x) + 5000 \quad (14)$$

We want to know the total cost for a production run of 2500 magnifying glasses. Setting $x = 2500$ in Equation (14), we find that

$$TC\big|_{x=2500} = 4(2500) - 3500 \ln[1 + 0.001(2500)] + 5000 \approx \$10,615.33$$

Rule of thumb:

If an integral involves a function with an argument, then set u equal to that argument.

In Examples 3 and 4, each of the indefinite integrals involved a quantity raised to a power, so we found it convenient to set u equal to that quantity when applying the method of substitution of variables. Another useful rule of thumb is: If an indefinite integral contains a function with an argument, then set u equal to that argument.

EXAMPLE 6

Evaluate $\int \sin(2x + 3)\, dx$.

Solution

In this indefinite integral, the argument of the sine function is $2x + 3$, so we set $u = g(x) = 2x + 3$, whereupon $du = g'(x)\, dx = 2\, dx$. The quantity $2\, dx$ is not part of the original indefinite integral, but this is easy to remedy:

$$\int \sin(2x + 3)\, dx = \int \left(\frac{2}{2}\right) \sin(2x + 3)\, dx$$

$$= \frac{1}{2} \int [\sin(2x + 3)]2\, dx$$

$$= \frac{1}{2} \int \sin u\, du = -\frac{1}{2}\cos u + c$$

using the sine rule of integration. Finally, replacing u with $g(x) = 2x + 3$, we have

$$\int \sin(2x + 3)\, dx = -\frac{1}{2}\cos(2x + 3) + c$$

The exponential function e^x is Euler's number e raised to the power specified by the variable x. Some books and all electronic spreadsheets write the exponential function e^x as $\exp(x)$, where the variable x is the argument of the natural exponential function. In a similar vein, we can think of the composite function $e^{g(x)}$ as the number e raised to the power specified by the output of $g(x)$, or we can write the function as $\exp(g(x))$ and think of $g(x)$ as the argument of the natural exponential function. This latter approach is taken whenever we use the method of substitution of variables to evaluate indefinite integrals of exponential functions.

EXAMPLE 7

Evaluate $\int e^{2x}\, dx$.

Solution

We set $u = g(x) = 2x$ as the argument of the exponential function and then use the chain rule to calculate $du = g'(x)\, dx = 2\, dx$. Then, using the natural exponential rule of integration, we get

$$\int e^{2x}\, dx = \int \left(\frac{2}{2}\right) e^{2x}\, dx = \frac{1}{2} \int (e^{2x})2\, dx$$

$$= \frac{1}{2} \int e^u\, du = \frac{1}{2} e^u + c$$

Now we replace u with $g(x) = 2x$ and we have

$$\int e^{2x}\, dx = \tfrac{1}{2}e^{2x} + c$$

The method of substitution of variables is a transformation method that replaces one indefinite integral by another, ideally a new indefinite integral that can be readily evaluated. In general, however, we retain a transformation as long as the new indefinite integral appears to be easier to evaluate than the original. If so, we subject the new indefinite integral to additional transformations.

We presented the method of substitution of variables in terms of variables u and x, but as always, we use other labels when appropriate.

EXAMPLE 8

Evaluate $\int \dfrac{\ln(5t^2)}{t}\, dt$.

Solution

The argument of this natural logarithm function is $5t^2$, so we set $u = g(t) = 5t^2$ and calculate $du = g'(t)\, dt = 10t\, dx$.

This expression for du is not part of the given indefinite integral so either we discard the proposed substitution or we see if we can algebraically insert du into the original indefinite integral. Although the multiplicative constant 10 can be embedded into the indefinite integral, like we did before, we cannot do that with the t-term: the constant multiple rule of integration deals only with constants, not with the independent variables that are being integrated. Instead, we rewrite the indefinite integral as follows:

$$\int \frac{\ln(5t^2)}{t}\, dt = \int \left(\frac{10t}{10t}\right)\frac{\ln(5t^2)}{t}\, dt$$

$$= \int \frac{1}{10}\frac{\ln(5t^2)}{t^2}\, 10t\, dt$$

$$= \frac{1}{2}\int \frac{\ln(5t^2)}{5t^2}\, 10t\, dt = \frac{1}{2}\int \frac{\ln u}{u}\, du$$

> *It is useless simply to set one variable equal to another when applying the method of substitution of variables.*

This last indefinite integral is certainly simpler in form than the original one, so we keep it. Now we have to figure out how to evaluate it.

We try another application of the method of substitution of variables, this time with a v-substitution since the label u has already been used. Neither rule of thumb applies to our latest indefinite integral: there are no powers, and the argument of the natural logarithm is simply u. The substitution $v = u$ is *never* recommended because it makes nothing simpler: the resulting indefinite integral will appear identical to the original indefinite integral except the label u will be replaced by the label v. It makes more sense to set v equal to a portion of the function we need to integrate. A simple choice is $v = \ln u$, giving us

$$dv = \left(\frac{d}{du}\ln u\right) du = \frac{1}{u}\, du$$

With these substitutions, we now have

$$\frac{1}{2}\int \frac{\ln u}{u}\, du = \frac{1}{2}\int \ln u\, \frac{du}{u} = \frac{1}{2}\int v\, dv$$

$$= \tfrac{1}{4}v^2 + c = \tfrac{1}{4}(\ln u)^2 + c$$

Thus

$$\int \frac{\ln(5t^2)}{t}\, dt = \frac{1}{2}\int \frac{\ln u}{u}\, du = \frac{1}{4}\ln^2(u) + c = \frac{1}{4}\ln^2(5t^2) + c$$

Occasionally an indefinite integral can be evaluated more than one way. For example, a second approach to solving Example 8 is suggested in Problem 40. If we are fortunate enough to recognize more than one approach, then we opt for the one that appears easiest.

IMPROVING SKILLS

Evaluate the indefinite integrals in Problems 1 through 39.

1. $\int 4x^3(1 + x^4)^6 \, dx$

2. $\int x^3(1 + x^4)^{0.4} \, dx$

3. $\int x^2(6 + 6x^3)^{30} \, dx$

4. $\int 8x\sqrt{4 - x^2} \, dx$

5. $\int (t + 1)\sqrt{t^2 + 2t} \, dt$

6. $\int \dfrac{t}{(t^2 - 3)^5} \, dt$

7. $\int \dfrac{x}{x^2 + 1} \, dx$

8. $\int \dfrac{x + 1}{x^2 + 2x - 1} \, dx$

9. $\int \dfrac{x}{(x^2 + 1)^2} \, dx$

10. $\int \dfrac{x + 1}{(x^2 + 2x - 1)^2} \, dx$

11. $\int \dfrac{3t}{\sqrt{2t^2 + 1}} \, dt$

12. $\int \dfrac{u^2}{\sqrt{u^3 - 6}} \, du$

13. $\int (y^2 + 2y)(y^3 + 3y^2) \, dy$

14. $\int \dfrac{y^2 + 2y}{y^3 + 3y^2} \, dy$

15. $\int \sin(5x) \, dx$

16. $\int 3 \sin(12\theta) \, d\theta$

17. $\int (\sin 5)x \, dx$

18. $\int \sin(\theta/2) \, d\theta$

19. $\int 4 \cos(0.1x) \, dx$

20. $\int \tfrac{1}{12} \cos(50u) \, du$

21. $\int \dfrac{\cos x}{\sin x} \, dx$

22. $\int \dfrac{\cos x}{\sin^2 x} \, dx$

23. $\int \dfrac{\sin(2t)}{\cos(2t)} \, dt$

24. $\int \dfrac{\sin(2t)}{\cos^4(2t)} \, dt$

25. $\int e^{3x} \, dx$

26. $\int e^{4x} \, dx$

27. $\int e^{-0.01t} \, dt$

28. $\int 2e^{0.3z} \, dz$

29. $\int 2xe^{x^2} \, dx$

30. $\int x^2 e^{2x^3} \, dx$

31. $\int (1 + y)e^{(y^2 + 2y)} \, dy$

32. $\int ve^{-2v^2} \, dv$

33. $\int (\sin \theta)e^{\cos \theta} \, d\theta$

34. $\int e^t e^{e^t} \, dt$

35. $\int 3e^{3x}(10 + e^{3x})^{2.1} \, dx$

36. $\int 3e^{3x}(10 + e^{4x}) \, dx$

37. $\int (3q^2 + 3e^q)(6 + q^3 + 3e^q) \, dq$

38. $\int (2t + 4e^{4t})(t^2 + e^{4t})^9 \, dt$

39. $\int \dfrac{e^\theta}{1 + e^\theta} \, d\theta$

40. Evaluate the indefinite integral in Example 8 by first observing that

$$\ln(5t^2) = \ln 5 + \ln t^2 = \ln 5 + 2 \ln t$$

CREATING MODELS

41. An entrepreneurial couple produces handcrafted chess sets in their home. The marginal cost (in dollars) to produce these sets is modeled by the function $f(n) = e^{0.9n}$, where *n* is the total number of chess sets produced each month. Find the total cost of producing five sets a month if the couple's monthly fixed cost is $500. (See Problem 33, Section 12.2.)

42. The marginal cost (in dollars) of producing shoes is modeled by the function $MC(s) = 5 + e^{0.01s}$, where *s* is the total number of pairs produced daily. What is the total cost of manufacturing 100 pairs of shoes each day if the daily fixed cost is $1000?

43. The Steelex Spring Company determines that the monthly marginal revenue for each case of springs is a function of the number *x* of cases sold as given by the formula $MR(x) = 40x/(2x^2 + 1)$. Determine the total monthly revenue for selling 1000 cases of springs.

44. The owner of a hang-glider company believes that its marginal revenue is given by the function $f(x) = 2 + 10/(1 + x)$, where *x* is the number of gliders (in thousands) offered for sale and revenue is measured in millions of dollars. What is the total revenue for a production run of 12,000 gliders? (See Problem 35, Section 12.2.)

45. A publishing company's yearly marginal revenue is given by the equation $MR(t) = 70e^{0.15t}$, where *t* denotes the number of years of operation and revenue is measured in thousands of dollars. Determine the total revenue earned by the company in its first 10 years in business. (See Problem 36, Section 12.2.)

46. A car dealership believes that its monthly marginal revenue is related to advertising expenditure *E* by the function $f(E) = 6000/(1 + 0.05E)$, where revenue is measured in dollars and advertising expenditures are measured in thousands of dollars. Without any advertising, the company will generate $150,000 in sales. What total revenue can the company expect if it budgets $45,000 for advertising? (See Problem 37, Section 12.2.)

47. The population $P(t)$ of a city is growing at the rate of

$$\frac{dP}{dt} = \frac{5000t}{1 + 2t^2}$$

where *t* is measured in years. Estimate the population of this city in 50 years if the current population is 250,000. (See Problem 38, Section 12.2.)

48. Scientists believe that the growth rate for a culture of a certain bacteria is modeled by the equation

$$\frac{dN}{dt} = 5e^{0.2t} \quad (0 \le t \le 12)$$

where time is measured in hours and *N* in the number of strands of bacteria. Based on this model, estimate the number of strands of bacteria after 10 hours if the culture is initialized with 100 strands at $t = 0$. (See Problem 39, Section 12.2.)

49. In calm waters, a grounded tanker with a ruptured hull produces a circular oil slick. The radius *r* of the slick is increasing at the rate of $dr/dt = 50(2t + 16)^{3/4}$ ft/min, where *t* is the number of minutes from the time the spill began. What is the surface area of the oil slick 30 minutes after the hull ruptured?

50. A study commissioned for a daily newspaper predicts that in *t* months, the growth rate of the number *S* of home delivery subscriptions will be $dS/dt = 34/(1 + t)^{0.7}$, where time is measured from the present time. How many subscriptions (rounded to the nearest hundred) will the company have in two years if home deliveries currently total 26,400? (See Problem 41, Section 12.2.)

51. A cup of 190°F coffee is poured and placed on a counter in a room that is at a constant temperature of 70°F. The rate of decrease of the temperature T of the coffee as it sits on the counter is given by the formula $dT/dt = -84e^{-0.7t}$, where t is measured in minutes from the time the coffee is poured. What is the temperature of the coffee after three minutes on the counter?

52. A glass of iced tea, chilled to a uniform temperature of 50°F, is placed outside on a hot (100°F) summer day. The rate of increase of the temperature T of the iced tea as it sits outdoors is given by the formula $dT/dt = 50e^{-0.8t}$, where t is measured in minutes from the time the tea is brought outdoors. What is the temperature of the tea after 5 minutes outdoors?

EXPLORING IN TEAMS

Evaluate the indefinite integrals in Problems 53 through 60.

53. $\int \dfrac{x}{x+1}\, dx$

54. $\int \dfrac{2x-1}{x+1}\, dx$

55. $\int \dfrac{1}{\sqrt{x}+1}\, dx$

56. $\int \dfrac{x}{\sqrt{x}+1}\, dx$

57. $\int \dfrac{\sqrt{x}-1}{\sqrt{x}+1}\, dx$

58. $\int \dfrac{e^{4x}}{1+e^{2x}}\, dx$

59. $\int [1/(x \ln x)]\, dx$

60. $\int x \sin(2x^2)\cos(2x^2)\, dx$

REVIEWING MATERIAL

61. (Section 1.5) Find the equation of the straight line that passes through the two points $(-10, 1000)$ and $(10, 4500)$.

62. (Section 5.3) A list of data has a mean of 100 and a standard deviation of 15. Use Chebyshev's theorem to determine the fraction of data that lie in the intervals: (a) [80, 120]; (b) [70, 130]; (c) less than 80 or more than 120.

63. (Sections 10.3 and 10.4) Find the derivative of the function

$$y = \frac{2x^4 \ln x}{3e^{-2x} - 5}$$

64. (Section 11.3) An orchard currently has 30 apple trees planted on each acre, with an average yield of 500 lb of apples per tree. New owners believe they can increase the total yield per acre by planting more trees, even though each additional tree planted per acre will reduce the yield per tree on that acre by 10 lb. How many additional trees should be planted on each acre to maximize total yield?

RECOMMENDING ACTION

65. Respond by memo to the following request:

MEMORANDUM

To: J. Doe Reader

From: Tony

Date: Today

Subject: **Help!!!**

I have to make an oral presentation on integration to my professor, and I have a problem with the indefinite integral $\int(x + 1)\, dx$. If I evaluate this indefinite integral directly, I get

$$\int(x + 1)\, dx = \int x\, dx + \int 1\, dx = \tfrac{1}{2}x^2 + x + c$$

whereas if I use the method of substitution of variables, with $u = x + 1$ and $du = dx$, I get

$$\int(x + 1)\, dx = \int u\, du = \tfrac{1}{2}u^2 + c = \tfrac{1}{2}(x + 1)^2 + c$$

How can I get two different functions from the same integral?

12.4   THE DEFINITE INTEGRAL

A region is bounded if it can be enclosed within a rectangle called a frame.

1 square mile is equal to 640 acres.

Determining area is a problem that has challenged decision-makers for thousands of years. Land is rented, leased, or sold by area, so it is imperative to both parties in such a transaction that they agree on the amount of area prior to establishing a price. Manufacturers design crates to ship goods and the cost of a container often depends on its surface area. People in the trades must make accurate estimates of areas before they set prices for painting a house, resurfacing a driveway, or fertilizing a lawn.

We have formulas for the areas of circles, triangles, rectangles, and other familiar geometrical shapes, but we need to address the more difficult problem of calculating the area of an irregularly shaped, bounded region, such as the shaded region in Figure 12.12. A region is *bounded* if it can be enclosed within a rectangle called a *frame*, which provides an upper bound for the area of the region. The window in Figure 12.12 is a frame for the Almond Hill Golf Course. This frame is 1/4 mile long in the vertical direction, 1/2 mile long in the horizontal direction, and the area is $A = \tfrac{1}{4}(\tfrac{1}{2}) = \tfrac{1}{8} = 0.125$ sq. miles. A square mile is equal to 640 acres, so the area of the golf course satisfies the inequality

$$0 \text{ acres} \le A \le 80 \text{ acres} \tag{15}$$

The frame in Figure 12.12 includes a coordinate system with the origin at the lower left corner. This is the preferred position for the origin because it places the region of interest in the first quadrant, where all coordinates are positive. Any rectangle larger than the one in Figure 12.12 can also be a frame for the golf course, as can some smaller rectangles. The one in Figure 12.12 was selected because it is reasonably "tight," with simple numbers for the lengths of the sides.

Inequality (15) is a rough estimate for the area and we can do better. First, though, we

FIGURE 12.12
The Almond Hill Golf Course Bounded by a Rectangle

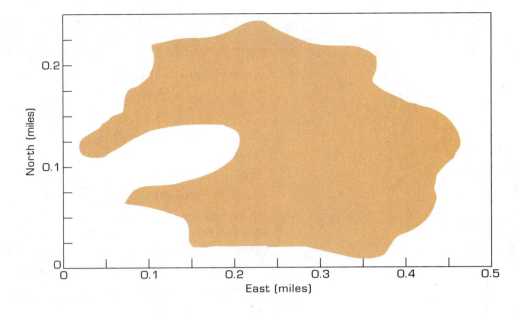

In most commercial transactions, area calculations are accepted within a specified interval of precision.

need to establish how precise an answer we seek. Do we require the exact area, or the area to the nearest acre, or to the nearest hundredth of an acre? In most commercial transactions, precision is allowed to be within a specified tolerance.

Inequality (15) establishes the area of the Almond Hill Golf Course at no more than 80 acres, which is probably not a good enough estimate for most purposes. We can improve the estimate by superimposing a grid over the region. The vertical length of the frame in Figure 12.12 is 0.25 miles, so if we divide this length by 5, there are tick marks every 0.05 miles. The horizontal length of the frame is 0.5 miles, so if we divide this length by 10, there are again tick marks every 0.05 miles. These subdivisions are shown in Figure 12.12 and the grid is created in Figure 12.13 by connecting the tick marks with horizontal and vertical lines.

FIGURE 12.13
The Almond Hill Golf Course Overlaid with a Grid of 0.05 × 0.05 Blocks

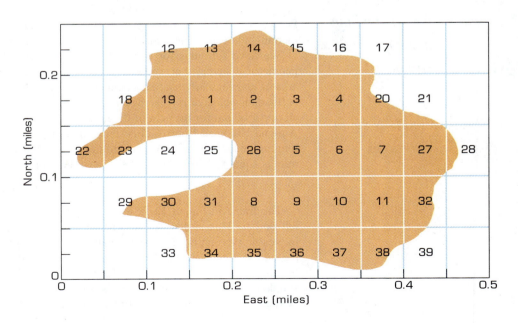

Each square in the grid has sides 0.05 miles long and area $(0.05)^2 = 0.0025$ sq. miles. We obtain an estimate for a *lower* bound of the area of the golf course by adding the areas of all of the squares that are completely inside the golf course, and we obtain an estimate for an *upper* bound of the area by adding the areas of all of the squares that overlap the golf course.

Squares 1 through 11 in Figure 12.13 are all completely embedded with the region of the golf course. Each of the 11 squares has an area of 0.0025 sq. miles, so the sum of their areas is $11(0.0025) = 0.0275$ sq. miles, or $0.0275(640) = 17.6$ acres. The golf course is at least this big. Squares 12 through 39 in Figure 12.13 are partially within the region defined by the golf course. If we combine these squares with the first 11, we have 39 squares that completely cover the golf course. The total area of these 39 squares is $39(0.0025) = 0.0975$ square miles, or $0.0975(640) = 62.4$ acres; this is an upper bound for the area of the Almond Hill Golf Course. Now we can say that the area of the Almond Hill Golf Course satisfies

$$17.6 \text{ acres} \leq A \leq 62.4 \text{ acres} \tag{16}$$

Inequality (16) is a more accurate estimate of the area of the golf course than Inequality (15), although neither is very precise. We can improve our latest estimate by halving the subdivisions on the vertical axis, which produces the finer grid shown in Figure 12.14. Now there is 0.025 miles between successive tick marks on the vertical axis, so the resulting grid is made up of rectangular boxes, each with area $(0.025)(0.05) = 0.00125$ sq. miles.

FIGURE 12.14

The Almond Hill Golf Course Overlaid with a Grid of 0.025 × 0.05 Blocks

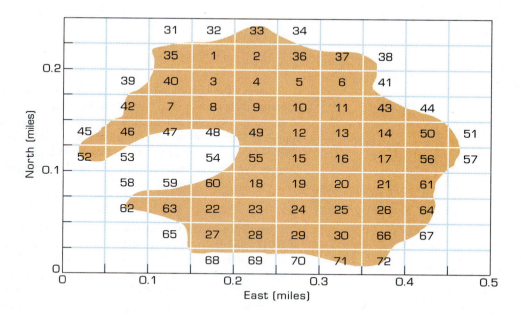

Squares 1 through 30 in Figure 12.14 are completely embedded within the region of the golf course, and their areas add to $30(0.00125) = 0.0375$ sq. miles or $(0.0375)(640) = 24$ acres. The golf course is at least this big. Squares 1 through 72 completely cover the golf course and their combined area is $72(0.00125) = 0.09$ sq. miles or

$(0.09)(640) = 57.6$ acres; this is an upper bound for the area of the Almond Hill Golf Course. We can refine our estimate of the area:

$$24 \text{ acres} \leq A \leq 57.6 \text{ acres} \tag{17}$$

We improve this estimate by using a still finer grid. The one in Figure 12.15 has 0.025 miles between tick marks on both axes. The resulting grid is a set of square boxes with area $(0.025)^2 = 0.000625$ sq. miles, which is half the area of the boxes in Figure 12.14.

FIGURE 12.15
The Almond Hill Golf Course Overlaid with a Grid of 0.025 × 0.025 Blocks

Squares 1 through 72 in Figure 12.15 lie completely inside the region of the golf course. Together their areas add to $72(0.000625) = 0.045$ sq. miles or $(0.045)(640) = 28.8$ acres, so the golf course is at least this big. Squares 1 through 135 completely cover the golf course and the combined area of these squares is $135(0.000625) = 0.084375$ sq. miles or $(0.084375)(640) = 54$ acres. This is an upper bound for the area of the Almond Hill Golf Course. We have a more refined estimate of the area of the Almond Hill Golf Course:

$$28.8 \text{ acres} \leq A \leq 54 \text{ acres} \tag{18}$$

We will now construct one last grid, yet a finer one. Figure 12.16 is a grid with spacings of 0.0125 miles on the vertical axis and 0.025 miles on the horizontal axis. This grid is made up of rectangular boxes, each with an area of $(0.0125)(0.025) = 0.0003125$ sq. miles. Squares 1 through 157 are completely within the golf course and cover a total area of $157(0.0003125) = 0.0490625$ sq. miles or $0.0490625(640) \approx 31.4$ acres, which is a lower bound for the area of the golf course. Squares 1 through 236 completely cover the golf course. Their combined area $236(0.0003125) = 0.07375$ sq. miles or $0.07375(640) = 47.2$ acres is an upper bound for the area of the golf course. Now we have the tightest estimate so far, namely,

$$31.4 \text{ acres} \leq A \leq 47.2 \text{ acres} \tag{19}$$

FIGURE 12.16
The Almond Hill Golf Course Overlaid with a Grid of 0.0125 × 0.025 Blocks

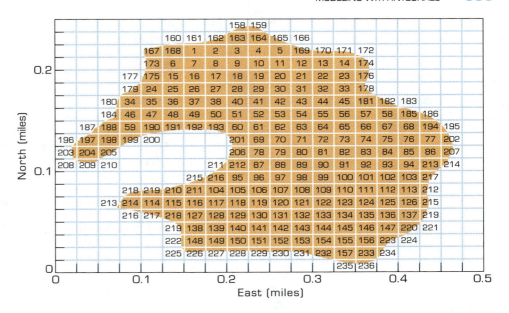

The procedure we have just used to calculate Inequalities (15) through (19) can be formalized by the following algorithm.

CALCULATING THE AREA OF AN IRREGULARLY SHAPED BOUNDED REGION

STEP 1 Enclose the region within a rectangle (the "frame").

STEP 2 On two adjoining sides of the frame, make tick marks that divide each side into equally spaced subintervals (which may differ for the two sides). Connect the tick marks with horizontal and vertical lines, creating a grid of identical rectangular blocks.

STEP 3 Count the number of rectangular blocks that are completely enclosed within the region and multiply this number by the area of each block to obtain a lower bound on the area of the region.

STEP 4 Count the number of rectangular blocks that overlap the region and multiply this number by the area of each block to obtain an upper bound on the area of the region.

STEP 5 If the bounds obtained from Steps 3 and 4 are sufficiently tight, stop; if not, continue.

STEP 6 Create a finer grid by dividing one or both sides of the frame into smaller, equally spaced subintervals and drawing lines through the resulting tick marks.

STEP 7 Return to Step 3.

Area estimates are improved by using finer grids at the cost of more work.

By following this procedure with the Almond Hill Golf Course, we sequentially generated Inequalities (15) through (19); we summarize that example in Table 12.1. Each inequality determines an interval that contains A, the area of the golf course. The midpoint of each interval is itself an estimate of the area. Because it is halfway from either endpoint, the maximum possible error for this midpoint is one-half the length of the interval. Using this method, our first estimate of the area is 40 acres, which has a maximum possible error of 40 acres. Our fifth estimate of the area is 39.3 acres, which has a maximum possible error of 7.9 acres. We can improve this last estimate to within any precision we want by using even finer grids. The more subdivisions we use, the better our estimate but the more work required to obtain that estimate.

TABLE 12.1   Area Estimates for the Almond Hill Golf Course

SUBDIVISION (miles)	INTERVAL ESTIMATE (acres)	MIDPOINT ESTIMATE (acres)	MAXIMUM ERROR (acres)
Frame	[0, 80]	40	40
0.05 × 0.05 blocks	[17.6, 62.4]	40	22.4
0.025 × 0.05 blocks	[24, 57.6]	40.8	16.8
0.025 × 0.025 blocks	[28.8, 54]	41.4	12.6
0.0125 × 0.025 blocks	[31.4, 47.2]	39.3	7.9

The algorithm we have just developed for calculating area is applicable to any bounded region that is defined by the graph of a continuous function.

EXAMPLE 1

Use the grid displayed in Figure 12.17 to estimate the area of the region that is below the graph of the equation $y = x^2 + 5$, above the x-axis, and between $x = 0$ and $x = 6$. (Both variables are in units of inches.)

FIGURE 12.17

A Region Overlaid with a Grid of 5 × 0.5 Blocks

FIGURE 12.18

Estimating the Area by Counting Blocks

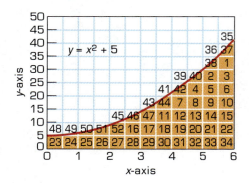

Solution

Each rectangular block in the grid is 5 in. high and 0.5 in. wide, with an area of $5(0.5) = 2.5$ sq. in. The region of interest is shaded in Figure 12.18. Blocks 1 through 34, which are completely embedded in the shaded region, have a total area of $34(2.5) = 85$ sq. in. This is a lower bound on the area of the region. Blocks 1 through 52 completely cover the region. Their combined area is $52(2.5) = 130$ sq. in.; this is an upper bound on the area of the shaded region. It follows that A, the area, is in the interval $[85, 130]$. The midpoint of this interval is 107.5 sq. in. If we take this as our estimate for the area, then the maximum possible error is 22.5 sq. inches, one-half of the length of the interval.

We will see in Section 12.5 that the true area of the shaded region is exactly 102 sq. in., so our current estimate of 107.5 sq. in. is much closer than 22.5 sq. in. to the actual area. To obtain an even better estimate, we would need a finer grid.

The shaded region in Figure 12.18 is bounded on three sides by straight line segments, which gives it more structure than the Almond Hill Golf Course. The additional structure allows us to obtain better area estimates with significantly less work.

To see how this works in general, suppose we have a function $y = f(x)$ with only nonnegative outputs, so that its graph never falls below the x-axis. We seek a method for calculating the area under the curve, above the x-axis, and between two values of x, say, $x = a$ and $x = b$, with $b > a$. Our approach is to subdivide the interval $[a, b]$ on the x-axis with equally spaced tick marks and to draw a vertical line through each tick mark. These vertical lines, including the lines $x = a$ and $x = b$, will be the sides of our approximating rectangles. The base of each rectangle is a line segment between two successive tick marks on the x-axis. If we take the top of each rectangle to be the *minimum* value of the function over the base, then the area of the rectangle is a lower bound for the area under that portion of the curve. If, on the other hand, we take the top of each rectangle to be the *maximum* value of the function over the base, we generate an upper bound for the area under the curve.

As an example, let us consider again the problem in Example 1, which asks us to estimate the area of the region that is below the graph of $y = x^2 + 5$, above the x-axis, and between $x = 0$ and $x = 6$, where both x and y are in inches. We begin by subdividing the interval $[0, 6]$ on the horizontal axis into one-inch segments and then drawing vertical lines through the tick marks. This is shown in Figure 12.19.

FIGURE 12.19

A Vertical Partition of the Area under the Graph of $y = x^2 + 5$

The function $f(x) = x^2 + 5$ is strictly increasing on the interval $[0, 6]$, so the lowest point on its graph on any interval always occurs at the left endpoint. To calculate a lower bound for the area under the curve, we take the height of each rectangle to be the value

FIGURE 12.20

A Lower Bound for the Area under the Graph of $y = x^2 + 5$

of the function at the left endpoint of the subdivision. These rectangles are shaded in Figure 12.20. The heights of the six rectangles are

$$f(0) = (0)^2 + 5 = 5$$
$$f(1) = (1)^2 + 5 = 6$$
$$f(2) = (2)^2 + 5 = 9$$
$$f(3) = (3)^2 + 5 = 14$$
$$f(4) = (4)^2 + 5 = 21$$
$$f(5) = (5)^2 + 5 = 30$$

and, since the base of each rectangle is 1 inch, the sum of their areas is

$$5(1) + 6(1) + 9(1) + 14(1) + 21(1) + 30(1) = 85 \text{ sq. in.}$$

Thus the area of interest is at least 85 sq. in.

The maximum value of the function $f(x) = x^2 + 5$ on any interval always occurs at the right endpoint. To calculate an upper bound for the area under the curve, we take the height of each rectangle to be the value of the function at the right endpoint of the sub-division. These rectangles are shaded in Figure 12.21. Again, the base of each rectangle is 1 inch, so the sum of the areas of the shaded rectangles is

$$1 \cdot f(1) + 1 \cdot f(2) + 1 \cdot f(3) + 1 \cdot f(4) + 1 \cdot f(5) + 1 \cdot f(6)$$
$$= 6 + 9 + 14 + 21 + 30 + 41 = 121 \text{ sq. in.}$$

Therefore the area of interest is no more than 121 sq. in.

FIGURE 12.21

An Upper Bound for the Area under the Graph of $f(x) = x^2 + 5$

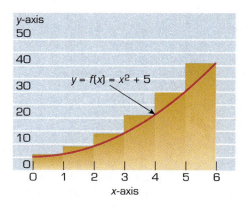

The area under the curve in Figure 12.19 is bounded by the inequalities

$$85 \text{ sq. in.} \le A \le 121 \text{ sq. in.}$$

This estimate is slightly better than the one obtained in Example 1 and it took significantly less work. To get yet a better estimate, we divide the interval [0, 6] into finer subdivisions. First, however, we can simplify this procedure for calculating area.

Often the most tedious part of calculating the areas of the approximating rectangles involves drawing the graph of the function and the rectangles themselves. It requires locating the points where the function attains its minimum and maximum values on each subdivision. This was easy for the function $f(x) = x^2 + 5$ because it is strictly increasing on the interval [0, 6], so the minimum occurs at the left endpoint of any subdivision and the maximum occurs at the right. Generally the work of identifying the global optimum on each partition can be extensive.

To avoid the graphing and locating of optima on numerous subdivisions, we can take the heights of the rectangles to be the values (outputs) of the function at the midpoint of each subdivision. This is illustrated by using the subdivisions in Figure 12.19 to construct the rectangles shown in Figure 12.22.

FIGURE 12.22

Estimating Area Using the Midpoint Method

The base of the leftmost rectangle is the interval [0, 1], with midpoint 0.5. We then can take the height of the first rectangle to be

$$f(0.5) = (0.5)^2 + 5 = 5.25$$

The base of the next rectangle is the interval [1, 2], with midpoint 1.5. We take the height of this second rectangle to be

$$f(1.5) = (1.5)^2 + 5 = 7.25$$

Continuing in this manner, we generate each of the rectangles shown in Figure 12.22, which have total area

$$
\begin{aligned}
1 \cdot f(0.5) + 1 \cdot f(1.5) &+ 1 \cdot f(2.5) + 1 \cdot f(3.5) + 1 \cdot f(4.5) + 1 \cdot f(5.5) \\
&= 5.25 + 7.25 + 11.25 + 17.25 + 25.25 + 35.25 \quad (20) \\
&= 101.5 \text{ sq. in.}
\end{aligned}
$$

We do not have to draw Figure 12.22 to generate the sum in Equation (20); all we need is the width of each rectangle and the location of the midpoint of its base. These are displayed compactly in Figure 12.23, where m_1 through m_6 denote the midpoints of the bases.

FIGURE 12.23
**Six Partitions of [0, 6],
Showing the Midpoint of
Each Base**

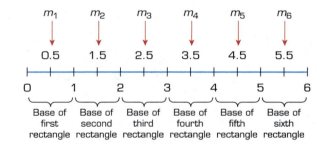

We formalize this method with an algorithm for estimating the area between the curve of a function and the x-axis, over the interval [a, b]. It is called the *midpoint method*. It is applicable only to graphs that do not lie below the x-axis.

THE MIDPOINT METHOD FOR ESTIMATING THE AREA BETWEEN THE GRAPH OF A NONNEGATIVE FUNCTION $f(x)$ AND THE x-AXIS, OVER [a, b]

STEP 1 Determine whether $f(x) \geq 0$ for all x in [a, b]. If not, stop—this algorithm does not apply; otherwise continue.

STEP 2 Select a value for the positive integer n and divide the interval [a, b] into n equal parts, called partitions. The width of each partition is

$$\Delta x = (b - a)/n$$

STEP 3 Locate the midpoint m_i of each partition using the formula

$$m_i = a + (i - \tfrac{1}{2}) \Delta x \quad (i = 1, 2, 3, \cdots, n)$$

STEP 4 Estimate the area by the sum

$$A \approx [f(m_1) + f(m_2) + f(m_3) + \cdots + f(m_n)] \Delta x$$

and use it as an estimate for the area.

EXAMPLE 2

Approximate the area between the curve of the function $f(x) = x^2 + 5$ and the x-axis, over the interval [0, 6] using ten partitions, where both x and $f(x)$ are in inches.

Solution

Here $n = 10$, so we divide the interval [0, 6] into 10 partitions, each with width $\Delta x = (6 - 0)/10 = 0.6$ in. These partitions, with their midpoints, are displayed in Figure 12.24. We approximate the area A under the curve as

$$A \approx [f(0.3) + f(0.9) + f(1.5) + f(2.1) + f(2.7) + f(3.3) + f(3.9) + f(4.5) + f(5.1) + f(5.7)] \Delta x$$
$$= [5.09 + 5.81 + 7.25 + 9.41 + 12.29 + 15.89 + 20.21 + 25.25 + 31.01 + 37.49](0.6)$$
$$= 101.82 \text{ sq. in.}$$

FIGURE 12.24

Ten Partitions of [0, 6] Showing the Midpoint of Each Base

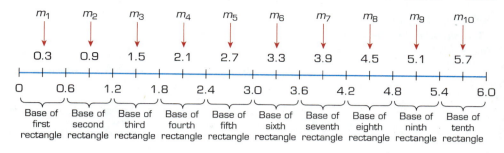

Using six rectangles ($n = 6$) to estimate the area between the graph of $f(x) = x^2 + 5$ and the x-axis, over the interval $[0, 6]$, we obtained 101.5 sq. in. as a first approximation. [See Equation (20).] This estimate is very close to the one obtained in Example 1, using many more rectangles. In Example 2, with $n = 10$, we obtained the estimate 101.82 sq. in. Still better estimates come with even finer subdivisions of the interval $[0, 6]$. Table 12.2 shows some of these results. It appears that the estimates are approaching 102 sq. in.

TABLE 12.2   Area Estimates over [0, 6] for $y = x^2 + 5$

NUMBER OF PARTITIONS n	BASE WIDTH $\Delta x = (6 - 0)/n$	AREA ESTIMATE (sq. in.)
6	1.0	101.5
10	0.6	101.82
20	0.3	101.955
40	0.15	101.98875
48	0.125	101.9921875
80	0.075	101.9971875

EXAMPLE 3

Approximate the area between the curve of the function $f(x) = e^{-2x}$ and the x-axis, over the interval $[-1, 1]$ using four partitions.

FIGURE 12.25

The Region between the Graph of $y = e^{-2x}$ and the x-axis, over [−1, 1]

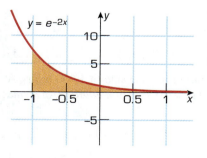

Solution

Units of linear measure are not specified, so we refer to them simply as "units" for length and "square units" for area.

The natural exponential function is always positive, so its graph lies above the x-axis. The area we wish to approximate is shaded in Figure 12.25. We will not use this graph to estimate the area; instead we apply the midpoint method directly to the function that defines the curve.

The number of partitions is $n = 4$, so we divide the interval $[-1, 1]$ into four subdivisions, each with width $\Delta x = [1 - (-1)]/4 = 0.5$ units. These subdivisions with their midpoints are displayed in Figure 12.26. We approximate the area under the curve as

$$A \approx [f(-0.75) + f(-0.25) + f(0.25) + f(0.75)]\, \Delta x$$
$$= [e^{-2(-0.75)} + e^{-2(-0.25)} + e^{-2(0.25)} + e^{-2(0.75)}](0.5)$$
$$\approx [4.48168907 + 1.64872127 + 0.60653066 + 0.22313016](0.5)$$
$$= 3.48003558 \text{ sq. units}$$

FIGURE 12.26

Midpoints of Four Partitions of [−1, 1]

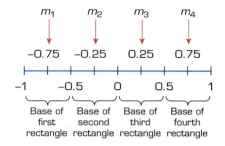

The midpoint method becomes more accurate as the widths of the partitions decrease.

One disadvantage to the midpoint method is that we have no handle on the precision of our answer. Using four partitions in Example 3, we obtained an estimate through eight decimal places. Nevertheless, we do not know whether it is *accurate* through eight decimal places, only through four decimal places, or even to the nearest unit.

In theory, the midpoint method becomes more accurate as the subdivisions get narrower, that is, as the number of partitions increases. Table 12.3, compiled with the aid of a computer, lists estimates for the area of the shaded region in Figure 12.25 with shrinking values for Δx. Based on these entries, we are confident that the first four digits have stabilized to 3.626 and that the next digit is most likely 8. Therefore we feel comfortable estimating the area as 3.627 square units, rounded to three decimal places. For greater precision we would need finer subdivisions.

TABLE 12.3 Area Estimates over [−1, 1] for $y = e^{-2x}$

NUMBER OF PARTITIONS n	WIDTH OF EACH BASE $\Delta x = [1 - (-1)]/n$ (units)	AREA ESTIMATE (square units)
4	0.5	3.48003558
8	0.25	3.58935429
10	0.2	3.60279370
16	0.125	3.61743265
20	0.1	3.62082269
32	0.0625	3.62450025
50	0.04	3.62589343
100	0.02	3.62661863
200	0.01	3.62679996
500	0.004	3.62685074
1000	0.002	3.62685799

The estimate in Step 4 of the midpoint method,

$$A = [f(m_1) + f(m_2) + f(m_3) + \cdots + f(m_n)] \Delta x \tag{21}$$

can involve many terms. Indeed, the last entry in Table 12.3 is the sum of 1,000 terms. A compact way to symbolically represent the sum in (21) uses *summation notation*, or *sigma notation*:

$$\sum_{i=1}^{n} f(m_i)\Delta x$$

The capital Greek letter sigma (Σ) denotes a sum, and the equation below the sigma establishes the smallest value for i. The sum starts by using that number to replace the i in the expression after the sigma. The subsequent terms in the sum are obtained by replacing the i in the expression with the consecutive integers after the first one, and the addition of terms is complete when i assume the value above the sigma. Thus

$$A \approx \sum_{i=1}^{n} f(m_i)\Delta x = f(m_1)\Delta x + f(m_2)\Delta x + \cdots + f(m_n)\Delta x$$

$$= [f(m_1) + f(m_2) + \cdots + f(m_n)]\Delta x \approx A$$

For examples and exercises on summation notation, see Appendix F.

In theory, the larger the value of n, the better our estimate for A. As the number of partitions becomes arbitrarily large, that is, as $n \to \infty$, the corresponding estimates approach the true value of the area and actually equal it in the limit. We write

$$A = \lim_{n \to \infty} \sum_{i=1}^{n} f(m_i)\Delta x \qquad (22)$$

Equation (22) is a formula for the *exact* area between the graph of a nonnegative continuous function and the x-axis, over a closed interval, as illustrated by the shaded region in Figure 12.27. Therefore it is an important mathematical quantity. We call the right side of Equation (22) the *definite integral* of $f(x)$ between $x = a$ and $x = b$ and denote it

$$\int_{a}^{b} f(x)\,dx = \lim_{n \to \infty} \sum_{i=1}^{n} f(m_i)\Delta x \qquad (23)$$

The integral sign, $\int$, is an elongated S, which helps us to recall the right side of Equation (23). The letter S stands for the word *sum*, and the elongation indicates a limiting operation, a "stretching-out," of the summation.

The terms on the right side of Equation (23) each involve a product of the function $f(x)$, evaluated at a specific input value, with Δx, representing a small interval of the x-axis.

Strictly speaking, the symbol $\int_{a}^{b} [\cdots]\,dx$ is a single entity, read "the definite interval of $[\cdots]$ with respect to the independent variable x." Nonetheless, many people associate the integral sign $\int$ with S for sum, and many associate the differential dx with an "infinitesimally" small piece of the x-axis to be multiplied by $f(x)$.

Although this latter interpretation is mathematically incorrect, it is a useful trick for remembering the right side of Equation (23). It is also helpful in figuring out the proper units for a definite integral. The terms being added on the right side are all of the form $f(x)\Delta x$. The partitions Δx are subintervals of the x-axis and so have the same units as the input variable x. Consequently, the units of the sum, and therefore of the definite integral, are the units of $f(x)$ times the units of x.

The notations are similar for the indefinite and the definite integrals of a continuous function $f(x)$, and they even have similar names, however, the two quantities are fundamentally different. The indefinite integral $\int f(x)\,dx$ denotes an *infinite set* of antiderivatives of $f(x)$ of the form $F(x) + c$, where c is an arbitrary constant and $F(x)$ satisfies the property $F'(x) = f(x)$. In contrast, the definite integral $\int_{a}^{b} f(x)\,dx$ is an area. As defined by Equation (23), the definite integral is a *number*, not a set.

All of our work so far in this section has dealt with the geometrical concept of area. The x- and y-axes involved units of length, such as miles for the Almond Hill Golf Course and inches in Example 1. As long as both the input units and the output units of a nonnegative continuous function $f(x)$ are measured in units of length, the resulting definite

FIGURE 12.27
The Area under the Graph of a Nonnegative Function over a Closed Interval

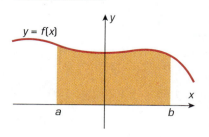

The definite integral of $f(x)$ between $x = a$ and $x = b$ is denoted $\int_{a}^{b} f(x)\,dx$. It is equal to $\lim_{n \to \infty} \sum_{i=1}^{n} f(m_i)\,\Delta x$.

integral will represent an area. Furthermore, we have seen nothing so far that allows $f(x)$ to be anything other than a function with length as an input and length as an output.

However, there is nothing about the right side of Equation (28) that prohibits us from considering functions for other types of input–output relationships. We can use our midpoint method algorithm to obtain purely numerical results, thereby extending the scope of definite integration. If these extensions have physical significance, then we have a new tool to solve other types of problems.

Consider the speed function $S(t) = \sqrt{t}$, with time measured in seconds and speed measured in feet/second. Table 12.4 lists a few corresponding speeds and times for this function, with all values rounded to two decimal places. At time $t = 0$, the object begins at rest, with a speed of $S(0) = \sqrt{0} = 0$ ft/sec. After four seconds, the object reaches a speed of $S(4) = \sqrt{4} = 2$ ft/sec.

TABLE 12.4 **Evaluations of $S(t) = \sqrt{t}$**

Time (sec)	0	0.5	1	1.5	2	2.5	3	3.5	4
Speed (ft/sec)	0	0.71	1	1.22	1.41	1.58	1.73	1.87	2

Let us see how we can use our knowledge of an object's speed to estimate how far it travels in four seconds. During those four seconds, the object begins at rest and accelerates to a speed of 2 ft/sec. Halfway through the trip, at $t = 2$, the object achieves a speed of 1.41 ft/sec. If we use this speed as an estimate for the average speed over the entire four-second interval, which we represent as Δt, then

$$\text{distance traveled} \approx S(2)\,\Delta t = (1.41)\cdot 4 = 5.64 \text{ ft} \qquad (24)$$

The speed function is graphed in Figure 12.28. The product in (24) is obtained by multiplying Δt, the width of the time interval $[0, 4]$, by the speed reached at its midpoint. This is *numerically* equal to the area of the shaded rectangle in Figure 12.28.

FIGURE 12.28

Estimating Distance with a Single Partition

We can improve our estimate for distance traveled by refining our estimates for average speed. Over the first two seconds, the object begins at rest and accelerates to a speed of 1.41 ft/sec. Halfway through this interval, at $t = 1$, the object achieves a speed of 1 ft/sec, which we can use as an estimate for the average speed over the first two-second interval. Over the last two seconds, the object accelerates from a speed of 1.4 ft/sec to a speed of

2 ft/sec. Halfway through this last two-second interval, at $t = 3$, the object achieves a speed of 1.73 ft/sec, which we can use as an estimate for the average speed over the last two-second interval.

For this second estimate, we have partitioned the interval [0, 4] into two subintervals, each of duration $\Delta t = 2$ sec. Using an average speed at 1 ft/sec over the first two-second partition and an average speed of 1.73 ft/sec over the last two-second partition, we estimate the total distance traveled over 4 sec to be

$$\text{distance traveled} \approx [S(1) + S(3)]\, \Delta t$$
$$= (1 + 1.73) \cdot 2 = 5.46 \text{ ft} \tag{25}$$

The estimate in (25) is obtained numerically by partitioning [0, 4] into two equal subintervals, multiplying the width of each subinterval by the output of the speed function at the midpoint, and then summing the results. This is numerically equal to the area of shaded portions of Figure 12.29.

FIGURE 12.29

Estimating Distance with Two Subintervals

To improve our estimate still further, we partition the interval [0, 4] into four equal subintervals, each of duration $\Delta t = 1$ sec. On each subinterval, we estimate the average speed over that partition by the speed at its midpoint. Our estimate for total distance traveled becomes

$$\text{distance traveled} \approx [S(0.5) + S(1.5) + S(2.5) + S(3.5)]\, \Delta t$$
$$= (0.71 + 1.22 + 1.58 + 1.87) \cdot 1 \tag{26}$$
$$= 5.38 \text{ ft}$$

The estimate in (26) is numerically equal to the area of shaded portions of Figure 12.30.

FIGURE 12.30

Estimating Distance with Four Subintervals

In general, we calculated the distance estimates by dividing the time interval $[0, 4]$ into n equal parts, each of width $\Delta t = (4 - 0)/n$, and then adding the terms:

$$\text{distance traveled} \approx [S(m_1) + S(m_2) + S(m_3) + \cdots + S(m_n)] \Delta t$$

where m_i denotes the midpoint of the ith partition ($i = 1, 2, 3, \cdots, n$) and $S(t)$ is the speed function. Using sigma notation, we write this estimate more compactly as

$$\text{distance traveled} \approx \sum_{i=1}^{n} S(m_i) \Delta t \qquad (27)$$

When $n = 4$, the estimate for the distance traveled is

$$\sum_{i=1}^{4} S(m_i) \Delta t = [S(m_1) + S(m_2) + S(m_3) + S(m_4)] \Delta t$$

$$= [S(0.5) + S(1.5) + S(2.5) + S(3.5)] \Delta t$$

which is Estimate (26). Table 12.5 displays these estimates for $n = 1, 2, 4, 8, 20, 100,$ and 1000. When $n = 1000$, Estimate (27) expands to become

$$\sum_{i=1}^{1000} S(m_i) \Delta t = [S(m_1) + S(m_2) + S(m_3) + \cdots + S(m_{1000})] \Delta t$$

$$= [S(0.002) + S(0.006) + S(0.008) + \cdots + S(3.998)] \Delta t$$

$$\approx 5.33$$

which is the last entry in Table 12.5, rounded to two places. It is our best estimate for the distance traveled by the object over the first four seconds.

TABLE 12.5 Distance Estimates on [0, 4]

NUMBER OF PARTITIONS (n)	WIDTH OF PARTITION $\Delta t = (4 - 0)/n$	DISTANCE TRAVELED (feet)
1	4	5.64
2	2	5.46
4	1	5.38
8	0.5	5.35
20	0.2	5.34
100	0.04	5.33
1000	0.004	5.33

Our estimates improve as the number n of subdivisions of equal width becomes larger. Our estimate with $n = 4$ in (27) is better than the one with $n = 2$. The estimate with $n = 100$ is better than the one with $n = 20$. As the number of subdivisions of $[0, 4]$ grows, our estimates get closer and closer to the actual answer. In the limit,

$$\text{distance traveled} = \lim_{n \to \infty} \sum_{i=1}^{n} S(m_i) \Delta t \qquad (28)$$

The right side of Equation (28) looks remarkably similar to Equation (22); the only difference is the labels. In Equation (28), we use S and t for the function and the independent variable, respectively, while the corresponding labels in Equation (22) are f and x. We can make the same label exchange in Equation (23), the definition of the definite integral. This, together with Equation (28), gives us

$$\text{distance traveled} = \lim_{n\to\infty} \sum_{i=1}^{n} S(m_i)\,\Delta t = \int_{a}^{b} S(t)\,dt$$

The area under the graph of a speed function over [a, b] is numerically equal to the distance traveled over that interval.

with $a = 0$ and $b = 4$ as the endpoints of the domain in our example.

We have done something very important and very powerful here: We adapted a procedure for solving area problems to a procedure for solving distance problems. The same procedure works for both problems because the mathematical models for both processes are identical except for the labels. The area under the graph of a speed function over a closed interval on the horizontal axis is numerically equal to the distance traveled over that interval, and both can be obtained by definite integration. The units are different but the numbers are the same.

IMPROVING SKILLS

In Problems 1 through 4, use the grids shown on each frame to estimate the areas of the shaded regions. Then refine each grid and calculate a tighter bound for each estimate by halving the partition widths on each axis.

1.

2.

3.

4.

In Problems 5 through 14, use the midpoint method to estimate the area under the curve of the given equation and over specified interval on the x-axis.

5. $y = 30 - x^2$; [0, 5] with 5 partitions
6. $y = 30 - x^2$; [0, 5] with 10 partitions
7. $y = x^2 + 3x + 6$; [−6, 6] with 3 partitions
8. $y = x^2 + 3x + 6$; [−6, 6] with 6 partitions
9. $y = x^2 + 3x + 6$; [−6, 6] with 12 partitions
10. $y = x^3 + 2$; [−1, 1] with 5 partitions
11. $y = x^3 + 2$; [−1, 1] with 10 partitions
12. $y = e^x$; [0, 4] with 2 partitions
13. $y = e^x$; [0, 4] with 4 partitions
14. $y = e^x$; [0, 4] with 8 partitions

15. Estimate $\int_0^2 (x^2 + 2x)\, dx$ using two partitions.

16. Estimate the integral in Problem 15 using four partitions.

17. Estimate $\int_1^3 \frac{1}{x}\, dx$ using three partitions.

18. Estimate the integral in Problem 17 using six partitions.

19. Estimate $\int_1^4 \frac{2}{x^2}\, dx$ using three partitions.

20. The following table lists the speedometer readings in an automobile at various times during a thirty-minute trip.

Time (min)	0	5	10	15	20	25	30
Speed (mph)	0	50	40	60	60	30	40

Estimate the distance traveled over this thirty-minute period using three partitions.
21. Solve Problem 20 using six partitions.
22. The speed of an object, s, in units of feet per minute, over the time interval $1 \le t \le 4$, measured in minutes, is given by the formula $s = 200 + 400/t$. Estimate how far the object travels over this time interval using three partitions.
23. Solve Problem 22 using six partitions.
24. The speed of an object, s, in units of miles per hour, is given by the formula $s = 20t^2 + 30t + 10$ where t is measured in hours. Estimate how far the object travels during the first hour using two partitions.
25. Solve Problem 24 using five partitions.

EXPLORING IN TEAMS

26. Modify the midpoint method so that it is applicable to all functions, not just nonnegative ones. Use your modified algorithm to estimate the area between the curve of the function $y = x^3$ and the interval [−1, 1].
27. The algorithm for calculating the area of an irregularly shaped bounded region can be made more efficient if at each stage we calculate the total area of all of the rectangles that fall completely within the region and then apply a finer grid to the remaining

region. For example, the first 11 rectangles in Figure 12.13 are all within the Almond Hill Golf Course so we know that their combined area is totally within the golf course and there is no need to subdivide this area any further. We delete it from further consideration and Figure 12.14 becomes Figure A; our concern is now with determining the area of the shaded region in this graph. Use this approach to recalculate Table 12.1. Is this approach really more efficient? Write an algorithm to execute this method.

FIGURE A

EXPLORING WITH TECHNOLOGY

28. Verify each line in Table 12.2.
29. Verify each line in Table 12.3.
30. Solve Problem 5 with 100 partitions.
31. Solve Problem 7, first with 50 partitions and then with 100 partitions.
32. Estimate to one decimal place the area under the graph of the equation $y = x^3 + 2$ between $x = -1$ and $x = 1$.
33. Estimate to one decimal place the area under the graph of the equation $y = \sin(\pi x)$ between $x = 0$ and $x = 1$.
34. Estimate to one decimal place the area under the graph of the equation $y = e^x$ between $x = 1$ and $x = 2$.
35. Estimate to one decimal place the area under the graph of the equation $y = xe^{-x}$ between $x = 1$ and $x = 3$.

REVIEWING MATERIAL

36. (Section 3.3) Solve the following linear programming problem:

$$\text{Minimize:} \quad z = 3x_1 + 5x_2$$
$$\text{subject to:} \quad 2x_1 + 7x_2 \geq 28{,}000$$
$$4x_1 + 3x_2 \geq 24{,}000$$
$$x_1 \leq 20{,}000$$
$$x_2 \leq 6{,}000$$

37. (Section 4.3) A corporation decides to create a sinking fund to finance the down payment on a new office building. They want the fund to have a value of $1,500,000 in 6 years. How much should they deposit monthly if interest rates are 8%/year, compounded monthly?

38. (Section 10.4 and 10.5) Find the derivative of the function $y = [(3e^{-2x} - 5)^{5.8}/ 2x^4 \ln x]$.

39. (Section 11.4) A company is increasing production of an item at the rate of 100 units/ day. The selling price, p (in dollars), is a function of the production level as modeled by the equation $p = 80 - (\sqrt{2n}/5)$, where n represents the number of items produced. How quickly is the price changing when daily production is at 20,000 items?

RECOMMENDING ACTION

40. Respond by memo to the following request.

MEMORANDUM

To: J. Doe Reader

From: Troy

Date: Today

Subject: **Definite Integrals**

I was explaining to Harvey over lunch the difference between definite and in-definite integrals. I mentioned that an indefinite integral is a function and that a definite integral is a number. Harvey said I was not technically correct, that a definite integral can also be a function. It all depends on the interval. Is Harvey wrong or am I?

12.5 APPLICATIONS OF THE DEFINITE INTEGRAL

FIGURE 12.31

The Area between the Graph of $f(x)$ and the x-axis, over the Interval $[a, x]$

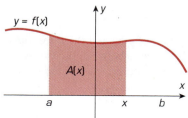

It is to our advantage to have a simple procedure for calculating the area of the region between the graph of a nonnegative continuous function $f(x)$ and the x-axis, over an interval $[a, b]$. Such a procedure is provided by the *fundamental theorem of calculus*, which acts as a bridge between indefinite integration and definite integration. We begin by developing a method for determining the area of the region under the graph of a nonnegative function $f(x)$ over the interval $[a, x]$ for *any* value of x, $a \leq x \leq b$, as shown in Figure 12.31. As the value of x changes, so does the domain $[a, x]$. The area of the shaded region in Figure 12.31 is therefore a function of x; we denote it by $A(x)$. We know very little about this area function and yet, surprisingly, we can find its derivative $A'(x)$. Then once we know $A'(x)$, we can use indefinite integration to identify $A(x)$.

From the definition of derivative, we have

$$A'(x) = \lim_{h \to 0} \frac{A(x + h) - A(x)}{h} \tag{29}$$

FIGURE 12.32

The Area between the Graph of $f(x)$ and the x-axis, over the Interval $[a, x + h]$

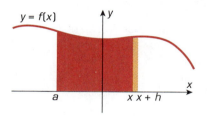

The numerator of the right side of Equation (29) is the difference of two areas: $A(x + h)$ is the area of the region between the graph of $f(x)$ and the interval $[a, x + h]$; it is the area of the entire shaded region in Figure 12.32, including both the area shaded in pink and the area shaded in orange. The output $A(x)$ is the area of the region between the graph of $f(x)$ and the interval $[a, x]$; it is shaded in pink in Figure 12.32. The difference, $A(x + h) - A(x)$, is the area of just the region shaded in orange.

If we extract and magnify the orange-shaded region in Figure 12.32, we obtain Figure 12.33 as a geometrical rendition of the difference $A(x + h) - A(x)$. We know from Equation (29) that we are interested in this difference as h tends to zero, so the base of the shaded region in Figure 12.33 becomes very narrow as $h \to 0$. Using the techniques developed in Section 12.4, we approximate the area of the shaded region in Figure 12.33 by a rectangle having width h and height equal to the function $f(x)$ at the midpoint m of the interval $[x, x + h]$. This point is $m = x + \frac{1}{2}h$, and the top of the approximation rectangle is shown with a dashed line in Figure 12.33. Thus we have

$$A(x + h) - A(x) \approx h \cdot f(x + \tfrac{1}{2}h)$$

$$\frac{A(x + h) - A(x)}{h} \approx \frac{h \cdot f(x + \tfrac{1}{2}h)}{h} = f(x + \tfrac{1}{2}h)$$

FIGURE 12.33

The Area of the Region between the Graph of $f(x)$ and the x-axis, over the Interval $[x, x + h]$

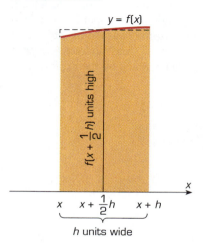

and so

$$A'(x) = \lim_{h \to 0} \frac{A(x + h) - A(x)}{h} = \lim_{h \to 0} f(x + \tfrac{1}{2}h) = f(x)$$

The function $A(x)$, which is the area between the graph of $f(x)$ and the x-axis, over the interval $[a, x]$, is an antiderivative of $f(x)$!

Now let $F(x)$ be an antiderivative of $f(x)$ obtained by indefinite integration. In the beginning of Section 12.2, we deduced the following fact: Any two antiderivatives of the same function must differ by a constant. Therefore it must be the case that

$$A(x) - F(x) = c \tag{30}$$

where c denotes a constant. Equation (30) is valid for all values of x in the interval $[a, b]$. In particular, when $x = a$, the interval $[a, x]$ collapses to the single point $x = a$ and, because the area of a line segment is zero, $A(a) = 0$. Thus when $x = a$, Equation (30) becomes

$$A(a) - F(a) = 0 - F(a) = c$$

or more simply, $-F(a) = c$. We can now rewrite Equation (30) as $A(x) - F(x) = -F(a)$, or

$$A(x) = F(x) - F(a) \qquad (31)$$

When $x = b$ in Equation (31), we find that

$$A(b) = F(b) - F(a) \qquad (32)$$

The output $A(b)$ is the area of the region between the graph of $f(x)$ and the x-axis, over the interval $[a, b]$. This is the entire region in Figure 12.31 that is between the two heavy lines. In Section 12.4 we found the area of this region to be equal to the definite integral of $f(x)$ between $x = a$ and $x = b$. Thus

$$A(b) = \int_a^b f(x)\, dx \qquad (33)$$

Together Equations (32) and (33) give us the fundamental theorem of calculus.

THE FUNDAMENTAL THEOREM OF CALCULUS

If $f(x)$ is continuous on the interval $[a, b]$ and if $F(x)$ is an antiderivative of $f(x)$ on this interval, then

$$\int_a^b f(x)\, dx = F(b) - F(a)$$

The area A between the graph of a nonnegative continuous function $f(x)$ and the x-axis, from $x = a$ to $x = b$, is

$$A = \int_a^b f(x)\, dx.$$

The fundamental theorem of calculus is a powerful method for evaluating a definite integral. It provides us with a robust approach to computing *exact* areas when two conditions are met: the function $f(x)$ must be nonnegative and continuous on the interval of interest.

The fundamental theorem of calculus is the justification for having similar notations for both the definite integral and the indefinite integral. To evaluate the definite integral of $f(x)$ from $x = a$ to $x = b$, we first use indefinite integration to obtain an antiderivative of $F(x) = \int f(x)\, dx$ and, for convenience, we choose zero for the constant of integration. Then we evaluate $F(x)$ at the two input values $x = b$ and $x = a$ and subtract $F(a)$ from $F(b)$ to get our answer. The difference $F(b) - F(a)$ is often denoted by the expression

$$F(x)\,\big|_{x=a}^{x=b} \text{ or } [F(x)]_{x=a}^{x=b}$$

EXAMPLE 1

Evaluate $\int_0^6 (x^2 + 5)\, dx$.

Solution
The indefinite integral of $f(x) = x^2 + 5$ is

$$\int (x^2 + 5)\, dx = \tfrac{1}{3}x^3 + 5x + c$$

where c is an arbitrary constant. Setting $c = 0$, we have as *one* antiderivative $F(x) = \frac{1}{3}x^3 + 5x$. The limits on the definite integral are $b = 6$ and $a = 0$, hence

$$\int_0^6 (x^2 + 5)\, dx = \left[\frac{x^3}{3} + 5x\right]_{x=0}^{x=6} = \left[\frac{(6)^3}{3} + 5(6)\right] - \left[\frac{(0)^3}{3} + 5(0)\right]$$

$$= 102 \text{ sq. units}$$

The function $f(x) = x^2 + 5$ in Example 1 is nonnegative and continuous over the interval $[0, 6]$, therefore the definite integral represents the shaded region in Figure 12.34. It follows from Example 1 that the area of the region between the graph of $y = x^2 + 5$ and the x-axis from $x = 0$ to $x = 6$ is 102 sq. units. If x and y are both measured in inches, then the area is 102 sq. in. Compare this approach with the work required to compile Table 12.2 in Section 12.4.

FIGURE 12.34

The Region between the Graph of $y = x^2 + 5$ and the x-axis, over the Interval [0, 6]

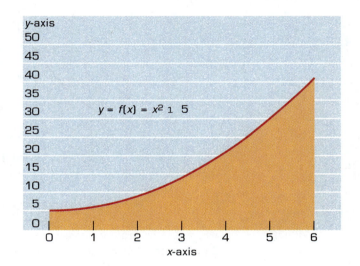

EXAMPLE 2

Evaluate $\int_{-1}^{1} e^{-2x}\, dx$.

Solution
The indefinite integral of $f(x) = e^{-2x}$ is

$$\int e^{-2x}\, dx = -\tfrac{1}{2}e^{-2x} + c$$

FIGURE 12.25

The Region between the Graph of $y = e^{-2x}$ and the x-axis, over [−1, 1]

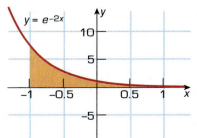

where c is an arbitrary constant. Setting $c = 0$, we have as *one* antiderivative $F(x) = -\tfrac{1}{2}e^{-2x}$. The limits on the definite integral are $b = 1$ and $a = -1$, hence

$$\int_{-1}^{1} e^{-2x}\, dx = -\tfrac{1}{2}e^{-2x}\Big|_{x=-1}^{x=1} = -\tfrac{1}{2}e^{-2(1)} - \left(-\tfrac{1}{2}e^{-2(-1)}\right) = \tfrac{1}{2}(e^2 - e^{-2})$$

$$= 3.62686041 \text{ sq. units}$$

rounded to eight decimal places.

Because $f(x) = e^{-2x}$ is nonnegative and continuous everywhere, the definite integral in Example 2 is the exact area of the shaded region in Figure 12.25. Compare this approach to the work necessary to compile Table 12.3 in Section 12.4.

EXAMPLE 3

Find the area of the region that is between the graph of $y = 2x/\sqrt{x^2 + 16}$ and the x-axis, over the interval $[2, 8]$, with both variables measured in miles.

Solution

We seek the area of the shaded region in Figure 12.35. Since $f(x) = 2x/\sqrt{x^2 + 16}$ is nonnegative and continuous on $[2, 8]$, it follows from Equation (33) that the area A is

$$A = \int_2^8 \frac{2}{\sqrt{x^2 + 16}} \, dx$$

FIGURE 12.35

The Region between the Graph of $y = 2x/\sqrt{x^2 + 16}$ and the x-axis, over [2, 8]

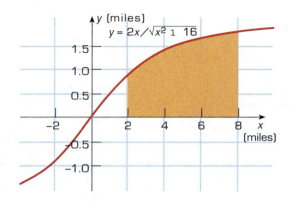

In Example 3 of Section 12.3, we used the method of substitution of variables to calculate the indefinite integral of $f(x) = 2x/\sqrt{x^2 + 16}$:

$$\int \frac{2x}{\sqrt{x^2 + 16}} \, dx = 2\sqrt{x^2 + 16} + c$$

where c is an arbitrary constant. Setting $c = 0$, we have as *one* antiderivative $F(x) = 2\sqrt{x^2 + 16}$. Here $b = 8$ and $a = 2$, hence

$$\int_2^8 \frac{2x}{\sqrt{x^2 + 16}} \, dx = 2\sqrt{x^2 + 16} \Big|_{x=2}^{x=8}$$
$$= 2\sqrt{(8)^2 + 16} - 2\sqrt{(2)^2 + 16}$$
$$= 4\sqrt{5} \approx 8.9443 \text{ sq. miles}$$

The indefinite integral in Example 3,

$$\int \frac{2x}{\sqrt{x^2 + 16}} \, dx = 2\sqrt{x^2 + 16} + c$$

represents infinitely many antiderivatives of $f(x) = 2x/\sqrt{x^2 + 16}$, one for each choice of the constant c. The fundamental theorem of calculus states that we can use *any* antiderivative of $f(x)$ to evaluate the definite integral, so we set $c = 0$ and took $F(x) = 2\sqrt{x^2 + 16}$. To see that we get the same answer regardless of our choice of c, let us redo Example 3 with $F(x) = 2\sqrt{x^2 + 16} + 7$:

$$\int_2^8 \frac{2x}{\sqrt{x^2 + 16}} \, dx = 2\sqrt{x^2 + 16} + 7 \Big|_{x=2}^{x=8}$$
$$= (2\sqrt{80} + 7) - (2\sqrt{20} + 7) = 4\sqrt{5}$$

The units associated with the definite integral of an input–output function are the product of the output units and the input units.

Notice that the constant 7 appears by itself in both $F(b)$ and $F(a)$ but when the difference $F(b) - F(a)$ is computed, the two sevens cancel. This will be the case for any arbitrary constant c, so we save work when evaluating a definite integral by the fundamental theorem of calculus by choosing the antiderivative with $c = 0$.

As we noted in Section 12.4, the units of measure for a definite integral are the units associated with $f(x)$ times the units associated with the independent variable x. Often our functions are models of input–output relationships, with the letters x, y, and even f replaced by more appropriate labels. The units associated with a definite integral of such a function are the product of the units of output and the units of input.

EXAMPLE 4

City planners believe that the growth rate of their population, in people per year, will increase as a function of time t according to the model

$$g(t) = 500 + 200 \sin(2\pi t)$$

Determine the units associated with the definite integral

$$\int_0^{10} g(t)\, dt$$

Solution

The function $g(t)$ is an input–output relationship with the input t measured in years and the output g measured in people/year. The units of this definite integral are

$$\left(\frac{\text{people}}{\text{year}}\right)\text{years} = \text{people}$$

Therefore

$$\int_0^{10} g(t)\, dt = \int_0^{10} [500 + 200 \sin(2\pi t)]\, dt$$

$$= \left[500t - \frac{100}{\pi} \cos(2\pi t) \right]_{t=0}^{t=10}$$

$$= 5{,}000 \text{ people}$$

EXAMPLE 5

The speed S of an object is modeled by the function $S(t) = \sqrt{t}$, with time t measured in seconds and speed measured in feet/second. Find the units associated with the definite integral

$$\int_0^4 S(t)\, dt$$

Solution

The function $S(t)$ is an input–output relationship with time as the input, in seconds, and speed as the output, in feet/second. The units of the definite integral are therefore

$$\left(\frac{\text{feet}}{\text{second}}\right)\text{seconds} = \text{feet}$$

and so

$$\int_0^4 S(t)\, dt = \int_0^4 t^{1/2}\, dt = \frac{2}{3} t^{3/2} \Big|_{t=0}^{t=4} = \frac{16}{3} \text{ feet}$$

EXAMPLE 6

The speed of a car (in mph) is given by the function

$$S(t) = 60(1 - e^{-t})$$

with time measured in hours. How far does the car go in its second hour of travel?

Solution

We want the distance traveled over the time interval [1, 2]. We know from our work in Section 12.4 that

$$\text{distance traveled} = \int_a^b S(t)\, dt$$

$$= \int_1^2 60(1 - e^{-t})\, dt = 60(t + e^{-t}) \Big|_{t=1}^{t=2} \approx 46.047$$

The input units for time are measured in hours and the output units of the speed function are measured in mph. Therefore the definite integral has units

$$\left(\frac{\text{miles}}{\text{hour}} \right) \text{hours} = \text{miles}$$

The car travels a distance of 46 miles in the second hour, rounded to the nearest mile.

Figure 12.36 is a graph of the speed function from Example 6. The area under this curve and over the horizontal axis from $t = 1$ to $t = 2$ is numerically equal to the distance traveled in the interval [1, 2].

FIGURE 12.36

The Area under the Graph of the Speed Function is Equal to the Distance Traveled

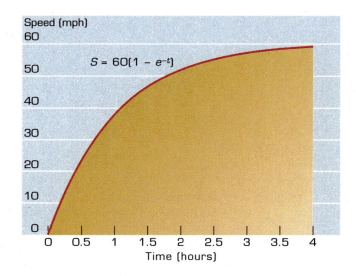

Speed and distance traveled are related concepts. Let us model the speed and distance of an object at any time t by the functions $S(t)$ and $D(t)$, respectively. We know from our work in Chapter 10 that speed is the instantaneous rate of change of distance, hence

$$S(t) = D'(t) \tag{34}$$

The distance traveled over a time interval $[a, b]$ is $D(b) - D(a)$. But then, using Equation (34),

$$D(b) - D(a) = \text{distance traveled} = \int_a^b S(t)\, dt = \int_a^b D'(t)\, dt \tag{35}$$

This equation is an immediate consequence of the fundamental theorem of calculus, with t replacing x as the independent variable, $D'(t)$ replacing $f(x)$, and $D(t)$ replacing $F(x)$. Equation (35) states that the distance traveled by an object from time $t = a$ to time $t = b$ can be calculated precisely by evaluating the definite integral of the speed function, which is the derivative of the distance function, over the interval $[a, b]$.

The total change in the output of $f(x)$ on the interval $[a, b]$ is equal to the definite integral of $f'(x)$ over $[a, b]$.

Now consider any differentiable function $f(x)$; its derivative is $f'(x)$. This tells us that $f(x)$ is an antiderivative of $f'(x)$. It follows from the fundamental theorem of calculus that

$$f(b) - f(a) = \int_a^b f'(x)\, dx \tag{36}$$

The difference $f(b) - f(a)$ is the total change in the output of $f(x)$ as x varies from $x = a$ to $x = b$. Thus Equation (36) states that the total change in the output of $f(x)$ on $[a, b]$ is equal to the definite integral of the derivative of $f(x)$ over the interval $[a, b]$.

The definite integral of a nonnegative continuous function can be interpreted numerically as the area of the region under the graph of that function. Thus if $f'(x)$ is a nonnegative continuous function, then the right side of Equation (36) is numerically equal to the area under the graph of $f'(x)$ from $x = a$ to $x = b$. These relationships are illustrated graphically in Figure 12.37.

FIGURE 12.37

The Area under the Graph of $y = f'(x)$ over $[a, b]$

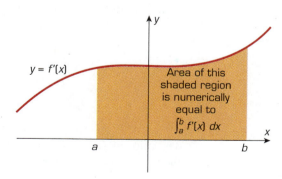

In Chapter 10, we defined marginal cost, MC, to be the derivative of the total cost function $TC = f(x)$, where x denotes the number of items produced:

$$MC = \frac{d}{dx}(TC) = \frac{d}{dx} f(x) = f'(x)$$

Since marginal cost is the derivative of total cost, it follows that total cost is the antiderivative of marginal cost. Therefore management has a model for marginal cost as a function of x; the change in total cost can be modeled by the equation

$$\begin{array}{l}\text{The change in total cost of}\\ \text{increasing production from}\\ a \text{ items to } b \text{ items}\end{array} = \int_a^b (\text{marginal cost function})\, dx \qquad (37)$$

EXAMPLE 7

A company manufactures magnifying glasses for stamp collectors and other hobbyists. The weekly marginal cost function for the production of x magnifying glasses is $MC(x) = 0.5 + 0.002x$, where $MC(x)$ is in units of dollars per item. What is the cost of beginning the business with a production run of 2500 magnifying glasses a week?

Solution

Starting the business means that the company is increasing its production from no magnifying glasses to 2500 magnifying glasses. We are given a formula for marginal cost and we are asked for the change in its antiderivative—the total cost—resulting from this increase in production. It follows from Equation (37) that

$$\begin{array}{l}\text{The change in total cost of}\\ \text{increasing production from}\\ 0 \text{ to } 2500 \text{ items}\end{array} = \int_0^{2500} (0.5 + 0.002x)\, dx$$

$$= [0.5x + 0.001x^2]_{x=0}^{x=2500} = \$7500$$

It is important to realize that the definite integral in Equation (37) gives the cost of moving from a production level of $x = a$ items to $x = b$ items; it is *not* the total cost of producing b items. To calculate the total cost of producing b units, management must add the cost of changing production from a units to b units to the actual cost of producing a units.

In a more general context, we can rewrite

$$f(b) - f(a) = \int_a^b f'(x)\, dx$$

as

$$f(b) = f(a) + \int_a^b f'(x)\, dx$$

The value of $f(x)$ at $x = b$ is the value of $f(x)$ at $x = a$ *plus* a definite integral that gives the total change in $f(x)$ over the interval $[a, b]$.

EXAMPLE 8

Determine the total cost of scheduling a production run of 2500 magnifying glasses for the company in Example 7 if fixed costs for its manufacturing facility are $5000/week.

Solution

Fixed costs are total costs when no items are produced, hence the total cost associated with a production run of zero magnifying glasses is $5000/week. We have from Example

7 that the incremental cost of moving from a production run of zero to a production run of 2500 is $7500. Therefore the total cost of a production run of 2500 magnifying glasses is $12,500. Compare this result with that of Example 2 of Section 12.2. There we solved the same problem using indefinite integration; here we used definite integration.

Total revenue is an antiderivative of marginal revenue, so if management has a model for marginal revenue as a function of x (the number of items produced), then

$$\begin{array}{l} \text{The change in revenue} \\ \text{from increasing production} \\ \text{from } a \text{ to } b \text{ items} \end{array} = \int_a^b [\text{marginal revenue function}]\, dx \qquad (38)$$

Furthermore, total revenue is usually zero when production is zero, because there is no income when there are no products available for sale.

EXAMPLE 9

A company knows from past experience that its marginal revenue is given by the function

$$MR(n) = 562.5 - 0.225n$$

where n is the number of items offered for sale and $MR(n)$ is in units of dollars/item. What is the total revenue for the production of 1200 items?

Solution

Here the label for the independent variable is n, not x. Consequently, Equation (38) becomes

$$\begin{array}{l} \text{Change in revenue from} \\ \text{increasing production} \\ \text{from 0 to 1200 items} \end{array} = \int_0^{1200} (562.5 - 0.225n)\, dn$$

$$= [562.5n - 0.1125n^2]_{n=0}^{n=1200} = \$513{,}000$$

This is the *additional* revenue the company can expect from increasing production from zero to 1200 items. Since no revenue is expected from no production, the total revenue expected from a production run of 1200 is simply $513,000. Compare this result with Example 3 of Section 12.2. There we solved the same problem using indefinite integration.

EXAMPLE 10

City planners believe that the growth rate of their population, in people per year, will increase as a function of time t (in years) according to the model

$$g(t) = 500 + 200 \sin(2\pi t)$$

Use this model to predict the population of the city in 10 years if the current population is 28,000 people.

Solution

Let $P(t)$ denote the population of the city at any time t. Then

$$\frac{dP}{dt} = g(t) = 500 + 200 \sin(2\pi t)$$

Using the results of Example 4, we have that

$$\text{Change in population over the next 10 years} = \int_0^{10} [500 + 200 \sin(2\pi t)] \, dt$$

$$= 5000 \text{ people}$$

The actual population at $t = 0$ is 28,000 people, therefore the expected population in 10 years is 33,000 people.

Definite integrals are evaluated by finding their corresponding indefinite integrals, hence many of the transformation rules associated with indefinite integration can be recast as rules for definite integrals. The constant multiple rule for any constant k, for instance, becomes

$$\int_a^b kf(x) \, dx = k \int_a^b f(x) \, dx$$

The sum and difference rules become

$$\int_a^b [f(x) + g(x)] \, dx = \int_a^b f(x) \, dx + \int_a^b g(x) \, dx$$

$$\int_a^b [f(x) - g(x)] \, dx = \int_a^b f(x) \, dx - \int_a^b g(x) \, dx \tag{39}$$

FIGURE 12.38

Area A is the Sum of Area A_1 and Area A_2

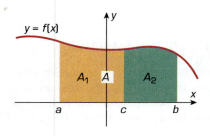

A fourth rule of definite integration has no counterpart in indefinite integration. Notice that the area A of the entire shaded region in Figure 12.38 is the sum of the areas A_1 and A_2 of the two differently shaded regions:

$$A = A_1 + A_2 \tag{40}$$

Since $f(x) \geq 0$ and continuous on $[a, b]$, we can use definite integration to calculate each of these areas:

$$A = \int_a^b f(x) \, dx$$

$$A_1 = \int_a^c f(x) \, dx \qquad A_2 = \int_c^b f(x) \, dx$$

Substituting these integrals into Equation (40), we obtain the following important result:

THE INTERMEDIATE VALUE RULE FOR DEFINITE INTEGRATION

$$\int_a^b f(x) \, dx = \int_a^c f(x) \, dx + \int_c^b f(x) \, dx$$

This rule was derived under the conditions that $f(x) \geq 0$ and that $a \leq c \leq b$, however, we will see that the intermediate value rule is still true when these conditions are removed.

FIGURE 12.39

The Region between the Graphs of Two Nonnegative Continuous Functions

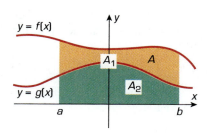

We can use our rules for definite integration to find the area A of a region between the graphs of two nonnegative continuous functions over a closed interval, such as the orange shaded region in Figure 12.39. The definite integral

$$\int_a^b f(x)\, dx$$

gives the area A_1 between the graph of the function $f(x)$ and the x-axis, over the interval $[a, b]$; this is the entire shaded region in Figure 12.39. The definite integral

$$\int_a^b g(x)\, dx$$

gives the area A_2 between the graph of the function $g(x)$ and the x-axis, over the interval $[a, b]$; this is the region shaded in green in Figure 12.39. If we subtract these two areas, we obtain the area $A = A_1 - A_2$, the region shaded in orange:

$$A = A_1 - A_2 = \int_a^b f(x)\, dx - \int_a^b g(x)\, dx$$

It now follows from Equation (39) that

$$A = \int_a^b [f(x) - g(x)]\, dx \qquad (41)$$

This formula yields the area of the region that lies between the graphs of the nonnegative continuous functions $f(x)$ and $g(x)$ over the interval $[a, b]$, where $f(x) \geq g(x) \geq 0$.

EXAMPLE 11

Find the area A of the region between the graphs of $y_1 = x$ and $y_2 = x^3$ over the interval $[0, 1]$.

Solution

The region of interest is shaded in Figure 12.40. Both functions are continuous on $[0, 1]$ and, on this interval, $x \geq x^3 \geq 0$, so we can use Equation (41) to calculate A:

$$A = \int_0^1 (x - x^3)\, dx = \left[\frac{1}{2} x^2 - \frac{1}{4} x^4 \right]_{x=0}^{x=1}$$

$$= [\tfrac{1}{2}(1)^2 - \tfrac{1}{4}(1)^4] - [\tfrac{1}{2}(0)^2 - \tfrac{1}{4}(0)^4] = \tfrac{1}{4}$$

FIGURE 12.40

The Region between the Graphs of $y_1 = x$ and $y_2 = x^3$ on [0, 1]

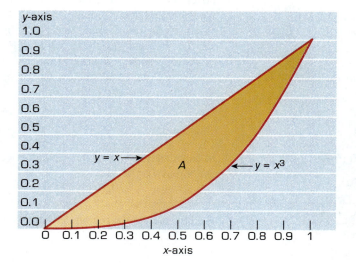

EXAMPLE 12

Find the area of the region between the graphs of $y_1 = x$ and $y_2 = x^3$ over the interval [1, 2].

Solution

Here the region of interest, over [1, 2], is shaded in Figure 12.41. Both functions are continuous on the interval [1, 2] but here $x^3 \geq x \geq 0$. We again use Equation (41) but this time we integrate $x^3 - x$:

$$A = \int_1^2 (x^3 - x)\, dx = \left[\frac{1}{4} x^4 - \frac{1}{2} x^2 \right]_{x=1}^{x=2}$$

$$= \left[\tfrac{1}{4}(2)^4 - \tfrac{1}{2}(2)^2 \right] - \left[\tfrac{1}{4}(1)^4 - \tfrac{1}{2}(1)^2 \right] = \tfrac{9}{4}$$

FIGURE 12.41

The Region between the Graphs of $y_2 = x^3$ and $y_1 = x$ on [1, 2]

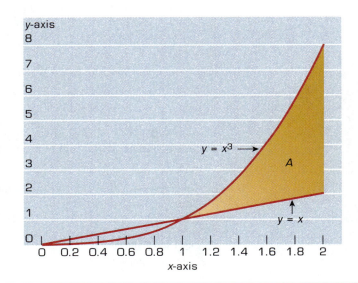

Both Examples 11 and 12 involved the same two functions but over different intervals. The interval in Example 11 was [0, 1] and over that interval $x \geq x^3$, so we integrated the difference $x - x^3$. The interval in Example 12 was [1, 2] and over that interval $x^3 \geq x$, so we integrated the difference $x^3 - x$. In both cases we constructed the difference $f(x) - g(x)$ to satisfy the requirement that $f(x) \geq g(x)$ over the interval of interest. In Example 4, we set $f(x) = x$ and $g(x) = x^3$, while in Example 5 we set $f(x) = x^3$ and $g(x) = x$.

As long as $f(x) \geq g(x)$ we have $f(x) - g(x) \geq 0$, and the graph of the difference $y = f(x) - g(x)$ never falls below the x-axis, even if one or both of the graphs of $f(x)$ or $g(x)$ do. Thus we can extend the formula

$$A = \int_a^b [f(x) - g(x)]\, dx$$

to calculate the area of the region between the graphs of any two functions $f(x)$ and $g(x)$, as long as $f(x) \geq g(x)$.

If $f(x) \geq g(x)$, then the area of the region between the graphs of the equations $y = f(x)$ and $y = g(x)$ over the interval [a, b] is

$$\int_a^b [f(x) - g(x)]\, dx.$$

EXAMPLE 13

Find the area of the region between the graphs of $y_1 = x^2 - 5$ and $y_2 = x^3$ over the interval [-1, 1].

Solution

The region of interest is shaded in Figure 12.42. Throughout the interval $[-1, 1]$, the graph of the function $y_2 = x^3$ is always above the graph of the function $y_1 = x^2 - 5$, so we use Equation (41) to calculate the area of the region:

$$A = \int_{-1}^{1} [x^3 - (x^2 - 5)] \, dx = \int_{-1}^{1} (x^3 - x^2 + 5) \, dx$$

$$= \left[\frac{1}{4} x^4 - \frac{1}{3} x^3 + 5x \right]_{x=-1}^{x=1}$$

$$= [\tfrac{1}{4}(1)^4 - \tfrac{1}{3}(1)^3 + 5(1)] - [\tfrac{1}{4}(-1)^4 - \tfrac{1}{3}(-1)^3 + 5(-1)]$$

$$= \tfrac{59}{12} - (-\tfrac{53}{12}) = \tfrac{28}{3}$$

FIGURE 12.42

The Region between the Graphs of $y_2 = x^3$ and $y_1 = x^2 - 5$ over $[-1, 1]$

Either of the functions $f(x)$ or $g(x)$ in Equation (41) can be the zero function, whose graph is the x-axis. If $g(x) = 0$ on the interval $[a, b]$, then the requirement $f(x) \geq g(x)$ reduces to $f(x) \geq 0$, and Equation (41) simplifies to Equation (32) for finding the area of the region under a nonnegative, continuous curve and over the x-axis. If, on the other hand, $f(x) = 0$ on the interval $[a, b]$, then the requirement $f(x) \geq g(x)$ reduces to $g(x) \leq 0$. The graph of such a function falls on or *below* the x-axis. The region of interest is the shaded portion of Figure 12.43, and Equation (41) simplifies to

The area of the region between the graph of a nonpositive function $g(x)$ and the x-axis, from $x = a$ to $x = b$, is

$$\int_a^b -g(x) \, dx.$$

$$A = \int_a^b -g(x) \, dx \tag{42}$$

We obtain the area of a region between the x-axis and the graph of a function that lies completely *below* the x-axis, from $x = a$ to $x = b$, by evaluating the definite integral of the *negative* of that function from a to b.

FIGURE 12.43

The Region between the x-axis and the Graph of $y = g(x) \leq 0$, on the Interval $[a, b]$

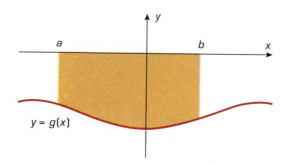

EXAMPLE 14

Find the area A of the region between the graph of $y = \sin x$ and the x-axis, over the interval $[\pi, 2\pi]$.

Solution

The region of interest is shaded in Figure 12.44. The function $g(x) = \sin x$ is continuous and nonpositive throughout the interval $[\pi, 2\pi]$, so we can use Equation (42) to calculate A:

$$A = \int_\pi^{2\pi} -\sin x \, dx = (\cos x)\Big|_{x=\pi}^{x=2\pi}$$

$$= \cos 2\pi - \cos \pi = 1 - (-1) = 2$$

FIGURE 12.44

The Region between the x-axis and the Graph of $y = \sin x$, from $x = \pi$ to $x = 2\pi$

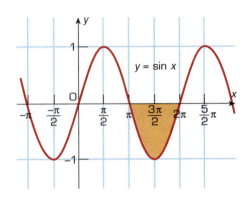

FIGURE 12.45

The Region between the Graph of $y = f(x)$ and the x-axis, where $f(x) \leq 0$ on $[a, c]$ and $f(x) \geq 0$ on $[c, b]$

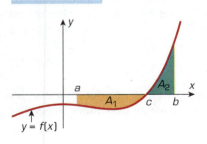

The shaded region in Figure 12.45 is interesting because it is bounded by two different curves (one is the x-axis) that cross at $x = c$. On the interval $[a, c]$, notice that $f(x) \leq 0$ (the function is nonpositive and its graph bounds the bottom of the region), while on the interval $[c, b]$, $f(x) \geq 0$ (the function is nonnegative and its graph bounds the top of the region). We can use our existing formulas to calculate the area A of the entire shaded region if we separate the area into two parts, $A = A_1 + A_2$.

On the interval $[a, c]$, the region with area A_1 is entirely above the graph of $f(x)$ and under the x-axis, so its area is given by the formula

$$A_1 = \int_a^c [0 - f(x)] \, dx = \int_a^c -f(x) \, dx$$

On the interval $[c, b]$, the region with area A_2 is entirely *under* the graph of $f(x)$ and *above* the x-axis, so its area is given by the formula

$$A_2 = \int_c^b [f(x) - 0] \, dx = \int_c^b f(x) \, dx$$

It now follows from Equation (37) that the entire shaded region has area

$$A = \int_a^c -f(x) \, dx + \int_c^b f(x) \, dx$$

To find the area of a region between the x-axis and the graph of a function that is partially below the x-axis and partially above, over an interval $[a, b]$, we find the area of the part that is below the x-axis, then the area of the part above the x-axis, and add the results.

EXAMPLE 15

Find the area of the region between the graph of the function $y = 2x - 1$ and the x-axis, over the interval $[-2, 2]$.

Solution

The region of interest is shaded in Figure 12.46. The two curves, $y = 2x - 1$ and $y = 0$, intersect at the point where their y-values are equal, that is, when

$$2x - 1 = 0$$

FIGURE 12.46

The Region between the Graph of $y = 2x - 1$ and the x-axis, over the Interval $[-2, 2]$

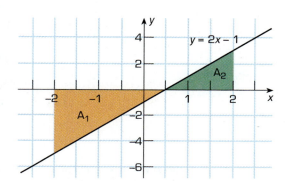

This occurs at $x = \frac{1}{2}$, which agrees nicely with a graphical estimate from Figure 12.46. On the interval $[-2, \frac{1}{2}]$ the function is nonpositive, so we calculate

$$A_1 = \int_{-2}^{1/2} -(2x - 1) \, dx = [-x^2 + x]_{x=-2}^{x=1/2} = 6.25$$

On the interval $[\frac{1}{2}, 2]$ the function is nonnegative, so we calculate

$$A_2 = \int_{1/2}^{2} (2x - 1) \, dx = [x^2 - x]_{x=1/2}^{x=2} = 2.25$$

Thus the area of the entire shaded region in Figure 12.46 is

$$A = A_1 + A_2 = 6.25 + 2.25 = 8.5 \text{ square units}$$

Notice that since both A_1 and A_2 are areas of triangles, we can check our answers using elementary geometry.

If the two curves cross, so that one is above the other over one part of the interval and then below the other over the rest of the interval, we handle each section separately. In each section, we subtract the function defining the lower curve *from* the function defining the upper curve, thus ensuring that the difference is nonnegative over the interval of integration. The sum of the areas over each section is the area of the entire region.

EXAMPLE 16

Find the area of the region between the graphs of $y = x$ and $y = x^3$ over the interval $[0, 2]$.

Solution

The region of interest is shaded in Figure 12.47. On part of the interval $[0, 2]$, the graph of $y = x^3$ is above the graph of $y = x$, and on the other part, the graph of $y = x^3$ is below

FIGURE 12.47

The Region between the Graphs of $y = x^3$ and $y = x$, over the Interval [0, 2]

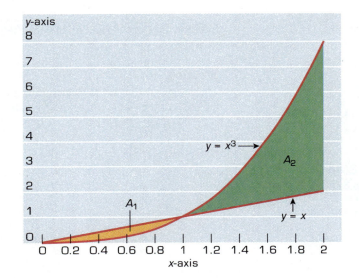

the graph of $y = x$. The two curves intersect at those points where their y-values are equal, or when $x^3 = x$. Solving for x, we find that

$$x^3 - x = x(x^2 - 1) = 0$$

$$x(x - 1)(x + 1) = 0$$

The two curves intersect at $x = 0$, $x = 1$, and $x = -1$. Since this last point is not in [0, 2], we ignore it. Within the interval [0, 2], the two curves intersect at $x = 0$ and $x = 1$, both of which agree nicely with graphical estimates from Figure 12.47.

On the interval [0, 1] we have $x \geq x^3$, so here we integrate the nonnegative difference $x - x^3$. Using the results of Example 11, the area above [0, 1] is

$$A_1 = \int_0^1 (x - x^3)\, dx = \frac{1}{4}$$

On the interval [1, 2] we have $x^3 \geq x$, so there we integrate the nonnegative difference $x^3 - x$. Now using the results of Example 12, the area of this region is

$$A_2 = \int_1^2 (x^3 - x)\, dx = \frac{9}{4}$$

Therefore the area of the entire shaded region in Figure 12.47 is

$$A = A_1 + A_2 = \frac{1}{4} + \frac{9}{4} = 2.5 \text{ square units.}$$

IMPROVING SKILLS

In Problems 1 through 26, evaluate each definite integral.

1. $\int_0^2 x^4\, dx$

2. $\int_{-1}^0 x^4\, dx$

3. $\int_{-3}^{2} x^4 \, dx$

4. $\int_{1}^{1.5} 2t^7 \, dt$

5. $\int_{-2}^{5} (3x - 22) \, dx$

6. $\int_{4}^{25} (\sqrt{x} - 5) \, dx$

7. $\int_{1}^{10} \left(x - \frac{1}{x} \right) dx$

8. $\int_{9}^{36} \frac{1}{\sqrt{Q}} \, dQ$

9. $\int_{0}^{\pi} 3 \sin \theta \, d\theta$

10. $\int_{-\pi}^{\pi} 3 \sin \theta \, d\theta$

11. $\int_{-\pi}^{\pi} (\tfrac{1}{3} \sin t + 3) \, dt$

12. $\int_{-10}^{0} (1 - 3 \cos \theta) \, d\theta$

13. $\int_{0}^{\pi/2} (\sin x - \cos x) \, dx$

14. $\int_{-1}^{1} (3 + 2e^x) \, dx$

15. $\int_{-4}^{-1} (e^u - 1) \, du$

16. $\int_{0}^{0.5} 4x^3 (1 + x^4)^6 \, dx$

17. $\int_{0}^{1} 8x\sqrt{4 - x^2} \, dx$

18. $\int_{0}^{9} \frac{x}{x^2 + 1} \, dx$

19. $\int_{3}^{6} \frac{u^2}{\sqrt{u^3 - 6}} \, du$

20. $\int_{0}^{\pi} \sin(\theta/2) \, d\theta$

21. $\int_{1}^{1.1} (\cos x)/(\sin x) \, dx$

22. $\int_{-0.3}^{0.5} e^{4x} \, dx$

23. $\int_{0}^{100} e^{-0.01t} \, dt$

24. $\int_{0.1}^{0.4} 2xe^{x^2} \, dx$

25. $\int_{-1}^{1} e^t e^{e^t} \, dt$

26. $\int_{0}^{1} \frac{e^\theta}{1 + e^\theta} \, d\theta$

In Problems 27 through 40, find the areas of the regions between the given boundaries.

27. $y = x^2 + 1$ and the x-axis; $x = 1$ to $x = 5$
28. $y = x^2 + 1$ and the x-axis; $x = -2$ to $x = 2$
29. $y_1 = x^2 + 1$ and $y_2 = 10$; $x = -2$ to $x = 2$
30. $y_1 = x^2 + 1$ and $y_2 = 10$; $x = 3$ to $x = 4$
31. $y_1 = x^2 + 1$ and $y_2 = x$; $x = 0$ to $x = 3$
32. $y_1 = x^2 - 2$ and $y_2 = x$; $x = -1$ to $x = 2$
33. $y_1 = x^2 - 2$ and $y_2 = x$; $x = 2$ to $x = 3$
34. $y = x/2$ and the x-axis; $x = -1$ to $x = 1$
35. $y_1 = x^2$ and $y_2 = 2 - x^2$; $x = -1$ to $x = 1$
36. $y = e^{-x}$ and the x-axis; $x = 0$ to $x = 1$
37. $y_1 = e^{-x}$ and $y_2 = -x$; $x = -1$ to $x = 1$
38. $y_1 = e^{-x}$ and $y_2 = e^x$; $x = -1$ to $x = 1$
39. $y = \cos x$ and the x-axis; $x = 0$ to $x = \pi$
40. $y = \cos x$ and $y = \sin x$; $x = 0$ to $x = \pi/2$

CREATING MODELS

Use definite integration to solve the following problems. Many of these problems are identical to problems in Section 12.2, where they were solved using indefinite integration.

41. A entrepreneurial couple produces handcrafted chess sets in their home. The marginal cost (\$/item) to produce these sets is modeled by the function $MC(n) = 4 - 2n +$

$6n^2$, where n is the total number of chess sets produced each month. Find the total cost of producing five chess sets a month if the couple's monthly fixed cost is $500.

42. What is the additional cost in Problem 41 if they decide to increase monthly production from five sets to eight?

43. What are the savings in Problem 41 if they decide to reduce monthly production from five sets to four?

44. A manufacturer of electric can openers has a daily fixed cost of $21,000 and knows that marginal cost is modeled by the equation $MC = \$5.50$/item. Find the total cost of producing 3200 can openers in one day.

45. What is the additional cost to the manufacturer in Problem 44 if production is increased from 3200 can openers to 3500?

46. A company's marginal revenue is given by the function $MR(n) = -0.008n^3 + 0.255n^2 - 2.3n + 5.87$, where n is the number of items (in thousands) offered for sale and total revenue is measured in millions of dollars. What is the total revenue from a weekly production of 12,000 items?

47. What is the additional revenue to the company in Problem 46 if they decide to increase weekly production from 12,000 items to 13,000 items?

48. Summer vacations taken by employees of the company in Problem 46 reduce weekly production capacity from 12,000 items to 10,000 items from June through September. What is the loss in revenue to the company during this period?

49. A firm's yearly marginal revenue is given by the equation $MR(t) = 3 + t$, where t denotes the number of years the firm has been in operation and total revenues are measured in millions of dollars. How much did their total revenues increase from year 1 to year 3?

50. How much total revenue did the firm in Problem 49 attract over its first three years in business?

51. A car dealership believes that its monthly marginal revenue is related to advertising expenditure, E, by the function $MR(E) = 6,000 - 50E$, where total revenue is measured in dollars and advertising expenditures are measured in thousands of dollars. Without any advertising, the company will generate $150,000 in sales. What total revenues can the company expect if it budgets $45,000 for advertising?

52. How much will revenues increase if the dealership in Problem 51 borrows an additional $5,000 and increases its advertising budget to $50,000?

53. Scientists believe that the growth rate in a certain bacterial culture is modeled by the equation $dN/dt = 0.005e^t$ ($0 \le t \le 12$), where time t is measured hours and N is measured in the number of strands of bacteria. Based on this model, estimate the number of strands of bacteria after ten hours if the culture is initialized at $t = 0$ with 100 strands.

54. How much does the population increase in Problem 53 from $t = 1$ to $t = 3$?

55. The population $P(t)$ of a city is declining at the rate $dP/dt = -4(t + 105)$, where t is in years. Estimate the population of this city in 50 years if the current population is 250,000 people.

56. How much will the city's population decrease in Problem 55 over the next five years?

57. A study commissioned for a daily newspaper finds that the growth rate of S, the number of subscribers to home delivery, is $dS/dt = 36\sqrt{t}$, with t measured in months. How many subscriptions (to the nearest hundred) will the company have in two years if home deliveries currently total 26,400?

58. How far will the object in Example 5 travel in ten seconds? How much distance will it cover between $t = 1$ and $t = 5$?

59. How far did the car in Example 6 travel during the first two hours? The next two hours?

EXPLORING IN TEAMS

60. Use the fundamental theorem of calculus to show that $\int_{-1}^{1} x^3 \, dx = 0$. Explain this result in terms of areas.

61. Use the fundamental theorem of calculus to show that $\int_{\pi}^{2\pi} \sin x \, dx = -2$. Explain this result in terms of areas.

62. Use the fundamental theorem of calculus to show that $\int_{-1}^{2} x^3 \, dx = 3.75$. Explain this result in terms of areas.

63. All of the definite integrals considered in this section and the previous one involved continuous functions. Explain how would you extend the process of definite integration to evaluate $\int_{0}^{4} f(x) \, dx$ for each of the following functions:

(**a**) $f(x) = \begin{cases} x, & 0 \le x < 2 \\ 3, & 2 \le x \le 4 \end{cases}$

(**b**) $f(x) = \begin{cases} x^2, & 0 \le x \le 1 \\ 5 - x, & 1 < x \le 4 \end{cases}$

(**c**) $f(x) = 1/\sqrt{x}$

(**d**) $f(x) = 1/x$

EXPLORING WITH TECHNOLOGY

64. Many graphing calculators have menus for evaluating a definite integral. Use such a calculator to reproduce the integration results in Examples 4 through 9.

REVIEWING MATERIAL

65. (Section 1.4) Construct a qualitative graphical model that relates the balance in a family's checking account (vertical axis) to time (horizontal axis) over a period of a month. Assume that all family members get paid at the end of each week.

66. (Section 4.1) An employee begins a new job at a salary of $22,500/yr and receives a 3% raise every six months. What will the salary of this employee be after ten years?

67. (Section 10.6) Determine where on the interval [0, 6] the following function is differentiable:

$$f(x) = \begin{cases} x, & 0 \le x < 2 \\ 4 - x, & 2 \le x < 4 \\ x^2 - 9x + 20, & 4 \le x \le 6 \end{cases}$$

68. (Section 11.5) Find the second derivative of the function

$$y = 3x^2 - 50x + 6 \sin 2x$$

RECOMMENDING ACTION

69. Respond by memo to the following request:

MEMORANDUM

To:	J. Doe Reader
From:	V.P. for Planning
Date:	Today
Subject:	**Figure 12.37**

I enjoyed your presentation, especially your graph Figure 12.37. As you displayed that graph, I wondered whether the curve for $f'(x)$ could ever be negative. If so, what would it represent? What area would we consider, and what would be the significance of that area?

12.6   PROBABILITY DENSITY FUNCTIONS

Definite integration is particularly useful for calculating probabilities associated with continuous processes. As we noted in Section 8.3, such processes allow for *any* outcome in some interval on the real line. For example, the time between bus arrivals at a station may be 8 minutes or 20 minutes or any real number between these two values. In contrast, the number of customers in a bank at any given time is a discrete process, not a continuous one. There, we may have an outcome of 2 customers or 5 customers, but not 3.7 customers.

In Section 8.3, we showed how to model a continuous process by using histograms and frequency polygons. In Sections 8.4 and 8.5, we refined some of our models by recognizing their frequency polygons as approximations to normal and exponential curves. Now we want to increase our pool of recognizable functions.

A frequency polygon is a graphical model of a sample. Different samples from the same population may lead to different frequency polygons, but all of them will have similar shapes, that is, they will all approximate the same curve. We call such a curve a *probability function,* and we use it as a graphical model for the underlying continuous process being sampled. Figures 12.48 and 12.49 are two frequency polygons that model a sample of male heights, first considered in Examples 3 and 5 of Section 8.3; each is an approximation to the same normal curve.

All frequency polygons are nonnegative, because the height of any bar in a histogram is the number of elements in a particular subset. Furthermore, since we create a frequency polygon by drawing a straight line between the midpoints of adjacent bars in a histogram, frequency polygons are continuous. Thus probability functions are nonnegative continuous functions.

FIGURE 12.48

A Frequency Polygon of Heights from 5′4″ to 6′4″ with Partitions of 2 Inches

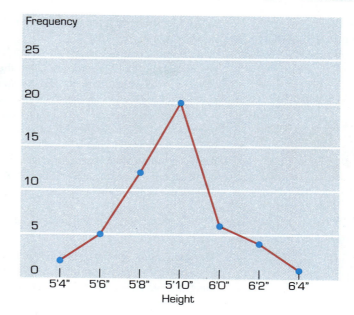

FIGURE 12.49

Frequency Polygon of Heights from 5′3″ to 6′3″ with Partitions of 2 Inches

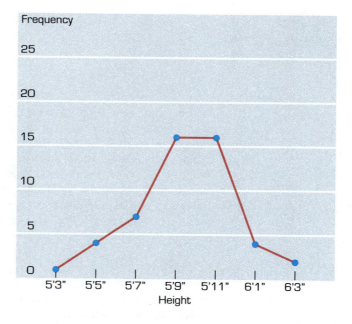

A nonnegative continuous function on a finite interval is a probability function.

The domain of a probability function is the set of all possible outcomes from a continuous process. In theory, the domain of a continuous process may be infinite, but as a practical matter, this does not happen. For example, the time between bus arrivals at a station can be very long, but it is not likely to be more than a day and almost certainly less than a week. If buses stop arriving at a scheduled stop for any protracted time period, customers will complain and the bus company will remedy the problem. Thus we begin by defining a probability function as a nonnegative continuous function on a finite interval

FIGURE 12.50

The Region under the Graph of a Probability Function with Domain [a, b]

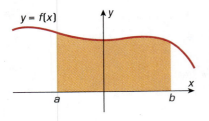

$y = f(x)$

FIGURE 12.51

Regions under the Graph of a Probability Function: over an Interval [a, b] and over a Subinterval [c, d]

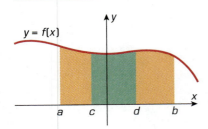

$y = f(x)$

[a, b]. Consequently, the area between the graph of such a curve and its domain is a finite number; see Figure 12.50.

The simplest types of probability calculations occur when processes have a finite number of equally likely outcomes. Then the probability $P(E)$ that an event E will happen is equal to the cardinal number $n(E)$ divided by the cardinal number $n(U)$ of the entire sample space U, that is, $P(E) = n(E)/n(U)$, which is Equation (1) of Section 6.1.

Surprisingly, we can extend this equation in a very straightforward manner to a continuous process with probability function $f(x)$. The sample space U remains the set of all possible outcomes, which, for the moment, we take to be the interval [a, b]. We let X denote a random outcome from such a process, and we consider an event E to be the set of all outcomes on an interval subset of [a, b], say, [c, d]. We then define the probability that an event E will occur to be

$$P(E) = P(c \le X \le d) = \frac{\text{area under } f(x) \text{ and over } [c, d]}{\text{area under } f(x) \text{ and over } [a, b]} \qquad (43)$$

The numerator on the right side of Equation (43) is represented by the green region in Figure 12.51. The denominator is the region shaded in both green and orange. In terms of integrals, we can rewrite Equation (43) as

$$P(c \le X \le d) = \frac{\displaystyle\int_c^d f(x)\, dx}{\displaystyle\int_a^b f(x)\, dx} \qquad (44)$$

EXAMPLE 1

International clients of an advertising agency may pay their invoices by transferring funds electronically through the World Wide Web. Based on historical data, it appears that the length of time (in days) between an invoice's being sent and its being paid is a continuous process modeled by the probability function $f(x) =$ $100x - x^2$ on [20, 80]: no client pays in less than 20 days and all clients pay within 80 days of receiving an invoice. Determine the probability that the next invoice will be paid between 50 and 60 days after the invoice is sent.

Solution

The function $f(x) = 100x - x^2$ is graphed in Figure 12.52 on the domain [20, 80], and we see that the function is nonnegative and continuous there. We want to determine the probability that an outcome X will fall in the interval [50, 60]. It follows from Equation (44) that

$$P(50 \le X \le 60) = \frac{\displaystyle\int_{50}^{60} (100x - x^2)\, dx}{\displaystyle\int_{20}^{80} (100x - x^2)\, dx} = \frac{\left. (50x^2 - \tfrac{1}{3}x^3) \right|_{50}^{60}}{\left. (50x^2 - \tfrac{1}{3}x^3) \right|_{20}^{80}}$$

$$= \frac{24{,}666\tfrac{2}{3}}{132{,}000} \approx 0.1869$$

FIGURE 12.52

The Probability Function $y =$ **$100x - x^2$ on the Interval [20, 80] with Subinterval [50, 60]**

This number is the ratio of the area of the region in Figure 12.52 shaded in green to the area of the entire region, which is shaded in green and orange.

EXAMPLE 2

Determine the probability that when the next invoice is sent by the ad agency in Example 1, it will be paid within 30 days.

Solution

We are interested in determining the probability that an outcome X is at most 30. We have from historical data that no outcome occurs in less than 20 days, hence we want the probability that $20 \le X \le 30$. It follows from Equation (44) that

$$P(20 \le X \le 30) = \frac{\displaystyle\int_{20}^{30} (100x - x^2)\, dx}{\displaystyle\int_{20}^{80} (100x - x^2)\, dx} = \frac{\left(50x^2 - \frac{1}{3}x^3\right)\Big|_{20}^{30}}{132{,}000}$$

$$= \frac{18{,}666\frac{2}{3}}{132{,}000} \approx 0.1414$$

This number is the ratio of the area of the region in Figure 12.53 shaded in green to the area of the entire region, which is shaded in green and orange.

FIGURE 12.53

The Probability Function $y =$ **$100x - x^2$ on the Interval [20, 80] with Subinterval [20, 30]**

The denominator in Equation (44),

$$\int_a^b f(x)\, dx = A \tag{45}$$

is the area A of the region under the graph of the probability function $f(x)$ and over the domain $[a, b]$ of the sample space. In Examples 1 and 2, when $a = 20$ and $b = 80$, we found the area under the curve to be $A = 132{,}000$. If we divide both sides of Equation (45) by A, we get

$$\frac{1}{A} \int_a^b f(x)\, dx = 1 \quad \text{or} \quad \int_a^b \frac{f(x)}{A}\, dx = 1$$

> *A probability density function $g(x)$ is a nonnegative continuous function over $[a, b]$ with*
> $$\int_a^b g(x)\, dx = 1.$$

This process of dividing a probability function $f(x)$ by A, the area of the region under the curve and over the domain, is called *normalizing* the function. It yields a new function, $g(x) = f(x)/A$, that is also nonnegative and continuous, with

$$\int_a^b g(x)\, dx = 1$$

The area of the region under the graph of $g(x)$ and over the domain $[a, b]$ is 1. A *probability density function* is a probability function that has been normalized.

EXAMPLE 3

Normalize the probability function from Example 1.

Solution

The probability function is $f(x) = 100x - x^2$ over the domain $[20, 80]$. Since

$$\int_a^b f(x)\, dx = \int_{20}^{80} (100x - x^2)\, dx = 132{,}000$$

the associated probability density function is

$$g(x) = \frac{1}{132{,}000}(100x - x^2)$$

We obtain a probability density function $g(x)$ by normalizing a probability function $f(x)$ over its domain. We can use $g(x)$ to simplify Equation (44):

$$P(E) = \frac{\int_c^d f(x)\, dx}{\int_a^b f(x)\, dx} = \frac{\int_c^d f(x)\, dx}{A} = \frac{1}{A} \int_c^d f(x)\, dx = \int_c^d \frac{f(x)}{A}\, dx$$

or

$$P(c \le X \le d) = \int_c^d g(x)\, dx \qquad (46)$$

This is the probability that a random outcome X will fall in a subset $[c, d]$ of the domain $[a, b]$.

EXAMPLE 4

A continuous process is modeled by the probability function $g(x) = 4x^3$ on the domain $[0, 1]$. What is the probability that a random outcome will be less than or equal to 3/4?

Solution

The function, which is graphed in Figure 12.54, is nonnegative and continuous on the interval $[0, 1]$, with

$$\int_0^1 4x^3\, dx = x^4 \Big|_0^1 = 1$$

FIGURE 12.54

The Probability Function $y = 4x^3$ on the Domain [0, 1]

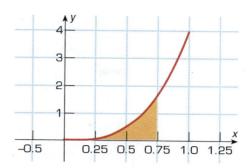

Therefore $g(x) = 4x^3$ is a probability density function on the specified domain and we can use Equation (46).

The domain is [0, 1], so for a random outcome to be less than or equal to 3/4, it must fall in the interval [0, 3/4]. Thus

$$P(0 \le X \le \tfrac{3}{4}) = \int_0^{3/4} 4x^3 \, dx = x^4 \Big|_0^{3/4} = \tfrac{81}{256}$$

This is the area of the shaded region in Figure 12.54.

EXAMPLE 5

What is the probability that an outcome in the continuous process described in Example 4 will fall in the interval [0.25, 0.75]?

Solution

$$P(0.25 \le X \le 0.75) = \int_{0.25}^{0.75} 4x^3 \, dx = x^4 \Big|_{0.25}^{0.75} = 0.3125$$

We calculate the probability of an event in a continuous process in one of two equivalent ways: either we use Equation (44) with a probability function $f(x)$, or we use Equation (46) with a probability density function $g(x)$. Since Equation (46) is simpler, most practitioners work with probability density functions by first normalizing a probability function when necessary.

Let us summarize how we use calculus to find probabilities associated with events in a continuous process. We begin by sampling the continuous process and constructing a frequency polygon from the sample. This frequency polygon is an approximation to a probability function on a domain [a, b]. If we recognize the probability function $f(x)$ or have a way to obtain it (see Problems 47 to 52), then we have a model for the underlying continuous process. We can normalize $f(x)$ to obtain a probability density function $g(x)$, which we can then use with Equation (46) to calculate probabilities.

Any sample space U associated with continuous processes can be contained within a finite interval, since U is a finite collection of observations. It will have a largest element, b, and a smallest element, a, so all outcomes in U will fall in the interval [a, b]. In theory, however, some sample spaces, such as the time it takes people to repay loans to a bank, can be infinitely large, and it is often convenient to model a continuous process with a function over an infinite domain, such as $(-\infty, \infty)$ or $[a, \infty)$, which represents the interval $a \le x < \infty$.

To calculate probabilities over an infinite domain $[a, \infty)$ or $(\infty, b]$, we must give meaning to integrals of the form

$$\int_a^\infty f(x)\, dx \quad \text{or} \quad \int_\infty^b f(x)\, dx \tag{47}$$

An integral with infinity for one of its "limits" is called an *improper integral*. None of our rules of integration so far, including the fundamental theorem of calculus, are applicable to improper integrals because those rules were developed for finite domains. We can, however, use the concept of a limit and adapt our integration rules to accommodate improper integrals.

If $f(x)$ is a nonnegative continuous function on $[a, \infty)$, then the left-hand integral in (47) represents the area between the graph of $f(x)$ and the x-axis for all values of $x \ge a$. We select a large number m and use our rules of integration to calculate the area under $f(x)$ between $x = a$ and $x = m$. Since both a and m are real numbers, the interval $[a, m]$ is finite and all our rules of integration apply. Geometrically, we are finding the area of the shaded region in Figure 12.55. We then let m approach infinity. If there is a finite number L such that the limit of this area exists and is equal to L, that is, if

$$\lim_{m \to \infty} \int_a^m f(x)\, dx = L$$

FIGURE 12.55

A Probability Function on the Domain $[a, m]$ as $m \to \infty$

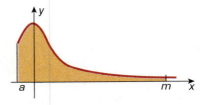

then we say that "the improper integral (47) *converges* to L" and we write

$$\int_a^\infty f(x)\, dx = \lim_{m \to \infty} \int_a^m f(x)\, dx = L$$

In such cases, we say that $f(x)$ is a probability function on $[a, \infty)$. If, in addition, $L = 1$, then we say that $f(x)$ is a probability density function on $[a, \infty)$. On the other hand, if there is no such finite number, then we say "the improper integral (47) *diverges*" and we cannot calculate probabilities with $f(x)$ on $[a, \infty)$.

EXAMPLE 6

Determine whether the integral $\int_0^\infty \frac{1}{2} e^{-x/2}\, dx$ converges.

Solution
The natural exponential function is nonnegative and continuous everywhere. Here

$$\int_0^\infty \frac{1}{2} e^{-x/2}\, dx = \lim_{m \to \infty} \int_0^m \frac{1}{2} e^{-x/2}\, dx = \lim_{m \to \infty} (-e^{-x/2}) \Big|_0^m$$

$$= \lim_{m \to \infty} (-e^{-m/2} + e^{0/2}) = \lim_{m \to \infty} \left(1 - \frac{1}{e^{m/2}}\right) = 1$$

and the integral converges to $L = 1$. The function $\frac{1}{2} e^{-x/2}$ is a probability density function on $[0, \infty)$.

The function in Example 6 is a probability density function for an exponential distribution (see Section 8.5). It has a maximum of 1/2 at $x = 0$, so it has a mean of 2. More generally, the probability density function for an exponential distribution with a mean of μ is

$$g(x) = \frac{1}{\mu} e^{-x/\mu} \quad (0 \le x < \infty) \tag{48}$$

Its graph is displayed in Figure 12.56.

The probability density function for an exponential distribution on $[0, \infty]$ with a mean of μ is $g(x) = \frac{1}{\mu} e^{-x/\mu}$.

FIGURE 12.56
The Probability Density Function for an Exponential Distribution with Mean μ

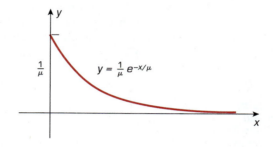

$$y = \frac{1}{\mu} e^{-x/\mu}$$

EXAMPLE 7

The interarrival time of cars to the drive-in window of a fast-food restaurant is exponentially distributed with a mean of 2 minutes. Determine the probability that the next car will arrive (*a*) between 0.5 and 3 minutes after the previous arrival; (*b*) in 3 minutes or less after the previous arrival.

Solution

(*a*) We seek the probability that a random event X will fall within the interval $[0.5, 3]$. The probability density function for this distribution is given by Equation (48) with $\mu = 2$:

$$P(0.5 \le X \le 3) = \int_{0.5}^{3} \frac{1}{2} e^{-x/2} \, dx = -e^{-x/2} \Big|_{0.5}^{3}$$

$$= -e^{-3/2} + e^{-0.5/2} \approx 0.5557$$

(*b*) Now we want the probability that a random event X will fall in the interval $[0, 3]$, so we use Equation (48) again but with a different lower limit:

$$P(0 \le X \le 3) = \int_{0}^{3} \frac{1}{2} e^{-x/2} \, dx = -e^{-x/2} \Big|_{0}^{3}$$

$$= -e^{-3/2} + e^{-0/2} \approx 0.7769$$

Compare this approach with that in Example 2 of Section 8.5.

The probability density function for the normal distribution on $[-\infty, \infty]$ with zero mean and standard deviation 1 is

$$g(x) = \frac{1}{\sqrt{2\pi}} e^{-x^2/2}.$$

The probability density function for the normal distribution (see Section 8.4) with zero mean and standard deviation 1 is

$$g(x) = e^{-x^2/2}/\sqrt{2\pi} \quad (-\infty < x < \infty)$$

Like any normal curve, $g(x)$ is symmetric around its mean (see Section 8.4). Consequently, for any positive number c,

$$\int_{-c}^{0} \frac{1}{\sqrt{2\pi}} e^{-x^2/2} \, dx = \int_{0}^{c} \frac{1}{\sqrt{2\pi}} e^{-x^2/2} \, dx$$

The graph of $g(x)$ is displayed in Figure 12.57. The area shaded in green is identical to the area shaded in orange. Thus, when working with a probability density function for a normal distribution, we need to evaluate only integrals with nonnegative limits.

FIGURE 12.57
The Probability Density Function for a Normal Distribution with Zero Mean and Standard Deviation 1

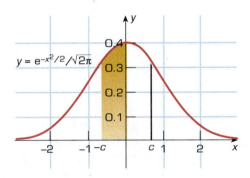

$y = e^{-x^2/2}/\sqrt{2\pi}$

EXAMPLE 8

Find the probability that a random outcome X from a normal distribution with zero mean and standard deviation 1 will fall into the interval $[-1, 0]$.

Solution

We can use Equation (46) here:

$$P(-1 \le X \le 0) = \int_{-1}^{0} \frac{1}{\sqrt{2\pi}} e^{-x^2/2} \, dx = \int_{0}^{1} \frac{1}{\sqrt{2\pi}} e^{-x^2/2} \, dx$$

$$= \frac{1}{\sqrt{2\pi}} \int_{0}^{1} e^{-x^2/2} \, dx$$

Unfortunately, we do not know an antiderivative for $f(x) = e^{-x^2/2}$, so we cannot use the fundamental theorem of calculus to evaluate this integral. We can, however, use the midpoint method developed in Section 12.4 to estimate this area. Doing so, we generate Table 12.6. We can feel confident that

$$P(-1 \le X \le 0) = P(0 \le X \le 1) = \frac{1}{\sqrt{2\pi}} \int_{0}^{1} e^{-x^2/2} \, dx = 0.3413$$

rounded to four decimal places.

TABLE 12.6 Estimates for $\dfrac{1}{\sqrt{2\pi}} \displaystyle\int_0^1 e^{-x^2/2}\,dx$

NUMBER OF PARTITIONS n	WIDTH OF EACH BASE $\Delta x = \dfrac{1 - (0)}{n}$	INTEGRAL ESTIMATE
10	0.1	0.34145
20	0.05	0.34137
50	0.02	0.34135
100	0.01	0.34135
200	0.005	0.34134
400	0.0025	0.34134

The answer in Example 8 is identical, to four decimal places, to the entry in Table 8.29 with $z = 1$. This is not a coincidence. Table 8.29 was compiled using numerical techniques, such as the midpoint method, to evaluate the integral

$$\frac{1}{\sqrt{2\pi}} \int_0^z e^{-x^2/2}\,dx$$

for different values of z. It is much easier to use such a table than to estimate integrals of the probability density function for a standard normal distribution as we did in Example 8.

Of course, most normally distributed processes do not have a zero mean and standard deviation 1. However, if X is a random outcome from a normally distributed process with mean μ and standard deviation σ, then the transformation

$$z = (X - \mu)/\sigma, \tag{49}$$

which is identical in form to Equation (13) in Section 8.4, converts X into a random outcome z selected from a normally distributed process with zero mean and standard deviation 1.

EXAMPLE 9

A sample space of blood pressures is normally distributed with a mean of 120 and a standard deviation of 10. What is the probability that a person chosen at random from this population will have a blood pressure between 115 and 132?

Solution

Using Equation (49) with $\mu = 120$ and $\sigma = 10$, we find that $X = 115$ corresponds to $z = (115 - 120)/10 = -0.5$, while $X = 132$ corresponds to $z = (132 - 120)/10 = 1.2$. Now we compute the probability for z:

$$P(115 \le X \le 132) = P(-0.5 \le z \le 1.2) = \int_{-0.5}^{1.2} \frac{1}{\sqrt{2\pi}} e^{-x^2/2}\,dx$$

Using symmetry, we rewrite this last equation as

$$\int_{-0.5}^{1.2} \frac{1}{\sqrt{2\pi}} e^{-x^2/2}\, dx = \int_{-0.5}^{0} \frac{1}{\sqrt{2\pi}} e^{-x^2/2}\, dx + \int_{0}^{1.2} \frac{1}{\sqrt{2\pi}} e^{-x^2/2}\, dx$$

$$= \int_{0}^{0.5} \frac{1}{\sqrt{2\pi}} e^{-x^2/2}\, dx + \int_{0}^{1.2} \frac{1}{\sqrt{2\pi}} e^{-x^2/2}\, dx$$

We can use the midpoint method to estimate either of these last two integrals, or we go to Table 8.29 and locate the estimates as tabular values associated with $z = 0.5$ and $z = 1.2$. This latter approach yields $0.19146 + 0.38493 = 0.57639$ as the probability that X will be between 115 and 132, rounded to five decimal places.

IMPROVING SKILLS

In Problems 1 through 20, determine whether the given function is a probability function over the specified domain. For those that are, construct a probability density function and then determine the probability that a random outcome from such a process will fall in the interval [2, 5].

1. $f(x) = x$; [0, 10]

2. $f(x) = x$ [0, 50]

3. $f(x) = x$; [−50, 50]

4. $f(x) = 2x$; [0, 10]

5. $f(x) = 2x + 10$; [0, 10]

6. $f(x) = 2x − 10$; [0, 10]

7. $f(x) = x^2$; [0, 10]

8. $f(x) = x^2$; [−10, 10]

9. $f(x) = 20$; [−10, 10]

10. $f(x) = 2$; [0, 20]

11. $f(x) = e^{2x}$; [0, 5]

12. $f(x) = e^{-2x}$; [0, 5]

13. $f(x) = e^{-2x}$; [−5, 5]

14. $f(x) = \begin{cases} 2x & 0 \le x \le 5 \\ 15 - x & 5 < x \le 15 \end{cases}$

15. $f(x) = \begin{cases} 2x & 0 \le x \le 5 \\ 15 - x & 5 < x \le 20 \end{cases}$

16. $f(x) = \begin{cases} 2x & 0 \le x \le 5 \\ 15 + x & 5 < x \le 20 \end{cases}$

17. $f(x) = \begin{cases} x + 5 & 0 \le x \le 4 \\ x^2 - 7 & 4 < x \le 10 \end{cases}$

18. $f(x) = \begin{cases} 15 & 0 \le x \le 5 \\ 25 - 2x & 5 < x \le 10 \end{cases}$

19. $f(x) = \begin{cases} 7 & 0 \le x \le 5 \\ 17 - 2x & 5 < x \le 10 \end{cases}$

20. $f(x) = \begin{cases} x^2 & 0 \le x \le 4 \\ 16 & 4 < x \le 10 \end{cases}$

21. Determine the probability that a randomly selected outcome will fall in the specified interval if the sample space is exponentially distributed with a mean of 50.
(**a**) [60, 70]   (**b**) [0, 20]
(**c**) [54, 68]   (**d**) [48, 55]

22. A sample space is exponentially distributed with a mean of 5. Find the probability that a randomly selected outcome will be (**a**) greater than 8; (**b**) greater than 4; (**c**) less than or equal to 7.

23. A sample space is exponentially distributed with a mean of 30. Find the probability that a randomly selected outcome will be (**a**) greater than 31; (**b**) greater than 20; (**c**) greater than 40.

24. Construct an integral for the probability that a randomly selected outcome will fall in the specified interval if the sample space is normally distributed with zero mean and standard deviation 1. Estimate the value of each integral.

(*a*) [0, 0.3] (*b*) [0, 2]
(*c*) [−1.5, 0] (*d*) [−0.2, 0.3]

25. Construct an integral for the probability that a randomly selected outcome will fall in the specified interval if the sample space is normally distributed with a mean of 22 and a standard deviation of 3. Estimate the value of each integral.

(*a*) [22, 24] (*b*) [22, 26]
(*c*) [19, 22] (*d*) [21, 25]

26. Construct an integral for the probability that a randomly selected outcome will fall in the specified interval if the sample space is normally distributed with a mean of 200 and a standard deviation of 15. Estimate the value of each integral.

(*a*) [200, 225] (*b*) [180, 210]
(*c*) [205, 208] (*d*) [185, 195]

27. Construct an integral for the probability that a randomly selected outcome will fall in the specified interval if the sample space is normally distributed with a mean of −50 and a standard deviation of 8. Estimate the value of each integral.

(*a*) [−60, −55] (*b*) [−56, −44]
(*c*) [−54, −40] (*d*) [−45, −40]

28. A sample space is normally distributed with a mean of 150 and a standard deviation of 22. Construct an integral for the probability that a randomly selected outcome will be (*a*) greater than 170; (*b*) greater than 200; (*c*) less than 100.

29. A sample space is normally distributed with a mean of 30 and a standard deviation of 0.3. Construct an integral for the probability that a randomly selected outcome will be (*a*) greater than 31; (*b*) greater than 29.5; (*c*) less than 29.

30. A sample space is normally distributed with a mean of −10 and a standard deviation of 1. Construct integrals for the probability that a randomly selected outcome will be (*a*) greater than −8; (*b*) less than −10; (*c*) less than −9.

In Problems 31 through 38, determine whether each improper integral converges.

31. $\int_{1}^{\infty} (1/x^3)\, dx$

32. $\int_{5}^{\infty} (1/x^2)\, dx$

33. $\int_{2}^{\infty} (1/x)\, dx$

34. $\int_{4}^{\infty} (1/\sqrt{x})\, dx$

35. $\int_{0}^{\infty} e^x\, dx$

36. $\int_{0}^{\infty} e^{-x}\, dx$

37. $\int_{0}^{\infty} [x/(1 + x^2)]\, dx$

38. $1/(x \ln x)\, dx$

CREATING MODELS

39. Salaries for employees of the Endurance Racing Bike Mfg. Company are exponentially distributed with a mean of $45,000. (*a*) What percent of all employees make between $25,000 and $35,000? (*b*) What percent make more than $60,000?

40. The sale of coffee beans at the Drop-In Coffee House is exponentially distributed with a mean of 2 lb. (*a*) What is the probability that the next sale will be for a pound or less? (*b*) Between 1 and 3 lb?

41. Rainfall in a region is exponentially distributed with an annual mean of 250 in. (**a**) What is the probability that the rainfall next year will exceed 300 inches? (**b**) 100 inches or less?

42. A standard IQ test is prepared so that scores will be normally distributed with a mean of 100 and a standard deviation of 15. What is the probability that a person will score between 100 and 130?

43. Scholastic Aptitude Tests are scaled so that scores are approximately normally distributed with a mean of 500 and a standard deviation of 100. What is the probability that a person will score 700 or higher?

44. Body weights of males at a large state university are normally distributed with a mean of 170 lb and a standard deviation of 20 lb. (**a**) What is the probability that the next male student who enters the infirmary will weigh between 160 and 180 lb? (**b**) Less than 140 lb?

45. The life span of a particular model of muffler is normally distributed with a mean of 40,000 miles and a standard deviation of 2,500 miles. What is the probability that such a muffler will wear out before 35,000 miles?

46. Salaries for technicians at the Classical Computer Company are normally distributed with a mean of $30,000 and a standard deviation of $5,500. (**a**) What percent of all technicians make between $25,000 and $35,000? (**b**) What percent make more than $40,000?

EXPLORING IN TEAMS

To determine a fair and reasonable price for a certain surgical procedure, an HMO reviewed the total cost of the same operation at many hospitals in the same geographical region. The results are tabulated in the following table, with all entries rounded to the nearest dollar.

8019	8028	8115	8153	8164	8200	8210
8265	8312	8321	8419	8452	8455	8512
8519	8543	8546	8573	8607	8612	8633
8666	8702	8719	8723	8744	8751	8813
8817	8828	8833	8855	8876	8917	8922
8923	8938	9002	9011	9012	9022	9034
9148	9156	9162	9165	9177	9183	9290
9218	9238	9279	9312	9329	9341	9352
9367	9368	9423	9438	9468	9473	9480
9493	9523	9550	9615	9627	9640	9652
9738	9752	9771	9785	9804	9829	9858
9914	9949	9981				

47. Construct a frequency table for this data, with ten categories.

48. Use two least-squares straight lines to fit the data from the frequency table obtained in Problem 47. Use one line to fit the increasing sequence of data points and a second line to fit the decreasing sequence of points.

49. Locate the point of intersection of the two lines found in Problem 48 and then create a probability function from these two lines.

50. Use the probability function from Problem 49 to estimate the probability that for the same procedure in the same geographical region, the next operation will cost $9100 or less.

51. Find a least-squares quadratic curve that best fits the data. *Hint*: See Problem 26 of Section 2.3.

52. Use the probability function from Problem 51 to estimate the probability that for the same procedure in the same geographical region, the next operation will cost $9100 or less. Compare your answer with the one in Problem 50.

EXPLORING WITH TECHNOLOGY

53. Use the midpoint method to estimate the integral

$$\int_0^m \frac{1}{\sqrt{2\pi}} e^{-x^2/2} \, dx$$

for $m = 3, 4, 5, 8, 10,$ and 20. Then use these results to verify that

$$\int_{-\infty}^{\infty} \frac{1}{\sqrt{2\pi}} e^{-x^2/2} \, dx = 2 \int_0^{\infty} \frac{1}{\sqrt{2\pi}} e^{-x^2/2} \, dx = 2 \lim_{m \to \infty} \int_0^m \frac{1}{\sqrt{2\pi}} e^{-x^2/2} \, dx$$

$$= 2(\tfrac{1}{2}) = 1$$

54. Many graphing calculators have menus for evaluating a definite integral. Use such a calculator to evaluate the integrals produced in Problems 24 through 30.

REVIEWING MATERIAL

55. (Section 3.3) A merchant has 100 lb of peanuts and 50 lb of cashews that she will use for making two types of mixes. The deluxe mix is 30% cashews and 70% peanuts and sells for $7.50/lb while the regular mix is 20% cashews and 80% peanuts and sells for $6/lb. How many pounds of each mix will maximize sales?

56. (Section 5.2) In 1997, the United Nations reported the following geographical distribution of HIV-infected adults: 20,800,000 in Sub-Saharan Africa; 210,000 in North Africa and the Middle East; 150,000 in Eastern Europe and Central Asia; 530,000 in western Europe; 1,300,000 in Latin America; 310,000 in the Caribbean; 860,000 in North America; 12,000 in Australia and New Zealand; 6,000,000 in South and Southeast Asia; and 440,000 in East Asia and Pacific countries. Draw a pie chart of this data.

57. (Section 6.1) Use the data in Problem 56 to estimate the probability that the next person to become infected with HIV will be from North America.

58. (Section 11.2) A couple has 100 ft of fencing to create *two* square plots of land: one for growing tomatoes and one for growing flowers. The two plots will be in different parts of their yard and do not have to be the same size. The sides of each plot cannot be greater than 20 ft or less than 3 ft. Find the dimensions of each plot if the combined enclosed area is to be as large as possible.

RECOMMENDING ACTION

59. Respond by memo to the following request:

MEMORANDUM

To: J. Doe Reader

From: Shoshanna

Date: Today

Subject: **Graphs**

I understand how to create a probability density function $g(x)$ from a probability function $f(x)$ by normalizing the latter. I was wondering, however, if there is any relationship between the graphs of these two functions. Do you know?

CHAPTER 12 KEYS

KEY WORDS

antiderivative (p. 845)
bounded region (p. 879)
definite integral (p. 891)
differential (p. 846)
frame (p. 879)
improper integral (p. 924)

indefinite integral (p. 846)
normalizing a function (p. 922)
probability density function (p. 922)
probability function (p. 918)
total cost (p. 859)
total revenue (p. 861)

KEY CONCEPTS

12.1 Antiderivatives and Indefinite Integrals

■ If $F(x)$ is an antiderivative of $f(x)$, then so too is the indefinite integral of $f(x)$. In fact, $F(x) + c$ is an antiderivative of $f(x)$ for every choice of an arbitrary constant c.

■ After an indefinite integral has been evaluated, the answer can be verified by differentiating it and checking that the result equals the function that was integrated.

12.2 Applications of the Indefinite Integral

■ Another variable label may be used instead of x in each rule of integration, as long as the new label replaces x throughout, including the x in the differential dx.

- The graphs of the functions from a family of indefinite integrals comprise a set of parallel curves.

- There is at most one function in a family of indefinite integrals that yields a given output value for a particular input value.

- The graph of a particular antiderivative is singled out from its set of parallel curves by specifying a point in the plane through which that graph must pass.

- Total cost is the indefinite integral of marginal cost. To evaluate the constant of integration, use the fact that total cost equals fixed cost when production is zero.

- Total revenue is the indefinite integral of marginal revenue. To evaluate the constant of integration, use the fact that total revenue is zero when production is zero.

- Position is an antiderivative of velocity; velocity is an antiderivative of acceleration.

12.3 Substitution of Variables

- Sometimes an integral that is difficult to evaluate can be rewritten algebraically as a combination of integrals that are easier to evaluate.

- The method of substitution of variables is a transformation method that replaces one integral by another, ideally one that is easier to evaluate.

- It accomplishes nothing to simply replace one variable with another (such as $v = u$ or $u = x$) when applying the method of substitution of variables.

- A rule of thumb for the method of substitution of variables is to set $u = g(x)$ when an integral involves a function $g(x)$ raised to a power.

- A rule of thumb for the method of substitution of variables is to set u equal to the argument of a function when an integral involves a function with that argument.

- The constant multiple rule of integration is used to transform the integral of a function with a multiplicative constant a into the integral of a function with multiplicative constant b, $b \neq a$.

12.4 The Definite Integral

- In most commercial transactions, area calculations are accurate only to a specified level of precision.

- Numerical techniques can be used to estimate areas of irregularly shaped regions.

- The midpoint method for estimating an area of a region between the graph of a non-negative continuous function and a closed interval on the x-axis becomes more accurate as the number of equally wide partitions of the interval increases. In the limit, as the number of partitions goes to infinity, these estimates equal that area.

- The units associated with the definite integral of an input–output function are the product of the input units and the output units.

- The area under the graph of a speed function and over a time interval on the horizontal axis is numerically equal to the distance traveled over that interval of time.

12.5 Applications of the Definite Integral

- The total change in the output of $f(x)$ over an interval $[a, b]$ is equal to the definite integral of $f'(x)$ over that interval.

■ The change in the total cost of increasing a production run from a items to b items is the definite integral of the marginal cost function over the interval $[a, b]$.

■ The change in the revenue from increasing a production run from a items to b items is the definite integral of the marginal revenue function over the interval $[a, b]$.

■ The fundamental characteristic of all of the formulas that use definite integration to find area is that the function being integrated must be nonnegative over the interval of integration.

12.6 Probability Density Functions

■ Any nonnegative continuous function on a finite interval can be used to assign probabilities.

■ The probability density function for an exponential distribution on $[0, \infty)$ with a mean of μ is

$$g(x) = \frac{1}{\mu} e^{-x/\mu}$$

■ The probability density function for the normal distribution on $(-\infty, \infty)$ with zero mean and standard deviation 1 is

$$g(x) = \frac{1}{\sqrt{2\pi}} e^{-x^2/2}$$

■ The transformation

$$z = (X - \mu)/\sigma$$

converts a random outcome X from a normally distributed process with mean μ and standard deviation σ into a random outcome z from a normally distributed process with zero mean and standard deviation 1.

■ Table 8.29 lists numerical estimates, precise to five decimal places, to the integral

$$\frac{1}{\sqrt{2\pi}} \int_0^z e^{-x^2/2} \, dx$$

for many values of z.

KEY FORMULAS

■ The power rule of integration: If n is a real number, $n \neq -1$, then

$$\int u^n \, du = \frac{u^{n+1}}{n+1} + c$$

where c is an arbitrary constant.

■ The indefinite integral of the power function $y = f(u) = u^{-1}$:

$$\int u^{-1} \, du = \int \frac{1}{u} \, du = \ln|u| + c$$

where c is an arbitrary constant.

- The natural exponential rule of integration:

$$\int e^u \, du = e^u + c$$

where c is an arbitrary constant.

- The sine rule of integration:

$$\int \sin u \, du = -\cos u + c$$

where c is an arbitrary constant.

- The cosine rule of integration:

$$\int \cos u \, du = \sin u + c$$

where c is an arbitrary constant.

- The constant multiple rule of integration: If $y = kf(u)$, where k is a known real number and the indefinite integral of the function $f(u)$ is known, then

$$\int kf(u) \, du = k \int f(u) \, du$$

A similar formula is applicable to definite integrals.

- The sum rule of integration: If $f(u)$ and $g(u)$ are two functions whose indefinite integrals are known, then the indefinite integral of their sum is

$$\int [f(u) + g(u)] \, du = \int f(u) \, du + \int g(u) \, du$$

A similar formula is applicable to definite integrals.

- The difference rule of integration: If $f(u)$ and $g(u)$ are two functions whose indefinite integrals are known, then the indefinite integral of their difference is

$$\int [f(u) - g(u)] \, du = \int f(u) \, du - \int g(u) \, du$$

A similar formula is applicable to definite integrals.

- The intermediate value rule for definite integration:

$$\int_a^b f(u) \, du = \int_a^c f(u) \, du + \int_c^b f(u) \, du$$

- The fundamental theorem of calculus: If $F(x)$ is an antiderivative of $f(x)$ and if $f(x)$ is continuous on the interval $[a, b]$, then

$$\int_a^b f(x) \, dx = F(b) - F(a)$$

- The area A of the region below the graph of a nonnegative function $f(x)$ and over the x-axis, between $x = a$ and $x = b$, is

$$A = \int_a^b f(x) \, dx$$

- The area A of the region above the graph of a nonpositive function $g(x)$ and below the x-axis, between $x = a$ and $x = b$, is

$$A = \int_a^b -g(x)\, dx$$

- If $f(x) \geq g(x)$, then the area A of the region between the graphs of the functions $y = f(x)$ and $y = g(x)$, and over the interval $[a, b]$, is

$$A = \int_a^b [f(x) - g(x)]\, dx$$

- If $f(x)$ is a probability function on $[a, b]$, then the probability that a random outcome X will fall in an interval $[c, d]$, where $a \leq c \leq d \leq b$, is

$$P(c \leq X \leq d) = \frac{\text{area under } f(x) \text{ and over } [c, d]}{\text{area under } f(x) \text{ and over } [a, b]}$$

$$= \frac{\displaystyle\int_c^d f(x)\, dx}{\displaystyle\int_a^b f(x)\, dx}$$

- If $g(x)$ is a probability density function on $[a, b]$, then the probability that a random outcome X will fall in an interval $[c, d]$, where $a \leq c \leq d \leq b$, is

$$P(c \leq X \leq d) = \int_c^d g(x)\, dx$$

KEY PROCEDURES

- The method of substitution of variables:

  Step 1 Set $u = g(x)$, where $g(x)$ is part of the function to be integrated.

  Step 2 Calculate $du = g'(x)\, dx$. If the quantity $g'(x)\, dx$ is not part of the original integral or cannot easily be made part of the original integral, then return to Step 1 and try another substitution; otherwise continue.

  Step 3 Use the substitutions $u = g(x)$ and $du = g'(x)\, dx$ to write the original integral completely in terms of the new independent variable, u. The resulting integral must *not* involve x. If this cannot be done, then return to Step 1 and try another substitution; otherwise continue.

  Step 4 If the resulting integral is simpler in structure than the original one, attempt to evaluate it; if this cannot be done, return to Step 1 and try another substitution.

  Step 5 After the integral has been evaluated, replace u by $g(x)$ to obtain the final answer in terms of x, the original independent variable.

- Calculating the area of an irregularly shaped, bounded region:

  Step 1 Enclose the region within a rectangle, the "frame."

  Step 2 Make tick marks on two adjoining sides of the frame by dividing each side into equally spaced subintervals, which can be different for the two sides. Connect the tick marks with horizontal and vertical lines, creating a grid of identical rectangular blocks.

Step 3 Count the number of rectangular blocks that are completely enclosed within the region of interest. Multiply this number by the area of each rectangular block to obtain a lower bound on the area of the region.

Step 4 Count the number of rectangular blocks that overlap or are within the region of interest. Multiply this number by the area of each rectangular block to obtain an upper bound on the area of the region.

Step 5 If the bounds obtained from Steps 3 and 4 are sufficiently tight, stop; if not, continue.

Step 6 Create a finer grid by dividing one side of the frame into smaller, equally spaced subintervals and drawing lines through the resulting tick marks.

Step 7 Return to Step 3.

■ The midpoint method for area of the region under the curve of $y = f(x)$ on $[a, b]$:

Step 1 Determine whether $f(x) \geq 0$ for all x in $[a, b]$. If not, stop, this algorithm does not apply; otherwise, continue.

Step 2 Select a positive integer n and then divide the interval $[a, b]$ into n equal partitions. The width of each partition is $\Delta x = (b - a)/n$.

Step 3 Locate the midpoints $m_1, m_2, m_3, \cdots, m_n$ of the n partitions, using the formula
$$m_i = a + (i - \tfrac{1}{2})\Delta x \quad (i = 1, 2, \cdots, n)$$

Step 4 Calculate the sum
$$[f(m_1) + f(m_2) + f(m_3) + \cdots + f(m_n)]\Delta x$$

and use it as an estimate for the area.

TESTING YOURSELF

Use the following problems to test yourself on the material in Chapter 12. The odd-numbered problems comprise one test, the even-numbered problems a second test.
   In Problems 1 through 10, evaluate the given integrals.

1. $\int_2^5 2x^2 \, dx$

2. $\int_{-2}^2 (6x^2 + 3x) \, dx$

3. $\int_{-1}^2 \frac{e^x}{2} \, dx$

4. $\int (z^{-3} + z^{-2}) \, dz$

5. $\int e^{-x/2} \, dx$

6. $\int x^2/(2x^3 + 9) \, dx$

7. $\int \frac{u^2 - 2}{u} \, du$

8. $\int_0^3 2/(u + 9) \, du$

9. $\int \sin(\theta/3) \, d\theta$

10. $\int_0^{1/2} x \sin(\pi x^2) \, dx$

11. Describe in words how to calculate the area of the region shown in Figure A.

FIGURE A

12. A manufacturer of pencils has a fixed cost of $2,000/day. The marginal cost (in cents/item) of producing x boxes of pencils is given by the formula

$$f(x) = 0.25x^{1/2} + 0.10$$

Find the total cost of producing 2500 boxes of pencils daily.

13. Use the midpoint method to *estimate* the area of the region between the graph of $y = \ln x$ and the x-axis, from $x = 1$ to $x = 2$, using four partitions.

14. Construct an integral for the probability that a randomly selected outcome will fall in the interval [25, 45] if the sample space is normally distributed with mean 20 and standard deviation 15.

15. Find the area of the region between the graphs of $y = x^3$ and $y = \sqrt{x}$, from $x = 0$ to $x = 1$.

16. Determine whether the integral $\int_1^{\infty} (1/x^{0.3})\, dx$ converges.

17. Find the area of the region between the graph of $y = 1 - 2x^2$ and the x-axis over the interval [1, 2].

18. Determine whether the function $f(x) = 2x^2$ is a probability function on the interval [−1, 1].

19. The marginal revenue function MR (in dollars/item), for a particular product, is $MR(n) = 5 + e^{-0.25n}$, where n denotes the number of units sold. Determine the total revenue obtained from the sale of 200 items.

20. Use the grid displayed in Figure B to estimate the area of the shaded region.

FIGURE B

21. A ball is thrown downward with an initial speed of 16 ft/sec from a helicopter hovering 2000 ft above the ground. Assume that air resistance is negligible and that the only force on the ball is due to gravity. How long will it take the ball to hit the ground and at what speed will it make contact?

22. Find the area of the region between the graphs of $y = -x^2$ and $y = 1 - x$, from $x = 0$ to $x = 2$.

23. A continuous process is modeled by the probability function $f(x) = 0.1x^2 + 2x + 5$ on the domain $[1, 4]$. Determine the probability that a random outcome X will fall in the interval $[2, 3]$.

24. Normalize the function given in Problem 23.

25. Determine the probability that a randomly selected outcome will fall in the interval $[25, 35]$ if the sample space is exponentially distributed with a mean of 30.

26. Use the midpoint method to *estimate* the area of the region under the graph of $y = x^4 + 1$ and over the x-axis from $x = -1$ to $x = 1$, using four partitions.

27. Construct an integral for the probability that a randomly selected outcome will fall in the interval $[25, 35]$ if the sample space is normally distributed with mean 30 and standard deviation 10.

28. Find the area of the region between the graph of $y = \ln x$ and the x-axis, from $x = 1$ to $x = 2$.

29. Determine whether the integral $\int_1^\infty (1/x^5) \, dx$ converges.

30. The demand D of a new brand of dog food is increasing at the rate $D'(t) = 10 - 7e^{-t}$, where demand is measured in thousands of cases and time t is measured in years from the moment the product was introduced to the market. Determine the total sales of this product over its first four years.

31. Create a probability density function on $[0, 5]$ from the function $f(x) = 2x^3 + 3$ and then use the result to determine the probability that a random outcome will fall in the interval $[1, 2]$.

32. Determine the probability that a randomly selected outcome will be greater than or equal to 40 if the sample space is exponentially distributed with a mean of 30.

13

MULTIVARIABLE
MODELS

13.1 FUNCTIONS OF MANY VARIABLES

All of the functions we have considered in the last four chapters are functions of a single input variable, modeling relationships of the form

$$f(input) = output$$

where the rule f transforms a single input into a single output.

Although the simplest functions involve a single input and a single output, many processes create a single output from multiple inputs. For example, paint, fabric, glass, brakes, fenders, and tires are only a few of the many inputs needed to manufacture a single automobile. Similarly, sugar, flour, flavorings, food coloring, and preservatives are all required to produce a single box of cake mix.

Many processes produce a single output from multiple inputs.

We have already seen many examples of functions with multiple inputs. The objective function for the linear programming problem developed in Example 3 of Section 3.3 is

$$P = 0.75x + 1.10y \tag{1}$$

The variables x and y denote inputs (the number of bags of regular fertilizer and the number of bags of deluxe fertilizer, respectively) while the variable P represents profit and is the output of the process. The dependent variable P is a function of the two independent variables, x and y. Once values of x and y are specified, the value of P is fully determined.

In Chapter 4, we developed the compound interest formula,

$$FV = PV\left(1 + \frac{r}{N}\right)^n$$

for the future value of a lump-sum deposit, PV. To calculate this amount, we need the initial deposit, PV; the nominal annual interest rate, r; the number N of conversion periods in a year; and the number n of conversion periods until the final payment is due. Therefore the output FV is a function of four input variables: PV, r, N, and n.

In Chapter 6, we developed the following formula for the expected value E:

$$E = d_1p_1 + d_2p_2 + \cdots + d_Np_N$$

where the d_i are the possible outcomes and the p_i are the corresponding probabilities. The output variable E appears to be a function of the $2N$ input variables $d_1, \cdots, d_N$ and $p_1, \cdots, p_N$. In fact, E is a function of only $2N - 1$ independent variables since $p_1 + \cdots + P_{N-1} + p_N = 1$: once $p_1, \cdots, p_{N-1}$ are specified, $p_N = 1 - (p_1 + \cdots + p_{N-1})$.

The notation

$$z = f(x, y)$$

denotes a function of two input variables (or "two variables," for short) that transforms the inputs x and y into an output z according to the rule of correspondence specified by f. This equation is shorthand notation for the formula

$$output = f(input\ one,\ input\ two)$$

which is symbolized in Figure 13.1. We can use other labels for the inputs, the output, or the rule of correspondence. In Equation (1), the output is denoted by P, the inputs by x and y, and the rule of correspondence is defined by the formula $P = f(x, y) = 0.75x + 1.10y$. In general, the relationship $P = g(a, b)$ indicates that an output P is a function of

FIGURE 13.1

A Schematic for a Function of Two Variables

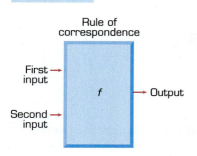

the two (input) variables a and b according to a rule of correspondence specified by g, yet to be defined.

The notation

$$z = f(x_1, x_2, x_3)$$

signifies that an output z is a function of the three input variables x_1, x_2, and x_3 according to the rule of correspondence specified by f. This relationship models a process in which three inputs are required to produce a single output. In a similar fashion, the notation

$$z = g(a, b, c, d)$$

indicates that the output z is a function of four (input) variables, a, b, c, and d, and the rule of correspondence is represented by g.

The domain of a function of two variables is a subset of a two-dimensional plane.

The natural domain of a function $z = f(x, y)$ of two variables is the set of all real-number pairs (x, y), where x and y are both input variables for which the rule of correspondence makes sense. In this case the domain is a subset of the xy-plane. The natural domain of a function of three variables is the set of all *triples* of real-valued inputs that are consistent with the rule of correspondence, and so on for any finite number of inputs. Unless stipulations are made to the contrary, we take the domain of a function to be its natural domain.

EXAMPLE 1

Determine the natural domain for the function $f(x, y) = \sqrt{x^2 + 4y^2 - 16}$.

Solution

The square root of a real number is meaningful only when its argument is nonnegative. Thus the domain of this function is restricted to the region in the xy-plane defined by the inequality $x^2 + 4y^2 - 16 \geq 0$, or $x^2 + 4y^2 \geq 16$. This domain is represented by the shaded region in Figure 13.2.

FIGURE 13.2
The Region of the *xy*-Plane Defined by the Inequality $x^2 + 4y^2 \geq 16$

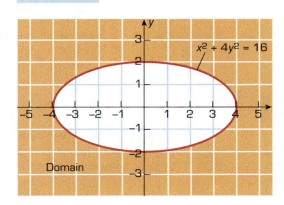

MODEL 13.1

Important Factors

P = profit ($)

x = number of bags of regular fertilizer

y = number of bags of deluxe fertilizer

Objective

Find x and y to maximize

$P = 0.75x + 1.10y$

Constraints

$3x + 4y \leq 7,400$

$7x + 6y \leq 14,100$

$x \geq 0$

$y \geq 0$

The function $P = f(x, y) = 0.75x + 1.10y$ [Equation (1)] is the objective function for Model 13.1. In Example 3 of Section 3.3, we found the feasible region for this linear programming problem to be the shaded region in Figure 13.3. This region, a subset of the *xy*-plane, is the domain for the function *P*. Points in the *xy*-plane that are not in this feasible region violate the constraints and are not allowable inputs for this process.

FIGURE 13.3

The Feasible Region for Model 13.1

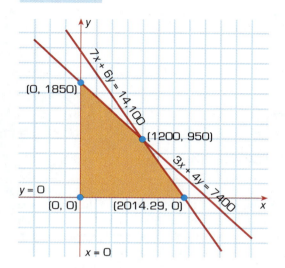

FIGURE 13.4

A Standard *xyz*-Coordinate System

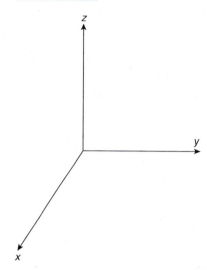

FIGURE 13.5

How to Get to the Point (2, 3, 4) from the Origin in the *xyz*-Coordinate System

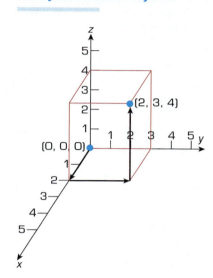

A three-dimensional coordinate system is needed to graph a function of two variables.

To graph a function of two variables of the form $z = f(x, y)$, we require a three-dimensional coordinate system, with an axis for each of the two input variables x and y, and a third axis for the output z. Such a coordinate system is shown in Figure 13.4. The positive z-axis points toward the top of the page from the origin, the positive y-axis points toward the right edge of the page, and the positive x-axis points out of the page, directly toward the reader.

Every point in a three-dimensional coordinate system is uniquely determined by an ordered triplet of the form (x, y, z). To graph the point $(2, 3, 4)$, which has inputs $x = 2$, $y = 3$, and output $z = 4$, we first measure a distance of 2 units along the positive x-axis; from there we measure a distance of 3 units parallel to the positive y-axis; then we measure a distance of 4 units parallel to the positive z-axis. This path is illustrated by the arrows in Figure 13.5, beginning at the origin and ending at the point $(2, 3, 4)$. The point $(2, 3, 4)$ is a graphical model of the functional relationship $f(2, 3) = 4$. In this model, the inputs are listed first and the output last.

EXAMPLE 2

Model the equation $g(50, 30) = -20$ on a three-dimensional coordinate system.

Solution

If we let x denote the first input variable, y the second, and z the output, then the equation $g(50, 30) = -20$ stipulates that the rule of correspondence g transforms the inputs $x = 50$ and $y = 30$ into the output $z = -20$. We represent this relationship by the point $(50, 30, -20)$, listing the inputs first and the output last. To plot $(50, 30, -20)$ on a three-dimensional coordinate system, we first measure a distance of 50 units along the positive x-axis; from there we measure a distance of 30 units parallel to the positive y-axis; and finally we measure a distance of 20 units parallel to the negative z-axis. This path is illustrated by the arrows in Figure 13.6, beginning at the origin and ending at the point $(50, 30, -20)$. This point is our graphical model of the relationship $g(50, 30) = -20$.

FIGURE 13.6

A Graphical Model of the Equation $g(50, 30) = -20$ on the xyz-Coordinate System

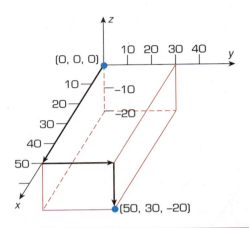

We graph the function $z = f(x, y)$ by plotting all of the points (x, y, z) that satisfy the equation. Each point in the domain is a point in the xy-plane. If $x = a$ and $y = b$ is such a point, then we evaluate the function at these inputs to obtain a corresponding output, say, $z = c$. That is, $c = f(a, b)$. We then plot the ordered triplet (a, b, c) on a three-dimensional coordinate system. Doing this for each point in the domain results in a surface over (or under or through) the domain, as shown in Figure 13.7. The shaded green rectangle

FIGURE 13.7

The Graph of the Function $f(x, y) = z$ is a Surface over (or under) Its Domain

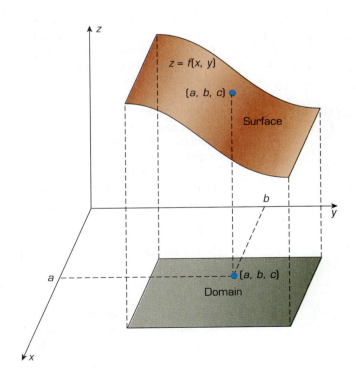

in the xy-plane is the domain of the function and the red surface above it represents the graph of the function. When the output z is positive, the corresponding point is above the domain, and we can think of the resulting three-dimensional surface as a roof over a portion of the floor (the domain) in the xy-plane. When the output z is negative, the corresponding point is under the domain.

EXAMPLE 3

Construct a graphical model of the function

$$z = f(x, y) = x^2 + y^2 \tag{2}$$

Solution

We begin by plotting individual points. When $x = y = 0$, the corresponding output is

$$z = f(0, 0) = (0)^2 + (0)^2 = 0$$

This relationship is modeled graphically by the point $(0, 0, 0)$ in a three-dimensional coordinate system. When $x = 3$ and $y = 4$, the corresponding output is

$$z = f(3, 4) = (3)^2 + (4)^2 = 25$$

This relationship is modeled graphically by the point $(3, 4, 25)$. If we were to continue in this manner, plotting a great many individual points, we would obtain the surface displayed in Figure 13.8. This surface is the shell of a paraboloid; its interior is empty and the surface continues upward through all values of positive z.

FIGURE 13.8

A Graph of the Paraboloid
$z = x^2 + y^2$

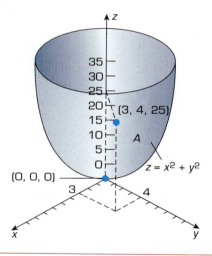

Because it is difficult to graph a three-dimensional surface by plotting individual points, we often begin by graphing a series of *level curves*. A level curve for a function $z = f(x, y)$ is the set of all points (inputs) in the xy-plane that yield the same value of the output variable z. To create a level curve, we fix the output z at a constant value, say $z = c$, and then graph the resulting equation $c = f(x, y)$ in the xy-plane.

A level curve for a function $z = f(x, y)$ is the set of all points in the xy-plane that yield the same value of the output z.

For example, using the function $z = f(x, y) = x^2 + y^2$ in Example 3 and setting $z = 25$, we obtain

$$25 = x^2 + y^2 \tag{3}$$

As we saw in Example 3 of Section 1.4, the graph of this equation is a circle of radius 5 centered at the origin in the xy-plane. This circle is the level curve of the function $z =$

FIGURE 13.9
Some Level Curves for
$z = x^2 + y^2$

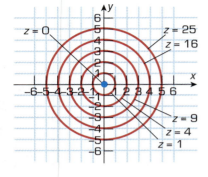

$x^2 + y^2$ at the height of $z = 25$. Similarly, if we set $z = 1$ in Equation (2), we obtain the equation

$$1 = x^2 + y^2 \tag{4}$$

which is the equation of a circle of radius 1 centered at the origin in the xy-plane. This circle is the level curve of the function $z = x^2 + y^2$ at the height of $z = 1$. By choosing other values for z we obtain other level curves, and some of these are shown in Figure 13.9.

A level curve is a tracing onto the xy-plane of the shape of a surface at a fixed height of the output variable z. Geometrically, this shape is the intersection of the surface $z = f(x, y)$ with a plane parallel to but a fixed number of units above or below the xy-plane. The intersection of the surface shown in Figure 13.8 with the plane $z = 25$ is the blue circle shown in Figure 13.10. Points (x, y, z) on this circle satisfy the two conditions $z = 25$ and $x^2 + y^2 = 25$. If we then project this circle onto the xy-plane (where z is always zero), we obtain the red circle shown in Figure 13.10. This red circle is the level curve associated with the blue circle in Figure 13.10. The red circle is a tracing or projection of the blue circle (fixed at a height of $z = 25$) onto the xy-plane.

FIGURE 13.10

The Intersection of the Plane
$z = 25$ **with the Surface**
$z = x^2 + y^2$

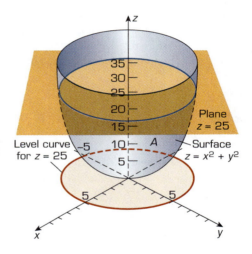

A contour map is a set of level curves for the same function.

A *contour map* for a function is a set of level curves for that function, all drawn on the same xy-coordinate system. Figure 13.9 is a contour map for the function described in Example 3. We sketch a surface from its contour map by telescoping out the various level curves to their corresponding heights and filling in the gaps between these curves.

EXAMPLE 4

Construct a contour map for the function $z = \sqrt{x^2 + 4y^2 - 16}$.

Solution

The output variable z is defined in terms of a positive square root, so z is always nonnegative. Starting with $z = 0$, the least possible output value, we get

$$0 = \sqrt{x^2 + 4y^2 - 16}$$

or, squaring both sides,

$$x^2 + 4y^2 = 16$$

This is the equation of an ellipse, which is graphed in Figure 13.11. Now setting $z = 3$, we get

$$3 = \sqrt{x^2 + 4y^2 - 16}$$

which gives us another ellipse:

$$x^2 + 4y^2 = 25$$

We include this ellipse in Figure 13.11. By choosing still other values of z, say, $z = 5$, $z = 8$, and $z = 10$, and adding them to Figure 13.11, we obtain a contour map of level curves for the function $z = \sqrt{x^2 + 4y^2 - 16}$.

FIGURE 13.11

A Contour Map for
$z = \sqrt{x^2 + 4y^2 - 16}$

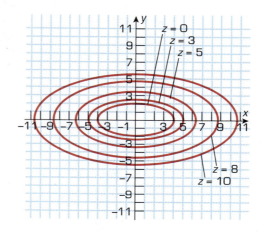

EXAMPLE 5

Graph the function $z = f(x, y) = \sqrt{x^2 + 4y^2 - 16}$.

Solution

A contour map for this function is displayed in Figure 13.11. The corresponding surface is shown in Figure 13.12.

FIGURE 13.12

The Graph of $z =$
$\sqrt{x^2 + 4y^2 - 16}$

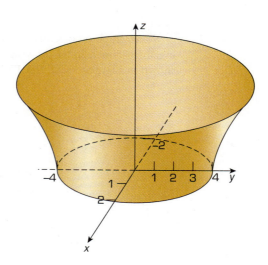

EXAMPLE 6

Graph the function $z = f(x, y) = 2x + 5y + 10$.

Solution

We begin by drawing level curves. Arbitrarily setting $z = 0$, we get the equation $0 = 2x + 5y + 10$. Rearranging terms, we have

$$2x + 5y = -10$$

which is the equation of a straight line. By choosing other values of z, say, $z = -20$, $z = -10$, $z = -5$, $z = 5$, and $z = 10$, we generate the equations of other straight lines, all of which are displayed on the contour map in Figure 13.13. The corresponding surface is a plane, part of which is shown in Figure 13.14.

FIGURE 13.14
The Graph of
$z = 2x + 5y + 10$

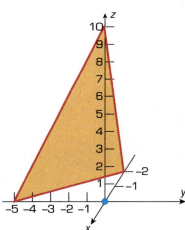

FIGURE 13.13
A Contour Map for
$z = 2x + 5y + 10$

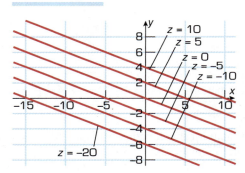

Some graphing calculators and computer software packages contain menus for rendering three-dimensional drawings of functions of two variables. Two such examples are shown in Figures 13.15 and 13.16.

FIGURE 13.15
The Graph of
$z = (x^2 + y^2)e^{1-x^2-y^2}$

FIGURE 13.16
The Graph of
$z = (x^2 + 8y^2)e^{1-x^2-y^2}$

IMPROVING SKILLS

In Problems 1 through 6, evaluate each of the given functions where indicated and find the outputs associated with the given inputs.

1. $f(x, y) = 2x^2 - 3y^2 + 8$:
   (a) $f(0, 0)$
   (b) $f(1, -2)$
   (c) $f(-2, 1)$

2. $f(x, y) = xy + 2y^2$:
   (a) $f(2, 3)$
   (b) $f(3, 2)$
   (c) $f(3, -2)$

3. $f(a, b) = a/b$:
   (a) $f(5, 15)$
   (b) $f(15, 3)$
   (c) $f(8, -5)$

4. $f(a, b, c) = a^2 + b^3 + c^4 + 5$:
   (a) $f(3, 2, 1)$
   (b) $f(4, 3, 2)$
   (c) $f(2, -1, -2)$

5. $f(u, v, w) = uw^2 + 2v - 100$:
   (a) $f(2, 5, 3)$
   (b) $f(2, -5, 3)$
   (c) $f(2, -5, -3)$

6. $f(a, b, c, d) = \dfrac{a + c - 1}{b - d}$:
   (a) $f(1, 2, 3, 4)$
   (b) $f(1, 2, 1, -2)$
   (c) $f(10, 20, 25, -10)$
   (d) $f(0, 8, -1, -3)$

In Problems 7 through 15, determine the natural domain for each of the given functions.

7. $f(x, y) = \sqrt{x^2 + 2y^2}$

8. $f(x, y) = \sqrt{x^2 - 2y^2}$

9. $f(a, b) = \sqrt{ab}$

10. $f(a, b) = a/b$

11. $f(a, b) = a/(1 + b^2)$

12. $f(P, Q) = 20P^2 - 30Q^2 + 45$

13. $f(u, v) = \ln(u + v)$

14. $f(u, v) = \sin u/\cos v$

15. $f(r, s) = 1 + 2e^{r-s}$

In Problems 16 through 25, plot each point on a three-dimensional coordinate system.

16. $(2, 3, -4)$

17. $(-2, 3, 4)$

18. $(3, 1, 5)$

19. $(3, -1, 5)$

20. $(100, 50, -100)$

21. $(-10, 20, 10)$

22. $(-2, 5, -2)$

23. $(-20, -20, 20)$

24. $(8, 1, 5)$

25. $(1000, 3000, 5000)$

26. Figure A is a topographical map of a mountain range. How many peaks are in this range? Which peak has a gully down one of its sides?

FIGURE A

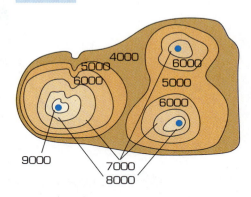

In Problems 27 through 32, sketch the surface corresponding to each contour map.

27.

28.

29.

30.

31.

32.

In Problems 33 through 41, construct a contour map of level curves for each of the given functions and then sketch the corresponding three-dimensional graph.

33. $z = x^2 + 2y^2$
34. $z = 4x^2 + y^2$
35. $z = 1 - x^2 - y^2$
36. $z = 1 + x^2 + y^2$
37. $z = \sqrt{16 - x^2 - y^2}$
38. $z = \sqrt{16 - x^2 - 4y^2}$
39. $z = 2x + 3y - 6$
40. $z = 2x + 3y + 6$
41. $z = 20x - 10y + 400$

CREATING MODELS

42. Write the equation for the volume of a rectangular box.
  (**a**) What are the input variables for this function and what is the output variable?
  (**b**) What is the domain of this function as a mathematical expression?
  (**c**) What is the domain of this function as a model for the volume of a box?
43. Write the equation for the volume of a mailing tube.
  (**a**) What are the input variables for this function and what is the output variable?
  (**b**) What is the domain of this function as a mathematical expression?
  (**c**) What is the domain of this function as a model for the volume of a tube?
44. Write the equation for a combination of selecting r objects from a set of n objects, $r \le n$. (See page 428.)
  (**a**) What are the input variables for this function and what is the output variable?
  (**b**) What is the domain of this function?
45. Write the equation for the future value of an ordinary annuity. (See page 282.)
  (**a**) What are the input variables for this function and what is the output variable?
  (**b**) What is the domain of this function?
46. A television manufacturer produces three different models, A, B, and C. Model A costs $210 to manufacture and sells for $320; model B costs $230 to manufacture and sells for $365, while model C costs the same as model B to manufacture but sells for $21 more. Model the profit function for this manufacturer, assuming the manufacturer can sell all televisions produced. How many input variables are there?
47. Solve Problem 46 if there is no relationship between the number of sets produced and the number sold. Unsold sets remain in inventory.

48. A company pays its workers a regular wage of $10.50/hr and an overtime rate of $15.75/hr. Model the total daily labor cost for this company. How many input variables are there?

49. A publisher prints a travel book that sells wholesale for $15, a cookbook that sells wholesale for $12, and a garden book that sells wholesale for $10. At these prices, all books published can be sold, up to the capacity of the printing presses, which is 20,000 copies a week. Model the total revenue function for this publisher. How many input variables are there? What is the domain of this function?

50. A company has an annual advertising budget of $675,000, which it divides between print advertising, local television advertising, direct mail marketing, and billboards. Print advertising costs $2,500/page, local television advertising costs $8,500/min, direct mail marketing costs 53¢ per piece of mail, and billboard advertising costs $2,300/billboard/month. Model the total advertising expenditures for this company. How many input variables are there? What is the domain of this function?

51. A new apartment building contains 30 one-bedroom units that rent at $500/month each, 20 two-bedroom units that rent at $850/month each, and 10 three-bedroom suites that rent at $1200/month each. Model the total monthly revenue function for this complex. How many input variables are there? What is the domain of this function?

52. Figure B is a weather map showing various temperature lines throughout the country. Is this a contour map? What is the function? What is the domain? What is the output variable?

FIGURE B

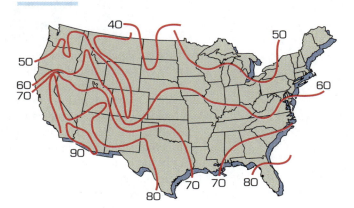

53. The Cobb–Douglas production function is used by economists to model the output P of a production process for which labor and capital investments (including equipment and buildings) are the inputs. If L denotes the number of units of labor and C the number of units of capital available for investment, then the Cobb–Douglas production function is

$$P = kL^{\alpha}C^{1-\alpha}$$

where k and α are constants, $0 < \alpha < 1$. Determine the number of input variables for this function.

EXPLORING IN TEAMS

The general equation of a plane in three dimensions is the linear equation

$$Ax + By + Cz = D$$

where A, B, C, and D are constants, with A, B, and C not all zero.

54. Graph the plane $x + 0y + 0z = 3$, generally written $x = 3$.
55. Graph the plane $x = 5$.
56. Graph the plane $0x + 0y + z = 3$, generally written $z = 3$.
57. Graph the plane $z = -3$.
58. Graph the plane $x + y + 0z = 1$, generally written $x + y = 1$.
59. What can you say about a plane with $D = 0$?

EXPLORING WITH TECHNOLOGY

In Problems 60 through 63, using a computer software package or a graphing calculator that supports three-dimensional graphing, draw surfaces corresponding to each of the given functions.

60. $z = x^2 + y^2$ (Compare with Figure 13.8.)
61. $z = \cos(x^2 + y^2)$
62. $z = xy(x^2 - y^2)/(x^2 + y^2)$
63. $z = \sin x \cos x$ (Note that y is arbitrary.)

REVIEWING MATERIAL

64. (Section 1.5) Find the equation of a line that passes through the point $(8, -250)$ with a slope of -10.
65. (Section 4.3) A village must repay the face value of a \$2 million bond in 30 years, and it decides to create a sinking fund to meet this obligation. Deposits will be made quarterly as taxes are received. How much must be deposited into this fund at the end of each quarter if the interest rate is 6%/yr, compounded quarterly?
66. (Section 11.5) Find d^2y/dx^2 for $y = 3 \sin^2(2x)$.
67. (Section 12.1) Evaluate $\int \left(\dfrac{2}{x^2} - \dfrac{3}{x} \right) dx$

RECOMMENDING ACTION

68. Respond by memo to the following:

MEMORANDUM

To: J. Doe Reader

From: Tanya

Date: Today

Subject: **Vertical Line Test**

I understand the vertical line test for determining whether a two-dimensional graph represents a function of a single input variable. I was wondering if a similar test exists for a function of two variables. What about a function of three variables?

13.2 PARTIAL DERIVATIVES

Previously we dealt with rates of change in a function of a single variable. In Chapter 10 we developed the concept of a derivative to measure the change of an output variable resulting from a small change in its input variable. In Chapter 11 we used the derivative to identify points where a differentiable function attains its maximum and minimum values. Now we want to extend this same analysis to functions of many variables.

We begin by modeling a cost function for a manufacturer of wooden chairs. Production of each chair requires 40 board–feet of lumber and 0.8 hours of an assembler's time together with 1.5 pints of paint and 0.3 hours of a decorator's time. The direct cost associated with this process is displayed in Model 13.2 as a function of four variables. It depends on the hourly wages of the assemblers and decorators and the supply costs for the wood and paint.

Paint currently costs the company \$3/pint and wood, \$1.15/board–foot. Intense competition has kept both supply costs stable, and no changes are expected for at least a year. Assemblers receive \$14.50/hr and their union contract holds for another six months. Decorators have been charging \$12.30/hr, but their contract has expired and a new one is under negotiation. Before agreeing to an increase in the decorator's hourly wage, the company wants to know the effect of such a change on its overall cost structure.

Over the next six months, all costs—with the exception of the decorators' hourly wage—are fixed, so $w = 1.15$, $p = 3$, $a = 14.50$, and the cost function in Model 13.2 simplifies to

$$C = 40(1.15) + 1.5(3) + 0.8(14.50) + 0.3d$$
$$= 62.10 + 0.3d \tag{5}$$

The cost function is now a function of the single input variable d, and we can take the derivative of this function to find the rate of change in C with respect to a unit change in d.

MODEL 13.2

Important Factors

C = cost of a chair (\$)

w = cost of wood (\$/board–foot)

p = cost of paint (\$/pint)

a = wage of an assembler (\$/hr)

d = wage of a decorator (\$/hr)

Objective

Find C.

Constraints

$C = 40w + 1.5p + 0.8a + 0.3d$

To emphasize that some variables are subject to change at the present time while others are not, we write in red those variables that might change. The cost function in Model 13.2 looks like this:

$$C = 40w + 1.5p + 0.8a + 0.3d \qquad (6)$$

The variables w, p, and a are presently fixed and act as the constants in this equation. If necessary, we can substitute numbers for each of the variables that are in black, as we did in Equation (5). For now, only the input variable d and the output variable C can vary.

Once labor negotiations are concluded, the decorators' hourly wage will remain fixed for some time, as will the cost function for each chair. In six months, however, the contract expires for the assemblers' hourly wage, and *their* rate may change. Then we will write the cost function in Model 13.2 as

$$C = 40w + 1.5p + 0.8a + 0.3d \qquad (7)$$

where now C and a are in red, to signify that they can vary, and d is in black because it is fixed for the time being. Later on the supply costs may also be subject to change, in which case w and p will be in red.

Equations (6) and (7) model the output variable C as a function of the four input variables, w, p, a, and d. In each equation, however, a different set of three variables is held constant, at least for awhile, and during that time period only the output variable and one of the input variables vary. Effectively, the output variable has become a function of a *single* input variable. That input variable is d in Equation (6) and a in Equation (7).

When we have a function f of many variables, we use the notation $\dfrac{\partial f}{\partial x}$ for the derivative with respect to just the input variable x, under the assumption that all other input variables are held constant. However, rather than use the word *derivative* in this case, we use a new term, *partial derivative*. Thus, $\dfrac{\partial f}{\partial y}$ is the partial derivative of the function f with respect to the variable y, assuming all other input variables to f are held constant. Similarly, $\dfrac{\partial g}{\partial t}$ is the partial derivative of the function g with respect to the variable t, assuming all other input variables to g are held constant. When we view a function of many variables as a function of a single input variable, we can find the partial derivative by applying any of the appropriate differentiation rules established in Chapter 10.

When we label the output of a function, such as $z = f(x, y)$, then alternate notations are

> *The partial derivative $\dfrac{\partial f}{\partial x}$ is the derivative of a function f of many variables with respect to the input variable x under the assumption that all other input variables have fixed values.*

$$\frac{\partial z}{\partial x} \text{ for } \frac{\partial f}{\partial x} \quad \text{and} \quad \frac{\partial z}{\partial y} \text{ for } \frac{\partial f}{\partial y}.$$

EXAMPLE 1

Determine $\dfrac{\partial z}{\partial x}$ and $\dfrac{\partial z}{\partial y}$ for the function $z = 2x + yx + 3y^3$.

Solution

To calculate $\dfrac{\partial z}{\partial x}$, we assume that the input variable y is held constant. Only the input variable x and the output variable z are allowed to vary, at least over the short run. We show this by writing z and x in red and y in black. The result is

$$z = 2x + yx + 3y^3$$

We treat y as a constant, in which case so is $3y^3$, and its derivative is zero:

$$\frac{\partial z}{\partial x} = 2 + y + 0$$

To calculate $\frac{\partial z}{\partial y}$, we assume that the input variable x is held constant. Only the input variable y and the output variable z are allowed to vary. We show this by writing z and y in red and x in black. The result is

$$z = 2x + yx + 3y^3$$

We are treating x as a constant, so the derivative of $2x$ is zero:

$$\frac{\partial z}{\partial y} = x + 9y^2$$

EXAMPLE 2

Determine $\frac{\partial w}{\partial x}$ and $\frac{\partial w}{\partial y}$ for the function $w = xe^{xy}$.

Solution

To calculate $\frac{\partial w}{\partial x}$, we assume that the input variable y is held constant. Only the input variable x and the output variable w are allowed to vary. We show this by writing w and x in red and y in black. This looks as follows:

$$w = xe^{xy}$$

Using the product rule of differentiation, we obtain

$$\frac{\partial w}{\partial x} = x(e^{xy}y) + 1 \cdot e^{xy} = (xy + 1)e^{xy}$$

To calculate $\frac{\partial w}{\partial y}$, we assume that the input variable x is fixed and treat it as a constant. We write in red those variables that can vary, so the given function now looks like this:

$$w = xe^{xy}$$

Therefore the partial derivative of w with respect to y is

$$\frac{\partial w}{\partial y} = x(e^{xy}x) = x^2e^{xy}$$

For simplicity, we can also adopt a subscript notation for partial derivatives, writing the input that can vary as a subscript of the output variable. When $z = f(x, y)$, we use f_x and z_x for $\frac{\partial f}{\partial x}$ and $\frac{\partial z}{\partial x}$, respectively; similarly, we use f_y and z_y for $\frac{\partial f}{\partial y}$ and $\frac{\partial z}{\partial y}$, respectively. For the function in Example 2, we have

$$w_x = (xy + 1)e^{xy} \qquad \text{and} \qquad w_y = x^2e^{xy}$$

For the function in Example 1, we have

$$z_x = 2 + y \qquad \text{and} \qquad z_y = x + 9y^2$$

Partial derivatives exist for functions of more than two input variables, under the same assumption: that all but one of the input variables can be held constant. Equation (6),

$$C = 40w + 1.5p + 0.8a + 0.3d$$

models the cost function C of manufacturing beach chairs under the assumption that each input variable is fixed except the hourly wage d of the decorators. Then C is a function of the single input variable d, and we can take its partial derivative with respect to this variable, obtaining

$$C_d = \frac{\partial C}{\partial d} = 0.3$$

As another approach, we can first substitute values for the input variables that are presently fixed into the original cost function, obtaining $C = 62.10 + 0.3d$. We arrive at the same result, however: the derivative of C with respect to d is 0.3. We also have from Equation (6) that

$$C_w = \frac{\partial C}{\partial w} = 40, \quad C_p = \frac{\partial C}{\partial p} = 1.5, \quad \text{and} \quad C_a = \frac{\partial C}{\partial a} = 0.8$$

EXAMPLE 3

Find all partial derivatives for the function $P = 2x^2y^3/(5y + t^4)$.

Solution

There are three input variables to this function, so there are three partial derivatives: $\partial P/\partial x$, $\partial P/\partial y$, and $\partial P/\partial t$, one for each independent variable.

To calculate $\partial P/\partial x$, we hold y and t fixed, allowing just the output P and the input x to vary. Writing these variables in red, we have

$$P = \frac{2x^2y^3}{5y + t^4} = \left(\frac{2y^3}{5y + t^4}\right)x^2$$

where all terms in black are treated as constants. Effectively, this function has the form of a constant (the terms in the parentheses) times x^2. The partial derivative of P with respect to x is

$$P_x = \frac{\partial P}{\partial x} = \left(\frac{2y^3}{5y + t^4}\right)2x = \frac{4xy^3}{5y + t^4}$$

To calculate $\partial P/\partial y$, we hold x and t fixed, allowing just P and y to vary. Writing these variables in red, we have $P = 2x^2y^3/(5y + t^4)$. Because y appears in both the numerator and denominator of this quotient, we must use the quotient rule to obtain the partial derivative with respect to y:

$$P_y = \frac{\partial P}{\partial y} = \frac{[(2x^2)(3y^2)](5y + t^4) - 2x^2y^3(5)}{(5y + t^4)^2}$$
$$= \frac{x^2y^2(20y + 6t^4)}{(5y + t^4)^2}$$

To calculate $\partial P/\partial t$, we hold x and y fixed (written in black), allowing just P and t (in red) to vary. Our original equation now looks like this:

$$P = 2x^2y^3(5y + t^4)^{-1}$$

Using the chain rule, we have

$$P_t = \frac{\partial P}{\partial t} = 2x^2y^3(-1)(5y + t^4)^{-2}(4t^3) = \frac{-8x^2y^3t^3}{(5y + t^4)^2}$$

If $z = f(x, y)$ is a function of two variables, we write $\left.\dfrac{\partial z}{\partial x}\right|_{(a,b)}$ to denote the value of the partial derivative of z with respect to x at the point $x = a$ and $y = b$ in the domain. One way to proceed is to first set $y = b$ and then differentiate the resulting function $z = f(x, b)$ with respect to the one input variable x that can still vary. Then we evaluate the resulting derivative at $x = a$. Easier, though, is to differentiate z with respect to x, keeping y fixed, and then evaluate both x and y at their respective values.

EXAMPLE 4

Evaluate $\left.\dfrac{\partial w}{\partial x}\right|_{(2,-1)}$ and $\left.\dfrac{\partial w}{\partial y}\right|_{(3,0)}$ for the function $w = f(x, y) = xe^{xy}$.

Solution
Using the results of Example 2, we have

$$\left.\frac{\partial w}{\partial x}\right|_{(2,-1)} = (xy + 1)e^{xy}\Big|_{\substack{x=2\\y=-1}}$$

$$= [(2)(-1) + 1]e^{(2)(-1)} = -e^{-2} \approx -0.1353$$

and

$$\left.\frac{\partial w}{\partial y}\right|_{(3,0)} = x^2 e^{xy}\Big|_{\substack{x=3\\y=0}} = (3)^2 e^{(3)(0)} = 9e^0 = 9$$

> *A partial derivative measures the rate of change of an output variable with respect to a unit change in one of its input variables.*

A partial derivative measures the rate of change in the output of a function of several variables with respect to a unit change in *one* of the input variables, assuming all of the other input variables remain constant. In Example 4, we found that $\left.\dfrac{\partial w}{\partial x}\right|_{(2,-1)} \approx -0.1353$. Thus if we increase x from 2 to 3 while holding y fixed at -1, then we expect the output w to decrease approximately 0.1353 units. Similarly, if we increase y from 0 to 1 while holding x at 3, we expect w to increase by 9 units, because $\left.\dfrac{\partial w}{\partial y}\right|_{(3,0)} = 9$.

EXAMPLE 5

FIGURE 13.17

A Storage Shed Attached to a Building

h = 5 ft
l = 4 ft
w = 6 ft

A storage shed will be attached to the side of a building so as to form one side of the shed. (See Figure 13.17.) This shed will be in the shape of a closed rectangular box with a base of 4 ft (parallel to the building) by 6 ft (perpendicular to the building) and with a height of 5 ft. The floor of the shed will be made from concrete at \$2/sq. ft, the three walls will be made from plywood at 50¢/sq. ft, and the roof will cost \$1.25/sq. ft. Prior to constructing the shed, the builder wants to know how the cost will change if any one of the three dimensions is altered.

Solution
The total cost C of the shed is $C(l, w, h) = 2wl + 0.5(2wh + lh) + 1.25wl$, or

$$C(l, w, h) = wh + 0.5lh + 3.25lw$$

from which we calculate the cost of the proposed design to be

$$C(4, 6, 5) = 6 \cdot 5 + 0.5(4)(5) + 3.25(4)(6) = \$118$$

Furthermore,

$$\left.\frac{\partial C}{\partial l}\right|_{(4,6,5)} = (0.5h + 3.25w)\Big|_{\substack{l=4\\w=6\\h=5}} = 0.5(5) + 3.25(6) = 22$$

The total cost will increase by approximately \$22 if the width remains at 6 ft, the height remains at 5 ft, and the length is increased from 4 ft to 5 ft. On the other hand,

$$\frac{\partial C}{\partial w}\bigg|_{(4,6,5)} = (h + 3.25l)\bigg|_{\substack{l=4 \\ w=6 \\ h=5}} = 5 + 3.25(4) = 18$$

and the total cost will increase by approximately \$18 if the length remains at 4 ft, the height remains at 5 ft, and the width is increased from 6 ft to 7 ft. Finally,

$$\frac{\partial C}{\partial h}\bigg|_{(4,6,5)} = (w + 0.5l)\bigg|_{\substack{l=4 \\ w=6 \\ h=5}} = 6 + 0.5(4) = 8$$

Thus the total cost will increase by approximately \$8 if the length remains at 4 ft, the width at 6 ft, and the height increases from 5 ft to 6 ft.

A second partial derivative is a partial derivative of a partial derivative.

A partial derivative of a function is again a function, so it can be differentiated again. The result is called a *second partial derivative*. A function of two variables, $z = f(x, y)$, has *two* partial derivatives associated with it, f_x and f_y, and *four* second partial derivatives, as follows: We can successively differentiate the original function twice with respect to x, obtaining $\frac{\partial}{\partial x}\left(\frac{\partial f}{\partial x}\right)$, which we denote as either $\frac{\partial^2 f}{\partial x^2}$ or f_{xx}. We can successively differentiate the original function twice with respect to y, obtaining $\frac{\partial}{\partial y}\left(\frac{\partial f}{\partial y}\right)$, which we denote as either $\frac{\partial^2 f}{\partial y^2}$ or f_{yy}. When we differentiate the original function first with respect to x and then that result with respect to y, we write $\frac{\partial}{\partial y}\left(\frac{\partial f}{\partial x}\right)$, which we denote as either $\frac{\partial^2 f}{\partial y \partial x}$ or f_{xy}. Finally, we can differentiate the original function first with respect to y and then with respect to x, obtaining $\frac{\partial}{\partial x}\left(\frac{\partial f}{\partial y}\right)$ which we denote as either $\frac{\partial^2 f}{\partial x \partial y}$ or f_{yx}. These last two second partial derivatives are called *mixed partials*, because they involve consecutive differentiations with respect to two different variables. When we use the subscript notation for mixed partial derivatives of second or higher degree, then the order in which the subscripts appear from left to right indicates the order in which the differentiations occur.

EXAMPLE 6

Find all second partial derivatives of the function $f(x, y) = x^2 \ln(2y)$.

Solution

The (first) partial derivatives of f are

$$f_x = 2x \ln(2y) \qquad \text{and} \qquad f_y = x^2/y$$

Then

$$f_{xx} = \frac{\partial}{\partial x}[2x \ln(2y)] = 2 \ln(2y)$$

$$f_{yy} = \frac{\partial}{\partial y}\left(\frac{x^2}{y}\right) = -\frac{x^2}{y^2}$$

$$f_{xy} = \frac{\partial}{\partial y}\left(\frac{\partial f}{\partial x}\right) = \frac{\partial}{\partial y}[2x \ln(2y)] = \frac{2x}{y}$$

$$f_{yx} = \frac{\partial}{\partial x}\left(\frac{\partial f}{\partial y}\right) = \frac{\partial}{\partial x}\left(\frac{x^2}{y}\right) = \frac{2x}{y}$$

Notice that the two mixed partials in Example 6 are identical. This is not coincidental. For all functions in this book, the order in which second partial derivatives are taken is immaterial, that is, $f_{xy} = f_{yx}$.

IMPROVING SKILLS

In Problems 1 through 30, find the first partial derivative with respect to each of the input variables.

1. $f(x, y) = 2x^5y^3$
2. $f(x, y) = 3x^5 + 8y^3 - 4xy$
3. $f(x, y) = 4x^2 - 2x^3y^3 + 5$
4. $f(x, t) = 5(xt)^3 + 3t - 1$
5. $f(x, t) = 2x^2 \sin t$
6. $f(x, t) = 2 \sin(x^2 t)$
7. $f(x, t) = x \sin(x^2 t)$
8. $f(x, t) = \sin^2(x^2 t)$
9. $f(t, y) = 5t^7 e^y$
10. $f(t, y) = 5ye^{t^7}$
11. $f(t, y) = e^{5yt^7}$
12. $f(t, y) = ye^{t^7 y^3}$
13. $f(u, v) = u/(u^2 + v^2)$
14. $f(u, v) = u^2/(uv + 1)$
15. $f(u, v) = (uv)/(u^2 + v^2)$
16. $f(u, v) = (uv^2 + 1)/u^2v^3 - 1)$
17. $g(x, y) = y^3\sqrt{1 - x^4}$
18. $g(x, y) = \sqrt{x^4 - 3x^3y + 2x^2y^3}$
19. $g(x, y) = (x^2y^5 + 4y^3)^{-0.5}$
20. $g(x, y) = (2.1x^{4.2}y^{-3.7} + 46)^{-1.5}$
21. $g(s, t) = \sqrt{(2s)/(1 + t^4)}$
22. $g(s, t) = \sqrt{(2s)/(1 + st)}$
23. $g(s, t) = \sqrt{(t \cos t)/(1 + st)}$
24. $g(s, t) = [(2e)^{3t}/(1 - 5e^{-st})]^{10}$
25. $f(x, y, z) = 3x^2y + 2y^3z - 4xz^4$
26. $f(x, y, z) = 3x^2y^3z^4 + 12$
27. $f(x, y, z) = x^{10}y^{-8}z^2 + 12xz^5$
28. $f(x, y, z) = \ln(3x - yz)$
29. $f(x, y, z) = x^2e^{3yz}$
30. $f(x, y, z) = x^2e^{xyz}$

In Problems 31 through 44, evaluate each partial derivative at the indicated point.

31. $f(x, y) = 3x^5y^3 - 4xy$; $\left.\dfrac{\partial f}{\partial x}\right|_{(2,0)}$

32. $f(x, y) = 3x^5y^3 - 4xy$; $\left.\dfrac{\partial f}{\partial y}\right|_{(1,2)}$

33. $f(x, y) = x^2 \cos(xt)$; $\left.\dfrac{\partial f}{\partial x}\right|_{(-1,1)}$

34. $f(x, t) = x^2 \cos(xt)$; $\left.\dfrac{\partial f}{\partial t}\right|_{(-1,1)}$

35. $g(x, y) = x^2ye^{2y}$; $\left.\dfrac{\partial g}{\partial x}\right|_{(1,2)}$

36. $g(x, y) = x^2ye^{2y}$; $\left.\dfrac{\partial g}{\partial y}\right|_{(1/2,1/4)}$

37. $g(t, w) = tw^3 \ln(2t)$; $\left.\dfrac{\partial g}{\partial t}\right|_{(2,3)}$

38. $g(t, w) = tw^3 \ln(2t)$; $\left.\dfrac{\partial g}{\partial w}\right|_{(2,3)}$

39. $f(x, y, z) = 3x^5y^2z^3 - 4x^2z$; $\left.\dfrac{\partial f}{\partial x}\right|_{(1,2,3)}$

40. $f(x, y, z) = 3x^5y^2z^3 - 4x^2z$; $\left.\dfrac{\partial f}{\partial y}\right|_{(1,2,3)}$

41. $f(x, y, z) = 3x^5y^2z^3 - 4x^2z$; $\left.\dfrac{\partial f}{\partial z}\right|_{(1,2,3)}$

42. $f(x, y, z) = 3xy^3e^{3xz} + 5$; $\left.\dfrac{\partial f}{\partial x}\right|_{(1,-1,-2)}$

43. $f(x, y, z) = 3xy^3e^{3xz} + 5$; $\left.\dfrac{\partial f}{\partial y}\right|_{(1,-1,1)}$

44. $f(x, y, z) = 3xy^3e^{3xz} + 5$; $\left.\dfrac{\partial f}{\partial z}\right|_{(2,2,-1)}$

In Problems 45 through 50, find f_{xy} and f_{yx} and show they are equal.

45. $f(x, y) = 4x^3y^8 - 4y^2$

46. $f(x, y) = xy^4 + 4xy$

47. $f(x, y) = x \sin(xt)$

48. $f(x, y) = xye^{-y}$

49. $f(x, y, z) = 3x^5y^2z^3 - 4x^2z$

50. $f(x, y, z) = xy^2/\ln(yz)$

51. Show that $z = x^2 + 6y$ satisfies the equation $3z_{xx} - z_y = 0$.

52. Show that $z = e^{-3y}\cos x$ is another solution to the equation in Problem 51.

53. Show that $z = e^{-(x+3y)}$ satisfies the equation $z_{xx} - 2z_x + z_y = 0$.

54. Show that $z = e^{x+2y}\sin x$ is another solution to the equation in Problem 53.

55. Show that $z = \sin(2x)\cos(4t)$ satisfies the equation $4z_{xx} = z_{tt}$.

CREATING MODELS

56. An economist believes there is a relationship between the number N of sport utility vehicles sold in a year and both the average cost, C, of these vehicles and the average price, p, of a gallon of gasoline. What can you say about the signs of the partial derivatives N_C and N_p?

57. The economist in Problem 56 also believes that the amount G of gasoline sold is related to both v, the number of vehicles on the road, and p, the average price of a gallon of gasoline. What can you say about the signs of the partial derivatives G_v and G_p?

58. The monthly payment m on a car loan depends on A, the amount borrowed, r, the nominal interest rate, and L, the length of the loan. What can you say about the signs of the partial derivatives m_A, m_r, and m_L?

59. A design for a closed rectangular carton with a square base calls for the top and bottom to be 4 sq. ft each and the height to be 3 ft. Estimate how the volume of the carton will change as either of its dimensions is altered.

60. For the carton described in Problem 59, estimate how the surface area changes as either of the box's dimensions is altered.

61. A design for a closed rectangular box calls for a base of 2 ft × 3 ft and a height of 4 ft. Estimate how the volume of the box changes as any one of its dimensions is altered.

62. A design for a closed cylindrical mailing tube calls for a radius of 2 in. and a length of 12 in. Estimate how the volume of this tube changes as either of its dimensions is altered.

FIGURE A

63. The volume V of the frustum of the cone shown in Figure A is given by the formula

$$V = \tfrac{1}{3}\pi h(R^2 + Rr + r^2)$$

Determine the rate of change of this volume with respect to a *decrease* in the base radius R when $R = 5$, $r = 4$, and $h = 3$.

64. A small factory produces two types of joints. The daily cost C (in dollars) of producing x Type 1 joints (in hundreds) and y Type 2 joints (in hundreds) is modeled by the equation

$$C = 5000 + 4x + 2y + \sqrt{xy}$$

What is the cost associated with a production run of 2000 Type 1 joints and 500 Type 2 joints? Estimate the rate at which this cost will change with respect to an increase in the production of either type of joint.

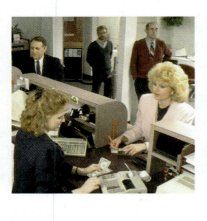

65. The volume V (in liters) of a particular gas in a canister is related to P, its pressure (in mm of mercury) and T, its temperature (in °C) by the equation

$$V = (8400 + 40T)/P$$

Current readings show $T = 22$ and $P = 810$. How quickly will the volume of the gas change with respect to a change in temperature? How quickly with respect to a change in pressure?

66. Mr. and Mrs. Johanssen are thinking of creating an ordinary annuity by depositing $20/month into an account at 6%/yr, compounded monthly, for the next 15 years. At what rate will the future value of this account grow if they increase their monthly deposits?

67. Christina's bank advertises three-year certificates of deposit that earn 5% a year, compounded quarterly, so she decides to invest $1000 there. Estimate how the future value of this investment will change as the interest rate changes between the present and the time when Christina actually deposits the money.

68. A *Cobb–Douglas model* relates P, the output of a production process to L, the number of units of labor, and C, the number of units of capital investment available to the production facility. This model is described by the equation

$$P = 1.2L^{0.75}C^{0.25}$$

for this process. Find the partial derivatives P_L, the *marginal productivity of labor*, and P_C, the *marginal productivity of capital*, when $L = C = 16$.

69. Solve Problem 68 when $L = 256$ and $C = 81$.

70. Find the marginal productivity of labor and the marginal productivity of capital (see Problem 68) for the Cobb–Douglas model described by

$$P = 5L^{0.6}C^{0.4}$$

when $L = 32$ and $C = 243$.

EXPLORING IN TEAMS

Two related products in the same marketplace are classified as *substitute products* if an increase in the sale of one of these products results in a decrease in the sale of the other. (Laundry detergents are a good example.) Let d_1 be the demand for one of these products at a unit price of p_1 and let d_2 be the demand for its substitute product at a unit price of p_2.

71. Explain how the demand for one of these products depends on both prices.

72. What can you say about the sign of the partial derivative $\partial d_1/\partial p_1$? What are the units of this partial derivative?

73. What can you say about the sign of the partial derivative $\partial d_1/\partial p_2$? What are the units of this partial derivative?

74. Use the information from Problems 71–73 to sketch a possible graph of the function $d_1 = f(p_1, p_2)$.

75. The demand for two related products is modeled by the equations

$$d_1 = 5000 - 2p_1 + 3p_2$$

$$d_2 = 8000 - 5p_1 - 4p_2$$

Does it follow from this model that the two products are substitute products?

76. The demand for two related products is modeled by the equations

$$d_1 = 5000 - 2p_1 + 3p_2$$
$$d_2 = 8000 + 5p_1 - 4p_2$$

Does it follow from this model that the two products are substitute products?

REVIEWING MATERIAL

77. (Section 2.3) *Mutual Funds* magazine reported that the total number of mutual funds was 3427 in 1991, 3850 in 1992, 4558 in 1993, 5357 in 1994, and 5761 in 1995. Determine a least-squares straight line fit for this data and then use it to estimate the total number of mutual funds that will exist in the year 2000.

78. (Section 10.2) Use the formula

$$f'(x_1) = \lim_{h \to 0} \frac{f(x_1 + h) - f(x_1)}{h}$$

to calculate $f'(2)$ for $f(x) = (2 + x)/x$.

79. (Section 11.2) Using the cost of materials listed in Example 5 of this section, find the dimensions of a storage shed similar to the one shown in Figure 13.7 but with a square base, that minimizes the total cost of materials while providing an interior volume of 120 cubic ft.

80. (Section 12.3) Evaluate the indefinite integral $\int 3e^{(1-5x)} \, dx$

RECOMMENDING ACTION

81. Respond by memo to the following:

MEMORANDUM

To: J. Doe Reader

From: Subhojit

Date: Today

Subject: **Definition of a Partial Derivative**

When we studied functions of a single input variable of the form $y = f(x)$, we developed the formula

$$f'(x_1) = \lim_{h \to 0} \frac{f(x_1 + h) - f(x_1)}{h}$$

for the derivative of the function at $x = x_1$. Is there a similar formula that we can use to define a partial derivative of the variable $z = f(x, y)$ at the point (x_1, y_1)?

13.3 OPTIMIZATION

We calculate partial derivatives by holding all but one of the input variables constant and then differentiating with respect to the single input that can vary. Partial derivatives have interesting geometrical interpretations when we restrict ourselves to functions of two input variables where graphing is possible.

Figure 13.18 is a graph of the function

$$z = f(x, y) = 4x - x^2 + 10y - y^2 + 50 \qquad (8)$$

Its partial derivatives at the point (3, 7) are

$$\frac{\partial z}{\partial x}\Big|_{(3,7)} = (4 - 2x)\Big|_{\substack{x=3 \\ y=7}} = -2 \qquad (9)$$

$$\frac{\partial z}{\partial y}\Big|_{(3,7)} = (10 - 2y)\Big|_{\substack{x=3 \\ y=7}} = -4 \qquad (10)$$

FIGURE 13.18
The Graph of $z = 4x - x^2 + 10y - y^2 + 50$

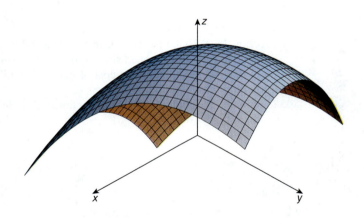

The partial derivative evaluated at (3, 7),
$\frac{\partial f}{\partial x}\Big|_{(3,7)}$ *is the slope of the tangent line to the curve $z = f(x, 7)$ at $x = 3$. This is the intersecting curve of the surface $z = f(x, y)$ with the plane $y = 7$.*

To calculate Equation (9), we hold y fixed and then differentiate Equation (8) with respect to x. Although we did not substitute a fixed value for y prior to differentiating, we could have. Setting $y = 7$ in Equation (8), we obtain

$$z = f(x, 7) = 4x - x^2 + 10(7) - (7)^2 + 50 = 4x - x^2 + 71 \qquad (11)$$

The derivative of this function of a single variable is $4 - 2x$, which equals -2 when x is 3.

The graph of the constraint $y = 7$ is a plane that is perpendicular to the y-axis and intersects it at $y = 7$. By stipulating that y must be 7, as we did to obtain Equation (11), we are restricting the surface $f(x, y)$ to its intersection with the $y = 7$ plane. Figure 13.19 shows both the surface defined by the function $z = 4x - x^2 + 10y - y^2 + 50$ and the $y = 7$ plane on the same coordinate system. The intersection of these two surfaces, designated by the thick curve in Figure 13.19, is the graph of the function $z = f(x, 7)$. This intersection is the graph of Equation (11).

Figure 13.20 is the graph of this intersection when we restrict ourselves to the $y = 7$ plane. The equation of this curve is the function $z = 4x - x^2 + 71$, and its derivative at

FIGURE 13.19

The Intersection of a Surface with the Plane y = 7

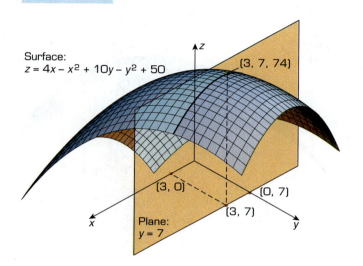

Surface:
$z = 4x - x^2 + 10y - y^2 + 50$

(3, 7, 74)

(3, 0)

(0, 7)

(3, 7)

x

Plane:
y = 7

z

y

FIGURE 13.20

The Surface $z = 4x - x^2 + 10y - y^2 + 50$ Restricted to the Plane y = 7, and the Tangent Line to the Curve at x = 3

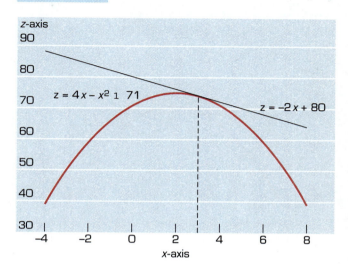

z-axis

$z = 4x - x^2$ 1 71

$z = -2x + 80$

x-axis

$x = 3$ is the slope of the tangent line to the curve at $x = 3$. This tangent line is also displayed in Figure 13.20, and its slope is -2. Thus $\dfrac{\partial z}{\partial x}\Big|_{(3,7)}$ is the slope of the tangent line to $z = f(x, 7)$ at $x = 3$.

EXAMPLE 1

Find $\dfrac{\partial z}{\partial y}\Big|_{(3,7)}$ and then give a geometrical interpretation of this partial derivative for the function in Equation (8):

$$z = f(x, y) = 4x - x^2 + 10y - y^2 + 50$$

Solution

To calculate $\dfrac{\partial z}{\partial y}\Big|_{(3,7)}$, we first fix $x = 3$ in Equation (8), obtaining

$$z = f(3, y) = 4(3) - (3)^2 + 10y - y^2 + 50 = 10y - y^2 + 53 \qquad (12)$$

The graph of this equation is shown in Figure 13.21.

Now we obtain $\dfrac{\partial z}{\partial y}\Big|_{(3,7)}$ by differentiating Equation (12) with respect to the input variable y, which gives us $\partial z/\partial y = 10 - 2y$, and then setting $y = 7$:

$$\frac{\partial z}{\partial y}\Big|_{(3,7)} = 10 - 14 = -4$$

This is the slope of the tangent line to the graph of Equation (12) at $y = 7$. Figure 13.21 also shows this tangent line.

FIGURE 13.21

The Curve $f(3, y) = 10y - y^2 + 53$ and the Tangent Line to the Curve at $y = 7$

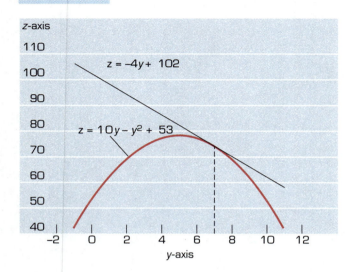

FIGURE 13.22

The Intersection of the Surface $z = 4x - x^2 + 10y - y^2 + 50$ with the Plane $x = 3$

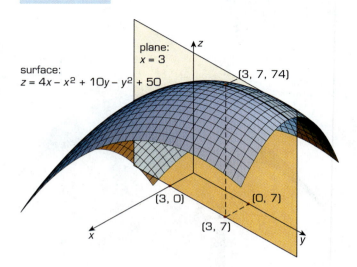

The graph of Equation (12) is the intersection of the graph of Equation (8) and that of the constraint equation $x = 3$. Both surfaces are shown in Figure 13.22, with their intersection highlighted by a thick curve. The graph of the equation $x = 3$ is a plane that is perpendicular to the x-axis and that intersects it at $x = 3$. By stipulating that x must be 3, we are restricting the output of the function $f(x, y)$ to the $x = 3$ plane. Figure 13.21 is the graph on the $x = 3$ plane of the intersection displayed in Figure 13.22.

We can use this geometrical interpretation of partial derivatives to develop a useful optimization algorithm for functions of many variables. Figure 13.22 is the graph of Equation (8),

$$z = f(x, y) = 4x - x^2 + 10y - y^2 + 50$$

when x is restricted to lie on the plane $x = 3$. The curve in Figure 13.21 attains its maximum at $y = 5$. This is where the tangent line to the curve has zero slope and where the derivative of the function $z = 10y - y^2 + 53$ is zero. The derivative of Equation (12) is the partial derivative of Equation (8) with x fixed at 3. Therefore, when we restrict ourselves to the plane $x = 3$, we obtain the largest value of z when

$$\left. \frac{\partial z}{\partial y} \right|_{x=3} = \frac{d}{dy}(10y - y^2 + 53) = 10 - 2y = 0$$

or when $y = 5$. The corresponding output is $z = f(3, 5) = 78$.

If we now set $y = 5$, Equation (8) reduces to

$$z = f(x, 5) = 4x - x^2 + 10(5) - (5)^2 + 50 = 4x - x^2 + 75 \qquad (13)$$

The graph of this equation is shown in Figure 13.23. It is the curve created by intersecting the surface defined by Equation (8) and the $x = 5$ plane, shown as a thick curve in Figure

FIGURE 13.23

The Graph of the Function
$z = 4x - x^2 + 75$

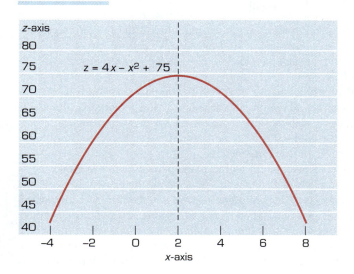

FIGURE 13.24

The Intersection of the Surface $z = 4x - x^2 + 10y - y^2 + 50$ with the Plane $y = 5$

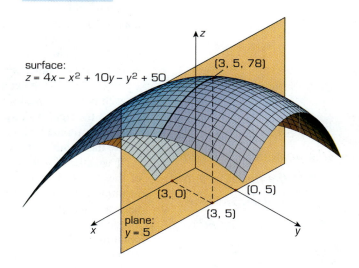

13.24. The function in Equation (13) attains its maximum where its derivative is zero. The derivative of the function $z = 4x - x^2 + 75$ is the partial derivative of Equation (8) with y fixed at 5:

$$\left.\frac{\partial z}{\partial x}\right|_{y=5} = \frac{d}{dx}(4x - x^2 + 75) = 4 - 2x$$

which is zero when $x = 2$. (Note the dashed line at $x = 2$ in Figure 13.23.) The corresponding output is $z = f(2, 5) = 79$.

Figure 13.25 shows a path moving through the domain from the point $(3, 7)$ through $(3, 5)$ and ending up at $(2, 5)$, as depicted by the dashed arrows. We trace the corresponding path on the surface, as indicated by the solid arrows in Figure 13.25. Our path on the surface gets higher and higher, while our movement in the domain is parallel to either the

FIGURE 13.25

The Surface $z = 4x - x^2 + 10y - y^2 + 50$ with a Path to a Maximum Point

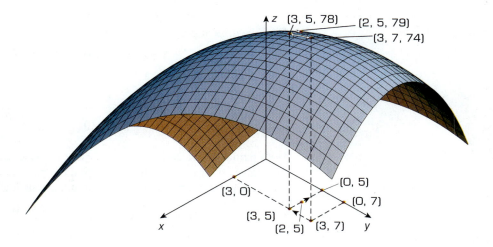

x- or the *y*-axis. We stop when we reach a point on the surface where all partial derivatives are zero. We could accelerate our search for the maximum by moving directly to the point on the surface where both partial derivatives are zero.

In Section 10.1, we defined the global maximum of a function as its largest output and the global minimum as its smallest output. On the graph of such a function, the global maximum occurs at the highest point and the global minimum at the lowest point. Furthermore, if a graph has more than one peak and/or pit, then we call the top of each peak a local maximum and the bottom of each pit a local minimum. These geometrical concepts extend easily to functions of two input variables.

Figure 13.26 is the graph of a function $z = f(x, y)$ with two peaks and one pit. The top of each peak is a local maximum, and the output variable *z* attains its largest value at the peak on the right, so this value is also the global maximum for this function. The global and local minima both equal the value of *z* at the bottom of the single pit, because that is where the smallest output occurs for this function.

If we slice a surface through a peak or pit with a plane perpendicular to either the *x*- or the *y*-axis, the curve on the plane will retain the peak or pit there. At this point, the resulting two-dimensional curve has a critical point: a point where either the curve has a tangent line with zero slope, or it has a cusp and its corresponding function does not have a derivative. Therefore we define a *critical point for a function of many variables* to be a point in the domain where either all of the function's first partial derivatives exist and are zero or at least one partial derivative does not exist. If a function of many variables has a peak or a pit at an interior point of its domain, then that point must be a critical point for the function.

> *A critical point for a function of many variables is a point in the domain where either all of the function's first partial derivatives are zero or one or more partial derivatives do not exist.*

FIGURE 13.26
A Surface $f(x, y)$ with Two Peaks and One Pit

EXAMPLE 2

Find the critical points for the function

$$z = 3x^2 + 2xy + 2y^2 - 10x - 10y + 30$$

Solution

We have

$$\frac{\partial z}{\partial x} = 6x + 2y - 10 \qquad \text{and} \qquad \frac{\partial z}{\partial y} = 2x + 4y - 10$$

Both partial derivatives exist everywhere because each can be evaluated at every pair of real values for *x* and *y*. Setting each of these partial derivatives equal to zero, we find that a critical point must satisfy the two simultaneous linear equations

$$6x + 2y = 10$$
$$2x + 4y = 10$$

The only solution to this set of equations is $x = 1$ and $y = 2$, therefore $(1, 2)$ is the only critical point. We see from Figure 13.27 that this critical point is the location of a local (global) minimum.

FIGURE 13.27

The Graph of $z = 3x^2 + 2xy + 2y^2 - 10x - 10y + 30$

EXAMPLE 3

Find the critical points for the function

$$z = x^2 - y^2$$

Solution

We have

$$\frac{\partial z}{\partial x} = 2x \quad \text{and} \quad \frac{\partial z}{\partial y} = 2y$$

Both partial derivatives exist everywhere because each can be evaluated at every pair of real values for x and y. Setting each of these partial derivatives equal to zero, we find that $x = y = 0$ is the only critical point.

A saddle point for a function occurs at a critical point where the function has neither a maximum nor a minimum.

A *saddle point* for a function occurs at a critical point where the function assumes neither a local maximum nor a local minimum. Near a saddle point there are points on the graph of the function that are higher than the saddle point and also points that are lower. Figure 13.28 is a graph of the function of Example 3. The critical point $(0, 0)$ is the location *in the domain* where this function attains the saddle point $(0, 0, 0)$.

A critical point can be the location of a local maximum, a local minimum, or a saddle point. We can identify the type of optimum at each critical point by graphing the function around the critical point, as we did in the last two examples. Graphing, however, is tedious, especially when software packages are not available. In such situations, the following algorithm for functions of two input variables may be useful. This algorithm is applicable only to critical points where all of the first derivatives exist and are zero.

FIGURE 13.28
The Graph of $z = x^2 - y^2$

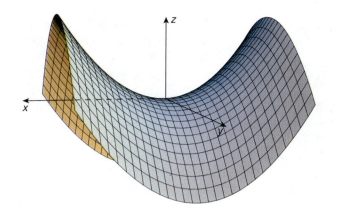

THE SECOND-DERIVATIVE TEST

Let (a, b) be a critical point for the function $z = f(x, y)$, defined in a circle centered at (a, b) with partial derivatives of all orders in this domain. Set

$$A = \frac{\partial^2 z}{\partial x^2}\bigg|_{(a,b)} \qquad B = \frac{\partial^2 z}{\partial x \, \partial y}\bigg|_{(a,b)} \qquad C = \frac{\partial^2 z}{\partial y^2}\bigg|_{(a,b)}$$

and

$$D = AC - B^2$$

1 If $D < 0$, then (a, b) is the location of a saddle point.

2 If $D > 0$ and $A < 0$, then (a, b) is the location of a local maximum.

3 If $D > 0$ and $A > 0$, then (a, b) is the location of a local minimum.

4 If $D = 0$, this test is inconclusive and other analysis is needed.

EXAMPLE 4

Apply the second-derivative test to the function from Example 2:

$$z = 3x^2 + 2xy + 2y^2 - 10x - 10y + 30$$

Solution
We showed in Example 2 that the only critical point for this function is $(1, 2)$, with

$$\frac{\partial z}{\partial x} = 6x + 2y - 10 \qquad \text{and} \qquad \frac{\partial z}{\partial y} = 2x + 4y - 10$$

Here

$$A = \frac{\partial^2 z}{\partial x^2}\bigg|_{(1,2)} = 6, \qquad B = \frac{\partial^2 z}{\partial x \partial y}\bigg|_{(1,2)} = 2, \qquad C = \frac{\partial^2 z}{\partial y^2}\bigg|_{(1,2)} = 4$$

and

$$D = AC - B^2 = 6(4) - (2)^2 = 20$$

Because $D = 20 > 0$ and $A = 6 > 0$, it follows directly from the second-derivative test that the critical point $(1, 2)$ is the location of a local minimum.

EXAMPLE 5

Use the second-derivative test to classify the critical points of the function

$$z = 4xy - x^4 - y^4 \qquad (14)$$

Solution

For this function,

$$\frac{\partial z}{\partial x} = 4y - 4x^3 \qquad \text{and} \qquad \frac{\partial z}{\partial y} = 4x - 4y^3$$

Here critical points occur only when both of these partial derivatives are zero, so the coordinates of a critical point must simultaneously satisfy the two equations

$$4y + 4x^3 = 0 \qquad \text{and} \qquad 4x - 4y^3 = 0$$

or, equivalently, $y = x^3$ and $x = y^3$. This last equation can also be written $y = x^{1/3}$.

We can solve these two nonlinear simultaneous equations by eliminating the x-variable between them and solving for y, as follows:

$$y = x^3 = (y^3)^3 = y^9 \quad \text{or} \quad y(y^8 - 1) = 0$$

The only real-valued solutions to the second equation are $y = 0$ and $y = \pm 1$. Since $x = y^3$, it follows that $x = 0$ when $y = 0$, $x = 1$ when $y = 1$, and $x = -1$ when $y = -1$. Thus the function in Equation (14) has three critical points, namely, $(-1, -1)$, $(0, 0)$, and $(1, 1)$. Alternatively, these points can be obtained by graphing the two equations $y = x^{1/3}$ and $y = x^3$ on the same coordinate system and locating their points of intersection, as shown in Figure 13.29.

FIGURE 13.29
The Graphs of $y = x^{1/3}$ and $y = x^3$ with Critical Points $(-1, -1)$, $(0, 0)$, and $(1, 1)$

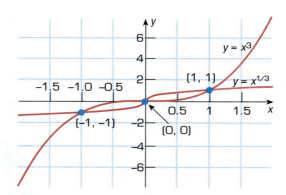

Taking second partial derivatives of the given function, we find that

$$\frac{\partial^2 z}{\partial x^2} = -12x^2, \qquad \frac{\partial^2 z}{\partial x \partial y} = 4, \quad \text{and} \quad \frac{\partial^2 z}{\partial y^2} = -12y^2$$

At the critical point $(-1, -1)$

$$A = \left.\frac{\partial^2 z}{\partial x^2}\right|_{(-1,-1)} = -12(-1)^2 = -12$$

$$B = \left.\frac{\partial^2 z}{\partial x \partial y}\right|_{(-1,-1)} = 4$$

$$C = \left.\frac{\partial^2 z}{\partial y^2}\right|_{(-1,-1)} = (-12)(-1)^2 = -12$$

$$D = AC - B^2 = (-12)(-12) - (4)^2 = 128$$

Because $D > 0$ and $A < 0$, it follows from the second-derivative test that the critical point $(-1, -1)$ is the location of a local maximum.

At the critical point (0, 0)

$$A = \frac{\partial^2 z}{\partial x^2}\bigg|_{(0,0)} = -12(0)^2 = 0$$

$$B = \frac{\partial^2 z}{\partial x \partial y}\bigg|_{(0,0)} = 4$$

$$C = \frac{\partial^2 z}{\partial y^2}\bigg|_{(0,0)} = -12(0)^2 = 0$$

$$D = AC - B^2 = 0 - (4)^2 = -16$$

Because $D < 0$, the critical point $(0, 0)$ is the location of a saddle point.

At the critical point (1, 1)

$$A = \frac{\partial^2 z}{\partial x^2}\bigg|_{(1,1)} = -12(1)^2 = -12$$

$$B = \frac{\partial^2 z}{\partial x \partial y}\bigg|_{(1,1)} = 4$$

$$C = \frac{\partial^2 z}{\partial y^2}\bigg|_{(1,1)} = -12(1)^2 = -12$$

$$D = AC - B^2 = (-12)(-12) - (4)^2 = 128$$

Because $D > 0$ and $A < 0$, the critical point $(1, 1)$ is the location of a local maximum.

   Equation (14) is graphed in Figure 13.30. This tooth-like surface has two local maxima with a saddle point between them.

FIGURE 13.30

The Graph of
$z = 4xy - x^4 - y^4$

EXAMPLE 6

A packaging company must design a cardboard box with a bottom, no top, and a partition in the middle that divides the box into two equal sections, each having a volume of 24 cu. in. Find the dimensions that use the least amount of cardboard.

FIGURE 13.31
A Bisected Cardboard Box

Solution

The box has the shape displayed in Figure 13.31. The surface area S of this box is

$$S = 3xz + 2yz + xy \tag{15}$$

Each compartment must have a volume of 24 cu. in., so we must satisfy the constraint $\frac{1}{2}xyz = 24$. Solving this equation for z, we find that $z = 48/xy$. Substituting this expression into Equation (15), we have

$$S = \frac{144}{y} + \frac{96}{x} + xy = 144y^{-1} + 96x^{-1} + xy \tag{16}$$

Thus we need to minimize a function of two input variables. It is impossible to construct a box with a side of zero length and positive volume, so neither the $x = 0$ nor $y = 0$ is part of the domain.

For this function,

$$\frac{\partial S}{\partial x} = -96x^{-2} + y \quad \text{and} \quad \frac{\partial S}{\partial y} = -144y^{-2} + x$$

These derivatives do not exist when either x or y is zero, but neither is in the domain. We need to know when both partial derivatives are zero, i.e., when

$$-96x^{-2} + y = 0 \quad \text{and} \quad -144y^{-2} + x = 0$$

Solving each equation for y, we obtain the two nonlinear equations

$$y = 96/x^2 \quad \text{and} \quad y = \pm 12/\sqrt{x}$$

We discard the negative quantity in the second equation because the dimensions of the box must be positive.

We can solve these equations by graphing them on the same coordinate system and locating their points of intersection. Alternatively, we can locate the intersection points algebraically by equating the y-values: $96/x^2 = 12/\sqrt{x}$, which reduces to $x^{3/2} = 8$ or $x = 4$. Then $y = 12/\sqrt{4} = 6$. Therefore the only critical point for Equation (16) is $(4, 6)$.

Taking second partial derivatives of the function in Equation (16), we find that

$$\frac{\partial^2 S}{\partial x^2} = 192x^{-3}, \quad \frac{\partial^2 S}{\partial x \partial y} = 1, \quad \frac{\partial^2 S}{\partial y^2} = 288y^{-3}$$

Therefore

$$A = \left. \frac{\partial^2 S}{\partial x^2} \right|_{(4,6)} = 192(4)^{-3} = 3$$

$$B = \left. \frac{\partial^2 S}{\partial x \partial y} \right|_{(4,6)} = 1$$

$$C = \left. \frac{\partial^2 S}{\partial y^2} \right|_{(4,6)} = 288(6)^{-3} = \frac{4}{3}$$

$$D = AC - B^2 = 3(\tfrac{4}{3}) - (1)^2 = 3$$

Because $D > 0$ and $A > 0$, the critical point $(4, 6)$ is the location of a local minimum.

In this case we can infer that the local minimum is also a global minimum. Because $(4, 6)$ is the location of a local minimum, the graph of S, the function in Equation (16), has a pit that bottoms out at $(4, 6)$ but increases around it. Furthermore, this graph never decreases anywhere else, even far from the critical point, because if it did we would have a local maximum or a saddle point, where the graph changes from an increasing function to a decreasing one. But there are no other local optima (S has only one critical point) so this function always increases as we move away from the point (4, 6).

With $x = 4$ and $y = 6$, it follows from the equation $z = 48/xy$ that $z = 2$. The box should have dimensions $4'' \times 6'' \times 2''$.

In Chapter 11 we found that a function of a single input variable on a closed interval attains its global maximum (and minimum) at a critical point or at an endpoint. A similar situation exists for a function of two input variables. If a domain is a bounded region and if the domain includes all points on the boundary of that region, then a function of two variables attains its global maximum (and minimum) either at a critical point or at a point on the boundary of the domain. For functions of many variables, the boundary of the domain presents difficulties because it generally contains infinitely many points, and such analysis is beyond the scope of this book.

Local and global optima for a function can occur at points where the function's partial derivatives do not exist. At such points, the second-derivative test is not applicable because this test is restricted to points where all of the first partial derivatives of a function are zero. Figure 13.32 is the graph of the function $z = \sqrt{x^2 + y^2}$ with partial derivatives

FIGURE 13.32
The Graph of $z = \sqrt{x^2 + y^2}$

$$\frac{\partial z}{\partial x} = \frac{x}{\sqrt{x^2 + y^2}} \quad \text{and} \quad \frac{\partial z}{\partial y} = \frac{y}{\sqrt{x^2 + y^2}}$$

These partial derivatives do not exist at $x = y = 0$, yet Figure 13.32 shows that the function is defined at $x = y = 0$ with $z = \sqrt{0^2 + 0^2} = 0 = 0$ and has a global minimum at the origin. Generally we must graph a function near points where at least one of its first derivatives does not exist to determine whether the function has a global maximum or minimum there.

IMPROVING SKILLS

For each function in Problems 1 through 27, locate the critical points and then classify them using the second-derivative test.

1. $z = x^2 - 4x + y^2 - 2y + 6$
2. $z = x^2 + 4x + y^2 + 2y + 6$
3. $z = x^2 - 4x - y^2 - 2y + 6$
4. $z = 2y - y^2 - x^2 - 2y$
5. $z = xy$
6. $z = (xy)^2$
7. $z = 2x^2 + xy - y^2 - 7x + 10$
8. $z = 2x^2 + xy + y^2 - 7x + 10$
9. $z = 2x^2 + 3xy + y^2 - 7x + 10$
10. $z = 4x^3 - 12x^2 - 36x + y^2 + 5$
11. $z = -3x^2 + 2xy - 4y^2 - 16x + 20y$
12. $z = 4x^2 - 8xy + 4y^2$
13. $z = 3x^2 - 3xy + y^2$
14. $z = 3x^2 + 3xy + y^2$
15. $z = 3x^2 + 4xy + y^2$
16. $z = x^3 + y^3 - 3xy + 1$
17. $z = 2x^2 + y^3 - 4xy - 2$
18. $z = 2y^3 - 3y^2 + 2x^2y - 3$
19. $z = x^3 + y^3 - 3xy + 4$
20. $z = y^4 - x^4$
21. $z = 2x^2 + y^4 - 4xy + 8$
22. $z = x^2 - 4x + 2y^3 - y^2 - 4y + 6$
23. $z = x^2 + 2y^4 - 12xy$
24. $z = e^{xy}$
25. $z = e^{x^2 - 2x + y^2}$
26. $z = e^{-(x^2 + y^2)}$
27. $z = \ln(xy) + 2x^2 - xy - 6x$
28. Show that the function $z = (x^2 + 8y^2)e^{1 - x^2 - y^2}$ has critical points at (1, 1), (1, −1), (−1, 1), (−1, −1), and (0, 0). Find all local maxima, local minima, and saddle points.

CREATING MODELS

29. Find three numbers x, y, and z whose sum is 10 and whose product x^2yz is a maximum.

30. Find three numbers whose sum is 10 and the sum of whose squares is a minimum.

31. A packaging company must design an open box with a volume of 144 ft³. The material for the bottom of the box costs \$2/ft² and the material for the sides costs \$1.50/ft². Find the dimensions of the box that minimize the total cost of materials.

32. A packaging company must design a closed box with a volume of 144 ft³. The material for the top and bottom of the box is twice as expensive as the material for its sides. Find the dimensions of the box that minimize the total cost of materials.

FIGURE A

33. A packaging company must design an open cardboard box with a total volume of 64 ft³. The box will have a bottom, no top, and two parallel partitions that divide the box into three equal sections (see Figure A). Find the dimensions of the box that will use the least amount of cardboard.

34. A packaging company must design an open cardboard box with a bottom, no top, and two perpendicular partitions that divide the box into four equal sections (see Figure B). Each section must have a volume of 18 ft³. Find the dimensions of the box that will use the least amount of cardboard.

35. A company manufactures two types of editing systems designed to run on personal computers, to be used by television stations. The company's profit P (in dollars) is modeled by the equation

$$P = 975x + 1575y + 15xy - 20x^2 - 10y^2$$

where x and y are the numbers of each type of system produced. Determine a production schedule that will maximize profit.

FIGURE B

36. A company manufactures incandescent and fluorescent light bulbs. The daily profit P (in dollars) of producing i incandescent bulbs and f fluorescent bulbs (both in hundred-unit lots) is modeled by the function

$$P = 0.1i^2 - 0.4if + 0.2f^2 + 20i + 40f - 2700$$

Determine a production schedule that will maximize daily profit.

37. A company produces two blends of coffee that it markets as regular (r) and gourmet (g). It costs the company \$1 to produce a pound of the regular blend and \$2 to produce a pound of the gourmet blend. The demands d_r and d_g for each type of coffee (in lb) are related to their respective prices p_r and p_g (in dollars/lb) by the equations

$$d_r = 40{,}000 - 4000p_r + 1000p_g$$

$$d_g = 30{,}000 + 1000p_r - 5000p_g$$

   (**a**) Write expressions for the total revenue, total cost, and total profit in terms of the two prices.

   (**b**) Determine a set of prices that will maximize profit.

38. A new-car dealership advertises regularly in two newspapers that serve substantially different markets. Each month, x dollars of advertising in the first paper yields approximately $40x/(x + 5)$ new-car sales, while y dollars of advertising in the second paper yields approximately $20y/(y + 10)$ new-car sales. On average, each sale earns the dealership \$1250 above the direct cost of the car. The monthly profit from advertising is the amount of money earned from sales minus the cost of the ads.

   (**a**) Write an expression for the monthly profit from advertising in terms of x and y.

   (**b**) Determine how much advertising in each newspaper will maximize this profit.

EXPLORING IN TEAMS

39. Review Section 2.3 on least-squares straight-line fit. Using the notation developed there, show that the sum-of-squares error in fitting data points (x_1, y_1), (x_2, y_2), $(x_3, y_3), \cdots, (x_N, y_N)$ with the line $y = mx + b$ is

$$E = (mx_1 + b - y_1)^2 + (mx_2 + b - y_2)^2$$
$$+ (mx_3 + b - y_3)^2 + \cdots + (mx_N + b - y_N)^2$$

40. The least-squares straight line estimate is the straight line that minimizes the sum-of-squares error. Explain why, for any fixed set of data points, the sum-of-squares error in Problem 39 is a function of only two variables. What are these two variables?

41. Find the partial derivatives of the sum-of-squares error E with respect to each of the two input variables identified in Problem 40.

42. Set each of the partial derivatives from Problem 41 equal to zero, thereby generating two equations for a critical point of E. Show that these equations reduce to System (16) in Section 2.3.

EXPLORING WITH TECHNOLOGY

43. Show that the second-derivative test is inconclusive for each of the following functions:

(**a**) $z = x^4 + y^4$ (**b**) $z = x^4 - y^4$ (**c**) $z = -x^4 - y^4$

Then use a computer software package or a graphing calculator that supports three-dimensional graphing to display the surface of each function near its critical point. Using these graphs, classify each critical point. Add a sentence to the end of the second-derivative test that explains why the test is inconclusive when $D = 0$.

44. Slice the surface shown in Figure 13.32 with the $x = 0$ plane and graph the intersection on a yz-coordinate system. Using a graphing calculator on this graph, zoom in on the origin and discuss the derivative from a geometrical perspective.

45. Using the results of Problem 44, give a geometrical condition for when a function that is well defined everywhere does not have partial derivatives at a point.

46. Graph the function

$$z = \left| |x| - |y| \right| - |x| - |y|$$

Make a geometrical case that this function has many points at which the partial derivatives do not exist, giving rise to saddle points.

REVIEWING MATERIAL

47. (Section 5.2) The *New York Times* reported that consumers spent $7.854 billion on ready-to-eat cereals over the 52-week period ending April 21, 1996. They purchased $2.84 billion worth of Kellogg's cereals, $2.073 billion worth of General Mills cereals, $1.006 billion worth of Post cereals, $0.598 billion worth of Quaker cereals, and $0.514 billion worth of store brands. Create a pie chart from this data.

48. (Sections 6.1 and 6.3) Using the data from Problem 47, determine the probability that a randomly selected cereal buyer is a Kellogg's customer. What is the probability that such a customer is a Kellogg's customer if it is known that the customer does not buy General Mills or Post cereals?

49. (Section 10.2) Determine the right- and left-hand limits at $x = 1, 2,$ and 3 for the function displayed in Figure C. Does this function have a limit at these points? Where is this function continuous?

FIGURE C
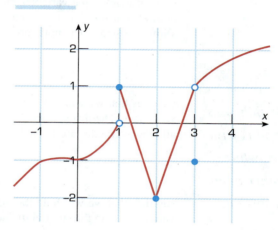

50. (Section 12.5) Evaluate the definite integral $\int_0^1 (x/\sqrt{1 + x^2})\, dx$

RECOMMENDING ACTION

51. Respond by memo to the following:

MEMORANDUM

To: J. Doe Reader

From: Head of Production

Date: Today

Subject: **Box Design**

Thank you for your proposal on the design of our new open box with one partition [see Example 6]. Since I am not a mathematician, I am unable to understand your analysis. I know that you took something called a partial derivative and set it equal to zero. Could you explain in plain English why this is the right thing to do? What does it mean?

13.4 ITERATED INTEGRALS

In Section 13.3, we extended our formulas for differentiation of functions of a single variable to functions of many variables by holding all but one input variable fixed and then differentiating with respect to the single input allowed to vary. The result is partial differentiation. We now want to reverse this process; we want to reconstruct a function from its partial derivative.

The process of reconstructing a function of a single variable from its derivative is called indefinite integration, which we developed in Chapter 12. We extend this process to functions of many input variables by the same approach we used to extend differentiation: we hold all but one input variable fixed and then integrate with respect to the single input allowed to vary. The notation

$$\int f(x, y)\, dx \tag{17}$$

The letter x in the differential dx indicates that integration is with respect to x, while holding all other input variables constant.

signifies that we will hold the input variable y fixed and then integrate with respect to the variable x. The differential dx indicates that x is the only variable that is allowed to vary; all other variables are treated as constants. Similarly, the notation

$$\int f(x, y)\, dy \tag{18}$$

signifies that we will hold the input variable x fixed and then integrate with respect to the variable y. The differential dy indicates that y is the only variable that is allowed to vary.

EXAMPLE 1

Evaluate the indefinite integral $\int (6x^2 + 8xy + 4x)\, dx$.

Solution

The differential dx indicates that x may vary and that all other variables, here y, remain fixed and are treated as constants. Thus $8y$, the coefficient of x in the second term, is treated the same way as 4, the coefficient of x in the third term. Applying first the constant multiple rule of integration and then the power rule of integration, we obtain

$$
\begin{aligned}
\int (6x^2 + 8xy + 4x)\, dx &= 6\int x^2\, dx + 8y\int x\, dx + 4\int x\, dx \\
&= 6(\tfrac{1}{3}x^3) + 8y(\tfrac{1}{2}x^2) + 4(\tfrac{1}{2}x^2) + c \tag{19} \\
&= 2x^3 + 4x^2y + 2x^2 + c
\end{aligned}
$$

Indefinite integration is created to reverse the effects of differentiation. Therefore we check integration by differentiating the answer to see if we get back the function we integrated. In Example 1, we integrated with respect to x; to check our answer, we differentiate the answer with respect to the same variable.

Check: $\qquad \dfrac{\partial}{\partial x}(2x^3 + 4x^2y + 2x^2 + c) = 6x^2 + 8xy + 4x$

which is indeed the function we integrated in Example 1.

> *A constant of integration depends on all of the input variables held fixed during integration.*

The letter c in Equation (19) represents a constant of integration. In this case the constant of integration can depend on the variable y because the integration was with respect to x and y was treated as a constant. To emphasize this dependence, we normally write Equation (19) as

$$\int (6x^2 + 8xy + 4x)\, dx = 2x^3 + 4x^2y + 2x^2 + C(y)$$

where $C(y)$ denotes any function of y. For such a function,

$$\frac{\partial C(y)}{\partial x} = 0$$

EXAMPLE 2

Evaluate the indefinite integral $\int (6x^2 + 8xy + 4x)\, dy$.

Solution

The differential dy indicates that we are integrating with respect to y, holding x fixed. Thus the first term, $6x^2$, and the third term, $4x$, as well as the coefficient $8x$, are all treated as constants. Therefore

$$\int (6x^2 + 8xy + 4x)\, dy = 6x^2 \int 1\, dy + 8x \int y\, dy + 4x \int 1\, dy$$
$$= 6x^2(y) + 8x(\tfrac{1}{2}y^2) + 4x(y) + C(x)$$
$$= 6x^2y + 4xy^2 + 4xy + C(x)$$

Check: We take the partial derivative of our answer with respect to y, the variable of integration:

$$\frac{\partial}{\partial y}\, [6x^2y + 4xy^2 + 4xy + C(x)] = 6x^2 + 8xy + 4x + 0$$

which is the function we integrated.

The extension of indefinite integration to functions of many variables is applicable also to definite integration.

EXAMPLE 3

Evaluate the definite integral $\int_{-1}^{1} xe^{x^2y}\, dy$.

Solution

The differential dy indicates that we are integrating with respect to y, holding x fixed. For example, if x is fixed at 2, the integral becomes $\int_{-1}^{1} 2e^{4y}\, dy$, which is similar to Example 7 of Section 12.3; if x is 5, the integral becomes $\int_{-1}^{1} 5e^{25y}\, dy$. We are not given a value for x, so we cannot actually substitute a number for it. Nevertheless, we can evaluate the given integral as if x is a constant, such as 2 or 5. Using the method of substitution of variables, we set $u = x^2y$, so that

$$du = \frac{\partial}{\partial y}\, (x^2y)\, dy = x^2\, dy$$

Then, since we are viewing x as if it is a constant, we have

$$\int xe^{x^2 y}\, dy = \int \frac{x^2}{x} e^{x^2 y}\, dy \qquad \text{(multiply by } x/x\text{)}$$

$$= \frac{1}{x} \int e^{x^2 y}(x^2\, dy) \qquad \text{(constant multiple rule of integration)}$$

$$= \frac{1}{x} \int e^u\, du = \frac{1}{x} e^u = \frac{1}{x} e^{x^2 y}$$

as an antiderivative of the function we need to integrate. Thus,

$$\int_{-1}^{1} xe^{x^2 y}\, dy = \frac{1}{x} e^{x^2 y}\Big|_{y=-1}^{y=1} = \frac{1}{x} e^{x^2(1)} - \frac{1}{x} e^{x^2(-1)} = \frac{1}{x}(e^{x^2} - e^{-x^2})$$

EXAMPLE 4

Evaluate the definite integral $\displaystyle\int_0^2 y\cos(xy)\, dx$.

Solution

The differential dx tells us that we are integrating with respect to x and treating y as a constant. Using the method of substitution of variables, we set $u = xy$, so that

$$du = \frac{\partial}{\partial x}(xy)\, dx = y\, dx$$

Then

$$\int y\cos(xy)\, dx = \int \cos(xy)(y\, dx)$$

$$= \int \cos u\, du = \sin u = \sin xy$$

as one antiderivative of the function we want to integrate. Thus

$$\int_0^2 y\cos(xy)\, dx = \sin(xy)\Big|_{x=0}^{x=2}$$

$$= \sin 2y - \sin 0 = \sin 2y$$

The answer to Example 4, $\sin(2y)$, can now be looked at as a function of the *variable* y, and it can be integrated, too. The *iterated integral*

$$\int_a^b \int_c^d f(x, y)\, dx\, dy = \int_a^b \left[\int_c^d f(x, y)\, dx \right] dy \qquad (20)$$

is a double integration: First we evaluate the integral inside the brackets with respect to x, treating y as a constant. The result is a function of y, which we now treat as a variable, and so we can integrate it. Similarly, the iterated integral

$$\int_c^d \int_a^b f(x, y)\, dy\, dx = \int_c^d \left[\int_a^b f(x, y)\, dy \right] dx$$

also is a double integration: Here we first evaluate the integral inside the brackets with respect to y, treating x as a constant, then we look at the result as a function of x, and we integrate that.

EXAMPLE 5

Evaluate the double integral $\int_{-\pi}^{\pi} \int_0^2 y \cos(xy)\, dx\, dy$.

Solution

This is an iterated integral in the form of Equation (20). Using the results of Example 4, we have

$$\int_{-\pi}^{\pi} \int_0^2 y \cos(xy)\, dx\, dy = \int_{-\pi}^{\pi} \left[\int_0^2 y \cos(xy)\, dx \right] dy$$

$$= \int_{-\pi}^{\pi} \sin(2y)\, dy = -\tfrac{1}{2}\cos(2y) \Big|_{y=-\pi}^{y=\pi}$$

$$= -\tfrac{1}{2}\cos(2\pi) + \tfrac{1}{2}\cos(-2\pi) = 0$$

EXAMPLE 6

Evaluate the iterated integral $\int_2^5 \int_1^3 x^3 y^2\, dx\, dy$.

Solution

$$\int_2^5 \int_1^3 x^3 y^2\, dx\, dy = \int_2^5 \left[\int_1^3 x^3 y^2\, dx \right] dy$$

$$= \int_2^5 \left[\frac{1}{4} x^4 y^2 \right]_{x=1}^{x=3} dy$$

$$= \int_2^5 \left[\frac{1}{4}(3)^4 y^2 - \frac{1}{4}(1)^4 y^2 \right] dy$$

$$= \int_2^5 20 y^2\, dy = \frac{20}{3} y^3 \Big|_{y=2}^{y=5}$$

$$= \tfrac{20}{3}(5)^3 - \tfrac{20}{3}(2)^3 = 780$$

EXAMPLE 7

Evaluate the iterated integral $\int_1^3 \int_2^5 x^3 y^2\, dy\, dx$.

Solution

$$\int_1^3 \int_2^5 x^3 y^2\, dy\, dx = \int_1^3 \left[\int_2^5 x^3 y^2\, dy \right] dx$$

$$= \int_1^3 \left[\frac{1}{3} x^3 y^3 \right]_{y=2}^{y=5} dx$$

$$= \int_1^3 \left[\frac{1}{3} x^3 (5)^3 - \frac{1}{3} x^3 (2)^3 \right] dx$$

$$= \int_1^3 39 x^3\, dx = \frac{39}{4} x^4 \Big|_{x=1}^{x=3}$$

$$= \frac{39}{4}(3)^4 - \frac{39}{4}(1)^4 = 780$$

It is not coincidental that the answers to Examples 6 and 7 are the same. It helps to look at iterated integrals from a geometrical point of view. The limits on the integrals in Equation (20),

FIGURE 13.33

A Rectangular Domain for the Function $f(x, y)$

$$\int_a^b \int_c^d f(x, y) \, dx \, dy = \int_a^b \left[\int_c^d f(x, y) \, dx \right] dy \qquad \text{(20 repeated)}$$

indicate that first x will vary between $x = c$ and $x = d$ and then y will vary between $y = a$ and $y = b$. These constraints limit the domain of $f(x, y)$ to the rectangle shown in Figure 13.33.

The graph of the function $f(x, y)$ is a surface over the xy-plane. For the moment, let us assume that $f(x, y) > 0$, so that this surface is actually a roof over the xy-plane, similar to the one shown in Figure 13.34. From our work in Section 13.3, we know that holding y fixed is equivalent to intersecting the surface $f(x, y)$ with a plane that is perpendicular to the y-axis and intersects it at some fixed value of y, say, $y = y_0$. The curve of intersection is shown in Figure 13.34 as the thick curve on the surface of $f(x, y)$.

Let us restrict our attention to the plane in Figure 13.34 and ask how we might calculate the area under the curve of intersection and above the xy-plane, with y fixed between a and b and x allowed to vary between c and d. With y fixed, $f(x, y)$ is effectively a function of the single variable x.

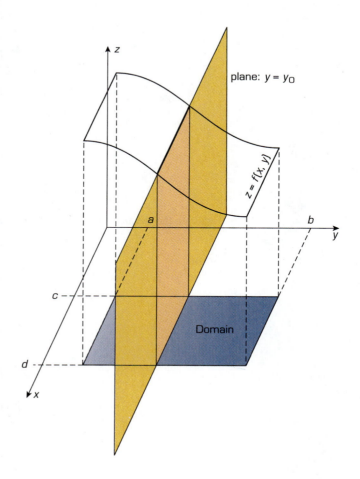

It follows from our work in Chapter 12 that the area we seek—the area represented by the dark orange shading in Figure 13.34—is

$$A(y) = \int_c^d f(x, y) \, dx \qquad (21)$$

which is the inside integral in Equation (20). Note that after we perform the integration in Equation (21), the variable x no longer appears. The area $A(y)$ is a function of the variable y. If we multiply $A(y)$ by a small width, say Δy, we have the volume of the slice shown in Figure 13.35.

FIGURE 13.35

A Slice of Width Δy Parallel to the x-Axis

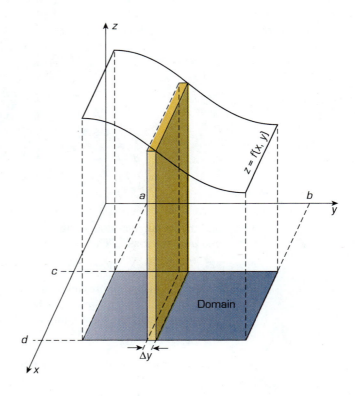

Let us divide the interval $[a, b]$ on the y-axis into four equal parts, each of width $\Delta y = (b - a)/4$. We define four numbers y_1, y_2, y_3, y_4 as follows:

$$y_1 = a + \Delta y$$

$$y_2 = y_1 + \Delta y = a + 2 \, \Delta y$$

$$y_3 = y_2 + \Delta y = a + 3 \, \Delta y$$

$$y_4 = y_3 + \Delta y = a + 4 \, \Delta y = b$$

The area under the surface in Figure 13.34 and above the xy-plane at each of these points is $A(y_1)$, $A(y_2)$, $A(y_3)$, and $A(y_4)$; these areas can be calculated using Equation (21). If we then multiply each of these areas by Δy, we generate the slices shown in Figure 13.36. The sum of these slices is an approximation of the volume V between the surface $f(x, y)$

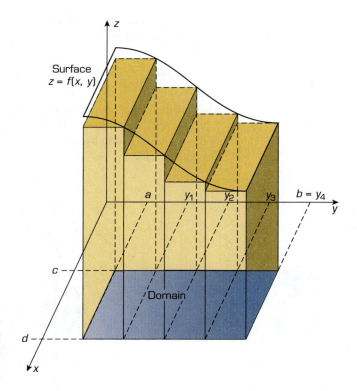

and the rectangular domain that is bounded between a and b on the y-axis and between c and d on the x-axis. That is,

$$V \approx A(y_1)\Delta y + A(y_2)\Delta y + A(y_3)\Delta y + A(y_4)\Delta y$$

We get a better approximation to the volume of the region under the surface in Figure 13.36 by dividing the y-axis into more than four equal parts. If we divide the interval $[a, b]$ into n equal parts, each of width Δy, then $\Delta y = (b - a)/n$ and we have the n points

$$y_1 = a + \Delta y, \qquad y_2 = a + 2\,\Delta y, \qquad y_3 = a + 3\,\Delta y, \cdots, y_n = a + n\,\Delta y$$

Our approximation to the volume now becomes

$$V \approx A(y_1)\Delta y + A(y_2)\Delta y + A(y_3)\Delta y + \cdots + A(y_n)\Delta y = \sum_{i=1}^{n} A(y_i)\Delta y$$

This approximation becomes better and better as n gets larger, so in the limit we have

$$V = \lim_{n \to \infty} \sum_{i=1}^{n} A(y_i)\Delta y$$

The right side of this equation is simply the definition of the definite integral (see Section 12.5) of $A(y)$ over the interval $[a, b]$. Therefore

$$V = \int_a^b A(y)\,dy$$

It now follows from Equation (21)

$$A(y) = \int_c^d f(x, y)\,dx \qquad\qquad \text{(21 repeated)}$$

that

$$V = \int_a^b \left[\int_c^d f(x, y) \, dx \right] dy \tag{22}$$

The iterated integral on the right side of Equation (22) is the volume between the surface defined by the function $f(x, y)$ and the rectangle in the xy-plane, provided the function is nonnegative over this domain.

We can repeat the same analysis but with slices of width Δx parallel to the y-axis, as shown in Figure 13.37. Using this procedure, the volume is given by

$$V = \int_c^d \left[\int_a^b f(x, y) \, dy \right] dx \tag{23}$$

FIGURE 13.37

A Slice of Width Δx Parallel to the y-Axis

> *The average value of the function $z = f(x, y)$ over a domain D is the area of D divided into an iterated integral of the function over D.*

Although Equations (22) and (23) are different, they both represent the same volume. Using Equation (22), we calculate volume by summing slices parallel to the x-axis; using Equation (23), we calculate volume by summing slices parallel to the y-axis. Since both approaches yield the volume of the same figure, both equations must be equal. This is the geometrical explanation of why Examples 6 and 7 have identical answers. In each of those examples, the integral is the volume under the graph of $z = x^3 y^2$ and over the rectangle in the xy-plane bounded between 2 and 5 on the y-axis and between 1 and 3 on the x-axis. On this domain, the function $z = x^3 y^2$ is positive.

The *average value* of the function $z = f(x, y)$ over a domain D is the value of (either) iterated integral of the function over D divided by the area of D. If D is a rectangle in the

xy-plane, bounded by the interval $[c, d]$ on the x-axis and by the interval $[a, b]$ on the y-axis, then the area of domain is simply $(b - a)(d - c)$ and the average value is either

$$\frac{1}{(b - a)(d - c)} \int_a^b \int_c^d f(x, y) \, dx \, dy \qquad \text{or} \qquad \frac{1}{(b - a)(d - c)} \int_c^d \int_a^b f(x, y) \, dy \, dx$$

EXAMPLE 8

A planned community with a sports complex at the center is created from a square tract of land measuring 6,000 feet on a side. Land values are modeled by the formula

$$p(x, y) = 5 - (0.0000002)(1.1x^2 + 1.2y^2)$$

where $p(x, y)$ is the price of land (in dollars/sq. ft) at the point (x, y), measured from the center of the community. The closer land is to the sports complex, the more expensive a plot is. What is the average price of land in this community?

FIGURE 13.38

A Domain Bounded above and below by Two Continuous Functions

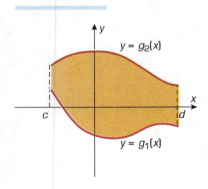

Solution

The area of the planned community is $(6,000)(6,000) = 36,000,000$ sq. ft. Placing the center of the community at coordinates $x = 0$ and $y = 0$ makes each side of the square range from $-3,000$ ft to $3,000$ ft. Therefore the average price AP of land is

$$
\begin{aligned}
AP &= \frac{1}{36,000,000} \int_{-3000}^{3000} \int_{-3000}^{3000} (5 - 0.00000022x^2 - 0.00000024y^2) \, dy \, dx \\
&= \frac{1}{36,000,000} \int_{-3000}^{3000} \left[5y - 0.00000022x^2 y - 0.00000024 \frac{y^3}{3} \right]_{y=-3000}^{y=3000} dx \\
&= \frac{1}{36,000,000} \int_{-3000}^{3000} (30,000 - 0.00132x^2 - 4320) \, dx \\
&= \frac{1}{36,000,000} \int_{-3000}^{3000} (25,680 - 0.00132x^2) \, dx \\
&= \frac{1}{36,000,000} \left[25,680x - 0.00132 \frac{x^3}{3} \right]_{x=-3000}^{x=3000} \\
&= \frac{1}{36,000,000} [154,080,000 - 23,760,000] \\
&= \$3.62/\text{sq. ft}
\end{aligned}
$$

FIGURE 13.39

Another Example of a Domain Bounded by Two Continuous Functions

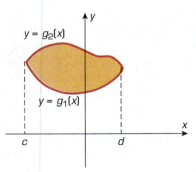

Most planned communities are not built on square tracts of land, as in Example 8, nor do most iterated integrals have rectangular domains. Some domains may be a region in the xy-plane bounded by two continuous functions $y = g_1(x)$ and $y = g_2(x)$, with

$$g_2(x) \geq g_1(x) \tag{24}$$

and x restricted to an interval $[c, d]$. Figures 13.38 and 13.39 illustrate two domains of this type, although there are many others. Note that we are showing only a domain in each figure. The corresponding surface associated with the function $z = f(x, y)$ appears above, below, or through the domain.

To integrate a function $z = f(x, y)$ over a domain that satisfies $c \le x \le d$ and the inequality (24), we use the iterated integral

$$\int_c^d \left[\int_{g_1(x)}^{g_2(x)} f(x, y) \, dy \right] dx \tag{25}$$

The limits for the integral (25) are determined by momentarily fixing x at an arbitrary point between c and d and concentrating on the limits for the integral inside the brackets. This inside integral involves integration with respect to y. Regardless of the value in $[c, d]$ chosen for x, the variable y will be somewhere between $g_1(x)$ and $g_2(x)$, which is why the limits on the inside integral in (25) are $g_1(x)$ and $g_2(x)$. This is the case for every choice of x between c and d, which is why the limits on the outside integral in (25) are c and d. This covers all of the points in the domain and, therefore, all values of y that lie between $g_1(x)$ and $g_2(x)$.

EXAMPLE 9

Determine the volume between the surface $z = x(1 + xy)$ and the region in the xy-plane bounded by the graphs of $y = x$ and $y = x^2$ with x restricted to the interval $[0, 1]$.

Solution

The domain is the shaded region in Figure 13.40, with a shape similar to that in Figure 13.39. Because the function $z = x(1 + xy) = x + x^2y$ is nonnegative over this domain, we obtain the volume V by evaluating an iterated integral of the function over its domain. Using the integral in (25), with $g_1(x) = x^2$ and $g_2(x) = x$ (note that $x^2 \le x$ on $[0, 1]$), we have

$$V = \int_0^1 \left[\int_{x^2}^x (x + x^2y) \, dy \right] dx$$

$$= \int_0^1 \left[xy + \frac{1}{2} x^2 y^2 \right]_{y=x^2}^{y=x} dx$$

$$= \int_0^1 \left[\left(x \cdot x + \frac{1}{2} (x^2)(x)^2 \right) - \left(x \cdot x^2 + \frac{1}{2} (x^2)(x^2)^2 \right) \right] dx$$

$$= \int_0^1 \left[x^2 - x^3 + \frac{1}{2} x^4 - \frac{1}{2} x^6 \right] dx$$

$$= \left[\frac{1}{3} x^3 - \frac{1}{4} x^4 + \frac{1}{10} x^5 - \frac{1}{14} x^7 \right]_{x=0}^{x=1}$$

$$= \tfrac{1}{3} - \tfrac{1}{4} + \tfrac{1}{10} - \tfrac{1}{14} \approx 0.112 \text{ cubic units}$$

FIGURE 13.40

The Domain for the Function
$z = x(1 + xy)$

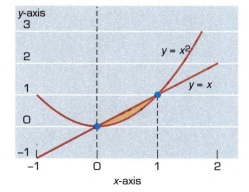

EXAMPLE 10

Integrate the function $z = xe^{-y}$ over the region bounded by the graphs of $y = 3 - x^2$, $y = -1$, $x = -1$, and $x = 2$.

Solution

The domain is the shaded region in Figure 13.41. Using the integral in (25), we have

$$\int_{-1}^{2} \left[\int_{-1}^{3-x^2} xe^{-y} \, dy \right] dx = \int_{-1}^{2} [-xe^{-y}]_{y=-1}^{y=3-x^2} \, dx$$

$$= \int_{-1}^{2} (xe - xe^{x^2-3}) \, dx$$

$$= e \int_{-1}^{2} x \, dx - \int_{-1}^{2} xe^{x^2-3} \, dx$$

$$= e \left[\frac{1}{2} x^2 \right]_{x=-1}^{x=2} - \left[\frac{1}{2} e^{x^2-3} \right]_{x=-1}^{x=2}$$

$$= e[\tfrac{1}{2}(2)^2 - \tfrac{1}{2}(-1)^2] - [\tfrac{1}{2}e^{(2)^2-3} - \tfrac{1}{2}e^{(-1)^2-3}]$$

$$= e + \tfrac{1}{2}e^{-2} \approx 2.7859$$

FIGURE 13.41

The Domain for the Function
$z = xe^{-y}$

IMPROVING SKILLS

In Problems 1 through 30, evaluate each integral.

1. $\int_{1}^{2} xy^2 \, dx$

2. $\int_{1}^{2} xy^2 \, dy$

3. $\int_{-1}^{1} xy^2 \, dx$

4. $\int_{-1}^{1} xy^2 \, dy$

5. $\int_{1}^{3} (x/y^2) \, dx$

6. $\int_{1}^{3} (x/y^2) \, dy$

7. $\int_{0}^{4} (3x^2 + y) \, dx$

8. $\int_{0}^{4} (3x^2 + y) \, dy$

9. $\int_{1}^{4} \left(\frac{y}{x} + 1 \right) dx$

10. $\int_{1}^{4} \left(\frac{y}{x} + 1 \right) dy$

11. $\int_{\pi}^{2\pi} x \sin(x^2 y) \, dx$

12. $\int_{\pi}^{2\pi} x \sin(x^2 y) \, dy$

13. $\int_{0}^{x^2} xy^3 \, dy$

14. $\int_{0}^{x^2} \left(1 - \frac{y}{x}\right) dy$

15. $\int_{1}^{e^x} (x/y) \, dy$

16. $\int_{x}^{2x} e^{xy} \, dy$

17. $\int_{x}^{x^2} (y^3/x^4) \, dy$

18. $\int_{x}^{\pi x} x \cos(y/x) \, dy$

19. $\int_{1}^{2} \int_{3}^{4} 2xe^y \, dy \, dx$

20. $\int_{1}^{2} \int_{3}^{4} 2xe^y \, dx \, dy$

21. $\int_{-1}^{1} \int_{0}^{2} (x^3 + 3y) \, dy \, dx$

22. $\int_{-1}^{1} \int_{0}^{2} (x^3 + 3y) \, dx \, dy$

23. $\int_{-1}^{1} \int_{0}^{2} 3x^3 y \, dx \, dy$

24. $\int_{0}^{3} \int_{0}^{1+x} 2x^4 y \, dy \, dx$

25. $\int_{0}^{3} \int_{x}^{1+x} x(x - y) \, dy \, dx$

26. $\int_{0}^{1} \int_{x^3}^{x} (xy^2)^2 \, dy \, dx$

27. $\int_{1}^{2} \int_{x}^{x^3} (xy^2)^2 \, dy \, dx$

28. $\int_{1}^{2} \int_{\sqrt{x}}^{x} (x + y) \, dy \, dx$

29. $\int_{0}^{1} \int_{0}^{\sqrt{1+x}} 4y \, dy \, dx$

30. $\int_{1}^{2} \int_{x^2}^{6-x} (x - 2y) \, dy \, dx$

In each of Problems 31 through 38, write (but do not evaluate) the iterated integral of the function $z = f(x, y)$ over the shaded domain.

31.

32.

33.

34.

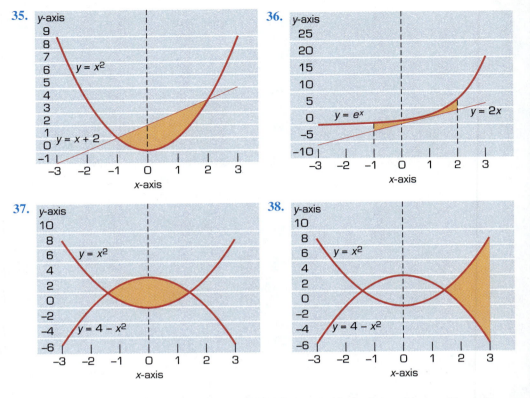

In Problems 39 through 44, sketch the domain associated with the iterated integral in each cited problem.

39. Problem 19

40. Problem 20

41. Problem 24

42. Problem 25

43. Problem 26

44. Problem 27

45. Find the volume of the region between the graph of the function $z = y(2 + \cos x)$ and the xy-plane, over the rectangle defined by the intervals $[0, 4]$ on the x-axis and $[1, 2]$ on the y-axis.

46. Find the volume of the region between the graph of the function $z = x^2 - 2(xy)^3$ and the fourth quadrant of the xy-plane, over the square with sides of length 2 and one corner at the origin.

47. Find the volume of the region between the graph of the function $z = 12/xy^2$ and the region in the xy-plane bounded by the graphs of $y = x^3$ and $y = x^2$, with x restricted to the interval $[1, 3]$.

CREATING MODELS

48. Solve Example 8 with $p(x, y) = 5 + (0.0000002)(1.1x^2 + 1.2y^2)$, so that the farther away a piece of property is from the noise and congestion of the sports complex, the more expensive it is.

49. Solve Problem 48 if the sports complex has coordinates $x = 0$ and $y = 0$ and is placed at the lower left corner of the square tract of land, which is $6000' \times 6000'$.

50. The developer in Example 8 has a second tract of land on which she wants to build a similar community. This tract is bounded between the graphs of $y = x$ and $y = x^3$, where x (in miles) is restricted to the interval $[0, 1]$. The sports complex will be placed at the lower left part of the property, with coordinates $x = 0$ and $y = 0$, and land will be priced by the formula $p(x, y) = 140 + 160xy$ dollars/sq. mile. What is the average price of land in this community? What is the average price in dollars/sq. ft?

51. A company manufactures two products. Its weekly profit P (in dollars) is modeled by the formula $P = 20x + 50y - 0.01xy - 60{,}000$, where x and y denote how many items of the first and second products, respectively, sold each week. The company sales each week are uniformly distributed between 4000 and 6000 items for the first product and between 1000 and 3000 items for the second product. What is the average weekly profit for this company?

52. Solve Problem 51 if

$$P = 90{,}000 - (x - 4800)^{1.5} - (y - 2500)^{1.75}$$

53. A Cobb–Douglas model relates the output P of a production process to two quantities: L, number of units of labor, and C, the number of units of capital investment available to the production facility. This model is described by the equation $P = 1.2L^{0.75}C^{0.25}$. Find the average output of this process if labor is uniformly distributed between 4 and 4.4 (thousand-hour) units and capital investment between 1 and 3 (million-dollar) units.

54. A city's population density d (in people/sq. mile) is modeled by the formula $d = 15{,}000e^{-0.1(x+y)}$, where the distance of the point (x, y) from the origin (the center of town) is measured in miles. What is the average population density inside the rectangular region defined by $-5 \le x \le 6$ and $-4 \le y \le 6$?

55. A town is in the shape of a rectangle that is 2 miles by 1 mile. At its center is an industrial plant that emits pollutants into the air. The concentration C of these pollutants is modeled by the formula

$$C = 120 - 4(x^2 + y^2)$$

where the distance of a point (x, y) from the town's center is measured in miles. What is the average concentration of pollutants in the air throughout the town?

56. Solve Problem 55 if the plant is located at one corner of the town on the shore of a lake that stretches along the two borders of the town that extend from the plant.

57. Solve Problem 55 if the plant has coordinates $x = 0$ and $y = 0$ and the town is the region between the graphs of $y = x^2 - 5$ and $y = x^3$, with x restricted to the interval $[-1, 1]$.

EXPLORING IN TEAMS

58. Solve Example 9 if x is restricted to the interval $[-1, 0]$.

59. Solve Example 9 if x is restricted to the interval $[1, 2]$.

60. Solve Example 9 if x is restricted to the interval $[0, 2]$.

61. Find the volume of the region between the graph of $z = x^2y$ and the triangle in the xy-plane defined by the three lines $y = x$, $y = 2 - x$, and $y = 0$.

62. Show that the iterated integral

$$\int_0^1 \int_0^2 (xy - 4) \, dy \, dx$$

is a negative. Give a geometrical explanation for this. How would you find the volume between the graph of $z = xy - 4$ and the rectangle in the xy-plane with $0 \le x \le 1$ and $0 \le y \le 2$?

63. Find the volume between the graph of $z = x^2y$ and the region in the xy-plane bounded between the graphs of $y = x$ and $y = x^3$, with x restricted to the interval $[-1, 0]$.

REVIEWING MATERIAL

64. (Section 3.3) Use graphical methods to solve the following linear programming model:

$$\text{Minimize:} \quad z = 3x - 2y$$
$$\text{subject to:} \quad y \le x$$
$$y \ge 2 - x$$
$$x \le 3$$

65. (Section 7.2) Find a stable distribution vector for the Markov chain with transition matrix

$$P = \begin{bmatrix} 0.5 & 0.5 & 0 \\ 0 & 0.3 & 0.7 \\ 0.1 & 0.4 & 0.5 \end{bmatrix}$$

66. (Section 12.7) A continuous random process on the interval $[1, 5]$ is modeled by the probability function $f(x) = 1 + 2x^2$. What is the probability that a random outcome from such a process has a value of at least 3?

RECOMMENDING ACTION

67. Respond by memo to the following:

MEMORANDUM

To: J. Doe Reader
From: Terry
Date: Today
Subject: **Area**

Can we use iterated integrals to calculate the area of a region R in the xy-plane? The graph of the function $z = 1$ is a plane parallel to the xy-plane and one unit above it. It seems to me that the volume of the space between this plane and R is numerically equal to the area of R, but with different units. Therefore can we also get the area of R by using iterated integration?

CHAPTER 13 KEYS

KEY WORDS

contour map (p. 946)
critical point (p. 968)
domain (p. 942)
function of two variables (p. 941)
iterated integral (p. 980)

level curve (p. 945)
local maximum (p. 969)
local minimum (p. 969)
partial derivative (p. 954)
saddle point (p. 969)

KEY CONCEPTS

13.1 Functions of Many Variables

■ The domain of a function of two variables is a subset of a two-dimensional plane.

■ A three-dimensional coordinate system is needed to graph a function of two variables.

■ Often it is easier to graph a function of two variables by first drawing a contour map for the function.

13.2 Partial Derivatives

■ The partial derivative of a function f of many variables, denoted $\partial f/\partial x$, is the derivative of f with respect to the input variable x, under the assumption that all other input variables have fixed values. The partial derivative of a function is itself a function.

■ A partial derivative measures the rate of change of an output variable with respect to a unit change in one of the input variables, with all other variables held fixed.

■ A second partial derivative is the partial derivative of a partial derivative, and it is again a function.

13.3 Optimization

■ A critical point of a function of many variables can be the location of either a local maximum, a local minimum, or a saddle point.

■ If the domain of a function of two variables is a bounded region that includes all of the points on its boundary, then the function attains its global maximum and/or minimum either at a critical point or at a point on the boundary of the domain.

13.4 Iterated Integrals

■ When evaluating an iterated integral of a function of x and y, the differential dx indicates that integration is with respect to x with y held constant. Similarly, the differential dy indicates that integration is with respect to y, holding x constant.

■ A constant of integration is a function that depends on all of the input variables held fixed during integration.

KEY FORMULAS

- $\dfrac{\partial}{\partial y}\left(\dfrac{\partial f}{\partial y}\right) = \dfrac{\partial^2 f}{\partial y^2} = f_{yy}; \qquad \dfrac{\partial}{\partial y}\left(\dfrac{\partial f}{\partial x}\right) = \dfrac{\partial^2 f}{\partial y \partial x} = f_{xy}$

- $\displaystyle\int_a^b \int_c^d f(x,\, y)\, dx\, dy = \int_a^b \left[\int_c^d f(x,\, y)\, dx\right] dy;$

  $\displaystyle\int_c^d \int_a^b f(x,\, y)\, dy\, dx = \int_c^d \left[\int_a^b f(x,\, y)\, dy\right] dx$

KEY PROCEDURES

- Let (a, b) be a critical point for the function $z = f(x, y)$, where f is defined in a circle centered at (a, b) with partial derivatives of all orders existing in this domain. Set

$$A = \left.\dfrac{\partial^2 z}{\partial x^2}\right|_{(a,b)} \qquad B = \left.\dfrac{\partial^2 z}{\partial x \partial y}\right|_{(a,b)} \qquad C = \left.\dfrac{\partial^2 z}{\partial y^2}\right|_{(a,b)}$$

  and

$$D = AC - B^2$$

  (*i*) If $D < 0$, then (a, b) is the location of a saddle point.
  (*ii*) If $D > 0$ and $A < 0$, then (a, b) is the location of a local maximum.
  (*iii*) If $D > 0$ and $A > 0$, then (a, b) is the location of a local minimum.
  (*iv*) If $D = 0$, then this test is inconclusive and other analysis is needed.

- To calculate the average value of the function $z = f(x, y)$ over a domain D, evaluate either iterated integral of the function over D and divide the result by the area of D.

- To integrate the function $z = f(x, y)$ over a domain in which $c \le x \le d$ and $g_1(x) \le y \le g_2(x)$ for two continuous functions $y = g_1(x)$ and $y = g_2(x)$, calculate the iterated integral

$$\int_c^d \left[\int_{g_1(x)}^{g_2(x)} f(x,\, y)\, dy\right] dx$$

TESTING YOURSELF

Use the following problems to test yourself on the material in Chapter 13. The odd-numbered problems comprise one test, the even-numbered problems a second test.

1. Plot the two points $(1, 2, 5)$ and $(2, -1, -3)$ on the same three-dimensional coordinate system.
2. Find three positive numbers whose sum is 48 and whose product is a maximum.
3. Construct a contour map for the function $z = 2x^2 + 4y^2$.
4. Evaluate the iterated integral $\displaystyle\int_0^1 \int_0^{x^2} (16xy + 10x^2)\, dy\, dx$
5. Find $\partial z/\partial x$ for each of the following functions:
   (**a**) $z = 3x^4 + 2x^2 y^3 - 5y^5 + 1$ \qquad (**b**) $z = xy/(3 + x^2 y)$
   (**c**) $z = \ln(x^3 + y^3 + 3)$

6. Find z_{xy} when $z = y \cos(x^2 y)$.

7. Find $\left.\dfrac{\partial z}{\partial y}\right|_{(2,\pi)}$ when $z = x^3 \sin(xy)$.

8. A company manufactures two products. Its weekly profit P (in dollars) is modeled by the formula $P = 80{,}000 - 500\sqrt{x} + 2y$, where x and y denote how many of the first and second products, respectively, are sold weekly. Each week, company sales are uniformly distributed between 500 and 800 items of the first product and between 500 and 1000 for the second product. What is the average weekly profit for this company?

9. Find $\partial^2 z/\partial x \partial y$ when $z = 2x + 3xy + y^2$.

10. Plot the two points $(50, 100, -50)$ and $(150, -100, 100)$ on the same three-dimensional coordinate system.

11. A sociologist believes that the amount of crime, C, is a function of P, the number of police, and U, the number of unemployed people. Assuming that the sociologist is correct, what can you say about the signs of $\partial C/\partial P$ and $\partial C/\partial U$?

12. A corporation manufactures two products, Model A and Model B. The manufacturing of Model A costs \$1000/unit and that of Model B costs \$700/unit. The company has a daily fixed cost of \$7000. The demands x for Model A and y for Model B depend on the prices p_A and p_B of each model, according to the equations

$$p_A = 40x - 20y + 2200$$

$$p_B = -60x + 30y + 1300$$

(**a**) Find an equation for the daily profit as a function of x and y.
(**b**) Find a production schedule that maximizes daily profit.

13. Find three numbers whose sum is 1 and the sum of whose squares is as small as possible.

14. Post office regulations require that the height plus girth of a parcel cannot exceed 108 in. Find the dimensions of a rectangular box that meets postal regulations and has maximum volume.

15. Show that for a given volume V, a closed rectangular box with minimum surface area is a cube.

16. Find $\partial z/\partial y$ for each of the following functions:
(**a**) $z = x^4 y + 2x^2 y^2 + xy^5$ (**b**) $z = (x + 3xy^2)^{3.5}$
(**c**) $z = ye^{2xy+1}$

17. An electronics company manufactures two types of computerized editing systems for television stations. The company's profit P (in dollars) is modeled by the equation $P = 975x + 1575y + 15xy - 20x^2 - 10y^2$, where x and y are the numbers of each type of system produced per day. Determine a production schedule that maximizes daily profit.

18. Find the volume between the graph of the function $z = 2x^3 y^5$ and the region in the xy-plane bounded by the graphs of the equations $y = x^4$ and $y = x^2$, $0 \le x \le 1$.

19. Evaluate the iterated integral $\displaystyle\int_{1}^{2}\int_{-1}^{1} (2x^2 y + 3xy^2)\, dx\, dy$.

20. Find $\left.\dfrac{\partial z}{\partial a}\right|_{(0,2,3)}$ when $z = 3ab^3 - 3b^2 c^3 + 4a^2 c - 5abc$.

21. Find the volume between the graph of the function $z = x^4 + x^2 y^2 + 1$ and the rectangle in the xy-plane defined by $0 \le x \le 3$ and $1 \le y \le 2$.

22. Construct a contour map for the function $z = 2x - 6y$.

APPENDIX

A EXPONENTS

For any real number a and any positive integer n, we define the *exponential* a^n as

$$a^n = \underbrace{a \cdot a \cdot a \cdots a}_{n \text{ times}} \tag{1}$$

The number a is called the *base*; the letter n is the *exponent*. Thus,

$$(-4)^3 = (-4)(-4)(-4) = -64$$

is an exponential with base 4 and exponent 3, and

$$(\tfrac{1}{3})^4 = (\tfrac{1}{3})(\tfrac{1}{3})(\tfrac{1}{3})(\tfrac{1}{3}) = \tfrac{1}{81}$$

is an exponential with base $\frac{1}{3}$ and exponent 4, while

$$2^{10} = \underbrace{(2 \cdot 2 \cdot 2 \cdot 2 \cdot 2 \cdot 2 \cdot 2 \cdot 2 \cdot 2 \cdot 2)}_{10 \text{ times}} = 1{,}024$$

is an exponential with base 2 and exponent 10.

An exponential can be decomposed into a product of two other exponentials, each with the same base. For example,

$$a^6 = aaaaaa = (aaaa)(aa) = a^4 a^2$$

and also

$$a^6 = aaaaaa = (aaa)(aaa) = a^3 a^3$$

In general,

$$a^{n+m} = a^n a^m \tag{2}$$

where n and m are positive integers.

EXAMPLE 1

Simplify the following products:　**(a)** $2^5 2^{10}$; **(b)** $(-3)^4(-3)^7$; **(c)** $(\tfrac{1}{2})^3(\tfrac{1}{2})^2$.

Solution
(a) $2^5 2^{10} = 2^{5+10} = 2^{15}$
(b) $(-3)^4(-3)^7 = (-3)^{4+7} = (-3)^{11}$
(c) $(\tfrac{1}{2})^3(\tfrac{1}{2})^2 = (\tfrac{1}{2})^{3+2} = (\tfrac{1}{2})^5$

Equation (2) is valid only when each exponential has the same base. We have from Example 1 that $2^5 2^{10} = 2^{15}$. Similarly, $3^5 3^{10} = 3^{15}$. However, Equation (2) is not applicable to the product $2^5 3^{10}$ because the two bases are different.

Equation (1),

$$a^n = \underbrace{a \cdot a \cdot a \cdots a}_{n \text{ times}}$$

is valid for any real number a, even if the base is itself expressed as an exponential:

$$(5^2)^4 = (5 \cdot 5)^4 = (5 \cdot 5)(5 \cdot 5)(5 \cdot 5)(5 \cdot 5) = 5^8 = 5^{2(4)}$$

and

$$\begin{aligned}
[(-\tfrac{1}{3})^3]^3 &= [(-\tfrac{1}{3})(-\tfrac{1}{3})(-\tfrac{1}{3})]^3 \\
&= [(-\tfrac{1}{3})(-\tfrac{1}{3})(-\tfrac{1}{3})][(-\tfrac{1}{3})(-\tfrac{1}{3})(-\tfrac{1}{3})][(-\tfrac{1}{3})(-\tfrac{1}{3})(-\tfrac{1}{3})] \\
&= (-\tfrac{1}{3})^9 = (-\tfrac{1}{3})^{3(3)}
\end{aligned}$$

In general, for any real number a and any positive integers n and m, we have

$$(a^n)^m = a^{nm} \tag{3}$$

EXAMPLE 2

Simplify the following expressions: (**a**) $(2^5)^2$; (**b**) $[(-3)^3]^4$; (**c**) $[(0.6)^6]^7$.

Solution
(**a**) $(2^5)^2 = 2^{5 \cdot 2} = 2^{10}$
(**b**) $[(-3)^3]^4 = (-3)^{3 \cdot 4} = (-3)^{12} = 3^{12}$
(**c**) $[(0.6)^6]^7 = (0.6)^{6 \cdot 7} = (0.6)^{42}$

It follows from Equation (1) and both the associative and commutative properties of multiplication that

$$\begin{aligned}
(2 \cdot 5)^4 &= (2 \cdot 5)(2 \cdot 5)(2 \cdot 5)(2 \cdot 5) \\
&= (2 \cdot 2 \cdot 2 \cdot 2)(5 \cdot 5 \cdot 5 \cdot 5) \\
&= 2^4 5^4
\end{aligned}$$

In general, if n is a positive integer and a and b are real numbers, then

$$(ab)^n = a^n b^n \tag{4}$$

Note that all of the exponents in Equation (4) are identical, even though the bases are different. It therefore follows that

$$(3)^2(4)^2 = (3 \cdot 4)^2 = (12)^2$$

$$(\tfrac{1}{2})^5(6)^5 = (\tfrac{1}{2} \cdot 6)^5 = (3)^5$$

$$(\tfrac{5}{12})^7(4)^7 = (\tfrac{5}{12} \cdot 4)^7 = (\tfrac{5}{3})^7$$

Equation (1) can be extended to negative exponents as follows: for any nonzero real number a and any positive integer n, we define

$$a^{-n} = \frac{1}{a^n} \tag{5}$$

For example,

$$5^{-2} = 1/5^2 = 1/25 = 0.04$$

$$(-\tfrac{1}{3})^{-4} = \frac{1}{(-\tfrac{1}{3})^4} = \frac{1}{\frac{1}{81}} = 81$$

Using Equation (5), Equations (2) through (4) can now be extended to negative exponents.

EXAMPLE 3

Simplify the expressions: (a) $2^7 2^{-3}$; (b) $[(-1/3)^{-4}]^{-5}$; (c) $(1/2^7)(1/5^7)$.

Solution

(a) $2^7 2^{-3} = 2^{7+(-3)} = 2^4 = 16$

(b) $[(-1/3)^{-4}]^{-5} = (-1/3)^{(-4)(-5)} = (-1/3)^{20} = (1/3)^{20}$

(c) $(1/2^7)(1/5^7) = 2^{-7} 5^{-7} = (2 \cdot 5)^{-7} = 10^{-7}$

Alternatively,

$$\left(\frac{1}{2^7}\right)\left(\frac{1}{5^7}\right) = \frac{1 \cdot 1}{2^7 \cdot 5^7} = \frac{1}{(2 \cdot 5)^7} = \frac{1}{10^7}$$

When $n = 1$, Equation (5) reduces to $a^{-1} = 1/a$. Consequently

$$\left(\frac{1}{a}\right)^n = (a^{-1})^n = a^{(-1 \cdot n)} = a^{-n} = \frac{1}{a^n}$$

We have shown that

$$\left(\frac{1}{a}\right)^n = \frac{1}{a^n} \tag{6}$$

For example,

$$\left(\frac{1}{7}\right)^3 = \frac{1}{7^3} \qquad \left(\frac{1}{5}\right)^{10} = \frac{1}{5^{10}} \qquad \left(\frac{1}{20}\right)^{15} = \frac{1}{(20)^{15}}$$

We can combine Equations (4) and (6) to deduce that

$$\left(\frac{b}{a}\right)^n = \left(b \cdot \frac{1}{a}\right)^n = b^n \left(\frac{1}{a}\right)^n = b^n \left(\frac{1}{a^n}\right) = \frac{b^n}{a^n}$$

We have just proved another general property:

$$\left(\frac{b}{a}\right)^n = \frac{b^n}{a^n} \tag{7}$$

Note that all of the exponents are the same. For example,

$$\left(\frac{3}{4}\right)^8 = \frac{3^8}{4^8} \qquad \text{and} \qquad \left(\frac{12}{5}\right)^3 = \frac{(12)^3}{5^3}$$

We can use our properties of exponents to give meaning to the concept of a *zero* exponent. Note that

$$1 = \frac{8}{8} = \frac{2^3}{2^3} = 2^3 2^{-3} = 2^{3+(-3)} = 2^0$$

and

$$1 = \frac{9}{9} = \frac{3^2}{3^2} = 3^2 3^{-2} = 3^{2+(-2)} = 3^0$$

so it is reasonable to define

$$a^0 = 1 \tag{8}$$

for every nonzero real number a. Accordingly,

$$5^0 = 1 \qquad (-4)^0 = 1 \qquad (-\tfrac{1}{3})^0 = 1$$

Using Equations (2) and (5) in tandem, then for any real number a and any (positive, negative, or zero) integers n or m, we have

$$\frac{a^n}{a^m} = a^n a^{-m} = a^{n+(-m)} = a^{n-m} \tag{9}$$

Note that in this case, each base is the same but the exponents can be different. For example,

$$\frac{5^4}{5^2} = 5^{4-2} = 5^2 = 25$$

$$\frac{(-2)^5}{(-2)^7} = (-2)^{5-7} = (-2)^{-2} = \frac{1}{(-2)^2} = \frac{1}{4}$$

EXAMPLE 4

Write the expression $\dfrac{8(512)}{(64)(1{,}024)}$ as an exponent to the base 2.

Solution

We begin by writing each factor as a power of 2:

$$\frac{8(512)}{(64)(1{,}024)} = \frac{2^3 2^9}{2^6 2^{10}} = \frac{2^{12}}{2^{16}} = 2^{12-16} = 2^{-4}$$

We also note that

$$2^{-4} = 1/2^4 = 1/16$$

EXAMPLE 5

Write the expression

$$\frac{81\left(\dfrac{1}{6561}\right)\left(\dfrac{1}{729}\right)}{\left(\dfrac{1}{243}\right)\left(\dfrac{1}{27}\right)\left(\dfrac{1}{9}\right)}$$

as an exponent to the base 3.

Solution

Here we first write each factor as a power of 3:

$$\frac{81\left(\dfrac{1}{6561}\right)\left(\dfrac{1}{729}\right)}{\left(\dfrac{1}{243}\right)\left(\dfrac{1}{27}\right)\left(\dfrac{1}{9}\right)} = \frac{3^4\left(\dfrac{1}{3^8}\right)\left(\dfrac{1}{3^6}\right)}{\left(\dfrac{1}{3^5}\right)\left(\dfrac{1}{3^3}\right)\left(\dfrac{1}{3^2}\right)} = \frac{3^4 3^{-8} 3^{-6}}{3^{-5} 3^{-3} 3^{-2}}$$

$$= \frac{3^4 3^{-14}}{3^{-10}} = \frac{3^{-10}}{3^{-10}} = 3^{-10-(-10)} = 3^0$$

which is 1.

In Equations (1) through (9), the exponents n and m are presumed to be integers. However, these properties generalize to rational exponents of the form n/m, where both n and m are integers. For example, the property $(a^n)^m = a^{nm}$ holds for all rational numbers n and m (including integers). To see this, let us begin with the special case $n = 1/m$, where m is a positive integer and a is a positive real number. For instance, when $m = 2$ and $a = 4$, we want

$$(4^{1/2})^2 = 4^{(1/2)(2)} = 4^1 = 4$$

Thus $4^{1/2}$ is a number whose square equals 4. When $m = 3$ and $a = 8$, we want

$$(8^{1/3})^3 = 8^{(1/3)(3)} = 8^1 = 8$$

so $8^{1/3}$ is a number whose third power is 8. When $m = 10$ and $a = 120$, we want

$$(120^{1/10})^{10} = 120^{(1/10)(10)} = 120^1 = 120$$

hence $120^{1/10}$ is a number whose tenth power is 120. Given these examples, it is reasonable to define $a^{1/m}$ as a number whose mth power is a when a is a real positive number and m is a positive integer. The exponential $a^{1/m}$ is often called the "mth root of a," sometimes written $\sqrt[m]{a}$. The second or square root of a is $a^{1/2} = \sqrt[2]{a}$, or just $\sqrt{a}$ for short.

EXAMPLE 6

Evaluate: (*a*) $\sqrt[5]{32}$; (*b*) $\sqrt[4]{625}$.

Solution

(*a*) The number $\sqrt[5]{32} = (32)^{1/5}$ has a fifth power of 32. Since $32 = 2 \cdot 2 \cdot 2 \cdot 2 \cdot 2 = 2^5$, it follows that $\sqrt[5]{32} = \sqrt[5]{2^5} = (2^5)^{1/5} = 2$.

(*b*) Here $\sqrt[4]{625} = (625)^{1/4}$ is a number whose fourth power is 625. Since $625 = 5 \cdot 5 \cdot 5 \cdot 5 = 5^4$, it follows that $\sqrt[4]{625} = \sqrt[4]{5^4} = (5^4)^{1/4} = 5$.

Most mth roots, such as $(120)^{1/10}$, are infinite decimals and not as easy to evaluate as those in Example 6. Generally we need a calculator to evaluate such roots. If a calculator has a y^x or an x^y key, then on many calculators, the sequence of keystrokes to evaluate $(120)^{1/10}$ is

The equivalent sequence of keystrokes on many graphing calculators is

Either way, the result is 1.614054 rounded to six decimal places. If we raise 1.614054 to the tenth power, we find that $(1.614054)^{10} \approx 119.9998$. This number is close to but not precisely 120 because we rounded our evaluation of $(120)^{1/10}$ to six decimals.

Sometimes a positive number a can have more than one root. For example, both 2 and -2 are square roots of 4 because both numbers raised to the second power equal 4. Each positive real number, however, has only one *positive* mth root; unless asked otherwise, that is the root we display when evaluating $a^{1/m}$. In particular, $y = \sqrt{a}$ denotes the positive square root of a. To denote the negative square root we write $-\sqrt{a}$.

We require the base a to be positive when defining $a^{1/m}$ because when m is an even number, the mth root of a negative number does not exist as a real number, and we want to work only with real numbers in this book. There is no real number, for example, whose square is -2. We can give $\sqrt{-2}$ meaning if we admit complex numbers, but we shall not do so here. All of the applications and models in this book are limited to real numbers.

Finally, if a is a positive real number and n and m are both positive integers, we define $a^{n/m}$ to be $(a^{1/m})^n$. First we calculate the mth root of a and then we raise the result to the nth power.

EXAMPLE 7

Evaluate: (**a**) $8^{2/3}$; (**b**) $(625)^{5/4}$.

Solution
(**a**) $8^{2/3} = (8^{1/3})^2 = 2^2 = 4$
(**b**) $(625)^{5/4} = [(625)^{1/4}]^5 = 5^5 = 3125$

Most exponentials with rational exponents are not as easy to calculate as those in Example 7 and we usually need a calculator to evaluate them. For example, consider $5^{2/3}$. If a calculator has a y^x or an x^y key, then on many calculators, the sequence of keystrokes to evaluate $5^{2/3}$ is

The equivalent sequence of keystrokes on many graphing calculators is

The result is 2.9240 rounded to four decimal places.

IMPROVING SKILLS

In Problems 1 through 54, write each expression to a single base raised to a single exponent.

1. $2^2 2^3$	**2.** $2^8 2^5$	**3.** $5^5 5^4$
4. $5^9 5^9$	**5.** $(-4)^3(-4)^{15}$	**6.** $(-7)^{16}(-7)^9$
7. $(0.3)^6(0.3)^3$	**8.** $(-\frac{1}{4})^{23}(-\frac{1}{4})^{34}$	**9.** $(2^3)^7$
10. $(2^4)^4$	**11.** $(7^5)^5$	**12.** $(8^3)^6$
13. $[(-5)^3]^2$	**14.** $[(-\frac{1}{5})^2]^4$	**15.** $[(-4)^3]^{10}$
16. $[(-\frac{1}{4})^3]^2$	**17.** $[(2.3)^3]^8$	**18.** $[(7.9)^5]^1$
19. $(7^{-2})^2$	**20.** $(3^{-3})^3$	**21.** $(2^4)^{-2}$
22. $(2^{-3})^{-2}$	**23.** $[(\frac{1}{2})^{-3}]^2$	**24.** $[(\frac{1}{2})^3]^{-5}$
25. $[(\frac{1}{3})^{-5}]^{-2}$	**26.** $[(\frac{1}{4})^{-2}]^2$	**27.** $[(\frac{1}{5})^{-3}]^{-5}$
28. $[(-\frac{1}{2})^{-4}]^2$	**29.** $(1/3^2)(1/5^2)$	**30.** $(1/2^3)(1/4^3)$
31. $7^6/7^9$	**32.** $3^2/3^5$	**33.** $3^8/3^3$
34. $2^5/2^7$	**35.** $2^9/2^4$	**36.** $2^9/4^2$
37. $(-5)^2/(-5)^3$	**38.** $(-2)^3/(-2)^4$	**39.** $(-5)^2/(-5)^{-3}$
40. $(-2)^{-3}/(-2)^4$	**41.** $5^{-2}/5^4$	**42.** $2^{-5}/2^{-4}$
43. $2^3(1/2^5)(1/2^4)$	**44.** $2^{-3}(1/2^6)(1/2^{-3})$	**45.** $(1/3^2)(1/3^{-6})(1/3^2)$
46. $(1/5^{-5})(1/5^{-6})(1/5^7)$	**47.** $(3^2/3^5)(3^3/3^4)$	**48.** $(3^{-2}/3^3)(3^4/3^{-4})$

49. $\dfrac{5^2(5^2/5^{-3})}{5^6/5^3}$

50. $\dfrac{2^3(2^{-2}/2^{-3})}{2^{-6}/2^3}$

51. $5^{-2}(5^2/5^{-4})(5^{-5}/5^{-3})$

52. $2^{-3}(2^{-4}/2^5)(2^{-7}/2^2)$

53. $\dfrac{3^3(3^{-2}/3^{-3})}{3^5/3^3}$

54. $\dfrac{3^{-8}(3^{-2}/3^4)}{3^{-9}/3^2}$

In Problems 55 through 64, evaluate each expression *without* using a calculator.

55. $9^{3/2}$

56. $(16)^{3/2}$

57. $9^{5/2}$

58. $(27)^{2/3}$

59. $(81)^{3/4}$

60. $(27)^{-2/3}$

61. $8^{-5/3}$

62. $(243)^{-3/5}$

63. $\left(\frac{1}{256}\right)^{5/4}$

64. $\left(\frac{1}{625}\right)^{3/4}$

In Problems 65 through 74, use a calculator to evaluate each expression; round your answer to four decimal places.

65. $7^{3/2}$

66. $(21)^{3/2}$

67. $(2.1)^{5/2}$

68. $(110)^{2/3}$

69. $(80)^{3/4}$

70. $(7)^{-2/3}$

71. $9^{-5/3}$

72. $(263)^{5/3}$

73. $(119)^{-5/3}$

74. $(1263)^{3/5}$

B POLYNOMIALS

A monomial in x is an expression of the form ax^n, where a is a real number and n is a nonnegative integer.

A *monomial* in x is an expression of the form ax^n, where a is a real number, called the *coefficient* of the monomial, and the exponent n is a nonnegative integer. Examples are

$$7x^5 \quad \text{with } a = 7 \text{ and } n = 5 \tag{10}$$

$$-3.1x^3 \quad \text{with } a = -3.1 \text{ and } n = 3 \tag{11}$$

$$\tfrac{1}{2}x^{84} \quad \text{with } a = \tfrac{1}{2} \text{ and } n = 84 \tag{12}$$

The letter x in the monomial is called a *variable,* and it may be replaced by any other label. The expression $7t^2$ is a monomial in the variable t with coefficient 7 and exponent 2, while $-3y^{121}$ is a monomial in the variable y with coefficient -3 and exponent 121.

The exponent of a monomial with a nonzero coefficient is the *degree* of that monomial. For example, the monomial in (10) is of degree 5, or it is a fifth-degree monomial, and the monomial in (11) is of degree 3, so it is a third-degree monomial, while the expression in (12) is an 84th-degree monomial.

Monomials in the same variable and with the same degree are called *like terms,* and they may be added or subtracted by simply adding or subtracting their coefficients. The expressions $8x^5$ and $3x^5$ are like terms because both are fifth-degree monomials in the variable x. Consequently, using the distributive law,

$$8x^5 + 3x^5 = (8 + 3)x^5 = 11x^5$$

$$8x^5 - 3x^5 = (8 - 3)x^5 = 5x^5$$

Similarly, the monomials $2y^{17}$ and $-6y^{17}$ are like terms because both are expressions in the variable y with exponent 17. Therefore

$$2y^{17} + (-6y^{17}) = [2 + (-6)]y^{17} = -4y^{17}$$

$$2y^{17} - (-6y^{17}) = [2 - (-6)]y^{17} = 8y^{17}$$

The product of a monomial with a real number k is obtained by multiplying the coefficient of the monomial by k. For example,

$$4(3x^5) = (4 \cdot 3)x^5 = 12x^5$$

$$\tfrac{1}{4}(2y^{17}) = (\tfrac{1}{4} \cdot 2)y^{17} = \tfrac{1}{2}y^{17}$$

$$-20(-5t^{101}) = [(-20)(-5)]t^{101} = 100t^{101}$$

Two monomials in the same variable are multiplied together by multiplying their coefficients and adding their exponents.

Two monomials in the same variable are multiplied together by multiplying their coefficients and adding their exponents. For example,

$$(3x^6)(5x^7) = (3 \cdot 5)(x^6 \cdot x^7) = 15x^{(6+7)} = 15x^{13}$$

$$(-8t^3)(9t^6) = (-8 \cdot 9)(t^3 \cdot t^6) = -72t^{(3+6)} = -72t^9$$

A *polynomial* in x is the sum of two or more monomials in that variable. The polynomial $3x^2 + 5x$ in the variable x is the sum of the two monomials $3x^2$ and $5x$. The polynomial $5y^4 - 3y^2 - 1$ in the variable y is the sum of three monomials $5y^4$, $-3y^2$, and $-y^0 = -1$. We may also consider this same polynomial $5y^4 - 3y^2 - 1$ to be the sum of the five monomials, $5y^4$, $0y^3$, $-3y^2$, $0y$, and $-y^0 = -1$. That is,

A polynomial is the sum of two or more monomials in the same variable.

$$5y^4 - 3y^2 - 1 = 5y^4 + 0y^3 - 3y^2 + 0y - y^0 \tag{13}$$

We can "pad" a polynomial by adding monomials with zero coefficients when convenient.

The *degree of a polynomial* is the highest degree of all of the monomials with nonzero coefficients comprising that polynomial. The polynomial in (13) is the sum of three monomials with degrees 4, 2, and 0, with nonzero coefficients. The highest of these degrees is 4, hence the degree of this polynomial is 4.

Arithmetic properties of polynomials are derived from those of monomials. Two polynomials in the same variable are added or subtracted by adding or subtracting like terms.

EXAMPLE 1

Find the sum and difference of the two polynomials

$$3x^4 + 5x^3 - 2x \quad \text{and} \quad -x^4 + 8x^3 - 6x$$

Solution

$(3x^4 + 5x^3 - 2x) + (-x^4 + 8x^3 - 6x)$

$$= [3 + (-1)]x^4 + [5 + 8]x^3 + [(-2) + (-6)]x$$
$$= 2x^4 + 13x^3 - 8x$$

$(3x^4 + 5x^3 - 2x) - (-x^4 + 8x^3 - 6x)$

$$= [3 - (-1)]x^4 + [5 - 8]x^3 + [(-2) - (-6)]x$$
$$= 4x^4 - 3x^3 + 4x$$

If one polynomial in a sum or difference has a monomial of a degree that is not in the other one, we add to it a monomial in that degree with a zero coefficient prior to the addition or subtraction.

EXAMPLE 2

Find the sum and difference of the two polynomials

$$-x^5 + 3x^4 - 5x \qquad \text{and} \qquad 2x^5 - 4x^3 + 6$$

Solution

$$(-x^5 + 3x^4 - 5x) + (2x^5 - 4x^3 + 6)$$
$$= (-x^5 + 3x^4 + 0x^3 - 5x + 0) + (2x^5 + 0x^4 - 4x^3 + 0x + 6)$$
$$= [(-1) + 2]x^5 + [3 + 0]x^4 + [0 + (-4)]x^3 + [(-5) + 0]x + (0 + 6)$$
$$= x^5 + 3x^4 - 4x^3 - 5x + 6$$

$$(-x^5 + 3x^4 - 5x) - (2x^5 - 4x^3 + 6)$$
$$= (-x^5 + 3x^4 + 0x^3 - 5x + 0) - (2x^5 + 0x^4 - 4x^3 + 0x + 6)$$
$$= [(-1) - 2]x^5 + [3 - 0]x^4 + [0 - (-4)]x^3 + [(-5) - 0]x + (0 - 6)$$
$$= -3x^5 + 3x^4 + 4x^3 - 5x - 6$$

We multiply a polynomial by a real number k by multiplying each coefficient in the polynomial by k.

EXAMPLE 3

Multiply by 1/2 each of the polynomials

$$10x^5 + 4x^4 - 8x^3 + 2x \qquad \text{and} \qquad -3t^8 + 6.1t^5 + 4.2t^3 - 5$$

Solution

$$\frac{1}{2}(10x^5 + 4x^4 - 8x^3 + 2x) = \frac{1}{2}(10x^5) + \frac{1}{2}(4x^4) + \frac{1}{2}(-8x^3) + \frac{1}{2}(2x)$$
$$= 5x^5 + 2x^4 - 4x^3 + x$$

$$\frac{1}{2}(-3t^8 + 6.1t^5 + 4.2t^3 - 5) = \frac{1}{2}(-3t^8) + \frac{1}{2}(6.1t^5) + \frac{1}{2}(4.2t^3) + \frac{1}{2}(-5)$$
$$= -1.5t^8 + 3.05t^5 + 2.1t^3 - 2.5$$

Two polynomials are multiplied together by multiplying each monomial in the first polynomial by every monomial in the second polynomial and then summing the results.

Two polynomials are multiplied together by multiplying each monomial in the first polynomial by every monomial in the second polynomial and then adding all of these products. This yields a new polynomial, which is then simplified by combining like terms.

EXAMPLE 4

Find the product of the polynomial $x + 4$ with the polynomial $2x^2 - 3x + 5$.

Solution

$$(x + 4)(2x^2 - 3x + 5)$$
$$= x(2x^2) + x(-3x) + x(5) + 4(2x^2) + 4(-3x) + 4(5)$$
$$= 2x^3 - 3x^2 + 5x + 8x^2 - 12x + 20$$
$$= 2x^3 + [(-3) + 8]x^2 + [5 + (-12)]x + 20$$
$$= 2x^3 + 5x^2 - 7x + 20$$

EXAMPLE 5

Find the product of the polynomial $3y^2 - 2y + 6$ with the polynomial $3y^3 - 5y^2 + y$.

Solution

$(3y^2 - 2y + 6)(3y^3 - 5y^2 + y)$

$$= 3y^2(3y^3) + 3y^2(-5y^2) + 3y^2(y) + (-2y)(3y^3) +$$
$$(-2y)(-5y^2) + (-2y)(y) + 6(3y^3) + 6(-5y^2) + 6(y)$$
$$= 9y^5 - 15y^4 + 3y^3 - 6y^4 + 10y^3 - 2y^2 + 18y^3 - 30y^2 + 6y$$
$$= 9y^5 - 21y^4 + 31y^3 - 32y^2 + 6y$$

IMPROVING SKILLS

In each of the following problems, perform the indicated operations and simplify. Determine the degree of the final polynomial.

1. $(2x^2 + 3x + 4) + (5x^2 + 6x + 7)$

2. $(t^2 - 8t + 6) + (2t^2 - 6t - 7)$

3. $(x^2 - 8x + 6) - (2x^2 - 6x - 7)$

4. $(y^2 - 3y + 7) + (2y^3 - 5y^2 - 4y)$

5. $(-u^2 - 2u + 10) + (3u^3 + 2u^2 - 2u + 5)$

6. $(-y^2 - 2y + 10) - (3y^3 + 2y^2 - 2y + 5)$

7. $(3t^5 + 8t^4 - 3t^2 + 1) + (4t^4 + 3t^3 - 2t^2 + t)$

8. $3(4x^4 + 8x^3 - 3x - 5)$

9. $4(t^5 + 3t^3 - 5t)$

10. $0.1(4m^3 + 3m^2 - 18m + 7)$

11. $\frac{1}{4}(5y^6 + 4y^4 - 3y^2 + 2)$

12. $-3(2t^2 + 4t - 9) + 2(t^2 - 3t + 8)$

13. $5(-r^3 + 2r^2 + 4r + 3) - 3(-4r^4 + 2r^2 - 8r)$

14. $\frac{1}{2}(3u^3 - 8u^2 - u + 6) + \frac{1}{4}(2u^2 + 5u - 1)$

15. $(x + 5)(x - 2)$

16. $(y - 3)(y + 3)$

17. $(2y + 5)(3y - 5)$

18. $(u^2 - 1)(u + 1)$

19. $(u^2 - 1)(u^2 + 1)$

20. $(2r^2 + 3r)(-r - 4)$

21. $(\frac{1}{2}m^2 + \frac{1}{4}m)(\frac{1}{2}m + \frac{1}{4})$

22. $(x^2 + 3x + 4)(x^2 + 6x + 7)$

23. $(m^2 - 8m + 6)(2m^2 - m - 1)$

24. $(t^2 - 6)(2t^2 + 3t + 1)$

25. $(y^2 - y + 7)(y^3 - 2y^2 - 3y)$

𝒞 FACTORING

The multiplication of two or more polynomials in the same variable yields a single polynomial as their product. *Factoring* is the reverse process, that of decomposing a single polynomial into the product of two or more quantities.

The simplest type of factoring is to decompose a single polynomial into the product of a real number k with another polynomial. This is possible when *every* coefficient of a polynomial is divisible by k. For example, the coefficients of the polynomial $6x^2 + 24x - 12$ are all divisible by the number 3, so we can write $6x^2 + 24x - 12 = 3(2x^2 + 8x - 4)$. Thus we have factored the original polynomial into the product of the real number 3 with the polynomial $2x^2 + 8x - 4$. All the coefficients of the polynomial $6x^2 +$

$24x - 12$ are also divisible by the number 2, so we can write $6x^2 + 24x - 12 = 2(3x^2 + 12x - 6)$ as another factorization of the same polynomial.

The largest positive integer divisor common to all of the coefficients in a polynomial is called its *greatest common factor*. For the polynomial $6x^2 + 24x - 12$ this number is 6, and we also have the factorization

$$6x^2 + 24x - 12 = 6(x^2 + 4x - 2)$$

In general, when we factor a number from a polynomial, we try to factor out its greatest common factor.

EXAMPLE 1

Factor a real number from each of the following polynomials:

(a) $30x^2 - 60x - 45$;   **(b)** $60t^4 + 120t^3 + 12t$.

Solution

(a) The coefficients of this polynomial in the variable x are 30, -60, and 45. Each is divisible by 3, 5, and 15. The largest of these is 15, so the greatest common factor is 15 and we write

$$30x^2 - 60x - 45 = 15(2x^2 - 4x - 3)$$

(b) The coefficients of this polynomial in the variable t are 60, 120, and 12. Each is divisible by 2, 3, 4, 6, and 12, of which the greatest common factor is 12. We write

$$60t^4 + 120t^3 + 12t = 12(5t^4 + 10t^3 + t)$$

The polynomial in Part (*b*) of Example 1 is interesting because the variable t to the first power is also common to each of the monomials. Therefore we can factor $12t$ from the original polynomial:

$$60t^4 + 120t^3 + 12t = 12t(5t^3 + 10t^2 + 1)$$

In general, when the same variable appears in *all* of the monomials of a polynomial, we factor out the variable to the highest power common to each monomial.

EXAMPLE 2

Factor the polynomials: **(a)** $2t^5 - 6t^4 + 10t^3$;   **(b)** $10u^6 + 50u^4 + 30u^2$.

Solution

(a) The monomials that comprise this polynomial in t are $2t^5$, $-6t^4$, and $10t^3$. Each contains the variable t, with 3 as the highest power of t common to all. The coefficients of this polynomial are 2, -6, and 10, with greatest common factor 2. Consequently we can factor $2t^3$ from the original polynomial:

$$2t^5 - 6t^4 + 10t^3 = 2t^3(t^2 - 3t + 5)$$

(b) The monomials that are the terms of this polynomial in u are $10u^6$, $50u^4$, and $30u^2$. Each term contains the variable u, with u^2 common to all. The coefficients of this polynomial are 10, 50, and 30, with greatest common factor 10, so we factor $10u^2$ from the original polynomial:

$$10u^6 + 50u^4 + 30u^2 = 10u^2(u^4 + 5u^2 + 3)$$

In Example 1 we factored constants from polynomials, while in Example 2 we factored monomials from polynomials. More complex still is the process of factoring a polynomial into a product of other polynomials. We begin by trying to factor a second-degree (i.e., *quadratic*) polynomial $x^2 + bx + c$, with 1 as the coefficient of x^2, into a product of first-degree (i.e., *linear*) polynomials, as follows:

$$x^2 + bx + c = (x + \alpha)(x + \beta) \tag{14}$$

$x^2 + bx + c$
$= (x + \alpha)(x + \beta)$
providing $\alpha + \beta = b$
and $\alpha\beta = c.$

Here the coefficients b, c, α (the Greek letter alpha), and β (the Greek letter beta) all denote integers. Since

$$(x + \alpha)(x + \beta) = x^2 + (\alpha + \beta)x + \alpha\beta$$

it follows that Equation (14) is true if and only if $\alpha + \beta = b$ and $\alpha\beta = c$.

EXAMPLE 3

Factor the following polynomials:

(**a**) $x^2 + 5x + 6$ (**b**) $x^2 + 8x + 15$
(**c**) $x^2 - 6x + 8$ (**d**) $x^2 - x - 6$
(**e**) $x^2 + x + 1$

Solution

(**a**) Here $b = 5$ and $c = 6$. We seek integers α and β having the properties $\alpha + \beta = 5$ and $\alpha\beta = 6$. Two such numbers are $\alpha = 2$ and $\beta = 3$. Therefore we have the factorization

$$x^2 + 5x + 6 = (x + 2)(x + 3)$$

(**b**) Here $b = 8$ and $c = 15$. We seek real integers α and β having the properties $\alpha + \beta = 8$ and $\alpha\beta = 15$. Two such numbers are $\alpha = 3$ and $\beta = 5$, so we have the factorization

$$x^2 + 8x + 15 = (x + 3)(x + 5)$$

(**c**) We rewrite this polynomial as

$$x^2 - 6x + 8 = x^2 + (-6)x + 8$$

so $b = -6$ and $c = 8$. We seek integers α and β such that $\alpha + \beta = -6$ and $\alpha\beta = 8$. Two numbers that work are $\alpha = -2$ and $\beta = -4$, giving us the factorization

$$x^2 - 6x + 8 = [x + (-2)][x + (-4)] = (x - 2)(x - 4)$$

(**d**) We rewrite this polynomial as

$$x^2 - x - 6 = x^2 + (-1)x + (-6)$$

Here $b = -1$ and $c = -6$, so we want integers α and β that meet the requirements $\alpha + \beta = -1$ and $\alpha\beta = -6$. Two such numbers are $\alpha = -3$ and $\beta = 2$, which gives us the factorization

$$x^2 - x - 6 = [x + (-3)][x + 2] = (x - 3)(x + 2)$$

(**e**) Here $b = c = 1$, so we need numbers α and β with the properties $\alpha + \beta = 1$ and $\alpha\beta = 1$. However, no such real numbers exist, so we are unable to factor this polynomial.

If the coefficient of x^2 in a second-degree output is a real number other than 1, then the polynomial has the form $ax^2 + bx + c$, where a, b and c are integers, $a \neq 1$. In this case we will factor such a polynomial into a product

$$ax^2 + bx + c = (\alpha_1 x + \beta_1)(\alpha_2 x + \beta_2) \tag{15}$$

When we multiply out the right side of Equation (15) we get

$$(\alpha_1 x + \beta_1)(\alpha_2 x + \beta_2) = (\alpha_1 \alpha_2)x^2 + (\alpha_1 \beta_2 + \alpha_2 \beta_1)x + (\beta_1 \beta_2)$$

It follows that Equation (15) is true if and only if

$$\alpha_1 \alpha_2 = a \tag{16}$$

$$\beta_1 \beta_2 = c \tag{17}$$

$$\alpha_1 \beta_2 + \alpha_2 \beta_1 = b \tag{18}$$

EXAMPLE 4

Factor the quadratic polynomial $2x^2 + 7x + 6$.

Solution
Here $a = 2$, $b = 7$, and $c = 6$, so Equation (15) becomes

$$2x^2 + 7x + 6 = (\alpha_1 x + \beta_1)(\alpha_2 x + \beta_2) \tag{19}$$

To satisfy Equation (16) we must have $\alpha_1 \alpha_2 = 2$. Only four integer pairs satisfy this requirement:

α_1	1	−1	2	−2
α_2	2	−2	1	−1

To satisfy Equation (17), we must have $\beta_1 \beta_2 = 6$. Many integer pairs satisfy this requirement:

β_1	1	−1	6	−6	2	−2	3	−3
β_2	6	−6	1	−1	3	−3	2	−2

All of these pairs of values of α_1, α_2 and β_1, β_2 make up our pool of test coefficients for Equation (19).

To satisfy Equation (18) we must have $\alpha_1 \beta_2 + \alpha_2 \beta_1 = 7$. It turns out that the only set of test coefficients that satisfies this equation is $\alpha_1 = 2$, $\alpha_2 = 1$ and $\beta_1 = 3$, $\beta_2 = 2$. With these values, Equation (19) becomes

$$2x^2 + 7x + 6 = (2x + 3)(x + 2)$$

If we multiply the two first-degree polynomials on the right side, we obtain the second-degree polynomial on the left.

In practice, we do not test Equation (18) directly. Instead, we generate a pool of values for α_1, α_2, β_1, and β_2, as we did in Example 4, and then substitute different pairs of values into the product $(\alpha_1 x + \beta_1)(\alpha_2 x + \beta_2)$, comparing each result to the given quadratic polynomial we are attempting to factor.

EXAMPLE 5

Factor the quadratic polynomial $16x^2 + 32x + 15$.

Solution

Here $a = 16$, $b = 32$, and $c = 15$, and Equation (15) becomes

$$16x^2 + 32x + 15 = (\alpha_1 x + \beta_1)(\alpha_2 x + \beta_2) \tag{20}$$

To satisfy Equation (16) we must have $\alpha_1 \alpha_2 = 16$. There are many integer pairs that satisfy this requirement:

α_1	1	-1	16	-16	2	-2	8	-8	4	-4
α_2	16	-16	1	-1	8	-8	2	-2	4	-4

To satisfy Equation (17) we must have $\beta_1 \beta_2 = 15$. The integer pairs that satisfy this requirement are:

β_1	1	-1	15	-15	3	-3	5	-5
β_2	15	-15	1	-1	5	-5	3	-3

Each of these pairs of values of α_1, α_2 and β_1, β_2 provide test coefficients for Equation (20).

To satisfy Equation (18) we must have $\alpha_1 \beta_2 + \alpha_2 \beta_1 = 32$. Of the various sets of test coefficients, the only sets that satisfy this equation are $\alpha_1 = \alpha_2 = 4$ and $\beta_1 = 3$, $\beta_2 = 5$. With these values, Equation (20) becomes

$$16x^2 + 32x + 15 = (4x + 3)(4x + 5)$$

EXAMPLE 6

Factor the quadratic polynomial $6x^2 + 7x - 5$.

Solution

We rewrite this polynomial as

$$6x^2 + 7x - 5 = 6x^2 + 7x + (-5)$$

with $a = 6$, $b = 7$, and $c = -5$. Then Equation (15) becomes

$$6x^2 + 7x + (-5) = (\alpha_1 x + \beta_1)(\alpha_2 x + \beta_2) \tag{21}$$

For this polynomial we need α_1 and α_2 such that $\alpha_1 \alpha_2 = 6$, and we need β_1 and β_2 such that $\beta_1 \beta_2 = -5$. The candidates are:

α_1	1	-1	6	-6	2	-2	3	-3
α_2	6	-6	1	-1	3	-3	2	-2

β_1	1	-1	5	-5
β_2	-5	5	-1	1

The only pairs of values for α_1, α_2 and β_1, β_2 that satisfy Equation (18), which is $\alpha_2 \beta_2 + \alpha_2 \beta_1 = 7$, are $\alpha_1 = 3$, $\alpha_2 = 2$ and $\beta_1 = 5$, $\beta_2 = -1$. With these values, Equation (21) becomes

$$6x^2 + 7x - 5 = (3x + 5)[2x + (-1)]$$
$$= (3x + 5)(2x - 1)$$

EXAMPLE 7

Factor the polynomial $18x^2 + 21x - 15$.

Solution

We note that we can factor 3 out of each term, obtaining

$$18x^2 + 21x - 15 = 3(6x^2 + 7x - 5)$$

Then, using the results of Example 6, we have

$$6x^2 + 21x - 15 = 3(3x + 5)(2x - 1)$$

Factoring can be laborious with no guarantee of success, as was the case in part (*e*) of Example 3. Fortunately, some factorizations occur with such regularity that they are very familiar. For any real number *a*:

The difference-of-squares rule	$x^2 - a^2 = (x + a)(x - a)$
The perfect-square rule	$x^2 \pm 2ax + a^2 = (x \pm a)^2$
The difference-of-cubes rule	$x^3 - a^3 = (x - a)(x^2 + ax + a^2)$
The sum-of-cubes rule	$x^3 + a^3 = (x + a)(x^2 - ax + a^2)$

EXAMPLE 8

Factor the polynomials: (*a*) $x^2 - 6x + 9$;   (*b*) $x^3 + 8$.

Solution

(*a*) By the perfect-square rule, with $a = 3$:

$$x^2 - 6x + 9 = x^2 - 2(3)x + 3^2 = (x - 3)^2$$

(*b*) By the sum-of-cubes rule, with $a = 2$ (since $8 = 2^3$):

$$x^3 + 8 = x^3 + 2^3 = (x + 2)(x - 2x + 4)$$

IMPROVING SKILLS

Factor the following polynomials, if possible.

1. $30x^2 - 60x - 45$
2. $8x^2 - 20x + 36$
3. $30t^3 + 35t^2 + 55$
4. $42y^5 + 54y^3 - 42$
5. $3t^3 + 5t^2 + 3t$
6. $y^5 + 12y^3 - 9y^2$
7. $4x^8 + 2x^7 + 6x^6$
8. $10p^5 - 14p^4 - 18p^2$
9. $2m^2 + 16$
10. $m^2 + 16$
11. $m^2 - 16$
12. $p^3 + 16$
13. $x^2 + 7x + 10$
14. $x^2 + 9x + 18$
15. $x^2 - 3x - 10$
16. $x^2 - 9x + 18$
17. $x^2 - 9x - 10$
18. $x^2 + 11x + 18$
19. $t^2 + 9t + 14$
20. $y^2 + 5y - 14$
21. $t^2 - 100$
22. $m^3 - 81m$
23. $y^2 - y + 1$
24. $p^2 - 2p + 1$
25. $2x^2 + 18x + 40$
26. $3x^2 - 3x - 60$
27. $m^4 + 13m^3 + 2m^2$

28. $2t^3 + 18t^2 - 44t$ **29.** $4x^2 - 1$ **30.** $9x^2 - 16$
31. $4x^2 + 6x + 2$ **32.** $2t^2 + 5t + 2$ **33.** $2m^2 - 3m - 2$
34. $3x^2 - 11x + 10$ **35.** $4x^2 + 4x + 1$ **36.** $6y^2 + 7y - 3$
37. $5t^2 + 21t + 4$ **38.** $4m^2 - 16m + 15$ **39.** $12p^2 - 7p - 12$
40. $12t^2 - 25t + 12$

ROOTS AND THE QUADRATIC FORMULA

A root of a polynomial equation in the variable x is a value for x that makes the equation true.

Let $p(x)$ denote polynomial in the variable x. A *root* of the polynomial equation

$$p(x) = 0 \qquad (22)$$

is a number that, when substituted for x, makes Equation (22) true. For example, $x = 2$ is a root of the polynomial equation $5x - 10 = 0$ because $5(2) - 10 = 0$, but $x = 3$ is *not* a root of this same equation because $5(3) - 10 = 5 \neq 0$. The value $x = 4$ is *not* a root of the polynomial equation $x^2 - 11x + 30 = 0$ because $(4)^2 - 11(4) + 30 = 16 - 44 + 30 = 2 \neq 0$, but both $x = 5$ and $x = 6$ are roots because $(5)^2 - 11(5) + 30 = 25 - 55 + 30 = 0$, and $(6)^2 - 11(6) + 30 = 36 - 66 + 30 = 0$.

Equation (22) is *linear* if the polynomial $p(x)$ is of degree 1. Such polynomial equations in the variable x have the general form

The only root of the linear equation $bx + c = 0$ is $x = -c/b$

$$bx + c = 0 \qquad (23)$$

where both b and c are known real numbers, $b \neq 0$. Equation (23) has the single root $x = -c/b$, obtained by solving Equation (23) for x.

EXAMPLE 1

Find the root of the linear equation $2x + 3 = 0$.

Solution

Here $b = 2$ and $c = 3$. Solving the linear equation for x, we obtain $x = -3/2$ as its root.

The polynomial equation $p(x) = 0$ is *quadratic* if the polynomial $p(x)$ is of degree 2. Quadratic equations in the variable x have the general form

$$ax^2 + bx + c = 0 \qquad (24)$$

where a, b, and c are real numbers, $a \neq 0$. The equation $2x^2 + 5x - 7 = 0$ is a quadratic equation in x with $a = 2$, $b = 5$, and $c = -7$ while the equation $7x^2 - 2x - 1 = 0$ is a quadratic equation in x with $a = 7$, $b = -2$, and $c = -1$. The equation $3t^2 + t = 0$ is a quadratic equation in t with $a = 3$, $b = 1$, and $c = 0$, while the equation $y^2 - 6.15 = 0$ is a quadratic equation in y with $a = 1$, $b = 0$, and $c = -6.15$.

One procedure for locating the roots of a quadratic equation is to factor the quadratic polynomial on the left side of Equation (24) into the product of two linear polynomials (see Appendix C):

$$ax^2 + bx + c = (\alpha_1 x + \beta_1)(\alpha_2 x + \beta_2)$$

Then we can rewrite Equation (24) as

$$(\alpha_1 x + \beta_1)(\alpha_2 x + \beta_2) = 0 \qquad (25)$$

Now, a product of two quantities is zero if and only if one or both of the quantities are zero, so Equation (25) is true if either

$$\alpha_1 x + \beta_1 = 0 \qquad \text{or} \qquad \alpha_2 x + \beta_2 = 0$$

A root of either of these linear equations is also a root of Equation (24).

EXAMPLE 2

Find the roots of the quadratic equation $2x^2 + 7x + 6 = 0$.

Solution

Using the results of Example 4 in Appendix C, we have

$$2x^2 + 7x + 6 = (2x + 3)(x + 2)$$

so we can rewrite our original equation in the factored form

$$(2x + 3)(x + 2) = 0$$

This last equation implies that either

$$2x + 3 = 0 \qquad \text{or} \qquad x + 2 = 0.$$

It follows from Example 1 that the root of the first linear equation is $x = -3/2$. A solution to the second equation is $x = -2$. Thus, the two roots of the quadratic equation $2x^2 + 7x + 6 = 0$ are $x = -3/2$ and $x = -2$.

EXAMPLE 3

Find the roots of the quadratic equation $6x^2 + 7x - 5 = 0$.

Solution

Using the results of Example 6 in Appendix C, we can rewrite our original equation in the factored form

$$6x^2 + 7x - 5 = (3x + 5)(2x - 1) = 0$$

This factored equation implies that either $3x + 5 = 0$ or $2x - 1 = 0$. The first equation is true when $x = -5/3$, and the second equation is true when $x = 1/2$. These are the two roots of the original quadratic equation.

EXAMPLE 4

Find the roots of the equation $x^2 - 6x + 9 = 0$.

Solution

Using the results of Example 8 in Appendix C,

$$x^2 - 6x + 9 = (x - 3)^2 = 0.$$

This implies that $x - 3 = 0$, or $x = 3$. This is the only root of the original quadratic equation.

The quadratic polynomial equation $ax^2 + bx + c = 0$ (with $a \neq 0$) can be rewritten $ax^2 + bx = -c$. Now dividing both sides of this equation by a, we obtain

$$x^2 + \frac{bx}{a} = -\frac{c}{a}$$

Then, adding the quantity $b^2/(4a^2)$ to both sides of this last equation, we have

$$x^2 + \frac{bx}{a} + \frac{b^2}{4a^2} = -\frac{c}{a} + \frac{b^2}{4a^2} \tag{26}$$

Using the perfect-square rule of factorization, we can factor the left side of Equation (26) as follows:

$$x^2 + \frac{bx}{a} + \frac{b^2}{4a^2} = x^2 + 2\left(\frac{b}{2a}\right)x + \left(\frac{b}{2a}\right)^2 = \left(x + \frac{b}{2a}\right)^2$$

The right side of Equation (26) can be simplified to

$$-\frac{c}{a} + \frac{b^2}{4a^2} = \frac{b^2 - 4ac}{4a^2}$$

This allows us to rewrite Equation (26) as

$$\left(x + \frac{b}{2a}\right)^2 = \frac{b^2 - 4ac}{4a^2}$$

Taking the square root of both sides of this last equation, we conclude that

$$x + \frac{b}{2a} = \pm\sqrt{\frac{b^2 - 4ac}{4a^2}} = \pm\frac{\sqrt{b^2 - 4ac}}{\sqrt{4a^2}}$$

Solving for x, we have

$$x = -\frac{b}{2a} \pm \frac{\sqrt{b^2 - 4ac}}{2a}$$

or

$$x = \frac{-b \pm \sqrt{b^2 - 4ac}}{2a} \qquad (27)$$

> *A root of the quadratic equation*
> $$ax^2 + bx + c = 0$$
> *must satisfy the quadratic formula*
> $$x = \frac{-b \pm \sqrt{b^2 - 4ac}}{2a}$$

Equation (27) is known as the *quadratic formula*. We can find the roots of a quadratic equation by substituting the coefficients of the polynomial into the quadratic formula and then simplifying.

EXAMPLE 5

Use the quadratic formula to find the roots of the equation $6x^2 + 7x - 5 = 0$.

Solution

This is a quadratic equation, having the form of Equation (24) with $a = 6$, $b = 7$, and $c = -5$. Substituting these values into the quadratic formula, we obtain

$$x = \frac{-7 \pm \sqrt{7^2 - 4(6)(-5)}}{2(6)} = \frac{-7 \pm \sqrt{49 + 120}}{12}$$
$$= \frac{-7 \pm \sqrt{169}}{12} = \frac{-7 \pm 13}{12}$$

Taking the plus sign, one root is

$$x = \frac{-7 + 13}{12} = \frac{6}{12} = \frac{1}{2}$$

Taking the minus sign, the other root is

$$x = \frac{-7 - 13}{12} = \frac{-20}{12} = -\frac{5}{3}$$

These are the same two roots we identified in Example 3 by factoring.

EXAMPLE 6

Find the roots of the equation $4y^2 - 2y - 5 = 0$.

Solution

This equation is quadratic in the variable y, with coefficients $a = 4$, $b = -2$, and $c = -5$. The polynomial on the left side of the equation is difficult to factor, so instead we use the quadratic formula:

$$y = \frac{-(-2) \pm \sqrt{(-2)^2 - 4(4)(-5)}}{2(4)} = \frac{2 \pm \sqrt{4 + 80}}{8} = \frac{-2 \pm \sqrt{84}}{8}$$

The plus sign yields the root

$$y = (-2 + \sqrt{84})/8 \approx 0.895644$$

while the minus sign gives us a second root,

$$y = (-2) - \sqrt{84})/8 \approx 1.395644$$

Example 6 illustrates the power of the quadratic formula. It allowed us to easily locate the roots of the polynomial equation $4y^2 - 2y - 5 = 0$, even though we could not readily factor its left side.

We identified two roots for each of the equations in Examples 5 and 6, one root corresponding to the plus sign in the quadratic formula and a second root corresponding to the minus sign. Occasionally, however, the quadratic formula yields only one root. If the coefficients a, b, and c satisfy the condition $b^2 = 4ac$, so that $b^2 - 4ac = 0$, then Equation (27) reduces to the single root

$$x = \frac{-b \pm \sqrt{b^2 - 4ac}}{2a} = \frac{-b \pm \sqrt{0}}{2a} = -\frac{b}{2a}$$

If $b^2 - 4ac < 0$, then its square root is not a real number and the corresponding quadratic equation has no real roots: they are complex numbers and not of interest to us in this book. All of our applications and models deal strictly with real numbers.

EXAMPLE 7

Find the roots of the equation $3t^2 - 6t + 3 = 0$.

Solution

This equation is quadratic in the variable t, with coefficients $a = 3$, $b = -6$, and $c = 3$. Using the quadratic formula, we find that

$$t = \frac{-(-6) \pm \sqrt{(-6)^2 - 4(3)(3)}}{2(3)} = \frac{6 \pm \sqrt{36 - 36}}{6} = \frac{6 \pm 0}{6} = 1$$

The only root of this equation is $t = 1$. We obtain the same result by factoring if we note that $3t^2 - 6t + 3 = 3(t^2 - 2t + 1) = 3(t - 1)^2$.

EXAMPLE 8

Find the roots of the equation $5z^2 + 8 = 0$.

Solution

This is a quadratic equation in the variable z, with $a = 5$, $b = 0$, and $c = 8$. Substituting these values into the quadratic formula, we obtain

$$z = \frac{-0 \pm \sqrt{0^2 - 4(5)(8)}}{2(5)} = \frac{\pm\sqrt{-160}}{10}$$

The square root of a negative number is not a real number, hence the given quadratic equation does not have any real roots.

IMPROVING SKILLS

Find all real roots of the following polynomial equations.

1. $30x - 60 = 0$	**2.** $8x - 20 = 0$	**3.** $3t + 36 = 0$
4. $4y + 18 = 0$	**5.** $5t + 8 = 0$	**6.** $7u = 0$
7. $-4x + 2 = 0$	**8.** $-10p - 14 = 0$	**9.** $x^2 + 7x + 10 = 0$
10. $x^2 + 9x + 18 = 0$	**11.** $x^2 - 3x - 10 = 0$	**12.** $x^2 - 9x + 18 = 0$
13. $t^2 - 9t - 10 = 0$	**14.** $y^2 + 11y + 18 = 0$	**15.** $2y^2 - 3y - 2 = 0$
16. $2p^2 + 6p - 4 = 0$	**17.** $c^2 - c - 1 = 0$	**18.** $m^2 + 6m + 9 = 0$
19. $y^2 - y - 2 = 0$	**20.** $3p^2 + 2p - 1 = 0$	**21.** $5t^2 - t - 1 = 0$
22. $x^2 - 8x + 16 = 0$	**23.** $m^2 + 2m = 0$	**24.** $2m^2 + 16 = 0$
25. $m^2 - 16 = 0$	**26.** $m^2 - 15 = 0$	**27.** $4x^2 - 1 = 0$
28. $9x^2 - 16 = 0$	**29.** $4x^2 + 6x + 1 = 0$	**30.** $4x^2 + 6x + 2 = 0$
31. $4x^2 + 6x + 3 = 0$	**32.** $2t^2 + 5t + 2 = 0$	**33.** $2t^2 + 5t + 1 = 0$
34. $2t^2 + 5t = 0$	**35.** $2t^2 + 5t + 3 = 0$	**36.** $3x^2 - 11x + 10 = 0$
37. $m^4 + 13m^3 + 2m^2 = 0$	**38.** $2t^3 + 18t^2 - 44t = 0$	**39.** $p^4 + 13p^3 = 0$
40. $y^3 + 8y^2 - y = 0$		

E COMMON LOGARITHMS

Multiplication and division are simplified greatly when each factor is expressed as a power of the same base (see Examples 4 and 5 in Appendix A). Bases of 2, 8, and 16 are particularly important in this age of computer technology, but the first base to receive widespread attention was 10, because it is the foundation of our basic number system.

Every positive number N can be expressed as 10 raised to some power p. That power, or exponent, p is called the *common logarithm of N,* denoted $\log N$. That is, if

If $N = 10^p$ then

$\log N = p$

$$N = 10^p \quad \text{then} \quad \log N = p \tag{28}$$

The number 100 is 10^2, so 2 is the common logarithm of 100. We write $\log 100 = 2$ because $100 = 10^2$. The number 0.001 can be written as an exponential in the base 10 with exponent -3:

$$0.001 = \frac{1}{1000} = \frac{1}{10^3} = 10^{-3}$$

Therefore -3 is the common logarithm of 0.001. That is, $\log 0.001 = -3$ because $0.001 = 10^{-3}$. Similarly,

$$\log 1 = 0 \quad \text{because} \quad 1 = 10^0$$

$$\log 10 = 1 \quad \text{because} \quad 10 = 10^1$$

$$\log 100 = 2 \quad \text{because} \quad 100 = 10^2$$

$$\log 10,000 = 4 \quad \text{because} \quad 10,000 = 10^4$$

Common logarithms are found most easily using a calculator with a [log] key. Depending on the type of calculator available, either first a number key is pressed and then the [log] key, or first the [log] key is pressed and then a number key. The common logarithm of the number is displayed after pressing the [=] key on most scientific calculators or the [ENTER] key on many graphing calculators.

EXAMPLE 1

Find log 8.27.

Solution

The sequence of keystrokes on most scientific calculators is

$$[8.27] \quad [log] \quad [=]$$

The equivalent sequence on many graphing calculators is

$$[log] \quad [8.27] \quad [ENTER]$$

Either way, the result is 0.91750551 rounded to eight decimal places. That is,

$$\log 8.27 \approx 0.91750551 \quad \text{because} \quad 8.27 \approx 10^{0.91750551}$$

For most numbers N, the number $\log N$ is an infinite decimal. Calculators can display only a part of such numbers, and we generally round the displayed numbers to the precision appropriate for our needs.

EXAMPLE 2

Find log 0.004163.

Solution

The sequence of keystrokes needed on many scientific calculators is

The equivalent sequence on many graphing calculators is

Either way, the result is -2.3806, rounded to four decimal places. That is,

$$\log 0.004163 \approx -2.3806 \quad \text{because} \quad 0.004163 \approx 10^{-2.3806}$$

If $N = 10^p$ for a positive number N and exponent p, then $\log N = p$. Substituting this value of p into $N = 10^p$, we obtain

$$10^{\log N} = N \tag{29}$$

This tells us that log N is the power to which 10 must be raised to obtain N. In other words, if we know the common logarithm of a number N, we can recapture the number N itself by raising 10 to the power log N. For example, if log $P = 5.4785997$, then $P = 10^{5.4785997} \approx 301{,}023$. If log $Q = -0.89279$, then $Q = 10^{-0.89279} \approx 0.128$.

We can reverse the procedure: if log $N = p$, where $N = 10^p$, then

$$\log 10^p = p \tag{30}$$

Equation (30) is simply the definition of the common logarithm.

For hundreds of years, common logarithms were used to facilitate multiplication and division. The technique used was to identify the logarithm of each factor of a product or quotient and then use Equation (29) to write it as an exponential in the base 10. The required products and quotients were then calculated using the properties of exponents. The next example illustrates this procedure; it is presented here only for a historical perspective.

EXAMPLE 3

Use common logarithms to calculate $N = \dfrac{8(511)}{(65)(1023)}$.

Solution

Using a calculator, we determine that

$$\log 8 \approx 0.903090 \quad \text{hence} \quad 8 \approx 10^{0.903090}$$

$$\log 511 \approx 2.708421 \quad \text{hence} \quad 511 \approx 10^{2.708421}$$

$$\log 65 \approx 1.812913 \quad \text{hence} \quad 65 \approx 10^{1.812913}$$

$$\log 1023 \approx 3.009876 \quad \text{hence} \quad 1023 \approx 10^{3.009876}$$

Then

$$N = \frac{8(511)}{(65)(1023)} \approx \frac{10^{0.903090}10^{2.708421}}{10^{1.812913}10^{3.009876}} = \frac{10^{0.903090+2.708421}}{10^{1.812913+3.009876}}$$

$$= \frac{10^{3.611511}}{10^{4.822789}} = 10^{3.611511-4.822789} = 10^{-1.211278}$$

$$\approx 0.061478$$

Today we do not use common logarithms to do calculations like the one in Example 3; rather we use a general-purpose calculator to perform the multiplications and divisions directly. However, prior to the widespread availability of electronic calculators in the 1970s, the procedure illustrated in Example 3 was the preferred method for calculating many products and quotients, using published tables to obtain the common logarithms of individual numbers. Now common logarithms are primarily used as a tool for evaluating data, and their properties are much more useful then the numbers themselves.

Common logarithms are exponents (of the number 10), so their properties are derived from those of exponents. For example, if A and B are positive numbers with log $A = a$ and log $B = b$, then it follows from Equation (29) that $A = 10^a$ and $B = 10^b$. Now using the properties of exponents (see Appendix A) and Equation (30), we have that

$$\log(AB) = \log(10^a 10^b) = \log(10^{a+b}) = a + b = \log A + \log B$$

We have just proven the following rule:

THE PRODUCT RULE OF LOGARITHMS

$$\log(AB) = \log A + \log B$$

Similarly,

$$\log(A/B) = \log(10^a/10^b) = \log(10^{a-b}) = a - b = \log A - \log B$$

THE QUOTIENT RULE OF LOGARITHMS

$$\log \frac{A}{B} = \log A - \log B$$

We can prove yet another rule:

$$\log(A^m) = \log([10^a]^m) = \log(10^{am}) = am = ma = m \log A$$

THE POWER RULE OF LOGARITHMS

$$\log(A^m) = m \log A$$

These rules of logarithms are particularly useful for solving equations in which the unknown quantity appears as an exponent.

EXAMPLE 4

Find y if $(2.5)^y = 8$.

Solution

Here the unknown quantity y appears as an exponent of 2.5. Taking common logarithms of both sides of the given equation, we have $\log(2.5)^y = \log 8$. It follows from the power rule of logarithms (with $a = 2.5$) that $y \log 2.5 = \log 8$. Solving for y, we obtain

$$y = \frac{\log 2.5}{\log 8} = 0.44064$$

rounded to five decimal places.

EXAMPLE 5

Find y if $5^{2y} = 17(4^{-3.1\,y})$.

Solution

Here the unknown quantity y appears *twice* as an exponent. Taking common logarithms of both sides of the given equation, we have

$$\log(5^{2y}) = \log[17(4^{-3.1\,y})]$$

Then

$$\log(5)^{2y} = \log 17 + \log(4^{-3.1\,y}) \qquad \text{(product rule of logarithms)}$$

$$\log(5)^{2y} = \log 17 + (-3.1y)\log 4 \qquad \text{(power rule of logarithms)}$$

$$2y \log 5 = \log 17 - 3.1y \log 4 \qquad \text{(power rule of logarithms)}$$

The quantity y no longer appears as an exponent, so we can solve for it directly.

$$2y \log 5 + 3.1y \log 4 = \log 17$$

$$y[2 \log 5 + 3.1 \log 4] = \log 17$$

and

$$y = \frac{\log 17}{2 \log 5 + 3.1 \log 4} = 0.376938$$

rounded to six decimal places.

IMPROVING SKILLS

In Problems 1 through 15, find the common logarithms, rounded to six decimal places, of the given numbers.

1. 1.26   **2.** 12.6   **3.** 126   **4.** 0.00126
5. 4.53   **6.** 4530   **7.** 0.453   **8.** 800
9. 74.3   **10.** 15.1   **11.** 15.15   **12.** 0.0112
13. 0.1012   **14.** 13,933   **15.** 145,982

In Problems 16 through 30, find the number that corresponds to each given common logarithm.

16. 2.683947   **17.** 0.683947   **18.** −4.39794   **19.** 1.262451
20. 2.262451   **21.** −0.737549   **22.** −1.737549   **23.** 1.790285
24. 2.790285   **25.** −1.045758   **26.** −2.045758   **27.** 0.011147
28. 1.757776   **29.** −2.110698   **30.** −0.30103

In Problems 31 through 50, use the properties of common logarithms to determine which of the following equalities are true. Then use a calculator to determine whether any of the other equalities are true.

31. $\log(2 \cdot 7) = (\log 2)(\log 7)$ **32.** $\log(3 \cdot 5) = \log 3 + \log 5$
33. $\log(-8) = -\log 8$ **34.** $\log 8 = 3 \log 2$
35. $\log(2 + 7) = \log 2 + \log 7$ **36.** $\log(2 + 7) = (\log 2)(\log 7)$
37. $\log(4 \cdot 9) = \log 4 + \log 9$ **38.** $\log(4 + 9) = \log 4 + \log 9$
39. $\log(4 \cdot 9) = (\log 4)(\log 9)$ **40.** $\log(4 \cdot 9) = 2 \log 6$
41. $\log(4 \cdot 5) = 2 \log 2 + \log 5$ **42.** $\log(4 \cdot 9) = 2 \log 2 + 3 \log 3$
43. $\log \dfrac{3}{5} = \dfrac{\log 3}{\log 5}$ **44.** $\log \dfrac{3}{5} = \log 3 - \log 5$
45. $\log(3 - 5) = \dfrac{\log 3}{\log 5}$ **46.** $\log \dfrac{3}{5} = \log(3 - 5)$

47. $\log \dfrac{1}{4} = -\log 4$

48. $\log \dfrac{1}{6} = -\log 6$

49. $\log \dfrac{9}{25} = 2 \log 3 - 2 \log 5$

50. $\log \dfrac{9}{25} = -\log \dfrac{25}{9}$

Solve the following equations for y.

51. $2^y = 7$
52. $7^y = 2$
53. $3^y = 7$
54. $5^y = 200$
55. $2^{y+3} = 20$
56. $7^{y-1} = 16$
57. $2^{3y} = 20$
58. $8^{-y} = 9$
59. $2^{0.1y} = 400$
60. $8^{-y} = 3$
61. $2^{4.5y} = 100$
62. $4(3^{2y}) = 5$
63. $3(2^{0.1y}) = 85$
64. $5.1(2.3^{-2y}) = 90$
65. $20(2^{4y}) = 3^{2y}$
66. $14(3.5^{3y}) = 8^{2.4y}$
67. $40(3^{5y}) = 30(2^{3y})$
68. $19(5^{3.2y}) = 6(7^{-3.7y})$

SIGMA NOTATION

The sum

$$a_1 + a_2 + a_3 + a_4 + \cdots + a_{99} + a_{100}$$

can be written more compactly as

$$\sum_{i=1}^{100} a_i$$

The capital Greek letter sigma (Σ) denotes a sum, and the equation below the sigma establishes the smallest value for the variable i. The sum starts by using that number, in this case $i = 1$, to replace the i in the expression following the sigma. The subsequent terms in the sum are obtained by replacing the i in the expression with consecutive integers starting with 2, and the addition of terms is complete when i assumes the value above the sigma, in this case 100.

The notation $\sum_{i=3}^{8} x_i$ indicates a sum that begins with x_3, since the quantity below the sigma is $i = 3$, and ends with x_8, since the quantity above the sigma is $i = 8$; and the sum includes all integer values of i between 3 and 8 in the expression x_i. Written out (or "expanded"), the sum is

$$\sum_{i=3}^{8} x_i = x_3 + x_4 + x_5 + x_6 + x_7 + x_8$$

Similarly,

$$\sum_{i=1}^{5} T_i = T_1 + T_2 + T_3 + T_4 + T_5$$

$$\sum_{i=2}^{4} y_i = y_2 + y_3 + y_4$$

$$\sum_{i=m}^{n} q_i = q_m + q_{m+1} + q_{m+2} + \cdots + q_{n-1} + q_n$$

EXAMPLE 1

Determine a numerical value for $\sum_{i=1}^{5} \dfrac{i}{i+1}$.

Solution

$$\sum_{i=1}^{5} \frac{i}{i+1} = \frac{1}{1+1} + \frac{2}{2+1} + \frac{3}{3+1} + \frac{4}{4+1} + \frac{5}{5+1}$$

$$= \frac{1}{2} + \frac{2}{3} + \frac{3}{4} + \frac{4}{5} + \frac{5}{6} = \frac{71}{20} = 3.55$$

EXAMPLE 2

Write the following sum in Σ notation:

$$\left(\frac{3^2}{2} + 5\right) + \left(\frac{4^2}{2} + 5\right) + \left(\frac{5^2}{2} + 5\right) + \cdots + \left(\frac{99^2}{2} + 5\right) + \left(\frac{100^2}{2} + 5\right)$$

Solution

Each term in the sum is of the form $(i^2/2) + 5$, where the variable i assumes integer values between 3 and 100, inclusive. The sum can be expressed compactly as

$$\sum_{i=3}^{100} \left(\frac{i^2}{2} + 5\right)$$

EXAMPLE 3

Each week the manager of a store reports the total sales for that week to corporate headquarters. Develop a formula for calculating yearly sales.

Solution

Denote weekly sales for the ith week of a year by W_i ($i = 1, 2, 3, \ldots,$ 52), where W_1 represents sales for the first week of the year, W_2 represents sales for the second week, and so on. Yearly sales, Y, are the sum of the individual weekly sales, so

$$Y = \sum_{i=1}^{52} W_i$$

Although we have been using the letter i to specify the bounds for a sum, any other label will do equally well, for example,

$$\sum_{k=5}^{9} k^2 = 5^2 + 6^2 + 7^2 + 8^2 + 9^2 = 255$$

and

$$\sum_{n=2}^{5} (x^n - n) = (x^2 - 2) + (x^3 - 3) + (x^4 - 4) + (x^5 - 5)$$

$$= x^5 + x^4 + x^3 + x^2 - 14$$

Many graphing calculators have menus for efficiently calculating sums expressed in sigma notation, and this technology should be used when it is available.

IMPROVING SKILLS

In Problems 1 through 8, expand each expression of terms without using Σ notation.

1. $\displaystyle\sum_{i=1}^{3} (x_i)^2$

2. $\displaystyle\sum_{i=3}^{11} 2x_i$

3. $\displaystyle\sum_{k=2}^{7} (x_k + y_k)$

4. $\displaystyle\sum_{j=99}^{105} (3M_j + 4)$

5. $\displaystyle\sum_{j=2}^{5} (3x_j)^2$

6. $\displaystyle\sum_{n=0}^{4} \sin(n\theta)$

7. $\displaystyle\sum_{n=0}^{4} \sin(n\theta_n)$

8. $\displaystyle\sum_{m=1}^{2} \ln(y/m)$

In Problems 9 through 18, determine a numerical value for each sum.

9. $\displaystyle\sum_{k=1}^{6} k$

10. $\displaystyle\sum_{k=1}^{6} k^2$

11. $\displaystyle\sum_{k=1}^{6} 2k$

12. $\displaystyle\sum_{j=3}^{7} (j - 1)$

13. $\displaystyle\sum_{j=1}^{5} (j + 1)$

14. $\displaystyle\sum_{i=0}^{6} (-1)^i$

15. $\displaystyle\sum_{i=1}^{5} (i)^{-1}$

16. $\displaystyle\sum_{i=1}^{5} (i)^i$

17. $\displaystyle\sum_{m=3}^{6} \frac{m + 2}{m + 3}$

18. $\displaystyle\sum_{p=7}^{14} (p - 10)^2$

19. Write the following expressions in Σ notation.
   (a) $3(2)^2 + 3(3)^2 + 3(4)^2 + 3(5)^2 + \cdots + 3(28)^2 + 3(29)^2$
   (b) $2(3)^2 + 3(3)^2 + 4(3)^2 + 5(3)^2 + \cdots + 28(3)^2 + 29(3)^2$
   (c) $2(3)^2 + 2(3)^3 + 2(3)^4 + 2(3)^5 + \cdots + 2(3)^{12} + 2(3)^{13}$
   (d) $3(2)^2 - 3(3)^2 + 3(4)^2 - 3(5)^2 + \cdots + 3(28)^2 + 3(29)^2$

20. Write the following sums in Σ notation.
   (a) $f(0.5) + f(1.5) + f(2.5) + f(3.5) + f(4.5)$
   (b) $f(1) + f(4) + f(9) + f(16) + f(25)$
   (c) $f(10) + f(12) + f(14) + f(16) + f(18)$
   (d) $f(10) + f(13) + f(16) + f(19) + f(22)$

21. Calculate the following sums for the data listed in Table A:

   (a) $\displaystyle\sum_{i=1}^{5} x_i$

   (b) $\displaystyle\sum_{j=1}^{5} y_j$

   (c) $\displaystyle\sum_{i=1}^{3} x_i$

   (d) $\displaystyle\sum_{k=1}^{5} (x_k + y_k)$

   (e) $\displaystyle\sum_{k=2}^{4} x_k + \sum_{i=1}^{4} y_i$

   (f) $\displaystyle\sum_{m=3}^{5} x_m y_m$

22. Calculate the following sums for the data listed in Table B:

   (a) $\displaystyle\sum_{i=1}^{6} x_i$

   (b) $\displaystyle\sum_{i=1}^{6} y_i$

   (c) $\displaystyle\sum_{j=1}^{3} (x_j)^2$

   (d) $\displaystyle\sum_{k=2}^{5} (x_k - 2)$

   (e) $\displaystyle\sum_{n=1}^{6} x_n y_n$

   (f) $\displaystyle\sum_{n=2}^{6} x_n y_{n-1}$

23. Determine whether any of the following equations are true.

   (a) $2\left(\displaystyle\sum_{i=1}^{n} x_i\right) = \sum_{i=1}^{n} 2x_i$

   (b) $\displaystyle\sum_{i=1}^{n} (x_i + y_i) = \sum_{i=1}^{n} x_i + \sum_{i=1}^{n} y_i$

   (c) $\displaystyle\sum_{i=1}^{m} x_i + \sum_{i=m+1}^{n} x_i = \sum_{i=1}^{n} x_i$

   (d) $\displaystyle\sum_{i=1}^{n} x_i y_i = \left(\sum_{i=1}^{n} x_i\right)\left(\sum_{i=1}^{n} y_i\right)$

TABLE A

i	x_i	y_i
1	0	6
2	1	7
3	2	8
4	-1	3
5	-4	-2

TABLE B

i	x_i	y_i
1	0	3
2	8	2
3	-2	6
4	5	9
5	-3	10
6	7	1

24. Derive formulas using Σ notation for the mean and variance of the set of grades $\{G_1, G_2, G_3, \ldots, G_{100}\}$.

25. Use Σ notation to write a system of normal equations for determining the equation of a linear least-squares straight line with data points (x_i, y_i), $i = 1, 2, \ldots, N$. [See System (*16*) of Section 2.3]

ANSWERS TO SELECTED PROBLEMS

CHAPTER 1

Section 1.1   Improving Skills

1. Discover design errors, develop estimates for fuel consumption and life-support requirements.
3. Study new traffic patterns and their impact on the community. Determine whether new access roads are needed. Estimate construction costs, expected use of the facility, and anticipated revenues.
5. Determine the immediate effect of each policy or a combination of policies on the different economic classes (low-, middle-, and high-income), on the national debt, on inflation, and on unemployment, and determine the long-run impact on the economy.
7. Minimize unnecessarily high inventory levels with their storage costs and avoid premature depletion of stock, which results in lost sales and customer dissatisfaction.
9. Estimate the amount of business that the new store will attract and its impact on each of the existing stores.
11. Current injuries or addictions, personality quirks that make the player either a distraction to or role model for other players, family or financial problems that can affect performance.
13. Provides no information on the exterior design and dimensions of the set.
15. Provides no information on how the plane will react to pilot commands, how much weight the plane can carry, or passenger comfort in the cabin.
17. Provides no information on the condition of tracks, trains, or switches, vandalism, weather conditions, or labor unrest.

Section 1.2   Creating Models

1. $722 = 80 + 0.15C$, C is total costs
3. $D = 0.10U$, D is the number of defective units, U is the number of units
5. $n = 200,000 + 5,000E$; n is the number of bottles, E is thousands of dollars spent on advertising
7. $A = 2B$; A is the number of votes candidate A receives, B is the number of votes candidate B receives
9. $D = 10Y$; D is depreciation in dollars, Y is the number of days
11. $2,856 = 8A + 5C$; A is the number of adult tickets, C is the number of children's tickets
13. $D = T + 1.5C$; D is total decoration time in hours, T is the number of tables, C is the number of chairs
15. $P = 10D + 8R$; P is total padding, D is the number of deluxe mattresses, R is the number of regular mattresses
17. $TC = 1.5n$, $TS = 3N$; TC is total cost, n is the number of calculators produced, TS is total sales, N is the number of calculators sold.
19. $x + y = 1,250$, $35x + 20y = 37,200$; x is the number of $35 seats, y is the number of $20 seats
21. $M = 100 + 59H$; M is total moving charges, H is the number of hours actually used to move
23. $2F + 3S = 20,000$; F is the number of books in the first job, S is the number of books in the second job
25. $M = 60J_1 + 50J_2$, $T = 5J_1 + 7J_2$; M is the number of seconds available for molding, T is the number of seconds available for threading, J_1 is the number of type-I joints, J_2 is the number of type-II joints

27. $H = 4{,}000d_1 + 3{,}000d_2$, $M = 7{,}000d_1 + 12{,}000d_2$, $L = 2{,}000d_1 + 6{,}000d_2$; H is the number of tons of high-grade ore, M is the number of tons of medium-grade ore, L is the number of tons of low-grade ore, d_1 is the number of days the first mine is in operation, d_2 is the number of days the second mine is in operation

29. $78.58 = 32(2) + 0.09M$; M is the number of miles driven

31. $630 = 200 + 0.05S$; S is the total amount of sales

33. $59{,}000 = 40{,}000 + 5r(40{,}000)$; r is the annual interest rate

35. $35r + 50d = 16{,}900$, $10r + 12d = 4{,}440$; r is the number of regular mattresses, d is the number of deluxe mattresses

37. $3C + 10F + L = 6{,}300$, $5C + 8F + L = 6{,}600$; C is the number of cabinets, F is the number of grandfather clock frames, L is the number of lamp bases

39. $75A + 50B = 3{,}125$, $6A + 2B = 206$; A is the number of days to operate plant A, B is the number of days to operate plant B

Section 1.3   Improving Skills

1. 18 inches
3. 13.5 inches
5. $167,500
7. a. 15 feet by $15\frac{5}{6}$ feet, b. 5 feet by $18\frac{1}{3}$ feet, c. $46\frac{2}{3}$ feet by 50 feet
9. $24'2''$ by $36'8''$
11. 4.35 inches
13. 2.7 inches
15. They are inverses of each other.
17. a. $A = kr^2$ b. $k \approx 3.1416$

Creating Models

19. 3 billion dollars
21. 53 feet 4 inches
23. $40,250
25. $1,110
27. a. 0.8464 cc b. 0.778688 cc
29. $\dfrac{\text{number of right-handed people}}{\text{total population}} = .82$; constant of proportionality $= .82$
31. 506,000
33. 320 ft/sec

Exploring with Technology

37. straight line
39. straight line

Section 1.4   Improving Skills

1. either every million or every 500,000 from 0 to 5 million
3. every 50 units from 5,900 to 6,400
5.

7. 4500 inches
9. 5 inches
11. a. 0 b. 0
13. They all have the same x coordinate.
15.

17.

19.

21.

23.

25.

27.

29.

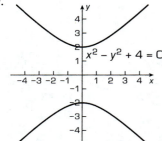

31. Generally, a straight line with two exceptions in 1980 and 1988

33. In general, an inverted U-shape; the unemployment rate increased from 1979 to 1982 and then decreased from 1983 to 1989.

35. a. 30% b. 24% c. 1971 d. 1986

37. a.

b. straight line

c. inverted-U

39.

somewhat similar to exponential growth

41.

There is no general trend in the annual precipitation in Baltimore from 1940 through 1955.

Creating Models

43.

45.

47.

49.

51.

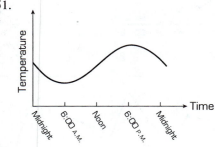

Exploring with Technology

53. $(-5, -27)$ By substituting this point of intersection into both equations one can conclude that it satisfies both equations.

55. There is no point of intersection, so there is no solution to this set of equations.

Section 1.5 Improving Skills

1. a. linear **b.** not linear **c.** not linear **d.** linear **e.** linear **f.** linear **g.** not linear **h.** linear **i.** not linear **j.** linear

3.

5.

7.

9.

11.

13.

15.

17.

19.

21.

23.

25. $\frac{5}{8}$ 27. -1

29. 30   31. $-\frac{40}{21}$

33. $\frac{20}{9}$ 35. $\frac{2}{3}$, (0, 2)

37. $-\frac{2}{3}$, (0, -2)   39. 2, (0, -10)

41. $\frac{1}{200}$, (0, 25)   43. 0, (0, 7)

45. 0, (0, -7)   47. 7.5, (0, -4000)

49. $\frac{2}{5}$, (0, 0)   51. 2, (0, 0)

53. $y = 5x - 5$ 55. $y = 3x + 1$

57. $y = -3x + 1$ 59. $y = 2x + 4$

61. $y = 40x - 312$ 63. $y = \frac{1}{3}x - \frac{1}{3}$

65. $y = -2x + 200$ 67. $x = 700$

69. $y = 2$ 71. $x = 8$

73. a. neither b. perpendicular c. parallel d. parallel e. perpendicular f. neither

75. $y = \frac{2}{3}x + \frac{170}{3}$

77. $y = -\frac{1}{2}x + 175$

79. $x = 100$

81.

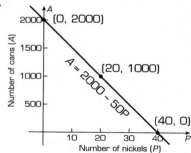

a. 50¢ is 10 nickels, so the number of cans is 1,500,

b. $1.00 is 20 nickels, so the number of cans is 1,000,

c. The price at which demand is zero.

83.

a. $546,000

b. 15.7 years

c. The value of the building when it was first constructed.

85.

Horizontal intercept—no significance; Vertical intercept—the base cost to a customer without any mileage used.

87. a. $I = \frac{1}{5}t + 15$ b. $I = -\frac{1}{5}t + 23$

Creating Models

89. $P = 2 + .15b$, where b and P are in dollars

91. $C = 630 + 40n$

93.

95.

97.

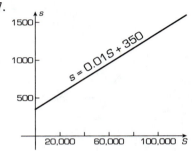

Section 1.6   Improving Skills

1. a. only (c)

3.

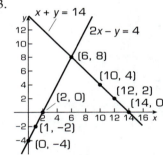

Estimated solution (6, 8) is the solution.

5.

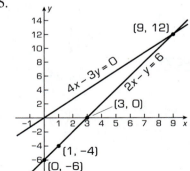

Estimated solution (9, 12) is the solution.

7.

$2x - y = -5$
and
$-4x + 2y = 10$

Infinitely many solutions.

9.

$(2.4, 0.8)$ is the solution.

11.

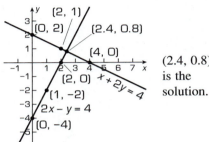

$(50, 0)$ is the solution.

13.

$(0, 0)$ is the solution.

15.

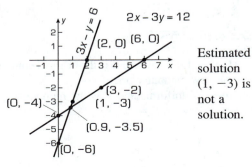

$2x - 3y = 12$

Estimated solution $(1, -3)$ is not a solution.

17.

$(600, 200)$ is a solution.

19.

No solution.

21. $C = \$5.50n + \$21,000$, $R = \$9.00n$, where n is the number of can openers

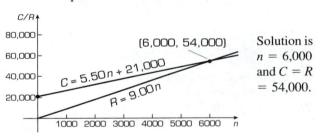

Solution is $n = 6,000$ and $C = R = 54,000$.

23. $C = 10n + 20,000$, $R = 22n$, where n is the number of books

$n = 1650$ is an estimate of the break-even point, not the actual point.

25. $C = 0.20n + \$15,000$, $R = 0.30n$

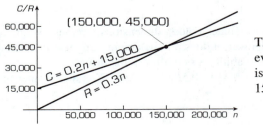

The break-even point is $n = 150,000$.

27. $C = 2n + 15,000$ $R = 3n$

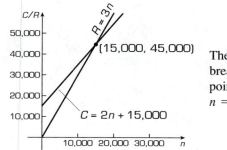

The break-even point is $n = 15,000$.

29. $C = 2n + 30,000$, $R = 4n$

The break-even point is $n = 22,500$.

Exploring with Technology

35. $x \approx 6019.23$ $y \approx 1489.23$
37. $x \approx 0.86$ $y \approx -3.43$
39. $x = -1$ $y = 0$

Testing Yourself

1. $R = 0.05W$ where R is the number of released workers and W is original work force

2.

3. Model includes colors, dimensions, options, weight, and horsepower; it does not include missing features, reliability, availability, or comfort to the driver.

4. $F = 0.5P$, $M = 0.5P$, $B = 0.6F$; F is the number of females, M is the number of males, P is the total population, B is the number of offspring

5. $\$500.50 = 1.50S + 2.50D$, $265 = S + D$; S is the number of single-scoop cones, D is the number of double-scoop cones

6.

7.

8. Model includes general layout of field, position of seats, permanent obstructions, and whether stadium is domed; it does not include number and locations of rest rooms, information on food servces, accessibility by public transportation, security, or any changes to stadium since picture was taken.

9. $C = \$200 + \$65n$, where n is the number of meals

10. 80 feet by 100 feet

11. $x = 30$, $y = 20$

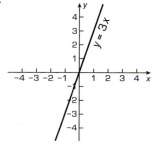

Estimate: (30, 20)

12. $P = C + 0.03C = 1.03C$, where C is the original computer price, P is the new computer price

13. I is the number of uniforms and P is the number of participants; $I = 180$ when $P = 12,000$

14. $T + G = 130$, $12T + 23G = 2077$; T is the number of tourist special dinners, G is the number of gourmet delight dinners

15.

16.

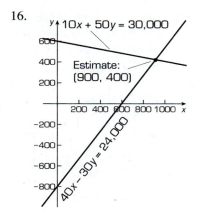

CHAPTER 2

Section 2.1   Improving Skills

1. linear
3. not linear
5. not linear
7. not linear
9. a. no b. no c. yes
11. $\begin{bmatrix} 2 & 3 & | & -4 \\ 5 & -6 & | & 7 \end{bmatrix}$
13. $\begin{bmatrix} 23 & -37 & | & 0 \\ 66 & -72 & | & 0 \end{bmatrix}$
15. $\begin{bmatrix} 1 & -2 & 3 & -1 & | & 4 \\ 2 & 0 & -1 & 4 & | & 8 \end{bmatrix}$
17. $\begin{bmatrix} 8 & 1 & 1 & | & 3 \\ 4 & 2 & -1 & | & 2 \end{bmatrix}$
19. $11y + 12x = 20$, $13y + 14x = 30$
21. $-8x + 2y = 0$, $3y = 1$
23. $-x = 3$, $2y = 4$
25. $x + 2y + 3z = 10$, $2x + 3y + 4z = 20$, $3x + 4y + 5z = 30$
27. $-A + 2B - 3C = 5$, $2A - 6B + 6C = -7$, $-3A + 4C = 0$
29. $x - 4y + \frac{1}{2}z = 2$, $y + \frac{1}{2}z = -3$, $0 = 0$
31. $x - 2y + 5w = 100$, $y - z + 4w = 150$, $z + 3w = 200$, $w = 250$
33. a. $\begin{bmatrix} 3 & 4 & | & 6 \\ 1 & 2 & | & 5 \end{bmatrix} R_1 \leftrightarrow R_2$
    b. $\begin{bmatrix} 1 & 2 & | & 5 \\ -6 & -8 & | & -12 \end{bmatrix} 2R_2 \rightarrow R_2$

c. $\begin{bmatrix} 1 & 2 & | & 5 \\ 7 & 12 & | & 26 \end{bmatrix} R_2 + 4R_1 \rightarrow R_2$
d. $\begin{bmatrix} 1 & 2 & | & 5 \\ 1 & \frac{4}{3} & | & 2 \end{bmatrix} \frac{1}{3}R_2 \rightarrow R_2$

35. a. $\begin{bmatrix} 1 & -5 & | & 8 \\ 0 & 18 & | & -18 \end{bmatrix} R_2 + (-3)R_1 \rightarrow R_2,$
    b. $\begin{bmatrix} 1 & -5 & | & 8 \\ 0 & 1 & | & -1 \end{bmatrix} \frac{1}{18}R_2 \rightarrow R_2$

37. a. $\begin{bmatrix} 1 & 2 & 3 & | & 10 \\ 0 & 1 & -6 & | & -20 \\ 0 & 1 & 4 & | & 30 \end{bmatrix} R_2 + (-4)R_1 \rightarrow R_2,$
    b. $\begin{bmatrix} 1 & 2 & 3 & | & 10 \\ 0 & 1 & -6 & | & -20 \\ 0 & 0 & 10 & | & 50 \end{bmatrix} R_3 + (-1)R_2 \rightarrow R_3,$
    c. $\begin{bmatrix} 1 & 2 & 3 & | & 10 \\ 0 & 1 & -6 & | & -20 \\ 0 & 0 & 1 & | & 5 \end{bmatrix} (0.1)R_3 \rightarrow R_3$

39. $x = -12$, $y = 8$
41. $x = y = 0$
43. $x = 39$, $y = -16$, $z = 6$
45. $x = -114$, $y = -16$, $z = -2$
47. $x = \frac{1}{6}$, $y = -\frac{4}{9}$, $z = -\frac{1}{3}$
49. $x = 36$, $y = 60$, $z = -11$, $w = 2$

51. Each of the systems of equations in Problems 39 through 50 is upper triangular since each corresponding augmented matrix has zero entries below the main diagonal:

39. $\begin{bmatrix} 1 & 2 & | & 4 \\ 0 & 1 & | & 8 \end{bmatrix}$

40. $\begin{bmatrix} 2 & -3 & | & -1 \\ 0 & 2 & | & 5 \end{bmatrix}$

41. $\begin{bmatrix} 11 & -8 & | & 0 \\ 0 & -1 & | & 0 \end{bmatrix}$

42. $\begin{bmatrix} 7 & -11 & | & 0 \\ 0 & 3 & | & 1 \end{bmatrix}$

43. $\begin{bmatrix} 1 & 2 & -1 & | & 1 \\ 0 & 1 & 3 & | & 2 \\ 0 & 0 & 1 & | & 6 \end{bmatrix}$

44. $\begin{bmatrix} 1 & -1 & -3 & | & 0 \\ 0 & 2 & 3 & | & 6 \\ 0 & 0 & 1 & | & 4 \end{bmatrix}$

45. $\begin{bmatrix} 1 & -8 & 7 & | & 0 \\ 0 & 1 & -8 & | & 0 \\ 0 & 0 & 1 & | & -2 \end{bmatrix}$

46. $\begin{bmatrix} 2 & 0 & -1 & | & 1 \\ 0 & 3 & 1 & | & 8 \\ 0 & 0 & 2 & | & 5 \end{bmatrix}$

47. $\begin{bmatrix} 2 & 3 & 0 & | & -1 \\ 0 & 3 & -4 & | & 0 \\ 0 & 0 & 3 & | & -1 \end{bmatrix}$

48. $\begin{bmatrix} 3 & 4 & -5 & | & 2 \\ 0 & 6 & 7 & | & 1 \\ 0 & 0 & 8 & | & 9 \end{bmatrix}$

49. $\begin{bmatrix} 1 & -1 & -2 & 1 & | & 0 \\ 0 & 1 & 5 & -2 & | & 1 \\ 0 & 0 & 1 & 8 & | & 5 \\ 0 & 0 & 0 & 1 & | & 2 \end{bmatrix}$

50. $\begin{bmatrix} 1 & 4 & -1 & 5 & | & 10 \\ 0 & 1 & 6 & -7 & | & 20 \\ 0 & 0 & 1 & 8 & | & 30 \\ 0 & 0 & 0 & 1 & | & 40 \end{bmatrix}$

Creating Models

53. $A + Y = 1{,}201$, $8A + 5Y = 8{,}510$
55. $c + f = 207$, $250c + 490f = 54{,}150$
57. $30C + 10D = 40$, $5C + 30D = 60$

Reviewing Material

62. $y = x^3$ is not linear—that's why the curve is not a straight line.

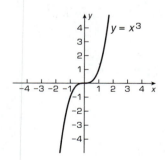

Section 2.2 Improving Skills

1. a. $\begin{bmatrix} 1 & 2 & | & 1 \\ 0 & 1 & | & -4 \end{bmatrix}$ b. $\begin{bmatrix} 1 & 0 & | & 9 \\ 0 & 1 & | & -4 \end{bmatrix}$

3. a. $\begin{bmatrix} 1 & 1 & | & 0 \\ 0 & 1 & | & 2 \end{bmatrix}$ b. $\begin{bmatrix} 1 & 0 & | & -2 \\ 0 & 1 & | & 2 \end{bmatrix}$

5. a. $\begin{bmatrix} 1 & 2 & | & 20 \\ 0 & 1 & | & 10 \end{bmatrix}$ b. $\begin{bmatrix} 1 & 0 & | & 0 \\ 0 & 1 & | & 10 \end{bmatrix}$

7. a. $\begin{bmatrix} 1 & \frac{1}{2} & | & 2 \\ 0 & 1 & | & 3 \end{bmatrix}$ b. $\begin{bmatrix} 1 & 0 & | & \frac{1}{2} \\ 0 & 1 & | & 3 \end{bmatrix}$

9. a. $\begin{bmatrix} 1 & 1 & | & 1 \\ 0 & 1 & | & \frac{3}{5} \end{bmatrix}$ b. $\begin{bmatrix} 1 & 0 & | & \frac{2}{5} \\ 0 & 1 & | & \frac{3}{5} \end{bmatrix}$

11. a. $\begin{bmatrix} 1 & 2 & | & 4 \\ 0 & 1 & | & \frac{15}{7} \end{bmatrix}$ b. $\begin{bmatrix} 1 & 0 & | & -\frac{2}{7} \\ 0 & 1 & | & \frac{15}{7} \end{bmatrix}$

13. a. $\begin{bmatrix} 1 & 2 & 3 & | & -2 \\ 0 & 1 & 2 & | & -\frac{8}{3} \end{bmatrix}$ b. $\begin{bmatrix} 1 & 0 & -1 & | & \frac{10}{3} \\ 0 & 1 & 2 & | & -\frac{8}{3} \end{bmatrix}$

15. a. $\begin{bmatrix} 1 & 3 & | & -1 \\ 0 & 1 & | & -1 \\ 0 & 0 & | & -1 \end{bmatrix}$ b. $\begin{bmatrix} 1 & 0 & | & 2 \\ 0 & 1 & | & -1 \\ 0 & 0 & | & -1 \end{bmatrix}$

17. a. $\begin{bmatrix} 1 & \frac{1}{2} & 6 & | & -10 \\ 0 & 1 & \frac{9}{4} & | & -24 \\ 0 & 0 & 1 & | & \frac{3}{16} \end{bmatrix}$ b. $\begin{bmatrix} 1 & 0 & 0 & | & \frac{139}{128} \approx 1.0859 \\ 0 & 1 & 0 & | & -\frac{1563}{64} \approx -24.422 \\ 0 & 0 & 1 & | & \frac{3}{16} = 0.1875 \end{bmatrix}$

19. a. $\begin{bmatrix} 1 & 2 & 6 & | & -2 \\ 0 & 1 & \frac{8}{3} & | & -\frac{4}{3} \\ 0 & 0 & 1 & | & 1 \end{bmatrix}$ b. $\begin{bmatrix} 1 & 0 & 0 & | & 0 \\ 0 & 1 & 0 & | & -4 \\ 0 & 0 & 1 & | & 1 \end{bmatrix}$

21. a. and b. $\begin{bmatrix} 1 & 0 & 1 & | & 4 \\ 0 & 1 & 0 & | & \frac{5}{2} \\ 0 & 0 & 0 & | & -\frac{1}{2} \end{bmatrix}$

23. a. $\begin{bmatrix} 1 & 1 & 1 & | & 1 \\ 0 & 1 & 2 & | & 3 \\ 0 & 0 & 1 & | & 1 \end{bmatrix}$ b. $\begin{bmatrix} 1 & 0 & 0 & | & -1 \\ 0 & 1 & 0 & | & 1 \\ 0 & 0 & 1 & | & 1 \end{bmatrix}$

25. a. $\begin{bmatrix} 1 & \frac{1}{2} & \frac{1}{2} & | & \frac{3}{2} \\ 0 & 1 & -\frac{5}{3} & | & \frac{5}{3} \\ 0 & 0 & 1 & | & \frac{1}{4} \end{bmatrix}$ b. $\begin{bmatrix} 1 & 0 & 0 & | & \frac{1}{3} \\ 0 & 1 & 0 & | & 2\frac{1}{12} \\ 0 & 0 & 1 & | & \frac{1}{4} \end{bmatrix}$

27. a. $\begin{bmatrix} 1 & 2 & 3 & | & 0 \\ 0 & 1 & 2 & | & 0 \\ 0 & 0 & 0 & | & 0 \end{bmatrix}$ b. $\begin{bmatrix} 1 & 0 & -1 & | & 0 \\ 0 & 1 & 2 & | & 0 \\ 0 & 0 & 0 & | & 0 \end{bmatrix}$

29. First interchange the first and third rows.

a. $\begin{bmatrix} 1 & 1 & 1 & | & -500 \\ 0 & 1 & -2 & | & 400 \\ 0 & 0 & 1 & | & -390 \end{bmatrix}$ b. $\begin{bmatrix} 1 & 0 & 0 & | & 270 \\ 0 & 1 & 0 & | & -380 \\ 0 & 0 & 1 & | & -390 \end{bmatrix}$

31. $x = 2\frac{2}{3}$, $y = \frac{1}{3}$
33. $a = \frac{11}{26}$, $b = -\frac{9}{26}$
35. $N = \frac{12}{5}$, $S = \frac{6}{5}$
37. $s = \frac{4369}{3777} \approx 1.1567$, $t = -\frac{320}{3777} \approx -0.0847$
39. $x = \frac{29}{18}$, $y = -\frac{13}{18}$
41. $x = 2.875$, $y = -2$, $z = -2.125$

43. infinitely many solutions; $x = \frac{5}{2}z + 4$, $y = -\frac{1}{2}z - 1$, z arbitrary
45. no solution
47. $N = S = E = 100$
49. $x = \frac{16}{35} \approx 0.457$, $y = \frac{26}{35} \approx 0.743$, $z = -\frac{126}{35} = -3.6$, $w = -\frac{124}{35} \approx -3.543$

Creating Models

51. $50d + 35r = 16{,}900$, $12d + 10r = 4{,}440$; d is the number of deluxe mattresses, r is the number of regular mattresses; solution: $d = 170$, $r = 240$
53. $A + Y = 1201$, $8A + 5Y = 8510$; A is the number of adult tickets, Y is the number of youth tickets; solution: $A = 835$, $Y = 366$
55. $C + F = 207$, $250C + 490F = 54{,}150$; C is the number of coach seats, F is the number of first-class seats; solution: $F = 10$, $C = 197$
57. $75a + 50b = 3125$, $12a + 5b = 384$, $6a + 2b = 206$; a is the number of days to operate plant A, b is the number of days to operate plant B; no solution
59. $0.7r + 0.6p + 0.55s = 180{,}000$, $0.3r + 0.4p + 0.45s = 120{,}000$; r is the number of gallons of regular, p is the number of gallons of plus, s is the number of gallons of supreme; infinitely many solutions: $r = \frac{1}{2}s$, $p = -1.5s + 300{,}000$, s is any real number between 0 and 200,000. If $s > 200{,}000$ then p is negative

Exploring with Technology

65. $x = \frac{1700}{7} \approx 242.857$, $y = -\frac{400}{7} \approx -57.143$

Reviewing Material

66. $x = 100$, $y = -\frac{1}{2}x + 85$; these two lines are not perpendicular.

Section 2.3 Improving Skills

1. 0.01193
3. 9
5. $N = 5$, $S_x = 0$, $S_{x^2} = 10$, $S_y = 63$, $S_{xy} = 10$; $5b + 0m = 63$, $0b + 10m = 10$; $b = 12.6$, $m = 1$, so $y = x + 12.6$; $E = 7.2$. This answer is smaller than that obtained in Problems 3 and 4.

7.

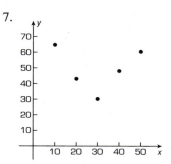

9.

x	y	x^2	xy
-2	60	4	-120
-1	70	1	-70
0	90	0	0
1	110	1	110
2	120	4	120
$S_x = 0$	$S_y = 450$	$S_{x^2} = 10$	$S_{xy} = 40$

$5b + 0m = 450$, $0b + 10m = 40$

11.

x	y	x^2	xy
$1950 \to -20$	1	400	-20
$1960 \to -10$	3	100	-30
$1970 \to 0$	6	0	0
$1980 \to 10$	8	100	80
$1990 \to 20$	10	400	200
$S_x = 0$	$S_y = 28$	$S_{x^2} = 1000$	$S_{xy} = 230$

$5b + 0m = 28$, $0b + 1000m = 230$

Creating Models

13. $y = \frac{193}{58}x - \frac{85}{58}$, or $y \approx 3.3276$, $x - 1.4655$; when $y = 25$, then $x \approx 7.95$ or nearly 8 weeks
15. code years; $y = 2.036x + 37.125$; for 1993, $x = 4.5$, $y \approx 46.3$ maltreated children for 1,000 U.S. children
17. code years; $y = 0.78x + 5.9167$; for 1994, $x = 3.5$, $y \approx 8.6$ billion
19. code years; $y = 0.161x + 74.457$; for 1990, $x = 7$, $y \approx 75.6$; for 1970 $x = -13$, $y \approx 72.4$
21. code years; $y = 0.243x + 24.357$; for 1990, $x = 7$, $y \approx 26.1$; for 1975, $x = -8$, $y \approx 22.4$
23. code years; $y = 11.29x + 190.74$; for 1992, $x = 8$ $y \approx 281.1$; for 1975, $x = -9$, $y \approx 89.1$
25. code years; $y = 4.999x + 66.181$; for 1965, $x = -15$, $y \approx -8.8$; for 2000, $x = 20$, $y \approx 166.2$

Reviewing Material

32.

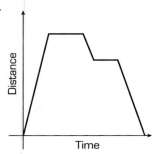

Section 2.4 Improving Skills

1. Entries in second row and first column do not match, so there is no solution.

3. $x = -10, y = 10$

5. **A** is 2×2, **B** is 2×2, **U** is 2×1, **C** is 2×3, **D** is 2×3, **V** is 2×1, **E** is 3×2, **F** is 3×2, **W** is 3×1, **G** is 3×3, **H** is 3×3, **x** is 1×4, **y** is 1×4,

7. $\begin{bmatrix} 12 & 1 \\ 1 & 4 \end{bmatrix}$

9. $\begin{bmatrix} 1 & 1 & 7 \\ 7 & 2 & 8 \end{bmatrix}$

11. $[-1 \quad 1 \quad 2 \quad -8]$

13. $\begin{bmatrix} -4 & 1 \\ -3 & 3 \\ 5 & -6 \end{bmatrix}$

15. $\begin{bmatrix} 2 & 2 & 4 \\ 5 & 2 & -1 \\ 3 & 3 & 1 \end{bmatrix}$

17. $\begin{bmatrix} 10 & -3 \\ 7 & 2 \end{bmatrix}$

19. $\begin{bmatrix} 6 & -3 \\ -1 & -3 \\ -3 & 2 \end{bmatrix}$

21. not defined

23. $[5 \quad -3 \quad 0 \quad 4]$

25. $\begin{bmatrix} 2 & 4 \\ -6 & 2 \end{bmatrix}$

27. $\begin{bmatrix} -3 \\ -21 \end{bmatrix}$

29. $\begin{bmatrix} -10 & -5 & -15 \\ -20 & -5 & -35 \end{bmatrix}$

31. $\begin{bmatrix} -21 & 4 \\ -11 & -5 \end{bmatrix}$

33. not defined

35. $[-8 \quad 5 \quad 1 \quad -10]$

37. $\begin{bmatrix} 19 & 5 \\ -29 & 6 \end{bmatrix}$

39. not defined

41. not defined

43. not defined

45. $\begin{bmatrix} 15 \\ 4 \end{bmatrix}$

47. not defined

49. $\begin{bmatrix} 0 \\ 9 \\ 4 \end{bmatrix}$

51. $\begin{bmatrix} -9 & 6 & 3 & -18 \\ 6 & -4 & -2 & 12 \end{bmatrix}$

53. not defined

55. not defined

57. $\begin{bmatrix} 7 & 4 & -6 \\ 11 & -1 & -5 \\ 6 & 3 & -1 \end{bmatrix}$

59. $\mathbf{M} = \begin{bmatrix} 9 & -5 \\ 10 & 1 \end{bmatrix}$

61. $\mathbf{AB} = \mathbf{AC} = \begin{bmatrix} 6 & 0 \\ 3 & 0 \end{bmatrix}$

Creating Models

63. a. 25 TV sets, 5 air conditioners, 7 refrigerators, 10 stoves, and 5 dishwashers

b. $[20 \quad 5 \quad 6 \quad 8 \quad 4]$ (20 TV sets, 5 air conditioners, 6 refrigerators, 8 stoves, and 4 dishwashers)

c. $[22 \quad 12 \quad 4 \quad 7 \quad 6]$ (22 TV sets, 12 air conditioners, 4 refrigerators, 7 stoves, and 6 dishwashers)

65. a. $\begin{bmatrix} 45 & 27 & 36 \\ 135 & 63 & 72 \\ 36 & 18 & 72 \end{bmatrix}$ b. $\begin{bmatrix} 25 & 30 & 40 \\ 150 & 49 & 80 \\ 38 & 19 & 76 \end{bmatrix}$

67. $\mathbf{pn} = [46, 480]$ is the total dollar value of the tickets purchased for a recent flight

69. a. $\mathbf{w} = \begin{bmatrix} 21 \\ 19 \\ 16 \end{bmatrix}$ b. $\mathbf{Cw} = \begin{bmatrix} 63 \\ 170 \end{bmatrix}$

c. $\dfrac{3.25}{59.75} \approx 5.4\%$ increase in labor costs for manufacturing television cabinets

$\dfrac{8.5}{161.50} \approx 5.3\%$ increase in labor costs for manufacturing grandfather clock frames

71. a. $\mathbf{C} = \begin{bmatrix} 5 & 8 \\ 4 & 10 \end{bmatrix}$ b. $\mathbf{d} = \begin{bmatrix} 800 \\ 1000 \end{bmatrix}$ $\mathbf{Cd} = \begin{bmatrix} 12,000 \\ 13,200 \end{bmatrix}$

Total cost if all items are shipped from Townsand is $1,200. Total cost if all items are shipped from Rockland is $13,200.

Reviewing Material

76. There is no solution that simultaneously satisfies all three equations.

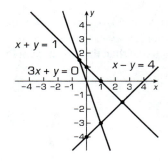

Section 2.5   Improving Skills

1. $\begin{bmatrix} -8 & 3 \\ 2 & -1 \end{bmatrix}$

3. $\begin{bmatrix} -4 & 3 \\ 3 & -2 \end{bmatrix}$

5. $\begin{bmatrix} 1 & -1 \\ -1 & 1.5 \end{bmatrix}$

7. $\begin{bmatrix} 0 & 1 \\ 1 & 0 \end{bmatrix}$

9. $\begin{bmatrix} \frac{1}{7} & -\frac{1}{7} \\ \frac{4}{7} & \frac{3}{7} \end{bmatrix}$

11. $\begin{bmatrix} 0 & 0 & 1 \\ 0 & 1 & 0 \\ 1 & 0 & 0 \end{bmatrix}$

13. $\begin{bmatrix} 1 & 0 & 0 \\ -2 & 1 & 0 \\ -6 & 3 & 1 \end{bmatrix}$

15. $\begin{bmatrix} 1 & 0 & 2 \\ -2 & 1 & -4 \\ 0 & 0 & 1 \end{bmatrix}$

17. $\begin{bmatrix} -1.5 & -1 & 3.5 \\ 1 & 1 & -2 \\ 0.5 & 0 & -0.5 \end{bmatrix}$

19. matrix does not have an inverse

21. a. $\begin{bmatrix} 1 & 3 \\ 2 & 7 \end{bmatrix}\begin{bmatrix} x \\ y \end{bmatrix} = \begin{bmatrix} 8 \\ 3 \end{bmatrix}$, $x = 47, y = -13$

b. $x = -19, y = 6$

c. $x = 250, y = 50$

d. $p = -19, q = 6$

23. $\begin{bmatrix} 1 & 2 \\ 2 & 5 \end{bmatrix}\begin{bmatrix} x \\ y \end{bmatrix} = \begin{bmatrix} 3 \\ 7 \end{bmatrix}$, $x = 1, y = 1$

25. $\begin{bmatrix} 4 & 2 \\ 7 & -3 \end{bmatrix}\begin{bmatrix} a \\ b \end{bmatrix} = \begin{bmatrix} 1 \\ 4 \end{bmatrix}$, $a = \frac{11}{26}, b = -\frac{9}{26}$

27. $\begin{bmatrix} 3 & 4 \\ 2 & 1 \end{bmatrix}\begin{bmatrix} N \\ S \end{bmatrix} = \begin{bmatrix} 12 \\ 6 \end{bmatrix}$, $N = 2.4, S = 1.2$

29. $\begin{bmatrix} 2.1 & -15 \\ 4 & 7.4 \end{bmatrix}\begin{bmatrix} s \\ t \end{bmatrix} = \begin{bmatrix} 3.7 \\ 4 \end{bmatrix}$, $s \approx 1.157, t \approx -0.085$

31. $\begin{bmatrix} 5 & 7 \\ 4 & 2 \\ 6 & 12 \end{bmatrix}\begin{bmatrix} x \\ y \end{bmatrix} = \begin{bmatrix} 3 \\ 5 \\ 1 \end{bmatrix}$

Coefficient matrix does not have an inverse

33. $\begin{bmatrix} 1 & 2 & -1 \\ 2 & 3 & -2 \\ 3 & -1 & 5 \end{bmatrix}\begin{bmatrix} x \\ y \\ z \end{bmatrix} = \begin{bmatrix} 1 \\ 4 \\ 0 \end{bmatrix}$

$x = 2.875, y = -2, z = -2.125$

35. $\begin{bmatrix} 1 & 1 & -2 \\ 2 & 4 & -3 \\ 1 & 3 & -1 \end{bmatrix}\begin{bmatrix} x \\ y \\ z \end{bmatrix} = \begin{bmatrix} 3 \\ 4 \\ 1 \end{bmatrix}$

no inverse

37. $\begin{bmatrix} 1 & 1 & -2 \\ 2 & 4 & -3 \\ 1 & 3 & -1 \end{bmatrix}\begin{bmatrix} x \\ y \\ z \end{bmatrix} = \begin{bmatrix} 3 \\ 4 \\ 0 \end{bmatrix}$

no inverse

39. $\begin{bmatrix} 2 & 3 & -2 \\ 2 & 4 & -3 \\ 4 & -3 & 2 \end{bmatrix}\begin{bmatrix} N \\ S \\ E \end{bmatrix} = \begin{bmatrix} 300 \\ 300 \\ 300 \end{bmatrix}$, $N = S = E = 100$

41. $\begin{bmatrix} 1 & 2 & -1 & 1 \\ -2 & -3 & -1 & -1 \\ 4 & 2 & -4 & 5 \\ -3 & -3 & 2 & -7 \end{bmatrix}\begin{bmatrix} x \\ y \\ z \\ w \end{bmatrix} = \begin{bmatrix} 2 \\ 4 \\ 0 \\ 14 \end{bmatrix}$

$x = \frac{16}{35}, y = \frac{26}{35}, z = -\frac{18}{5}, w = -\frac{124}{35}$

43. $\mathbf{S} = \mathbf{AM} = \begin{bmatrix} 1 & 2 \\ 3 & 4 \end{bmatrix}\begin{bmatrix} 9 & 1 & 0 & 5 \\ 0 & 13 & 13 & 0 \end{bmatrix}$

$= \begin{bmatrix} 9 & 27 & 26 & 5 \\ 27 & 55 & 52 & 15 \end{bmatrix}$

45. $\mathbf{S} = \mathbf{AM} = \begin{bmatrix} 2 & 3 \\ -4 & 5 \end{bmatrix}\begin{bmatrix} 12 & 15 & 4 & 9 & 0 & 5 & 5 \\ 12 & 25 & 0 & 19 & 8 & 18 & 0 \end{bmatrix}$

$= \begin{bmatrix} 60 & 105 & 8 & 75 & 24 & 64 & 10 \\ 12 & 65 & -16 & 59 & 40 & 70 & -20 \end{bmatrix}$

47. $\mathbf{M} = \mathbf{A}^{-1}\mathbf{S} = \begin{bmatrix} -2 & 1 \\ 1.5 & -0.5 \end{bmatrix}\begin{bmatrix} 22 & 45 & 5 & 44 & 23 \\ 56 & 91 & 15 & 102 & 69 \end{bmatrix}$

$= \begin{bmatrix} 12 & 1 & 5 & 14 & 23 \\ 5 & 22 & 0 & 15 & 0 \end{bmatrix}$

LEAVE NOW

Creating Models

49. a. $8A + 5C = 1{,}356, A + C = 240$; $A = 52$ adult tickets, $C = 188$ children tickets

b. $8A + 5C = 1{,}477, A + C = 188$; $A = 179$ adult tickets, $C = 9$ children tickets

51. a. $i + t \geq 1200$; $10i + 15t \leq 13{,}800$

b. if an exact amount of time is needed

53. $h + m = 50{,}000$, $0.14h + 0.0825m = 0.105\,(50{,}000) = 5{,}250$; $h = \$19{,}565.22$ in high-risk investment, $y = \$30{,}434.78$ in moderate-risk investment

55. $A + B = 142$, $400{,}000 + 80{,}000A = 600{,}000 + 75{,}000B$ or $80{,}000A - 75{,}000B = 200{,}000$; $A = 70$ (buses produced at plant A, $B = 72$ (buses produced at plant B)

57. $C = 0.02\,(200{,}000 - S)$, $S = 0.08\,(200{,}000 - C)$, $F = 0.2\,(200{,}000 - S - C)$; $C = \$3{,}685.90$ city taxes, $S = \$15{,}705.13$ state taxes, $F = \$36{,}121.79$ federal taxes

59. $50d + 40r + 35e = 105{,}000$, $12d + 10r + 9e = 26{,}100$, $10d + 10r + 10e = 25{,}500$, $d = 300 + 0.5e$ (number of deluxe mattresses), $r = 2{,}250 - 1.5e$ (number of regular mattresses), e is any real number of economy mattresses less than or equal to 1,500

Reviewing Material

68. $P = 2t - 3975$

Testing Yourself

1. $x = 1, y = 1, z = -2$
2. $x = 9, y = 3, z = 2$
3. $\begin{bmatrix} 1 & 2 & -1 & | & 4 \\ 0 & 1 & 1.5 & | & -6 \\ 0 & 0 & 1 & | & -\frac{46}{9} \end{bmatrix}$
4. code years with 1980 as $x = 0$; $y = 3081.95x + 16,369$; in 1981, $x = 1$ and $y \approx 19,451$
5. a. $x = 5z, y = -3z + 4, z$ is any real number b. no solution
6. $\begin{bmatrix} 1 & 2 & -1 & | & 3 \\ 0 & 1 & \frac{3}{2} & | & -4 \\ 0 & 0 & 1 & | & -\frac{38}{17} \end{bmatrix}$
7. a. $\begin{bmatrix} 3 & 0 \\ 5 & -2 \\ 2 & 9 \end{bmatrix}$ b. not defined c. $\begin{bmatrix} 0 & -6 \\ 5 & -5 \\ -1 & 6 \end{bmatrix}$ d. not defined
   e. $\begin{bmatrix} -2 & 4 & -2 \\ 1 & 8 & -1 \\ 26 & -7 & 17 \end{bmatrix}$

8. a. not defined b. $\begin{bmatrix} 1 & -1 & -4 \\ 4 & 2 & -1 \end{bmatrix}$ c. $\begin{bmatrix} 5 & -1 & 12 \\ 8 & -2 & 7 \end{bmatrix}$
   d. not defined e. $\begin{bmatrix} -10 & -2 & -2 \\ -9 & -6 & -3 \end{bmatrix}$

9. $\begin{bmatrix} 1 & -0.2 \\ -0.5 & 0.2 \end{bmatrix}$
10. a. no solution b. $x = 0, y = 4, z = 0$
11. $x = 5, y = -6, z = 8$
12. $x = 5, y = -6, z = 8$
13. $y \approx 0.173x + 146$
    a. m and b determine the salary; m is the average percent tip (17.3%) left by customers; b is the base salary of \$146 per week.
    b. no
14. $\begin{bmatrix} 1 & -1 \\ -2 & 2.5 \end{bmatrix}$

CHAPTER 3

Section 3.1   Improving Skills

1. $x < 44$
3. $N \leq \frac{9}{2}$
5. $x \geq 12$
7. $t \leq 9$
9. $x < 0$
11. $s \geq -12.5$
13. $1 \leq x \leq 6$
15. $4 < y < 6$
17. $-6 \leq y < -1$
19. $3 \leq z \leq 8$
21. $1 \leq x < 3$
23. $4 < z < 9$
25. $N > 15$
27. $p \leq 4\frac{8}{11}$
29. $p \geq \frac{11}{26}$

Creating Models

31. $200w + 500s \geq 70,000$; w is the number of wing chairs, s is the number of sofas
33. $9s + 5l \leq 300, s \geq 10$; s is the number of sofas, l is the number of love seats
35. $60J_1 + 50J_2 \leq 1,800,000, J_1 \geq 100, J_2 \geq 200$; J_1 is the number of Type I joints, J_2 is the number of Type II joints

37. $50d + 35r \leq 17,000, 12d + 10r \leq 5,000, r \geq 2d$; d is the number of deluxe mattresses, r is the number of regular mattresses
39. $2t + \frac{1}{2}b \leq 6,000, 3t + \frac{3}{4}b \leq 6,500, b \geq 5,000$; t is the number of television cabinets, b is the number of lamp bases
41. $a + 0.05b \geq .2(a + b), a + b \geq 500$; a is the number of gallons of Drink A, b is the number of gallons of Drink B
43. $5f_1 + 10f_2 \leq 7(x + y)$; f_1 is the number of pounds of Feed I, f_2 is the number of pounds of Feed II
45. $20f_1 + 5f_2 \geq 10(f_1 + f_2)$; f_1 is the number of pounds of Feed I, f_2 is the number of pounds of Feed II
47. $5t + 10g + b \leq 10,000, 3t + 8g + 2b \leq 12,000$; t is the number of television cabinets, g is the number of grandfather clock frames, b is the number of lamp bases
49. $20f_1 + 5f_2 + 15f_3 + 3f_4 \geq 10(f_1 + f_2 + f_3 + f_4)$; f_1 is the number of pounds of Feed I, f_2 is the number of pounds of Feed II, f_3 is the number of pounds of Feed III, f_4 is the number of pounds of Feed IV

Reviewing Material

53.

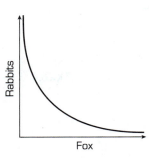

54. $s = -1,500, p = 2,000$

Section 3.2 Improving Skills

1.

3.

5.

7.

9.

11.

13.

corner point: (2, 1)

15.

corner point: (4, 0)

17.

corner point:
$(142\frac{6}{7}, 142\frac{6}{7})$

19.

corner point:
$(5, -\frac{2}{3})$

21.

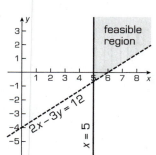

corner point:
$(7, 2)$

23.

corner points:
$(0, 3), (0, 1),$
$(6, 1)$

25.

corner points:
$(0, 0), (0, 5),$
$(4, 0)$

27.

corner points:
$(-1, -6.25), (6.4, 3)$

29.

corner points:
$(-1\frac{2}{3}, 2), (6, 2),$
$(\frac{1}{4}, -3\frac{3}{4})$

31.

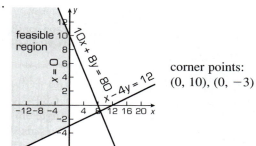

corner points:
$(0, 10), (0, -3)$

33.

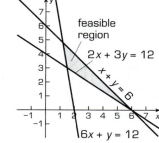

corner points:
$(6, 0), (1.2, 4.8),$
$(1.5, 3)$

35.

corner points:
$(5\frac{25}{37}, \frac{35}{37})$,
$(5, 0)$,
$(2\frac{1}{2}, -2\frac{1}{2})$

37.

corner points:
$(0, 0)$, $(0, 5)$,
$(1\frac{7}{33}, 3\frac{31}{33})$,
$(2, 0)$

Creating Models

39.

41.

43.

45.

47.

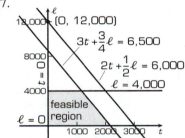

Reviewing Material

50. $-\frac{1}{4}$

51. $x = \frac{3}{5}, y = -8$

Section 3.3   Improving Skills

1. a. $x = -3, y = 5$ b. $x = 5, y = -3$
   c. $x = 5, y = -3$ d. $x = -3, y = 5$
3. a. $x = 1.316, y = 0.368$ b. $x = 0, y = 3$
   c. $x = 1.316, y = 0.368$ d $x = 5$, and $y = 0$
5. $z = 20$ when $x = 2$ and $y = 0$
7. $z = 5\frac{2}{3}$ when $x = \frac{16}{9}$ and $y = \frac{35}{9}$
9. $z = 16$ when $x = 4$ and $y = 0$
11. $z = 54$ when $x = 4$ and $y = 2$
13. $z = 200$ when $x = 10$ and $y = 0$
15. $z = 24,600$ when $x = 2600$ and $y = 8400$
17. $z = 340,000$ when $x = 25$ and $y = 20$ Superfluous constraints are $x \geq 0$ and $y \geq 0$.
19. There is no feasible region, so there is no optimal solution.

21. $z = 42,000$ when $x = 0$ and $y = 2800$
23. There is no feasible region, so there is no optimal solution.
25. $z = 1120$ when $x = 28$ and $y = 0$
27. $z = -200$ when $x = 8$ and $y = -22$
29. $z = 495$ when $x = 5$ and $y = 15$
31. $z = 13,230$ when $x = 9$ and $y = 60$ First constraint is superfluous
33. $z = 1700$ when $x = 1200$ and $y = 2000$

Creating Models

35. x is the number of TV cabinets, y is the number of lamp bases; maximize $z = 7x + 4y$ subject to: $3x + 2y \le 14,000$, $5x + y \le 14,000$, $x \ge 0$, $y \ge 0$; $z = \$30,000$ when $x = 2000$ and $y = 4000$
37. x is the number of Type I joints; y is the number of Type II joints; maximize $z = 0.03\,x + 0.02y$ subject to: $60x + 10y \le 240,000$, $2x + 5y \le 47,200$, $x \ge 0$, $y \ge 0$; $z = \$246$ when $x = 2600$ and $y = 8400$
39. x is the number of days for mine I, y is the number of days for mine II; maximize $z = 8000x + 7000y$ subject to: $4x + 3y \ge 160$, $6x + 10y \ge 350$, $4x + 5y \ge 0$, $x \ge 0$, $y \ge 0$ $x \le 31$, $y \le 31$; $z = 340,000$ when $x = 25$ and $y = 20$
41. x is the number of Sleepers, y is the number of Johnny Firms; maximize $z = 15x + 30y$ subject to: $0.4x + 0.8y \le 2,240$, $0.6x + 0.4y \le 1,440$, $x \ge 0$, $y \ge 0$; $z = \$84,000$ when $x = 0$ and $y = 2800$ or when $x = 800$ and $y = 2400$
43. x is the number of crates from the first manufacturer, y is the number of crates from the second manufacturer; maximize $z = 80x + 95y$ subject to: $60x + 40y \le 8,000$, $100x + 120y \le 12,000$, $x \ge 0$, $y \ge 0$ $z = 9,600$ when $x = 120$ and $y = 0$
45. x is the number of regional participants, y is the number of national participants; maximize $z = x + y$ subject to: $500x + 800y \le 63000$, $y \ge 2x$, (or $2x - y \le 0$), $x \ge 0$, $y \ge 0$; $z = 90$ when $x = 30$ and $y = 60$
47. x is the number of swings, y is the number of ladders; maximize $z = 2x + 3y$ subject to: $12x + 14y \le 14,952$, $x + 4y \le 1,892$, $0.03x + 0.12y \le 72$, $x \ge 0$, $y \ge 0$; $z = 2644$ for $x = 980$ and $y = 228$
49. x is the number of pounds of hamburger, y is the number of pounds of picnic patties; maximize $z = 0.4x + 0.40y$ subject to: $0.2x \le \frac{1}{2}(160)$, $0.6x + 0.5y \le \frac{1}{2}(600)$, $0.3y \le \frac{1}{2}(300)$, $x \ge 0, y \ge 0$; $z = \$233.33$ with $x = 83\frac{1}{3}$ and $y = 500$
51. x is the number of dogs, y is the number of cats; maximize $z = 80x + 50y$ subject to: $x + y \le 50$, $2x + y \le 80$, $x \ge 0$, $y \ge 0$; $z = 3400$ for $x = 30$ and $y = 20$
53. x is the number of gallons of fuel oil, y is the number of gallons of gasoline; maximize $z = 1.05x + 0.90y$ subject to: $x \le 130$, $x \ge 60$, $y \le 110$, $y \ge 50$; $z = 235.5$ for $x = 130$ and $y = 110$
55. x is the number of regular mattresses, y is the number of extra-firm mattresses, z is the number of economy mattresses; maximize $z = 40x + 60y + 35z$ subject to: $30x + 50y + 20z \le 80,000$, $10x + 5y + 14z \le 14,000$, $4x + 4y + 3z \le 7,200$, $x \ge 0$, $y \ge 0$, $z \ge 0$
57. x is the number of TV cabinets, y is the number of grandfather clock frames, z is the number of lamp bases; maximize $z = 7x + 22y + 4z$ subject to: $3x + 10y + 2z \le 30,000$, $5x + 8y + z \le 40,000$, $0.2x + 0.6y + 0.1z \le 1,600$, $x \ge 0$, $y \ge 0$, $z \ge 0$
59. x is the amount invested in the low-risk stock, y is the amount invested in the medium-risk stock, z is the amount invested in high-risk stock; maximize $z = 0.04x + 0.06y + 0.09z$ subject to: $x + y + z \le 10,000$, $x + y \ge 6000$, $-x + z \le 2000$, $x \ge 0$, $y \ge 0$, $z \ge 0$

Section 3.4   Reviewing Material

66. 182
67. $$\left[\begin{array}{ccc|c} 1 & 5 & -5 & 40 \\ 0 & 1 & 3 & -8 \\ 0 & 0 & 1 & -\frac{8}{3} \end{array}\right]$$

Section 3.4   Improving Skills

1.
$$\begin{array}{c} \\ s_1 \\ s_2 \\ \\ \end{array}\begin{array}{cccccc} x_1 & x_2 & s_1 & s_2 & z & \\ \left[\begin{array}{ccccc|c} 60 & 10 & 1 & 0 & 0 & 240,000 \\ 2 & 5 & 0 & 1 & 0 & 47,200 \\ \hline -3 & -2 & 0 & 0 & 1 & 0 \end{array}\right] \end{array}$$

3.
$$\begin{array}{c} \\ s_1 \\ s_2 \\ \\ \end{array}\begin{array}{cccccc} x_1 & x_2 & s_1 & s_2 & z & \\ \left[\begin{array}{ccccc|c} 1 & -3 & 1 & 0 & 0 & 3 \\ 2 & 4 & 0 & 1 & 0 & 18 \\ \hline -2 & -7 & 0 & 0 & 1 & 0 \end{array}\right] \end{array}$$

5.
$$\begin{array}{c} \\ s_1 \\ s_2 \\ \\ \end{array}\begin{array}{cccccc} x_1 & x_2 & s_1 & s_2 & z & \\ \left[\begin{array}{ccccc|c} 1 & 2 & 1 & 0 & 0 & 2 \\ -3 & 4 & 0 & 1 & 0 & 7 \\ \hline -2 & 1 & 0 & 0 & 1 & 0 \end{array}\right] \end{array}$$

7.
$$\begin{array}{c} \\ s_1 \\ s_2 \\ \\ \end{array}\begin{array}{ccccccc} x_1 & x_2 & x_3 & s_1 & s_2 & z & \\ \left[\begin{array}{cccccc|c} 6 & 5 & 3 & 1 & 0 & 0 & 26 \\ 4 & 2 & 5 & 0 & 1 & 0 & 8 \\ \hline -15 & -10 & -14 & 0 & 0 & 1 & 0 \end{array}\right] \end{array}$$

9.
$$\begin{array}{c} \\ s_1 \\ s_2 \\ \\ \end{array}\begin{array}{cccccccc} x_1 & x_2 & x_3 & x_4 & s_1 & s_2 & z & \\ \left[\begin{array}{ccccccc|c} 3 & -4 & 5 & -6 & 1 & 0 & 0 & 90 \\ 7 & 7 & 8 & -9 & 0 & 1 & 0 & 135 \\ \hline -5 & -6 & -7 & -8 & 0 & 0 & 1 & 0 \end{array}\right] \end{array}$$

11.

	x_1	x_2	x_3	s_1	s_2	s_3	z	
s_1	0.3	0.4	0.3	1	0	0	0	9000
s_2	0.2	0.8	0	0	1	0	0	5500
s_3	0.9	0	0.1	0	0	1	0	73000
	0	0	−1	0	0	0	1	0

13.

	x_1	x_2	x_3	s_1	s_2	s_3	s_4	s_5	z	
s_1	4	4	5	1	0	0	0	0	0	0.03
s_2	2	3	8	0	1	0	0	0	0	0.07
s_3	1	1	3	0	0	1	0	0	0	0.01
s_4	2	1	1	0	0	0	1	0	0	0.08
s_5	4	3	5	0	0	0	0	1	0	0.12
	−9	−9	−1	0	0	0	0	0	1	0

15.

	x_1	x_2	x_3	x_4	s_1	s_2	s_3	s_4	z	
s_1	2	3	3	4	1	0	0	0	0	20
s_2	3	4	4	5	0	1	0	0	0	35
s_3	5	5	4	3	0	0	1	0	0	35
s_4	3	3	2	1	0	0	0	1	0	20
	0	−1	0	−2	0	0	0	0	1	0

17. Both constraints have equality signs, and the first contraint has a negative right side.

19. x_2 can be negative.

21. The objective function is minimized.

23. First constraint has a negative right side.

25. Both contrainsts involve $\geq$ inequalities.

27. maximize: $z = 3x_1 - 4x_2 + 5x_3$ subject to: $-x_1 + 2x_2 - 2x_3 \leq 0$, $-3x_1 - 4x_2 + 4x_3 \leq 0$, assuming x_1, x_2, and x_3 are nonnegative

29. maximize $z = 5x_1 + 6x_2 + 7x_3 + 8x_4$ subject to: $-3x_1 + 4x_2 - 5x_3 + 6x_4 \leq 90$, $-7x_1 - 7x_2 + 8x_3 + 9x_4 \leq 0$, assuming x_1, x_2, x_3, and x_4 are nonnegative

Creating Models

31. x_1 is the numbe r of regional participants, x_2 is the number of national participants, x_3 is the number of international participants; maximize $z = x_1 + x_2 + x_3$ subject to: $500x_1 + 800x_2 + 900x_3 \leq 63{,}000$, $2x_1 - x_2 \leq 0$, $x_1 - x_3 \leq 0$, assuming x_1, x_2, x_3 are nonnegative

33. x_1 is the amount invested in the low-risk fund, x_2 is the amount invested in the medium-risk fund, x_3 is the amount invested in the high-risk fund; maximize $z = 0.04x_1 + 0.06x_2 + 0.09x_3$ subject to $x_1 + x_2 + x_3 \leq 10{,}000$, $x_1 - x_2 + x_3 \leq 0$, assuming x_1, x_2, x_3 are nonnegative

35. x_1 is the number of cookbook titles x_2 is the number of garden book titles x_3 is the number of self-help titles maximize $z = 8{,}500x_1 + 8{,}000x_2 + 9{,}000x_3$ subject to $x_1 + x_2 + x_3 \leq 24$, $-x_1 + x_2 + x_3 \leq 0$, assuming x_1, x_2, x_3 are nonnegative

Section 3.5 Reviewing Materials

40.

41.
$$\begin{bmatrix} 2 & 1 & 1 \\ 5 & 4 & -3 \\ 4 & 3 & -2 \end{bmatrix}$$

Section 3.5 Improving Skills

1.

	x_1	x_2	s_1	s_2	z	
s_1	−0.5	0	1	−0.5	0	4.5
x_2	0.75	1	0	0.25	0	2.75
	−0.5	0	0	1.5	1	16.5

a second iteration is necessary; 0.75 is the pivot.

3.

	x_1	x_2	s_1	s_2	z	
x_2	0.5	1	0.5	0	0	500
s_2	1	0	−2	1	0	1600
	37.5	0	17.5	0	1	17,500

optimal solution: $x_2 = 500$, $s_2 = 1600$, $x_1 = s_1 = 0$, $z = 17{,}500$,

5. The problem has no solution.

7.

	x_1	x_2	s_1	s_2	s_3	z	
s_1	0	2.32	1	−0.44	0	0	290.6
x_1	1	0.88	0	0.04	0	0	15.4
s_3	0	8.84	0	−0.28	1	0	412.2
	0	−0.72	0	1.24	0	1	477.4

a second iteration is necessary; 0.88 is the pivot.

9. Either column 1 or 2 can be taken as the first work column. Using column 1:

	x_1	x_2	s_1	s_2	s_3	s_4	z	
s_1	0	2.2	1	−0.4	0	0	0	13.08
x_1	1	0.4	0	0.2	0	0	0	3.66
s_3	0	3.8	0	−0.6	1	0	0	24.72
s_4	0	2.8	0	−0.6	0	1	0	18.62
	0	−0.6	0	0.2	0	0	1	3.66

a second iteration is necessary; 2.2 is the pivot.

11. $z = 8$; $x_1 = 0$, $x_2 = 4$, $s_1 = 0$

13. $z = \frac{51}{9} \approx 5.67$; $x_1 = \frac{16}{9} \approx 1.78$, $x_2 = \frac{35}{9} \approx 3.89$, $s_1 = s_2 = 0$

15. $z = 24{,}600$; $x_1 = 2600$, $x_2 = 8400$, $s_1 = s_2 = 0$

17. $z = 1120$; $x_1 = 28$, $x_2 = 0$, $s_1 = 0$, $s_2 = 20$

19. $z = \$75{,}520$; $x_1 = 0$, $x_2 = 9440$, $s_1 = s_2 = 0$

21. $z = 31.5$; $x_1 = 0$, $x_2 = 4.5$, $s_1 = 16.5$, $s_2 = 0$

23. $z = 4$; $x_1 = 2$, $x_2 = 0$, $s_1 = 0$, $s_2 = 13$

25. $z = 35,777\frac{7}{9} \approx 35,777.78$; $x_1 = 1022\frac{2}{9} \approx 1022.22$, $x_2 = 0$, $s_1 = 1322\frac{2}{9} \approx 1322.22$, $s_2 = 844\frac{4}{9} \approx 844.44$, $s_3 = 0$

27. $z = 287.5$; $x_1 = 0$, $x_2 = 0$, $x_3 = 57.5$, $s_1 = 122.5$, $s_2 = 0$

29. $z = 23,000$; $x_1 = 1000$, $x_2 = 800$, $x_3 = 0$, $s_1 = 10,000$, $s_2 = s_3 = 0$

31. $z = 22,750$; $x_1 = 1,050$, $x_2 = 0$, $x_3 = 500$, $s_1 = 28,500$, $s_2 = s_3 = 0$, $s_4 = 1,300$

33. $z = 52.5$; $x_1 = 0$, $x_2 = 12.5$, $x_3 = 0$, $x_4 = 2.5$, $s_1 = s_3 = 0$, $s_2 = 52.5$

Creating Models

35. x_1 is the number of regular mattresses, x_2 is the number of extra-firm mattresses; maximize $z = 40x_1 + 60x_2$ subject to: $30x_1 + 50x_2 \leq 80,000$, $10x_1 + 5x_2 \leq 14,000$, $4x_1 + 4x_2 \leq 7,200$, $x_1 \geq 0$, $x_2 \geq 0$; $z = 98,000$ when $x_1 = 500$, $x_2 = 1300$

37. x_1 is the number of Type I joints, x_2 is the number of Type II joints; maximize $z = 0.03\ x_1 + 0.02x_2$ subject to: $60x_1 + 10x_2 \leq 240,000$, $2x_1 + 5x_2 \leq 47,200$, $x_1 \geq 0$, $x_2 \geq 0$; $z = \$246$ when $x_1 = 2600$, $x_2 = 8400$

39. x_1 is the number of crates from the first manufacturer, x_2 is the number of crates from the second manufacturer; maximize $z = 80x_1 + 95x_2$ subject to: $60x_1 + 40x_2 \leq 8000$, $100x_1 + 120x_2 \leq 12000$, $x_1 \geq 0$, $x_2 \geq 0$; $z = 9600$, $x_1 = 120$, $x_2 = 0$

41. x_1 is the number of TV cabinets, x_2 is the number of grandfather clock frames, x_3 is the number of lamp bases; maximize $z = 7x_1 + 22x_2 + 4x_3$ subject to: $3x_1 + 10x_2 + x_3 \leq 30,000$, $5x_1 + 8x_2 + x_3 \leq 40,000$, $0.1x_1 + 0.6x_2 + 0.1x_3 \leq 120$, $x_1 \geq 0$, $x_2 \geq 0$; $z = 8400$, $x_1 = 1200$, $x_2 = x_3 = 0$

43. x_1 is the number of TV cabinets, x_2 is the number of grandfather clock frames, x_3 is the number of lamp bases; maximize $z = 7x_1 + 22x_2 + 4x_3$ subject to: $3x_1 + 10x_2 + x_3 \leq 30,000$, $5x_1 + 8x_2 + x_3 \leq 40,000$, $0.1x_1 + 0.6x_2 + 0.1x_3 \leq 3,000$, $x_1 \geq 0$, $x_2 \geq 0$, $x_3 \geq 0$; $z = \$120,000$ with $x_1 = x_2 = 0$, $x_3 = 30,000$

45. x_1 is the number of dogs cared for by regular plan, x_2 is the number of cats cared for by kennel, x_3 is the number of dogs cared for by economy plan; maximize $z = 80x_1 + 50x_2 + 25x_3$ subject to: $x_1 + x_2 + x_3 \leq 50$, $2x_1 + x_2 \leq 80$, $x_1 \geq 0$, $x_2 \geq 0$, $x_3 \geq 0$; $z = \$3450$ with $x_1 = 40$, $x_2 = 0$, $x_3 = 10$

47. x_1 is the number of steak knives, x_2 is the number of pocket knives, x_3 is the number of chef's knives; maximize $z = 9x_1 + 22x_2 + 50x_3$ subject to $4x_1 + 10x_2 + 5x_3 \leq 180$, $x_1 + x_2 + 2.5x_3 \leq 30$, $x_1 \geq 0$, $x_2 \geq 0$, $x_3 \geq 0$; $z = \$630$ with $x_1 = 0$, $x_2 = 15$, $x_3 = 6$

49. x_1 is the number of regional participants, x_2 is the number of national participants, x_3 is the number of international participants; maximize $z = x_1 + x_2 + x_3$ subject to $500x_1 + 800x_2 + 900x_3 \leq 63,000$, $2x_1 - x_2 \leq 0$, $x_1 - x_3 \leq 0$, $x_1 \geq 0$, $x_2 \geq 0$, $x_3 \geq 0$; $z = 84$ with $x_1 = 21$, $x_2 = 42$, $x_3 = 21$

51. x_1 is the amount invested in the 4% fund, x_2 is the amount invested in the 6% fund, x_3 is the amount invested in the 9% fund; maximize $z = 0.04x_1 + 0.06x_2 + 0.09x_3$ subject to $x_1 + x_2 + x_3 \leq 10,000$, $-x_1 - x_2 + x_3 \leq 0$, $x_3 \leq 4,000$, $x_1 \geq 0$, $x_2 \geq 0$, $x_3 \geq 0$; $z = \$720$ with $x_1 = 0$, $x_2 = 6000$, $x_3 = 4000$

Reviewing Material

64.

$$y = \frac{1}{x^2}$$

65. $y = -2.05x + 152.96$ for $x = 45$, $y = 61$

Section 3.6 Improving Skills

1. a.

	x_1	x_2	s_1	s_2	z	
s_1	-60	-10	1	0	0	$-240,000$
s_2	2	5	0	1	0	$47,200$
	-3	-2	0	0	1	0

b.

	x_1	x_2	s_1	s_2	z	
x_1	1	$\frac{1}{6}$	$-\frac{1}{60}$	0	0	$4,000$
s_2	0	$\frac{14}{3}$	$\frac{1}{30}$	1	0	$39,200$
	0	$-\frac{3}{2}$	$-\frac{1}{20}$	0	1	$12,000$

c. $z = 70,800$ with $x_1 = 23,600$, $x_2 = 0$

3. a.

	x_1	x_2	s_1	s_2	z	
s_1	-60	-10	1	0	0	$-240,000$
s_2	-2	-5	0	1	0	$-47,200$
	3	2	0	0	1	0

b.

	x_1	x_2	s_1	s_2	z	
x_1	1	$\frac{1}{6}$	$-\frac{1}{60}$	0	0	$4,000$
s_2	0	$\frac{14}{3}$	$-\frac{1}{30}$	1	0	$-39,200$
	0	$\frac{3}{2}$	$\frac{1}{20}$	0	1	$-12,000$

c. $z = 24600$ with $x_1 = 2600$, $x_2 = 8400$

5. a.

$$\begin{array}{c}\\ s_1 \\ s_2 \\ \\ \end{array}\begin{array}{c}x_1 \quad\; x_2 \quad\; s_1 \;\; s_2 \;\; z \\ \left[\begin{array}{ccccc|c} -0.4 & -0.8 & 1 & 0 & 0 & -2240 \\ -0.6 & -0.4 & 0 & 1 & 0 & -1440 \\ \hline 15 & 30 & 0 & 0 & 1 & 0 \end{array}\right]\end{array}$$

b.

$$\begin{array}{c}\\ x_2 \\ s_2 \\ \\ \end{array}\begin{array}{c}x_1 \;\; x_2 \quad\; s_1 \;\; s_2 \;\; z \\ \left[\begin{array}{ccccc|c} 0.5 & 1 & -1.25 & 0 & 0 & 2{,}800 \\ -0.4 & 0 & -0.5 & 1 & 0 & -320 \\ \hline 0 & 0 & 37.5 & 0 & 1 & -84{,}000 \end{array}\right]\end{array}$$

z = 84,000 with $x_1 = 800$, $x_2 = 2400$

7. a.

$$\begin{array}{c}\\ s_1 \\ s_2 \\ s_3 \\ \\ \end{array}\begin{array}{c}x_1 \quad\; x_2 \quad s_1 \;\; s_2 \;\; s_3 \;\; z \\ \left[\begin{array}{cccccc|c} 0.4 & 0.8 & 1 & 0 & 0 & 0 & 2240 \\ -0.4 & -0.8 & 0 & 1 & 0 & 0 & -2240 \\ 0.6 & 0.4 & 0 & 0 & 1 & 0 & 1440 \\ \hline 15 & 30 & 0 & 0 & 0 & 1 & 0 \end{array}\right]\end{array}$$

b.

$$\begin{array}{c}\\ s_1 \\ x_2 \\ s_3 \\ \\ \end{array}\begin{array}{c}x_1 \;\; x_2 \;\; s_1 \quad\;\; s_2 \quad s_3 \;\; z \\ \left[\begin{array}{cccccc|c} 0 & 0 & 1 & 1 & 0 & 0 & 0 \\ 0.5 & 1 & 0 & -1.25 & 0 & 0 & 2800 \\ 0.4 & 0 & 0 & 0.5 & 1 & 0 & 320 \\ \hline 0 & 0 & 0 & 37.5 & 0 & 1 & -84000 \end{array}\right]\end{array}$$

c. z = 84,000 with $x_1 = 0$, $x_2 = 2800$

9. a.

$$\begin{array}{c}\\ s_1 \\ s_2 \\ s_3 \\ \\ \end{array}\begin{array}{c}x_1 \;\; x_2 \;\; s_1 \;\; s_2 \;\; s_3 \;\; z \\ \left[\begin{array}{cccccc|c} -1 & 3 & 1 & 0 & 0 & 0 & -3 \\ 1 & -3 & 0 & 1 & 0 & 0 & 3 \\ 2 & 4 & 0 & 0 & 1 & 0 & 18 \\ \hline -2 & -7 & 0 & 0 & 0 & 1 & 0 \end{array}\right]\end{array}$$

b.

$$\begin{array}{c}\\ x_1 \\ s_2 \\ s_3 \\ \\ \end{array}\begin{array}{c}x_1 \;\; x_2 \;\; s_1 \;\; s_2 \;\; s_3 \;\; z \\ \left[\begin{array}{cccccc|c} 1 & -3 & -1 & 0 & 0 & 0 & 3 \\ 0 & 0 & 1 & 1 & 0 & 0 & 0 \\ 0 & 10 & 2 & 0 & 1 & 0 & 12 \\ \hline 0 & -13 & -2 & 0 & 0 & 1 & 6 \end{array}\right]\end{array}$$

c. z = 21.6 with $x_1 = 6.6$, $x_2 = 1.2$

11. a.

$$\begin{array}{c}\\ s_1 \\ s_2 \\ s_3 \\ \\ \end{array}\begin{array}{c}x_1 \;\; x_2 \;\; s_1 \;\; s_2 \;\; s_3 \;\; z \\ \left[\begin{array}{cccccc|c} -1 & 3 & 1 & 0 & 0 & 0 & -3 \\ 1 & -3 & 0 & 1 & 0 & 0 & 3 \\ -2 & -4 & 0 & 0 & 1 & 0 & -18 \\ \hline -2 & -7 & 0 & 0 & 0 & 1 & 0 \end{array}\right]\end{array}$$

b.

$$\begin{array}{c}\\ s_1 \\ s_2 \\ x_2 \\ \\ \end{array}\begin{array}{c}x_1 \quad\; x_2 \;\; s_1 \;\; s_2 \quad\; s_3 \;\; z \\ \left[\begin{array}{cccccc|c} -2.5 & 0 & 1 & 0 & 0.75 & 0 & -16.5 \\ 2.5 & 0 & 0 & 1 & -0.75 & 0 & 16.5 \\ 0.5 & 1 & 0 & 0 & -0.25 & 0 & 4.5 \\ \hline 1.5 & 0 & 0 & 0 & -1.75 & 1 & 31.5 \end{array}\right]\end{array}$$

c. There is no optimal solution.

13. a.

$$\begin{array}{c}\\ s_1 \\ s_2 \\ \\ \end{array}\begin{array}{c}x_1 \quad\; x_2 \quad s_1 \;\; s_2 \;\; z \\ \left[\begin{array}{ccccc|c} -50 & -27 & 1 & 0 & 0 & -170 \\ -8 & -25 & 0 & 1 & 0 & -80 \\ \hline 180 & 0 & 0 & 0 & 1 & 0 \end{array}\right]\end{array}$$

b.

$$\begin{array}{c}\\ x_1 \\ s_2 \\ \\ \end{array}\begin{array}{c}x_1 \quad\;\; x_2 \quad\;\; s_1 \;\; s_2 \;\; z \\ \left[\begin{array}{ccccc|c} 1 & 0.54 & -0.02 & 0 & 0 & 3.4 \\ 0 & -20.68 & -0.16 & 1 & 0 & -52.8 \\ \hline 0 & -97.2 & 3.6 & 0 & 1 & 612 \end{array}\right]\end{array}$$

c. z = 0 with $x_1 = 0$, $x_2 = 6.2963$

15. a.

$$\begin{array}{c}\\ s_1 \\ s_2 \\ \\ \end{array}\begin{array}{c}x_1 \;\; x_2 \;\; s_1 \;\; s_2 \;\; z \\ \left[\begin{array}{ccccc|c} -1 & 2 & 1 & 0 & 0 & 2 \\ 3 & -4 & 0 & 1 & 0 & 7 \\ \hline -2 & 1 & 0 & 0 & 1 & 0 \end{array}\right]\end{array}$$

b. No preprocessing step required.

c. z = 15.5 with $x_1 = 11$, $x_2 = 6.5$

17. a.

$$\begin{array}{c}\\ s_1 \\ s_2 \\ s_3 \\ s_4 \\ \\ \end{array}\begin{array}{c}x_1 \quad x_2 \quad x_3 \;\; s_1 \;\; s_2 \;\; s_3 \;\; s_4 \;\; z \\ \left[\begin{array}{cccccccc|c} -6 & -5 & -3 & 1 & 0 & 0 & 0 & 0 & -26 \\ 6 & 5 & 3 & 0 & 1 & 0 & 0 & 0 & 26 \\ -4 & 2 & 5 & 0 & 0 & 1 & 0 & 0 & 8 \\ 4 & -2 & -5 & 0 & 0 & 0 & 1 & 0 & -8 \\ \hline -15 & -10 & -14 & 0 & 0 & 0 & 0 & 1 & 0 \end{array}\right]\end{array}$$

b.

$$\begin{array}{c}\\ x_1 \\ s_2 \\ s_3 \\ s_4 \\ \\ \end{array}\begin{array}{c}x_1 \;\; x_2 \;\; x_3 \;\; s_1 \;\; s_2\, s_3\, s_4\, z \\ \left[\begin{array}{cccccccc|c} 1 & \frac{5}{6} & \frac{1}{2} & -\frac{1}{6} & 0 & 0 & 0 & 0 & \frac{13}{3} \\ 0 & 0 & 0 & 1 & 1 & 0 & 0 & 0 & 0 \\ 0 & \frac{16}{3} & 7 & -\frac{2}{3} & 0 & 1 & 0 & 0 & \frac{76}{3} \\ 0 & -\frac{16}{3} & -7 & \frac{2}{3} & 0 & 0 & 1 & 0 & -\frac{76}{3} \\ \hline 0 & \frac{5}{2} & -\frac{13}{2} & -\frac{5}{2} & 0 & 0 & 0 & 1 & 65 \end{array}\right]\end{array}$$

c. z ≈ 88.52 with $x_1 \approx 2.52$, $x_2 = 0$, $x_3 \approx 3.62$

19. a.

$$\begin{array}{c}\\ s_1 \\ s_2 \\ \\ \end{array}\begin{array}{c}x_1 \quad x_2 \quad x_3 \;\; s_1 \;\; s_2 \;\; z \\ \left[\begin{array}{cccccc|c} -6 & -5 & -3 & 1 & 0 & 0 & -26 \\ -4 & -2 & -5 & 0 & 1 & 0 & -8 \\ \hline 15 & 10 & 14 & 0 & 0 & 1 & 0 \end{array}\right]\end{array}$$

b.

$$\begin{array}{c}\\ x_1 \\ s_2 \\ \\ \end{array}\begin{array}{c}x_1 \;\; x_2 \;\; x_3 \;\; s_1 \;\; s_2 \;\; z \\ \left[\begin{array}{cccccc|c} 1 & \frac{5}{6} & \frac{1}{2} & -\frac{1}{6} & 0 & 0 & \frac{13}{3} \\ 0 & \frac{4}{3} & -3 & -\frac{2}{3} & 1 & 0 & \frac{28}{3} \\ \hline 0 & -\frac{5}{2} & \frac{13}{2} & \frac{5}{2} & 0 & 1 & -65 \end{array}\right]\end{array}$$

c. z = 52 with $x_1 = 0$, $x_2 = 5.2$, $x_3 = 0$

21. a.

$$\begin{array}{c}\\ s_1 \\ s_2 \\ \\ \end{array}\begin{array}{c}x_1 \;\; x_2 \;\; x_3 \;\; x_4 \;\; s_1 \;\; s_2 \;\; z \\ \left[\begin{array}{ccccccc|c} 3 & -4 & 5 & -6 & 1 & 0 & 0 & 90 \\ -7 & -7 & -8 & 9 & 0 & 1 & 0 & 135 \\ \hline -5 & -6 & -7 & -8 & 0 & 0 & 1 & 0 \end{array}\right]\end{array}$$

c. There is no optimal solution.

23. a.

	x_1	x_2	x_3	x_4	s_1	s_2	z	
s_1	-3	4	-5	6	1	0	0	-90
s_2	-7	-7	-8	9	0	1	0	-135
	5	6	7	8	0	0	1	0

b.

	x_1	x_2	x_3	x_4	s_1	s_2	z	
s_1	1.375	8.375	0	0.375	1	-0.625	0	-5.625
x_3	0.875	0.875	1	-1.125	0	-0.125	0	16.875
	-1.125	-0.125	0	15.875	0	0.875	1	-118.125

c. $z = 126$ with $x_1 = 0$, $x_2 = 0$, $x_3 = 18$, $x_4 = 0$

25. a.

	x_1	x_2	x_3	s_1	s_2	s_3	s_4	z	
s_1	0.3	0.4	0.3	1	0	0	0	0	$9{,}000$
s_2	-0.3	-0.4	-0.3	0	1	0	0	0	$-9{,}000$
s_3	-0.2	0.8	0	0	0	1	0	0	$5{,}500$
s_4	0.9	0	0.1	0	0	0	1	0	$7{,}300$
	0	0	1	0	0	0	0	1	0

b.

	x_1	x_2	x_3	s_1	s_2	s_3	s_4	z	
s_1	0	0	0	1	1	0	0	0	0
x_2	0.75	1	0.75	0	-2.5	0	0	0	$22{,}500$
s_3	-0.8	0	-0.6	0	2	1	0	0	$-12{,}500$
s_4	0.9	0	0.1	0	0	0	1	0	$7{,}300$
	0	0	1	0	0	0	0	1	0

c. $z \approx 11{,}760.87$ with $x_1 \approx 6804.35$, $x_2 \approx 8576.09$, $x_3 \approx 11{,}760.87$, $x_4 = 0$

27. a.

	x_1	x_2	x_3	s_1	s_2	s_3	s_4	s_5	s_6	z	
s_1	-4	-4	-5	1	0	0	0	0	0	0	-0.03
s_2	-2	3	8	0	1	0	0	0	0	0	0.07
s_3	-1	-1	-3	0	0	1	0	0	0	0	-0.01
s_4	2	1	1	0	0	0	1	0	0	0	0.08
s_5	-2	-1	-1	0	0	0	0	1	0	0	-0.08
s_6	4	3	5	0	0	0	0	0	1	0	0.12
	9	9	1	0	0	0	0	0	0	1	0

b.

	x_1	x_2	x_3	s_1	s_2	s_3	s_4	s_5	s_6	z	
s_1	0	-2	-3	1	0	0	0	-2	0	0	0.13
s_2	0	4	9	0	1	0	0	-1	0	0	0.15
s_3	0	-0.5	-2.5	0	0	1	0	-0.5	0	0	0.03
s_4	0	0	0	0	0	0	1	1	0	0	0
x_1	1	0.5	0.5	0	0	0	0	-0.5	0	0	0.04
s_6	0	1	3	0	0	0	0	2	1	0	-0.04
	0	4.5	-3.5	0	0	0	0	4.5	0	1	-0.36

c. There is no optimal solution.

29. a.

	x_1	x_2	x_3	x_4	s_1	s_2	s_3	s_4	s_5	s_6	z	
s_1	2	3	3	4	1	0	0	0	0	0	0	20
s_2	-2	-3	-3	-4	0	1	0	0	0	0	0	-20
s_3	3	4	4	5	0	0	1	0	0	0	0	35
s_4	-3	-4	-4	-5	0	0	0	1	0	0	0	-35
s_5	5	5	4	3	0	0	0	0	1	0	0	35
s_6	3	3	2	1	0	0	0	0	0	1	0	20
	0	-1	0	-2	0	0	0	0	0	0	1	0

b.

	x_1	x_2	x_3	x_4	s_1	s_2	s_3	s_4	s_5	s_6	z	
x_1	-0.4	-0.2	-0.2	0	1	0	0	0.8	0	0	0	-8
s_2	0.4	0.2	0.2	0	0	1	0	-0.8	0	0	0	8
s_3	0	0	0	0	0	0	1	1	0	0	0	0
s_4	0.6	0.8	0.8	1	0	0	0	-0.2	0	0	0	7
s_5	3.2	2.6	1.6	0	0	0	0	0.6	1	0	0	14
s_6	2.4	2.2	1.2	0	0	0	0	0.2	0	1	0	13
	1.2	0.6	1.6	0	0	0	0	-0.4	0	0	1	14

c. There is no optimal solution.

Creating Models

31. x_1 is the number of days for mine I, x_2 is the number of days for mine II; $z = 340{,}000$ with $x_1 = 25$, $x_2 = 20$

33. x_1 is the number of days for mine I, x_2 is the number of days for mine II; x_3 is the number of days for mine III; $z = 340{,}000$ with $x_1 = 25$, $x_2 = 20$, $x_3 = 0$

35. x_1 is the amount of supplement I, x_2 is the amount of supplement II, x_3 is the amount of supplement III; $z = \$0.33$ with $x_1 = 1$, $x_2 = 2$, $x_3 = 2$

37. x_1 is the number of pounds of hamburger, x_2 is the number of pounds of picnic patties; $z = \$233.33$ with $x_1 = 83\frac{1}{3}$, $x_2 = 500$

39. x_1 is the number of compacts, x_2 is the number of mid-size cars; $z = 50{,}000$ with $x_1 = 25$, $x_2 = 0$

41. x_1 is the number of cookbooks, x_2 is the number of garden book titles, x_3 is the number of self-help titles; $z = \$206{,}000$ with $x_1 = 12$, $x_2 = 4$, $x_3 = 8$

43. x_1 is the number of cars shipped from A to C, x_2 is the number of cars shipped from A to D, x_2 is the number of cars shipped from B to C, x_4 is the number of cars shipped from B to D; $z = 724$ with $x_1 = 15$, $x_2 = 0$, $x_3 = 3$, $x_4 = 19$

Reviewing Material

50. 2160 chairs

51. $\begin{bmatrix} 4 & -6 \\ 5 & -5 \\ 2 & 2 \end{bmatrix}$

Testing Yourself

1.

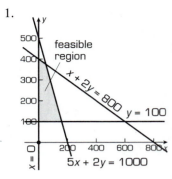

feasible region

$x + 2y = 800$

$y = 100$

$5x + 2y = 1000$

$x = 0$

2. x_1 is the number of standard tents, x_2 is the number of deluxe tents; max $z = 5x_1 + 7x_2$ subject to $3x_1 + 4x_2 \le 9600$, $2x_1 + 3x_2 \le 4000$, with $x_1 \ge 0, x_2 \ge 0$

3. $(0, 60)$ maximizes $z = 10x + 23y$ for a value of $1{,}380$

4.

	x_1	x_2	s_1	s_2	z	
x_2	0	1	0.2	1	0	2
x_1	1	0	-0.4	1	0	8
	0	0	.8	1	1	12

$x_1 = 8, x_2 = 2, s_1 = 0, s_2 = 0$,
$z = 12$; this solution is optimal.

5.

	x_1	x_2	x_3	x_4	s_1	s_2	z	
s_1	3	-4	5	-6	1	0	0	90
s_2	7	7	8	-9	0	1	0	135
	-5	-6	-7	-8	0	0	1	0

6.

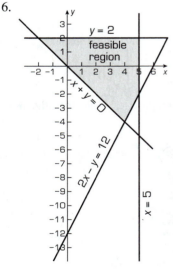

$y = 2$

feasible region

$x + y = 0$

$2x - y = 12$

$x = 5$

7.

	x_1	x_2	x_3	s_1	s_2	s_3	z	
s_1	1	-2	1	1	0	0	0	12
s_2	-1	2	-1	0	1	0	0	-12
s_3	3	1	-2	0	0	1	0	-5
	2	-3	-1	0	0	0	1	0

8. Either $(0, 4)$ or $(3, 6)$ minimizes $z = 100x - 150y$ for a value of -600.

9.

	x_1	x_2	x_3	s_1	s_2	z	
x_2	0.5	1	0.5	-0.5	0	0	50
s_2	-2	0	2	4	1	0	100
	2.5	0	-0.5	2.5	0	1	650

$x_1 = 0, x_2 = 50, s_1 = 0, s_2 = 100, z = 650$; this solution is not optimal.

10. x_1 is the number of model SE, x_2 is the number of model LE; min $z = 2600x_1 + 2300x_2$ subject to $x_1 + x_2 \ge 9$, $16{,}000x_1 + 20{,}000x_2 \ge 160{,}000$, $x_2 \ge 2$, $x_1 \le 8$, with $x_1 \ge 0$

11.

	x_1	x_2	s_1	s_2	z	
x_2	-0.5	1	-0.5	0	0	500
s_2	3	0	0	1	0	3,600
	2.5	0	-17.5	0	1	17,500

a. $x_1 = 0, x_2 = 500, s_1 = 0, x_2 = 3{,}600, z = 17{,}500$
b. The solution is feasible.
c. The optimal solution does not exist.

12.

	x_1	x_2	x_3	s_1	s_2	s_3	z	
s_1	0	-1	4	1	1	0	0	-30
x_1	1	-1	-3	0	-1	0	0	20
s_3	0	4	5	0	2	1	0	-45
	0	35	45	0	10	0	1	-200

a. $x_1 = 20, x_2 = 0, x_3 = 0, s_1 = -30, s_2 = 0, s_3 = -45$, $z = -200$
b. The solution is not feasible.
c. There is no solution. There are no negative elements in the third row except in last column.

13. x_1 is the number of cartons of regular soda, x_2 is the number of cartons of diet soda; max $z = 0.2x_1 + 0.18x_2$ subject to $1.10x_1 + 1.20x_2 \le 5400$, $x_1 + x_2 \le 6000$, with $x_1 \ge 0, x_2 \ge 0$

14.

	x_1	x_2	x_3	s_1	s_2	s_3	z	
s_1	4	-4	8	1	0	0	0	1225
s_2	1	2	1	0	1	0	0	850
s_3	3	0	2	0	0	1	0	1100
	-3	-2	-5	0	0	0	1	0

15. x_1 is the number of days the Maryland plant should operate, x_2 is the number of days the Utah plant should operate; min $z = x_1 + x_2$ subject to $x_1 - x_2 \ge 0$, $3000x_1 + 4000x_2 \ge 30{,}000$, $4000x_1 + 3000x_2 \ge 24{,}000$, with $x_1 \ge 0, x_2 \ge 0$

16.

	x_1	x_2	x_3	s_1	s_2	s_3	s_4	z	
s_1	-2	-2	1	1	0	0	0	0	-100
s_2	1	-3	1	0	1	0	0	0	500
s_3	3	2	-1	0	0	1	0	0	350
s_4	-3	-2	1	0	0	0	1	0	-350
	6	3	-5	0	0	0	0	1	0

CHAPTER 4

Section 4.1   Improving Skills

1. a. $240 b. $1,240
3. a. $45 b. $1,045
5. a. $6,000 b. $31,000
7. a. $8,437.50 b. $33,437.50
9. a. $5,460 b. $12,285
11. $10,000
13. a. $1,806.11 b. $806.11
15. a. $1,819.40 b. $819.40
31.

17. a. $8,705.12 b. $3,705.12
19. a. $4,005.43 b. $605.43
21. a. $4,009.59 b. $609.59
23. a. $6,135.81 b. $1,385.81
25. a. $24,063.80 b. $4,063.80
27. a. $809.00 b. $129.00
29. a. $2,786.35 b. $561.35

The interest earned on $7,500 invested at 10% for 12 years			
TYPE OF COMPOUNDING	MATURITY	FORMULA FOR VALUE AT MATURITY	VALUE AT MATURITY
none	12 years	$A = 7{,}500\,(1 + 0.10(12))$	$16,500.00
annual	12 periods	$P_{12} = 7{,}500\,(1 + 0.10)^{12}$	$24,538.21
semiannual	$2 \times 12 = 24$ periods	$P_{24} = 7{,}000\left[1 + \dfrac{0.10}{2}\right]^{24}$	$24,188.25
quarterly	$4 \times 12 = 48$ periods	$P_{48} = 7{,}500\left[1 + \dfrac{0.10}{4}\right]^{48}$	$24,536.17
monthly	$12 \times 12 = 144$ periods	$P_{144} = 7{,}500\left[1 + \dfrac{0.10}{12}\right]^{144}$	$24,777.37
weekly	$52 \times 12 = 624$ periods	$P_{624} = 7{,}500\left[1 + \dfrac{0.10}{52}\right]^{624}$	$24,872.20
daily	$365 \times 12 = 4380$ periods	$P_{4380} = 7{,}500\left[1 + \dfrac{0.10}{365}\right]^{4380}$	$24,896.78

33. a. 12.55% b. 12.68% c. 12.73% d. 12.75%

Creating Models

35. $200
37. $193.75
39. $130.21
41. $27,455.71
43. $2,610.17
45. $6,910.86
47. $28,594.88
49. Take $10,000 now
51. 6.10% compounded semiannually
53. 5% a year interest, compounded quarterly
55. $27,901.32
57. $311,545.48
59. $525,869.55

Exploring with Technology

65. The compound interest graph rises faster than the simple interest.

67.

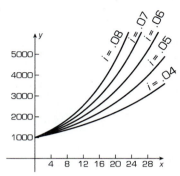

Reviewing Material

69. a. 870 ft by 1305 ft

   b. Yes, the areas are directly proportional with a constant of proportionality of 1 inch to $(87)^2$ inches, that is, 1 inch to 7569 inches.

70. $z = -2, x = 1, y = 1$

71.

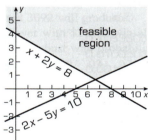

Section 4.2   Improving Skills

1. $8,741.35
3. $8,083.50
5. $1,134.60
7. $717.47
9. $12,264.48
11. $1,788.49
13. $2,560.48
15. $16,224.63
17. $1,945.64
19. $1,179.26

21. option 2
23. option 2
25. option 2
27. option 1
29. option 1
31. 8.14%
33. 10.26%
35. 8.61%
37. 14.11%
39. 24.57%

Creating Models

41. $6,131.14
43. $42,618.46
45. No, the advisor is not correct, because her money would yield $28,640.89 in a bank account.
47. buyer A
49. buyer A
51. the second opportunity
53. paying after 10 years

Reviewing Material

60. $y = -\frac{1}{10}x + 26$
61. $a = -17, b = 9.5$
62. $z = 12$ with either $x = 0$ and $y = 3$ or $x = 2\frac{2}{11}$ and $y = 1\frac{4}{11}$

Section 4.3   Improving Skills

1. a. $5,426.79   b. $5,508.19
3. a. $19,377.20 b. $19,764.74
5. a. $7,059.69   b. $7,065.60
7. a. $50,353.35 b. $51,486.30
9. a. $29,943.48 b. $30,916.64
11. a. $40,883.80 b. $42,825.78
13. a. $24,015.59 b. $24,040.53
15. a. $30,082.21 b. $30,608.65
17. $143.60   19. $108.35   21. $4,120.57
23. $20,071.04   25. $59.39

Creating Models

27. $532.96
29. $39,012.62
31. $66,298.98
33. $816.12
35. 56,641.61
37. $34.09
39. $101.19
41. $75.64
43. $67,831.82
45. $5,789.11
47. $3,800.42
49. $195.27

Reviewing Material

56.

57. Code years with 1990 as $x = 0$; $y = -1.26x + 11.5$; for 1993 $x = 3$ and the trade deficit is 7.72 billions dollars.

58. $z = 100,000$ with either $x_1 = 0$ and $x_2 = 4000$ or $x_1 = \frac{10000}{7}$ and $x_2 = \frac{24,000}{7}$

Section 4.4 Improving Skills

1. $2,991.58
3. $3,974.45
5. $6,471.69
7. $33,735.62
9. $9,468.04
11. $7,691.41
13. $10,688.02

15. $3,751.28
17. $193.60
19. $187.88
21. $729.79
23. $1,005.78
25. $1,375.51

27.

Amortization Schedule for a $1000 Loan for 6 Months at 12%

PAYMENT NUMBER	PAYMENT AMOUNT	PAYMENT ON INTEREST	PAYMENT ON DEBT	OUTSTANDING DEBT
0	$1000.00			
1	$172.55	$10.00	$162.55	$837.45
2	172.55	8.37	164.18	673.27
3	172.55	6.73	165.82	507.45
4	172.55	5.07	167.48	339.97
5	172.55	3.40	169.15	170.82
6	172.53	1.71	170.82	0

29.

Amortization Schedule for a $4000 Loan for 3 Years at 10%

PAYMENT NUMBER	PAYMENT AMOUNT	PAYMENT ON INTEREST	PAYMENT ON DEBT	OUTSTANDING DEBT
0				$4,000.00
1	$129.07	$33.33	$95.74	3,904.26
2	129.07	32.54	96.53	3,807.73
3	129.07	31.73	97.34	3,710.39

$646.52 interest

31.

Amortization Schedule for a $30,000 Loan for 10 Years at 8%

PAYMENT NUMBER	PAYMENT AMOUNT	PAYMENT ON INTEREST	PAYMENT ON DEBT	OUTSTANDING DEBT
0				$30,000.00
1	$363.98	$200.00	$163.98	29,836.02
2	363.98	198.91	165.07	29,670.95
3	363.98	197.81	166.17	29,504.78

$13,677.60 interest

33.

| \multicolumn{5}{l}{**Amortization Schedule for a $150,000 Loan for 25 Years at 7%**} | | | | |
PAYMENT NUMBER	PAYMENT AMOUNT	PAYMENT ON INTEREST	PAYMENT ON DEBT	OUTSTANDING DEBT
0				$150,000.00
1	$1,060.17	$875.00	$185.17	149,814.83
2	1,060.17	873.92	186.25	149,628.58
3	1060.17	872.83	187.34	149,441.24

$168,050.64 interest

Creating Models

35. $R = \$277.22$, retire loan with $8,978.18
37. $R = \$141.01$, retire loan with $2,133.27
39. $R = \$1,502.53$, retire loan with $164,532.05
41. $355.08
43. $11,941.18
45. $216,283.49
47. $49,090.74
49. The more attractive opportunity is the first one—invest $20,000 and receive $2,300 at the end of each quarter for the next three years.

Reviewing Material

62.

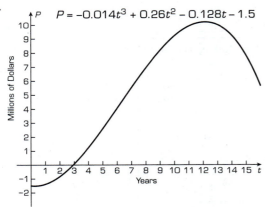

$P = -0.014t^3 + 0.26t^2 - 0.128t - 1.5$

$t \approx 7.5$ when $P = 6\frac{1}{2}$ million.

63. There is no solution to this system.
64. $z = 100,000$ when $x_1 = \frac{30000}{7}$ and $x_2 = \frac{16000}{7}$ or when $x_1 = 8000$ and $x_2 = 800$

Testing Yourself

1. $11,239.86
2. $7,762.85
3. $10,409.69
4. a. $3,362.67
   b. $59.48
5. 9.06%

6.

| \multicolumn{5}{l}{**Amortization Table for a $50,000 Loan Over 20 Years at $8\frac{1}{4}$% a Year**} | | | | |
PAYMENT NUMBER	PAYMENT AMOUNT	PAYMENT ON INTEREST	PAYMENT ON DEBT	OUTSTANDING DEBT
0				$50,000.00
1	$426.03	$343.75	$82.28	49,917.72
2	426.03	343.18	82.85	49,834.87

7. The first buyer's offer with present value $9,157.30 is better than the second offer with present value $8,823.03
8. $56,928.26
9. a. $132,991.86
   b. $300.77

10. $2,161.40
11. $2,612.01
12. $5,272.25

13.

Amortization Schedule for a $2,000 Loan Over 3 Years at 6%

PAYMENT NUMBER	PAYMENT AMOUNT	PAYMENT ON INTEREST	PAYMENT ON DEBT	OUTSTANDING DEBT
0				$2,000.00
1	$60.84	$10.00	$50.84	1,949.16
2	60.84	9.75	51.09	1898.07

14. 6.17%

CHAPTER 5

Section 5.1   Improving Skills

1. true 3. false 5. true
7. false 9. true 11. false
13. true 15. false
17. φ, {1}, {2}, {3}, {1, 2}, {1, 3}, {2, 3}, {1, 2, 3}
19. a. {1, 2, 3, 5, 6} b. {2, 5, 6} c. {3, 4} d. {1, 4}
    e. {3} f. {4} g. {1, 3, 4} h. {2, 3, 5, 6}
21. a. {Nashville, Tampa} b. φ c. {Cleveland}
    d. {Buffalo, Richmond, Cleveland, Nashville, Tampa}
    e. {Cleveland, Buffalo, Nashville, Tampa}
    f. {Buffalo, Nashville, Tampa}
23. a. B' b. $A \cap B$ c. $A \cap C$ d. $B \cap C$ e. $B' \cap C$
    f. $A' \cap B$ g. C' h. $A \cap C'$

Creating Models

25. a. 13.9 b. 112.7 c. 70.9 d. 169.7 e. 8.5 f. 104.2
27. a. 54.3 b. 6.7 c. 6.8 d. 11.3 e. 15.8 f. 14.2 g. 71.9 h. 5.2
29. 40
31.

a. 7 b. 13

33.

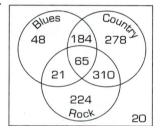

a. 48 b. 249 c. 184 d. 20

35.

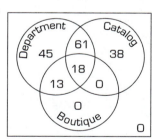

a. 31 b. 0 c. 18 d. 0

Reviewing Material

41. 625 miles
42. There is no solution to this system.
43. $z = 800$ with $x_1 = 100$, $x_2 = 0$, $x_3 = 200$
44. $5,013.71

Section 5.2 Creating Models

1. a.

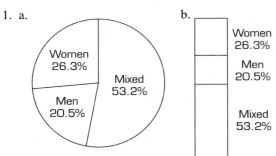

b.

Women
26.3%

Men
20.5%

Mixed
53.2%

3. a.

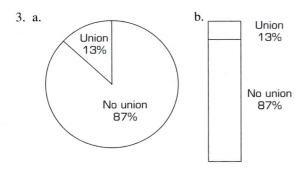

b.

Union
13%

No union
87%

5. a.

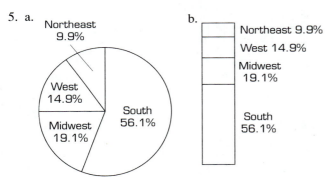

b.

Northeast 9.9%

West 14.9%

Midwest
19.1%

South
56.1%

7. a.

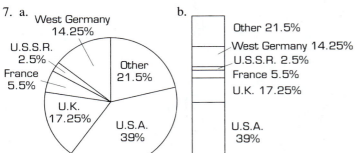

b.

Other 21.5%

West Germany 14.25%

U.S.S.R. 2.5%

France 5.5%

U.K. 17.25%

U.S.A.
39%

9. a.

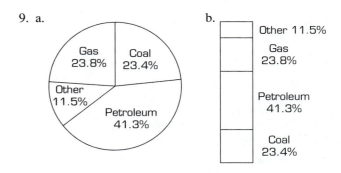

b.

Other 11.5%

Gas
23.8%

Petroleum
41.3%

Coal
23.4%

11. a.

b.

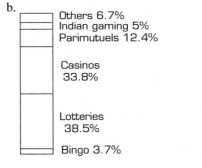

Others 6.7%
Indian gaming 5%
Parimutuels 12.4%

Casinos
33.8%

Lotteries
38.5%

Bingo 3.7%

13. a.

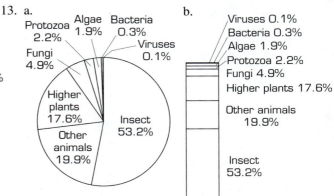

b.

Viruses 0.1%
Bacteria 0.3%
Algae 1.9%
Protozoa 2.2%
Fungi 4.9%
Higher plants 17.6%

Other animals
19.9%

Insect
53.2%

15. a.

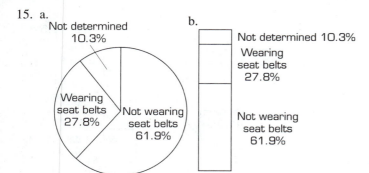

b.

Not determined 10.3%

Wearing seat belts 27.8%

Not wearing seat belts 61.9%

17. a.

b.

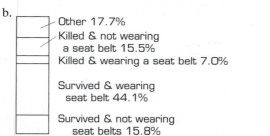

Other 17.7%

Killed & not wearing a seat belt 15.5%

Killed & wearing a seat belt 7.0%

Survived & wearing seat belt 44.1%

Survived & not wearing seat belts 15.8%

19.

21.

23.

25.

27.

29.

31.

33.

35.

37.

39.

41.

Reviewing Material

45.

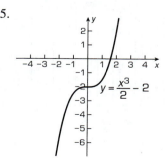

46. code years with 1986.5 as $x = 0$; $y = 2.39 - 0.08667x$, for 1995, $x = 8.5$ and $y \approx 1.7$

47. $z = 1254\frac{6}{11}$ with $x_1 = 190\frac{10}{11}$ and $x_2 = 290\frac{10}{11}$

48. 7.46%

Section 5.3   Improving Skills

1. a. 12 b. 10 c. 5 d. 10.3636 e. 1.6667
3. a. 9 and 12 b. 10 c. 5 d. 10.2727 e. 1.7104
5. a. 9 b. 9 c. 4 d. 9.7778 e. 1.4741
7. a. no mode b. 5 c. 8 d. 5 e. 2.8284
9. a. no mode b. 7 c. 10 d. 7 e. 3.4157

11. a. 2 b. 2 c. 898 d. 101.7778 e. 282.21

13. a. no mode b. 1 c. 0 d. 1 e. 0

15. a. at least $\frac{3}{4}$ b. at least $\frac{8}{9}$ c. at least $\frac{15}{16}$ d. at least $\frac{24}{25}$
    e. at most $\frac{1}{4}$ f. at most $\frac{1}{9}$

Creating Models

17. mean ≈ 19.28   standard deviation ≈ 4.01

19. mean ≈ 184.46   standard deviation ≈ 34.07

21. mean ≈ 279.5   standard deviation ≈ 69.20

23. mean ≈ 60.46   standard deviation ≈ 4.64

Reviewing Material

29. $y = -\frac{5}{4}x - \frac{30}{4}$

30.
$$\mathbf{A}^2 = \begin{bmatrix} 1 & 1 & 3 \\ 2 & 1 & 1 \\ 1 & 3 & 4 \end{bmatrix} \quad \mathbf{A}^3 = \begin{bmatrix} 2 & 4 & 7 \\ 3 & 2 & 4 \\ 4 & 7 & 9 \end{bmatrix}$$

31. 20 skirts and 4 trousers

32. $5,794.93

Section 5.4   Creating Models

1.

3.

5.

7.

9.

11.

13.

15.

17. 350
19. 1,575
21. 252
23. 21,060
25. 24,024
27. 129,600
29. 15,120
31. 10,000

Reviewing Material

36. 80 adult females, 40 adult males
37. Using elementary row operations, the matrix reduces to
$$\begin{bmatrix} 1 & 2 & 3 \\ 0 & 1 & 1 \\ 0 & 0 & 0 \end{bmatrix}$$, which has a row of all zero elements.
38. $z = 100$ with $x_1 = x_3 = 0$, $x_2 = 50$
39.

Amortization Schedule for a $6000 Loan over 4 Years at 9.6%			
PAYMENT NUMBER	PAYMENT AMOUNT	PAYMENT ON INTEREST	PAYMENT ON DEBT	OUTSTANDING DEBT
0				$6,000.00
1	$151.03	$48.00	$103.03	5,896.97
2	151.03	47.18	103.85	5,793.12
3	151.03	46.34	104.69	5,688.43

Testing Yourself

1. a. {B, F, G} b. {A, F} c. {A, C, D, E, F}
   d. There are 32 possibilities from ϕ to {A, B, D, F, G}.
   e. {A, B, E}
2.

Consumption of eggs

3. a. $A \cap B$ b. A' c. $A' \cap B$ d. $B \cup C$ e. $A \cap C$
   f. $A' \cap B'$

4. a.

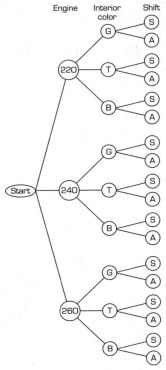

   b. $4(9)(3) = 108$

5.

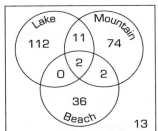

   a. 36 b. 201 c. 125 d. 13
6. a. {0, 6, 7, 8, 9} b. {2, 4} c. {0, 1, 2, 4, 6, 7, 8, 9}
   d. {0, 1, 2, 3, 4, 5, 6, 7, 8, 9} e. {7, 8, 9} f. There are
   many possible answers. {3, 4}, {3, 4, 5}, etc.
7. Commercials 28.9%, Olympic events 47.2%, Features 23.9%

8. a. three-month moving averages:
   March 1.3, April 1.37, May 1.5, June 1.53, July 1.47,
   August 1.33, September 1.37, October 1.47, November
   1.53, December 1.6
   b. mean = 1.45, mode = 1.5, median = 1.45, range =
   0.06, standard deviation ≈ 0.180

9.

10.

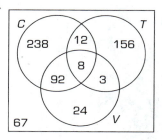

a. 127 b. 27 c. 156 d. 421

11. mean ≈ 2.91, modes are 1 and 2, median $= 2$, standard deviation ≈ 1.83, range $= 6$

12.

13. a.

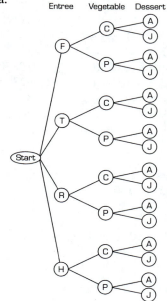

b. $(8)(10)(5) = 400$

14.

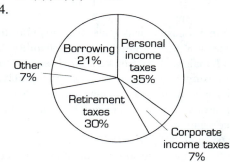

CHAPTER 6

Section 6.1   Improving Skills

1. *a* and *c* are equally likely

3. *c* and *d* are equally likely

5. a. The sum is 2. b. The sum is 2 or 12. c. The faces on both dice are either a 1 or 2. d. The faces on the two dice are the same. e. The faces on both dice are either a 1 or 2 or the faces are both 6. f. The faces on the two dice are not the same number.

7. a. All the children are boys. b. The family does not consist of three boys. c. The family consists of exactly two girls. d. The family consists of exactly two boys. e. The family consists of exactly two girls or two boys.

9. a. {HHH, HHT, HTH, HTT, THH, THT, TTH, TTT}
   b. {HHT, HTH, THH}
   c. {HTT, THT, TTH, TTT}
   d. $\frac{3}{8}$
   e. $\frac{4}{8} = \frac{1}{2}$

11. Distinguish ties by color and number.
   a. Take sample space to be all listings of two ties, with tie chosen first listed first.
   b. $3(4) + 2(5) + 2(5) = 32$
   c. $\frac{32}{42} = \frac{16}{21}$

13. 31

15. 8

Creating Models

17. (a) $\frac{4}{6} = \frac{2}{3}$ (b) $\frac{3}{6} = \frac{1}{2}$

19. $\frac{6}{36} = \frac{1}{6}$ 21. $\frac{8}{400} = 0.02$

23. a. 0.45 b. 0.73 c. 0.31

25. a. $\frac{875}{1535} \approx 0.5700$ b. $\frac{136}{1535} \approx 0.0886$ c. $\frac{594}{1535} \approx 0.3870$

27. a. $\frac{890}{992} \approx 0.8972$ b. $\frac{55}{992} \approx 0.0554$ c. $\frac{92}{992} \approx 0.0927$

29. $\frac{1}{3}$ 31. $\frac{9}{30} = 0.3$ 33. 0.0888

35. 0.0536   37. 0.0150   39. 0.05

Reviewing Material

45. a. 3.6 years b. 8 years c. 2.5 years and 7 years

46. code years; $y = 7.395x + 140.12$
    a. 132.7 b. 162.3

47. $128,025.29

48. a. {0, 4, 5, 6} b. {2} c. {0, 1, 2, 4}
    d. ϕ

Section 6.2 Improving Skills

1. yes   3. no   5. no

7. yes   9. yes   11. no

13. no   15. no

17. a. probability that a customer is male b. probability that a customer does not make a purchase c. probability that a customer is female and does make a purchase d. probability that a customer is a female or makes a purchase e. probability that a customer is a female or does not make a purchase f. probability that a customer is a female and does not make a purchase

19. a. 0.75 b. 0.65 c. 0 d. 0.60 e. 0.75 f. 0.40

21. The probabilities of the three simple outcomes do not sum to 1.

23.

Face value	1	2	3	4	5	6
Probability	0.237	0.168	0.180	0.103	0.165	0.147

(a) 0.418 (b) 0.312 (c) 0.763

25.

Age group	18–23	24–29	30–35	36–41	42–53	over 53
Probability of smoking	$\frac{30}{93}$	$\frac{47}{142}$	$\frac{46}{145}$	$\frac{44}{133}$	$\frac{70}{203}$	$\frac{68}{316}$

27. 1 to 1   29. 1 to 5   31. $\frac{3}{5}$

33. $\frac{8}{17}$ 35. $\frac{2}{5}$ 37. $\frac{28}{41}$

39.
$$\frac{P(E)}{P(E')} = \frac{a}{b}$$
$$aP(E') = bP(E)$$
$$aP(E') = b(1 - P(E'))$$
$$aP(E') = b - bP(E')$$
$$P(E')(a + b) = b$$
$$P(E') = \frac{b}{a + b}$$

41. yes

Creating Models

43. 0.38   45. a. 0.79 b. 0.21

47. 0.94   49. 0.19

51. $\frac{3}{4}$ 53. $\frac{4}{7}$

55. a. $\frac{1}{7}$ b. $\frac{2}{7}$

Reviewing Material

58. $x = -10$, $y = 10$, $z = 0$

59. no solution.

60. the second offer is better

61.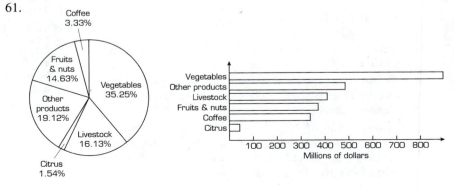

Section 6.3   Improving Skills

1. $P(E/F) = 0.2$ $P(F/E) = 0.25$
3. $P(E/F) \approx 0.3333$ $P(F/E) = 0.5$
5. $P(E/F) \approx 0.42857$ $P(F/E) \approx 0.64286$
7. $P(E/F) \approx 0.3333$ $P(F/E) = 0.5$
9. $P(E/F) = 0.25$ $P(F/E) = 0.4$
11. 1. not independent 2. independent 3. not independent
    4. independent 5. not independent 6. not independent
    7. not independent 8. independent 9. independent 10. not
    independent
13. a. $\frac{136}{1535}$ b. $\frac{70}{875}$ c. $\frac{66}{660}$ d. $\frac{66}{136}$ e. $\frac{805}{1399}$ f. $\frac{594}{660}$
15. a. $\frac{102}{992} \approx 0.10288$ b. $\frac{92}{937} \approx 0.0982$ c. $\frac{10}{55} \approx 0.1818$
    d. $\frac{10}{102} \approx 0.098$ e. $\frac{45}{55} \approx 0.8182$ f. $\frac{845}{890} \approx 0.9494$
17. a. $\frac{5}{8}$ b. $\frac{3}{8}$ c. $\frac{1}{3}$ d. $\frac{2}{3}$ e. $\frac{5}{12}$ f. $\frac{3}{8}$ g. $\frac{7}{12}$ h. $\frac{5}{12}$
19. $\frac{1}{4}$ 21. 0.72
23. 0.05   25. 0.88

Creating Models

27.

$P(Y) = 0.0586$

29.

$P(N/NB) = 0.9444$

31.

a. $P(E'/A) = \dfrac{P(E' \cap A)}{P(A)}$
$= 0.1111$
b. $P(A/E') = 0.1852$

33.

a. $P(S \cap C) = 0.104$
b. $P(C/N) = 0.0125$

35. 0.5789
37. $\frac{1}{9}, \frac{2}{11}$
39. $P(E) = \frac{7}{8}$ $P(E/F) = 1$
    These events are not independent.
41. $\frac{1}{2}$
43. 0.9409
45. 0.81, 0.729
47. 0.9409

Reviewing Material

54. $141,166.67
55. 9 steak knives and 16 pocket knives
56. The better offer is the second offer
57. mean $\approx$ 75.07, standard deviation $\approx$ 5.50, median = 75.9,
    range 17.7; there is no mode.

Section 6.4   Improving Skills

1. $\frac{1}{6}$ 3. $\frac{4}{9}$
5. $\frac{5}{14}$ 7. $\frac{9}{14}$
9. $\frac{40}{247}$ 11. $\frac{45}{247}$
13. $\frac{27}{31}$ 15. 0
17. 0.60435   19. 0.26842
21. 0.01965   23. 0.03837

Creating Models

25. 0.63067   27. 0.56853
29. 0.40678   31. a. 0.6842 b. 0.3158
33. a. 0.9313 b. 0.1892   35. 0.059
37. 0.353

Reviewing Material

45. $\mathbf{x} = \begin{bmatrix} -4\frac{2}{3} \\ \frac{1}{3} \end{bmatrix}$
46. x = the number of lbs of beef, y = the number of lbs of
    pork, minimize $z = 1.05x + 1.20y$ subject to $0.72x + 0.80y \geq .75(x + y)$, with $x \geq 0, y \geq 0$
47. $13,553.10

48. (a) 8 million
   (b) 6.4 trillion

Section 6.5   Improving Skills

1. 362,880	3. 720	5. 42
7. $\frac{6}{5}$	9. 720	11. 120
13. 6,720	15. 56	17. 20
19. 20	21. 492,960	23. 82,160

Creating Models

25. 117,600	27. 51,875,417,160
29. 2,184	31. 20,160
33. 265,182,525	35. 28
37. 4,060	39. 2,002
41. 28	43. 65,780
45. 1,677,106,640	47. 25,200
49. 2,520	51. 640,550,877,120
53. 132,300	55. 56
57. $\frac{1}{28}$	59. 0.01124
61. 0.2143	63. 0.0333
65. 0.0002401	67. 0.0000215
69. 0.002641	71. 0.467
73. $\frac{1}{80}$	

Exploring with Technology

79. true	81. true	83. true

Reviewing Material

84.

85. $y = -48x + 1130$
86. There is no solution to this system.
87. $84,848.96

Section 6.6   Improving Skills

1. 0.1700	3. 0.1943
5. 0.0683	7. 0.00000006
9. 0	

Creating Models

11. 0.375	13. 0.1875
15. 0.0625	17. 0.4202
19. 0.0879	21. 0.1719
23. 0.0914	25. a. 0.2725 b. 0.8948
27. 0.00013	29. 0.9139
31. 0.0916	33. 0.2658
35. 0.1426	

Reviewing Material

39.

40. $\begin{bmatrix} -5 & -2 & 2 \\ -3 & -1 & 1 \\ -3 & -1 & 0 \end{bmatrix}$

41. $z = 150$ when $x_1 = 0$, $x_2 = 18.75$, $x_3 = 43.75$
42. mean ≈ 55.16, standard deviation ≈ 4.68, mode $= 53$, median $= 53$, range $= 16$

Section 6.7   Improving Skills

1. 2.5	3. 2	5. 2.9
7. -2.3	9. 3.3	11. 14.25
13. 7.7	15. 2.45	17. 29.9
19. -0.135	21. 3.0733	23. -0.07
25. 11.04	27. $2.50	29. $2
31. $3.50	33. $7.70	

Creating Models

35. 1.7 days	37. $1,250	39. 3.72
41. $-$0.50	43. 1.5	45. 94 cents
47. 1.875	49. 0.6923	51. 0.5
53. $-$0.26	55. $-$0.26	57. 3.5
59. -0.556	61. $-$0.3333	

Reviewing Material

69. $x = 0.58$ $y = 0.1$
70. $x_1 = 66.6667$, $x_2 = 233.333$, $x_3 = 0$, $z = 833.333$
71. 0.4508
72. 0.349

Testing Yourself

1. a. {G₁G₂, G₁G₃, G₁R, G₂G₁, G₂G₃, G₂R, G₃G₁, G₃G₂, G₃R, RG₁, RG₂, RG₃}

   b. $\frac{1}{2}$

   c. 1

2. 0.4414

3. $P(A \mid B) = 0.4762$
   $P(B \mid A) = 0.5714$

4. a. $\frac{31}{39}$
   b. $\frac{1}{13}$
   c. $\frac{3}{16}$

5. a. not independent
   b. 0.75

6. 0.88757   7. $\frac{8}{17}$ 8. 210

9. −0.0769   10. 5.5 doors

11. a. 0.2
    b. 0.76

12. a. 0.2
    b. 0.7
    c. 0
    d. 0.7

13. a. $\frac{509}{714}$
    b. $\frac{235}{714}$
    c. $\frac{73}{347}$

14. 0.942755   15. 0.717949

16. 0.007128   17. 0.68

18. a. {F₁, F₂, F₃, S₁, S₂, J₁, J₂, J₃, J₄, J₅, J₆, SN₁}
    b. $\frac{7}{12}$
    c. $\frac{2}{3}$

19. 6,375,600

20. 8 to 17   21. 0.216346

22. a. $\frac{3}{4}$
    b. $\frac{1}{4}$
    c. $\frac{1}{6}$
    d. $\frac{4}{7}$
    e. $\frac{3}{7}$

23. $\frac{1}{1,000}$ 24. 0.0161   25. 0.0857

Chapter 7

Section 7.1   Improving Skills

1. In matrices a, b, d, and f, not all of the rows sum to 1. In c and e, not all the entries are between 0 and 1.

3. a. 2 b. 1 c. 0.6 d. 0.4

5. a. 4 b. 0 c. 0.2 d. 0 e. 0.1

7.
$$\mathbf{P}^2 = \begin{bmatrix} 0.4 & 0.6 \\ 0.24 & 0.76 \end{bmatrix}$$

   a. 0.4 b. 0.6 c. 0.76 d. 0.24

9.
$$\mathbf{P}^2 = \begin{bmatrix} 0.15 & 0.75 & 0.10 \\ 0.18 & 0.78 & 0.04 \\ 0.12 & 0.72 & 0.16 \end{bmatrix}$$

   a. 0.10 b. 0.75 c. 0.16 d. 0.04

11.
$$\mathbf{P}^2 = \begin{bmatrix} 0.19 & 0.23 & 0.04 & 0.54 \\ 0.06 & 0.14 & 0.04 & 0.76 \\ 0.16 & 0.32 & 0.09 & 0.43 \\ 0 & 0 & 0 & 1 \end{bmatrix}$$

   a. 0 b. 0.04 c. 0.14 d. 0

13. $p_{11}^{(2)} = \frac{40}{63}$

15. $p_{12}^{(3)} = 0.84$

17. $p_{31}^{(2)} = 0.24$

19. $p_{21}^{(2)} = 0.37$

21.

23.

Creating Models

25.

	Democrat	Republican
Democrat	0.45	0.55
Republican	0.4	0.6

27.

	dry	wet
dry	0.20	0.80
wet	0.65	0.35

29.

	regular	not regular
regular	0.98	0.02
not regular	0.04	0.96

31.

	regular user	occasional user	nonuser
regular	0.85	0.10	0.05
occasional user	0	1	0
nonuser	0.1	0.15	0.75

33.

	A	B	C	D
A	0	$\frac{1}{3}$	$\frac{1}{3}$	$\frac{1}{3}$
B	$\frac{1}{2}$	0	$\frac{1}{4}$	$\frac{1}{4}$
C	$\frac{1}{2}$	$\frac{1}{4}$	0	$\frac{1}{4}$
D	$\frac{1}{2}$	$\frac{1}{4}$	$\frac{1}{4}$	0

Exploring with Technology

41.
$$\begin{bmatrix} 0.5714285714 & 0.4285714286 \\ 0.5714285714 & 0.4285714286 \end{bmatrix} = P^{32} = P^{64} =$$
$$P^{128} = P^{256} = P^{512} = P^{1024}$$

43.
They approach
$$\begin{bmatrix} 0.25 & 0.125 & 0.625 \\ 0.25 & 0.125 & 0.625 \\ 0.25 & 0.125 & 0.625 \end{bmatrix}$$

Reviewing Material

44. $a = \frac{1}{4}, b = \frac{1}{8}, c = \frac{5}{8}$

45. 10.85%

46.

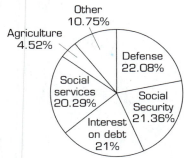

Other 10.75%
Agriculture 4.52%
Defense 22.08%
Social services 20.29%
Social Security 21.36%
Interest on debt 21%

47. $\frac{1}{7}$

Section 7.2   Improving Skills

1. a., c., e., f., i., j., k.
3. [0.35 0.65]
5. [0.235 0.487 0.278]
7. $[\frac{1}{4} \quad \frac{1}{4} \quad \frac{1}{4} \quad \frac{1}{4}]$
9. $d_1 = [0.1 \quad 0.9], d_2 = [0.64 \quad 0.36], d_3 = [0.316 \quad 0.684]$
11. $d_1 = [0.34 \quad 0.66], d_2 = [0.496 \quad 0.504],$
    $d_3 = [0.4024 \quad 0.5976]$
13. $d_1 = [0 \quad 1], d_2 = [0.4 \quad 0.6], d_3 = [0.24 \quad 0.76]$
15. $d_1 = [0.1 \quad 0.9], d_2 = [0.36 \quad 0.64], d_3 = [0.256 \quad 0.744]$
17. $d_1 = [0.38 \quad 0.62], d_2 = [0.252 \quad 0.748],$
    $d_3 = [0.2008 \quad 0.7992]$
19. $d_1 = [0.2857 \quad 0.7143], d_2 = [0.2993 \quad 0.7007],$
    $d_3 = [0.3000 \quad 0.7000]$
21. $d_1 = [0.15 \quad 0.75 \quad 0.1], d_2 = [0.165 \quad 0.765 \quad 0.07],$
    $d_3 = [0.1695 \quad 0.7695 \quad 0.061]$
23. $d_1 = [0.07 \quad 0.03 \quad 0.9], d_2 = [0.097 \quad 0.003 \quad 0.9],$
    $d_3 = [0.0997 \quad 0.0003 \quad 0.9]$

25. $d_1 = [0.1 \quad 0.3 \quad 0.6], d_2 = [0.1 \quad 0.35 \quad 0.55],$
    $d_3 = [0.1 \quad 0.345 \quad 0.555]$
27. regular   29. not regular
31. regular   33. not regular
35. regular   37. $s = [0.4375 \quad 0.5625]$
39. $s = [\frac{1}{6} \quad \frac{5}{6}]$ 41. $s = [\frac{10}{19} \quad \frac{9}{19}]$
43. $s = [\frac{11}{110} \quad \frac{38}{110} \quad \frac{61}{110}]$ 45. $s = [\frac{1}{2} \quad \frac{1}{4} \quad \frac{1}{4}]$
47. $s = [\frac{1}{22} \quad \frac{9}{22} \quad \frac{9}{22} \quad \frac{3}{22}]$

Creating Models

49. a. 0.5665 b. $\frac{22}{37}$
51. a. 51.2% overweight b. $\frac{22}{27} \approx 81.5\%$ overweight
53. 53.4% residential, 31.1% commercial, 15.5% public land
55. a. 44.4875% high-risk, 22.9625% medium-risk, 32.55% low-risk
    b. 68.7% high risk, 10.9% medium-risk, 20.4% low-risk

Exploring with Technology

63. $s = [\frac{4}{7} \quad \frac{3}{7}]$
65. $s = [\frac{1}{4} \quad \frac{1}{8} \quad \frac{5}{8}]$
    The rows of the powers of the transition matrix approach the stabile distribution vector.

Reviewing Material

66. $5,000 in the stock fund, $6,000 in the real estate fund, and $9,000 in the bank
67. $64,810.91
68. There is no mode. Range = 5,904 mean = 90524600, standard deviation $\approx$ 2080.7, median = 89,659
69. 0.2784

Section 7.3   Improving Skills

1. absorbing Markov chain; absorbing state: state 2
3. absorbing Markov chain; absorbing states: states 1 and 2
5. absorbing Markov chain; absorbing states: states 2 and 3
7. not an absorbing Markov chain—an object in state 1 does not remain there
9. not an absorbing Markov chain—not possible for objects in nonabsorbing states 1 and 2 to enter absorbing states 3 and 4
11. not an absorbing Markov chain—not possible for objects in nonabsorbing states to enter absorbing states
13. not an absorbing Markov chain; not possible for objects in states 3 or 4 to enter state 2
15. $R = \begin{bmatrix} 0.3 \\ 0.2 \end{bmatrix}$ $F = \begin{bmatrix} \frac{20}{7} & \frac{5}{7} \\ \frac{10}{7} & \frac{20}{7} \end{bmatrix}$
17. $R = \begin{bmatrix} 0.7 \\ 0.5 \end{bmatrix}$ $F = \begin{bmatrix} \frac{10}{9} & \frac{4}{9} \\ 0 & 2 \end{bmatrix}$

19.
$$R = \begin{bmatrix} 0.7 \\ 0 \end{bmatrix} \qquad F = \begin{bmatrix} \frac{10}{7} & \frac{4}{7} \\ \frac{10}{7} & \frac{18}{7} \end{bmatrix}$$

21.
$$R = \begin{bmatrix} 0 \\ 0.1 \end{bmatrix} \qquad F = \begin{bmatrix} \frac{50}{3} & 10 \\ 15 & 10 \end{bmatrix}$$

23.
$$R = \begin{bmatrix} 0.3 & 0.5 \\ 0 & 0.5 \end{bmatrix} \quad F = \begin{bmatrix} \frac{35}{33} & \frac{10}{33} \\ \frac{10}{33} & \frac{50}{33} \end{bmatrix}$$

25.
$$R = \begin{bmatrix} 0.25 \\ 0.25 \\ 0.25 \end{bmatrix} \qquad F = \begin{bmatrix} 2 & 1 & 1 \\ 1 & 2 & 1 \\ 1 & 1 & 2 \end{bmatrix}$$

Creating Models

27. 58 discharged eventually, 22 will die

29. Over the long run, all offspring must be of the dominant genotype.

31. Over the long run, 100% of the land will be public.

33. 239   **35.** 243

37. Approximately 1038 will leave college without graduating and 3092 will graduate.

39. Player with \$2 will win all the money with probability $\frac{2}{3}$ and go bust with probability $\frac{1}{3}$.

43.
$$P^2 = \begin{bmatrix} 0 & 0 & .03 & .97 & 0 \\ 0 & 0 & 0 & .96 & .04 \\ 0 & 0 & 0 & .6 & .4 \\ 0 & 0 & 0 & 1 & 0 \\ 0 & 0 & 0 & 0 & 1 \end{bmatrix}$$

$$P^4 = P^8 = P^{16} = \begin{bmatrix} 0 & 0 & 0 & .988 & .012 \\ 0 & 0 & 0 & .96 & .04 \\ 0 & 0 & 0 & .6 & .4 \\ 0 & 0 & 0 & 1 & 0 \\ 0 & 0 & 0 & 0 & 1 \end{bmatrix}$$

The entries in the upper right section of P^2, P^4, P^8, and P^{16} approach the entries in the product **FR**.

45. As the exponent n increases, the power P^n will have entries in its upper right section that approach the entries in the product **FR**.

Reviewing Material

46.

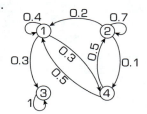

47. \$11,916.16

Testing Yourself

1.

$$\begin{array}{cc} & \begin{array}{cc} M & P \end{array} \\ \begin{array}{c} M \\ P \end{array} & \begin{bmatrix} 0.98 & 0.02 \\ 0.5 & 0.95 \end{bmatrix} \end{array}$$

Objects: commuting residents; Period is two years; States: mass transport (M) and private transport (P)

2.
$$P^2 = \begin{bmatrix} 1 & 0 & 0 \\ 1 & 0 & 0 \\ 1 & 0 & 0 \end{bmatrix} = P^3 = P^4$$

All objects end up in state 1 after two time periods.

3.

Chapter 8

Section 8.1   Improving Skills

1. If a number is in the interval $[0, 0.2)$, then show one finger; if a number is in the interval $[0.2, 0.7)$, then show two fingers; and if a number is in the interval $[0.7, 1)$, then show three fingers.

3. If a number is in the interval $[0, 0.65)$, then show one finger; if a number is in the interval $[0.65, 0.8)$, then show two fingers; and if a number is in the interval $[0.8, 1)$, then show three fingers.

5. If a number is in the interval $[0, \frac{3}{7})$, then show one finger; if a number is in the interval $[\frac{3}{7}, \frac{5}{7})$, then show two fingers; and if a number is in the interval $[\frac{5}{7}, 1)$, then show three fingers.

7. If a number is in the interval $[0, 0.05)$, then show one finger; there are no numbers in the interval; $[0.05, 0.05)$, so the player will never show two fingers; and if a number is in the interval $[0.05, 1)$, then show three fingers.

9.

TOSS	RANDOM NUMBER	COIN TOSS	PAYOFF	TOTAL WINNINGS	TIMES EVEN
1	0.5348	T	−1	−1	0
2	0.9804	T	−1	−2	0
3	0.5953	T	−1	−3	0
4	0.4058	H	1	−2	0
5	0.4321	H	1	−1	0
6	0.9734	T	−1	−2	0
7	0.5812	T	−1	−3	0
8	0.2624	H	1	−2	0
9	0.8329	T	−1	−3	0
10	0.2282	H	1	−2	0

11.

WEEK	DELIVERY	RETURNED DELIVERY	RESTOCK FEE	BEG. STOCK	RANDOM NUMBER	DEMAND	STOCK AT END	SHORTFALL	LOST PROFIT
1				6000	0.024390	2000	4000	0	0
2	4500	2500	375	6000	0.736792	5000	1000	0	0
3	4500	0	0	5500	0.945015	6000	0	500	55
4	4500	0	0	5000	0.100406	3000	2000	0	0
5	4500	500	75	6000	0.282385	4000	2000	0	0
6	4500	500	75	6000	0.514565	4000	2000	0	0
7	4500	500	75	6000	0.253310	4000	2000	0	0
8	1500	500	75	6000	0.610121	4000	2000	0	0
9	4500	500	75	6000	0.621430	5000	1000	0	0
10	4500	0	0	5500	0.627292	5000	500	0	0
			750						55

Total lost revenue = 750 + 55 = $805.

13.

WEEK	DELIVERY	RETURNED DELIVERY	RESTOCK FEE	BEG. STOCK	RANDOM NUMBER	DEMAND	STOCK AT END	SHORTFALL	LOST PROFIT
1				6000	0.099377	3000	3000	0	0
2	4000	1000	150	6000	0.418348	4000	2000	0	0
3	4000	0	0	6000	0.933511	6000	0	0	0
4	4000	0	0	4000	0.345862	4000	0	0	0
5	4000	0	0	4000	0.261539	4000	0	0	0
6	4000	0	0	4000	0.969422	6000	0	2000	220
7	4000	0	0	4000	0.952613	6000	0	2000	220
8	4000	0	0	4000	0.572245	4000	0	0	0
9	4000	0	0	4000	0.110237	3000	1000	0	0
10	4000	0	0	5000	0.450540	4000	1000	0	0
			150						440

Total lost revenue = 150 + 440 = $590

Creating Models

15. Use the first two columns of random numbers. Your answers may vary if you used different numbers.

TOSS	RANDOM NUMBER	COIN TOSS	MATCH PREVIOUS?	PAYOFF	TOTAL WINNINGS
1	0.92630	T	not applicable	0	0
2	0.79445	T	yes	2	2
3	0.59654	T	yes	2	4
4	0.31524	H	no	−1	3
5	0.06348	H	yes	2	5
6	0.78240	T	no	−1	4
7	0.78735	T	yes	2	6
8	0.71966	T	yes	2	8
9	0.49587	H	no	−1	7
10	0.76938	T	no	−1	6

17. Use columns 1 and 2 of the random number table. Your answers may vary if you used different numbers.

PLAY	RANDOM NUMBER	COLOR	PAYOFF	TOTAL WINNINGS
1	0.92630	black	−1	−1
2	0.79445	black	−1	−2
3	0.59654	black	−1	−3
4	0.31524	red	1	−2
5	0.06348	red	no wager	−2
6	0.78240	black	no wager	−2
7	0.78735	black	−1	−3
8	0.71966	black	−1	−4
9	0.49587	black	−1	−5
10	0.76938	black	−1	−6

19. Use random numbers in columns 3 and 4. Your answers may vary if you used different numbers.

PERIOD	BEGINNING BALANCE	PAYMENT	RANDOM NUMBER	INTEREST RATE	ENDING BALANCE
1	$0	$50,000	none	6%	$P_n = 50{,}000.00(1 + \frac{0.06}{4})^1 = \$50{,}750.00$
2	$50,750	$50,000	0.19267	6%	$P_n = 100{,}750.00(1 + \frac{0.06}{4})^1 = \$103{,}772.50$
3	$103,772.50	$50,000	0.71549	5%	$P_n = 153{,}772.50(1 + \frac{0.05}{4})^1 = \$157{,}616.81$
4	$157,616.81	$50,000	0.27386	7%	$P_n = 207{,}616.81(1 + \frac{0.07}{4})^1 = \$214{,}883.40$
5	$214,883.40	$50,000	0.76612	5%	$P_n = 264{,}883.40(1 + \frac{0.05}{4})^1 = \$271{,}505.49$
6	$271,505.49	$50,000	0.90379	4%	$P_n = 321{,}505.49(1 + \frac{0.04}{4})^1 = \$327{,}935.60$
7	$327,935.60	$50,000	0.95457	4%	$P_n = 377{,}935.60(1 + \frac{0.04}{4})^1 = \$385{,}494.31$
8	$385,494.31	$50,000	0.44843	7%	$P_n = 435{,}494.31(1 + \frac{0.07}{4})^1 = \$450{,}736.61$
9	$450,736.61	$50,000	0.50004	8%	$P_n = 500{,}736.61(1 + \frac{0.08}{4})^1 = \$520{,}766.07$
10	$520,766.07	$50,000	0.39789	7%	$P_n = 570{,}766.07(1 + \frac{0.07}{4})^1 = \$590{,}742.88$

21. Use random numbers in columns 1 and 2 for determining the number of houses that are sold, and the numbers in columns 3, 4, and 5 (if necessary) to determine the commission. Your answers may vary if you used different numbers.

WEEK	RANDOM NUMBER (HOUSE SALE)	RANDOM NUMBER (COMMISSION)	NUMBER OF HOUSES	WEEKLY COMMISSION	TOTAL EARNINGS
1	0.92630	0.19267	0	0	$0
2	0.79445	0.71549	1	$6.000	$6.000
3	0.59654	0.27386	0	0	$6,000
4	0.31524	0.76612	0	0	$6,000
5	0.06348	0.90379	0	0	$6,000
6	0.78240	0.95457	1	$8,000	$14,000
7	0.78735	0.44843	1	$5,000	$19,000
8	0.71966	0.50004	1	$5,000	$24,000
9	0.49587	0.39789	0	0	$24,000
10	0.76938	0.51392	1	$5,000	$29,000

23. Use random numbers in columns 7 and 8. Your answers may vary if you used different numbers.

PERIOD	TIME UNTIL OPENING	RANDOM NUMBER	TIME TO NEXT CUSTOMER ARRIVAL	TOTAL CUSTOMERS
1	15 min.	0.37708	2	1
2	13 min.	0.00701	1	2
3	12 min.	0.29281	1	3
4	11 min.	0.59483	3	4
5	8 min.	0.09209	1	5
6	7 min.	0.79862	3	6
7	4 min.	0.34986	2	7
8	2 min.	0.18544	1	8
9	1 min.	0.60680	3	9
10	open	—	—	9

25. Use columns 1 and 2 for Window 1, columns 3 and 4 for Window 2, and columns 5 and 6 for Window 3. Your answers may vary if you used different values.

PULL OF SLOT	RANDOM NUMBER (WINDOW 1)	RANDOM NUMBER (WINDOW 2)	RANDOM NUMBER (WINDOW 3)	WINDOW 1	WINDOW 2	WINDOW 3	WIN
1	0.92630	0.19267	0.53497	watermelon	apple	orange	−1
2	0.79445	0.71549	0.26104	watermelon	watermelon	apple	−1
3	0.59654	0.27386	0.05358	orange	apple	apple	−1
4	0.31524	0.76612	0.13537	apple	watermelon	apple	−1
5	0.06348	0.90379	0.55887	apple	watermelon	orange	−1
6	0.78240	0.95457	0.23894	watermelon	watermelon	apple	−1
7	0.78735	0.44843	0.67318	watermelon	orange	watermelon	−1
8	0.71966	0.50004	0.94031	watermelon	orange	watermelon	−1
9	0.49587	0.39789	0.48086	orange	orange	orange	10
10	0.76938	0.51392	0.71015	watermelon	orange	watermelon	−1

27. Use columns 3 and 4 for the Hygienist time, columns 5 and 6 for the Dentist time, and columns 7 and 8 for the Dentist reduction time. Your answers may vary if you used different values. n/a = not applicable.

PATIENT	RANDOM NUMBER (HYGIENIST)	RANDOM NUMBER (DENTIST 20)	RANDOM NUMBER (DENTIST 30)	HYGIENIST TIME	DENTIST TIME IF HYGIENIST TAKES 20	DENTIST REDUCTION IF HYGIENIST TAKES 30	TOTAL TIME
1	0.19267	0.53497	0.37708	20	10	n/a	30
2	0.71549	0.26104	0.00701	20	10	n/a	30
3	0.27386	0.05358	0.29281	20	10	n/a	30
4	0.76612	0.13537	0.59483	20	10	n/a	30
5	0.90379	0.55887	0.09209	30	10	−1	39
6	0.95457	0.23894	0.79862	30	10	0	40
7	0.44843	0.67318	0.34986	20	15	n/a	35
8	0.50004	0.94031	0.18544	20	15	n/a	35
9	0.39789	0.48086	0.60680	20	10	n/a	30
10	0.51392	0.71015	0.79157	20	15	n/a	35
							334

Reviewing Material

33. Number of Absences

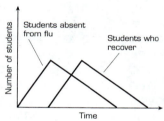

34. $x = 14{,}782.61$
    $y = 2{,}086.96$
    $z = 35{,}652.21$

35. 0.0975

Section 8.2   Improving Skills

1.

RANDOM NUMBER (r)	R = 5r
0.53479	2.674
0.81115	4.056
0.98036	4.902
0.12217	0.611
0.59526	2.976
0.40238	2.012
0.40577	2.029
0.39351	1.968
0.03172	0.159
0.69255	3.463

3.

RANDOM NUMBER (r)	R = 5r + 3
0.53479	5.674
0.81115	7.056
0.98036	7.902
0.12217	3.611
0.59526	5.976
0.40238	5.012
0.40577	5.029
0.39351	4.968
0.03172	3.159
0.69255	6.463

5.

RANDOM NUMBER (r)	R = 9r − 6
0.53479	−1.19
0.81115	1.30
0.98036	2.82
0.12217	−4.90
0.59526	−0.64
0.40238	−2.38
0.40577	−2.35
0.39351	−2.46
0.03172	−5.71
0.69255	0.23

7.

RANDOM NUMBER (r)	R = 2400r + 2500
0.53479	3783.5
0.81115	4446.8
0.98036	4852.9
0.12217	2793.2
0.59526	3928.6
0.40238	3465.7
0.40577	3473.8
0.39351	3444.4
0.03172	2576.1
0.69255	4162.1

9. New simulation of gasoline station

WEEK	DELIVERY	RETURNED DELIVERY	RESTOCK FEE	STOCK AT BEGINNING	RANDOM NUMBER	DEMAND	STOCK AT END	SHORTFALL	LOST PROFIT
1				6,000.00	0.819499	5,278.00	722.00	0.00	0.00
2	4500	0.00	0.00	5,222.00	0.059492	2,237.97	2,984.03	0.00	0.00
3	4500	1,484.04	222.61	6,000.00	0.649792	4,599.17	1,400.83	0.00	0.00
4	4500	0.00	0.00	5,900.83	0.826910	5,307.64	593.19	0.00	0.00
5	4500	0.00	0.00	5,093.19	0.252296	3,009.18	2,084.01	0.00	0.00
6	4500	584.01	87.60	6,000.00	0.390961	3,563.84	2,436.16	0.00	0.00
7	4500	936.16	140.42	6,000.00	0.812566	5,250.26	749.74	0.00	0.00
8	4500	0.00	0.00	5,249.74	0.976262	5,905.05	0.00	655.31	72.08
9	4500	0.00	0.00	4,500.00	0.813032	5,252.13	0.00	752.13	82.73
10	4500	0.00	0.00	4,500.00	0.166790	2,667.16	1,832.84	0.00	0.00
			$450.63						$154.82

Total lose revenue = 450.63 + 154.82 = $605.45.

11.

BUS NUMBER	RANDOM NUMBER	INTERARRIVAL TIME	PEOPLE WAITING	RANDOM NUMBER	PEOPLE ON BUS	PEOPLE REJECTED
1	0.819499	16.92	17	0.819499	46	3
2	0.059492	12.36	15	0.216748	38	0
3	0.649792	15.90	16	0.900191	47	3
4	0.826910	16.96	20	0.042029	36	0
5	0.252296	13.51	14	0.184579	38	0
6	0.390961	14.35	14	0.806528	46	0
7	0.812566	16.88	17	0.967748	48	5
8	0.976262	17.86	23	0.562969	43	6
9	0.813032	16.88	23	0.997746	48	11
10	0.166790	13.00	24	0.848450	46	11
						39

13. Enclose region in a rectangle with sides $x = 1$ to $x = 4$ and $y = 0$ to $y = 9$.

TRIAL POINT	RANDOM NUMBER	x-VALUE	RANDOM NUMBER	y-VALUE	$y = 2x + 1$	ACCEPT POINT	TOTAL ACCEPTED
1	0.00149	1.00447	0.53250	4.79250	3.00894	no	0
2	0.84745	3.54235	0.73200	6.58800	8.08470	yes	1
3	0.63222	2.89666	0.84066	7.56594	6.79332	no	1
4	0.50533	2.51599	0.59620	5.36580	6.03198	yes	2
5	0.50159	2.50477	0.61009	5.49081	6.00954	yes	3
6	0.60433	2.81299	0.38542	3.46878	6.62598	yes	4
7	0.04822	1.14466	0.05758	0.51822	3.28932	yes	5
8	0.49577	2.48731	0.06178	0.55602	5.97462	yes	6
9	0.89049	3.67147	0.80193	7.21737	8.34294	yes	7
10	0.16162	1.48486	0.26466	2.38194	3.96972	yes	8

area of region $\approx (\frac{8}{10})27 = 21.6$

15. Enclose region in a rectangle with sides $x = 0$ to $x = 3$ and $y = 0$ to $y = 5$.

TRIAL POINT	RANDOM NUMBER	x-VALUE	RANDOM NUMBER	y-VALUE	$y = 5 - x$	ACCEPT POINT	TOTAL ACCEPTED
1	0.00149	0.00447	0.53250	2.66250	4.99553	yes	1
2	0.84745	2.54235	0.73200	3.66000	2.45765	no	1
3	0.63222	1.89666	0.84066	4.20330	3.10334	no	1
4	0.50533	1.51599	0.59620	2.98100	3.48401	yes	2
5	0.50159	1.50477	0.61009	3.05045	3.49523	yes	3
6	0.60433	1.81299	0.38542	1.92710	3.18701	yes	4
7	0.04822	0.14466	0.05758	0.28790	4.85534	yes	5
8	0.49577	1.48731	0.06178	0.30890	3.51269	yes	6
9	0.89049	2.67147	0.80193	4.00965	2.32853	no	6
10	0.16162	0.48486	0.26466	1.32330	4.51514	yes	7

area of region $\approx (\frac{7}{10})15 = 10.5$

17. Enclose the region in a rectangle with sides $x = -2$ to $x = 2$ and $y = 0$ to $y = 4$.

TRIAL POINT	RANDOM NUMBER	x-VALUE	RANDOM NUMBER	y-VALUE	$y = x^2$	ACCEPT POINT	TOTAL ACCEPTED
1	0.00149	−1.99404	0.53250	2.13000	3.9762	no	0
2	0.84745	1.38980	0.73200	2.92800	1.9315	yes	1
3	0.63222	0.52888	0.84066	3.36264	0.2797	yes	2
4	0.50533	0.02132	0.59620	2.38480	0.0005	yes	3
5	0.50159	0.00636	0.61009	2.44036	0.0000	yes	4
6	0.60433	0.41732	0.38542	1.54168	0.1742	yes	5
7	0.04822	−1.80712	0.05758	0.23032	3.2657	no	5
8	0.49577	−0.01692	0.06178	0.24712	0.0003	yes	6
9	0.89049	1.56196	0.80193	3.20772	2.4397	yes	7
10	0.16162	−1.35352	0.26466	1.05864	1.8320	no	7

area of region $\approx (\frac{7}{10})16 = 11.2$

19. Enclose the region in a rectangle with sides $x = -3$ to $x = 3$ and $y = -1$ to $y = 5$.

TRIAL POINT	RANDOM NUMBER	x-VALUE	RANDOM NUMBER	y-VALUE	$y = 5 - x^2$	ACCEPT POINT	TOTAL ACCEPTED
1	0.00149	−2.99106	0.53250	2.19500	−3.94644	no	0
2	0.84745	2.08047	0.73200	3.39200	0.65403	no	0
3	0.63222	0.79332	0.84066	4.04306	4.37064	yes	1
4	0.50533	0.03198	0.59620	2.57720	4.99898	yes	2
5	0.50159	0.00954	0.61009	2.66054	4.99991	yes	3
6	0.60433	0.62598	0.38542	1.31252	4.60815	yes	4
7	0.04822	−2.71068	0.05758	−0.65452	−2.34779	no	4
8	0.49577	−0.02538	0.06178	−0.62932	4.99936	yes	5
9	0.89049	2.34294	0.80193	3.81158	−0.48937	no	5
10	0.16162	−2.03028	0.26466	0.58796	0.87796	yes	6

area of region $\approx (\frac{6}{10})36 = 21.6$

Creating Models

Answers for Problems 21 through 33 will vary depending on which random numbers are chosen.

21.

PERIOD	BEGINNING AMOUNT	RANDOM NUMBER	INTEREST RATE	ENDING AMOUNT
1	$1,000.00		0.0400	$P_n = 1,000.00(1 + 0.0400/2) = \$1,020.00$
2	$1,020.00	0.15025	0.0373	$P_n = 1,020.00(1 + 0.0373/2) = \$1,039.00$
3	$1,039.00	0.95610	0.0493	$P_n = 1,039.00(1 + 0.0493/2) = \$1,064.63$
4	$1,064.63	0.09026	0.0364	$P_n = 1,064.63(1 + 0.0364/2) = \$1,083.98$
5	$1,083.98	0.81431	0.0472	$P_n = 1,083.98(1 + 0.0472/2) = \$1,109.57$
6	$1,109.57	0.21431	0.0382	$P_n = 1,109.57(1 + 0.0382/2) = \$1,130.78$
7	$1,130.78	0.20237	0.0380	$P_n = 1,130.78(1 + 0.0380/2) = \$1,152.28$
8	$1,152.28	0.08030	0.0362	$P_n = 1,152.28(1 + 0.0362/2) = \$1,173.14$
9	$1,173.14	0.40378	0.0411	$P_n = 1,173.14(1 + 0.0411/2) = \$1,197.22$
10	$1,197.22	0.99955	0.0500	$P_n = 1,197.22(1 + 0.0500/2) = \$1,227.15$

23.

WEEK	RANDOM NUMBER (HOUSE SALE)	NUMBER OF HOUSES SOLD	RANDOM NUMBER (COMMISSION)	COMMISSION	WEEKLY COMMISSION
1	0.15025	0			$0.00
2	0.95610	3	0.63386	$7,803.16	
			0.81469	$8,888.14	
			0.05731	$4,343.86	$21,035.16
3	0.09026	0			$0.00
4	0.81431	1	0.52462	$7,147.72	$7,147.72
5	0.21431	0			$0.00
6	0.20237	0			$0.00

continues on next page

WEEK	RANDOM NUMBER (HOUSE SALE)	NUMBER OF HOUSES SOLD	RANDOM NUMBER (COMMISSION)	COMMISSION	WEEKLY COMMISSION
7	0.08030	0			$0.00
8	0.40378	0			$0.00
9	0.99955	3	0.58627	$7,517.62	
			0.71122	$8,267.32	
			0.91066	$9,463.96	$25,248.90
10	0.59335	0			$0.00
					$53,431.78

25.

PERIOD	TIME UNTIL OPENING (IN SECONDS)	RANDOM NUMBER	TIME TO NEXT CUSTOMER ARRIVAL (IN SECONDS)	TOTAL CUSTOMERS
1	900	0.94982	87	1
2	813	0.01224	31	2
3	782	0.31669	49	3
4	733	0.21240	43	4
5	691	0.41018	55	5
6	636	0.32324	49	6
7	587	0.28097	47	7
8	540	0.42605	56	8
9	484	0.65952	70	9
10	415	0.85100	81	10

27.

TRIAL	RANDOM NUMBER (BALL)	RANDOM NUMBER (BOX)	BALL SIZE	BOX SIZE	FIT?	NUMBER ROUTED BACK
1	0.63386	0.06620	11.99016	11.95662	no	1
2	0.81469	0.88857	12.04441	12.03886	no	2
3	0.05731	0.74298	11.81719	12.02430	yes	2
4	0.52462	0.97332	11.95739	12.04733	yes	2
5	0.58627	0.65484	11.97588	12.01548	yes	2
6	0.71122	0.07415	12.01337	11.95742	no	3
7	0.91066	0.56583	12.07320	12.00658	no	4
8	0.55128	0.49196	11.96538	11.99920	yes	4
9	0.67667	0.69808	12.00300	12.01981	yes	4
10	0.94822	0.09641	12.08447	11.95964	no	5

29.

STORE	RANDOM NUMBER (SITE)	RANDOM NUMBER (CONTRACT)	RANDOM NUMBER (CONSTRUCTION)	TIME FOR SITE SELECTION	TIME FOR CONTRAST NEGOTIATIONS	TIME FOR CONSTRUCTION	TOTAL NUMBER OF DAYS
1	0.63386	0.71122	0.06620	119	44	254	417
2	0.81469	0.91066	0.88857	124	48	303	476
3	0.05731	0.55128	0.74298	102	41	295	437
4	0.52462	0.67667	0.97332	116	44	308	468
5	0.58627	0.94822	0.65484	118	49	289	456

31. Measure time (in minutes) from when the shop opens. Time 0 is 9:00 a.m.

TIME	RANDOM NUMBER (ARRIVALS)	MINUTES TO NEXT ARRIVAL	RANDOM NUMBER (HAIRCUT)	TIME FOR NEXT HAIRCUT	NEXT ARRIVAL TIME	NEXT SERVICE COMPLETION	BARBER STATUS	PEOPLE WAITING
0.000	0.63386	21.339	0.06620	18.26	21.339	39.603	idle	0
21.339	0.81469	23.147	0.88857	21.55	44.486	39.603	busy	0
39.603					44.486	61.158	idle	0
44.486	0.05731	15.573	0.74298	20.97	60.059	61.158	busy	0
60.059	0.52462	20.246	0.97332	21.89	80.305	61.158	busy	1
61.158					80.305	82.130	busy	0
80.305	0.58627	20.863	0.65484	20.62	101.168	82.130	busy	1
82.130					101.168	104.023	busy	0
101.168	0.71122	22.112	0.07415	18.30	123.280	104.023	busy	1
104.023					123.280	124.642	busy	0
123.280	0.91066	24.107	0.56583	20.26	147.386	124.642	busy	1
124.642					147.386	142.939	busy	0
142.939					147.386	167.650	idle	0

33.

ARRIVING VEHICLE	TIME UNTIL OPENING (IN MINUTES)	RANDOM NUMBER	TIME TO NEXT CUSTOMER ARRIVAL (IN MINUTES)	NUMBER WAITING	NUMBER WHO MUST LEAVE
1	15	0.94982	3	1	0
2	12	0.01224	2	2	0
3	10	0.31669	2	3	0
4	8	0.21240	2	4	0
5	6	0.41018	2	5	0
6	4	0.32324	2	5	1
7	2	0.28097	2	5	2
8	0	0.42605	2	5	3

Reviewing Material

42. $y = -240x + 32,000$; (200, 2000) is not on this line.

43.

Amortization Schedule for a $2000 Loan Over 24 Months at 10%				
PAYMENT NUMBER	PAYMENT AMOUNT	PAYMENT ON INTEREST	PAYMENT ON DEBT	OUTSTANDING DEBT
0				$2000.00
1	$92.29	$16.67	$75.62	1924.38
2	92.29	16.04	76.25	1848.13
3	92.29	15.40	76.89	1771.24

44. a. 11 b. 140 c. 158 d. 9

Section 8.3   Improving Skills

1. $\bar{x} = 10.1$, standard deviation = 1.595
3. $\bar{x} = 10.5$, standard deviation = 1.7159
5. $\bar{x} = 10$, standard deviation = 1.8856
7. $\bar{x} = 5$, standard deviation = 3.1623
9. $\bar{x} = 7$, standard deviation = 3.7417

11. yes
15. yes
19. yes

13. no
17. no

21. [10.5 12.5), [12.5, 14.5), [14.5, 16.5), [16.5, 18.5), [18.5, 20.5), [20.5, 22.5)
23. [80, 84), [84, 88), [88, 92), [92, 96), [96, 100)

Creating Models

25. not uniformly distributed

27.

29.

31.

Not uniformly distributed

Exploring With Technology

35. 24. $\bar{x} \approx 132.8$, s.d. ≈ 9.8
25. $\bar{x} \approx 3.374$, s.d. ≈ 1.079
26. $\bar{x} \approx 8.032$, s.d. ≈ 1.834
27. $\bar{x} \approx 103.878$, s.d. ≈ 73.363
28. $\bar{x} \approx 39529$, s.d. ≈ 6122
29. $\bar{x} \approx 56,781$, s.d. ≈ 6259
30. $\bar{x} \approx 6.69$, s.d. ≈ 0.45
31. $\bar{x} \approx 3.083$, s.d. ≈ 3.324

36.
$$\mathbf{AB} = \begin{bmatrix} -2 & 18 & -12 \\ 10 & -21 & 21 \\ -10 & 22 & -20 \end{bmatrix} = \mathbf{BA}$$

A and **B** commute under the operation of multiplication.
37. $41.18
38. 120,200,080

Section 8.4 Improving Skills

1. 0.24857	3. 0.34134	5. 0.21204
7. 0.45254	9. 0.22868	11. 0.49714
13. 0.39435	15. 0.29017	17. 0.15870

19. a. 0.18141 b. 0.0116 c. 0.0116
21. a. 0.02275 b. 0.5 c. 0.84134

23.

TRIAL	RANDOM NUMBER	RANDOM NUMBER	SAMPLE
1	0.209729	0.517216	8.0862
2	0.672490	0.251967	7.7669
3	0.983139	0.957622	8.0878
4	0.364483	0.146088	7.6123
5	0.987746	0.496411	8.3540

Creative Models

25. 0.47725
27. a. 0.63718
    b. 0.03438
29. 0.02275
31. 0.93.3%
33. 51.65 inches triggers a flood condition
35. 0.043%
37. 0.07353
39. 0.90508
41. effectively zero
43. $P (0.5 \leq X < 10.5) = 0.56314$
45. 0.0268
47. 0.23576

Exploring with Technology

55. Random numbers from Table 8.32. If you used different numbers, your answer may be different.

PERIOD	TIME UNTIL OPENING (IN SECONDS)	RANDOM NUMBER	RANDOM NUMBER	NUMBER OF SECONDS TO NEXT ARRIVAL	TOTAL CUSTOMERS
1	900.00	0.312175	0.516666	166.87	1
2	733.13	0.511149	0.529704	146.26	2
3	586.87	0.814643	0.341134	197.38	3
4	389.48	0.630753	0.846835	168.22	4
5	211.26	0.946331	0.964791	187.58	5
6	33.68	0.502784	0.124030	118.71	6

57.

ITEM	RANDOM NUMBER (PROD. COST)	RANDOM NUMBER	PRODUCTION	RANDOM NUMBER (SHIPPING)	SHIPPING COST	RANDOM NUMBER (ADMIN COST)	RANDOM NUMBER	ADMIN COST	TOTAL COST
1	0.312175	0.516666	33.25	0.481757	4.41	0.547500	0.205303	12.90	50.56
2	0.511149	0.529704	30.50	0.169053	2.85	0.493285	0.764916	15.81	49.15
3	0.814643	0.341134	37.32	0.460438	4.30	0.453595	0.963627	17.22	58.84
4	0.630753	0.846835	33.43	0.864601	6.32	0.989448	0.584736	21.10	60.85
5	0.946331	0.964791	36.01	0.718323	5.59	0.904337	0.709617	20.05	61.65
6	0.502784	0.124030	26.83	0.227727	3.14	0.356356	0.983102	17.66	47.62
7	0.444183	0.513289	30.66	0.871400	6.36	0.251226	0.501083	17.97	54.99
8	0.363832	0.213480	30.39	0.905983	6.53	0.351937	0.585693	16.15	53.07
9	0.939675	0.623328	45.19	0.689715	5.45	0.305809	0.561010	16.89	67.53
10	0.174745	0.326635	37.73	0.949443	6.75	0.889902	0.937984	18.83	63.30
									$567.50

59.

PERIOD	PAYMENT	BEGINNING AMOUNT	RANDOM NUMBER	RANDOM NUMBER	INTEREST RATE	ENDING AMOUNT
1	$50,000.00	$50,000.00			0.0600	Pn = 50,000.00(1 + 0.0600/4) = $ 50,750.00
2	$50,000.00	$100,750.00	0.511149	0.529704	0.0515	Pn = 100,750.00(1 + 0.0515/4) = $102,047.29
3	$50,000.00	$152,047.29	0.814643	0.341134	0.0720	Pn = 152,047.29(1 + 0.0720/4) = $154,782.36
4	$50,000.00	$204,782.36	0.630753	0.846835	0.0603	Pn = 204,782.36(1 + 0.0603/4) = $207,868.80
5	$50,000.00	$257,868.80	0.946331	0.964791	0.0680	Pn = 257,868.80(1 + 0.0680/4) = $262,254.63
6	$50,000.00	$312,254.63	0.502784	0.124030	0.0405	Pn = 312,254.63(1 + 0.0405/4) = $315,415.11
7	$50,000.00	$365,415.11	0.444183	0.513289	0.0520	Pn = 365,415.11(1 + 0.0520/4) = $370,164.10
8	$50,000.00	$420,164.10	0.363832	0.213480	0.0512	Pn = 420,164.10(1 + 0.0512/4) = $425,539.08
9	$50,000.00	$475,539.08	0.939675	0.023328	0.0956	Pn = 475,539.08(1 + 0.0956/4) = $486,900.15
10	$50,000.00	$536,900.15	0.174745	0.326635	0.0732	Pn = 536,900.15(1 + 0.0732/4) = $546,722.11

63.

TELEVISION	RANDOM NUMBER (T.V.)	RANDOM NUMBER	NUMBER OF MINUTES TO NEXT T.V.	RANDOM NUMBER (INSPECT)	RANDOM NUMBER	LENGTH OF INSPECTION	TIME NEXT T.V. ARRIVES	TIME INSPECTION ENDS	IDLE TIME
1	0.312175	0.516666	18	0.40445	0.39640	8	0:18	0:26	0:00
2	0.511149	0.529704	14	0.92959	0.31046	15	0:32	0:47	0:00
3	0.814643	0.341134	23	0.88184	0.25468	15	0:55	1:10	0:00
4	0.630753	0.846835	18	0.64768	0.96129	10	1:13	1:23	0:00
5	0.946331	0.964791	21	0.86943	0.25829	14	1:34	1:48	0:00
6	0.502784	0.124030	10	0.72017	0.61232	10	1:44	1:58	0:04
7	0.444183	0.513289	15	0.03796	0.40050	15	1:59	2:14	0:00
8	0.363832	0.213480	14	0.43127	0.70103	9	2:13	2:23	0:01
9	0.939675	0.023328	33	0.78843	0.09042	13	2:46	2:59	0:00
10	0.174745	0.326635	23	0.55228	0.38574	7	3:09	3:16	0:00
									0:05

Reviewing Material

65. $x = 10$, $y = 8$, $z = 11$

66.

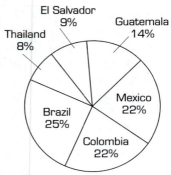

El Salvador 9%
Guatemala 14%
Thailand 8%
Mexico 22%
Brazil 25%
Colombia 22%

67. 1 to 3   68. 0.8718

Section 8.5   Improving Skills

1. 0.0320	3. 0.0537	5. 0.1935
7. 0.0432	9. 0.2387	11. 0.0814
13. 0.0546	15. 0.0829	17. 0.3297

19. a. 0.2019 b. 0.4493 c. 0.7534

21. a. .7408 b. .3679 c. 0

23.

TRIAL	RANDOM NUMBER	EXPONENTIAL VALUE
1	0.293296	9.8126
2	0.966885	0.2694
3	0.459493	6.2211
4	0.776727	2.0213
5	0.646329	3.4916
6	0.080450	20.1610

Creating Models

25. a. 11.4% b. 26.4%

27. a. 0.3012 b. 0.3297

29. Given $\mu = 2$ pounds.

SALES	RANDOM NUMBER	COFFEE BEANS (IN POUNDS)
1	0.650582	0.86
2	0.213081	3.09
3	0.069642	5.33
4	0.679353	0.77
5	0.534584	1.25
6	0.185438	3.37
7	0.962169	0.08
8	0.744978	0.59
9	0.842627	0.34
10	0.793525	0.46

31.

	CURRENT TIME	NEXT ARRIVAL	SERVER STATUS	NEXT SERVICE COMPLETION	CUSTOMERS IN LINE
status update 1	0.00	7.58	idle	—	none
status update 2	7.58	8.33	busy	10.15	none
status update 3	8.33	9.82	busy	10.15	#2
status update 4	9.82	9.97	busy	10.15	#2 and #3
status update 5	9.97	15.25	busy	10.15	#2, #3, and #4
status update 6	10.15	15.25	busy	14.52	#3 and #4
status update 7	14.52	15.25	busy	18.75	#4
status update 8	15.25	33.36	busy	18.75	#4
status update 9	18.75	33.36	busy	21.59	none
status update 10	21.59	33.36	idle	—	none
status update 11	33.36	—	busy	35.62	none

Exploring with Technology

37. Simulation of Television Inspector. Random numbers from Table 9.38 were used for this simulation. If you used different numbers, then your answers may vary.

TELEVISION	RANDOM NUMBER (T.V.)	NUMBER OF MINUTES TO NEXT T.V.	RANDOM NUMBER (INSPECT)	LENGTH OF INSPECTION	TIME NEXT T.V. ARRIVES	TIME INSPECTION ENDS
1	0.650582	4	0.185438	9	0:04	0:13
2	0.213081	15	0.962169	12	0:19	0:31
3	0.069642	27	0.744978	11	0:46	0:57
4	0.679353	4	0.842627	11	0:50	1:08
5	0.534584	6	0.793525	11	0:56	1:19

39.

CUSTOMER	ARRIVAL TIME	RANDOM NUMBER (INTERARRIVAL)	NUMBER OF MINUTES TO INTERARRIVAL	RANDOM NUMBER (SERVICE)	TIME FOR SERVICE	TIME SERVICE ENDS	CUSTOMERS WAITING
1	0	0.650582	3.44	0.589031	6.18	6.18	#2
2	3.44	0.213081	12.37	0.088126	5.18	11.36	none
3	15.81	0.069642	21.32	0.695422	6.39	22.20	none
4	37.12	0.679353	3.09	0.950441	6.90	44.02	#5
5	40.22	0.534584	5.01	0.492928	5.99	50.01	#6
6	45.23	0.185438	13.48	0.255914	5.51	55.52	none
7	58.71	0.962169	0.31	0.538624	6.08	64.79	#8
8	59.01	0.744978	2.36	0.036984	5.07	69.86	#9
9	61.37	0.842627	1.37	0.430644	5.86	75.72	n/a

Reviewing Material

41. $x = 10$, $y = 8$, $z = 11$

42. $218,108.43

43.

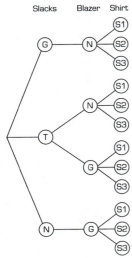

44. 0.3672

Chapter 8 Test

1.

TOSS	RANDOM NUMBER	COIN FACE
1	0.771	tails
2	0.035	heads
3	0.189	heads
4	0.992	tails
5	0.501	tails
6	0.530	tails
7	0.627	tails
8	0.100	heads
9	0.726	tails
10	0.334	head

2.

TRIAL	RANDOM NUMBER	EXPONENTIAL NUMBER
1	0.771	3.2508
2	0.035	41.9051
3	0.189	20.8251
4	0.992	0.1004
5	0.501	8.6394

3. Use a random number generator as follows: number of washers for a random number in the interval: in [0, 0.2), 0 washers; in [0.2, 0.55), 1 washer; in [0.55, 0.85), 2 washers; in [0.85, 1), 3 washers. Number of dryers for a random number in the interval: in [0, 0.55), 0 dryers; in [0.55, 0.75), 1 dryer; in [0.75, 0.90), 2 dryers; in [0.9, 1), 3 dryers.

WEEK	RANDOM NUMBER (WASHER)	WASHER SALE	RANDOM NUMBER (DRYER)	DRYER SALE
1	0.771	2	0.530	0
2	0.035	0	0.627	1
3	0.189	0	0.100	0
4	0.992	3	0.726	1
5	0.501	1	0.334	0

4. mean $= \frac{20}{10} = 2$ standard deviation $= \sqrt{\frac{45}{9}} = 2.4495$

5. Sales: for a random number in the interval: [0, 0.25), 1; in [0.25, 0.50), 2; in [0.50, 0.75), 3; in [0.75, 1), 4.

WEEK	BEGINNING STOCK	RANDOM NUMBER	CUSTOMERS	LOST SALES	INVENTORY AT END
1	3	0.771	4	1	0
2	4	0.035	1	0	3
3	3	0.189	1	0	2
4	2	0.992	4	2	0
5	4	0.501	3	0	1
6	4	0.530	3	0	1
7	4	0.627	3	0	1
8	4	0.100	1	0	3
9	3	0.726	3	0	0
10	4	0.334	2	0	2

6.

7.

TRIAL	RANDOM NUMBER	UNIFORM NUMBER
1	0.771	2.668
2	0.035	−3.220
3	0.189	−1.988
4	0.992	4.436
5	0.501	0.508

8. For a random number in the interval [0, 0.12), red; in [0.12, 0.4), green; in [0.40, 1), white.

TRIAL	RANDOM NUMBER	COLOR
1	0.771	white
2	0.035	red
3	0.189	green
4	0.992	white
5	0.501	white
6	0.530	white
7	0.627	white
8	0.100	red
9	0.726	white
10	0.334	green

9.

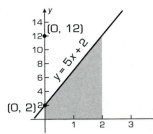

Enclose the region
in a rectangle with
sides $x = 0$ to $x = 2$
and $y = 0$ to $y = 12$.

TRIAL POINT	RANDOM NUMBER	x-VALUE	RANDOM NUMBER	y-VALUE	y = 5x + 2	ACCEPT POINT	TOTAL ACCEPTED
1	0.771	1.542	0.530	6.360	9.71	yes	1
2	0.035	0.070	0.627	7.524	2.35	no	1
3	0.189	0.378	0.100	1.200	3.89	yes	2
4	0.992	1.984	0.726	8.712	11.92	yes	3
5	0.501	1.002	0.334	4.008	7.01	yes	4

10. $P(x > 70) = P(z > +1) = 0.15866$

11. $\bar{x} = 12.5$, $s.d. = 3.338$

12.

TRIAL	RANDOM NUMBER	UNIFORM NUMBER
1	0.771	431.3
2	0.035	210.5
3	0.189	256.7
4	0.992	497.6
5	0.501	350.3
6	0.530	359.0
7	0.627	388.1
8	0.100	230.0
9	0.726	417.8
10	0.334	300.2

13.

Approximately exponentially distributed

14.

TRIAL	AMOUNT BET	RANDOM NUMBER	CARD	WINNINGS PER GAME	TOTAL WINNINGS
1	1	0.771	not ace	−1	−1
2	1	0.035	ace	10	9
3	2	0.189	not ace	−2	7
4	2	0.992	not ace	−2	5
5	2	0.501	not ace	−2	3

15. .91044

16.

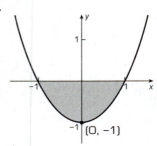

(0, −1)

Enclose the region
in a rectangle with
sides $x = -1$ to $x = 1$
and $y = -1$ to $y = 0$.

TRIAL POINT	RANDOM NUMBER	x-VALUE	RANDOM NUMBER	y-VALUE	$y = x^2 - 1$	ACCEPT POINT	TOTAL ACCEPTED
1	0.771	0.542	0.530	−0.470	−0.706	yes	1
2	0.035	−0.930	0.627	−0.373	−0.135	no	1
3	0.189	−0.622	0.100	−0.900	−0.613	no	1
4	0.992	0.984	0.726	−0.274	−0.032	no	1
5	0.501	0.002	0.334	−0.666	−0.749	yes	2

17.

TRIAL	RANDOM NUMBER	EXPONENTIAL NUMBER
1	0.771	1.30
2	0.035	16.76
3	0.189	8.33
4	0.992	0.04
5	0.501	3.46
6	0.530	3.17
7	0.627	2.33
8	0.100	11.51
9	0.726	1.60
10	0.334	5.48

18.

DAY	BEGINNING STOCK	RANDOM NUMBER	DAILY DEMAND	ENDING STOCK	LOST SALES
1	100	0.771	104	0	4
2	100	0.035	52	48	0
3	100	0.189	63	37	0
4	100	0.992	119	0	19
5	100	0.501	85	15	0
6	100	0.530	87	13	0
7	100	0.627	94	6	0
8	100	0.100	57	43	0
9	100	0.726	101	0	1
10	100	0.334	73	27	0
					24

The average daily lost sales based on this simulation is $\frac{24}{10} = 2.4$ pounds.

Chapter 9

Section 9.1   Improving Skills

1. function. Range = {10, 20, 30}
3. not a function.
5. not a function.
7. function. Range = {−2, 0, 2}
9. not a function.
11. function. Range = {2}
13. function. Range = {2, 3, 4}
15. function. Range = {1, 2, 3, 4, 5, 6}
17. function. Range = {1, 2, 3, 4, 5, 6}
19. function.   Natural domain = all real numbers.
21. not a function.
23. not a function.
25. not a function.
27. function.   Natural domain = all real numbers between −3 and 3, inclusive.
29. function.   Natural domain = all real numbers.
31. all real numbers.
33. all real numbers.
35. all real numbers.
37. all real numbers except $x = 2$ and $x = 3$.
39. all real numbers except $x = \sqrt{6}$ and $x = -\sqrt{6}$.
41. a. 3
    b. 5
    c. 5
    d. 3.5
43. a. −3
    b. 9
    c. $-6a - 3$
    d. $-3a - 3b + 3$
45. a. $-\dfrac{3}{2}$

    b. $\dfrac{1}{6}$

    c. $\dfrac{a + 1}{2a - 2}$

    d. $\dfrac{a + b - 2}{2a + 2b - 4}$

Creating Models

47. Revenue, R, is a function of the number of $8 ticket sales, x: $R = f(x) = 8x$. R is the output and x is the input. Natural domain for x = all nonnegative integers (cannot sell negative or fractional tickets).

49. Revenue, R, is a function of the number of $8 tickets, t, and the number of $2.50 bags of popcorn sold, p: $R = f(t, p) = 8t + 2.5p$. R is the output and t and p are the inputs. Natural domains for t and p = all nonnegative integers (cannot sell negative or fractional tickets or bags of popcorn)

51. Interest per quarter, I, is a function of the amount in the account earning 6% a year compounded quarterly, x: $I = f(x) = 0.015x$. I is the output, and the amount in the account at the beginning of the quarter, x, is the input. Natural domain for x = all nonnegative rational numbers (basically, any dollar amount).

53. The amount of interest earned on a $1,000 investment, I, is a function of the interest rate, r: $I = f(r) = 1000r$. The output is I and the input is r. Natural domain for r = all nonnegative rational numbers between 0 and 1 (interest rates will be decimal values between 0 and 1 [0% and 100%]).

55. The perimeter of a square, P, is a function of the length of the side, s: $P = f(s) = 4s$. The output is P, and the input is s. Natural domain of s = all positive real numbers.

57. The perimeter of a rectangle, P, depends on its height, h, and width, w: $P = f(h, w) = 2h + 2w$. The output is P, and the inputs are h and w. Natural domains for h and w: all positive real numbers.

59. The area of a square, A, is dependent on the length of its side, s: $A = f(s) = s^2$. The output is A, and the input is s. Natural domain for s = all positive real numbers.

61. The area of a rectangle, A, with one side length 4 in. depends on the length of the other side, x: $A = f(x) = 4x$. The output is A and the input is x. Natural domain for x = all positive real numbers.

63. The area of a circle, A, is dependent on the radius, r: $A = f(r) = \pi r^2$. The output is A and the input is r. Natural domain for r = all positive real numbers.

65. The area of a circle, A, depends on the diameter, d: $A = f(d) = \pi\left(\dfrac{d}{2}\right)^2$. The output is A, and the input is d. Natural domain for d = all positive real numbers.

67. The area of a triangle, A, with base 4 in. depends on its height, h: $A = f(h) = 2h$. The output is A and the input is h. Natural domain for h = all positive real numbers.

69. The area of a triangle, A, standing no higher than 5 in. depends both on its height, h, and the length of its base, b: $A = f(h, b) = \frac{1}{2}(h)(b)$. The output is A and the inputs are h and b. Natural domain for b = all positive real numbers; natural domain for h = all positive real numbers less than or equal to 5 inches.

Exploring with Technology

75.

Adding a positive constant to the function moves the graph up vertically by that amount.

80. $x \approx 14{,}782.61$
$y \approx 2.086.96$
$z \approx 35.65$

81. 0.0975.

82. expected number of houses = 0.6
expected profit per house = \$5,700
expected weekly profit = \$3,420

Section 9.2   Improving Skills

1. polynomial of degree 2   3. polynomial of degree 2
5. polynomial of degree 0   7. polynomial of degree 0
9. polynomial of degree 5   11. *not* a polynomial
13. *not* a polynomial   15. *not* a polynomial
17. *not* a polynomial
19. 11, 13, 16, 17, 18
   sum of the degrees of each polynomial in the product
23. If f(x) is a polynomial of degree n, then $f(0) = a_n(0)^n -$
   $\ldots + a_2(0)^2 + a_1(0) + a_o = 0 + a_o$. The point $(0, a_o)$ is
   the y-intercept.

Creating Models

25. Cost = cost per book + fixed costs, so let $C = 10x + 20{,}000$, where x = number of books manufactured.
27. Let x = number of loaves of bread made and sold. Then the cost function is $C = 120{,}000 + 0.23x$, the revenue function is $R = 0.55x$, and the profit function is $P = R - C = (0.55x) - (120{,}000 - 0.23x) = 0.32x - 120{,}000$.
29. a. 1,536 feet   b. 3,600 feet   c. 50 seconds
31. $V = s^3$ is a polynomial of degree 3.
33. $V = \frac{4}{3}\pi r^3$ is a polynomial of degree 3.

Exploring with Technology

Exploring with Technology

37.

$x \approx 25$

39. A 20-foot piece of rope is cut into two pieces, one with length x and the other with length $20 - x$. The sum of the squares of each length is $y = (x)^2 + (20 - x)^2$. Its graph on [0, 30] and [0, 500], and zoomed in is

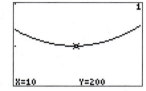

$x \approx 10, 20 - x \approx 10$

41. Four examples of fourth degree polynomials, all on $[-5, 5]$ and $[-10, 10]$
$y = 3x^4 - 4x^3 - x^2 - 1$ $y = -x^4 + 5x^2 - 1$

$y = 0.5x^4 - 2$ $y = x^4 - 4x^3 - 6x^2 - 4x + 1$

The maximum number of peaks and pits in a fourth degree polynomial is three, but there could be also be one peak/pit. The number of times the graph can cross the x-axis is zero, one, two, three, or a maximum of four.

Reviewing Material

43. $y = -240x + 32{,}000$
The point (200, 2,000) is not on the line.

44.

PAYMENT NUMBER	PAYMENT AMOUNT	PAYMENT ON INTEREST	PAYMENT ON DEBT	OUTSTANDING DEBT
0				$2,000.00
1	$92.29	$16.67	$75.62	$1,924.38
2	$92.29	$16.04	$76.25	$1,848.12
3	$92.29	$15.40	$76.89	$1,771.23

45. a. 11
   b. 140
   c. 158
   d. 9

Section 9.3   Improving Skills

1. Yes, $b = 7$
3. No
5. Yes, $b = 2$
7. $7^{2x} = (7^2)^x = 49^x$. Yes, $b = 49$
9. Yes, $b = \pi$
11. Yes, $b = e$
13. 18^x
15. $\left(\dfrac{6}{5}\right)^x$
17. $\frac{1}{64}4^x$
19. $2(4^x)$
21. a. 7.46496
   b. 4.56287
   c. 5.31967
   d. 1.73611
   e. 1.97244
   f. 3.18798
23. a. -34.8917
   b. -4.68794
   c. -0.25811
   d. -1.53703
   e. -0.78723
   f. -1.16307

Creating Models

31. $27,542.56
33. $2,610.23
35. $22,074.83
37. $28,098.96
39. approximately 279 strands
41. 35.36 million

Exploring with Technology

49. window: $[-10, 10]$ and $[-10, 10]$:

The graph is moved horizontally to the right by the c value in $x - c$.

25.

TYPE OF COMPOUNDING	FORMULA FOR VALUE AT MATURITY	VALUE AT MATURITY
Annually	$\$1\left(1 - \dfrac{0.09}{1}\right)^1$	1.09
Semiannually	$\$1\left(1 - \dfrac{0.09}{2}\right)^2$	1.092025
Quarterly	$\$1\left(1 - \dfrac{0.09}{4}\right)^4$	1.093083319
Monthly	$\$1\left(1 - \dfrac{0.09}{12}\right)^{12}$	1.093806898
Weekly	$\$1\left(1 - \dfrac{0.09}{52}\right)^{52}$	1.094089166
Daily	$\$1\left(1 - \dfrac{0.09}{365}\right)^{365}$	1.094162145
Hourly	$\$1\left(1 - \dfrac{0.09}{8760}\right)^{8760}$	1.094173778
Every minute	$\$1\left(1 - \dfrac{0.09}{525,600}\right)^{525,600}$	1.094174275
Every second	$\$1\left(1 - \dfrac{0.09}{31,536,000}\right)^{31,536,000}$	1.094174286

27. 4.08%
29. 8.55%

53. window: $[-10, 10]$ and $[-10, 10]$:

When x is replaced by $x + c$, the graph is moved horizontally to the left by c units.

Reviewing Material

55. $\mathbf{AB} = \mathbf{BA} = \begin{bmatrix} -2 & 18 & -12 \\ 10 & -21 & 21 \\ -10 & 22 & -20 \end{bmatrix}$

The matrices commute under multiplication.

56. $41.18

57. $C(80, 1)\, C(79, 4) = 120,200,080.$

Section 9.4 Improving Skills

1. a. 0.91629   b. -1.13943 c. 4.60517
   d. 3.80666   e. undefined   f. 8.08178
3. a. 6.15   b. -2.04 c. -18
   d. 8.28   e. 0.14   f. 5.25
5. $\ln(0.25) = \ln(\tfrac{1}{4}) = \ln(4)^{-1} = -1\,\ln(4)$
7. $\ln(16) = \ln(4)^2 = \ln(\tfrac{1}{4})^{-2} = -2\,\ln(\tfrac{1}{4})$
9. $\ln(64) = \ln(4)^3 = \ln(4^2)^{3/2} = \ln(16)^{3/2} = \tfrac{3}{2}\ln(16) = 1.5\,\ln(16)$
11. $\ln(6) = \ln(2 \cdot 3) = \ln(2) - \ln(3)$
13. $\ln(18) = \ln(9 \cdot 2) = \ln(9) + \ln(2) = \ln(3)^2 + \ln(2) = 2\,\ln(3) - \ln(2)$
15. $\ln(9.5) = \ln(\tfrac{19}{2}) = \ln(19) - \ln(2)$
17. $\ln(72) = \ln(8 \cdot 9) = \ln(8) + \ln(9) = \ln(2)^3 + \ln(3)^2 = 3\,\ln(2) + 2\,\ln(3)$
19. $\ln(9) = \ln(\tfrac{81}{9}) = \ln(81) - \ln(9) = \ln(81) - \ln(3)^2 = \ln(81) - 2\,\ln(3)$
21. 0.23105   23. 0.231   25. 0.69528
27. 0.79543   29. 1.24568

Creating Models

31. 15.338 years   33. 15.83 years
35. 14.11%   37. 10.94 years
39. a. $N(t) = 100e^{0.02789t}$, t in hours
    b. approximately 403 strands
41. approximately 62 strands

43. Let $t = 0$ in 1984
    $N(t) = 424e^{1.0564k}$
    In 1985, $t = 1$ and $N(1) = 1,219.42$.
    In 1987, $t = 3$ and $N(3) = 10,086.3$.
    In 1988, $t = 4$ and $N(4) = 29.008.19$.
45. a. $1.14
    b. 19.8 years
47. 73.6%
49. 54.6%
51. 1,593.44 years
53. 9%

Exploring with Technology

57. When $n < 1$, $\ln(x)$ grows faster.
    When $n > 1$, x^n grows faster.

Reviewing Material

58. $v = 931.44x + 11.908.5$
    The estimated average tuition charges in 2000 will be $y = 931.44(12.5) + 11,908.5 = 23,547.38$.

59.

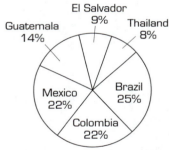

60. Level 1 (choice of slack color). Level 2 (choice of blazer color). Level 3 (choice of shirt color).

61. 0.8716

Section 9.5   Improving Skills

1. a. 0.384 radians
    b. 1.3963 radians
    c. 2.9147 radians
    d. 7.1558 radians
    e. −0.2443 radians
    f. −2.4086 radians
    g. −11.781 radians
    h. 0.0548 radians

3.

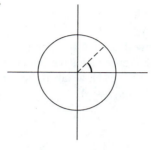

Both sine and cosine are positive.

5.

The sine will be positive and the cosine will be negative.

7.

Both sine and cosine are positive.

9.

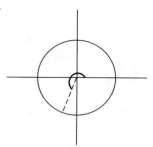

Both sine and cosine are negative.

11. −0.5366   13. 0.8415

15. 0.335   17. 0.342

19. 0.9569   21. −0.7818

23. $b = 2$ and $A = 2$ 25. $b = 4$ and $A = \frac{1}{2}$

27. $b = \pi$ and $A = 5$

29. $y = 3 \sin \theta$

31. $y = 3 + 3 \sin \theta$

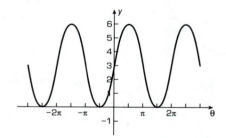

33. $y = \sin 2\theta$

35. $y = 2 \sin 2\theta$

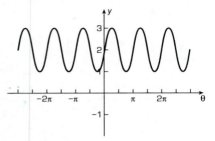

37. $y = 7 \sin(\theta/4)$

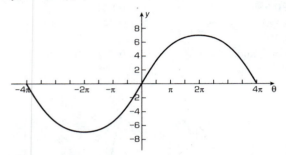

39. $y = -2 + 7 \sin(\theta/4)$

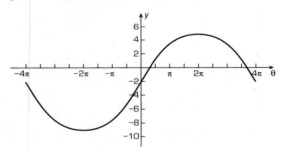

41. $y = \sin \pi\theta$

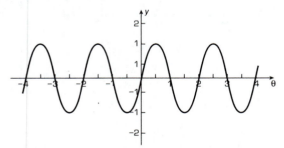

43. $y = 100 + \sin \pi\theta$

45. $y = \sin(\pi\theta/4)$

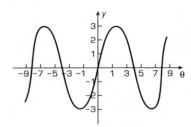

47. $b = 2$, $k = 1$, and $A = 1$

49. $b = 4$, $k = 7.5$, and $A = 2.5$

51. $b = \pi$, $k = 1.5$, and $A = 3.5$

53. On the unit circle, we know that $\sin \theta = y_o$ and $\cos \theta = x_o$. The right triangle in the Figure yields the Pythagorean Theorem in the form $x_o^2 + y_o^2 = 1$. Now substitute $(\sin \theta)^2 - (\cos \theta)^2 = 1$, or $\sin^2 \theta + \cos^2 \theta = 1$.

55. Consider the picture:

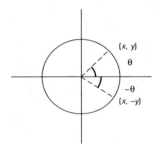

Since $\cos \theta = x$, note that $\cos(-\theta) = x = \cos \theta$. Therefore, $\cos(-\theta) = \cos \theta$.

Creating Models

57. $y = 78.5 - 21.5 \cos(\pi\theta/6)$ with $\theta = 0$ representing July (the high)

59. 9 am

61. The distance from the Earth to the sun is irrelevant. We put time on the x-axis, and angle on the y-axis. Over the course of a year, the angle increases as time increases. On January 1, the angle returns to zero, and the passage begins again.

Exploring with Technology

67. We obtain a line as we zoom in.

69. In general, you should find that as you zoom in, the graph looks like a line.

Reviewing Material

71. $x = 10$, $y = 8$, and $z = 11$
72. $58.275.67
73. 0.0506

Testing Yourself

1. a. *not* a function. The domain value 4 is assigned to two elements in the codomain.
    b. *not* a function. The domain element 21 is not assigned to any element in the codomain.
2. a. $y = 1.5 \sin(2\theta)$
    b. $y = \pi \sin(\pi\theta)$
3. a. 0   b. 3   c. $-\frac{3}{4}$ d. $1/z^6 - 1$
4. a. $N(t) = 100e^{0.3t}$
    b. 817 strands
5. a. 0.1225   b. 3.178
    c. 0.034   d. -0.9589
    e. -1.9378 f. 1.0838
6. 593.65
7. a. This graph is periodic with period $= 2\pi/2 = \pi$.

b. This graph is not periodic.

8. a. function
    b. not a function
9. $6,254.04
10. 4.084 radians
11. -5.307
12. 4,222 years old
13. approximately 6.25 years

14. a.

b.

15. 80.21°
16. a. polynomial function
    b. power function
    c. polynomial function
    d. power function
    e. polynomial function
    f. exponential function
    g. rational function
    h. polynomial function
17. a. The amplitude is 2. The period is $2\pi/3$.

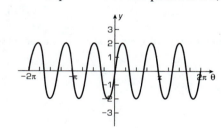

b. The amplitude is 7. The period is $2\pi/(\pi/3) = 2\pi(3/\pi) = 6$.

18. Value of the machinery in month t $= V(t) = 15,000 - 200(t)$. The input is the month ($0 \le t \le 24$) and the output is the value in the given month.

Chapter 10

Section 10.1 Improving Skills

1. 5

3. $\dfrac{-1}{3}$

5. $\dfrac{-1}{4}$

7. $\sin 1 \approx 0.841$

9. 0.535

11. 2

13. -3

15.

x_1	x_2	$f(x_1)$	$f(x_2)$	$\dfrac{f(x_2) - f(x_1)}{x_2 - x_1}$
1	1.5	2	4.375	4.75
1	0.5	2	1.125	1.75
1	1.2	2	2.728	3.64
1	0.8	2	1.512	2.44
1	1.1	2	2.331	3.31
1	0.9	2	1.729	2.71
1	1.05	2	2.157625	3.1525
1	0.95	2	1.857375	2.8525
1	1.01	2	2.030301	3.0301
1	0.99	2	1.970299	2.9701
1	1.001	2	2.003003001	3.003001
1	0.999	2	1.997002999	2.997001
1	1.0001	2	2.00030003	3.00030001
1	0.9999	2	1.99970003	2.99970001
1	1.00001	2	2.00003	3.00003
1	0.99999	2	1.99997	2.99997

The instantaneous rate of change at $x_1 = 2$ appears to be 3.

17. -2

19. -4

21. approximately 0.7682

23. 1

25. 1.250

27. $y = -2x + 4$

29. $y = 2$

31. $y = 7.389x - 7.389$

33. c, a, b.

35. a. 0.69 million people/year
b. 1.919 million people/year

37. 2.281671 million people/year

39. 54250.52 dollars/year

41. a. 0.145 million/year
b. 0.1895 million/year

43. (Answers may vary.)

a.

b.

c.

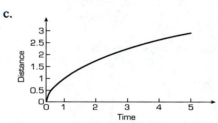

45. 64 ft./sec.

47. a. 6.5 million dollars
b. 10 million dollars per year
c. $80 million
d. $75.5 million

Reviewing Material

53. $x = 10; y = 8; z = 11$

54. 1 to 3

55. 3.8% are in zone 1, 34% are in zone 2, and 27.2% are in zone 3

56. a, b, c, and e

Section 10.2 Improving Skills

1.

x	LIMIT FROM LEFT	LIMIT FROM RIGHT	LIMIT	FUNCTION VALUE	CONTINUOUS AT x?
1	-1	-2	None	-1	No
2	1	2	None	1	No
3	2	-1	None	2	No

3.

x	LIMIT FROM LEFT	LIMIT FROM RIGHT	LIMIT	FUNCTION VALUE	CONTINUOUS AT x?
1	1	1	1	-1	No
2	1	1	1	-1	No
3	1	1	1	1	Yes

5.

x	LIMIT FROM LEFT	LIMIT FROM RIGHT	LIMIT	FUNCTION VALUE	CONTINUOUS AT x?
1	1	1	1	2	No
2	-1	-2	None	-2	No
3	0	-2	None	0	No

7.

x	LIMIT FROM LEFT	LIMIT FROM RIGHT	LIMIT	FUNCTION VALUE	CONTINUOUS AT x?
1	1	1	1	1	Yes
2	2	2	2	2	Yes
3	-1	-1	-1	-1	Yes

9.

x	LIMIT FROM LEFT	LIMIT FROM RIGHT	LIMIT	FUNCTION VALUE	CONTINUOUS AT x?
1	-1	-2	None	-1	No
2	1	2	None	1	No
3	2	-1	None	2	No

11.

x	LIMIT FROM LEFT	LIMIT FROM RIGHT	LIMIT	FUNCTION VALUE	CONTINUOUS AT x?
1	-1	1	None	1	No
2	1	-1	None	1	No
3	-2	0	None	-2	No

13.

x	LEFT POLY.	LEFT LIMIT	RIGHT POLY.	RIGHT LIMIT	LIMIT	FUNCTION VALUE	CONTINUOUS AT x?
1	x	1	1	1	1	1	Yes
2	1	1	$x + 1$	3	None	3	No
3	$x + 1$	4	$x + 1$	4	4	4	Yes

15.

x	LEFT POLY.	LEFT LIMIT	RIGHT POLY.	RIGHT LIMIT	LIMIT	FUNCTION VALUE	CONTINUOUS AT x?
1	x	1	3	3	None	3	No
2	3	3	$x + 1$	3	3	3	Yes
3	$x + 1$	4	$x + 1$	4	4	4	Yes

17.

x	LEFT POLY.	LEFT LIMIT	RIGHT POLY.	RIGHT LIMIT	LIMIT	FUNCTION VALUE	CONTINUOUS AT x?
1	x^3	1	x^3	1	1	1	Yes
2	x^3	8	$6 + x$	8	8	6	No
3	$6 + x$	9	$2 + x$	5	None	5	No

19.

x	LEFT POLY.	LEFT LIMIT	RIGHT POLY.	RIGHT LIMIT	LIMIT	FUNCTION VALUE	CONTINUOUS AT x?
1	$2x - 1$	1	1	1	1	1	Yes
2	1	1	1	1	1	1	Yes
3	1	1	$4 - x$	1	1	1	Yes

21. 5 **23.** $\dfrac{1}{2}$ **25.** 1

27. 2 **29.** 1 **31.** no limit

33. no limit **35.** 0 **37.** no limit

39. 1

47.

$$C = \begin{cases} 12, & 0 < t \le 1 \\ 18, & 1 < t \le 2 \\ 18 + 3(t - 2), & 2 < t \le 4 \\ 24, & 4 < t \le 24 \end{cases}$$

The cost for one full day is $24.00. The function is discontinuous at 1, 2, and 3.

49. Let C = cost of coffee, and x = amount, in pounds.

$$C = \begin{cases} 4x, & 0 \le x < 50 \\ 200 + 3.80(x - 50), & 50 \le x \le 100 \\ 390 + 3.50(x - 100), & x > 100 \end{cases}$$

The function is continuous and is defined as follows.

51. In all currently marketed calculators $1 + \dfrac{1}{10^{50}}$ will yield 1 because the calculator would need to carry 50 places to do the addition correctly. The computation $1^{10^{50}}$ gives 1. This is wrong because of the earlier error in adding 1 and $\dfrac{1}{10^{50}}$.

Creating Models

41. The cost for a 4.5 minute call is 50 cents.

$$c = \begin{cases} 0 & t = 0 \\ 0.10 & 0 < t \le 1 \\ 0.20 & 1 < t \le 2 \\ 0.30 & 2 < t \le 3 \\ 0.40 & 3 < t \le 4 \\ 0.50 & 4 < t \le 5 \end{cases}$$

where c = cost, and t = time, in minutes. The function is discontinuous at all integral multiples of one minute.

43. Let t = time (in years), and $f(t)$ = the balance in the account, then

$$f(t) = \begin{cases} P, & 0 \le t < 0.5 \\ P + 0.05/2P = P(1 + 0.025), & 0.5 \le t < 1 \\ P(1 + 0.025)^2, & 1 \le t < 1.5 \\ \vdots \\ P(1 + 0.025)^6, & 2.5 \le t < 3 \end{cases}$$

The function has discontinuities at all integral multiples of six months. At the end of three years, the balance is $1000(1.025)^6 = \$1159.93$.

45. Daily salary $s = 50 + 0.02S$, where s is sales. This function is continuous.

Reviewing Material

53. Yes. $A = 3B$, and $B = 5C$, so $A = 3B = 3(5C) = 15C$. The constant is 15.

54.

$$\begin{bmatrix} 1 & \frac{-1}{2} & 1 \\ 1 & \frac{-1}{2} & 2 \\ \frac{-1}{3} & \frac{1}{3} & \frac{-2}{3} \end{bmatrix}$$

55. $f(x) = 4 + 6 \sin 0.2\pi x$

Section 10.3 Improving Skills

1. $6x^5$
3. $15x^5$
5. $-8x^{-9}$
7. $-40x^{-9}$
9. $0.2x^{-0.8}$
11. $-0.2x^{-1.2}$
13. $6x^5 + 5x^4$
15. $7x^6 - 12x^{11}$
17. $12.2x^{5.1} - 18.3x^{-7.1}$
19. $\pi x^{\pi-1}$
21. 0
23. $5x^4 - 21x^2 + 4$
25. $7x^6 + 18x^2 - 8x$
27. $16x^7 + 30x^4 - 9x^2 + 9$
29. $16x^{-5} - 18x^2 - 4$
31. $x^4 - x^2 - x$
33. $7x^6 + 15x^4$
35. $4t^3 - 3$
37. $8.4N^{3.2} + 3$
39. $4\theta^3 - 4\theta$
41. $4r$
43. $-1/t^2$
45. a. -5
   b. 352
47. -17 49. 96 51. 0.25
53. a. Positive slope on the intervals $(-12, -6)$, $(1, 6)$, and $(9, 12)$.
   b. Negative slope on the intervals $(-6, 1)$, and $(6, 9)$.
   c. Derivative is zero at the points $x = -6$, $x = -1$, and $x = 9$.

55. $6x$ 57. $4x^3$ 59. $\dfrac{-1}{x^2}$

Creating Models

61. $\frac{9}{5}$ degrees F/degree C 63. 64 ft./sec.
65. 10 million dollars
67. a. $C = 21000 + 5.50N$
   b. 5.50
   c. The cost increases at a constant rate of $5.50 for each additional can opener produced.
69. a. 0.23 dollars/loaf
   b. 0.55 dollars/loaf
71. $2\pi r$ 73. $P/8$

Reviewing Material

80. mean = 5.6
   mode = 7; median = 6
   The standard deviation is $\sqrt{5.817} \approx 2.412$.
81. 0.04   82. a, b, c, d, f

Section 10.4 Improving Skills

1. $xe^x + e^x$
3. $x^4e^x + 4x^3e^x$
5. $(x^3 + 3x^2 + x + 1)e^x$
7. $(z^{0.8} - z^{-0.8})e^z + e^z(0.8z^{-0.2} + 0.8z^{-1.8})$
9. $e^x(-\sin x + e^x \cos x)$
11. $x + 2x \ln x$

13. $\dfrac{1 - 2 \ln t}{t^3}$

15. $x^2 - 5x + (3x^2 - 10x) \ln x$

17. $\dfrac{-1}{4}\left(\dfrac{\cos \theta}{\theta} - \ln \theta \cdot \sin \theta\right)$

19. $x \cos x + \sin x$

21. $q^{-6} \cos q + (-6q^{-7}) \sin q$

23. $\cos x$

25. $\sqrt{x}e^x + \dfrac{e^x}{2\sqrt{x}}$

27. $\dfrac{-3x^2 - 10x - 3}{(x^2 - 1)^2}$

29. $\dfrac{(3x + 4)(2x + 2)}{(3x + 4)^2}$

31. $\dfrac{(x^2 - 5x + 1)(2x + 2) - (x^2 + 2x - 3)(2x - 5)}{(x^2 - 5x + 1)^2}$

33. $\dfrac{7t^2 - 56}{(t^2 + 8)^2}$

35. $\dfrac{-1}{\sqrt{x}(\sqrt{x} - 1)^2}$

37. $\dfrac{-8e^x}{(e^x - 3)^2}$

39. $\dfrac{\ln x - 1}{(\ln x)^2}$

41. $\dfrac{1 - 3 \ln t}{t^4}$

43. $\dfrac{-1}{\sin^2 \theta}$

45. $\dfrac{(e^x + 1)(1 + \ln x) - xe^x \ln x}{(e^x + 1)^2}$

47. $\dfrac{\theta \cos \theta + \sin \theta + 2\theta^2 \cos \theta}{(1 + 2\theta)^2}$

49. $\dfrac{t^2e^t\left(t^2 - \dfrac{1}{t} + 3t^2 \ln t\right) - (t^3 - 1) \ln t(t^2e^t + 2te^t)}{(t^2e^t)^2}$

51. a. -0.3012
   b. -1
53. -0.7358 55. 147.32 57. $y = \frac{2}{45}x + \frac{7}{45}$.

Creating Models

59. a. 0.8 doll houses/day
   b. 0.102 doll houses/day
61. The population either stays the same or increases, but does not decrease over time.
63. $E' < 0$ over the interval (1950, 1973), and $E' > 0$ over the inverval (1973, present).

65. a. 11.0806 foxes/yr.
    b. 4.2715 foxes/yr.
67. 1500   69. -250
71. $R'(3) = -1775$, meaning that an increase from the current price of $3 results in a revenue decrease, so we would not recommend increasing price.

Exploring with Technology

79. The selling price should be about $10.00, at which point revenue no longer increases when price increases. After $10.00, the marginal revenue becomes negative, indicating that the revenue function is decreasing.

81.

The largest value of y (0.36787944) occurs when $x = 2.7182823$. At this point the derivative is equal to 0. This is not surprising, since, at its highest point, the function has a horizontal tangent, whose slope is 0.

Reviewing Material

82. $4015.46
83. 94.12% of the infected will be cured, and 5.88% will be dead.
84. a. graph A
    b. graph C
    d. graph B
    g. graph D

Section 10.5   Improving Skills

1. $(2x + 1)^2$
3. $(2x + 1)^2 + 1$
5. $\sin(x^2 + 2x - 5)$
7. $e^{2\sin x^2}$
9. $e^{\sin^4 x}$
11. $\dfrac{-2x^3}{\sqrt{1 - x^4}}$
13. $-x(x^2 + 4)^{-1.5}$

15. $\dfrac{-2x^3\sqrt{1 + x^4}}{(1 + x^4)^2}$
17. $\dfrac{-1}{(x - 1)^{3/2}(x + 1)^{1/2}}$
19. $\dfrac{-20480}{(x^2 - 5)^{11}}$
21. $2e^{2x}$ 23. $-3e^{-3x}$
25. $6xe^{3x^2-5}$ 27. $\cos x e^{\sin x}$
29. $\dfrac{-5e^{x/x-5}}{(x - 5)^2}$
31. $\dfrac{1}{x}$ 33. $-\dfrac{1}{x}$
35. $2\cos 2x$ 37. $2\sin x \cos x$
39. $-(2x + 2)\sin(x^2 + 2x + 3)$
41. $2 + 6\cos 2x$ 43. $2xe^{2x} + e^{2x}$
45. $2x^2e^{2x} + 2xe^{2x}$ 47. $2x^3 + 2xe^{x^2}$
49. $\dfrac{2x\ln x - x}{(\ln x)^2}$
51. $6x\sin^2(x^2)\cos(x^2)$
53. $\dfrac{5}{2}(1 + e^{x/2})^4 e^{x/2}$
55. $\dfrac{-15}{(x + 3)^2}\cos^4\!\left(\dfrac{x}{x + 3}\right)\sin\!\left(\dfrac{x}{x + 3}\right)$
57. $\dfrac{-1}{4\sqrt{x}\sqrt{6 - \sqrt{x}}}$
59. $\dfrac{7}{x}\cos(\ln x^7)$

Creating Models

61. (Answers may vary.)

63. (Answers may vary.)

65. $1877.62/year   67. $169.74/year
69. a. -2.383 mg/hr
    b. -2.144 mg/hr
71. -0.008627 g/yr.
73. a. $N(t) = 19.3e^{0.03853t}$
    b. The rate of increase in 1970 is 1.33 cents/year. The rate of increase in 1990 is 2.86 cents/year.
75. a. $f(t) = 78.5 + 21.5\cos(\pi/6t)$.
    b. 0 deg./mo.

Exploring with Technology

85. a. The graph shows that sales are seasonal for the first few years, then stabilize.

b. $t = 0.25$ years, or at the three month mark.
c. 0

87. a. The graph shows that sales increase linearly with time.

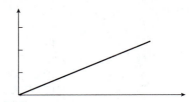

b. The sales function has no maximum value.
c. The sales function has no maximum value.

Reviewing Material

88. $z = 400$, when $x_1 = 100$, $x_2 = 0$, $x_3 = 0$, $s_1 = 200$.
89. $118,067.77
90. a. 220
    b. 60
    c. 0.273
91. a and d

Section 10.6   Improving Skills

1. $y = 27x - 52$
3. $y = 2x + 1$
5. $y = -ex$
7. $y = 2x - 1$; $(1.1)^2 \approx 1.2$; Actual value $= 1.21$; Error $= 0.01$
9. $y = \frac{1}{4}x + 1$; $\sqrt{4.15} \approx 2.0375$; Actual value $= 2.0371549$; Error $= -0.0003451$
11. $y = x$; $\sin 0.02 \approx 0.02$; Actual value $= 0.0199987$; Error $= -0.0000013$
13. a. $y = \frac{1}{2}x + \frac{1}{2}$; $\sqrt{1.23} \approx 1.115$
    b. $y = \frac{1}{6}x + \frac{3}{2}$; $\sqrt{8.95} \approx 2.9916667$
    c. $y = \frac{1}{12}x + 3$; $\sqrt{36.4} \approx 6.0333333$

15. The table below gives the values for e^x and $1 + x$ difference $e^x - (x + 1)$, for values of x close to 0. Notice that as $x \to 0$, the difference approaches 0.

x	e^x	$x + 1$	$e^x - (x + 1)$
-0.1	0.9048374	0.9	0.004837418
-0.09	0.9139312	0.91	0.003931185
-0.08	0.9231163	0.92	0.003116346
-0.07	0.9323938	0.93	0.00239382
-0.06	0.9417645	0.94	0.001764534
-0.05	0.9512294	0.95	0.001229425
-0.04	0.9607894	0.96	0.000789439
-0.03	0.9704455	0.97	0.000445534
-0.02	0.9801987	0.98	0.000198673
-0.01	0.9900498	0.99	4.98337E-05
0	1	1	0
0.01	1.0100502	1.01	5.01671E-05
0.02	1.0202013	1.02	0.00020134
0.03	1.0304545	1.03	0.000454534
0.04	1.0408108	1.04	0.000810774
0.05	1.0512711	1.05	0.001271096
0.06	1.0618365	1.06	0.001836547
0.07	1.0725082	1.07	0.002508181
0.08	1.0832871	1.08	0.003287068
0.09	1.0941743	1.09	0.004174284
0.1	1.1051709	1.1	0.005170918

Creating Models

17. a. 10 million dollars/year
    b. 10 million dollars
    c. $y(8) - y(7) = 10.5$ million dollars
19. a. -1224 items/dollar
    b. $-$1224
    c. $n(4) - n(3) = -$1071
21. a. 223,029 people/yr.
    b. 223,029 people
    c. $N(11) - N(10) = 266,625$ people

Exploring with Technology

23. differentiable at both $x = 0$ and $x = 1$
25. no, at $x = 0$; yes, at $x = 1$
27. no, at $x = 0$; yes, at $x = 1$
29. no, at $x = 0$; no, at $x = 1$
31. yes, at $x = 0$; yes, at $x = 1$
33. yes, at $x = 0$; yes, at $x = 1$
35. yes, at $x = 0$; yes, at $x = 1$
37. yes, at $x = 0$; yes, at $x = 1$

Reviewing Material

41. (100, 116.667)

42.

Yearly Rental Cost

43. a. 0.9   b. 0   c. 0.63
44. a. $N(t) = 20e^{0.8047t}$
    b. approximately 224 strands
    c. 0.861 days

Testing Yourself

1. $\dfrac{-1}{2}$

2. $= 0.1667$ million/yr.

3. The values in the table below suggest that $\lim\limits_{x \to 1} f(x)$ exists and is equal to 1.

$x \to 1$ FROM THE LEFT	FUNCTION VALUE
0.9	0.62
0.99	0.9602
0.999	0.996002
0.9999	0.9996
0.99999	0.99996
0.999999	0.999996
$x \to 1$ FROM THE RIGHT	FUNCTION VALUE
1.1	1.4641
1.01	1.040604
1.001	1.004006
1.0001	1.0004001
1.00001	1.00004
1.000001	1.000004

4. yes, at $x = 1$; yes, at $x = 2$
5. The graph below shows that the function is continuous everywhere.

6. no, at $x = -1$; yes, at $x = 0$ and $x = 1$

7. a. $f'(2) = \lim\limits_{h \to 0} \dfrac{f(2+h) - f(2)}{h}$. Using the following table, it appears that the limit exists and is equal to 12.

$h \to 0$ FROM LEFT	$\dfrac{f(2+h) - f(2)}{h}$
−0.1	11.7
−0.01	11.97
−0.001	11.997
−0.0001	11.9997
−0.00001	11.99997
−0.000001	11.999997
−0.0000001	11.9999997
$h \to 0$ FROM RIGHT	
0.1	12.3
0.01	12.03
0.001	12.003
0.0001	12.0003
0.00001	12.00003
0.000001	12.000003
0.0000001	12.00000028

b. $f'(x) = 6x$; $f'(2) = 6(2) = 12$
8. $y = \frac{2}{3}x + \frac{1}{3}$; $(1.08)^{2/3} \approx 1.053333$
9. $5x^4$
10. $\dfrac{-5e^{2x}}{x} - 10e^{2x} \ln(1.3x)$
11. $\dfrac{-6x^8 - 35x^4 - 5}{(1 - 3x^4)^2}$
12. $-12\pi \sin(\pi x) \cos(\pi x)e^{2 \cos(\pi x)}$
13. $(x^5 - 6x^3 + x^{12})(-4x^{-5} - 4x^3) + (x^{-4} - x^4)(5x^4 - 18x^2 + 12x^{11})$
14. 0
15. $\dfrac{(4 + \cos x)(1 - e^x) - (x - e^x)(-\sin x)}{(4 + \cos x)^2}$
16. $\dfrac{3x^2 - 3}{2(-3x + x^3)^{-1/2}}$
17. $e^{-2x} \cos x - 2e^{-2x} \sin x$
18. $\dfrac{2}{2x + 7}$
19. $12x^2 \cos^3(1 - x^3) \sin(1 - x^3)$
20. $56x^6 - 21x^2$
21. $S'(1) = 0.73576$ million items/million dollars. For each additional million spent in advertising, sales increases by only 0.73576 million items, that is, 735,760 items. The question of advisability of increasing the advertising budget cannot be answered with the given information, since revenue depends on selling price. If the item sells for $1.00, for instance, it would not be advisable to increase

the advertising budget, since an increase of $1 million in advertising would result in a revenue increase of only $0.73576 million in revenue. However, if the item sells for $100, then an increase of $1 million in advertising results in an increase of $73.576 million in revenue.

22. $\dfrac{3}{\pi}$

23. $y = -\frac{1}{4}x + \frac{1}{2}$

24.

x_1	-1	0	1	2	3	4
$\lim\limits_{x \to x_1^-} f(x)$	-1	1	-2	3	2	3
$\lim\limits_{x \to x_1^-} f(x)$	-1	1	-1	-2	2	3
Continuous	yes	yes	no	no	no	yes
Differentiable	yes	no	no	no	no	no

25. The table below gives the values of $\ln x$, $x - 1$, and the difference $[\ln x - (x - 1)]$ near $x = 1$. As $x \to 1$, the difference approaches 0.

x	$\ln x$	$1 - x$	$\ln x - (x - 1)$
0.9	-0.105360516	-0.1	-0.005360516
0.99	-0.010050336	-0.01	$-5.03359\text{E-}05$
0.999	-0.0010005	-0.001	$-5.00334\text{E-}07$
0.9999	-0.000100005	$-1\text{E-}04$	$-5.00033\text{E-}09$
0.99999	$-1.00001\text{E-}05$	$-1\text{E-}05$	$-5.00003\text{E-}11$
0.999999	$-1\text{E-}06$	$-1\text{E-}06$	$-5\text{E-}13$
1.1	0.09531018	0.1	-0.00468982
1.01	0.009950331	0.01	$-4.96691\text{E-}05$
1.001	0.0009995	0.001	$-4.99667\text{E-}07$
1.0001	$9.9995\text{E-}05$	$1\text{E-}04$	$-4.99967\text{E-}09$
1.00001	$9.99995\text{E-}06$	$1\text{E-}05$	$-4.99997\text{E-}11$
1.000001	$9.99999\text{E-}07$	$1\text{E-}06$	$-5\text{E-}13$
1.0000001	$1\text{E-}07$	$1\text{E-}07$	$-5\text{E-}15$

26. a. $f'(-1) = \lim\limits_{h \to 0} \dfrac{f(-1 + h) - f(-1)}{h}$. The table below gives values of the difference quotient as $h \to 0$ from the right and from the left. It appears that the limit exists and is equal to 4.

$x \to -1$ FROM LEFT	
-0.1	3.47107438
-0.01	3.940790119
-0.001	3.99400799
-0.0001	3.99940008
-0.00001	3.999940001
-0.000001	3.999994
-0.0000001	3.999999403

$x \to -1$ FROM RIGHT	
0.1	4.691358025
0.001	4.00600801
0.0001	4.00060008
0.00001	4.000060001
0.000001	4.000006
0.0000001	4.000000597

b. $f'(x) = \dfrac{-4}{x^3}$; $f'(-1) = \dfrac{-4}{(-1)^3} = 4$

27. $V(x) \approx V(1000) + V'(1000)(x - 1000)$
$V(1001) = 0.89315$ dollars

28. $x = -1$: not continuous and not differentiable
$x = 0$: both continuous and differentiable
$x = 1$: both continuous and differentiable

Chapter 11

Section 11.1 Improving Skills

1. Global maximum is -75, at $x = -2$.
Global minimum is -375, at $x = 2$ and $x = 8$.
Local maximum is -125, at $x = 4$ and $x = 6$.
Local minimum is -225, at $x = 5$.

3.

	x	$y = 3x^2 - 24x - 5$	REMARKS
endpoints	0	-5	global maximum
	6	-41	
critical point	4	-53	global minimum

3. critical point at $x = 4$, global maximum is -5 at $x = 0$, global minimum is -53 at $x = 4$

5. no critical points, global maximum is 480 at $x = 5$, global minimum is 25 at $x = 0$

7. no critical points, global maximum is 99 at $x = 50$, global minimum is 49 at $x = 25$

9. critical points at $x = -2$ and $x = 2$, global maximum is 23 at $x = -2$, global minimum is -9 at $x = 2$

11. critical points at $x = -2$ and $x = 2$, global maximum is 72 at $x = 5$, global minimum is -58 at $x = -5$

13. critical point at $x = 2$, global maximum is 7 at $x = 0$, global minimum is -9 at $x = 2$

15. critical point at $x = 3$, global maximum is 2908 at $x = 6$, global minimum is 2692 at $x = 3$

17. critical point at $x = 4$, global maximum is 57 at $x = 0$, global minimum is -71 at $x = 4$

19. critical points at $x = 0$ and $x = 8$, no global maximum, global minimum is -220 at $x = 8$

21. critical points at $x = 1$ and $x = 3$, global maximum is 10 at $x = 1$ and $x = 4$, global minimum is 6 at $x = 3$

23. critical points at $x = 1$ and $x = 5$, global maximum is -3 at $x = 1$ and $x = 7$, global minimum is -35 at $x = -1$ and $x = 5$

25. critical points at $x = \frac{2}{3}$ and $x = 6$, global maximum is 130 at $x = 10$, global minimum is -62 at $x = 6$

27. critical points at $x = 1$ and $x = 2$, no global maximum, no global minimum

Exploring with Technology

29. Global maximum: (4, 257); Global minimum: (0, 0)

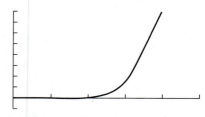

31. Global maximum: None; Global minimum: (1, 0)

33. Global maximum: (6.449, 6.361); Global minimum: (3.446, -3.288)

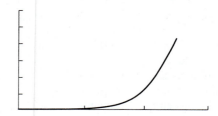

35. Global maximum: (6.573, 41.403); Global minimum: (3.638, -11.638)

37. Global maximum: (0.751, 0.322); Global minimum: (3.927, -0.014)

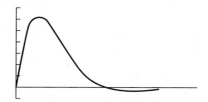

39. Global maximum: (0.751, 0.322); Global minimum: None

41. Global maximum: None; Global minimum: (1.829, -0.707)

43. Global maximum: (1, 6); Global minimum: (4, 1)

45. Global maximum: None; Global minimum: None

47. Global maximum: (1, 1); Global minimum: (0, 0)

49. Global maximum: (2, 8); Global minimum: (0, 0)

51. Global maximum: (0, 1); Global minima: $(-2, 0.135)$, $(2, 0.135)$

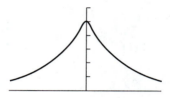

53. Global maximum: (0, 1); Global minimum: None

55. Global maximum: (0, 1); Global minimum: $(2, -1)$

Reviewing Material

61. $x = 6.5$ and $y = 6$
62. $12,145.88
63. 0.650
64. There is no limit at 0, so $f'(0)$ does not exist.

Section 11.2   Creating Models

1. 75 ft. by 150 ft. with area of 11,250 sq. ft.
3. square corral: sides of 228.79 ft.; rectangular corral: 7.07 ft. by 14.14 ft.

5. (Answers may vary.) 8 ft. by 150 ft. With four foot aisles, plus two feet on each side for merchandise, neither dimension should be less than eight feet.
7. 8.11 in. by 12.17 in.
9. $r = \dfrac{28}{\pi} \approx 8.9$ in. and $h = 28$ in.
11. base: 1.145 ft. by 1.145 ft.; height: 1.526 ft.
13. $l \approx 2.6$ ft., $w \approx 1.73$ ft., $h \approx 2.08$ ft.
15. rectangular base: 0.56 m. by 1.121 m.; semicircle radius: 0.56 m.
17. r^2
19. $0.1667x + 0.125y = 1$.

Reviewing Material

27.

28.

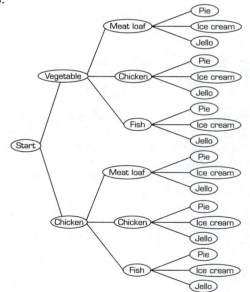

18 different dinners

29. 0.143

30. $1.34

Section 11.3 Creating Models

1. maximum profit is $17,500, when 150 sets per week are produced
3. maximum profit is $225,000, when 25,000 items are produced each month
5. The maximum profit is $4,750, when 100 units are produced with a total cost of $25,000
7. 105 items
9. 5000 copies at a loss of $21,566
11. $43,820
13. 30 trees
15. 60
17. 57
19.

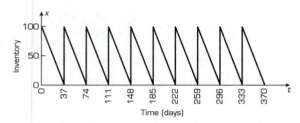

21. $EOQ = 90$ Each order should be placed every $36\frac{1}{2}$ days.
23. $EOQ = 200$ Each order should be placed every 365/8 $\approx$ 46 days.

Exploring with Technology

29.

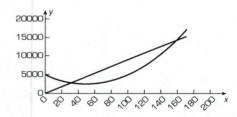

The break-even points are approximately 31.56 and 158.44. They are not interesting: 31.56 is neither a maximum nor a minimum, and 158.44 is outside the domain, beyond the capacity of the plant.

31. $C(95) = 4525$; $I(95) = 8550$; $C'(95) = 90$; $I'(95) = 90$
The derivatives are the same at $x = 95$. If the derivative P' exists at the point where profit is a maximum, then it is zero there. But profit $P = C - I$, so $P' = C' - I'$. $P' = 0 \Rightarrow C' - I' = 0 \Rightarrow C' = I'$.

Reviewing Material

33. Yes, if A is directly proportional to $5B$, then $A = k(5B) = (5k)B$. Let $c = 5k$. Then $A = cB \Rightarrow A$ is directly proportional to B.

34. 80 sofas and no love seats or chairs. 120 upholstery labor hours, and 280 decorator labor hours are unused.
35. [0.36066, 0.47541, 0.16393].
36. $6x - e^{-2x} + 2xe^{-2x}$

Section 11.4 Improving Skills

1. a. $y = 1 = 2t^2 + 5t + 3$
$$\frac{dy}{dt} = 4t + 5$$

b. $\dfrac{dy}{dt} = \dfrac{dy}{dx} \cdot \dfrac{dx}{dt} = (4x - 3)(1) = 4x - 3$

$x = t + 2 \Rightarrow \dfrac{dy}{dt} = 4t + 5$

3. a. $y = t^2 - 8t + 15$
$$\frac{dy}{dt} = 2t - 8$$

b. $\dfrac{dy}{dt} = \dfrac{dy}{dx} \cdot \dfrac{dx}{dt} = (2x - 2)(1)$

$x = t - 3 \Rightarrow \dfrac{dy}{dt} = 2t - 8$

5. a. $y = (3t + 6)^3 + 3(3t^2 + 6)$
$$\frac{dy}{dt} = 18t(9t^4 + 36t^2 + 36) + 18t$$

b. $\dfrac{dy}{dt} = \dfrac{dy}{dx} \cdot \dfrac{dx}{dt} = (3x^2 + 3)(6t)$

$x = 3t^2 + 6 \Rightarrow \dfrac{dy}{dt} = 18t(9t^4 + 36t^2 + 36) + 18t$

7. a. $y = \cos\dfrac{1}{t}$

$$\frac{dy}{dt} = \frac{\sin\dfrac{1}{t}}{t^2}$$

b. $\dfrac{dy}{dt} = \dfrac{dy}{dx} \cdot \dfrac{dx}{dt} = -\sin x(-t^{-2}) = \dfrac{\sin x}{t^2}$

$x = \dfrac{1}{t} \Rightarrow \dfrac{dy}{dt} = \dfrac{\sin\dfrac{1}{t}}{t^2}$

9. a. $y = \ln\left(\dfrac{1}{2t}\right)$

$$\frac{dy}{dt} = \frac{-1}{t}$$

b. $\dfrac{dy}{dt} = \dfrac{dy}{dx} \cdot \dfrac{dx}{dt} = \dfrac{1}{x} \cdot \dfrac{1}{2}(-t^{-2}) = \dfrac{-1}{2xt^2}$

$x = \dfrac{1}{2t} \Rightarrow \dfrac{dy}{dt} = \dfrac{-1}{t}$

11. $\dfrac{dy}{dt} = \dfrac{dy}{dx} \cdot \dfrac{dx}{dt} = 48x - 72;$ $\left.\dfrac{dy}{dt}\right|_{x=10} = 408$

13. $\dfrac{dy}{dt} = \dfrac{dy}{dx} \cdot \dfrac{dx}{dt} = -6x^2 + 6;$ $\left.\dfrac{dy}{dt}\right|_{x=3} = -48$

15. $\dfrac{dy}{dt} = \dfrac{dy}{dx} \cdot \dfrac{dx}{dt} = \dfrac{120}{(3x - 100)^2}$

$\left.\dfrac{dy}{dt}\right|_{x=45} = 0.09796$

17. $\dfrac{dy}{dt} = \dfrac{dy}{dx} \cdot \dfrac{dx}{dt} = 15 \cos 3x;$ $\left.\dfrac{dy}{dt}\right|_{x=1} = -14.845$

19. $\dfrac{dy}{dt} = \dfrac{dy}{dx} \cdot \dfrac{dx}{dt} = \dfrac{6}{x};$ $\left.\dfrac{dy}{dt}\right|_{x=0.05} = 120$

Creating Models

21. \$312,500
23. 0.9091 items/month
25. -0.01 items/year
27. -0.00064 cases/day
29. 0.407 ft/min, independent of depth
31. -512π cm/hr
33. 2261.95 in^2/min
35. 0.000796 cm/sec
37. 17.25 dollars/hour

Reviewing Material

43. The system has no solution
44. 0.013

45. The average rate of change is 39.542. The instantaneous rate of change at $t = 10$, estimated from the table below is 52.175 dollars/year.

t_1	t_2	$V(t_1)$	$V(t_2)$	$[V(t_2) - V(t_1)]/(t_2 - t_1)$
10	10.200000	895.4238483	905.9199632	52.48057457
10	10.100000	895.4238483	900.6566158	52.32767572
10	10.010000	895.4238483	895.945754	52.19057391
10	10.001000	895.4238483	895.4760252	52.17689008
10	10.000100	895.4238483	895.4290658	52.17552196
10	10.000010	895.4238483	895.42437	52.17538515
10	10.000001	895.4238483	895.4239004	52.17537129
10	9.700000	895.4238483	879.9072527	51.72198534
10	9.800000	895.4238483	885.0493428	51.87252733
10	9.900000	895.4238483	890.2214828	52.02365467
10	9.990000	895.4238483	894.9022466	52.16017189
10	9.999000	895.4238483	895.3716744	52.17384988
10	9.999900	895.4238483	895.4186307	52.17521794
10	9.999990	895.4238483	895.4233265	52.17535476
10	9.999999	895.4238483	895.4237961	52.17536856

Section 11.5   Improving Skills

1. $20x^3 + 8$
   28
   168
3. 0   0   0
5. 6   6   6
7. -2 -2 -2
9. $\dfrac{-2}{x^3}$ -2

   $\dfrac{-1}{4}$
11. $-\sin x$
    -0.8415 -0.9093

13. $-4x \cos 2x - 4 \sin 2x$
    -1.9726
    8.256
15. $\dfrac{-2}{x^2}$

    -2 $\dfrac{-1}{2}$
17. $4e^{x^2}(x^2 + 1)$
    21.7463   1091.963
19. $2e^x \cos x$
    2.9374
    -6.1499
21. $y' > 0$ and $y'' < 0$ on $(0, 4)$

23. $y' > 0$ on $(-3, 0)$, $y' < 0$ on $(0, 3)$, y' does not exist at $x = 0$;
    $y'' > 0$ on $(-3, 0)$ and $(0, 3)$
25. $y' > 0$ on $(-2, 1)$, $y' < 0$ on $(-4, -2)$ and $(1, 4)$, $y' = 0$ at $x = -2$ and $x = 1$;
    $y'' > 0$ on $(-4, -\frac{1}{2})$, $y'' < 0$ on $(-\frac{1}{2}, 4)$, $y'' = 0$ at $x = -\frac{1}{2}$
27. $y' < 0$ on $(-3, 0)$ and $(0, 3)$, $y' = 0$ at $x = 0$;
    $y'' > 0$ on $(-3, 0)$, $y'' < 0$ on $(0, 3)$, $y'' = 0$ at $x = 0$
29. $y' > 0$ on $(-3, -1.7)$ and $(0, 0.9)$, $y' < 0$ on $(-1.7, 0)$ and $(0.9, 2)$, $y' = 0$ at $x = -1.7$ and $x = 0.9$;
    $y'' > 0$ on $(-1.2, 0.6)$, $y'' < 0$ on $(-3, -1.2)$ and $(0.6, 2)$, $y'' = 0$ at $x = -1.2$ and $x = 0.6$.
31. strictly increasing when $x > 0$, strictly decreasing when $x < 0$, concave up everywhere
33. strictly increasing when $x > 1$, strictly decreasing when $x < 1$, concave up everywhere
35. strictly increasing everywhere, concave up when $x > 2$, concave down when $x < 2$
37. strictly increasing and concave up everywhere
39. strictly increasing and concave down everywhere
41. strictly decreasing everywhere, concave up when $x > 0$, concave down when $x < 0$
43. $(0, 10)$ is a local maximum.
45. $(6, -36)$ is a local minimum.
47. $(-2, 16)$ is a local maximum.
    $(2, -16)$ is a local minimum.
49. $(0, 0)$ is a local minimum.
    $(2, 4)$ is a local maximum.
51. $(1, -2)$ is a local minimum.
    $(-2, 25)$ is a local maximum.
53. $(1.816, -4.089)$ is a local minimum.
    $(0.184, -1.911)$ is a local maximum.
55. $(-1, -1)$ is a local maximum.
    $(0, 0)$ is a local maximum.
    $(1, -1)$ is a local minimum.
57. $(0, 1)$ is a local maximum.
59. $\left(-1, -\frac{1}{2}\right)$ is a local minimum; $\left(1, \frac{1}{2}\right)$ is a local maximum.

Exploring with Technology

65.

x	$f(x)$	$f'(x)$	$f''(x)$
0	0	1	0
−1	−0.5	0	0.5
1	0.5	0	−0.5
−2	−0.4	−0.12	−0.032
2	0.4	−0.12	0.032

At $x = 0$ the first derivative is positive, so the function is increasing. The second derivative is zero at $x = 0$, negative to the left of $x = 0$, and positive to the right of $x = 0$, so $x = 0$ is an inflection point of the graph.

The points $x = -1$ and $x = 1$ are critical points, since the first derivative is zero there. At the critical point $x = -1$ the second derivative is positive, so the point $(-1, -0.5)$ is a local minimum. At the critical point $x = 1$ the second derivative is negative, so the point $(1, 0.5)$ is a local maximum.

At the point $x = -2$ the first derivative is negative, so the function is decreasing there, and concave down, since the second derivative is negative at $x = -2$. At the point $x = 2$ the first derivative is negative, so the function is decreasing there, and concave up, since the second derivative is positive at $x = 2$.

Reviewing Material

67. $2,994.92
68. a. 65.6%
    b. 60.7%
    c. 58.6%
69. $A = 40$, $b = \dfrac{2\pi}{5}$, $k = 60$

Testing Yourself

1.

	x	y	REMARKS
endpoints	5	1950	global minimum
	25	2450	global maximum
critical point	10	2450	global maximum
	20	1950	global minimum

2. local minimum at $(20, 1950)$; local maximum at $(10, 2450)$
3. 10 and 10
4. The first derivative is positive for $x > 0$, so the function is strictly increasing everywhere in its domain. There are no critical points, so there are no local or global optima. The second derivative is negative for $x > 0$, so the graph is concave down everywhere in the domain.
5. $1.50
6. a. $18x + 4$
    b. $60x^4 - 30x + 4$
    c. $-4\cos 2x - 4\sin 2x$
7. 6.3662 cm./sec.

8. (Answers may vary.)

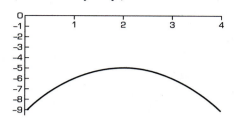

9. 50

10.

	x	y	REMARKS
endpoints	−1	−6	
	3	10	global maximum
critical points	0	1	
	2	−15	global minimum

11. local maximum at $(0, 1)$, local minimums at $(-2, -15)$ and $(2, -15)$

12. a. $(-5, 5)$ b. $(-10, -5)$ and $(5, 12)$ c. $(-10, 0)$ d. $(0, 12)$

13. The points $(-10, 0)$ and $(5, 0)$ are local and global maxima. The point $(-5, -500)$ is a local minimum. The point $(12, -1000)$ is a global minimum, assuming that the domain of the function is the closed interval $[-10, 12]$.

14. 3145.93 ft.^{2}/hr.

15. (Answers may vary.)

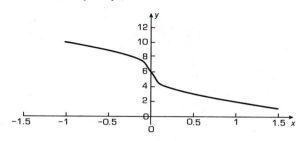

16. Each piece should be 10 ft. long.

17. a. $(-\infty, 1)$ and $(3, \infty)$ b. $(1, 3)$ c. $(2, \infty)$ d. $(-\infty, 2)$

18. $r = \dfrac{2R}{3}$

19. a. 42   b. −0.3679

20. order size of 13, placed every $365/123 = 3$ days.

Chapter 12

Section 12.1   Improving Skills

1. $\dfrac{1}{5}x^5 + c$

3. $1.1x^5 + c$

5. $\dfrac{1}{4}x^8 + c$

7. $\frac{1}{5}x^5 - \frac{1}{8}x^8 + c$

9. $2x + c$

11. $\frac{3}{2}x^2 - 22x + c$

13. $\frac{2}{3}x^{3/2} - 5x + c$

15. $\frac{1}{3}x^3 + \frac{1}{2}x^2 + x + c$

17. $-\frac{1}{4}x^{-4} + \frac{1}{3}x^{-3} - \frac{1}{2}x^{-2} + c$

19. $-x^{-1} + \frac{1}{3}x^3 + 2x + c$

21. $x - \ln|x| + c$

23. $\frac{23}{42}x^{4.2} - \frac{1}{8}x^{-4.8} + c$

25. $2\sqrt{x} + c$

27. $\frac{3}{4}x^{4/3} + \frac{8}{5}x^{5/4} - \frac{9}{2}x^{2/3} + c$

29. $-3 \cos x + c$

31. $\frac{5}{7} \sin x + c$

33. $-\cos x - \sin x + c$

35. $3x + 2e^x + c$

37. $e^x - x + c$

39. $e^x - \ln|x| + c$

Exploring with Technology

53. The graph below is Figure 12.2 reproduced, with tangent lines at $x = -2$. The tangent lines are parallel and each has the same slope of 58. This pattern will hold for any x-value you choose.

55. Below are the graphs of $y = -3 \ln|x| + 5$, $y = -3 \ln|x| + 1$, and $y = -3 \ln|x| - 2$ on $[-5, 5]$ and $[-7, 7]$. The tangent line is drawn at $x = 1$. The tangent lines are parallel slope $= -3$. Notice that $f'(x) = -3/x$ and that $f'(1) = -3$.

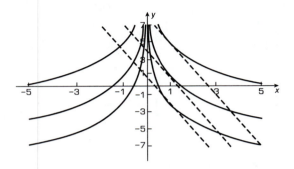

Reviewing Material

56. a. 0.3333   b. 0.2927
57. The graph in Figure A is not a function. The graph in Figure B is a function.
58. $-6e^{-2x}$
59. The global minimum is $y = -144$ at endpoint $x = -3$, and the global maximum is $y = 360$ at endpoint $x = 3$.

Section 12.2   Improving Skills

1. $u^3 - 4u^2 + 6u + c$
3. $\frac{10}{13}u^{1.3} + \frac{10}{7}u^{0.7} + c$
5. $\frac{2}{3}z^{3/2} + c$
7. $2\sqrt{3n} + c$
9. $2\sqrt{5}/3y^{3/2} - \sqrt{5}y + c$
11. $2.55t^2 - 0.2778t + c$
13. $1/\pi t + c$
15. $2e^n - 4n + c$
17. $2e\,y + c$
19. $2e^\theta + c$
21. $5 \sin u + c$
23. $-\frac{503}{75} \cos \theta + c$
25. $-2 \cos t - 3 \sin t + c$
27. $\frac{1}{2}y^2 - 8 \sin y + c$
29. $te^{\cos 5} + c$
31. a. $c = 1$ b. $c = 2$
    c. $c = -1$ d. $c = 1$
    e. $c = -1$ f. $c = 0$

Creating Models

33. $745   35. $10.248 million
37. $1,635,000

39. approximately 210 strands
41. 29,200 subscriptions
43. $v(t) = 32t - 16$
    $x(t) = 16t^2 - 16t$
    $0 \le t \le 7.5887$
45. 277.12 feet per second.
47. 4.29 seconds

Reviewing Material

51. $\mathbf{A}^{-1} = \begin{bmatrix} -2.6 & 1.8 & -0.6 \\ 0.4 & -0.2 & 0.4 \\ 0.8 & -0.4 & -0.2 \end{bmatrix}$

52. The amplitude is 3 and the period is $2\pi/4 = \pi/2$.
53. $2x^3 + 8x^3 \ln x$
54. 27.5 feet by 27.5 feet.

Section 12.3   Improving Skills

1. $\frac{1}{7}(1 + x^4)^7 + c$
3. $\frac{1}{558}(6 + 6x^3)^{31} + c$
5. $\frac{1}{3}(t^2 + 2t)^{3/2} + c$
7. $\frac{1}{2} \ln|x^2 + 1| + c$
9. $\frac{-1}{2(x^2 + 1)} + c$
11. $\frac{3}{2}\sqrt{2t^2 + 1} + c$
13. $\frac{1}{6}(y^3 + 3y^2)^2 + c$
15. $-\frac{1}{5} \cos(5x) + c$
19. $40 \sin(0.1x) + c$
23. $-\frac{1}{2} \ln|\cos(2t)| + c$
31. $\frac{1}{2}e^{(y^2 + 2y)} + c$
35. $\frac{10}{31}(10 + e^{3x})^{3.1} + c$
37. $\frac{1}{2}(6 + q^3 + 3e^q)^2 + c$
39. $\ln|1 + e^\theta| + c$

Creating Models

41. $598.91   43. $145.09
45. $1,624.79 thousand   47. 260,647 people
49. 108,625.16 sq. ft.   51. 84.69°F

Reviewing Material

61. $y = 175x + 2,750$
62. a. 55.6%   b. 75%   c. 44.4%
64. 10 more trees per acre

Section 12.4   Improving Skills

1. 10,000 sq ft $\le A \le$ 22,500 sq ft.
3. 0.75 sq mi $\le A \le$ 2.05 sq mi.
5. 108.75   7. 200   9. 215
11. 4   13. 51.4284   15. 6.5
17. 1.0833   19. 1.372154   21. 1400 miles
23. 1150.78 feet   25. 31.6 miles

Exploring with Technology

31. 215.9424 with 50 partitions
    215.9856 with 100 partitions
33. 0.6   35. 0.5

Reviewing Material

36. $x_1 \approx 3{,}818.18$
    $x_2 \approx 2{,}909.1$
    $z = 26{,}000$
37. $16,299.86

38.
$$\frac{(-69.6x^4 e^{-2x} \ln x)(3e^{-2x} - 5)^{4.8} - (3e^{-2x} - 5)^{5.8}(2x^3 + 8x^3 \cdot \ln x)}{(2x^4 \ln x)^2}$$
39. decreasing by 10 cents per day

Section 12.5   Improving Skills

1. 6.4	3. 55	5. -122.5
7. 47.1974	9. 6	11. 6π
13. $2\sqrt{2} - 2$	15. -2.6504	17. 14
19. 6.755	21. 0.0574	23. 63.212
25. 13.7096	27. 45.3333	29. 30.667
31. 7.5	33. 1.8333	35. 2.667
37. 2.35	39. 2	

Creating Models

41. $745	43. $-$117
45. $1,650	47. $1.335 million
49. $10 million	51. $369,375

53. approximately 210 strands
55. 224,000 people
57. 29,200 subscriptions
59. 68.122 miles, 112.979 miles

Reviewing Material

65. Answers will vary depending on how much money the family spends each week!

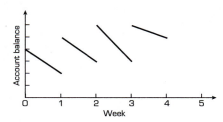

66. $30,304.24
67. everywhere on the interval $(0, 6)$, except at $x = 2$
68. $6 - 24 \sin 2x$

Section 12.6   Improving Skills

1. yes, 0.21	3. no
5. yes, 0.17	7. yes, 0.117
9. yes, 0.15	11. yes, 0.9976
13. yes, 0.0000008295	15. no
17. yes, 0.00895	19. no

21. a. 0.0546   b. 0.3297   c. 0.0829   d. 0.05
23. a. 0.3558   b. 0.5134   c. 0.2636
25. a. 0.24857   b. 0.40824   c. 0.34134   d. 0.47064
27. a. 0.1587   b. 0.54674   c. 0.58581   d. 0.1587

29. a. $\displaystyle \lim_{n \to \infty} \int_{3.333}^{n} \frac{1}{\sqrt{2\eth}} e^{-x^2/2} \, dx$

    b. $\displaystyle \lim_{n \to \infty} \int_{-1.667}^{n} \frac{1}{\sqrt{2\eth}} e^{-x^2/2} \, dx$

    c. $\displaystyle \lim_{n \to \infty} \int_{3.333}^{n} \frac{1}{\sqrt{2\eth}} e^{-x^2/2} \, dx$

31. yes	33. no
35. no	37. no

Creating Models

39. a. 11.43%	b. 26.36%
41. a. 0.3012	b. 0.3297
43. 0.02275	45. 0.02275

Exploring with Technology

53. $\displaystyle \lim_{m \to \infty} \int_{0}^{m} \frac{1}{\sqrt{2\eth}} e^{-x^2/2} \, dx = 0.5.$

Reviewing Material

55. 142.86 lbs of deluxe mix and no regular mix
56.

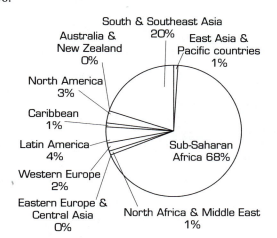

57. 0.0281

58. one plot is 3 ft × 3 ft and the other is 22 ft × 22 ft

Testing Yourself

1. 14

2. 32.25

3. 3.5106

4. $-\frac{1}{2}z^{-2} - z^{-1} + c$

5. $-2e^{-x/2} + c$

6. $\frac{1}{6}\ln|2x^3 + 9| + c$

7. $\frac{1}{2}u^2 - 2\ln|u| + c$

8. 0.5754

9. $-3\cos(\theta/3) + c$

10. 0.0466

11. (See the algorithm in Section 12.4.)

12. $23,083.33

13. 0.387588

14. $\int_{0.33}^{1.67} \frac{1}{\sqrt{2\eth}} e^{-x^2/2} \, dx$

15. 0.4167

16. no

17. 3.667

18. yes

19. $1,004

20. 0.8 sq mi $\leq A \leq$ 5.8 sq mi.

21. 10.7 seconds

22. 2.667

23. 0.3313

24. $g(x) = \dfrac{1}{32.1}(0.1x^2 + 2x + 5)$

25. 0.1232

26. 2.320313

27. $\int_{-0.5}^{0.5} \frac{1}{\sqrt{2\eth}} e^{-x^2/2} \, dx$

28. approximately 0.3865

29. yes

30. 33,128 cases

31. $g(x) = \dfrac{1}{327.5}(2x^3 + 3)$, 0.0321

32. 0.2636

Chapter 13

Section 13.1   Improving Skills

1. a. 8   b. -2 c. 13

3. a. $\frac{1}{3}$ b. 5   c. -1.6

5. a. -72 b. -92 c. -92

7. entire xy plane

9. points in the first quadrant, the third quadrant, and the points on the coordinate axes.

11. entire ab plane

13. $\{(u, v)\,|\,v > -u\}$

15. entire rs plane

17.

19.

21.

23.

25.

27.

29.

35.

31.

33.

37.

41.

39.

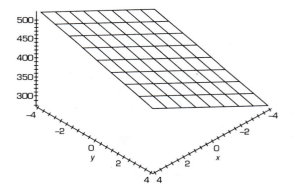

Creating Models

43. Volume $V = \pi r^2 h$, where r = radius of the tube, and h = length of the tube.
   a. Input variables: r, h
      Output variable: V
   b. all pairs (r, h) of real numbers.
   c. all pairs (r, h) of real numbers, with $r \geq 0$ and $h \geq 0$.

45. $FV = R\left[\dfrac{(1 + i)^n - 1}{i}\right]$

   a. Input variables: R, i, and n;   Output variable: FV
   b. all triples (R, i, n) of real numbers, with $R \geq 0$, $i \geq 0$, and $n \geq 0$.

47. Let a = number of model A televisions produced, b = number of model B produced, and c = number of model C produced. Let s_1 = the number of model A televisions sold, s_2 = the number of model B televisions sold, and s_3 = the number of model C televisions sold. Profit P = Revenue − Cost = $320a + 365b + (365 + 21)c - (210s_1 + 230s_2 + 230s_3)$. There are six input variables.

49. Let T = number of travel books sold, F = number of cookbooks sold, and G = number of garden books sold. Since the capacity of the press is 20,000 books, $T + F + G = 20000 \Rightarrow G = 20000 - F - T$. Total revenue R is given by $R = 15T + 12F + 10(2000 - F - T)$. There are two input variables. The domain of the function is all pairs of integers (T, F) such that T, F, and $(2000 - F - T)$, are all nonnegative.

51. Let x_1 = number of one-bedroom units rented, x_2 = number of two-bedroom units rented, and x_3 = number of three-bedroom units rented. Total monthly revenue $R = 500x_1 + 850x_2 + 1200x_3$. There are three input variables. The domain is the set of all triples (x_1, x_2, x_3) of integers, with $0 \leq x_1 \leq 30$, $0 \leq x_2 \leq 20$, $0 \leq x_3 \leq 10$.

53. L and C.

Exploring with Technology

61.

63.

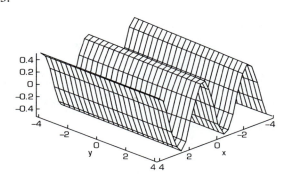

Reviewing Material

64. $y = -10x - 170$

65. \$5,947.82

66. $24[\cos^2(2x) - \sin^2(2x)]$

67. $\dfrac{-1}{x^2} - 3 \ln|x| + C$

Section 13.2 Improving Skills

1. $\dfrac{\partial f}{\partial x} = 10x^4y^3$; $\dfrac{\partial f}{\partial y} = 6x^5y^2$

3. $\dfrac{\partial f}{\partial x} = 8x - 6x^2 \cdot y^3$; $\dfrac{\partial f}{\partial y} = 4x^2 - 6x^3y^2$

5. $\dfrac{\partial f}{\partial x} = 4x \sin t$; $\dfrac{\partial f}{\partial t} = 2x^2 \cos t$

7. $\dfrac{\partial f}{\partial x} = 2x^2t \cos(x^2t) + \sin(x^2t)$

   $\dfrac{\partial f}{\partial t} = x^3 \cos(x^2t)$

9. $\dfrac{\partial f}{\partial t} = 35t^6e^y$; $\dfrac{\partial f}{\partial y} = 5t^7e^y$

11. $\dfrac{\partial f}{\partial t} = 35yt^6e^{5yt^7}$; $\dfrac{\partial f}{\partial y} = 5t^7e^{5yt^7}$

13. $\dfrac{\partial f}{\partial u} = \dfrac{-u^2 + v^2}{(u^2 + v^2)^2}$; $\dfrac{\partial f}{\partial v} = \dfrac{-2uv}{(u^2 + v^2)^2}$

15. $\dfrac{\partial f}{\partial u} = \dfrac{v^3 - u^2v}{(u^2 + v^2)^2}$

   $\dfrac{\partial f}{\partial v} = \dfrac{u^3 - uv^2}{(u^2 + v^2)^2}$

17. $\dfrac{\partial g}{\partial x} = \dfrac{-2y^3x^3}{\sqrt{1 - x^4}}$; $\dfrac{\partial g}{\partial y} = 3y^2\sqrt{1 - x^4}$

19. $\dfrac{\partial g}{\partial x} = -xy^5(x^2y^5 + 4y^3)^{-1.5}$

   $\dfrac{\partial g}{\partial y} = -\dfrac{5x^2y^4 + 12y^2}{2(x^2y^5 + 4y^3)^{1.5}}$

21. $\dfrac{\partial g}{\partial s} = \dfrac{1}{(1 + t^4)\sqrt{\dfrac{2s}{1 + t^4}}}$

$\dfrac{\partial g}{\partial t} = \dfrac{-4st^3}{(1 + t^4)^2\sqrt{\dfrac{2s}{1 + t^4}}}$

23. $\dfrac{\partial g}{\partial s} = \dfrac{-t^2 \cos t}{2(1 + st)^2\sqrt{\dfrac{t \cos t}{1 + st}}}$

$\dfrac{\partial g}{\partial t} = \dfrac{1}{2\sqrt{\dfrac{t \cos t}{1 + st}}}\left(\dfrac{-t \sin t - st \sin t + \cos t}{(1 + st)^2}\right)$

25. $\dfrac{\partial f}{\partial x} = 6xy - 4z^4; \qquad \dfrac{\partial f}{\partial y} = 3x^2 + 6y^2z;$

$\dfrac{\partial f}{\partial z} = 2y^3 - 16xz^3$

27. $\dfrac{\partial f}{\partial x} = 10x^9y^{-8}z^2 + 12z^5; \qquad \dfrac{\partial f}{\partial y} = -8x^{10}y^{-9}z^2;$

$\dfrac{\partial f}{\partial z} = 2x^{10}y^{-8}z + 60xz^4$

29. $\dfrac{\partial f}{\partial x} = 2xe^{3yz}; \qquad \dfrac{\partial f}{\partial y} = 3x^2ze^{3yz}; \qquad \dfrac{\partial f}{\partial z} = 3x^2ye^{3yz}$

31. 0   33. -0.2391

35. 218.39   37. 64.43

39. 1596   41. 320

43. 180.77   45. $96x^2y^7$

47. $-x^2y \sin(xy) + 2x \cos(xy)$

49. $30x^4yz^3$

51. $\dfrac{\partial z}{\partial x} = 2x; \quad \dfrac{\partial z}{\partial y} = 6; \quad z_{xx} = \dfrac{\partial}{\partial x}\left(\dfrac{\partial z}{\partial x}\right) = \dfrac{\partial}{\partial x}(2x) = 2;$

$3z_{xx} - z_y = 3(2) - 6 = 6 - 6 = 0$

53. $\dfrac{\partial z}{\partial x} = -e^{-(x+3y)}; \qquad \dfrac{\partial z}{\partial y} = -3e^{-(x+3y)};$

$z_{xx} = \dfrac{\partial}{\partial x}\left(\dfrac{\partial z}{\partial x}\right) = \dfrac{\partial}{\partial x}(-e^{-(x+3y)}) = e^{-(x+3y)};$

$z_{xx} - 2z_x + z_y = e^{-(x+3y)} - 2(-e^{-(x+3y)}) - 3e^{-(x+3y)} = e^{-(x+3y)} + 2e^{-(x+3y)} - 3e^{-(x+3y)} = 0$

55. $\dfrac{\partial z}{\partial x} = 2 \cos(2x) \cos(4t); \qquad \dfrac{\partial z}{\partial t} = -4 \sin(2x) \sin(4t);$

$z_{xx} = -4 \sin(2x) \cos(4t)$

$z_{tt} = -16 \sin(2x) \cos(4t)$

$4z_{xx} = 4(-4 \sin(2x) \cos(4t)) = -16 \sin(2x) \cos(4t) = z_{tt}$

Creating Models

57. $\dfrac{\partial G}{\partial v} > 0$—the more vehicles on the road, the more gas sold.

$\dfrac{\partial G}{\partial p} < 0$—the higher the price, the less gas sold

59. $\left.\dfrac{\partial V}{\partial l}\right|_{\substack{l=2 \\ h=3}} = 12 \text{ ft}^3/\text{ft.}$

$\left.\dfrac{\partial V}{2h}\right|_{\substack{l=2 \\ h=3}} = 4 \text{ ft}^3/\text{ft}$

61. $\left.\dfrac{\partial V}{\partial l}\right|_{\substack{l=2 \\ w=3 \\ h=4}} = 12 \text{ ft}^3/\text{ft.}$

$\left.\dfrac{\partial V}{\partial w}\right|_{\substack{l=2 \\ w=3 \\ h=4}} = 8 \text{ ft}^3/\text{ft.}$

$\left.\dfrac{\partial V}{\partial h}\right|_{\substack{l=2 \\ w=3 \\ h=4}} = 6 \text{ ft}^3/\text{ft.}$

63. -14π 65. -0.014 liters/mm

67. 3439.27 dollars/100%

69. $\left.\dfrac{\partial P}{\partial L}\right|_{P_{(256.81)}} = 0.675$

$\left.\dfrac{\partial P}{\partial C}\right|_{P_{(256.81)}} = 0.711$

Reviewing Material

77. $y = 617.5x + 4590.6$, where $x = \text{year} - 1993$

78. $f(2 + h) - f(2) = \dfrac{2 + (2 + h)}{2 + h} - \dfrac{2 + 2}{2} = \dfrac{-2h}{4 + 2h}$

$f'(2) = \lim_{h \to 0} \dfrac{f(2 + h) - f(2)}{h} = \lim_{h \to 0} \dfrac{\dfrac{-2h}{4 + 2h}}{h} =$

$\lim_{h \to 0} \dfrac{-2h}{(4 + 2h)h} = \lim_{h \to 0} \dfrac{-2}{4 + 2h} = -\dfrac{1}{2}$

79. Base and roof are 3.33×3.33, walls are 3.33×10.82.

Section 13.3   Improving Skills

1. $(2, 1)$ is a local minimum.

3. $(2, 1)$ is a saddle point.

5. $(0, 0)$ is a saddle point.

7. $\left(\dfrac{14}{9}, \dfrac{14}{18}\right)$ is a saddle point.

9. $(-14, 21)$ is a saddle point.

11. $(-2, 2)$ is a local maximum.

13. $(0, 0)$ is a local minimum.

15. $(0, 0)$ is a saddle point.

17. (0, 0) is a saddle point.
    $(\frac{4}{3}, \frac{4}{3})$ is a local minimum.
19. (0, 0) is a saddle point.
    (1, 1) is a local minimum.
21. (−1, −1) is a local minimum.
    (0, 0) is a saddle point.
    (1, 1) is a local minimum.
23. (0, 0) is a saddle point.
25. (1, 0) is a local minimum.
27. $(\frac{3}{2}, \frac{2}{3})$ is a saddle point.

Creating Models

29. 5, 5/2, and 5/2.
31. square base of side 6 ft. and height $h = 4$ ft.
33. base 8 ft. × 4 ft. and height $h = 2$ ft.
35. $x = 75$, $y = 135$
37. a. Revenue $= p_r d_r + p_g d_g$;   Cost $= 1d_r + 2d_g$
       Profit $= -4000p_r^2 + 6000p_r + 2000p_r p_g - 5000p_g^2 + 39000p_g - 64000$
    b. $p_r = \dfrac{69}{38}$ $p_g = \dfrac{162}{38}$

Exploring with Technology

43. a. There is an absolute minimum at (0, 0).
    b. There is a saddle point at (0, 0).
    c. There is an absolute maximum at (0, 0).
45. If the intersection of the graph of f with a vertical plane through (x_0, y_0) and perpendicular to the x-axis has a sharp corner at y_0, then the partial derivative with respect to y does not exist there. A similar statement can be made for x.

Reviewing Material

47.
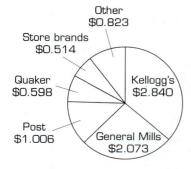

Cereal Sales (billions)

48. 0.5948

49.

x	LEFT-HAND LIMIT	RIGHT-HAND LIMIT	LIMIT	CONTINUOUS
1	0	1	no	no
2	−2	−2	−2	yes
3	1	1	1	no

50. $\sqrt{2} - 1 \approx 0.4142$

Section 13.4   Improving Skills

1. $1.5y^2$
3. 0
5. $4y^{-2}$
7. $64 - 4y$
9. $y \ln 4 + 3$
11. $\dfrac{1}{2y}(\cos(\pi^2 y) - \cos(4\pi^2 y))$
13. $\dfrac{1}{4}x^9$
15. x^2
17. $\dfrac{1}{4}(x^4 - 1)$
19. $3e^3(e - 1)$
21. 12
23. 0
25. −9
27. 817.09
29. 3
31. $\displaystyle\int_0^1 \int_0^{2x} f(x, y)\, dy\, dx$
33. $\displaystyle\int_0^2 \int_{-x-1}^{2x} f(x, y)\, dy\, dx$
35. $\displaystyle\int_{-1}^2 \int_{x^2}^{x+2} f(x, y)\, dy\, dx$
37. $\displaystyle\int_{-\sqrt{2}}^{\sqrt{2}} \int_{x^2}^{4-x^2} f(x, y)\, dy\, dx$
39. The region for Problem 19 is

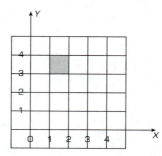

41. The region for Problem 24 is

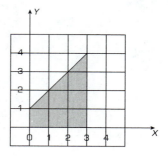

43. The region for Problem 26 is

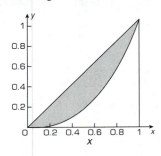

45. 10.86

47. 1.48

49. 10.52

51. $40,000

53. 4.1515 units

55. $118\frac{1}{3}$

57. 89.28

59. 7.3881 cubic units

61. $\frac{11}{30}$ cubic units

63. $\frac{2}{45}$

Reviewing Material

64. $x = 3, y = -1, z = 11$

65. [0.1045   0.373   0.522]

66. 0.777

Testing Yourself

1.

2. 16, 16, 16

3.

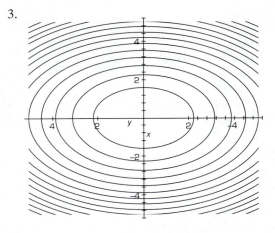

4. $\dfrac{10}{3}$

5. a. $12x^3 + 4xy^3$ b. $\dfrac{y}{3 + x^2y}$ c. $\dfrac{3x^2}{x^3 + y^3z + 3}$

6. $-2x^3y \cos(x^2y) - 2x \sin(x^2y)$

7. 16

8. $56,873.63

9. 3

10.

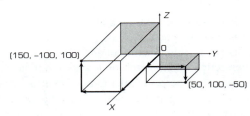

11. $\dfrac{\partial C}{\partial P} < 0$—the larger the number of police, the smaller the amount of crime

$\dfrac{\partial C}{\partial U} > 0$—the larger the number of unemployed, the larger the amount of crime

12. Profit $= 40x^2 - 80xy + 1200x + 30y^2 + 600y - 7000$

$x = 75, y = 90$

13. -1, 1, and 1

14. $18 \times 18 \times 36$

16. a. $x^4 + 4x^2y + 5xy^4$

b. $21xy(x + 3xy^2)^{2.5}$

c. $2xye^{2xy+1} + e^{2xy+1}$

17. $x = 75, y = 135$

18. $\dfrac{1}{112}$

19. 2

20. -6

21. 72.6

22.

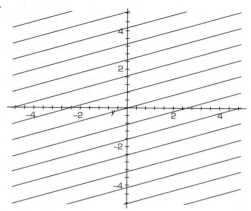

Appendix A   Exponents

1. 2^5 3. 5^9 5. $(-4)^{18}$
7. $(0.3)^9$ 9. 2^{21} 11. 7^{25}
13. $(-5)^6$ 15. $(-4)^{30}$ 17. $(2.3)^{24}$
19. 7^{-4} 21. 2^{-8} 23. 2^6
25. 3^{10} 27. 5^{-15} 29. 15^{-2}
31. 7^{-3} 33. 3^5 35. 2^5
37. $(-5)^{-1}$ 39. $(-5)^5$ 41. 5^{-6}
43. 2^{-6} 45. 3^2 47. 3^{-4}
49. 5^4 51. 5^2 53. 3^2
55. 27   57. 243   59. 27
61. $\frac{1}{32}$ 63. $\frac{1}{1024}$ 65. 18.5203
67. 6.3907   69. 26.7496   71. 0.0257
73. 0.0003473

Appendix B   Polynomials

1. $7x^2 + 9x + 11$
3. $-x^2 - 2x + 13$
5. $3u^3 + u^2 - 4u + 15$
7. $3t^5 + 12t^4 - 5t^2 + t + 1$
9. $4t^5 + 12t^3 - 20t$
11. $\frac{5}{4}y^6 + y^4 - \frac{3}{4}y^2 + \frac{1}{2}$
13. $12r^4 - 5r^3 + 4r^2 + 44r + 15$
15. $x^2 + 3x - 10$
17. $6y^2 + 5y - 25$
19. $u^4 - 1$
21. $\frac{1}{4}m^3 + \frac{1}{4}m^2 + \frac{1}{16}m$
23. $2m^4 - 17m^3 + 19m^2 + 2m - 6$
25. $y^5 - 3y^4 - 6y^3 - 11y^2 - 21y$

Appendix C   Factoring

1. $15(2x^2 - 4x - 3)$ 3. $5(6t^3 + 7t^2 + 11)$
5. $t(3t^2 + 5t + 3)$ 7. $2x^6(2x^2 + x + 3)$
9. $2(m^2 + 8)$ 11. $(m + 4)(m - 4)$
13. $(x + 5)(x + 2)$ 15. $(x - 5)(x + 2)$
17. $(x - 10)(x + 1)$ 19. $(t + 2)(t + 7)$
21. $(t + 10)(t - 10)$ 23. not factorable
25. $2(x + 4)(x + 5)$ 27. $m^2(m^2 + 13m + 2)$
29. $(2x + 1)(2x - 1)$ 31. $2(2x + 1)(x + 1)$
33. $(2m + 1)(m - 2)$ 35. $(2x + 1)^2$
37. $(5t + 1)(t + 4)$ 39. $(4p + 3)(3p - 4)$

Appendix D   Roots and the Quadratic Formula

1. $x = 2$ is a root   3. $t = -12$ is a root
5. $t = -\frac{8}{5}$ is a root   7. $x = \frac{1}{2}$ is a root
9. $x = -5$ and $x = -2$ are roots
11. $x = 5$ and $x = -2$ are roots
13. $t = 10$ and $t = -1$ are roots
15. $y = -\frac{1}{2}$ and $y = 2$ are roots
17. $c = (1 + \sqrt{5})/2 \approx 1.618034$ and
$c = (1 - \sqrt{5})/2 \approx -0.618034$ are roots
19. $y = 2$ and $y = -1$ are roots
21. $t = (1 + \sqrt{21})/10 \approx 0.55826$ and
$t = (1 - \sqrt{21})/10 \approx -0.35826$ are roots
23. $m = 0$ and $m = -2$ are roots
25. $m = -4$ and $m = 4$ are roots
27. $x = -\frac{1}{2}$ and $x = \frac{1}{2}$ are roots

29. $x = (-6 + \sqrt{20})/8 \approx -0.19098$ and
$x = (-6 - \sqrt{20})/8 \approx -1.30902$ are roots

31. Since we cannot take the square root of a negative number, there are no real roots.

33. $t = (-5 + \sqrt{17})/4 \approx -0.21922$ and
$t = (-5 - \sqrt{17})/4 \approx -2.28078$ are roots

35. $t = -\frac{3}{2}$ and $t = -1$ are roots

37. $m = 0$ and $m = -0.155711$ and
$m = -12.84429$ are all roots

39. $p = 0$ and $p = -13$ are roots

Appendix E   Common Logarithms

1. 0.100371
3. 2.100371
5. 0.656098
7. −0.343902
9. 1.870989
11. 1.180413
13. −0.994819
15. 5.164299
17. 4.829999
19. 18.299996
21. 0.183
23. 61.699977
25. 0.09
27. 1.025999
29. 0.00775

31. This equality is false. This can also be verified by calculating:
$\log(2 \cdot 7) = \log(14) = 1.14613 \neq (\log 2)(\log 7) = 0.2544$

33. This equality is false.

35. This equality is false. This can also be verified by calculating:
$\log(2 + 7) = \log 9 = 0.954243 \neq \log 2 + \log 7 = 1.146128$

37. This equality is true.

39. This equality is false. This can also be verified by calculating:
$\log(4 \cdot 9) = \log(36) = 1.556303 \neq (\log 4)(\log 9) = 0.574511$

41. This equality is true.

43. This equality is false. This can also be verified by calculating:
$\log(\frac{3}{5}) = -0.221849 \neq \log 3/\log 5 = 0.682606$

45. This equality is false.

47. This equality is true.
49. This equality is true.
51. $y = 2.807355$
53. $y = 1.771244$
55. $y = 1.321928$
57. $y = 1.44064$
59. $y = 86.438562$
61. $y = 1.476412$
63. $y = 48.24428$
65. $y = -5.206672$
67. $y = -0.084275$

Appendix F   Sigma Notation

1. $(x_1)^2 + (x_2)^2 + (x_3)^2$

3. $(x_2 + y_2) + (x_3 + y_3) + (x_4 + y_4) + (x_5 + y_5) + (x_6 + y_6)$
$+ (x_7 + y_7)$

5. $(3x_2)^2 + (3x_3)^2 + (3x_4)^2 + (3x_5)^2$

7. $\sin(0) + \sin(\theta_1) + \sin(2\theta_2) + \sin(3\theta_3) + \sin(4\theta_4)$

9. 21
11. 42
13. 20

15. $\frac{137}{60}$
17. 3.4544

19. a. $\sum_{i=2}^{29} 3(i)^2$
b. $\sum_{i=2}^{29} i(3)^2$

c. $\sum_{i=2}^{13} 2(3)^i$
d. $\sum_{i=2}^{29} 3(-1)^i(i)^2$

21. a. -2
b. 22
c. 3
d. 20
e. 26
f. 21

23. a. Yes, $2\sum_{i=1}^{n} x_i = \sum_{i=1}^{n} 2x_i$.

b. Yes, $\sum_{i=1}^{n} (x_i + y_i) = \sum_{i=1}^{n} x_i + \sum_{i=1}^{n} y_i$.

c. Yes, $\sum_{i=1}^{m} x_i + \sum_{i=m+1}^{n} x_i = \sum_{i=1}^{n} x_i$.

d. No, $\sum_{i=1}^{n} x_i y_i \neq \sum_{i=1}^{n} x_i \sum_{i=1}^{n} y_i$.

25. System (16) becomes:

$$(N)b + \left(\sum_{i=1}^{N} x_i\right)m = \sum_{i=1}^{N} y_i$$

$$\left(\sum_{i=1}^{N} x_i\right)b + \left(\sum_{i=1}^{N} x_i^2\right)m = \sum_{i=1}^{n} x_i y_i$$

INDEX

PHOTO CREDITS